Handbook of Research on Modern Optimization Algorithms and Applications in Engineering and Economics

Pandian Vasant
Universiti Teknologi Petronas, Malaysia

Gerhard-Wilhelm Weber
Middle East Technical University, Turkey

Vo Ngoc Dieu
Ho Chi Minh City University of Technology, Vietnam

A volume in the Advances in Computational Intelligence and Robotics (ACIR) Book Series

ENGINEERING SCIENCE REFERENCE
An Imprint of IGI Global

Published in the United States of America by
Engineering Science Reference (an imprint of IGI Global)
701 E. Chocolate Avenue
Hershey PA, USA 17033
Tel: 717-533-8845
Fax: 717-533-8661
E-mail: cust@igi-global.com
Web site: http://www.igi-global.com

Library of Congress Cataloging-in-Publication Data

Names: Vasant, Pandian, editor. | Weber, Gerhard-Wilhelm, editor. | Dieu, Vo Ngoc, 1971- editor.
Title: Handbook of research on modern optimization algorithms and applications in engineering and economics / Pandian Vasant, Gerhard-Wilhelm Weber, and Vo Ngoc Dieu, editors.
Description: Hershey : Information Science Reference, 2016. | Includes bibliographical references and index.
Identifiers: LCCN 2015037648| ISBN 9781466696440 (hardcover) | ISBN 9781466696457 (ebook)
Subjects: LCSH: Mathematical optimization. | Algorithms.
Classification: LCC QA402.5 .H3655 2016 | DDC 330.01/5196--dc23 LC record available at http://lccn.loc.gov/2015037648

This book is published in the IGI Global book series Advances in Computational Intelligence and Robotics (ACIR) (ISSN: 2327-0411; eISSN: 2327-042X)

British Cataloguing in Publication Data
A Cataloguing in Publication record for this book is available from the British Library.

All work contributed to this book is new, previously-unpublished material. The views expressed in this book are those of the authors, but not necessarily of the publisher.

For electronic access to this publication, please contact: eresources@igi-global.com.

Advances in Computational Intelligence and Robotics (ACIR) Book Series

Ivan Giannoccaro
University of Salento, Italy

ISSN: 2327-0411
EISSN: 2327-042X

Mission

While intelligence is traditionally a term applied to humans and human cognition, technology has progressed in such a way to allow for the development of intelligent systems able to simulate many human traits. With this new era of simulated and artificial intelligence, much research is needed in order to continue to advance the field and also to evaluate the ethical and societal concerns of the existence of artificial life and machine learning.

The **Advances in Computational Intelligence and Robotics (ACIR) Book Series** encourages scholarly discourse on all topics pertaining to evolutionary computing, artificial life, computational intelligence, machine learning, and robotics. ACIR presents the latest research being conducted on diverse topics in intelligence technologies with the goal of advancing knowledge and applications in this rapidly evolving field.

Coverage

- Computational Logic
- Algorithmic Learning
- Cognitive Informatics
- Natural Language Processing
- Machine Learning
- Synthetic Emotions
- Artificial Intelligence
- Fuzzy Systems
- Heuristics
- Agent technologies

IGI Global is currently accepting manuscripts for publication within this series. To submit a proposal for a volume in this series, please contact our Acquisition Editors at Acquisitions@igi-global.com or visit: http://www.igi-global.com/publish/.

The Advances in Computational Intelligence and Robotics (ACIR) Book Series (ISSN 2327-0411) is published by IGI Global, 701 E. Chocolate Avenue, Hershey, PA 17033-1240, USA, www.igi-global.com. This series is composed of titles available for purchase individually; each title is edited to be contextually exclusive from any other title within the series. For pricing and ordering information please visit http://www.igi-global.com/book-series/advances-computational-intelligence-robotics/73674. Postmaster: Send all address changes to above address.

Titles in this Series

For a list of additional titles in this series, please visit: www.igi-global.com

Handbook of Research on Modern Optimization Algorithms and Applications in Engineering and Economics
Pandian Vasant (Universiti Teknologi Petronas, Malaysia) Gerhard-Wilhelm Weber (Middle East Technical University, Turkey) and Vo Ngoc Dieu (Ho Chi Minh City University of Technology, Vietnam)
Engineering Science Reference • copyright 2016 • 684pp • H/C (ISBN: 9781466696440) • US $325.00 (our price)

Emerging Technologies in Intelligent Applications for Image and Video Processing
V. Santhi (VIT University, India) D. P. Acharjya (VIT University, India) and M. Ezhilarasan (Pondichery Engineering College, India)
Information Science Reference • copyright 2016 • 518pp • H/C (ISBN: 9781466696853) • US $235.00 (our price)

Handbook of Research on Design, Control, and Modeling of Swarm Robotics
Ying Tan (Peking University, China)
Information Science Reference • copyright 2016 • 854pp • H/C (ISBN: 9781466695726) • US $465.00 (our price)

Handbook of Research on Emerging Perspectives in Intelligent Pattern Recognition, Analysis, and Image Processing
Narendra Kumar Kamila (C.V. Raman College of Engineering, India)
Information Science Reference • copyright 2016 • 477pp • H/C (ISBN: 9781466686540) • US $255.00 (our price)

Handbook of Research on Advanced Hybrid Intelligent Techniques and Applications
Siddhartha Bhattacharyya (RCC Institute of Information Technology, India) Pinaki Banerjee (Goldstone Infratech Limited, India) Dipankar Majumdar (RCC Institute of Information Technology, India) and Paramartha Dutta (Visva-Bharati University, India)
Information Science Reference • copyright 2016 • 653pp • H/C (ISBN: 9781466694743) • US $285.00 (our price)

Research Advances in the Integration of Big Data and Smart Computing
Pradeep Kumar Mallick (Institute for Research and Development, India)
Information Science Reference • copyright 2016 • 374pp • H/C (ISBN: 9781466687370) • US $210.00 (our price)

Innovative Research in Attention Modeling and Computer Vision Applications
Rajarshi Pal (Institute for Development and Research in Banking Technology, India)
Information Science Reference • copyright 2016 • 456pp • H/C (ISBN: 9781466687233) • US $200.00 (our price)

Handbook of Research on Swarm Intelligence in Engineering
Siddhartha Bhattacharyya (RCC Institute of Information Technology, India) and Paramartha Dutta (Visva-Bharati University, India)
Engineering Science Reference • copyright 2015 • 743pp • H/C (ISBN: 9781466682917) • US $335.00 (our price)

IGI GLOBAL
DISSEMINATOR OF KNOWLEDGE
www.igi-global.com

701 E. Chocolate Ave., Hershey, PA 17033
Order online at www.igi-global.com or call 717-533-8845 x100
To place a standing order for titles released in this series, contact: cust@igi-global.com
Mon-Fri 8:00 am - 5:00 pm (est) or fax 24 hours a day 717-533-8661

Editorial Advisory Board

List of Reviewers

List of Contributors

Table of Contents

Detailed Table of Contents

With the rise of large city and the need of large civil engineering structures and city planning, surveying industry improves continuously their instruments/software in order to get cm accuracy position anywhere. Moreover, since the boom of mobile phones in the late 90s, location has become very valuable information for security, emergency and commercial applications. Depending of the application, the location technologies vary based on the accuracy of the location and the price of the system, which delivers the location information to the user. For outdoor applications, Global Navigation Satellite System is the main candidate, whereas if the user/mobile node is indoors or in a narrow street other technologies will be preferred such as the ones based on Wi-Fi or radio-frequency signal.This chapter provides an overview of different positioning technologies used in geo-location together with their limits/advantages. This chapter studies also a number of algorithms developed to estimate the position coordinates of a static or mobile user or target.

This chapter is intended as an overview of robust optimization models related to optimization problems subject to uncertain data, with special focus on the case when uncertainty impacts the right-hand side coefficients in the constraints. Two-stage as well as multistage models are addressed, emphasizing links with applications and computational complexity issues. A class of multistage robust optimization problems for which exact optimal strategies can be efficiently computed (via a robust dynamic programming recursion) is discussed. An application to a multiperiod energy production planning problem is presented into detail, and computational results are reported.

Supply chain network design is one of the most important strategic issues in operations management. The main objective in designing a supply chain is to keep the cost as low as possible. However, the modelling of a supply chain requires more than single-objective such as lead-time minimization, service level maximization, and environmental impact maximization among others. Usually these objectives may cause conflicts such as increasing the service level usually causes a growth in costs. Therefore, the aim should be to find trade-off solutions to satisfy the conflicting objectives. The aim of this chapter is to propose a new method based on a hybrid version of the Bees Algorithm with Slope Angle Computation and Hill Climbing Algorithm to solve a multi-objective supply chain network design problem. A real case from the literature has been selected and solved in order to show the potentiality of the proposed method in solving a large scale combinatorial problem.

Over the years the supply chain industry has been transforming to improve the end-to-end (production to delivery) process. Supply chain management (SCM) allows various industries to oversee and better handle how their product is manufactured and delivered. It allows them to track and identify the location of the product and to be more efficient in delivery. Integrating total asset visibility (TAV) technology into the supply chain structure can provide excellent visibility of a product. This kind of visibility complemented with various packaging schemes can assist in accommodating optimization strategies for visualizing the movement of a product throughout the entire supply chain pipeline. The chapter will define SCM, discuss TAV, review how transportation as well as optimization impacts SCM and TAV, and examine the role of packaging in the context of SCM and TAV.

In this article it is studied the application of a genetic algorithm in the problem of variable selection for multiple linear regression, minimizing the least squares criterion. The algorithm is based on a chromosomic representation of variables that are considered in the least squares model. A binary chromosome indicates the presence (1) or absence (0) of a variable in the model. The fitness function is based on the adjusted square R, proportional to the fitness for chromosome selection in a roulette wheel model selection. Usual genetic operators, such as crossover and mutation are implemented. Comparisons are performed with benchmark data sets, obtaining satisfying and promising results.

In power systems, we still lack the existence of standardized test systems that can be used to benchmark the performance and solution quality of proposed optimization techniques. It is therefore necessary to develop new methods for design of test cases for economic analysis in power systems. We compared two methods to generate test systems: time series and a method for simulating stable random variables that can be used to multiperiod unit commitment based on the use of Chambers-Mallows-Stuck. Hence, after comparing both methods, we describe the method for simulating stable random variables in the generation of test systems for economic analysis in power systems. A study focused on generating test electrical systems through fat tail model for unit commitment problem in electrical power systems is presented. Usually, the instances of test systems in Unit Commitment are generated using normal distribution, but in this work, simulations data are based on alpha-stable distribution. Numerical results illustrate the applicability of the proposed method.

In real-world most of manufacturing systems are large, complex, and subject to uncertainty. This is mainly due to events as random demands, breakdowns, repairs of production machines, setup and cycle times, inventory fluctuations and more. If items move too quickly, workers may work too hard. If items move too slowly, workers may have great leisure times. However, must make decisions here and now regarding the operation of the system optimally and quickly. In practice, these decisions are based on recent statistics of the system behavior, in the experience of the analyst and the urgency of the solution. In this chapter, we present a real problem associated with the production of individual parts in metalworking industry for the refrigerators production. We develop a model based on the Markov Decision Process to study the dynamics of the trajectory of end products in a manufacturing line that works by process. Then, we propose a measure of the average production rate of the line by using the Monte Carlo method. We illustrate our proposal using a numerical example with real data obtained in situ.

This chapter addresses the family of problems known in the literature as Capacitated Vehicle Routing Problems (CVRP). A procedure is introduced for the optimization of a version of the generic CVRP. It generates feasible clusters and, in a first step, yields a coding of their ordering. The next stage provides this information to a genetic algorithm for its optimization. A selective pressure process is added in order to improve the selection and subsistence of the best candidates. This arrangement allows improving the performance of the algorithm. We test it using Van Breedam and Taillard's problems, yielding similar results as other algorithms in the literature. Besides, we test the algorithm on real-world problems, corresponding to an Argentinean company distributing fresh fruit. Four instances, with 50, 100, 150 and 200 clients were examined, giving better results than the current plans of the company.

In this chapter, the Artificial Bee Colony (ABC) algorithm and its variants are presented and applied to spiral antennas design for RFID tag application at the UHF band. The ABC variants include the Improved ABC (I-ABC), which is an improved version of the original ABC algorithm. The I-ABC introduces the best-so-far solution, inertia weight and acceleration coefficients to modify the search process. Furthermore, another ABC variant is the Gbest ABC (ABC), which includes global best (gbest) solution information into the search equation to improve the exploitation. These algorithms are applied to antenna design where the optimization goals are antenna size minimization, gain maximization, and conjugate matching. The algorithms performance is compared with other popular evolutionary algorithms. The optimization results produced show that the ABC family of algorithms is a powerful tool that can be efficiently applied to antenna design problems.

In this chapter the application of a nature-inspired technique in conjunction with simulation models to optimize the siting of concentration nodes in a traffic network of urban area is presented. The solutions of this optimization problem involve the redesign of the network by adding nodes and arcs to the current traffic network. The problem is a sub-problem of the most general "Traffic Network Design Problem". For this resolution a Genetic Algorithm approach was design and development. The popular Simulator of Urban Mobility (SUMO) is use as traffic simulator in order to evaluate the solutions obtained. This chapter contains the conceptual models and the results of the Optimization via Simulation technique proposed.

This chapter proposes a Hybrid Cuckoo search algorithm to determine optimal location and sizing of Static VAR Compensator (SVC). Hybrid Cuckoo search algorithm is a simple combination of the Cuckoo search algorithm (CSA) and Teaching-learning-based optimization (TLBO), where the learner phase of TLBO is added to improve performance of Cuckoo eggs. The proposed method is applied for optimizing location and sizing of SVC in electric power system. This problem is a kind of discrete and combinatorial problem. The objective function considers loss power, voltage deviation and operational cost of SVC and other operating constraints in power system. Numerical results from three various tested systems show that the proposed method is better than the conventional CSA and TLBO in finding the global optimum solutions and its performance is also high than others.

In cases where the size and colour of cable are changed, the cable industry is classified as a multi-product, mass production system. The paper provides a mixed integer linear programming model based on continuous time representation for a case study on the scheduling problem of the cable industry to minimize the total cost including setup cost, operating cost, and inventory holding cost. As the solution methodology, three grouping policies are proposed while Xpress solver could not give any feasible solution for the model. Cables of the same size and the same colour, respectively, of the different types of cable are grouped together. A metaheuristic based on a simulated annealing algorithm is applied to minimize the total cost of proposed solutions. Finally the solution with the smallest total cost is selected as the production schedule of the study case.

The purpose of this chapter is to present a set of algorithms and their efficiency for the consistency based Multiple Sequence Alignment (MSA) problem. Based on the strength and adaptability of the Genetic Algorithm (GA) two approaches are developed depending on the MSA type. The first approach, for the non related sequences (no consistency), involves a Hybrid Genetic Algorithm (GA_TS) considering also Tabu Search (TS). The Traveling Salesman Problem (TSP) is also applied determining MSA orders.

Recently, renewable power sources such as WTG and PV have become viable economic options for generating sustainable energy. However, WTG and PV have an inconstant power production problem. To solve this problem, multi-state models have been proposed. The electricity generated from these units varies with different time scales: hourly, daily and seasonally. Since planning an optimal size generates cost losses to the customer, three models should be built: a load model, generation model, and service cost model. Loss of energy expectation (LOEE) and loss of load expected (LOLE) are calculated for the load and generation model. The reliability value is calculated to determine the number of required renewable generators. As a result, the system is constructed to have sufficient capacity, and the utility cost became the main objective of the total service costs.

Nature has always been a source of inspiration for human beings. Large numbers of complex optimization problems have been solved by the techniques inspired by nature. Software modularization is one of such complex problems that have been encountered by software engineers. It is the process of organizing modules of a software system into optimal clusters. In this chapter, some bio-inspired algorithms such as bat, artificial bee colony, black hole and firefly algorithm have been proposed for the cause of software modularization. The hybrid of these algorithms with crossover and mutation operators of the genetic algorithm has also been proposed. All the algorithms along with their hybrids are tested on seven benchmark open source software systems. It has been evaluated from the results thus obtained that the hybrid of these algorithms proved to optimize better than the existing genetic and hill-climbing approaches.

Electrification of Transportation has undergone major modifications since the last decade. Success of combining smart grid technology and renewable energy exclusively depends upon the large-scale participation of Plug-in Hybrid Electric Vehicles (PHEVs) towards reach the desired pollution-free transportation industry. One of the key Performance pointers of hybrid electric vehicle is the State-of-Charge (SoC) which needs to be enhanced for the advancement of charging station using computational intelligence methods. In this Chapter, authors applied Hybrid Particle swarm and gravitational search Optimization (PSOGSA) technique for intelligently allocating energy to the PHEVs considering constraints such as energy price, remaining battery capacity, and remaining charging time. Computational experiment results attained for maximizing the highly non-linear fitness function estimates the performance measure of both the techniques in terms of best fitness value and computation time.

A project where activity times are characterized by intervals or triangular fuzzy numbers is considered in this chapter. Classical project evaluation and review technique (PERT) cannot be directly applied to such projects. Comparison of two intervals or two triangular fuzzy numbers plays an important role in this problem. Ranking methods based on some fuzzy mathematical techniques are discussed and two algorithms for finding the critical path for such a project are given. An illustrative example is also provided.

Multi objective (MO) optimization is an emerging field which is increasingly being implemented in many industries globally. In this work, the MO optimization of the extraction process of bioactive compounds from the Gardenia Jasminoides Ellis fruit was solved. Three swarm-based algorithms have been applied in conjunction with normal-boundary intersection (NBI) method to solve this MO problem. The gravitational search algorithm (GSA) and the particle swarm optimization (PSO) technique were implemented in this work. In addition, a novel Hopfield-enhanced particle swarm optimization was developed and applied to the extraction problem. By measuring the levels of dominance, the optimality of the approximate Pareto frontiers produced by all the algorithms were gauged and compared. Besides, by measuring the levels of convergence of the frontier, some understanding regarding the structure of the objective space in terms of its relation to the level of frontier dominance is uncovered. Detail comparative studies were conducted on all the algorithms employed and developed in this work.

In the present chapter, we give an overview of iterative methods for linear complementarity problems (abbreviated as LCPs). We also introduce these iterative methods for the problems based on fixed-point principle. Next, we present some new properties of preconditioned iterative methods for solving the LCPs. Convergence results of the sequence generated by these methods and also the comparison analysis between classic Gauss-Seidel method and preconditioned Gauss-Seidel (PGS) method for LCPs are established under certain conditions. Finally, the efficiency of these methods is demonstrated by numerical experiments. These results show that the mentioned models are effective in actual implementation and competitive with each other.

In this chapter, we propose to perform a numerical technique based on genetic algorithms (GAs) to identify the electrical parameters (Is, Iph, Rs, Rsh, and n) of photovoltaic (PV) solar cells and modules. The one diode type approach is used to model the I–V characteristic of the solar cell. To extract electrical parameters, the approach is formulated as optimization problem. The GAs approach was used as a numerical technique in order to overcome problems involved in the local minima in the case optimization criteria. Compared to other methods, we find that the GAs is a very efficient technique to estimate the electrical parameters of photovoltaic solar cells and modules. Compared with other parameter extraction techniques, based on statistical study, results indicate the consistency and uniformity of method in terms of the quality of final solutions. In parallel, the simulated data with the extracted parameters of method base with GAs are in very good agreement with the experimental data in all cases.

In this chapter, two approaches are developed to solve the integrated production planning and maintenance problem. Moreover, Some Propositions and mathematical properties were suggested and applied in the proposed heuristics to solve the problem. The first heuristic developed is based on Dantzig-Wolfe decomposition. The Dantzig-Wolfe Decomposition principle reformulates the original model and Column generation is then used to deal with the huge number of variables of the reformulated model. A simple rounding heuristic and a smoothing procedure are finally carried out in order to obtain integer solutions. The second heuristic is based on Lagrangean relaxation of the capacity constraints and sub-gradient optimization. At every step of sub-gradient method, feasibility and improvement procedures are applied to the solution of the Lagrangean problem. Computational experiments are carried out to show the results obtained by our approaches and compared to those of commercial solver.

The aim of this chapter is to investigate the multiple attribute decision making problems to a selected project with generalized intuitionistic fuzzy information in which the information about weights is completely known and the attributes values are taken from the generalized intuitionistic fuzzy environment. Here, we extend the technique for order performance by similarity to ideal solution (TOPSIS) for the generalized intuitionistic fuzzy data. In addition, obtained the concept of possibility degree of generalized intuitionistic fuzzy numbers and used to solve ranking alternative in multi-attribute decision making problems.

The optimization of the maintenance decision making can be defined as an attempt to resolve the conflicts of decision situation in such a way that variable under the control of the decision maker take their best possible value. One of the most important controllable parameters is the time interval between maintenance. Most of the researchers have kept the fact that whenever the suitable maintenance interval is reached, the system is replaced with the original one. However the improvement of a system life not only depends on the replacement of deteriorated components, but also on the effectiveness of the maintenance. Taking care about this fact, the effects of maintenance of a multi-component system by combining the three main different PM actions, namely (1a), (1b) and (2p)-maintenance actions. Thus, the main purpose of an effective maintenance program is to present a technique for finding the optimal maintenance interval for the system by considering the multiple goals of the organization viz. maximum availability, minimum maintenance cost.

A new optimization algorithm, specifically, the cuckoo search algorithm (CSA), which inspired by the unique breeding strategy of cuckoos, has been developed recently. Preliminary studies demonstrated the comparative performances of the CSA as opposed to genetic algorithm and particle swarm optimization, however, with the competitive advantage of employing fewer control parameters. Given enough computation, the CSA is guaranteed to converge to the optimal solutions, albeit the search process associated to the random-walk behavior might be time-consuming. Moreover, the drawback from the fixed step size searching strategy in the inner computation of CSA still remain unsolved. The adaptive cuckoo search algorithm (ACSA), with the effort in the aspect of integrating an adaptive search strategy, was attached in this study. Its beneficial potential are analyzed in the benchmark test function optimization, as well as engineering optimization problem. Results showed that the proposed ACSA improved over the classical CSA.

During the last decades, several metaheuristics have been developed to solve complex engineering optimization problems which most of them have been inspired by natural phenomena and swarm behaviors. Metaheuristics are the most selected techniques to find optimal solution intelligently in many areas of scheduling, space allocation, decision making, pattern recognition, document clustering, control objectives, image processing, system and filter modeling, etc. These algorithms have promised better solutions in single and multi-objective optimization. Gravitational search algorithm (GSA) is one of the recent created metaheuristic search algorithms, which is inspired by the Newtonian laws of gravity and motion. GSA was first proposed by Rashedi et al. and in the short time it became popular among the scientific community and researchers resulting in a lot of variants of the basic algorithm with improved performance. This chapter book presents a detailed review of the basic concepts of GSA and a comprehensive survey of its advanced versions. We propose a number of suggestions to the GSA community that can be undertaken to help move the area forward.

This paper focuses on a problem of minimizing total weighted tardiness of jobs in a real-world single batch-processing machine (SBPM) scheduling in existence of fuzzy due date. In this paper, first a fuzzy mixed integer linear programming model is developed. Then, due to the complexity of the problem, which is NP-hard, we design two hybrid metaheuristics called GA-VNS and VNS-SA applying the advantages of genetic algorithm (GA), variable neighborhood search (VNS) and simulated annealing (SA) frameworks. Besides, we propose three fuzzy earliest due date heuristics to solve the given problem. Through computational experiments with several random test problems, a robust calibration is applied on the parameters. Finally, computational results on different-scale test problems are presented to compare the proposed algorithms.

This paper considers issues in the design and construction of a fuzzy logic system to model complex (nonlinear) systems. Several important applications are considered and methods for the decomposition of complex systems into hierarchical and multi-layered fuzzy logic sub-systems are proposed. The learning of fuzzy rules and internal parameters is performed using evolutionary computing. The proposed method using decomposition and conversion of systems into hierarchical and multi-layered fuzzy logic sub-systems reduces greatly the number of fuzzy rules to be defined and improves the learning speed for such systems. However such decomposition is not unique and may give rise to variables with no physical significance. This can raise then major difficulties in obtaining a complete class of rules from experts even when the number of variables is small. Application areas considered are: the prediction of interest rate, hierarchical control of the inverted pendulum, robot control, feedback boundary control for a distributed optimal control system and image processing.

In recent years the bi-level programming problem (BLPP) is interested by many researchers and it is known as an appropriate tool to solve the real problems in several areas such as economic, traffic, finance, management and so on. Also it has been proved that the general BLPP is an NP-hard problem. The literature shows a few attempts for using approximate methods. In this chapter we attempt to develop an effective approach based on Taylor theorem to obtain an approximate solution for the non-linear BLPP. In this approach using the Karush-Kuhn–Tucker, the BLPP has been converted to a non-smooth single problem, and then it is smoothed by the Fischer – Burmeister function. Finally the smoothed problem is solved using an approach based on Taylor theorem. The presented approach achieves an efficient and feasible solution in an appropriate time which is evaluated by comparing to references and test problems.

Over the years, all parts of a commercial refrigerator, such as the compressor, heat exchangers, refrigerant, and packaging, have been improved considerably due to the extensive research and development efforts carried out by academia and industry. However, the achieved and anticipated improvement in conventional refrigeration technology are incremental since this technology is already nearing its fundamentals limit of energy efficiency is described is ‘magnetic refrigeration’ which is an evolving cooling technology. The word ‘green’ designates more than a colour. It is a way of life, one that is becoming more and more common throughout the world. An interesting topic on ‘sustainable technologies for a greener world’ details about what each technology is and how it achieves green goals. Recently, conventional chillers using absorption technology consume energy for hot water generator but absorption chillers carry no energy saving. With the aim of providing a single point solution for this dual purpose application, a product is launched but can provide simultaneous chilling and heating using its vapour absorption technology with 40% saving in heating energy. Using energy efficiency and managing customer energy use has become an integral and valuable exercise. The reason for this is green technology helps to sustain life on earth. This not only applies to humans but to plants, animals and the rest of the ecosystem. Energy prices and consumption will always be on an upward trajectory. In fact, energy costs have steadily risen over last decade and are expected to carry on doing so as consumption grows. This article discusses the potential for such integrated systems in the stationary and portable power market in response to the critical

Foreword

The main aim of this book is to offer a state-of-the-art chapters' collection that cover themes related to modern optimization algorithms and applications in engineering and economics that could be applied for different hard real problems solving.

Efficient solving of various real problems by computers may require different problem solving algorithms, adapted on the specific of the problems type, complexity and the known information about the problems. Computational Intelligence is one of the most successful paradigms used for different computational hard problems solving. But only a few chapters in the book use classical algorithms of Computational Intelligence, like Genetic Algorithms, Fuzzy Logic, or Swarm Intelligence. Most of them use a hybridization with other algorithms, in order to obtain their combined advantages. There are many other chapters that use recently proposed new algorithms, like Artificial Bee Colony Algorithm, Cuckoo Search Algorithm, or Gravitational Search Algorithm.

The book covers a large range of applications in the fields of power generation and distribution, manufacturing and supply chain management, vehicle routing problems, RFID applications, traffic networks design, local and global positioning, multiple sequence alignment, various mathematical problems or chemical models. In my opinion, this book is a must read for all the engineers and economists interested in the field of optimization.

Like other books edited or written by Prof. Pandian M. Vasant, this book is a delight for practitioners in the field of optimization and we must bring deep thanks to the Editorial Board, to the authors and to the publisher that have made possible the coming out of this book.

Rustem Popa
"Dunarea de Jos" University of Galati, Romania

Rustem Popa *(PhD 1999) is an Associate Professor at the Department of Electronics and Telecommunications of the "Dunarea de Jos" University in Galati, Romania ("Low Danube" University in Galati). He was a Scientific Researcher at the Research and Design Institute for Shipbuilding in Galati, and since 1990, is with "Dunarea de Jos" University. His research interests include computational intelligence, evolvable hardware, digital signal processing, medical electronics. He is an author/coauthor/editor of 6 books, 2 journals and more than 50 journal and conference papers.*

Foreword

It has been a long time since simplex method was invented, in period when computers actually did not exist. Main value of that pioneer, well-known method was silent revolution that results in something which book *Modern Optimization Algorithms and Applications in Engineering and Economics* talks about.

From this, very first beginning of optimization algorithms, where simplex method has been calculated by usage of desk calculator, increasing computer power gives us opportunity for development and usage of modern complex optimization algorithms.

Those algorithms are often inspired by scientific fields like mathematics, physics, as well by biological (evolutionary) systems and living things like bees, ants, bats and birds.

Each author of the chapter gives personal contribution as personal vision how to efficient apply some of the optimization algorithms, or /and personal contribution to new solutions in this area.

Important aspect is applicability and usability of optimization algorithms on everyday problems, for which this book has adequate answer.

Book Modern Optimization Algorithms and Applications in Engineering and Economics consolidate theory and practice, where optimization algorithms plays key role in solving real problems in domain of engineering and economics.

Many innovative approaches of using well know optimization techniques, as well as new methods which design has been based upon real problems are shown through book chapters.

Innovation is visible through area of application, as well as an adoption of well-known solutions for particular problem or situation.

All that witnesses on optimization algorithms, as on fruitful area for future research activities, which consolidate different disciplines like engineering, economy, finance and many others, which from optimization algorithms can harvest additional benefits, leaned on interdisciplinary approach.

Researchers will find in book many inspiring ideas how to observe well known methods from some other perspective for achieving additional value to commonly used methods. Also, book for the researchers can be used as a milestone how to apply some existing optimization algorithms in economic or engineering.

Practitioners can deepen knowledge about basic concept of optimization algorithms used in the book, as well as to learn some new algorithmic solutions. Most valuable feature for practitioners is detailed explanations and case studies described in some chapters.

Prof. Vasant as an editor chose an excellent subject for the book, interesting to vide auditorium from researchers, theoretician and practitioners interested in developing, using, and innovation of optimization algorithms.

Taking in consider excellent researches consolidated within book chapters, book will be for sure interesting literature for anyone interested in area which book covers.

Goran Klepac
Raiffeisenbank Austria d.d., Croatia

Goran Klepac, *PhD, University College Professor works as a head of Strategic development in Raiffeisenbank Austria d.d., Croatia, Europe. In several universities in Croatia, he lectures subjects in domain of data mining, predictive analytics, decision support system, banking risk, risk evaluation models, expert system, database marketing and business intelligence. As a team leader, he successfully finished many data mining projects in different domains like retail, finance, insurance, hospitality, telecommunications, and productions. He is an author/coauthor of several books published in Croatian and English in domain of data mining. www.goranklepac.com*

Foreword

The objective of this volume, *Modern Optimization Algorithms and Applications in Engineering and Economics,* is to provide readers with the cases studies of the applications of the state of the art optimization algorithms based on both combinations of heuristics and mathematically fully justified methods.

During of the XX century, Optimization Methods paved a road for modern development of various areas of computer science, inverse problems, bio and life sciences gaining the attention of engineers working in the fields of industrial mathematics, emerging analytics, algorithms, and information technologies, providing a strong impact on all areas of engineering, economics, finance, and social sciences. In many real world applications practitioner faces with various challenges caused by justified model absence, noisy and incomplete data, and mathematically such problems are related to inverse ill-posed problems. In order to tackle with such challenges, the various regularization methods are usually employed. From the other hand, meta-heuristics have become very successful in solving various real world problems. The purpose of this book is to fill the gap between the model-based and model-free streams of modern industrial mathematics.

This volume applications areas covers wide area of industrial mathematics including traffic networks design, power generation and distribution, manufacturing and supply chain management and other applications. The authors of this book chapters are experienced scientists and engineers from vibrant south Asia countries, Americas and Europe.

We extend sincere appreciation and thanks to all the authors for having shared their knowledge and bright ideas with the academic and engineering community. We are grateful to the IGI Global publishing house, Editor Prof. Dr. Pandian Vasant and the Editorial Advisory Board, for having provided the opportunity for authors to publish their newest results.

Denis N. Sidorov
Energy Systems Institute of Russian Academy of Sciences, Russia

Denis N. Sidorov *(PhD 1999, DrSc 2014) is Senior Researcher at Energy Systems Institute of Russian Academy of Sciences. He was involved in European research projects at Trinity College Dublin and Université de Technologie de Compiègne. His research interests cover various areas of industrial mathematics including nonlinear dynamical systems, integral equations, DSP, forecasting and classification. He gained his industrial experience at ASTI Holdings Ltd Pte (Singapore). He also serves as associate professor of Irkutsk State University and Irkutsk National Research Technical University. Dr. Sidorov was visiting professor at Tampere University of Technology (2002), Isaac Newton Institute of Mathematical Sciences (2014), Institute of Information Theory and Automation of Czech Academy of Sciences (2008) and University of Siegen (2013).*

Preface

Modern optimization approaches have attracted many research scientists, decision makers and practicing researchers in recent years as powerful intelligent computational techniques for solving unlimited numbers of complex real-world problems, particularly related to the research area of optimization. Under the unpredictable and turbulent environment, classical and traditional approaches are struggling to obtain a holistic solution with a greater level of satisfaction for the real world application problems on optimization. Therefore, new global optimization methods are required to handle these issues seriously. One such method must be the modern optimization technique with a generic, flexible, robust, and versatile framework for solving complex problems of global optimization in real the world.

Modern optimization techniques are robust, less time-consuming, dependable, high solution quality, and efficient productive for solving the nonlinear real-world problems in engineering, technology, science and economy in an uncertain environment. The modern optimization techniques developed in this book are user-friendly, easy-to-use and can serve as a teaching and research tool, besides being useful for practicing scientists in the areas of engineering, technology and economy. In this book, the editors aim to give a valuable enrichment to state-of-the-art scientific discourse in theory, methods, and applications; to strengthen interdisciplinary ties; and to refresh the given and initiate new joint ventures in research and education, for improvements of the industries and economies of the world and for better living conditions and perspectives of its people.

This book will be very useful for the undergraduate students, postgraduate students, decision makers and researchers in private sectors, universities, and industries in the fields of various sciences and management such as mathematics/applied mathematics, physics, chemistry, computer science, management, business, economics and finance, or wherever one wants to model their uncertain practical and real life problems. It is well known that uncertainty is inevitable in every field of engineering, management and science. This book aims to become significant and very helpful for humankind.

The following research topics are well covered in this book:

- Wind turbine generator
- Supply chain management
- Gravitational search algorithm
- Silicon solar cells
- Robust Programming
- Transportation
- Unit commitment
- Single batch-processing machine

- Water cycle algorithm
- Markov decision processes
- Teaching-learning-based optimization
- Hydrothermal scheduling
- Cable industry
- Smart grid
- Antenna design
- Traffic Simulation
- Bat algorithm
- Global Positioning System

The book is organized into 32 chapters. A brief description of each of the chapters is as follows:

Chapters 1: This chapter provides an overview of different positioning technologies used in geo-location together with their limits/advantages. This chapter studies also a number of algorithms developed to estimate the position coordinates of a static or mobile user or target.

Chapter 2: This chapter is intended as an overview of robust optimization models related to optimization problems subject to uncertain data, with a special focus on a case where the uncertainty has impacts on the right-hand side coefficients in the constraints. An application to a multi-period energy production planning problem is presented into detail, and computational results are reported.

Chapter 3: The aim of this chapter is to propose a new method based on a hybrid version of the bees algorithm with slope angle computation and hill climbing algorithm to solve a multi-objective supply chain network design problem. A real case from the literature has been selected and solved in order to show the potentiality of the proposed method in solving a large-scale combinatorial problem.

Chapter 4: This chapter defines SCM, discusses TAV, reviews how transportation as well as optimization impacts of SCM and TAV, and examines the role of packaging in the context of SCM and TAV.

Chapter 5: In this chapter, it is studied the application of a genetic algorithm in the problem of variable selection for multiple linear regressions to minimize the least squares criterion. Comparisons are performed on benchmark data sets with the obtained satisfying and promising results.

Chapter 6: This chapter focuses on generating test electrical systems through fat tail model for unit commitment problem in electrical power systems. Usually, the instances of test systems in unit commitment are generated using the normal distribution. However, the simulation data is based on alpha-stable distribution in this study. Numerical results illustrate the applicability of the proposed method.

Chapter 7: The authors have developed a model based on the Markov decision process to study the dynamics of the trajectory of end products in a manufacturing line that works by process. Then, they have proposed a measure of the average production rate of the line by using the Monte Carlo method. They illustrate their proposal using a numerical example with real data obtained in situ.

Chapter 8: This chapter addresses the family of problems known in the literature as Capacitated Vehicle Routing Problems (CVRP). A procedure is introduced for the optimization of a version of the generic CVRP. It generates feasible clusters and, in a first step, yields a coding of their ordering. The next stage provides this information to a genetic algorithm for its optimization.

Chapter 9: In this chapter, the artificial bee colony (ABC) algorithm and its variants are presented and applied to the spiral antennas design for RFID tag application at the UHF band. The algorithms' performance is compared with other popular evolutionary algorithms. The obtained optimal results have shown that the ABC algorithms are a powerful tool that can be efficiently applied to antenna design problems.

Chapter 10: In this chapter, the application of a nature-inspired technique in conjunction with simulation models to optimize the siting of concentration nodes in a traffic network of urban area is presented. The popular simulator of urban mobility (SUMO) is used as traffic simulator in order to evaluate the obtained solutions. This chapter contains the conceptual models and the optimal results by the proposed simulation technique.

Chapter 11: This chapter proposes a hybrid cuckoo search algorithm to determine optimal location and sizing of static VAR compensator (SVC). The hybrid cuckoo search algorithm is a simple combination of the cuckoo search algorithm (CSA) and teaching-learning-based optimization (TLBO) where the learner phase of TLBO is added to improve the performance of the cuckoo search algorithm. Numerical results from three various tested systems have shown that the proposed method is better than the conventional CSA and TLBO in finding the global optimum solutions with higher performance than others.

Chapter 12: This chapter provides a mixed-integer linear programming model based on the continuous time representation for a case study on the scheduling problem of the cable industry to minimize the total cost including setup cost, operating cost, and inventory holding cost. A metaheuristic based on a simulated annealing algorithm is applied to minimize the total cost of proposed solutions. Finally, the solution with the smallest total cost is selected as the production schedule of the study case.

Chapter 13: The purpose of this chapter is to present a set of algorithms and their efficiency for the consistency based multiple sequence alignment (MSA) problem. The proposed GA (GA_TS_VS) also uses TS but it works with partitions. In conclusion, GAs is one of the stochastic approaches those are proved very beneficial for MSA in terms of their performance.

Chapter 14: This chapter proposes a cuckoo search algorithm (CSA) and a modified cuckoo search algorithm (MCSA) for solving short-term hydrothermal scheduling (ST-HTS) problem. The CSA method is a new meta-heuristic algorithm inspired from the obligate brood parasitism of some cuckoo species by laying their eggs in the nests of other host birds of other species for solving optimization problems. The robustness and effectiveness of the CSA and MCSA are tested on several systems with different objective functions of thermal units. The results obtained by the CSA and MCSA are analyzed and compared to those from other methods. The result comparison has shown that the two methods are favorable for solving short-term hydrothermal scheduling problems.

Chapter 15: In this chapter, a general strategy is recommended to solve variety of linear and nonlinear ordinary differential equations (ODEs) with boundary value conditions. The optimization task is carried out using different optimizers including the particle swarm optimization and the water cycle algorithm. The obtained optimal results have shown that the proposed method equipped with meta-heuristic algorithms can be successfully applied for approximate solving of different types of ODEs.

Chapter 16: Recently, renewable power sources such as wind turbine generator (WTG) and photovoltaic (PV) have become viable economic options for generating sustainable energy. However, WTG and PV have an inconstant power production problem. To solve this problem, multi-state models have been proposed in this chapter. The electricity generated from these units varies with different time scales: hourly, daily and seasonally. As a result, the system is constructed to have sufficient capacity, and the utility cost has become the main objective of the total service costs.

Chapter 17: In this chapter, some bio-inspired algorithms such as bat, artificial bee colony, black hole and firefly algorithm have been proposed for the cause of software modularization. The hybrid of these algorithms with crossover and mutation operators of genetic algorithm has also been proposed. All the algorithms along with their hybrids are tested on seven benchmark open source software systems. It has been evaluated from the obtained results that the hybrid of these algorithms has proved to find better optimal solution than the existing genetic and hill climbing approaches.

Chapter 18: In this chapter, the authors applied a hybrid particle swarm and gravitational search optimization (PSOGSA) technique for intelligently allocating energy to the PHEVs considering constraints such as energy price, remaining battery capacity, and remaining charging time. The computational experiment results have attained for maximizing the highly non-linear fitness function estimates the performance measure of both the techniques in terms of best fitness value and computation time.

Chapter 19: A project where activity times are characterized by intervals or triangular fuzzy numbers is considered in this chapter. Ranking methods based on some fuzzy mathematical techniques are discussed and two algorithms for finding the critical path for such a project are given.

Chapter 20: In this chapter, the multiobjective (MO) optimization of the extraction process of bioactive compounds from the Gardenia Jasminoides Ellis fruit was solved. Three swarm-based algorithms have been applied in conjunction with normal-boundary intersection (NBI) method to solve this MO problem. The gravitational search algorithm (GSA) and the particle swarm optimization (PSO) technique were implemented in this work.

Chapter 21: In this chapter, the authors give an overview of iterative methods for linear complementarity problems (LCPs). They also introduce these iterative methods for the problems based on fixed-point principle. They has presented some new properties of preconditioned iterative methods for solving the LCPs. Convergence results of the sequence generated by these methods have been compared to classic Gauss-Seidel method and preconditioned Gauss-Seidel (PGS) method for the LCPs under certain conditions.

Chapter 22: In this chapter, the author performs a numerical technique based on genetic algorithms (GAs) to identify the electrical parameters (Is, Iph, Rs, Rsh, and n) of photovoltaic (PV) solar cells and modules. The GAs approach has been used as a numerical technique in order to overcome problems involved in the local minima in the case optimization criteria. Consequently, the simulated data with the extracted parameters of method based on GAs is very good agreement with the experimental data in all cases.

Chapter 23: In this chapter, two approaches are developed to solve the integrated production planning and maintenance problem. Computational experiments are carried out to show the results obtained by the authors' approaches and compared to those of commercial solvers.

Chapter 24: The aim of this chapter is to investigate the multiple attribute decision making problems to select projects with generalized intuitionistic fuzzy information, in which the information about weights is completely known and the attribute values are taken from the generalized intuitionistic fuzzy environment. In addition, the concept of possibility degree of generalized intuitionistic fuzzy numbers is obtained and used to solve ranking alternative in multi-attribute decision making problems.

Chapter 25: The optimization of the maintenance decision making can be defined as an attempt to resolve the conflicts of decision situation in such a way that the variable under the control of the decision maker takes their best possible value. Thus, the main purpose of this chapter is to present a technique for finding the optimal maintenance interval for the system by considering the multiple goals of the organization viz. maximum availability and minimum maintenance cost.

Chapter 26: The adaptive cuckoo search algorithm (ACSA) with the effort in the aspect of integrating an adaptive search strategy is presented in this chapter. Its beneficial potential is analyzed in the benchmark test functions as well as engineering optimization problems. The obtained results have showed that the proposed ACSA is more efficient the classical CSA.

Chapter 27: This book chapter presents a detailed review of the basic concepts of gravitational search algorithm (GSA) and a comprehensive survey of its advanced versions. The authors have proposed a number of suggestions to the GSA community that it can be undertaken to help moving the area forward to the next level of research enlightenment.

Chapter 28: This chapter focuses on a problem of minimizing total weighted tardiness of jobs in a real-world single batch-processing machine (SBPM) scheduling in existence of fuzzy due date. The authors have designed two hybrid metaheuristics called GA-VNS and VNS-SA applying to the advantages of genetic algorithm (GA), variable neighborhood search (VNS) and simulated annealing (SA) frameworks. The computational results on different-scale test problems are presented to compare to the proposed algorithms.

Chapter 29: This chapter considers issues in the design and construction of a fuzzy logic system to model complex (nonlinear) systems. Several important applications are considered and methods for the decomposition of complex systems into hierarchical and multi-layered fuzzy logic sub-systems are proposed. The real world application problems such as the prediction of interest rate, hierarchical control of the inverted pendulum, robot control, feedback boundary control for a distributed optimal control system and image processing were considered in this chapter.

Chapter 30: In this chapter, the authors attempt to develop an effective approach based on Taylor theorem to obtain an approximate solution for the non-linear BLPP. In this approach using the Karush-Kuhn–Tucker, the BLPP has been converted to a non-smooth single problem, and then it is smoothed by the Fischer – Burmeister function. The presented approach achieves an efficient and feasible solution in an appropriate time which is evaluated by comparing to references and test problems.

Chapter 31: This chapter provides an interesting topic on "sustainable technologies for a greener world". The details about what each technology is and how it achieves green goals are presented.

Chapter 32: This chapter addresses the problem of time-varying nonlinear modeling and monitoring of a continuously stirred tank reactor (CSTR) process using state estimation techniques. The estimation performance is evaluated on a synthetic example in terms of estimation accuracy, root mean square error and execution times.

This book will be a wonderful reference for the global research scholars across the planet in the research areas of soft computing, computational intelligence, computer science, engineering, management, finance, economics, medicine, and hotel industry.

Firstly, we would like to thank to all the authors from Malaysia, UK, Vietnam, France, USA, Columbia, Argentina, Mexico, South Korea, Turkey, Greece, Japan, Australia, Qatar, Algeria, India and Iran for their valuable contribution to the book in submitting their book chapters. Secondly, thanks to all the referees' for their valuable time and great effort in reviewing all the book chapters. Lastly but not least, we also thank to the authorities of Universiti Teknologi PETRONAS (UTP), Middle East Technical University (METU) and Ho Chi Minh City University of Technology (HCMCUT) for their marvelous support and encouragement in making this book project very successful.

Pandian Vasant
Universiti Teknologi PETRONAS, Malaysia

Gerhard-Wilhelm Weber
Middle East Technical University, Turkey

Vo Ngoc Dieu
Ho Chi Minh City University of Technology, Vietnam

Acknowledgment

The book editors are very grateful to the editorial team and entire staff of *IGI Global*, for their confidence, interest, continuous guidance and support at all levels of preparation of this book's preparation.

We Editors wish all the readers a pleasant and enjoyable, insightful and inspiring lecture of the contributions of this *IGI Global* book. In fact, we cordially hope that this special issue will present and value *IGI Global* as a premium publishing house in science, engineering, economics and finance, which strongly fosters very much needed intellectual advances and their contributions to humanity and mankind in all over the world.

Pandian Vasant
Universiti Teknologi PETRONAS, Malaysia

Gerhard-Wilhelm Weber
Middle East Technical University, Turkey

Vo Ngoc Dieu
Ho Chi Minh City University of Technology, Vietnam

Chapter 1
Optimization Algorithms in Local and Global Positioning

Jean-Philippe Montillet
Central Washington University, USA

Lukasz Kosma Bonenberg
The University of Nottingham, UK

Kegen Yu
Wuhan University, China

Gethin Wyn Roberts
The University of Nottingham – Ningbo, China

ABSTRACT

With the rise of large city and the need of large civil engineering structures and city planning, surveying industry improves continuously their instruments/software in order to get cm accuracy position anywhere. Moreover, since the boom of mobile phones in the late 90s, location has become very valuable information for security, emergency and commercial applications. Depending of the application, the location technologies vary based on the accuracy of the location and the price of the system, which delivers the location information to the user. For outdoor applications, Global Navigation Satellite System is the main candidate, whereas if the user/mobile node is indoors or in a narrow street other technologies will be preferred such as the ones based on Wi-Fi or radio-frequency signal.This chapter provides an overview of different positioning technologies used in geo-location together with their limits/advantages. This chapter studies also a number of algorithms developed to estimate the position coordinates of a static or mobile user or target.

INTRODUCTION

Location and tracking plays an increasingly important role in civilian, military applications and services (e.g. location based services, surveying, and car navigation). Global Navigation Satellite Systems (GNSS) typically can provide satisfactory positioning, navigation and timing services in an open outdoor environment. However, conventional GNSS receivers do not work well in an indoor environment (e.g. office and warehouse) or in a built-up area (e.g. urban canyons) because the satellite signals are blocked by roofs, walls or nearby buildings. To overcome this indoor positioning problem a number of ground-based positioning systems have been developed by researchers and engineers such as those based on Wi-Fi, ultra-wideband (UWB), radio frequency identification (RFID), and inertial measurement unit

DOI: 10.4018/978-1-4666-9644-0.ch001

(IMU). For example, some systems are dedicated to locate a user in a building at a particular floor and room (e.g. PLUS - Precision Location Ultra-wideband System, MIT Cricket indoor positioning system - e.g. (Kolodziej & Hjelm, 2006)). Also, to enhance satellite-based positioning performance in harsh propagation environments, assisted-GNSS, RTK GNSS, and data fusion methods can be employed. In particular, many optimization algorithms have been used to solve local and global positioning problems to achieve superior positioning performance in severe multipath and Non-Line-Of-Sight (NLOS) environments (Wang, Groves & Ziebart, 2013).

This chapter deals with both local and global positioning, especially with a focus on algorithms. Since there are so many positioning and localization algorithms reported in the literature, either iterative or non-iterative algorithms (Yu et al., 2006), it is impractical to provide a thorough study on this issue. In particular, mathematical descriptions are provided for a number of algorithms including the optimization theory for localization in centralized networks (e.g. Levenberg-Marquadt, Broyden family of algorithms), convex optimization in sensor ad-hoc networks, or the (nonlinear) least-squares theory for satellite positioning. Note that interested readers in localization, navigation and tracking can refer to Kolodziej and Hjelm (2006) and Yu, Sharp and Guo (2009) for more background information related to local and global positioning.

The remainder of the chapter is organized in sections, providing a short summary on the history of radio positioning and its importance. Section II is a brief introduction to local and global positioning. Section III deals with satellite positioning system and the challenges in outdoor applications (i.e. urban canyons). It also includes a comprehensive description of the theory of adjustment (i.e. least-squares), which is used to triangulate the position of GNSS receivers. The section also discusses the models used to compensate tropospheric and ionospheric delays. Latest advances include the terrestrial based positioning systems (e.g. Locata), which is currently a hot research topic for the integration with GNSS receivers in order to solve various challenges (i.e. lack of satellite availability in urban canyons). Local positioning is discussed in Section IV. Several algorithms are reviewed using least-squares, optimization theory methods and data fusion approach with a special emphasis in the NLOS scenario. This scenario is a major challenge for office localization and more generally indoor applications (i.e. consumer applications in shopping malls, locating goods in warehouses). This chapter ends with the application of adaptive filters, which are suited to integrate multiple technologies (measurements) for positioning. The role of the stochastic models is emphasized (e.g. fractional Brownian motion) in order to integrate different time series of receivers' position in multi-sensors application.

BACKGROUND

Positioning and localization has been an active and important area of research especially in electrical engineering, computer engineering and civil engineering with a particular emphasis in surveying. Galileo Galilei's work of triangulating the position of a ship using the positions of the stars in the 17th century can be seen as a pioneering work in this area (Singer, 1941). Large efforts were focused on emerging radio navigation with the first antennas manufactured in the late 1800. Subsequent improvements in radio communications and the discovery of radar systems in the 1930's accentuated further the need to locate ships and planes (Balanis, 2012). In 1978, a second revolution started with the advent of Global Positioning System (GPS) motivated by the US army to get a position of military vehicles and ships anywhere at anytime on Earth thanks to the emerging satellite technology (Strang & Borre, 1997).

Since the end of the 1990s with the explosion of new technologies and the wireless consumer market, a third revolution is now driving the research work. Localization is not only important for military applications or maritime navigation, but various markets have also emerged with many new applications such as using localization for game purposes via mobile phone positioning, tracking object in a warehouse using Radio Frequency IDentification (RFID) or localizing victims of accidents for emergency services (Kolodziej & Hjelm, 2006). All these applications belong to the topic of Location Based Services (LBS). At the software level, many positioning algorithms have been developed based on the requirements of users or the potential targeted markets. For example, tracking objects in a warehouse can be generally realized using RFID technology. However, if an ad-hoc sensor network is utilized for localizing people in an office or players in a sky tunnel, Ultra-Wide Band (UWB) technology is one possible candidate (Yu et al., 2006). Moreover, applications using localization of mobile phones generally use cell-ID or triangulation of GSM signals via multiple base stations. Note that some applications are just using the position estimated with the GPS embedded in devices such as the mobile phone. The number of positioning systems developed in the last 20 years is tremendous and generally gathered under the name of ubiquitous positioning (Meng et al., 2007). All these applications can be categorized as local positioning applications as the infrastructures or network requested for positioning is generally installed in the vicinity of the user. Here, "vicinity" means the distance between the user and a sensor is less than a few kilometers.

On the other hand, there are many applications using GPS. One in particular is surveying in civil engineering. It includes a variety of applications such as building deformation, open-pit mining, or assets monitoring. By setting up permanent GPS stations with geodetic grade of GPS receivers, one can monitor up to a few millimeters level displacement of the ground due to geophysical phenomena (e.g. land-slide, tectonic movement), which requires accuracy better than one centimeter (Strang & Borre, 1997). To obtain such accuracy, the GPS stations have to be well calibrated (i.e. taking care of ionospheric and tropospheric delays, multipath interference). Note that these receivers are also used in assisting airport control for the landing and takeoff of planes. Generally, the number of satellites in view with the GPS receivers (on board of the plane) is augmented (artificially) with Pseudolites for a better coverage and thus a better accuracy of the plane location. The idea of augmenting artificially the number of satellites in view with the receiver to improve the accuracy of the rover/target has been reused to solve the challenge of precise positioning in urban canyons. Urban canyons localization is one of the hot topics of the 21st century in the location and tracking research when using GPS and other satellite constellations (i.e. GLONASS). With more and more skyscrapers in cities around the world, many streets are constrained with "limited access to the sky" creating many challenges (i.e. signal acquisition, robust multipath, triangulation) not only in Location Based Services (LBS), but also for surveying applications such as building monitoring and precise localization of assets (i.e. buried pipes in cities) (Montillet et al., 2007). These challenges have been addressed differently depending of the accuracy requested for the application. In the LBS market, the accuracy for the user is better than meter and thus using a combination of various technologies (i.e. GPS + Wi-Fi) may be a solution (Yu et al., 2009). However, surveying applications need an accuracy of one centimeter for mapping assets and even better in monitoring building deformation. In the last 15 years, one evolution of Pseudolites technology has been the development of Locata technology. It was shown that Locata technology is a possible solution to solve the challenges of surveying in urban canyons (Montillet et al., 2009).

Last but not least, the latest research in global positioning is the integration of GPS with other concurring constellations such as the Russian GLONASS, the European Galileo and the Chinese BeiDou.

This multiple Global Navigation Satellite System (or multi GNSS) can help to solve some challenges in places where multiple constellations are available (Montillet & Yu, 2012; Hofmann-Wellenhof, Lichtenegger & Collins, 1993). Specifically, the Japanese Quazi Zenith Satellite System (QZSS), as well as the Chinese BeiDou system incorporate inclined geospatial satellite orbits in order to overcome the lack of satellites in the zenith. The BeidDou system also includes GEostationary Orbits (GEO), in addition to Medium Earth Orbits (MEO). In these parts of the world, there are additional positioning satellites at very high inclination angles, meaning that in built up cities there is more likelihood of being able to gain a positional fix. Currently, QZSS is transmitting on one satellite, and it is planned that QZSS will have four satellites in orbit. BeiDou currently has 6 GEO and 5 IGSO satellites in its constellation, all based nominally over China and the surrounding area.

POSITION ESTIMATION IN GLOBAL POSITIONING SYSTEM

1. Global Positioning: Overview and Characteristics

Satellite-based positioning was an idea first emerged in the 1960s. GPS was designed as a dual-use system, with the primary purpose of meeting the military's needs for positioning and timing. The network is owned and operated by the U.S. government. The first generation of satellite (Block I) was designed with life-time expectancy of 8-9 years. These were experimental satellites, used to prove the concepts of GPS. 11 satellites were launched between 1978 and 1985 into three 63 degrees inclined orbital planes. With the evolution of this technology, more satellites (e.g. Block IIA, IIR (R means replenishment), IIF (F denotes follow on) and III (future generation of GPS satellites with first one expected to be launched in April 2016)) have been launched to orbit around the earth to cope with the needs of the market and new features have been added as the technology evolves constantly (e.g. more reliable and more precise navigation solution) (Hofmann-Wellenhof et al., 1993). After the first generation, GPS satellites orbit at 55 degrees with the equator line (US Army, 1996).

GPS consists of three major segments: the Control Segment, the Space Segment and the User Segment. The Control Segment consists of five ground based control stations equipped to monitor the satellites and update the information transmitted by the satellites. The Space Segment consists of the satellites themselves and the User Segment comprises an unlimited number of users equipped with GPS receivers capable of tracking the signals transmitted by the satellites (UoN, 2014).

The signals transmitted by the satellites consist of two carrier waves (L1 and L2), which are modulated with timing codes (C/A and P). The C/A code is only modulated on the L1 carrier wave at 1.57542 GHz. The P code is modulated on both L1 and L2 carrier waves. L2 carrier wave has a center frequency of 1.2276 GHz. The chipping rate of the C/A code is 1.023 MHz with a period of 1 ms, whereas P(Y) code is 10.23 MHz with a period of 38 weeks. It is implied that the length of the C/A and P codes are respectively 293 m and 29.3m. The accuracy of the GPS technology comes from the atomic clocks embedded in the satellites and transmitting continuously the codes on L1 and L2 frequencies (US Army, 1996). The clocks may be considered as perfect in terms of clock drift, but on the surface of the earth the GPS receiver uses a much inaccurate clock than the one embedded in a satellite for navigation. Thus the clock drift will depend on the grade/price of the product. For example, it is known that the accuracy of the position calculated using a GPS hand-held receiver is around 10 m in an open area, whereas with a GPS device used for Geodesy applications the accuracy may be better than 5 cm depending on the

Figure 1. Overview of GPS segments (Montillet, 2008)

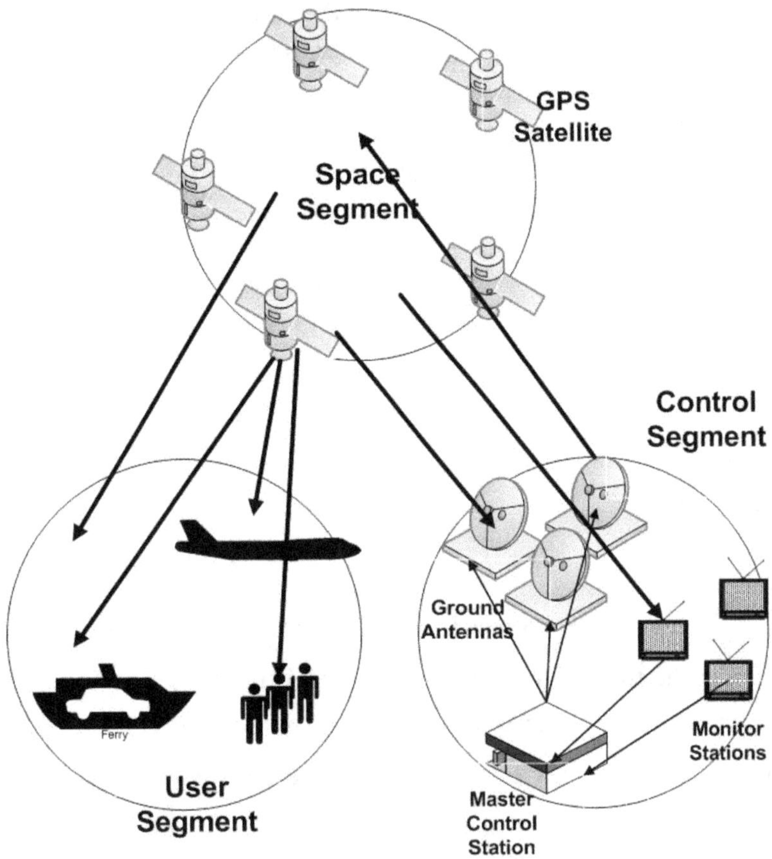

measurement scenario (Schwieger, 2003; Leick, 2004). Some new signals are already on test and transmitted by the latest version of GPS satellites (i.e. L5 signal with a chipping rate of 1.17645 GHz). These additional signals will help to improve the anti-jamming capability, the protection against anti-spoofing, other services (i.e. safety signal, resistance to interferences), and the estimation of the ionosphere corrections (Hofmann-Wellenhof et al., 1993; Montillet, 2008).

2. Theory of Satellite Positioning

a. Measurement Model

Pseudoranges and carrier-phase measurements are the two types of GPS observables used for positioning. They can be used together or separately to compute the position of the receiver. As previously underlined, the carrier-phase observations are used for precise positioning at the centimetre level accuracy. During the last three decades, processing techniques have been developed to constantly improve the navigation solution using those observables. An example is the double and triple difference algorithms developed to detect some issues related to the carrier-phase measurements such as cycle slips and integer ambiguity resolution (Hofmann-Wellenhof et al., 1993). Moreover, GPS receiver architecture includes a great deal of advanced signal processing hardware and software to extract the observables from the satellites' signals. It is well known that the signal power transmitted from the satellites is severely attenuated (-135dBm), when travelling through the different atmospheric layers of the Earth (e.g. Troposphere and Ionosphere).

i. Pseudoranges Ranging

Let us consider positioning based only on the Pseudoranges. According to (Strang & Borre, 1997), the Pseudoranges between the i-th GPS satellite and the ground receiver at the observation time t can be described

$$P_r^{(i)}(t) = \rho_r^{(i)}(t) + c\delta t_r(t) - c\delta t^{(i)}\left(t - \tau_r^{(i)}(t)\right) + I_r^{(i)}(t) + T_r^{(i)}(t) + \varepsilon_{Pr}^{(i)}(t) \tag{1}$$

Where

$$\rho_r^{(i)}(t) = \left\| x^{(i)}\left(t - \tau_r^{(i)}(t)\right) - x_r(t) \right\|$$

$\rho_r^{(i)}(t) = x^{(i)}\left(t - \tau_r^{(i)}(t)\right) - \boldsymbol{x}_r(t)$ is the geometric range between a receiver and a satellite; $x^{(i)}\left(t - \tau_r^{(i)}(t)\right)$ is the satellite position at the time of transmission; $\tau_r^{(i)}(t)$ is the time duration of signal propagation; $x_r(t)$ is the receiver position; **c** is the speed of light in vacuum; $\delta t_r(t)$ is the receiver clock error; $\delta t^{(i)}\left(t - \tau_r^{(i)}(t)\right)$ is the satellite clock error; $I_r^{(i)}(t)$ is the ionospheric delay; $T_r^{(i)}(t)$ is the tropospheric delay; and $\varepsilon_{\mathrm{Pr}}^{(i)}(t)$ is the unmodeled errors including measurement noise and multipath effect. The delays due to Ionosphere and Troposphere are corrected in the observations using well-known models such as the Klobuchar model and the Ifadis mapping function (wet tropospheric delay, see (Hofmann-Wellenhof et al., 1993)) and Saastomoinen (dry tropospheric delay) model (Saastomoinen, 1973). After compensating the tropospheric and ionosperic delays and ignoring the unmodeled errors, the Pseudoranges model from Equation 1 is simplified as

$$P_r^{(i)} \approx \rho_r^{(i)} + c\delta t^{(i)}, i = 1,2,\ldots,m \tag{2}$$

where it is assumed that the ground receiver can detect signals transmitted from m GPS satellites. However, with single-frequency receiver it is not possible to mitigate completely the tropospheric and ionospheric delays. These additional errors propagate in the estimation of the receiver coordinates, which ends up in the accuracy delivered by the receiver (i.e. 10-20 m for a cheap hand-held receiver).

ii. Carrier-Phase Measurements Ranging

Carrier-phase measurements (CP) are the second type of observable provided by the GPS technology. This observable contains less noise than the Pseudoranges, so that it is possible to locate a GPS receiver with a higher precision. When calculating the carrier-phase measurement, the order of magnitude of the accuracy is about 0.01 cycles, which gives the receiver's position at the millimetre level. The equation of the carrier-phase signal is described in:

$$\varphi_r^{(i)}(t)=\frac{1}{\lambda}\left(\rho_r^{(i)}(t)+c\delta t_r(t)-c\delta t^{(i)}\left(t-\tau_r^{(i)}(t)\right)+I_r^{(i)}(t)+T_r^{(i)}(t)\right)+N_r^{(i)}+\varepsilon_{\varphi r}^{(i)}(t) \tag{3}$$

With $\varphi_r^{(i)}(t)$ the carrier-phase measurement in cycles at the epoch t; $\delta t_r(t)$, $I_r^{(i)}(t)$ and $T_r^{(i)}(t)$ are respectively the receiver's clock drift, the ionospheric and tropospheric delays. $\varepsilon_{\varphi r}^{(i)}(t)$ is the carrier-phase signal error in cycles. It is common knowledge that the signal transmitted from the i-th satellite is reflected on the ground or on the scatterers before the GPS's antenna received it, which results in some shadowing and multipath effects (Strang & Borre, 1997). The error $\varepsilon_{\varphi r}^{(i)}(t)$ is much smaller than $\varepsilon_{\varphi r}^{(i)}(t)$, which is one reason why carrier-phase measurement is more accurate than Pseudoranges. Finally $N_r^{(i)}$ is the integer ambiguity between the receiver and the i-th satellite. When a GPS receiver locks onto a given satellite's signal, it is then possible to measure how many cycles there are between the receiver and the satellite. The ambiguity value is the unknown number of cycles accumulated before the receiver locked the signal. This number should remain constant through the measurements until the receiver losses to the satellite's signal. To calculate this unknown variable, techniques differ depending on whether the GPS receiver is single or dual frequency (i.e. Stop-and-Go technique for single frequency receiver, wide lane technique using linear combinations of L1 and L2 frequency, Three-Carrier Ambiguity Resolution (TCAR)) (Hofmann-Wellenhof et al., 1993). Finally, the ambiguities must be converted from float to integer. Several techniques have been developed the past two decades such as On-The-Fly (OTF) ambiguity (Roberts, 1997) and the Least-Squares AMBiguity Decorrelation Adjustment (LAMBDA) (Teunissen, 1995). Note that it has been recently shown that fractional ambiguity can also be estimated for rapid integer ambiguity resolution (Geng et al., 2012).

b. The Linear LS Iterative Estimator in Stand-Alone Positioning (Pseudoranges)

The geometric distance can be expressed in terms of the satellite position (x_i, y_i, z_i) and the receiver position (x_r, y_r, z_r) as

$$\rho_r^{(i)}=\sqrt{(x_r-x_i)^2+(y_r-y_i)^2+(z_r-z_i)^2} \tag{4}$$

By expanding the above geometric distance in a Taylor series around a position (x_0, y_0, z_0), which is in the vicinity of the receiver position, the Pseudoranges displayed in Equation 2 becomes

$$P_r^{(i)}-\rho_0^{(i)}\approx-\frac{x_0-x_i}{\rho_0^{(i)}}(x_r-x_0)-\frac{y_0-y_i}{\rho_0^{(i)}}(y_r-y_0)-\frac{z_0-z_i}{\rho_0^{(i)}}(z_r-z_0)+c\delta t_r-c\delta t^{(i)} \tag{5}$$

where $\rho_0^{(i)}$ is the Euclidean (or geometric) distance between the i-th satellite and the position point (x_0, y_0, z_0), which is calculated with respect to the approximate values of receiver position and receiver clock error (Strang & Borre, 1997). Equation 5 can be rewritten as $\Delta\rho_i \approx -\frac{x_0-x_i}{\rho_0^{(i)}}\Delta x-\frac{y_0 y_i}{\rho_0^{(i)}}\Delta y-\frac{z_0-z_i}{\rho_0^{(i)}}\Delta z+c\delta t_r$ (6)where $\Delta\rho_i \approx P_r^{(i)}-\rho_0^{(i)}+c\delta t_r$ is the difference between the observed and estimated Pseudoranges, subtracted by the range offset induced by the satellite clock error; and $\Delta x = x_r - x_0$, $\Delta y = y_r - y_0$, and $\Delta z = z_r - z_0$. Note that GPS satellite clock error can be estimated using a model as described in Montillet and Yu (2012). Clearly, the system described by Equation 6 is resolvable provided that m is greater than 3. Equation 5 can be written in a compact form as

$$\Delta \boldsymbol{\rho} \approx \boldsymbol{A\delta} \tag{7}$$

where

$$\Delta \rho = \begin{bmatrix} \Delta\rho_1 \\ \vdots \\ \Delta\rho_m \end{bmatrix}, A = \begin{bmatrix} \frac{x_1 - x_0}{\rho_0^{(1)}} & \frac{y_1 - y_0}{\rho_0^{(1)}} & \frac{z_1 - z_0}{\rho_0^{(1)}} & c \\ \frac{x_2 - x_0}{\rho_0^{(2)}} & \frac{y_2 - y_0}{\rho_0^{(2)}} & \frac{z_2 - z_0}{\rho_0^{(2)}} & c \\ \vdots & \vdots & \vdots & \vdots \\ \frac{x_N - x_0}{\rho_0^{(m)}} & \frac{y_N - y_0}{\rho_0^{(m)}} & \frac{z_N - z_0}{\rho_0^{(m)}} & c \end{bmatrix}, \delta = \begin{bmatrix} \Delta x \\ \Delta y \\ \Delta z \\ \delta t_r \end{bmatrix} \tag{8}$$

Clearly, there are four unknown parameters in the linear system as described by Equations 7 and 8. Thus, when the GPS receiver can detect signals transmitted from at least four GPS satellites, the system is resolvable and the well-known linear least-squares (LS) estimator can be applied to estimate the unknowns. For instance, the ordinary LS solution to Equation 7 is given by

$$\hat{\delta} \approx \left(\mathbf{A}^T\mathbf{A}\right)^{-1} \mathbf{A}^T\Delta\rho \tag{9}$$

The coordinates of the receiver position and the transmission time can be iteratively determined. Initially, the unknown parameters $\boldsymbol{\theta} = \left[x_r, y_r, z_r, t_r\right]^T$ are assigned with rough initial values. Then, the variation vector $\hat{\delta}$ is determined by Equation 9 and the parameter vector is updated accordingly. The process continues until the variations are sufficiently small. It can be seen that in Equation 9 all the measurements are treated on an equal basis. To achieve better accuracy, weighting coefficients can be applied to weight measurements of different quality. Thus, the weighted LS estimator can be applied so that Equation 9 becomes

$$\hat{\delta} \approx \left(\mathbf{A}^T\mathbf{Q}^{-1}\mathbf{A}\right)^{-1} \mathbf{A}^T\mathbf{Q}^{-1}\Delta\rho \tag{10}$$

where $\boldsymbol{Q}$ is the variance-covariance matrix of the observation $\Delta\boldsymbol{\rho}$. Accordingly, the unknown position and transmission time can be estimated iteratively until $\hat{\delta}$ is significantly small or when the number of iterations is greater than a predefined threshold. The variance-covariance matrix $\boldsymbol{Q}$ can be approximated to be a diagonal matrix as

$$\mathbf{Q} = \text{diag}\left\{q_1^2, q_2^2, \ldots, q_m^2\right\} \tag{11}$$

where the diagonal components can be modelled as the elevation angle dependent weighting function which is inversely proportional to the sinusoidal function of the elevation angle, multiplied by the predicted standard deviation of the error for a given satellite, which is also called User Range Accuracy (URA) (US Army, 1996). That is,

$$q_i = \frac{\sigma_i}{\sin \alpha_i} \tag{12}$$

Where σ_i is the URA value and α_i is the elevation angle of the i-th satellite. Note that the predicted URA values are rather imprecise and they are not available in some circumstances. This weighted linear iterative LS estimator is generally called LS estimator (Leick, 2004). The position is generally estimated in a procedure such as described in the block diagram shown in Figure 2 (Montillet & Yu, 2012). In the block diagram, there is possibility to exclude some satellites based on a chosen cut-off angle. The input data are the GPS ephemeris and the field measurement data file. The output data are the estimated parameters namely, 3D coordinates of the receiver, receiver's clock drift and the residuals. In the block diagram, the Pseudoranges compensation means compensation for the ionospheric and tropospheric delays. No differential GPS (DGPS) technology is employed to make further Pseudoranges' corrections. Finally, the position of the receiver is expressed in the World Geodetic System 84 (WGS 84). The LS algorithm is iterative, and thus the chosen criterion to stop the iteration satisfies the relation:

$$\| \tilde{\theta}_{i+1} - \tilde{\theta}_i \| \leq Th \tag{13}$$

where $\dot{\boldsymbol{\theta}}_i = \left[x_r^{(i)}, y_r^{(i)}, z_r^{(i)}\right]^T$ is the position vector estimate, at the i-th iteration, Th is a threshold level which is typically around 10^{-5}, and $||.||$ denotes the Euclidean distance. The small threshold is chosen to ensure a good convergence of the LS algorithm (Strang & Borre, 1997).

c. Positioning with Carrier-Phase Measurements

It has been shown that the use of CP measurements requires the estimation of the integer ambiguities. CP measurements are used frequently in baseline estimation between two GPS receivers. It is possible with one receiver to use CP measurements only with a dual frequency (L1 and L2) GPS receiver using Precise Point Positioning (PPP) and precise ephemerides. Alternatively, On-The-Fly (OTF) ambiguity resolution can be used, which requires either a dual frequency base station (in order to send corrections to the targeted station) or a network Real Time kinematic (i.e. SmartNet in the United Kingdom and

Figure 2. Block Diagram of the LS iterations (Montillet & Yu, 2012)

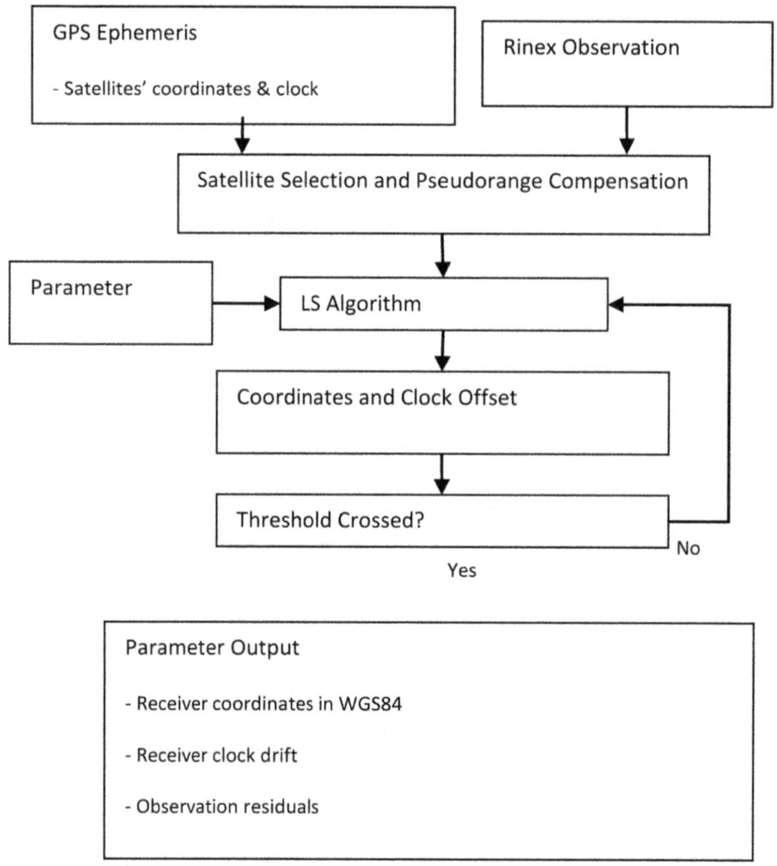

Ireland). With new frequency bands (i.e. L5, L2C), which will be soon available to users, there will be more frequency combinations. This could then speed up the time for fixing the ambiguities as integer values. With a single frequency receiver, the user can start on a known location (e.g. Stop-and-Go technique). This is generally a time-consuming process (Leick, 2004).

Depending on the baseline length, one can make hypothesis on the double difference measurements. When using double difference, one can cancel out the receiver's clock drift. For example, for short baseline (less than 10km), the Troposphere delay and Ionosphere delay are approximately the same at both receivers. Thus, one can then cancel them out by double differencing. Readers interested in this topic are referred to Hofmann-Wellenhof, Lichtenegger and Collins (1993), Strang and Borre (1997) and Bonenberg (2014) for more detailed information.

3. GPS Limitations: Challenges and Possible Solutions

To recall the previous section, a wireless node (rover or user) is able to get its location if enough satellites are available to do the triangulation. In deep urban canyons (e.g. between two skyscrapers), the line-of-sight is often blocked between the satellite and the receiver. This scenario also occurs when users are surrounded by dense foliage (e.g. forest) or in an indoor scenario. The limitation of the number of satellites available to perform the triangulation also influences the network geometry. Thus, the fluctuation of the values for the Dilution of Precision (DOP) introduces additional bias in the estimated position (Strang & Borre, 1997). Finally, one has to take into account phenomena linked to signal propagation such as multipath, scattering and shadowing (Leick, 2004; Montillet et al., 2009).

This section underlines the ongoing work on integrating new satellite constellations with GPS, and the development of pseudo satellite technology (ground based transmitters), which are foreseen to solve partially actual challenges that GPS technology alone is facing.

a. Multi-Constellation and Measurements Integration

In the previous section, the GPS was described, which is the best-known and the most advanced one of all the GNSS. Russia's GLONASS has been fully operational since 1995, but the number of operational satellites was diminishing. Efforts have been dedicated to the system since 2001, and a restoration scheme was made priority by the Russian Government. In October 2011, the full orbital constellation was restored to 24 satellites. The development of GNSS, and future generations of GNSS has been taking place for a number of reasons, including the ownership and sovereignty of systems that are becoming key and critical to many every day activities. In addition, the number of applications using GNSS is growing at a rapid pace. This means that countries and states can launch specific satellites into orbit to enhance existing GNSS operational capability over their territory, such as Japan's QZSS. It also means that revenue can be made through ownership of specific GNSS. The next generation GNSS will include the U.S.'s modernized GPS and planned GPS-III, Russia's revitalised GLONASS, Europe's Galileo system, and China's BeiDou system. Furthermore, a number of Space Based Augmentation Systems (SBAS) and Regional Navigation Satellite Systems (RNSS) will add extra satellites and signals to the multi-constellation GNSS/RNSS 'mix'. The modernisation of such systems doesn't only include new satellites broadcasting enhanced and new signals, but it also includes improvements at the ground monitoring stations, as well as the GNSS receivers themselves. There are many challenges when integrating all these systems together (e.g. different frequencies, multiple-access protocol, and ambiguity resolution). The advantages include improved precisions and accuracies, as well as availability. The theory of least-squares adjustment will be the basis for the triangulation of the receiver. Going through a comprehensive review of this topic is not the focus of this chapter. Readers interested in multi-constellations and future positioning applications are referred to (Roberts et al., 1999; Rizos, 2008; Odolinski et al., 2014; Teunissen, Odolinski & Odijk, 2014).

b. Pseudolites and Locata Technology

i. Pseudolites: Overview and Limitations

Since the earliest days of the GPS concept with the first satellite launched in 1978, ground-based transmitters of GPS signals have been under development to complete a satellites constellation. These ground based transmitters were called Pseudo-Satellites quickly renamed as Pseudolites. They have been used to test GPS system elements and to enhance GPS in certain applications by providing better accuracy, integrity and availability through the use of Pseudolite signals in addition to the GPS signals. Figure 3 shows three Pseudolites.

Figure 3. The location of the three Pseudolites, with their antennas mounted vertically facing the receiver locations, on the footpaths alongside the River Trent (Cosser et al., 2003)

Pseudolites have been a promising technology for providing positioning in high multipath environments where GPS signals are generally unavailable or severely attenuated and of questionable quality. Thus, they are generally used for both indoor and outdoor positioning applications transmitting a GPS-like signal either augmented with the GPS satellites' constellation or in stand-alone (Cobb, 1997). Pseudolites work in an unsynchronized mode and double differencing is used to eliminate the Pseudolites internal clock bias and receiver clock bias. In the mid 1990s growing interest foresaw Pseudolites technology as the next big thing. Since then, numerous Pseudolites applications have flourished such as Local Area Augmentation Systems (LAAS) (Cobb, 1997), plane landing, and civil engineering applications (bridge deformation monitoring, open pit-mining, reducing streetworks) (Van Dierendonck, 1997; Barnes et al., 2002; Roberts et al., 2004; Cosser, 2004; Montillet, 2008; Bonenberg, 2014).

However, there are some fundamental issues that limit the effectiveness of a Pseudolite system using C/A code on L1/L2. These include the legality of transmitting on L1/L2, cross-correlation between Pseudolites and GPS signals (GPS jamming), saturation of GPS receiver front-ends, and the limited multipath mitigation offered by C/A code. When combined with other problems inherent to all Pseudolite systems such as near- far, multipath, and synchronisation, the issues in using L1/L2 C/A code Pseudolite systems further complicate the design and deployment of such systems and places limits on their operational effectiveness (Kanli, 2005). Locata was created with sole aim of solving those problems.

ii. Locata: a New Ground-based Positioning Technology

Locata's terrestrial positioning technology which has been developed since 1995 in Australia, offers cm level positioning accuracy. Similar to Pseudolite, it uses a GPS-like signal structure. Terrestrial systems need precise time synchronization, which is limited by multipath and near-far effect. Solutions to these problems set the Locata technology apart from previous technologies such as the Pseudolites. The main features are:

- Utilization of the 2.4 GHz licence-free Industrial, Scientific and Medical (ISM) radio band, preventing any transmitting restrictions but also opening system to in-band interference.
- TimeLoc procedure, allowing Locata network time synchronisation at nanosecond level.
- Multipath and near-far effect is mitigated by the use of Time and Code Division Multiple Access (TDMA/CDMA), pulsing digital signal and spatial separation of transmitting antennas.

Successful trilateration requires visibility to at least four Locata transceivers or LocataLites (e.g. Figure 4), and unlike GPS there is no need for a base station (only single differencing is required) (Montillet et al., 2009).

Locata technology can provide position on its own. Successful implementations include the partial coverage of the University of New South of Wales and the University of Nottingham for research purposes (Barnes et al., 2007; Montillet et al., 2009); White Sands, Boddington Gold Mine and Sydney Bay for the specific commercial applications (Harcombe, 2012; Craig & Corporation, 2012; Barnes et al., 2007). The transmitting range of the Locata technology is about 10km according to the ISM band

Figure 4. A Locata transceiver (or LocataLite) antenna array unit (Bonenberg, 2014)

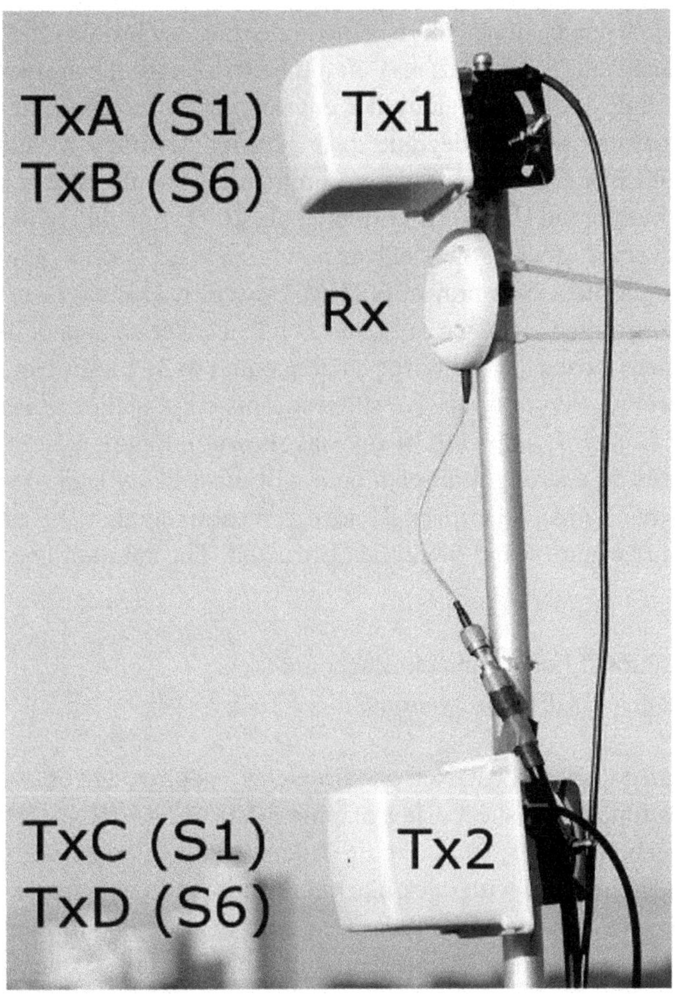

regulations. Locata was initially developed to tackle the problem of positioning in scenarios where GPS is not working properly such as in urban canyons. Technology can also be used for centimeter level indoor positioning, using dedicated antenna and positioning algorithm, described in (Rizos at al, 2010).

The deployment of the Locata transceivers (or LocataLites) should be tailored to the application and the combination of topography, buildings, landscape and other environmental issues in the survey area should be carefully considered. Those can severely degraded rover position, mostly due to scattering/diffraction and multipath of Locata signal. As rule of thumb, there should always be a good visibility between the transceivers (master and slaves) (Montillet et al., 2009; Bonenberg et al., 2012).

A network of Locata transceivers ('LocataNet') consists of two types of devices: static transceivers; transmitting and receiving signal at known fixed position; and moving receiver, known as a rover. Network transceivers consist of single master, providing a network time (or synchronization to GPS time) and number of slaves, able to maintain a constant time frame by synchronizing with the master transceivers. However, when it is not be possible, the slave can synchronize with another slave, which is also called cascade synchronization. When synchronizing with another slave, the synchronization of this transmitter with the whole network decreases. This miss synchronization can be given with the formula $\sigma_{timeloc}\sqrt{n}$, where n is number of cascades and $\sigma_{TimeLoc}$ is the synchronization time between a master and a slave. If one assumes an averaged synchronization time of 0.12 nanoseconds, no more than four hoops or cascaded TimeLoc can be done before being under the synchronization threshold. The deployment of the Locata transceivers (or LocataLites) should be tailored to the application and the combination of topography, buildings, landscape and other environmental issues in the survey area should be carefully considered. They can severely degrade rover position accuracy, mostly due to scattering/diffraction and multipath of Locata signal. As rule of thumb, there should always be a good visibility between the transceivers (master and slaves) (Montillet et al., 2009; Bonenberg et al., 2012).

From a mathematical point of view, the ideal shape for standard Locata network (with 10 units due to the limitation of the multiple access protocol (TDMA) used by Locata) is an equally spaced grid in order to offer maximum coverage as shown in Figure 5. It can cover an area of 400 km^2 if it is assumed that on average the master LocataLite can cover an area equal to 314 km^2. Note that the firmware 6.0 version allows the interconnection of up to five subnets. This combination covers an area of 3120 km^2 almost twice the size of Greater London (1570 km^2), as shown in Figure 6.

Deploying a LocataNet in a strong multipath area will most likely lead to signal obstructions due to buildings and the surroundings of the rover. Thus, it is most likely that at least one of four signals in any given square shown in Figure 5 and 6 will be obstructed. The solution is generally a combination of these two methods:

- Introduction of additional Locata units in a specific area.
- Integration of Locata and GPS observations.

To demonstrate how to deploy a LocataNet, one simulates a deployment of Locata network in Canary Wharf area of London's financial district using mapping data from EDINA (2000) and the simulator tool developed in (Bonenberg et al., 2012). Figure 7 demonstrate a proposed Locata deployment for the whole area (building are marked with thick black line and five Locata transmitters marked as green dots). For simplification one presents a worse case trajectory (a red line in Figure 7), with largest sky coverage and Locata signal obstruction. As presented in top of Figure 8, Locata on its own offer only

Figure 5. The implementation of single Locata network (Bonenberg, 2014)

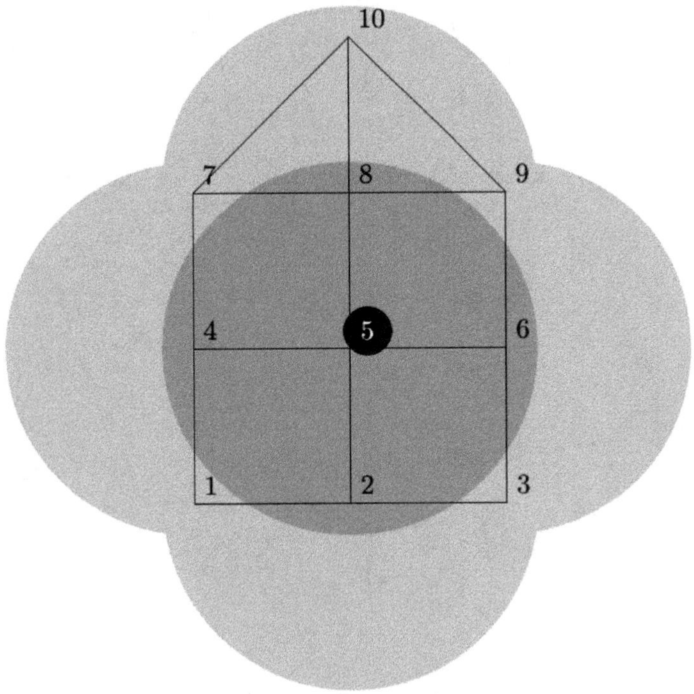

Figure 6. The implementation of five Locata sub-networks (Bonenberg, 2014)

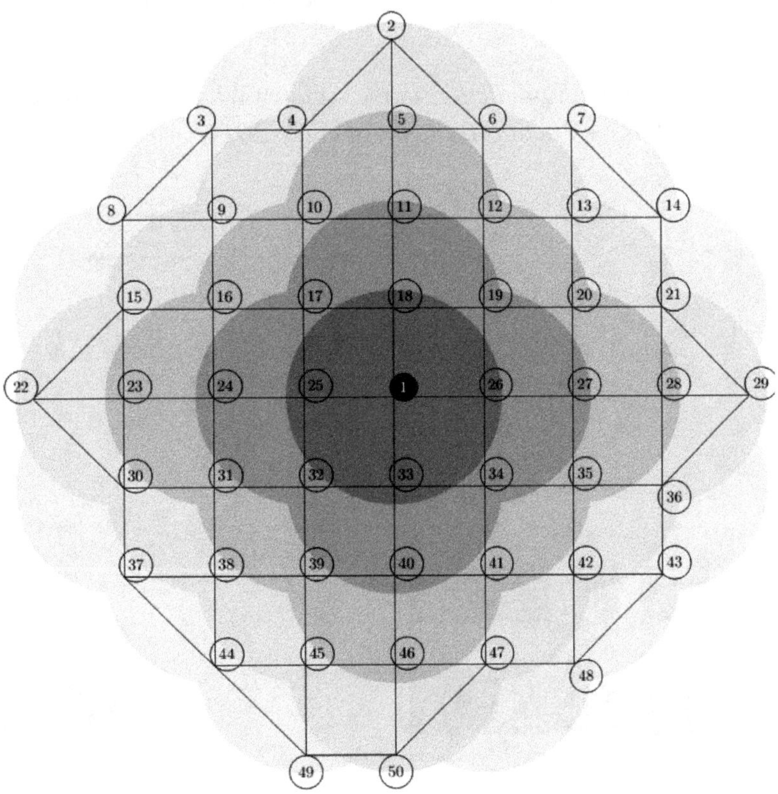

Figure 7. Canary Warf scenario (Bonenberg et al., 2012)

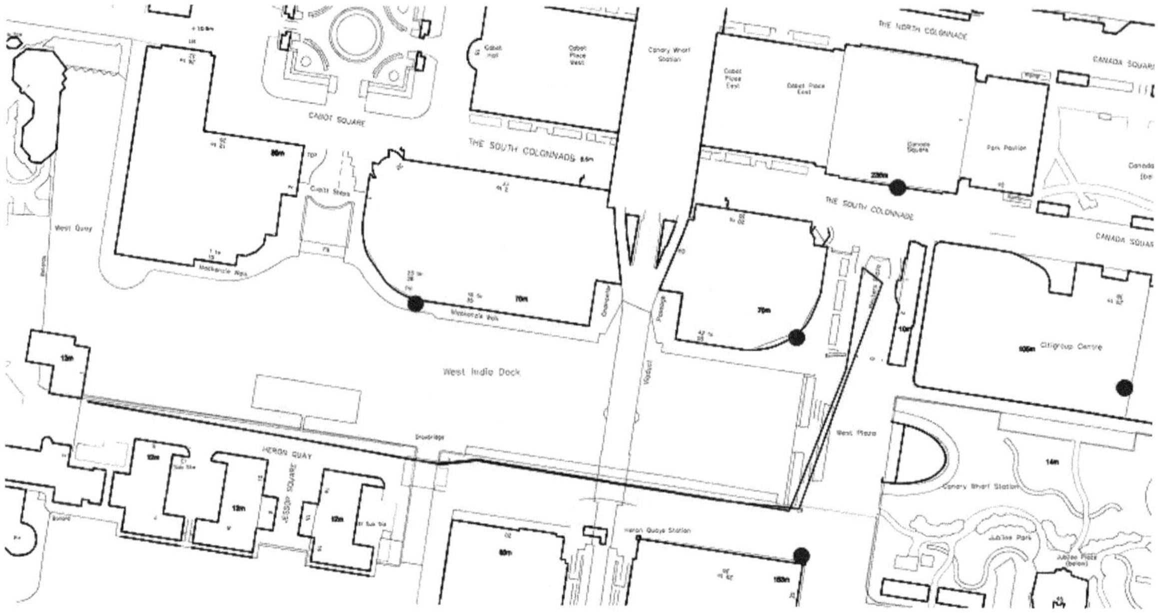

marginal improvement over GPS, with combination of both systems almost doubling the positional coverage (51% vs 30%).

Equally important is the accuracy of the position, and in an urban environment the multipath (reception of the same signal from more then one direction, due to reflection) is the largest limiting factor

Figure 8. The visibility and multipath prediction in the Canary Wharf scenario. Simulations with GPS alone and GPS, Locata and GPS + Locata (Bonenberg et al., 2014)

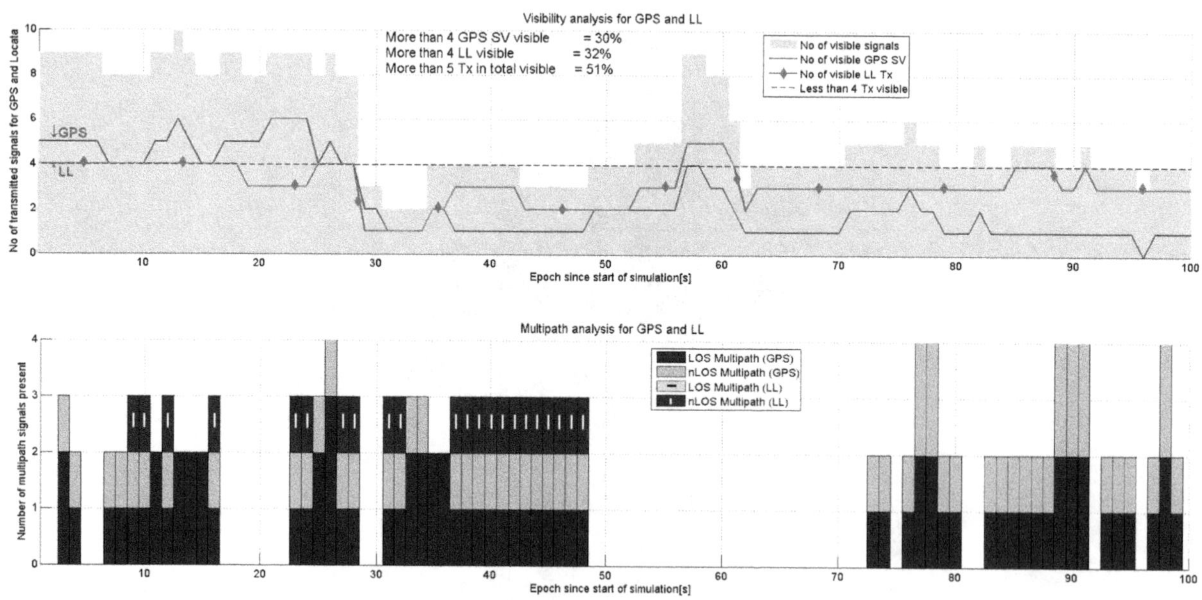

(Bonenberg at al., 2012). Two types of multipath exist – line of sight (LOS) one, when direct signal can also be recorded by the rover, and non line of sight (NLOS), when direct signal is not available. The latter can lead to big positioning errors (Wang, Groves & Ziebart, 2013).

Multipath simulation, shown in the bottom of Figure 8, shows better performances for Locata than GPS both for LOS and NLOS multipath. Locata is able to maintain better accuracy in the urban environment, and its combination with GPS would allow eliminating all NLOS due to extra observations present.

The simulation demonstrates, that even in the most problematic scenario combination of Locata and GPS can offer a large quality improvement. Further discussion can be found in Section III (bridge monitoring scenario) and in Bonenberg (2014).

LOCAL POSITIONING SYSTEMS

This section first discusses location based services for outdoor and indoor environments and indicates the related challenging problems. Next, an overview on the local positioning systems especially for indoor positioning and early pioneering systems is provided. Then, one positioning algorithm is described with details.

1. Location Based Services

With the advent of advanced technologies such as wireless networks, internet, geographical information systems (GIS) and GPS, all future mobile wireless nodes can be localized or self-localized such as using an embedded GPS receiver. User location information is important for a wide range of applications and services including advertisement, games, healthcare and security. For example, it would be possible to use the online yellow pages to know the location of a departmental store or restaurant and at the same time to get the direction to the destination based on user location. Some applications will use the location of a user to broadcast advertisements specific to the local area. These applications, which are based on the location of a user or a wireless node, define the LBS market (Abulleif & Al-Dossary, 2008). The LBS market is the sum of all the applications on top of the positioning systems and technologies.

Outdoor and indoor LBS are usually viewed as two separate domains, basically because outdoor positioning systems generally do not work well indoors. There are several factors that make outdoor LBS an ideal environment for value added application development:

- GPS works virtually everywhere outdoors. In urban environments it is enhanced with GSM and Wi-Fi positioning to provide satisfactory positioning, navigation and timing services. This means no new infrastructure is necessary for new applications to emerge.
- There is a well-defined software distribution model. All major mobile operating systems have online software marketplaces, such as Apple Store and Google Play that are instantly available to smartphones.
- There is a well-defined content generation model. Business owners are interested in their details being available to popular LBSs such as assistance for locating cafes and restaurants, or booking hotels. This means that a sufficiently popular service does not need to update their database as it will be done for them.

- Modern smartphones have high definition touchscreens and are able to deliver a large variety of multimedia content in the form of text, audio, and video.

While there are many public buildings that could greatly benefit from indoor LBS, such as museums, airports, hospitals, and universities, there are several factors that make the emergence of such services problematic. For example, to enable indoor LBS, an alternative positioning system to GPS must first be installed. This is generally an expensive investment.

Because different applications require different levels of accuracy, LBS require large infrastructure such as application servers, which connect different positioning technologies to continuously track a user. LBS components are user devices, position technologies, wireless network, application server and geographical data server. All these components communicate with each other to produce a service for the mobile user. For example, if a user dials the emergency service (e.g. E911) number, it requires wireless carriers to identify the caller's location for a specified percentage of 911 calls within a range of 50 to 150 meters for carriers that use handset-based technology (Kolodziej & Hjelm, 2006). GPS/GNSS can locate precisely a user outdoor, but in built-up areas and in indoor environments the services will require getting the location information using other positioning technologies such as those to be discussed next.

2. Overview of Local Positioning System

Various wireless technologies are used for indoor location. They may be classified based on: 1) the measurement of signal parameter such as Time of Arrival (ToA), Time Difference of Arrival (TDoA), Received Signal Strength (RSS), Angle of Arrival (AoA), or a combination of these; 2) the physical layer or location sensor infrastructure (i.e. the wireless technology/network used to enable communication between a mobile device and a base station or between two mobile devices). Medium to long range communication networks include GSM, 3G, and 4G cellular network, while short to medium range networks include ultra-wide band (UWB), Wireless Local Area Network (WLAN, especially using Wi-Fi), infrared, Bluetooth (Rodriguez et al., 2005), ZigBee, and ultrasound (Yu, Sharp & Guo, 2009).

Positioning systems can be classified into four different system topologies (Al Nuaimi & Kamel, 2011; Niculescu, 2004). The first one is the remote positioning topology in which the signal transmitter is the mobile and several fixed measuring units are the receiver. The measurements from all measuring units are collected, and the location of the transmitter is determined by a master unit or station. The second is self-positioning in which the measuring unit is the mobile such as in an ad-hoc network. The mobile receives the signals transmitted by a number of stations in known locations, and it has the capability to compute its location based on the forwarded measurements. If a wireless data link is provided between the transmitter and receiver in a positioning system, it is possible to send the measurements from a self-positioning measuring unit to a remote processing center, and this is called indirect remote positioning, which is the third system topology. If the measurements or useful information is sent from a base station or processing center to a mobile unit via a wireless data link to aid position determination, this case is named indirect self-positioning, which is the fourth system topology (Liu, et al., 2007; Yu et al., 2006).

It is worth mentioning a number of first local indoor positioning systems developed in the 1990s and early 2000s. Readers are referred to (Caffery, 2000) for information about early cellular network based positioning systems especially for emergency services. Cambridge University and AT&T developed the Active Badge System using infrared signals transmitted at a frequency of about 10-15 Hz (Want et al.,

1992). Currently, there are about 1500 such devices used in the United States and Europe. Later, AT&T developed BAT positioning system based on TOA measurement of ultrasonic signals with centimeter accuracy (Harter et al., 1999). The first Wi-Fi based positioning system RADAR was developed by Microsoft Research using RSS measurements at access points with accuracy around 2-3 meters (Bahl & Padmanabhan, 2000). The MIT's Cricket system was also among the systems developed in the early 2000s, which makes use of measurement of TDOA between a radio frequency signal and an ultrasonic signal with position accuracy about 1-3 centimeters (Priyantha, Chakraborty & Balakrishnan, 2000). Then UCLA developed the AHLos system, which also uses radio frequency and ultrasonic signals and position determination is realized in a iterative way, intended for positioning in a distributed network (Savvides et al., 2001). Due to ultra-low power transmission and high time resolution, UWB is a competing technology for short to medium range communications and positioning (Fontana, 2004). CSIRO developed the WASP system based on round trip time measurement of the ISM frequency band signal with sub-meter position accuracy in severe multipath propagation environments (Hedley, Humphrey & Ho, 2008; Sharp, Yu & Sathyan, 2012).

In the remainder of this section, an optimization based approach for positioning in cellular networks is studied. The position determination is based on data fusion of measurements of distance, speed and direction (Yu & Dutkiewicz, 2012).

3. Data Fusion for Cellular Network-Based Mobile Tracking

In cellular network based mobile positioning, there are a number of hurdles affecting positional accuracy. One such hurdle is the very limited number, typically three or four, of base stations, which the mobile station can communicate with. Another hurdle is the relatively large distance estimation errors, which are often caused by multipath and NLOS propagations especially in urban areas where there are rich high-rise buildings and structures. This section describes two positioning algorithms, which can reduce multipath and NLOS effect considerably.

a. Two-Step Position Determination Method

Distance measurement is taken at each base station, typically by measuring the round trip time of the radio frequency signal. In the first step, single-shot position determination is performed by applying optimization with a technique to mitigate the NLOS effect. In the second step, basic mobile track geometry, mobile velocity and heading direction measurements are utilized to adjust the initial position estimates produced in the first step.

i. Optimization with NLOS Mitigation

The estimate of the distance between the mobile and the *i*th base station at time instant t_j may be simply described as

$$r_{i,j} = d_{i,j} + n_{i,j} \tag{14}$$

where $n_{i,j}$ is the distance measurement error and

$$d_{i,j} = \sqrt{(x_{m,j} - x_{b,i})^2 + (y_{m,j} - y_{b,i})^2} \tag{15}$$

where $(x_{m,j}, y_{m,j})$ are the mobile position coordinates at time instant t_j, whereas $(x_{b,i}, y_{b,i})$ are the position coordinates of the ith base station. Position determination can be performed in different ways. Here the optimization based nonlinear minimization is considered. The cost function at time instant t_j is defined as

$$\Lambda(\theta) = \sum_{i=1}^{N} w_i \varepsilon_{i,j}^2(\theta) \tag{16}$$

where it is assumed that there are N base stations within radio range of the mobile so that distance measurement can be taken via communication. The parameter vector is defined as

$$\theta = \left[x_{m,j}, y_{m,j}, \eta\right]^T \tag{17}$$

where the parameter η is included in the error term $\varepsilon_{i,j}(\theta)$ which is defined as

$$\varepsilon_{i,j}(\theta) = r_{i,j} - d_{i,j} - \eta \tag{18}$$

The weights .. are selected to emphasize the contribution of smaller error terms among $\{\varepsilon_{i,j}\}$, which can be chosen to be inversely proportional to the variances of the distance measurement errors. In the event of unknown or very similar statistics of $\{\varepsilon_{i,j}\}$, equal weights can be simply used. Note that in minimization or optimization, when the weights are properly selected with respect to the quality of each measurement, better estimation results will be expected. When the optimal weights are employed, optimal estimation accuracy would be produced. In the simulation equal weights are used. Minimizing the cost function produces the estimate of the unknown parameter vector:

$$\hat{\theta} = \arg\min_{\theta} \Lambda(\theta) \tag{19}$$

Note that the η parameter is introduced to model the error caused by the NLOS propagation. It usually varies as the vehicle reaches different positions on the track. More specifically, it represents the mean of the NLOS-induced distance measurement errors of all base stations at one specific position point. That is, when the iterative process is at the steady state, the parameter η will approach the mean of the distance measurement errors. Accordingly, the NLOS-induced distance error will be greatly mitigated. Although this mitigation technique cannot completely suppress the NLOS effect, it is simple and reliable.

Intuitively, it would be difficult to predict how the NLOS parameter varies since it will be dependent on the specific propagation environments. It can be correlated with previous measurements; however, the NLOS induced bias errors are usually assumed random and mutually independent. Also, researchers are more interested in the distribution of the bias errors especially through field measurements in one or more areas of interest. It is certainly worth further investigating the correlation issue in future work.

There are a number of practical non-linear minimization algorithms that can be applied to solve the positioning problem. In the simulation, the Levenberg-Marquardt algorithm will be employed. To improve the accuracy, constraints can be applied. For instance, the constraints can be defined as

$$x_L \le x_{m,j} \le x_U, y_L \le y_U, 0 \le \eta \le \eta_{\max} \tag{20}$$

where (x_L, y_L) and (x_U, y_U) are the lower and upper bounds of $(x_{m,j}, y_{m,j})$, respectively. They can be selected based on the dimensions of the monitored area. The parameter $\eta_{\max}$ can be chosen according to the severity of the NLOS propagation. In the simulation, $\eta_{\max}$ is set at 0.3. To start the iterative minimization process, initial values of the three parameters are required. Since η is zero in LOS propagation, the initial value for η can be simply set at zero. The initial estimate of $(x_{m,j}, y_{m,j})$ can be obtained by using a non-iterative algorithm such as the linearization based method. The linearization based method used in the simulation is briefly described as follows. First, one squares both sides of each distance measurement equation. Then, one introduces an extra unknown parameter $R_{m,j} = x^2_{m,j} + y^2_{m,j}$ so that the estimate equation becomes linear with respect to the three unknown parameters. Finally, a linear least-squares solution will be produced. It is worth mentioning that the NLOS variable is typically assumed to have a stationary distribution regardless of the path, with distribution parameter(s) suited to specific location areas. Such an assumption is useful especially when deriving theoretical results that are mathematically tractable.

ii. Geometry and Motion Based Data Adjustment

The position estimates produced by the algorithm described above are then adjusted by making use of motion characteristics and basic track geometry. Figure 9 shows an example of a track segment relative to the base station in a two dimensions coordinate system. The vehicle moves along the track from at P_1 time t_1 and reaches P_k at time t_k. The track segment between each pair of neighboring time instants is assumed linear; however, the whole track can be in any type of form. This assumption would be reasonable since such a short track segment between two neighboring time instants would be straight or very close to straight in general.

Denote the mobile position and velocity at point P_j as $(x_{m,j}, y_{m,j})$ and v_j, respectively. Typically, a vehicle is equipped with a speedometer so that the velocity estimate at any time instant is available. Assume that the acceleration of the mobile is either zero or a non-zero constant during two neighboring time instants. Then, the acceleration between time instants t_{j-1} and t_j can be calculated by

$$\alpha_j = \frac{v_j - v_{j-1}}{\Delta t} \tag{21}$$

where Δt is the sampling time period, i.e. the position estimation updating period. Accordingly, the distance between time instants t_{j-1} and t_j can be described as

Figure 9. Illustration of the geometric relationship between a base station and a track segment (Yu and Dutkiewicz, 2012)

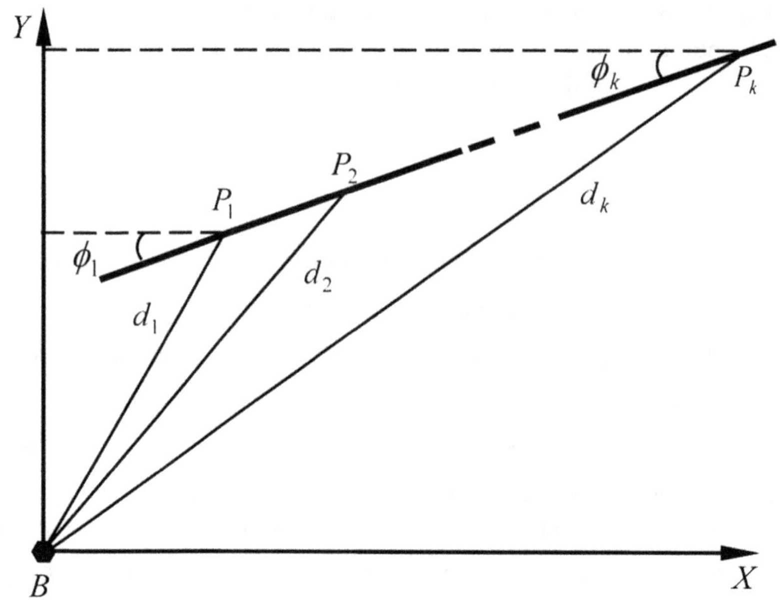

$$\delta d_j = v_{j-1}\Delta t + \frac{1}{2}(\Delta t)^2 \alpha_j = \frac{1}{2}(v_{j-1} + v_j)\Delta t \tag{22}$$

Then, the mobile position coordinates at time t_j can be expressed as

$$x_{m,j} = x_{m,1} + \delta_{x,j}, \quad y_{m,j} = y_{m,1} + \delta_{y,j}, j \geq 2 \tag{23}$$

where

$$\delta_{y,j} = \sum_{i=2}^{j} \delta d_i \sin\phi_i, \quad \delta_{x,j} = \sum_{i=2}^{j} \delta d_i \cos\phi_i \tag{24}$$

where ϕ_i is the heading angle at time instant t_i, which is defined as the angle between the motion direction and the x coordinate axis. The heading direction can be estimated by a heading sensor. For instance, the Honeywell HMR3200 digital compass provides heading information with accuracy of 0.5 degree.

The purpose is to employ Equation 23 to continuously adjust the position estimate $\left(x_{m,j}, y_{m,j}\right)$ to achieve better position accuracy as described below.

Suppose that the current time instant is t_k. Then, in the presence of errors in initial position, velocity, and heading direction estimates, Equation 23 can be written as

$$x_{m,1} \approx \hat{x}_{m,i} - \hat{\delta}_{x,i}$$
$$y_{m,1} \approx \hat{y}_{m,i} - \hat{\delta}_{y,i}, \quad i = 2, 3, \cdots, k \tag{25}$$

where $\left(\hat{x}_{m,i}, \hat{y}_{m,i}\right)$ are the initial position estimates produced by the method described earlier, and

$$\hat{\delta}_{y,i} = \sum_{j=2}^{i} \delta\hat{d}_j \sin\hat{\phi}_j$$
$$\hat{\delta}_{x,i} = \sum_{j=2}^{i} \delta\hat{d}_j \cos\hat{\phi}_j \tag{26}$$

where

$$\hat{\delta d}_j = \frac{1}{2}\left(\hat{v}_{j-1} + \hat{v}_j\right)\Delta t \tag{27}$$

From Equation 25, it can be seen that the estimates of $\left(x_{m,1}, y_{m,1}\right)$ at time t_k can be obtained by solving

$$\tilde{x}_{m,1}^{(k)} = \frac{1}{k-1}\sum_{i=2}^{k}(\hat{x}_{m,i} - \hat{\delta}_{x,i})$$
$$\tilde{y}_{m,1}^{(k)} = \frac{1}{k-1}\sum_{i=2}^{k}(\hat{y}_{m,i} - \hat{\delta}_{y,i}) \tag{28}$$

Therefore, the adjusted position estimate at point P_k can be calculated by

$$\tilde{x}_{m,k} = \tilde{x}_{m,1}^{(k)} + \hat{\delta}_{x,k}$$
$$\tilde{y}_{m,k} = \tilde{y}_{m,1}^{(k)} + \hat{\delta}_{y,k} \tag{29}$$

Clearly, the strategy is that the parameter measurements or estimates at multiple time instants, including $\left\{\hat{x}_{m,i}, \hat{y}_{m,i}, \hat{\delta}_{x,i}, \hat{\delta}_{y,i}\right\}$, are utilized to produce an accurate estimate of $\left\{x_{m,1}, y_{m,1}\right\}$, which are then employed to adjust current position estimate using velocity and heading angle information. From Equations 26 and 28, it can be seen that when calculating $\left\{\hat{\delta}_{x,i}, \hat{\delta}_{y,i}\right\}$ and $\left\{\tilde{x}_{m,1}^{(k)}, \tilde{y}_{m,1}^{(k)}\right\}$ an averaging process is actually involved. As a consequence, the accuracy of the adjusted position estimates would be enhanced significantly especially when the number of measurements is large.

Note that in the proposed method as described above, the initial position estimates, the velocity and heading angle estimates at current and all previous data points are required to produce a position estimate at current time instant. Clearly, the utilization of the full history of measurements is not due to whether or not the measurements are correlated. Instead, it makes use of the path geometry information to integrate all the measurements to achieve better position estimates.

As the mobile moves, the number of data employed for position adjustment is getting bigger. However, the required data storage and computation at each time instant are quite small. This is illustrated by the recursive calculation described as follows. At time instant t_{k+1}, the first line of Equation 28 becomes

$$\begin{aligned}\tilde{x}_{m,1}^{(k+1)} &= \frac{1}{k}\sum_{i=2}^{k+1}(\hat{x}_{m,i} - \hat{\delta}_{x,i}) \\ &= a\tilde{x}_{m,1}^{(k)} + \frac{1}{k}(\hat{x}_{m,k+1} - \hat{\delta}_{x,k+1})\end{aligned} \tag{30}$$

where

$$a = \frac{k-1}{k} \tag{31}$$

Then, at time instant t_{k+1}, Equation 29 can be written as

$$\tilde{x}_{m,k+1} = a(\tilde{x}_{m,1}^{(k)} + \hat{\delta}_{x,k}) + \frac{a}{2}\left(\hat{\upsilon}_{k+1} + \hat{\upsilon}_{k}\right)\Delta t \cos\hat{\phi}_{k+1} + \frac{1}{k}\hat{x}_{m,k+1} \tag{32}$$

In the same way, one obtains

$$\tilde{y}_{m,k+1} = a(\tilde{y}_{m,1}^{(k)} + \hat{\delta}_{y,k}) + \frac{a}{2}\left(\hat{\upsilon}_{k+1} + \hat{\upsilon}_{k}\right)\Delta t \sin\hat{\phi}_{k+1} + \frac{1}{k}\hat{y}_{m,k+1} \tag{33}$$

Thus, at time instant t_k, after adjusting the position estimate, one only needs to store five numbers, $\left\{\tilde{x}_{m,1}^{(k)}, \tilde{y}_{m,1}^{(k)}, \hat{\delta}_{x,k}, \hat{\delta}_{y,k}, \tilde{\upsilon}_{k}\right\}$. At time instant t_{k+1}, the five numbers are called, new velocity and heading angle estimates are collected, the single-shot optimization based position estimate is produced, and finally the mobile position coordinates are adjusted according to Equations 32 and 33.

Since this work focuses on vehicle tracking and vehicles are typically equipped with a speedometer, it is assumed that the velocity at any time instant is available. In the case of tracking pedestrians, a speedometer is generally not available and thus different approaches need to be developed.

As mentioned earlier, the proposed method presented in this section and the proposed one-step method in the next section do not require any knowledge of the road structure in the monitored area as long as the small track segment between each pair of neighboring time instants is linear or close to linear. Accordingly, they are insensitive to the frequent change in road direction during tracking. In the case where the vehicle moves along a circle with a relatively small radius such as a roundabout, the assumption of a piecewise linear track segment would be still appropriate. This is because the vehicle velocity will be rather small in such a circumstance so that the position variation during a sampling period over the circle/roundabout will be small too. As a consequence, it is still reasonable to approximate the track segment between two neighboring time instants as piecewise linear. Such an approximation would not cause a significant error due to the small dimensions of the roundabouts.

b. One-step Position Determination Method

In the method studied above the step involving optimization based initial position determination with NLOS mitigation and the step involving track geometry and motion based adjustment are performed separately. The position determination scheme which jointly handles the two processes in one step is described below.

Let the current time instant be k and define the cost function as

$$\Lambda(x_{m,1}, y_{m,1}, \eta) = \sum_{i=1}^{N} \sum_{j=1}^{k} w_i (r_{i,j} - \tilde{d}_{i,j} - \eta)^2 \tag{34}$$

where the true distance

$$d_{i,j} = \sqrt{\left(x_{m,1} + \delta_{x,j} - x_{b,i}\right)^2 + \left(y_{m,1} + \delta_{y,j} - y_{b,i}\right)^2} \tag{35}$$

is approximated by

$$\tilde{d}_{i,j} = \sqrt{\left(x_{m,1} + \hat{\delta}_{x,j} - x_{b,i}\right)^2 + \left(y_{m,1} + \hat{\delta}_{y,j} - y_{b,i}\right)^2} \tag{36}$$

where $(\hat{\delta}_{x,j}, \hat{\delta}_{y,j})$ are defined by Equation 26. As a result, all the parameters except for $x_{m,1}$, $y_{m,1}$ and η in the cost function are known. Minimizing the cost function with respect to the three unknown parameters produces

$$(\hat{x}_{m,1}, \hat{y}_{m,1}, \hat{\eta}) = \arg\min_{x_{m,1}, y_{m,1}, \eta} \Lambda(x_{m,1}, y_{m,1}, \eta) \tag{37}$$

To enhance the estimation performance, constraints can be applied in the same way as discussed in Section 2.1. Once the estimates of $x_{m,1}$ and $y_{m,1}$ are produced, the current mobile position can be adjusted as discussed earlier.

To start the nonlinear minimization process defined by Equation 37, initial estimates of the unknown parameters are required. The initial value for parameter η can be simply set at zero. The initial position estimate can be obtained as follows. By ignoring the measurement noise and the NLOS caused error, the distance measurement equations can be approximated as

$$r_{i,j} \approx \sqrt{(x_{m,1} + \hat{\delta}_{x,j} - x_{b,i})^2 + (y_{m,1} + \hat{\delta}_{x,j} - y_{b,i})^2} \tag{38}$$

Squaring both sides of the above equation and then re-arranging the equation produces

$$x_{m,1}^2 + y_{m,1}^2 + 2c_{i,j}x_{m,1} + 2s_{i,j}y_{m,1} \approx \rho_{i,j} \tag{39}$$

where

$$\begin{aligned} \rho_{i,j} &= r_{i,j}^2 - c_{i,j}^2 - s_{i,j}^2 \\ c_{i,j} &= \hat{\delta}_{x,j} = x_{b,i} \\ s_{i,j} &= \hat{\delta}_{y,j} = y_{b,i} \end{aligned} \tag{40}$$

By treating $x_{m,1}^2 + y_{m,1}^2$ as one variable, Equation 40 becomes a linear equation, which can be written in a compact form as

$$\boldsymbol{C\theta} \approx \boldsymbol{\rho} \tag{41}$$

where

$$\boldsymbol{C} = \begin{bmatrix} 1 & 2c_{1,1} & 2s_{1,1} \\ \vdots & \vdots & \vdots \\ 1 & 2c_{1,k} & 2s_{1,k} \\ \vdots & \vdots & \vdots \\ 1 & 2c_{N,1} & 2s_{N,1} \\ \vdots & \vdots & \vdots \\ 1 & 2c_{N,k} & 2s_{N,k} \end{bmatrix}, \boldsymbol{\theta} = \begin{bmatrix} x_{m,1}^2 + y_{m,1}^2 \\ x_{m,1} \\ y_{m,1} \end{bmatrix}, \boldsymbol{\rho} = \begin{bmatrix} \rho_{1,1} \\ \vdots \\ \rho_{1,k} \\ \vdots \\ \rho_{N,1} \\ \vdots \\ \rho_{N,k} \end{bmatrix} \tag{42}$$

The ordinary LS solution for Equation 42 is simply given by

$$\hat{\boldsymbol{\theta}} = (\mathbf{C}^T\mathbf{C})^{-1}\mathbf{C}^T\boldsymbol{\rho} \tag{43}$$

That is, the initial estimates of the position coordinates are produced as

$$\hat{x}_{m,1} = \hat{\theta}_{2,1}, \hat{y}_{m,1} = \hat{\theta}_{3,1} \tag{44}$$

Clearly, when k is large, the size of the matrix **C** and vector **ρ** will become large. Accordingly, relatively large computational complexity and data storage are involved, which might be a disadvantage for some applications. To avoid such a drawback, an alternative method can be used to generate the initial position estimate at time t_k. That is, the adjusted position estimate at the previous time instant t_{k-1} is utilized to serve as the initial position estimate at time t_k. Thus, the possible large computational complexity and data storage involved in the calculation in Equation 43 will be avoided. Suppose that a vehicle moves at a speed of 72 km/h and the data sampling frequency is 1 Hz. Then, the distance between two sampling points is 20 m, which would be around the accuracy of the adjusted position estimates as shown later. Therefore, such a way to obtain the initial position estimate would be acceptable.

Unlike the two-step method, there is no simple way to significantly reduce the computational complexity involved in the minimization in Equation 37 where the cost function consists of $N * k$ terms. Thus, as k is getting larger, the complexity of the one-step approach would be greater than that of the two-step method.

c. Simulation Results

Figure 10 shows the placement of four base stations, which are denoted by small squares and the mobile track with three segments for simulation. Note that when the vehicle goes outside of the boundary defined by the four base stations, it is likely that one or more other neighboring base stations can communicate with the vehicle. Thus, it is reasonable to assume that the vehicle is always surrounded by a number of base stations. The base station coordinates and the coordinates of the four points on the track are all in units of kilometers. For simulation simplicity the mobile is assumed to move at a constant velocity along the track from point P_1 to P_4 without considering the possible velocity variations around the two turning points in practice. Clearly, the simple path geometry is chosen for simplicity. Certainly, it would be more realistic if the paths are randomly drawn with some variety. However, similar observations would be expected when different path geometries are used for simulation. The distance measurements are made simultaneously once per second at all base stations. Since a NLOS propagation environment is considered, the distance measurement error is modeled as the sum of a zero-mean Gaussian noise variable and an exponential NLOS induced bias variable.

Figure 11 shows how an imperfect heading angle measurement affects the position accuracy of the proposed two algorithms. It can be seen that when the standard deviation (STD) of the heading angle error goes from 1 to 10 deg, the accuracy of the two-step method decreases by about 25%, whereas the accuracy of the one-step method degrades by around 30%. Clearly, the performance degradation is significant. On the other hand, when the STD of the heading measurement error is less than 4 deg, the accuracy degradation is really marginal, around 2% for the two-step method and 3% for the one-step method. As mentioned earlier, the accuracy of the heading angle measurements provided by the off-the-shelf sensors is typically high such as on the order of one degree. As a result, the impact of the heading angle measurement errors on the position accuracy will be minor in general.

Figure 12 shows the effect of the mobile velocity estimation error on the position accuracy. The maximum STD of the velocity estimation error is set at 12 km/h, which is 20% of the actual mobile velocity. It can be seen that over the range of velocity estimation errors, the accuracy of the two-step method decreases by about 21%, whereas the accuracy of the one-step method degrades by around 36%. When the STD of the velocity error is up to around 3 km/h, the impact of velocity error on the position accuracy is negligible. As the velocity error increases, the one-step method is more sensitive to the velocity error.

Figure 10. Base station configuration and mobile track for simulation (Yu and Dutkiewicz, 2012)

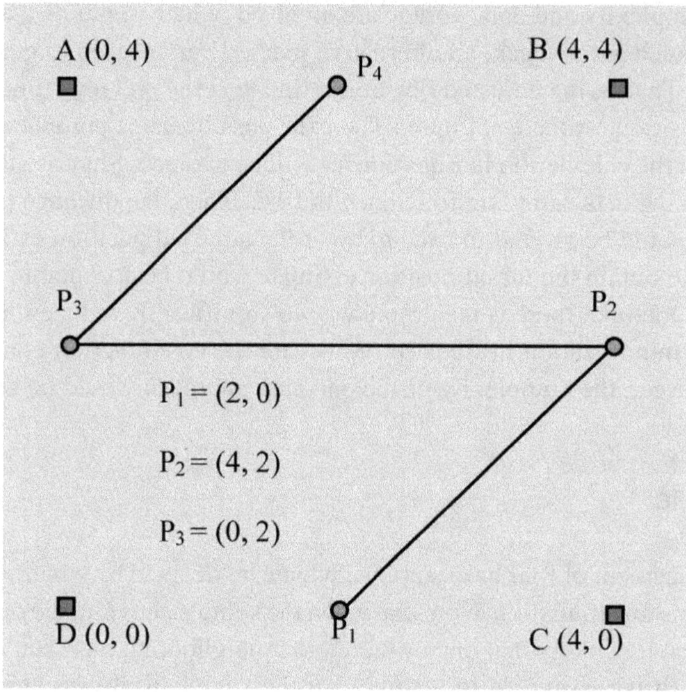

Figure 11. Effect of heading angle error on position accuracy (Yu and Dutkiewicz, 2012)

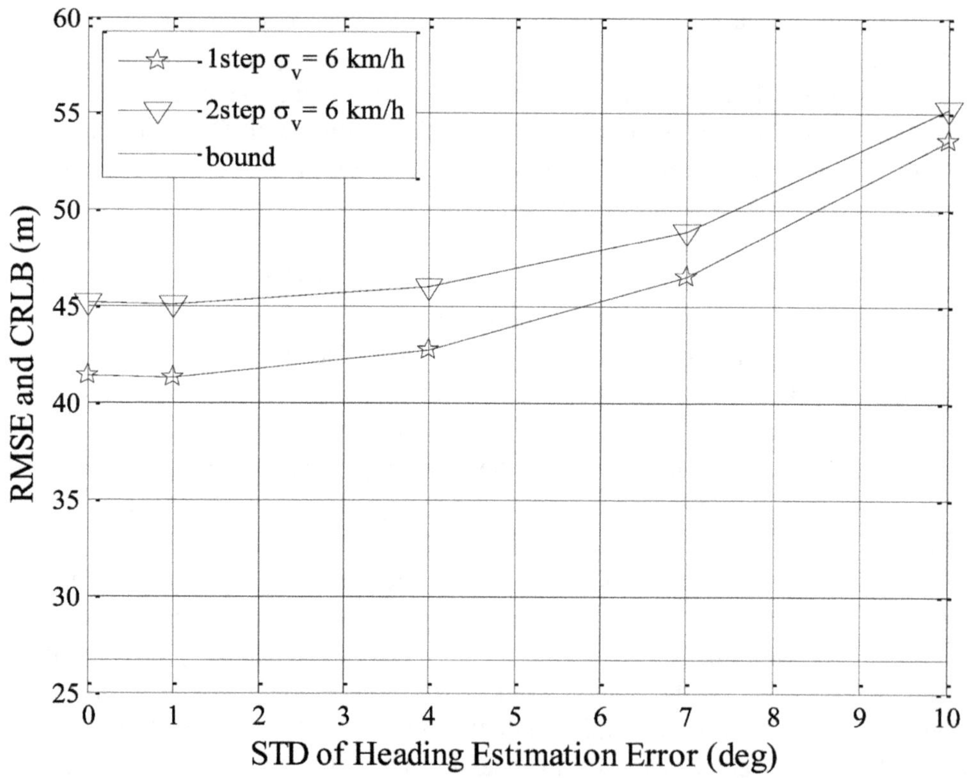

Figure 12. Effect of mobile velocity estimation error on position accuracy (Yu and Dutkiewicz, 2012)

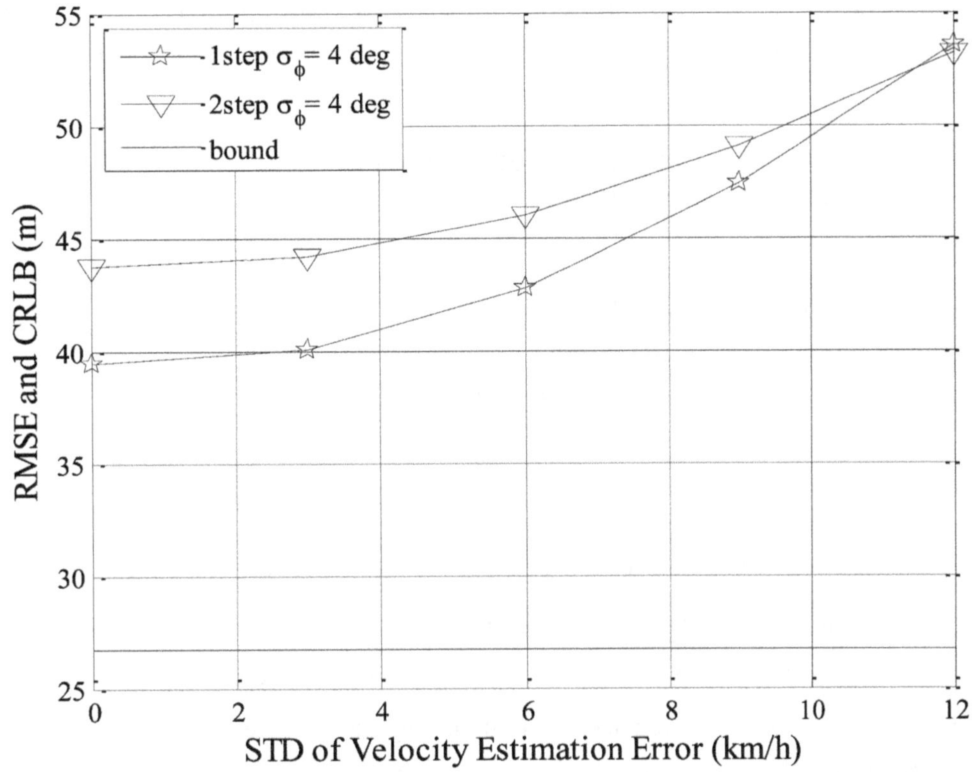

COMBINING VARIOUS MEASUREMENTS AND TECHNOLOGIES

Finally, this section deals with two examples. The first example is the augmentation of GPS constellation with Pseudolites in order to monitor the deformation of a bridge. The second example shows how a least mean squares algorithm can be used to track the position of a static receiver. The stochastic model of the noise is described using the fractional Brownian motion. Then, the theory is evaluated using GNSS measurements.

1. Integration of GPS and Pseudolites: a Practical Example with Monitoring Bridge Deformation.

The case study is the bridge monitoring applied to the Millennium Bridge in London prior to the refurbishment of the structure in November 2000. The Millennium Bridge field tests consisted of placing a number of survey grade (CP and Pseudoranges) GPS receivers on the bridge, being positioned relative to a reference receiver placed on top of an adjacent building (Roberts et al., 2006). After the data were processed in OTF kinematic, the authors noticed that there seemed to be larger movements in the longitudinal axis of the bridge than in the lateral, which was not as expected. It turned out that as the bridge's longitudinal axis in orientated in an almost North-South direction. It was deduced that the errors in the resulting positional solution due to the geometry of the location of the satellites used, introduced a larger error in the North-South direction than the East-West. This error resulted from the fact that GPS satellites have an inclination angle of 55 degrees, which means that there has never been a satellite tracked in the North direction in latitudes such as the UK. Following this issue, the real data were used to create simulated data, with simulated Pseudolite observations inserted, in order to evaluate the potential of using Pseudolites in such a circumstance (Roberts et al., 2006).

a. Integration of Pseudolites and GPS Measurements in the Least-Squares Triangulation

Recalling Section 3, the equation of Pseudoranges between the i-th GPS satellite and the ground receiver at the observation time t can be described by

$$P_r^{(i)}(t) = \rho_r^{(i)}(t) + c\delta t_r(t) - c\delta t^{(i)}\left(t - \tau_r^{(i)}(t)\right) + I_r^{(i)}(t) + T_r^{(i)}(t) + \varepsilon_{Pr}^{(i)}(t) \tag{45}$$

where $\rho_r^{(i)}(t) = x^{(i)}\left(t - \tau_r^{(i)}(t)\right) - \boldsymbol{x}_r(t)$ is the geometric range between a receiver and a satellite; $x^{(i)}\left(t - \tau_r^{(i)}(t)\right)$ is the satellite position at the time of transmission; $\tau_r^{(i)}(t)$ is the time duration of signal propagation; $\boldsymbol{x}_r(t)$ is the receiver position; c is the speed of light in vacuum; $\delta t_r(t)$ is the receiver clock error; $\delta t^{(i)}\left(t - \tau_r^{(i)}(t)\right)$ is the satellite clock error; $I_r^{(i)}(t)$ is the ionospheric delay; $T_r^{(i)}(t)$ is the tropospheric delay; and $\varepsilon_{Pr}^{(i)}(t)$ is the unmodeled errors including measurement noise and multipath effect. It is possible to remove partially the Ionosphere and Troposphere delays using some models (i.e. Klobuchar and Saastomoinen (Hofmann-Wellenhof, Lichtenegger and Collins, 1993)). It is underlined that due to the ultra stable clock embedded in the satellites (e.g. atomic clock), the clock error is averaged with time and it becomes a small bias compared to the clock error of the receiver (Strang and Borre, 1997). With these assumptions, Equation 45 can be re-written as:

$$P_r^{(i)} \approx \rho_r^{(i)} + c\delta t_r, i = 0,1,2,\ldots,m \tag{46}$$

Similarly, the Pseudorange measurements of the k-th Pseudolite can be written as:

$$p_r^{(k)}(t) = \rho_r^{(k)}(t) + c\delta t_r(t) - c\delta t_P^{(k)}\left(t - \tau_r^{(k)}\right) + \varepsilon_{Pr}^{(k)}(t) \tag{47}$$

With $\rho_r^{(k)}(t)$ the geometric distance between the receiver and the k-th Pseudolite; $\delta t_r(t)$ is the receiver clock error; $\varepsilon_{Pr}^{(k)}(t)$ is the error due to the signal propagation (i.e. multipath). $\delta t_P^{(k)}$ is the clock error bias of the k-th Pseudolite. It is generally not common to have an ultra stable clock in a Pseudolite, because of the low-cost of the system for commercial applications (Cosser, 2004). Note that some military applications have been using expensive Pseudolites with embedded atomic clock (Cobb, 1997). In the case of a low-cost Pseudolite, one can approximate Equation 46 as:

$$p_r^{(k)} \approx \rho_r^{(k)} + c\delta t_r - c\delta t_p^{(k)}, k = 1,2,\ldots,h \tag{48}$$

Clearly, there is an additional variable per Pseudolite when estimating the position of the receiver. One way to solve this issue is to estimate the position of the receiver using (single difference) Differential GPS (or DGPS, see (Roberts, 1997)). The idea is to subtract the Pseudoranges between a reference satellite ($P_r^{(0)}$) and the other satellites and Pseudolites in range with the receiver. Note that a Pseudolite can also be used as reference. As a result, the clock error of the receiver (δt_r) is then eliminated from Equation 48 as:

$$\Delta P_r^{0i} = P_r^{(0)} - P_r^{(i)} \approx \rho_r^{(0)} - \rho_r^{(i)}, i = 1,2,\ldots,m$$

$$\Delta p_r^{0k} = P_r^{(0)} - p_r^{(k)} \approx \rho_r^{(0)} - \rho_r^{(k)} + c\delta t_p^{(k)}, k = 1,2,\ldots,h \tag{49}$$

In this way, 4 satellites will be needed to estimate the coordinates of the receiver. As for the k-th Pseudolite, one needs to estimate the clock error bias $\delta t_p^{(k)}$, and the usefulness of introducing Pseudolites in this manner is to improve the geometry of the network (i.e. geometry dilution of precision in Leick (2004)). In this case, the derivatives of the Pseudoranges function of the coordinates of the receiver [x_r, y_r, z_r] and the clock error of the Pseudoiltes [$\delta t_p^{(1)}, \delta t_p^{(2)}, \ldots, \delta t_p^{(h)}$] are (Cobb, 1997):

$$\mathrm{d}\Delta P_r^{0i} \approx \left(-\frac{x_{st} - x_0}{\rho_r^{(0)}} + \frac{x_{st} - x_i}{\rho_r^{(i)}} \right) \Delta x + \left(-\frac{y_{st} - y_0}{\rho_r^{(0)}} + \frac{y_{st} - y_i}{\rho_r^{(i)}} \right) \Delta y + \left(-\frac{z_{st} - z_0}{\rho_r^{(0)}} + \frac{z_{st} - z_i}{\rho_r^{(i)}} \right) \Delta z, i = 1,2,\ldots,m \tag{50}$$

$$\mathrm{d}\Delta p_r^{0k} \approx \left(-\frac{x_{st} - x_0}{\rho_r^{(0)}} + \frac{x_{st} - x_k}{\rho_r^{(k)}} \right) \Delta x + \left(-\frac{y_{st} - y_0}{\rho_r^{(0)}} + \frac{y_{st} - y_k}{\rho_r^{(k)}} \right) \Delta y + \left(-\frac{z_{st} - z_0}{\rho_r^{(0)}} + \frac{z_{st} - z_k}{\rho_r^{(k)}} \right) \Delta z + c\delta t_p^{(k)}, k = 1,2,\ldots,h$$

which can be written as a compact from as:

$$\Delta \rho \approx \mathbf{G}\delta \tag{51}$$

where

$$\Delta \rho = \begin{bmatrix} \mathrm{d}\Delta P_r^{01} \\ \vdots \\ \mathrm{d}\Delta P_r^{0m} \\ \mathrm{d}\Delta p_r^{01} \\ \vdots \\ \mathrm{d}\Delta p_r^{0h} \end{bmatrix},$$

$$\mathbf{G}=\begin{bmatrix} & & & 0,0 & \cdots & 0,0 \\ & & & \vdots & \ddots & \vdots \\ -\frac{x_{st}-x_0}{\rho_r^{(0)}}+\frac{x_{st}-x_i}{\rho_r^{(i)}} & -\frac{y_{st}-y_0}{\rho_r^{(0)}}+\frac{y_{st}-y_i}{\rho_r^{(i)}} & -\frac{z_{st}-z_0}{\rho_r^{(0)}}+\frac{z_{st}-z_i}{\rho_r^{(i)}} & \vdots & \cdots & \vdots \\ \vdots & \vdots & \vdots & & \vdots & \\ -\frac{x_{st}-x_0}{\rho_r^{(0)}}+\frac{x_{st}-x_i}{\rho_r^{(i)}} & -\frac{y_{st}-y_0}{\rho_r^{(0)}}+\frac{y_{st}-y_i}{\rho_r^{(i)}} & -\frac{z_{st}-z_0}{\rho_r^{(0)}}+\frac{z_{st}-z_i}{\rho_r^{(i)}} & 0,0 & \cdots & 0,0 \\ -\frac{x_{st}-x_0}{\rho_r^{(0)}}+\frac{x_{st}-x_k}{\rho_r^{(k)}} & -\frac{y_{st}-y_0}{\rho_r^{(0)}}+\frac{y_{st}-y_i}{\rho_r^{(i)}} & -\frac{z_{st}-z_0}{\rho_r^{(0)}}+\frac{z_{st}-z_i}{\rho_r^{(i)}} & c,0 & \cdots & 0,0 \\ \vdots & \vdots & \vdots & \vdots & \ddots & \vdots \\ -\frac{x_{st}-x_0}{\rho_r^{(0)}}+\frac{x_{st}-x_k}{\rho_r^{(k)}} & -\frac{y_{st}-y_0}{\rho_r^{(0)}}+\frac{y_{st}-y_i}{\rho_r^{(i)}} & -\frac{z_{st}-z_0}{\rho_r^{(0)}}+\frac{z_{st}-z_i}{\rho_r^{(i)}} & \vdots & \cdots & \vdots \\ & & & & \vdots & \\ & & & 0,0 & \cdots & 0,0 \end{bmatrix}$$

$$\delta=\begin{bmatrix}\Delta x & \Delta y & \Delta z & \delta t_p^{(1)} & \ldots & \delta t_p^{(h)}\end{bmatrix}^T$$

and the least-squares solution is simply given by:

$$\hat{\delta}\approx\left(\mathbf{G}^T\mathbf{G}\right)^{-1}\mathbf{G}^T\Delta\rho \tag{52}$$

The position point (x_{st}, y_{st}, z_{st}) is the starting point of the receiver position (i.e. a rough apriori estimation); the receiver position (x_r, y_r, z_r); and $\Delta x = x_r - x_{st}$, $\Delta y = y_r - y_{st}$, and $\Delta z = z_r - z_{st}$. As described in Section 3, the vector of the estimates $\hat{\delta}$ is iterated until it reaches a certain threshold (Cobb, 1997; Montillet & Yu, 2012).

This algorithm is based on single difference measurements. It is possible to estimate the position of the receiver with Pseudolites and GPS using double difference or triple difference CP measurements (Ford et al., 1997; Leick, 2004). Note that the estimates in Equation 52 can also be formulated using a weighting matrix in a similar way as explained in Section 3 in order to achieve a better accuracy (e.g. BLUE theory in (Kay, 1993)). Here, the algorithm is only shown using Pseudoranges in order to understand the basic principle of the integration of the measurements of the two technologies (GPS satellites and Pseudolites) in the positioning algorithm. However, one can use carrier-phase measurements with the estimation of the float ambiguities as additional variables following Section 3.

b. Experimental Method Applied to the Millennium Bridge Deformation

It is known that the GPS constellation usually causes the vertical positioning accuracy to be two to three times worse than the horizontal accuracy. This is due to all satellites being located above the horizon. For bridge monitoring applications the vertical component is the most important component and so improving the accuracy in this component is a research aim (Meng et al., 2004). For the purpose of creating a more accurate bridge deformation monitoring system and overcoming some of the deficiencies of the current GPS satellite constellation, the use of Pseudolites has been investigated (Roberts et al., 2006).

Pseudolites can be used as another ranging source, which can improve the DOP values and the overall quality of the solution (Cosser, 2004).

The Millennium Bridge in London was completed in 2000, linking St Paul's cathedral with Tate Modern, the Bankside Gallery and Globe theater. The footbridge is a shallow suspension bridge with cables below the deck allowing pedestrians unobstructed views of the surrounding areas. The bridge has spans of 81 m, 144 m and 108 m. During its opening on the 10 June 2000, a sponsored walk was organized during which a maximum of 2,000 people were walking over the bridge at any instance. It is thought that due to the people walking close to each other, they would naturally walk in step. This large additional weight and force on the bridge resulted in greater than expected movements.

The bridge was subsequently closed for the installation of dampers to prevent such future movements. Today, the Bridge is fully operational and is a major tourist attraction. During this closure, and prior to the refurbishment, staff from Nottingham Geospatial Institute (NGI) and former Institute of Engineering, Surveying and Space Geodesy (IESSG) were allowed to conduct trials upon the bridge in order to assess whether GPS could in fact be used to monitor the deflections of such a structure (Roberts et al., 2004).

The viability trials conducted on the 22 to 24 November 2000 consisted of clamping three Leica AT504 choke ring antennas, connected to SR530 dual frequency GPS receivers, upon the bridge at pre-defined locations. A reference Leica SR530 GPS receiver with an AT 504 choke ring antenna was placed upon an adjacent building. The GPS data were recorded at a rate of 10Hz for post processing. The reference station's coordinates were established through processing all the data gathered relative to Ordnance Survey active stations located around London.

In addition to the GPS receivers, a tri-axial accelerometer (Kistler) was also used at one of the Bridge GPS locations, gathering data at a rate of 200 Hz. One of the Bridge's GPS receivers was co-located with the accelerometer throughout the trials, whilst the other two GPS receivers spent their time between 4 further locations upon the Bridge. The locations and sessions are illustrated in Figure 13 and Table 1. All the data were collected and periodically downloaded for post processing.

In all, 11 hours of GPS data were recorded at a rate of 10Hz, and 4 hours of accelerometer data at a rate of 200Hz. It resulted in 1.98 million points and 2.88 million points respectively. During the trials there were sporadic passengers upon the bridge, consisting of construction workers and the researchers. However, there was no significant pedestrian loading applied to the Millennium Bridge during the trials. The GPS post processing was carried out using both the commercial software (Leica's SKI-PRO) and the in-house software called IESSG's Kinpos (Cosser, 2004). Note that the software KinPos integrates the carrier-phase measurements in the positioning algorithm (in addition to the Pseudoranges), in order to do the estimation of the bridge deformations at the sub-centimeter level.

Once the GPS data were post processed, the initial task was to convert the results from GPS coordinates into bridge coordinates. This was conducted by knowing the orientation of the bridge in WGS84, derived from the coordinates of two GPS stations along the length of the bridge. The Millennium Bridge runs in an almost North-South direction. The lateral direction is that across the bridge's span and the longitudinal is that along its length. In Roberts et al. (2006), the displacement of the bridge (laterally and longingly) is estimated by using a moving average filter (Proakis, 2000) on the coordinate time series of each GPS receiver (e.g., A, B, C, D, E in Table 1) shown in Figure 14. The results are displayed in Figure 14 using a 5 seconds filter.

However in the initial analysis in Roberts et al. (2006), it was suggested that the longitudinal deflections of the Bridge were in fact larger than those in the lateral direction. But the analysis of the 24 hour sky plot (orbit of the GPS satellites) for the millennium Bridge, showed that there was never a GPS

Figure 13. The locations of the Bridge's Monitoring Stations (Roberts et al., 2006)

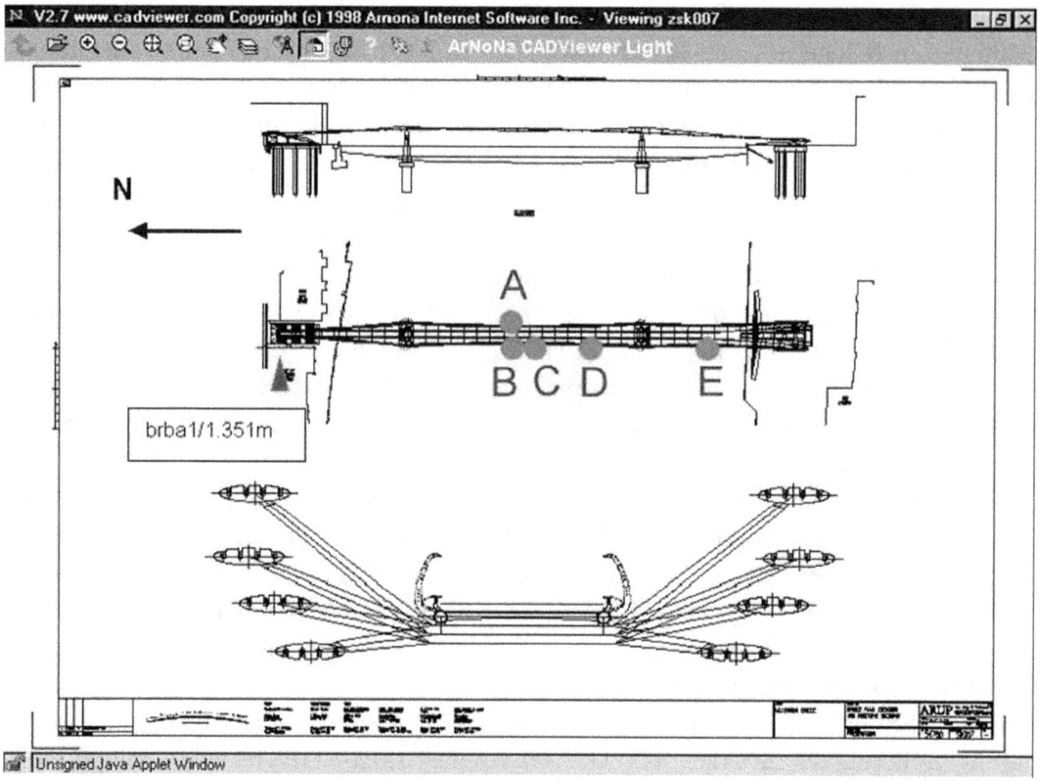

Figure 14. The coordinate time series of the two midspan GPS receivers's deflections together with a 5 seconds filter (Roberts et al., 2006)

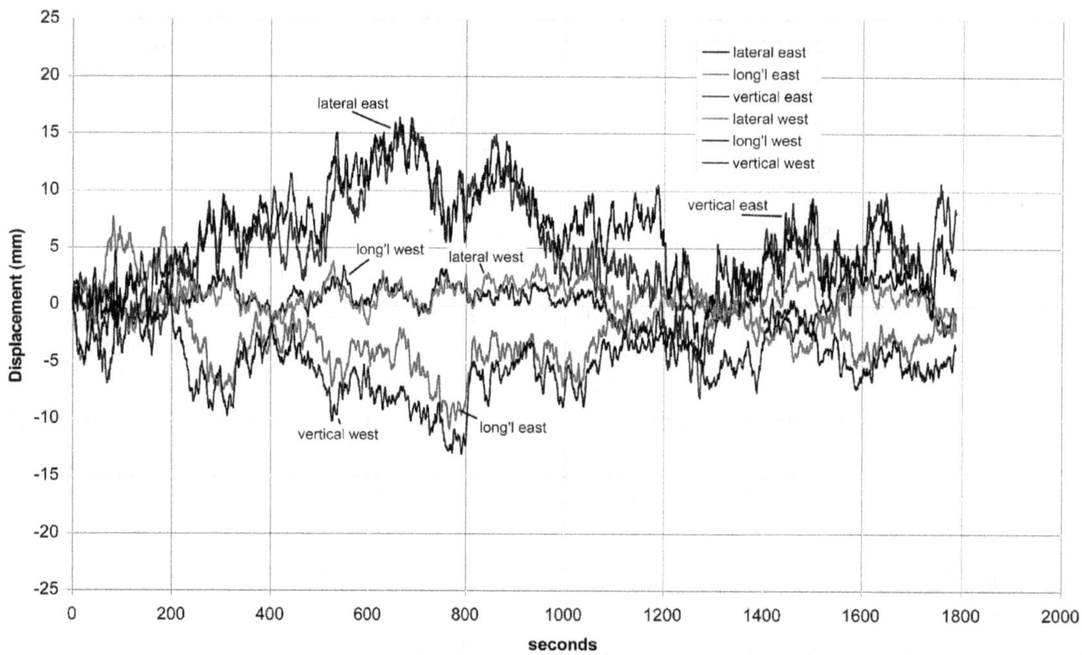

Table 1. Data recording session times

Date	Session Number	Start Time	End Time	Antenna Location
22/11/2000	1	14.35.42	16.25.52	A
	1	14.36.03	16.24.58	B
	1	14.35.52	16.25.26	C
	1	15.08.12	16.08.14	B (accelerometer)
23/11/2000	2	09.51.23	12.13.26	B
	2	09.38.25	12.13.37	C
	2	09.38.58	12.13.22	D
	3	12.37.24	14.05.50	B
	3	12.39.24	14.50.21	C
	3	12.37.57	14.50.13	A
	3	12.42.08	13.42.10	B (accelerometer)
	3	13.46.26	14.46.28	B (accelerometer)
	4	15.14.24	16.47.12	B
	4	15.16.23	16.47.17	C
	4	15.15.24	16.47.28	A
24/11/2000	5	08.56.09	12.06.13	B
	5	09.05.15	12.06.36	D
	5	09.15.28	12.07.20	E
	5	09.04.35	10.04.38	B (accelerometer)

satellite due north of the survey site. This result implied that the North DOP value was high, resulting in the East-West component being of the highest quality, followed by North- South then the vertical component. In other words, the bad geometry of the satellites position during the measurements created spurious signals and higher noise in the time series of the GPS coordinates (especially the North and vertical components) (Meng et al., 2004).

Finally, the results were augmented using (one or two) simulated Pseudolites to investigate how the problems of bad satellites geometry could be addressed. The NGI developed a GPS data simulator that will output RINEX GPS data for a receiver at a given location and time. The simulator was used to re-create the original bridge data, which compared well with the real bridge data (Cosser, 2004). The simulated longitudinal results were larger than the lateral results. Figure 15 illustrates the simulated results.

The next step introduced code and carrier data from a stationary satellite located at a pre defined location within the simulation like a Pseudolite. All the data produced from the simulator, in RINEX format, were then processed using Kinpos (Hide, 2003; Cosser, 2004). At this stage no weighting was used for the Pseudolites within the processing. The simulation was used purely as a means to evaluate whether Pseudolites could be a feasible way to improve such a scenario. Figure 16 illustrates the Millennium Bridge data with the introduction of one Pseudolite.

The introduction of the Pseudolite dramatically improves the results in both the North direction as well as in the vertical direction.

Figure 15. The GPS only simulated results using the GPS data simulator (Roberts et al., 2006)

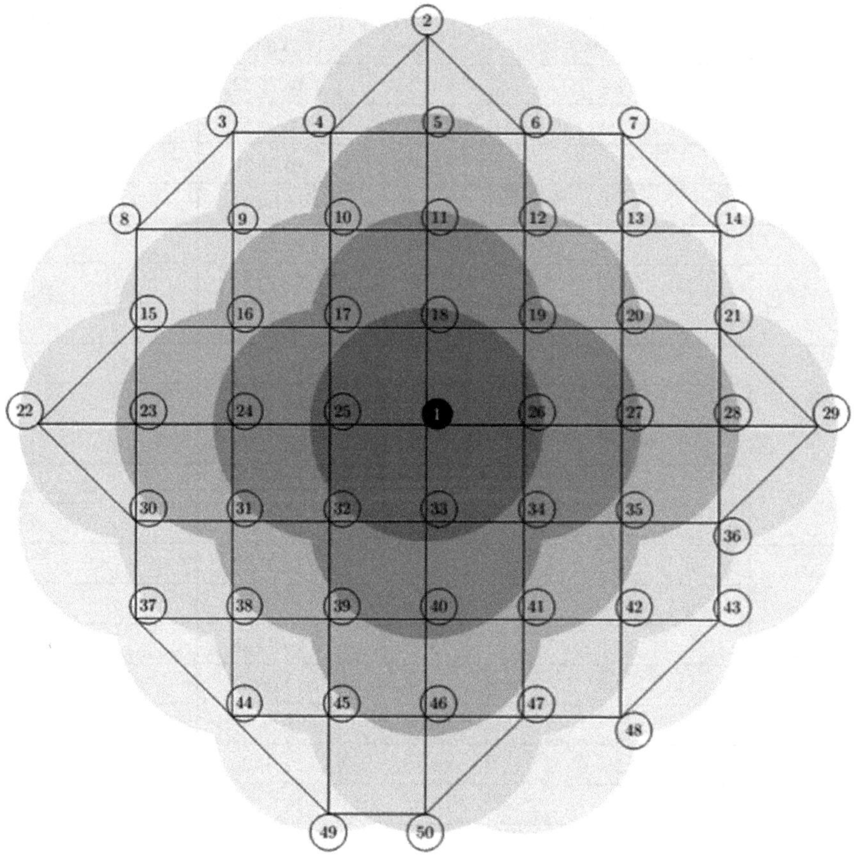

Figure 16. Simulated coordinates of a GPS receiver augmented with one Pseudolite (Roberts et al., 2006)

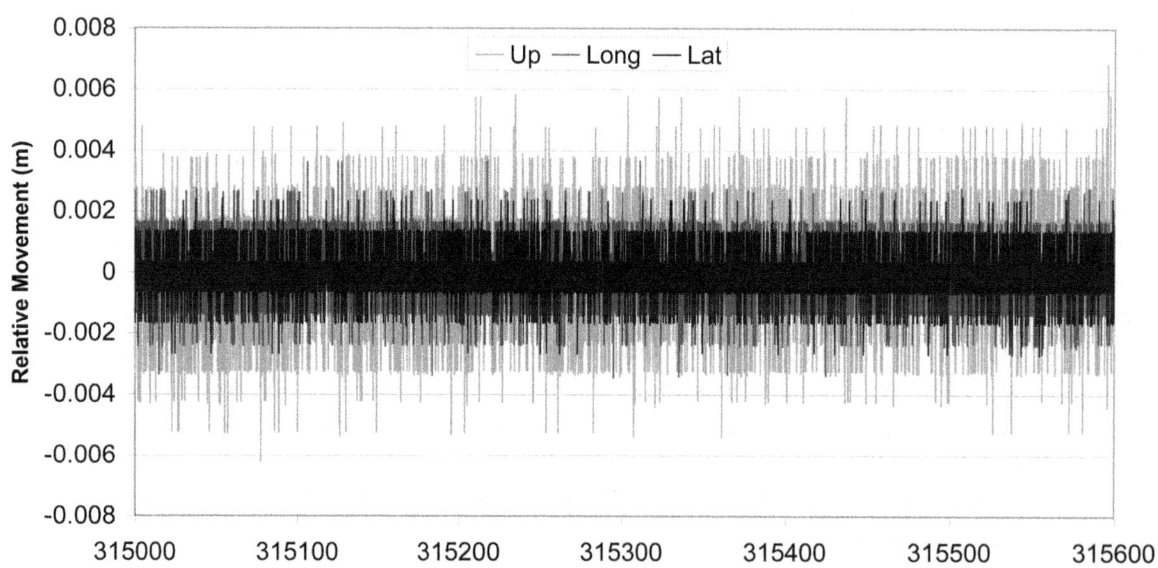

2. Combining Measurements of a Static Rover with a Modified LMS Filter

Following the recent work of Montillet and Yu (2011) and Montillet and Yu (2014b), it has been shown that the stochastic properties of the coordinate time series of a rover in stand-alone positioning or in RTK positioning can be modeled by the fractional Brownian motion (fBm). To some extend, the stochastic model of the daily GPS time series can also be modeled with the fBm model due to the coloured noise contained in the time series (Montillet et al., 2012). The Hurst parameter plays a key role to determine the degree of fitting of the fBm model to the time series of the receiver coordinates. This section introduces the fBm model and its application to time series modeling and recalls the fundamentals about Least Mean Squares (LMS) adaptive filter. Furthermore, based on the work of Montillet and Yu (2011) and Montillet and Yu (2014b), a modified adaptive filter is developed in order to couple measurements recorded in a static position from different technologies (i.e. GPS RTK, Locata) with the assumption that the measurement noise experiences a similar stochastic model.

a. Overview of the Fractional Brownian Motion and Estimation of the Hurst Parameter

The Brownian motion is a well-known model to describe the movement of a particle in a liquid (Papoulis & Unnikrishna Pillai, 2002). Due to its stochastic properties, the model is used in many research topics such as in financial market it is used to model the randomness of the price of the stokes (Wooldridge, 2010) and in telecommunications to model the signal noise (Norros, 1995). Mandelbrot and Van Ness (1968) developed the fractional Brownian motion (fBm) based on the Hurst parameter (*H*) to model long-run non-periodic statistical dependences of time series. This fBm model has been successfully applied in chemistry with the time series of the turbulences of the pressure signal of an airlift reactor (Scheffer & Marciel Filho, 2001) and in telecoms for the study of specific networks such as in Norros (1995). When it comes to model long time series of a phenomenon, the limit process is never Brownian motion, nor is it necessarily Gaussian. The fBm model gives some degree of freedom to characterize the process noise of the time series thanks to the Hurst parameter (*H*). Note that this model is generally referred in the literature as first or second order fBm (Gupta & Joshi, 2008). Recent developments of the fBm model have shown its generalization to the m-th order ($m \geq 2$), which gives interesting properties for the associated covariance matrix with some approximations for long time series (Montillet & Yu, 2015).

The Hurst parameter was introduced in an early work from Hurst, Black, and Sinaika (1965) where the authors studied the long-run non-periodic statistical dependences of time series in hydrology. This parameter is in theory in the interval [0,1]. Unlike the Brownian motion, the fBm has long-range dependency property when $H \sim 0.5$. If $H < 0.5$ the increments of the process are negatively correlated and the statistics of the process noise are more Gaussian, whereas if $H > 0.5$ the increments are positively correlated and the statistics of the process noise are more Brownian. Note that the standard Brownian motion can be obtained from the fBm formula when H is only equal to 0.5. The first challenge before applying the fBm model is to estimate the H parameter from the time series. The research community has been investigating the fractional Brownian motion for more than a couple of decades. Taqqu, Teverovsky, and Willinger (1996) summarized the empirical studies to calculate the H parameter. Early works were based on the regression of the time series, which uses the property of self-similarities to compute the H parameter (e.g. (Hurst et al., 1965)) with high standard deviation as a drawback of the methodology. By windowing the time series, Hosking (1981) and Taqqu et al. (1996) developed the auto-regressive moving average (ARMA) and fractional auto-regressive integrated moving average (FARIMA) methods to compute the Hurst parameter. In Montillet and Yu (2011), the authors developed a technique similar to the aggregated variance method developed by Taqqu et al. (1996), but they reformulated the solution in block matrices. Consequently, the Hurst parameter estimation performance reaches the minimum variance unbiased estimator (MVU) using a Gauss-Markov model for fluctuations of *H*. Note that in Montillet and Yu (2014b), the authors applied the ARMA and FARIMA model to daily position GPS time series in order to model the coloured noise. All those methods previously cited are generally referred as aggregated variance method. More recently wavelet theory has been applied to this engineering problem (Stoev & Taqqu, 2003). The fBm was introduced as a model of the phenomena in hydrology or fluctuations in solids (Mandelbrot & Van Ness, 1968). This work was based on an early work from Hurst et al. (1965), which derived an empirical formula from various time series with a main emphasis on hydrology (i.e. long-term storage capacity of reservoirs). The fBm model is defined as:

$$B_H\left(t,\omega\right)-B_H\left(0,\omega\right)=\frac{1}{\Gamma\left(\mathrm{H}+0.5\right)}\left\{\begin{array}{l}\int_{-\infty}^{0}\left[\left(t-\mathrm{s}\right)^{\{\mathrm{H}-1/2\}}-\left(-\mathrm{s}\right)^{\{\mathrm{H}-1/2\}}\right]\mathrm{dB}\left(\mathrm{s},\omega\right)+\int_{0}^{t}\left[\left(t-\mathrm{s}\right)^{\{\mathrm{H}-1/2\}}\right.\\ \left.-\left(-\mathrm{s}\right)^{\{\mathrm{H}-1/2\}}\right]\mathrm{dB}\left(\mathrm{s},\omega\right)\end{array}\right\} \tag{53}$$

With *H* the Hurst parameter, $B_H(0, \omega)$ a starting value. Note that *t* has to be positive in the general expression of the fBm. Finally, many algorithms estimating the Hurst parameter are based on the self-similarity property (Taqqu et al., 1996; Montillet & Yu, 2011).

b. Modified Leaky LMS Algorithm for Smoothing Position Estimates

i. Statistical Models in Adaptive Filtering

The time series of the focus here is the coordinates of a receiver (in WGS 84) triangulated with GPS Pseudoranges in single point positioning. For a fixed receiver, it is interesting to model the long-term and short-term correlation of these time series (Hosking, 1981) when recording the rover's position during hours and more. The choice of this model is to give some degree of freedom instead of using the conventional Gaussian model, which is generally used in satellite-based positioning theory (Strang & Borre, 1997). One way to constrain the variance of the estimated receiver position is generally with the use of a Kalman filter in backward filtering (Strang & Borre, 1997). Here, a simpler approach with a LMS filter is studied as the case study is limited to a static receiver. The scientific community has already investigated the application of the adaptive filters to constrain the error of time-of-arrival measurements in sensor localization (Montillet, Yu & Oppermann, 2007). The authors in Weinbach, Raziq and Collier (2009) developed a LMS filter to reduce the recurrent multipath error due to satellite geometry on the double difference observations for a static rover. As a result, the double difference residuals have smaller variances and the rover's coordinates are steadier.

The LMS filter is an iterative algorithm which adapts a stationary random process (*v(n)*) to a desired output (*d(n)*) with the same stochastic properties. Let us define the discrete-time processes:

$$\mathbf{v}_{\mathrm{k}}=\mathbf{x}_{\mathrm{k}}-\mathrm{x}_0$$

$$\boldsymbol{d}_k=\mathbf{g}_{\mathrm{k}}-\mathrm{x}_0 \tag{54}$$

with the time series of the receiver's coordinates in one dimension equal to $\mathbf{x}_k = \left[x(k), x(k+1), \ldots, x(k+N)\right]^T$. $\mathbf{x}_k$, $\mathbf{v}_k$, $\mathbf{g}_k$ and $\mathbf{d}_k$ are in $\mathbb{R}^N$. As v_k is the time series on which to adapt a desired time series $\boldsymbol{d}_k$, it follows that $\sigma_d^2 < \sigma_v^2$. In other words, it can be seen that $\boldsymbol{d}_k$ is the coordinate time series of the same rover, but filtered (i.e. moving average) or coming from another sensor if integrated together (i.e. Inertial Measurement Unit – (Moore et al., 2007)). x_0 is the true rover's coordinate. Due to propagation errors on the satellite measurements, $\mathbf{v}_k$ and $\boldsymbol{d}_k$ are not necessary zero-mean Gaussian distributed, but instead follow a fractional Brownian motion model such as:

$$\mathbf{v}_k \sim \text{fbm}\left(H_1, \sigma_v^2, \mu_v\right)$$

$$\boldsymbol{d}_k \sim \text{fbm}\left(H_2, \sigma_d^2, \mu_d\right) \quad (55)$$

H_1 and H_2 are the Hurst parameter and satisfy $\left|H_1 - H_2\right| < \xi_1$ where ξ_1 is a very small positive constant. To guarantee the convergence of the LMS filter and suppress the unwanted bias of the solution, it is important to have a strong hypothesis on the means such as $\left|\mu_v - \mu_d\right| < \xi_2$ with ξ_2 very small.

The simpler form of the LMS algorithm described in Haykin (2001) reduces the cost function based on the mean square error $e_k = d_k - y_k$ with $y_k = \boldsymbol{w}_k^T \mathbf{v}_k$ as:

$$E\left\{e_k^2\right\} = E\left\{d_k^2\right\} - 2\mathbf{p}^T\mathbf{w}_k + \mathbf{w}_k{}^T\mathbf{R}\mathbf{w}_k \quad (56)$$

with *E{.}* the expectation operator, $\boldsymbol{p}$ @$E\left\{d_k \mathbf{v}_k\right\}$ and $\boldsymbol{R} \triangleq \mathbf{v}_k^T \mathbf{v}_k$. The LMS filter is an iterative algorithm which updates the weight vector $\boldsymbol{w}_k$ according to:

$$\mathbf{w}_{k+1} = \mathbf{w}_k + 2\mu e_k \mathbf{v}_k \quad (57)$$

where μ is a selectable parameter. It has been shown that in the case of uncorrelated Gaussian data, $\boldsymbol{w}_k$ converges to the Wiener solution with $0 < \mu < 1/\left(3tr\left(\boldsymbol{R}\right)\right)$ (Kamenetsky & Widrow, 2004). The time delay for the weight vector to relax to its optimal value is proportional to the n-th eigenvalue of $\boldsymbol{R}$. However, the eigenvalue spread $\frac{\lambda_{max}}{\lambda_{min}}$ plays a critical role in converging time of the LMS filter. Figure 17 shows the basic principle of the LMS adaptive filter.

Figure 17. The Least Mean Squares (LMS) adaptive filter applied to smooth position estimates $\mathbf{v}_k$ using a reference time series $\mathbf{d}_k$ with the output of the filter $\boldsymbol{y}_k$. $\boldsymbol{b}_1$ is the tap weights (Montillet & Yu, 2014b)

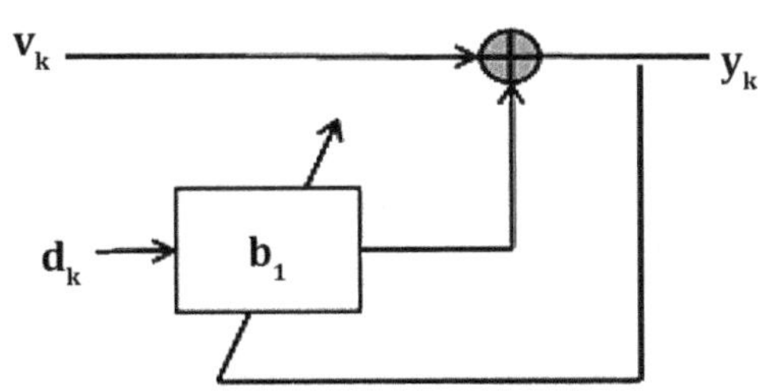

ii. Modified Leaky LMS

Let us consider two fBm processes with Hurst parameters H_1 and H_2 such as H_1<0.5 and H_2>0.5, with respectively covariance matrix $\boldsymbol{R}_1$ and $\boldsymbol{R}_2$. Then, the eigenvalues spread satisfies the equation:

$$\left(\frac{\lambda_{max}}{\lambda_{min}}\right)_{H_1} < \left(\frac{\lambda_{max}}{\lambda_{min}}\right)_{H_2} \tag{58}$$

Gupta and Joshi (2006) demonstrated the following theorem: the $N \times \mathrm{N}$ auto-covariance matrix $\boldsymbol{R}(\mathrm{n})$ ofadiscretefractionalBrownianmotion,withHurstparameter,diagonalizableas $\boldsymbol{R}(\mathrm{n}) = \boldsymbol{Q}(\mathrm{n})\Lambda(\mathrm{n})\boldsymbol{Q}(\mathrm{n})^T$ can be approximated as $\hat{\boldsymbol{R}}(\mathrm{n}) = \boldsymbol{Q}\hat{\Lambda}(\mathrm{n})\boldsymbol{Q}^T$ for large n where $\boldsymbol{Q}$ is a constant orthogonal matrix and $\hat{\Lambda}(n) = diag\ \{\lambda_1,\ \lambda_2, ..., \lambda_N\ (\mathrm{n})\}$. Thus only the largest eigenvalue is time dependent for large *n* with:

$$\lambda_N(\mathrm{n}) \infty \frac{1}{\Gamma(2\mathrm{H}+1)|\sin \mathrm{H}|}\left[2\mathrm{Nn}^{2\mathrm{H}} + 2\mathrm{Hn}^{2\mathrm{H}-1}(\mathrm{N}-1)\mathrm{N} - \frac{2}{\mathrm{N}}\sum_{\mathrm{i}=1}^{\mathrm{N}-1}\mathrm{i}(\mathrm{N}-\mathrm{i})^{2\mathrm{H}}\right]$$

if $N \gg 1$ and $n ?\ 0$, one can approximate the largest eigenvalue such as

$$\lambda_N(\mathrm{n}) \infty \frac{1}{\Gamma(2\mathrm{H}+1)|\sin \mathrm{H}|}\left[2\mathrm{Nn}^{2\mathrm{H}} + 2\mathrm{Hn}^{2\mathrm{H}-1}(\mathrm{N}-1)\mathrm{N}\right] \tag{59}$$

Equation 59 is the dividend of $\frac{P(2,N,H)}{f(H)}$, with *P* a polynomial of degree 2, and *f* a function of the Hurst parameter. It is easy to see that *f* is slowly increasing and bounded (with [0, 3.5]) as *H* is in [0, 1], whereas by definition *P* increases much quicker regards to *N* and *H*. It then follows for $H_1 < H_2$, $N?$ 1 and $n \gg 0$:

$$\lambda_{\mathrm{N}}(n) < \lambda_{\mathrm{N}}(n+1)$$

$$\lambda_{\mathrm{N}}(H_1) < \lambda_{\mathrm{N}}(H_2) \tag{60}$$

and thus,

$$\left(\frac{\lambda_{\mathrm{N}}(n)}{\lambda_1}\right)_{H_1} < \left(\frac{\lambda_{\mathrm{N}}(n)}{\lambda_1}\right)_{H_2}$$

$$\left(\frac{\lambda_{max}}{\lambda_{min}}\right)_{H_1} < \left(\frac{\lambda_{max}}{\lambda_{min}}\right)_{H_2} \tag{61}$$

When the eigenvalue spread increases, Kamenetsky & Widrow, (2004) showed that the rate of convergence of the LMS algorithm decreases. In this case study, the eigenvalue spread increases as *H* is close to 1. Thus, the cost function of the leaky LMS (or L-LMS) is

$$\mathrm{J}_{\mathrm{k}} = \mathrm{e}_{\mathrm{k}}^{2} + \gamma \mathbf{w}_{\mathrm{k}}{}^{\mathrm{T}} \mathbf{w}_{\mathrm{k}} \tag{62}$$

with γ is called the "leak" parameter and it is selected by the user ($\gamma \geq 0$). The name stems from the fact that, when the input is turned off, the weight vector of the regular LMS algorithm stalls. A full description of the properties of the L-LMS is described in Kamenetsky and Widrow, (2004). One of the important properties is that the leaky algorithm decreases the eigenvalue spread. If the eigenvalue spread is defined by λ_{max} and λ_{min}, then according to Kamenetsky and Widrow, (2004),

$$\left(\frac{\lambda_{\max} + \gamma}{\lambda_{\min} + \gamma}\right) \leq \left(\frac{\lambda_{\max}}{\lambda_{\min}}\right), \ \gamma \geq 0 \tag{63}$$

Similarly to the cost function defined in the variable leaky LMS algorithm, Montillet and Yu (2014b) define:

$$\mathrm{J}_{\mathrm{k}}(\mathrm{H}) = \mathrm{e}_{\mathrm{k}}^{2} + \mathrm{f}(\mathrm{H}) \mathbf{w}_{\mathrm{k}}{}^{\mathrm{T}} \mathbf{w}_{\mathrm{k}} \tag{64}$$

With the function $f(H)$ defined as:

$$\mathrm{f}(\mathrm{H}) = \begin{cases} \mathrm{H}^{\beta} \ \textit{if} \ H \leq 0.5, \\ (\mathrm{H} - 0.5)^{\beta} \ \textit{if} \ H > 0.5, \\ \beta \geq 2 \end{cases} \tag{65}$$

Clearly the *H* parameter is directly used to adjust the cost function. In Montillet & Yu (2011), the authors used the cost function $f(H) = H$. Thus, the tap weight vector is updated by:

$$\mathbf{w}_{\mathrm{k+1}} = \left(1 - 2\mu \mathrm{f}(\mathrm{H})\right) \mathbf{w}_{\mathrm{k}} + 2\mu \mathrm{e}_{\mathrm{k}} \mathbf{v}_{\mathrm{k}} \tag{66}$$

In other words, the cost function in Equation 63 shows the trade-off between the optimal LMS algorithm when the noise properties experience Gaussian increments (*H<0.5*) or Brownian increments (*H>0.5*). Note that the literature also refers these properties to persistent and anti-persistent. This can be shown by reformulating the eigenvalue spread following Equation 64:

$$\left(\frac{\lambda_{\max} + \gamma}{\lambda_{\min} + \gamma}\right) \leq \left(\frac{\lambda_{\max} + \mathrm{f}(\mathrm{H})}{\lambda_{\min} + \mathrm{f}(\mathrm{H})}\right) \leq \left(\frac{\lambda_{\max}}{\lambda_{\min}}\right), \ \gamma \geq 0 \tag{67}$$

Thus, the proposed Modified L-LMS (ML-LMS) in Montillet and Yu (2011) and Montillet and Yu (2014b) includes a degree of freedom when the fBm noise changes abruptly (i.e. persistent and anti-persistent effects in the time series).

iii. Results on Experimental Data

Finally, this example ends with some of the results published in Montillet and Yu (2014b), which are based on real time series. The early results using simulations can be found in Montillet and Yu (2011). The work developed in Montillet and Yu (2014b) particularly underlined the comparison in terms of performances between the L-LMS proposed in Kamenetsky and Widrow (2004).

The adaptive filters are applied in two real case scenarios. The first case study is a GNSS station in the middle of London in a built-up environment. The position was recorded during 1780 epochs using Real Time Kinematic, without ($\mathbf{v}_k$) and with a choke ring antenna ($\mathbf{d}_k$). The position is initially triangulated in WGS84 reference frame and then translated in a local East, North, Up reference frame. The figure at the top of Figure 18 shows the time series of the East coordinate. The remaining figures display the Minimum Square Error (MSE) when adapting the time series of $\mathbf{v}_k$ with $\mathbf{d}_k$ using the L-LMS and ML-LMS. The statistics of the MSE when applying the adaptive filters on the East and North coordinates are shown in Table 2. The error is generally very small as the amplitude of the time series coordinates of the GNSS receiver is in the order of magnitude of the centimetre. In particular, one can see that the statistics of the MSE corresponding to the ML-LMS are the lowest among all the adaptive filters tested here.

The second case study is based on the GNSS data recorded at the station in Nottingham (UK) in open sky environment, when using Pseudoranges to triangulate the position of the GNSS receiver ($\mathbf{v}$). One can also use the carrier-phase smoothing technique to improve the accuracy of the position of the receiver estimates ($\boldsymbol{d}$). The statistics of the position are calculated over 1800 epochs. The statistics of the MSE when applying the adaptive filters on the East and North coordinates, are shown in the Table 3. In this case, the standard deviation of the time series of the receiver coordinates $\mathbf{v}_k$ or $\boldsymbol{d}_k$ is large. One can then see that the mean and standard deviation of the MSE is large compared to the first case study. Overall, the results are similar to the one before. The ML-LMS outperforms the other algorithms.

To conclude, the modified LMS filter can couple two coordinates time series if the stochastic properties of these time series are well modeled. Beyond the application to satellite positioning systems and surveying, this work can also be applied in order to couple different kind of sensors (i.e. Ultra Wide band technology, Wi-Fi).

FUTURE RESEARCH DIRECTIONS

Regarding satellite based positioning systems, one future research direction would focus on the integration of multi constellations of satellites, enabling more reliable and more accurate position estimation such as for car navigation and surveying (especially in urban canyons and other areas with limited view of the sky). Another future research focus would be on improving models in signal propagation (e.g. troposphere and ionosphere model with new available frequencies) such as by exploiting meteorology. A joint study of GNSS and meteorology are beneficial to both fields, enhancing GNSS positioning accuracy and weather forecast reliability.

Figure 18. East coordinates of a GNSS receiver with and without using a choke ring antenna (top), and Mean Square Error (MSE) when using either L-LMS, or ML-LMS adaptive filter (Montillet & Yu, 2014b)

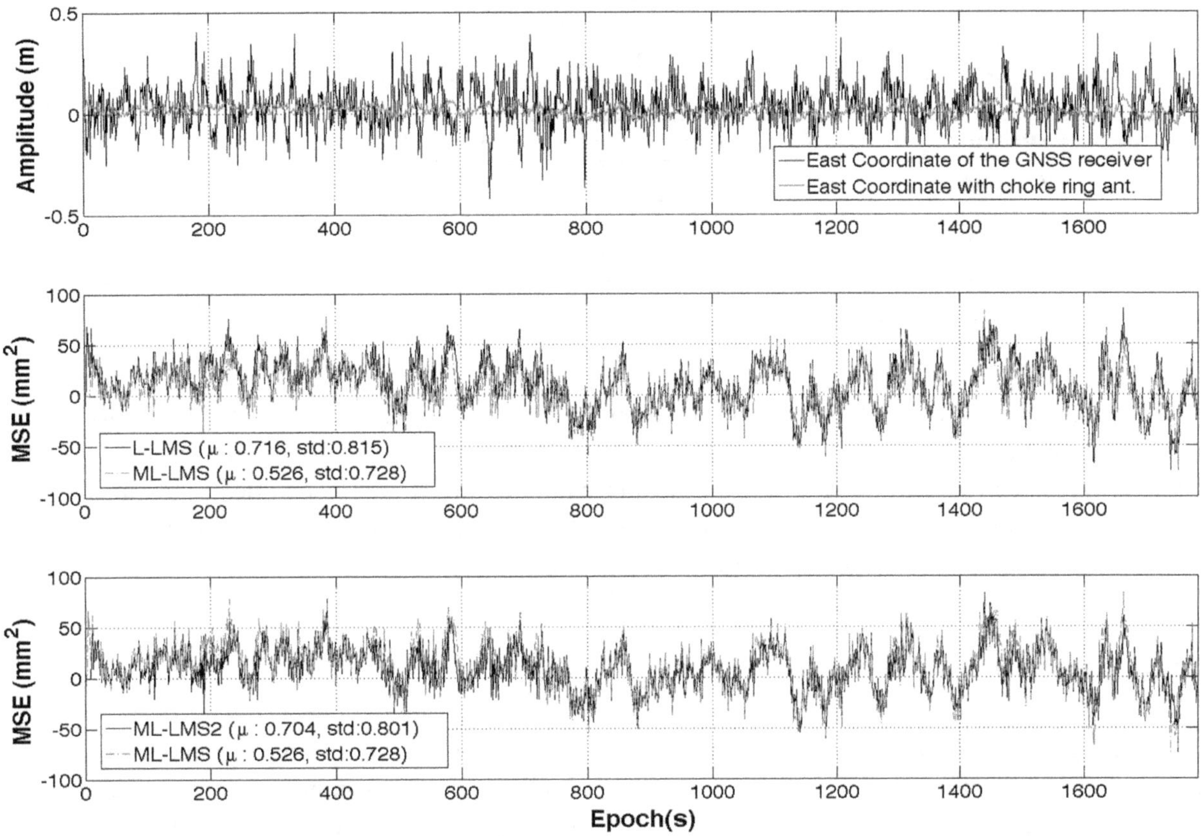

Table 2. Mean Square Error (MSE) and Standard Deviation (std) for the different adaptive filters when applied to the East and North coordinate of the GNSS station following the first case study

MSE (m^2)	East		North	
	H=0.14		H=0.22	
	μ	Std	μ	std
Original	0.033	0.107	0.043	0.124
Reference	0.034	0.040	0.042	0.073
LMS (0.001)	0.625	0.955	1.824	2.377
L-LMS (0.001)	0.716	0.815	1.786	2.410
ML-LMS (0.001)	0.526	0.728	1.484	2.043

Table 3. Mean Square Error (MSE) and Standard Deviation (std) for the different adaptive filters when applied to the East and North coordinate of the GNSS station following the second case study

MSE (m^2)	East		North	
	H=0.32		H=0.45	
	μ	Std	μ	Std
Original	-1.168	2.095	-0.628	3.129
Reference	-0.211	1.047	-0.534	1.564
LMS	0.306	0.571	0.981	1.752
L-LMS	0.377	0.537	1.169	1.679
ML-LMS	0.129	0.328	0.791	1.579

Currently multi-sensor data fusion for positioning (e.g. combining measurements from GNSS and local positioning systems) is an active research area. Multi-sensors typically would provide a better solution to locate in real time a user or to survey in rich multipath environments such as urban canyons. In addition, this multi-sensors system will improve the coverage of LBS in areas where at the moment services are limited due to the accuracy of the positioning ability (i.e. indoor shopping mall).

CONCLUSION

Local and global positioning are two areas of fast development and high impact research, which has already found many applications in the industry, army and medical environment. Some of theses applications are already revolutionizing some daily tasks (i.e. locating stores or chemists closest to the user's position). Another intense research area is car navigation. Car manufacturers are developing self-driving cars with the availability of cheap and reliable positioning systems.

This chapter has studied concepts, methods and techniques of local and global positioning systems. In particular, detailed descriptions of a number of positioning algorithms have been provided. In the first part, satellite positioning systems, such as GNSS and the future multi-constellations, have been described together with an overview of the least-squares theory which is the main algorithm to estimate the position of a GNSS receiver. In the second part, Pseudolites including their latest development (e.g. Locata) were studied, which are an alternative to perform precise positioning (centimeter level accuracy) in urban canyons or dense multipath areas. Then, the focus was on the discussion of estimating indoor location using technologies relying on GSM or Wi-Fi signals. Several well-known algorithms were briefly mentioned, which can achieve accuracy from centimeters to a few meters, depending on the technology and the environment. Details of a network based positioning algorithm for locating a vehicle were studied. This algorithm makes use of fusion of distance, speed and heading data to mitigate multipath and NLOS propagation.

Finally in satellite based positioning systems, the integration of multi constellations enables more reliable receivers with applications such as in car navigation system and surveying measurements (especially in urban canyons and other areas with limited view of the sky). Non-linear least-squares method is used for the triangulation of the receiver position. Some research is constantly on going in improving models in signal propagation (e.g. troposphere and ionosphere model with new available frequencies), fixing integer ambiguities for precise positioning with satellites from various constellations, and general improvement on receivers hardware (i.e. low-cost, size, capability of tracking various channels). However, there is a great deal of research work motivated by the need of more precise position in order to generate new applications for Location Based Services .One of the key challenges is to integrate in the location algorithms terrestrial based localization technologies and satellite positioning systems. Some algorithms will use optimization algorithms in order to refine the position of a user together with continuous stream of data. Last but not least, the number of users using LBS is rapidly growing, which means that new challenges such as the storage and handling of high volume of data generated by users are appearing. This will certainly impact the infrastructure of the networks supporting LBS and the design of future location algorithms. Time consuming optimization algorithms may not be the best choice when dealing with large number of users.

Note that the industrial and research communities are also motivated for the need of very precise surveying accuracy (millimeter level) for geodesy and geophysics applications such as to refine terrestrial reference frame and the estimation of the deformation of the Earth after strong events (i.e. Earthquakes). However this topic is beyond the scope of this chapter, which aimed at reviewing the topics in navigation, tracking and precise positioning for industrial applications.

REFERENCES

Abulleif, T., & Al Dossary, A. (2008). *Location Based Services (LBS)*. Report from Surveying Services Division, Saudi Aramco Dhahran, Saudi Arabia. Retrieved from: www.saudigis.org/FCKFiles/File/.../79_E_ThamerAbulleif_KSA.pdf

Al Nuaimi, K., & Kamel, H. (2011). A survey of indoor positioning systems and algorithms. In *Proceedings of the IEEE International Conference on Innovation in Information Technology (IIT)*. doi:10.1109/INNOVATIONS.2011.5893813

Army, U. S. (1996). *Navstar GPS - user equipment introduction*. Technical Report released by the U.S. Government. Retrieved from www.navcen.uscg.gov/pubs/gps/gpsuser/gpsuser.pdf

Bahl, P., & Padmanabhan, V. N. (2000). RADAR: An in-building RF-based user location and tracking system. In *Proceedings of the nineteenth Annual Joint Conference of the IEEE Computer and Communications Societies (INFOCOM 00)* (vol. 2, pp. 775-784). doi:10.1109/INFCOM.2000.832252

Balanis, C. A. (Ed.). (2012). Theory, analysis and design (3rd ed.). John Wiley & Sons.

Barnes, J. B., Rizos, C., Pahwa, A., Politi, N., & Van Cranenbroeck, J. (2007). The potential of Locata technology for structural monitoring applications. *Journal of Global Positioning Systems*, *6*(2), 166–172. doi:10.5081/jgps.6.2.166

Barnes, J. B., Rizos, C., Wang, J., Small, D., Voigt, G., & Gambale, N. (2003). LOCATA: A new positioning technology for high precision indoor and outdoor positioning. *Proceedings of the 16th International Technical Meeting of the Satellite Division of The Institute of Navigation (ION GPS/GNSS 2003)*. Portland, OR.

Barnes, J. B., Wang, J., Rizos, C., & Tsujii, T. (2002). The performance of a Pseudolite based positioning system for deformation monitoring. In *Proceedings of the 2nd Symposium on Geodesy for Geotechnical and Structural Engineering*. Berlin, Germany.

Bonenberg, L. K. (2014). *Closely-coupled integration of Locata and GPS for engineering applications.* (doctoral dissertation). The University of Nottingham.

Bonenberg, L. K., Roberts, G. W., & Hancock, C. M. (2012). Using Locata to augment GNSS in a kinematic urban environment. *Archives of Photogrammetry. Cartography and Remote Sensing*, *12*, 63–74.

Caffery, J. J. (2000). *Wireless location in CDMA cellular radio systems.* Springer.

Cobb, S. H. (1997). *GPS Pseudolites: Theory, design, and applications.* (doctoral dissertation). Stanford University.

Cosser, E. (2004). *Bridge deformation monitoring with single frequency GPS augmented by Pseudolites.* (Doctoral Dissertation). The University of Nottingham.

Cosser, E., Roberts, G. W., Meng, X., & Dodson, A. H. (2003). The comparison of single and dual frequency GPS for bridge deflection and vibration monitoring. In *Proceedings of the 11th International Symposium on Deformation Measurements, International Federation Surveyors (FIG), Commission 6 - Engineering Surveys, Working Group 6.1.* Santorini, Greece.

Craig, D. & Locata Corporation. (2012). Truth on the range. *Inside GNSS*, *7*(3), 37-48.

EDINA, University of Edinburgh. (2000). *EDINA Digimap Ordinance Survey Service.* Retrieved from http://digimap.edina.ac.uk/digimap/

Fontana, R. J. (2004). Recent system applications of short-pulse ultra-wideband (UWB) technology. *IEEE Transactions on Microwave and Techniques*, *52*(9), 2087–2104. doi:10.1109/TMTT.2004.834186

Ford, T., Neumann, J., Toso, N., Petersen, W., Anderson, C., Fenton, P., . . . Barltrop, K. (1997, September). HAPPI- a high accuracy Pseudolite/GPS positioning integration. In *Proceedings of ION/GPS 1997, the 10th Technical Meeting of the Satellite Division of the Institute of Navigation* (pp. 1719-1728).

Geng, J., Shi, C., Ge, M., Dodson, A., Lou, H. Y., Zhao, Q., & Liu, J. (2012). Improving the estimation of fractional-cycle biases for ambiguity resolution in precise point positioning. *Journal of Geodesy*, *86*(8), 579–589. doi:10.1007/s00190-011-0537-0

Gupta, A., & Joshi, S. (2006). A novel Least Mean Squares algorithm for tracking a discrete-time fBm Process. In *Proceedings of the IEEE Indian Conference 2006 (INDCON'06).* doi:10.1109/INDCON.2006.302790

Gupta, A., & Joshi, S. (2008). Some studies on the structure of covariance matrix of discrete-time fBm. *IEEE Transactions on Signal Processing*, *56*(10), 4635–4650. doi:10.1109/TSP.2008.927467

Harcombe, P. (2012). *NSW Sydney satellite.* Oral Presentation from the NSW State Government. Retrieved from http://www.locata.com/wp-content/uploads/2012/12/

Harter, A., Hopper, A., Steggles, P., Ward, A., & Webster, P. (1999). The anatomy of a context-aware application. In *Proceedings of the 5th annual ACM/IEEE international conference on Mobile computing and networking (MobiCom '99)*. New York. doi:10.1145/313451.313476

Haykin, S. (Ed.). (2001). *Adaptive filter theory* (4th ed.). Prentice Hall Upper Saddle River.

Hedley, M., Humphrey, D., & Ho, P. (2008). System and algorithms for accurate indoor tracking using low-cost hardware. In *Proceedings of the IEEE/ION Position, Location And Navigation Symposium (PLANS)*. doi:10.1109/PLANS.2008.4569983

Hide, C. (2003). *Integration of GPS and low cost INS measurements*. (doctoral dissertation). University of Nottingham, Nottingham, UK.

Hofmann-Wellenhof, B., Lichtenegger, H., & Collins, J. (Eds.). (1993). *Global positioning system - theory and practice*. Springer-Verlag.

Hosking, J. R. M. (1981). Fractional differencing. *Biometrika*, *68*(1), 165–176. doi:10.1093/biomet/68.1.165

Hurst, H. E., Black, R. P., & Sinaika, Y. M. (1966). Long term storage in reservoirs: An experimental study. *Journal of the Royal Statistical Society. Series A (General)*, *129*(4), 591–593. doi:10.2307/2982267

Kamenetsky, M., & Widrow, B. (2004). A variable leaky LMS adaptive algorithm. In *Proceedings of the 38th Asilomar Conference on Signals, Systems and Computers*. doi:10.1109/ACSSC.2004.1399103

Kanli, M. O. (2005). Limitations of pseudolites systems using off-the-shelf GPS receivers. *Journal of Global Positioning Systems*, *3*(1-2), 154–166.

Kay, S. (Ed.). (1993). Fundamentals of statistical signal processing: Vol. 1. *Estimation theory* (1st ed.). Prentice Hall Signal Processing Series, Prentice Hall PTR Upper Saddle River.

Kolodziej, K. W., & Hjelm, J. (Eds.). (2006). *Local Positioning Systems: LBS applications and services*. Taylor & Francis Group.

Leick, A. (Ed.). (2004). *GPS satellite surveying* (3rd ed.). New York: John Wiley & Sons.

Liu, H., Darabi, H., Banerjee, P., & Liu, J. (2007). Survey of wireless indoor positioning techniques and systems. *IEEE Transactions on Systems, Man, and Cybernetics (Part C)*, *37*(6), 1067-1079.

Mandelbrot, B. B., & Van Ness, J. W. (1968). Fractional Brownian motions, fractional noises and applications. *RE:view*, *10*(4), 422–437.

Meng, X., Dodson, A. H., Moore, T., & Roberts, G. W. (2007). Ubiquitous positioning: anyone, anything: anytime, anywhere. *GPS World*, *18*(6), 60–65.

Meng, X., Roberts, G. W., Dodson, A. H., Cosser, E., Barnes, J., & Rizos, C. (2004). Impact of GPS satellite and pseudolite geometry on structural deformation monitoring: Analytical and empirical studies. *Journal of Geodesy*, *77*(12), 809–822. doi:10.1007/s00190-003-0357-y

Montillet, J.-P. (2008). *Precise positioning in urban canyons: Applied to the localisation of buried assets*. (doctoral dissertation). University of Nottingham.

Montillet, J.-P., Roberts, G. W., Hancock, C., Meng, X., Ogundipe, O., & Barnes, J. (2009). Deploying a Locata network to enable precise positioning in urban canyons. *Journal of Geodesy*, *83*(2), 91–103. doi:10.1007/s00190-008-0236-7

Montillet, J.-P., Taha, A., Meng, X., & Roberts, G. W. (2007). Buried assets: Testing GPS and GSM in urban canyons. *GPS World*, *18*(3), 39–43.

Montillet, J.-P., Tregoning, P., McClusky, S., & Yu, K. (2012). Extracting white noise statistics in GPS coordinate time series. *IEEE Geoscience and Remote Sensing Letters*, *10*(3), 563–567. doi:10.1109/LGRS.2012.2213576

Montillet, J.-P., & Yu, K. (2011). Leaky LMS algorithm and fractional Brownian motion model for GNSS receiver position estimation. In *Proceedings of the IEEE Vehicular Technology Conference (VTC'11 fall)*. doi:.10.1109/VETECF.2011.6092850

Montillet, J.-P., & Yu, K. (2012). On a new weighting matrix to enhance the accuracy of stand-alone GNSS positioning. In *Proceedings of the IEEE International Symposium on Communications and Information Technologies (ISCIT 12)*. Gold Coast, Australia. doi:10.1109/ISCIT.2012.6380845

Montillet, J.-P., & Yu, K. (2014). Modified leaky LMS algorithms applied to satellite positioning. In *Proceedings of the IEEE Vehicular Technology Conference (VTC'14 fall)*. Vancouver, Canada.

Montillet, J.-P., & Yu, K. (2014b). Modelling geodetic processes with Levy alpha-Stable distribution and FARIMA. *Mathematical Geosciences*. doi:.10.1007/s11004-014-9574-6

Montillet, J.-P., & Yu, K. (2015). Covariance Matrix Analysis for Higher Order Fractional Brownian Motion Time Series. In *Proceedings of the 28th IEEE Canadian Conference on Electrical and Computer Engineering (CCECE)*. Halifax, Canada. doi:10.1109/CCECE.2015.7129488

Montillet, J.-P., Yu, K., & Oppermann, I. (2007). Location performance enhancement with recursive processing of time-of-arrival measurements. In *Proceedings of the IEEE 18th International Symposium on Personal, Indoor and Mobile Radio Communications (PIMRC 2007)*. doi:10.1109/PIMRC.2007.4394529

Moore, T., Hill, C., Hide, C., Ochieng, W., Feng, S., Aguado, E., & Lau, L. et al. (2007). *End-to-End testing of an integrated centimetric positioning test-bed. In Proceedings of the ION GNSS 2007*. Fort Worth, TX: Fort Worth Convention Center.

Niculescu, D. (2001) Ad-hoc Positioning System (APS). In *Proceedings of the IEEE Global telecommunications Conference, 2001 (GLOBECOM'01)* (vol. 5, pp. 2926-2931). doi:10.1109/GLOCOM.2001.965964

Niculescu, D. (2003) Ad-hoc Positioning System (APS) using AoA. In *Proceedings of the twenty-second Annual Joint Conference of the IEEE Computer and Communications Societies (INFOCOM03)* (vol. 3, pp. 1734-1743). doi:10.1109/INFOCOM.2003.1209196

Niculescu, D. (2004). Positioning in Ad Hoc Sensor Networks. *IEEE Network*, *18*(4), 24–29. doi:10.1109/MNET.2004.1316758

Norros, I. (1995). On the use of fractional Brownian motion in the theory of connectionless networks. *IEEE Journal on Selected Areas of Telecommunications*, *13*(6). doi:10.1109/49.400651

Odolinski, R., Teunissen, P. J. G., & Odijk, D. (2014). Combined BDS, Galileo, QZSS and GPS single-frequency RTK. *GPS Solutions*, *19*(1), 151–163. doi:10.1007/s10291-014-0376-6

Papoulis, A., & Unnikrishna Pillai, S. (Eds.). (2002). *Probability, random variables, and stochastic processes* (4th ed.). Mc Graw Hill.

Priyantha, N. B., Chakraborty, A., & Balakrishnan, H. (2000). The Cricket location-support system. In *Proceedings of the 6th ACM International Conference on Mobile Computing and Networking (ACM MOBICOM)*. doi:10.1145/345910.345917

Proakis, J. G. (Ed.). (2000). Digital communications (4th ed.). Mc-Graw Hill.

Rizos, C. (2008). Multi-constellation GNSS/RNSS from the perspective of high accuracy users in Australia. *Journal of Spatial Sciences*, *53*(2), 29–63. doi:10.1080/14498596.2008.9635149

Roberts, G. W. (1997). *Real Time on-the-fly Kinematic GPS*. (Doctoral Dissertation). The University of Nottingham.

Roberts, G. W., Ashkenazi, V., & Dodson, A. H. (1999). Monitoring the height deflections of the Humber Bridge by GPS, GLONASS and finite element modeling. In *Proceedings of the International Association of Geodesy (IAG) Symposia: Geodesy Beyond 2000, the challenges of the first decade.*

Roberts, G. W., Meng, X., & Brown, C. (2004, July). From St Paul's to the Tate Modern; Overcoming problems in monitoring bridge deflections using GPS. In *Proceedings of the 1st FIG (Federation Internationale des Geometres – International Federation of Surveyors) International Symposium on Engineering Surveys for Construction Works and Structural Engineering, session 8: Frequencies and Cyclic Monitoring of Structures.*

Roberts, G. W., Meng, X., Brown, C. J., & Dallard, P. (2006). GPS measurements on the London Millennium Bridge. *The Institution of Civil Engineers: Bridge Engineering*, *159*(4), 153–161. doi:10.1680/bren.2006.159.4.153

Rodriguez, M., Pece, J. P., & Escudero, C. J. (2005) In-building location using Bluetooth. In *Proceedings of the International Workshop on Wireless Ad-hoc Networks (IWWAN)*. doi:10.1.1.126.2802

Saastamoinen, J. (1973). Contribution to the theory of atmospheric refraction. *Bulletin Géodésique*, *107*(1), 13–34. doi:10.1007/BF02522083

Savvides, A., Han, C.-C., & Srivastava, M. (2001). Dynamic fine-grained localization in ad-hoc networks of sensors. In *Proceedings of the seventh Annual international Conference on Mobile Computing and Networking (ACM MOBICOM)*. Rome, Italy.

Scheffer, R., & Maciel Filho, R. (2001). The fractional Brownian motion as a model for an industrial airlift reactor. *Chemical Engineering Science*, *56*(2), 707–711. doi:10.1016/S0009-2509(00)00279-7

Schwieger, V. (2003). Using handheld GPS receivers for precise positioning. In *Proceedings of the 2nd International Federation of Surveyors (FIG) Regional Conference*. Marrakech, Morocco.

Sharp, I., Yu, K., & Sathyan, T. (2012). Positional accuracy measurement and error modeling for mobile tracking. *IEEE Transactions on Mobile Computing*, *11*(6), 1021–1032. doi:10.1109/TMC.2011.119

Singer, C. (1941). *A Short History of Science to the Nineteenth Century.* Clarendon Press.

Stoev, S., & Taqqu, M. S. (2003). Lecture Notes in Physics: Vol. 621. *Wavelet estimation for the Hurst parameter in stable processes.* doi:10.1007/3-540-44832-2_4

Strang, G., & Borre, K. (1997). *Linear algebra, geodesy, and GPS.* Wellesley, MA: Wellesley Cambridge Press.

Taqqu, M. S., Teverovsky, V., & Willinger, W. (1996). Estimators for long-range dependence: An empirical study. *Fractals, 3*(4), 785–798. doi:10.1142/S0218348X95000692

Teunissen, P. (1995). The Least-Squares AMbiguity Decorrelation Adjustment: A method for fast GPS integer ambiguity estimation. *Journal of Geodesy*, *70*(1-2), 65–82. doi:10.1007/BF00863419

Teunissen, P. J. G., Odolinski, R., & Odijk, D. (2014). Instantaneous BeiDou plus GPS RTK positioning with high cut-off elevation angles. *Journal of Geodesy*, *88*(4), 335–350. doi:10.1007/s00190-013-0686-4

University of Nottingham (UoN). (2014). *Lecture notes in space geodesy. Material retrieved from lectures in GNSS.* Surveying and Space Geodesy at the University of Nottingham.

Van Dierendonck, A. (Ed.). (1997). *Satellite radio navigation, in avionic navigation sytems.* John Wiley & Sons.

Wang, L., Groves, P., & Ziebart, M. K. (2013). GNSS shadow matching: Improving urban positioning accuracy using a 3D city model with optimized visibility scoring scheme. *Journal of Navigation*, *60*(3), 195–207. doi:10.1002/navi.38

Want, R., Hopper, A., Falcao, V., & Gibbons, J. (1992). The active badge location system. *ACM Transactions on Information Systems*, *10*(1), 91–102. doi:10.1145/128756.128759

Weinbach, U., Raziq, N., & Collier, P. (2009). Mitigation of periodic GPS multipath errors using a normalised least mean square adaptive filter. *Journal of Spatial Science*, *54*(1), 1–13. doi:10.1080/144 98596.2009.9635162

Wooldridge, J. M. (Ed.). (2010). *Econometric analysis of cross section and panel data.* MIT Press.

Yu, K., & Dutkiewicz, E. (2012). Geometry and motion based positioning algorithms for mobile tracking in NLOS environments. *IEEE Transactions on Mobile Computing*, *11*(2), 254–263. doi:10.1109/ TMC.2011.24

Yu, K., Sharp, I., & Guo, J. Y. (2009). *Ground-based wireless positioning.* Wiley-IEEE Press. doi:10.1002/9780470747988

Yu, K., Saarnisaari, H., Montillet, J.-P., Rabbachin, A., Oppermann, I., & Thadeu Freitas de Abreu, G. (2006). Localization. In *Ultra-wideband wireless communications and networks.* John Wiley & Sons.

ADDITIONAL READING

Garg, H. (2013). Predicting uncertain behavior of complex repairable industrial systems using PSO and fuzzy confidence interval based methodology. In P. Vasant (Ed.), *Handbook of research on novel soft computing intelligent algorithms: Theory and practical applications* (pp. 414–449). IGI Global USA.

Garg, H. (2013). An approach for analyzing fuzzy system reliability using particle swarm optimization and intuitionistic fuzzy set theory. *Journal of Multiple-Valued Logic and Soft Computing*, *21*(3-4), 335–354.

Garg, H. (2013). Performance analysis of complex repairable industrial systems using PSO and fuzzy confidence interval based lambda-tau methodology. *ISA Transactions*, *52*(2), 171–183. doi:10.1016/j.isatra.2012.09.010 PMID:23098922

Garg, H. (2014). A hybrid GA-GSA algorithm for optimizing the performance of an industrial system by utilizing uncertain data. In P. Vasant (Ed.), *Handbook of research on artificial intelligence techniques and algorithms* (pp. 625–659). USA: IGI Global.

Garg, H. (2014). Solving structural engineering design optimization problems using an artificial bee colony algorithm. *Journal of Industrial and Management Optimization*, *10*(3), 777–794. doi:10.3934/jimo.2014.10.777

Garg, H. (2015). An approach for analyzing the reliability of industrial system using fuzzy Kolmogorov's differential equations. *The Arabian Journal for Science and Engineering*, *40*(3), 975–987. doi:10.1007/s13369-015-1584-2

Garg, H., & Rani, M. (2013). An approach for reliability analysis of industrial systems using PSO and IFS technique. *ISA Transactions (Elsevier)*, *52*(6), 701–710. doi:10.1016/j.isatra.2013.06.010 PMID:23867122

Garg, H., Rani, M., & Sharma, S. P. (2015). An approach for analyzing the reliability of industrial systems using soft computing based technique. *Expert systems with Applications (Elsevier)*, *41*(2), 489–501. doi:10.1016/j.eswa.2013.07.075

Garg, H., & Sharma, S. P. (2013). Multi-objective reliability-redundancy allocation problem using particle swarm optimization. *Computer and Industrial Engineering (Elsevier)*, *64*(1), 247–255. doi:10.1016/j.cie.2012.09.015

Ghaderi, A. (2012). Investigating of hybrid meta-heuristics to solve the large-scale multi-source Weber problems and performance measuring of them with statistical tests. In P. Vasant (Ed.), *Handbook meta-heuristics optimization algorithms in engineering, business, economics, and finance* (pp. 171–197). USA: IGI Global.

Gupta, A., & Joshi, S. (2013). DCT and Eigenvectors of Covariance of 1st and 2nd order Discrete fractional Brownian motion. Retrieved from ArXiv: http://arxiv.org/pdf/1302.5556.pdf

Hui, X., & Yan, Q. (2010). Multi-algorithm hybrid location model based on data fusion. *In Proceedings of the International Conference on Communications and Mobile Computing (CMC)*, 1, 408-412. doi:10.1109/CMC.2010.105

Litvinchev, I., & Ozuna, E. L. (2012). Lagrangian bounds and a heuristic for the two-stage capacitated facility location problem. [IJEOE]. *International Journal of Energy Optimization and Engineering*, *1*(1), 59–71. doi:10.4018/ijeoe.2012010104

Mukhopadhyay, P., Dutta, S., & Roy, P. K. (2015). Optimal location of TCSC using opposition teaching learning based optimization. [IJEOE]. *International Journal of Energy Optimization and Engineering*, *4*(1), 85–101. doi:10.4018/ijeoe.2015010106

Nechaev, I. A., & Palamarchuk, S. (2014). Generation scheduling based on two-level optimization problem. [IJEOE]. *International Journal of Energy Optimization and Engineering*, *3*(1), 1–11. doi:10.4018/ijeoe.2014010101

Polprasert, J., Ongsakul, W., & Dieu, V. N. (2013). A new improved particle swarm optimization for solving nonconvex economic dispatch problems. [IJEOE]. *International Journal of Energy Optimization and Engineering*, *2*(1), 60–77. doi:10.4018/ijeoe.2013010105

Purnomo, H. D., & Wee, H.-M. (2013). Metaheuristic methods for configuration of assembly lines: A survey. In P. Vasant (Ed.), *Handbook of research on novel soft computing intelligent algorithms: Theory and practical applications* (pp. 165–199). IGI Global USA.

Savsani, V. (2013). Multi-objective design optimization of rolling element bearings using ABC, AIA and PSO techniques. [IJEOE]. *International Journal of Energy Optimization and Engineering*, *2*(3), 102–125. doi:10.4018/ijeoe.2013070107

Singh, B. K. (2013). Evaluation of generic algorithm as learning system in rigid space interpretation. In P. Vasant (Ed.), *Handbook of research on novel soft computing intelligent algorithms: Theory and practical applications* (pp. 475–510). IGI Global USA.

Taylan, P., Weber, G.-W., & Beck, A. (2007) New approaches to regression by generalized additive models and continuous optimization for modern applications in finance, science and technology, *Optimization: A Journal of Mathematical Programming and operations Research*, 56 (5-6), 675-698. doi:.10.1080/02331930701618740

Wang, W., Huang, T., Liu, H., & Pang, F. (2009). Localization algorithm based on SVM-data fusion in wireless sensor networks. *In Proceedings of the 3rd International Conference on Genetic and Evolutionary Computing (WGEC '09)*, 447-450. doi:10.1109/WGEC.2009.16

Weber, G.-W., Batmaz, I., Köksal, G., Taylan, P., & Yerlikaya-Özkurt, F. (2011). CMARS: A new contribution to nonparametric regression with multivariate adaptive regression splines supported by continuous optimization. *Inverse Problems in Science and Engineering*, *20*(3), 371–400. doi:10.1080/17415977.2011.624770

KEY TERMS AND DEFINITIONS

Adaptive Algorithm: algorithm that changes its behavior based on the information available at the time it is running. This field includes the families of least mean squares, recursive least mean squares, and Kalman filters.

Global Navigation Satellite System (GNSS): generic name about the various satellite constellations: Global Positioning System (USA), GLONASS (Russia), BeiDou (China) and Galileo (European Union).

Localization: generic term which is used to describe the algorithms and technologies to estimate the position of a user or node. The field of localization is broad and includes satellite systems, navigation systems, indoor and outdoor terrestrial based technologies.

Location Based Services (LBS): general class of computer program-level services that use location data. It gathers user applications linked to various contexts such as health, indoor object search, entertainment, work, and personal life.

Navigation: filed of study that focuses on the process of monitoring and controlling the movement of a craft or vehicule from one place to another. This field is broad and includes land, marine, aeronautic and space navigation (i.e. satellite systems).

Positioning Accuracy: error on the estimated position of a receiver related to the true position. The positioning accuracy is the results of various phenomena (e.g., signal propagation through different layers of the atmospheres, NLOS between transmitters and receiver, multipath, geophysical processes, …). For many applications in satellite based precise positioning and especially in the application to geodesy, the study of positioning accuracy includes the broad field of time series analysis (e.g. parameters estimation, stochastic modeling, …).

Tracking: In the field of navigation and tracking, it is the action of a system to keep track of vehicles or more generally mobile nodes in real-time. This action involves location technologies (i.e. terrestrial based and/or satellite-based technologies) and GIS.

Chapter 2
Robust Two-Stage and Multistage Optimization:
Complexity Issues and Applications

Michel Andre Minoux
University P. and M. Curie, France

ABSTRACT

This chapter is intended as an overview of robust optimization models related to optimization problems subject to uncertain data, with special focus on the case when uncertainty impacts the right-hand side coefficients in the constraints. Two-stage as well as multistage models are addressed, emphasizing links with applications and computational complexity issues. A class of multistage robust optimization problems for which exact optimal strategies can be efficiently computed (via a robust dynamic programming recursion) is discussed. An application to a multiperiod energy production planning problem is presented into detail, and computational results are reported.

1. INTRODUCTION

The problem of taking the best possible decisions to operate a complex system in an uncertain environment arises in a huge variety of applications in Engineering, Management Science, Operations Research. A rich body of literature related to the question of how to take uncertainty into account within optimization problems has been developed over the years providing a wide variety of models and algorithms, including scenario-based stochastic programming (e.g. Birge, 1985), stochastic dynamic programming (Ross, 1983), chance constraint programming (Charnes & Cooper, 1959).

A well-known limitation of these approaches is that they assume that the uncertain parameters of the problems addressed can be considered as random variables with *perfectly known probability distributions*; however such an assumption is often quite unrealistic because of the lack of input data or the absence of systematic prior statistical analysis.

In order to overcome the above-mentioned limitations, the field of *robust optimization* has emerged and developed in the last 15 years or so in order to handle situations in which probability distributions

DOI: 10.4018/978-1-4666-9644-0.ch002

are essentially unknown, and when it is felt acceptable to optimize the system against the most adverse scenarios taken in a specified set of scenarios, called the *uncertainty set*; some basic references are Ben Tal and Nemirovski (1999, 2000, 2002), Ben Tal *et al.* (2004), Bertsimas and Sim (2003, 2004), Iyengar (2005)

The main focus of this chapter is on robust optimization models and solution algorithms, and it is intended to provide both a *review of already known results* concerning two-stage and multistage robust optimization models (sections 2 and 3), and *new results* concerning: a) comparison of robust optimization models and stochastic optimization models in their capability of taking risk concerns into account (section 2.2); b) a class of efficiently solvable special cases of multistage robust optimization models (section 4). Central to the definition of this class is the concept of *state-space representable uncertainty* set; it includes several types of uncertainty sets already proposed in the literature, and, as an original contribution to the present work, it is shown here to provide a compact representation of uncertainty sets related to Markov chains (section 4.4). Links with applications will also be emphasized throughout the chapter and computational results will be discussed (section 5) in connection with a multiperiod energy production planning problem.

2. BACKGROUND: ROBUST 2-STAGE (LINEAR) OPTIMIZATION PROBLEMS

In this section we fist introduce a generic model for the class of robust 2-stage (linear) optimization problems featuring right-hand side uncertainty (RHS-uncertainty). We next discuss links with related stochastic models and illustrate how the two approaches differ in their capability of taking risk concerns into account. Finally, we provide an overview of possible applications, emphasizing known complexity (NP-hardness) results and polynomial-time solvable special cases.

2.1 Robust 2-Stage Optimization with RHS Uncertainty

For the sake of simplicity in the presentation, we restrict here to the case of *linear* programs with uncertain right-hand side vectors. Of course extensions to nonlinear programming models are possible (an example of such extensions can be found e.g. in (Minoux, 2014)).

Also we note that problems involving nonlinear but *convex* constraints and/or objective function can be transformed (using tangential approximation), up to any prescribed level of accuracy, into LP problems; thus assuming linearity in the generic model considered here is less restrictive than it might appear at first sight.

The generic model of 2-stage robust LP problems discussed in the sequel is based on the deterministic LP problems in variables *x, y*:

$$(\mathrm{I}) \qquad \begin{cases} Minimize\ \gamma^T x + q^T y \\ subject\ \ to: \\ Fx + Hy = b \\ \chi \in x, y \geq 0 \end{cases} \tag{2.1}$$

where $x \in \mathbb{R}^n$, $y \in \mathbb{R}^p$, $\gamma \in \mathbb{R}^n$, $q \in \mathbb{R}^p$; F and G are given matrices of dimensions m × n and m x p respectively ; $b \in \mathbb{R}^m$ is the right-hand side vector ;

X is the positive orthant $\mathbb{R}^n_+$, or a given polyhedron included in $\mathbb{R}^n_+$ (in the latter case, a linear description of X can be assumed to be available).

Clearly, when the right-hand side vector b is supposed to be deterministically known and given, (I) above is just a regular linear program. However we consider here the case when b is *uncertain* and only known to belong to some specified subset $B \subseteq \mathbb{R}^m$ called the *uncertainty set* for *b*. (It is also sometimes referred to as the *scenario set* for *b*).

In the sequel, B will be assumed to be either a finite subset in $\mathbb{R}^m$, or a nonempty bounded (compact) subset (e.g. a polyhedron).

Now, a close look at the basic model (I) reveals that two distinct sets of variables are involved, namely:

- the x variables, which will be called the *first-stage* (or "here and now") variables;
- the y variables, which will be called the *second-stage* (or "wait-and-see") variables.

The role of the second stage variables y is, once the first stage variables have been fixed, to provide some flexibility to adapt the various possible realizations of the right-hand side in order to satisfy the constraints (2.1). This is the analogue to the role played by the *recourse variables* in stochastic programming (see e.g. Birge, 1985 ; Shapiro, 2006 ; Shapiro & Nemirovski, 2006).

When the structure of problem (I) is such that, for all $x \in X$, and all $b \in B$ there is $y \geq 0$ verifying (2.1), then we say that we have *complete recourse*. When this condition is not fulfilled, then an optimal first-stage solution x can only belong to the subset $\overline{X} \subset \mathbb{R}^n_+$ defined as:

$$\overline{X} = \{\, x \in \mathbb{R}^n_+ : \forall\, b \in B, \exists\, y \geq 0 : Hy = b - Fx \}$$

It is easy to check that $\overline{X}$ is a polyhedron. Moreover, it has been shown in Minoux (2011) that, for any given $x = \hat{x}$, testing whether $\hat{x} \in \overline{X}$ can be done by solving the following *separation problem* where minimization is to be carried out with respect to both *u* and *b*:

$$(SEP) \quad \begin{cases} z^* = Min \quad u^T b - u^T F\hat{x} \\ subject\ to: \\ u^T H \geq 0 \\ \|u\|_\infty \leq 1 \\ b \in \mathcal{B} \end{cases}$$

If $z^* \geq 0$ then $\hat{x} \in \overline{X}$, if $z^* < 0$ then $\hat{x} \notin \overline{X}$. (Note that the above problem has bilinear objective since $u^T b = \sum_{i=1}^{m} u_i b_i$ involves products of u_i variables and b_i variables).

When $\mathscr{B}$ is a finite scenario set $\{b^1, b^2, \ldots, b^k\}$ of "reasonably small" cardinality K, (SEP) can be solved by solving K linear programs in *u* variables only, one for each possible b^k in $\mathscr{B}$.

However, when $\mathscr{B}$ is an arbitrary polyhedron, it has been shown in Minoux (2011) that (SEP) is *strongly NP-hard*, hence practically difficult to solve in general.

For every $x \in X \cap \overline{X}$ and $b \in \mathscr{B}$, the minimum second stage cost achievable by adjusting the second-stage variables is equal to $\Phi\,(b - Fx) = \text{Min}\,\{q^T y \,/\, Hy = b - Fx,\ y \geq 0\,\}$ where $\Phi: \mathbb{R}^m \to \mathbb{R}$ (so-called *value function* or *perturbation function*) is defined, for any $\beta \in \mathbb{R}^m$ as the optimal value to the linear program

$$(II)\qquad \begin{cases} Min\, q^T y \\ s.t. \\ Hy = \beta \\ y \geq 0 \end{cases}$$

Therefore, for any given x, the corresponding worst-case total cost is:

$$\phi(x) = \gamma^T x + \underset{b \in \mathscr{B}}{Max}\left\{\Phi(b - Fx)\right\} \tag{2.2}$$

The *robust counterpart* to problem (I), given the uncertainty set $\mathscr{B}$ for the right-hand side b, is therefore:

$$RC\,(I)\begin{cases} Minimize\ \ \phi(x) \\ subject\ \ to: \\ x \in X \cap \overline{X} \end{cases}$$

Remark

The model (I) and its robust counterpart RC(I) include other formulations as special cases. For instance, if instead of (I) we consider:

$$(I)'\qquad \begin{cases} Min\quad \gamma^T x + q^T y \\ s.t: \\ Fx + Hy \leq b \\ x \in X,\ y \geq 0 \end{cases}$$

with $b \in \mathscr{B}$. (Thus (I)' is deduced from (I) by replacing equality constraints with inequalities).

Let $\tilde{H} = \left[H, I\right]$ where I is the m × m identity matrix and $\tilde{y} = \begin{pmatrix} y \\ \alpha \end{pmatrix}$ where $\alpha \in \mathbb{R}^m$, then it is easily checked that the problem (I)″ below (which has the same structure as (I)) is equivalent to (I)' for every occurrence of b in $\mathscr{B}$:

$$(I)'' \quad \begin{cases} Min\ \gamma^T x + (q^T, 0)\begin{pmatrix} y \\ \alpha \end{pmatrix} \\ s.\ t.: \\ Fx + \tilde{H}\begin{pmatrix} y \\ \alpha \end{pmatrix} = b \\ x \geq 0,\ y \geq 0, \alpha \geq 0 \end{cases}$$

Then a robust counterpart for (I)″ is readily deduced from the robust counterpart RC(I)' for (I)'. ||

2.2 Robust Optimization vs. Stochastic Optimization

Before going further into the description of the robust 2-stage model introduced in § 2.1, we find it interesting to provide some additional motivation via a comparison with the standard approach of optimization problems under uncertainty, namely, the *stochastic optimization approach*.

A first, straightforward, observation is that, properly stating a stochastic programming version of problem (I) for random right-hand side b, requires that a m-variate probability distribution of b over $\mathbb{R}^m$ be known and given. Under this assumption, the resulting stochastic optimization problem to be solved can be formulated as minimizing over $x \in X$ the function:

$$\theta(x) = \gamma^T x + \mathbb{E}\left\{\Phi(b - Fx)\right\} \tag{2.3}$$

where: $\mathbb{E}$ denotes the expectation with respect to the probability distribution of b in $\mathbb{R}m$; and Φ: $\mathbb{R}m \rightarrow \mathbb{R}$ is the value function of problem (II) introduced in section 2.1.

For the sake of simplicity, we will assume in this section, that $-\infty < \Phi(b - Fx) < +\infty$ for all $x \in X$ and b in the support S of its probability distribution (these assumptions will be fulfilled in the small example considered in this section).

Now, if we take $\mathscr{B} = S$, it is seen that the expression of θ(x) in (2.3) differs from the expression φ(x) in (2.2) by only replacing the Max operator by an expectation.

Indeed the stochastic optimization approach minimizes over $x \in X$ the expected value of total cost (first stage + second stage costs); whereas the robust optimization approach minimizes over $x \in X$ the worst-case value of total cost with respect to all possible occurrences of b in the uncertainty set.

One argument which is sometimes mentioned as a potential drawback of the stochastic optimization approach is that the expected cost criterion does not capture any notion of risk: a solution which minimizes the expectation may lead to high variance in the resulting cost, when subject to random occurrences of b (drawn according to the given probability distribution). By contrast, and in line with this argument, there is hope that an optimal robust solution provides global hedging against the most unfavorable realizations of uncertainty, thereby taking risk concerns into account.

We now show, by means of a small instance of the two-stage LP problem (I) introduced in section 2.1, that there are indeed situations for which the above statements can be justified, and the differences between the two approaches, precisely evaluated on a quantitative basis.

In this instance n = 1, m = 1, p = 2, so there is one first-stage ("here and now") variable $x \geq 0$, and two second-stage ("wait and see") variables $y_1 \geq 0$ $y_2 \geq 0$ and there is one constraint:

$$x + y_1 - y_2 = b.$$

where $b \in \mathbb{R}_+$ is the right-hand side value subject to uncertainty, x, y_1, y_2, are subject to nonnegativity constraints, and the objective function (to be minimized) is:

$x + q_1y_1 + q_2y_2$ with $q_1 > 1 > q_2 > 0$. (so, $\gamma = 1$ in this case)

In the stochastic version of the problem, b is supposed to be a random real variable, uniformly distributed on a given interval $[b^-, b^+] \subset \mathbb{R}_+$, hence with probability density $\mu(\alpha) = \frac{1}{B}$ if $\alpha \in [b^-, b^+]$, (we denote $B = b^+ - b^-$) ; and $\mu(\alpha) = 0$ if $\alpha \notin [b^-, b^+]$.

In the robust 2-stage version, the probability distribution is not assumed to be available, b is only known to belong to the uncertainty set $\mathcal{B} = [b^-, b^+]$.

It is easily seen that, for any fixed $x \in \mathbb{R}$, the LP problem (II) reduces in this case to:

$$\begin{cases} \text{Minimize } q_1y_1 + q_2y_2 \\ \text{subject to :} \\ y_1 - y_2 = b - x \\ y_1 \geq 0 \text{ , } y_2 \geq 0 \end{cases}$$

which, in view of fact that $q_1 + q_2 > 0$, has optimal solution value:

$$\Phi(b - x) = q_1(b - x)^+ + q_2(x - b)^+$$

(where $(b - x)^+ = \text{Max}\{0, b - x\}$, $(x - b)^+ = \text{Max}\{0, x - b\}$)

Let us first determine the optimal solution to the robust 2-stage version of the problem, i.e x_R^* minimizing $\varphi(x) = x + \underset{b \in [b^-, b^+]}{Max}\left\{ q_1(b - x)^+ + q_2(x - b)^+ \right\}$ (2.4)

As the pointwise maximum of the sum of two functions convex in b, and for fixed x, $\Phi(b - x)$ above is convex in b, and therefore the maximum w.r.t. b is necessarily attained at $b = b^-$ or $b = b^+$, hence (2.4) can be rewritten:

$$\varphi(x) = x + \text{Max}\{q_1(b^- - x)^+ + q_2(x - b^-)^+ ; q_1(b^+ - x)^+ + q_2(x - b^+)^+\}$$

From this it is seen that:

$\varphi(x) > \varphi(b^+)\ \forall\ x > b^+$; and $\varphi(x) > \varphi(b^-)\ \forall\ x < b^-$

therefore x_R^* necessarily lies in $[b^-, b^+]$ and is such that:

$q_2(x - b^-) = q_1(b^+ - x)$

So:

$$x_R^* = \frac{q_1 b^+ + q_2 b^-}{q_1 + q_2} \tag{2.5}$$

Now let us consider the stochastic version of the problem.
The expected second-stage cost for any fixed $x \geq 0$ is:

$c(x) = \mathbb{E}\{q1_(b - x)+ + q2_(x - b)+^\}$, and therefore:

$$c(x) = \int_{b^-}^{b^+} q_1(\alpha - x)\frac{d\alpha}{B} = -q_1 x + q_1\left(\frac{b^+ + b^-}{2}\right) \text{ if } x < b^-$$

$$c(x) = \int_{b^-}^{b^+} q_2(x - \alpha)\frac{d\alpha}{B} + \int_{b^-}^{b^+} q_1(\alpha - x)\frac{d\alpha}{B}$$

$$= \frac{q_1(b^+)^2 + q_2(b^-)^2}{2B} - \frac{(q_2 b^- + q_1 b^+)}{B}x + \frac{(q_1 + q_2)}{2B}x^2 \text{ if } b^- \leq x \leq b^+$$

$$c(x) = \int_{b^-}^{b^+} q_2(x - \alpha)\frac{d\alpha}{B} = q_2 x - q_2\frac{(b^+ + b^-)}{2} \text{ if } x > b^+$$

It is easily seen that $c(x)$ is convex in $x \in \mathbb{R}_+$, as soon as $q_1 + q_2 > 0$, and if we assume $q_1 > 1 > q_2 > 0$, the minimum value of $\theta(x) = x + c(x)$ is attained on $[b^-, b^+]$ at:

$$x_S^* = \frac{q_1 b^+ + q_2 b^- - B}{(q_1 + q_2)} = b^- + \frac{(q_1 - 1)B}{q_1 + q_2} \tag{2.6}$$

For instance, if $b^- = 5$, $b^+ = 10$, $q_1 = 1.5$; $q_2 = 0.4$, we obtain: $x_R^* = 8.94$, $x_S^* = 6.31$.

It can be seen from this example, that the two optimal solutions can be significantly different; but it can be shown that they also significantly differ in terms of variance of the total cost, i.e. $\sigma^2(Z(x))$ where $Z(x)$ is the random variable:

Table 1. Comparison of optimal stochastic solution (x_S^*) and optimal robust solution (x_R^*) in terms of variance and confidence level on total cost.

	Optimal Stochastic Solution x_S^* (min expected cost)	Optimal Robust Solution x_R^* (min worst-case cost)	Variances of cost $\sigma^2(Z_S^*) / \sigma^2(Z_R^*)$	Target values for total cost c_i (i = 1...4)	Confidence levels for x_S^* P{ $x_S^* \le c_i$} (i = 1...4)	Confidence levels for x_R^* P{ $x_R^* \le c_i$} (i = 1...4)
q_1 = 1.2 q_2 = 0.4	5.62 (7.93)	8.75 (10.25)	2.80/0.19	c_1= 9.95 c_2 = 10.10 c_3 = 10.17 c_4 = 10.22	0.84 0.87 0.88 0.89	0.8 0.9 0.95 0.98
q_1 = 1.5 q_2 = 0.2	6.46 (8.38)	9.41 (10.30)	3.07/0.07	c_1= 10.12 c_2 = 10.2 c_3 = 10.25 c_4 = 10.27	0.78 0.792 0.797 0.8	0.8 0.9 0.95 0.98
q_1 = 1.5 9_2 = 0.4	6.31 (8.42)	8.94 (10.53)	3.21/0.21	c_1= 10.22 c_2 = 10.37 c_3 = 10.45 c_4 = 10.5	0.78 0.80 0.81 0.82	0.8 0.9 0.95 0.98
9_1 = 1.5 q_2 = 0.8	6.11 (8.47)	8.23 (10.88)	3.4/0.58	c_1= 10.35 c_2 = 10.61 c_3 = 10.75 c_4 = 10.82	0.78 0.82 0.84 0.85	0.8 0.9 0.95 0.98
q_1 = 2 q_2 = 0.8	6.76 (9.10)	8.57 (11.45)	3.94/0.7	c_1= 10.85 c_2 = 11.15 c_3 = 11.29 c_4 = 11.37	0.76 0.79 0.80 0.81	0.8 0.9 0.95 0.98
q_1 = 3 q_2 = 0.8	7.62 (9.86)	8.94 (12.12)	4.01/0.86	c_1= 11.47 c_2 = 11.8 c_3 = 11.95 c_4 = 12.04	0.78 0.8 0.81 0.82	0.8 0.9 0.95 0.98

$Z(x) = x + q_1(b - x)^+ + q_2(x - b)^+$

Indeed, in the above example (q_1 = 1.5, q_2 = 0.4) for the optimal solution to the stochastic model, $\sigma^2\left(Z(x_S^*)\right) = 3.21$; while for the optimal robust solution, a much smaller variance is obtained: $\sigma^2\left(Z(x_R^*)\right)$ = 0.21 (more results of the same kind are shown in Table 1).

Such a result is clearly in line with the above-stated idea that robust optimal solutions may lead to reduced risk as opposed to optimal stochastic solutions. However, in order to confirm this intuition, a more systematic quantitative analysis is needed. To that aim, we are going to compare the optimal solutions of both the robust version and the stochastic version of the problem in terms of *confidence levels* corresponding to various target values on total cost. More precisely, for a given solution $x = \overline{x} \in \left[b^-, b^+\right]$, the confidence level corresponding to a target value c on total cost is defined as the probability $\mathbb{P}\left\{Z(\overline{x}) \le c\right\}$, where $Z(\overline{x})$ is the random variable $Z(\overline{x}) = \overline{x} + q_1(b - \overline{x})^+ + q_2(\overline{x} - b)^+$.

The probability distribution of the random variable $Z(\bar{x})$ is easily deduced from the probability distribution of b in [b⁻, b⁺], and assuming the uniform distribution with probability density $\frac{1}{B}$, it can be shown that:

$$\mathbb{P}\left\{Z(\bar{x}) \leq c\right\} = \frac{\alpha_{max} - \alpha_{min}}{B}$$

where:

$$\alpha_{max} = Min\left\{b^{+},\ \bar{x} + \frac{c - \bar{x}}{q_1}\right\}\ ;\ \alpha_{min} = Min\left\{b^{-},\ \bar{x} - \frac{c - \bar{x}}{q_2}\right\}$$

When $\bar{x} = x_S^*$, the optimal solution to the stochastic problem, the random variable Z (x_S^*) will be denoted Z_S^* ; and when $\bar{x} = x_R^*$, $Z(x_R^*)$ will be denoted Z_R^*

Table 1 compares for various choices for the parameters q_1 and q_2, the confidence levels $\mathbb{P}\left\{Z_S^* \leq c\right\}$ and $\mathbb{P}\left\{Z_R^* \leq c\right\}$ for various values of c chosen in order that at least one of the resulting probabilities lie in the range [0.8, 1]. (in practice, confidence levels less than 0.8 are of little interest). From the results displayed, it is seen that for most cases the optimal robust solution yields significantly higher confidence levels (the unique exception being for $q_1 = 1.2$, $q_2 = 0.5$ and $c = c_1$).

From the results in Table 1, it is clear that optimal robust solutions are to be preferred when a reduced risk level is required.

2.3 Overview of Applications and Complexity Results

The purpose of the present section is to show that the class of robust 2-stage optimization problems featuring right-hand side uncertainty introduced in § 2.1 is related to many potentially interesting applications. In most cases, the resulting robust counterparts (RC(I)) lead to computationally hard problems, however some related polynomial-time solvable special cases will be pointed out.

Most examples mentioned in this section correspond to what is referred to as polyhedral uncertainty (PU), i.e. when the uncertainty sets under consideration are polytopes (= bounded polyhedra). We will also refer to an important special case of polyhedral uncertainty, namely the so-called "knapsack constrained uncertainty" (KCU). This corresponds to uncertainty sets which are in 1-1 correspondence with the solution set of a given (continuous) knapsack problem.

To illustrate, suppose that we have n parameters d_i (i = 1, …, n) each of them can take any value in a given interval $\left[d_i^0, d_i^0 + \varepsilon\delta_i\right]$ (where d_i^0 is the nominal value and $d_i^0 + \varepsilon\delta_i$ is the worst-case value); ε is fixed, either equal to 1 or equal to –1. Then knapsack-constrained uncertainty corresponds to uncertainty sets $\mathcal{D}$ composed of all $d = (d_i)_{i=1,\ldots,n}$ of the form $d_i = d_i^0 + \varepsilon\sigma_i\delta_i (i = 1,\ldots, n)$ for all $\sigma = (\sigma_i)_{i=1,\ldots,n}$

in the solution set $\sum$ defined by the (continuous) knapsack constraints: $\sum_{i=1}^{n} \sigma_i \leq \Gamma, 0 \leq \sigma \leq \hat{\sigma}$ for given upper bounds $\hat{\sigma}_i = 1$ (i = 1, . . ., n) and $\Gamma \in [1, n]$.

If each of the uncertain parameters, d_i, can take any value in an interval of the form $\left[d_i^0 - \delta_i, d_i^0 + \delta_i\right]$ (where $d_i^o - \delta_i$ and $d_i^0 + \delta_i$ are the extreme values and the nominal value is the central value d_i^0) then we can define a (KCU)-type uncertainty set $\mathcal{D}$' composed of all $d = (d_i)_{i=1,\ldots,n}$ of the form: $d_i = d_i^0 + \sigma_i^+ \delta_i - \sigma_i^- \delta_i$ for all $\sigma = (\sigma^+, \sigma^-)$ in the set:

$$\Sigma' = \left\{(\sigma^+, \sigma^-) \,/\, \sum_{i=1}^{n} (\sigma_i^+ + \sigma_i^-) \leq \Gamma, \quad 0 \leq \sigma_i^+ \leq 1, 0 \leq \sigma_i^- \leq 1\right\}$$

again recognized as the solution set of a continuous knapsack problem.

The parameter Γ involved in the above definitions is called the *budget of uncertainty*. Note that when Γ is an integer, the extreme points of $\sum$ (or $\sum$') are integer-valued, and the conditions $d \in \mathcal{D}$ (or $d \in \mathcal{D}$') can be interpreted as accepting as possible scenarios only those $d = (d_i)_{i=1,\ldots,n}$ such that the number of components which can take on extreme values does not exceed Γ. This is why we refer to this special case of (KCU) as *cardinality constrained uncertainty* (CCU). We note here that (KCU) uncertainty is the basic model of uncertainty considered by Bertsimas and Sim (2003, 2004).

2.3.1 Robust PERT Scheduling

PERT scheduling is a very classical problem in Operations Research, at least when the task durations are deterministically known. An instance of this problem is specified by a directed circuitless graph G in which the nodes correspond to the various tasks i = 1, …, n of a project, and there is an arc (i, j) whenever there is a precedence constraint stating that processing of task j should not start before completion of task i. In the deterministic case the duration d_i of each task i is fixed and given, and the problem is to determine the minimum total duration of the project (the total time spent between the beginning of the initial task and the end of the terminal task). This problem is efficiently solved (in polynomial time O(m) where m is the number of arcs) as a longest path problem in the given circuitless graph G, when each arc (i, j) has length d_i (the duration of task i). The extension of this problem to the case of uncertain task durations has been investigated in Minoux (2007) assuming a polyhedral uncertainty set of (KCU) type. In this extended version of PERT scheduling, a minimum robust project duration t* is looked for, i.e. the minimum value of t guaranteeing that the project can be completed in no more than t time steps for all possible d in the uncertainty set. The special case of *cardinality constrained uncertainty* (CCU) is shown to be polynomially solvable in time O(mn2) (where m is the number of arcs in G). Other more general versions of the robust PERT scheduling problem leading to *pseudo-polynomial algorithms* are described in Minoux (2007), including the robust *time-dependent* PERT scheduling problem.

2.3.2. Robust Network Flow Problems

The robust maximum flow problem under uncertain arc capacities is defined as follows. Given a directed graph G = [V, U] where V is the vertex set and U the arc set, $s \in V$, $t \in V$ two specified vertices in G, and $C \subset R_{+}^{|U|}$ a set of possible scenarios of capacities associated with the arcs, we want to determine a *maximum robust s – t flow value in G*: in other words, the maximum value of z such that, for all c ∈ C, there exists a s – t flow of value at least z. This problem has been investigated in Minoux (2009) and shown to be strongly NP-hard even in the case of an uncertainty set of the (KCU) type. The special case when the graph G' (deduced from G by adding the return arc (t, s)) is *planar* and the uncertainty set of type (CCU) is polynomially solvable.

Another type of single-commodity network flow problems featuring multiple sources and sinks with *uncertain requirements* at nodes in the set of sinks has been investigated in Atamtürk and Zhang (2007), Minoux (2010).

The problem is then to determine how much extra capacity x_u has to be added on each arc u ∈ U in order to guarantee the existence of a feasible flow from the source nodes S to the sink nodes T for any scenario of requirements in the uncertainty set, while minimizing total cost. It has been shown in Minoux (2010) that this problem class is strongly NP-hard, even in the single-source case (|S| = 1) and uncertainty sets of type (KCU). On the other hand, polynomially solvable special cases have been proposed in Ordoñez and Zhao (2007) and Minoux (2010). In the former reference, polynomial solvability is obtained assuming that the uncertainty set for requirements is a polytope defined as a *product of intervals*. In the latter reference, it is obtained by assuming:

(a) a single source node (|S| = 1),
(b) *planarity* of the graph G' deduced from G by adding an extra node t and |T | arcs of the form (j, t) for all j ∈ T ;
(c) the uncertainty set for requirements if of the (CCU) type.

2.3.3 Robust Inventory Management under Uncertain Customer Requirements

Restricting to the single- product case, we consider a given time horizon discretized into T consecutive time periods t = 1, 2, …, T . In the deterministic case (i.e. when customer requirements are assumed to be exactly known in each time period of the horizon) this classical problem consists in determining the amount y_t to be provisioned in each time period t = 1, …, T so as to meet customer requirements, while minimizing total procurement costs and keeping the total amount in store below a given capacity limit C.

It can be formulated as a linear program in which the customer requirements d appear in the right-hand side coefficients. Now, when the customer requirements $d = (d_t)_{t=1,\dots,T}$ are uncertain, and only known to belong to a given uncertainty set $\mathcal{D} \subset \mathbb{R}^{T}$ a 2-stage robust linear programming formulation is obtained by considering recourse variables representing the possibility of having s*tockout.* The recourse variable corresponding to period t is zero if the inventory level It in period t is nonnegative; it is equal to –It (the amount of product missing to meet customer requirements) in time period t in case It < 0 (stockout). A penalty term involving the recourse variables with high penalty costs is added to the objective, in order that stockout situations be avoided, as far as possible, in an optimal robust solution. In the case when the uncertainty set $\mathcal{D}$ is a product of intervals, a pseudo-polynomial algorithm, based on a dynamic programming approach, has been described in Minoux (2011) to solve this 2-stage robust version of inventory management. The multistage version of the inventory management problem under uncertainty is investigated in Minoux (2008).

3. ROBUST MULTISTAGE OPTIMIZATION PROBLEMS

In this section we discuss the extension of the robust 2-stage model to decisions problems under uncertainty featuring several ($K \geq 2$) steps – or stages – each stage k involving a set of "here and now" decision variables $x^{(k)}$ and a set of "wait-and-see" variables $y^{(k)}$.

3.1. Proposed Generic Model

In each stage $k > 1$, it is assumed that the values of the uncertain parameters $\omega^{(1)}\ \omega^{(2)} \ldots \omega^{(k-1)}$ related to the previous stages 1, 2, … $k - 1$ have been revealed, and the optimal values of the $x^{(k)}$ variables have to be fixed « here and now », i.e. before knowing which occurrence of the $\omega^{(k)}$ (from a given uncertainty set) will arise.

Then, once $x^{(k)}$ has been fixed and $\omega^{(k)}$ has been revealed, the "wait-and-see" variables $y^{(k)}$ for stage k, have to be determined, before proceeding to stage k+1.

Clearly, in such a process, the optimal values for $x^{(k)}$ (resp: $y^{(k)}$) are fonctions of $\omega^{(1)}, \omega^{(2)}, \ldots \omega^{(k-1)}$, (resp: $\omega^{(1)}, \omega^{(2)}, \ldots \omega^{(k)}$), and the concept of an optimal solution is no longer appropriate.

Indeed, solving a multistage robust otpimization problem consists in determining an optimal "strategy", specifying, for each stage k:

a. The best $\overline{x}^{(k)}$ values for $x^{(k)}$ as a function of the previous decisions $\overline{x}^{(1)}, \ldots \overline{x}^{(k-1)}$, and the realizations of $\omega^{(1)}, \omega^{(2)}, \ldots \omega^{(k-1)}$ observed in the previous stages ;
b. The best $\overline{y}^{(k)}$ values for $y^{(k)}$ as a function of: $\overline{x}^{(1)}, \overline{x}^{(2)}, \ldots \overline{x}^{(k)}$ and the observed realizations of $\omega^{(1)}, \omega^{(2)}, \ldots \omega^{(k)}$.

Again, restricting, for the sake of simplicity in the presentation, to the case of linear programs with uncertain right-hand side vector, a possible generic multistage robust optimization model can be stated as follows:

$$(MO)\begin{cases} Minimize: \sum_{k=1}^{K} \left(\gamma^{(k)}\right)^{T} x^{(k)} + \left(q^{(k)}\right)^{T} y^{(k)} \\ Subject\ to: \\ \sum_{t=1}^{k} F^{(t,k)} x^{(k)} + \sum_{t=1}^{k} H^{(t,k)}\ y^{(k)} = b^{(k)} \qquad (k = 1 \ldots K) \\ x^{(k)} \in X^{(k)} \qquad (k = 1 \ldots K) \\ y^{(k)} \geq 0 \end{cases}$$

Here $b^{(k)}$, $k = 1\ldots K$, are right-hand sides subject to uncertainty, and such that the global right-hand side vector:

$$b = \begin{pmatrix} b^{(1)} \\ b^{(2)} \\ . \\ . \\ b^{(K)} \end{pmatrix}$$

is known to belong to a given uncertainty set $\mathscr{B}$. $F^{(t,k)}$ and $H^{(t,k)}$ (for all k = 1, … K, and t = 1, … k) are given matrices of appropriate dimensions.

Observe that, is most applications of practical interest, $\mathscr{B}$ should not be restricted to consist of a product of K uncertainty sets (one for each $b^{(k)}$, k = 1, …K) ; considering more general structures for $\mathscr{B}$ allows one to describe dependence among uncertainties occurring in different stages, a phenomenon commonly observed in practice.

It is also worth noting that, given a multistage optimization model in the form (MO), there is an associated 2-stage version of the same problem, which is just obtained by grouping all the decision variables $x^{(1)}$, $x^{(2)}$, … $x^{(K)}$ into a single "here and now" decision vector ; and grouping all the decision variables $y^{(1)}$, $y^{(2)}$, …$y^{(K)}$ into a single "wait-and-see" decision vector.

Clearly an optimal robust solution to this 2-stage version of the problem can be taken as an approximation to the exact optimal multistage strategy. The computational experiments discussed in section 5 below will illustrate the fact that the differences in terms of optimal cost values between a given multistage problem and its corresponding 2-stage version can be quite reduced.

3.2 Example of Application

As a typical illustration of the above generic multistage model, we show in Table 2, the matrix structure corresponding to the multiperiod energy production planning problem considered in the computational experiments reported in section 5. (Table 2 corresponds to an instance with 4 time periods hence K = 4 in this case).

In this example, (x_k, z_k) (k = 1, … 4) are the "here and now" decision variables and (y_k^+, y_k^-) are the "wait-and-see" decision variables. x_k represents the production level to be set for stage k, and z_k the corresponding production cost for stage k. The y_k^+, y_k^- variables represent the adjustments which are necessary to achieve balance between production and demand in stage k.

ρ_k (k = 2, … 4) are given vectors all with the same dimension ν, and e denotes the vector of all ones of dimension ν.

The vectors $\eta_1 \in \mathbb{R}^{\nu}$ and $\eta_k \in \mathbb{R}^{\nu+2}$ (k = 2, …4) which appear as part of the right-hand side, are deterministic and given ; the only components of the right-hand side subject to uncertainty are $d_k \in \mathbb{R}_+$ (k = 1, … 4), they represent customer load requirements and are assumed to belong to some specified uncertainty set $\mathfrak{D} \subset \mathbb{R}^4_+$.

Table 2. illustration of the matrix structure of a multistage robust linear program arising from the optimal energy planning problem discussed in §5 (case of K = 4 stages). d_k (k = 1,...4) are uncertain right-hand side coefficients representing customer load requirements in the various time periods.

0	**1**	**α_1**	**$-\beta_1$**	**0**	**1**	**α_2**	**$-\beta_2$**	**0**	**1**	**α_3**	**$-\beta_3$**	**0**	**1**	**α_4**	**$-\beta_4$**		
x_1	**z_1**	**y_1^+**	**y_1^-**	**x_2**	**z_2**	**y_2^+**	**y_2^-**	**x_3**	**z_3**	**y_3^+**	**y_3^-**	**x_4**	**z_4**	**y_4^+**	**y_4^-**		
ρ_1	$-e$	0	0			0				0				0		$\le$	η_1
1	0	0	-1													$=$	d_1
0	0	0	0	ρ_2	$-e$	0	0									$\le$	η_2
-1	0	0	0	1	0	0	0			0				0			
1	0	0	0	-1	0	0	0										
0	0	0	0	1	0	1	-1									$=$	d_2
				0	0	0	0	ρ_3	$-e$	0	0					$\le$	η_3
	0			-1	0	0	0	1	0	0	0			0			
				1	0	0	0	1	0	0	0						
				0	0	0	0	1	0	1	-1					$=$	d_3
								0	0	0	0	ρ_4	$-e$	0	0	$\le$	η_4
	0					0		-1	0	0	0	1	0	0	0		
								1	0	0	0	-1	0	0	0		
								0	0	0	0	1	0	1	-1	$=$	d_4

The cost coefficients are $\gamma^{(k)} = \begin{pmatrix} 0 \\ 1 \end{pmatrix}$ (k = 1, … 4) and $q^{(k)} = \begin{pmatrix} \alpha_k \\ -\beta_k \end{pmatrix}$ the unit prices of energy, in case of supply shortage (α_k) or excess production ($-\beta_k$).

In this case it is seen that, ∀ k = 2, … 4:

$$F^{(k-1,k)} = \begin{bmatrix} 0 & 0 \\ -1 & 0 \\ 1 & 0 \end{bmatrix} H^{(k-1,k)} = \begin{bmatrix} 0 & 0 \\ 0 & 0 \\ 0 & 0 \end{bmatrix}$$

$$F^{(k,k)} = \begin{bmatrix} \rho_k & -e \\ 1 & 0 \\ -1 & 0 \\ 1 & 0 \end{bmatrix} H^{(k,k)} = \begin{bmatrix} 0 & 0 \\ 0 & 0 \\ 0 & 0 \\ 1 & -1 \end{bmatrix}$$

and $F^{(t,k)} = 0$, $H^{(t,k)} = 0$ ∀ t < k-1.

4. STATE-SPACE REPRESENTABLE ROBUST MULTISTAGE PROBLEMS AND THEIR EFFICIENT SOLUTION

In view of the various complexity (NP-hardness) results mentioned in section 2 for the case K = 1 (i.e 2-stage LP), it is clear that multistage robust optimization problems fitting the generic model (MO) presented in section 3.1 form a class of essentially intractable problems when exact optimal strategies are required. In this respect, identifying subclasses of instances featuring special structures and amenable to efficient computation of exact optimal robust strategies is of special interest. Indeed, if a given (a priori intractable) multistage problem can be sufficiently closely approximated, in some way, as an instance of such an efficiently solvable subclass, then exact optimal robust strategies for the latter can be expected to yield good approximate robust strategies for the former.

The purpose of the present section is to describe a special class of robust multistage optimization problems for which exact optimal strategies can be efficiently computed. There are two basic assumptions underlying this class of problems, namely:

a) The evolution over successive stages (time periods) of the system under consideration can be represented in terms of transitions between successive states taken in some finite set of states of "reasonably small" cardinality;
b) The uncertainty process describing the possible occurrences of uncertainty from one stage to the next is also state-space representable e.g. by means of a given sequential graph featuring one layer per stage. Again "reasonably small" cardinality of the set of states has to be assumed. (Note that what is meant here by "reasonably small", will become more precise in light of the complexity result in Proposition 4.2 below.)

4.1 General Statement of Robust Multistage Optimization in State-Space Representation

We consider a dynamic system evolving over a given time horizon [0, T] decomposed into a given finite number T of time periods. For any $t \in \{1, 2, \ldots, T\}$, the time period numbered t corresponds to the time interval between instants t – 1 and t. In this setting, there are K = T stages, each stage corresponding to a time period.

At each time instant t = 1, …, T, the system can be in any possible state in a discrete finite set of states $\mathcal{S} = \{1, \ldots, N\}$. The state S0 of the system at time t = 0 is supposed to be known. In a classical (deterministic) dynamic programming model (see e.g. Bertsekas, 2005), in each time period t, when the system is in state S at time t – 1, its evolution over the tt^{h} time period (i.e. between time t – 1 and time t) is described by a *state transition equation* providing the new state S1 of the system at time t as a function of:

- The state S at time t – 1;
- The decision x_t taken at the beginning of the tth time period.

Moreover there is a reward associated with such a transition which also depends on the state S at time t – 1 and on the decision x_t. The set X (t, S) of possible decisions starting from state S at the beginning of time period t is supposed to be a known finite discrete set.

The resulting classical (deterministic) dynamic optimization problem consists in determining an optimal sequence of decisions maximizing the total reward obtained over the T time periods. This is achieved through application of the classical dynamic programming recursion.

To define a robust version of such a problem we now consider that both the state transition function and the reward function in time period t depend on one or several uncertain parameters ω_t taken in some given finite uncertainty set. The structure of the uncertainty sets considered in the present section will be described in more detail in § 4.2 below.

In view of preserving the genericity of the model as much as possible, we will consider that each ω_t is a q-component vector, some of the components of ω_t influencing the state transition function, and some of the components influencing the reward function. Observe that the set of components of ω_t influencing the state transition function and the set of components of ω_t influencing the reward function are not assumed to be disjoint, but they are not assumed to be identical either. The model investigated here is thus capable of representing all kinds of situations which may be viewed as intermediates between full dependence and full independence of the uncertainty factors acting on the state transition function and on the reward function.

We will denote

$$S_1 = F(t, S, x_t, \omega_t) \in \mathcal{S} \tag{4.1}$$

the state reached by the end of time period t when starting from $S \in \mathcal{S}$ at the beginning of time period t after taking the decision xt$\in$ X (t, S) and for the value ωt of the uncertainty vector. The corresponding reward will be denoted

$$R(t, S, x_t, \omega_t) \in \mathbb{R}. \tag{4.2}$$

We assume that the decision x_t for period t has to be taken prior to the occurrence (or: realization) of an uncertainty vector ω_t. Therefore it will be assumed that the set of possible decisions X (t, S) is defined in such a way that for any $x_t \in X(t, S)$ and any ω_t in the uncertainty set, the value $F(t, S, x_t, \omega_t)$ is well defined and leads to a feasible state $S_1 \in \mathcal{S}$ by the end of time period t.

4.2 State Space Representable Uncertainty Sets

To complete the description of the class of robust multistage optimization problems considered in the present section, we now define precisely the concept of *state space representable uncertainty sets*.

In each time period t, let ω_t denote the uncertain parameter(s) influencing the evolution of the system considered. Depending on the context of application, each possible realization of ω_t may correspond to one or several numerical values. For instance in a single-product inventory management problem where uncertainty impacts both customer requirements and procurement prices, a possible realization of ω_t will correspond to two numerical values, a required amount of product, and a procurement price. In a multi-product version of the same problem, any possible realization of ω_t would correspond to a number of numerical values equal to twice the number of products (for each product, a customer requirement and a procurement price).

In the proposed model of uncertainty, we assume that:

a. in each time period t, there is a given finite number of possible realizations of ω_t, each of which corresponding to one or several *modalities*. We denote K_t the total number of modalities for all the possible realizations of ω_t, and for each modality k = 1, …, K_t, we denote v(t, k) the corresponding value of ω_t (please note that the same value of ω_t can arise from different modalities, therefore, as a general rule, there will be several k' ≠ k such that v(t, k') = v(t, k)). In the sequel, if the kth modality of realization of ω_t occurs during time period t, we will say that the process of uncertainty is in state (t, k) for time period t.
b. The set of possible scenarios (i.e. sequence of realized values for $\omega_1, \omega_2, \ldots, \omega_T$) are in one-to- one correspondence with the set of directed paths in a given directed sequential graph G = $[\mathcal{V}, \mathcal{U}]$ with vertex set $\mathcal{V}$ such that $\mathcal{V} = V1 \cup V2 \cup \ldots \cup VT \cup \{0\}$ where, for each t = 1, …, T, the vertices in Vt correspond to all pairs of the form (t, k) for k = 1, …, Kt, i.e. each vertex corresponds to a possible uncertainty state. The vertex numbered 0 represents the common origin of all the paths, and is connected to vertices of V1 only. $\mathcal{U}$ is a given set of directed edges (arcs) of the form [(t, k)(t + 1, k')] with (t, k) ∈ Vt, (t+ 1, k') ∈ Vt + 1; it is assumed that $\mathcal{U}$ contains all the arcs of the form [0, (1, k)] for k = 1, …, K1. For each vertex (t, k) we will denote succ(t, k) the set of direct successors of vertex (t, k) in the given graph G. As already mentioned, succ(0) = V1.

Clearly, how to build the sequential graph G in practice essentially depends on the application considered. Several examples are presented and discussed into detail in Minoux (2015). A new example will be discussed in section 4.4 below where it is shown that the concept of state-space representable uncertainty set introduced above turns out to be well suited to the (frequently encountered) case of an uncertainty process deriving from a discr*ete finite Markov chain model.*

We are now going to investigate how to exploit the above-defined structure of state-space representable uncertainty set in order to efficiently solve the multistage robust dynamic problem, i.e. to compute exac*t optimal robust strategies.*

4.3 Efficient Computation of Exact Optimal Robust Strategies

Consider a robust multistage optimization problem given by specifying:

a. the state space $\mathcal{S}$ used to represent the evolution over time of the dynamic system to be optimized, and S0 ∈ $\mathcal{S}$ the initial state at time 0 (i.e. at the beginning of the first time period);
b. the state transition function F defined in (4.1) and the reward function R defined in (4.2);
c. the sequential graph G = $[\mathcal{V}, \mathcal{U}]$ describing the state space representation of the uncertainty set, with the special vertex 0 in $\mathcal{V}$ corresponding to the initial state of the process representing uncertainty. We recall that $\mathcal{V} = \{0\} \cup V1 \cup V_2 \cup \ldots \cup VT$, and for any t ∈ [1, T], the vertices in Vt represent the Kt possible modalities of realization of ωt.

In the same way as for the case of stochastic multiperiod optimization problems, a solution to the above stated problem does not correspond to some well-defined sequence of decisions (x1, x_2, … , xT) leading to maximum reward (in the worst situation created by uncertainty).

Indeed a solution corresponds to a str*ategy (a*lso sometimes referred to as "closed loop" solution) which is specified by associating with each state S, at each time instant t – 1, the value xt o$_f$ an optimal decision to be taken during period t, given the current state of the uncertainty process at time t – 1. Clearly the state of the uncertainty process at time t – 1 is perfectly characterized by specifying which vertex in the graph G has been reached at time t – 1, as a result of the successive realizations of ω1, $_ω$2, .., ωt – $_{1.}$

Indeed, the knowledge of which vertex (t – 1, k) in Vt $_{-1\ h}$as been reached in the partial realization of the uncertainty process from time period 1 to time period t – 1 provides the necessary and sufficient information on the past occurrences of uncertainty, in view of properly restricting the future possible scenarios to be taken into account to only those paths in the graph G originating from vertex (t – 1, k).

In view of this, a s*trategy* corresponding to our robust multistage optimization problem will be defined by associating a decision xt ∈ X (t, S) with each pair (S, (t – 1, k)) where S is the state of the dynamic system reached by the end of time period t – 1, and (t – 1, k) is the vertex in Vt$_{-1}$ defining the state of the uncertainty process reached by the end of time period t – 1. Thus, for a given strategy φ, we will denote φ(t – 1, S, k) ∈ X (t, s) the decision to be taken during time period t, when the dynamic system is in state S at time t – 1, and the uncertainty process is in the state corresponding to vertex (t – 1, k) at time t – 1.

An *optimal robust strategy* φ* is a strategy such that, for each t > 1, for each S ∈ $\mathcal{S}$, and k ∈ [1, Kt$_{-1]}$, φ* (t – 1, S, k) is the decision leading to the maximum reward over the time periods t, t + 1, …, T in the worst possible scenario in the set of scenarios corresponding to all the paths in G originating at vertex (t – 1, k).

Let us denote Z*(t – 1, S, k) the maximum reward which results from using the optimal strategy φ* when starting from state S in the beginning of time period t, with an *uncertainty state* (t – 1, k).

We now show that the Z* and φ* functions can be determined by the following backward dynamic programming type recursion:

$$\begin{cases} Z^*(t-1,S,k) = \underset{x_t \in X(t,S)}{Max} \left\{ \Psi(t,S,k,x_t) \right\} \\ \text{where:} \\ \Psi(t,S,k,x_t) = \underset{k' \in succ(t-1,k)}{Min} \left\{ R(t,\, S,\, x_t,\, v(t,\, k')) + Z^*\,(t,\, F(t,\, S,\, x_t,\, v(t,\, k')),\, k') \right\} \\ \text{and } \varphi^*(t-1,S,k) = \underset{x_t \in X(t,S)}{\arg\max} \left\{ \Psi(t,\, S,\, k,\, x_t) \right\} \end{cases} \qquad (4.3),\ (4.4)$$

The above backward recursion is initialized by specifying the values of Z* (T + 1, S, k) for all S ∈ $\mathcal{S}$, k ∈ [1, KT$_]$ ("end of game values"). A standard way of initializing, if nothing is known about the future (beyond time T), is to set Z* (T + 1, S, k) = 0 for all S ∈ $\mathcal{S}$ and k = 1, …, KT. We can then state:

PROPOSITION 4.1

The Z* (t – 1, S, k) and φ* (t – 1, S, k) values computed from the backward recursion (4.3)-(4.4), starting with given Z* (T + 1, S, k) "end of game" values, define the reward and decision functions associated with an optimal robust strategy over [0, T]. In particular, the best (robust) decision to be taken in the first time period starting with the system in state S_0 at time 0, and the uncertainty state corresponding to node 0 in the graph representation of uncertainty is:

$$\varphi^* (0, S_0, 0) = \underset{x_1 \in X(1, S_0)}{\arg\max} \{\Psi(t, S_0, k, x_1)\}$$

and the corresponding optimal (worst-case) reward is

$$Z^* (0, S_0, 0) = \underset{x_1 \in X(1, S_0)}{Max} \{\Psi(t, S_0, k, x_1)\}.$$

Proof. Consider the dynamic system in state S at t – 1 while the uncertainty state at t – 1 is (t – 1, k). Assume that all optimal values Z* (τ, S', k') have already been computed for all τ ≥ t, S' ∈ $\mathcal{S}$, k' ∈ [1, Kτ_1. The possible decisions which can be taken during time period [t – 1, t] are those in X (t, S). Suppose a specific xt∈ X (t, S) is chosen. Then, depending on which realization of uncertainty occurs during time period [t – 1, t], i.e. depending on which state of uncertainty (t, k') is reached during this period, the state of the system by the end of time period [t – 1, t] is S' = F (t, S, xt v(t, k')) and the reward obtained during this period is R (t, S, xt v(t, k')). In view of the definition of the graph G representing the process of uncertainty, the only possible uncertainty states during time period [t – 1, t] correspond to those vertices (t, k') which are the direct successors of the vertex (t – 1, k).

Since the value of the best worst-case reward which can be obtained over time periods t + 1, t + 2, …, T, starting at time t with the system in state S' and the uncertainty process in state (t, k'), is known and equal to Z* (t, S', k'), then the total reward over the periods t, t + 1, …, T under decision x_t is R (t, S, x_t, v(t, k')) + Z* (t, S', k'). Now, the worst-case situation given that decision x_t has been taken for time period t, is obtained by taking the minimum reward over the set of possible uncertainty states during time period t, i.e. over the set of vertices (t, k') ∈ succ(t – 1, k). The resulting value is exactly ψ (t, S, k, x_t) given by (4.4).

Now the best possible decision x_t is obtained by taking the maximum of ψ with respect to x_t ∈ X (t, S) according to (4.3). This completes the proof. □

The following result settles the time complexity of solving the robust multistage optimization problem using the recursion (4.3)-(4.4).

PROPOSITION 4.2

Let |$\mathcal{U}$| denote the number of arcs in the graph G = [$\mathcal{V}$, $\mathcal{U}$] representing the process of uncertainty. Then the time complexity of solving the recursion (4.3)-(4.4) is:

$\mathcal{O}$ (|S| ×| X| × |$\mathcal{U}$|)

where |S| is the cardinality of the state space representing the dynamic system, and |X| the maximum cardinality of the decision sets X (t, S).

Proof. For each t = 1, …, T, and for given S ∈ $\mathcal{S}$ and xt ∈X (t, S), we have to compute all values ψ (t, S, k, xt) for all vertices (t – 1, k) ∈ Vt–1 of G. Each of these values is computed in time proportional to | succ(t – 1, k)|, thus for each t, S ∈ S and xt ∈X (t, S) the computation time is proportional to the number of arcs of G which have their initial endpoint in Vt–1. so the overall computational effort for given S ∈ S and xt ∈X (t, S) is $\mathcal{O}$(|$\mathcal{U}$|), and the claimed complexity result follows. □

It is of interest to compare the above result with the complexity of the standard dynamic programming algorithm when applied to a problem of the same size in terms of cardinality of the state space |S| and decision space |X|, but without uncertainty, namely $\mathcal{O}(T \times |S| \times |X|)$. It is seen that the use of the state-space representation provided by the graph G to handle uncertainty increases complexity by a factor $\frac{|\mathcal{U}|}{T}$ (the average number of arcs per time period in G) which is typically considerably smaller than the number of scenarios implicitly represented by the graph G (the latter being of the order of magnitude of the number of paths in G).

Also worth noting as a consequence of Proposition 4.2 is that, due to the exponential growth of the number of paths of G as a function of its size, the complexity of our procedure depends only log*arithmically on* the number of paths, i.e. on the number of scenarios in the uncertainty set.

By contrast, in multistage linear programming models, the size of the deterministic equivalent LP grows lin*early* with the number of scenarios.

4.4 State-Space Representable Uncertainty Sets for Markov Chains

It has been shown in Minoux (2015) that the concept of state-space representable uncertainty extends several types of uncertainty sets previously investigated in the literature, including (KCU) and the so-called "parsimonious" uncertainty model from Minoux (2008).

Here we are going to show that state space representable uncertainty sets are also well-suited to represent *in a compact way* sets of most likely events arising from finite state stationary Markov processes. In this case, it will be seen that the set of sequences of transitions which compose such uncertainty sets are in 1-1 correspondence with the various paths satisfying a well-defined *weight constraint* in the graph associated with the probability transition matrix of the given Markov chain. A state space representation is then easily deduced via a standard *graph expansion procedure*.

Consider n real parameters $a_1, a_2, \ldots, a_n$ which are subject to uncertainty, each of which can take a finite number m of possible values. For each $j \in [1, n]$, denote $v_{j,1}, v_{j,2}, \ldots, v_{j,m}$ the corresponding set of possible values.

Assume that the possible sequences of values $\left[v_{1,i_1}, v_{2,i_2}, \ldots, v_{n,i_n}\right]$ are the result of a random process corresponding to a finite state *stationary and ergodic* Markov chain with given $m \times m$ probability transition matrix $P = (p_{ij})_{\substack{i=1,\ldots,m \\ j=1,\ldots,m}}$ (for an overview of some useful basic results on Markov chains, refer to Appendix at the end of the chapter). We also assume (as is often the case in applications) that *the initial state s of the Markov process is known and given.*

Alternatively, the set of possible sequences of values can be viewed as corresponding to the set of directed paths of length $n - 1$ originating from node (1, s) in the sequential graph G defined as follows. The set of nodes corresponds to all (step, state) pairs of the form (k, i) where $k \in [1, n]$ and $i \in [1, m]$; assuming $1 \leq k \leq n - 1$, there is an arc joining (k, i) to (k + 1, j) whenever $p_{ij} > 0$. Moreover in the sequel, with each such arc will be associated a weight $w_{ij} = -\ln(p_{ij})$.

According to the definition of a stationary Markov chain with transition matrix P, the probability of occurrence of a sequence of states $\Sigma = (i_1, i_2, \ldots, i_n)$ (with $i_1 = s$) is

$$p(\Sigma) = p_{i_1,i_2} \times p_{i_2,i_3} \times \ldots \times p_{i_{n-1},i_n}. \tag{4.5}$$

Thus

$$\ln(p(\Sigma)) = w(\Sigma) = \sum_{k=2}^{n} w_{i_{k-1}\, i_k} \tag{4.6}$$

is the total weight of the path in G corresponding to the sequence S.

In view of the above, it can be seen that the set of possible combinations of values for $a = (a_1, a_2, \ldots, a_n)^T$ is a discrete set endowed with the discrete probability distribution defined by (4.5).

Now, a natural way of constructing an uncertainty set $\mathcal{U}$ for robust optimization purposes is to consider the subset of sequences obtained by selecting t*he most likely sequences,* in such a way that the probability of occurrence of sequences not belonging to the subset is less than a prescribed value $\alpha < 1$. Clearly such a set $\mathcal{U}$ can be defined as the set of all sequences Σ such that $p(\Sigma)$, or equivalently $\ln(p(\Sigma))$, is less than a given threshold θ, for an appropriate value of θ. Before discussing how to choose the right value for θ, we observe that the set of sequences in $\mathcal{U}$ will correspond to the set of paths originating in node (1, s) in G having total weight $\leq \theta$. Now, it is always possible to approximate (as closely as desired) the values $w_{ij} = -\ln(pij_)$ as integer multiples $\tilde{w}_{ij}$ of some (sufficiently small) real value $\delta > 0$, where $\tilde{w}_{ij} = \left\lfloor \frac{w_{ij}}{\delta} \right\rfloor$. Similarly θ can be approximated as $\tilde{\theta} = \left\lfloor \frac{\theta}{\delta} \right\rfloor$.

Then the set of sequences in $\mathcal{U}$ can be represented as the set of paths in another "expanded" graph $\tilde{G}$, where the nodes correspond to triples (k, i, z) where $z \in \left[0, \tilde{\theta}\right]$ and there is an arc joining (k, i, z) to $(k + 1, j, z + \tilde{w}_{ij})$ whenever $z + \tilde{w}_{ij} \leq \tilde{\theta}$. Only those paths in $\tilde{G}$ originating at node (1, s, 0) are considered.

In this way it is realized that a state-space representation of the uncertainty set $\mathcal{U}$ is obtained. The number of nodes in $\tilde{G}$ is $m \times n\,(\tilde{\theta} + 1)$, it thus depends on the accuracy looked for.

Now we turn to the issue of choosing the right threshold value θ. For this, we use the results in Appendix, in particular Theorem A.2 and Corollary A.3 providing the exact expressions of mean and covariance matrix of the *transition counts* $\tau = (\tau_{ij})$ for n-step realizations of stationary and ergodic Markov processes (started at a given state s), and justifying the fact that, for n large enough, the distributions are asymptotically normal. Denoting U the set $\{(i, j)/p_{ij} > 0\}$, the weight of a path originating at s in G in terms of the weights $w_{ij} = -\ln(p_{ij})$ is equal to $Z = \sum_{(i,j)\in U} w_{ij}\tau_{ij}$ where τ_{ij} is the corresponding transition count (the number of times a transition (i, j) is activated when traversing the path) and Corollary A.3 provides the expressions of mean E(Z) and variance σ^2 (Z) of the random variable Z. Since Z is asymptotically normal, assuming n large enough, θ can thus be chosen equal to:

$$\theta = E(Z) + \gamma\sigma(Z)$$

where γ is the $(1-\alpha)\%$ quantile of the standard normal distribution $\mathcal{N}(0, 1)$.

Table 3. The Q matrix corresponding to the 3x3 transition matrix P.

	(1.1)	(1.2)	(2.1)	(2.2)	(2.3)	(3.2)	(3.3)
(1.1)	0.6	0.4	0	0	0	0	0
(1.2)	0	0	0.3	0.5	0.2	0	0
(2.1)	0.6	0.4	0	0	0	0	0
(2.2)	0	0	0.3	0.5	0.2	0	0
(2.3)	0	0	0	0	0	0.5	0.5
(3.2)	0	0	0.3	0.5	0.2	0	0
(3.3)	0	0	0	0	0	0.5	0.5

Let us illustrate the above construction of state-space representations of uncertainty sets associated with finite state Markov processes on a small 3-state example (m = 3) with probability transition matrix:

$$P = \begin{bmatrix} 0.6 & 0.4 & 0 \\ 0.3 & 0.5 & 0.2 \\ 0 & 0.5 & 0.5 \end{bmatrix}$$

For several values of n, we consider n-step realizations of the process started at s = 1 (sequences of n consecutive states, beginning with s = 1). The P matrix defined as above is irreducible and the corresponding stationary probability distribution is $\pi = (0.348, 0.465, 0.186)^T$.

The set of transitions having nonzero probability in P is U = {(1, 1), (1, 2), (2, 1), (2, 2), (2, 3), (3, 2), (3, 3)} and q = |U| = 7.

The Q matrix (see Appendix, Corollary A.3) is the 7 × 7 matrix shown in Table 3.

This matrix is also irreducible, and the stationary distribution of the corresponding Markov chain is

$\tilde{\pi} = (0.209, 0.139, 0.139, 0.232, 0.093, 0.093, 0.093)^T$.

It can be checked that $\tilde{\pi}_{ij} = \pi_i \times p_{ij}$ for all (i, j) ∈ U. For instance: $\tilde{\pi}_{11} = \pi_i \times p_{11} = 0.348 \times 0.6 = 0.2088$; $\tilde{\pi}_{23} = \pi_2 \times p_{23} = 0.465 \times 0.2 = 0.093$, etc.

The fact that s = 1 is taken as initial state of the first (3-state) Markov process translates into the initial probability vector: $\hat{p} = [0.6, 0.4, 0, 0, 0, 0, 0]$ for the second (7-state) Markov process.

For all (i, j) ∈ U, the weight vector w with components $w_{ij} = -\ln(p_{ij})$ is

$w = [0.510, 0.916, 1.203, 0.693, 1.609, 0.693, 0.693]^T$.

We know from Corollary A.4 that, for sufficiently large n, the distribution of Z, the weight of (n – 1)- step realizations of the latter Markov process, is approximately normal with mean

$$E(Z) = \left(\widehat{p}^T \sum_{k=1}^{n-1} Q^{(k-1)} \right) w \tag{4.7}$$

and variance

$$\sigma^2(Z) = (n-1) w^T \; C(Q, \widehat{p}, n-1) w \tag{4.8}$$

where the matrix C(Q, $\widehat{p}$, n – 1) is defined as in Appendix, eq. (a.6), using $\widehat{p}$ instead of $\overline{p}$, and Q instead of P.

Table 4 indicates for various values of n the corresponding values of mean and standard deviation deduced from (4.7) and (4.8), and the values of the threshold θ for various values of α (10%, 5%, 1%). Column 5 provides the observed frequency of Z > θ based on simulations corresponding to 10^5 realizations of the Markov process started at s = 1. The figures shown appear to be reasonably close to the target frequencies and this is especially true for larger values of n, which is what might be expected: as n increases, the gaussian approximation becomes more and more accurate. The last column in Table 2 shows empirical estimates (derived from the same simulation) of $\overline{\theta}$, the exact value guaranteeing Prob{Z > $\overline{\theta}$ } = α. It is observed that the $\overline{\theta}$ values are extremely close to the corresponding θ values, thus validating the proposed model of uncertainty.

5. APPLICATION TO A MULTIPERIOD ENERGY PRODUCTION PROBLEM UNDER UNCERTAINTY

5.1 Problem Description

In this section we discuss an application of the robust multistage optimization model and solution approach presented in section 4 above to a problem arising in the context of multiperiod production planning in energy management. This problem may be viewed as an aggregate and simplified version of the well-known Unit Commitment problem which consists in determining the best possible production schedules of a given set of electrical power production units over a given multi-period horizon (in the extensive body of literature devoted to this problem, we refer to Carpentier *et al.* (2012), Frangioni *et al.* (2008) and Padhy (2004)).

In the problem discussed here, we consider an aggregate representation of the various production units into a single "super-unit" capable of delivering in each time period t of a given T -period horizon, any power level between 0 and $P_t^{\max}$, where $P_t^{\max}$ represents the sum of the maximum power which can be produced by all the units which can be activated during time period t (in practice, $P_t^{\max}$ will actually depend on t, e.g. to take into account the necessary outage periods for maintenance of some units). To fix ideas, if we consider a daily planning problem, the number of time periods will typically lie between 48 and 96, each time period corresponding to half an hour. Among the most important constraints which have to be taken into account for each individual production unit are the so-called *ramping constraints* i.e. constraints limiting the speed of variation of the power delivered from one time period to the next. We will therefore assume that a simplified aggregate counterpart of all these individual constraints can

Table 4. Experimental validation of the proposed threshold values θ

n	E (Z) σ (Z)	α	Threshold θ	Observed frequency of Z > θ	Value of $\bar{\theta}$: Prob {Z > $\bar{\theta}$ } = α
10	$\begin{bmatrix} 7.36 \\ 0.95 \end{bmatrix}$	0.1	8.57	10.1%	8.57
		0.05	8.93	3.9%	8.84
		0.01	9.57	0.5%	9.21
20	$\begin{bmatrix} 15,79 \\ 1.33 \end{bmatrix}$	0.1	17.5	9%	17.45
		0.05	18	3.5%	17.9
		0.01	18.9	0.45%	18.7
40	$\begin{bmatrix} 32.64 \\ 1.88 \end{bmatrix}$	0.1	35.05	9.6%	35.03
		0.05	35.74	4.2%	35.65
		0.01	37.02	0.57%	36.75
100	$\begin{bmatrix} 83.2 \\ 2.97 \end{bmatrix}$	0.1	87	9.8%	86.9
		0.05	88.1	4.5%	87.9
		0.01	90.1	0.76%	89.8
200	$\begin{bmatrix} 167.46 \\ 4.19 \end{bmatrix}$	0.1	172.83	9.8%	172.5
		0.05	174.38	4.4%	174.1
		0.01	177.24	0.77%	177

be expressed in the form of a gradient-type constraint on the global production of the "super-unit". More precisely, denoting x_t the total power produced during time period t, these constraints can be expressed as:

$$\begin{cases} \delta_t^- \leq x_t - x_{t-1} \leq \delta_t^+ & (t = 2, ..., T) \\ 0 \leq x_t \leq P_t^{\max} & (t = 1, ..., T) \end{cases} \qquad (5.1), (5.2)$$

where δ_t^+ and δ_t^- (t = 2, …, T) are given values.

The objective function (to be minimized) is the sum of two parts:

a. The sum over all periods t = 1, …., T, of the production costs $\gamma_t(x_t)$; ∀ t, γ_t is supposed to be a given convex polyhedral function: $\mathbb{R}_+ \rightarrow \mathbb{R}_+$, defined as the pointwise maximum of a finite number of given affine functions. We note here that such a representation of the production costs is a good approximation of reality since, as power load requirements increase, those units having minimum marginal production cost are activated first; next, units corresponding to the second best marginal production cost are activated, and so on. As a result, a convex piecewise affine production cost function is obtained.

b. The second part of the objective function is intended to account for the possible extra costs incurred either in case of supply shortage or in case of excess production. More precisely, when $d_t > x_t$ (case of supply shortage) the missing power $d_t - x_t$ has to be bought on the market at a given unit price α_t which usually exceeds the marginal production cost $\gamma(x_t + 1) - \gamma(x_t)$; when $d_t < x_t$ (case of production in excess), the extra power $x_t - d_t$ can be sold on the market at a given unit price β_t but usually at the expense of a reduced benefit (due to market behavior, one has $\beta_t < \alpha_t$; moreover β_t is often less than the marginal production cost). Thus, for a given production plan $x = (x_t)_{t=1,\dots,T}$ and given power requirements $d = (d_t)_{t=1,\dots,T}$, the second part of the objective reads:

$$H(x, d) = \sum_{t=1}^{T}\left(\alpha_t\left[y_t\right]^- - \beta_t\left[y_t\right]^+\right) \tag{5.3}$$

where $y_t = x_t - d_t$, $[y_t]^+ = \text{Max}\{0, x_t - d_t\}$, $[y_t]^- = \text{Max}\{0, d_t - x_t\}$.

Now, assuming that the uncertainty set $\mathcal{D}$ for load requirements is *state-space representable,* it is realized that the problem of determining an optimal strategy for power production over the time periods 1, 2, …, T, can be solved using the procedure described in Section 4, provided that the set of possible decisions (production levels) in each time period is discretized. In the computational results reported in Section 5.2, this will indeed be the case (for each t, the set of production levels will correspond to all integer values between 0 and $P_t^{\max}$). Also note that, in order to satisfy the constraints (5.1)-(5.2) above, the set $X(t, x_{t-1})$ (the set of allowed decisions for period t, given that the production level was x_{t-1} in period t – 1) is simply defined as the set of integer values in the interval $[0, P_t^{\max}] \cap \left[x_{t-1} - \delta_t^-, x_{t-1} + \delta_t^+\right]$.

Also observe that the uncertainty on load requirement only impacts the cost function (to be minimized) as follows: if production level x_t has been decided for period t and customer load d_t occurs during that period, then the resulting cost for period t is $\alpha_t\left[y_t\right]^- - \beta_t\left[y_t\right]^+$ (where $y_t = x_t - d_t$). Of course the only possible values for d_t are those which are consistent with the structure of the given uncertainty set, so they depend on which state has been reached by the uncertain process during period t – 1; or, equivalently, on which node (t – 1, k) has been reached in the underlying graph G. These are actually all the values v (t, k') for all (t, k') ∈ succ(t – 1, k) (the set of direct successors of node (t – 1, k) in G). In view of this, an optimal strategy will be obtained by applying the robust dynamic programming recursion (4.3)-(4.4) in Section 4, just replacing Max by Min in (4.3) and Min by Max in (4.4), because the problem addressed is to minimize a cost (instead of maximizing a revenue).

Before presenting the computational experiments, we also have to define what we refer to as the "2-stage version" of the above dynamic optimization problem. This is obtained when withdrawing the possibility for the decision-maker to adjust his/her decision in each time period, depending on the realizations of uncertainty observed so far. In other words, in the 2-stage version, all the production levels $x_1, x_2, \dots, x_T$ have to be chosen beforehand, prior to getting any information about which scenario for load requirements will arise.

It is easily seen that solving the 2-stage version of the problem consists in finding $x = (x_1, x_2, \dots, x_T)$ minimizing the cost function:

$$\sum_{t=1}^{T}\gamma_t(x_t) + \underset{d \in \mathcal{D}}{Max}\left\{H(x, d)\right\} \tag{5.4}$$

subject to (5.1)-(5.2) together with integrality of x_t (t = 1, ..., T).

It has been shown in Minoux (2014) that the continuous version of the above 2-stage robust optimization problem can be solved efficiently (i.e. in polynomial time) when the uncertainty set $\mathscr{D}$ for load requirements is state-space representable. When integrality conditions on x are imposed, of course the problem is no longer polynomially solvable, but close-to-optimal integer solutions can be deduced by simply rounding the optimal fractional solution obtained.

In addition to illustrating the efficiency of the robust dynamic programming approach proposed in Section 4 on the energy management problem, the computational results presented in the next subsection provide an experimental basis for comparing, in terms of objective function values, optimal robust multistage strategies and robust 2-stage solutions.

5.2 Computational Experiments, Results and Discussion

We describe here a set of computational experiments carried out on a series of randomly generated test instances of the above-described problem for values of T ranging from 30 to 40 and state-space representable uncertainty sets with underlying graphs featuring up to 287 nodes per time period.

The instances of the power production planning problem considered have been drawn at random as follows.

1. For each t = 1, …, T, the cost functions γ_t (x) do not depend on t and are convex piecewise affine, given as:

$$\gamma(x) = \text{Max}\{15x;\ 450 + 20(x - 30);\ 1050 + 25(x - 60);\ 1800 + 40(x - 90)\}.$$

2. $P^{max} = 100$, and $\forall$ t, the δ_t^+ and δ_t^- values are all equal to the same value $\Delta = 30$.
3. For each t = 1, …, T, the α_t and β_t values are randomly chosen in a given interval $[v_{min}, v_{max}]$ while complying with the condition $\alpha_t > \beta_t$ for all t.
4. The graph providing the state-space representation of uncertain load requirements over time is constructed as explained in Minoux (2014). The main parameters in this construction are:
    - The interval $[\mu^-, \mu^+]$ in which the average values μ_t for load requirements are randomly selected;
    - The parameter $\lambda \in [0, 1]$ representing the dispersion of load requirements around their mean values μ_t ;
    - The number K of possible values for load requirements in each time period t. These K values, denoted $\theta_t^1, \theta_t^2, \ldots, \theta_t^K$, are chosen equally spaced between the extreme values: $\theta_t^1 = \mu_t(1 - \lambda)$ and $\theta_t^K = \mu_t(1 + \lambda)$ (in all our computational experiments, the same value K = 7 has been used);
    - The (integer) value W_{max} corresponding to the maximum allowed path weights in the graph (for details, refer to the Appendix in Minoux (2014)). Note that this value W_{max} plays a role analogue to the one played by the parameter θ in the construction of the graph representing uncertainty in section 4.4.

The results obtained on a series of 21 randomly generated instances are displayed in Tables 5, 6, 7, and 8.

Table 5. $T = 30$, $K_{max} = 7$, $W_{max} = 20$ (the corresponding graph has 147 nodes and 1337 arcs per time period), μ_t randomly chosen in [40, 100], $\alpha(t)$ and $\beta(t)$ randomly chosen in [15, 35] and such that $\alpha(t) > \beta(t)$.

Instance #	Dispersion of demand around average demand (λ)	Cost of average production plan	Cost optimal strategy (1st decision)	Cost optimal 2-stage production plan (1st decision)	Relative difference 2-stage/optimal strategy
P1	± 5%	39787	39339 (69)	39341 (69)	$5\ 10^{-5}$
P2	± 5%	46978	46714 (90)	46724 (90)	$4.3\ 10^{-5}$
P3	± 5%	37057	36803 (62)	36816 (62)	$1.6\ 10^{-4}$
P4	± 10%	40732	40273 (69)	40285 (69)	$8\ 10^{-4}$
P5	± 10%	47825	47578 (90)	47592 (90)	$2\ 10^{-4}$
P6	± 10%	37769	37498 (62)	37518 (62)	$1.6\ 10^{-4}$
P7	± 30%	44484	43998 (69)	44010 (69)	$2\ 10^{-4}$
P8	± 30%	51845	51572 (90)	51593 (90)	$3.5\ 10^{-4}$
P9	± 30%	41163	40805 (63)	40834 (63)	$6\ 10^{-4}$

Table 6. $T = 30$, $K_{max} = 7$, $W_{max} = 40$ (the corresponding graph has 287 nodes and 1405 arcs per time period), μ_t randomly chosen in [40, 100], $\alpha(t)$ and $\beta(t)$ randomly chosen in [15, 35] and such that $\alpha(t) > \beta(t)$ (the only change as compared with Table 4 concerns the value Wmax. The instances considered in Table 6 are the same as for Table 4, the only change is in the increased value of Wmax. This results in a bigger uncertainty set and the cost function values are increased accordingly.

Instance #	Dispersion of demand around average demand (λ)	Cost of average production plan	Cost optimal strategy (1st decision)	Cost optimal 2-stage production plan (1st decision)	Relative difference 2-stage/optimal strategy
P1	± 5%	40456	39945 (70)	39964 (69)	$7\ 10^{-5}$
P2	± 5%	47758	47420 (90)	47439 (90)	$6\ 10^{-5}$
P3	± 5%	37711	37355 (62)	37376 (62)	$1.3\ 10^{-4}$
P4	± 10%	42173	41573 (71)	41599 (69)	$2\ 10^{-4}$
P5	± 10%	49502	49089 (90)	49116 (90)	$2.6\ 10^{-4}$
P6	± 10%	39113	38633 (62)	38675 (62)	$3.4\ 10^{-4}$
P7	± 30%	48917	48138 (78)	48202 (69)	$1.1\ 10^{-3}$
P8	± 30%	56817	56303 (90)	56340 (90)	$4.6\ 10^{-4}$
P9	± 30%	45432	44454 (63)	44537 (63)	$1.5\ 10^{-3}$

Table 7. $T = 36$, $K_{max} = 7$, $W_{max} = 30$ (the corresponding graph has 217 nodes and 1319 arcs per time period), μ_t randomly chosen in [20, 100], $\alpha(t)$ and $\beta(t)$ randomly chosen in [10, 40] and such that $\alpha(t) > \beta(t)$.

Instance #	Dispersion of demand around average demand (λ)	Cost of average production plan	Cost optimal strategy (1st decision)	Cost optimal 2-stage production plan (1st decision)	Relative difference 2-stage/optimal strategy
P10	± 10%	58949	58653 (69)	58664 (74)	$1.3\ 10^{-4}$
P11	± 10%	45041	44348 (60)	44362 (70)	$7\ 10^{-5}$
P12	± 10%	60111	59280 (90)	59282 (62)	$3.3\ 10^{-5}$
P13	± 20%	62294	61962 (74)	61973 (74)	$1.6\ 10^{-5}$
P14	± 20%	47833	47087 (60)	47127 (71)	$4.4\ 10^{-4}$
P15	± 20%	63243	62418 (90)	62418 (99)	$5\ 10^{-5}$

Table 8. $T = 40$, $K_{max} = 7$, $W_{max} = 25$ (the corresponding graph has 182 nodes and 1313 arcs per time period), μ_t randomly chosen in [30, 90], $\alpha(t)$ and $\beta(t)$ randomly chosen in [10, 50] and such that $\alpha(t) > \beta(t)$

Instance #	Dispersion of demand around average demand (λ)	Cost of average production plan	Cost optimal strategy (1st decision)	Cost optimal 2-stage production plan (1st decision)	Relative difference 2-stage/optimal strategy
P16	± 10%	50274	48662 (69)	48759 (60)	$5\ 10^{-4}$
P17	± 10%	49566	47367 (42)	47409 (42)	$1.5\ 10^{-4}$
P18	± 10%	49090	47304 (60)	47388 (60)	$9\ 10^{-4}$
P19	± 20%	53419	51506 (60)	51593 (60)	$6\ 10^{-4}$
P20	± 20%	52453	49847 (42)	49938 (42)	$1.1\ 10^{-3}$
P21	± 20%	51938	49739 (61)	49930 (61)	$2.8\ 10^{-3}$

In each table the columns show successively:

- The instance identifier;
- The value of λ (dispersion of demand around average demand) corresponding to the instance;
- The cost of the "average production plan", which corresponds to fixing $x_t = \mu_t$ ($\forall$ t), i.e. in each time period, the production level is set equal to the average demand;
- The cost of the optimal robust strategy obtained using the dynamic programming based procedure described in Section 4; next to each cost function value, the value of the first decision to be taken under the optimal strategy (i.e. the production level in time period 1) is shown in parenthesis;
- The cost of the robust 2-stage production plan obtained using the approach described in Minoux (2014) (as already mentioned, the cost figures shown correspond to the values of (5.4) obtained after componentwise rounding of the optimal 2-stage solutions); next to each cost function value, the value of the first decision (the production level in time period 1) is shown in parenthesis;

- The relative difference in terms of cost between the optimal robust strategy and the rounded optimal robust 2-stage solution;

The main observations which can be derived from the results of Tables 4 to 7 are the following:

a. The relative differences between the cost of the optimal strategy and the cost of the "average production plan" are significant, typically up to 4%, with an average 1,5%. This confirms the benefit of considering a robust optimization model as opposed to the crude approach consisting in converting the problem into a deterministic one with the average values of the uncertain parameters (load requirements).
b. As could be expected, for each instance the cost of the optimal strategy is smaller than the cost of the rounded optimal 2-stage solution. However the differences in cost appear to be quite small, typically less than 0.2% (the only exception is P21 with a difference of 0.28%).
c. In a majority of instances, the first decisions (i.e. the production level to be set in the first time period) corresponding either to the rounded optimal 2-stage solution or to the optimal strategy turn out to be *almost identical* (the only cases where significant differences are observed are P7 in Table 5 and P12 and P14 in Table 6). This observation suggests that the two-stage robust optimization model investigated here for power production planning under uncertainty is a good candidate to serve as a basis for a "rolling horizon" approach capable of closely approximating optimal strategies.
d. Concerning computational efficiency, the solution procedure based on (4.3)-(4.4) solves all the instances considered in less than 1 minute CPU (with a 2.5 GHz Intel core i5 and 4GB RAM in a MATLAB environment). This clearly leaves open the possibility of solving somewhat larger problems, either in terms of number of states, or in terms of number of time periods.

6. CONCLUSION

In this chapter, two-stage and multistage robust optimization models have been discussed emphasizing complexity issues (NP-hardness results) and polynomially solvable special cases, in connection with various applications. A class of multistage robust optimization problems for which exact optimal strategies can be computed efficiently (using a dynamic programming type recursion) has been introduced. Central to the definition of this class is the concept of state-space representable uncertainty set, which includes various models of uncertainty previously investigated in the literature and, as an original contribution of the present work, is shown here to provide a relevant model for building uncertainty sets associated with discrete finite state Markov processes. To illustrate the class of problems investigated here, a multiperiod energy production planning problem under uncertain load requirements has been discussed, and computational results on instances featuring up to 40 time periods have been reported. From the point-of-view of practical application, the main conclusions drawn from the results obtained are that optimal robust strategies can be closely approximated by considering the robust two-stage version of the problem, which is computationally somewhat easier to solve.

7. FUTURE RESEARCH DIRECTIONS

The results discussed in the present work have clearly shown the possibility of by-passing the inherent computational hardness of the class of two-stage and multistage robust optimization problems under right-hand side uncertainty, when properly exploiting special structures underlying the definition of the uncertainty sets. In this respect, exhibiting further types of special structures (other than those corresponding to perfect independence of uncertain parameters related to distinct time periods) is a potentially fruitful direction for future research. Another issue deserving further investigations concerns the search for new possible links between the robust optimization models which have been discussed here, and other alternative approaches for handling uncertainty in optimization problems, in particular Stochastic Programming models and Chance Constraint Programming models. As a first step in this direction, extending the discussion presented in section 2.2 of this chapter to the multistage case would be worth considering.

ACKNOWLEDGMENT

Two reviewers are gratefully acknowledged for their constructive comments on a first version of this chapter. We are also indebted to A. Ben-Abbes, S. Charousset, R. Zorgati and W. van Ackooij for stimulating discussions concerning the application to the energy production planning problem which has been the focus of our computational experiments.

REFERENCES

Atamtürk, A., & Zhang, M. (2007). Two-stage robust network flow and design under demand uncertainty. *Operations Research*, *55*(4), 662–667. doi:10.1287/opre.1070.0428

Ben-Tal, A., Goryashko, A., Guslitzer, E., & Nemirovski, A. (2004). Adjustable robust solutions of uncertain linear programs. *Mathematical Programming*, *99*(2), 351–376. doi:10.1007/s10107-003-0454-y

Ben-Tal, A., & Nemirovski, A. (1999). Robust solutions of uncertain linear programs. *Operations Research Letters*, *25*(1), 1–13. doi:10.1016/S0167-6377(99)00016-4

Ben-Tal, A., & Nemirovski, A. (2000). Robust solution of linear programming problems contaminated with uncertain data. *Mathematical Programming*, *88*(3), 411–424. doi:10.1007/PL00011380

Ben-Tal, A., & Nemirovski, A. (2002). Robust Optimization-Methodology and applications. *Mathematical Programming*, *92*(3), 453–480. doi:10.1007/s101070100286

Bertsekas, D. P. (2005). *Dynamic programming and optimal control* (2 vols.). Athena Scientific Massachussets.

Bertsimas, D. P., & Sim, M. (2003). Robust discrete optimization and network flows. *Math. Prog. B*, *98*(1-3), 49–71. doi:10.1007/s10107-003-0396-4

Bertsimas, D. P., & Sim, M. (2004). The price of robustness. *Operations Research, 52*(1), 35–53. doi:10.1287/opre.1030.0065

Billingsley, P. (1961). Statistical methods in Markov chains. *Annals of Mathematical Statistics, 32*(1), 12–40. doi:10.1214/aoms/1177705136

Birge, J. R. (1985). Decomposition and partitioning methods for multistage stochastic linear programs. *Operations Research, 33*(5), 989–1007. doi:10.1287/opre.33.5.989

Carpentier, P., Cohen, G., Culioli, J. C., & Renaud, A. (2012). Stochastic optimization of unit commitment: A new decomposition framework. *IEEE Transactions on Power Systems, 11*(2), 1067–1073. doi:10.1109/59.496196

Charnes, A., & Cooper, W. W. (1959). Chance Constrained Programming. *Management Science, 6*(1), 73–79. doi:10.1287/mnsc.6.1.73

Derriennic, Y., & Lin, M. (2003). The Central Limit Theorem for Markov Chains started at a point. *Probability Theory and Related Fields, 125*(1), 73–76. doi:10.1007/s004400200215

Frangioni, A., Gentile, C., & Lacalandra, F. (2008). Solving unit commitment problems with general ramp constraints. *International Journal of Electrical Power & Energy Systems, 30*(5), 316–326. doi:10.1016/j.ijepes.2007.10.003

Iyengar, G. N. (2005). Robust dynamic programming. *Mathematics of Operations Research, 30*(2), 257–280. doi:10.1287/moor.1040.0129

Minoux, M. (2007). Models and Algorithms for Robust PERT Scheduling with Time-Dependent Task Durations. *Vietnam Journal of Mathematics, 35*(4), 387–398.

Minoux, M. (2008). Solving some multistage robust decision problems with huge implicitly-defined scenario trees. *Algorithmic O.R., 4*(1), 1–18.

Minoux, M. (2009). On robust maximum flow with polyhedral uncertainty sets. *Optimization Letters, 3*(3), 367–376. doi:10.1007/s11590-009-0116-y

Minoux, M. (2010). Robust network optimization under polyhedral demand uncertainty is NP-Hard. *Discrete Applied Mathematics, 158*(5), 597–603. doi:10.1016/j.dam.2009.09.025

Minoux, M. (2011). On 2-stage robust LP with RHS uncertainty: Complexity results and applications. *Journal of Global Optimization, 49*(3), 521–537. doi:10.1007/s10898-010-9645-2

Minoux, M. (2014). Two-stage robust optimization, state-space representable uncertainty and applications. *RAIRO Oper. Res., 48*(4), 455–475. doi:10.1051/ro/2014017

Minoux, M. (2015). (to appear). Efficient Robust Multistage Optimization with State-Space Representation of Uncertainty and Applications. *IMA Journal of Management Mathematics.*

Ordon˜ez, F., & Zhao, J. (2007). Robust capacity expansion of network flows. Networks, 50(2), 136–145.

Padhy, N. P. (2004). Unit Commitment - A bibliographical survey. *IEEE Transactions on Power Systems, 19*(2), 1196–1205. doi:10.1109/TPWRS.2003.821611

Ross, S. M. (1983). *Introduction to stochastic dynamic programming*. New York, NY: Academic Press.

Shapiro, M. (2006). On complexity of multistage stochastic programs. *O.R. Letters*, *34*, 1–8.

Shapiro, M., & Nemiroski, A. (2006). On complexity of stochastic programming problems. In V. Jeyakumar & A. Rubinov (Eds.), Continuous Optimization: Current Trends and Applications, (pp. 111-144). Berlin: Springer Verlag.

ADDITIONAL READING

Barmish, B.R., & Lagoa, C.M. (1997). The uniform distribution: a rigorous justification for the use in robustness analysis, *Math. Control, Signals, Systems 10*, .203–222.

Ben-Tal, A., El Ghaoui, L., & Nemirovski, A. (2009). *Robust Optimization*. Princeton, NJ: Princeton University Press. doi:10.1515/9781400831050

Ben-Tal, A., Goryashko, A., Guslitzer, E., & Nemirovski, A. (2003). Adjustable robust solutions of uncertain linear programs. *Mathematical Programming*, *99*(2), 351–376. doi:10.1007/s10107-003-0454-y

Ben-Tal, A., & Nemirovski, A. (1998). Robust convex optimization. *Mathematics of Operations Research*, *23*(4), 769–805. doi:10.1287/moor.23.4.769

Bertsekas, D. P., & Tsitsiklis, J. N. (1996). *Neuro-Dynamic Programming*. Nashua, NH: Athena Scientific.

Bertsimas, D., & Brown, D. (2009). Constructing uncertainty sets for robust linear optimization. *Operations Research*, *57*(6), 1483–1495. doi:10.1287/opre.1080.0646

Bertsimas, D., Brown, D. B., & Caramanis, C. (2011). Theory and applications of robust optimization. *SIAM Review*, *53*(3), 464–501. doi:10.1137/080734510

Bertsimas, D., Goyal, V., & Lu, B. Y. (2015). A tight characterization of the performance of static solutions in two-stage adjustable robust linear optimization. *Mathematical Programming*, *150*(2), 281–319. doi:10.1007/s10107-014-0768-y

Bertsimas, D., Iancu, D., & Parrilo, P. (2009). Optimality of affine policies in multi-stage robust optimization, in *Proceedings of the 48th Conference on Decision and Control* (pp. 1131-1138).Red Hook, NY: Curran Associates Inc. doi:10.1109/CDC.2009.5399533

Birge, J., & Louveaux, F. (1997). *Introduction to Stochastic Programming*. New York, NY: Springer-Verlag.

Calafiore, G., & Campi, M. C. (2006). The scenario approach to robust control design. *IEEE Transactions on Automatic Control*, *51*(5), 742–753. doi:10.1109/TAC.2006.875041

Cerbakova, J. (2010). Information input for multistage stochastic programs. *IMA J. of Management Math.*, *21*(2), 89–109. doi:10.1093/imaman/dpp006

Chekuri, C., Oriolo, G., Scutella, M. G., & Shepherd, F. B. (2007). Hardness of robust network design. *Networks*, *50*(1), 50–54. doi:10.1002/net.20165

Chen, X. M., Sim, M., & Sun, P. (2007). A robust optimization perspective to stochastic programming. *Operations Research*, *55*(6), 1058–1071. doi:10.1287/opre.1070.0441

Delage, E., & Mannor, S. (2010). Percentile optimization for Markov decision processes with parameter uncertainty. *Operations Research*, *58*(1), 203–213. doi:10.1287/opre.1080.0685

Dyer, M., & Stougie, L. (1997). Computational complexity of stochastic programming problems. *Mathematical Programming*, *106*(3), 423–432. doi:10.1007/s10107-005-0597-0

Erdogan, E., & Iyengar, G. (2006). Ambiguous chance constrained problems and robust optimization. *Mathematical Programming*, *107*(1-2), 37–61. doi:10.1007/s10107-005-0678-0

Garey, M. R., & Johnson, D. S. (1979). *Computers and intractability. A guide to the theory of NP-Completeness*. San Francisco, CA: W.H Freeman and Co.

Gartska, S. J., & Wets, R. J.-B. (1974). On decisions rules in stochastic programming. *Mathematical Programming*, *7*(1), 117–143. doi:10.1007/BF01585511

Goldfarb, D., & Iyengar, G. (2003). Robust portfolio selection problems. *Mathematics of Operations Research*, *1*(1), 1–38. doi:10.1287/moor.28.1.1.14260

Goulart, P. J., Kerrigan, E. C., & Ralph, D. (2007). Efficient robust optimization for robust control with constraints. *Math. Program. Series A*, *114*, 1–33.

Hansen, L., & Sargent, T. (2001). Robust control and model uncertainty. *The American Economic Review*, *91*(2), 60–66. doi:10.1257/aer.91.2.60

Iyengar, G., & Erdogan, E. (2006). Ambiguous chance constrained problems and robust optimization. *Math. Progr. Series B*, *107*(1-2), 37–61. doi:10.1007/s10107-005-0678-0

Kall, P., & Wallace, S. (1994). *Stochastic Programming*. Chichester, UK: John Wiley and Sons.

King, A. J., & Wallace, S. (2012). *Modeling with stochastic programming*. New York, NY: Springer Verlag. doi:10.1007/978-0-387-87817-1

Kouvelis, P., & Yu, G. (1997). *Robust Discrete Optimization and Its Applications*. Norwell, MA: Kluwer Academic. doi:10.1007/978-1-4757-2620-6

Laguna, M. (1998). Applying robust optimization to capacity expansion of one location in telecommunications with demand uncertainty. *Management Science*, *44*(11), 5101–5110.

Malcom, S., & Zenios, S. (1994). Robust optimization for power systems capacity expansion under uncertainty. *J. ORSA*, *45*(9), 1–55.

Mulvey, J. M., & Ruszczyński, A. (1995). A new scenario decomposition method for large-scale stochastic optimization. *Operations Research*, *43*(3), 477–490. doi:10.1287/opre.43.3.477

Natarajan, K., Pachamanova, D., & Sim, L. M. (2009). Constructing risk measures from uncertainty sets. *Operations Research*, *57*(5), 1129–1141. doi:10.1287/opre.1080.0683

Nemirovski, A., & Shapiro, A. (2006). Convex approximations of chance constrained programs. *SIAM Journal on Optimization*, *17*(4), 969–996. doi:10.1137/050622328

Nilim, A., & El Ghaoui, L. (2005). Robust Markov decision processes with uncertain transition matrices. *Operations Research, 53*(5), 780–798. doi:10.1287/opre.1050.0216

Prékopa, A. (1995). *Stochastic Programming*. Dordrecht, The Netherlands: Kluwer Academic. doi:10.1007/978-94-017-3087-7

Puterman, M. L. (1994). *Markov Decision Processes: Discrete Stochastic Dynamic Programming*. New York, NY: John Wiley & Sons. doi:10.1002/9780470316887

Ruszczyński, A., & Shapiro, A. (Eds.). (2003). *Stochastic Programming*. Amsterdam, The Netherlands: North Holland Publishing Co.

Shapiro, A. (2008). Stochastic programming approach to optimization under uncertainty. *Math. Program., Series B, 112*(1), 183–220. doi:10.1007/s10107-006-0090-4

Shapiro, A. (2011). Dynamic Programming approach to adjustable robust optimization. *O.R. Lett., 39*(2), 83–87. doi:10.1016/j.orl.2011.01.001

Shapiro, A., Dentcheva, D., & Ruszczyński, A. (2009). *Lectures on Stochastic Programming: Modeling and Theory*. Philadelphia, PA: SIAM. doi:10.1137/1.9780898718751

Shapiro, A., & Nemirovski, A. (2005). On complexity of stochastic programming problems. In V. Jeyakumar & A. M. Rubinov (Eds.), *Continuous Optimization: Current Trends and Applications* (pp. 111–144). New York, NY: Springer. doi:10.1007/0-387-26771-9_4

Soyster, A. (1973). Convex programming with set-inclusive constraints and applications to inexact linear programming. *Operations Research, 21*(5), 1154–1157. doi:10.1287/opre.21.5.1154

Soyster, A. (1979). Inexact Linear programming with generalized resource sets. *EJOR, 3*(4), 316–321. doi:10.1016/0377-2217(79)90227-3

Zhu, L., Li, C., & Xu, W. (2012). Affinely Adjustable Robust optimization model for multi-period production and inventory system under risk preference. *Journal of Appl. Anal. and Comp., 2*(3), 341–350.

KEY TERMS AND DEFINITIONS

Linear Programming Problem: A linear programming (LP) problem is an optimization problem which consists in finding the maximum or the minimum value of a given linear function of n real decision variables subject to a given set of linear equality or inequality constraints.

Multistage Robust or Stochastic Optimization: This refers to an optimization problem with uncertain or random parameters, the realizations of which are revealed progressively over successive time periods (or stages). For a k-stage instance, the set of decision variables is decomposed into $(x_1, y_1, x_2, y_2, \ldots, x_k, y_k)$; in any stage t, the decision variables x_t have to be fixed knowing the realizations of the uncertain or random parameters in stages 1, 2, … t-1, but prior to getting any information about the realization of the uncertain or random parameters in stage t and in subsequent stages; the decision variables y_t can be adjusted after observing the realized values of the uncertain or random parameters in stages 1, 2, .., t.

NP-Hard Problem: In complexity theory, this refers to a class of combinatorial problems which are considered as computationally difficult to solve in the sense that no efficient algorithm (= running in time polynomial in the size of the instance) is known for solving them. The formal definition of the class is that it contains any problem which can be shown to be at least as difficult as some NP-complete problem (as typical examples of NP-complete problems we can mention: Satisfiability of a logical expression, the existence of a hamiltonian cycle in a graph, the vertex coloring problem in graphs, the stable set problem in graphs, the 0-1 knapsack problem; for a more complete list, refer to the book by Garey and Johnson (1979) see Additional Reading Section).

Polyhedral Uncertainty Set: This corresponds to the special case (frequently arising in applications) of an uncertainty set defined as the set of solutions to a given linear equality/inequality system.

Right-Hand Side Uncertainty: An optimization problem featuring right-hand side uncertainty corresponds to the situation when the only uncertain parameters correspond to all (or part of) the right-hand side coefficients in the constraints.

Robust Optimal Solution: We define this here for a two-stage minimization problem with decision variables x, recourse variables y, and depending on u, a p-component vector the components of which represent the various parameters subject to uncertainty; assuming that u is known to belong to some given (finite or compact) uncertainty set U in p-dimensional space, the value h(x,v) of any solution x in case u=v (with v an element of U) is the minimum objective function value achievable by adjusting the recourse variables y. The *worst-case value* w(x) of any given solution x is then defined as the *maximum* taken over v in U of h(x,v), and an optimal robust solution x* is a solution minimizing w(x). It is thus realized that such a x* is obtained as the result of a Min-Max optimization process. The problem of minimizing w(x) is called the *deterministic equivalent* to the robust optimization problem under consideration.

Stochastic Optimization Problem: We define this here for a two-stage minimization problem with decision variables x, recourse variables y, and depending on u, a p-component vector the components of which represent the various random parameters assumed to follow a given probability distribution in p-dimensional space; the value h(x,v) of any solution x in case u=v arises, is the minimum objective function value achievable by adjusting the recourse variables y. The expected value of any given solution x is then defined as the expectation of h(x,v) with respect to the probability distribution of v, and an optimal solution to the stochastic optimization problem is a solution x* minimizing this expected value.

Two-Stage Robust or Stochastic Optimization: This refers to an optimization problem with uncertain or random parameters, in which the set of decision variables (x, y) is decomposed into two distinct subsets: the set of primary (or 'here and now') decision variables x, and the set of secondary (or 'wait-and-see') variables y; the former have to be fixed prior to getting any information about the realization of the uncertain or random parameters, while the latter can be adjusted after observing the realized values of the uncertain or random parameters.

Uncertainty Set: This is one of the basic input data which is required to specify an optimization problem under uncertainty on some of the parameters such as coefficients in the constraints or in the objective function. Given p uncertain real parameters, the uncertainty set (in p-dimensional real space) is the set of values which can possibly be taken by the p parameters considered. This set may feature finite cardinality, or may contain infinitely many elements (compactness being assumed in this case).

APPENDIX. SOME USEFUL PROBABILISTIC RESULTS ON DISCRETE MARKOV CHAINS

Many stochastic processes encountered in applications can be represented in terms of the basic *Markov chain model*: sequences of random variables $x_1, x_2, \ldots, x_t, \ldots$ taking values in a given set X (referred to as the *set of states* of the Markov process) such that for any given sequence $(a_1, a_2, \ldots, a_t) \in X^T$, the probability: $\mathbb{P}\{x_t = a_t / (x_1, x_2, \ldots, x_{t-1}) = (a_1, a_2, \ldots, a_{t-1})\}$ only depends on the k last values taken by the process, i.e. only depends on $a_{t-1}, a_{t-2}, \ldots, a_{t-k}$ (where k is a given integer). The case of $k = 1$ corresponds to what is referred to as a *simple* Markov chain.

Here we will restrict to considering discrete simple Markov chains, i.e. assuming finite state space X, with *stationary* transition probabilities:

$$\mathbb{P}\{x_t = i/x_{t-1} = j\} = p_{i,j} \ (i \in X, j \in X)$$

(stationarity means that the $p_{i,j}$ values do not depend on t).
Such Markov chains are completely defined by the m×m transition matrix $P = (p_{i,j})_{\substack{i=1,\ldots,m \\ j=1,\ldots,m}}$ where $|X| = m$ is the cardinality of the associated state space.
The transition matrix P is called a *stochastic matrix*, i.e. it has nonnegative entries and all the row sums are equal to 1.
The transition matrix P of a discrete Markov chain is called *irreducible* if its associated graph G(P) = [X,Y] is *strongly connected.* (G(P) has vertex set X, and the arc set Y contains all pairs $(i, j) \in X^2$ such that $p_{i,j} > 0$). Thus irreducibility means that for any pair of states (i, j), if $x_t = i$ for some t, then there exists t' > t such that $\mathbb{P}\left\{x_{t'} = j\right\} > 0$. Each state in such a Markov chain is thus *recurrent.*
A state $i \in X$ of a discrete stationary Markov chain with probability transition matrix $P = (p_{i,j})$ is called aperiodic iff the g.c.d. of the lengths of all the circuits of G(P) containing vertex i is equal to 1.
It can be shown that if an irreducible Markov chain has one aperiodic state, then all states are *aperiodic*, and such a chain is called *ergodic*. A sufficient condition for a discrete stationary Markov chain with associated probability transition matrix P to be *ergodic* is that there exists an integer n > 0 such that $G(P^n)$ is the complete graph on the vertex set X (with a loop at each vertex).
A basic property of a discrete stationary ergodic Markov chain is that it admits a *stationary probability distribution* $\pi > 0$ on the set of states which corresponds to the unique eigenvector of the transposed matrix P^T associated with the eigenvalue 1. Thus π is defined as the unique solution to the linear system:

$$\begin{cases} P^T \pi - \pi = 0 \\ \sum_{i=1}^{m} \pi_i = 1. \end{cases}$$

A basic result for ergodic Markov chains is that the sequence of powers $P^2, \ldots, P^k, \ldots$ of the probability transition matrix converges to the matrix:

$$w = \begin{bmatrix} \pi^T \\ \pi^T \\ \vdots \\ \pi^t \end{bmatrix}$$

, the matrix featuring m identical rows, each of which is π^T.

We now recall in Lemma A.1 below a basic result on the behavior of Markov chains in terms of "state counts" which can be obtained from n-step realizations of the corresponding random process, i.e. from sequences of n successive states obtained by simulating the process. It is worth pointing out that this result assumes that the initial state is randomly selected according to the stationary probability distribution π. For any given n, Lemma A.1 provides the exact analytic expression of the expectation and covariance matrix of the n-component random vector ζ (directly related to the state count vector $f = (f_i)_{i=1,\dots,m}$). Moreover, it also recalls the well-known Central Limit Theorem for Markov chains and provides the exact expression of the covariance matrix of the limit gaussian distribution obtained for $n \to \infty$.

LEMMA A.1 (Billingsley, 1961) Consider a m-state stationary ergodic Markov chain with (irreducible) transition matrix $P = (p_{ij})_{\substack{i=1,\dots,m\\ j=1,\dots,m}}$. Let $\pi = (\pi_i)_{i=1,\dots,m}$, $\pi > 0$, denote the stationary probability distribution on the set of states. For given integer $n > 1$, consider the random vector $f = (f_i)_{i=1,\dots,m}$, the components of which represent the "state counts", i.e. the number of times the various states i = 1, …, m are encountered in n-step realizations of the Markov process. Let $\zeta = (\zeta_i)_{i=1,\dots,m}$ be the random m-vector defined as:

$\zeta_i = (f_i - n\pi_i) / \sqrt{n}$ (i = 1, …, m).

Then

$$\text{(i)} \begin{cases} \mathbb{E}\{\zeta\} = 0 \\ cov(\zeta,\zeta) = S(P,n) = diag(\pi) - \pi\pi^T + \Delta + \Delta^T \end{cases} \qquad \text{(a.1)}$$

where: $\Delta = \frac{1}{n}\sum_{k=1}^{n-1}(n-k)\left[diag(\pi)\times P^k - \pi\pi^T\right]$.

(ii) For $n \to \infty$ the distribution of ζ converges to a m-variate normal distribution with mean 0 and covariance matrix

$$\bar{S}(P) = diag(\pi) - \pi\pi^T + \sum_{k=1}^{\infty}\left(diag(\pi)\times P^k - \pi\pi^T\right) + \sum_{k=1}^{\infty}\left((P^k)^T \times diag(\pi) - \pi\pi^T\right) \qquad \text{(a.2)}$$

(in the above, diag(π) denotes the m × m diagonal matrix, the diagonal terms of which are the components of π).

Proof. The proof of (i) can be found in Billingsley (1961) (lemma 3.2), (a.1) being an equivalent in matrix notation to equation (3.10) of this reference which gives the expression of $cov\{\zeta_i, \zeta_j\} = E\{\zeta_i, \zeta_j\}$. Morever, (ii), the convergence in distribution to a normal distribution (central limit theorem) follows from Billingsley (1961), Theorem 3.3. □

However, for the purpose of our analysis in Section 4.4, a related but somewhat different result will be needed. Actually it is observed that Lemma A.1 is obtained with the underlying assumption that each state i of the process is a possible initial state and the probability for i to be selected as initial state in a realization of the process is exactly π_i. The following result addresses the more general case when the probability for i to be selected as initial state in a realization of the process is given and equal to $\overline{p}_i$, and $\overline{p} \neq \pi$ (we assume $(\sum_{i=1}^{m} \overline{p}_i = 1)$. Note that a special case corresponds to choosing $\overline{p}_s = 1$ and $\overline{p}_j = 0$, $\forall$ $j \neq s$; this arises when all the realizations of the process have to start from the same initial state s. Theorem A.2 below provides the exact expressions of the expectation and covariance matrix corresponding to the state counts in this case. To the best of our knowledge, these expressions are given here for the first time.

THEOREM A.2 Let $P = (p_{ij})$ be the transition matrix of a given m-state stationary and ergodic Markov process. Let $f = (f_i)_{i=1,\dots,m}$ denote the random vector whose components are the state counts of n-step realizations of the process, assuming that the initial state i is randomly chosen with given probability $\overline{p}_i$ ($\sum_{i=1}^{m} \overline{p}_i = 1$ is assumed). For any $k \geq 1$, and $i \in [1, m]$, let us denote $P_{\bullet,i}^{(k-1)}$ the ith column of $P^{(k-1)}$ (P raised at the power (k – 1)). Let $\zeta = (\zeta_i)_{i=1,\dots,m}$ denote the random m-vector defined as:

$$\zeta_i = (f_i - \mathbb{E}(f_i)) / \sqrt{n} \tag{a.4}$$

where:

$$\mathbb{E}(f_i) = \sum_{k=1}^{n} \overline{p}^T P_{\bullet,i}^{(k-1)}. \tag{a.5}$$

Then $\mathbb{E}(\zeta) = 0$ and the (i, j) entry of the corresponding covariance matrix reads:

$$\mathbb{E}(\zeta_i, \zeta_j) = \frac{1}{n} \sum_{k=1}^{n} \sum_{\ell=1}^{n} \left[\psi_{i,j}(k, \ell) - \overline{p}^T P_{\bullet,i}^{(k-1)} \overline{p}^T P_{\bullet,j}^{(\ell-1)} \right] \tag{a.6}$$

where for any $(k, \ell) \in [1, n] \times [1, n]$:

$$\psi_{i,j}(k, \ell) = \begin{cases} \overline{p}^T P_{\bullet,i}^{(k-1)} \times P_{ij}^{(\ell-k)} \text{ if } k < \ell \\ \overline{p}^T P_{\bullet,i}^{(k-1)} \text{if } k = \ell \text{ and } i = j \\ 0 \text{ if } k = \ell \text{ and } i = j \\ \overline{p}^T P_{\bullet,j}^{(\ell-1)} \times P_{ji}^{(k-\ell)} \text{ if } k > \ell \end{cases} \tag{a.7),(a.8),(a.9),(a.10}$$

Proof. (a.5) is obtained by observing that in the 1-step process, (case of k = 1), the probability of the process to be in state i is $\overline{p}_i$. So (a.5) holds for n = 1. Now for $k \geq 2$, the probability of the process to be in state i in the kth-step, given that it is in some given state s at step 1 is $P_{s,i}^{(k-1)}$. Since each possible initial state s ∈ [1, m] has probability $\bar{p}_s$, then the probability for the process to be in state i in the kth-step is:

$$\sum_{s=1}^{m} \bar{p}_s P^{(k-1)}_{s,i} = \bar{p}^T \; P^{(k-1)}_{\bullet,i}$$

(we recall that $P^{(k-1)}_{s,i}$ denotes the (s, i) entry of the matrix $P^{(k-1)}$, i.e. P raised at the power (k −1)). Now let z_k (i) denote the random variable which is equal to 1 if the process is in state i in step k, and equal to 0 otherwise. We can write:

$$f_i = \sum_{k=1}^{n} z_k(i)$$

and E $(z_k(i)) = \bar{p}^T \; P^{(k-1)}_{\bullet,i}$. From this it follows that:

$$\mathrm{E}\ (\zeta_i, \zeta_j) = \frac{1}{n} \sum_{k=1}^{n} \sum_{\ell=1}^{n} \mathrm{E}\left\{\left(z_k(i) - \bar{p}^T \; P^{(k-1)}_{\bullet,i}\right)\left(z_\ell(j) - \bar{p}^T \; P^{(\ell-1)}_{\bullet,j}\right)\right\}$$

$$= \frac{1}{n} \sum_{k=1}^{n} \sum_{\ell=1}^{n} \mathrm{E}\left[\left\{z_k(i)\; z_\ell(j)\right\} - \bar{p}^T \; P^{(k-1)}_{\bullet,i} \times \bar{p}^T \; P^{(\ell-1)}_{\bullet,j}\right]$$

To obtain expression (a.6) it thus remains to show that: $\mathbb{E}\ \left\{z_k(i)\; z_\ell(j)\right\} == \psi_{i,j}(k, \ell)$.
Suppose first that k < *l*. Then the probability for having both $z_k(i) = 1$ and $z_\ell(j) = 1$ is exactly $\bar{p}^T \; P^{(k-1)}_{\bullet,i}$ (the probability for the process to be in state i in step k), multiplied by $P^{(\ell-k)}_{ij}$ (the probability for the process to be in state j in step l, given that it was in state i in step k). Thus (a.7) is obtained. For the case l < k, (a.10) is deduced via a similar argument.
Now, for the case k = l, (a.9) obviously holds (the process cannot be in two distinct states at the same time). And if k = l and i = j, the probability is just $\bar{p}^T \; P^{(k-1)}_{\bullet,i}$, according to (a.8). This completes the proof.
In the sequel, given P and the initial probability distribution $\bar{p}$, we will denote C(P, $\bar{p}$, n) the covariance matrix defined by (a.6) in theorem A.2.
Now, in Section 4.4, we actually need the following corollary which concerns the *transition counts* of a given stationary ergodic Markov chain, rather than the *state counts*.

As before, P denotes the m × m probability transition matrix, and given any n-step realization of the process corresponding to n − 1 successive transitions from a given initial state s, for any (i, j) ∈ [1, m] × [1, m], τ_{ij} denotes the number of times the transition (i, j) is activated. Let U = {(i, j)/p_{ij} > 0} |U| = q, and consider the random vector θ ∈ $\mathbb{R}^U$ defined as:

$$\theta_{ij} = \left(\tau_{ij} - (n-1)\;\pi_i\; p_{ij}\right) / \sqrt{n-1}. \qquad \text{(a.11)}$$

We can state:

COROLLARY A.3 Let Q denote the q × q stochastic matrix, the entries of which correspond to pairs $[(i,j),(k,l)] \in U \times U$; the value of the entry $[(i,j),(k,l)]$ being equal to $\delta_{jk} p_{k\ell} (= p_{k\ell}$ if $j = k, 0$ otherwise$)$.

Let $\hat{p}$ (initial probability vector) denote the vector in $\mathbb{R}^U$ defined as: $\hat{p}_{(i,j)} = 0$ for all (i, j) ∈ U such that i ≠ s ; $\hat{p}_{(s,j)} = p_{sj}$ for j = 1, …, m. Then:

$$\begin{cases} \mathbb{E}(q) = 0 \\ cov(\theta,\theta) = C(Q,\hat{p},n-1) \end{cases}$$

where C(Q, $\hat{p}$, n – 1) is the covariance matrix obtained as in Theorem A.2, (a.6), P, being replaced by Q and $\bar{p}$ replaced by $\hat{p}$.

Moreover, for n → ∞, θ converges in distribution to a multivariate normal distribution with mean 0 and covariance matrix $\bar{S}(Q)$, ($\bar{S}(Q)$ is defined as in Lemma A.1, (a.2), substituting matrix Q for matrix P).

Proof. The above corollary is obtained by considering the so-called *multiple Markov* process deduced from the given initial Markov process as follows. The set of states of the new process corresponds to the transitions (i, j) ∈ U . From any state of the new process, say (i, j), there is a possible transition to any other state of the form (j, k) ∈ U, the corresponding probability being p_{jk}.

It is easily checked that this new Markov process is ergodic and the expressions of E(θ) and cov(θ, θ) are deduced from Theorem A.2. Finally, the convergence to a normal distribution (Central Limit Property) can be deduced from the results in Derrienic and Lin (2003).

COROLLARY A.4 Consider n-step realizations of a m-state stationary ergodic Markov process with (irreducible) probability transition matrix P started with initial probability distribution $\bar{p} = (\bar{p}_i)_{i=1,\dots,m}$.

Let Q denote the probability transition matrix of the multiple Markov process described in Corollary A.3 and $\hat{p}$ the initial probability distribution defined as: $\hat{p}_{(i,j)} = \bar{p}_i \times p_{ij}$ (for all (i, j) ∈ U). Then the following holds:

(i) For any weight vector $\mu = (\mu_i)_{i=1,\dots,m}$ associated with the states i = 1, …, m, consider the random variable

$$Y = \sum_{i=1}^{m} \mu_i f_i = \sum_{i=1}^{m} \mu_i \; (\mathrm{E}(f_i) + \sqrt{n}\zeta_i)$$

with $\mathbb{E}$ (fi) = $\sum_{k=1}^{n} p^T \; P^{(k-1)}_{\bullet,i}$. Then

$$\begin{cases} \mathbb{E}(Y) = \sum_{i=1}^{m} \mu_i E(f_i) \\ \sigma^2(Y) = \; n\mu^T C(P,\bar{p},n)\mu \end{cases}$$

and, for n $\rightarrow \infty$, $\frac{Y}{\sqrt{n}}$ tends to a normal (univariate) distribution with variance $\mu^T \overline{S}(P)\mu$.

(ii) For any weight vector w = $\left(w_{ij}\right)_{(i,\,j)\in U}$ associated with the transitions of the process, consider the random variable:

$$Z = \sum_{(i,\,j)\in U} w_{ij}\,\tau_{ij} = \sum_{(i,\,j)\in U} w_{ij}\left(\mathrm{E}(\tau_{ij}) + \sqrt{n-1}\,\theta_{ij}\right)$$

with $\mathbb{E}(\tau_{ij}) = \sum_{k=1}^{n-1} \hat{p}^T Q^{(k-1)}_{\bullet,(i,j)}$. Then

$$\begin{cases} \mathbb{E}(Z) = \sum_{(i,\,j)\in U} w_{ij}\mathbb{E}(\tau_{ij}) \\ \sigma^2(Z) = (n-1)w^T C(Q,\,\hat{p},n-1)w \end{cases}$$

and for n $\rightarrow \infty$, $\frac{Z}{\sqrt{n-1}}$ tends to a normal (univariate) distribution with variance .

Proof. The expression of $\mathbb{E}$(Y) and $\mathbb{E}$(Z) directly follow from the linearity of the expectation. The expressions of σ2(Y) and σ2(z) are obtained by using the following observation: if v ~ $\mathcal{N}$ (0, Σ) is a random vector in ℝv with covariance matrix Σ, then for any weight vector α ∈ ℝv, αTv is a normal (univariate) random variable with variance αT Σα. The expressions of the variances of the limit distributions for n → ∞ follow from the fact when n → + ∞, $C(P,\,\overline{p},\,n) \rightarrow \overline{S}(P)$ and $C(Q,\,\hat{p},\,n-1) \rightarrow \overline{S}(Q)$.

Chapter 3
Supply Chain Network Design Using an Enhanced Hybrid Swarm-Based Optimization Algorithm

Baris Yuce
Cardiff University, UK

Ernesto Mastrocinque
University of London, UK

ABSTRACT

Supply chain network design is one of the most important strategic issues in operations management. The main objective in designing a supply chain is to keep the cost as low as possible. However, the modelling of a supply chain requires more than single-objective such as lead-time minimization, service level maximization, and environmental impact maximization among others. Usually these objectives may cause conflicts such as increasing the service level usually causes a growth in costs. Therefore, the aim should be to find trade-off solutions to satisfy the conflicting objectives. The aim of this chapter is to propose a new method based on a hybrid version of the Bees Algorithm with Slope Angle Computation and Hill Climbing Algorithm to solve a multi-objective supply chain network design problem. A real case from the literature has been selected and solved in order to show the potentiality of the proposed method in solving a large scale combinatorial problem.

INTRODUCTION

The supply chain (SC) is a complex system (Mastrocinque et al., 2014) aiming to move products or services from suppliers to customers and involves people, technologies, information, materials among others. At the strategic level, supply chain network design is a crucial decision affecting the future success of the business (Lambiase et al., 2013). A supply network can have different configuration depending on the criteria used during the design stage.

DOI: 10.4018/978-1-4666-9644-0.ch003

Optimization is a powerful tool employed by decision-makers to calculate the optimal value of decision variables in order to minimize/maximize certain objective functions. It has been widely used for solving production related problems (Vasant, 2010; Vasant & Barsoum, 2009; Vasant & Basoum, 2010). The most common supply chain optimization model is based on the evaluation of a single objective function, usually an economic index such as total cost of the supply chain or profit. However, reality can be more complex than reduce the modelling of a supply network to a single-objective In fact other objectives might be considered when it comes to design the configuration of a supply chain such as lead-time minimization, environmental impact minimization, service level maximization, among others (Voudouris, 1996). Typically, these objectives may be in contrast between them such as increasing the service level usually causes a growth in costs. Therefore, the aim of a multi-objective supply network design should be to find trade-off solutions in order to satisfy the conflicting objectives.

In multi-objective optimization there is no single optimum solution, but there are different solutions forming the so-called Pareto set. Pareto solutions are a set of trade-offs between different objectives and they are non-dominated solutions which means there is no other solution which would improve an objective without causing a worsening in at least one of the other objectives (Deb, 2001).

In literature, several methods have been proposed to solve supply chain design problems to get the Pareto optimal solutions. Especially, evolutionary and meta-heuristics algorithms such as genetic algorithms (Altiparmak et al., 2006), particle swarm optimization algorithm (Mahnam et al., 2009), ant colony algorithm (Moncayo-Martinez & Zhang, 2011), the Bees Algorithm (Mastrocinque et al., 2013) among others, have proven to be a valid tool in order to solve the multi-objective supply chain network design.

The Bees Algorithm has proven to be a powerful method for multi-objective supply network design (Yuce et al., 2015). Recently an enhanced version of the Bees Algorithm (BA) using Slope Angle Computation and Hill Climbing Algorithm (SACHCA) has been proposed (Yuce et al., 2015a) and successfully applied for single-objective problems such as continuous type functions and single-objective job shop scheduling.

In this chapter the authors apply the SACHCA-based Bees Algorithm to a multi-objective optimization problem such as supply chain design in order to test its performance on a large scale combinatorial problem and give to the decision maker an alternative tool. In order to explore the behavior of the SACHCA-based Bees Algorithm, it has been used to solve a real-world problem such as the notebook supply network design (Graves & Willems, 2005) considering two conflicting objectives such as minimizing the total cost and the total lead time of the network.

This chapter is organized as follows: the background section revises the literature on the multi-objective supply chain optimization and the Bees Algorithm. In section 1, the enhanced version of the Bees Algorithm with Slope Angle Computation and Hill Climbing Algorithm is presented. In section 3 the notebook supply chain case study from the literature is explained. In section 4 the optimization results are presented and discussed. Finally future research direction and conclusion are given.

BACKGROUND

In literature, several multi-objective supply chain design models have been proposed. In (Altiparmak et al., 2006) a facility location problem of a four echelons supply chain (suppliers, plants, distribution centers-DCs and customers) have been proposed, where the objectives are to minimize the total cost, maximize customer services and the capacity utilization balance for DCs using a genetic algorithm based approach. In Che and Chiang (2010) a supply chain design is based on the supplier selection, product assembly and distribution system using a modified Pareto genetic algorithm to minimize the total cost, the delivery time and maximize the quality. In (Liao et al., 2011) a multi-objective location–inventory problem has been investigated using a multi-objective evolutionary algorithm based on the non-dominated sorting genetic algorithm II (NSGAII) in order to minimize total costs, maximize the volume fill rate and the responsiveness level . In Yeh and Chuang (2011) a planning model for green partner selection involving four objectives such as cost, time, product quality and green appraisal score, has been developed; this model employed two types of genetic algorithms to solve the multi-objectives problem by means of the weighted sum approach to find the set of Pareto optimal solutions. In Pishvaee and Razm (2012) authors have developed a multi-objective fuzzy mathematical programming model for a forward/reverse supply chain minimizing the total cost and the environmental impact. A multi-product, multi-stage, and multi-period scheduling model is proposed in (Chen & Lee, 2004) to deal with multiple goals for a multi-echelon supply chain network with uncertain market demands and product prices; a two-phase fuzzy decision-making method is presented to maximize the participants' expected profits, average safe inventory levels, average customer service levels and robustness of selected objectives to demand uncertainties. A bi-objective optimization approach to designing and planning of supply chain is proposed in (Pinto-Varela et al., 2011) in order to maximize the annual profit and minimize the environmental impact. In (Sazvar et al., 2014) a bi-objective stochastic programming model for a centralized green supply chain with deteriorating products have been proposed in order to minimize total supply chain costs and environmental impacts at the same time. In (Sarrafha et al., 2015) a supply chain design model minimizing the total SC costs as well as minimizing the average tardiness of product to DCs have been proposed and solved with a multi-objective biogeography based optimization (MOBBO).In (Pasandideh et al., 2015) a bi-objective optimization of a multi-product multi-period three-echelon supply chain network consisting of manufacturing plants, distribution centers each with uncertain services, and customer nodes is proposed. The two objectives are minimization of the total cost while maximizing the average number of products dispatched to customers.

Other models have also been proposed based on swarm inspired algorithm to solve the supply chain design problem. In (Mahnam et al., 2009) an inventory model for an assembly supply chain network with fuzzy demand has been proposed considering total cost and fill rate and solved using a hybridization of multi-objective particle swarm optimization and simulation optimization. In (Che, 2012) an optimization mathematical model integrating cost and time criteria has been solved using a modified particle swarm optimization method (MEDPSO) for solving a multi-echelon unbalanced supply chain planning problem. The results indicated that the MEDPSO method can obtain a better quality solution compared to classical GA and PSO. Furthermore, another swarm based optimization model based on the Ant Colony optimization method has been developed in order to configure a supply network with different options for each stage minimizing the total cost and the total lead time of the supply chain (Moncayo-Martinez & Zhang, 2011). The same problem has been solved using the Bees Algorithm and its enhanced version based on adaptive neighborhood search and abandonment strategy (Yuce et al., 2014).

The Bees Algorithm has been used successfully to solve many problems including mechanical design, job shop scheduling, supply chain optimization, robot path planning, chemical engineering problems, assembly problems and several other applications (Ang et al., 2013a; Ang et al., 2013b; Castellani et al., 2012; Pham et. al., 2008; Pham et. al., 2007a; Pham et. al., 2007b; Pham et. al., 2007c; Pham et. al., 2007d; Pham et. al., 2007e and Pham et. al., 2007f; Mastrocinque et. al., 2013; ; Lien and Cheng, 2014; Yuce et. al., 2014). Despite these several successful applications of the Bees Algorithm, a careful examination shows that the algorithm needs to be more robust and sensitive to a particular problem. Several attempts have been made in the past to improve the Bees Algorithm using methods such as early neighborhood search and efficiency based recruitment (Pham et al., 2012), adaptive neighborhood search and site abandonment strategy (Yuce et. al., 2013), combination of adaptive enlargement and reduction of the search neighborhood (Azfanizam et al., 2014) developing new local search strategies (Ahmad, 2012), tuning the algorithms parameters (Otri, 2011), introducing new parameters (Pham and Koc, 2011; Pham and Ghanbarzadeh, 2007), combining with other optimization algorithms (Packianather et al., 2014; Sholedolu, 2009,) novel initialization based on the patch concept and Levy flight distribution (Hussein et al., 2014).

Recently another hybrid version of the Bees Algorithm (BA) using Slope Angle Computation and Hill Climbing Algorithm (SACHCA) has been proposed (Yuce et al., 2015a). The aim of the SACHCA is to strengthen the neighborhood search. The SACHCA-base BA has been successfully applied for single-objective problems such as continuous type functions and single-objective job shop scheduling.

Since there is no algorithm which can find the best solution for all types of optimization problems according to the no-free lunch theorem (Wolpert & Macready, 1997) the aim of this chapter is to use the SACHCA-based BA to solve a bi-objective supply chain design problem minimizing the total cost and total lead time of a well-known problem such as the notebook supply network (Graves & Willems, 2005).

THE BEES ALGORITHM AND SLOPE ANGLE COMPUTATION AND HILL CLIMBING ALGORITHM

The Bees Algorithm

The Bees Algorithm (BA) is a swarm based algorithm proposed by (Pham et. al. 2005; 2006a). Swarm based algorithms emulate the behavior of animals in swarm such as Ant Colonies, Bee Colonies, Bird Flocking, Animal Herding, Fish Schooling among others (Beni, 1989).

The BA has been inspired from the foraging behavior of the honey bees. A swarm of honey bees consists of a queen bee and thousands of worker bees which are allocated to do different works in a colony such as cleaning their nest, foraging and constructing comb (Seeley, 1995). When a scout bee finds a food source, it returns to the hive and performs special ritual which is a communicational movement, known as the waggle dance in order to inform the colony (Seeley, 1995, Gould and Gould, 1988). The waggle dance contains information about direction, distance and quality of the flower patch found by the bee (Talbi, 2009). After the waggle dance, a number of bees is allocated for that particular food source. Promising patches should have more bees than the others. The recruited bees evaluate the related patch and look around of the patch as well as share this information with their peers in the hive. The same process happens for all patches emulating local search. After the recruitment stage the scout bees will continue searching for other promising patches in a random way emulating the global search and whenever they find another patch this information will be given to the colony randomly (Von Frisch, 1955). The Bees Algorithm involves several parameters given in Table 1.

The steps of the BA are: placing the "n" scout bees on the search space; in step 2 fitness values of the visited patches are evaluated. Step 3 regards the selection of the best patches with respect to their fitness value where these selected best patches will be split into two groups containing more scout bees to the elite patches 'e', and less scout bees to the non-elite best patches 'm-e' in step 4. Step 5 is the

Table 1. Bees Algorithm parameters

Parameter	Symbol
Number of scout bees in the selected patches n	*n*
Number of best patches out of the selected patches	*m*
Number of elite patches out of the selected best patches	*e*
Number of recruited bees in the elite patches	*nep*
Number of recruited bees in the non-elite best patches	*nsp*
The size of neighborhood for each patch	*ngh*
Number of iterations MaxIter	*MaxIter*
Difference between fitness values of consecutive iterations	*Error*

neighborhood search in the patches given in steps 3 and 4. In step 6, the remainder bees, generated in the initial population, will be recruited for the random search to find better random solutions. In step 7 the random patches' fitness values will be evaluated and this process will continue until one of the stopping criteria is met.

The procedure described above equips the algorithm to combine exploitative neighborhood search with explorative global search enabling effective location of the globally optimal solution to a problem (Yuce, 2012). Due to the powerful search capability of the algorithm, the basic BA was successfully utilized to solve different types of optimization problems such as job scheduling (Pham et. al., 2007a), forming manufacturing cells (Pham et. al., 2007b), data clustering (Pham et. al., 2007c), tuning a fuzzy logic controller (Pham et. al., 2007d), training neural networks (Pham et. al., 2006b), finding the multiple feasible solutions to preliminary design problems (Pham et. al., 2007e) and design of mechanical components (Pham et. al., 2007f). The improvements made to the basic BA with slope angle computation and hill climbing algorithm is given in the following section.

The Bees Algorithm with Slope Angle Computation and Hill Climbing Algorithm

The enhancement to the basic Bees Algorithm (BA) based on the Slope Angle Computation and Hill Climbing Algorithm (SACHCA) was recently proposed by (Yuce et al., 2015) in order to enhance the blind local search process (Yuce, 2012). The Hill Climbing Algorithm (HCA) is an iterative single element-based local search algorithm, also known as Gradient Ascent / Descent algorithm. Although the local minimum of an optimization problem can be found by the HCA the global optimum is not guaranteed (Grosan and Abraham, 2011).

The SACHCA-based improved BA deals with location process of the best sites. The inclination of the current sites is evaluated by slope angle computation. In general, the promising location will be far from the current position if the slope angle is close to 90°, while the current location will be close to a local optimum point if the slope angle is close to 0°. The direction of the local optimum can be determined according to the slope angle orientation.

The slope angle is computed using the first order numerical derivation. The numerical derivation of each site is calculated from its neighborhood. The two end points in the vicinity of the neighborhood are used to compute the numerical derivation. The central difference method is utilized for numerical derivation (see Equation 1).

Figure 1. Slope angle for three selected sites (adapted from Yuce et al., 2015)

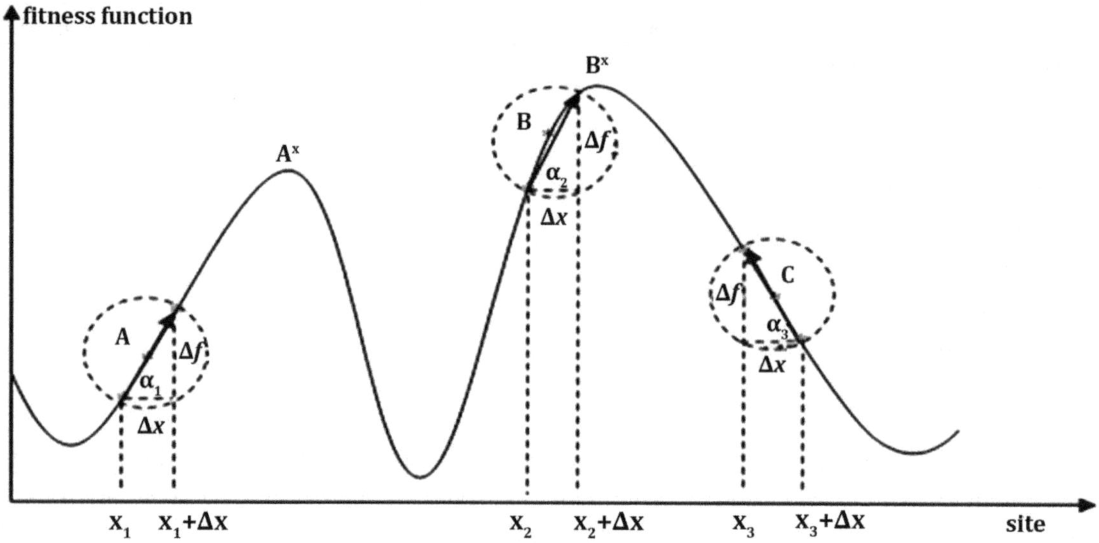

$$Slope\,angle = f'\left(x\right) = \frac{f\left(x+\dfrac{\Delta x}{2}\right)-f\left(x-\dfrac{\Delta x}{2}\right)}{\Delta x} \tag{1}$$

If the slope angle is very steep, then the promised location will be more likely to be far from the selected site but when the slope angle is close to zero, the promised location is very close to the selected site, as shown in Figure 1.

Figure 1 shows three selected sites. Around site A, the angle direction is towards site A^x, which is a local optimum. The direction of sites B and C are towards the site B^x, which is another local optimum. At the end of the searching process, all the local optima will be sorted and the biggest will be selected as the global optimum which in Figure 1 is site B^x .The local search process was accomplished with the use of the Hill Climbing Algorithm, according with Equation 2.

$$x\left(i+1\right) = x\left(i\right) + h\nabla f \tag{2}$$

where i is the iteration number, $x(i)$ is the current position, $x(i+1)$ is the next position, h is the incremental size, ∇f is the gradient of the current position. The pseudo-code of the algorithm is the following (Yuce et al., 2015):

Generate the initial population size as n, set the best patch size as m, set the elite patch size as e, set the number of forager bees around of elite as nep, set the number of forager bees around of non-elite best patches as nsp, set the step size for HC algorithm as h, set the angle limit as $angle\_\lim it$, set the "number of the waiting time for HC" algorithm as $HC\_time\_\lim it$, set the neighborhood size as ngh, set the maximum iteration number as $MaxIter$, and set the error limit as $Error$.

i= 0, $time = 0$; $slope\_angle(1:m) = 0$; Generate initial population; Evaluate Fitness Value of initial population;

Sort the initial population based on the fitness result.

While $i \le MaxIter$ or $FitnessValue_i - FitnessValue_{i-1} \ge Error$

i= i+1;

Select the elite patches and non-elite best patches for neighborhood search.

Recruit the forager bees to the elite patches and non-elite best patches.

Evaluate the fitness value of each patch; Sort the results based on their fitness.

For k=1:m

Calculate ..

While $slope\_angle(k) > angle\_\lim it$ *and* $time \le HC\_time\_\lim it$

$$x(i+1,k) = x(i,k) + h\nabla f\left(x(i,k)\right)$$

Evaluate Fitness value for each position

End

Record all the local optimum sites found and sort them (end of the neighbourhood search).

Allocate the rest of the bees for global search to the non-best locations;

Evaluate the fitness value of non-best patches;

Sort the fitness values and positions;

Run the algorithm until termination criterions are met.

End

SUPPLY NETWORK DESIGN CASE STUDY AND METHODOLOGY

The notebook supply chain network design case analyzed in this chapter is based on the Graves and Willems (2005) case, readapted by Moncayo-Martinez and Recio (2014). The supply chain in this case, has been represented as a Bill of Materials (BOM) where each node could represent a supplying manufacturing, assembly, delivery and so on. Each of this stage could be performed by different options. In fact a part could be supplied by more than a supplier, an assembly operation could be performed in different plants, a manufacturing operations could use different technologies and a different transportation modes can be used chosen to reach the customers. Moreover, each of this option for each stage, has a different cost and needs different time to be performed. Usually, cost and time are conflicting objective, because if an option has a higher cost than another, usually it need less time to be performed. For instance, if I want to be delivered a part or raw material in a shorter amount of time, I might have to pay more.

The notebook supply network is composed by several stages going from the raw materials supply, to assembly of finished products and delivery to the markets. The structure of the network is shown in Figure 2. The network encompasses 17 stages and each of them can be performed by different options n. A cost and a lead time are associated with each option (Table 2). cost and lead time associated to each

Figure 2. Notebook supply chain structure

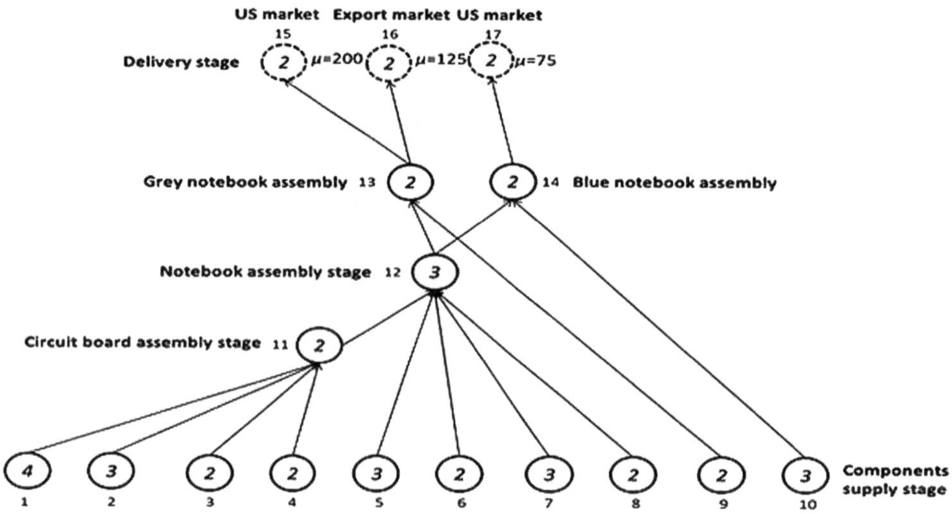

stage is are shown in Table 1. The number of possible different configuration of the supply network is given by $\prod_{s=1}^{S} n_s$ which is equal to 1990656 in the proposed case.

The aim of the optimization is to find the optimal configuration of the notebook supply chain minimizing the total cost and the total lead time of the network given by Equations 3 and 4 respectively, where d_i represents the demand of products for each stage, Θ the period of time considered, C_{sn} is the cost of option n for stage s, T_{sn} the lead time of option n for stage s and finally x_{sn} is the binary decision variable equal to 1 if option n for the stages s is selected, 0 otherwise. Only one option for each stage can be selected, according to the constraint in Equation 5. In the case analyzed, Θ=360 days, $d_{15} = 200, d_{16} = 125,\ d_{17} = 75$ per day.

$$TCost = \Theta \sum_{s=1}^{S} \left(d_s \sum_{n=1}^{N_s} C_{sn} x_{sn} \right) \tag{3}$$

$$TLeadtime = \max_{s=15,16,17} LT_s \tag{4}$$

with $LT_s = \sum_{n=1}^{N_s} T_{sn} x_{sn} + \max_{k=input\ nodes\ to\ stage\ s} \left(LT_k \right)$

$$\sum_{n=1}^{N_s} x_{sn} = 1 \quad for\ s \in S \tag{5}$$

Table 2. Cost and lead time values for each stage option in the supply chain

Stage s	*Option n*	*C*	*T*
1	1 2 3 4	130.00 133.25 134.91 136.59	40 20 10 0
2	1 2 3	200.00 202.50 205.03	20 10 0
3	1 2	155.00 156.93	10 0
4	1 2	204.00 200.00	10 0
5	1 2 3	300.00 325.00 350.00	60 10 5
6	1 2	225.00 240.00	70 30
7	1 2 3	40.00 45.00 55.00	60 20 10
8	1 2	200.00 205.00	30 10
9	1 2	5.00 5.50	40 15
10	1 2 3	5.00 5.50 7.50	40 15 5
11	1 2	120.00 150.00	20 5
12	1 2 3	120.00 132.00 140.00	5 2 1
13	1 2	20.00 30.00	2 1
14	1 2	20.00 30.00	2 1
15	1 2	12.00 20.00	5 1
16	1 2	15.00 30.00	15 2
17	1 2	12.00 20.00	5 1

The basic and the SACHCA-based Bees Algorithm have been applied to solve the bi-objective notebook supply chain design. Four different combinations of the BA parameters (Table 3) are tested in order to obtain the Pareto front line. All the experiments have been carried out with the number of iterations fixed to 1000, ngh=2, sc=1, h=1, angle limit=0.5, HC time limit=1000. In all tests the algorithm

Table 3. Bees Algorithm parameters values for the four combinations tested

	Parameters					
Combination	*n*	*m*	*e*	*nep*	*nsp*
1	10	5	1	3	2
2	25	10	2	6	4
3	50	15	5	12	8
4	100	30	10	24	16

is run 100 times and results are updated according to previous Pareto solutions after each run in order to increase robustness. The weighted sum approach has been used to generate the Pareto solutions. The bi-objective problem has been reduced to a single objective problem aiming to find the solution that minimize with the fitness function expressed by Equation 6.

$$fitness = w_1 normTCost + w_2 normTLeadtime \tag{6}$$

where normTCost and normTLeadtime are the normalized solutions and w_1 and w_2 are the weights where the summation is equal to 1. For this case a w_1=w_2=0.5.

In order to evaluate the Pareto frontline, two evaluation criteria (Zitzler et al., 2000; Tan et al., 2002) as coverage of two sets and spacing are considered. The coverage of two sets (Mallipeddi et al., 2010): let X', $X' \subseteq X$ be two sets of decision vectors. The function C maps the ordered pairs $\left(X', X''\right)$ to the interval [0, 1] as follows: $C\left(X',X''\right) > \frac{\left|\{a'' \in X''; a' \in X' : a' \leq a''\}\right|}{\left|X''\right|}$. The value $C\left(X', X''\right) = 1$ means that all solutions in X'' are dominated by or equal to solutions in X'. The opposite, $C\left(X', X''\right) = 0$, represents the situations where none of the solutions in X'' are covered by the set X'. Note that both $C\left(X', X''\right)$ and $C\left(X'', X'\right)$ have to be considered since $C\left(X', X''\right)$ is not necessarily equal to $1 - C\left(X'', X'\right)$. Spacing (Zhao & Suganthan, 2010) is a measure of relative distance between consecutive solutions in a non-dominated set Q as follows: $S = \sqrt{\frac{1}{|Q|}\sum_{i=1}^{|Q|}\left(d_i - \bar{d}\right)^2}$ with Q size of the set and $\bar{d}$ average of the d_i, measuring the distance between the solution i and the nearest solution in the objective space: $d_i = \min\left\{\sum_{m=1}^{M}\left(f_m^i - f_m^k\right)^2 \mid k \in Q\, k \neq i\right\}$. A value of this metric equal to 0 would mean that each point of the Pareto front is equally spaced from the other.

RESULTS

As shown from the Pareto frontlines in Figure 3 and the results summarized in Table 4, SACHCA BA found more solutions than the BA for all the tested combinations of parameters. Both the algorithms found 34 day as minimum total lead time, while the minimum value of the total cost 187834680 was found by the SACHCA BA. The minimum total cost value found by the SACHCA BA was always lower than the BA one for all the parameters combinations.

When it comes to the spacing metric, only in combination 3 the SACHCA BA Pareto frontline had a lower value compared to the BA, which means the BA Pareto solutions were better distributed in the front. However, for all the parameters combinations, C(SACHCA,BA) had always a much higher value than the C(BA,SACHCA), which means the SACHCA BA solutions dominated the BA ones.

Overall both the algorithms performed very well on the complex combinatorial problem selected, and there was no notable differences in the results obtained with both algorithms for each parameters combination tested.

Figure 3. Pareto front lines obtained with the two algorithms tested on the notebook supply chain problem for combination 1 (a), 2 (b), 3 (c) and 4 (d) of the parameters

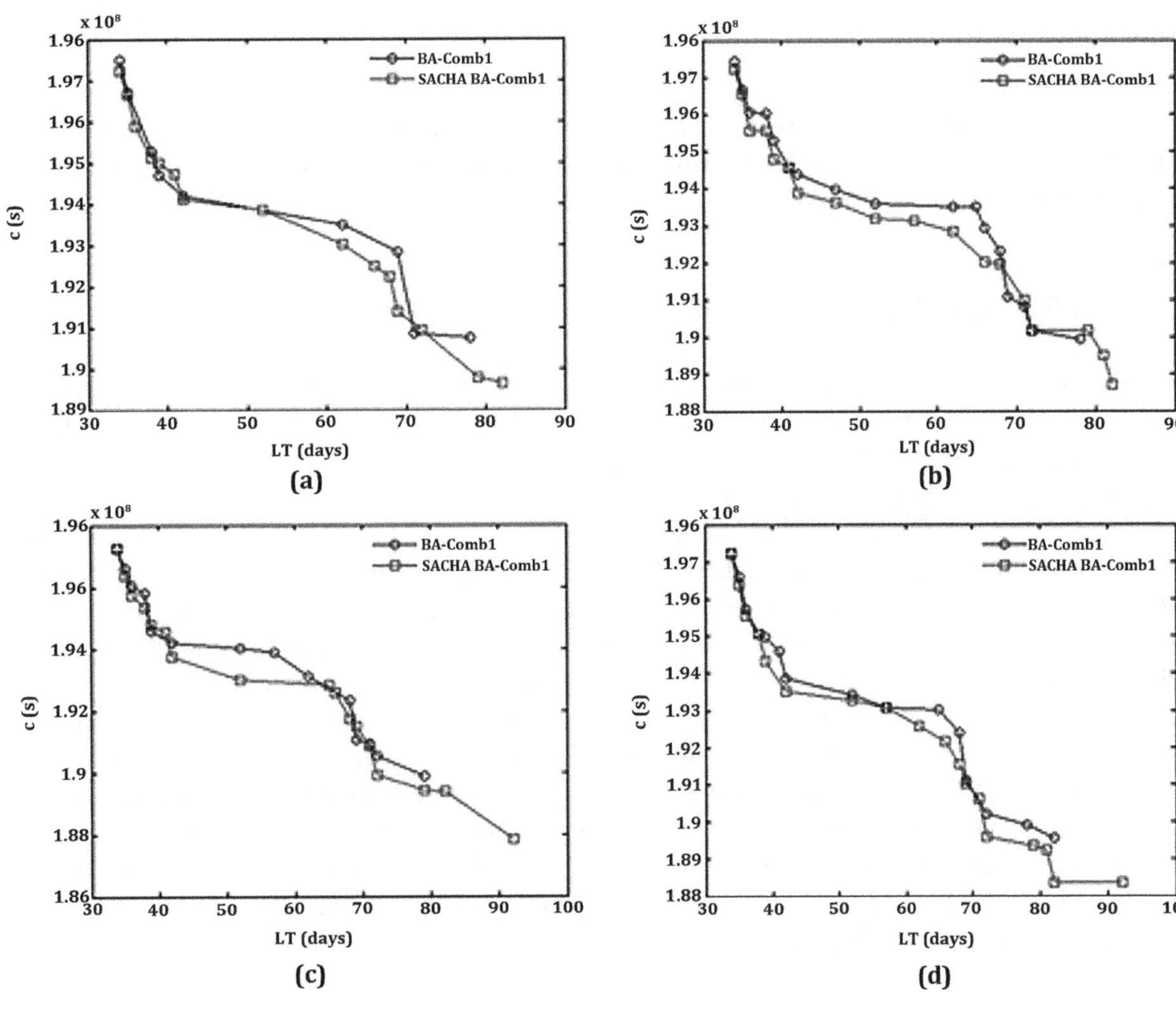

Table 4. Results and performance of the two algorithm tested on the notebook supply chain case

Comb	Algorithm	No of Pareto solutions	Min TLeadtime	Min TCost	Spacing	Coverage of two sets
1	Basic BA	9	34	190756440	0.037	**C(BA,SACHCA) 0.067**
	SACHCA BA	15	34	189638280	0.042	**C(SACHCA,BA) 0.667**
2	Basic BA	17	34	189943740	0.031	**C(BA,SACHCA) 0.055**
	SACHCA BA	18	34	188716500	0.025	**C(SACHCA,BA) 0.705**
3	Basic BA	14	34	189867780	0.043	**C(BA,SACHCA) 0.118**
	SACHCA BA	17	34	187834680	0.055	**C(SACHCA,BA) 0.571**
4	Basic BA	15	34	189554940	0.031	**C(BA,SACHCA) 0**
	SACHCA BA	18	34	188369280	0.031	**C(SACHCA,BA) 0.600**

From a managerial perspective, since the SACHCA BA allowed to find more Pareto solutions dominating the BA solutions, it gives to the decision maker more and better options to choose from regarding the supply network design problem.

FUTURE RESEARCH DIRECTIONS

The work presented in this chapter can be developed for future research from different perspectives. One possible development regards testing the proposed SACHCA BA on a more complex supply chain optimization problem considering uncertainty in the parameters such as for instance demand or lead time in order to make the model more fitting to reality.

Another possible future research would consist in comparing the SACHA BA and the BA to other well-known optimization algorithms such as genetic, ant colony, particle swarm and so on. In fact, the SACHA BA has already shown to provide better performance than other well-known algorithm when applied to solve continuous type optimization functions. However it would be interesting to compare their performance on a multi-objective combinatorial problem such as the supply chain design.

Also it would be interesting to use other hybrid version of the Bees Algorithm in order to benefit from the strength of different approaches. Recently, a hybrid Bees Algorithm with Genetic Algorithm has been proposed in order to benefit from a strong random search form the GA and local search from the BA, reducing the inconvenient of stacking in local minima. This hybrid algorithm was applied to solve the single-machine job shop scheduling problem with Earliness and Tardiness optimization (Packianather et al, 2014). The results were very promising and it would be interesting in future to test it on the combinatorial supply chain problem.

Finally, as future research the SACHCA BA could be tested on other well-known operational research problems such as for instance the traveling salesman problem.

CONCLUSION

In this chapter, an enhanced version of the Bees Algorithm with Slope Angle Computation and Hill Climbing Algorithm has been proposed to solve a bi-objective supply chain design problem. The complex combinatorial problem deals with the configuration of a notebook supply network in order to minimize two objective functions simultaneously such as the total cost and total lead-time of the network. The SACHCA-BA and the basic BA have been applied to solve the above mentioned problem with four different combinations of the algorithm parameters in order to find the Pareto solutions. The results showed the efficiency of the proposed algorithm as a more powerful tool to find more and better Pareto solution for the supply chain problem compared with the basic Bess Algorithm. In fact the minimum total cost found by the SACHCA-BA was lower than the one found by the BA for all the parameters combinations tested. Moreover, the coverage of two set C showed how the Pareto solutions obtained with the SACHCA BA dominate the ones obtained with the basic BA. This allows the decision maker to select a supply chain configuration where the tradeoff between cost of final products and delivery time is more balanced. However, the basic BA showed to produce solutions being better distributed in the Pareto front line as confirmed by the lower values of the spacing metric.

REFERENCES

Ahmad, S. A. (2012). *A Study of Search Neighborhood in the Bees Algorithm.* (PhD Thesis). Cardiff University.

Altiparmak, F., Gen, M., Lin, L., & Paksoy, T. (2006). A genetic algorithm approach for multi-objective optimization of supply chain networks. *Computers & Industrial Engineering*, *51*(1), 196–215. doi:10.1016/j.cie.2006.07.011

Ang, M. C., Ng, K. W., & Pham, D. T. (2013b). Combining the Bees Algorithm and shape grammar to generate branded product concepts. *Proceedings of the Institution of Mechanical Engineers. Part B, Journal of Engineering Manufacture*, *227*(12), 1860–1873. doi:10.1177/0954405413494922

Ang, M. C., Ng, K. W., Pham, D. T., & Soroka, A. (2013a). Simulations of PCB Assembly Optimization Based on the Bees Algorithm with TRIZ-Inspired Operators. In Advances in Visual Informatics (pp. 335-346).

Azfanizam, A. S., Pham, D. T., & Faieza, A. A. (2014). Combination of Adaptive Enlargement and Reduction in the Search Neighborhood in the Bees Algorithm. *In Applied. Mechanics of Materials*, *564*, 614–618. doi:10.4028/www.scientific.net/AMM.564.614

Beni, G., & Wang, J. (1989). Swarm Intelligence in Cellular Robotic Systems. *Advanced Workshop on Robots and Biological Systems*, Tuscany, Italy.

Castellani, M., Pham, Q. T., & Pham, D. T. (2012). Dynamic optimization by a modified bees algorithm. *Proceedings of the Institution of Mechanical Engineers. Part I, Journal of Systems and Control Engineering*, *226*(7), 956–971. doi:10.1177/0959651812443462

Che, Z. H. (2012). A particle swarm optimization algorithm for solving unbalanced supply chain planning problems. *Applied Soft Computing*, *12*(4), 1279–1287. doi:10.1016/j.asoc.2011.12.006

Che, Z. H., & Chiang, C. J. (2010). A modified Pareto genetic algorithm for multi-objective build-to-order supply chain planning with product assembly. *Advances in Engineering Software*, *41*(7), 1011–1022. doi:10.1016/j.advengsoft.2010.04.001

Chen, C. L., & Lee, W. C. (2004). Multi-objective optimization of multi-echelon supply chain networks with uncertain product demands and prices. *Computers & Chemical Engineering*, *28*(6), 1131–1144. doi:10.1016/j.compchemeng.2003.09.014

Deb, K. (2001). *Multi-objective optimization using evolutionary algorithms* (Vol. 16). John Wiley & Sons.

Gould, J. L., & Gould, C. G. (1988). *The Honey Bee*. New York: Scientific American Library.

Graves, S. C., & Willems, S. P. (2005). Optimizing the supply chain configuration for new products. *Management Science*, *51*(8), 1165–1180. doi:10.1287/mnsc.1050.0367

Grosan, C., & Abraham, A. (2011). *Intelligent Systems: A Modern Approach*. Berlin: Springer. doi:10.1007/978-3-642-21004-4

Hussein, W. A., Sahran, S., & Abdullah, S. N. H. S. (2014). Patch-Levy-based Initialization Algorithm for Bees Algorithm. *Applied Soft Computing*, *23*, 104–121. doi:10.1016/j.asoc.2014.06.004

Lambiase, A., Mastrocinque, E., Miranda, S., & Lambiase, A. (2013). Strategic planning and design of supply chains: A literature review. *Int. J. Eng. Bus. Manag*, *5*(49), 1–11. doi:10.5772/56858

Liao, S. H., Hsieh, C. L., & Lai, P. J. (2011). An evolutionary approach for multi-objective optimization of the integrated location–inventory distribution network problem in vendor-managed inventory. *Expert Systems with Applications*, *38*(6), 6768–6776. doi:10.1016/j.eswa.2010.12.072

Lien, L. C., & Cheng, M. Y. (2014). Particle bee algorithm for tower crane layout with material quantity supply and demand optimization. *Automation in Construction*, *45*, 25–32. doi:10.1016/j.autcon.2014.05.002

Mahnam, M., Yadollahpour, M. R., Famil-Dardashti, V., & Hejazi, S. R. (2009). Supply chain modeling in uncertain environment with bi-objective approach. *Computers & Industrial Engineering*, *56*(4), 1535–1544. doi:10.1016/j.cie.2008.09.038

Mallipeddi, R., Mallipeddi, S., & Suganthan, P. N. (2010). Ensemble strategies with adaptive evolutionary programming. *Information Sciences*, *180*(9), 1571–1581. doi:10.1016/j.ins.2010.01.007

Mastrocinque, E., Yuce, B., Lambiase, A., & Packianather, M. S. (2013). A multi-objective optimization for supply chain network using the Bees Algorithm. *Int. J. Eng. Bus. Manage*, *5*, 1–11. doi:10.5772/56754

Mastrocinque, E., Yuce, B., Lambiase, A., & Packianather, M. S. (2014). A System of Systems Approach to Supply Chain Design. *Applied Mechanics and Materials*, *496*, 2807–2814. doi:10.4028/www.scientific.net/AMM.496-500.2807

Moncayo-Martínez, L. A., & Recio, G. (2014). Bi-criterion optimization for configuring an assembly supply chain using Pareto ant colony meta-heuristic. *Journal of Manufacturing Systems*, *33*(1), 188–195. doi:10.1016/j.jmsy.2013.12.003

Moncayo-Martínez, L. A., & Zhang, D. Z. (2011). Multi-objective ant colony optimization: A meta-heuristic approach to supply chain design. *International Journal of Production Economics*, *131*(1), 407–420. doi:10.1016/j.ijpe.2010.11.026

Otri, S. (2011). *Improving The Bees Algorithm for Complex Optimization Problems.* (PhD Thesis). Cardiff University.

Packianather, M. S., Yuce, B., Mastrocinque, E., Fruggiero, F., Pham, D. T., & Lambiase, A. (2014, August). Novel Genetic Bees Algorithm applied to single machine scheduling problem. In *World Automation Congress (WAC)*, (pp. 906-911). IEEE. doi:10.1109/WAC.2014.6936194

Pasandideh, S. H. R., Niaki, S. T. A., & Asadi, K. (2015). Optimizing a bi-objective multi-product multi-period three echelon supply chain network with warehouse reliability. *Expert Systems with Applications*, *42*(5), 2615–2623. doi:10.1016/j.eswa.2014.11.018

Pham, D. T., Afify, A. A., & Koç, E. (2007b). *Manufacturing cell formation using the Bees Algorithm.* 3rd International Virtual Conference on Intelligent Production Machines and Systems (IPROMS 2007), Whittles, Dunbeath, Scotland.

Pham, D. T., Castellani, M., & Ghanbarzadeh. (2007e). Preliminary design using the Bees Algorithm. In *Proc Eighth International Conference on Laser Metrology, CMM and Machine Tool Performance, LAMDAMAP*.

Pham, D. T., Castellani, M., & Fahmy, A. A. (2008). Learning the inverse kinematics of a robot manipulator using the Bees Algorithm. *Proceedings of INDIN*, *2008*, 493–498.

Pham, D. T., Darwish, A. H., Eldukhri, E. E., & Otri, S. (2007d). Using the Bees Algorithm to tune a fuzzy logic controller for a robot gymnast. In *Proceedings of IPROMS 2007 Conference*.

Pham, D. T., Ghanbarzadeh, A., Koc, E., Otri, S., Rahim, S., & Zaidi, M. (2005). *The Bees Algorithm. Technical Report: MEC 0501*. Cardiff, UK: Manufacturing Engineering Centre, Cardiff University.

Pham, D. T., Ghanbarzadeh, A., Koc, E., Otri, S., Rahim, S., & Zaidi, M. (2006a). The bees algorithm-a novel tool for complex optimization problems. In *Proceedings of the 2nd Virtual International Conference on Intelligent Production Machines and Systems (IPROMS 2006)* (pp. 454-459). doi:10.1016/B978-008045157-2/50081-X

Pham, D. T., Koc, E., Lee, J. Y., & Phrueksanant, J. (2007a). *Using the Bees Algorithm to schedule jobs for a machine*. LAMDAMAP, 8th international Conference on Laser Metrology, CMM and Machine Tool Performance, Cardiff, Euspen, UK.

Pham, D. T., Otri, S., Afify, A., Mahmuddin, M., & Al-Jabbouli, H. (2007c). *Data clustering using the Bees Algorithm*. 40th CIRP International Manufacturing Systems Seminar, Liverpool, UK.

Pham, D. T., Packianther, M. S., Imaguliyev, A., & Yuce, B. (2012). Early Neighborhood Search and Efficiency Based Improved The Bees Algorithm. IMS 2012 Conference, Sakarya, Turkey.

Pham, D. T., Soroka, A. J., Koç, E., Ghanbarzadeh, A., & Otri, S. (2007f). Some applications of the Bees Algorithm in engineering design and manufacture. In *Proc Int. Conference on Manufacturing Automation (ICMA 2007)*, Singapore.

Pinto-Varela, T., Barbosa-Póvoa, A. P. F., & Novais, A. Q. (2011). Bi-objective optimization approach to the design and planning of supply chains: Economic versus environmental performances. *Computers & Chemical Engineering*, *35*(8), 1454–1468. doi:10.1016/j.compchemeng.2011.03.009

Pishvaee, M. S., & Razmi, J. (2012). Environmental supply chain network design using multi-objective fuzzy mathematical programming. *Applied Mathematical Modelling*, *36*(8), 3433–3446. doi:10.1016/j.apm.2011.10.007

Sarrafha, K., Rahmati, S. H. A., Niaki, S. T. A., & Zaretalab, A. (2015). A bi-objective integrated procurement, production, and distribution problem of a multi-echelon supply chain network design: A new tuned MOEA. *Computers & Operations Research*, *54*, 35–51. doi:10.1016/j.cor.2014.08.010

Sazvar, Z., Mirzapour Al-e-hashem, S. M. J., Baboli, A., & Akbari Jokar, M. R. (2014). A bi-objective stochastic programming model for a centralized green supply chain with deteriorating products. *International Journal of Production Economics*, *150*, 140–154. doi:10.1016/j.ijpe.2013.12.023

Seeley, T. D. (1995). *The Social Physiology of Honey Bee Colonies: The Wisdom of the Hive*. London: Hardvard University Press.

Sholedolu, M. O. (2009). *Nature-inspired Optimization: Improvements to the Particle Swarm Optimization Algorithm and the Bees Algorithm*. (PhD Thesis). Cardiff University.

Talbi, E. G. (2009). *Metaheuristics: From Design to Implementation*. John Wiley and Sons. doi:10.1002/9780470496916

Tan, K. C., Lee, T. H., & Khor, E. F. (2002). Evolutionary algorithms for multi-objective optimization: Performance assessments and comparisons. *Artificial Intelligence Review*, *17*(4), 251–290. doi:10.1023/A:1015516501242

Vasant, P. (2010). Hybrid simulated annealing and genetic algorithms for industrial production management problems. *International Journal of Computational Methods*, *7*(02), 279–297. doi:10.1142/S0219876210002209

Vasant, P., & Barsoum, N. (2009). Hybrid genetic algorithms and line search method for industrial production planning with non-linear fitness function. *Engineering Applications of Artificial Intelligence*, *22*(4), 767–777. doi:10.1016/j.engappai.2009.03.010

Vasant, P., & Barsoum, N. (2010). Hybrid pattern search and simulated annealing for fuzzy production planning problems. *Computers & Mathematics with Applications (Oxford, England)*, *60*(4), 1058–1067. doi:10.1016/j.camwa.2010.03.063

Von Frisch, K. (1955). *The dancing bees: An Account of the Life and Senses of the Honey Bee*. New York: Harcourt.

Voudouris, V. T., & Consulting, A. (1996). Mathematical programming techniques to debottleneck the supply chain of fine chemical industries. *Computers & Chemical Engineering*, *20*, 1269–1274. doi:10.1016/0098-1354(96)00219-0

Wolpert, D. H., & Macready, W. G. (1997). No free lunch theorems for optimization. *IEEE Transactions on Evolutionary Computation*, *1*(1), 67–82. doi:10.1109/4235.585893

Yeh, W. C., & Chuang, M. C. (2011). Using multi-objective genetic algorithm for partner selection in green supply chain problems. *Expert Systems with Applications*, *38*(4), 4244–4253. doi:10.1016/j.eswa.2010.09.091

Yuce, B. (2012). *Novel Computational Technique for Determining DepthUsing Bees Algorithmand Blind Image Deconvolution*. (PhD Thesis). Cardiff University.

Yuce, B., Mastrocinque, E., Lambiase, A., Packianather, M. S., & Pham, D. T. (2014). A multi-objective supply chain optimization using enhanced Bees Algorithm with adaptive neighborhood search and site abandonment strategy. *Swarm and Evolutionary Computation*, *18*, 71–82. doi:10.1016/j.swevo.2014.04.002

Yuce, B., Mastrocinque, E., Packianather, M. S., Lambiase, A., & Pham, D. T. (2015). The Bees Algorithm and Its Applications. In P. Vasant (Ed.), *Handbook of Research on Artificial Intelligence Techniques and Algorithms* (pp. 122–151). Hershey, PA: Information Science Reference; doi:10.4018/978-1-4666-7258-1.ch004

Yuce, B., Packianather, M. S., Mastrocinque, E., Pham, D. T., & Lambiase, A. (2013). Honey bees inspired optimization method: The Bees Algorithm. *Insects*, *4*(4), 646–662. doi:10.3390/insects4040646 PMID:26462528

Yuce, B., Pham, D. T., Packianather, M. S., & Mastrocinque, E. (2015a). An enhancement to the Bees Algorithm with slope angle computation and Hill Climbing Algorithm and its applications on scheduling and continuous-type optimization problem. *Production & Manufacturing Research*, *3*(1), 3–19. doi:10.1080/21693277.2014.976321

Zhao, S. Z., & Suganthan, P. N. (2010). Multi-objective evolutionary algorithm with ensemble of external archives. *International Journal of Innovative Computing, Information, & Control*, *6*(1), 1713–1726.

Zitzler, E., Deb, K., & Thiele, L. (2000). Comparison of multiobjective evolutionary algorithms: Empirical results. *Evolutionary Computation*, *8*(2), 173–195. doi:10.1162/106365600568202 PMID:10843520

KEY TERMS AND DEFINITIONS

Bees Algorithm: A honey bees foraging behavior inspired optimization algorithm.

Foraging Behavior: Behavior regarding searching food sources.

Hill Climbing Algorithm: A local search optimization method.a local search optimization method.

Honey Bee: Insect specialized in production and storage of honey.

Multi-Objective Optimization: Optimization problem involving more than one objective function to be optimized simultaneously.

Supply Chain Management: Management of the whole productive chain from suppliers to customers.

Swarm-Based Optimization: Artificial intelligence technique based on the collective behavior of self-organized systems.

Waggle Dance: Method used by the scout bees to communicate information about the food source to the rest of the colony.

Chapter 4
Real World Applications:
Using Technology to Improve Supply Chain Management and Total Asset Visibility (TAV)

Ibibia K. Dabipi
University of Maryland – Eastern Shores, USA

Judy A. Perkins
Prairie View A&M University, USA

Tierney Moore
Prairie View A&M University, USA

ABSTRACT

Over the years the supply chain industry has been transforming to improve the end-to-end (production to delivery) process. Supply chain management (SCM) allows various industries to oversee and better handle how their product is manufactured and delivered. It allows them to track and identify the location of the product and to be more efficient in delivery. Integrating total asset visibility (TAV) technology into the supply chain structure can provide excellent visibility of a product. This kind of visibility complemented with various packaging schemes can assist in accommodating optimization strategies for visualizing the movement of a product throughout the entire supply chain pipeline. The chapter will define SCM, discuss TAV, review how transportation as well as optimization impacts SCM and TAV, and examine the role of packaging in the context of SCM and TAV.

INTRODUCTION

In an effort to improve the supply chain and total asset visibility (TAV), new technological advancements in shipping, transporting, tracking, and packaging have created significant changes to the role transportation and optimization will impact the overall end-to-end process. Globalization of the world economy and the corresponding supply chains has advanced rapidly such that a dynamic rapid response system has to be developed in order to cope with the ever-changing world economy and regulations as barriers

DOI: 10.4018/978-1-4666-9644-0.ch004

to trade and interdependence are reduced across the globe. Growth in world trade, and corresponding cargo container movements, continues to substantially exceed overall economic growth, meaning that trade volumes are doubling every 5-7 years. With that being said, a typical supply chain flow would have a set of common elements consisting of the following:

- Consumers
- Retailers
- Wholesalers
- Manufacturers
- Suppliers

Supply chain management is a critical component of any company's viability in a global economy ruled by fluidity of unpredictable events such as wars, economic and natural disasters. The ability to navigate and mitigate these events from a global perspective minimizes the negative impact both in profits and in company image. The most successful companies in this regard all have the following attributes in common namely exceptional capabilities in supplier development, collaborative product development, demand shaping and demand sensing capabilities, lean manufacturing and innovative retailer practices. For most industries and for many companies, the ability to achieve and sustain global supply chain excellence will be a substantial determinant of their overall corporate success. More importantly, the technology support for global logistics is at an immature stage for all but a small number of companies. This lack of technology support contributes strongly to the challenges in controlling costs and executing effectively. Most of those available in the past were designed to solve very specific problems, rather than providing a more comprehensive set of solutions. Today, while there is still room for continued advancement in global logistics and trade management solutions, there are systems available right now that can automate and improve a significant portion of the global logistics challenge. The robustness and functionality of the software is consistently improving, in tandem with the growing customer demand for global logistics solutions. This chapter describes how transportation, optimization, and packaging technologies impact current supply chain management and TAV processes and procedures. It will discuss a few of the difficulties associated with accepting new technologies that support improving logistical practices. Also, key elements of transportation, optimization, and packaging initiatives that address the gaps in supply chain methods used by an array of organizations to advance the future of logistics and different types of case studies will be presented.

BACKGROUND

In the transportation arena, there are several cutting-edge technological advancements such as RFID (radio-frequency identification), 2D Codes, QR Codes, integrated 2D & QR Code, RTLS (real-time locating systems), remote management, and sustainable energy (Lu & Swaminathan, 2013) using a minimal carbon "footprint" are being used to support management of the supply chain and total asset visibility (TAV). RFID technology use tags and readers to easily track goods that are being shipped. The tags carry information on them that states how much product is shipped and the destination of the shipment. This allows for a smoother transition from production to delivery at a store and/or distribution center. The tags are beneficial in reducing the time it takes to locate an item or pallet, improves visibility, and offers real-time movement and location (RFID Arena 2013). Real-time locating systems are used to pinpoint packages and their location by using transactional data. That information is then stored and viewed on a dashboard (a real-time user interface that displays graphical information on product status). A North American Nissan assembly plant utilizes the RFID and RTLS technologies to track vehicle components and cars throughout their manufacturing process. Each vehicle has a RFID tag that is associated with the vehicle identification number (VIN) (Selko, 2006). This allows the company to easily locate parts and store or disperse them as needed. Vehicles and parts will retain their tags until the parts are assembled and the vehicles have been shipped.

Even though Nissan manufactures their vehicles in the United States, their parts are still shipped in from other countries and that is where remote management comes into play. Remote management's goal is to produce transparency from order to delivery. This is seen mostly in e-commerce where the product is being shipped globally. The challenges that arise with global e-commerce are achieving uniformity amongst operations and systems, quality of service (delivery dates, lack of communication), and reducing waste ("defective products, inventory, idle capacity due to failures") (Gunasekaran & Ngai, 2004). The key factor in remote management is utilizing RFID and RTLS technology to help track a product and management of the inventory/ assets. Along with RFID tags, Codes, and QR Codes mechanisms are used to track product information. One of the new widely used type of coding is QR tags. QR tags are used to store information and can be read by any smart phone or device that has the capability or a QR reader. The tags are used for informative information such as the background and origin of a product compared to just storing information about the quantity and tracking information. Unlike RFID tags, QR tags are not rewritable and tend to remain with a product after the product has been delivered to a customer or reaches the store shelve. The 2D Codes such as data matrix codes are used by manufactures for data tracking. The 2D coding tags are mostly used in the aerospace and automotive industry (Totally Integrated Automation 2011). While the QR Code is widely used throughout the retail community, companies such as Wal-Mart have turned to sustainable energy technology to assist in transporting their goods. Their goal is reduce the amount of waste that is created by the end-to-end process. They have also implemented incentives for suppliers to comply with their goals.

Moreover, the term "Big Data" was coined to reference the magnitude of data that is being managed each day in the supply chain industry. Companies such like Quintiq and JDA have created software that assists in optimizing the process and processing data. The information that is collected form tags some type of system has to be able to process it and categorize it. Companies such as GE have even gone as far as creating their own system that streamlines the process of production to delivery. Software products are created to better plan and execute the process of delivering goods. Industries such as e-commerce, automotive, and other industries that handle high volumes of traffic are affected by the Big Data transition. It allows them to easily plan shipments, better communicate with the customer, transport items, and keep track of products. The main goal of the optimization tools are to better plan and forecast, track inventory, manage orders, manage transportation, and manage distribution (Manhattan Associates 2008). In bettering supply chain management transportation and optimization advancements work hand-in-hand to improve the end-to-end process.

Lastly, TAV is the ability to identify the location of a product while it is in transit from the location of production to delivery. It is used to streamline the supply process and automate the tracking process. Currently, the military is an avid user of TAV because of the capability to provide well-timed and precise product status updates (Department of Defense, Office of Inspector General, 2001). As a result of using TAV, customers will receive their shipment in a timely manner. Undoubtedly, there are the levels of complexity in the decision making process that are required to stay competitive within the global economy.

WHAT IS SUPPLY CHAIN MANAGEMENT?

Definition of Supply Chain

Supply chain management (SCM) is the process in which an item is produced, shipped, and delivered to the customer in a timely matter. Or as Investopedia (2015) defines supply chain management as the "streamlining of a business' supply-side activities to maximize customer value and to gain a competitive advantage in the marketplace." The emphasis is on the process itself. How will the product be transported, how long will it take for it be delivered, how will the product be tracked, what type of care needs to be taken in to consideration, are there any factors that will or may serve as an obstacle throughout the process, or can the product be tracked while in route?

Components of Supply Chain

The key players that are involved in supply chain management are suppliers, distributors, and the customers. Perfected coordination between the players is a result from streamlining the planning, coordination and collaboration amongst all aspects. The four components of supply chain management, which are consistent throughout the transportation industry, (Perez, 2013) are industry framework, value proposal, managerial focus, and internal processes. Industry framework refers to how a supply chain process is established to better serve a demand in an industry. It is the "interaction of suppliers, customers, technological developments, and economic factors that affect competition (Perez, 2013)." Demand, what is the need for a product and how long a consumer will continue to want or need the product, is what drives production, efficiency and costs. A higher demand requires more production, more product, heavier workload, more costs, and efficiency to ensure that the right amount of product is delivered in a timely manner. As a result, market fluctuation needs to be taken into consideration. When there is too much supply and not enough demand, product liquidations or discounts occur. When there is more demand than supply, sales are lost. Another factor in the industrial framework is product lifecycle. How long will a product be in demand, how long will there be a need for the product? Due to new breakthroughs technology many products do not last long in the market. Products are constantly changing and becoming more efficient. This puts a strain on determining market costs and demand predictions. The last factor is the correlation between asset costs and total cost. How much is the consumer paying for a product compared to how much it cost the company to manufacture the product including operation costs (labor, resources, company utilities, etc.). If there is a large company with a great deal of labor and resources the company has to produce and sale product or constantly develop new product to stay afloat. However, a smaller company will focus on responsiveness to the demand. An example of industrial framework establishment, is when an electronics company, E-Best, is releasing a new gaming system. The demand is high because the product provides a new cutting edge technology that no other gaming system on the market has. In order to prepare for the release, E-Best must fulfill orders for each store that has ordered the product. Each order has a minimum order of 1,000 systems. They have also determined that air, rail, and ground transportation will be the best way to transport product and ensure that product will be there by Black Friday, E-Best's release date. Because they are a company that has high utilization rates (they have funds and labor), they are able to produce the gaming systems in mass quantities. The other gaming systems that are about to become obsolete, their prices will began to lower and will mark the beginning of new development and liquidation of assets. The coordination of the process is very detrimental to the supply cycle.

Before E-Best actually decided upon a release date and before stores began to place orders, they had to conduct a consumer trial run. The trial run allowed them to predict the demand for the product and to drum up conversation and notable reviews about the product. By E-Best providing test products to stores and conducting trial tests it added value and drummed up the consumer demand for the product. Even though E-Best is a reputable company and has a highly valued reputation for producing quality electronics and gaming systems, if they did not have an extra boost to increase the value of the product that they are trying to release then the sales and demand will not be as high. This puts them at a competitive advantage in the industry. It is also important to make sure that there are professionals that manage the system. The mismanagement and non-collaboration between business strategy and supply chain can lead to the failure of your supply chain. And lastly, internal processes ensures the quality of a product, as well as, the time efficiency in which the product is produced. Products must also be able to be customized to customer's standards and still be able to produce and deliver in a timely manner. Nothing should be able to change in the order process and delay the time that it takes for a product to be produced and delivered with the upmost efficiency and quality.

Industries that Use Supply Chain

Supply chain management can be seen in industries such as automotive, technological, pharmaceuticals, food, and beverages. Each have their own set of internal processes but they all use some type of supply chain management to deliver their goods to meet the consumer's demand. In the automotive industry vehicles are assembled on an assembly line then shipped by rail or boat then ground to dealerships. Pharmaceuticals have a system that produces medications then they have a transportation system that delivers the medicine to hospitals and drug stores. The technological industry has assembly lines that assemble electronics and computer hardware that are then shipped by boat then ground transportation takes over and delivers the products to perspective locations. The food and beverage industry has a number of systems and processes that occur in order to produce the product then the product is transported to stores and restaurants by ground transportation. Supply chain management allows various industries to oversee and better handle how their product is manufactured and delivered. It allows them to track and identify the location of the product and to be more efficient in delivery.

TOTAL ASSET VISIBILITY

What is Total Asset Visibility?

Total asset visibility (TAV) has been referred to as a capability as well as an initiative within the military community for over four decades. According to the US Government Accounting Office (GAO) National Security and International Affairs Division (NSIAD) (1999), "obtaining total asset visibility has been a

Department of Defense (DoD) goal since at least 1972." Specifically, TAV is an operational term used by the military community to describe the oversight that is applied to identifying the location of equipment, supplies, or personnel during in-processing, while in transit or in storage (US Army Field Manual (FM)101-5-1/US Marine Corps Reference Publication (MCRP) 5-2A, Operational Terms and Graphics, 1997). TAV's subordinate components (i.e., in-storage, in-transit, and in-process) provides the following: (1) visibility of static inventory, (2) information on the asset as it moves through the logistical pipeline, and (3) visibility of the asset during maintenance, calibration, or acquisition (US Army Field Manual (FM) 55-80, Army Container Operations, 1997). When implementing these three components under a single umbrella, the supply chain end-to-end process is streamlined and the end-user gains timely, accurate, and comprehensive knowledge about their asset.

Achieving Total Asset Visibility

Over the years, the supply chain industry has been transforming to improve the end-to-end (production to delivery) process. Achieving TAV within the supply chain process requires overcoming significant challenges similar to those found in the military and gaps that were shown consistently across an array of industries (i.e., the military, e-commerce, and retail). The gaps were in the areas of physical and information flow. Anderson (2001) identified the challenges that the military needed to address to be "(1) material, (2) training, (3) leader development, (4) organizations, and (5) doctrine." Specific lessons learned from challenges faced by the military can provide a framework that must incorporate at a minimum the following Anderson's (2001):

- TAV systems must have access to uninterrupted communications that enable them operate in an interactive, semi-interactive, or remote mode.
- TAV automated systems must be able to transmit near-real-time data to common data bases.
- Communications networks must be robust enough to handle Intranet and Internet requirements.
- Personnel having to operate TAV systems must be trained on how to obtain accurate information rapidly.
- Training programs for personnel operating TAV systems must meet the needs of the organizations responsible for providing TAV information.
- Leaders must learn how to obtain useful data and be able to use the data in their decision-making process.
- Leaders must learn how to evaluate the effectiveness and efficiency of TAV systems
- Organizations may have to be restructured based on their respective TAV responsibilities as it relates to the supply chain (end-to-end) process in order to provide a unified approach to effectively managing assets.
- Formalized guidance should be published to explain how TAV processes will work at levels of the organization

In regards to closing the physical and information flow gaps prevalent to TAV, there has been a wave of advancements in technology that has automated the tracking process in a manner that feeds information into a computer system for authorized users to access and use for making real-time decisions. Cutting-edge technological advancements such as RFID (radio-frequency identification), 2D Codes, QR Codes, integrated 2D & QR Code, RTLS (real-time locating systems), remote management, and sustainable energy (Lu & Swaminathan, 2013) that uses minimal carbon "footprint" are being used to support management of the supply chain and TAV. RFID technology use tags and readers to easily track goods that are being shipped. The tags carry information on them that states how much product is shipped and the destination of the shipment. This allows for a smoother transition from production to delivery at a store and/or distribution center. The tags are beneficial in reducing the time it takes to locate an item or pallet, improves visibility, and offers real-time movement and location (RFID Arena 2013). Real-time locating systems are used to pinpoint packages and their location by using transactional data. That information is then stored and viewed on a dashboard (a real-time user interface that displays graphical information on product status). A North American Nissan assembly plant utilizes the RFID and RTLS technologies to track vehicle components and cars throughout their manufacturing process. Each vehicle has a RFID tag that is associated with the vehicle identification number (VIN) (Selko, 2006). This allows the company to easily locate parts and store or disperse them as needed. Vehicles and parts will retain their tags until the parts are assembled and the vehicles have been shipped.

Even though Nissan manufactures their vehicles in the United States, their parts are still shipped in from other countries and that is where remote management comes into play. Remote management's goal is to produce transparency from order to delivery. This is seen mostly in e-commerce where the product is being shipped globally. The challenges that arise with global e-commerce are achieving uniformity amongst operations and systems, quality of service (delivery dates, lack of communication), and reducing waste ("defective products, inventory, idle capacity due to failures") (Gunasekaran & Ngai, 2004). The key factor in remote management is utilizing RFID and RTLS technology to help track a product and management of the inventory/ assets. Along with RFID tags, Codes, and QR Codes mechanisms are used to track product information. One of the new widely used type of coding is QR tags. QR tags are used to store information and can be read by any smart phone or device that has the capability or a QR reader. The tags are used for informative information such as the background and origin of a product compared to just storing information about the quantity and tracking information. Unlike RFID tags, QR tags are not rewritable and tend to remain with a product after the product has been delivered to a customer or reaches the store shelve. The 2D Codes such as data matrix codes are used by manufactures for data tracking. The 2D coding tags are mostly used in the aerospace and automotive industry (Totally Integrated Automation 2011). While the QR Code is widely used throughout the retail community, companies such as Wal-Mart have turned to sustainable energy technology to assist in transporting their goods. Their goal is reduce the amount of waste that is created by the end-to-end process. They have also implemented incentives for suppliers to comply with their goals.

Benefits and Limitations Total Asset Visibility Bring to Supply Chain Operations

As a result of TAV implementation, customers or the end-user can expect to receive their shipment in a timely manner. The TAV system will enable the shipment to be monitored, from the time of order until the time of delivery. With a concomitant need to employ TAV enabling technology to improve supply chain operations, there are some limitations associated with technologies that support improving logistical practices.

One such limitation is that TAV technology can experience periods of non-communication, thus permitting visibility of the asset to be lost. Chisholm and Hammons (2006) states that TAV is dependent on communication and location technology to provide visibility into the transportation phases of the logistics supply chains and depending on location, combined satellite and cellular communication capabilities offer the ability for optimized communications.TAV devices need to incorporate dual technologies for determination and reporting of position information. Conceptually, Global Positioning System (GPS) provides the primary location information; however, when GPS information is unavailable, the Inertial Navigation Unit (INU) continues to provide location information (from the last known GPS latitude/longitude). This then suggest that TAV incorporate both internal and external power sources to "enable communication over the complete transportation path" (Chisholm & Hammons, 2006) in order to better accommodate periods of non-communication.

Another limitation is the security of information systems linking TAV technology to data providers and users. While security breaches can and do occur, the protection of information systems that could be subjected to US Government Accounting Office (GAO) National Security and International Affairs Division (NSIAD) (1999) "data loss, destruction, or unwanted browsing" should become part of an organization's TAV implementation plan.

Clearly, TAV has gained increased attention over the years as supply chain efforts have become focused on providing the correct assets at the right time and place. TAV can only be an improvement over existing systems provided the organization deploying the concept are able to fully incorporate, at a minimum, the aforementioned challenges and gaps.

TRANSPORTATION IMPACTING SUPPLY CHAIN MANAGEMENT AND TOTAL ASSET VISIBILITY

Previous Transportation Methods

Before there were airplanes there were ocean liners and before ocean liners were coach boats. The first form of transportation that is documented is foot transportation (walking, running). Once humans discovered that animals could be used as a form of transportation, the next (Lambert, 2014) forms of transportation were horses, donkeys, and camels. All early forms of transportation have been dated back to 4,000 BC to 2,000 BC. In 2000 BC wheels with spokes were born while in 3100 BC the earliest form of sail boats was created. As time progresses, the earliest methods of transportation became obsolete while more modern versions were created. Animal transportation evolved into horses and donkeys pulling wagons and covered stage coaches. In the 17th Century the first turnpike was created and that was the first toll road established. The Egyptians developed a pulley system constructed by rope and knots while the Romans perfected the water transportation vehicles. Following the 17th Century, the 18th Century's transportation advancements included the construction of canals and the invention of the hot air balloon. But the type of transportation that would revolutionize the industry was created in the 19th Century and that mode of transportation was the railroad followed by the underground railway and steamships. The 20th century had one of the most profound impacts of all time. The airplane and ferry were created during that time period. Animals, wood (sailboats), stone (wheels) and human feet were all stepping stones to the modern day transportation systems that are currently used today.

Solutions to Closing the Gap

New Technologies and Uses

With all of the new forms of technology that are surfacing a few of the most widely used are: (1) radio frequency identification (RFID), (2) real-time locating systems (RTLS), (3) 2D barcoding, and (4) electronic supply chain (ESC). RFID is used when transporting goods from supplier to distributor by tracking the shipment and provide information on each package entailing what the package consists and the quantity. RFID technology requires tags and readers. The information is stored on a passive or active tag which is attached to a pallet or box. Once the tag is in the vicinity of a reader, the reader reads the tag then sends information to a computer, storage bank, or the reader itself which can have a built-in function to display the information. RTLS works in collaboration with RFID in order to track products. The real-time locating system pinpoints products using active tags that are constantly read in real-time. 2D barcodes are not your typical barcodes, (Totally Integrated Automation, 2011) they do not need to be registered and they are able to store more characters than barcodes. Another advantage of using the 2D barcodes is that they have a unique number that is assigned to the 2D codes that last until the code is destroyed. They are also read by a 2D barcode reader. Electronic supply chain, also known as remote management, is a system where a customer can order a product off line and it will be sent to them. The inventory database is automatically updated and notifies the customer immediately if an item is in stock or not. The item is then shipped once it is shipped off (Gunasekaran & Ngai, 2004) and tracked until it reaches its destination.

Example #1

Waltmans, a high-quality retail store, uses RFID and RTLS to track their goods. Their tracking system is used once the products are packed. Once the item is packed and ready for shipping, every box will receive a RFID tag which will enable anyone who scans the tag to identify the box content and quantity. The scanners that are above the port doors enables the distribution manager to verify the information in the computer with the information that the readers scanned. While the shipment is in route to specified destination, the package can be tracked by scanning the tag as well. Once the package arrives, a store manager scan the packages to ensure that the shipment is correct. After the product has been checked, each item receives a RFID barcode tag or RFID sensor that allows the store to track the inventory in real-time.

How New Technology will be Applied to Transportation

Applying new technology to the transportation aspect of supply chain management is a very valuable investment. When RFID is implemented to a supply chain process, fixed readers will be placed at designated openings where containers and pallets will pass through so that the readers are able to scan and read the tags. The tags are placed in an area where there is no interference. Once the packages leave the suppliers and are delivered to the distributors, the tags are scanned again when they cross over the threshold opening of the loading dock. The tags are also useful inventory control mechanism when resources first enter a manufacturing site. Tracking products with RTLS is an extra amenity. It allows suppliers, distributors, and manufactures to trace products in real-time in order to keep a better handle on the products. Another advancement in RFID technology is electronic supply chain (ESC). ESC is mostly seen in international as well as Internet commerce when shipping, identifying inventory as well as tracking inventory from supplier to the customer. RFID and RTLS is a part of the system that allows suppliers, distributors, and manufactures to control inventory remotely. The manufacturer may be in Hong Kong and the supplier is in China while the customer is in London. 2D barcoding is a new type of barcoding that that does not require multiple systems to register a product and track the items in the system. Each item has its own code that is used for tracking and identification purposes.

Example #2

A beverage company, Sippers, imports their bottles from a manufacturing company in China. The bottles are branded with the new barcoding system. It is a part of Sippers new initiative to remain at the top of the technology curve. The codes will not have to be changed once they are programmed. Once the bottles are filled, packaged, and registered, each pallet is given an RFID tag that identifies which beverage is on the pallet and the amount of units that are on the pallet. Once the pallet leaves the warehouse the readers scan the active RFID tags. While in route to the delivery truck, the RTLS system consistently updates the system on how many pallets exist and the position of a given pallet. An electronic system is keeping track of where each shipment is located and if it has reached its final destination. Once the shipment reaches the distributor and they began to scan the bottles into the system, the information is automatically uploaded and no other registration steps are taken. The transportation methods that were taken, was one that reduced the amount of environmental impact.

Feasibility with Growing Economy and Demand

As stated in the previous section, implementing new technology is very feasible option to pursue. According to RFID Journal (2015), tags cost $0.05 per tag, readers range in cost from $500-$2,000, antennas cost anywhere from $200 and up. The costs of implementing a RFID and RTLS system benefits businesses, especially when it comes to the aspect of pilfering. Tags are now equipped to be read on various materials. They are even able to be read on metals and near water, which they were not able to a few years ago. With the new advancements in technology relating to RFID, many industries have already invested in the new technology.

Example #3

The pharmaceutical industry is one that is known to have a high rate of theft and fraud. One company, namely, Healthy Body has taken new measures in order to reduce those rates. This specific company has adopted the new radio frequency identification (RFID) tracking concept. Healthy Body Industries utilizes real-time locating system (RTLS) and RFID technology to track their inventory. Tags are placed on inventory that displays the quantity and name of the medications that are stored in the cabinets. Readers are also placed at the opening of the cabinets to ensure that only authorized personnel can access that area. RTLS allows the company to identify where products are and when they have left their designated area in real-time. If any personnel is removing items for an authorized reason, they must scan the designated tag and edit the material. The information on the tag and the inventory data must also mirror the companies records. If they do not, decision-makers of the company are made aware of three possible scenarios, which are: (1) the medicine was illegally removed from the area, (2) personnel made a mistake in recording the transaction and/or deposits correctly, or (3) there was a system malfunction and the information failed to record.

After one year of implementation of Healthy Body Industries' RFID/RTLS system, they were able to achieve 25% return on investment. Their tags cost them $0.05 a piece. The badge readers costs $700 dollars. The inventory scanners cost just under $7,000 for the purchase of two readers. Also, Healthy Body Industries gained improvements to their security breaches, better inventory management, and an increase shipping and tracking reliability.

If there are companies or industries that have the resources to upgrade facilities, implement, and maintain the system, it will. The end result will be more beneficial than not investing. The same tags that are used internal to the facility can also be used external to the facility (shipping and store use). Industries such as aerospace, manufacturing, packaged goods, retail, and financial systems all use RFID. Examples can be seen by toll authorities, tap and go financial transactions at gas stations and stores, tracking retail within stores by incorporating RFID tags into sensors and barcodes, real-time locating systems with hospitals and pharmaceutical companies, and they also can be seen in the automotive industry by tracking parts and keeping track of which ones leave the lot/yard and which do not. The feasibility of incorporating such technology is quite high. With the advances in being able to read tags in various environments, the feasibility of utilizing tags in various industries have increased. Ultra-high frequency tags can be read as far as 20 to 30 feet depending on the situations. If the antenna is in an ideal place then the tags have a stronger reading range, but if the antennas are blocked the read range drastically goes down. The RFID Journal has also stated that "most governments restrict the output of readers so that they don't interfere with other RF devices." When the strongest signals are needed more antennas and readers are needed to achieve, but in some cases they are not mandatory.

OPTIMIZATION IMPACTING SUPPLY CHAIN MANAGEMENT AND TOTAL ASSET VISIBILITY

Detailed Description of Optimization

Supply chain management (SCM) can be described as the monitoring of materials, information, and finances as they progress through the system namely from supplier to manufacturer to wholesaler to retailer to consumer. This process encompasses the coordination and integration of flows both within and among companies in real time. The ultimate goal of any effective supply chain management system is to reduce inventory without sacrificing customer demand expectations. Supply chain management flows can be divided into three main flows:

- Product flow
- Information flow
- Finances flow

The product flow includes the movement of goods from a supplier to a customer, as well as any customer returns or service needs. The information flow involves transmitting orders and updating the status of delivery. The financial flow consists of credit terms, payment schedules, and consignment and title ownership arrangements. According to Doran (2004), robust concentration on core competencies is essential as enterprises are forced to streamline their efforts in order to achieve significant, comprehensive cost reductions; reduce their utilization of resources, shorten the cycle times and reduce inventories all while improving their service simultaneously. These tasks translate into immense challenges for corporate management, which requires the optimization of frictions in the production and logistics processes, as well as the establishment of continuous materials and information flows along the value-added process.

Previous Optimization Strategy

Optimization of the supply chain management can therefore be examined within the context of the three main flows. Alex Brisbourne, President and Chief Operating Officer (COO), KORE Telematics, stated in his article "The Current State Of The Internet Of Things" that the term "Internet of Things, (Field Technologies Online Magazine, 2013)" refers to uniquely identifiable objects (things) and their virtual representations in an Internet-like structure. The definition is, for the most part, analogous to descriptions of machine-to-machine (M2M) communications and connected devices. With the complexity of the supply chain in a global economic environment, lapses in real-time information updates, can be harmful when different time zones are taken into account. Thus connectivity and optimization of the process can mean the difference between success and failure of the business. This has been recognized by more shipping companies who are now using wireless connectivity to remotely monitor their mobile assets. This connectivity enables M2M-based Internet of Things functionality, such as the ability to report on the location and route of a truck or the humidity, temperature, and/or pressure of its cargo. The added intelligence provided by the Internet of Things plays a critical role in ensuring regulatory compliance and the safety of our global food and pharmaceutical supply chains. The optimization of this process tracks the product flow within the system.

The information flow occurs through a network that requires active monitoring. Network optimization becomes the challenge of the Supply Chain Organization when it expands in a global context. The more the network expands, the more complex it is to assess the efficacy of the network. Optimizing this process therefore requires evaluating the information technologies in the context of Transactional IT as well as Analytical IT. While Transactional IT address the acquisition, processing and communicating raw data about the company's past and current supply chain operations such as general ledger systems, quarterly sales reports, and e-commerce systems, Analytical IT address the development and application of systems for evaluating and disseminating decisions based on models constructed from supply chain decision databases (e.g., production scheduling systems, forecasting systems, supply chain network optimization systems). These models could be descriptive such as forecasting, data mining, activity-based costing, performance metrics, simulation, and systems dynamics or prescriptive such as mathematical programming combined with heuristic methods.

The financial flow is mostly governed by behavioral realities underlying organizational decision-making. Functionally under the Transactional IT domain, the financial flow as well as its decision-making processes are derived from, transactional database where optimization models implicitly combine structural and numerical elements that are based on aggregations of product families, customers into markets, vendors into suppliers

Gaps and Issues in Optimization

With customer satisfaction as the driver, studies have shown that there exists a void between customer expectations and industry delivery of products. This is guided by the inevitable vulnerability of the global market place as a result of fluctuations in the supply and demand of goods and services. The recent fluctuation in oil pricing is an example of how vulnerable the entire supply chains can be impacted by a single cost-driver. Incidents such as natural disasters, security breaches, spikes in product demand, technology failures, political unrest and economic instability, shipping delays, and other disruptive events further complicate the efficiency of the global supply chains. Mitigating these circumstances with a measure of normalcy in the operational domain becomes a key target for supply chain management especially in evaluating what if scenarios associated with the recovery process and the time to recover assessment. Thus optimization can be viewed in two broad categories (supply chain capacity and supply chain capability) defining the evaluation process to address output concerns. The supply chain capacity relates more to a volume measure (i.e. the volume of throughput can you achieve across your supply chain) while supply chain capability, reflects on how robust and agile the supply chain is to disruptions. The issue of capacity addresses the ability to meet the demand from customers and is predicated on the inventory model and statistical analysis of previous demand cycles and growth projections. On the other hand, the uncertainty in disruption from either global economy meltdown, terrorist acts or transportation disasters present a different set of problems to be addressed. Further questions such as are there bottlenecks in the system and are there redundancies built in to mitigate them arise. Within this context, the whole approach to optimization could center on prioritization on which category presents the higher risk for the company.

Solutions for Gaps and Issues

There are currently various software packages developed by various companies addressing specific aspects of optimization within the industry. Their degrees of success ascertained by the developers of these software packages can be categorized as supply chain capacity (volume measure) relating how much throughput can be achieved across the supply chain versus supply chain capability, which reflects on how robust and agile the supply chain is to disruptions described above. These should take into account the planning, controlling and coordination of the value-added processes and the service activities along the entire supply chain. For the purpose of achieving the common targets, specific inter-organizational planning, controlling and coordination systems are required. This will address the capacity optimization issue thus providing a reasonable predictor to the level of capacity maintenance within the entire organization. The risk of failure from a capability gap could be more disruptive to the production or output plan than a reduction of throughput of the supply chain. From a purely risk management perspective, the potential for the greatest failure exists if the supply chain capability is inadequate. In 1996, Foxmeyer was the second largest wholesale drug distributor in the U.S., with sales over $5 billion dollars in a highly competitive industry. Tens of millions of dollars in unrecoverable shipping errors later, the company filed for bankruptcy and the main operating division of the $5 billion company was sold to its larger rival, McKesson, for only $80 million. Starting in 1993, athletic shoe and gear maker adidas tried to implement first one and then a second warehouse management system in its Spartanburg, SC, distribution center. The adidas 1996 Warehouse Meltdown was a result of adidas insisting that the vendor's Unix-based system be ported to fault tolerant Stratus computers. This eventually led to the company (Integrated Software Logistics Engineering) going under in mid-project.

Some examples are: Supplier Management Software by Intelex Technologies, a web-based Supplier Management Software that allows organizations to effectively track supplier performance, perform evaluations, assign ratings and drive continual performance improvement; Collaborative Supply Portal by Ultriva, which provides real-time visibility and collaboration across the supplier network. The product consumption based replenishment supplements forecast based replenishment. It includes electronic kanban, vendor managed inventory, buyer and supplier dashboard, and RFID tagging. Another example is the VISCO Software by VISCO for importers and wholesale distributors. VISCO is an ERP software designed specifically for companies that are importing and distributing. Uniquely designed to handle profit by shipment, inventory visibility (including overseas and on the water), document management (Bill of Lading, Packing List, Commercial Invoices) as well as foreign currencies and unit of measure conversions.

This is not an endorsement by the authors of these packages. There are far more software packages that can be enumerated here. This is just an introduction to the sample of packages designed to mitigate the supply chain gap issues some of which are customized to specific industry needs.

PACKAGING, SUPPLY CHAIN MANAGEMENT, AND TOTAL ASSET VISIBILITY

Packaging plays a critical role in supply chain management (SCM) and total asset visibility (TAV). A number of supply chain operations within the commercial industry (Naval Studies Board, 1997) use intermodal (truck, rail, and ship) shipping of containerized cargo for all goods except bulk commodities such as wheat and coal. Currently, packaging of containerized cargo has become more modular in nature and is shipped by a variety of containers (e.g., 20-foot International Standards Organization (ISO), double containers (BICONS), triple containers (TRICONS), and quadruple containers (QUADCONS)) in "shelf-ready" or "rack-ready" condition so that the cargo can be easily moved to its final destination (i.e., end-user). Also, per a variety of specifications and standards, packaging involves having to address a wide range of areas in order to properly transport as well as protect the cargo from getting damaged, deteriorating, or both.

Lambert, Stock, and Ellram (1998) indicated that packaging performs six functions (e.g., containment, protection, apportionment, unitization, convenience, and communication) and there are several types of packaging (e.g., corrugated containers, foam-in-place, stretch wrapping, and strapping). While all six of the packaging functions affects SCM, unitization is the one function that impacts both SCM and TAV the most. Unitization ultimately reduces the number of times a product will be handled. This specific function permits a package to be placed inside of another package for shipping purposes. The shipping container, packaging aids used in the container's interior, and TAV technology establishes a framework for providing ready and/or selective access to cargo while retaining protected and secure storage. Additionally, the unitization shipping configuration scheme addresses two key issues commonly associated with packaging, and they are added space and weight.

Moreover, when this logistics management practice is employed by commercial industry it drives SCM performance and efficiency responsiveness within the supply chain structure. Without question, aspects of packaging, SCM, and TAV are interrelated and it's this relationship that effectively link together the various phases of the supply chain.

LIMITATION OF THIS RESEARCH WORK

This chapter is limited to information that has been published or tested by researchers and industry developers. The information on total asset visibility was limited only to the amount of information that the military was willing to release due to the sensitivity. As far as the industries use, limitations were subjective to only the industries who openly shared information, as well as, those within the industries that were willing to divulge their information.

FUTURE RESEARCH DIRECTIONS

When most people think about transportation and managing supply chains, they think about vehicles and goods. The tracking industry is headed in the direction of going past tracking vehicles and goods. Researchers are exploring the idea of tracking humans in work environments and hospitals. The next step in search regarding supply chain management and total asset visibility is incorporating other forms of technology and expanding its usability to every industry. Retailers will be able to locate items using RFID and RTLS.

CONCLUSION

Supply chain management (SCM) has come of age since its initial conception in the mid-eighties and early nineties. The growing dependence by the transportation industry in an ever-changing and unstable global economy requires that industries take advantage of advances in technology to better optimize their network throughput as well as customer satisfaction. Several aspects of SCM performance and adaptive responsiveness to the prevailing circumstances have been examined with a view of reviewing solutions that are used to mitigate disruptions in the SCM chain. Technologies such as RFID, GPS and M2M-based Internet of Things functionality are enabling technologies for addressing both total asset visibility (TAV) as well as optimization within the context of Transactional IT as well as Analytical IT. Supply chain flexibility is defined as the supply chain's promptness and the degree to which it can adjust its supply chain speed, destinations and volumes in response to changes in customer demand. Flexible supply chains, have the potential of improving customer satisfaction, lower inventories, provide positive relationship between each node of the supply chain and supply chain flexibility, and provide positive relationship between IT-systems and supply chain flexibility. There is the need to understand the transparency of a system determined by the degree of knowledge of the system elements in all the categories (information, products/services, processes, resources and partners). Since these are all connected to the success of the company, having a clear understanding of the interconnectivity of these elements and the necessary optimization strategy will avert the types of failures examined in this chapter. For a variety of reasons, companies have moved into global sourcing and exporting programs without fully understanding the complexity of the planning and execution requirements, and with insufficient technology support. This lack of technology support contributes strongly to the challenges in controlling costs and executing effectively. Companies must therefore within of context capacity or capability, evaluate the whole approach to optimization and prioritizing on which category presents the higher risk for the company.

CASE STUDIES

Case Study #1: Finding a Tracking Inventory Solution

In the oil and gas industry delivery is very important. Lost time equates to money lost and the oil and gas industry is about profit. One company, namely, Oil Seekers Express want to see the profit, but is being negatively affected by one of their processes. It is slowing down other processes and systems which is in return affecting business relationships, overall profit, transportation and the delivery schedule.

Moreover, Oil Seekers Express has a motto service you can depend on, but that is not quite true. Oil Seekers Express, is having an issue with keeping track of which tankers in the yard are filled, partially filled, empty, and which tankers are expected to arrive and those that supposed to be departing. Because of this issue, deliveries are late and there is a significant amount of time that is wasted on identifying which tanks transport certain materials. The tankers transport various hazardous materials ranging from solids to chemicals. Oil Seekers Express now need a system that will allow them to do the following: (1) track tanker cars, (2) be able to identify which ones are filled, partially filled, or empty, (3) be able to identify which cars have arrived and which cars are leaving the yard, and (4) identify which cars transport specific hazardous chemicals without the hassle of using a manual process.

Case Study #2: Developing a Supply Chain

A new hospital is in the process of being built. They are a new cutting edge hospital with all of the new and future technology that cures their patients. The hospital includes laboratories, research facilities, etc. The hospital wants to fix the unfixable, which is they hold dear to their hearts.

Healing Wounds, the new hospital, does not have a system, as of yet, that tracks patients, medicine, and equipment. The Operations Manager does not have any idea as to how he is going to simultaneously keep track of inventory, patients, and keep the hospital running as efficiently as possible. The Operations Manager has recently read that many hospitals are using some type of new technology known as radio frequency identification (RFID) and is interested in it as an option. The hospital needs a system that will allow them to track inventory once it is shipped and until it reaches its destination, track equipment upon entering the hospital, track which beds are occupied, and how much medicine is in their inventory. Another requirement is that the information must be in real-time and they will also need to perform a cost analysis of each option.

DISCUSSION TOPICS

1. In your own words, define supply chain. List the four components of the supply chain and explain how the components work together. List five industries which utilize supply chain management (SCM).
2. Explain what total asset visibility (TAV) is and how it works. Can anyone, besides the military, use total asset visibility? If so, which industries would best benefit from using total asset visibility? List the pros and cons of total asset visibility for those industries.
3. Give an example of how a company implemented one or more of the new technologies that were highlighted in the chapter. Then explain how it affected the company's supply chain process.

4. How are the three supply chain management's flows divided? Give an example of how all three are intertwined.
5. How can packaging be a hindrance in an automated supply chain process? What might be some possible solutions?
6. Packaging performs six functions, provide a definition of each functions. Describe how each packaging function impacts both SCM and TAV.
7. What are the limitations of active and passive radio frequency identification (RFID) technology? How do these limitations affect the visibility of a product that moves through the supply chain?
8. In regards to optimization software packages, which software allows organizations to effectively track supplier performance, perform evaluations, assign ratings and drive continual performance improvement? Which software provides real-time visibility and collaboration across the supplier network.?
9. Provide an example of the VISCO software and describe how it impacts each of the three main supply chain management (SCM) flows.
10. Based on your understanding of the topics covered in this chapter, provide an example layout of how to optimize the supply chain as well as incorporate total asset visibility into the an existing logistics process.

REFERENCES

Anderson, N. J. (2001). Challenges of Total Asset Visibility. *Army Logistician, 33*(1). Retrieved from http://www.alu.army.mil/alog/issues/JanFeb01/MS602.htm

Brisbourne, A. (2013, February 25). The Current State Of The Internet Of Things. *Field Technologies Online Magazine*.

Chisholm, G., & Hammons, M. J. (2006). Enabling Total Asset Visibility. *Defense Transportation Journal*. Retrieved from http://www.ndtahq.com/images/pdf/total_asset_visibility.pdf

Doran, D. (2004). Rethinking the Supply Chain: An Automotive Perspective. *Supply Chain Management: An International Journal, 9*(1), 102 - 109.

Gunasekaran, A., & Ngai, E. W. T. (2004). Article. *Virtual Supply-Chain Management, 15*(6), 584–595. Retrieved from http://www.tandf.co.uk/journals

Investopedia. (2015). *Supply Chain Management-SCM*. http://www.investopedia.com/terms/s/scm.asp

Lambert, D. M., Stock, J. R., & Ellram, L. M. (1998). *Fundamentals of Logistics Management*. Irwin/McGraw-Hill.

Lambert, T. (2014). *A Brief History of Transport*. Retrieved from http://www.localhistories.org/transport.html

Lu, L. X., & Swaminathan, J. M. (2013). *Advances in Supply Chain Management*. Retrieved from http://public.kenan-flagler.unc.edu/faculty/lul/LuSwaminathan_SupplyChainManagement_2013.pdf

Naval Studies Board. (1997). *Technology for the United States Navy and Marie Corps, 2000-2035. Becoming a 21st-Century Force*. Washington, DC: National Academy Press.

Perez, H. D. (2015). *Supply Chain Strategies: Which One Hits The Mark?*. Retrieved from http://www.supplychainquarterly.com/topics/Strategy/20130306-supply-chain-strategies-which-one-hits-the-mark/

Porter, M. E. (2008). *Competitive Strategy: Techniques for Analyzing Industries and Competitors*. New York: The Free Press.

Selko, A. (2006). Nissan To Use RFID Enabled RTLS To Automate Supply Chain. *Industry Week: Advancing the Business of Manufacturing*. Retrieved from http://www.industryweek.com/planning-amp-forecasting/nissan-use-rfid-enabled-rtls-automate-supply-chain

Totally Integrated Automation. (2011). *Benefits and Latest Advances of 2D Bar Codes for Consumer Packaged Goods Manufacturers*. Retrieved from http://www.totallyintegratedautomation.com/2011/06/benefits-and-latest-advances-of-2d-bar-codes-for-consumer-packaged-goods-manufacturers/

US Army. (1997). *Army Container Operations, Field Manual (FM) 55-80*. Retrieved from http://archive.org/stream/ost-military-doctrine-fm55_80/fm55_80_djvu.txt

US Army, & Us Marine Corps. (1997). *Field Manual (FM)101-5-1/Marine Corps Reference Publication (MCRP)5-2A, Operational Terms and Graphics*. Retrieved from http://fas.org/man/dod-101/army/docs/fm101-5-1/f545con.htm

US Government Accounting Office (GAO) National Security and International Affairs Division. (1999). *NSIAD*. Washington, DC: Defense Inventory DOD Could Improve Total Asset Visibility Initiative With Results Act Framework.

Vella, D. (2015). *Supply Chain Management: A Needed and Growing Opportunity*. http://www.inboundlogistics.com/cms/article/supply-chain-management-a-needed-and-growing-opportunity/

ADDITIONAL READING

Bragdon, C. (2011). *Transportation Security*. Butterworth-Heinenmann.

Journal, R. F. I. D. (2015). RFID Journal. http://www.rfidjournal.com/faq/show?139

Perez, H. D. (2015). *Supply Chain Strategies: Which One Hits The Mark?* Retrieved from http://www.supplychainquarterly.com/topics/Strategy/20130306-supply-chain-strategies-which-one-hits-the-mark/

Porter, M. E. (2008). *Competitive Strategy: Techniques for Analyzing Industries and Competitors*. New York: The Free Press.

Vella, D. (2015). *Supply Chain Management: A Needed and Growing Opportunity*. http://www.inboundlogistics.com/cms/article/supply-chain-management-a-needed-and-growing-opportunity/

KEY TERMS AND DEFINITIONS

Big Data: Extremely large data sets that may be analyzed computationally to reveal patterns, trends, and associations, especially relating to human behavior and interactions.

End-to-End Process: A term used in many business arenas referring to the beginning and end points of a method or service.

Inventory Management: The process of ensuring that a company always has the products it needs on hand and that it keeps costs as low as possible.

Machine-2-Machine: This describes technology that enables networked devices to exchange information and perform actions without the manual assistance of humans.

Optimization: A process or methodology of creating a system as fully functional as possible.

Supply Chain Management: The management of the flow of goods and services.

Sustainable Development: This refers to economic development that is conducted without depletion of natural resources.

Total Asset Visibility: A capability to provide users with timely and accurate information on the location, movement, status, and identity of units, personnel, equipment, material, and supplies.

Transportation: The act or process of moving people or things from one place to another.

NOMENCLATURE/ABBREVIATION FOR ACRONYMS

BICONS: Double Containers
COO: Chief Operating Officer
DoD: Department of Defense
ESC: Electronic Supply Chain
FM: Field Manual
GAO: Government Accounting Office
GPS: Global Positioning System
INU: Inertial Navigation Unit
ISO: International Standards Organization
IT: Information Technology
M2M: Machine-2-Machine
MCRP: Marine Corps Reference Publications
NSIAD: National Security and International Affairs Division
QUADCONS: Quadruple Containers
RFID: Radio Frequency Identification
RTLS: Real-time Locating Systems
SCM: Supply Chain Management
TAV: Total Asset Visibility
TRICONS: Triple Containers
VIN: Vehicle Identification Number

Chapter 5
Variable Selection in Multiple Linear Regression Using a Genetic Algorithm

Javier Trejos
University of Costa Rica, Costa Rica

Mario A. Villalobos-Arias
University of Costa Rica, Costa Rica

Jose Luis Espinoza
Technological Institute of Costa Rica, Costa Rica

ABSTRACT

In this article it is studied the application of a genetic algorithm in the problem of variable selection for multiple linear regression, minimizing the least squares criterion. The algorithm is based on a chromosomic representation of variables that are considered in the least squares model. A binary chromosome indicates the presence (1) or absence (0) of a variable in the model. The fitness function is based on the adjusted square R, proportional to the fitness for chromosome selection in a roulette wheel model selection. Usual genetic operators, such as crossover and mutation are implemented. Comparisons are performed with benchmark data sets, obtaining satisfying and promising results.

INTRODUCTION

Model selection is a very important task in Statistics, as it can be viewed in several ways: a dimension reduction procedure, an important feature selection task. In any way, it simplifies the situation modeled and gives sense to the analysis of the data.ds

In multiple linear regression, a numerical variable Y is modeled by a linear combination of explanatory numerical variables $X_1, X_2, \ldots, X_p$. For minimizing the sum of squares of the differences between Y and this linear combination, it is well known (Draper & Smith (1968), Tomassone, Audrain, Lesquoy, & Millier (1992), Venables & Ripley (1994)) that a solution can be obtained if the X_j are not linearly

DOI: 10.4018/978-1-4666-9644-0.ch005

dependent. In the case of collinearity between these variables, several approaches have been proposed for overcoming the problem: (i) Stepwise regression, selecting the most explanatory variables in a forward or backward greedy procedure; (ii) regularization via principal component analysis, using the independent principal components; (iii) use of metaheuristics for selecting the best explanatory subset of X_j.

There are many other types of regression. In recent years, the tree based algorithms have become very popular, such as CART (Breiman, Friedman, Ohlsen, & Stone, 1984; Gordini & Veglio, 2014). Also, fuzzy procedures are useful in some cases (de los Cobos, 2011). PLS regression is a generalization for the case of several variables to be explained. In the section of additional readings the authors have put several references to these and other situations.

Nonlinear regression is appropriate in cases where the linear model does not explain correctly the variable Y. The most well known method for non linear regression is the Gauss-Newton method, based on a first-order Taylor approximation, and iteratively approximating the solution. Another method is the gradient one, that looks for the steepest descent at each point. And the Marquardt method is some kind of combination of the preceding methods. Even if these approaches are reasonable, there is no guarantee of reaching the best least squares solution, moreover, in some cases the iterations may not converge at all. For this, somewhere else the authors have also applied metaheuristics with very good results in nonlinear regression, using simulated annealing and tabu search (Villalobos & Trejos, 2000; Villalobos, Trejos, & de los Cobos, 2006).

The problem of selecting explanatory variables in linear regression was tackled by using a genetic algorithm (Holland, 1975), a metaheuristic that has shown to behave properly in many difficult optimization methods (Vasant, 2013). For this, it is necessary to define an appropriate fitness function. In our case, a balance between two conflicting objectives is needed: (i) want to include all variables that have legitimate predicting skill; (ii) want to exclude any redundant or sample-specific variables. Of course, there is no single definition of "best", and it is well known that different algorithms may produce different solutions and, in linear regression, problems are magnified by correlation among predictors.

Among the different criteria that may be used, in this article it is used one such that it increases only if new variables included add significantly to the model since it is not good to add too many explanatory variables that do not seem to contribute much to the model. Of course, this is not the case of the determination coefficient, since it increases with the number of variables in the model.

There are several criteria that may be used to tackle this problem: the Adjusted R square, the Mallow's statistic, the Press statistic, the Akaike information criterion (AIC), the Bayesian information criterion (BIC), and so on. In this investigation, it was used the Adjusted R square, which can decline in value if the contribution to the explained deviation by the additional variable is less than the impact on the degrees of freedom. In the background section these criteria are developed in more detail.

BACKGROUND

In this section the concepts needed to present our regression approach are developed. First, details on linear regression are presented, illustrate it and present several different ways to do this kind of regression. Particularly, some other approaches for variable selection in linear regression are presented. Second, different numerical criteria for measuring the quality of the selected variables in the regression model are presented. Finally, general genetic algorithms are developed.

Linear Regression

Given a set of p numerical explanatory variables $X_1, X_2, \ldots, X_p$ and another numerical variable to be explained Y, general regression models aim to find a way to explain Y from the explanatory variables. This explanation could be for descriptive purposes or in a predictive goal. Also, it could be parametric, as it is the case showed in this chapter, or non parametric.

In classical multiple linear regression one seeks for a set of p coefficients $\beta_1, \beta_2, \ldots, \beta_p$ such that Y can be written as a linear combination of $X_1, X_2, \ldots, X_p$:

$$Y = \beta_1 X_1 + \beta_2 X_2 + \cdots + \beta_p X_p + \varepsilon$$

where ε is an error term. The goal is to find suitable values of $\beta_1, \beta_2, \ldots, \beta_p$ so that it is minimized the sum of squares criterion:

$$\left\| Y - \sum_{j=1}^{p} \beta_j X_j \right\|^2 = \sum_{i=1}^{n} \left(y_i - \vec{x}_i^{\,t} \vec{\beta} \right)^2 \rightarrow (\beta_1, \ldots, \beta_p).$$

Solution of the problem, that is values for $\beta_1, \beta_2, \ldots, \beta_p$, is called *regression coefficients.*

Graphically, the situation can be illustrated by Figure 1. Variable Y is intended to be explained by explanatory variables ... Eventually, the user might want to predict the value of Eventually, the user might want to predict the value of Y for a new observation.

One of the variables $X_1, X_2, \ldots, X_p$ might be a constant, in that case it is said that the model has an intercept. In matricial terms, linear regression can be illustrated by Figure 2, where X is the matrix that contains variables X_j in its columns.

Figure 1. Illustration of the explanatory situation tackled by regression models

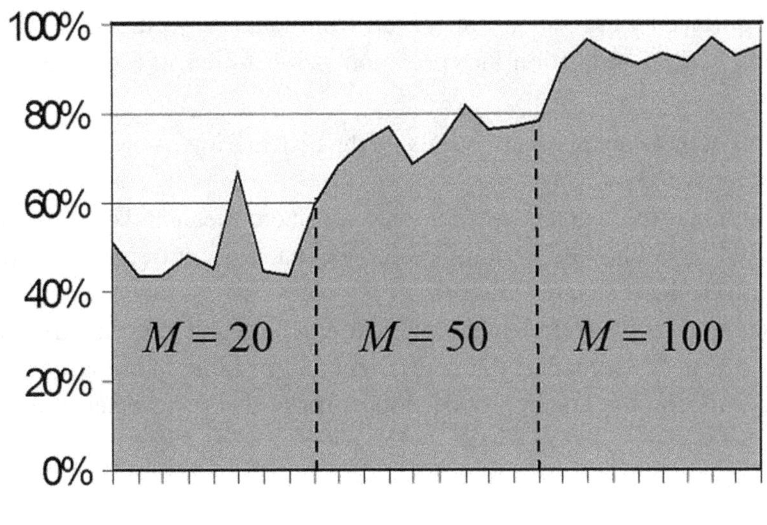

Figure 2. Illustration linear regression in matricial terms

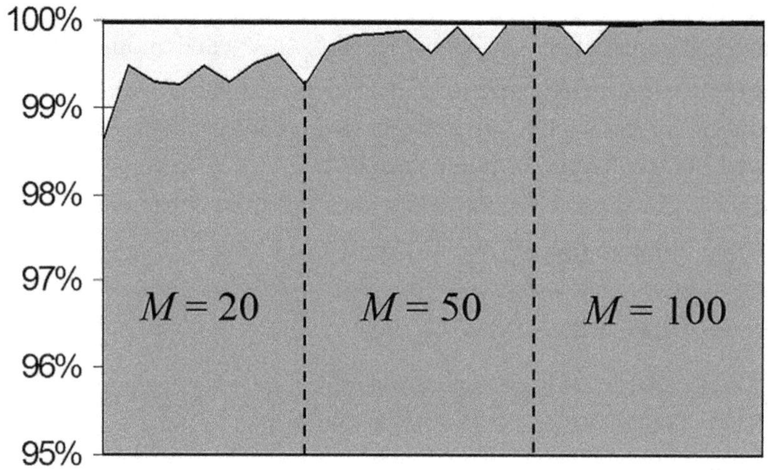

If all explanatory variables are included in the model, then it is well known (Draper & Smith, 1968) that the solution to this problem is given by:

$$(\beta_1,\ldots,\beta_p)^t = (X^tX)^{-1}X^tY.$$

This definition supposes, of course, that the inverse of matrix X^tX exists, that is, the explanatory variables are not collinear, or -in statistical terms- they are not highly correlated. It is said that Y is modeled by $\hat{Y}$ where

$$\hat{Y} = \beta_1X_1 + \beta_2X_2 + \cdots + \beta_pX_p.$$

Linear regression can be seen as a projection method, since explanatory variables generate a vector subspace in $\mathbb{R}^n$, and the problem can be stated as the projection of vector Y in that subspace. Regression coefficients are then defined by the coefficients of the orthogonal projection of Y in the subspace generated by $X_1, X_2,\ldots,X_p$. This projection interpretation is illustrated in Figure 3, for two explanatory variables.

It is worth noting that there are no assumptions on the distribution of the errors ε for minimizing the least squares criterion given above.

However, if one assumes these errors are Gaussian with zero mean and constant standard deviation, then, using the maximum likelihood criterion, it is proved that the solution is given by the same regression coefficients as for the least squares criterion.

In practice not all variables are interesting, or there might be highly correlated explanatory variables, so the numerical problem of estimating the coefficients is unstable. That is the reason why several methods of variable selection have been proposed to estimate the best subset of explanatory variables (see Figure 4).

Figure 3. Illustration of the linear regression model as a projection method

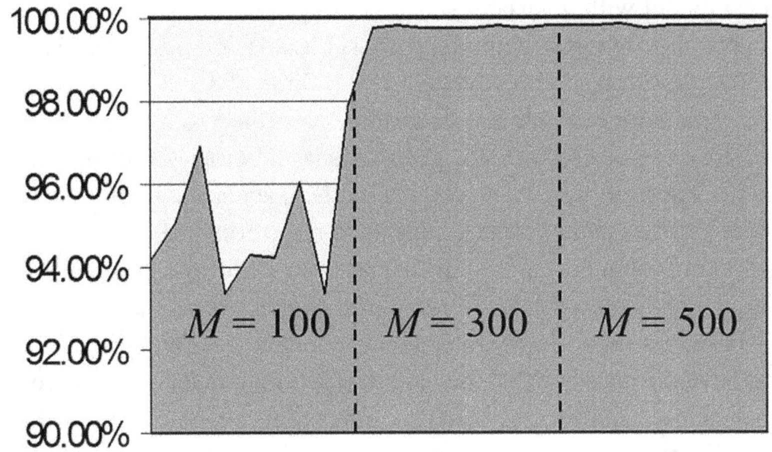

Figure 4. Scheme for variable selection

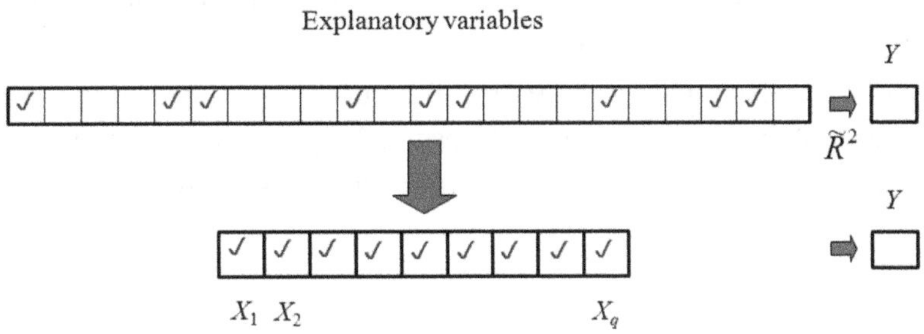

One of the most popular methods for doing so are the stepwise methods, forward or backward. In the former, the first step calculates linear ordinary regression of Y on all single explanatory variables, and the one with the best determination coefficient is selected; in the next step, a new variable is introduced into the model if the multiple correlation coefficient is maximum, among all possible pairs of explanatory variables such that the first variable chosen is always in the model. The iterations stop according to a criterion, and a linear combination of the q explanatory variables introduced in the model is presented as the model.

Backward stepwise regression is similar, but extracting the worst variable in each step. At the beginning of the procedures, all the variables enter in the model, and each one is deleted, one-by-one, That is, p model are built, each with p -1 explanatory variables. Criterion is calculated and the best model is maintained. Again, p - 1 models are built deleting one variable in each one and the best one is retained. The procedure continues until a stop criterion ir reached, usually related to a hypothesis test based on the Fisher distribution, that is, when no predictor produces the F-ration statistic greater than a threshold. One disadvantage of the backward stepwise regression is that it supposes that, at the beginning, the regression coefficients with the complete model can be computed, which is not the case when there is strong collinearity between the explanatory variables.

These stepwise methods have a problem, which is that they are greedy methods (Cormen, Leiserson, Rivest, & Stein., 2009), that is, the optimum is a local optimum since the optimization is made without revising the variables that are already in the model. That is, the model with q explanatory variables is not necessarily the best model with q variables.

It is important to note also that tree methods, such as classification and regression tree (CART), are usually based on greedy algorithms (Breiman et al., 1984). Indeed, CART builds a tree of classification, by selecting the most explanatory variable at each step, according to a variance related criterion. The algorithm does not revise its choices, even if it prunes the tree before reporting a solution. The method is very popular at the present time, particularly in the medical sciences, mainly because it does not need special conditions to be verified and interpretations are straightforward.

Regularized regression is another approach that tries to avoid the problem of collinearity in the explanatory variables, which leads to undefinition of the coefficients. One of the most popular regularization procedures is based on the use of principal components (Joliffe, 2002; Jackson, 1991). Indeed, principal components have the property that they are independent, that is, non correlated. The procedure first calculates these principal components, and then regression coefficients are calculated over the principal components. Finally, regression coefficients of the regression of Y over the original explanatory variables are calculated using the principal components equations based on the eigenvectors that served to define the principal components.

This regularization is not a variable selection procedure but can be used for the case that singularities of X^tX arise and the user wants a complete model.

Penalization methods are also popular procedures for variable selection in regression. Among these, ridge regression has shown to have good behavior (Hoerl, 1962). The idea of ridge regression is to shrink the estimated coefficients towards zero. A tuning parameter $\lambda \geq 0$ is introduced in the penalty term, so the criterion to be minimized is now

$$\left\| Y - \sum_{j=1}^{p} \beta_j X_j \right\|_2^2 + \lambda \left\| \vec{\beta} \right\|_2^2 = \sum_{i=1}^{n} \left(y_i - \vec{x}_i^t \vec{\beta} \right)^2 + \lambda \sum_{j=1}^{p} \beta_j^2 \rightarrow \left(\beta_1, \ldots, \beta_p \right)^{\text{ridge}}.$$

It is proved that the solution is given by:

$$\left(\beta_1, \ldots, \beta_p \right)^{\text{ridge}} = \left(X^tX + \lambda I \right)^{-1} X^tY.$$

Note that the inclusion of λ makes problem to be non-singular even if X^tX is not invertible.

Solution is very similar to the ordinary least squares (OLS) one given above, but with the addition of a "ridge" down the diagonal, which has the following effects:

- $\left(\beta_1, \ldots, \beta_p \right)^{\text{ridge}}$ tends to the OLS coefficients $\left(\beta_1, \ldots, \beta_p \right)$ as λ tends to 0.
- $\left(\beta_1, \ldots, \beta_p \right)^{\text{ridge}}$ tends to (0,...,0) as λ tends to ∞.

In the case of an orthonormal design matrix, then OLS and ridge regression coefficients are related by the following equation that illustrated the feature of shrinking that has ridge regression:

$$\hat{\beta}_j^{\text{ridge}} = \frac{\hat{\beta}_j^{\text{OLS}}}{1+\lambda}.$$

An alternative to ridge regression is the lasso (Tibshirani, 1996), using the L_1 norm instead of the L_2 metric used in ridge regression. That is, the minimization criterion is:

$$\left\| Y - \sum_{j=1}^{p} \beta_j X_j \right\|_1 + \lambda \left\| \vec{\beta} \right\|_1 = \sum_{i=1}^{n} \left| y_i - \vec{x}_i^t \vec{\beta} \right| + \lambda \sum_{j=1}^{p} \left| \beta_j \right| \to \left(\beta_1, \ldots, \beta_p \right)^{\text{lasso}}.$$

Lasso means *least absolute selection and shrinkage operator*, hence it has also the property of shrinking as in ridge regression. The tuning parameter λ has the same effects in the lasso as in ridge regression, whether it tends to 0 or to infinity. However, when using the L_1 norm the penalty causes some coefficients to be shrunken to zero exactly. Therefore, lasso is more suitable than ridge regression for variable selection. In practice, parameter λ is tuned via cross-validation.

Criteria

There are several ways of measuring the quality of a given regression model. Determination coefficient does not serve, since it only grows with the inclusion of new variables. The explanation for this is very simple from the geometric point of view; indeed, the space of projection is "bigger" with more explanatory variables, hence the error measured with R^2 is lesser as more variables are included in the model.

A first idea is then a modification of R^2. These modifications are called *adjusted R square* and noted $\bar{R}^2$; however, there is no single definition of adjusted R square (Leach & Henson, 2007).

A measure of quality of a model given by a subset of q explanatory variables, is the maximization of the adjusted R square, defined by:

$$\bar{R}^2 = 1 - \left(\frac{n-1}{n-q} \right) \left(1 - R_q^2 \right),$$

where R_q^2 is the determination coefficient of the linear regression of Y on the q selected variables; as above n is the sample size. This formulation of the adjusted R square was proposed by Wherry, as indicated in Leach & Henson (2007). These authors cite 13 more formulations for the adjusted R square:

Smith,

$$\bar{R}^2 = 1 - \left(\frac{n}{n-q} \right) \left(1 - R_q^2 \right).$$

Ezekiel (also called McNemar),

$$\bar{R}^2 = 1 - \left(\frac{n-1}{n-q-1}\right)\left(1 - R_q^2\right).$$

Olkim & Pratt,

$$\bar{R}^2 = R^2 - \left(\frac{q-2}{n-q-1}\right)\left(1 - R_q^2\right) - \left(\frac{2(n-3)}{(n-q-1)^2}\right)\left(1 - R_q^2\right).$$

Pratt,

$$\bar{R}^2 = 1 - \left(\frac{(n-3)(1-R_q^2)}{n-q-1}\right)\left(1 + \frac{2(1-R_q^2)}{n-q-2.3}\right).$$

Claudy (1),

$$\bar{R}^2 = \left(2\rho - R_q^2\right)^2$$

where ρ is estimated with the square root of the Ezekiel value.

6. Claudy (2),

$$\bar{R}^2 = 1 - \left(\frac{n-1}{n-q-1}\right)\left(\frac{n-2}{n-q-2}\right)\left(\frac{n-1}{n}\right)\left(1 - R_q^2\right).$$

7. Claudy (3),

$$\bar{R}^2 = 1 - \left(\frac{(n-4)(1-R_q^2)}{n-q-1}\right)\left(1 + \frac{2(1-R_q^2)}{n-q+1}\right).$$

8. Lord (1),

$$\bar{R}^2 = 1 - \left(\frac{n+q-1}{n-q-1}\right)\left(1 - R_q^2\right).$$

9. Lord (2),

$$\bar{R}^2 = 1 - \left(\frac{(n+q+1)(n-1)}{(n-q-1)n} \right) \left(1 - R_q^2\right).$$

10. Darlington & Stein,

$$\bar{R}^2 = 1 - \left(\frac{n-1}{n-q-1} \right) \left(\frac{n-2}{n-q-2} \right) \left(\frac{n+1}{n} \right) \left(1 - R_q^2\right).$$

11. Browne,

$$\bar{R}^2 = \frac{(n-q-3)\rho^4 + \rho^2}{(n-2q-2)\rho^2 + \rho}.$$

12. Rozeboom (1),

$$\bar{R}^2 = 1 - \left(\frac{n+q}{n-q} \right) \left(1 - R_q^2\right).$$

13. Rozeboom (2),

$$\bar{R}^2 = \rho^2 \left[1 + \left(\frac{q}{n-q-2} \right) \left(\frac{1-\rho^2}{\rho^2} \right) \right]^{-1}.$$

Wherry's formula is used since, in our opinion, it makes a balance between explanatory power of variables in the model and the quantity of variables in it; also, it is a simple formula to implement and to understand. According to Field, Miles, & Field. (2012), this is the formula used in R software.

As it was said, there many other quality criteria for variable selection in linear regression (Zhang, Li, & Tsai, 2010). Among the most popular are Akaike Information Criterion (*AIC*) and Bayesian Information Criterion (*BIC*), both with the same first goodness-of-fit term but with a different penalty term, and defined as:

$$AIC = -2\ln(L_q) + 2q,$$
$$BIC = -2\ln(L_q) + q\ln(n),$$

where L_q is the maximum of the likelihood function of the model with the selected q explanatory variables. The penalty term of *BIC* is more stringent than the penalty term of *AIC*, since for $n \geq 8$, $q\ln(n)$ exceeds $2q$.

According to Zhang, et al. (2010), *AIC* is more efficient since it selects the model so that its average squared error is asymptotically equivalent to the minimum offered by the candidate models when the true model is approximated by a family of candidate models, and *BIC* is more consistent since it identifies the true model with probability that approaches to 1 in large samples when a set of candidate models contains the true model. Generally, *BIC* tends to favor smaller models than *AIC*.

PRESS in another selection criterion (Liu, Weiss, Jennrich, & Wenger,1999). Let $\hat{\beta}_{(i)}$ be the estimate of β when the ith observation is deleted and $\hat{y}_{(i)} = x_i^t \hat{\beta}_{(i)}$ be the case deleted predicted value of the response when the independent variable has value x_i. Then *PRESS* is defined as:

$$PRESS = \sum_{i=1}^{n} \left(y_i - \hat{y}_{(i)} \right)^2 .$$

PRESS can be seen as a measure of how well observation y_i is predicted by the given model and the data $Y_{(i)}$, with $Y_{(i)} = (y_1, \ldots, y_{i-1}, y_{i+1}, \ldots, y_n)^t$.

Genetic Algorithms

Among the most popular metaheuristcs for combinatorial optimization, genetic algorithms (Holland (1976) have a special place (Reeves (1995)). Since their introduction in the mid seventies, they have shown to be very effective to solve complex optimization problems in Engineering, Informatics, Economics, etc. (Chis, 2010) Particularly, for Statistics, they have been used for clustering, dimension reduction, and classification, for instance. Very good results were obtained in clustering numerical (Trejos, Murillo, & Piza (1998); Trejos, Villalobos, Murillo, Chavarría, & Fallas, (2014)) and binary variables (Trejos & Piza (2000)), and in constructing Pawlak's rough sets (Pawlak (1991)) for discrimination as well (Espinoza & Trejos (2002)). Mathar (1999) and Groenen, Mathar, & Trejos (2000) report good results in multidimensional scaling, and Bozdogan (2004) in the determination of the number of clusters.

Among the attractive features that genetic algorithms have, is the asymptotic convergence to the global optimum of a combinatorial optimization problem, proved by Rudolph (1994). This proof, based on a homogeneous Markov chain modeling, suggests several hints for the implementation of genetic algorithms, which were followed here. For instance, the best solution present in the chromosomic population must be maintained for the next iteration, that is, the random selection based on a roulette wheel cannot make this best solution disappear.

Originally, Goldberg (1989) proposed a schemata analysis for determining that genetic algorithms were able to find good solutions. This analysis was based on the fact that, from one iteration to the following, some fixed positions in chromosomes repeated, and only some allele changed. These fixed positions were considered stable and may represent features searched on the good solutions. However, there was no proof of convergence. Rudolph's theorem was the first demonstration that genetic algorithms, as simulated annealing, converge asymptotically to the overall solution of the problem.

Flow chart, or block diagram, of the basic genetic algorithm is showed in Figure 5.

Some authors have already used genetic algorithms for variable selection. For instance, Leardi, Boggia, & Terrile (1992) in the context of feature selection.

Figure 5. Flow chart of the basic genetic algorithm

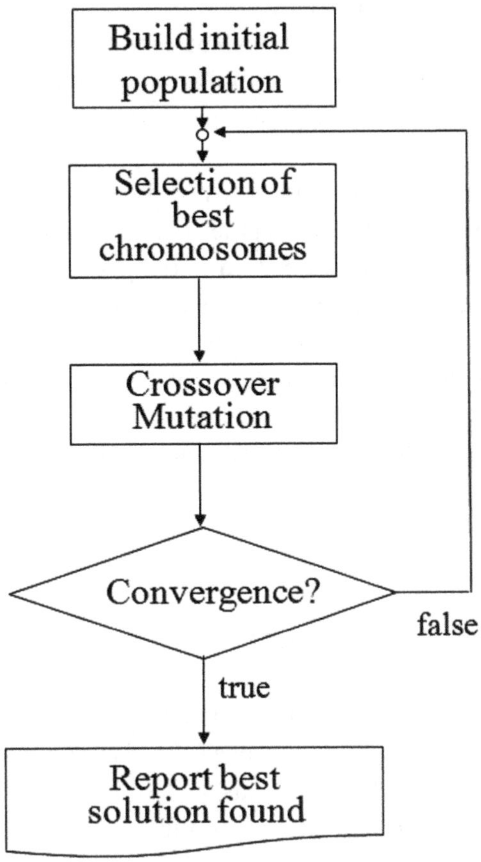

In the last thirty years, several metaheuristics have arose for solving combinatorial optimization problems, many of them with good results. Among these heuristics, simulated annealing (Kirkpatrick el al., 1983), tabu search (Glover, 1989), ant colonies (Dorigo, Maniezzo, & Colorni, 1996), and particle swarms (Kennedy & Eberhart, 2000) are very popular and effective (Reeves, 1995). The first two are based on concepts of neighborhoods, and the latter on populations of solutions, as for genetic algorithms.

Simulated annealing is based on the annealing physical procedure, introducing control parameters that play the role of temperature, that controls the probability of acceptation of new states of the optimization problem that have a worst cost function than the current state. Cooling is made slowly, and it is proved (Laarhoven & Aarts, 1988), with a Markov chain model, that it converges asymptotically to the global optimum. Efficient implementations lead to very good results (Aarts & Korst, 1990). In statistical problems, the authors have applied it to clustering in the numerical (Trejos et al, 1998), binary (Piza et al, 2000) or two-mode case (Trejos & Castillo, 2000), non linear regression (Villalobos & Trejos, 2000b) and multidimensional scaling (Trejos, Castillo, González, & Villalobos, 2000; Groenen et al., 2000).

Tabu search looks for the best state in the neighborhood of the current state, handling a tabu list; the best state is chosen unless it is in the tabu list. This list usually contains encodings of movements made during the iterations, and the intention is to avoid previously visited stated, in order to cover the maximum of the search space. The researchers have also used it for numerical clustering (Trejos et al,

1998), two-mode clustering (Castillo & Trejos, 2002), multidimensional scaling (Villalobos & Trejos, 2000a) and non linear regression (Villalobos et al., 2006).

Ant colony optimization handles a population of solutions, related to artificial ants, and the main idea is that good solutions reinforce the probability of forming these solutions, using a simulated pheromone trail that controls this probability. In numerical clustering very good results (Trejos, Murillo, & Piza, 2004) were obtained.

Finally, particle swarm optimization (PSO) uses the principle of moving in a multidimensional space a set of solutions, called particles. Moving in the space is ruled by three components: imitation to the overall best particle, imitation to the best position in the past of each particle, and inertia from last movement. In numerical clustering, very good results were also obtained (Trejos & Villalobos, 2007). PSO has been used for multiple regression (Chandra-Satapathy, et al., 2009) in a prediction context.

In Trejos et al. (2014) several comparisons are made of the above four metaheuristics in the numerical clustering problem for minimizing the intra-variance criterion, with very good results with them all, much better than classical methods such as Ward's hierarchical clustering or *k*-means partitioning.

METHODOLOGY

Genetic Algorithm for Variable Selection

In the genetic algorithm model presented here, the explanatory variable indicators are coded as chromosomes that define the presence (coded as 1) or absence (coded as 0) of that variable in the linear model. That is, a chromosome is a p binary vector. As it was said above, the fitness of each chromosome will be defined as a function of the adjusted R squared. In the implementation it is used a roulette wheel proportional to the fitness as the chromosome selection procedure.

Genetic operators are mutation and crossover. A mutation is done selecting (with a uniform distribution) a chromosome with probability p_{m1} and an allele with probability p_{m2}; hence, mutation probability is $p_m = p_{m1}p_{m2}$. A mutation consists on shifting the values of 0 into 1 or vice-versa. On the other hand, crossover is made with probability p_c and it is defined as the combination of the selected parents that interchange their alleles in one single cross point.

Iterations are performed with a population of chromosomes initialized randomly, until a stop criterion is satisfied, that is, stop if the variance of the population's fitness is less than a given threshold, or a maximum number of iterations is attained. A value of 0.00001 was used as the stop threshold.

Parameters to be tuned are the population size M, probabilities of crossover p_c and of mutation p_m (it was used $p_{m1} = p_{m2}$, hence $p_m = 2p_{m1}$). It is also used a tolerance number for non singular matrices of 100 000 000.

Evaluation of the Method

Several evaluations have been performed of the method. In this article, it is presented an experiment on the most sensible parameters: M, p_c and of mutation p_m, using some standard benchmark data sets.

Table 1. Longley data set

Employed Y	GNP deflator X_1	GNP X_2	Unemployed X_3	Armed Forces X_4	Population X_5	Year X_6
60323	83.0	234289	2356	1590	107608	1947
61122	88.5	259426	2325	1456	108632	1948
60171	88.2	258054	3682	1616	109773	1949
61187	89.5	284599	3351	1650	110929	1950
63221	96.2	328975	2099	3099	112075	1951
63639	98.1	346999	1932	3594	113270	1952
64989	99.0	365385	1870	3547	115094	1953
63761	100.0	363112	3578	3350	116219	1954
66019	101.2	397469	2904	3048	117388	1955
67857	104.6	419180	2822	2857	118734	1956
68169	108.4	442769	2936	2798	120445	1957
66513	110.8	444546	4681	2637	121950	1958
68655	112.6	482704	3813	2552	123366	1959
69564	114.2	502601	3931	2514	125368	1960
69331	115.7	518173	4806	2572	127852	1961
70551	116.9	554894	4007	2827	130081	1962

Longley Data

Longley (1967) studied labour data in the period 1947-1962. Variables are X_1: prices, deflated with base the year 1954; X_2: GNP or gross national product; X_3: number of unemployed; X_4: number of people in the armed forces; X_5: 'noninstitutionalized' population over 14 years of age; X_6: year; Y: number of people employed. Explanatory variables are known to be highly collinear, so the dataset is often used for testing linear regression methods. Complete data are presented in Table 1.

There are 126 possible combinations of selecting at least one explanatory variable (considering models that may include the constant). By an exhaustive exploration it was found that the best linear combination is defined by X_0, X_2, X_3, X_4 and X_6. It has an $\bar{R}^2$ of 0.9937.

In 100 applications of the program with a population of $M = 10$, 300 as the maximum of iterations, $p_m = 0.02$, and $p_c = 0.5$, it was found the optimal solution in 88% of the times.

CPU-Machine Data

The data table (Ein-Dor & Feldmesser (1987)) concerns Relative CPU Performance Data and has 6 explanatory, integer variables observed on 209 instances. The variables are MYCT = X_1: machine cycle time in nanoseconds; MMIN = X_2: minimum main memory in kilobytes; MMAX = X_3: maximum main memory in kilobytes; CACH = X_4: cache memory in kilobytes; CHMIN = X_5: minimum channels in units; CHMAX = X_6: maximum channels in units; PRP = Y: published relative performance of the CPU.

As for the Longley data set, there are 126 possible combinations of parameters (including the constant). By an exhaustive of all these combinations, the best one is the model without X_5, associated to an $\overline{R}^2$ of 0.8615.

Executing 100 times the program with $M = 10$, and the rest of parameters as for the Longley data set, it was found the best solution defined by X_0, X_1, X_2, X_3, X_4 and X_6 in 92% of the cases.

Triazines Data

With 60 continuous variables observed on 186 cases, the data set intends to predict the activity of triazines from the descriptive structural attributes (Hisrt, King, & Sternberg, (1994), King, Hirst, & Sternberg, (1995)). There are around 10^{18} combinations of explanatory variables, so it is not reasonable to exhaustively explore all these combinations.

Executing the program 100 times with $M = 40$, a maximum number of iterations of 400, $p_m = 0.02$, and $p_c = 0.5$, the best solution founded has an $\overline{R}^2$ of 0.3356. This solution contains X_4, X_6, X_8, X_9, X_{10}, X_{11}, X_{23}, X_{32}, X_{33}, X_{34}, X_{36}, X_{37}, X_{47}, X_{48}, X_{49}, and X_{56}, and it was found in 45% of the times.

Sensibility to Some Critical Factors

It is well known that most optimization metaheuristics are sensible to parameters tuning, see for example Ries, Beullens, & Wang (2013).

For studying the sensibility of the method to the clue factors, three factors were considered at different levels:

- three sizes of the genetic population M (20, 50 and 100);
- three values for the mutation probability p_m (0.01, 0.1 and 0.2);
- three values for the crossover probability p_c (0.3, 0.5 and 0.7).

The attraction rate τ was measured of the suboptimal states found by the method in 100 executions, for each of the 27 combinations. For numerical reasons, it was added 0.1 to all the values of fitness reported in this section.

Sensibility for the Longley Data

Table 2 show the results for Longley data, with a table for each population size M; rows and columns correspond to the probabilities p_m and p_c, respectively.

Table 2. Attraction rate for Longley dataset for different values of the 3 factors

M =20				M =50				M =100			
p_m	p_c			p_m	p_c			p_m	p_c		
	0.3	0.5	0.7		0.3	0.5	0.7		0.3	0.5	0.7
0.01	42	32	34	0.01	63	60	64	0.01	87	93	83
0.1	38	33	50	0.1	53	56	58	0.1	83	82	79
0.2	28	26	30	0.2	53	62	57	0.2	84	86	86

Mean values of the attraction rates for the three population sizes are:

$$\bar{\tau}_{M=20} = 34.8,$$
$$\bar{\tau}_{M=50} = 58.4,$$
$$\bar{\tau}_{M=100} = 84.8.$$

It was performed an analysis of variance (ANOVA), obtaining an F statistic of 1905 with a p-value less than 0.0001, hence the null hypothesis is rejected: it seems that the population size is has an influence on the quality of results, this means that for the three values of M, the best is $M = 100$.

When studying the quality of results by the mean value of fitness in 100 executions for each choice of the three parameters, there were obtained the results shown in Table 2. The ANOVA on factor M, gives an F statistic of 137.1 with a p-value < 0.0001, which leads to rejecting the null hypothesis. Again, it is clear that the population size has an influence on the results.

Since the best value found is 1.099767, it was compared the mean values of fitness of Table 3 with this best value, for calculating the percentage of its value which is found for several levels. This leads to the tables shown in Table 4, associated to Figure 6.

Sensibility for the CPU-Machine data

As for the Longley data, it was performed the same experiment with 3 factors, with 3 levels each. For each one, it was made 100 executions of the method, with a maximum of 200 iterations. The best fitness found was 0.917388.

Table 3. Mean values of fitness for Longley data, for the different levels of the three factors

$M=20$				$M=50$				$M=100$			
p_m	p_c			p_m	p_c			p_m	p_c		
	0.3	0.5	0.7		0.3	0.5	0.7		0.3	0.5	0.7
0.01	0.56	0.48	0.48	0.01	0.75	0.81	0.85	0.01	0.10	1.06	1.02
0.1	0.53	0.50	0.74	0.1	0.76	0.80	0.90	0.1	1.00	1.03	1.01
0.2	0.49	0.48	0.65	0.2	0.84	0.85	0.86	0.2	1.07	1.02	1.05

Table 4. Percentage of the best value of fitness for the mean fitness average for different levels of the 3 factors

$M=20$				$M=50$				$M=100$			
p_m	p_c			p_m	p_c			p_m	p_c		
	0.3	0.5	0.7		0.3	0.5	0.7		0.3	0.5	0.7
0.01	51%	44%	44%	0.01	68%	74%	77%	0.01	91%	96%	93%
0.1	48%	45%	67%	0.1	69%	73%	82%	0.1	91%	94%	92%
0.2	45%	44%	60%	0.2	76%	77%	78%	0.2	97%	93%	95%

Figure 6. Longley data - Illustration of the percentage of the best value of fitness for the mean fitness average for three levels of the population size

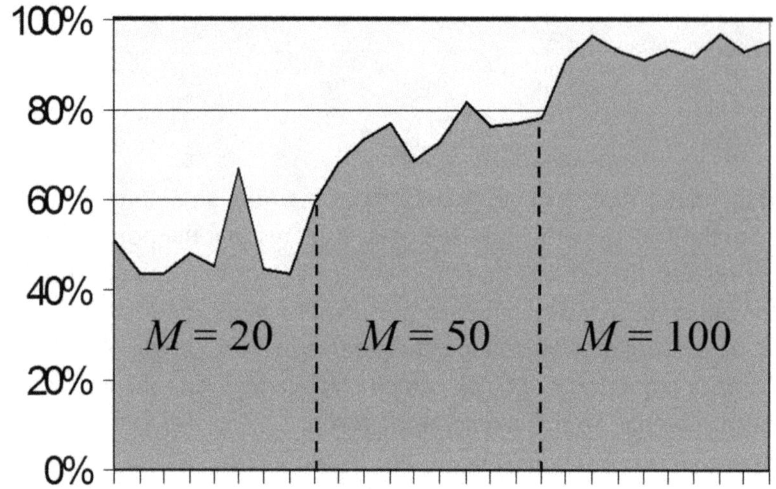

The attraction rates of the optima, for each level of the parameters, are shown in Table 5.

Mean values of the attraction rates for the three population sizes are:

$$\overline{\tau}_{M=20} = 34.9,$$
$$\overline{\tau}_{M=50} = 63.2,$$
$$\overline{\tau}_{M=100} = 83.6.$$

For comparing these means, ANOVA gives $F = 73.37$ with p-value < 0.0001, hence the null hypothesis is rejected again; it seems that the population size also has an influence in the quality of results, and hence the biggest value of M is the best.

The average of fitness (see Table 6), in all cases, is near the best value of 0.917388. However, there is a difference when $M = 100$, since the ANOVA gives $F = 25.25$ with p-value < 0.0001.

Figure 7 shows the percentage of approximation of the value in table 5 to the best value 0.917388.

Table 5. Attraction rates for CPU-Machine data, for different levels of 3 factors

$M=20$				$M=50$				$M=100$			
p_m	p_c			p_m	p_c			p_m	p_c		
	0.3	0.5	0.7		0.3	0.5	0.7		0.3	0.5	0.7
0.01	24	31	32	0.01	50	59	58	0.01	70	74	83
0.1	34	36	44	0.1	57	69	66	0.1	76	86	89
0.2	32	40	41	0.2	54	80	76	0.2	84	95	89

Table 6. Mean values of fitness for CPU-Machine data, for the different levels of the three factors

M =20				*M* =50				*M* =100			
p_m	p_c			p_m	p_c			p_m	p_c		
	0.3	**0.5**	**0.7**		**0.3**	**0.5**	**0.7**		**0.3**	**0.5**	**0.7**
0.01	.9048	.9127	.9109	0.01	.9148	.9159	.9160	0.01	.9170	.9140	.9169
0.1	.9107	.9127	.9108	0.1	.9162	.9139	.9166	0.1	.9170	.9173	.9174
0.2	.9128	.9139	.9106	0.2	.9138	.9171	.9171	0.2	.9172	.9174	.9174

Figure 7. CPU Machine data - Illustration of the percentage of the best value of fitness for the mean fitness average for three levels of the population size

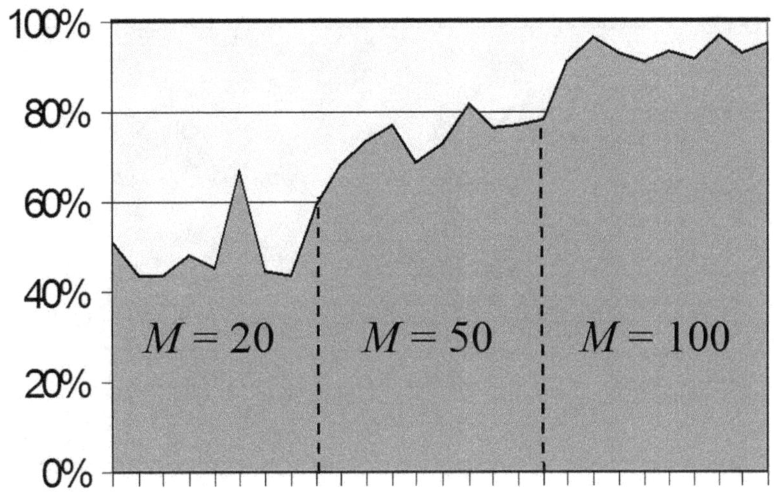

Sensibility for the Triazines data

Again, it was applied the method 100 times on each of the 27 combinations of parameters, with a maximum of 200 iterations. Parameter levels were:

- size of genetic population *M*: 100, 300 and 500;
- the same values for mutation probability p_m: 0.01, 0.1 and 0.2;
- the same values for crossover probability p_c: 0.3, 0.5 and 0.7.

Table 7 shows the best fitness values, highlighting with boldface the better ones.

For each of the 27 cases, the mean value of fitness has a percentage of approximation with respect to the best fitness value, in 100 executions, as shown in Table 8. An ANOVA obtains $F = 78.70$ and *p*-value < 0.0001, so again population size has an influence in the quality of results. Figure 8 illustrates this situation.

Table 7. Best fitness values for the triazines data set

M =100				M =300				M =500			
p_m	p_c			p_m	p_c			p_m	p_c		
	0.3	0.5	0.7		0.3	0.5	0.7		0.3	0.5	0.7
0.01	1.0521	**1.0525**	1.0519	0.01	1.0528	1.0526	1.0528	0.01	1.05249	1.05266	1.05255
0.1	1.0521	1.0520	1.0520	0.1	**1.0530**	105263	1.0525	0.1	1.05295	1.05241	1.05279
0.2	1.0517	1.0521	1.0522	0.2	1.0522	1.0524	1.0525	0.2	1.05231	**1.05302**	1.05251

Figure 8. Triazines data - Illustration of the percentage of the best value of fitness for the mean fitness average for three levels of the population size

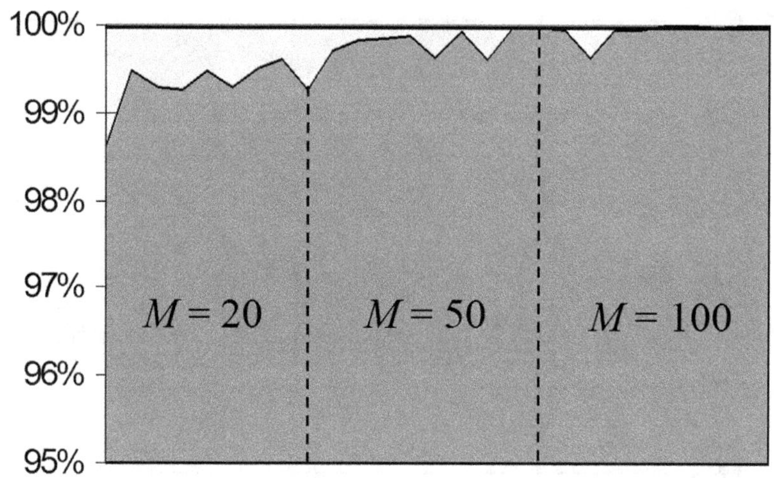

Table 8. Percentage of approximation to the best value of fitness for the triazines data

M =100				M =300				M =500			
p_m	p_c			p_m	p_c			p_m	p_c		
	0.3	0.5	0.7		0.3	0.5	0.7		0.3	0.5	0.7
0.01	94.15	95.06	96.90	0.01	99.75	99.81	99.80	0.01	99.80	99.84	99.86
0.1	93.35	94.27	94.24	0.1	99.75	99.79	99.80	0.1	99.79	99.86	99.84
0.2	96.04	93.36	97.87	0.2	99.82	99.80	99.80	0.2	99.81	99.79	99.85

DISCUSSION

It was implemented the above method for selecting variables in linear regression for testing its performance. The method is based on a classical genetic algorithm respecting the Rudolph's asymptotic convergence conditions, with binary chromosomes that correspond to the presence/absence of explanatory variables in the linear model. Parameters to be tuned are the population size, crossover and mutation probabilities, a tolerance for the condition number of the variance matrix, and the convergence threshold.

According to the results, the population size seems to be critic, since larger populations give better results, until an optimum size. Of course, probabilities are also important, but in the sensibility analysis they seem not to be critical.

Results obtained so far are promising, giving better models than popular stepwise regression or regularization by principal components.

FUTURE RESEARCH DIRECTIONS

For this first investigation, a version of the adjusted R square was used. In further research, other criteria, like AIC and BIC, will be used in order to assess the capability of the method to select the variables that correspond to a good model.

A Monte Carlo experiment with some controlled factors on the data sets generated, will also be performed. Factors will be: the number n of objects, the number of explanatory variables q in the model, the number of noise explanatory variables $p - q$, correlations, or degree of correlation, between some explanatory variables.

Also, it can experimented with some hybridization of the method with other metaheuristics, as it is used in some applications (Yin, 2013). For instance, metaheuristics methods based in neighborhoods may be used with our genetic algorithm by defining a neighbor solution by a mutation (swap between 0 and 1 in the chromosomic representation) and using the Metropolis rule for deciding on the acceptation of the new solution.

CONCLUSION

The genetic algorithm presented here has shown to be able to solve the model selection problem in multiple linear regression. Some parameters seem to be critical, such as the size of the genetic population.

ACKNOWLEDGMENT

Authors would like to express their gratitude for the valuable remarks made by seven referees, who helped to substantially improve this chapter. Research was partially funded by University of Costa Rica, project CIMPA 821-B1-121, and by the Costa Rica Institute of Technology.

REFERENCES

Aarts, E., & Korst, J. (1990). *Simulated Annealing and Boltzmann Machines. A Stochastic Approach to Combinatorial Optimization and Neural Computing*. Chichester, UK: John Wiley & Sons.

Bozdogan, H. (2004). Intelligent statistical data mining with information complexity and genetic algorithms. In H. Bozdogan (Ed.), *Statistical Data Mining & Knowledge Discovery* (pp. 15–56). Boca Raton, FL: Chapman & Hall/CRC.

Breiman, L., Friedman, J. H., Ohlsen, R. A., & Stone, C. J. (1984). *Classification and Regression Trees*. Belmont: Wadsworht.

Castillo, W., & Trejos, J. (2002). Two-mode partitioning: review of methods and application of tabu search. In K. Jajuga, A. Sokolowski, & H.-H. Bock (Eds.), *Classification, Clustering and Data Analysis* (pp. 43–51). Berlin: Springer. doi:10.1007/978-3-642-56181-8_4

Chandra-Satapathy, S., Murthym, J. V. R., Prasad-Reddy, P. V. G. D., Misra, B. B., Dash, P. K., & Panda, G. (2009). Particle swarm optimized multiple regression linear model for data classification. *Applied Soft Computing*, *9*(2), 470–476. doi:10.1016/j.asoc.2008.05.007

Chis, M. (2010). Introduction: A survey of the evolutionary computation techniques for software engineering. In M. Chis (Ed.), *Evolutionary Computation amd Optimization Algorithms in Software Engineering: Applications and Techniques* (pp. 1–12). Hershey, PA: Information Science Reference; doi:10.4018/978-1-61520-809-8.ch001

Cormen, T. H., Leiserson, C. E., Rivest, R. L., & Stein, C. (2009). *Introduction to Algorithms (3rd ed.)*. Cambridge, MA: The MIT Press.

De los Cobos, S., Goddard, J., & Gutiérrez, M. A. (2011). Fuzzy regression vs. ordinary least squares regression: Case study. *Revista de Matemática: Teoría y Aplicaciones*, *18*(1), 33–48.

Dorigo, M., Maniezzo, V., & Colorni, A. (1996). Ant System: Optimization by a Colony of Cooperating Agents. *IEEE Transactions on Systems, Man, and Cybernetics. Part B, Cybernetics*, *26*(1), 29–41. doi:10.1109/3477.484436 PMID:18263004

Draper, N. R., & Smith, H. (1968). *Applied Regression Analysis*. New York: John Wiley & Sons.

Ein-Dor, P., & Feldmesser, J. (1987). Attributes of the performance of central processing units: A relative performance prediction model. *Communications of the ACM*, *4*(4), 308–317. doi:10.1145/32232.32234

Espinoza, J. L. (2002). Obtaining reducts with a genetic algorithm. In K. Jajuga, A. Sokolowski, & H. H. Bock (Eds.), *Classification, Clustering and Analysis* (pp. 2119–2226). Berlin: Springer. doi:10.1007/978-3-642-56181-8_24

Field, A., Miles, J., & Field, Z. (2012). *Discovering Statistics Using R*. Thousand Oaks, CA: Sage Publications.

Glover, F. (1989). Tabu search - Part I. *ORSA Journal on Computing, 1*, 190-206.

Goldberg, D. E. (1989). *Genetic Algorithms in Search, Optimization and Machine Learning*. Reading, MA: Addison-Wesley.

Gordini, N., & Veglio, V. (2014). Customer relationship management and data mining: a classification tree to predict customer purchasing behavior in global market. In P. M. Vasant (Ed.), *Handbook of Research on Novel Soft Computing Intelligent Algorithms: Theory and Practical Applications* (pp. 1–40). Hershey, PA: Information Science Reference; doi:10.4018/978-1-4666-4450-2.ch001

Groenen, P., Roosmalen, J., Trejos, J., & Castillo, W. (2008). Optimization strategies in two-mode partitioning. *Journal of Classification*, *26*(2), 155–181.

Groenen, P. J. F., Mathar, R., & Trejos, J. (2000). Global optimization methods for multidimensional scaling applied to mobile communications. In W. Gaul, O. Opitz, & M. Schader (Eds.), *Data Analysis. Scientific Modeling and Practical Application* (pp. 459–469). Berlin: Springer. doi:10.1007/978-3-642-58250-9_37

Hirst, J. D., King, R. D., & Sternberg, M. J. (1994). Quantitative structure-activity relationships by neural networks and inductive logic programming: 2. The inhibition of dihydrofolate reductase by triazines. *Journal of Computer-Aided Molecular Design*, *8*(4), 405–420. doi:10.1007/BF00125375 PMID:7815092

Hoerl, A. E. (1962). Application of ridge analysis to regression problems. *Chemical Engineering Progress*, *58*, 54–59.

Holland, J. H. (1975). *Adaptation in Natural and Artificial Systems.* Ann Arbor, MI: The University of Michigan Press.

Jackson. (1991) *A User's Guide to Principal Components.* New York: John Wiley & Sons.

Joliffe, I. T. (2002). *Principal Component Analysis* (2nd ed.). New York: Springer.

Kennedy, J., & Eberhart, R. C. (2000). *Intelligent Swarm Systems.* New York: Academic Press.

King, R. D., Hirst, J. D., & Sternberg, M. J. (1995). Comparison of artificial intelligence methods for modeling pharmaceutical QSARS. *Applied Artificial Intelligence an International Journal*, *9*(2), 213–233. doi:10.1080/08839519508945474

Kirkpatrick, S., Gelatt, D., & Vecchi, M. P. (1983). Optimization by simulated annealing. *Science*, *220*(4598), 671–680. doi:10.1126/science.220.4598.671 PMID:17813860

Laarhoven, P., & Aarts, E. (1988). *Simulated Annealing: Theory and Applications.* Dordrecht: Kluwer Academic Publishers.

Leach, L. F., & Henson, R. K. (2007). The use and impact of adjusted R^2. Effects in published regression research. *Multiple Linear Regression Viewpoints*, *33*(1), 1–11.

Leardi, R., Boggia, R., & Terrile, M. (2002). Genetic algorithms as a strategy for feature selection. *Journal of Chemometrics*, *6*(5), 267–281. doi:10.1002/cem.1180060506

Liu, H., Weiss, R. E., Jennrich, R. I., & Wenger, N. S. (1999). PRESS model selection in repeated measures data. *Computational Statistics & Data Analysis*, *30*(2), 169–184. doi:10.1016/S0167-9473(98)00088-7

Longley, J. W. (1967). An appraisal of least-squares programs from the point of view of the user''. *Journal of the American Statistical Association*, *62*, 819–841. doi:10.1080/01621459.1967.10500896

Mathar, R. (1997). A hybrid global optimization algorithm for multidimensional scaling. In R. Klar & O. Opitz (Eds.), *Classification and Knowledge Organization* (pp. 63–71). Berlin, Heidelberg: Springer. doi:10.1007/978-3-642-59051-1_7

Pawlak, Z. (1991). *Rough Sets. Theoretical Aspects of Reasoning About Data.* Dordrecht: Kluwer Publ. Co.

Piza, E., Trejos, J., & Murillo, A. (2000). Clustering with non-Euclidean distances using combinatorial optimisation techniques. In J. Blasius, J. Hox, E. de Leeuw, & P. Schmidt (Eds.), Science Methodology in the New Millennium. Academic Press.

Reeves, C. (Ed.). (1995). *Modern Heuristic Techniques for Combinatorial Problems*. London: McGraw-Hill.

Ries, J., Beullens, P., & Wang, Y. (2013). Instance-specific parameter tuning for meta-heuristics. In P. N. Vasant (Ed.), *Meta-Heuristics Optimization Algorithms in Engineering, Business, Economics, and Finance* (pp. 136–170). Hershey, PA: IGI Global; doi:10.4018/978-1-4666-2086-5.ch005

Rudolph, G. (1994). Convergence analysis of canonical genetic algorithms. *IEEE Transactions on Neural Networks*, *5*(1), 96–101. doi:10.1109/72.265964 PMID:18267783

Steel, S. J., & Uys, D. W. (2006). Influential data cases when the *Cp* criterion is used for variable selection in multiple linear regression. *Computational Statistics & Data Analysis*, *50*(7), 1840–1854. doi:10.1016/j.csda.2005.02.003

Tibshirani, R. (1996). Regression shrinkage and selection via the Lasso. *Journal of the Royal Statistical Society. Series B. Methodological*, *58*(1), 267–288.

Tomassone, R., Audrain, S., Lesquoy, E., & Millier, C. (1992). *La Régression. Des Nouveaux Regards sur une Ancienne Méthode Statistique*. Paris: Masson.

Trejos, J. (1993). A simulated annealing implementation for oblique varimax rotations. In J. Janssen & C. H. Skiadas (Eds.), Applied Stochastic Models and Data Analysis (vol. 2, pp. 981-989). Singapore: World Scientific.

Trejos, J., & Castillo, W. (2000). Simulated annealing optimization for two-mode partitioning. In W. Gaul & R. Decker (Eds.), *Classification and Information at the Turn of the Millenium* (pp. 133–142). Berlin: Springer. doi:10.1007/978-3-642-57280-7_15

Trejos, J., Castillo, W., González, J., & Villalobos, M. (2000). Application of simulated annealing in some multidimensional scaling problems. In H. A. L. Kiers, J.-P. Rasson, P. J. F. Groenen, & M. Schader (Eds.), *Data Analysis, Classification and Related Methods* (pp. 297–302). Berlin: Springer. doi:10.1007/978-3-642-59789-3_47

Trejos, J., Murillo, A., & Piza, E. (1998). Global stochastic optimization for partitioning. In A. Rizzi, M. Vichi, & H. H. Bock (Eds.), *Advances in Data Science and Classification* (pp. 185–190). Berlin: Springer. doi:10.1007/978-3-642-72253-0_25

Trejos, J., Murillo, A., & Piza, E. (2004). Clustering by ant colony optimization. In D. Banks, L. House, F. R. McMorris, P. Arabie, & W. Gaul (Eds.), *Classification, Clustering, and Data Mining Applications* (pp. 25–32). Berlin: Springer. doi:10.1007/978-3-642-17103-1_3

Trejos, J., & Piza, E. (2000). Partitioning of binary data applying a genetic algorithm (in Spanish). Technical Report, CIMPA, University of Costa Rica.

Trejos, J., & Villalobos, M. (2007). Partitioning by particle swarm optimization. In P. Brito, P. Bertrand, G. Cucumel, & F. de Carvalho (Eds.), *Selected Contributions in Data Analysis and Classification* (pp. 235–244). Berlin: Springer. doi:10.1007/978-3-540-73560-1_22

Trejos, J., Villalobos, M., Murillo, A., Chavarria, J., & Fallas, J. J. (2014). Evaluation of optimization metaheuristics in clustering. In C. Travieso, J. Arroyo & M. Ramírez (Eds.), *IEEE 3rd International Conference and Workshop on Bioinspired Intelligence* (pp. 154-161). doi:10.1109/IWOBI.2014.6913956

Vasant, P. M. (2013). *Meta-Heuristics Optimization Algorithms in Engineering, Business, Economics, and Finance*. Hershey, PA: IGI Global; doi:10.4018/978-1-4666-2086-5

Venables, W. N., & Ripley, B. D. (1994). *Modern Applied Statistics with S-Plus*. New York: Springer-Verlag. doi:10.1007/978-1-4899-2819-1

Villalobos, M., & Trejos, J. (2000a). Metric multidimensional scaling using tabu search [in Spanish]. *Revista de Matemática: Teoría y Aplicaciones*, *7*(1-2), 71–76.

Villalobos, M., & Trejos, J. (2000b). Simulated annealing optimization in nonlinear regression: Algorithm and software [in Spanish]. *Investigación Operacional*, *21*(3), 236–246.

Villalobos, M., Trejos, J., & de los Cobos, S. (2006). Application of tabu search in non linear regression [in Spanish]. *Revista de Matemática: Teoría y Aplicaciones*, *13*(1), 79–92.

Yin, P. Y. (2013). *Trends in Developing Metaheuristics, Algorithms, and Optimization Approaches*. Hershey, PA: IGI Global; doi:10.4018/978-1-4666-2145-9

Zhang, Y., Li, R., & Tsai, C. L. (2010). Regularization parameter selections via generalized information criterion. *Journal of the American Statistical Association*, *105*(489), 312–323. doi:10.1198/jasa.2009.tm08013 PMID:20676354

ADDITIONAL READING

Abdi, H. (2003). Partial least squares (PLS) regression. In M. Lewis-Beck, A. Bryman, & T. Futing (Eds.), *Ecyclopedia of Social Sciences Research Methods* (pp. 792–795). Thousand Oaks, CA: Sage.

Albert, A. (1972). *Regression and the Moore-Penrose Pseudoinverse*. New York: Academic Press.

Antoniadis, A., Berruyer, J., & Carmona, R. (1992). *Régression Non Linéaire et Applications*. Paris: Economica.

Bard, Y. (1974). *Nonlinear Parameter Estimation*. New York: Academic Press.

Bates, D. M., & Watts, D. G. (1988). *Nonlinear Regression Analysis and its Applications*. New York: John Wiley & Sons. doi:10.1002/9780470316757

Billard, L., & Diday, E. (2000). Regression analysis for interval-valued data. In H. A. L. Kiers, J.-P. Rasson, P. J. F. Groenen, & M. Schader (Eds.), *Data Analysis, Classification, and Related Methods* (pp. 369–374). Berlin: Springer. doi:10.1007/978-3-642-59789-3_58

Billard, L., & Diday, E. (2007). *Symbolic Data Analysis: Conceptual Statistics and Data Mining*. New York: Wiley.

Bock, H. H., & Diday, E. (2000). *Analysis of Symbolic Data*. Berlin: Springer. doi:10.1007/978-3-642-57155-8

Borowiak, D. S. (1989). *Model Discrimination for Nonlinear Regression Models*. Boca Raton: CRC Press, Taylor & Francis.

Bry, X. (1996). *Analyses Factorielles Multiples*. Paris: Economica.

Cadenas, J. M., Garrido, M. C., Muñoz, E., Cruz-Corona, C., Pelta, D. A., & Verdegaray, J. L. (2009). Different Approaches for Cooperation with Metaheuristcs. In J. Rabuñal Dopico, J. Dorado, & A. Pazos (Eds.), *Encyclopedia of Artificial Intelligence* (pp. 480–487). Hershey, PA: Information Science Reference; doi:10.4018/978-1-59904-849-9.ch073

Chabanon, C., Lechevallier, Y., & Milleman, S. (1992). Proposition d'une construction efficace d'un réseau de neurones à partir d'un arbre de décision. In E. Diday & Y. Kodratoff (Eds.), *III Journées Symbolique-Numérique*. Paris: Université Paris-Dauphine.

Chatterjee, S., Hadi, A. S., & Price, B. (2000). *Regression Analysis by Example*. New York: John Wiley & Sons.

Chipman, H. A., George, E. I., & McCulloch, R. E. (2010). BART: Bayesian Additive Regression Trees. *The Annals of Applied Statistics*, *4*(1), 266–298. doi:10.1214/09-AOAS285

Daniel, C., & Wood, F. S. (1971). *Fitting Equations to Data: Computer Analysis of Multifactor Data for Scientists and Engineers*. New York: John Wiley & Sons.

Davis, L., & Steenstrup, M. (1987). Genetic algorithms and simulated annealing: an overview. In L. Davis (Ed.), *Genetic Algorithms and Simulated Annealing*. London: Pitman.

Diday, E., & Noirhomme-Fraiture, M. (2008). *Symbolic Data Analysis and the SODAS Software*. New York: Wiley.

Dréo, J., Pétrowski, A., Siarry, P., & Taillard, E. (2006). *Metaheuristics for Hard Optimization. Simulated Annealing, Tabu Search, Evolutionary and Genetic Algorithms, Ant Colonies, ... Methods and Case Studies*. Berlin: Springer.

Emiliano, P. C., Vivanco, M. J. F., & de Menezes, F. S. (2014). Information criteria: How do they behave? *Computational Statistics & Data Analysis*, *69*, 141–153. doi:10.1016/j.csda.2013.07.032

Foucart, T. (1996). Analyse de la collinéarité. Classification de variables. *Revue de Statistique Appliquee*, *44*(4), 41–57.

Foucart, T. (2006). Colinéarité et régression linéaire. *Mathematiques et Sciences Humaines*, *173*(1), 5–25.

Fox, J., & Weisberg, S. (2010). *An R Companion to Applied Regression*. Thousand Oaks, CA: Sage Publications.

Galton, F. (1886). Regression towards mediocrity in hereditary stature. *The Journal of the Anthropological Institute of Great Britain and Ireland*, *15*, 246–263. doi:10.2307/2841583

García Torres, M. (2009). Metaheuristics in Data Mining. In J. Wang (Ed.), *Encyclopedia of Data Warehousing and Mining* (2nd ed., pp. 1200–1206). Hershey, PA: Information Science Reference; doi:10.4018/978-1-60566-010-3.ch187

Glover, F., Taillard, E., & de Werra, D. (1993). A user's guide to tabu search. *Annals of Operations Research*, *41*(1), 3–28. doi:10.1007/BF02078647

Hastie, T., Tibshirani, R., & Friedman, J. (2009). *The Elements of Statistical Learning. Data Mining, Inference, and Prediction* (2nd ed.). New York: Springer.

Hoerl, A. E., & Kennard, R. W. (1970). Ridge regression: Biased estimation for nonorthogonal problems. *Technometrics*, *12*(1), 55–67. doi:10.1080/00401706.1970.10488634

Hosmer, D. W., Lemeshow, S., & Sturdivant, R. X. (2013). *Applied Logistic Regression* (3rd ed.). New York: Wiley. doi:10.1002/9781118548387

Irizarry, R. (2005). Classification and regression trees. In: *Statistical Learning: Algorithmic and Non-parametric Approaches*, Johns Hopkins University. Consulted in: http://www.statmethods.net/advstats/cart.html

Ivarsson, A., Andersen, M. B., Johnson, U., & Lindwall, M. (2013). To adjust or not adjust: Nonparametric effect sizes, confidence intervals, and real-world meaning. *Psychology of Sport and Exercise*, *14*(1), 97–102. doi:10.1016/j.psychsport.2012.07.007

Lima Neto, E. A., & De Carvalho, F. A. T. (2008). Center and range methods for fitting a linear regression model to symbolic interval data. *Computational Statistics & Data Analysis*, *52*(3), 1500–1515. doi:10.1016/j.csda.2007.04.014

Lima Neto, E. A., & De Carvalho, F. A. T. (2008). Nonlinear regression model to symbolic interval-valued variable. In: *2008 IEEE International Conference on Systems, Man, and Cybernetics*, vol. 1 (pp. 1247-1252), Singapore. doi:10.1109/ICSMC.2008.4811454

Lima Neto, E. A., & De Carvalho, F. A. T. (2010). Constrained linear regression models for symbolic interval-valued variable. *Computational Statistics & Data Analysis*, *54*(2), 333–347. doi:10.1016/j.csda.2009.08.010

Linhart, H., & Zucchini, W. (1986). *Model selection*. New York: Wiley.

Loh, W. Y. (2002). Regression trees with unbiased variable selection and interaction detection. *Statistica Sinica*, *12*, 361–386.

Marquardt, D. W. (1963). An algorithm for least squares estimation of nonlinear parameters. *Journal of the Society for Industrial and Applied Mathematics*, *2*(2), 431–441. doi:10.1137/0111030

Michalewicz, Z. (1992). *Genetic Algorithms + Data Structures = Evolution Proggrams*. Berlin: Springer. doi:10.1007/978-3-662-02830-8

Michalewicz, Z., & Michalewicz, M. (1999). Teaching Evolutionary Algorithms. In: P.J. Angeline, Z. Michalewicz, M. Schoenauer, X. Yao, & A. Zalzala (Eds.), *Proceedings of the Congress of Evolutionary Computation*, vol. 3 (pp. 1702-1706). Piscataway, NJ: IEEE Press.

Miller, A. J. (1990). *Subset selection in regression.* London: Chapman and Hall. doi:10.1007/978-1-4899-2939-6

Murtagh, F. (1991). Multilayer perceptrons for classification and regression. *Neurocomputing, 2*(5-6), 183–197. doi:10.1016/0925-2312(91)90023-5

Ratkowsky, D. A. (1983). *Nonlinear Regression Modeling. A Unified Practical Approach.* New York: Marcel Dekker.

Ross, G. (1990). *Nonlinear Estimation.* New York: Springer-Verlag. doi:10.1007/978-1-4612-3412-8

Senvar, O., Turanoglu, E., & Kahraman, C. (2013). Usage of Metaheuristics in Engineering: A Literature Review. In P. Vasant (Ed.), *Meta-Heuristics Optimization Algorithms in Engineering, Business, Economics, and Finance* (pp. 484–528). Hershey, PA: Information Science Reference; doi:10.4018/978-1-4666-2086-5.ch016

Tibshirani, R., & Wasserman, L. (2014). *Nonparametric regression. Statistical Machine Learning Lectures, Spring 2014.* Carnagie Mellon University PA.

Vasant, P. M. (2014). Handbook of Research on Novel Soft Computing Intelligent Algorithms: Theory and Practical Applications (Vols. 1–2). pp. 1–1018). Hershey, PA: IGI Global; doi:10.4018/978-1-4666-4450-2

Young, D. S. (2014). *Regression Methods.* University Park, PA: Penn State University Press.

Zucchini, W. (2000). An introduction to model selection. *Journal of Mathematical Psychology, 44*(1), 41–61. doi:10.1006/jmps.1999.1276 PMID:10733857

KEY TERMS AND DEFINITIONS

Genetic Algorithm: An iterative meta-heuristic based on the evolution of species, that handles a population of solutions of the optimization problem that have a survive probability proportional to the quality of the respective solution, and makes combinations of solutions based on crossover and mutation operators.

Linear Regression: A regression model where explanatory variables model one variable to be explained, by means of a linear combination.

Metaheuristics: Optimization methods that search for good solutions of a problem, usually based on iterative steps that approximate the solution; the "meta" term refers to the fact that a basic frame is used to many different situations.

Regression: A set of methods that model one or more variables to be explained, usually numerical, by means of a set of explanatory variables, also usually numerical.

Variable Selection: Procedures that select the most explanatory variables in a regression model, according to some numerical criterion.

APPENDIX: NOMENCLATURE

X: data matrix n x p, with observed explanatory variables in the columns.
Y: column variable of dimension n with the observed variable to be explained.
$\hat{Y}$: linear combination of the explanatory variables retained in the model.
n: number of cases or objects.
p: number of observed explanatory variables.
q: number of explanatory variables selected in the model.
β: vector of regression coefficients.
R^2 or R square: determination coefficient, that is, squared correlation between Y and $\hat{Y}$.
$\bar{R}^2$: adjusted R square.
AIC: Akaike Information Criterion
BIC: Bayesian Information Criterion
M: size of the genetic population, that is, number of chromosomes in the genetic algorithm.
p_c: probability of crossover, applied over the chromosomes.
p_m: probability of mutation, applied over the genes.
t: iteration number of the genetic algorithm.
τ: attraction rate of the best solution obtained in a multistart experiment.

Chapter 6
An Application of Alpha-Stable Distributions for the Economic Analysis of Unit Commitment

Jose Antonio Marmolejo
Anahuac University, Mexico

Román Rodríguez
National Polytechnic Institute, Mexico

ABSTRACT

In power systems, we still lack the existence of standardized test systems that can be used to benchmark the performance and solution quality of proposed optimization techniques. It is therefore necessary to develop new methods for design of test cases for economic analysis in power systems. We compared two methods to generate test systems: time series and a method for simulating stable random variables that can be used to multiperiod unit commitment based on the use of Chambers-Mallows-Stuck. Hence, after comparing both methods, we describe the method for simulating stable random variables in the generation of test systems for economic analysis in power systems. A study focused on generating test electrical systems through fat tail model for unit commitment problem in electrical power systems is presented. Usually, the instances of test systems in Unit Commitment are generated using normal distribution, but in this work, simulations data are based on alpha-stable distribution. Numerical results illustrate the applicability of the proposed method.

1. INTRODUCTION

The optimization problems for electrical power systems have been studied for more than five decades. A definition of Economic Dispatch is the operation of generation facilities to produce energy at the lowest cost to reliably serve consumers, recognizing any operational limits of generation and transmission facilities.

In the typical Unit Commitment (Wood & Wollenberg, 1996) the problem consists as determining the mix of generators and their estimated output level to meet the expected demand of electricity over a

DOI: 10.4018/978-1-4666-9644-0.ch006

given time horizon (a day or a week), while satisfying the load demand, spinning reserve requirement and transmission network constraints. An electric network consists of many generation nodes with various generating capacities and cost functions, lines of transmission and nodes of power demand (Vasant et al. 2012).

The application of optimization models for electrical power systems is marked by constant development for new algorithms like exact methods, metaheuristics and hybrid strategies (Vasant, 2013). However, to benchmark the performance and solution quality for any solution technique, it is necessary to have variety of electrical test systems (Vasant, 2014).

Nowadays, we still lack the existence of standardized test systems that can be used to benchmark the performance and solution quality of proposed techniques. Many papers consider different test systems, which make it very difficult to perform a proper comparison between methods that have been proposed (Diniz, 2010).

In Zhang and Schaffner (2010), they refer that the existing IEEE test systems developed are mainly used for reliability, power flows and stability analysis but not for economic analysis. Short time ago, some panels focused on the development of standard test systems of transmission and distribution systems for economic analysis has emerged. In 2007 the IEEE Working group (WG) on Test Systems for Economic Analysis was created, sponsored by IEEE System Economics subcommittee.

Finding an appropriate method for generating test cases of a specific electricity network is not an easy task. In this sense, this paper compares two methods to generate test systems: time series models and a method for simulating stable random variables that can be used to multiperiod unit commitment based on the use of Chambers-Mallows-Stuck algorithm.

There are some methods to find the best estimation of load forecasting. The main methods include time series such as exponential smoothing or Autoregressive Integrated Moving Average models (ARIMA). However, the electricity demand pattern becomes more complex to forecast because it has a particular behavior.

Lévy and Aleksander Khintchine (1924) first developed the stable distributions theory in the 20's of last century. Since then, this distribution has been applied in different areas of knowledge, such as economics, physics, engineering and hydrology. The reason is that some phenomena of nature, like electrical demand, cannot be described assuming Normal distribution as they present observations with extreme values, which characterizes the instability of the series and denotes the presence of heavy tails, an effect known as impulsivity.

Usually electrical demand presents a greater degree of impulsivity that the Normal distribution cannot describe due to the presence of peaks in the series during the hours of the day and seasons of high-energy demand.

For this, we propose use Chambers-Mallows and Stuck algorithm for simulating alpha stable random variables characterizing demand patterns of real electrical systems. The use of Chambers-Mallows-Stuck method for simulating stable random variables provides a new way to generate test systems widely used in power systems research. By modeling the demand through the use of alpha stable distribution can catch the real behavior of the electrical demand and build possible extreme scenarios, each scenario corresponds to a price-elastic demand curve. The simulations are based on real observations of demand for different reliability test systems. Electrical network data are taken from the 24 (Charman et al. 1979) and 118 bus IEEE test systems (Christie, 1993), and a portion of electric energy system of Mainland Spain (Alguacil & Conejo, 2000). Cost functions of the thermal plants data are taken from the literature.

One of the main advantages of the proposed method to generate test instances is that once obtained the distribution parameters can create as many instances as needed without losing robustness in the estimates. This represents a great advantage over other techniques demand forecast.

After the elaboration of these test cases, we tested a mixed integer non-linear formulation of Unit Commitment problem based on these cases, in order to obtain estimation about the performance of these new test systems.

Additionally, we tested the Lagrangian relaxation of the same problem to verify numerical stability of the generated test systems.

2. STABLE DISTRIBUTIONS

Paul Lévy and Aleksander Khintchine (Lévy, 1924) first developed the stable distributions theory. This distribution has been applied in different areas of knowledge, such as economics, physics, hydrology, and signal processing. However, it was not until the work of Mandelbrot in 60's that the stable distributions were popularized. Mandelbrot proposed a revolutionary theory based on this distribution to solve the problem of price fluctuations, later shown that many other economic variables follow a stable distribution.

Significantly, the stable distribution meets the Central Limit Theorem and the stability property (which denotes that stable distributions are isomorphic) that also contains the normal distribution as a special case of this. On the other hand, there is some complexity in working with such distributions, since they lack, in general, of an analytical expression.

Definition. A random variable X has stable distribution having the following characteristic function (Nolan, 2005):

$$\Phi(w)=\begin{Bmatrix} \exp\left\{-|\gamma \mathrm{w}|^{\alpha}\left[1-\mathrm{isign}(\mathrm{w})\beta\tan\left(\frac{\pi\alpha}{2}\right)\right]+\mathrm{i}\delta\mathrm{w}\right\},(\alpha\neq 1) \\ \exp\left\{-|\gamma \mathrm{w}|\left[1+\mathrm{isign}(\mathrm{w})\beta\frac{2}{\pi}\log(|\mathrm{w}|)\right]+\mathrm{i}\delta\mathrm{w}\right\},(\alpha=1) \end{Bmatrix} \tag{1}$$

Where $sign(w)=\frac{w}{|w|}$ and á $\in(0,2]$. Whose parameters are defined as follows, α, represents the characteristic exponent, which controls the degree of impulsiveness of the random variable *W*. Moreover, the parameter $\beta\in[-1,+1]$ controls the symmetry of distribution (*β=0*, stable distribution symmetrical, *β=1* and *β=-1* to the family of stable distributions positive and negative respectively). While *γ>0* is a scale parameter, also called dispersion, and δ is the position parameter.

Remarkably, if the expression of the characteristic function parameter, α=2, then the parameter β becomes meaningless, since ² tanÀ= 0. In this case, the characteristic function becomes:

$$\varphi(w) = \exp\left\{-|\gamma w|^2 + i\delta w\right\} \tag{2}$$

The above expression is the characteristic function of a Gaussian random variable with mean δ and varianceó ^{2} =2ã ^{2}. Therefore, from the definition above, we also can show that the normal distribution is a particular case of stable distribution. Given the properties of stable distribution above, it follows that its use is justified in the same way as the Gaussian distributions and not only that, but the Gaussian distribution is a particular case of stable and therefore the range of application of stable distributions is even wider than the normal distribution.

This is mainly due the existence and continuity of the probability density function of stable distribution, but with a few exceptions, it cannot be expressed in a compact way. In other words, the integral with respect to *(w)* of the characteristic function (1), only have an analytical solution for the described cases, denoting the stable distribution by four parameters $f_{á,â(.|ã,ä)}$, (Nolan, 2005).

A α-stable distribution with parameters:

- $f_{2,0}(\cdot|\gamma,\delta)$ is a Gaussian distribution with mean δ and variance $2\gamma^2$

$$f_{2,0}(x|\gamma,\delta) = N(x \mid \delta, 2\gamma^2) = \frac{1}{2\gamma\sqrt{\pi}} \exp\left\{\frac{(x-\delta)^2}{4\gamma^2}\right\} \tag{3}$$

- $f_{1,0}(\cdot|\gamma,\delta)$ is a Cauchy distribution with density

$$f_{1,0}(x|\gamma,\delta) = pCauchy(x \mid \delta,\gamma) = \frac{\gamma}{\pi\left((x-\delta)^2 + \gamma^2\right)} \tag{4}$$

- $f_{\frac{1}{2},1}(\cdot|\gamma,\delta)$ is a Levy distribution with density

$$f_{\frac{1}{2},1}(x|\gamma,\delta) = pLévy(x \mid \delta,\gamma) = \left(\frac{\gamma}{2\pi}\right)^{\frac{1}{2}} \frac{1}{(x-\delta)^{\frac{3}{2}}} \exp\left\{-\frac{\gamma}{2(x-\delta)}\right\} \tag{5}$$

In this research, we will use the Nolan (2005) algorithm to estimate the alpha stable distribution parameters. Nolan (1997) algorithm is based on the Zotolarev (1986) representation of the characteristic function. The software performs the numerical integration of a set of splits of the random variable domain. These splits are based on the sign change of the trigonometric functions (sinus and cosines) that result of the transformation of Zotolarev's (1986) representation into the complex plane.

It is important to point out that there are other methodologies for estimating the parameters of stable distribution, like those presented in Belov (2005) proposing a combination of Gaussian and Laguerre quadratures; or Mittnik et al. (1999) presenting an algorithm that applies the fast Fourier transform. Finally, DuMouchel (1973) using a Bergström series expansion on the Zotolarev's characteristic function representation for approximating a stable cumulative distribution.

The current computational developments and their power applied to the distribution parameter calculations of stable distributions had been a key element in the recent use of such distributions in many areas.

The statistical significance of the estimated parameters can be contrasted by different statistical tests, the hypothesis to be tested is:

H0: *the series is distributed as* α-stable *distribution*
H1: *the series is not distributed* as α-stable distribution

The Anderson Darling test showed greater efficiency in the case of heavy tailed series, its relevance and use has been generally accepted for the analysis of stable series (Belov et al. 2006).

2.1 Stable Random Variable

The Chambers-Mallows-Stuck (Chambers et al. 1976) method can generate a random variable X with distribution $f_{\pm,^2}(X \mid 1,0)$ from a non-linear transformation of two random variables independent, one uniform (V) and another exponential (W) using the following theorem:

Theorem.

Let (V) a uniform random variable in the interval $\left(\frac{-\pi}{2},\frac{\pi}{2}\right)$ $(\frac{-ð}{2},\frac{ð}{2}$ and (W) an exponential random variable with mean equal one. If V and W are independent, then:

$$X=\left(S_{\alpha,\beta}\frac{\sin\left(\propto\left(V+B_{\propto,\beta}\right)\right)}{\left(cos V\right)^{\frac{1}{\propto}}}\right)\left(\frac{\cos\left(\left(1-\propto\right)V+B_{\propto,\beta}\alpha\right)}{W}\right)^{\frac{1-\propto}{\propto}}, \tag{6}$$

X follows a stable distribution with $f_{\propto,\beta}(X \mid 1,0)$, where:

$$B_{\alpha,\beta}=\frac{\arctan\left(\beta\tan\left(\frac{\pi\alpha}{2}\right)\right)}{\alpha}, \tag{7}$$

$$S_{\alpha,\beta}=(1+\beta^2 tan^2\frac{\pi\alpha}{2})^{\frac{1}{2\alpha}}. \tag{8}$$

Once you get the variable (X), is generated a variable with stable distribution for any value of the parameters $\alpha,\beta,\sigma\, and\, \mu$. If $X \sim f_{\propto,\beta}(X \mid 1,0)$ then:

$$f_{\propto,\beta}(\gamma,\mu) \sim \gamma X + \mu\, si\, \alpha \neq 1, and \tag{9}$$

$$f_{\propto,\beta}(\gamma,\mu) \sim \gamma X + \frac{2}{\pi}\beta\gamma ln\gamma + \mu\, if\, \alpha = 1. \tag{10}$$

Some examples of the application of Chambers-Mallows-Stuck are presented in Rodríguez (2013), Scalas (2007) and Belov et al. (2006) among others works.

3. THE MULTIPERIOD UNIT COMMITMENT MODEL

In this work we addresses a Multiperiod Unit Commitment based on Marmolejo et al. (2011) notation, where network constraints are represented through a DC model (Alguacil & Conejo, 2000). The following notation is used in the mathematical model:

Constants:

A_j Startup cost of power plant j.
B_{nm} Susceptance of line $n-m$.
C_{nm} Transmission capacity limit of line $n-m$.
D_{nk} Load demand at node n during period k.
$E_j(t_{jk})$ Nonlinear function representing the operating cost of power plant j as a function of its power output in period k.
E_{j1} Linear coefficient of operating cost for the plant j.
E_{j2} Quadratic coefficient of operating cost for the plant j.
F_j Fixed cost of power plant j.
K_{nm} Conductance of line $n-m$.
R_k Spinning reserve requirement during period k.
$\overline{T_j}$ Maximum power output of plant j.
$\underline{T_j}$ Minimum power output of plant j.
nr Reference node with angle cero.

Variables:

t_{jk} Power output of plant j in period k.
v_{jk} Binary variable, which is equal to one when plant j is committed in period k.
y_{jk} Binary variable, which is equal to one when plant j is started up at the beginning of period k.
δ_{nk} Angle of node n in period k.

Sets:

J Set of indices of all plants.
K Set of period indices.
N Set of indices of all nodes.
Λ_n Set of indices of the power plants j at node n.
Ω_n Set of indices of nodes connected and adjacent to node n.

The objective is minimizing a function that includes fixed cost, start up cost and operating cost. A second order polynomial describes the variable costs as a function of the electric power.

$$MinZ = \sum_{k\in K}\sum_{j=J}[F_j v_{jk} + A_j y_{jk} + E_j(t_{jk})] \tag{11}$$

There is a power balance constraint per node and time period. In each period, the production has to satisfy the demand and losses in each node. Line losses are modeled through cosine approximation and it is assumed that the demand for electric energy is known and is discretized into t periods.

$$\sum_{j\in\Lambda_n} t_{jk} + \sum_{m\in\Omega_n} B_{nm}[\delta_{mk} + \delta_{nk}] - \sum_{m\in\Omega_n} K_{nm}[1-\cos(\delta_{mk} - \delta_{nk})] = D_{nk} \\ \forall n \in N, \forall k \in K \tag{12}$$

Spinning reserve requirements are modeled. In each period the running units have to be able to satisfy the demand and the prespecified spinning reserve.

$$\sum_{j\in J}\overline{T}_j v_{jk} \geq \sum_{n\in N} D_{nk} + R_k \forall k \in K \tag{13}$$

Each unit has a technical lower and upper bound for the power production.

$$\overline{T}_j v_{jk} \leq t_{jk\leq}\overline{T}_j v_{jk} \forall j \in J, \forall k \in K \tag{14}$$

Transmission capacity limits of lines avoid dynamic stability system problems.

$$-C_{nm} \leq B_{nm}[\delta_{mk} - \delta_{nk}] \leq C_{nm} \forall n \in N, \forall k \in K, \forall m \in \Omega_n \tag{15}$$

This constraint holds the logic of running, start-up and shut-down of the units. A running unit cannot be started-up.

$$y_{jk} \geq v_{jk} - v_{jk-1} \forall j \in J, \forall k \in K \tag{16}$$

Angle in all buses has a lower and upper bound.

$$-\pi \le \delta_{nk} \le \pi \quad \forall n \in N / \{nr\}, \forall k \in K \tag{17}$$

4. SOLVING STRATEGIES

LR decomposes the original problem into a Dual Master Problem and makes better and easily manageable Dual Subproblems to be solved separately. Lagrange Multipliers that are added to the master problem to yield a dual problem link the sub problems. The dual problem has lower dimensions than the primal problem and is easier to solve. The multipliers are updated through different methods. In this work, we use subgradient method. The major difficulty of this method is associated with obtaining solution feasibility because of the dual nature of the algorithm.

Dual Master Problem:

$$\begin{aligned} &\underset{\lambda_j(k),\mu_i(k),\gamma(k),\beta(k)}{Max} g^v = \sum_{k\in K} \sum_{j=J} [L_n^v] \\ &subject\ to: \\ &\lambda_j(k), \mu_i(k), \gamma(k), \beta(k) \ge 0 \end{aligned} \tag{18}$$

Dual Subproblem:

$$\begin{aligned} &L_n(t_{jk}, v_{jk}, y_{jk}, \delta_{nk}, \lambda_k, \mu_k, \gamma_k, \beta_k) = \\ &\min\{\sum_{k\in K} \sum_{j\in\Lambda_n} [F_j v_{jk} + A_j y_{jk} + E_j(t_{jk})] \\ &+\sum_{k\in K} \lambda_k [D_{nk} - \sum_{j\in\Lambda_n} t_{jk} - \sum_{m\in\Omega_n} B_{nm}(\delta_{mk} - \delta_{nk}) + \sum_{m\in\Omega_n} K_{nm}(1-\cos(\delta_{mk} - \delta_{nk})] \\ &+\sum_{k\in K} \mu_k [D_{nk} + R_k - \sum_{j\in J} \overline{T}_j v_{jk}] + \sum_{k\in K} \gamma_k [C_{nm} - \sum_{m\in\Omega_n} B_{nm}(\delta_{mk} - \delta_{nk})\ ] \\ &+\sum_{k\in K} \beta_k [-\sum_{m\in\Omega_n} B_{nm}(\delta_{mk} - \delta_{nk}) - C_{nm}\ ]\ \} \end{aligned} \tag{19}$$

Subject to eq. (14, 16, 17).

Generated test systems can be used to benchmark the performance and solution quality in several solving strategies. In this work, we compare the direct solution through GAMS and Lagrangian Relaxation (LR) of the problem. Through LR, we can obtain a lower bound on the value of the objective function of the original problem. The instances estimated in the next section are used as inputs to resolve the problem posed.

5. TEST SYSTEMS GENERATION

In this work, a method for calculating test systems (instances) based on the generation of random numbers that follow a stable alpha distribution is presented. Considering a set of parameters estimated based on the behavior of real electrical power systems.

Such instances are used to test the estimates made by the multiperiod unit commitment model. The results obtained with the test instances generated through adjustment of stable alpha distribution and Chambers-Mallow-Stuck method for generating random stable variables are contrasted with a traditionally method used to generate demand forecasts (the autoregressive integrated moving average models).

To generate new instances by the methodology proposed we worked with three standardized test systems:

- **System I:** Based on 104-bus electric energy system of Mainland Spain with 104 nodes, 62 thermal units and 160 transmission lines (Alguacil and Conejo, 2000).
- **System II:** Based on IEEE-24 bus test system with 24 nodes, 24 thermal units and 38 transmission lines (Charman et l. 1979).
- **System III:** Based on IEEE-118 bus test system with 118 nodes, 54 thermal units and 186 transmission lines (Christie, 1993).

All instances consider a 24-hour planning horizon with one period per hour. We use stable distribution to model the demand from the original systems analyzed. After a quantile analysis, we conclude that the three systems do not follow a normal distribution and show the presence of extreme values and fat tails, see Figures 1, 2, 3.

Figure 1. Quantile-Quantile normal plot IEE-104 bus system (Source: Author's own elaboration)

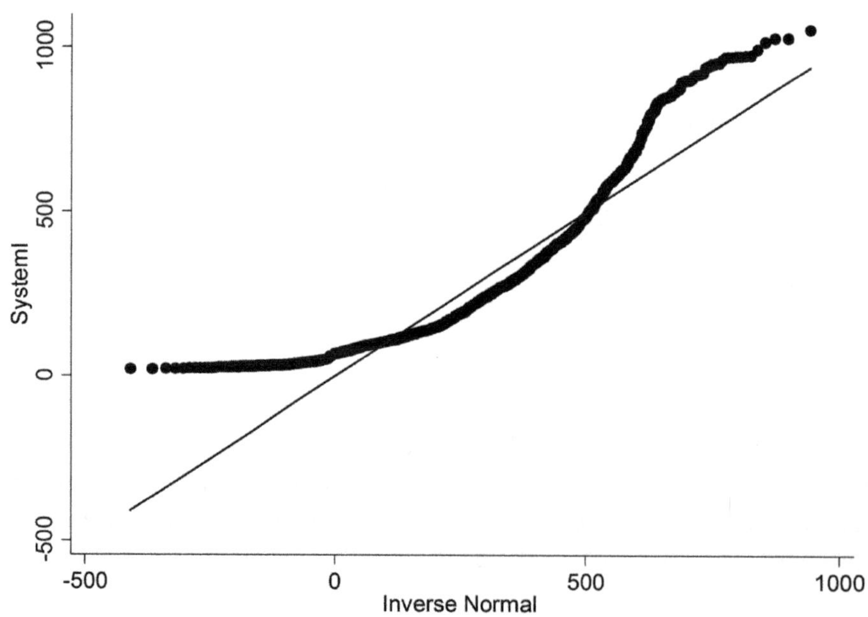

Figure 2. Quantile-Quantile normal plot IEEE-24 bus system (Source: Author's own elaboration)

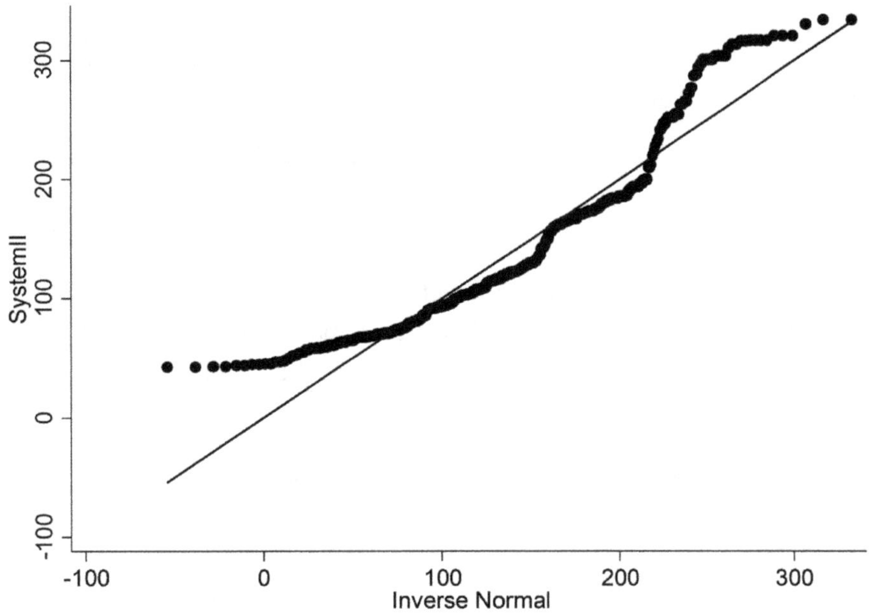

Figure 3. Quantile-Quantile normal plot IEEE-118 bus system (Source: Author's own elaboration)

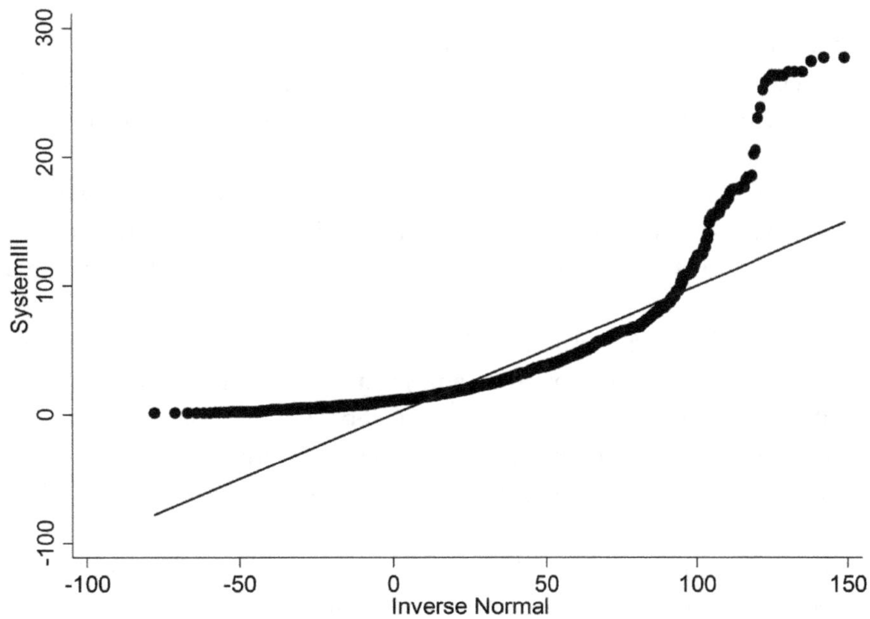

Table 1. Shapiro-Wilk Normal test (Source: Author's own elaboration)

Variable	Obs	w	v	z	P-value 5%
SystemI	1541	0.86224	128.857	12.239	0.0000
SystemII	408	0.91322	24.319	7.600	0.0000
SystemIII	2376	0.72011	388.189	15.259	0.0000

Normal test show statistically corroborated that three systems do not follow a normal distribution, and for the nature of the series have a fat tails characteristics, see Table 1.

The utilization of stable distribution to model energy demand it is proposed. It was felt that S_1 parameterization is usually used (Samorodnitsky, 1994) for modeling stable data.

$$\varphi(k) = lnfc(k) = \begin{cases} i\delta k - \gamma^{\alpha}|k|^{\alpha}\left\{1 - i\beta \text{sign}(k)\tan\frac{\pi\alpha}{2}\right\}, & (\alpha \neq 1) \\ i\delta k - \gamma|k|\left\{1 + i\beta \text{sign}(k)\frac{2}{\pi}\right\}, & (\alpha = 1) \end{cases}, \tag{20}$$

where $sign(k) = \frac{k}{|k|}$.

The parameters characterizing the stable distribution were obtained through the three methods generally accepted: a) maximum likelihood, b) quantile and c) regression (Nolan, 1999). In this case, the regression method used since it is the method generally used in the analysis of stable series that has a better adjustment of the tails of distribution.

The parameters for the stable distribution show a parameter alpha different to two (normal case) that means the data has impulsivity. To test the fit of the distribution to data, many authors agree with the utility of Anderson-Darling test (Stephens, 1977) for data with fat tails, in all cases the null hypothesis was not rejected.

The parameter alpha in three systems shows the presence of impulsivity in the series, all series are symmetric positive, and has high dispersion, see Tables 2 and 3. That shows that demand in the 24 hours of day and the habits of consumption. Figures 4, 5 and 6 show the fit of data with alpha stable distribution through Quantile-Quantile stable plots. The tails of the distributions have been captured. Moreover, this information will be considered into the simulation of random alpha stable variables.

The parameter alpha shows the impulsivity of data, to simulate the random alpha stable variables. Three scenarios with $\alpha = 1.50, \alpha = 1.69 and \alpha = 1.84$ were considered. The three proposed scenarios considered the impulsivity (alpha parameter) estimated for three sets of standardized electrical power systems used as input for modeling. The objective of presenting three scenarios is precisely cover a wider range of possible values for the instances generated in a range of likely values observed in real electrical power systems. Another parameters used to simulate random alpha stable variables, are the parameters of symmetry, dispersion and localization because that help to simulate extreme values observed in the real data. In the Figure 7, we can see the effect to the parameter alpha in the distribution function.

Table 2. Alpha Stable parameters with three methods of estimation (Source: Author's own elaboration)

IEE-104 bus system	α	β	γ	δ
Maximum likelihood	1.2000	1.0000	84.3954	423.414
Quantile	1.6132	1.0000	121.3540	291.3820
Regression	1.6938	1.0000	113.085	227.7650
IEE-24 bus system	**α**	**β**	**γ**	**δ**
Maximum likelihood	1.4000	1.0000	35.4172	160.6370
Quantile	1.6045	1.0000	41.7026	151.6880
Regression	1.8430	1.0000	42.8166	121.925
IEE-118 bus system	**α**	**β**	**γ**	**δ**
Maximum likelihood	1.2000	1.0000	10.7543	54.0993
Quantile	1.7539	1.0000	15.1875	32.7828
Regression	1.5092	1.0000	13.3914	35.9108

Table 3. Alpha Stable parameters (Source: Author's own elaboration)

System	α	β	γ	δ	Anderson Darling statistical	P-value 5%
I	1.6938	1.0000	113.085	227.765	0.623263	0.62625
II	1.8430	1.0000	42.8166	121.925	3.22121	0.02116
III	1.5092	1.0000	13.3914	35.9108	1.40198	0.20170

Figure 4. Quantile-Quantile stable plots 104-bus system (Source: Author's own elaboration)

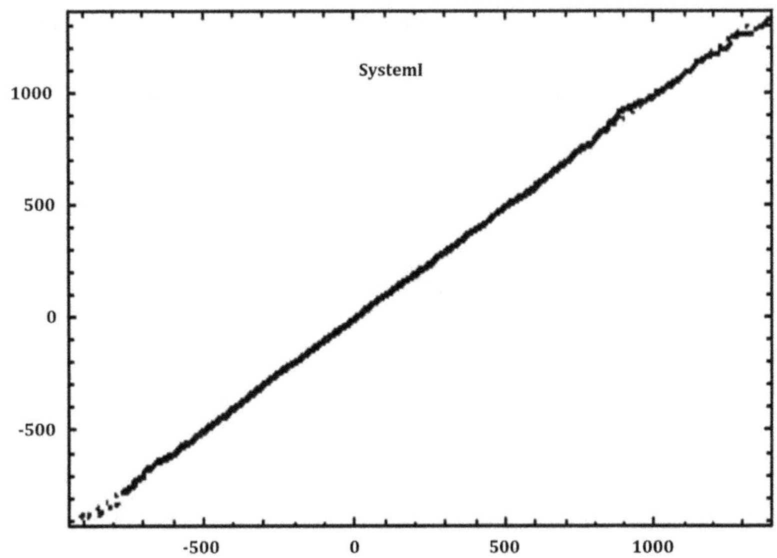

Figure 5. Quantile-Quantile stable plots IEEE-24 bus system (Source: Author's own elaboration)

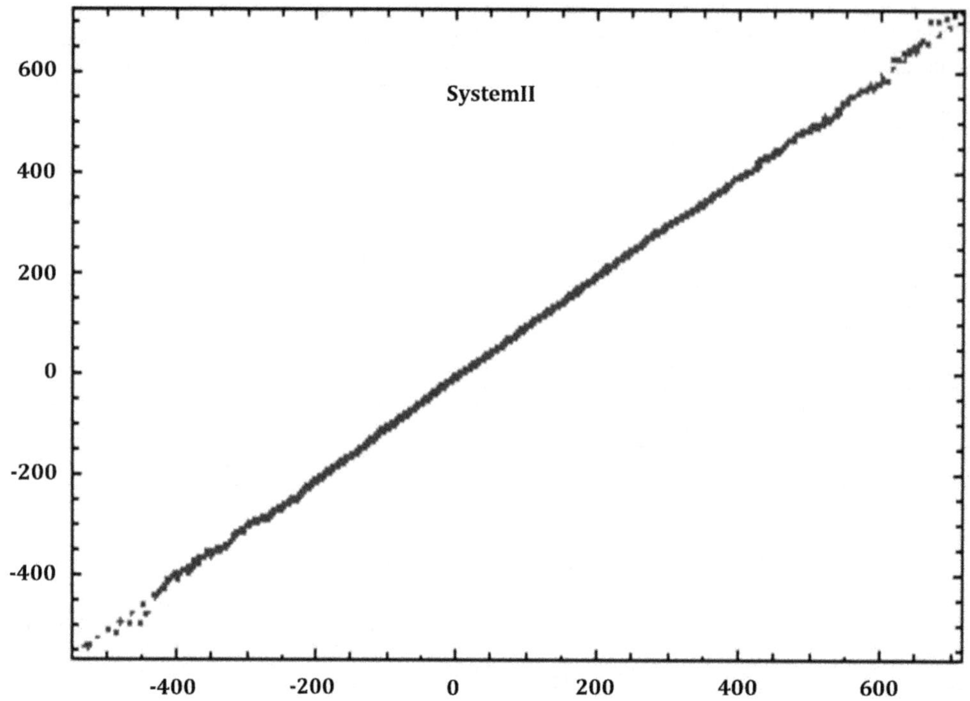

Figure 6. Quantile-Quantile stable plots IEEE-118 bus system (Source: Author's own elaboration)

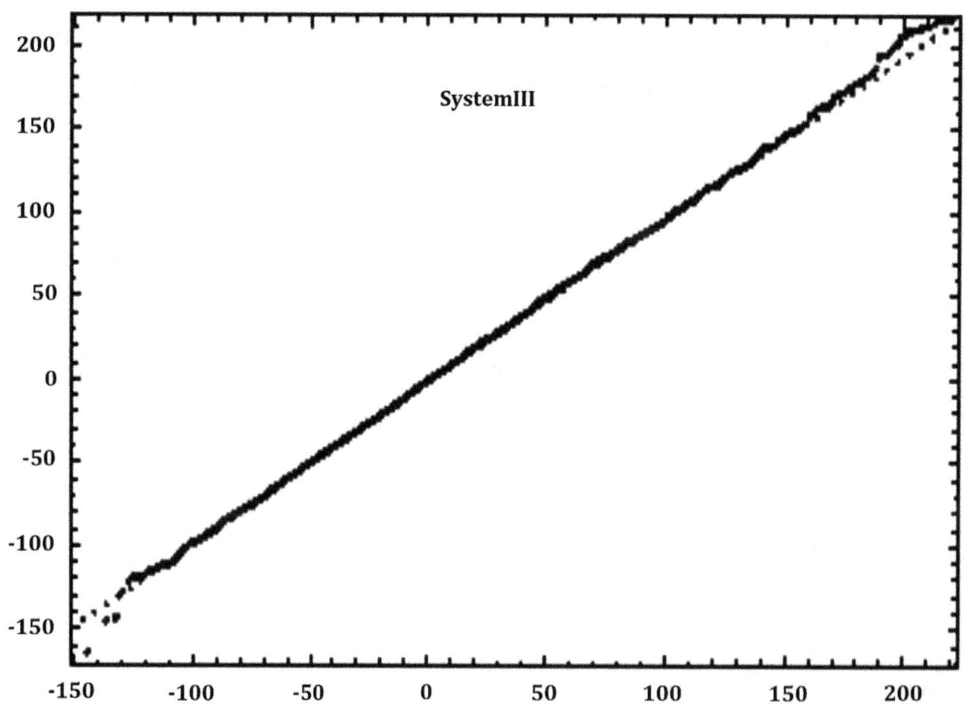

Figure 7. Stable distribution in three scenarios of alpha parameter $(\alpha = 0.5, \alpha = 1 and \alpha = 2)$ *(Source: Author's own elaboration)*

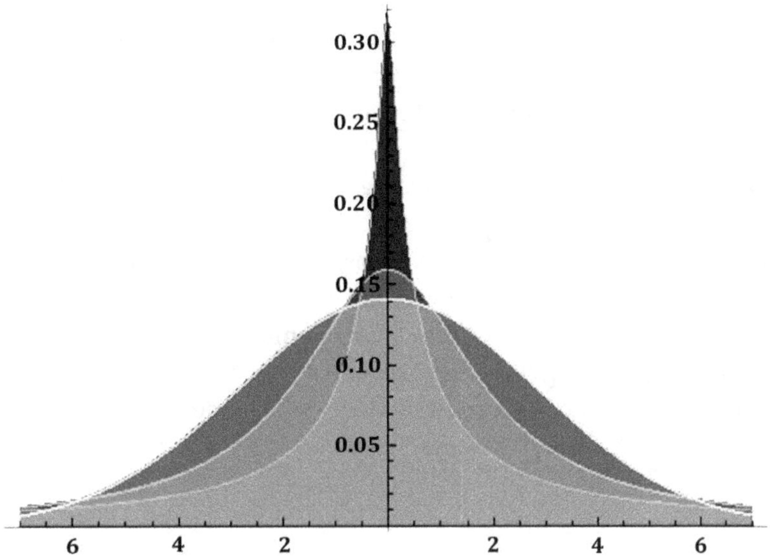

With the data the three scenarios considered to simulate the random variables, the next set of parameterswereused: $\alpha = (1.59, 1.69, 1.84)$, $\beta = 1.0$, $\gamma = (13.3, 42.8, 113.08)$and $\delta = (35.9, 121.9, 227.76)$. The effect of parameter alpha in the distribution function fitted is showed in the Figure 8. The parameter alpha measure the impulsivity of data that is the reason because is one of the most important parameters for this analysis.

Figure 8. Stable distribution for simulations in three scenarios of alpha parameter observed $(\alpha = 1.59, \alpha = 1.69 and \alpha = 1.84)$ *(Source: Author's own elaboration)*

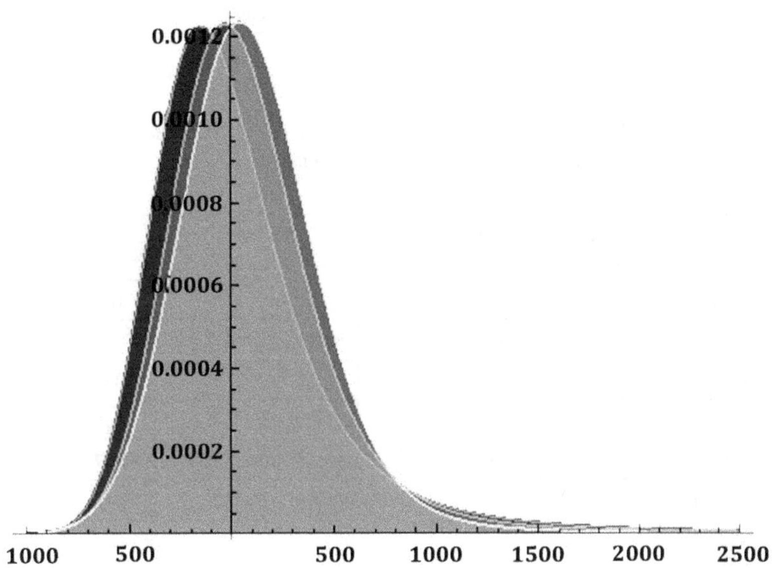

After last analysis and by Chambers-Mallows-Stuck method, we created 300 instances for each original test system according to the original energy demand ranges. The Numerical behaviors of some of these instances are showed in next section.

Additionally, three ARIMA models were estimated to compare two methods to generate demand data for the three systems analyzed. Dickey Fuller test was performed for stationary test series. SYS-104 system data series is stationary in levels.IEEE-24 system is stationary in first difference and IEEE-118 is stationary in second difference. See Table 4.

The autoregressive components and moving average shown the characterization of models estimated see Tables 5, 6 and 7.

Based on the time series models for the three systems, were forecast 50 observations of each one. One of the main limitations of using ARIMA models for generating test systems is that each model is a specific case so it is not possible to generalize results and the number of observations forecasted are limited.

From the initial values of (p) and (q) definition, several alternative models with various combinations AR (p) and MA (q) are proposed. The proposed models are compared to each other using the value of the coefficients and criteria of Akaike (AIC) and Schwarz (SC).

Table 4. Dicky Fuller test (Source: Author's own elaboration)

	Lags (n)	Test statistic	5% Critical value	Mackinnon approximate p-value for Z(t) at 95%
Sys-104	0	-38.910	-2.860	0.0000
IEEE-24	1	-2.573	-2.874	0.0101
IEEE-118	2	-3.029	-2.874	0.0323

Table 5. Results of estimations for ARIMA (3, 0, 1)-Sys-104 (Source: Author's own elaboration)

Variable	Coefficient	Standard error	P-value (5%)
Constant	265.5036	9.086378	0.0000
AR(1)	-0.860231	0.046913	0.0000
AR(2)	0.085455	0.034277	0.0130
AR(3)	0.150288	0.027799	0.0000
MA(1)	0.886205	0.027199	0.0080
Wald Chi2	608.6600		
Prob. > Chi2	0.000000		

Table 6. Results of estimations for ARIMA (2, 1, 0)-IEEE-24 (Source: Author's own elaboration)

Variable	Coefficient	Standard error	P-value (5%)
Constant	-0.008357	1.665174	0.0090
AR(1)	0.241515	0.024658	0.0000
AR(2)	0.010278	0.047456	0.0300
Wald Chi2	123.5100		
Prob. > Chi2	0.000000		

Table 7. Results of estimations for ARIMA (2, 2, 1)-IEEE-118(Source: Author's own elaboration)

Variable	Coefficient	Standard error	P-value (5%)
Constant	-0.000015	0.003098	0.0096
AR(1)	0.133620	0.008207	0.0000
AR(2)	0.084121	0.012665	0.0000
MA(1)	-0.102000	0.038950	0.0060
Wald Chi2	857.4416		
Prob. > Chi2	0.000000		

6. COMPUTATIONAL RESULTS

In this section, we present the results to the computational experiments that we carried out to evaluate the performance of the generated test systems by the proposed method (See Figures 9-14). All simulations and mathematical models were carried out on an AMD Phenom ™ II N970 Quad-Core with a 2.2 GHz processor and 4 GB RAM. Table 8 shows the results of the Unit Commitment for 15 generated instances by DICOPT solver.

The mathematical model of Multiperiod Unit Commitment was implemented in GAMS (Brooke et al. 1998). For Lagrangian Relaxation we use the solver DICOPT (Brooke et al. 2010) for solving the MINLP problems (dual subproblems). CONOPT for solving the NLP problems (primal sub problems) and CPLEX (Brooke et al. 2010) for MIP problems (master primal problem).

Concerning the computational times, we note that they are similar than original systems. One must highlight that the computational times do not exceeded more than 30 minutes. In general, the generated instances by the proposed method do not have numerical problems in the optimization process, which normally occurs in Unit Commitment Problem.

Figure 9. Bounds and computational time in IEEE-24 bus system by DICOPT Solver (Source: Author's own elaboration)

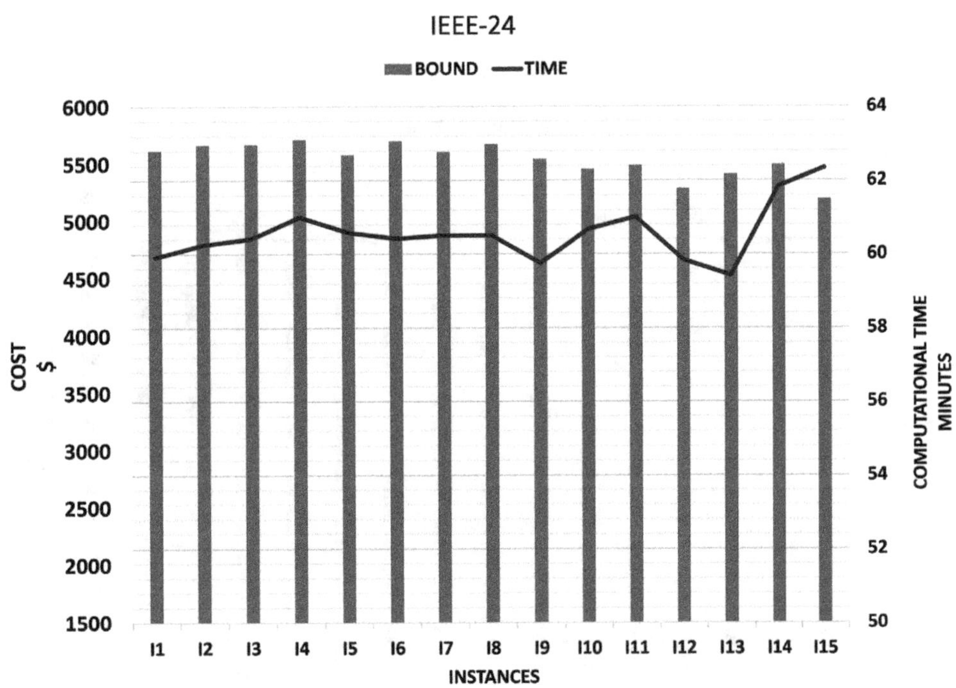

Figure 10. Bounds and computational time in Sys-104 bus system by DICOPT Solver (Source: Author's own elaboration)

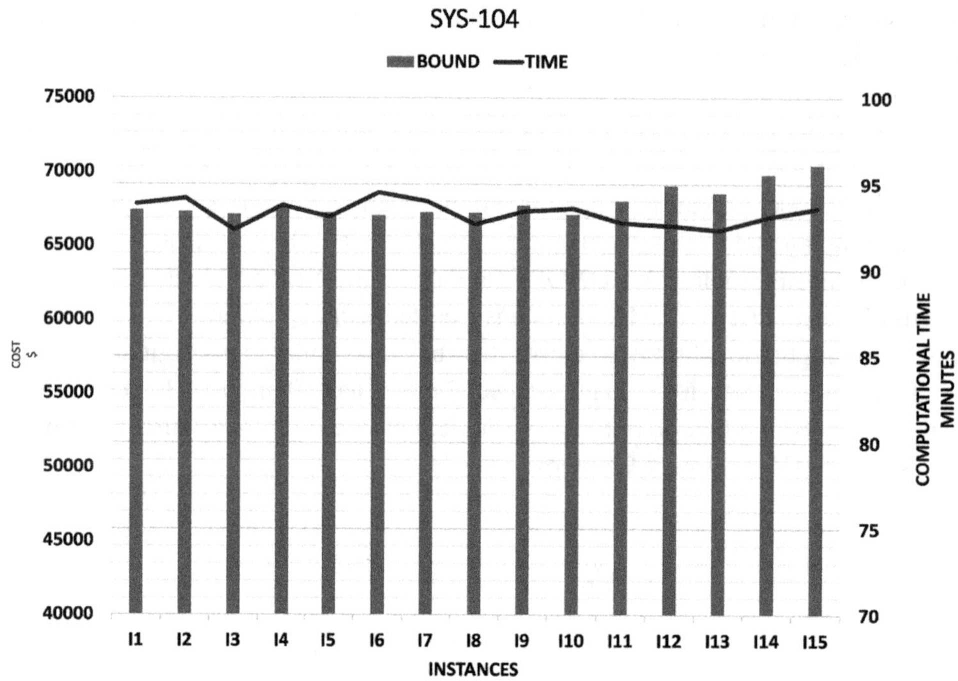

Figure 11. Bounds and computational time in IEEE-118 bus system by DICOPT Solver (Source: Author's own elaboration)

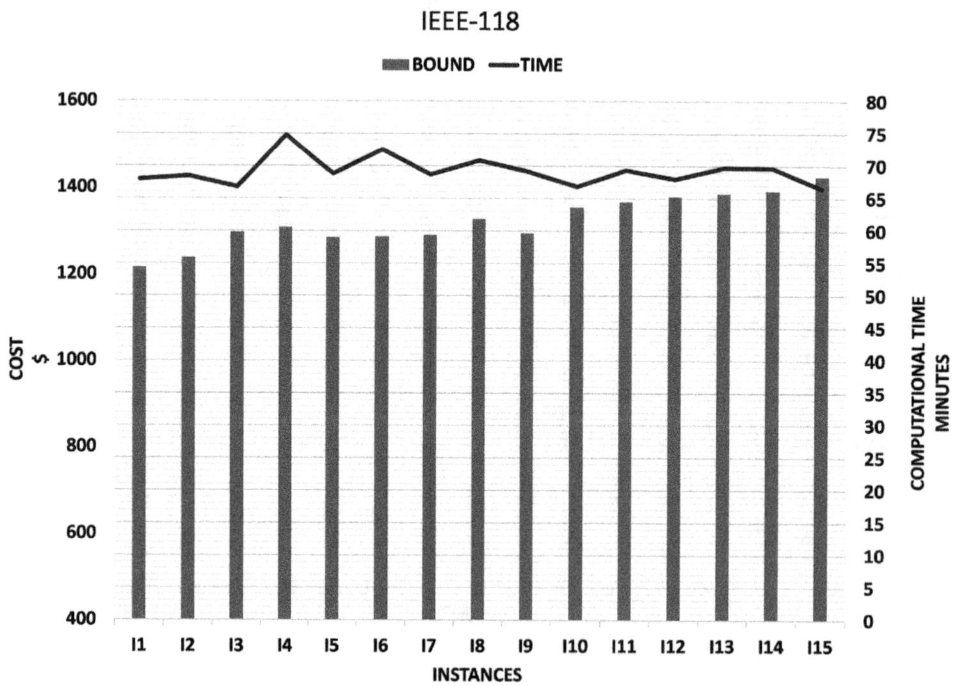

Figure 12. Lagrangian bound of IEEE-24 (Source: Author's own elaboration)

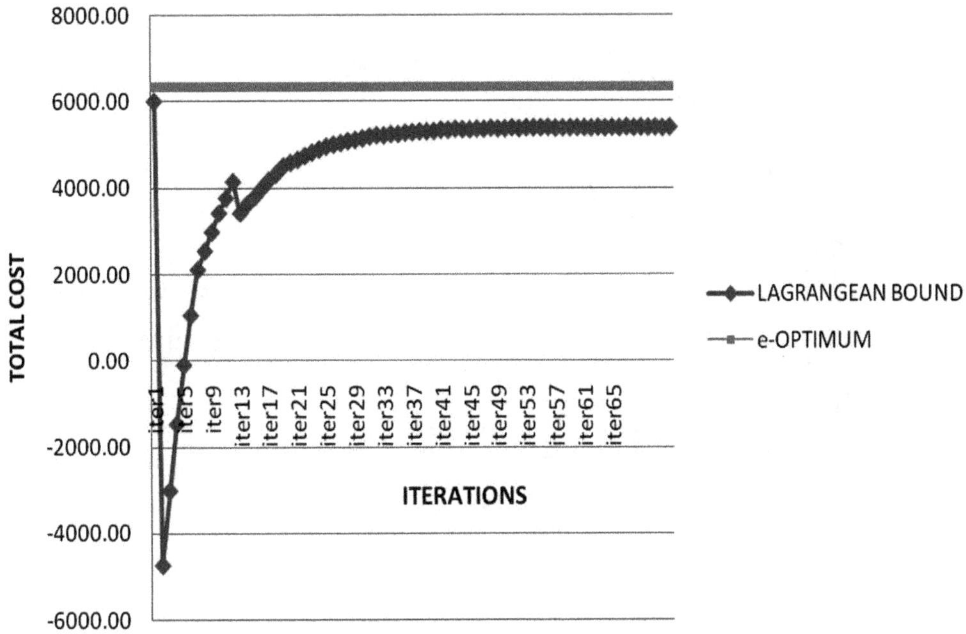

Figure 13. Lagrangian bound of SYS 104 (Source: Author's own elaboration)

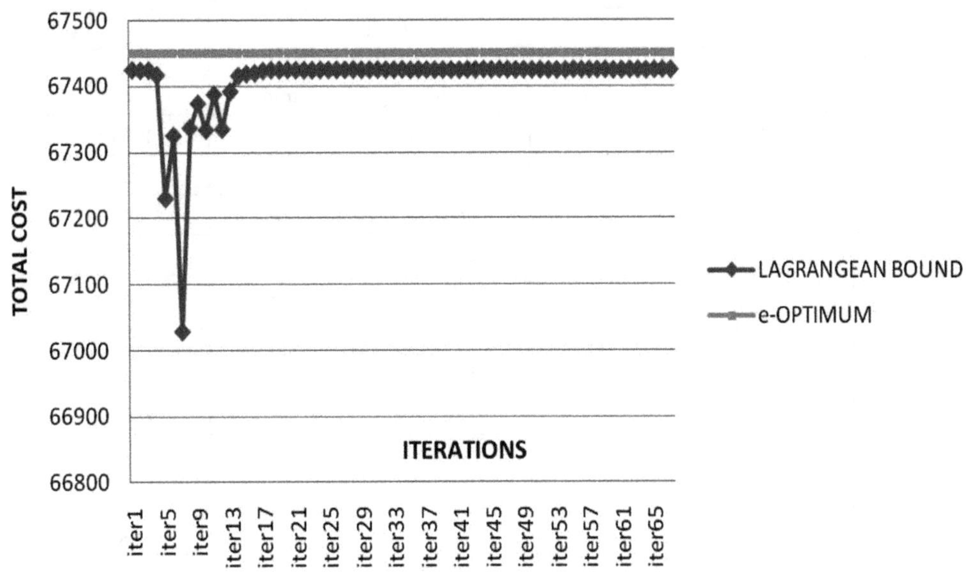

Figure 14. Lagrangian bound of IEEE-118 (Source: Author's own elaboration)

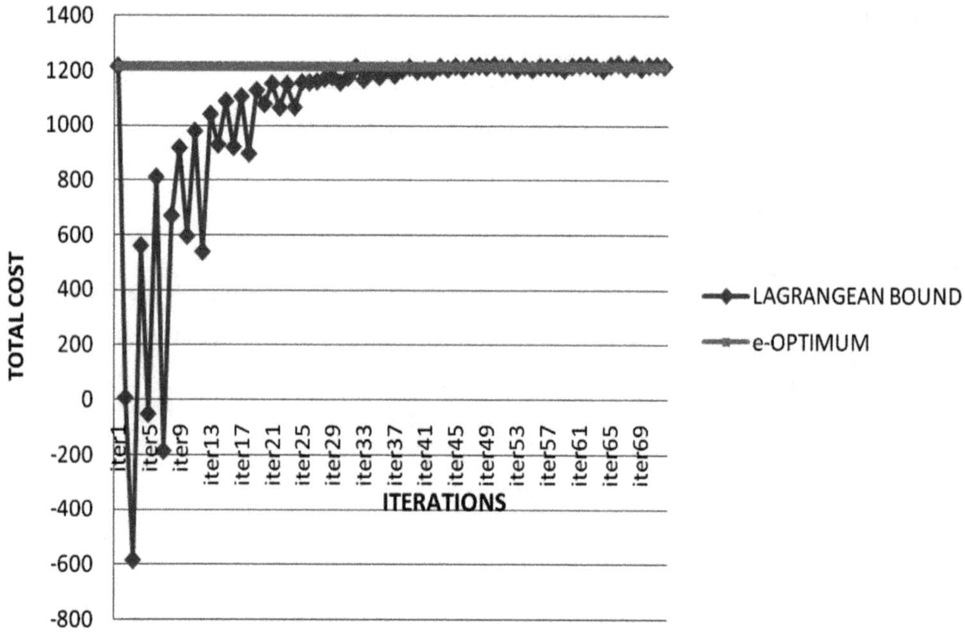

Table 8. Results obtained by DICOPT Solver

	SYS-104		IEEE-24		IEEE-118	
INSTANCE	**BOUND $**	**TIME min**	**BOUND $**	**TIME min**	**BOUND $**	**TIME min**
I1	67398.22	34	5616.41	18	1215.18	23
I2	67281.37	34	5665.61	18	1237.5	23
I3	67104.12	32	5669.13	18	1296.74	22
I4	67623.65	34	5711.97	19	1307.85	30
I5	67207.35	33	5579.78	19	1284.37	24
I6	67027.43	35	5699.76	18	1286.55	28
I7	67250.75	34	5603.32	19	1290	24
I8	67211.65	33	5670.61	19	1327.28	26
I9	67708.45	33	5543.96	18	1294.47	24
I10	67088.38	34	5453.76	19	1354.32	22
I11	68018.36	33	5484.37	19	1366.83	24
I12	69096.21	33	5283.34	18	1379.43	23
I13	68526.73	32	5408.32	17	1385.47	25
I14	69827.1	33	5491.87	20	1392.12	25
I15	70466.12	34	5192.34	20	1425.01	22

7. FUTURE RESEARCH DIRECTIONS

As future research is necessary to explore the presence of long memory in the series of demand for electric energy which could allow the use of fractal analysis techniques for generating test systems in combination with alpha stable distributions.

It is convenient to compare the results generated by the proposed methodology with other more sophisticated forecasting techniques such as genetic algorithms, evolutionary computation and neural networks, among other techniques. Previously evaluating the feasibility of its use in modeling the demand for electricity.

It is convenient to explore the combination of the above techniques and their usefulness for solving algorithms and their application in other optimization models (Vasant, 2015). With the aim of generating a wealth of knowledge focused on the generation of test systems much more attached to reality that can generate more efficient optimization algorithms.

8. CONCLUSION

Nowadays, we still lack the existence of standardized test systems that can be used to benchmark the performance and solution quality of proposed techniques.

Several authors report that the electric load pattern is very complex. It is therefore necessary to develop new methods. Therefore, we compared two methods to generate test systems. However, we found that the time series methods perform poorly.

The existence of herd behavior and complex microeconomic structures in the different electric load pattern produce the existence of impulsiveness of electrical demand series. These characteristics may lead to incorrect statistical analysis if the normality assumptions is not hold. The proposed methodology gives the researchers a measurement of the accuracy of such assumption.

This paper showed empirical evidence that the electrical demand presents fat tails distributions and the normality assumption made in the mainstream demand analysis are not always fulfilled, making them of little use to decision making or prognosis, especially in markets as power generation. Stable distributions can be an option to model power systems and generate test systems. The results show that the methods of stable distributions generate better test systems than those generated by time series models.

By introducing Chambers-Mallows- Stuck method, we presented a new way to generate electrical test systems for economic analysis. We show that the electrical demand is an alpha stable random variable, so we use it to generate several instances for the original test systems.

The generation of three scenarios through parameters alpha stable distribution estimates for the observed electrical power systems allowed cover a wider range of possible outcomes for simulated test systems. These parameters allowed simulate impulsivity, symmetry, location and dispersion of the observed data, and unlimited number of instances. Being the alpha parameter most important because it is what determines the impulsiveness of the series and therefore the presence of outliers in the data generated.

Electrical demand presents a greater degree of impulsivity due to the presence of peaks in the series during the hours of the day and seasons of high-energy demand. By modeling the demand through the use of alpha stable distribution can catch the real behavior of the electrical demand and build possible extreme scenarios, each scenario corresponds to a price-elastic demand curve.

The comparison of the results generated with forecasts of demand trough ARIMA models and data generated by the simulation of random variables alpha stable. Shows that the data generated by the ARIMA models are specific to each system power generated not be so in the case of alpha stable random variables, as shown in the generation of demands in the range of parameters set by the stable alpha distributions fitted. Therefore, the number of instances generated through simulation stable alpha allows generate more consistent and robust data that following the behavior of the real systems. This makes it possible to generate algorithms that are more efficient.

We test all generated instances in the Multiperiod Unit Commitment Problem and the Lagrangian relaxation. The results show that the proposed methodology is relevant, obtaining feasible solutions with GAMS solver in the same way of the original systems.

This work contributes to have standardized test systems that can be used to benchmark the performance of many proposed techniques.

ACKNOWLEDGMENT

The authors are very grateful to all the collaborators of the editorial, for their welcoming and approval of this chapter and for their guidance, help, and close advice at every stage. We wish all the readers an enjoyable, instructive, and inspiring study of the contributions of this chapter.

REFERENCES

Alguacil, N., & Conejo, A. (2000). Multiperiod optimal power flow using benders decomposition. *IEEE Transactions on Power Systems*, *15*(1), 196–201. doi:10.1109/59.852121

Belov, I., Kabasinskas, A., & Sakalauskas, L. (2006). A study of stable models of stock markets. *Information Technology and Control.*, *35*(1), 34–56.

Belov, I. A. (2005). On the computation of the probability density function of α-stable distributions. Mathematical Modelling and Analysis. *Proceedings of the 10th International Conference MMA 2005.*

Brooke, A., Kendrick, D., & Meeraus, A. (1998). *GAMS: A User's Guide*. The Scientific Press.

Brooke, A., Kendrick, D., & Meeraus, A. (2010a). *GAMS / Cplex 12. User Notes*. GAMS Development Corporation.

Brooke, A., Kendrick, D., & Meeraus, A. (2010b). *GAMS / DICOPT. User Notes*. GAMS Development Corporation.

Chambers, J. M., Mallows, C. L., & Stuck, B. W. (1976). A Method for Simulating Stable Random Variables. *JASA*, *71*(354), 340–344. doi:10.1080/01621459.1976.10480344

Charman, P., Bhavaraju, P., & Billington, R. (1979). IEEE reliability test system. *IEEE Transactions on Power Apparatus and Systems*, *98*(6), 2047–2054.

Christie, R. (1993). *IEEE 118 bus test case*. College of Engineering, Electric Engineering, University of Washington. Available in: http://www.ee.washington.edu/research/pstca

Diniz, A. L. (2010). *Test Cases for Unit Commitment and Hydrothermal Scheduling Problems*. IEEE Power and Energy Society General Meeting, Minneapolis, MN.

DuMouchel, W. H. (1973). On the asymptotic normality of the maximum-likelihood estimate when sampling from a stable distribution. *Annals of Statistics*, *1*(5), 948–957. doi:10.1214/aos/1176342516

Lévy, P. (1924). Théorie des erreurs la loi de Gauss et les lois exceptionelles. *Bulletin Soc. France.*, *52*, 49–85.

Marmolejo, J. A., Litvinchev, I., Aceves, R., & Ramirez, J. M. (2011). Multiperiod optimal planning of thermal generation using cross decomposition. *Journal of Computer and Systems Sciences International*, *50*(5), 793–804. doi:10.1134/S1064230711050145

Mittnik, S., Doganoglu, T., & Chenyao, D. (1999). Computing the probability density function of the stable Paretian distribution. *Mathematical and Computer Modelling*, *29*(10), 235–240. doi:10.1016/S0895-7177(99)00106-5

Nolan, J. P. (1997). Numerical calculation of stable densities and distribution functions. *Communications in Statistics. Stochastic Models*, *13*(4), 759–774. doi:10.1080/15326349708807450

Nolan, J. P. (1999). *Fitting Data and Assessing Goodness-of-fit with Stable Distributions*. Department of Mathematics and Statistics, American University. Available in: http://academic2.american.edu/~jpnolan/

Nolan, J. P. (2005). *Modeling financial data with stable distributions*. Department of Mathematics and Statistics, American University. Available in http://academic2.american.edu/~jpnolan/

Rodríguez, R. & Ake, S. (2013). Valuación de opciones de tipo de cambio asumiendo distribuciones α-estables. *Contaduría y Administración*, *58*(4).

Samorodnitsky, G., & Taqqu, M. (1994). *Stable Non-Gaussian Random Processes: Stochastic models with infinite variance*. New York: Chapman and Hall.

Scalas, E., & Kim, K. (2007). The art of Fitting Financial Time Series with Lévy Stable Distributions. *Journal of the Korean Physical Society*, *50*(1).

Stephens, M. A. (1977). Goodness of Fit for the Extreme Value Distribution. *Biometrika*, *64*(3), 583–588. doi:10.1093/biomet/64.3.583

Vasant, P. (2015). *Handbook of Research on Artificial Intelligence Techniques and Algorithms*. Hershey, PA: IGI Global; doi:10.4018/978-1-4666-7258-1

Vasant, P., Barsoum, N., & Webb, J. (2012). *Innovation in Power, Control, and Optimization: Emerging Energy Technologies*. Hershey, PA: IGI Global; doi:10.4018/978-1-61350-138-2

Vasant, P. M. (2013). *Meta-Heuristics Optimization Algorithms in Engineering, Business, Economics and Finance*. Hershey, PA: IGI Global; doi:10.4018/978-1-4666-2086-5

Vasant, P. M. (2014). *Handbook of Research on Novel Soft Computing Intelligent Algorithms: Theory and Practical Applications*. Hershey, PA: IGI Global; doi:10.4018/978-1-4666-4450-2

Wood, A. J., & Wollenberg, B. F. (1996). *Power generation, operation and control*. John Wiley & Sons.

Zhang, X.-P., & Schaffner, C. S. (Eds.). (2010). *Minutes of the Working Group meeting on Test Systems for Economic Analysis, 2007, 2008, 2009 and 2010*. Available at http://www.ieee.tea.ethz.ch/wiki/WG_meetings

Zolotarev, V. M. (1986). One-dimensional Stable Distributions, Volume 65 of Translations of mathematical monographs. American Mathematical Society.

ADDITIONAL READING

Benders, J. F. (1962). Partitioning Procedures for Solving Mixed-variables Programming Problems. *Numerische Math*, *4*(1), 238–252. doi:10.1007/BF01386316

Cohen, A., & Yoshimura, X. (1983). A Branch and Bound Algorithm for Unit Commitment. *IEEE Transactions on Power Apparatus and Systems*, *102*(2), 444–451. doi:10.1109/TPAS.1983.317714

Conejo, A. J. Castillo, & Minguez, E.R. et al. (2006). Decomposition Techniques in Mathematical Programming. Springer.

Deb, K. (2009). *Optimization for Engineering Design: Algorithms and Examples*. PHI Learning Pvt. Ltd.

Dieu, V. N., & Schegner, P. (2012). Real Power Dispatch on Large Scale Power Systems by Augmented Lagrange Hopfield Network. *International Journal of Energy Optimization and Engineering*, *1*(1), 19–38. doi:10.4018/ijeoe.2012010102

Dillon, T., Edwin, K., Kochs, D., & Taud, R. J. (1978). Integer Programming Approach to the Problem of Optimal Unit Commitment with Probabilistic Reserve Determination. *IEEE Transactions on Power Apparatus and Systems*, *97*(6), 2154–2166. doi:10.1109/TPAS.1978.354719

Drud, A. S. (1994). CONOPT—A Large Scale GRG Code. *ORSA J. Computing*, *6*(2), 207–216. doi:10.1287/ijoc.6.2.207

Eynard, J., Grieu, S., & Polit, M. (2011). Wavelet-based multi-resolution analysis and artificial neural networks for forecasting temperature and thermal power consumption. *Engineering Applications of Artificial Intelligence*, *24*(3), 501–516. doi:10.1016/j.engappai.2010.09.003

Goldsmid, H. J. (2009). The thermoelectric and Related Effects, pp. 1-6. In introduction to Thermoelectricity. Springer.

Gozde H. & Taplamacioglu M.C. (2011). J. Automatic generation control application with craziness based particle swarm optimization in a thermal power system. J. of Electrical Power & Energy Systems.

Granville, S., Pereira, M., Dantzig, G. B., ... (1998). Mathematical Decomposition Techniques for Power Systems. EPRI Tech. Rep. 2473–2476.

Hongchao, Y., & Wenzhi, D. (2010). Optimal Operational Planning of Steam Power Systems Using an IPSOSA algorithm. *Journal of Computer and Systems Sciences International*, *49*(5), 750–756. doi:10.1134/S1064230710050096

Kazarlis, S., Bakirtzis, A., & Petridis, V. (1996). A Genetic Algorithm Solution to the Unit Commitment Problem. *IEEE Transactions on Power Systems*, *11*(1), 83–92. doi:10.1109/59.485989

Khanh, D. V. K., Vasant, P., Elamvazuthi, I., & Dieu, V. N. (2014). Optimization of thermo-electric coolers using hybrid genetic algorithm and simulated annealing. *Archives of Control Sciences*, *24*(2), 155–176. doi:10.2478/acsc-2014-0010

Kou, H.-S., Lee, J.-J., & Chen, C.-W. (2008). Optimum thermal performance of microchannel heat sink by adjusting channel width and height. *International Communications in Heat and Mass Transfer*, *35*(5), 577–582. doi:10.1016/j.icheatmasstransfer.2007.12.002

Kurbastky, V., Sidorov, D., Tomin, N., & Spirayev, V. (2014). Optimal Training of Artificial Neural Networks to Forecast Power System State Variables. *International Journal of Energy Optimization and Engineering*, *3*(1), 65–82. doi:10.4018/ijeoe.2014010104

Lee, K. Y., & El-Sharkawi, M. A. (2008). *Modern Heuristic Optimization Techniques: Theory and Applications to Power Systems*. Wiley. doi:10.1002/9780470225868

Mantaway, A., Abdel-Magid, Y., & Selim, S. (1998). A Simulated Annealing Algorithm for Unit Commitment. *IEEE Transactions on Power Systems*, *13*(1), 197–204. doi:10.1109/59.651636

Mantaway, A., Abdel-Magid, Y., & Selim, S. (1998). Unit Commitment by Tabu Search. IEEE Proc. Generation, Transmission and Distribution, *145*(1), 56–54. doi:10.1049/ip-gtd:19981681

Pereira, M., & Pinto, L. (1983). Application of Decomposition Techniques to the Mid and Short Term Scheduling of Hydrothermal Systems. *IEEE Transactions on Power Apparatus and Systems*, *102*(11), 3611–3618. doi:10.1109/TPAS.1983.317709

Pezzini, P., Gomis-Bellmunt, O., & Sudri-Andreu, A. (2011). Optimization techniques to improve energy efficiency in power systems. *Renewable & Sustainable Energy Reviews*, *15*(4), 2028–2041. doi:10.1016/j.rser.2011.01.009

Ruzic, S., & Rajakovic, N. (1991). A New Approach for Solving Extended Unit Commitment Problem. *IEEE Trans. PWR.*, *6*(1), 269–275.

Som, T., & Chakraborty, N. (2013). Cost Minimization of Power Delivery Systems by Optimal Power Generation from Distributed Energy Resources Using Real Valued Cultural Algorithm. *International Journal of Energy Optimization and Engineering*, *2*(2), 44–61. doi:10.4018/ijeoe.2013040104

Tseng, C. L. (1996). On Power System Generation Unit Commitment Problems, Ph. D. Dissertation. UC Berkeley, December 1996.

Vasant, P., Weber, G., Barsoum, N., & Dieu, V. N. (2015). Power, control, and optimization. *TheScientificWorldJournal*, *2015*, 1–2. doi:10.1155/2015/964581 PMID:25789339

Virmani, S., Imhof, K., & Mukherjee, S. (1989). Implement of a Lagrange Relaxation Based Unit Commitment Problem. *IEEE Transactions on Power Systems*, *4*(4), 1373–1379. doi:10.1109/59.41687

Xinghui, Z., Jianshe, K., Lei, X., & Jianmin, Z. (2015). Alpha Stable Distribution Based Morphological Filter for Bearing and Gear Fault Diagnosis in Nuclear Power Plant. *Science and Technology of Nuclear Installations*. doi:10.1155/2015/460131

Xu, X., & Wang, S. (2007). Optimal simplified thermal models of building envelope based on frequency domain regression using genetic algorithm. *Energy and Building*, *39*(5), 525–536. doi:10.1016/j.enbuild.2006.06.010

Yu, G., Li, C., & Zhang, J. (2013). A new statistical modeling and detection method for rolling element-bearing faults based on alpha-stable distribution. *Mechanical Systems and Signal Processing*, *41*(1-2), 155–175. doi:10.1016/j.ymssp.2013.08.015

KEY TERMS AND DEFINITIONS

Economic Dispatch: The operation of generation facilities to produce energy at the lowest cost to reliably serve consumers, recognizing any operational limits of generation and transmission facilities.

Electric Network: Consists of many generation nodes with various generating capacities and cost functions, lines of transmission and nodes of power demand.

Heavy Tails: Heavy tail means that there is a larger probability of getting extreme values in a series.

Stable Distribution: Are a class of probability distributions which allowing skewness and heavy tails. They are described by and index of impulsiveness, and skewness parameter, a scale parameter, and a location parameter.

Unit Commitment Problem: The problem of unit commitment involves finding the least-cost dispatch of available generation resources to meet the electrical load.

Chapter 7
Stochastic Optimization of Manufacture Systems by Using Markov Decision Processes

Gilberto Pérez Lechuga
Universidad Autónoma del Estado de Hidalgo, Mexico

Francisco Venegas Martínez
Instituto Politécnico Nacional, Mexico

Elvia Pérez Ramírez
Universidad Nacional Autónoma de México, Mexico

ABSTRACT

In real-world most of manufacturing systems are large, complex, and subject to uncertainty. This is mainly due to events as random demands, breakdowns, repairs of production machines, setup and cycle times, inventory fluctuations and more. If items move too quickly, workers may work too hard. If items move too slowly, workers may have great leisure times. However, must make decisions here and now regarding the operation of the system optimally and quickly. In practice, these decisions are based on recent statistics of the system behavior, in the experience of the analyst and the urgency of the solution. In this chapter, we present a real problem associated with the production of individual parts in metalworking industry for the refrigerators production. We develop a model based on the Markov Decision Process to study the dynamics of the trajectory of end products in a manufacturing line that works by process. Then, we propose a measure of the average production rate of the line by using the Monte Carlo method. We illustrate our proposal using a numerical example with real data obtained in situ.

1 INTRODUCTION

Manufacturing is the process of converting raw materials, components, or parts into finished goods that meet a customer's expectations or specifications. Manufacturing commonly employs a man-machine setup with division of labor in a large scale production to make a good with tools and/or machines by effecting chemical, mechanical, or physical transformation of materials, substances, or components, or by simulating natural processes, usually repeatedly and on a large scale with a division of labor. Manufacturing flow

DOI: 10.4018/978-1-4666-9644-0.ch007

line systems consist of material work and storage areas. Material flows from work area to storage area to work area visiting each station exactly once in a fixed sequence. At the beginning of the production line there is an entity named raw material warehouse, and at the end there is another warehouse named of finished product. The time that parts spend in work areas can be random; this randomness may be due to random processing times, random failures and repair events or both. Storage areas can hold only a finite amount of material and the machines are always working. The work areas are called the shop floor and the storage areas are called buffers. Thus, a manufacturing system is a method of organizing production and it involves flow lines, job shops, transfer lines, flexible manufacturing systems, flexible assembly systems, cellular systems and more (Buzacot & Shatikumar, 1993).

Stochastic manufacturing is the analysis of the manufacturing systems from a randomly approach. Models for production planning which do not recognize the uncertainty can be expected to generate inferior planning decisions as compared to models that explicitly account for the uncertainty. In practice, there are many forms of uncertainty that affects the production line: a) The environmental uncertainty and b) System uncertainty. The first concept includes the uncertainty of customer demands and the uncertainty of suppliers, the second is related with the uncertainties within the production process such as operation yield uncertainty, production lead time uncertainty, quality uncertainty, failure machines and changes to product structure see (Mula, García Sabater, & Lario, 2006).

A manufacturing system therefore may be looked upon as an independent group of sub-systems, each sub-system performing a distinct function. Different sub-systems may perform different functions, yet they are inter-related and require to be unified to achieve overall objectives of the organization.

The manufacturing systems can be classified in intermittent and continuous systems (Balduzi, Giua, & Seatzu, 2001). Intermittent production systems are those where the production facilities are flexible enough to handle a wide variety of products and sizes. In the intermittent system the goods are manufactured specially to fulfill orders made by customers rather than for stock. Here, the flow of material is intermittent. These can be used to manufacture those products where the basic nature of inputs changes with the change in the design of the product and the production process requires continuous adjustments. Considerable storage between operations is required, so that individual operations can be carried out independently for further utilization of men and machines. In continuous systems the items are produced for the stocks and not for specific orders. Before planning manufacturing to stock, a sales forecast is made to estimate likely demand of the product and a master schedule is prepared to adjust the sales forecast according to past orders and level of inventory. In this systems the inputs are standardized and a standard set of processes and sequence of processes can be adopted. After setting of master production schedule, a detailed planning is carried on. Basic manufacturing information and bills of material are recorded. Information for machine load charts, equipment, personnel and material needs is tabulated.

In order to model these systems, many conventional techniques and intuitive approaches used for years have become satisfactory (Vitanov, Julkaa, Bainesb, Tjahjonob, & Lendermanna, 2007). Mathematical model and methods together with other methodologies such as simulation (Mula, García Sabater, & Lario, 2006), (Smith, 2003), (Liu & Song, 2012), (Fowler & Rose, 2015), (Anglania, Griecoa, Pacellaa, & Toliob, 1996), (Metropolis, Rosenbluth, Teller, & Teller, 1953), Petri nets (Jonsson P, 2003), (Popescu, Soto, & Lastra, 2012), (Hsieh & Lin, 2014), (Liu & Song, 2012), (Leitao, Mendez, Bepperling, Cachapa, Colombo, & Reestivo, 2012), and more have been used to study, analyze and to control these systems. Production planning problems have been formulated and solved as optimization problems since the early 1950's, and an extensive literature has been developed. The widespread use of Enterprise Resource Planning (ERP) systems and developments in information technology and scientific computing have opened the way for even wider use of these techniques in industry. For a review, and models developed of stochastic methods for production planning see see (Mula, García Sabater, & Lario, 2006), (Goetschalckxa, Vidalb, & Dognac, 2002), (Christou, Lagodimos, & Lycopoulou, 2007), (Schoenherr, Tummala, & Harrison, 2008), (Sethi, Yan, Zhang, & Zhang, 2002), (Zanjania, Nourelfatha, & Kadia, 2010), (Hindawi Publishing Corporation (undated), Pandian Vasant Universiti Teknologi Petronas, Malaysia 0000-0002-0755-1046 Scholarly Contributions [Data Provided by scopus], 2015), (Hindawi Publishing Corporation (undated), Pandian Vasant Universiti Teknologi Petronas, Malaysia 0000-0002-0755-1046 Scholarly Contributions [Data Provided by scopus], 2015), (Lodree Jr & Uzochukwub, 2008).

Hierarchical analysis of decomposition is refered in (Gerchad & Henig, 1989), (Nakashima, Arimitsua, Nosea, & Kuriyamab, 2004), (Sipper & Bulfin Jr., 1997), (Lee, Jeong, & Moon, 2002), (Yildirim, Tan, & Karaesmen, 2005), (Bayazit, 2005), (Smith & Ball, 2012), (Rubenstein, Y, 1980). For a discussion of manufacturing planning and control see (Günter H-O & Veek P V, 2003), (Flug, 1996), (Wang & Lin, 2009), (Gustavsson & Wänström, 2009), (Olhager & Selldin, 2007), and for advanced planning and scheduling systems see (Kilger, Staedtler, & Meyr, 2008), (Xiadong, Fernandez-G, & Fu, 2004), (Chen & Ji, 2007), (Hvolby & Jensen, 2010), (Soman, Van Donk, & Gaalman, 2004), (Fletena & Kristoffersen, 2008).

Other authors model production systems with focus on queue lines. They use an infinite-server model to represent the component plants; that is, the production process at each plant is modeled as a $M/G/\infty$ queuing system, for more information on this topic see (Eynan & Rosenblat, 1996), (Moona & Seob, 2005), (Gallien & Wein, 2001), (Song & Yao, 2002), (Son & Venkateswaran, 2007), (Cochran, Arinez, Duda, & Linck, 2002).

2 BACKGROUND MODEL

A Markov chain is a random process with the property that the next state depends only on the current state. It is a Markov model, named for Andrey Markov, for a particular type of Markov process in which the process can only be in a finite or countable number of states. Markov chains are useful as tools for statistical modeling in almost all fields of modern applied mathematics. Andrey Markov produced the first results (1906) for these processes, purely theoretically. A generalization to countably infinite state spaces was given by Kolmogorov (1936). Markov chains are related to Brownian motion and the ergodic hypothesis, two topics in physics, which were important in the early years of the twentieth century, but Markov appears to have pursued this out of a mathematical motivation, namely the extension of the law of large numbers to dependent events.

Regarding the issue of modeling of manufacturing systems and similar areas have been written a lot of articles about it. For example, in (Riddalls, Bennett, & Tipi, 2000) the authors present a comprehensive discussion about the mathematical methods used to model and analyze supply chains by using the dynamics of the supply chains. Their method is appraised from a system dynamics perspective, the authors categorize their work in: continuous time differential equation models, discrete time difference equation models, discrete event simulation models and operational research techniques.

About the use of Markov decision processes to model certain parts of manufacture systems there are many references on it. In (Ryua & Jung, 2003), the document explains how to formulate and solve models of production planning and inventory management for manufacturing systems under uncertainties. For modeling of manufacturing systems in closed loop see (Csjáji & Monostor, 2005). In (Boute,

Disney, Lambrecht, & Houdt, 2007) the writers modeling a two-stage make-to-order supply chain as a production-inventory system, where the retailer's inventory replenishment lead times are endogenously determined by the manufacturer's production facility. They approach the problem like a queueing system i.e., the placing of the production orders is viewed as a queue that evolves like a markovian process. In the same direction, (Chao, 2013) addresses a supply chain with one manufacturer, one retailer, and some online customers. Through the Markov decision process, the document describes the optimal production and availability policy for a manufacturer to determine whether to produce one more unit of products and whether to indicate "in stock" or "out of stock" on website. The autor also simulate the production and availability policy via a myopic method. Another important authors about this theme can be found in (Buzacot & Shatikumar, Stochastic Models of Manufacturing Systems, 1993), (Puterman, 2005), (Yin, Yin, & Liu, 2004), (Doshi, Goodwin, Akkiraju, & Verma, 2004), (Benjaar & Elhafs, 2006), (Kyriakidis & Dimitrakos, 2006), (Ilgina & Guptab, 2010), (Gayon, Benjaafar, & Vricourt, 2008)

In this proposal, we develop a Markov decision process to illustrate the operation of an intermittent manufacturing system. In this, the different assemblies used to the construction of a finished product, visit once each workstation in an order of production pre-specified. The workstations represent the states of the system. We suggest the modeling of the randomness of the trajectories of products through transition probability matrices. In our approach, we use an initial set of transition probabilities obtained through the historical information from the operation of each machine in the manufacturing system. Subsequently, it is suggested to obtain the steady state distributions of the probability matrices through the simulation of these. The expected amount of production is obtained by counting the number of visits to the final station of the manufacturing process.

3 PROBLEM DEFINITIONS

Consider a manufacturing line that operates intermittently producing metal pieces to be used as assemblies in refrigerators for commercial use. Here, the production process is characterized in that each piece should to visit a sequence of workstations at once. The path taken depends on the refrigerator model to be fabricated, the type of work piece, the physical and chemical characteristics of the material to be processed, the processing speed of each machine, as well as the volume produced.

At the beginning of the line there is a warehouse of raw materials, and at the end of it, a warehouse of finished products. Typically, there is also a temporary buffer B between each workstation that is used to decouple the flow between two consecutive workstations (Pérez L, Gress, Karelyn, & Orán, 2009), (Figure 1).

In this configuration there are workstations that cannot be visited on a specific trajectory due that the product does not require of that manufacturing process. Assume that the plant is able to manufacture N different assemblies, the inventory of raw material for manufacturing is infinite and the process moves always forward.

Once the product to make is determined, we proceeds to select the material to do it, the machines that will be involved, their processing speeds and the amount of raw materials needed for manufacturing. Then, the material is taken from the warehouse of raw materials and begins its trajectory of machining according to the specifications of the Industrial Engineering Department. The processing times at each workstation are considered random and exponentially distributed. The model also assumes that, given a

Figure 1. Workstation, buffers and sequencing process

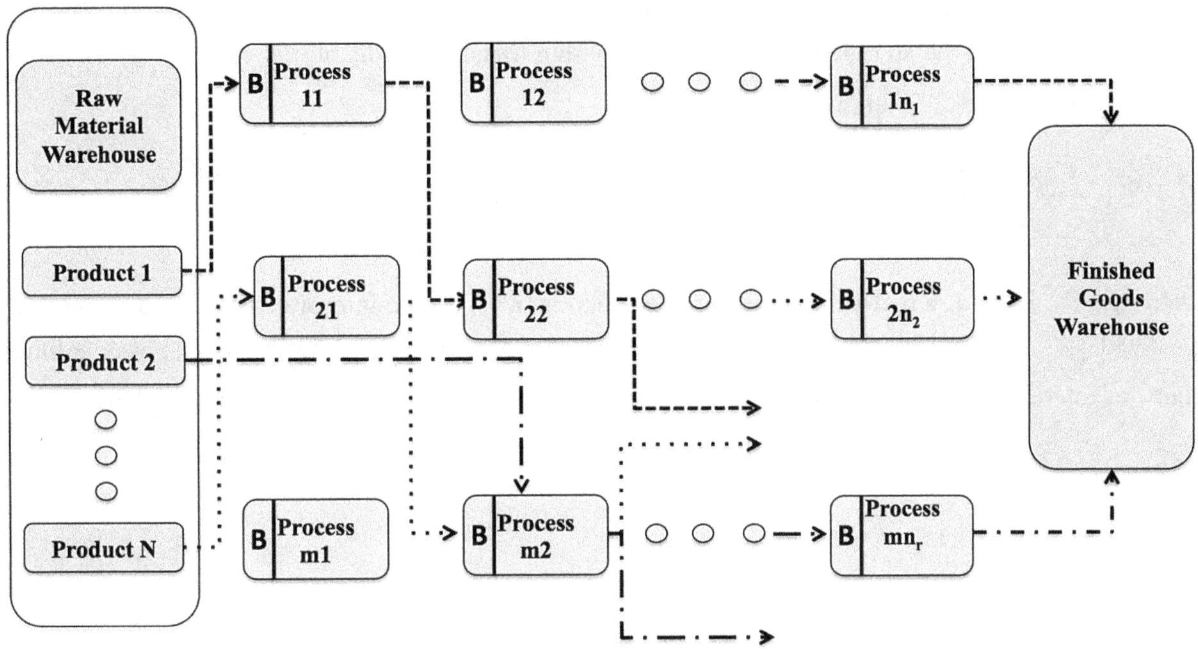

machining process, it can be done by one or more similar machines (Which may differ in capacity and operating conditions) whose nominal capacity is known.

3.1 The Mathematical Model

From the above description note that, each state of the manufacturing process is visited regardess the history of the process trajectory, *i.e.*, the next state to visit depends only on the state in which the process is currently. For this, the manufacturing process described above, can be modeled as a homogeneous Markov chain in discrete time with transition probabilities matrix given by $\mathcal{P}$. Formally, we define the stochastic process with discrete time $[M(n)]_{n=0,1,...}$ finite state space $I = \{z_1, ..., z_d\}$, where for all $n \in \mathbb{N}$ an all $\omega_0, \ldots, \omega_n \in I$ it holds that (Flug, 1996):

$$\mathbf{P}\left[M\left(0\right) = \omega_0, M\left(1\right) = \omega_1, \ldots, M\left(n\right) = \omega_n\right] =$$

$$\mathbf{P}\left(M\left(0\right) = \omega_0\right)\Pi_{i=1}^{n}\mathbf{P}\left(M\left(i\right)\omega_1 \middle| M\left(i-1\right) = \omega_{i-1}\right),$$

and $\mathbf{P}\left(M\left(n\right) = z_j \middle| M\left(n-1\right) = z_j\right) = p_{ij}$ for all n.

In the remainder of this chapter we assume that the Markov chain used to model the manufacturing process is positive-recurrent, and aperiodic, *i.e.*, ergodic. We also assume that each subassembly must pass through of at most d manufacturing processes. The idea is to generate the corresponding expected paths of the total process for each subassembly θ, θ = 1, ..., N

Let $\{X_n^\theta : n \geq 0\}$ be an ergodic Markov chain with a transition probabilities matrix $\{\pi_{kj}^\theta : k, j\epsilon I\}$ and state space $I \epsilon \mathbb{R}^d$. Then, we define

$$g_\theta = E\left[f\left(X^\theta\right)\right] = \sum_{j\epsilon I} f^\theta\left(j\right)P\left(X^\theta = j\right) = \sum_{j\epsilon I} f^\theta\left(j\right)p_j^\theta, \tag{3.1}$$

where $f : \mathbb{R}^d \to \mathbb{R}$ is a real-valued measurable function, $f^\theta(j)$ can be interpreted as a paid at the state $j, p_j^\theta = \mathrm{P}\left(X^\theta = j\right) = \lim_{m\to\infty} P\left(X_m^\theta = j\right)$ is the steady-state distribution. In our proposal, possible functions f of interest may be the following (Rubenstein, Y, 1980): $f(j) = 1$, if $j = k$ otherwise; then

$$\mathrm{E}\left[f\left(X\right)\right] = p_k \tag{3.2}$$

$f(j) = c_j$, value of being in state j; then

$$\mathrm{E}\left[f\left(X\right)\right] = \sum_{j\in I} c_j\mathrm{P}\left(x = j\right). \tag{3.3}$$

Here, Equation (3.3) represents the expected value per period. Then, in first instance we would be interested into obtain the smallest value of W_i, *i.e.*,

$$h^* = \min_{1\leq\theta\leq N} W_\theta = \min_{1\leq\theta\leq N} \mathrm{E}\left[f\left(X^\theta\right)\right], \tag{3.4}$$

that is, the minimum average cost of producing of any of the N subassemblies, and the path that they follows in the line of manufacturing.

Given the transition probabilities matrix $\mathcal{P}, X_0 = i,$ and $k \geq 0$, note that, Equation (3.7), is a Bernoulli random variable with probability of success $p_{ij}^{(k)} = \mathrm{E}\left[f^k\left(j\right)\right]$ It follows that, the number of visits $\upsilon_{ij}^{(n)}$, made to state j in first n steps (step 0 to step $n - 1$) from state i initially is $\upsilon_{ij}^{(n)} = \sum_{k=0}^{n-1} f^{(k)}\left(j\right)$ and that the expected number of visits is given by, $\mathrm{E}\left[\upsilon_{ij}^{(n)}\right] = \sum_{k=1}^{n-1} p_{ij}^{(k)}$, or in matrix form

$$V^{(n)} = \mathrm{E}\left[\left[\upsilon_{ij}^{(n)}\right]\right]_{i,j\in I} = \sum_{\xi=0}^{n-1} \mathcal{P}^{(\xi)} = p^\xi \tag{3.5}$$

Given that the chain is initially in state i, the expected number of visits to state j throughout the lifetime of the chain is $E\left[\upsilon_{ij}\right] = \lim_{n\to\infty} E\left[\upsilon_{ij}^{(n)}\right]$, or in matrix form

$$V^{\infty} = V = \left[E\left[\upsilon_{ij}\right]\right]_{i,j\in I} = \sum_{\xi=0}^{\infty} \mathcal{P}^{\xi} \tag{3.6}$$

In this proposal we are interested in to calculate $\left[E\left[\upsilon_{ij}\right]\right]_{i,j\epsilon I}$, when i and j are the initial and end nodes of the manufacturing system.

3.2 The Dynamic Programming and Markov Decision Processes Approach

Consider a discrete time Markov decision processes with a finite space state $I = \{1, 2, ..., N\}$ and a finite action set D. A policy s is a map assigning to each state i an action $s\left(i\right)\epsilon D$.

Let p_{ij}^{d} be the transition probability from state i to state j if the action $d \; \epsilon \; D$ is taken. In general, under this approach, the decision to be made is whether the subassembly θ will visit to the station i from the station j. The payment we should do when the state i is observed, and the action d is taken will be denoted as f_{t}^{d}. The time interval between two transitions is called a stage and will be denoted by the Greek letter ρ.

An optimal policy is defined as a rule that minimizes some predefined objective function. The optimization technique used depends on the form of the objective function. Some optimality criteria commonly used in a finite planning horizon are the following (Kristensen, 1996), (Rubenstein, Y, 1980):

The total production cost after the product visited M stages (the total expected net returns over a fixed number of stages is maximized).

$$g\left(s_1,\dots,s_M\right) = \mathrm{E}\left(\sum_{m=1}^{M} f_{I(m)}^{s^m}\right) \tag{3.7}$$

where s^m is the policy at the stage m, and I is the unknown state at stage m.

If we want to get the production cost at any point in the evolution of the chain when the lengths of the intervals are equal we will use the expected discounted payments

$$g\left(s_1,\dots,s_M\right) = \mathrm{E}\left(\sum_{m=1}^{M} \beta^{m-1} f_{I(m)}^{s^m}\right) \tag{3.8}$$

Where $\beta < 1$ is a discount factor defined by the interest rate at the end of the stage length. Note that, equation (3.7) is a particular case of Equation (3.8) when $\beta = 1$.

To solve these instances, often an iterative method is used through of the following functional

$$J_i\left(n\right) = \min_{d}\left\{f_i^d + \beta\sum_{j=1}^{u} p_{ij} J\left(n-1\right), i = 1,\dots\right\} \tag{3.9}$$

The functional $J_i(n)$ represents the total expected discounted payment from the process when it starts from state i and will operate for k stages before termination. The boundary conditions of the functionalmean that $J_i(0)$ is the salvage value of the system when it is in state i.

Due to the nature of the problem under consideration here, would be wise to consider it as an infinite horizon problem. That is, we can assume that the number of stages of the model is unknown. In this case the minimization of the expected average payment per unite of time can be used. Thus, if all stages are of equal length the objective could be (Kristensen, 1996)

$$g(s) = \sum_{i \in I} \pi_i^s f_i^s \tag{3.10}$$

where π_i^s is the limiting state probability under the policy s. An optimal policy that minimizes the expected cost of producing N units is given by (Howard, 1971)

$$J_i(n) = \min_d \left\{ \alpha \left(n \frac{f_i^d}{z_i^d} + j_i(0) \right) + (1-\alpha) \left(f_i^d + \sum_{j \in I} p_{ij}^d J_j \left(n - {}_i^d \right) \right) \right\}, n = 1 \ldots \tag{3.11}$$

Where $\alpha = 1$ if $z_i^d \geq n$, and $\alpha = 0$ if $z_i^d < n$. Here, $\left(f_i^d / z_i^d \right)$ represents the rate (production-cost/output), this value remains constant throughout each stage. In this proposal z_i^d can be viewed as a physical output (e.g. number of parts produced per period).

3.3 Implementation Strategies

The problem (3.4) can be viewed as a regenerative process whose solution is well known through the equivalent problem defined as [57].

$$\underset{p}{\text{Minimize}} \left\{ \sum_{q=1}^{Q} \mathrm{E}\left\{ f\left(X_q \right) \right\} p_q = 1, \;\; p_q \geq 0, \;\; q = 1, \ldots, Q \right\} \tag{3.12}$$

where Q is the number of stochastic processes obtained by simulation from $\left\{ X_n^\theta : n \geq 0 \right\}$, and its solution is given by a vector p^* with a single non-zero component. In our proposal, we will address the problem by using an estimate of the value of the process path and then, selecting the minimum value thereof. Formally, we have the following discrete-time Markov chain simulation problem: Given a transition matrix π, initial distribution μ, and real-valued function f, calculate

$$\xi = \mathrm{E}_\pi f\left(X_0, X_1, \ldots \right) \tag{3.13}$$

where $X = \{X_n: n \geq 0\}$ is the Markov chain associated with π and μ, ($E\pi(\cdot)$, denotes the expectation on the path space on X associated with π and μ).

The dilemma of simulating the transition probabilities matrix is strongly linked to the problem of the Monte Carlo integration to evaluate the integral

$$R = \int_{\Omega} f(x)\, d\mu(x) \tag{3.14}$$

using random sampling. In its basic form, this is done by independently sampling N points, X_1, ..., X_N according to some convenient density function g, and then computing the estimate

$$\hat{R} = \frac{I}{N}\sum_{i=1}^{N}\frac{f(X_i)}{g(X_i)} \tag{3.15}$$

Where X is a random variable with density g. Here, the importance sampling derives from the fact that we can chose g to be large in the regions that are most important namely where $|f|$ is large. In this formulation, Equation (3.15) is named the Horvitz-Thompson estimator, and in our approach, we use a generalization of this technique to the simulation of ergodic Markov Chains.

An initial way of formulating the problem is as follows.

Let $\mu = (\mu(i) : i \in I)$ and $\pi = (\pi(i, j) : i, j \in I)$ be, where $I = \{1, 2, ...\}$ is the state space X. Also assume that f depends on the trajectory of X only up to the deterministic time n, *i.e.*, $f(X_0, X_1, ...,) = f(X_0, X_1, ..., X_n)$. Then, it follows that

$$\mathrm{E}_{\pi} f(X_0, X_1, \ldots, X_n) = \sum_{i_0, i_1, \ldots, i_n} f(i_0, i_1, \ldots, i_n)\mu(i_0)\prod_{k=0}^{n-1}\pi(i_k, i_{k+1}) \tag{3.16}$$

As in Equation (3.15), to select the function g we can proceed as follows. Let $g : I^{n+1} \rightarrow \mathbb{R}$ be a Probability Density Function (PDF) on I^{n+1} such that $g(i_0, i_1, ..., i_n) > 0$. Then, we have that (Glynn & Iglehar, 1989).

$$\begin{aligned} R &= \sum_{i_0, i_1, \ldots, i_n} f(i_0, i_1, \ldots, i_n)\frac{\mu(i_0)\prod_{k=0}^{n-1}\pi(i_k, i_{k+1})}{g(i_0, i_1, \ldots, i_n)} g(i_0, i_1, \ldots, i_n) \\ &= \mathrm{E}_g f(X_0, X_1, \ldots, X_n)\frac{\mu(X_0)\prod_{k=1}^{n-1}\pi(X_k, X_{k+1})}{g(X_0, X_1, \ldots, X_n)} \end{aligned} \tag{3.17}$$

where $E_g f(\cdot)$ is the expectation over trajectories X having a PDF given by g.

Although this approach is practical for the purposes of this analysis, there is a more afiordable alternative to achieve the simulation of the Markov chain, one is, the Metropolis-Hasting Algorithm due to Metropolis et al. (Metropolis, Rosenbluth, Teller, & Teller, 1953), which we will discuss later. Regardless of the selected alternative, the simulation of the Markov chain model requires an initial distribution, μ and a transition probabilities matrix, $\tilde{\pi}_\theta$, obtained from historical data information generated in the manufacturing plant.

Once cleared the above, we will simulate this initial matrix to obtain the steady state matrix of transition probabilities by using the Metropolis-Hasting Algorithm, for approximately generating samples from the arbitrary distribution. The aim is to simulate a Markov chain such that the stationary distribution of this chain coincides with the target distribution (Leung & Wub, 2004).

Later, we will generate a random variable *X* for each *θ*, taking values in *I* according to a target distribution $\{\mu_i\}$, where

$$\mu_i^\theta = \frac{\alpha_i}{\sum_{i\in I}\alpha_i}, \alpha_i > 0, i\in I, \theta = 1,2,\ldots,N$$

The Metropolis-Hasting Algorithm, suggest to construct a Markov chain $\{X_n: n \geq 0\}, n\in I, I\in \mathbb{R}^d$ whose evolution on the horizon is controlled by an arbitrary transition matrix $Q = q_{ij}$. The generic algorithm is as follows (Rubenstein & Kroese, 2007).

1. If $X_n = i$, generate a random variable *Y* such that $P(Y = j) = q_{ij}, j\in I$ and PDF given by the *d* points obtained from the *ith* row of *Q*.
2. If $Y = j$, then

$$X_{n+1} = \begin{cases} j, & \text{with probability } \hat{p}_{ij} \\ i, & \text{with probability } 1 - \hat{p}_{ij} \end{cases} \tag{3.18}$$

where

$$\hat{p}_{ij} = \min\left[1, \frac{\mu_j q_{ji}}{\mu_i q_{ij}}\right] = \min\left[1, \frac{\alpha_j q_{ji}}{\alpha_i q_{ij}}\right]$$

In the limit it can be ensure that (Rubenstein & Kroese, 2007), the one step transition probabilities matrix $\pi = \left(\pi(i,j) : i, j\in I\right)$ is given by

$$\pi(i,j) = \begin{cases} q_{ij}\hat{p}_{ij} & \text{when } i \neq j \\ 1 - \sum_{k\neq i} q_{ik}\hat{p}_{ik} & \text{when } i = j, k \in I \end{cases}$$

After obtaining the stationary distribution of the process, we proceed to obtain the expected number of visits to each node in the chain using the Equation (3.6).

Figure 2. Layout and example of path for the product θ

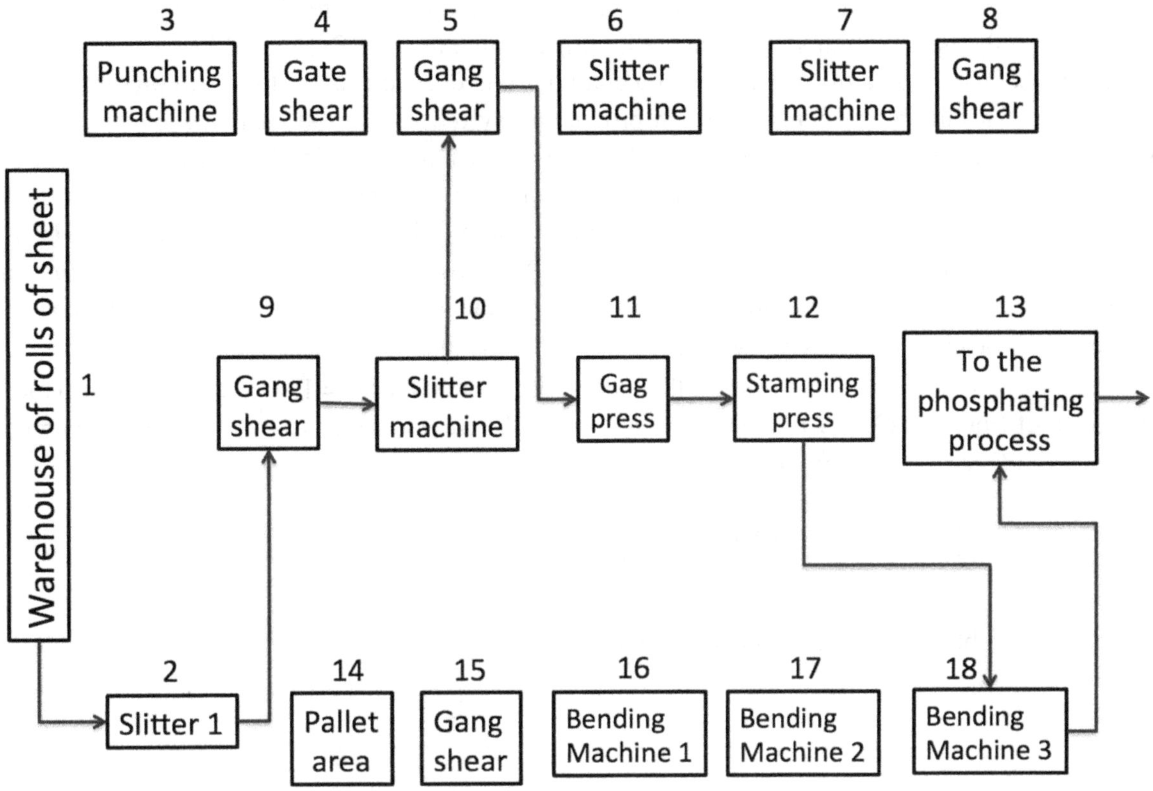

4 NUMERICAL EXAMPLE

Consider the following actual manufacturing system, which produces assemblies for refrigerator. (Figure 2) schematically shows the layout and an example of route taken by a piece. In this, the product θ made the following route: $1 \rightarrow 2 \rightarrow 9 \rightarrow 10 \rightarrow 5 \rightarrow 11 \rightarrow 12 \rightarrow 18 \rightarrow 13$.

Here, each number represents a workstation that the assembly should visit. In this case, the machines number 8, 9 and 15 can do the same manufacturing operation although at different speeds and capacities. Deprives the same situation in the group of machines number 6, 7, 10 and in the group of machines 16, 17 and 18. When more than equipment can do the same operation, its election is random if the corresponding workstation is available. Otherwise we choose the machine that is close to being vacated.

For this example, the numbers of states of the chain is 18 and correspond to the number of possible operations that the manufacturing line has ($d = 18$). We will consider a total of 4 different products to produce ($N = 4$).

As in [57], we also assume that the values of $f^{\theta}(j) = c_j$, are known but the transition matrices $\left\{\pi^{\theta}_{kj} : k, j, \in I, \theta = 1, \ldots, N\right\}$ are unknown, and therefore, g_{θ}, (Equation (3.1)), cannot be found analytically because p^{θ}_{j} depends on π^{θ}_{kj}. Here, we asume that $c_{j+1} > c\,j$ and $\Delta_j = c_{j+1} - c_j$, represents the increase of the cost of the product when it goes from state j to state $j + 1$ for $j \epsilon I$.

In practice, information about, μ and π can be obtained by evaluation of the past-periods data. This information is usually used as a basis to forecasting the future data or trends. In plant engineering, a system commonly used to do this, is called the Intelligent Maintenance System (IMS). This is a system that utilizes the history data by facilitating data analytics and decisión support tools to predict and prevent the potential failure of the machines. In production systems, database management involves the monitoring, administration, and maintenance of all machines and production process. Today, there is a lot of specialized software that recovers production statistics, occupancy time, machine fault count, estimate average times of failure, and more, see for example (ORACLE ENGINEERED SYSTEMS, 2015), (SAP para Pymes!-SAP Business One (undated), 2015)). In this proposal we assume that this information is available and, therefore, can be used to our analysis. To illustrate our example, consider the following 4 original matrices $\tilde{\pi}_\theta$, where $I = \{1, 2, ..., 18\}$ and $\theta = 1, 2, 3, 4$.

For simplicity our proposal, we assume that $\Delta_p = 500$, $c_p = 200$ for $\rho = 1, 2, ...$ Here, we use a modified version of Equation (3.3), *i.e.*, $f(\rho) = c_\rho = c_{\rho-1} + \Delta_\rho$, Where ρ is the position that keeps the process on the horizon (Figure 3). Therefore

$$E\left[f\left(X\right)\right] = g_\theta = \sum_{j\in I}\left(c_{j-1} + 500\right)\left[P\left(X = j\right)\right]$$

Thus, h^* can be obtained through Equations (3.4) or (3.12).

Table 1. Transitíon probabilities matrix $\tilde{\pi}_1$

1	2	3	4	5	6	7	8	9	10	11	12	13	14	15	16	17	18
2	0	1	0	0	0	0	0	0	0	0	0	0	0	0	0	0	0
3	0	0	0.19	0.26	0	0	0	0	0.34	0	0	0	0	0.21	0	0	0
4	0	0	0	0.12	0.17	0.09	0	0	0.11	0.3	0	0	0	0.09	0.12	0	0
5	0	0	0	0	0.1	0.03	0	0	0	0.02	0.14	0	0	0.12	0.19	0.4	0
6	0	0	0	0	0	0.1	0.5	0	0	0	0.31	0	0	0	0	0.09	0
7	0	0	0	0	0	0	0.1	0.33	0	0	0	0.1	0.23	0	0	0	0.2
8	0	0	0	0	0	0	0	0.17	0	0	0	0.15	0.19	0	0	0	0.04
9	0	0	0	0	0	0	0	0	0	0	0	0	0.75	0	0	0	0
10	0	0	0	0	0.28	0.2	0	0	0	0.19	0.02	0	0	0.12	0.19	0	0
11	0	0	0	0	0.26	0.19	0	0	0	0	0.15	0	0	0	0.2	0.2	0
12	0	0	0	0	0	0	0.2	0.35	0	0	0	0.28	0	0	0	0	0.12
13	0	0	0	0	0	0	0.2	0.15	0	0	0	0	0.4	0	0	0	0
14	0	0	0	0	0	0	0	0	0	0	0	0	1	0	0	0	0
15	0	0	0	0	0.1	0.23	0	0	0	0	0.12	0.22	0	0	0.15	0.18	0
16	0	0	0	0	0	0.02	0	0	0	0	0.4	0.08	0	0	0	0.5	0
17	0	0	0	0	0	0	0.2	0.18	0	0	0	0.31	0	0	0	0	0.31
18	0	0	0	0	0	0	0	0.3	0	0	0	0	0.2	0	0	0	0

Table 2. Transitíon probabilities matrix $\tilde{\pi}_2$

1	2	3	4	5	6	7	8	9	10	11	12	13	14	15	16	17	18
2	0	1	0	0	0	0	0	0	0	0	0	0	0	0	0	0	0
3	0	0	0.28	0.29	0	0	0	0	0.18	0.05	0	0	0	0.2	0	0	0
4	0	0	0	0.12	0.17	0	0	0	0.11	0.3	0	0	0	0.1	0.2	0	0
5	0	0	0	0	0.15	0.08	0	0	0	0.21	0.14	0	0	0.12	0.3	0	0
6	0	0	0	0	0	0.1	0.41	0	0	0	0.17	0.14	0	0	0	0.09	0.09
7	0	0	0	0	0	0	0.12	0.15	0	0	0	0.2	0.3	0	0	0	0.19
8	0	0	0	0	0	0	0	0.21	0	0	0	0.18	0.12	0	0	0	0.08
9	0	0	0	0	0	0	0	0	0	0	0	0	0.45	0	0	0	0
10	0	0	0	0.34	0.24	0.2	0	0	0	0.12	0.02	0	0	0.03	0.05	0	0
11	0	0	0	0	0.13	0.32	0	0	0	0	0.21	0	0	0	0.02	0.32	0
12	0	0	0	0	0	0	0.25	0.28	0	0	0	0.14	0	0	0	0	0.25
13	0	0	0	0	0	0	0.13	0.21	0	0	0	0	0.32	0	0	0	0
14	0	0	0	0	0	0	0	0	0	0	0	0	1	0	0	0	0
15	0	0	0	0	0.14	0.18	0	0	0	0	0.21	0.18	0	0	0.12	0.17	0
16	0	0	0	0	0	0	0.18	0	0	0	0.3	0.02	0	0	0	0.5	0
17	0	0	0	0	0	0	0.2	0.18	0	0	0	0.31	0	0	0	0	0.31
18	0	0	0	0	0	0	0	0.21	0	0	0	0	0.17	0	0	0	0

Table 3. Transitíon probabilities matrices $\tilde{\pi}_3$

1	2	3	4	5	6	7	8	9	10	11	12	13	14	15	16	17	18
2	0	1	0	0	0	0	0	0	0	0	0	0	0	0	0	0	0
3	0	0	0.16	0.32	0	0	0	0	0.09	0.15	0	0	0	0.28	0	0	0
4	0	0	0	0.15	0.21	0	0	0	0.18	0.26	0	0	0	0.1	0.1	0	0
5	0	0	0	0	0.15	0.08	0	0	0	0.21	0.14	0	0	0.12	0.3	0	0
6	0	0	0	0	0	0.1	0.5	0	0	0	0.11	0.11	0	0	0	0.09	0.09
7	0	0	0	0	0	0	0.12	0.15	0	0	0	0.2	0.3	0	0	0	0.19
8	0	0	0	0	0	0	0	0.21	0	0	0	0.18	0.12	0	0	0	0.08
9	0	0	0	0	0	0	0	0	0	0	0	0	0.45	0	0	0	0
10	0	0	0	0.31	0.21	0.2	0	0	0	0.14	0.02	0	0	0.08	0.04	0	0
11	0	0	0	0	0.13	0.32	0	0	0	0	0.21	0	0	0	0.02	0.32	0
12	0	0	0	0	0	0	0.25	0.28	0	0	0	0.14	0	0	0	0	0.25
13	0	0	0	0	0	0	0.12	0.19	0	0	0	0	0.35	0	0	0	0
14	0	0	0	0	0	0	0	0	0	0	0	0	1	0	0	0	0
15	0	0	0	0	0.14	0.18	0	0	0	0	0.21	0.18	0	0	0.12	0.17	0
16	0	0	0	0	0	0	0.18	0	0	0	0.3	0.02	0	0	0	0.5	0
17	0	0	0	0	0	0	0.2	0.18	0	0	0	0.31	0	0	0	0	0.31
18	0	0	0	0	0	0	0	0.28	0	0	0	0	0.11	0	0	0	0

Table 4. Transitíon probabilities matrices $\tilde{\pi}_4$

1	2	3	4	5	6	7	8	9	10	11	12	13	14	15	16	17	18
2	0	1	0	0	0	0	0	0	0	0	0	0	0	0	0	0	0
3	0	0	0.19	0.26	0	0	0	0	0.34	0	0	0	0	0.21	0	0	0
4	0	0	0	0.12	0.19	0.17	0	0	0.11	0.3	0	0	0	0.02	0.09	0	0
5	0	0	0	0	0.1	0.03	0	0	0	0.02	0.14	0	0	0.12	0.19	0.4	0
6	0	0	0	0	0	0.1	0.5	0	0	0	0.31	0	0	0	0	0.09	0
7	0	0	0	0	0	0	0.17	0.25	0	0	0	0.09	0.24	0	0	0	0.21
8	0	0	0	0	0	0	0	0.17	0	0	0	0.15	0.19	0	0	0	0.04
9	0	0	0	0	0	0	0	0	0	0	0	0	0.75	0	0	0	0
10	0	0	0	0	0.28	0.2	0	0	0	0.19	0.02	0	0	0.12	0.19	0	0
11	0	0	0	0	0.28	0.21	0	0	0	0	0.2	0	0	0	0.12	0.19	0
12	0	0	0	0	0	0	0.2	0.35	0	0	0	0.28	0	0	0	0	0.12
13	0	0	0	0	0	0	0.2	0.15	0	0	0	0	0.4	0	0	0	0
14	0	0	0	0	0	0	0	0	0	0	0	0	1	0	0	0	0
15	0	0	0	0	0.08	0.58	0	0	0	0	0.12	0.04	0	0	0.02	0.16	0
16	0	0	0	0	0	0.02	0	0	0	0	0.4	0.08	0	0	0	0.5	0
17	0	0	0	0	0	0	0.2	0.12	0	0	0	0.25	0	0	0	0	0.43
18	0	0	0	0	0	0	0	0.3	0	0	0	0	0.2	0	0	0	0

Table 5. The optimal results obtained

θ	**ψ_ε**	**h***	**Optimal path**
1	2	2643	$1 \Rightarrow 2 \Rightarrow 14 \Rightarrow 6 \Rightarrow 12 \Rightarrow 13$
2	1.6	1909	$1 \Rightarrow 2 \Rightarrow 9 \Rightarrow 15 \Rightarrow 12 \Rightarrow 13$
3	1.8	3154	$1 \Rightarrow 2 \Rightarrow 14 \Rightarrow 5 \Rightarrow 6 \Rightarrow 12 \Rightarrow 13$
4	2	3638	$1 \Rightarrow 2 \Rightarrow 4 \Rightarrow 6 \Rightarrow 8 \Rightarrow 13$

Using the Metropolis-Hasting Algorithms we obtained π_{ij}^{θ}, for $\theta = 1, 2, 3, 4$. Then, using Equation (3.6) we obtain the expected value of the production ψ_ε (for $i = 1$ and $j = 13$) by measuring the number of visits made to the final state of the process, the process number 13. Table 5 summarizes the results obtained, and Appendix 1 shows the results of the runs (Tables 6 and 7).

The expected number of visits in each node i is shown in (Figure 4). The results suggest that the nodes 13 and 18 are best visited with a maximum of 7 times per node.

Table 6. Some results of simulations runs products 1 and 2 (similar trajectories).

θ_1							θ_2						
i	j	ρ	p_{ij}	c_ρ	$f(\rho)$	$E[f(X)]$	i	j	ρ	p_{ij}	c_ρ	$f(\rho)$	$E[f(X)]$
1	2	1	1.00	200	700	700	1	2	1	1.00	200	700	700
2	3	2	0.19	200	1200	228	2	3	2	0.28	200	1200	336
3	4	3	0.12	200	1700	204	3	4	3	0.12	200	1700	204
4	6	4	0.03	200	2200	66	4	6	4	0.08	200	2200	176
6	8	5	0.33	200	2700	891	6	8	5	0.15	200	2700	405
8	13	6	0.75	200	3200	2400	8	13	6	0.45	200	3200	1440
					Total	**4489**						**Total**	**3261**
1	2	1	1.00	200	700	700	1	2	1	1.00	200	700	700
2	4	2	0.26	200	1200	312	2	4	2	0.29	200	1200	348
4	5	3	0.10	200	1700	170	4	5	3	0.15	200	1700	255
5	6	4	0.10	200	2200	220	5	6	4	0.10	200	2200	220
6	8	5	0.33	200	2700	891	6	8	5	0.15	200	2700	405
8	13	6	0.75	200	3200	2400	8	13	6	0.45	200	3200	1440
					Total	**4693**						**Total**	**3368**
1	2	1	1.00	200	700	700	1	2	1	1.00	200	700	700
2	9	2	0.34	200	1200	408	2	9	2	0.18	200	1200	216
9	5	3	0.28	200	1700	476	9	5	3	0.24	200	1700	408
5	11	4	0.31	200	2200	682	5	11	4	0.17	200	2200	374
11	8	5	0.35	200	2700	945	11	8	5	0.28	200	2700	756
8	13	6	0.75	200	3200	2400	8	13	6	0.45	200	3200	1440
					Total	**5611**						**Total**	**3894**
1	2	1	1.00	200	700	700	1	2	1	1.00	200	700	700
2	14	2	0.21	200	1200	252	2	14	2	0.20	200	1200	240
14	6	3	0.23	200	1700	391	14	6	3	0.18	200	1700	306
6	8	4	0.33	200	2200	726	6	8	4	0.15	200	2200	330
8	13	5	0.75	200	2700	2025	8	13	5	0.45	200	2700	1215
					Total	**4094**						**Total**	**2791**
1	2	1	1.00	200	700	700	1	2	1	1.00	200	700	700
2	9	2	0.34	200	1200	408	2	9	2	0.18	200	1200	216
9	11	3	0.02	200	1700	34	9	11	3	0.02	200	1700	34
11	12	4	0.28	200	2200	616	11	12	4	0.14	200	2200	308
12	13	5	0.40	200	2700	1080	12	13	5	0.32	200	2700	864
					Total	**2838**						**Total**	**2122**
1	2	1	1.00	200	700	700	1	2	1	1.00	200	700	700
2	14	2	0.21	200	1200	252	2	14	2	0.20	200	1200	240
14	6	3	0.23	200	1700	391	14	6	3	0.18	200	1700	306
6	12	4	0.10	200	2200	220	6	12	4	0.20	200	2200	440

continued on following page

Table 6. Continued

θ_1							θ_2						
i	j	ρ	P_{ij}	c_ρ	$f(\rho)$	$E[f(X)]$	i	j	ρ	P_{ij}	c_ρ	$f(\rho)$	$E[f(X)]$
12	13	5	0.40	200	2700	1080	12	13	5	0.32	200	2700	864
					Total	**2643**						**Total**	**2550**
1	2	1	1.00	200	700	700	1	2	1	1.00	200	700	700
2	9	2	0.34	200	1200	408	2	9	2	0.18	200	1200	216
9	15	3	0.19	200	1700	323	9	15	3	0.05	200	1700	85
15	12	4	0.08	200	2200	176	15	12	4	0.02	200	2200	44
12	13	5	0.40	200	2700	1080	12	13	5	0.32	200	2700	864
					Total	**2687**						**Total**	**1909**

Table 7. Some results of simulations runs products 2 and 3 (different trajectories).

i	j	ρ	P_{ij}	c_ρ	$f(\rho)$	$E[f(X)]$	i	j	ρ	P_{ij}	c_ρ	$f(\rho)$	$E[f(X)]$
1	2	1	1.00	200	700	700	1	2	1	1.00	200	700	700
2	3	2	0.16	200	1200	192	2	4	2	0.26	200	1200	312
3	5	3	0.21	200	1700	357	4	6	3	0.03	200	1700	51
5	7	4	0.50	200	2200	1100	6	8	4	0.25	200	2200	550
7	8	5	0.21	200	2700	567	8	13	5	0.75	200	2700	2025
8	13	6	0.45	200	3200	1440						**Total**	**3638**
					Total	**4356**	1	2	1	1.00	200	700	700
1	2	1	1.00	200	700	700	2	3	2	0.19	200	1200	228
2	4	2	0.32	200	1200	384	3	4	3	0.12	200	1700	204
4	5	3	0.15	200	1700	255	4	5	4	0.10	200	2200	220
5	7	4	0.50	200	2200	1100	5	6	5	0.10	200	2700	270
7	12	5	0.18	200	2700	486	6	8	6	0.25	200	3200	800
12	8	6	0.19	200	3200	608	8	13	7	0.75	200	3700	2775
8	13	7	0.45	200	3700	1665						**Total**	**5197**
					Total	**5198**	1	2	1	1.00	200	700	700
1	2	1	1	200	700	700	2	3	2	0.19	200	1200	228
2	10	2	0.15	200	1200	180	3	9	3	0.12	200	1700	204
10	5	3	0.13	200	1700	221	9	5	4	0.11	200	2200	242
5	7	4	0.5	200	2200	1100	5	11	5	0.31	200	2700	837
7	12	5	0.18	200	2700	486	11	12	6	0.28	200	3200	896
12	8	6	0.19	200	3200	608	12	13	7	0.4	200	3700	1480
8	13	7	0.45	200	3700	1665						**Total**	**4587**
					Total	**4960**	1	2	1	1.00	200	700	700
1	2	1	1.00	200	700	700	2	3	2	0.19	200	1200	228

continued on following page

Table 7. Continued

i	j	ρ	p_{ij}	c_ρ	$f(\rho)$	$E[f(X)]$	i	j	ρ	p_{ij}	c_ρ	$f(\rho)$	$E[f(X)]$
2	9	2	0.09	200	1200	108	3	6	3	0.17	200	1700	289
9	4	3	0.31	200	1700	527	6	12	4	0.09	200	2200	198
4	6	4	0.08	200	2200	176	5	11	5	0.31	200	2700	837
6	7	5	0.12	200	2700	324	11	12	6	0.28	200	3200	896
7	8	6	0.21	200	3200	672	12	13	7	0.14	200	3700	518
8	13	7	0.45	200	3700	1665						**Total**	**3666**
					Total	**4172**	1	2	1	1.00	200	700	700
1	2	1	1.00	200	700	700	2	9	2	0.34	200	1200	408
2	10	2	0.15	200	1200	180	9	5	3	0.28	200	1700	476
10	11	3	0.21	200	1700	357	5	16	4	0.09	200	2200	198
11	7	4	0.25	200	2200	550	16	7	5	0.20	200	2700	540
7	12	5	0.18	200	2700	486	7	12	6	0.15	200	3200	480
12	13	6	0.35	200	3200	1120	12	13	7	0.40	200	3700	1480
					Total	**3393**						**Total**	**4282**
1	2	1	1.00	200	700	700	1	2	1	1.00	200	700	700
2	14	2	0.28	200	1200	336	2	4	2	0.26	200	1200	312
14	5	3	0.14	200	1700	238	4	15	3	0.19	200	1700	323
5	6	4	0.10	200	2200	220	15	16	4	0.50	200	2200	1100
6	12	5	0.20	200	2700	540	16	8	5	0.20	200	2700	540
12	13	6	0.35	200	3200	1120	8	13	6	0.75	200	3200	2400
					Total	**3154**						**Total**	**5375**

CONCLUSION

Manufacturing's role is changing. The way it contributes to the economy shifts as nations mature: in today's advanced economies, manufacturing promotes innovation, productivity, and trade more than growth and employment. In these countries, manufacturing also has begun to consume more services and to rely more heavily on them to operate.

The influence of manufacturing goes far beyond the direct contribution to national product and employment. Manufacturing is a global business underpinning all economic activity. Spending on goods accounts for more than half of all consumer expenditure, whilst manufacturing goods makes up around two-thirds of the developed countries.

That is why, the applications of the mathematical modelling to the manufacturing systems is vital to strengthen its comprehension and improve its efficiency. The main difficulty here is that no two-production systems are identical. Even if they were designed identically, numerous changes and adjustments, introduced in the course of time by engineering and equipment maintenance personnel, force them to evolve so that they become fundamentally different. Thus, there are, practically speaking, infinitely many

Figure 3. Location of the ρ values in the process

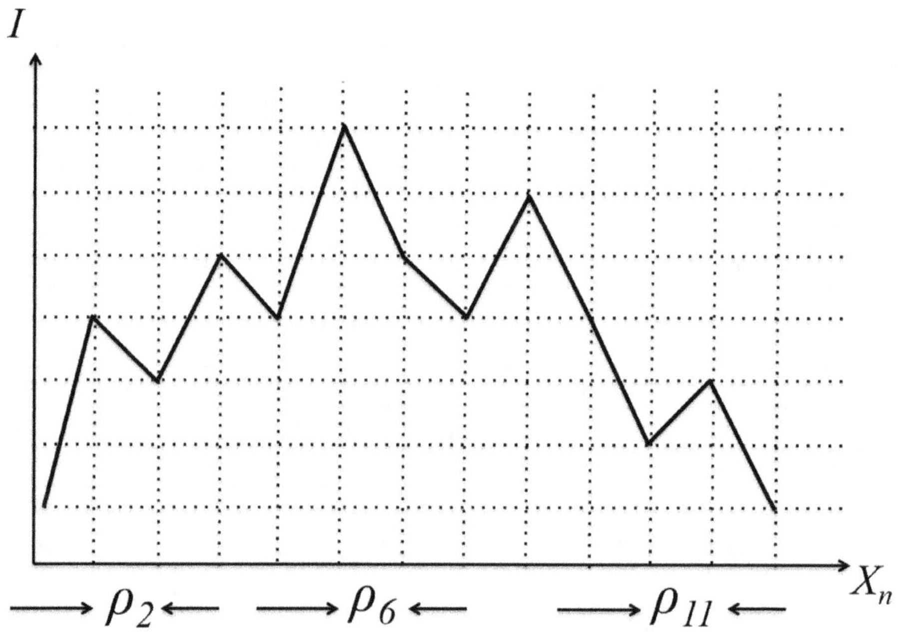

Figure 4. The expected number of visits from node i to node 13

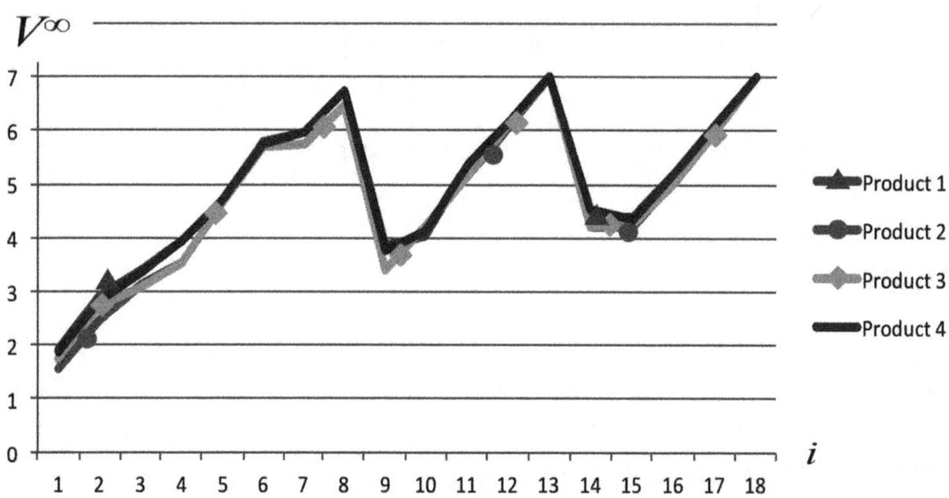

different production systems and therefore, infinite ways to model them. Nevertheless, it is possible to introduce a small set of standard models to which every production system may be reduced, perhaps at the expense of sacrificing some fidelity of the description.

In practice, the mathematical model of a production system is defined by one or more of the following five components: a) type of a production system, b) models of the machines, c) models of the material handling devices, d) rules of interactions between the machines and material handling devices, e) the performance measures. The combination of one or more of them makes the model very complex to be solved in a closed form, forcing to the analyst to eliminate surplus information that contributes little to an understanding thereof.

The approach given in this document to the problem of estimating the minimum cost of a production line by process has been through the Markov decision processes. This method provides an excellent tool to describe the trajectory of a product along the manufacturing line. However, in this as in other models made with this instrument we must face the problem of dimensionality, i.e., the various phenomena that arise when analyzing and organizing data in high-dimensional spaces. The problem of dimensionality often transforms a viable solution into an impractical solution, including in an untreatable solution. Therefore, models that require large state space dimensions must be treated carefully and search for alternative techniques to address them. The study of dimensionality in large-scale problems is a fertile field to study problems like those here mentioned. Dimensionality treatment is undoubtedly one of the future researches in the area of mathematical modelling by using the Markov decision processes.

Regarding the issue here treated, the proposed problem can be extended to include in it the inventories generated at each stage of the process, the cost of these, the starvation periods, the installed plant capacity, the efficiency of machines and the attention level to the customer as a probability measure. To give more realism, it is recommended the inclusion in the model of intermediate buffers between each operation to decouple the line. Assess the bottlenecks, the idle time and the equipment availability as a function of their failure and repair rates. Finally, the inclusion of the reliability of the system operation would come to round the proposal.

Given the enormous difficulty that these assumptions could generate in the construction of the model, it is recommended the use of other methods such as, the stochastic mathematical programming, the dynamic stochastic programming and/or the Monte Carlo simulation method. And to solve these instances, the use of evolutionary algorithms or other alternatives see (Hindawi Publishing Corporation (undated), Pandian Vasant Universiti Teknologi Petronas, Malaysia 0000-0002-0755-1046 Scholarly Contributions [Data Provided by scopus], 2015), (Hindawi Publishing Corporation (undated), Pandian Vasant Universiti Teknologi Petronas, Malaysia 0000-0002-0755-1046 Scholarly Contributions [Data Provided by scopus], 2015). In any case, the use of mathematical models to represent manufacturing systems is an exciting activity for scientists and engineers interested in the topic.

ACKNOWLEDGMENT

Our deepest gratitude to the anonymous referees for their valuable comments to improve the presentation and content of this work.

REFERENCES

Anglania, A., Griecoa, A., Pacellaa, M., & Toliob, T. (1996). Object-oriented modeling and simulation of flexible manufacturing systems: A rule based procedure. *Simulation Modeling Practice and Theory, 10*(3-4), 209-234.

Balduzi, F., Giua, A., & Seatzu, C. (2001). Modelling and simulation of manufacturing systems with first order hybrid Petri nets. *International Journal of Production Research, 39*(2).

Bayazit, O. (2005). Use of AHP in decision-making for flexible manufacturing systems. *Journal of Manufacturing Technology Management*, *16*(7), 808–819. doi:10.1108/17410380510626204

Benjaar, S., & Elhafs, M. (2006). Production and inventory control of a single product assemble to order system with multiple customer classes. *Management Science*, *52*(12), 1896–1912. doi:10.1287/mnsc.1060.0588

Boute, R., Disney, S., Lambrecht, M., & Houdt, B. (2007). An integrated production and inventory model to dampen upstream demand variability in the supply chain. *European Journal of Operational Research*, *178*(1), 121–142. doi:10.1016/j.ejor.2006.01.023

Buzacot, J. A., & Shatikumar, J. (1993). *Stochastic Models of Manufacturing Systems*. Prentice Hall.

Chao, G. (2013). Production and availability policies through the Markov Decision Process and myopic methods for contractual and selective orders. *European Journal of Operational Research*, *225*(3), 383–392. doi:10.1016/j.ejor.2012.10.005

Chen, K., & Ji, P. (2007). A mixed integer programming model for advanced planning and scheduling (APS). *European Journal of Operational Research*, *181*(1), 515–522. doi:10.1016/j.ejor.2006.06.018

Christou, A., Lagodimos, A., & Lycopoulou, D. (2007). Hierarchical production planning for multi-product lines in the beverage industry: Production Planning and Control. *The Management of Operations, 18*(5).

Cochran, D., Arinez, J., Duda, J., & Linck, J. (2002). A decomposition approach for manufacturing system design. *Journal of Manufacturing Systems*, *20*(6), 371–389. doi:10.1016/S0278-6125(01)80058-3

Csjáji, B., & Monostor, L. (2005). Stochastic Reactive Production Scheduling by Multi-agent Based Asynchronous Approximate Dynamic Programming. *CEEMAS LNAI*, 388-397.

Doshi, P., Goodwin, R., Akkiraju, R., & Verma, K. (2004). Dynamic workflow composition using Markov decision processes. *IEEE International Conference on*, (pp. 576-582). doi:10.1109/ICWS.2004.1314784

Eynan, A., & Rosenblat, M. (1996). Component commonality effects on inventory costs. *HE Transactions*, *28*(2), 93–104.

Fletena, S., & Kristoffersen, T. (2008). Short-term hydropower production planning by stochastic programming. *Computers & Operations Research*, *35*(8), 2656–2671. doi:10.1016/j.cor.2006.12.022

Flug, P. (1996). *Optimization of Stochastic Models: The Interface Between Simulation and Optimization*. Kluwer Academic Publishers. doi:10.1007/978-1-4613-1449-3

Fowler, J., & Rose, O. (2015). *Grand Challenges in modeling and simulation.* Retrieved from http://sim.sagepub.com/content/80/9/469.short

Gallien, L., & Wein, L. (2001). A simple and effective component procurement policy for stochastic assembly systems. *Queueing Systems*, *38*(2), 221–248. doi:10.1023/A:1010914600116

Gayon, J., Benjaafar, S., & Vricourt, F. (2008). Using imperfect advance demand information in production-inventory systems with multiple customer classes. *Manufacturing & Service Operations Management*, *11*(1), 128–143. doi:10.1287/msom.1070.0201

Gerchad, Y., & Henig, M. (1989). An inventory model with component commonality. *Naval Research Logistics*, 61-68.

Glynn, P., & Iglehar, D. (1989). Importance sampling for stochastic simulations. *Management Science*, *35*(11), 1367–1392. doi:10.1287/mnsc.35.11.1367

Goetschalckxa, M., Vidalb, C., & Dognac, K. (2002). Modeling and design of gloibal logistic systems: A review of integrated estrategic and tactical models and design algorithms. *European Journal of Operational Research*, *143*(1).

Günter H-O, & Veek P V. (2003). Advanced planning and scheduling in process industry. *Advances Planning and Scheduling Solutions in Process Industry*, 1-9.

Gustavsson, M., & Wänström, C. (2009). Assessing information quality in manufacturing planning and control processes. *International Journal of Quality & Reliability Management*, *26*(4), 325–340. doi:10.1108/02656710910950333

Hindawi Publishing Corporation. (n.d.a). Retrieved from http://www.hindawi.com/28921463/

Hindawi Publishing Corporation. (n.d.b). Retrieved from http://www.hindawi.com/journals/tswj/2015/964581/

Howard, R. (1971). *Dynamic probabilistic systems. Volume II: Semi Markov and decisión process*. John Wiley & Sons.

Hsieh, F., & Lin, J. (2014). A dynamic scheme for scheduling complex tasks in manufacturing systems based on collaboration of agents. *Applied Intelligence*, *41*(2), 366–382. doi:10.1007/s10489-014-0521-5

Hvolby, H., & Jensen, K. (2010). Technical and industrial issues of Advanced Planning and Scheduling (APS) systems. *Computers in Industry*, *61*(9), 845–851. doi:10.1016/j.compind.2010.07.009

Ilgina, M., & Guptab, S. (2010). Environmentally conscious manufacturing and product recovery (ECMPRO): A review of the state of the art. *Journal of Environmental Management*, *91*(3), 563–592. doi:10.1016/j.jenvman.2009.09.037 PMID:19853369

Jonsson, P., & Mattsson, S.-A. (2003). The implications of fit between planning environments and manufacturing planning and control methods. *International Journal of Operations & Production Management*, *23*(8), 872–900. doi:10.1108/01443570310486338

Kilger, C., Staedtler, H., & Meyr, H. (2008). *Supply Chain Management and Advanced Planning, Concepts, Models, Software, and Case Studies* (4th ed.). Springer.

Kristensen, A. (1996). *Textbook notes of herd management: Dynamic programming and Markov decision processes*. Retrieved from http://www.academia.edu/4892288/Dina,Notat,No.,49

Kyriakidis, E., & Dimitrakos, T. (2006). Optimal preventive maintenance of a production system with an intermediate buffer. *European Journal of Operational Research*, *168*(1), 86–99. doi:10.1016/j.ejor.2004.01.052

Lee, Y., Jeong, C., & Moon, C. (2002). Advanced planning and scheduling with outsourcing in manufacturing supply chain. *Computers & Industrial Engineering*, *43*(1-2), 351-374.

Leitao, P., Mendez, J., Bepperling, S., Cachapa, D., Colombo, A., & Reestivo, F. (2012). Integration of virtual and real environments for engineering service-oriented manufacturing systems. *Journal of Intelligent Manufacturing*, *23*(6), 2551–2563. doi:10.1007/s10845-011-0591-8

Leung, S., & Wub, Y. (2004). A robust optimization model for stochastic aggregate production planning. *Production Planning and Control*, *15*(5), 502–514. doi:10.1080/09537280410001724287

Liu, F., & Song, J. (2012). God and bad news about the (S,T) policy. *Manufacturing and Service Operations Management*, *14*(1), 42–49. doi:10.1287/msom.1110.0353

Lodree, J. Jr, & Uzochukwub, B. (2008). Production planning for a deteriorating ítem with stochastic demand and consumer choice. *International Journal of Production Economics*, *116*(2), 219–232. doi:10.1016/j.ijpe.2008.09.010

Metropolis, M., Rosenbluth, A., Teller, A., & Teller, E. (1953). Equations of state calculation by fast computing machines. *Journal of Chemical Physics, 21*, 1087-1092.

Moona, C., & Seob, Y. (2005). Evolutionary algorithm for advanced process planning and scheduling in a multi-plant. *Computers & Industrial Engineering*, *48*(2), 311–325. doi:10.1016/j.cie.2005.01.016

Mula, P. J., García Sabater, J., & Lario, F. (2006). Models for production planning under uncertainty: A review. *International Journal of Production Economics*, *103*(1), 271–285. doi:10.1016/j.ijpe.2005.09.001

Nakashima, K., Arimitsua, H., Nosea, T., & Kuriyamab, S. (2004). Optimal control of a remanufacturing system. *International Journal of Production Research*, *42*(17), 3619–3625. doi:10.1080/00207840410001721772

Olhager, J., & Selldin, J. (2007). Manufacturing planning and control approaches: Market alignment and performance. *International Journal of Production Research*, *45*(6), 1469–1484. doi:10.1080/00207540600635250

Oracle Engineered Systems. (2015). Retrieved from http://www.oracle.com/index.html

Pérez, L. G., Gress, E., Karelyn, A., & Orán, M. (2009). Some efficiency measures in the operation of flexible manufacturing systems: A stochastic approach. *Proceedings of the 2009 SIAM Conference on "Mathematics for Industry"*.

Popescu, C., Soto, M., & Lastra, J. (2012). A Petri net-based approach to incremental modeling of flow and resources in service oriented manufacturing systems. *International Journal of Production Research*, *50*(2), 325–343. doi:10.1080/00207543.2011.561371

Puterman, L. (2005). *Markov Decision Processes: Discrete Stochastic Dynamic Programming*. John Wiley & Sons.

Riddalls, C., Bennett, S., & Tipi, N. (2000). Modelling the dynamics of supply chains. *International Journal of Systems Science*, *31*(8), 969–976. doi:10.1080/002077200412122

Rubenstein, Y. (1980). Selecting the best stochastic stable system. Stochastic Processes and their Applications, 10, 75-85.

Rubenstein, Y., & Kroese, P. (2007). *Simulation and the Monte Carlo Method.* Wiley Series in Probability and Statistics. doi:10.1002/9780470230381

Ryua, K., & Jung, M. (2003). Agent-based fractal architecture and modelling for developing distributed manufacturing systems. *International Journal of Production Research, 41*(17), 4233–4255. doi:10.1080/0020754031000149275

SAP para Pymes!-SAP Business One. (n.d.). Retrievedfrom http://evolve-it.com.mx/sap-para-pymes/?gclid=CKrKwLfetMMCFYM8aQod9XMAGw

Schoenherr, T., Tummala, V., & Harrison, T. (2008). Assessing supply chain risks with the analytic hierarchy process: Providing decision support for the offshoring decision by a US manufacturing company. *Journal of Purchasing and Supply Management, 14*(2), 100–111. doi:10.1016/j.pursup.2008.01.008

Sethi, S., Yan, H., Zhang, H., & Zhang, Q. (2002). Optimal and hierarchical controls in dynamic stochastic manufacturing systems: A review. *Manufacturing & Service Operations Management, 4*(2), 133–170. doi:10.1287/msom.4.2.133.281

Sipper, D., & Bulfin, R. Jr. (1997). *Production Planning Control and Integration.* New York: McGraw Hill.

Smith, J. (2003). Survey on the use of simulation fr manufacturing systems design and operation. *Journal of Manufacturing Systems, 22*(2), 157–171. doi:10.1016/S0278-6125(03)90013-6

Smith, L., & Ball, P. (2012). Steps towards sustainable manufacturing through modelling material, energy and waste flows. *International Journal of Production Economics, 140*(1), 227–238. doi:10.1016/j.ijpe.2012.01.036

Soman, C., Van Donk, D., & Gaalman, G. (2004). Combined make-to-order and make-tostock in a food production system. *International Journal of Production Economics, 90*(2), 223-235.

Son, Y., & Venkateswaran, J. (2007). Hierarchical supply chain planning architecture for integrated analysis of stability and performance. *Int. J. Simulation and Process Modelling, 3*(3).

Song, J., & Yao, J. (2002). Performance Analysis and Optimization of Assemble-to-Order Systems with Random Leadtimes. *Operations Research, 50*(5), 889–903. doi:10.1287/opre.50.5.889.372

Vitanov, V., Julkaa, N., Bainesb, T., Tjahjonob, B., & Lendermanna, P. (2007). A review of multi-factor capacity expansion models for manufacturing plants: Searching for a holistic decisions aid. *International Journal of Production Economics, 106*(2), 607–621. doi:10.1016/j.ijpe.2006.08.002

Wang, L., & Lin, S. (2009). A multi-agent based agile manufacturing planning and control system. *Computers & Industrial Engineering, 57*(2), 620–640. doi:10.1016/j.cie.2009.05.015

Xiadong, Y., Fernandez-G, E., & Fu, M. (2004). Optimal preventive maintenance scheduling in semiconductor manufacturing. *Semiconductor Manufacturing. IEEE Transactions on, 17*(3), 345–356.

Yildirim, I., Tan, B., & Karaesmen, F. (2005). A multiperiod stochastic production planning and sourcing problem with service level constraints. *OR-Spektrum, 27*(2-3), 471–489. doi:10.1007/s00291-005-0203-0

Yin, K., Yin, G., & Liu, H. (2004). Stochastic modeling for inventory and production planning in the paper industry. *AIChE Journal. American Institute of Chemical Engineers*, *50*(11), 2877–2890. doi:10.1002/aic.10251

Zanjania, M., Nourelfatha, M., & Kadia, D. (2010). A multi-stage stochastic programming approach for production planning with uncertainty in the quality of raw materials and demand. *International Journal of Production Research*, *48*(16).

Chapter 8
A Genetic Algorithm's Approach to the Optimization of Capacitated Vehicle Routing Problems

Mariano Frutos
Universidad Nacional del Sur, Argentina & CONICET, Argentina

Fernando Tohmé
Universidad Nacional del Sur, Argentina & CONICET, Argentina

Fabio Miguel
Universidad Nacional de Río Negro, Argentina

ABSTRACT

This chapter addresses the family of problems known in the literature as Capacitated Vehicle Routing Problems (CVRP). A procedure is introduced for the optimization of a version of the generic CVRP. It generates feasible clusters and, in a first step, yields a coding of their ordering. The next stage provides this information to a genetic algorithm for its optimization. A selective pressure process is added in order to improve the selection and subsistence of the best candidates. This arrangement allows improving the performance of the algorithm. We test it using Van Breedam and Taillard's problems, yielding similar results as other algorithms in the literature. Besides, we test the algorithm on real-world problems, corresponding to an Argentinean company distributing fresh fruit. Four instances, with 50, 100, 150 and 200 clients were examined, giving better results than the current plans of the company.

INTRODUCTION

The competitive pressures faced by companies pushes for the development of new techniques for optimal decision-making. These methods are also intended to be fast enough to achieve highly competitive services. One of the main issues that such techniques should address is the design of plans for the distribution of products. The high complexity of these problems constitutes a roadblock for the creation of

DOI: 10.4018/978-1-4666-9644-0.ch008

methods able to yield optimal outputs for real-world cases in reasonable computation times (Bartolini et al., 2013; Ambrosino & Sciomachen, 2007). Despite the lack of appropriate methodologies, firms have to keep making short, median and long-term decisions with respect to transportation.

The economic importance of moving products requires if not optimal, at least good strategies to handle the problems fast enough, providing competitive solutions (Bodin & Golden, 1981; Kumar & Panneerselvam, 2012). Distribution, for instance, involves different phases, covering from the moment the products leave the firm until they reach the end consumers. The configuration and topology of the ensuing distribution network can be obtained as a solution of the so-called Vehicle Routing Problem (VRP) (Basu, 2012; Laporte & Osman, 1995). The VRP is a combinatorial optimization problem in the NP-Complete class. The academic and industrial motivation for seeking reasonable solutions is related to the hardness to find optimal ones (Pollaris et al., 2015; Mota et al., 2007).

VRP amounts to find the shortest routes of delivery of goods to a set of clients, whose placement in space and demand are known. A single point of departure (a depot) and a given fleet of vehicles are assumed. To this basic setting different operational conditions have been added in time. An immediate incorporation was a constraint on the capacity of the fleet of vehicles (CVRP, Capacitated Vehicle Routing Problem) (Kao & Chen, 2013; Wu et al., 2013; Frutos & Tohmé, 2012; Alba & Dorronsoro, 2008). In this chapter we address specifically the CVRP. To extend our approach in (Frutos & Tohmé, 2012), we add a selective pressure process that improves the performance of the individuals that are selected.

The chapter is organized as follows. First we present the VPR and establish its operational conditions. Then we characterize the CVRP as well as an alternative version. We solve Van Breedam and Taillard's problems using our improved algorithm and compare with already known results. Later we apply the algorithm to a real-world case of an Argentinean fresh fruit delivering firm. Finally, we present the conclusions and discuss possible lines of future research.

ADDITIONAL READINGS

As mentioned above, different extensions to the basic VRP have been developed in the last decade. So for instance, time windows for client attention were added to the basic problem yielding the VRPTW (Vehicle Routing Problem with Time Windows) (Wang, 2012; Desrochers et al., 2002; Khebbache-Hadji et al., 2013). Another way in which VRP has been enlarged is with multiple depots (MDVRP, Multi-Depot Vehicle Routing Problem) (Escobar et al., 2014). Other extensions involve the possibility of returning some delivered goods to the deposit (VRPPD, Vehicle Routing Problem with Pick-Up and Delivering) (Çatay, 2009; Thangiah et al., 2007), or that different vehicles can deliver to a single client (SDVRP, Split Delivery Vehicle Routing Problem) (Aleman & Hill, 2010; Aleman et al., 2010; Belenguer et al., 2010; Derigs et al., 2009; Nagao & Nagamochi, 2007; Moreno et al., 2010; Bolduc et al., 2010). Another addition is that some values (number of clients, their demands, servicing time or in-route time) can be random (SVRP, Stochastic Vehicle Routing Problem) (Zhang et al., 2013) and that some deliveries can be performed in certain dates (PVRP, Periodic Vehicle Routing Problem) (Kurz & Zäpfel, 2013). Finally, some problems combine some of these operational conditions (Belfiore & Yoshizaki, 2009; Nowak et al., 2009).

Algorithms have been developed, solving some of the aforementioned problems in reasonable time. Kao and Chen (2013) solve an instance of CVRP using a hybrid algorithm combining Particle Swarm Optimization (PSO) and Simulated Annealing (SA). The authors' approach is to use the technique known as "cluster first and route second". Frutos et al. also solve an instance of CVRP, using a Genetic Algorithm (GA) that groups routes drawn from a feasible set obtained in a pre-initialization process. Wang (2012) solves a VRPTW with a hybrid algorithm that incorporates an Ant Colony Optimization (ACO) algorithm and an Iterated Local Search (ILS) procedure. This algorithm also minimizes the total number of vehicles required for delivering the goods. Escobar et al. (2014) solve a MDVRP using a Hybrid Granular Tabu Search (HGTS) algorithm and diversification strategies that improve the initial solution. Çatay (2009), solves a VRPPD using an ACO algorithm while Aleman et al. (2010) solve a SDVRP with a constructive algorithm that uses a "route angle control measure", Furthermore, this approach is extended using adaptive memory. Zhang et al. (2013) solve a SVRP by means of the ITS algorithm and stochastic programming. Finally, Kurz and Zäpfel (2013) solve an instance of PVRP with a Tabu Search (TS) algorithm running on a bi-objective model.

Finally, updated discussions on VRP, complementing our point of view, can be found in Sitek and Wikarek, 2015, Nguyen et al. (2012) and Perboli et al. (2011). In these papers a variant called Two-Echelon Capacitated Vehicle Routing Problem is postulated. This version of the problem introduces a higher complexity, which can be framed appropriately in the light of Gonzalez-Feliu (2013).

DEFINITION OF THE CVRP

The Capacitated Vehicle Routing Problem (CVRP) is one of the variants of the Vehicle Routing Problem (VRP), which is salient in the field of product delivery, having started its treatment in the 1960s (Wu et al., 2008). Since the 1990s significant theoretical and practical inroads have been made, introducing new approaches to its solution (Rivera et al., 2015; Yu et al., 2008; Jin et al., 2007). The CVRP, in particular, involves the search for a class of routes of the shortest length allowing a fleet of vehicles, with a cap in their capacity, deliver the goods demanded by the clients of the firm (Tavakkoli-Moghaddam et al., 2007) (Belenguer et al., 2010). Moreover, all the routes have to start and end at a single depot. The capacity of the vehicles can be limited in several ways, like for instance weight and volume. All the clients have to be catered and their individual demands cannot be fractioned. This means that the load to be delivered to each client must be transported in a single vehicle. Further operational conditions on the clients, depots and vehicles can be added, giving rise to a variety of particular problems, briefly discussed in the Introduction.

A problem that can be conceived as an instance of the CVRP arises when a provider has to pick up goods from its production plants and carry them to a single depot. The same happens with garbage collection in urban environments: routes have to be chosen in terms of the amounts of garbage to be picked up. We could also consider the case in which the goods are not initially in a depot, but distributed in different sites that have to be visited before starting the delivery (Sitek, 2014; Sitek et al., 2014).

A common feature in these problems is that each vehicle has associated a variable cost, proportional to the distance that it has to cover. When the attributes of the vehicle (capacity, cost, etc.) are the same for all the fleet, the problem is called of homogeneous fleet. Otherwise it is said to involve a heterogeneous fleet. The amount of vehicles could be either a constant or a decision variable (Lee et al., 2006). The usual statement amounts to minimize the distance covered satisfying the demands of the clients, using the least number of vehicles (Belfiore & Yoshizaki, 2009; Wilck IV & Cavalier, 2012). Legal regulations impose constraints on the maximum time a vehicle can be circulating. It may happen that certain zones are forbidden for delivery trucks. In some cases it is asked that the time in route of the vehicles must not be too different (Oyola & Løkketangen, 2014; Shen et al., 2009). It is assumed in general that each vehicle covers a single route, although some models have been lately developed in which a vehicle can cover more than one route.

In this work we focus on slight variants of the CVRP. Consider a first version of the problem. Let N^C be the class of client nodes, endowed with a linear order <, N^S the set of storage nodes and $N=N^C \cup N^S$. C_{ij} is the cost (distance) of going from node i to j for any pair $(i,j) \in N \times N$. X_{ij} is a binary variable, valued 1 if a vehicle goes from i to j without intermediate stops. Otherwise $X_{ij}=0$. C_{max} is the maximum capacity of a vehicle and d_i the demand at $i \in N^C$. In turn, S_j is the class of nodes in a path that starts and ends at node $j \in N^S$. Then, the problem is:

$$\min \sum_{i \in N^c} \sum_{j \in N^s} C_{ij} X_{ij}$$

S.t.

(i) $\sum_{i \in N^c \cap S_j} d_i \leq C_{max}$

(ii) For every pair of paths S_j and S_k, $(S_j \cap S_k) \cap N^c = \varnothing$.

That is, the goal is the minimization of the total distance traversed by all vehicles, satisfying the demands on every route for a partition of client nodes in clusters, each one covered by a single vehicle. Constraint (ii) puts the problem in a complexity class above that of the linear programming problem.

A second version of the problem can be stated as follows: let $M = \{S_j\}_{j \in N^s}$ for some allocation of routes, $C_m = \sum_{(k,l) \in m} C_{kl}$ the sum of distances over route m (which is, for some $j \in N^S$, equal to S_j), X_m is a binary variable such that $X_m=1$ if route m is used and $X_m=0$ otherwise and e_{im} also a binary variable (1 if $i \in N^C$ belongs to route m and 0 otherwise). Then, the problem is:

$$min \sum_{m \in M} C_m X_m$$

S.t.

(iii) For each $i \in N^c, \sum_{m \in M} e_{im} X_m = 1$.

This means that this second version is analogous to the first one when (ii) is already satisfied.

METHODOLOGY

An algorithm for 2 can be divided in two stages. The first one generates feasible clusters, classifying and ordering them. The second stage solves the minimization problem by means of a Genetic Algorithm (Goldberg, 1989).

First Stage: The algorithm generates the clusters that will be elements of M. We take each $i \in N^C$ and consider all the clusters S_j such that $i \in S_j$. The inductive construction starts with $i=1$ (the lowest in the

$<$ order) building all the clusters of client nodes that include 1, satisfying condition (i). Then, if all possible clusters up to $i=k$ have been obtained, consider $i=k+1$ (where, of course, $k<k+1$). All routes starting and ending in a node of N^S that go through client nodes in $Nc=\{1,\ldots,k\}$, including node $k+1$, satisfying condition (i), are constructed. For each $i \in N^C$ its class of clusters, denoted S^i, is endowed with a weak order $\preceq_i$ (reflexive, antisymmetric and transitive, i.e. allowing ties) according to the number of clients in each cluster. That is, given $S_j, S_k \in S^i$, $S_j \preceq_i S_k$ if and only if $|S_j| \leq |S_k|$. The highest rank in $\preceq_i$ corresponds to routes covering i as well as the largest possible number of other client nodes in the upper level of order $<$. As an example, consider the case of 10 client nodes, each one demanding 50 units and a vehicle capacity of 100 units. Table 1 shows the distances among nodes, being 0 the storage site from which departs the vehicle. Table 2 shows the result of the first stage of the procedure. Notice that, since the capacity of the vehicle is 100 and each client node demands 50, a route that clusters some clients together can only cater two of them. So, for instance the group corresponding to node 9 includes only two possible clusters, {9} and {9,10}, because all other nodes have already been assigned to other clusters. Cluster {9,10} has a higher order than {9} because it includes two instead of just one clients.

Second Stage: At this point a Genetic Algorithm (Goldberg, 1989) can be applied on the clusters obtained in the first stage. An individual is coded as a chromosome satisfying conditions (ii) and (iii) and therefore are feasible under problem 2. The coding operates on a lexicographic basis. Each $i \in N^C$ is given a number, starting with $i=1$, and following the process according to order $<$. So, take all the clusters ordered according to $\preceq_1$. This order can be represented by the linear numerical order $1, \ldots, |S^1|$. So, if the k-th cluster in the ordering is chosen, the corresponding gene in the chromosome is k. For $i=k$, the order $\preceq_k$ is applied to a subset of S^k obtained by eliminating all the clusters that include nodes already present in clusters corresponding to genes $i=1,\ldots,k-1$. The new ordering is denoted $\preceq_k$. Then the k-th gene in the chromosome is the rank in $\preceq_k$ of the cluster that is chosen. If for a given i the domain obtained is $\varnothing$ (because all possible nodes belong to clusters corresponding to genes 1 to $i-1$), any number between 1 and $|S^i|$ can be assigned to the i-th gene. In this way, a chromosome is defined, each corresponding to a feasible solution to problem 2. The Genetic Algorithm seeks to find the optimal one.

Table 1. Distances (Case: 10 nodes / Demand per node: 50 units / Vehicle capacity: 100 units)

Nodes	0	1	2	3	4	5	6	7	8	9	10
0	0,00	5,00	14,14	7,07	11,18	15,81	14,14	5,00	11,18	11,18	18,03
1	5,00	0,00	11,18	11,18	15,81	20,62	18,03	10,00	7,07	10,00	22,36
2	14,14	11,18	0,00	15,81	25,00	25,50	20,00	18,03	5,00	20,62	32,02
3	7,07	11,18	15,81	0,00	11,18	10,00	7,07	5,00	15,00	18,03	18,03
4	11,18	15,81	25,00	11,18	0,00	11,18	15,00	7,07	22,36	15,81	7,07
5	15,81	20,62	25,50	10,00	11,18	0,00	7,07	11,18	25,00	25,00	15,00
6	14,14	18,03	20,00	7,07	15,00	7,07	0,00	11,18	20,62	25,00	20,62
7	5,00	10,00	18,03	5,00	7,07	11,18	11,18	0,00	15,81	14,14	14,14
8	11,18	7,07	5,00	15,00	22,36	25,00	20,62	15,81	0,00	15,81	29,15
9	11,18	10,00	20,62	18,03	15,81	25,00	25,00	14,14	15,81	0,00	20,00
10	18,03	22,36	32,02	18,03	7,07	15,00	20,62	14,14	29,15	20,00	0,00

Table 2. Clusters (Case: 10 nodes / Demand per node: 50 units / Vehicle capacity: 100 units)

Nodes	Clusters									
0	-	-	-	-	-	-	-	-	-	-
1	{0,1}	{0,1,2}	{0,1,3}	{0,1,4}	{0,1,5}	{0,1,6}	{0,1,7}	{0,1,8}	{0,1,9}	{0,1,10}
2	{0,2}	{0,2,3}	{0,2,4}	{0,2,5}	{0,2,6}	{0,2,7}	{0,2,8}	{0,2,9}	{0,2,10}	-
3	{0,3}	{0,3,4}	{0,3,5}	{0,3,6}	{0,3,7}	{0,3,8}	{0,3,9}	{0,3,10}	-	-
4	{0,4}	{0,4,5}	{0,4,6}	{0,4,7}	{0,4,8}	{0,4,9}	{0,4,10}	-	-	-
5	{0,5}	{0,5,6}	{0,5,7}	{0,5,8}	{0,5,9}	{0,5,10}	-	-	-	-
6	{0,6}	{0,6,7}	{0,6,8}	{0,6,9}	{0,6,10}	-	-	-	-	-
7	{0,7}	{0,7,8}	{0,7,9}	{0,7,10}	-	-	-	-	-	-
8	{0,8}	{0,8,9}	{0,8,10}	-	-	-	-	-	-	-
9	{0,9}	{0,9,10}	-	-	-	-	-	-	-	-
10	{0,10}	-	-	-	-	-	-	-	-	-

We can see some examples of how chromosomes encode clusters in Table 3, (3–3–2–1–4–2–3–2–1–1) and Table 4 (5–2–1–5–1–4–2–2–1–1): all routes start and end at node 0 and go through two client nodes. In gray we see the filtered out clusters and in red the selection. Consider for instance the second gene in Table 3: it has to discard (slashed) all routes that go through node 3, since it has already been assigned to the first gene. Number 3 assigned to this gene indicate that the route selected is the third in the list of non-filtered routes. On the other hand, a number assigned to a gene with a row with all routes filtered out does not have any effect. The initial population consists of individuals whose chromosomes are constituted by genes selected at random. Once established a chromosome, it is decoded and evaluated. This evaluation is performed according to problem 2. For the computation of each cluster we use Lin-Kerninghan's algorithm that yields the minimal distance among its nodes. The following operators have been chosen in the light of some previous evaluations. The cross-over operator is the two-point one (Table 5, Parent 1: 3–3–2–*–*–*–*–2–1–1, Parent 2: *–*–*–5–1–4–2–*–*–*, Off-spring: 3–3–2–5–1–4–2–2–1–1) and the mutation operator alters gene values at 10% of the chromosome (Mutation (Table 6, Offspring*: 3–3–2–5–1–2–2–2–1–1)). Finally, the selection is done by ranking the candidates.

The following pseudo-code details the steps towards a sound implementation of the algorithm.

```
Generate and sort the set of feasible routes
Randomly generate an initial population P(0) and initialize t=0
Repeat
         Evaluate the fitness of each individual of P(t)
         Select the fittest individuals of P(t) to act as parents
         Apply the genetic operators over the parents, generating P(t+1)
Do t=t+1
Until the stopping criterion is fulfilled
```

Table 3. Example of a Chromosome (3-3-2-1-4-2-3-2-1-1) that will act as Parent 1 in the Cross-Over of Table 5

Nodes	Chr.	Clusters									
1	3	{0,1}	{0,1,2}	**{0,1,3}**	{0,1,4}	{0,1,5}	{0,1,6}	{0,1,7}	{0,1,8}	{0,1,9}	{0,1,10}
2	3	{0,2}	~~{0,2,3}~~	{0,2,4}	**{0,2,5}**	{0,2,6}	{0,2,7}	{0,2,8}	{0,2,9}	{0,2,10}	-
3	2	~~{0,3}~~	~~{0,3,4}~~	~~{0,3,5}~~	~~{0,3,6}~~	~~{0,3,7}~~	~~{0,3,8}~~	~~{0,3,9}~~	~~{0,3,10}~~	-	-
4	1	**{0,4}**	~~{0,4,5}~~	{0,4,6}	{0,4,7}	{0,4,8}	{0,4,9}	{0,4,10}	-	-	-
5	4	~~{0,5}~~	~~{0,5,6}~~	~~{0,5,7}~~	~~{0,5,8}~~	~~{0,5,9}~~	~~{0,5,10}~~	-	-	-	-
6	2	{0,6}	**{0,6,7}**	{0,6,8}	{0,6,9}	{0,6,10}	-	-	-	-	-
7	3	~~{0,7}~~	~~{0,7,8}~~	~~{0,7,9}~~	~~{0,7,10}~~	-	-	-	-	-	-
8	2	{0,8}	**{0,8,9}**	{0,8,10}	-	-	-	-	-	-	-
9	1	~~{0,9}~~	~~{0,9,10}~~	-	-	-	-	-	-	-	-
10	1	**{0,10}**	-	-	-	-	-	-	-	-	-

Unlike the procedure in (Frutos & Tohmé, 2012) we incorporate here a selective pressure process that generates a new ranking. This is in order to address an issue arising after a number of rounds, which is that fitness values come too close and there is a risk of losing good solutions. The selective pressure is based on assigning new fitness values based on the position that the candidate has reached. More precisely, the new ranking obeys to the following expression:

$$R_q = (2 - SP) + 2(SP - 1)(Q_q - 1/P - 1)$$

Where R_q indicates the new rank of solution q, Q_q is its position, SP the selective pressure and P is the size of the population. SP is a number between 1 and 2, where in case $SP=1$, all solutions get the same rank.

Before testing the algorithm over problems found in the literature we run some experiments with this example. Figures 1 to 3 (Appendix 1) show the routes generated in several generations and the minimal total distance covered by the selected routes. For instance, according to the routes depicted in Figures 1 to 3, in Figure 4 (Appendix 1) we can see how new fitness degrees are assigned to s_1, s_2, s_3, s_4, s_5 and s_6. We can also the results with different values of *SP*.

EXPERIMENTS AND COMPARISONS

Van Breedam (1994) and Taillard (1993) present a large number of instances of features and values. They suggested also a reduced set of representative cases to be solved by different heuristics. Some of them involve 100 nodes with a total demand of 1000 units. We follow here this convention and use the author's nomenclature for the cases with the same capacity of vehicles. The problems tested range from Bre-1 to Bre-15. From Bre-1 to Bre-6 only the capacity of vehicles is restricted. The demand on each client node is 10 units, while the vehicle capacity is 100 units in Bre-1 and Bre-2, 50 units in Bre-3 and Bre-4, while it is of 200 units in Bre-5 and Bre-6. Problems Bre-7 and Bre-8 have not been considered because they involve both deliveries and pickups. In most of these problems a single type of good is as-

Table 4. Example of a Chromosome (5-2-1-5-1-4-2-2-1-1) that will act as Parent 2 in the Cross-Over of Table 5

Nodes	Chr.	Clusters									
1	5	{0,1}	{0,1,2}	{0,1,3}	{0,1,4}	**{0,1,5}**	{0,1,6}	{0,1,7}	{0,1,8}	{0,1,9}	{0,1,10}
2	2	{0,2}	**{0,2,3}**	{0,2,4}	~~{0,2,5}~~	{0,2,6}	{0,2,7}	{0,2,8}	{0,2,9}	{0,2,10}	-
3	1	~~{0,3}~~	~~{0,3,4}~~	~~{0,3,5}~~	~~{0,3,6}~~	~~{0,3,7}~~	~~{0,3,8}~~	~~{0,3,9}~~	~~{0,3,10}~~	-	-
4	5	{0,4}	~~{0,4,5}~~	{0,4,6}	{0,4,7}	{0,4,8}	**{0,4,9}**	{0,4,10}	-	-	-
5	1	~~{0,5}~~	~~{0,5,6}~~	~~{0,5,7}~~	~~{0,5,8}~~	~~{0,5,9}~~	~~{0,5,10}~~	-	-	-	-
6	4	{0,6}	{0,6,7}	{0,6,8}	~~{0,6,9}~~	**{0,6,10}**	-	-	-	-	-
7	2	{0,7}	**{0,7,8}**	~~{0,7,9}~~	~~{0,7,10}~~	-	-	-	-	-	-
8	2	~~{0,8}~~	~~{0,8,9}~~	~~{0,8,10}~~	-	-	-	-	-	-	-
9	1	~~{0,9}~~	~~{0,9,10}~~	-	-	-	-	-	-	-	-
10	1	~~{0,10}~~	-	-	-	-	-	-	-	-	-

Table 5. Cross-Over (Offspring: 3-3-2-5-1-4-2-2-1-1)

Nodes	Chr.	Clusters									
1	3	{0,1}	{0,1,2}	**{0,1,3}**	{0,1,4}	{0,1,5}	{0,1,6}	{0,1,7}	{0,1,8}	{0,1,9}	{0,1,10}
2	3	{0,2}	~~{0,2,3}~~	{0,2,4}	**{0,2,5}**	{0,2,6}	{0,2,7}	{0,2,8}	{0,2,9}	{0,2,10}	-
3	2	~~{0,3}~~	~~{0,3,4}~~	~~{0,3,5}~~	~~{0,3,6}~~	~~{0,3,7}~~	~~{0,3,8}~~	~~{0,3,9}~~	~~{0,3,10}~~	-	-
4	5	{0,4}	~~{0,4,5}~~	{0,4,6}	{0,4,7}	{0,4,8}	**{0,4,9}**	{0,4,10}	-	-	-
5	1	~~{0,5}~~	~~{0,5,6}~~	~~{0,5,7}~~	~~{0,5,8}~~	~~{0,5,9}~~	~~{0,5,10}~~	-	-	-	-
6	4	{0,6}	{0,6,7}	{0,6,8}	~~{0,6,9}~~	**{0,6,10}**	-	-	-	-	-
7	2	{0,7}	**{0,7,8}**	~~{0,7,9}~~	{0,7,10}	-	-	-	-	-	-
8	2	~~{0,8}~~	~~{0,8,9}~~	~~{0,8,10}~~	-	-	-	-	-	-	-
9	1	~~{0,9}~~	~~{0,9,10}~~	-	-	-	-	-	-	-	-
10	1	~~{0,10}~~	-	-	-	-	-	-	-	-	-

Table 6. Mutation (Offspring: 3-3-2-5-1-4-2-2-1-1)*

Nodes	Chr.	Clusters									
1	3	{0,1}	{0,1,2}	**{0,1,3}**	{0,1,4}	{0,1,5}	{0,1,6}	{0,1,7}	{0,1,8}	{0,1,9}	{0,1,10}
2	3	{0,2}	~~{0,2,3}~~	{0,2,4}	**{0,2,5}**	{0,2,6}	{0,2,7}	{0,2,8}	{0,2,9}	{0,2,10}	-
3	2	~~{0,3}~~	~~{0,3,4}~~	~~{0,3,5}~~	~~{0,3,6}~~	~~{0,3,7}~~	~~{0,3,8}~~	~~{0,3,9}~~	~~{0,3,10}~~	-	-
4	5	{0,4}	~~{0,4,5}~~	{0,4,6}	{0,4,7}	{0,4,8}	**{0,4,9}**	{0,4,10}	-	-	-
5	1	~~{0,5}~~	~~{0,5,6}~~	~~{0,5,7}~~	~~{0,5,8}~~	~~{0,5,9}~~	~~{0,5,10}~~	-	-	-	-
6	2	{0,6}	**{0,6,7}**	{0,6,8}	~~{0,6,9}~~	{0,6,10}	-	-	-	-	-
7	2	~~{0,7}~~	~~{0,7,8}~~	~~{0,7,9}~~	~~{0,7,10}~~	-	-	-	-	-	-
8	2	{0,8}	~~{0,8,9}~~	**{0,8,10}**	-	-	-	-	-	-	-
9	1	~~{0,9}~~	~~{0,9,10}~~	-	-	-	-	-	-	-	-
10	1	~~{0,10}~~	-	-	-	-	-	-	-	-	-

sumed. Exceptions are problems Bre-9 to Bre-11, in which demands are heterogeneous. Finally Bre-12 to Bre-15 are problems involving time windows, and therefore are not included in our experiments. On the other hand we run some of the problems of Taillard. For instance, Tai-75a and Tai-75b which have 75 client nodes and an heterogeneous demand. On the other hand, we run also problems Tai-100a and Tai-100b, with 100 clients and also with an heterogeneous demand. Table 7 shows the parameters under which the tests are run and Table 8 presents their results. The best known solution, drawn from (Van Breedam, 1994; Alba & Dorronsoro, 2008), is presented as well as information on the solutions under our procedure (best solution, success and average running times).

With respect to the processing time, some instances took only minutes while others required hours, although in average the best solutions took not much longer than an hour to be found. The algorithm, as stated in (Frutos & Tohmé, 2012), just reached a 33.33% of the best solutions known to Van Breedam problems. Through the application of selective pressure a 66.66% was achieved while the average running time decreased a 15.02%. In the case of Taillard's problem a 50% of the times the best known solution was reached.

A REAL-WORLD APPLICATION

In order to apply the algorithm we tested it on data from an Argentinean firm specialized in producing and distributing fresh fruit. We focus our analysis on the distribution in Buenos Aires and its surrounding areas. The company has a refrigerated depot in the Central Market of the city. The fleet of 5 refrigerated trucks, each with a legally bound capacity of 300 Kg., departs from there. In case it cannot satisfy

Table 7. Parameters and characteristics (Van Breedam and Taillard's problems)

Problems	Bre-1 to Bre-6	Bre-9 to Bre-11	Tai-75a and Tai-75b	Tai-100a and Tai-100b
Nodes or Clients	100	100	75	100
Initialization	Random	Random	Random	Random
Representation	Integer	Integer	Integer	Integer
Chromosome Size	100	100	75	100
Population Size	200	200	200	200
Type of Selection	Ranking	Ranking	Ranking	Ranking
Selective Pressure	1.3	1.3	1.3	1.3
Type of Crossing	2-point	2-point	2-point	2-point
Crossing Probability	0.75	0.75	0.75	0.75
Type of Mutation	Altering Gene	Altering Gene	Altering Gene	Altering Gene
Mutated Gene	10	10	8	10
Mutation Probability	0.1	0.1	0.1	0.1
Objective Functions	1	1	1	1
Executions	20	20	20	20
Termination Criteria	10 G unchanged	10 G unchanged	10 G unchanged	10 G unchanged
Notebook Used	Processor: 2.1 GHz AMD Turion X2 Ultra Dual-Core, Memory: 4 GB 800 MHz DDR2			

Table 8. Results (Van Breedam and Taillard's problems)

Instance	Best known solution	Best solution found					
		Best solution*	Success (%)*	Best solution**	Success (%)**	Average time*	Average time**
Bre-1	1106.00	1157.00	42.00	1157.00	70.00	25.16	19.23
Bre-2	**1506.00**	1587.00	18.00	**1506.00**	50.00	24.11	20.26
Bre-3	**1751.00**	**1751.00**	100.00	**1751.00**	100.00	11.2	9.01
Bre-4	**1470.00**	**1470.00**	100.00	**1470.00**	100.00	19.68	17.89
Bre-5	950.00	985.00	76.00	985.00	90.00	21.66	19.25
Bre-6	**969.00**	979.00	62.00	**969.00**	65.00	31.72	27.12
Bre-9	1690.00	1790.00	17.00	1790.00	25.00	38.52	33.54
Bre-10	**1026.00**	**1026.00**	52.00	**1026.00**	65.00	49.45	42.87
Bre-11	**1128.00**	1187.00	100.00	**1128.00**	100.00	73.47	62.35
Tai-75a	**1618.36**	-	-	**1618.36**	100.00	-	51.78
Tai-75b	**1344.64**	-	-	**1344.64**	100.00	-	62.21
Tai-100a	2041.33	-	-	2112.35	95.00	-	75.82
Tai-100b	1940.61	-	-	1978.51	95.00	-	73.45
*Results from (Frutos & Tohmé, 2012), **Results applying selective pressure. Average time in minutes.							

Table 9. Parameters and characteristics (50_300, 100_300, 150_300 and 200_300)

Problems	50_300	100_300	150_300	200_300
Nodes or Clients	50	100	150	200
Initialization	Random	Random	Random	Random
Representation	Integer	Integer	Integer	Integer
Chromosome Size	50	100	150	200
Population Size	200	200	200	200
Type of Selection	Ranking	Ranking	Ranking	Ranking
Selective Pressure	1.3	1.3	1.3	1.3
Type of Crossing	2-point	2-point	2-point	2-point
Crossing Probability	0.75	0.75	0.75	0.75
Type of Mutation	Altering Gene	Altering Gene	Altering Gene	Altering Gene
Mutated Gene	5	10	15	20
Mutation Probability	0.1	0.1	0.1	0.1
Objective Functions	1	1	1	1
Executions	20	20	20	20
Termination Criteria	10 G unchanged	10 G unchanged	10 G unchanged	10 G unchanged
Notebook Used	Processor: 2.1 GHz AMD Turion X2 Ultra Dual-Core, Memory: 4 GB 800 MHz DDR2			

Table 10. Results (50_300, 100_300, 150_300 and 200_300)

Instance	Best solution found		
	Best solution	Success (%)	Average time (min)
300_50	**1740.68**	95.00	9.12
100_300	**3201.96**	80.00	27.54
150_300	**4470.20**	70.00	42.81
200_300	**5821.89**	65.00	61.51

Table 11. Routes of the best solution (50_300)

Distance	By route	Load	Route	Sequence of deliveries
1740.68	519.76	289.00	1	0-44-27-14-48-13-41-25-29-7-37-6-34-0
	334.99	300.00	2	0-9-46-1-11-40-23-33-39-24-12-36-38-0
	481.80	294.00	3	0-32-45-18-49-30-31-26-35-3-5-4-28-0
	317.68	283.00	4	0-22-43-15-8-16-47-2-20-17-19-21-0
	86.45	50.00	5	0-42-10-0

Table 12. Routes of the best solution (100_300)

Distance	By route	Load	Route	Sequence of deliveries
3201.96	341.75	282.00	1	0-17-86-29-75-6-99-54-14-70-26-42-61-0
	336.19	300.00	2	0-27-62-5-68-21-71-31-67-72-82-13-78-0
	330.54	292.00	3	0-56-96-52-74-81-15-65-93-18-36-89-44-0
	566.36	290.00	4	0-50-2-23-85-63-64-97-51-73-88-34-32-0
	335.77	293.00	5	0-24-1-38-40-84-94-11-76-58-90-47-8-0
	309.63	281.00	6	0-20-22-16-60-83-4-77-45-19-48-39-0
	186.79	164.00	7	0-46-91-55-53-69-25-0
	296.60	284.00	8	0-33-41-79-49-9-87-28-10-3-57-98-0
	498.35	286.00	9	0-92-95-30-7-35-66-43-12-59-80-37-0

all the clients the firm contracts a delivery service. In any case it has a maximum of 200 clients a day. Each one receives between 50 Kg. and the full load of a truck. 4 situations, involving 50, 100, 150 and 200 clients (denoted 0_300, 100_300, 150_300 and 200_300, respectively), were considered. Table 9 shows the parameters under which the cases 50_300 (Appendix 2, Table 15), 100_300 (Appendix 2, Table 16), 150_300 (Appendix 2, Table 17) and 200_300 (Appendix 2, Table 18) were run, while Table 10 presents the results.

Tables 11 to 14 show the details of the best routes found (distance, load and sequence of deliveries). In turn, Figures 5 to 8 (Appendix 1), graph the routes.

For the 50_300 instance 5 routes were found to cater 50 clients. Each vehicle was used in a 81.06% of its capacity. In the current plan of the firm, these 50 clients are also provided through 5 routes with

Table 13. Routes of the best solution (150_300)

Distance	By route	Load	Route	Sequence of deliveries
4470.20	316.34	300.00	1	0-13-125-141-91-61-73-23-128-25-84-136-139-0
	249.28	240.00	2	0-46-69-89-33-109-58-19-142-7-0
	492.45	290.00	3	0-74-111-137-80-22-5-140-8-133-49-35-95-0
	382.27	297.00	4	0-24-97-76-17-124-117-20-123-85-126-4-81-0
	324.51	300.00	5	0-42-113-144-129-145-83-15-108-27-132-26-104-0
	321.42	300.00	6	0-55-120-116-75-131-110-98-134-18-36-96-135-0
	334.23	297.00	7	0-79-146-59-48-100-101-52-82-9-12-66-118-0
	326.55	297.00	8	0-88-38-56-45-41-6-51-57-32-147-54-53-0
	292.74	279.00	9	0-77-143-34-107-72-119-105-70-103-78-28-0
	311.73	289.00	10	0-30-64-90-1-99-21-65-50-86-148-63-0
	295.60	299.00	11	0-2-112-121-43-115-149-40-3-67-11-31-0
	491.10	282.00	12	0-10-29-44-127-14-47-138-102-130-87-68-0
	331.98	299.00	13	0-71-39-60-16-37-93-62-114-94-106-122-92-0

Table 14. Routes of the best solution (200_300)

Distance	By route	Load	Route	Sequence of deliveries
5821.89	321.86	300.00	1	0-88-60-58-1-54-140-4-121-39-45-47-99-0
	302.33	295.00	2	0-81-26-11-94-146-85-144-104-21-97-7-0
	308.27	288.00	3	0-105-109-102-136-142-14-96-20-24-150-28-0
	320.93	297.00	4	0-116-117-10-13-143-114-111-108-95-107-112-6-0
	385.64	288.00	5	0-184-186-182-170-168-139-129-165-32-130-38-36-0
	322.80	292.00	6	0-98-148-137-82-3-23-71-16-19-100-34-37-0
	520.29	295.00	7	0-133-8-167-178-180-187-179-141-198-196-197-110-0
	322.28	296.00	8	0-40-84-145-18-118-120-101-93-2-25-124-152-0
	332.05	299.00	9	0-73-22-174-176-74-72-67-63-62-68-66-69-0
	307.33	284.00	10	0-147-122-123-27-127-92-138-153-89-125-75-0
	487.64	280.00	11	0-199-195-194-191-192-189-183-128-177-181-171-0
	333.94	295.00	12	0-134-91-17-115-12-135-103-15-132-113-106-76-0
	329.63	296.00	13	0-29-126-56-156-41-59-65-160-61-158-159-161-0
	312.47	287.00	14	0-164-172-193-188-163-9-80-83-86-173-166-0
	289.64	299.00	15	0-77-162-64-87-57-157-155-51-131-149-169-0
	304.20	300.00	16	0-70-55-48-44-35-33-30-151-90-31-154-42-0
	292.16	296.00	17	0-175-190-79-52-46-50-5-49-43-119-53-78-0
	28.46	23.00	18	0-185-0

a similar degree of use of each vehicle. For the 100_300 case, 9 routes deliver to 100 clients and the capacity of the vehicles was used in a 91.55%. The firm is currently covering the demand of these 100 clients with 11 routes, using less than 85% of each vehicle. In the 150_300 case we got 13 routes for 150 clients and a 96.64% use of the trucks. The company does this with 15 routes and a 75% usage of capacity. Finally, for 200_300, the algorithm generated 18 routes for the 200 clients and a 92.77% usage. The real-world company does this with 22 routes and using less than 70% of the capacity. It is clear that the algorithm improves the allocation of resources over the current practices of the firm.

FUTURE RESEARCH DIRECTIONS

Our plan for further work starts with the consideration of the Multi-Depot Vehicle Routing Problem (MDVRP). In any real-world analysis a firm will own at least two depots and the distribution system will arise as a solution of this enlarged problem (Vidal et al., 2012) (Surekha & Sumathi, 2011). The fleet of trucks will also be assumed as having a larger capacity than 500 Kg., which involves a further modification in the operational conditions of the problem (Salhia et al., 2014). We also intend to evaluate the Two-Echelon Capacitated Vehicle Routing Problem (2E-CVRP) (Sitek & Wikarek, 2014) (Nguyen et al., 2012) (Perboli et al., 2011). The goal is to address the design of distribution systems with Cross-Docking facilities (Gonzalez-Feliu, 2013). An important aspect to analyze is the environmental aspects of distribution systems (Soysal et al., 2014). Last but not least, the increasing complexity of these problems requires the expansion of the toolbox of analysts, including new and improved meta-heuristics as well as alternative algorithms.

CONCLUSION

We presented a new algorithm for the CVRP (Capacitated Vehicle Routing Problem), distinguishing between the standard and its alternative formulation. While the outcomes are similar to other procedures in the literature, their running times, tested on Van Breedam and Taillard's problems, are not. We improved the performance of the algorithm adding selective pressure on the solutions of the genetic algorithm. This procedure yields a new ranking and warrants the survival of the best solutions. We also addressed a real-world problem. Running four instances in which 50, 100, 150 and 200 nodes had to be visited we found that the results obtained under the algorithm improved over the current plants enacted by the firm. Running the algorithm with more than 200 nodes the processing times increased, concentrated in the first stage of the algorithm.

ACKNOWLEDGMENT

The authors are thankful to the National Research Council of Argentina (CONICET), the Universidad Nacional del Sur (PGI 24/J056) and the Universidad Nacional de Río Negro.

REFERENCES

Alba, E., & Dorronsoro, B. (2008). A hybrid cellular genetic algorithm for the capacitated vehicle routing problem. *Studies in Computational Intelligence*, *82*, 379–422. doi:10.1007/978-3-540-75396-4_14

Aleman, R., & Hill, R. (2010). A tabu search with vocabulary building approach for the vehicle routing problem with split demands. *International Journal of Meta-heuristics*, *1*, 55–80.

Aleman, R., Zhang, X., & Hill, R. (2010). An adaptive memory algorithm for the split delivery vehicle routing problem. *Journal of Heuristics*, *16*(3), 441–473. doi:10.1007/s10732-008-9101-3

Ambrosino, D., & Sciomachen, A. (2007). A food distribution network problem: A case study. *IMA Journal of Management Mathematics*, *18*, 35–53.

Bartolini, E., Cordeau, J. F., & Laporte, G. (2013). Improved lower bounds and exact algorithm for the capacitated arc routing problem. *Mathematical Programming*, *137*(1-2), 409–452. doi:10.1007/s10107-011-0497-4

Basu, S. (2012). Tabu search implementation on the traveling salesman problem and its variations: A literature survey. *American Journal of Operations Research*, *2*(02), 163–173. doi:10.4236/ajor.2012.22019

Belenguer, J., Benavent, E., Labadi, N., Prins, C., & Reghioui, M. (2010). Split delivery capacitated arc routing problem: Lower bound and metaheuristic. *Transportation Science*, *44*(2), 206–220. doi:10.1287/trsc.1090.0305

Belfiore, P., & Yoshizaki, H. (2009). Scatter search for a real-life heterogeneous fleet vehicle routing problem with time windows and split deliveries in Brazil. *European Journal of Operational Research*, *199*(3), 750–758. doi:10.1016/j.ejor.2008.08.003

Bodin, L., & Golden, B. (1981). Classification in vehicle routing and scheduling. *Networks*, *11*(2), 97–108. doi:10.1002/net.3230110204

Bolduc, M., Laporte, G., Renaud, J., & Boctor, F. (2010). A tabu search heuristic for the split delivery vehicle routing problem with production and demand calendars. *European Journal of Operational Research*, *202*(1), 122–130. doi:10.1016/j.ejor.2009.05.008

Çatay, B. (2009). Ant colony optimization and its application to the vehicle routing problem with pickups and deliveries. In R. Chiong & S. Dhakal (Eds.), *Natural Intelligence for Scheduling, Planning and Packing Problems, Studies in Computational Intelligence* (Vol. 250, pp. 219–244). Heidelberg: Springer. doi:10.1007/978-3-642-04039-9_9

Derigs, U., Li, B., & Vogel, U. (2009). Local search-based metaheuristics for the split delivery vehicle routing problem. *The Journal of the Operational Research Society*, *61*(9), 1356–1364. doi:10.1057/jors.2009.100

Desrochers, M., Desrosiers, J., & Solomon, M. (2002). A new optimization algorithm for the vehicle routing problem with time windows. *Operations Research*, *40*(2), 342–354. doi:10.1287/opre.40.2.342

Escobar, J. W., Linfati, R., Toth, P., & Baldoquin, M. G. (2014). A hybrid granular tabu search algorithm for the multi-depot vehicle routing problem. *Journal of Heuristics*, *20*(5), 483–509. doi:10.1007/s10732-014-9247-0

Frutos, M., & Tohmé, F. (2012). A new approach to the optimization of the CVRP through Genetic Algorithms. *American Journal of Operations Research*, *2*(04), 495–501. doi:10.4236/ajor.2012.24058

Goldberg, D. (1989). *Genetic Algorithms in Search, Optimization, and Machine Learning*. Addison Wesley.

Gonzalez-Feliu, J. (2013). Vehicle Routing in Multi-Echelon Distribution Systems with Cross-Docking: A Systematic Lexical-Metanarrative Analysis. *Computer and Information Science*, *6*(3), 28–47. doi:10.5539/cis.v6n3p28

Jin, M., Liu, K., & Bowden, R. (2007). A two-stage algorithm with valid inequalities for the split delivery vehicle routing problem. *International Journal of Production Economics*, *105*(1), 228–242. doi:10.1016/j.ijpe.2006.04.014

Kao, Y., & Chen, M. (2013). Solving the CVRP Problem Using a Hybrid PSO Approach. In K. Madani, A. Dourado, A. Rosa, & J. Filipe (Eds.), *Computational Intelligence, Studies in Computational Intelligence* (Vol. 465, pp. 59–67). Heidelberg: Springer. doi:10.1007/978-3-642-35638-4_5

Khebbache-Hadji, S., Prins, C., Yalaoui, A., & Reghioui, M. (2013). Heuristics and memetic algorithm for the two-dimensional loading capacitated vehicle routing problem with time windows. *Central European Journal of Operations Research*, *21*(2), 307–336. doi:10.1007/s10100-011-0204-9

Kumar, S., & Panneerselvam, R. (2012). A survey on the vehicle routing problem and its variants. *Intelligent Information Management*, *4*(03), 66–74. doi:10.4236/iim.2012.43010

Kurz, A., & Zäpfel, G. (2013). Modeling cost-delivery trade-offs for distribution logistics by a generalized PVRP model. *Journal of Business Economics*, *83*(7), 705–726. doi:10.1007/s11573-013-0678-0

Laporte, G., & Osman, I. (1995). Routing problems: A bibliography. *Annals of Operations Research*, *61*(1), 227–262. doi:10.1007/BF02098290

Lee, C., Epelman, M., White, C. III, & Bozer, Y. (2006). A shortest path approach to the multiplevehicle routing problem with split pick-ups. *Transportation Research*, *40*(4), 265–284. doi:10.1016/j.trb.2004.11.004

Moreno, L., Poggi de Aragao, M., & Uchoa, E. (2010). Improved lower bounds for the split delivery vehicle routing problem. *Operations Research Letters*, *38*(4), 302–306. doi:10.1016/j.orl.2010.04.008

Mota, E., Campos, V., & Corbéran, A. (2007). A new metaheuristic for the vehicle routing problem with split demands. In J. Van Hemert & C. Cotta (Eds.), *Evolutionary Computation in Combinatorial Optimization* (pp. 121–129). Heidelberg: Springer. doi:10.1007/978-3-540-71615-0_11

Nagao, Y., & Nagamochi, H. (2007). A DP-based heuristic algorithm for the discrete split delivery vehicle routing problem. *Journal of Advanced Mechanical Design, Systems and Manufacturing*, *1*(2), 217–226. doi:10.1299/jamdsm.1.217

Nguyen, V. P., Prins, C., & Prodhon, C. (2012). Solving the two-echelon location routing problem by a GRASP reinforced by a learning process and path relinking. *European Journal of Operational Research, 216*(1), 113–126. doi:10.1016/j.ejor.2011.07.030

Nowak, M., Ergun, O., & White, C. III. (2009). An empirical study on the benefit of split loads with the pickup and de-livery problem. *European Journal of Operational Research, 198*(3), 734–740. doi:10.1016/j.ejor.2008.09.041

Oyola, J., & Løkketangen, A. (2014). GRASP-ASP: An algorithm for the CVRP with route balancing. *Journal of Heuristics, 20*(4), 361–382. doi:10.1007/s10732-014-9251-4

Perboli, G., Tadei, R., & Vigo, D. (2011). The Two-Echelon Capacitated Vehicle Routing Problem: Models and Math-Based Heuristics. *Transportation Science Archive, 45*(3), 364-380.

Pollaris, H., Braekers, K., Caris, A., Janssens, G. K., & Limbourg, S. (2015). Vehicle routing problems with loading constraints: State-of-the-art and future directions. *OR-Spektrum, 37*(2), 297–330. doi:10.1007/s00291-014-0386-3

Rivera, J. C., Afsar, H. M., & Prins, C. (2015). A multistart iterated local search for the multitrip cumulative capacitated vehicle routing problem. *Computational Optimization and Applications, 61*(1), 159–187. doi:10.1007/s10589-014-9713-5

Salhia, S., Imranb, A., & Wassana, N. A. (2014). The multi-depot vehicle routing problem with heterogeneous vehicle fleet: Formulation and a variable neighborhood search implementation. *Computers & Operations Research, 52*, 315–325. doi:10.1016/j.cor.2013.05.011

Shen, Z., Dessouky, M., & Ordóñez, F. (2009). A two-stage vehicle routing model for large scale bioterrorism emergencies. *Networks, 54*, 255–269. doi:10.1002/net.20337

Sitek, P. (2014). A Hybrid Approach to the Two-Echelon Capacitated Vehicle Routing Problem. In R. Szewczyk, C. Zieliński, & M. Kaliczyńska (Eds.), *Recent Advances in Automation, Robotics and Measuring Techniques, Advances in Intelligent Systems and Computing* (Vol. 267, pp. 251–263). Heidelberg: Springer. doi:10.1007/978-3-319-05353-0_25

Sitek, P., & Wikarek, J. (2014). A novel integrated approach to the modelling and solving of the Two-Echelon Capacitated Vehicle Routing Problem. *Production & Manufacturing Research: An Open Access Journal, 2*(1), 326–340.

Sitek, P., & Wikarek, J. (2015). A Hybrid Approach to the Optimization of multi-echelon Systems. *Mathematical Problems in Engineering, 2015*, 1–12. doi:10.1155/2015/925675

Sitek, P., Wikarek, J., & Grzybowska, K. (2014). A Multi-Agent Approach to the Multi-Echelon Capacitated Vehicle Routing Problem. In J. M. Corchado, J. Bajo, J. Kozlak, & P. Pawlewski (Eds.), *Highlights of Practical Applications of Heterogeneous Multi-Agent Systems, The PAAMS Collection, Communications in Computer and Information Science* (Vol. 430, pp. 121–132). Heidelberg: Springer. doi:10.1007/978-3-319-07767-3_12

Soysal, M., Bloemhof-Ruwaard, J. M., & Bektas, T. (2014). The time-dependent two-echelon capacitated vehicle routing problem with environmental considerations. *International Journal of Production Economics*, 1-13.

Surekha, P., & Sumathi, S. (2011). Solution To Multi-Depot Vehicle Routing Problem Using Genetic Algorithms. *World Applied Programming*, *1*(3), 118–131.

Taillard, E. (1993). *VRP benchmarks*. Retrieved from http://mistic.heigvd.ch/taillard/problemes.dir/vrp.dir/ vrp.html

Tavakkoli-Moghaddam, R., Safaei, N., Kah, M., & Rabbani, M. (2007). A new capacitated vehicle routing problem with split service for minimizing fleet cost by simulated annealing. *Journal of the Franklin Institute*, *344*(5), 406–425. doi:10.1016/j.jfranklin.2005.12.002

Thangiah, S., Fergany, A., & Awan, S. (2007). Real-time split-delivery pickup and delivery time window problems with transfers. *Central European Journal of Operations Research*, *15*(4), 329–349. doi:10.1007/s10100-007-0035-x

Van Breedam, A. (1994). *An Analysis of the Behavior of Heuristics for the Vehicle Routing Problem for a Selection of Problems with Vehicle Related, Customer-related, and Timerelated Constraints*. (PhD thesis). University of Antwerp, Belgium.

Vidal, T., Crainic, T. G., Gendreau, M., Lahrichi, N., & Rei, W. (2012). A hybrid genetic algorithm for multi-depot and periodic vehicle routing problems. *Operations Research*, *60*(3), 611–624. doi:10.1287/opre.1120.1048

Wang, Y. (2012). A hybrid approach based on ant colony system for the VRPTW. In Wu, Y. (Ed.), *Advanced Technology in Teaching, Proceedings of WTCS 2009, Advances in Intelligent and Soft Computing* (*vol. 117*, pp. 327-333). Springer. doi:10.1007/978-3-642-25437-6_46

Wilck, I. V. J. IV, & Cavalier, T. (2012). A construction heuristic for the split delivery vehicle routing. *American Journal of Operations Research*, *2*(02), 153–162. doi:10.4236/ajor.2012.22018

Wu, B., Lin, J., & Dong, M. (2013). Artificial Bee Colony Algorithm for three-dimensional loading capacitated vehicle routing problem. In E. Qi, J. Shen, & R. Dou (Eds.), *Proceedings of 20th International Conference on Industrial Engineering and Engineering Management* (pp. 815-825). Springer. doi:10.1007/978-3-642-40063-6_81

Yu, Y., Chen, H., & Chu, F. (2008). A new model and hybrid approach for large scale inventory routing problems. *European Journal of Operational Research*, *189*(3), 1022–1040. doi:10.1016/j.ejor.2007.02.061

Zhang, J., Lam, W. H. K., & Chen, B. Y. (2013). A stochastic vehicle routing problem with travel time uncertainty: Trade-off between cost and customer service. *Networks and Spatial Economics*, *13*(4), 471–496. doi:10.1007/s11067-013-9190-x

KEY TERMS AND DEFINITIONS

Crossover Operator: A genetic operator that generates diversity by combining fragments of the chromosomes of possible solutions yielding a new candidate.

Decision-Making: The process of choosing an alternative among a class of them, with the aim of solving a given problem.

Genetic Algorithm: A series of steps to allow the evolution of solutions to specific problems. It is inspired in biological evolution and particularly its genetic-molecular basis.

Genetic Operator: A function applied in a genetic algorithm in order to ensure the diversity of a population.

Meta-Heuristics: A search procedure intended to solve computational problems, by seeking appropriate solution methods by adjusting parameters and specifications.

Mutation Operator: A genetic operator that generates diversity by changing the chromosome of a candidate solution yielding an alternative candidate.

Optimization: The process of finding the maxima or minima of either a single (mono-objective) or of several (multi-objective) functions over a set of a constrained set of alternatives.

Selection Process: Procedure to choose some individuals out from a population in order to apply genetic operators to them and so generating a new population.

Vehicle Routing Problem: A class of transportation optimization problem. It seeks to find the minimal length routes of provision to different clients.

ACRONYMS USED IN THE TEXT

2E-CVRP: Two-Echelon Capacitated Vehicle Routing Problem.
ACO: Ant Colony Optimization.
CVRP: Capacitated Vehicle Routing Problem.
GA: Genetic Algorithm.
HGTS: Hybrid Granular Tabu Search.
ILS: Iterated Local Search.
MDVRP: Multi-Depot Vehicle Routing Problem.
PSO: Particle Swarm Optimization.
PVRP: Periodic Vehicle Routing Problem.
SA: Simulated Annealing.
SDVRP: Split Delivery Vehicle Routing Problem.
SVRP: Stochastic Vehicle Routing Problem.
TS: Tabu Search.
VRP: Vehicle Routing Problem.
VRPPD: Vehicle Routing Problem with Pick-Up and Delivering.
VRPTW: Vehicle Routing Problem with Time Windows.

APPENDIX 1: FIGURES

Figure 1. Solutions for generations 10 and 20

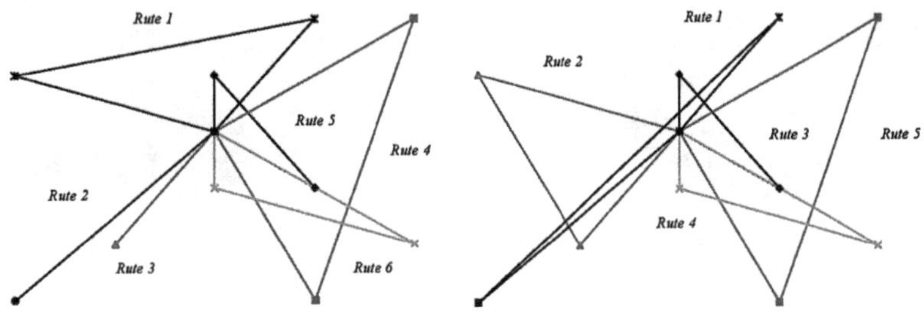

Generation 10, Distance: 205.61 *Generation 20, Distance: 205.56*

Figure 2. Solutions for generations 45 and 75

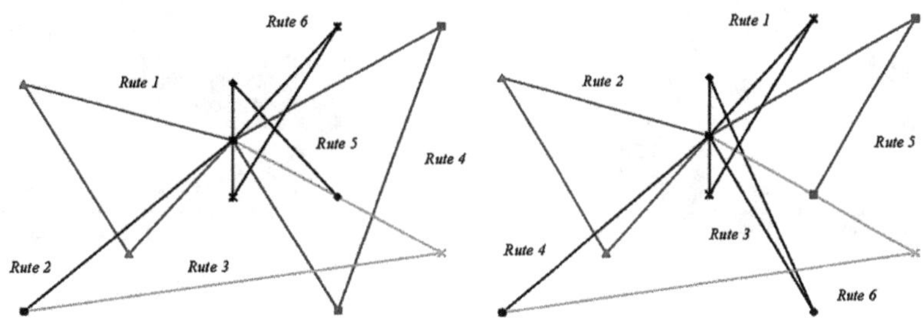

Generation 45, Distance: 201.65 *Generation 75, Distance: 201.40*

Figure 3. Solutions for generations 100 and 150

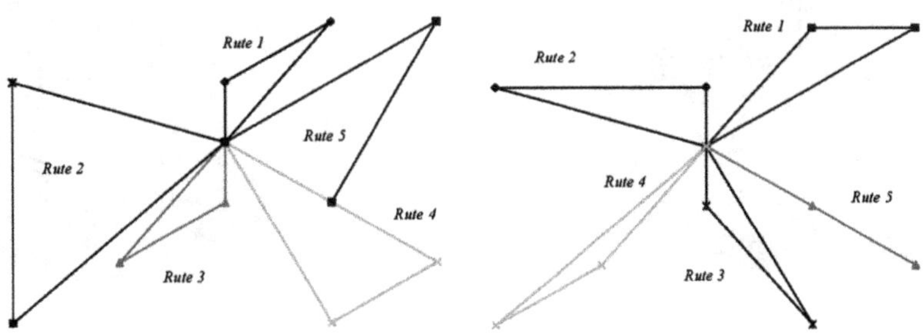

Generation 100, Distance: 169.76 *Generation 150, Distance: 153.06*

Figure 4. Selective pressure

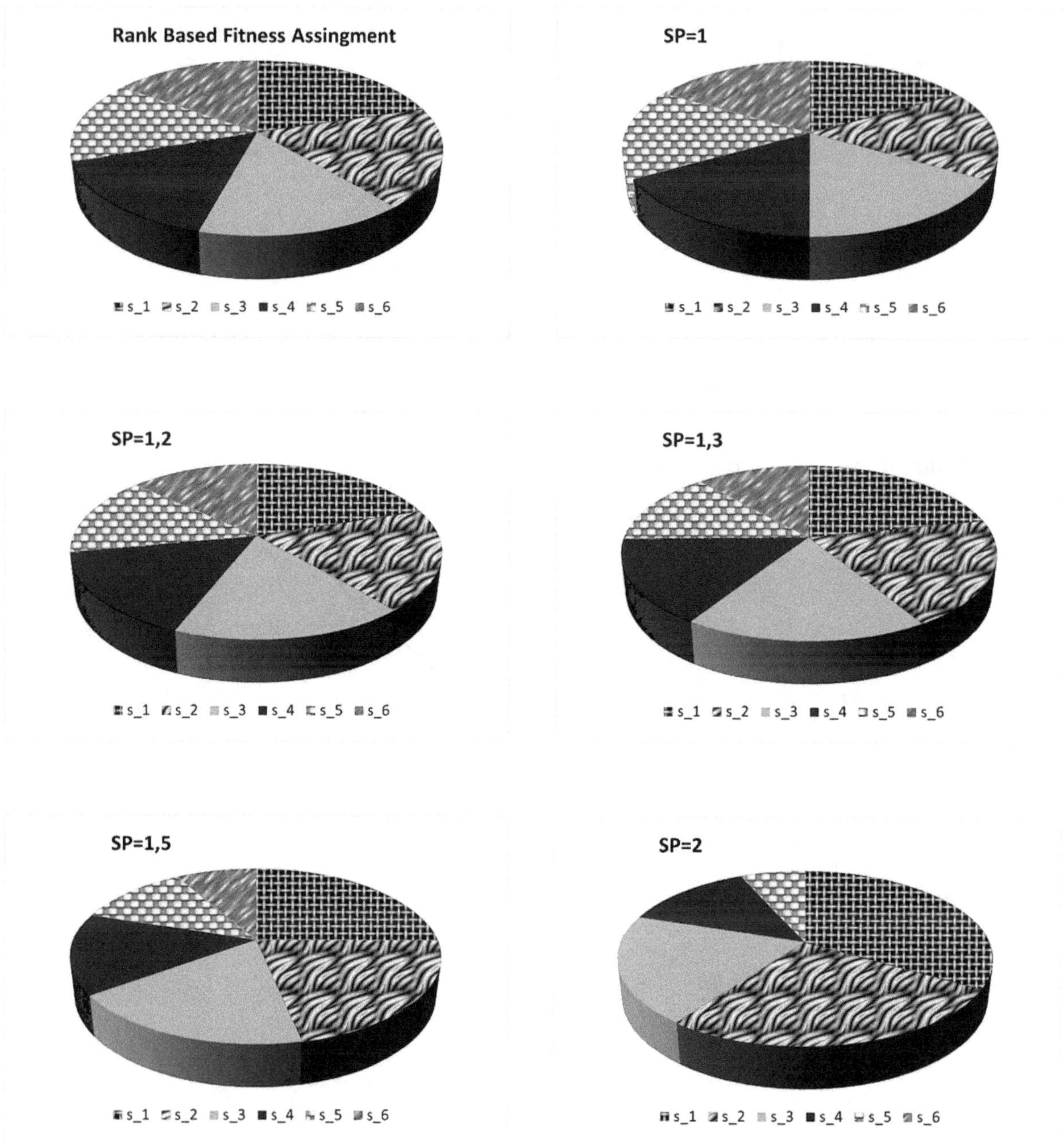

Figure 5. Routes of the best solution (50_300)

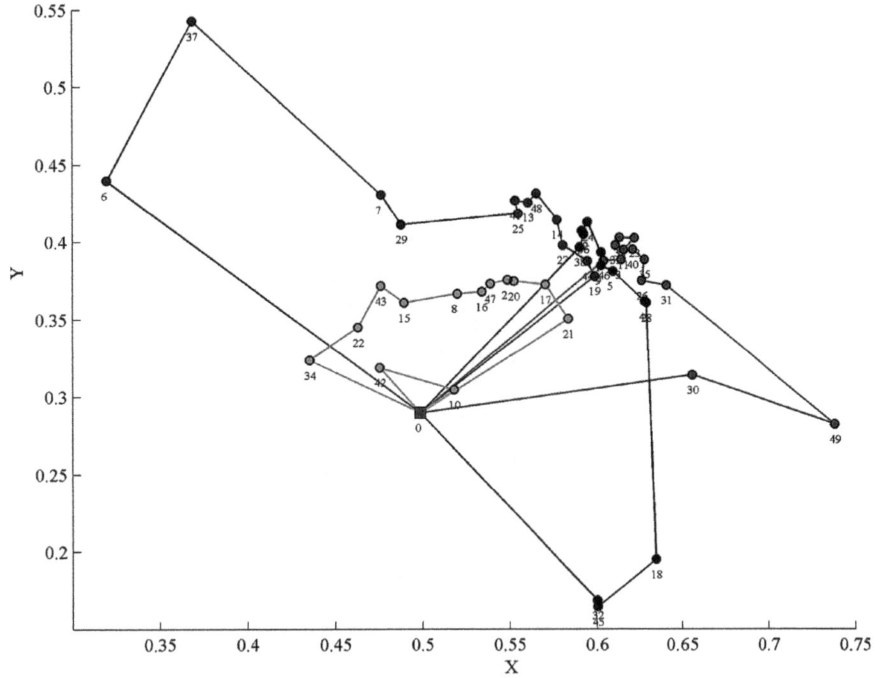

Figure 6. Routes of the best solution (100_300)

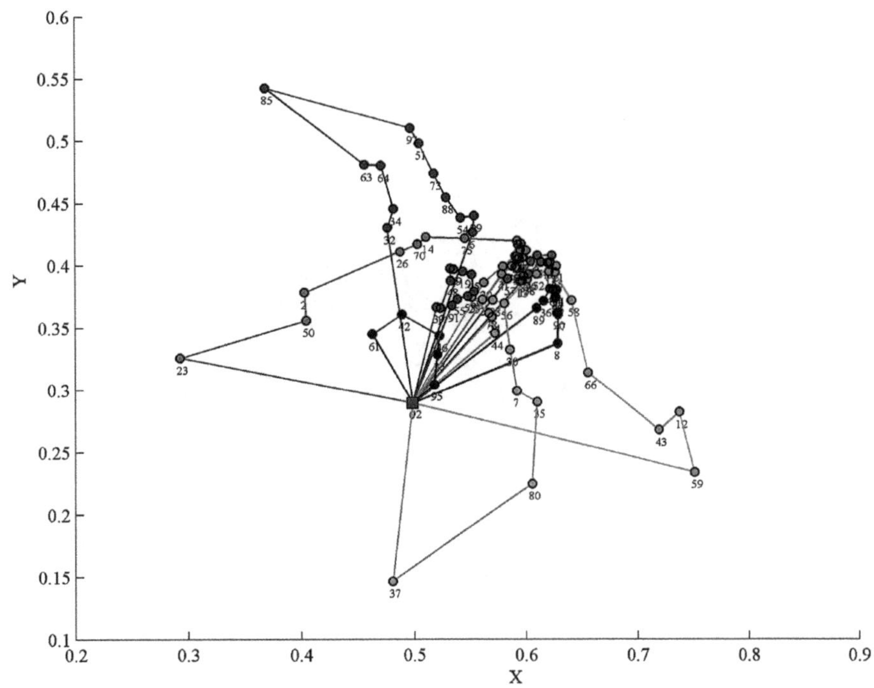

Figure 7. Routes of the best solution (150_300)

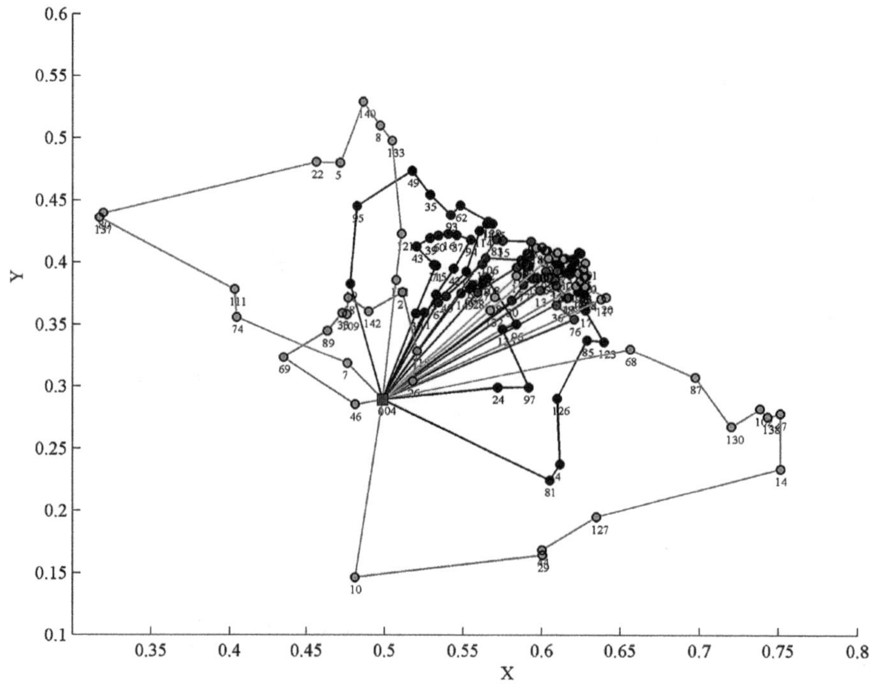

Figure 8. Routes of the best solution (200_300)

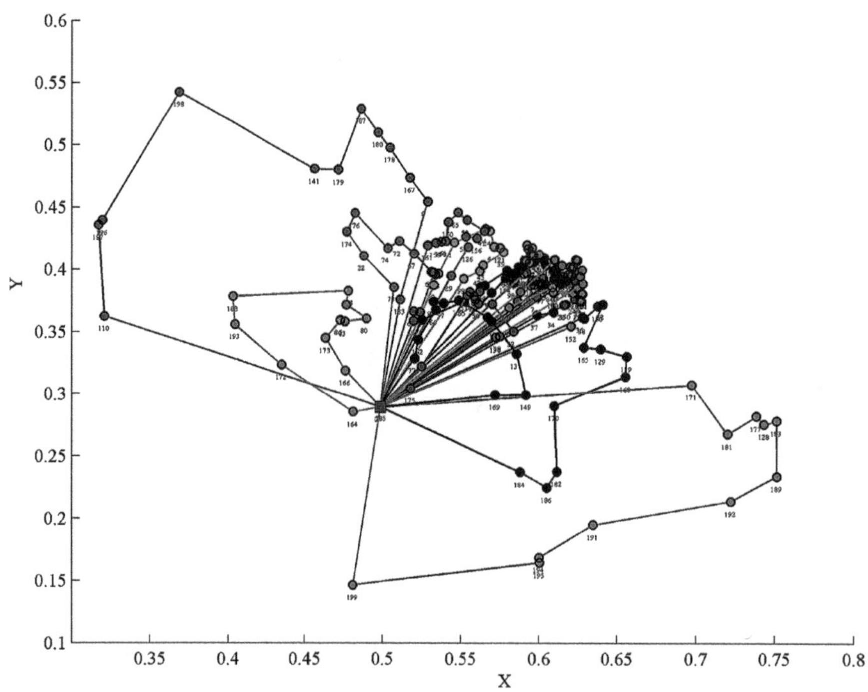

APPENDIX 2: INPUT DATA FROM THE REAL-WORLD PROBLEM

Table 15. Input data from problem 50_300

Nodes	x	y	Demand	Nodes	x	y	Demand	Nodes	x	y	Demand
0	0.499	0.290	0	17	0.570	0.372	20	34	0.435	0.324	30
1	0.602	0.393	24	18	0.635	0.195	27	35	0.628	0.389	21
2	0.548	0.375	30	19	0.599	0.378	25	36	0.592	0.405	27
3	0.614	0.389	22	20	0.552	0.375	24	37	0.369	0.543	20
4	0.628	0.362	21	21	0.584	0.350	29	38	0.590	0.397	21
5	0.609	0.381	25	22	0.463	0.345	28	39	0.611	0.398	21
6	0.320	0.440	21	23	0.622	0.402	28	40	0.621	0.395	27
7	0.476	0.430	22	24	0.595	0.413	25	41	0.553	0.427	20
8	0.520	0.367	23	25	0.555	0.418	26	42	0.475	0.319	27
9	0.602	0.385	27	26	0.626	0.375	21	43	0.476	0.372	25
10	0.518	0.305	23	27	0.580	0.398	24	44	0.595	0.387	27
11	0.615	0.395	27	28	0.629	0.361	24	45	0.600	0.164	27
12	0.591	0.407	24	29	0.488	0.411	23	46	0.604	0.388	21
13	0.560	0.425	21	30	0.655	0.314	27	47	0.539	0.373	23
14	0.577	0.414	27	31	0.641	0.372	26	48	0.565	0.431	28
15	0.489	0.361	26	32	0.600	0.169	23	49	0.738	0.282	30
16	0.534	0.368	30	33	0.613	0.403	28	-	-	-	-

Table 16. Input data from problem 100_300

Nodes	x	y	Demand	Nodes	x	y	Demand	Nodes	x	y	Demand
0	0.499	0.290	0	34	0.482	0.445	22	68	0.623	0.408	23
1	0.597	0.388	27	35	0.610	0.291	29	69	0.554	0.379	27
2	0.403	0.378	26	36	0.616	0.372	23	70	0.503	0.417	25
3	0.592	0.397	25	37	0.481	0.147	27	71	0.621	0.401	30
4	0.593	0.417	27	38	0.601	0.389	30	72	0.604	0.403	25
5	0.610	0.408	22	39	0.520	0.367	23	73	0.518	0.474	29
6	0.553	0.427	20	40	0.621	0.381	25	74	0.613	0.403	25
7	0.592	0.299	25	41	0.579	0.393	30	75	0.546	0.422	24
8	0.629	0.338	25	42	0.489	0.361	26	76	0.627	0.374	25
9	0.597	0.406	23	43	0.720	0.268	23	77	0.592	0.420	25
10	0.591	0.399	30	44	0.573	0.346	22	78	0.567	0.362	28
11	0.628	0.380	24	45	0.552	0.393	23	79	0.587	0.399	24
12	0.738	0.282	30	46	0.523	0.344	30	80	0.605	0.225	21

continued on following page.

Table 17. Continued

Nodes	x	y	Demand	Nodes	x	y	Demand	Nodes	x	y	Demand
13	0.595	0.387	27	47	0.629	0.361	24	81	0.622	0.402	28
14	0.511	0.423	23	48	0.533	0.388	26	82	0.598	0.391	23
15	0.621	0.395	27	49	0.591	0.407	24	83	0.595	0.418	30
16	0.590	0.407	24	50	0.404	0.356	21	84	0.624	0.380	23
17	0.524	0.366	25	51	0.505	0.498	20	85	0.369	0.543	20
18	0.626	0.375	25	52	0.609	0.393	22	86	0.533	0.398	21
19	0.544	0.395	29	53	0.548	0.375	30	87	0.594	0.406	30
20	0.563	0.386	21	54	0.542	0.439	26	88	0.529	0.455	26
21	0.627	0.400	22	55	0.539	0.373	23	89	0.609	0.366	23
22	0.580	0.399	28	56	0.581	0.370	26	90	0.628	0.362	21
23	0.293	0.326	21	57	0.584	0.389	23	91	0.534	0.368	30
24	0.570	0.359	20	58	0.641	0.372	26	92	0.499	0.290	23
25	0.552	0.375	24	59	0.751	0.234	29	93	0.626	0.375	21
26	0.488	0.411	23	60	0.595	0.413	25	94	0.627	0.381	23
27	0.561	0.373	23	61	0.463	0.345	28	95	0.518	0.305	23
28	0.592	0.405	27	62	0.600	0.412	28	96	0.602	0.393	24
29	0.536	0.397	21	63	0.456	0.481	30	97	0.497	0.510	27
30	0.586	0.333	29	64	0.471	0.480	26	98	0.570	0.372	20
31	0.620	0.402	21	65	0.626	0.394	26	99	0.554	0.440	20
32	0.476	0.430	22	66	0.655	0.314	27	-	-	-	-
33	0.521	0.329	28	67	0.613	0.403	28	-	-	-	-

Table 17. Input data from problem 150_300

Nodes	x	y	Demand	Nodes	x	y	Demand	Nodes	x	y	Demand
0	0.499	0.290	0	50	0.602	0.410	22	100	0.621	0.401	30
1	0.610	0.400	28	51	0.604	0.403	25	101	0.627	0.400	22
2	0.512	0.376	30	52	0.624	0.407	26	102	0.738	0.282	30
3	0.533	0.374	26	53	0.602	0.393	24	103	0.584	0.396	30
4	0.612	0.238	20	54	0.609	0.393	22	104	0.499	0.290	23
5	0.471	0.480	26	55	0.608	0.386	20	105	0.589	0.398	25
6	0.592	0.405	27	56	0.587	0.402	29	106	0.565	0.404	27
7	0.475	0.319	27	57	0.611	0.397	29	107	0.590	0.397	21
8	0.497	0.510	27	58	0.476	0.372	25	108	0.565	0.387	22
9	0.610	0.408	22	59	0.614	0.403	25	109	0.475	0.358	28
10	0.481	0.147	27	60	0.534	0.422	26	110	0.626	0.389	27

continued on following page.

Table 17. Continued

Nodes	x	y	Demand	Nodes	x	y	Demand	Nodes	x	y	Demand
11	0.525	0.360	22	61	0.627	0.381	27	111	0.403	0.378	26
12	0.584	0.389	23	62	0.548	0.446	21	112	0.507	0.386	29
13	0.599	0.378	25	63	0.593	0.417	27	113	0.552	0.393	23
14	0.751	0.234	29	64	0.604	0.388	21	114	0.560	0.425	21
15	0.575	0.418	23	65	0.603	0.409	29	115	0.531	0.398	30
16	0.540	0.423	30	66	0.554	0.379	27	116	0.623	0.391	28
17	0.628	0.362	21	67	0.534	0.368	30	117	0.638	0.370	20
18	0.616	0.372	23	68	0.656	0.330	22	118	0.521	0.329	28
19	0.477	0.383	24	69	0.435	0.324	30	119	0.591	0.399	30
20	0.641	0.372	26	70	0.587	0.399	24	120	0.615	0.392	23
21	0.609	0.408	25	71	0.533	0.398	21	121	0.511	0.423	23
22	0.456	0.481	30	72	0.592	0.397	25	122	0.562	0.399	27
23	0.628	0.380	24	73	0.627	0.381	23	123	0.639	0.336	28
24	0.572	0.299	30	74	0.404	0.356	21	124	0.627	0.374	25
25	0.626	0.375	21	75	0.626	0.394	26	125	0.609	0.382	20
26	0.518	0.305	23	76	0.620	0.354	27	126	0.610	0.291	29
27	0.562	0.382	28	77	0.588	0.382	26	127	0.635	0.195	27
28	0.559	0.376	23	78	0.563	0.386	21	128	0.626	0.375	30
29	0.600	0.164	27	79	0.601	0.388	25	129	0.566	0.433	24
30	0.581	0.370	26	80	0.320	0.440	21	130	0.720	0.268	23
31	0.520	0.359	26	81	0.605	0.225	21	131	0.628	0.389	21
32	0.618	0.396	25	82	0.623	0.408	23	132	0.567	0.362	28
33	0.472	0.359	24	83	0.571	0.419	20	133	0.505	0.498	20
34	0.597	0.388	27	84	0.626	0.375	25	134	0.621	0.381	25
35	0.529	0.455	26	85	0.629	0.338	25	135	0.575	0.346	29
36	0.609	0.366	23	86	0.600	0.412	28	136	0.623	0.376	30
37	0.546	0.422	24	87	0.697	0.308	20	137	0.317	0.436	21
38	0.587	0.402	26	88	0.570	0.372	20	138	0.743	0.276	29
39	0.529	0.420	23	89	0.463	0.345	28	139	0.617	0.372	27
40	0.539	0.373	23	90	0.609	0.399	30	140	0.486	0.529	21
41	0.591	0.407	24	91	0.624	0.380	23	141	0.609	0.381	25
42	0.544	0.395	29	92	0.556	0.382	27	142	0.489	0.361	26
43	0.520	0.413	30	93	0.542	0.439	26	143	0.595	0.387	27
44	0.600	0.169	23	94	0.555	0.418	26	144	0.565	0.431	28
45	0.590	0.407	24	95	0.482	0.445	22	145	0.569	0.431	29
46	0.480	0.286	28	96	0.584	0.350	29	146	0.613	0.403	25
47	0.751	0.279	25	97	0.592	0.299	25	147	0.617	0.391	22
48	0.620	0.402	21	98	0.624	0.385	26	148	0.596	0.412	25
49	0.518	0.474	29	99	0.613	0.403	28	149	0.548	0.375	30

Table 17. Continued

Table 18. Input data from problem 200_300

Nodes	x	y	Demand	Nodes	x	y	Demand	Nodes	x	y	Demand
0	0.499	0.290	0	67	0.520	0.413	30	134	0.588	0.382	26
1	0.546	0.422	24	68	0.524	0.366	25	135	0.621	0.401	30
2	0.626	0.375	21	69	0.520	0.359	26	136	0.622	0.391	28
3	0.602	0.410	22	70	0.525	0.360	22	137	0.591	0.399	30
4	0.571	0.419	20	71	0.603	0.407	24	138	0.592	0.420	25
5	0.559	0.376	23	72	0.511	0.423	23	139	0.656	0.330	22
6	0.614	0.389	22	73	0.507	0.386	29	140	0.569	0.431	29
7	0.599	0.378	25	74	0.503	0.417	25	141	0.456	0.481	30
8	0.529	0.455	26	75	0.590	0.407	24	142	0.623	0.391	28
9	0.476	0.372	25	76	0.628	0.389	21	143	0.613	0.403	25
10	0.610	0.400	28	77	0.521	0.329	28	144	0.602	0.393	24
11	0.597	0.388	27	78	0.525	0.322	24	145	0.609	0.382	20
12	0.620	0.402	21	79	0.575	0.346	29	146	0.595	0.391	30
13	0.612	0.403	23	80	0.489	0.361	26	147	0.590	0.397	21
14	0.626	0.389	27	81	0.586	0.375	30	148	0.589	0.398	25
15	0.623	0.408	23	82	0.597	0.406	23	149	0.592	0.299	25
16	0.604	0.403	25	83	0.475	0.358	28	150	0.617	0.372	27
17	0.610	0.408	22	84	0.608	0.386	20	151	0.587	0.402	26
18	0.621	0.381	25	85	0.592	0.405	27	152	0.620	0.354	27
19	0.604	0.388	21	86	0.472	0.359	24	153	0.593	0.417	27
20	0.624	0.380	23	87	0.533	0.374	26	154	0.579	0.393	30
21	0.601	0.388	25	88	0.533	0.388	26	155	0.567	0.362	28
22	0.488	0.411	23	89	0.595	0.413	25	156	0.560	0.425	21
23	0.603	0.409	29	90	0.587	0.399	24	157	0.548	0.375	30
24	0.623	0.376	30	91	0.609	0.408	25	158	0.538	0.423	26
25	0.626	0.375	25	92	0.595	0.418	30	159	0.534	0.422	26
26	0.595	0.387	27	93	0.626	0.375	30	160	0.542	0.439	26
27	0.596	0.412	25	94	0.598	0.391	23	161	0.529	0.420	23
28	0.616	0.372	23	95	0.614	0.393	24	162	0.523	0.344	30
29	0.544	0.395	29	96	0.627	0.381	23	163	0.477	0.383	24
30	0.587	0.402	29	97	0.602	0.385	27	164	0.480	0.286	28
31	0.584	0.396	30	98	0.584	0.389	23	165	0.629	0.338	25
32	0.641	0.372	26	99	0.552	0.393	23	166	0.475	0.319	27
33	0.580	0.398	24	100	0.609	0.381	25	167	0.518	0.474	29
34	0.609	0.366	23	101	0.628	0.380	24	168	0.655	0.314	27
35	0.580	0.399	28	102	0.621	0.395	27	169	0.572	0.299	30
36	0.628	0.362	21	103	0.622	0.402	28	170	0.610	0.291	29
37	0.599	0.364	22	104	0.601	0.389	30	171	0.697	0.308	20

continued on following page.

Table 18. Continued

Nodes	x	y	Demand	Nodes	x	y	Demand	Nodes	x	y	Demand
38	0.629	0.361	24	105	0.615	0.395	27	172	0.435	0.324	30
39	0.577	0.414	27	106	0.626	0.394	26	173	0.463	0.345	28
40	0.581	0.370	26	107	0.615	0.392	23	174	0.476	0.430	22
41	0.565	0.431	28	108	0.611	0.397	29	175	0.518	0.305	23
42	0.570	0.382	20	109	0.618	0.396	25	176	0.482	0.445	22
43	0.562	0.382	28	110	0.321	0.363	24	177	0.738	0.282	30
44	0.565	0.387	22	111	0.611	0.398	21	178	0.505	0.498	20
45	0.565	0.404	27	112	0.617	0.391	22	179	0.471	0.480	26
46	0.570	0.372	20	113	0.627	0.400	22	180	0.497	0.510	27
47	0.562	0.399	27	114	0.613	0.403	28	181	0.720	0.268	23
48	0.563	0.386	21	115	0.614	0.403	25	182	0.612	0.238	20
49	0.558	0.379	21	116	0.609	0.393	22	183	0.751	0.279	25
50	0.561	0.373	23	117	0.609	0.399	30	184	0.588	0.237	25
51	0.570	0.359	20	118	0.624	0.385	26	185	0.499	0.290	23
52	0.584	0.350	29	119	0.556	0.382	27	186	0.605	0.225	21
53	0.554	0.379	27	120	0.627	0.381	27	187	0.486	0.529	21
54	0.566	0.433	24	121	0.575	0.418	23	188	0.403	0.378	26
55	0.552	0.375	24	122	0.592	0.397	25	189	0.751	0.234	29
56	0.553	0.427	20	123	0.594	0.406	30	190	0.573	0.346	22
57	0.539	0.373	23	124	0.627	0.374	25	191	0.635	0.195	27
58	0.536	0.397	21	125	0.591	0.407	24	192	0.722	0.214	20
59	0.554	0.440	20	126	0.555	0.418	26	193	0.404	0.356	21
60	0.534	0.396	29	127	0.600	0.412	28	194	0.600	0.169	23
61	0.540	0.423	30	128	0.743	0.276	29	195	0.600	0.164	27
62	0.533	0.398	21	129	0.639	0.336	28	196	0.320	0.440	21
63	0.531	0.398	30	130	0.638	0.370	20	197	0.317	0.436	21
64	0.534	0.368	30	131	0.586	0.333	29	198	0.369	0.543	20
65	0.548	0.446	21	132	0.624	0.407	26	199	0.481	0.147	27
66	0.520	0.367	23	133	0.512	0.376	30	-	-	-	-

Chapter 9
Application of Artificial Bee Colony Algorithms to Antenna Design Problems for RFID Applications

Sotirios K. Goudos
Aristotle University of Thessaloniki, Greece

Katherine Siakavara
Aristotle University of Thessaloniki, Greece

John N. Sahalos
Aristotle University of Thessaloniki, Greece

ABSTRACT

In this chapter, the Artificial Bee Colony (ABC) algorithm and its variants are presented and applied to spiral antennas design for RFID tag application at the UHF band. The ABC variants include the Improved ABC (I-ABC), which is an improved version of the original ABC algorithm. The I-ABC introduces the best-so-far solution, inertia weight and acceleration coefficients to modify the search process. Furthermore, another ABC variant is the Gbest ABC (ABC), which includes global best (gbest) solution information into the search equation to improve the exploitation. These algorithms are applied to antenna design where the optimization goals are antenna size minimization, gain maximization, and conjugate matching. The algorithms performance is compared with other popular evolutionary algorithms. The optimization results produced show that the ABC family of algorithms is a powerful tool that can be efficiently applied to antenna design problems.

INTRODUCTION

Radiofrequency identification (RFID) has gained considerable growth in the last decade and remains in the front end of the general research and development sector concerning the remotely receiving and transmitting data using RF waves.

DOI: 10.4018/978-1-4666-9644-0.ch009

Nowadays the RFID technology providing automated wireless identification and tracking capability and being more robust than the barcode system, has shown a commercial worldwide deployment following frequency allocation in the UHF band, ranging from 860 MHz to 950MHz (Rao, Nikitin, & Lam, 2005). An ordinary RFID system comprises of at least, a reader (Interrogator) with a reader antenna, tags (transponders) which are microchips combined with an antenna in a compact package, a host computer and middleware including software and data base. An overview of criteria for RFID tag antenna design and an analysis of practical application aspects can be found in (Foster & Burberry, 1999; Rao et al., 2005).

One of major challenges in today' s RFID technologies that could potentially impede their practical implementation is the design of small size tag antennas with high efficiency and effective impedance matching to ICs with typically capacitive reactance. These antenna requirements are essential to optimize the RFID system power performance, especially for passive configurations where the only energy source is the incoming reader energy. Several RFID design cases for both passive and active tags can be found in the literature. Among others, these include covered slot antenna design (S. Y. Chen & Hsu, 2004), circular patch antenna analysis (Padhi, Karmakar, Law, & Aditya, 2003), planar inverted F-antenna (Hirvonen, Pursula, Jaakkola, & Laukkanen, 2004), folded dipole antenna (Xianming & Ning, 2004), U-shaped antenna (Alhawari, Ismail, Jalal, Raja Abdullah, & Rasid, 2013), compact strip dipole (Jiun-Peng & Powen, 2013), and patch antennas (Huang, Yang, Chew, & Ye, 2010; Zhongbao, Shaojun, Shiqiang, & Shouli, 2011).

The most commonly used shapes for UHF tags are those of meander line (Abdulhadi & Abhari, 2011; Bjorninen et al., 2008; Calabrese & Marrocco, 2008; Paredes, Zamora, Herraiz-Martinez, Martin, & Bonache, 2011), and spiral line (Alarco et al., 2010; Bin, Jianhua, Baiqiang, & Weiming, 2008) due to the characteristics of high gain, omni-directionality, planarity and relatively small surface size. Additionally, printed fractal antenna configurations exhibit similar attributes and recently several types of them have been proposed as efficient tag antennas (G. Monti, L. Catarinucci, & Tarricone, 2009; Lei, Shu, & Hanhua, 2009).

The evolutionary algorithms (EAs) that have emerged in the past decade mimic biological entities behavior and evolution. EAs are suitable optimization tools for solving the above-described design problem. Genetic algorithms (GA) (Holland, 1975) and Ant Colony Optimization (ACO) have been applied successfully to RFID antenna design (Calabrese & Marrocco, 2008; Marrocco, 2003; Randall, Lewis, Galehdar, & Thiel, 2007). Another popular EA is Differential evolution (DE), which is used in (S. K. Goudos, Siakavara, Samaras, Vafiadis, & Sahalos, 2011) for antenna and microwave filter design. Swarm intelligence (SI) can be defined as the collective behavior of decentralized and self-organized swarms. Among others Artificial Bee Colony (ABC) (Karaboga & Basturk, 2007) is a recently proposed SI algorithm, which has been applied to several real world engineering problems. The ABC algorithm models and simulates the honeybee behavior in food foraging. In ABC algorithm, a potential solution to the optimization problem is represented by the position of a food source while the nectar amount of a food source corresponds to the quality (objective function fitness) of the associated solution. The ABC algorithm has been applied successfully to RFID tag design (S. K. Goudos, Siakavara, & Sahalos, 2013a, 2013b, 2013c; S. K. Goudos, K. Siakavara, & J. N. Sahalos, 2014; Sotirios K. Goudos, Siakavara, & Sahalos, 2015).

Several ABC variants have been proposed in the literature. Among others, the Gbest-guided ABC (GABC) (Zhu & Kwong, 2010) is an improved ABC algorithm, which modifies the solution search equation by applying the global best (gbest) solution to guide the search of new candidate solutions in order to improve the exploitation. The Improved ABC (I-ABC) (Li, Niu, & Xiao, 2012) is another ABC variant that introduces the best-so-far solution, inertia weight and acceleration coefficients to modify the search process.

The purpose of this chapter is to briefly describe the above algorithms and present their application to RFID tag antenna design. This chapter presents results from design cases using the initial ABC, the GABC and the I-ABC. These include different shaped antennas like asymmetric spiral, load-ended spiral and an antenna with meander lines and loads. The chapter is supported with an adequate number of references.

This chapter is subdivided into four sections. The second section presents the different ABC algorithms. The next section describes the design cases and presents the numerical results. Finally, the last section contains the discussion about the advantages of using an ABC-based approach and the conclusion.

BACKGROUND

Genetic algorithms is a class of search techniques inspired by evolutionary biology that were introduced by J.H. Holland (Holland, 1975). They are based on Darwin's theory of evolution and the concept of natural selection. They were applied to several real world problems by Goldberg (Goldberg, 1989). Most commonly, they use selection, crossover, and mutation operators (Mitchell, 1996). Additionally, they could employ elitism (i.e. the best solution found survives to the next generation and replaces the worst solution in the current generation). GAs is the most widely used evolutionary algorithm in the literature (Vasant, 2011, 2012).

Particle Swarm Optimization (PSO) is a popular evolutionary algorithm, which is based on the intelligence and movement of swarms (birds, fishes, bees, etc.), and resembles their behavior (Kennedy & Eberhart, 1995). The most common algorithms include the classical Inertia Weight PSO (IWPSO) and the Constriction Factor PSO (CFPSO) (Clerc, 1999). The PSO algorithm is inherently used only for real-valued problems but can easily expand to discrete-valued problems. A simple modification of the real-valued PSO called binary PSO (binPSO) has been introduced by Kennedy & Eberhart (Kennedy & Eberhart, 1997). Several PSO variants have been introduced in the literature in order to further improve performance. Among others, Comprehensive Learning Particle Swarm Optimizer (CLPSO) (Liang, Qin, Suganthan, & Baskar, 2006), which utilizes a new learning strategy, and Boolean PSO (Afshinmanesh, Marandi, & Shahabadi, 2008), which introduces a new discrete PSO variant. PSO has been applied successfully to several real world problems (Arindam Majumder & Majumder, 2015).

The ABC (Karaboga & Basturk, 2007) algorithm models and simulates the honey bee behavior in food foraging. The ABC algorithm has been compared with GA, DE, PSO, and ES algorithms on a set of test functions in several cases (Karaboga & Akay, 2009; Karaboga & Basturk, 2008). The authors in (Karaboga & Akay, 2009) have shown that the performance of the ABC algorithm is better than or similar to that of above algorithms. Moreover, ABC uses less control parameters. Therefore, it can be efficiently applied to multimodal and multidimensional optimization problems. Among others ABC has been applied to neural network training (Karaboga, Akay, & Ozturk, 2007), image processing (Akay, 2013), wireless sensor networks (Gu, Xu, Du, Hou, & Qian, 2013), and the frequency assignment problem (da Silva Maximiano, Vega-Rodríguez, Gómez-Pulido, & Sánchez-Pérez, 2013). A detailed survey of ABC algorithms and its applications can be found in (Karaboga, Gorkemli, Ozturk, & Karaboga, 2014).

In ABC algorithm, a potential solution to the optimization problem is represented by the position of a food source while the nectar amount of a food source corresponds to the quality (objective function fitness) of the associated solution. For population-based optimization algorithms, both exploration and exploitation are necessary. The exploration refers to the ability to investigate the various unknown regions in the solution space to discover the global optimum, while, the exploitation refers to the ability to apply the knowledge of the previous good solutions to find better solutions (Trelea, 2003).

In order to find the best solution the algorithm defines three classes of bees: employed bees, onlooker bees and scout bees. The employed bee searches for the food sources, the onlooker bee makes a decision to choose the food sources by sharing the information of employed bee, and the scout bee is used to determine a new food source if a food source is abandoned by the employed bee and onlooker bee.

Artificial Bee Colony Algorithm (ABC)

Each employed bee corresponds to one food source (i.e. the number of the employed bees is equal to the number of solutions). The employed bees search for new neighbor food source near their hive. A new position of the $\overline{x}_i$ solution $\overline{x}_i = \left(x_{i,1},..,x_{i,j},..,x_{i,D}\right)$ solution, where D is the problem dimension, is generated using is generated using:

$$u_{i,j} = x_{i,j} + \varphi_{i,j}\left(x_{i,j} - x_{k,j}\right) \tag{1}$$

where k is a randomly chosen index from the population $\{1,2,\ldots,SN\}$different than i, $j$is a randomly chosen index from $\{1,2,\ldots,D\}$, SN is the number of food sources, and, $\varphi_{i,j}$ is a uniformly distributed random number within [-1,1]. ABC uses a greedy selection operator, which for minimization problems is defined by:

$$\overline{x}_i' = \begin{cases} \overline{u}_i, & \text{if } f(\overline{u}_i) < f(\overline{x}_i) \\ \overline{x}_i, & \text{otherwise} \end{cases} . \tag{2}$$

where $\overline{x}_i'$ is the new position of the food source.

An onlooker bee chooses a food source depending on the probability value associated with that food source, p_i, given by:

$$p_i = \frac{fit_i}{\sum_{m=1}^{SN} fit_m} \tag{3}$$

where fit_i is the fitness value of the ith solution which is proportional to the nectar amount of the food source in the ith position. If a solution representing a food source cannot be improved by a predetermined number of trials, which is a parameter called "limit", it means that the associated food source has been exhausted by the bees and then the employed bee of this food source becomes a scout bee. The position of the abandoned food source (solution) is then replaced with a randomly produced solution given by:

$$x_{i,j} = rand_{j(0,1)}\left(x_{j,U} - x_{j,L}\right) + x_{j,L} \quad j = 1,2,...,D \tag{4}$$

where $x_{j,L}$ and $x_{j,U}$ are the lower and upper bounds of the jth dimension respectively and $rand_{j(0,1)}$ is a uniformly distributed random number within (0,1).

Figure 1. ABC Flowchart

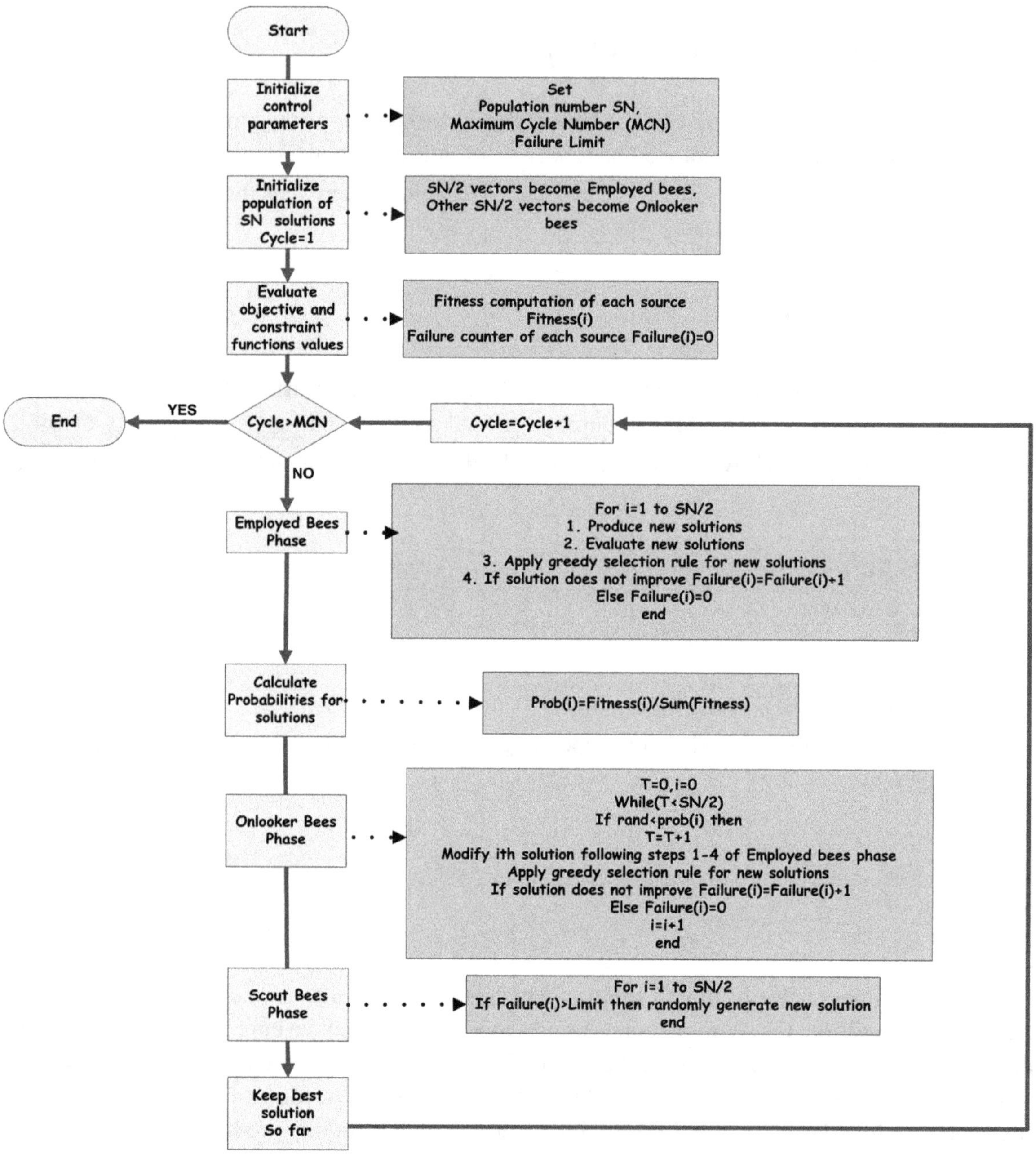

The flowchart of the ABC algorithm is given in Figure 1. The ABC algorithm is outlined as follows.

1. Initialize the ABC parameters. That is, set the maximum cycle number (*MCN*), the number *SN* of food sources (population size), and the limit parameter.
2. Evaluate the objective function value for every vector of the population.
3. Employed bees phase. For each employed bee. Produce a new solution vector u_i using (1). Evaluate new solutions and apply greedy selection rule for new solutions of (2).
4. Calculate the probability values p_i
5. Onlooker bees phase. For each onlooker bee depending on probability p_i, produce a new solution vector u_i using (1) from the selected solution x_i. Evaluate new solutions and apply greedy selection rule for new solutions of (2).
6. Scout bees Phase. If the limit number is reached for a solution randomly generate another using.
7. Keep the best solution so far.
8. If the maximum number of cycles *MCN* is reached, then exit; otherwise increment cycle number and go to step 4 for the next cycle.

Improved-ABC (I-ABC)

The ABC algorithm is a very effective technique, which can handle efficiently arbitrary optimization problems (see Figure 1). The main advantages of using ABC are not only the fact that it is a high performance optimizer but also that it is an easy to understand and implement algorithm that requires little computational bookkeeping. However, ABC could converge slowly and sometimes may trap in local optima.

The authors in (Li et al., 2012) propose three major changes by introducing the best-so-far solution, inertia weight and acceleration coefficients to modify the search process. In addition, the original solution generation equation of ABC is good at exploration but poor at exploitation (Li et al., 2012) Therefore, they improve the exploitation by differentiation of the modification forms of the employed bees and the onlooker ones in the second acceleration coefficient. The new proposed improved ABC algorithm is called as I-ABC. The I-ABC differs from the original ABC algorithm in new solution generation equation. A new position of the $\overline{x}_i$ solution is generated using:

$$u_{i,j} = x_{i,j} w_{i,j} + 2\left(rand_{1[0,1]} - 0.5\right)\left(x_{i,j} - x_{k,j}\right)\Phi_1 + rand_{2[0,1]}\left(x_{best,j} - x_{k,j}\right)\Phi_2 \qquad (5)$$

where $k \in \{1,2,\ldots,SN\}$, $k \neq i$, $j \in \{1,2,\ldots,D\}$ are randomly chosen indices, where *SN* is the number of food sources, $w_{i,j}$ is the inertia weight that controls impacts of the previous solution $x_{i,j}$, $x_{best,j}$ is the j-th parameter of the global best solution so far, $rand_{1[0,1]}$, $rand_{2[0,1]}$ are uniformly distributed random numbers in [0,1], and Φ_1, Φ_2 are the acceleration constants, that is positive control parameters that control the maximum step size. These along with the inertia weight are given by:

$$w_{i,j} = \Phi_1 = \frac{1}{1+e^{-fit_i/ap}}$$

$$\Phi_2 = \begin{cases} 1 & \text{for employed bees} \\ \dfrac{1}{1+e^{-fit_i/ap}} & \text{for onlooker bees} \end{cases} \qquad (6)$$

where *ap* is the fitness of the global best in the first iteration, and fit_i is the fitness value of the i-th solution. To balance the processes of the exploration and the exploitation, the modification forms of the employed bees and the onlooker ones differ in the acceleration coefficient Φ_2.

Gbest-Guided ABC (GABC)

The GABC differs from the original ABC algorithm in new solution generation equation. A new position of the $\bar{x}_i$ solution is generated using:

$$u_{i,j} = x_{i,j} + \varphi_{i,j}\left(x_{i,j} - x_{k,j}\right) + \psi_{i,j}\left(y_j - x_{i,j}\right)$$

where $k \in \{1,2,\ldots,SN\}$, $k \neq i$, $j \in \{1,2,\ldots,D\}$ are randomly chosen indices, where *SN* is the number of food sources, $\varphi_{i,j}$ is a uniformly distributed random number within [-1,1], y_j is the jth parameter of the global best solution, and $\psi_{i,j}$ is a uniform random number in [0,C], where C is non-negative constant.

More details about GABC can be found in (Zhu & Kwong, 2010).

RESULTS AND DISCUSSIONS

Several geometrical configurations have been employed for the design of RFID tag antennas and have been proved effective in meeting the operating requirements (Rao et al., 2005). In the present work the spiral shaped antenna printed on a dielectric slab was selected because this configuration depends on several geometrical parameters such as the number and the length of the spirals, the distance between them, the width of the printed lines etc. Thus, it is expected that among the large number of combinations of these parameter values at least one would lead to a tag antenna with not simply satisfactory performance but with really a very good one, especially when the design follows an optimization procedure.

Our motivation was to explore the potential of a spiral shaped antenna to meet the requirements of operation of a RFID tag. A theoretical study was conducted first. A spiral of ordinary shape was initially considered. Its input impedance was calculated approximately by a hybrid combination of the Transmission Line Model and the low of electromotive force induced to the spiral. To apply this model it was supposed that the spiral comes from an open terminated transmission line of two conductors bend around the input terminals of the line at which a generator or, in the case of a tag, the IC element is connected conductively. At this point also the input impedance is calculated. The radiated field and the gain were calculated via the ordinary method of Green functions.

Based on the above theory, the performance of the spiral antenna versus the number and the length of its turns, the distance between them, and the modification of the shape of its segments in meander configuration, instead of straight lines, was studied. The results showed that dense spirals of small size and with large number of turns have larger gain than that of less dense spirals with smaller number of turns and larger size. It was also ascertained that two modifications of the layout of the ordinary spirals would lead to further reduction of their size without downgrading of their performance. The first one was the formation of the spiral branches by the meander shape. In this way, the spirals fill the area they occupy more effectively and as they acquire larger length while the space they take up would be smaller. The second modification concerns to the loading of the spirals at their open ends via parts of transmission printed strips with characteristic impedance different from that of the spirals. By applying the Transmission Line model and describing the current approximately as that of an ordinary load-terminated line, it was found that the spirals resonate at lower frequencies. The resonance frequencies depend on the length of the loads, on the width of the strips etc. Therefore, in order for a loaded spiral to resonate at a frequency equal to the respective unloaded, the size of the loaded one had to be reduced. Concluding, this procedure would work as a minimization process.

It was considered that both the above techniques when applied simultaneously would lead to spirals with small size and satisfactory performance. To obtain this, the trial and error method would be inadequate as such spirals are geometrically defined by several parameters and additionally conjugate matching to the IC had to be achieved. To solve this problem an optimization algorithm had to be used. The optimization algorithm as a robust and effective process gave us the potential to modify further the spirals inserting loads not only at their terminals but also along their branches. In the following, we present the details of the spirals' shape geometrical description.

RFID Antenna Optimization Problem Description

The first difficulty in designing a tag antenna arises from the requirement to be conjugate matched to the IC chip, which has input impedance with relatively low real part and high capacitive imaginary part. Therefore, the antenna's input impedance must have an inductive imaginary part of equal value. Moreover, the demand for the tag antenna to be of small size is also opposed to the desired attribute of relatively high gain.

Therefore, the two optimization goals for RFID tag design are gain maximization and conjugate matching. We have designed tags with one objective function combing both the above goals. It is described by:

$$F_1(\overline{x}) = -G(\overline{x}) + \Xi \times \left|\max\left\{0, \left|\Gamma\right| - 0.3\right\}\right| \tag{7}$$

where $\overline{x}$ is the vector of the antenna geometry, G is the antenna Gain calculated, Ξ is a very large number, and Γ is the (load dependent) reflection coefficient of the tag antenna-load system which is computed by:

$$\Gamma = \frac{Z_{in,chip} - Z^{*}_{a}}{Z_{in,chip} + Z_{a}} \tag{8}$$

where $Z_{in,chip} = R_{in,chip} + jX_{in,chip}$ is the chip input impedance and $Z_a = R_a + jX_a$ the tag input impedance respectively.

An important parameter of the RFID system performance, as mentioned before, is the read range that is the maximum distance at which RFID reader can detect the backscattered signal from the tag. As reader sensitivity is typically high in comparison with tag, the read range is defined by the tag response. The read range is expressed using Friis free-space formula as (Rao et al., 2005):

$$R = \frac{\lambda}{4\pi}\sqrt{\frac{P_{EIRPrd}\, e D_{tag}\, p_{loss}\, \tau}{P_{in,chip}}} \tag{9}$$

where P_{EIRPrd} is the effective isotropically radiated power by the reader, D_{tag} is the tag directivity, *e* is the tag efficiency, p_{loss} is the polarization loss factor (represents the loss of EM power because of polarization mismatch $0 \leq p_{loss} \leq 1$), $P_{in,chip}$ is the power absorbed by the chip given by:

$$P_{in,chip} = \left(1 - |\Gamma|^2\right) P_{tag} \tag{10}$$

where P_{tag} is the available power at the input of the tag antenna, and τ is the power transmission coefficient given by:

$$\tau = \frac{4R_a R_{in,chip}}{\left|Z_{in,chip} + Z_a\right|^2} \tag{11}$$

All the RFID tags presented are designed for operation in Europe, where the minimum EIRP requirement is P_{EIRPrd}=3.28W. The power threshold for the chip to run on is -18dBm(15.84μW). All of the spirals are structured with cooper strips of 1mm width, printed on dielectric substrate with ε_r=1.046 and thickness h=3.17mm. The tags are assumed to be in free space.

Asymmetric Spiral RFID Tag Design Using the ABC Algorithm

The first example is that of an asymmetric spiral shaped antenna (S. K. Goudos et al., 2013c). The geometry is depicted in Figure 2. The optimization parameters are the first spiral length *SL*, the spacing between the two spirals D_0, the first branch number *m1* (i.e. the number of the horizontal and vertical sections of each spiral branch), and the second branch number *m2*. Additionally, we consider the design case of an asymmetric spiral with the last horizontal lengths randomly reduced. In this case, we need two additional design parameters, the percentage length reduction of the last horizontal length of the first *per1* and the second spiral *per2* respectively. For this case, the number of unknowns is six. Therefore, in this case the vector of the antenna geometry is $\bar{x} = (SL, D_0, m1, m2, per1, per2)$.

The length of a vertical spiral branch is given in this case by the following recursive formula:

$$\begin{aligned} &L_{vi}(m) = L_{vi}(m-1) + 2D_0 \quad for\, m > 2 \\ &with\, L_{v1}(1) = SL, \\ &\quad L_{v1}(2) = SL + D_0, \\ &\quad L_{v2}(1) = SL + D_0, \\ &\quad L_{v2}(2) = SL + 3D_0 \end{aligned} \tag{12}$$

where *m* is the branch number.

Figure 2. Asymmetric spiral antenna geometry.

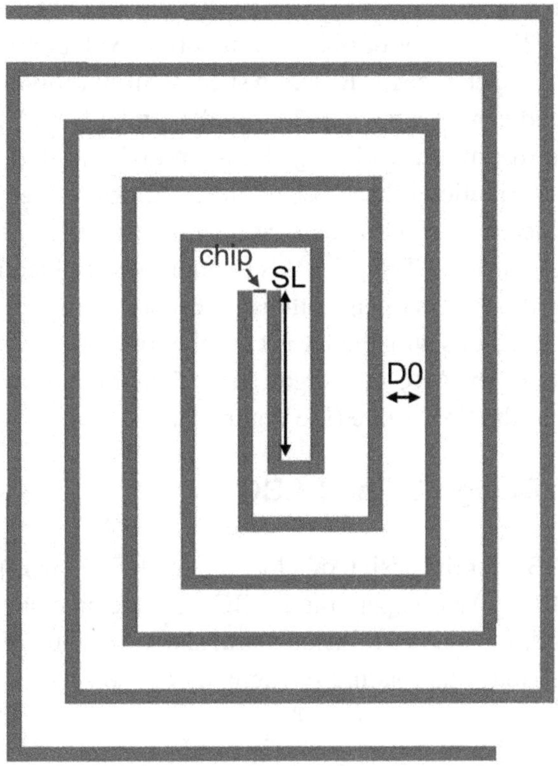

The length of a horizontal spiral branch is given by the following recursive formula:

$$
\begin{aligned}
&L_{hi}(m) = L_{hi}(m-1) + 2D_0 \quad for\, m > 2 \\
&with\, L_{h1}(1) = D_0, \\
&\quad L_{h1}(2) = 2r_0 + 2D_0, \\
&\quad L_{h2}(1) = 2r_0 + 2D_0, \\
&\quad L_{h2}(2) = 2r_0 + 3D_0
\end{aligned}
\tag{13}
$$

In this case, we compare the ABC algorithm with a binary-coded GA (), PSO (), and SADE (). The control parameters for these algorithms are those that commonly perform well regardless of the characteristics of the problem to be solved. For the GA the crossover probability and the mutation probability are set to 0.9 and 0.1 respectively. In the PSO algorithm, the cognitive learning factor and the social learning factor are both set equal to 2.05. The inertia weight was linearly decreased from 0.9 to 0.4. For the ABC the *limit* parameter is set to 100. SADE requires the adjustment of only population size and iteration number. For all algorithms we selected a population of 20 vectors and the total number of iterations was set to 1000. The algorithms are run for 20 independent trials. The best results are compared. The best value, the worst value, the mean, and the standard deviation of the last generation computed by each algorithm are presented here. The stopping criterion was the iteration number. Additionally, if each algorithm has not improved the objective function value for 200 iterations then it stops to avoid stagnation. The objective function was the one given by (7). We also obtain the minimum FE_{min}, maximum FE_{max}, and the average FE_{avg} number of objective-function evaluations for each algorithm and the success rate (S_r) which is the number of each algorithm successful runs divided by the total runs number.

The best RFID tag obtained is with a tag gain of 1.96dBi and |Γ|=0.27 at 867MHz. The tag size is 29.24mmX35.63mm and the maximum read range between 865 and 869.8MHz is 11.63m. Figure 3 presents the surface current distribution of the best spiral found, while Figure 4 shows the 3D radiation pattern. Table 1 shows the objective function value comparative results. It is evident that ABC obtains better results than the other algorithms. Table 2 holds the number of objective function evaluations results with the success rate for this case. All algorithms succeeded in finding feasible tag designs for all runs, but ABC has obtained the results with the fewest number of objective function evaluations. Figure 5 shows the convergence rate graph. ABC converges faster than the other algorithms and requires less than 50 iterations to reach near the final value (see Figure 3).

Load-Ended Antennas Design Using I-ABC

An example case of using I-ABC for the design of a load-ended RFID tag is presented in this subsection (Sotirios K. Goudos et al., 2015). We compare the I-ABC with the original ABC algorithm.

As it was above-mentioned, the load terminated spiral shape was one of the modifications selected for the RFID tag implementation. The loading of the ends of the spiral contributes to the conjugate

Figure 3. Asymmetric Spiral surface current distribution at 867MHz

Figure 4. Asymmetric Spiral 3D radiation pattern at 867MHz

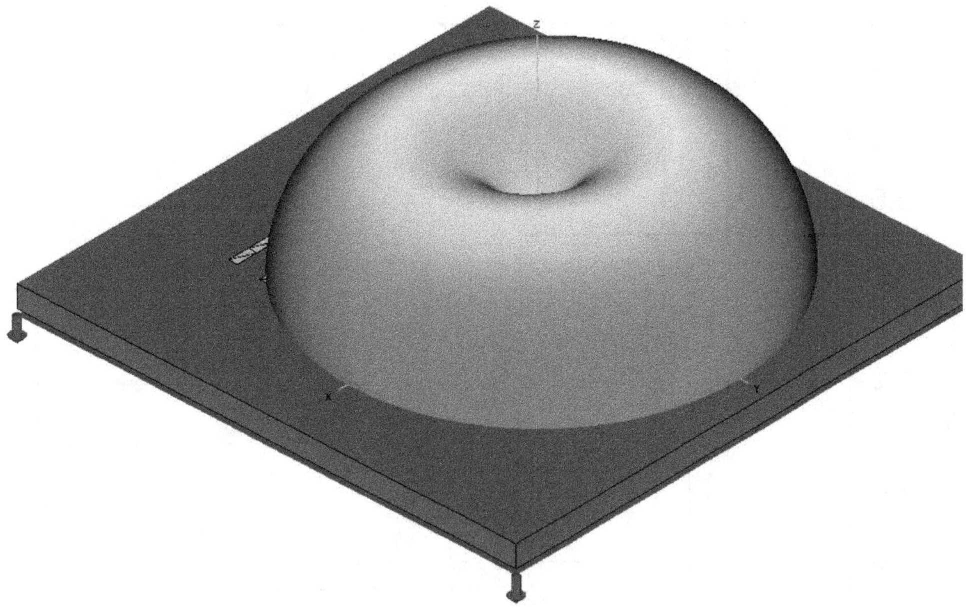

Table 1. Objective function values comparative results for the asymmetric spiral design case. The minimum value in each column is in bold

Algorithm	Best	Worst	Mean	St. Dev.
GA	-1.20	-0.35	-0.95	**0.25**
PSO	-1.39	-0.39	-1.01	0.28
SADE	-1.88	-0.40	-1.32	0.53
ABC	**-1.96**	**-0.41**	**-1.35**	0.52

Table 2. Number of objective function evaluations Comparative results for the asymmetric spiral design case. The minimum value in each column is in bold

Algorithm	Sr	FEmin	FEmax	FEavg
GA	1	6456	14881	7244
PSO	1	5403	15006	7152
SADE	1	5679	14332	7112
ABC	1	**4765**	**13279**	**6915**

Figure 5. Convergence rate plot for the asymmetric spiral case

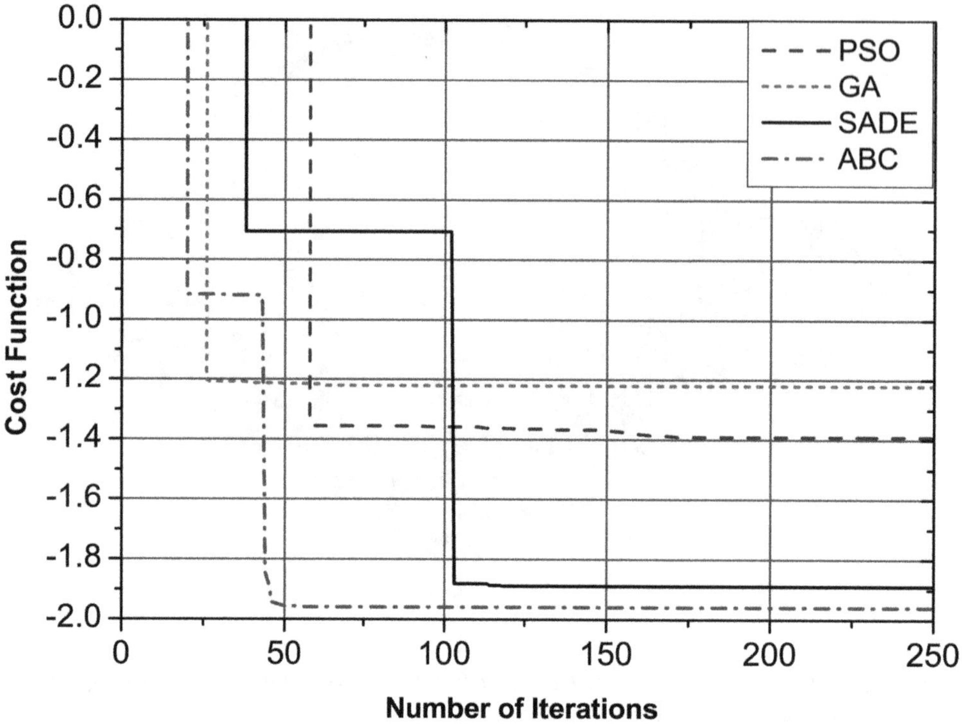

matching of the chip impedance. Therefore, we have designed a spiral antenna type with loaded ends. Figure 6 shows the antenna geometry. The optimization parameters are again the first six of the previous case with the addition of the horizontal load length *hl*, and the vertical load additional width *wl*. Thus, we have a total of eight design parameters. Hence, the vector of the antenna geometry is $\overline{x} = \left(SL, D_0, m1, m2, per1, per2, hl, wl\right)$.

The length of a vertical and a horizontal spiral branch is given by the formulas (12) and (13) respectively.

We compare I-ABC with the original ABC algorithm on the load-ended spiral case. The *limit* parameter is set to 100 for both algorithms. Additionally, if each algorithm has not improved the objective function value for 200 iterations, then it stops to avoid stagnation. The algorithms' statistical results are compared.

We have obtained using the I-ABC algorithm a tag gain of 2.15dBi and $|\Gamma|=0.24$ at 867MHz while ABC best results has a tag gain value of 2.00dBi and $|\Gamma|=0.29$ at 867MHz . The tag size is 66.82mm x65.82mm, the maximum read range between 865MHz and 869.8MHz is 14.19m. The conjugate matching bandwidth is about 2.4MHz. Figure 7 depicts the surface current distribution of the best spiral found and Figure 8 shows the 3D radiation pattern. We notice that the direction of the maximum radiation is perpendicular to the tag surface.

Table 3 reports the algorithms comparative results. The I-ABC algorithm seems to perform better than the ABC algorithm. Regarding computational time for this case both algorithms require about 4 hours for each algorithm run. ABC is slightly faster than I-ABC Table 4 holds the number of objective function evaluations results with the success rate. We notice that the I-ABC algorithm has achieved the highest success rate. The ABC obtained the smaller average number of objective-function evaluations.

Figure 6. Load-ended antenna geometry

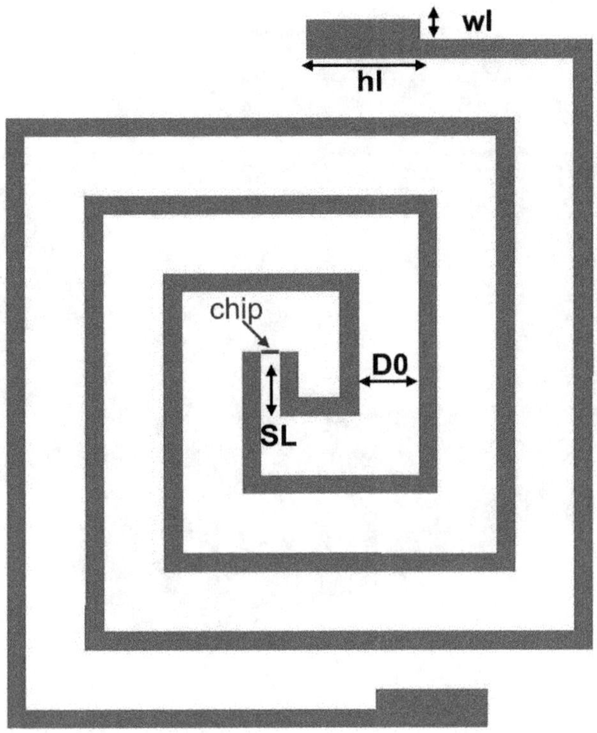

Figure 7. Load-ended spiral antenna surface current distribution at 867MHz

Figure 8. Load-ended spiral antenna 3D radiation pattern at 867MHz

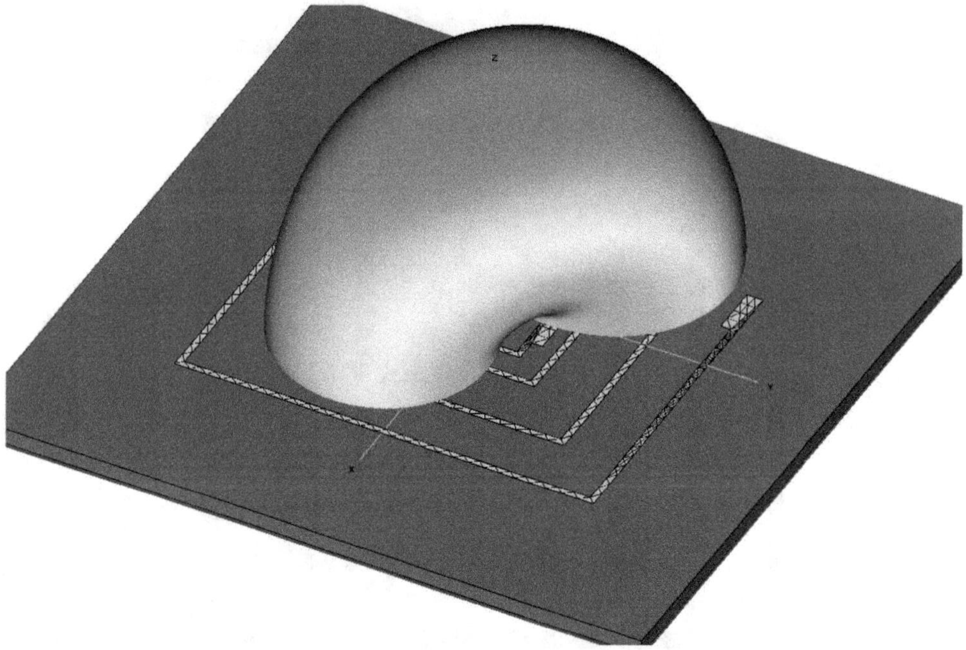

The obtained results show that both algorithms are powerful optimizers that can be efficiently applied to tag antenna design problems.

Modified Antennas Design Using GABC

This subsection addresses the problem of modified RFID antennas with meander lines and loads (Sotirios K. Goudos, Katherine Siakavara, & John N. Sahalos, 2014). Figure 9 shows the spiral geometry. The chip is placed between the two first vertical spiral branches as it is indicated in Figure 9 with red color. The horizontal spacing between the two first vertical branches is fixed and denoted as $2r_0$.The meander lines are constructed at the middle of each spiral branch. Each spiral branch is either meandered (modeled by binary "0") or loaded (modeled by binary "1"). Thus, the number of unknown variables is $2m1+2m2$ for this case, where $m1$, and $m2$ is the branch number (i.e. the number of the horizontal and vertical sections) of spiral one and two respectively. The other optimization parameters include the first

Table 3. Comparative results of the algorithms for the load-ended antenna case. The smaller values are in bold

Algorithm	Best	Worst	Mean	Standard Deviation	Avg. Computation Time (h)
I-ABC	**-2.15**	**-1.05**	**-1.60**	**0.37**	4.3
ABC	-2.00	-0.88	-1.51	0.51	**4.1**

Table 4. Number of objective function evaluations comparative results for the load-ended antenna case. The smaller values are in bold

Algorithm	Sr	FEmin	FEavg
I-ABC	0.8	**4455**	5788
ABC	0.7	4557	**5609**

vertical spiral length, *SLv*, the first horizontal spiral length, *SLh*, and the spacing between the two spirals, D_0. The use of two additional design parameters is required due to the meander lines. These are the vertical to the spiral branch length, *Df*, and the parallel to the spiral branch length, *sf*, which is defined as a percentage of the total branch length. Therefore, the total number of optimization variables is 7+2 $m1+2$ $m2$ while the vector of the antenna geometry becomes $\overline{x}=\left(SLv, SLh, D_0, m1, m2, Df, sf, b_1^1, \ldots. b_k^1 \ldots b_{2m1}^1, b_1^2, \ldots. b_n^2 \ldots b_{2m2}^2\right)$, where b_k^1 is the k-th bit of spiral one k-branch, and b_n^2 is the n-th bit of spiral two n-branch.

The length of a vertical spiral branch is given in this case by the following recursive formula:

$$\begin{aligned} &L_{vi}\left(m\right)=L_{vi}\left(m-1\right)+2D_0 \quad for\, m>2 \\ &with\, L_{v1}\left(1\right)=SL_v, \\ &\quad L_{v1}\left(2\right)=SL_v-D_0, \\ &\quad L_{v2}\left(1\right)=SL_v+D_0, \\ &\quad L_{v2}\left(2\right)=SL_v+3D_0 \end{aligned} \tag{14}$$

where *m* is the branch number.

The length of a horizontal spiral branch is given by the following recursive formula:

Figure 9. Modified antenna with meander lines and loads geometry

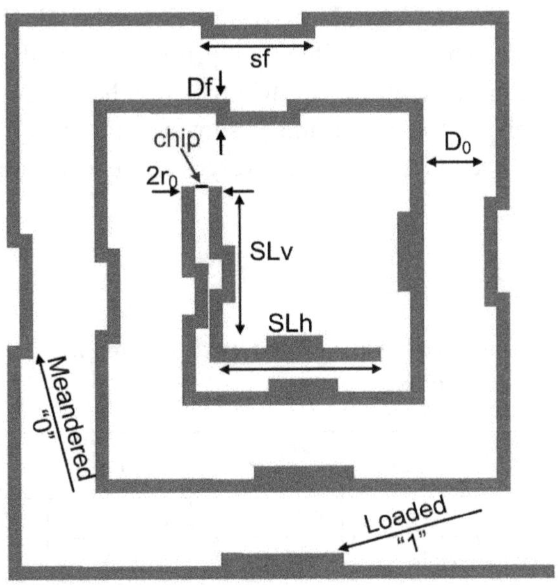

$$\begin{aligned} & L_{hi}\left(m\right) = L_{hi}\left(m-1\right) + 2D_0 \quad for\, m > 2 \\ & with\, L_{h1}\left(1\right) = SL_h, \\ & \quad L_{h1}\left(2\right) = 2r_0 + SL_h + D_0, \\ & \quad L_{h2}\left(1\right) = 2r_0 + SL_h + D_0, \\ & \quad L_{h2}\left(2\right) = 2r_0 + SL_h + 3D_0 \end{aligned} \tag{15}$$

This tag is structured with copper strips of 1mm width, printed on dielectric substrate with ε_r=3.25 and thickness h=0.5mm. We compare GABC with Particle Swarm Optimization (PSO) (Dieter, Fischer, & Menzel, 2011), a binary-coded Genetic Algorithm (GA) (Marrocco, 2003), and the original ABC algorithm. For the GA we set the crossover probability to 0.9 and the mutation probability to 0.1. In For the PSO algorithm, the cognitive learning factor and the social learning factor are both set equal to 2.05. The inertia weight was linearly decreased from 0.9 to 0.4. For both ABC algorithms, the limit parameter is set to 100. In the GABC algorithm, the C parameter is set to one. For both ABC algorithms, the limit parameter is set to 100. In the GABC algorithm, the C parameter is set to one. All the algorithms ran for 20 independent trials. A population of 20 vectors was selected and the total number of iterations was set to 400. The stopping criterion was the iteration number. The best value, the worst value, the mean, and the standard deviation of the last generation computed by each algorithm are presented here. The same initial conditions are used for all algorithms. Additionally the antenna size, in the optimization process was restricted to 50X50mm.GABC obtained a tag with gain of 2.06dBi and $|\Gamma|$=0.05 at 867MHz . The maximum read range between 865MHz and 869MHz is 7.5m. The tag size is 43.75mmX41.11mm. Figure 10 depicts the geometry and the surface current distribution of the best tag found by the GABC at 867MHz. The 3D radiation pattern at 867MHz is shown in Figure 11. We notice that again the maximum direction is perpendicular to the tag level. Table 5 presents the statistical results of the objective-function values for each algorithm and the success rate percentage. We define the success rate percentage as the percentage of the ratio successful runs/total runs. The GABC algorithm seems to perform better than the other algorithms. GABC obtains the best objective-function value, the highest mean value, and the smallest standard deviation value. The ABC original algorithm obtains the second best result close to that of GABC. Additionally we notice that the GABC algorithm has achieved the highest success rate of 70%. The convergence rate plot of the algorithms is shown in Figure 12. GABC and ABC converge at similar speeds faster than PSO and GA. GABC converges at the lower objective-function value.

Table 5. Comparative results of the algorithms for the modified antenna with meander lines and loads antenna case. The smaller values are in bold

Algorithm	Best	Worst	Mean	St. Dev.	Success Rate %
GA	-1.55	-0.24	-1.21	0.50	30
PSO	-1.77	-0.37	-1.13	0.59	40
ABC	-1.84	-0.55	-1.36	0.51	60
GABC	**-2.06**	**-1.12**	**-1.47**	**0.34**	70

Figure 10. Modified antenna with meander lines and loads surface current distribution at 867MHz

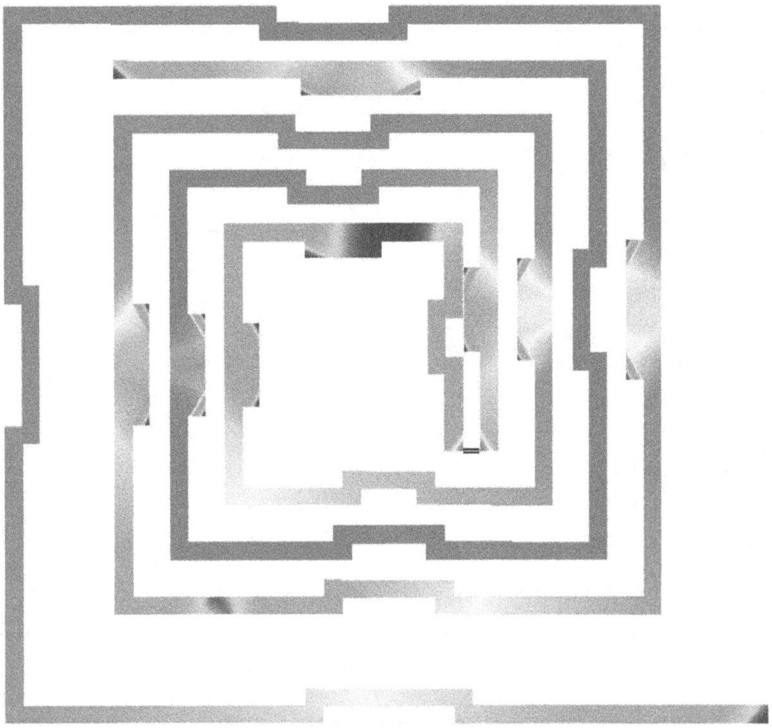

Figure 11. Modified antenna with meander lines and loads 3D radiation pattern at 867MHz

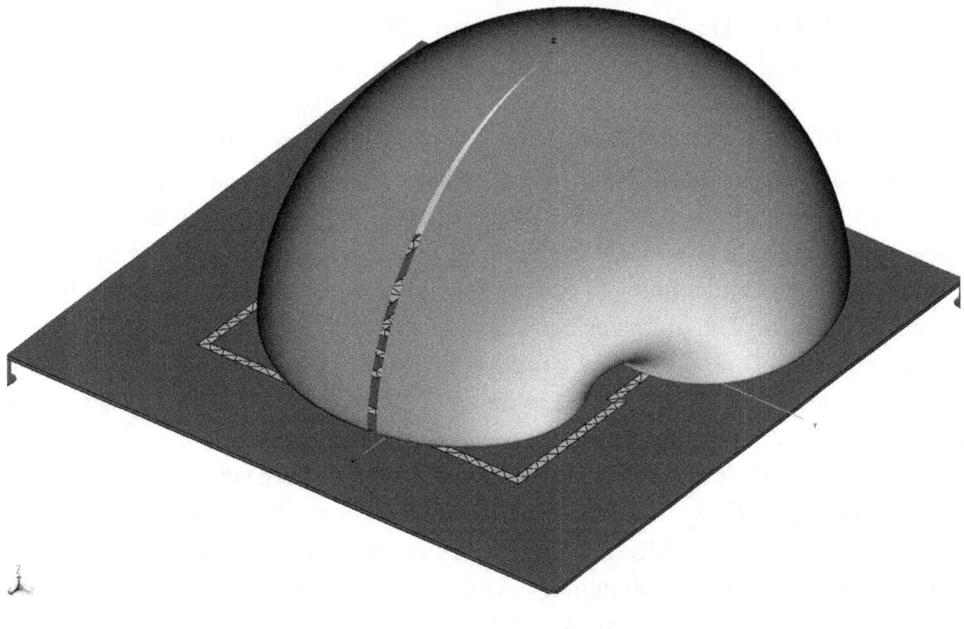

Figure 12. Convergence rate plot for the modified antenna with meander lines and loads case

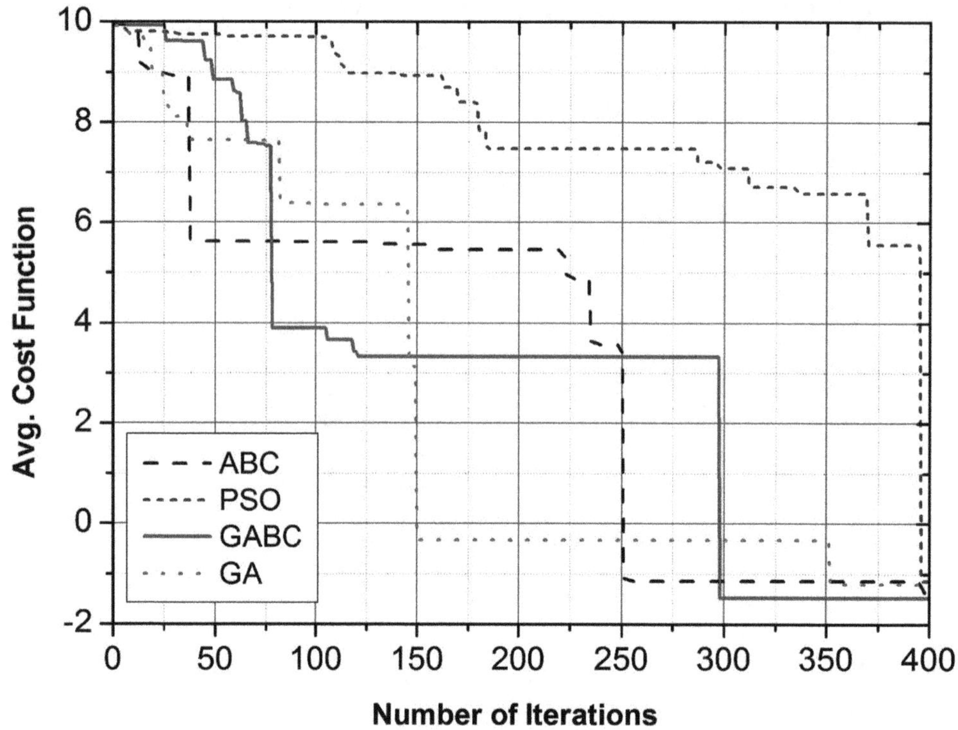

DISCUSSION

Similar to the other (EAs) such as GAs, and PSO, in the ABC approach there is a way of sharing information between solutions. This is accomplished using the employed and the scout bees. Diversity in ABC is provided by scout production. The ABC algorithm compared with GAs presents similarities and differences. In general, ABC does not have any evolution operators like crossover and mutation. Compared to GAs, ABC has fewer parameters to adjust and is easier to implement in any programming language. ABC is also computationally more efficient than a GA with the same population size. In ABC algorithm a stochastic selection scheme based on the fitness (nectar) values is used which is similar to ''roulette wheel selection'' in a GA. Additionally, ABC uses a greedy selection scheme as in DE. This scheme is used by onlookers and employed bees to select between the source position in their memory and the new source position. The onlooker bees use the solutions with the higher fitness values more often than those with lower fitness values to produce new trial solutions. This indicates that the promising regions of the search space are searched in detail. This type of selection process is similar to the natural selection employed in GAs.

Compared with PSO, ABC is quite similar and simple. They are both swarm intelligence-based algorithms. Both the PSO and ABC algorithms are efficient optimizers. However, ABC has a fewer control parameters that must be predefined. ABC requires the selection of only three parameters; population size, maximum iteration number and the limit number. ABC has been found to outperform PSO on numerical benchmark problems in (Karaboga & Basturk, 2008).

The way the algorithms keep the best solution found so far in the population is another issue. In GA (when elitism is employed) and PSO the best solution is always kept in the population. Therefore, it can be used for producing new solutions. However, in ABC, the best solution found in an iteration could be replaced with a randomly produced solution by a scout.

Although ABC has great potential, it was clear that some modifications to the original structure are required to improve significantly the performance. It is proposed in (Karaboga et al., 2014) that new neighbor production mechanisms can be introduced to ABC in order to improve performance in terms of convergence. An ABC drawback is the fact that it does not use a crossover operator like in GAs or DE. The distribution of good information between solutions is not at a required level. This may cause slow convergence (Karaboga et al., 2014). Additional improvements to the original ABC algorithm may provide performance improvements in that area.

FUTURE RESEARCH DIRECTIONS

The research domain of evolutionary algorithms and especially swarm algorithms is growing rapidly. Specific ABC optimizers that work well for a given problem are introduced. A current and growing research trend in evolutionary algorithms is their hybridization with local optimizers. These algorithms are called Memetic Algorithms (MAs). MAs are inspired by Dawkins' notion of meme. The advantage of such an approach is that the use of a local search optimizer ensures that specific regions of the search space can be explored using fewer evaluations and good quality solutions can be generated early during the search. Furthermore, the global search algorithm generates good initial solutions for the local search. MAs can be highly efficient due to this combination of global exploration and local exploitation. So far there some ABC-based MAs have been proposed in order to overcome some of the drawbacks and to improve the performance of common ABC optimizers (Ali, Hassanien, & Snasel, 2014; Bansal, Sharma, Arya, & Nagar, 2013; Fister, Fister Jr, Brest, & Žumer, 2012; Kumar, Kumar, Sharma, & Sharma, 2014; Kumar, Sharma, & Kumari, 2014).

Hybrid algorithms of ABC and other algorithms is also a growing research trend (Vasant, 2014). Among others recent papers use hybrid ABC-PSO to train neural networks (Ghanem & Jantan, 2014), hybrid of ant colony optimization (ACO) and ABC (Kefayat, Lashkar Ara, & Nabavi Niaki, 2015) for optimal placement and sizing of distributed energy resources, hybrid of ABC and evolution strategies (ES) (Mollinetti, Souza, & Teixeira, 2014), and hybrid ABC-DE in (Xiang, Ma, & An, 2014).

The application of multi-objective ABC algorithms to real world problems introduces new challenges regarding performance and computational cost (H. Chen, Zhu, Ma, & Niu, 2014; da Silva Maximiano et al., 2013; Ma, Hu, Zhu, & Chen, 2014; Mohammadi, Feizi Derakhshi, & Akbari, 2012; Novais, Batista, Rocha, & Guimaraes, 2013; Rubio-Largo, Vega-Rodríguez, Gómez-Pulido, & Sánchez-Pérez, 2013; Zhong, Xiang, & Liu, 2014). Therefore, several new research directions have to be explored.

LIMITATION

The antenna cases presented in the previous sections could be modified when fabricated. Therefore, differences between the theoretical predicted values and the measured ones may occur. This is due to the following reasons. During the fabrication process, a layer of adhesive will be used to glue the copper strip of the antenna to the dielectric substrate and a silicon layer will be added to cover the metal surface to prevent potential copper rusting. The above parameters were not taken into account during simulations. Additionally, all the antenna tags present linear polarization. However, there are commercially available antenna readers with circular polarization. In such a case, a polarization loss factor of 0.5 should be taken into account in read range calculation.

CONCLUSION

The ABC algorithm and two of its variants were presented in this chapter. The I-ABC algorithm is an improved ABC variant that uses the best-so-far solution, inertia weight and acceleration coefficients to modify the search process, while the GABC includes global best solution information into the search equation to improve the exploitation. The application of these algorithms to design of different spiral antennas for passive RFID tags showed that all were proven quite efficient. Compared with other popular algorithms they produced better results. The results obtained with I-ABC and GABC are better than those obtained with the original ABC algorithm. Overall, all three algorithms have obtained design configurations with small size and high gain. The spiral shape geometrical parameters simultaneously offer the potential to control the input impedance and thus the conjugate matching of the structure. The designed tags exhibit satisfactory large read distances and the antennas ensure the necessary input power to the chip. Among others a tag design with the higher gain is that with 2.15dBi gain and 14.19m read range. The smaller design has dimensions less than 35mm.

ACKNOWLEDGMENT

This research has been co-financed by the European Union (European Social Fund – ESF) and Greek national funds through the Operational Program "Education and Lifelong Learning" of the National Strategic Reference Framework (NSRF) - Research Funding Program: THALES: Reinforcement of the interdisciplinary and/or inter-institutional research and innovation.

REFERENCES

Abdulhadi, A. E., & Abhari, R. (2011, 3-8 July 2011). *Dual printed meander monopole antennas for passive UHF RFID tags.* Paper presented at the Antennas and Propagation (APSURSI), 2011 IEEE International Symposium on.

Afshinmanesh, F., Marandi, A., & Shahabadi, M. (2008). Design of a Single-Feed Dual-Band Dual-Polarized Printed Microstrip Antenna Using a Boolean Particle Swarm Optimization. *IEEE Transactions on Antennas and Propagation*, *56*(7), 1845–1852. doi:10.1109/TAP.2008.924684

Akay, B. (2013). A study on particle swarm optimization and artificial bee colony algorithms for multilevel thresholding. *Applied Soft Computing*, *13*(6), 3066–3091. doi:10.1016/j.asoc.2012.03.072

Alarco, X. N,.J., Deleruyelle, T., Pannier, P., & Egels, M. (2010). *A new spiral antenna for passive UHF RFID tag on different substrates.* Paper presented at the Antennas and Propagation (EuCAP).

Alhawari, A. R. H., Ismail, A., Jalal, A. S. A., Raja Abdullah, R. S. A., & Rasid, M. F. A. (2013). *U-shaped inductively coupled feed RFID tag antenna for gain enhancement.* Paper presented at the RFID-Technologies and Applications (RFID-TA), 2013 IEEE International Conference on.

Ali, A. F., Hassanien, A. E., & Snasel, V. (2014). Memetic artificial bee colony for integer programming. *Communications in Computer and Information Science*, *488*, 268–277. doi:10.1007/978-3-319-13461-1_26

Bansal, J. C., Sharma, H., Arya, K. V., & Nagar, A. (2013). Memetic search in artificial bee colony algorithm. *Soft Computing*, *17*(10), 1911–1928. doi:10.1007/s00500-013-1032-8

Bin, L., Jianhua, Z., Baiqiang, Y., & Weiming, X. (2008). *A dual-frequency ceramic spiral antenna with rectangle PBG structure array used for modern RFID system.* Paper presented at the Wireless, Mobile and Multimedia Networks (ICWMMN 2008), IET 2nd International Conference on.

Bjorninen, T., Nikkari, M., Ukkonen, L., Fan, Y., Elsherbeni, A., Sydanheimo, L., & Kivikoski, M. (2008). *Design and RFID signal analysis of a meander line UHF RFID tag antenna.* Paper presented at the Antennas and Propagation Society International Symposium.

Calabrese, C., & Marrocco, G. (2008). Meandered-Slot Antennas for Sensor-RFID Tags. *IEEE Antennas and Wireless Propagation Letters*, *7*, 5–8. doi:10.1109/LAWP.2007.914123

Chen, H., Zhu, Y., Ma, L., & Niu, B. (2014). Multiobjective RFID network optimization using multiobjective evolutionary and swarm intelligence approaches. *Mathematical Problems in Engineering*, 2014.

Chen, S. Y., & Hsu, P. (2004). CPW-fed folded-slot antenna for 5.8 GHz RFID tags. *Electronics Letters*, *40*(24), 1516–1517. doi:10.1049/el:20046780

Clerc, M. (1999). *The swarm and the queen: towards a deterministic and adaptive particle swarm optimization.* Paper presented at the 1999 Congress on Evolutionary Computation, Washington, DC. doi:10.1109/CEC.1999.785513

da Silva Maximiano, M., Vega-Rodríguez, M. A., Gómez-Pulido, J. A., & Sánchez-Pérez, J. M. (2013). A new Multiobjective Artificial Bee Colony algorithm to solve a real-world frequency assignment problem. *Neural Computing & Applications*, *22*(7-8), 1447–1459. doi:10.1007/s00521-012-1046-7

Demarcke, P., Rogier, H., Goossens, R., & De Jaeger, P. (2009). Beamforming in the Presence of Mutual Coupling Based on Constrained Particle Swarm Optimization. *Antennas and Propagation. IEEE Transactions on*, *57*(6), 1655–1666. doi:10.1109/TAP.2009.2019923

Dieter, S., Fischer, C., & Menzel, W. (2011). Analysis of a folded reflect-array antenna using particle swarm optimization. *International Journal of Microwave and Wireless Technologies*, *3*(3), 267–272. doi:10.1017/S1759078711000456

Fister, I., Fister, I., Jr., Brest, J., & Žumer, V. (2012). *Memetic artificial bee colony algorithm for large-scale global optimization.* Paper presented at the 2012 IEEE Congress on Evolutionary Computation, CEC 2012. doi:10.1109/CEC.2012.6252938

Foster, P. R., & Burberry, R. A. (1999). *Antenna problems in RFID systems.* Paper presented at the RFID Technology (Ref. No. 1999/123), IEE Colloquium on.

Ghanem, W. A. H. M., & Jantan, A. (2014). Using hybrid artificial bee colony algorithm and particle swarm optimization for training feed-forward neural networks. *Journal of Theoretical and Applied Information Technology*, *67*(3), 664–674.

Goldberg, D. E. (1989). *Genetic Algorithms in Search, Optimization and Machine Learning*. New York: Addison Wesley.

Goudos, S. K., Siakavara, K., & Sahalos, J. N. (2013a). *Modified spiral RFID tag antenna optimal design using Artificial Bee Colony optimization.* Paper presented at the Microwave Conference (EuMC), 2013 European.

Goudos, S. K., Siakavara, K., & Sahalos, J. N. (2013b). *Novel spiral antennas design using swarm intelligence for passive UHF RFID tags.* Paper presented at the 7th European Conference on Antennas and Propagation, EuCAP 2013, Gothenburg.

Goudos, S. K., Siakavara, K., & Sahalos, J. N. (2013c). *Novel spiral antennas design using swarm intelligence for passive UHF RFID tags.* Paper presented at the Antennas and Propagation (EuCAP), 2013 7th European Conference on.

Goudos, S. K., Siakavara, K., & Sahalos, J. N. (2014). Novel spiral antenna design using artificial bee colony optimization for UHF RFID Applications. *IEEE Antennas and Wireless Propagation Letters*, *13*, 528–531. doi:10.1109/LAWP.2014.2311653

Goudos, S. K., Siakavara, K., & Sahalos, J. N. (2014). *Optimizing meandered spiral antennas for RFID tags using gbest-guided Artificial Bee Colony algorithm.* Paper presented at the Antennas and Propagation (EuCAP), 2014 8th European Conference on.

Goudos, S. K., Siakavara, K., & Sahalos, J. N. (2015). Design of load-ended spiral antennas for RFID UHF passive tags using improved artificial bee colony algorithm. *AEÜ. International Journal of Electronics and Communications*, *69*(1), 206–214. doi:10.1016/j.aeue.2014.09.008

Goudos, S. K., Siakavara, K., Samaras, T., Vafiadis, E. E., & Sahalos, J. N. (2011). Self-adaptive differential evolution applied to real-valued antenna and microwave design problems. *IEEE Transactions on Antennas and Propagation*, *59*(4), 1286–1298. doi:10.1109/TAP.2011.2109678

Gu, Y., Xu, X., Du, J., Hou, R., & Qian, H. (2013). Anycast routing protocol for wireless sensor networks based on artificial bee colony. *Chinese Journal of Sensors and Actuators*, *26*(4), 564–569. doi:10.3969/j.issn.1004-1699.2013.04.024

Hirvonen, M., Pursula, P., Jaakkola, K., & Laukkanen, K. (2004). Planar inverted-F antenna for radio frequency identification. *Electronics Letters*, *40*(14), 848–850. doi:10.1049/el:20045156

Holland, J. H. (1975). *Adaptation in Natural and Artificial Systems*. Ann Arbor, MI: The University of Michigan Press.

Huang, J. Z., Yang, P. H., Chew, W. C., & Ye, T. T. (2010). A novel broadband patch antenna for universal UHF RFID tags. *Microwave and Optical Technology Letters*, *52*(12), 2653–2657. doi:10.1002/mop.25560

Jiun-Peng, C., & Powen, H. (2013). A Compact Strip Dipole Coupled Split-Ring Resonator Antenna for RFID Tags. *IEEE Transactions on Antennas and Propagation*, *61*(11), 5372–5376. doi:10.1109/TAP.2013.2278481

Karaboga, D., & Akay, B. (2009). A comparative study of Artificial Bee Colony algorithm. *Applied Mathematics and Computation*, *214*(1), 108–132. doi:10.1016/j.amc.2009.03.090

Karaboga, D., Akay, B., & Ozturk, C. (2007) Artificial Bee Colony (ABC) optimization algorithm for training feed-forward neural networks. Vol. 4617 *LNAI. 4th International Conference on Modeling Decisions for Artificial Intelligence, MDAI 2007* (pp. 318-329). Kitakyushu. doi:10.1007/978-3-540-73729-2_30

Karaboga, D., & Basturk, B. (2007). A powerful and efficient algorithm for numerical function optimization: Artificial bee colony (ABC) algorithm. *Journal of Global Optimization*, *39*(3), 459–471. doi:10.1007/s10898-007-9149-x

Karaboga, D., & Basturk, B. (2008). On the performance of artificial bee colony (ABC) algorithm. *Applied Soft Computing*, *8*(1), 687–697. doi:10.1016/j.asoc.2007.05.007

Karaboga, D., Gorkemli, B., Ozturk, C., & Karaboga, N. (2014). A comprehensive survey: Artificial bee colony (ABC) algorithm and applications. *Artificial Intelligence Review*, *42*(1), 21–57. doi:10.1007/s10462-012-9328-0

Kefayat, M., Lashkar Ara, A., & Nabavi Niaki, S. A. (2015). A hybrid of ant colony optimization and artificial bee colony algorithm for probabilistic optimal placement and sizing of distributed energy resources. *Energy Conversion and Management*, *92*, 149–161. doi:10.1016/j.enconman.2014.12.037

Kennedy, J., & Eberhart, R. (1995). *Particle swarm optimization.* Paper presented at the IEEE International Conference on Neural Networks, Piscataway, NJ.

Kennedy, J., & Eberhart, R. C. (1997). *Discrete binary version of the particle swarm algorithm.* Paper presented at the IEEE International Conference on Systems, Man and Cybernetics. doi:10.1109/ICSMC.1997.637339

Kumar, S., Kumar, A., Sharma, V. K., & Sharma, H. (2014). *A novel hybrid memetic search in Artificial Bee Colony algorithm.* Paper presented at the 2014 7th International Conference on Contemporary Computing, IC3 2014. doi:10.1109/IC3.2014.6897149

Kumar, S., Sharma, V. K., & Kumari, R. (2014). *Memetic search in Artificial Bee Colony algorithm with fitness based position update.* Paper presented at the International Conference on Recent Advances and Innovations in Engineering, ICRAIE 2014. doi:10.1109/ICRAIE.2014.6909301

Lei, C., Shu, Y., & Hanhua, Y. (2009). *Study and design of a modified fractal antenna for RFID applications.* Paper presented at the Computing, Communication, Control, and Management, 2009. CCCM 2009. ISECS International Colloquium on.

Li, G., Niu, P., & Xiao, X. (2012). Development and investigation of efficient artificial bee colony algorithm for numerical function optimization. *Applied Soft Computing*, *12*(1), 320–332. doi:10.1016/j.asoc.2011.08.040

Liang, J. J., Qin, A. K., Suganthan, P. N., & Baskar, S. (2006). Comprehensive learning particle swarm optimizer for global optimization of multimodal functions. *IEEE Transactions on Evolutionary Computation, 10*(3), 281–295. doi:10.1109/TEVC.2005.857610

Ma, L., Hu, K., Zhu, Y., & Chen, H. (2014). Cooperative artificial bee colony algorithm for multi-objective RFID network planning. *Journal of Network and Computer Applications, 42*, 143–162. doi:10.1016/j.jnca.2014.02.012

Majumder, A. (2015). Application of Standard Deviation Method Integrated PSO Approach in Optimization of Manufacturing Process Parameters In *Handbook of Research on Artificial Intelligence Techniques and Algorithms* (pp. 536-563). IGI Global.

Marrocco, G. (2003). Gain-optimized self-resonant meander line antennas for RFID applications. *IEEE Antennas and Wireless Propagation Letters, 2*(1), 302–305. doi:10.1109/LAWP.2003.822198

Mitchell, M. (1996). *An Introduction to Genetic Algorithms*. MIT Press.

Mohammadi, S. A., Feizi Derakhshi, M. R., & Akbari, R. (2012). *An Adaptive Multi-Objective Artificial Bee Colony with crowding distance mechanism.* Paper presented at the AISP 2012 - 16th CSI International Symposium on Artificial Intelligence and Signal Processing. doi:10.1109/AISP.2012.6313720

Mollinetti, M. A. F., Souza, D. L., & Teixeira, O. N. (2014). *ABC+ES: A novel hybrid Artificial Bee Colony Algorithm with evolution strategies.* Paper presented at the GECCO 2014 - Companion Publication of the 2014 Genetic and Evolutionary Computation Conference.

Monti, G., Catarinucci, L., & Tarricone, L. (2009). Compact microstrip antenna for RFID applications. *Progress In Electromagnetics Research Letters, 8*, 191–199. doi:10.2528/PIERL09042803

Novais, F. T., Batista, L. S., Rocha, A. J., & Guimaraes, F. G. (2013). *A multiobjective estimation of distribution algorithm based on artificial bee colony.* Paper presented at the 1st BRICS Countries Congress on Computational Intelligence, BRICS-CCI 2013. doi:10.1109/BRICS-CCI-CBIC.2013.75

Padhi, S. K., Karmakar, N. C. Sr, Law, C. L., & Aditya, S. Sr. (2003). A dual polarized aperture coupled circular patch antenna using a C-shaped coupling slot. *IEEE Transactions on Antennas and Propagation, 51*(12), 3295–3298. doi:10.1109/TAP.2003.820947

Paredes, F., Zamora, G., Herraiz-Martinez, F. J., Martin, F., & Bonache, J. (2011). Dual-Band UHF-RFID Tags Based on Meander-Line Antennas Loaded With Spiral Resonators. *IEEE Antennas and Wireless Propagation Letters, 10*, 768–771. doi:10.1109/LAWP.2011.2162716

Randall, M., Lewis, A., Galehdar, A., & Thiel, D. (2007). *Using ant colony optimisation to improve the efficiency of small meander line RFID antennas.* Paper presented at the 3rd IEEE International Conference on E-Science and Grid Computing, Bangalore. doi:10.1109/E-SCIENCE.2007.82

Rao, K. V. S., Nikitin, P. V., & Lam, S. F. (2005). Antenna design for UHF RFID tags: A review and a practical application. *IEEE Transactions on Antennas and Propagation, 53*(12), 3870–3876. doi:10.1109/TAP.2005.859919

Rubio-Largo, Á., Vega-Rodríguez, M. A., Gómez-Pulido, J. A., & Sánchez-Pérez, J. M. (2013). A multiobjective approach based on artificial bee colony for the static routing and wavelength assignment problem. *Soft Computing*, *17*(2), 199–211. doi:10.1007/s00500-012-0887-4

Trelea, I. C. (2003). The particle swarm optimization algorithm: Convergence analysis and parameter selection. *Information Processing Letters*, *85*(6), 317–325. doi:10.1016/S0020-0190(02)00447-7

Vasant, P. (2011). *Solving fuzzy optimization problems of uncertain technological coefficients with genetic algorithms and hybrid genetic algorithms pattern search approaches. In Innovation in Power, Control, and Optimization: Emerging Energy Technologies* (pp. 344–368). IGI Global.

Vasant, P. (2012). *Hybrid linear search, genetic algorithms, and simulated annealing for fuzzy nonlinear industrial production planning problems. In Meta-Heuristics Optimization Algorithms in Engineering, Business, Economics, and Finance* (pp. 87–109). IGI Global.

Vasant, P. (2014). *Hybrid Optimization Techniques for Industrial Production Planning: A Review. In Handbook of Research on Novel Soft Computing Intelligent Algorithms: Theory and Practical Applications* (pp. 41–68). IGI Global.

Xiang, W., Ma, S., & An, M. (2014). HABCDE: A hybrid evolutionary algorithm based on artificial bee colony algorithm and differential evolution. *Applied Mathematics and Computation*, *238*, 370–386. doi:10.1016/j.amc.2014.03.055

Xianming, Q., & Ning, Y. (2004). *A folded dipole antenna for RFID.* Paper presented at the Antennas and Propagation Society International Symposium, 2004. IEEE.

Zhong, Y. B., Xiang, Y., & Liu, H. L. (2014). A multi-objective artificial bee colony algorithm based on division of the searching space. *Applied Intelligence*, *41*(4), 987–1011. doi:10.1007/s10489-014-0555-8

Zhongbao, W., Shaojun, F., Shiqiang, F., & Shouli, J. (2011). Single-Fed Broadband Circularly Polarized Stacked Patch Antenna With Horizontally Meandered Strip for Universal UHF RFID Applications. *IEEE Transactions on Microwave Theory and Techniques*, *59*(4), 1066–1073. doi:10.1109/TMTT.2011.2114010

Zhu, G., & Kwong, S. (2010). Gbest-guided artificial bee colony algorithm for numerical function optimization. *Applied Mathematics and Computation*, *217*(7), 3166–3173. doi:10.1016/j.amc.2010.08.049

ADDITIONAL READING

Abdulhadi, A. E., & Abhari, R. (2012). Design and Experimental Evaluation of Miniaturized Monopole UHF RFID Tag Antennas. *IEEE Antennas and Wireless Propagation Letters*, *11*, 248–251. doi:10.1109/LAWP.2012.2187632

Attaran, A., Rashidzadeh, R., & Muscedere, R. (2014). Chipless RFID tag using RF MEMS switch. *Electronics Letters*, *50*(23), 1720–1722. doi:10.1049/el.2014.3075

Bashri, M. S. R., Ibrahimy, M. I., & Motakabber, S. M. A. (2014). Design of a planar wideband patch antenna for UHF RFID tag. *Microwave and Optical Technology Letters*, *56*(7), 1579–1584. doi:10.1002/mop.28389

Chen, K. H., Yen, J. H., Tarng, J. H., & Yang, C. F. (2014). A low-profile dipole RFID tag antenna with lateral omni-directional radiation properties for cold-chain logistics applications. Paper presented at the IEEE Antennas and Propagation Society, AP-S International Symposium (Digest). doi:10.1109/APS.2014.6905087

Chiu, C. W., Tsai, M. C., Wang, H. C., & Wu, T. F. (2014). Broadband T-matching loop tag antenna design for on-body UHF RFID applications. *Microwave and Optical Technology Letters*, *56*(5), 1194–1200. doi:10.1002/mop.28291

Choo, J., & Ryoo, J. (2014). UHF RFID tag applicable to various objects. *IEEE Transactions on Antennas and Propagation*, *62*(2), 922–925. doi:10.1109/TAP.2013.2290051

Huang, J. Z., Yang, P. H., Chew, W. C., & Ye, T. T. (2010). A novel broadband patch antenna for universal UHF RFID tags. *Microwave and Optical Technology Letters*, *52*(12), 2653–2657. doi:10.1002/mop.25560

Kim, Y. (2013). Design of near omnidirectional UHF RFID tag with one-off seal function for liquid bottles. *Microwave and Optical Technology Letters*, *55*(2), 375–379. doi:10.1002/mop.27285

Lehto, A., Nummela, J., Ukkonen, L., Sydanheimo, L., & Kivikoski, M. (2009). Passive UHF RFID in Paper Industry: Challenges, Benefits and the Application Environment. *IEEE Transactions on Automation Science and Engineering*, *6*(1), 66–79. doi:10.1109/TASE.2008.2007269

Li, Y., Rongwei, Z., Staiculescu, D., Wong, C. P., & Tentzeris, M. M. (2009). A Novel Conformal RFID-Enabled Module Utilizing Inkjet-Printed Antennas and Carbon Nanotubes for Gas-Detection Applications. *IEEE Antennas and Wireless Propagation Letters*, *8*, 653–656. doi:10.1109/LAWP.2009.2024104

Mahmoud, K. R. (2010). Design optimization of a bow-tie antenna for 2.45 GHz RFID readers using a hybrid BSO-NM algorithm. *Progress in Electromagnetics Research*, *100*, 105–117. doi:10.2528/PIER09102903

Mitatha, S., Udomariyasap, P., Noppanakeepong, S., & Yupapin, P. P. (2012). Highly capacity RFID generated by a soliton pulse within a panda ring resonator. *Microwave and Optical Technology Letters*, *54*(5), 1339–1342. doi:10.1002/mop.26779

Pan, T., Zhang, S., & He, S. (2014). Compact RFID tag antenna with circular polarization and embedded feed network for metallic objects. *IEEE Antennas and Wireless Propagation Letters*, *13*, 1271–1274. doi:10.1109/LAWP.2014.2333365

Park, I. Y., & Kim, D. (2014). Artificial magnetic conductor loaded long-range passive RFID tag antenna mountable on metallic objects. *Electronics Letters*, *50*(5), 335–336. doi:10.1049/el.2013.2671

Pouzin, A., Vuong, T. P., Tedjini, S., & Perdereau, J. (2011). Automated bench test for UHF RFID tags measurement in operational environment. *International Journal of Microwave and Wireless Technologies*, *3*(4), 415–422. doi:10.1017/S1759078711000511

Randall, M., Lewis, A., Galehdar, A., & Thiel, D. (2007). Using ant colony optimisation to improve the efficiency of small meander line RFID antennas. Paper presented at the 3rd IEEE International Conference on E-Science and Grid Computing, Bangalore. doi:10.1109/E-SCIENCE.2007.82

Santiago, A. G., Costa, J. R., & Fernandes, C. A. (2013). Broadband UHF RFID passive tag antenna for near-body applications. *IEEE Antennas and Wireless Propagation Letters*, *12*, 136–139. doi:10.1109/LAWP.2013.2243400

Tang, T., & Chen, X. (2014). UWB UHF RFID TAG antenna with branch-line slots for metallic objects. *Microwave and Optical Technology Letters*, *56*(7), 1689–1691. doi:10.1002/mop.28419

Tang, Z.-J., Wu, X.-F., Zhan, J., & Hu, S.-G. (2014). Broadband UHF RFID tag antenna design matched with different RFID chips. *Microwave and Optical Technology Letters*, *56*(1), 55–57. doi:10.1002/mop.28046

Zhang, J., & Long, Y. (2014). A novel metal-mountable electrically small antenna for RFID tag applications with practical guidelines for the antenna design. *IEEE Transactions on Antennas and Propagation*, *62*(11), 5820–5829. doi:10.1109/TAP.2014.2354412

Zi Long, M., Li Jun, J., Jingtian, X., & Ye, T. T. (2012). A Single-Layer Compact HF-UHF Dual-Band RFID Tag Antenna. *IEEE Antennas and Wireless Propagation Letters*, *11*, 1257–1260. doi:10.1109/LAWP.2012.2225821

Ziai, M. A., & Batchelor, J. C. (2014). Multi-state logging freeze detection passive RFID tags. *IEEE Transactions on Antennas and Propagation*, *62*(12), 6406–6411. doi:10.1109/TAP.2014.2359474

KEY TERMS AND DEFINITIONS

Ant Colony Optimization (ACO): An algorithm, which is inspired from the foraging behavior of some ant species. These ants deposit pheromone on the ground in order to mark some favorable path that should be followed by other members of the colony.

Antenna Gain: An antenna performance figure, which combines the antenna's directivity and electrical efficiency. For a transmitting antenna, the gain describes how well the antenna converts input power into radio waves headed in a specified direction. For a receiving antenna, the gain describes how well the antenna converts radio waves arriving from a specified direction into electrical power.

Genetic Algorithm: A stochastic population-based global optimization technique that mimics the process of natural evolution.

Omnidirectional Antenna: An antenna that radiates equally in all directions.

Particle Swarm Optimization (PSO): An evolutionary algorithm that mimics the swarm behavior of bird flocking and fish schooling.

Radio-Frequency Identification (RFID): The use of electromagnetic fields to transfer data in order to automatically identify, and track tags attached to objects.

Self-Adaptive Differential Evolution (SADE): A population-based stochastic global optimization algorithm, which requires the setting of two parameters; population size and maximum iteration number.

Ultra High Frequency (UHF): The term for radio frequencies from 300 MHz to 3 GHz, also known as the decimetre band.

ACRONYMS USED IN THE TEXT

ABC: Artificial Bee Colony.
ACO: Ant Colony Optimization.
DE: Differential Evolution.
GA: Genetic Algorithm.
GABC: Gbest-guided Artificial Bee Colony.
I-ABC: Improved Artificial Bee Colony.
PSO: Particle Swarm Optimization.
RFID: Radio-frequency identification.
SADE: Self-adaptive Differential Evolution.
SI: Swarm intelligence.
UHF: Ultra High Frequency.

APPENDIX: INPUT DATA FOR THE RFID DESING CASES

Table 6. Input Data for the asymmetric spiral design case

Parameter	Min value	Max value
SL	2.5mm	4.5mm
D_0	1mm	4mm
m1	1	9
m2	1	9
per1	0.01	0.5
per2	0.01	0.5

Table 7. Input Data for the load-ended antenna design case

Parameter	Min value	Max value
SL	2.5mm	4.5mm
D_0	1mm	4mm
m1	2	9
m2	2	9
per1	0.01	0.5
per2	0.01	0.5
hl	2mm	15mm
wl	1mm	2mm

Table 8. Input Data for the modified antenna with meander lines and loads design case

Parameter	Min value	Max value
SL_v	2.5mm	4.5mm
SL_h	2.5mm	4.5mm
D_0	1mm	4mm
$m1$	1	7
$m2$	1	7
Df	0.90	0.93
sf	0.1	0.2
$b_1^1 \ldots\ldots b_{2m1}^1$	0	1
$b_1^2 \ldots\ldots b_{2m1}^2$	0	1

Chapter 10
Optimization of Traffic Network Design Using Nature-Inspired Algorithm: An Optimization via Simulation Approach

Enrique Gabriel Baquela
Universidad Tecnológica Nacional, Argentina

Ana Carolina Olivera
Universidad Nacional de la Patagonia Austral, Argentina & CONICET, Argentina

ABSTRACT

In this chapter the application of a nature-inspired technique in conjunction with simulation models to optimize the siting of concentration nodes in a traffic network of urban area is presented. The solutions of this optimization problem involve the redesign of the network by adding nodes and arcs to the current traffic network. The problem is a sub-problem of the most general "Traffic Network Design Problem". For this resolution a Genetic Algorithm approach was design and development. The popular Simulator of Urban Mobility (SUMO) is use as traffic simulator in order to evaluate the solutions obtained. This chapter contains the conceptual models and the results of the Optimization via Simulation technique proposed.

INTRODUCTION

These days, urban expansion has become a special issue due to the impact in ecosystem, biodiversity and human activities (Bhatta, Saraswati & Bandyopadhyay, 2010; He, Tian, Shi, & Hu, 2011; Haregeweyn, Fikadu, Tsunekawa, Tsubo, & Meshesha, 2012; Poelmans & Rompaey, 2010). However, the absence of urbanization planning policies around the world has turned this expansion process chaotic and disorganized A very common phenomenon is the urban sprawl, which means that a city has spread into several suburbs connected with the commercial and economic center of the town through highways, resulting thus in vehicle dependency (He, Okada, Zhang, Shi, & Li, 2010).

DOI: 10.4018/978-1-4666-9644-0.ch010

When planning to expand a city, it is necessary to achieve a sustainable use of the land, to exploit the roads and highways with low traffic, and to reorganize the urbanization process in order to guarantee a minimization of the environmental impact and a reduction of traffic problems (He, Tian, Shi, & Hu, 2011).

In this context, in recent years a lot of work has focused on redefining expansion areas in urban settlements (Ferreira & Condessa, 2012), and on employing Geographical Information Systems with sensors to test the impact of urban planning policies (Stanilov & Batty, 2011; Xiao et al., 2006). If investing in new infrastructure (sensors, data recover stations, etc.) is not possible, then the simulation and optimization heuristics become the alternatives to analyze the impact of urban expansion models on the landscape (Poelmans & Rompaey, 2010). In this chapter it is proposed to use a framework based on Optimization via Simulation (OvS) which combines Genetic Algorithms (GA) and Traffic Simulation in order to evaluate the organization urban planning in the cities, taking into account its impact on the traffic network.

The objectives of this chapter are:

- Introduce the traffic optimization problem.
- Introduce the "Origin-Destiny Traffic Assignment Problem".
- Explain how to model the "Origin-Destiny Traffic Assignment Problem".
- Show the main aspects of Genetic Algorithms and Optimization via Simulation.
- Introduce the traffic simulation.
- Explain how to solve the "Origin-Destiny Traffic Assignment Problem" with Genetic Algorithms and Optimization via Simulation.
- Test the effects of parameters selection over the solutions found by OvS.

BACKGROUND

City growth reflects a progressive dynamic that is hardly ever planned, resulting in a decrease of the general performance of its sub-systems (Xiao et al., 2006). In the case of the traffic sub-system, a population growth means more vehicles in the system, higher probability of traffic jams, longer travel times etc. (He et al., 2010). Even though this is not easy to control, it is possible to establish policies related to urbanization permits, to create industrial centers and to encourage activities in strategic zones, which will allow us to control the way the city grows and its associated traffic sub-system (Xiao et al, 2006). In this line, the traffic simulation (Li & Sun, 2012; Nagel, & Schreckenberg, 1992) revels itself as a useful tool to quantify the impact of a certain traffic network topology and the corresponding population assignment (Garcia-Nieto, Olivera & Alba, 2013).

In this chapter, the authors present the problem of deciding where to foment the installation of urban, industrial and/or commercial complexes so that it has a minimum effect over the total traffic system: considering the network already has a certain structure and a dynamic flow restricting possible configurations. The authors define this proposed decision problem like "Restatement of The Traffic Network" and the underlying optimization problem "Origin-Destiny Traffic Assignment Problem" (ODTAP). The ODTAP is an underlying problem of the "Network Traffic Design Problem" (NTDP). The NTDP is the NP-Hard problem of building a traffic network in such a way to minimize the performing function of the system, i.e., the mean travel time in general. In this context, several literature works exist in relation to NTDP, some of them related to the ODTAP issue and urban planning (Friesz, 1981; Current & Min, 1986; Yang & Bell, 1997; Yang & Bell, 1998; Current & Marsh, 1993; Baquela & Olivera, 2014).

The metaheuristic approach has proven to be useful for different types of optimization problems (Romeijn & Smith, 1994; Hu & Wang, 2011; Vasant, 2013). For the NTDP, there are some works for which the solution is obtained by heuristics and bio-inspired algorithms. In this sense, Gallo, D'Acierno and. Montella (2012) optimize a cost function in base of the ant colony algorithm. Caggiani and Ottomanelli (2011) study the traffic network design problem in its equilibrium point by molding with fuzzy variables and optimizing by the "Sequential Quadratic Programming" method. Friesz, Cho, Mehta, Tobin, and Anandalingam (1992) analyze the use of Simulated Annealing for the design of traffic networks; Zhang, Lu and Xiang (2008) use Genetic Algorithms (GA) to optimize a railroad design. Dinu and Bordea (2011) implement a GA to remodel a transport network; Ceylan and Bell (2005) apply GA for the coordination of traffic lights in traffic jam areas. Ukkusuri, Mathew and Waller (2007) use a GA in order to obtain robust solutions for the NDTP, evaluating each solution in multiple scenarios. And Miandoabchi and Farahani (2011) implement a hybrid algorithm which combines GA and Simulated Annealing to assign a shape to traffic roads and to expand lanes to already existing roads. The development of resolution techniques for the NDTP generally follows the way of the optimized systems analytically defined. Although some previously mentioned works give good solutions for NDTP, their use in the ODTAP resolution has not been addressed. In this paper, we explore the Optimization via Simulation (OvS) capacity to obtain good solutions for ODTAP considering that this tool has proved to be useful to solve a lot of problems that, due to their complexity, are difficult to model analytically. For example, Madireddy, Medeiros and Kumara (2011) use a simulation version with agents to evaluate evacuation policies. Fu (2002) presents a more extended state of the art in relation of OvS development. With respect to problems associated to the traffic system, Garcia-Nieto et al., (2013) apply OvS optimizing with Particle Swarm Optimization and simulating with the SUMO-Simulated of Urban Mobility tool. Behrisch, Bieker, Erdmann and Krajzewicz (2011) use OvS to solve the problem to programming system of traffic lights cycles. Kalganova, Russell and Cumming (1999) use GA in an OvS framework to solve the traffic lights synchronizing problem. Finally, Horvat and Tošic (2012) apply OvS together with GA for the design of non-restriction traffic networks in preexisting networks.

In this chapter, an OvS approach based in GA is presented in order to establish where foment the installation of urban, industrial and/or commercial complexes with a minimum effect over the traffic system.

ORIGIN-DESTINY TRAFFIC ASSIGNMENT PROBLEM (ODTAP)

Introduction

Given a population growth scenario, one of the urban planning system goals is to achieve a traffic network restructure in such a way that this network can absorb the increase of vehicles, trying to change least possible the total system performance.

The traffic network context is a complex system consisting of a traffic network, traffic demand, control instruments and a set of policies governing the interaction between all elements of the system. The traffic network is modeled using an oriented graph in which the arcs represent the traffic lanes and nodes constitute changes in these lanes (intersections, bridges, extensions, traffic sources, traffic sinks). The traffic demand is the set of all vehicles in the traffic net and their routes.

For a given time window, it is possible to classify the nodes of a traffic network as nodes of *origin*, *destination* and *transfers*. With this classification of traffic network nodes, it is possible to use the for-

Figure 1. Given a traffic network A, where traffic origins and destinies correspond to the highlighted nodes, it is possible to associate this with a traffic and assignment problem (B) between origin nodes (B left) and destiny nodes (B right)

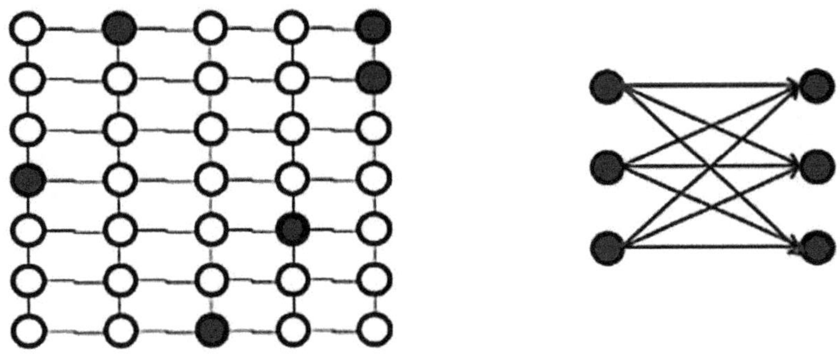

A B

malism of the traffic and assignment problem in its dual version (in which the variable decisions will be the capacities of the origin and destiny nodes). This approach is shown in Figure 1.

ODTAP: Problem Definition

Given a traffic network *T*, with a group of vehicles *C*, each one of them has been assigned a route between two network nodes, for each time instant t, is possible to obtain, by simulation or sampling of the real system, a function of mean travel time in that network. This is a random function and it depends, besides of the network itself, on the individual routes of each vehicle. Now, suppose there is an increase in the number of vehicles, in such a way that the new group *C'* > *C*. We want to determine which traffic network nodes are possible to restate, so they could act as origins or destinies of these new vehicles routes, changing least possible the rest of the network.

With the "Restatement of The Traffic Network" we make reference to define incentive policies for the installation of, for example, industrial parks and commercial centers, or investing in infrastructure that allows accommodation and a major vehicle flow circulation. The practical effect of these policies is to modify the network structure either by adding nodes and arcs, or by increasing the capacities associated to them.

Mathematical Formulation of ODTAP

ODTAP can be mathematically defined as a variation of an Assignment and Transport Problem, with two major differences:

- It's difficult to assign directly the vehicle´s trip. It's better to determinate the amount of vehicles that will leave from each origin and the amount of vehicles that will arrive to each destiny.
- Travel times between origins and destinies – in a real scenario – are not constant, but they follow a probability function that depends on the time and the global vehicle assignment to origins and destinies.

Mathematically, it is possible to formulate the ODTAP as follows:

$$Min\,Z = \sum_i \sum_j \left(t_{ij} \times X_{ij} \right) \tag{1}$$

Subject to

$$\sum_i X_{ijp} = PO_i \tag{2}$$

$$\sum_j X_{ijp} = PD_j \tag{3}$$

$$\sum X_{ijp} = \sum_i PO_i = \sum_j PD_j \tag{4}$$

$$PO_i \geq \text{Actual population at origin } i \tag{5}$$

$$PD_j \geq \text{Actual population at origin } j \tag{6}$$

where i is an origin node, j a destiny node and p is one of all routes that connect i with j, PO_i is the number of vehicles travel from i, PD_j is the number of vehicles travel to j, X_{ijp} is the number of vehicles that has to travel from the i node to j through p, and t_{ijp} is the average time that take from going from i to j by the path p. X and t are functions of PD and PO. The function to be optimized is the total travel time (Equation 1) of all vehicles in the system. Equation 2 ensures that all cars assigned to an origin i have an associated route. Equation 3 ensures that all cars assigned to a destination j have an associated route. Equation 4 ensures that vehicles are neither created nor destroyed. And Equations 5 and 6 restrict PD and PO to its current capacity.

It is worth highlighting that, from the total network nodes, not all are possible for restating. Most of them, as a result of the new traffic, will function as transfer nodes, which will take part in the route of the new vehicle, but will not function as origin or destiny of the new routes.

Therefore, the problem of defining which nodes to restate can be solved by the use of a formalism of an assignment problem, in which we have to assign a vehicle to a pair of origin-destiny nodes so as to cover the capacities and the demands of the nodes, minimizing as well the total travel time (Ozdemir, Basligil, & Sarsenov, 2012). If there is a significant increase in the size of the group of vehicles, the travel time of each trip will be unknown (since they depend on how many new vehicles will go through each route). As the assignment of the vehicle cannot be done in an origin-destiny level, our decision variables are the capacities to increase (*PO* and *PD*), being the random assigned routes to each vehicle according to the defined capacities of each network node.

Modeling the Network Expansion and Node Restatement

The previous ODTAP formulation means restating pre-existing nodes in the traffic network. But if there is a significant population growth, urbanizing or industrializing non populated areas might be needed. This is equal to expanding the original graph with new nodes and arcs. The treatment of this problem can be continued using the assignment formalism, taking into account that the options the planner has are boundless. Therefore, it is possible to model this new situation using a traffic network *T'* so that *T* is a sub-network, the new nodes having initial capacity zero, and associated to them, there is a group of binary restrictions allowing or not, intermediate arcs in terms of populated or not settlements in the nodes. The network updating process and the nodes restating can be seen in Figure 2.

Modeling Traffic System Behavior

The best way to represent the behavior of traffic systems is simulating them. There are four classes of traffic simulation models used in traffic research (Figure 3). Microscopic models simulate the movement of every single vehicle on the street. Macroscopic models simulate traffic at an aggregate level, through traffic flow equations. Mesoscopic models are an intermediate type of model which simulate the system using queue approaches and discrete events simulations. Lastly, submicroscopic models simulate at level of substructures of individual vehicles, like brake system and engine. Although accurate, submicroscopic models require too much computational effort. Moreover, macro and mesoscopic models obtain information too aggregated, unable to model some dynamic effects as spontaneous traffic

Figure 2. Current traffic network (full line) with an extended traffic network (new nodes and arcs with dotted line)

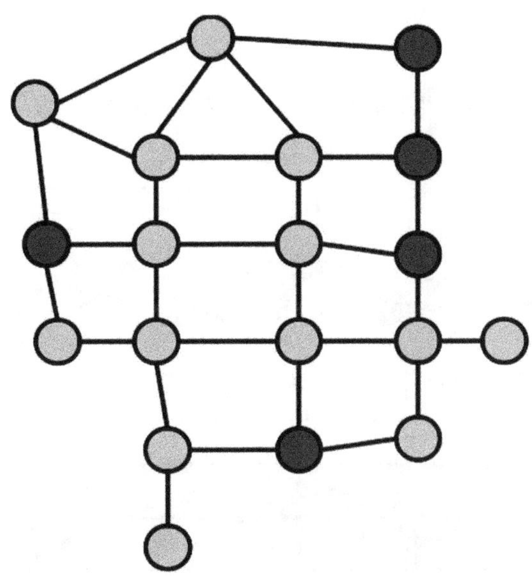

jams. Microscopic models are one of the most studied tools in recent years, due to the fact that their computational cost is comparatively low and their level of granularity allows researchers to appreciate the complexity of the system.

There are two groups of microscopic simulations: discrete- space simulations and continuous-space simulations, both with discrete-time (Figure 4).

In discrete-space simulations, the traffic network is discretized into a set of cells whose states reflect whether it is occupied by a vehicle (or part thereof) or empty (Li & Sun, 2012). In a model of a one way, the representation is done by a graph of a single path, which is modeled by a one-dimensional automaton (Figure 5). In the case of networks with more than one way, the model is embodied on a two-dimensional automaton, wherein the cells have an additional state indicating those in which the circulation of vehicles is not practicable (Figure 6).

Figure 3. Types of traffic simulations: macroscopic (A), mesoscopic (B), microscopic (C) and sub-microscopic (D)

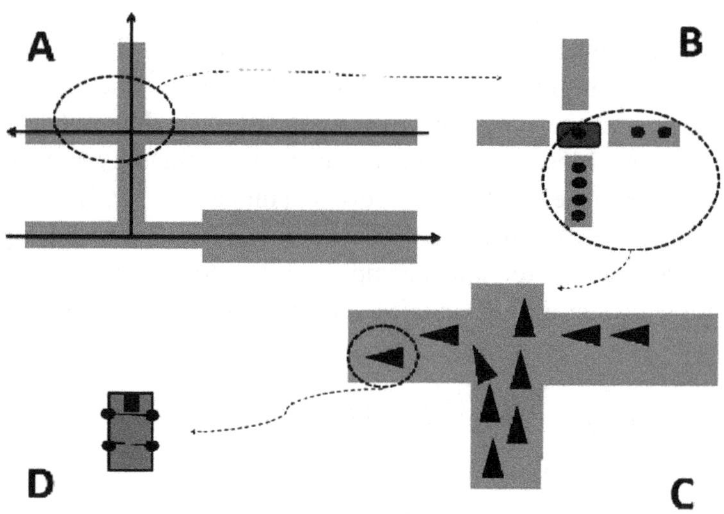

Figure 4. Difference between continuous-space (A) and discrete-space (B) simulations

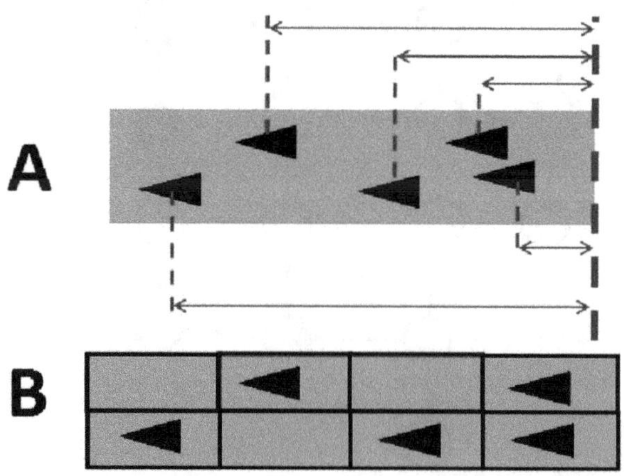

Figure 5. Unidimensional cellular automaton for traffic flow modelling

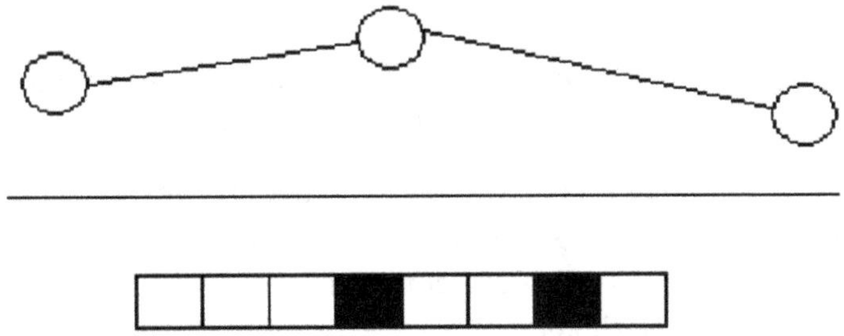

Figure 6. Bidimensional cellular automaton for traffic flow modelling

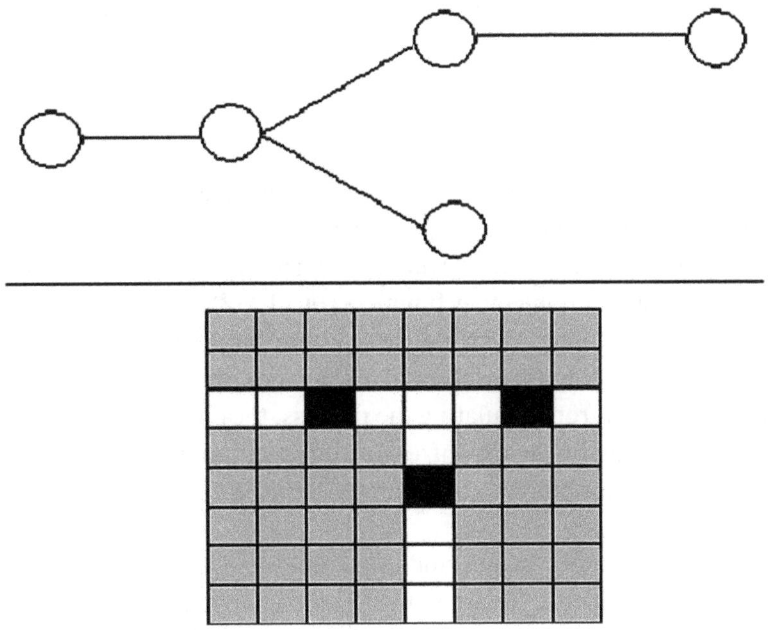

In continuous-space simulations, the space is not discretized, and each vehicle is an object with references to a particular path in the graph together with the distance to the origin of the road. In such models, Car-Following techniques are usually applied (Krauss, 1998). Behavior is governed by an equation of the stimulus- response type, in which the determining factor of each vehicle's behavior is the vehicle in front of it (the path to follow is fixed since the beginning of the simulation). Based on the individual information of each vehicle, it is possible to obtain aggregate variables. In the traffic flow theory, there is an essential relationship between the vehicle density k, the speed v, and the flow q (Equation 7)

$$q = k \times v \quad (7)$$

It can be empirically determined that the flow and the velocity are dependent of the vehicle density in the network, increasing the flow when vehicles are being added to the system until a specific amount

(k_c), when the vehicles begin to interfere between them until the velocity low up to zero (in a traffic jam) at the density (k_{jam}).

Even though differential equations could model the system, they do not allow us to represent spontaneous jams (kw_w) that occur with $w \neq \{jam, c\}$ when $w < jam$ due to differences in speed between vehicles and random variations of that speed. Microscopic models can model this type of phenomenon.

In this paper, a microscopic continuous-space and discrete-time simulator is used, however, the results can be extrapolated to many others type of microscopic simulators.

To solve the ODTAP, our approach should use the average travel time. This variable is also dependent on the density and it is correlated with the flow, tending to infinity when density tends to k_{jam}. In our problem, the density is constant (the number of vehicles to assign to origins and destinies is fixed), thus, solve the ODTAP consists in evaluate which distribution of vehicles has the lowest average time for a fixed density.

Optimization via Simulation Process

In order to solve the ODTAP, we development a novel technique based on Optimization via Simulation (OvS) . In the OvS, the solutions are generated by heuristics or metaheuristics, but the evaluation of the goodness of these solutions is done by the execution of a simulator, which functions as the generator of necessary values to calculate the objective function (Fu, 1994). Moreover, most of the restrictions are defined implicitly in the structure of the simulated system. Both the optimizer and the simulator are independent from each other, relating themselves in a black box way, having a system functioning as a controller between the two. Apart from regulating the process, the controller has the function of translating the resulting solutions brought by the optimizer to configuration parameters for the simulator, at the same time it converts the simulator outputs (data samples) into a value which the optimizer can accept as an aptitude or fitness function (Figure 7).

Traffic systems are very complex systems, and modeling them analytically means losing a lot of information on their behavior. In general, for urban traffic, the analytical models oversimplify the system studied, decreasing their utility considerably. OVS allows us to solve this problem because it allows evaluating the temporal dynamics of these systems.

Figure 7. OvS process diagram

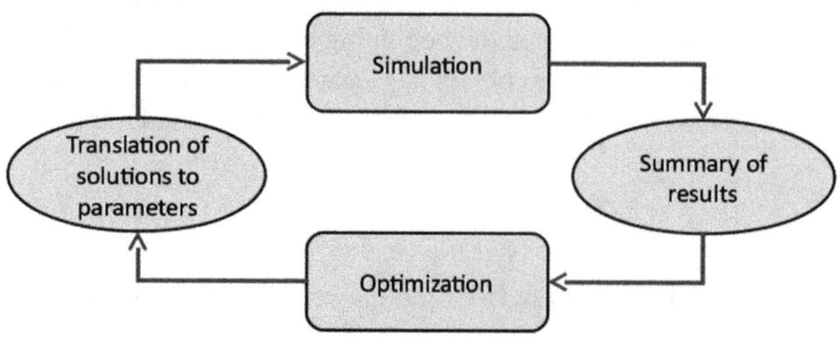

Figure 8. Complete OvS process diagram

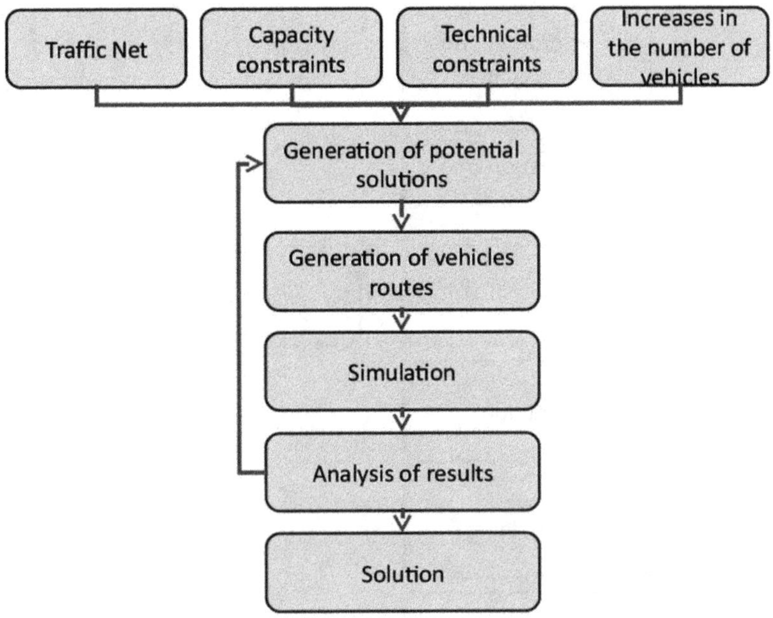

As you can see in Figure 8, the process input data are the extended traffic network, the capacity restrictions for each node, the possible technical restrictions and the increase in the number of vehicles. According to the inputs, the optimizer generates solutions (increase capacities of each node), which are translated in a route for every vehicle, stating from which and to which node they are going. In terms of this information, the system simulation is run and, new solutions to be evaluated are generated after analyzing the results, repeating the process until the quality of them does not present significant variations (the chosen stopping criteria).

Optimization by Genetic Algorithms

Genetic Algorithms (GA) have been proved to be good for the resolution of different optimization problems (Mahalakshmi, Kalaivani & Kiruba Nesamalar, 2013; Keshavarz Ghorabaee, Amiri & Azimi, 2015). Optimization by GA gives the advantage of exploring in depth the solution space (Luke, 2013).

Unlike other algorithms, the GA works over a group of potential solutions. This group of size *N* is called population and each solution is called individual. The individuals are coded in chains of binary or real numbers, referred to chromosomes. In each iteration, every individual is tested in order to determine its relative goodness with respect to the rest. As long as the population individuals do not fulfill a specific convergence criteria, a sequence consisting of selecting the subgroup M<N of individuals with the best fitness values and of crossover the individuals to generate a new populations of size N, is repeated. In this crossover, every new chromosome is composed of sections extracted from pairs of the subgroup M chromosomes, after this, new chromosomes can be submitted to a mutation, i.e., random changes in the chromosomes. The best are selected for the next population and the process is repeated until the convergence criteria are achieved (Figure 9). The crossover and mutation operations can be observed in Figures 10 and 11.

Figure 9. General diagram of a genetic algorithm

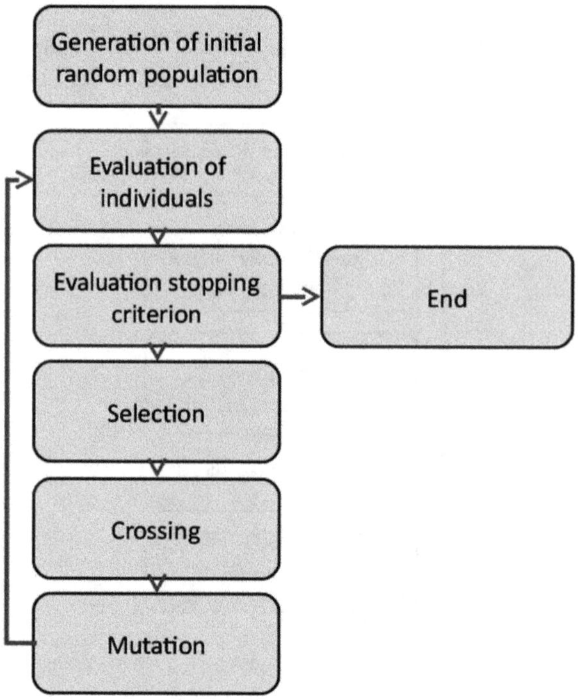

Figure 10. Chromosome crossover: in this figure is shown how from two individuals (P1 and P2) one of the possible descendants is generated (D1), combining chromosome segments from the parents (Luke, 2013)

P1	0	1	1	0	0	0	1	1	1	1	0	0	1
P2	0	0	0	0	1	1	1	0	0	0	0	1	1
D1	0	1	1	0	0	0	1	1	1	0	0	1	1

Figure 11. Chromosome mutation: in this figure is shown how a particular chromosome is mutated by a random change in one of its bits

D1	0	1	1	0	0	0	1	1	1	0	0	1	1
D1'	0	1	1	0	1	0	1	1	1	0	0	1	1

Figure 12. An example of the codification of a whole variable value to a binary chromosome of a length 4

$$X = 3 \longrightarrow \boxed{0\,|\,0\,|\,1\,|\,1}$$

It should be noted that every OvS iteration using GA implies simulate as many solutions as the number of individuals the generated populations have, therefore, the basic algorithm was modified by adding a consolidation of solution stage, so that if there were duplicated individuals, they would be simulated only once.

Regarding solutions, GA uses binary encoding. So, decision variable values are encoding in bit chains (of a fixed length in our case, Figure 12). This allows that the crossover and mutation do not take part in a decision variable level but in an inferior structure level.

For the ODTAP, it is decided to model the solutions by binary representation. Mainly for of two reasons, binary encoding implies a better individuals combinations for this problem and, in this point the accuracy of real variables due to the nature of the problem is not necessary.

Evaluation of Potential Solutions

With the objective to obtain a real evaluation of traffic journey associated with land use urbanization planning, estimation through microscopic simulation is proposed. In a microscopic evaluation of the traffic network, each vehicle is examined separately. In order to get a representative view SUMO software lets one run simulations that emulate vehicular mobility at a microscopic level. It is open source software, highly portable and capable of supporting full maps of real cities, like those obtained through "Google Maps"[1] or "Open Street Map"[2]. It is possible to define properties by means of SUMO, such as vehicular acceleration and deceleration, driver's ability, maximum vehicular speed, and all the information of street directions, roads, roundabouts, highways, etc. SUMO models the traffic network as a graph with nodes that represent the intersection of the roads and beginnings/endings of traffic lanes. These nodes are, in ODTAP, the transport points or origin-destiny nodes. Besides, each arc of the graph symbolizes a traffic lane with several properties: capacity, velocity limit, etc. In each intersection, SUMO allows define the cycles scheduling of traffic lights. Figure 13 shows how SUMO works; receiving nets, cars and routed as inputs, and generating performance measures as outputs.

To model the ODTAP, the DUAROUTER application (part of the SUMO package) was used with the objective to assigning to each pair origin-destiny, traffic flow in intervals of time. Taking into account an origin and a destination, DUAROUTER calculates the nodes sequence following to the vehicles during the simulation. In order to perform this task, DUAROUTER uses the Dijkstra's shortest path's algorithm (Dijkstra, 1959).

The flow between two nodes *i* and *j* can be defined as follow:

$$Vol\left(F_{ij}\right) = \frac{PO_i \times PD_j}{\sum_i PO_i} \tag{8}$$

Figure 13. SUMO as a black box. Net, cars and routes definitions are used to generate performance measures of traffic systems

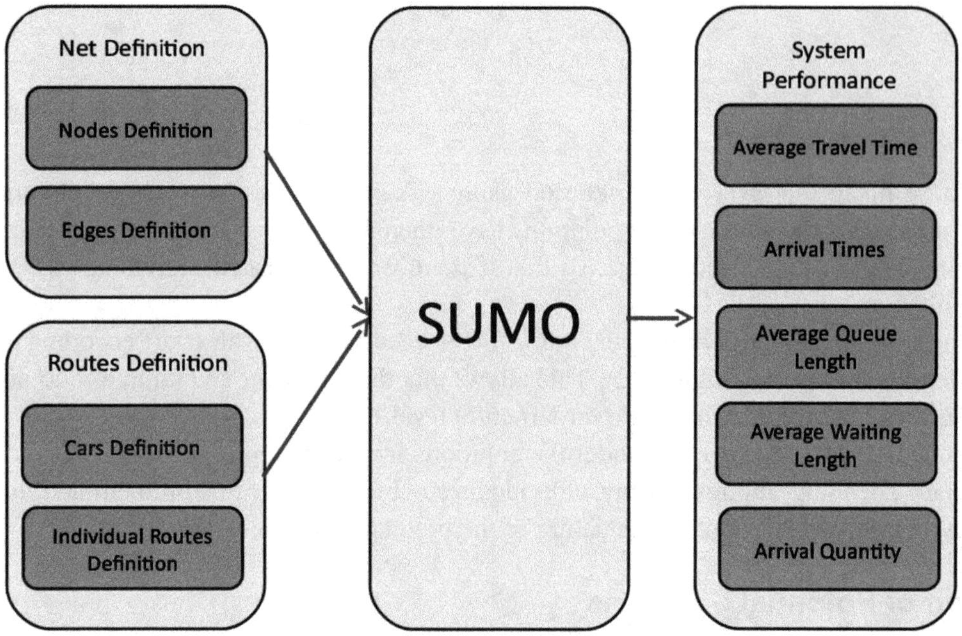

subject to

$$\sum X_{ijp} = \sum_i PO_i = \sum_j PD_j \quad (9)$$

OvS and Horizon of Simulations

While the solutions obtained by OvS are good, the processing time is high. This is inherent in the simulations of traffic itself. The implementation of OvS is highly demanding of computational time.

Figures 15, 16 and 17 illustrates the distribution of computational times of spider-style, grid-style and random-style networks simulations respectively; for a simulation horizon of 3500 seconds, with over 3000 vehicles in the system, randomly distributed over the entire network. It is important to consider that the computational effort depends on both, the simulation horizon and the number of vehicles (called density, constant for the ODTAP).

The horizon of simulation can be reduced in order to shorten the computational time. Then, if the horizon under study is h_{max}, it is possible to take other simulation horizon $h_g < h_{max}$ and infer results based on this simulation bounded. The authors have studied how the average time is distributed along the simulation time. The Figure 18 shows the accumulated mean travel time is distributed along the simulation time, over 540 simulations of three topologies (near 162,000 vehicles).

It can be concluded that the travel time is stabilized quickly if the density of vehicles in the system is constant.

Figure 14. Distribution of computation times for random simulations

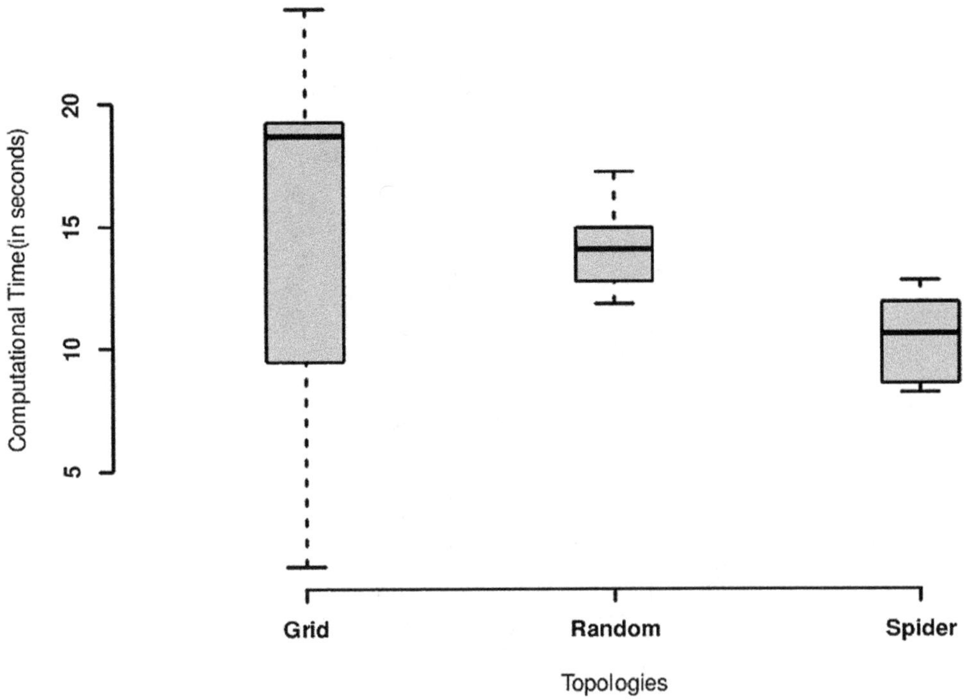

Figure 15. Grid style net

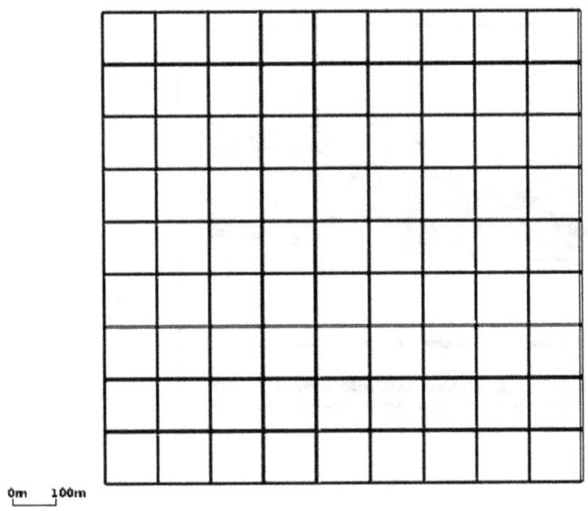

Figure 16. Spider style net

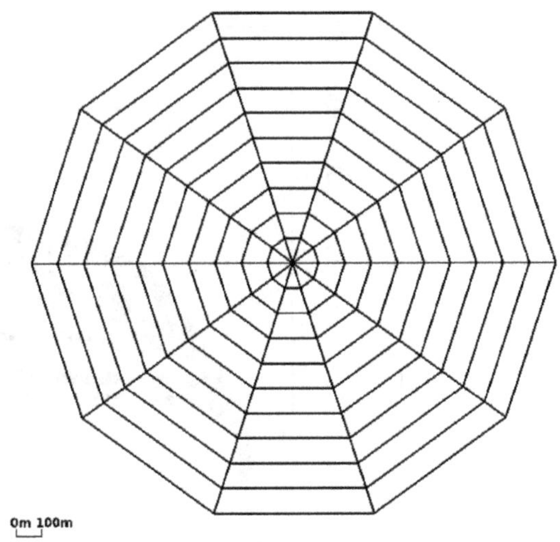

Figure 17. Random style net

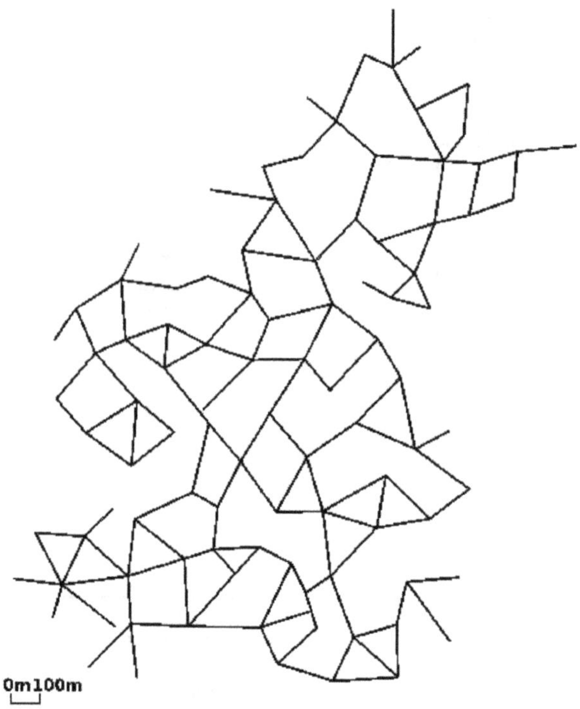

Figure 18. Mean travel time vs. simulated time

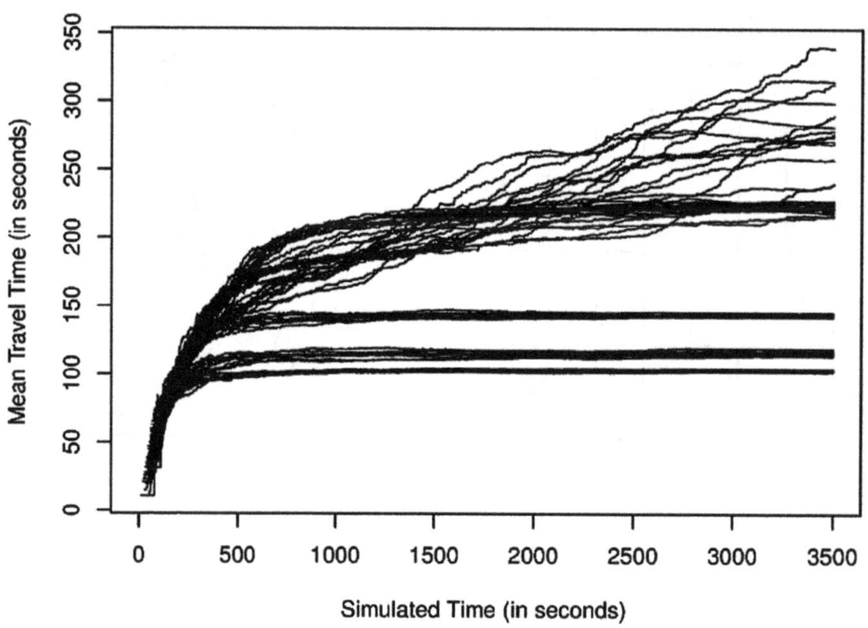

Experiments

To study the proposed algorithm, it was necessary to define fixed scenarios that serve as a base for the experiments. To do this, five networks with a Grid format with 100, 400, 900, 1600 and 2500 total nodes, from which 10% of them were randomly taken as origin and destiny nodes, were generated. The rest of the nodes were used as transfer nodes and remain encapsulated within the simulation loop.

For each net, 50 random traffic flow patterns were defined for a vehicle population equal to the number of total nodes in each net. The time range simulated comprised 3600 seconds (1 hour). The increase in the number of vehicles circulating through the system was of the 20% in each case.

Genetic Algorithms were tested using schemes based on binary chromosomes. For each scheme, scenarios with different parameter combinations were defined. The probability of one individual to be selected for mutation was established with 0.10, 0.15 and 0.20 values; the mutation procedure selected was the "Bit String Mutation"; the crossover type selected was "One-Point Crossover"; the population size fixed in 50; and the number of individuals selected in each generation in 5, 10 and 15. The selection of best solutions is made only in the offspring subset.

For their implementation, the optimization process involves a script in the language programming R (R Core Team, 2013), the evaluation of each proposed solution is done by the SUMO simulator (Behrisch et al., 2011) and the optimization and simulation stages correlation by the RSUMO package[3].

Results

Regarding the configurations used to test the OvS with GA, the average variation of travel times seen in each configuration can be observed in Table 1:

Table 1 shows the increase of average mean travel time for all combinations of mutation rate and number of parents selected, after the increase of cars population. Due to the existence of multiple minimums premises, it is expected that the GA works better when the exploration of the solution group is

Table 1. Results of experiments

N° Configuration	Mutation Rate	Number of Parents Selected	Average Increase of "Mean Travel Time"
1	0.05	5	1.22
2	0.10	5	1.14
3	0.15	5	1.14
4	0.20	5	1.14
5	0.05	10	1.17
6	0.10	10	1.13
7	0.15	10	1.07
8	0.20	10	1.14
9	0.05	15	1.15
10	0.10	15	1.11
11	0.15	15	1.06
12	0.20	15	1.08

Figure 19. Increase in the average travel time with four mutation rates

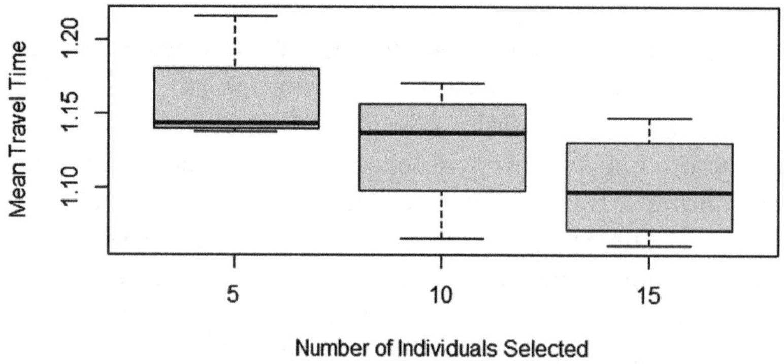

Figure 20. Increase in the average travel time for three selection sizes

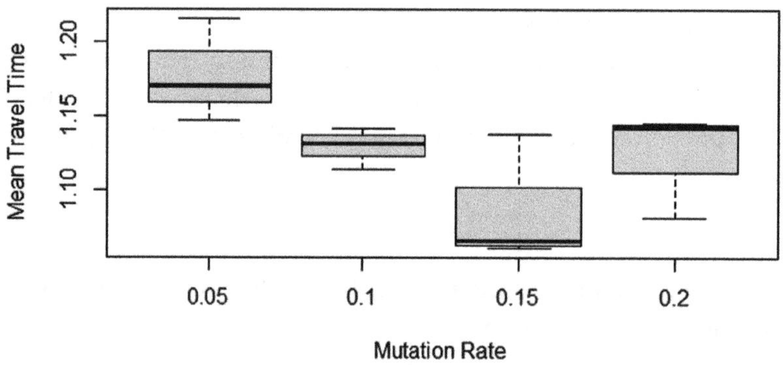

bigger. This can be inferred by contrasting the system performance against different mutation rates and solution group size. In Figure 19, we can analyze the evolution of the reached solutions in relation to the mutation rate, and in Figure 20 in relation to the size of the selection group.

As seen in Table 1, the selection of parameters has a strong impact on the performance of the algorithm. In the case of the tested examples, the solutions range from an increase in travel time of 6% for the best case and 22% for the worst. It can be appreciated how, at the same time that the mutation rate increases, the quality of the solutions found gets better in average, until a certain mutation level is reached, in which the selection effect of the best solutions seems to be lost to generate the following solutions. Regarding this selection, we can see that as allowing more solutions to generate descendants, mean travel times decrease (a linear behavior in our appreciation that a bigger sampling implies a bigger probability of better solutions).

The samples were analyzed with the Kruskal-Wallis test, under the null hypothesis that GA parameters do not affect the quality of the found solutions. With 2 degrees of freedom, the Kruskal-Wallis test yielded a "p-value" of 0.001. It shown that the differences in the results obtained by different configurations of GA were not random variations. Based on this result, we can infer that the framework works best when GA parameters allow further exploration in every generation.

LIMITATIONS

The main limitation of our approach is the high computational cost associated with running the simulations. Any OvS algorithm involves running hundreds or thousands of simulations to assess the quality of potential solutions. This paper proposes an approach to reduce the time required for each simulation, which, while not completely solve the problem, significantly reduces the time required for the execution of the algorithm.

CONCLUSION

In this chapter, an Optimization via Simulation procedure for Origin-Destiny Traffic Assignment Problem was introduced in order to improve the urbanization expansion. In this context, SUMO was used for simulation purpose and a Genetic Algorithm for the optimization part.

The combination of Traffic Simulation and Genetic Algorithms resulted in an interesting approach that merges the power of a good generation of potential solutions and a realistic way to estimate the fitness of these solutions.

Although, a priori, the implementation of OvS for this type of problem was computationally expensive, methodologies were developed, based on knowledge about the simulated system, for significantly reduce the computational cost, without sacrificing quality.

Lastly, the performance of GA may get better as the exploration of space solution increases, instead of focusing in a specific zone.

FUTURE RESEARCH DIRECTIONS

Future research activities are aimed primarily to reducing the computational cost of the algorithm presented, test another meta-heuristics for optimization and assess the optimization model on real scenarios.

The results of tests were very good. But, the algorithm was tested in depth with artificial networks. Is a future objective measure its performance using real traffic nets.

In addition to the use of Genetic Algorithms, we will evaluate the OvS scheme with other meta-heuristics like Simulated Annealing.

ACKNOWLEDGMENT

The work of E. G. Baquela was supported by the Universidad Tecnológica Nacional, under PID TVUTN-SN0003605. Ana C. Olivera thanks to CONICET (www.conicet.gov.ar) for grants RD 1901-14 and PCB-I CONICET-ANII. Authors thanks to Agencia Nacional de Promoción Científica y Técnológica for Grant PICT 2014-0430.

REFERENCES

Baquela, E. G., & Olivera, A. C. (2014, October). *Combining Genetic Algorithms with Traffic Simulations for restructuring traffic networks subject to increases in population.* Paper presented in the International Conference on Metaheuristics and Nature Inspired Computing. Marrakech, Morocco.

Behrisch, M., Bieker, L., Erdmann, J., & Krajzewicz, D. (2011). Sumo - simulation of urban mobility: An overview. In *SIMUL 2011, The Third International Conference on Advances in System Simulation* (pp. 63-68), Barcelona, Spain.

Bhatta, B., Saraswati, S., & Bandyopadhyay, D. (2010). Urban sprawl measurement from remote sensing data. *Applied Geography (Sevenoaks, England)*, *30*(4), 731–740. doi:10.1016/j.apgeog.2010.02.002

Caggiani, L., & Ottomanelli, M. (2011). Traffic equilibrium network design problem under uncertain constraints. *Procedia: Social and Behavioral Sciences*, *20*, 372–380. doi:10.1016/j.sbspro.2011.08.043

Ceylan, H., & Bell, M. (2005). Genetic algorithm solution for the stochastic equilibrium transportation networks under congestion. *Transportation Research Part B: Methodological*, *39*(2), 169–185. doi:10.1016/j.trb.2004.04.001

Current, J., & Marsh, M. (1993). Multiobjective transportation network design and routing problems: Taxonomy and annotation. *European Journal of Operational Research*, *65*(1), 4–19. doi:10.1016/0377-2217(93)90140-I

Current, J., & Min, H. K. (1986). Multiobjective design of transportation networks: Taxonomy and annotation. *European Journal of Operational Research*, *26*(2), 187–201. doi:10.1016/0377-2217(86)90180-3

Dijkstra, E. W. (1959). A note in two problems in connexion with graphs. *Numerische Mathematik*, *1*(1), 269–271. doi:10.1007/BF01386390

Dinu, S. & Bordea, G. (2011). A new genetic approach for transport network design and optimization. *Bulletin of the Polish Academy of Sciences - Technical Sciences, 59*, 263-272.

Ferreira, J. A., & Condessa, B. (2012). Defining expansion areas in small urban settlements an application to the municipality of Tomar (Portugal). *Landscape and Urban Planning*, *107*(3), 283–292. doi:10.1016/j.landurbplan.2012.06.008

Friesz, T. (1981). Multiobjective Optimization in Transportation: The Case of Equilibrium Network Design. *Lecture Notes in Economics and Mathematical Systems*, *190*, 116–127. doi:10.1007/978-3-642-45527-8_10

Friesz, T., Cho, H., Mehta, N., Tobin, R., & Anandalingam, G. (1992). A Simulated Annealing Approach to the Network Design Problem with Variational Inequality Constraints. *Transportation Science*, *26*(1), 18–26. doi:10.1287/trsc.26.1.18

Fu, M. C. (1994). Optimization via Simulation: A Review. *Annals of Operations Research*, *53*(1), 199–248. doi:10.1007/BF02136830

Fu, M. C. (2002). Optimization for simulation: Theory vs. practice. *INFORMS Journal on Computing*, *14*(3), 192–215. doi:10.1287/ijoc.14.3.192.113

Gallo, M., D'Acierno, L., & Montella, B. (2012). A Meta-heuristic Algorithm for Solving the Road Network Design Problem in Regional Contexts. *Procedia: Social and Behavioral Sciences*, *54*, 84–95. doi:10.1016/j.sbspro.2012.09.728

Garcia-Nieto, J., Olivera, A., & Alba, E. (2013). Optimal Cycle Program of Traffic Lights with Particle Swarm Optimization. *IEEE Transactions on Evolutionary Computation*, *17*(6), 823–839. doi:10.1109/TEVC.2013.2260755

Haregeweyn, N., Fikadu, G., Tsunekawa, A., Tsubo, M., & Meshesha, D. T. (2012). The dynamics of urban expansion and its impacts on land use/land cover change and small-scale farmers living near the urban fringe: A case study of Bahir Dar, Ethiopia. *Landscape and Urban Planning*, *106*(2), 149–157. doi:10.1016/j.landurbplan.2012.02.016

He, C., Okada, N., Zhang, Q., Shi, P., & Li, J. (2007). Modelling dynamic urban expansion processes incorporating a potential model with cellular automata. *Landscape and Urban Planning*, *86*(1), 79–91. doi:10.1016/j.landurbplan.2007.12.010

He, C., Tian, J., Shi, P., & Hu, D. (2011). Simulation of the spatial stress due to urban expansion on the wetlands in Beijing, China using a GIS-based assessment model. *Landscape and Urban Planning*, *101*(3), 269–277. doi:10.1016/j.landurbplan.2011.02.032

Horvat, A., & Tošic, A. (2012). Optimization of traffic networks by using genetic algorithms. *Elektrotehniški Vestnik*, *79*, 197–200.

Hu, J., & Wang, C. (2011). Discrete Optimization Via Approximate Annealing Adaptative Search With Stochastic Averaging. In *Proceedings of the 2011 Winter Simulation Conference* (pp. 4206-4216). Phoenix, AZ.

Kalganova, T., Russell, G., & Cumming, A. (1999). Multiple Traffic Signal Control Using A Genetic Algorithm. In Artificial Neural Nets and Genetic Algorithms (pp. 220-228). Roanne, France.

Keshavarz Ghorabaee, M., Amiri, M., & Azimi, P. (in press). Genetic algorithm for solving bi-objective redundancy allocation problem with k-out-of-n subsystems. *Applied Mathematical Modelling*.

Krajzewicz, D., Erdmann, J., Behrisch, M., & Bieker, L. (2012). Recent Development and Applications of "SUMO - Simulation of Urban Mobility". *International Journal On Advances in Systems and Measurements*, *5*(4), 128–138.

Krauss, S. (1998). *Microscopic Modeling of Traffic Flow: Investigation of Collision Free Vehicle Dynamics*. (Unpublished doctoral dissertation). Hauptabteilung Mobilitt und Systemtechnik des DLR. Koln.

Li, Y., & Sun, D. (2012). Microscopic car-following model for the traffic flow: The state of the art. *Journal of Control Theory and Applications*, *10*(2), 133–173. doi:10.1007/s11768-012-9221-z

Li, Y., Zhu, X., Sun, X., & Wang, F. (2010). Landscape effects of environmental impact on bay-area wetlands under rapid urban expansion and development policy: A case study of lianyungang, china. *Landscape and Urban Planning*, *94*(34), 218–227. doi:10.1016/j.landurbplan.2009.10.006

Luke, S. (2013). *Essentials of Metaheuristics*. Retrieved November 10, 2014, from http://cs.gmu.edu/~sean/book/metaheuristics/

Madireddy, M., Medeiros, D. J., & Kumara, S. (2011). An agent based model for evacuation traffic management. In *Proceedings of the 2011 Winter Simulation Conference* (pp. 222-233). doi:10.1109/WSC.2011.6147753

Mahalakshmi, M; Kalaivani, P; Kiruba Nesamalar, E. (2013). A Review on Genetic Algorithm and its Applications. *International Journal of Computing Algorithm, 2*, 415-423.

Miandoabchi, E., & Farahani, R. Z. (2011). Optimizing Reserve Capacity of Urban Road Networks in a Discrete Network Design Problem. *Advances in Engineering Software*, *42*(12), 1041–1050. doi:10.1016/j.advengsoft.2011.07.005

Nagel, K., & Schreckenberg, M. (1992). A cellular automaton model for freeway traffic. *Journal de Physique. I*, *2*(2), 2221–2229. doi:10.1051/jp1:1992277

Ozdemir, Y., Basligil, H., & Sarsenov, B. (2012). A Large Scale Integer Linear Programming to the Daily Fleet Assignment Problem: A Case Study in Turkey. *Procedia: Social and Behavioral Sciences*, *62*, 849–853. doi:10.1016/j.sbspro.2012.09.143

Poelmans, L., & Rompaey, A. V. (2010). Complexity and performance of urban expansion models. *Computers, Environment and Urban Systems*, *34*(1), 17–27. doi:10.1016/j.compenvurbsys.2009.06.001

R Core Team. (2013). *R: A language and environment for statistical computing*. R Foundation for Statistical Computing, Vienna, Austria. Retrieved from http://www.R-project.org/

Romeijn, H., & Smith, L. (1994). Simulated annealing and adaptive search in global optimization. *Probability in the Engineering and Informational Sciences*, *8*(04), 571–590. doi:10.1017/S0269964800003624

Stanilov, K., & Batty, M. (2011). Exploring the historical determinants of urban growth patterns through cellular automata. *Transactions in GIS*, *15*(3), 253–271. doi:10.1111/j.1467-9671.2011.01254.x

Ukkusuri, S. V., Mathew, T. V., & Travis Waller, S. (2007). Robust Transportation Network Design Under Demand Uncertainty. *Computer-Aided Civil and Infrastructure Engineering*, *22*(1), 6–18. doi:10.1111/j.1467-8667.2006.00465.x

Vasant, P. M. (2013). *Meta-Heuristics Optimization Algorithms in Engineering, Business, Economics, and Finance*. IGI Global. doi:10.4018/978-1-4666-2086-5

Xiao, J., Shen, Y., Ge, J., Tateishi, R., Tang, C., Liang, Y., & Huang, Z. (2006). Evaluating urban expansion and land use change in Shijiazhuang, China, by using GIS and remote sensing. *Landscape and Urban Planning*, *75*(12), 69–80. doi:10.1016/j.landurbplan.2004.12.005

Yang, H. & Bell, M.G. (1997). Traffic restraint, road pricing and network equilibrium. *Transportation Research, 31*(B), 303-314.

Yang, H., & Bell, M. G. (1998). Models and algorithms for road network design: A review and some new developments. *Transport Reviews*, *18*(3), 25–278. doi:10.1080/01441649808717016

Zhang, G., Lu, J., & Xiang, Q. (2008). Application of Genetic Algorithm to Network Design Problem. *2008 International Conference on Intelligent Computation Technology and Automation (ICICTA) 1*, (pp. 26-29). doi:10.1109/ICICTA.2008.252

ADDITIONAL READING

Barceló, J. (2010). *Fundamentals of Traffic Simulation - International Series in Operations Research & Management Science (145)*. Springer.

Forrester, J. W. (1969). *Urban Dynamics*. Pegasus Communications.

Fu, M. C. (1994). Optimization via Simulation: A Review. *Annals of Operations Research, 53*(1), 199–248. doi:10.1007/BF02136830

Garvin, A. (1995). *The American City: What Works and What Doesn't* (1st ed.). McGraw-Hill.

Luke, S. (2013). *Essentials of Metaheuristics*, Lulu press. Retrieved November 10, 2014, from http://cs.gmu.edu/~sean/book/metaheuristics/

O'Flaherty, C. A. (1996) - Transport Planning and Traffic Engineering, A Butterworth-Heinemann, 4th Edition, ISBN: 978-0340662793

Treiber, M., & Kesting, A. (2013). *Traffic Flow Dynamics. Data, Models and Simulation*. Springer. doi:10.1007/978-3-642-32460-4

KEY TERMS AND DEFINITIONS

Genetic Algorithm: Canonical optimization method that emulate the evolution and inheritance.

Optimization via Simulation: Optimization framework in which the evaluation of the objective function is made in a simulator.

Traffic Demand: Set of all vehicles in a traffic systems, with their associated routes.

Traffic Destination: Sink of traffic flow in the traffic system.

Traffic Flow: Set of all vehicles going from node A to node B in the same time window.

Traffic Jam: Stop of vehicle's circulation cause by a high density of vehicles in the system or difference of speed between a set of vehicles.

Traffic Net: Oriented graph that represents the physical structure of a traffic system.

Traffic Origin: Source of traffic flow in the traffic system.

Traffic Simulation: Dynamic representation of a traffic system achieved by building a computer model and moving it through time.

Traffic System: An specific traffic net with its traffic demand and its control policies.

ENDNOTES

1 http://maps.google.com

2 http://www.openstreetmaps.com

3 Rsumo can be downloaded from http://www.modelizandosistemas.com.ar/p/rsumo.html

Chapter 11
Hybrid Cuckoo Search Algorithm for Optimal Placement and Sizing of Static VAR Compensator

Khai Phuc Nguyen
Shibaura Institute of Technology, Japan

Dieu Ngoc Vo
Ho Chi Minh University of Technology, Vietnam

Goro Fujita
Shibaura Institute of Technology, Japan

ABSTRACT

This chapter proposes a Hybrid Cuckoo search algorithm to determine optimal location and sizing of Static VAR Compensator (SVC). Hybrid Cuckoo search algorithm is a simple combination of the Cuckoo search algorithm (CSA) and Teaching-learning-based optimization (TLBO), where the learner phase of TLBO is added to improve performance of Cuckoo eggs. The proposed method is applied for optimizing location and sizing of SVC in electric power system. This problem is a kind of discrete and combinatorial problem. The objective function considers loss power, voltage deviation and operational cost of SVC and other operating constraints in power system. Numerical results from three various tested systems show that the proposed method is better than the conventional CSA and TLBO in finding the global optimum solutions and its performance is also high than others.

INTRODUCTION

This chapter presents a hybrid version of the Cuckoo search algorithm (HCSA) and its application for determining optimal placement and sizing of Static VAR compensator (SVC) devices in the electric power system. The Cuckoo search algorithm is a nature-inspired optimization technique, which works on

DOI: 10.4018/978-1-4666-9644-0.ch011

the parasitic behavior of Cuckoo species. The conventional method includes two probability-generating stages. In the first stage, The Lévy flight generates random Cuckoo eggs, and then these eggs will be laid into the neighbors' nests to create new solutions. The better solutions will carry over the next generation. The second stage describes the action of the host birds to abandon Cuckoo eggs in their nests. In this proposed Hybrid Cuckoo search algorithm, we combine the learner stage of the Teaching-learning-based optimization to improve performances of Cuckoo eggs instead of abandoning them. The proposed method is applied for optimizing location and sizing of SVC devices to enhance the voltage profile of the power system. This problem combines discrete and continuous unknowns with many equal and unequal constraints. The discrete unknowns are natural numbers representing the location of installed SVC devices, and the continuous unknowns represent injected reactive power of SVC devices. The proposed method is evaluated on three IEEE standard systems and compared with both of conventional Cuckoo search algorithm and Teaching-learning-based optimization. Numerical results show that the proposed Hybrid Cuckoo search algorithm gives better solutions and has higher performance than other methods.

This chapter divides into seven sections. Section 2 gives the literature review of the Cuckoo search algorithm and the problem of optimal placement and sizing of SVC devices. Section 3 proposes the Hybrid Cuckoo search algorithm. Section 4 describes the structure of a SVC device and the objective function. In Section 5, the implement of Hybrid Cuckoo search algorithm for the problem has been discussed. Section 6 shows numerical results on three benchmark problems. Finally, conclusions are made.

BACKGROUND

Following the development of population-based optimization methods in recent years such as Particle swarm optimization (Kennedy & Ebemart, 1995) or Firefly algorithm (Yang, 2010a, 2010b), Yang and Deb proposed a powerful optimization technique with few controlling parameters; its name is Cuckoo search algorithm (Yang & Deb, 2009). At the first works to develop this method, Yang and Deb show that the Cuckoo search algorithm is better than particle swarm optimization and genetic algorithm in finding optimal solutions for 10 tested functions. After that, they applied the Cuckoo search algorithm for the Spring design optimization and the Welded beam design to proof that their method is favorable for engineering design problems (Yang & Deb, 2010). Then, many improvements and applications of the Cuckoo search were published. On a short survey of E. F. Shair et al, the number of journal papers concerning with the Cuckoo search algorithm increase rapidly from 2010 to 2013 (Shair et al., 2011). For example, E. Valian et al proposed an improved Cuckoo search algorithm and applied it for training neural network (Valian et al., 2011). In the improved Cuckoo search algorithm, they gave a strategy to change the step length α and probability rate p_a. At the beginning of search, p_a is a small value while α is a large one. These controlling parameters were adjusted by the change of iteration. The value of p_a is larger and the value of α is smaller by each generation. With the idea that makes the step size change following the iteration number, S. Walton et al. proposed a Modified CSA, where the step size α is depend of root square of iteration number (Walton et al., 2011). Furthermore, they used the golden ratio to create a new Cuckoo egg, which lay between the two top eggs. Their proposed method is better than the conventional CSA in some evaluated functions. Besides, few researchers tried to use different methods to generate the Cuckoo eggs instead of the Lévy flight. For instance, H. Zheng and Y. Zhou replace the Lévy flight by Gauss distribution (Zheng & Zhou, 2012). Results show that the Gauss distribution Cuckoo search algorithm is better than conventional one in seven benchmarks. Furthermore, modern optimization techniques, P. Civicioglu and E. Besdok made a deep survey to compare the effects of the conventional Cuckoo search algorithm with other three evolutionary methods (Civicioglu & Besdok, 2013). After obtaining 50 mathematical functions, they conducted that differential evolution and the Cuckoo search are quite better than PSO and artificial bee colony algorithm. In addition, A. Ouaarab et al proposed a fraction for smart Cuckoo eggs to improve the standard Cuckoo search algorithm for discrete problems (Ouaarab et al., 2014). They successfully applied it for the symmetric traveling salesman problem. On summary, there are many improvements of the original Cuckoo search algorithm, but no method is clearly more effective than the conventional one. Thus, the requirement to develop a new version of Cuckoo search algorithm increasingly continues.

In addition, in 2011, R.V. Rao et al. developed a new nature-inspired algorithm, its name is Teaching-learning-based algorithm (Rao & Patel, 2012; Rao et al., 2011, 2012). The algorithm is based on behaviors of teachers and students in class. Basing on numerical results, they conducted that the teaching-learning-based algorithm had better performance than other modern optimization techniques, such as genetic algorithm, particle swarm optimization, harmony search, etc. In electric power system field, many researchers successfully applied this method for various problems. For example, P.K. Roy used it to solve short-term hydrothermal scheduling problem with non-smoothing fuel cost functions (Roy, 2013, pp. 10-19). P.K Roy, B. Mandal and S. Bhui also proposed teaching-learning-based algorithm (Mandal & Roy, 2013) for optimal reactive power dispatch and economic emission load dispatch (Roy & Bhui, 2013). In addition, J.A. Martin Garcia and A.J. Gil Mena used a modified version of this method to identify optimal location and size of distributed generators in distribution grid (Martín García & Gil Mena, 2013).

On another hand, Flexible AC transmission system (FACTS) devices provide many benefits to improve the quality of power system control. The Static VAR Compensator, which is a shunt FACTS device, is widely used because of its low cost and good performance. As other FACTS devices, the problem of determining location and sizing of SVC in power system is one of the most important studies. To obtain the improvement of voltage profile when installing FACTS in power system, Y. Del Valle et al. apply PSO to find out size and location to set up a static synchronous compensator (STATCOM) in the Brazilian power system (Del Valle et al.,2006). After connecting the STATCOM into the grid, the voltage deviation is reduced and voltages at four buses are increased up to 0.95pu. To obtain both of technical and economic objectives, Pisica et al. propose a multi-objective function which combines two technical components and the investment cost (Pisica et al., 2009). They apply Genetic algorithm to determine location and size of a SVC. Basing on idea of Pisica et al., Reza Sirjani et al. successfully employ the Improved Harmony Search (IHS) to optimize allocation and size of many SVC devices in power system (Sirjani et al., 2011). The Improved Harmony search is compared with Particle swarm optimization in the IEEE 57-bus system. The numerical result shows that the IHS is slightly better than the PSO. They continue proposing another concept of the Harmony search, which is called global harmony search, and expanding the multi-objective function into four components, where the forth part obtains eigenvalue (Sirjani et al., 2012). Their proposed method is better than both of IHS and PSO in the IEEE 57-bus tested case.

Both of Cuckoo search algorithm and Teaching-learning-based optimization have two different procedures to generate random generations in their process. In this paper, we propose a hybrid version of the Cuckoo search algorithm, where the Cuckoo eggs will learn together to improve themselves as the learner stage of the Teaching-learning-based optimization. We apply it for solving the multi-objective function for optimal SVC devices in the electric power system. This chapter also gives a comparison between the proposed method and two conventional ones. Three systems of IEEE tested cases are obtained to find out the effect of these methods when increasing search space. The first benchmark is the IEEE 30-bus system with five candidate SVC devices. The second case study is the IEEE 57-bus system with six candidate SVC devices and the last case study is the IEEE 118-bus with 10 candidates SVC devices.

PROPOSED HYBRID CUCKOO SEARCH ALGORITHM

Teaching-Learning-Based Optimization

In 2011, R.V. Rao et al. developed the Teaching-learning- based optimization, a kind of population-based method. This method simulates communications between the teacher and learners in a class. A good teacher can transfer his knowledge to learners better than another average-level teacher can. It leads his learners get better marks. On the hand, learners in a class can exchange their knowledge together to improve themselves. Basing on these basic ideas, R.V. Rao et al. proposed the method with two stages in its process of working. The first stage is namely "Teacher phase" and the other is "Learner phase". The advantage of the teaching-learning-based optimization is that it is a parameter-less algorithm. Hence, it is very easy to apply for solving complex problems.

In teacher phase, the recent best solution plays role as a good teacher to move the mean value M_i toward the higher level. A factor, named teaching factor, T_F decides the changes of mean value. The teaching factor can be 1 or 2 and is decided randomly. Follow equations show the probable value of the teaching factor, the change of mean value and updated values for solutions:

$$T_F = round\left(1 + rand()\right) \tag{1}$$

$$D\_mean = rand().\left[M_{best} - T_F.M_i\right] \tag{2}$$

$$X_{new,i} = X_{old,i} + D\_mean \tag{3}$$

where:

- *rand*() is a probability function, returns a random number in the range [0, 1]
- M_{best} is the current best solution
- M_i is the mean value of populations
- $X_{old,i}$, $X_{new,i}$ are the existing and updated solutions, respectively.

In learner phase, a learner selects randomly another one in his class to exchange knowledge. He may learn something new from his friend if the friend is better than he is. Otherwise, he will help his friend improve knowledge of his friend. Figure 1 shows the pseudo code of the learner phase:

Figure 1. Pseudo code of the learner phase

```
for i = 1 to n
        Randomly select another learner Xj, i≠j
        if f(Xi) < f(Xj)
                X_new,i = X_old,i + rand().(X_i − X_j)
        else
                X_new,i = X_old,i + rand().(X_j − X_i)
        end if
end for
```

Cuckoo Search Algorithm

Basing on the parasitic reproduction strategy of Cuckoo species in nature, Cuckoo search algorithm simulates the actions of the female Cuckoo bird to lay her egg into the neighbor's nest. This method also considers the probability that the host bird finds out and abandons the Cuckoo egg. The process of Cuckoo search algorithm includes two separate stages. In the first stage, Cuckoo eggs are created and laid into the host bird's nest. Yang and Deb used the Levy flight to create the Cuckoo eggs. The other stage is the probability of abandonment of Cuckoo eggs. Using the Levy flight for Cuckoo search algorithm is the key contribution of Yang and Deb. The Lévy flight provides a random walk while the random step length is drawn from the Lévy distribution. The Lévy distribution is a continuous probability distribution for non-negative random variable. The formula of the Lévy distribution is below and Figure 2 shows the cumulative of the Lévy distribution with various values of c and μ is zero.

$$f(x;\mu,c)=\sqrt{\frac{c}{2\pi}}\,\frac{e^{-\frac{c}{2(x-\mu)}}}{(x-\mu)^{3/2}} \tag{4}$$

where μ is the location parameter and c is the scale parameter.

However, it is too tricky to generate the step length for the Levy flight. One of good strategies to generate the step length is the Mantegna's equations (Mantegna, 1994, pp. 4677). Following equations formulate the Mantegna's algorithm to generate the step length for Lévy flight:

$$step=\frac{u}{v^{\frac{1}{\beta}}}; \tag{5}$$

$$u=rand().\sigma;\ v=rand() \tag{6}$$

$$\sigma=\left(\frac{\Gamma(1+\beta).\sin\left(\frac{\pi.\beta}{2}\right)}{\Gamma\left(\frac{1+\beta}{2}\right).\beta.2^{\frac{\beta-1}{2}}}\right)^{\frac{1}{\beta}};\beta=\frac{3}{2} \tag{7}$$

where $\Gamma()$ is the gamma function.

Figure 2. Cumulative of the Lévy distribution

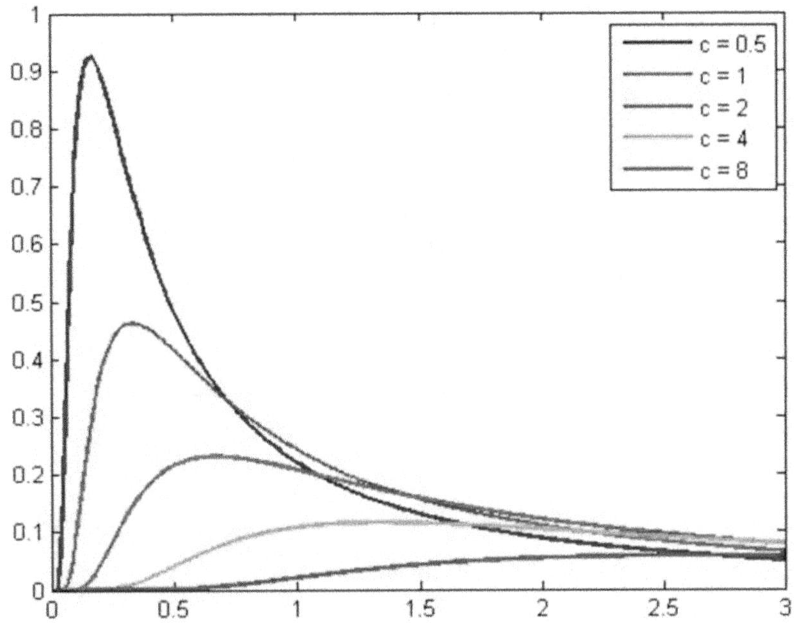

Hybrid Cuckoo Search Algorithm

The Hybrid Cuckoo search algorithm is a simple combination of the standard Cuckoo search algorithm and the Teaching-learning-based optimization. In the discovering alien eggs of the conventional version, new solutions are created from populations by subtracting two random nests. However, in the learner phase of the Teaching-learning-based optimization, the authors proposed a gradient process to generate new solutions, where worse solutions have to follow better ones. Basing on our experiment on this method, the gradient has both of advantages and disadvantages in search engine. With this gradient process, Teaching-learning-based optimization rapidly converges to optimum point and it may give better solutions. Nevertheless, the gradient might easily let the search space to local optimum points. Thus, we propose a hybrid factor p_h to combine the discovering-alien-eggs stage of the Cuckoo search algorithm and the learner phase of the Teaching-learning-based optimization. In this research, we select the hybrid factor p_h basing on our experience. If the hybrid factor p_h is larger, Cuckoo eggs could follow the best solution faster. However, there is risk that Cuckoo eggs may fall into local optima. When the hybrid factor p_h is in range [0.4, 0.8], the proposed method gives better solutions. Figure 3 shows the general pseudo code of the Hybrid Cuckoo search algorithm.

In the first stage, from equations (5) – (7), a random step is created via Lévy flight. The equations (8) and (9) describe the process of laying Cuckoo eggs into nests.

$$newNest_i = Nest_i + rand().\Delta X_i \tag{8}$$

$$\Delta X_i = step\left(Nest_{best} - Nest_i\right) \tag{9}$$

Figure 3. Pseudo code of Hybrid Cuckoo search algorithm

```
Initialize a population of n host nests xi (i = 1,2, …, n);
while (stop criterion) do
        Lay random cuckoo eggs into nests by Lévy flight;
        Evaluate the fitness Fi;
        Replace the host nests by new better ones;
        Randomly create a number ran_A in range [0;1];
        if ran_A < the hybrid factor ph
                Abandon probably alien eggs
        else
                Improve alien eggs by the gradient
        end if
end while
```

where:

- *step* is the flying step generated by Lévy flight in (5)
- $Nest_{best}$ is the best solution.

In the discovering-alien-eggs stage, a new nest is generated randomly from populations. There is a probability rate p_a to discover alien eggs. The new solutions can be found out as following way:

$$newNest_i = Nest_i + K.\Delta X_i^{dis} \tag{10}$$

$$K = \begin{cases} 1, \text{if } rand() < p_a \\ 0, \text{ otherwise} \end{cases} \tag{11}$$

$$\Delta X_i^{dis} = rand()\left[randperm(Nest_i) - randperm(Nest_i)\right] \tag{12}$$

where:

- $randperm(Nest_i)$ is the random perturbation for positions of nests.

In improving-alien-eggs state, a new nest is generated randomly from populations with a comparison of fitness value. Following equations describe this stage:

$$newNest_i = Nest_i + rand().\Delta X_i^{improve} \tag{13}$$

$$\Delta X_i^{improve} = \begin{cases} Nest_i - Nest_j, \text{if } f\left(Nest_i\right) < f\left(Nest_j\right) \\ Nest_j - Nest_i, \text{otherwise} \end{cases} \tag{14}$$

where $f(x)$ is the fitness function.

OPTIMAL PLACEMENT AND SIZING OF SVC DEVICES

Static VAR Compensator

Static VAR Compensator is a type of shunt connected Flexible AC transmission system (FACTS) device. A SVC device includes a combination of Thyristor controlled shunt capacitors and reactors. The most popular configuration for continuously controlled SVC is the combination of fixed capacitor and Thyristor controlled reactor. Figure 4 shows a schematic of the SVC device. The SVC operates in inductive region to absorb the reactive power and operates in capacitive region to supply the reactive power to the system. By injecting or absorbing reactive power, SVC devices are used to control and make the voltage more stable. In real power system, the voltage profile is controlled to operate around the specified value that is 1.0 per unit (pu). Thus, we approximate the nodal voltage magnitude at connected bus is always maintained at 1.0 pu. On summary, a reactive power represents each installed SVC device. The active power can be injected if it is a positive number or absorbed if it is a negative value.

Objectives and Operational Constraints

Objectives

The problem of placement and sizing of SVC is described by a multi-objective function. This function considers both of technical and economic targets to determine optimal location and sizing of SVC devices. With technical approach, the power loss needs to be minimized and the voltage profile has to be improved. However, the investment cost of SVC device is also obtained. The more SVC devices are

Figure 4. Schematic of the SVC device

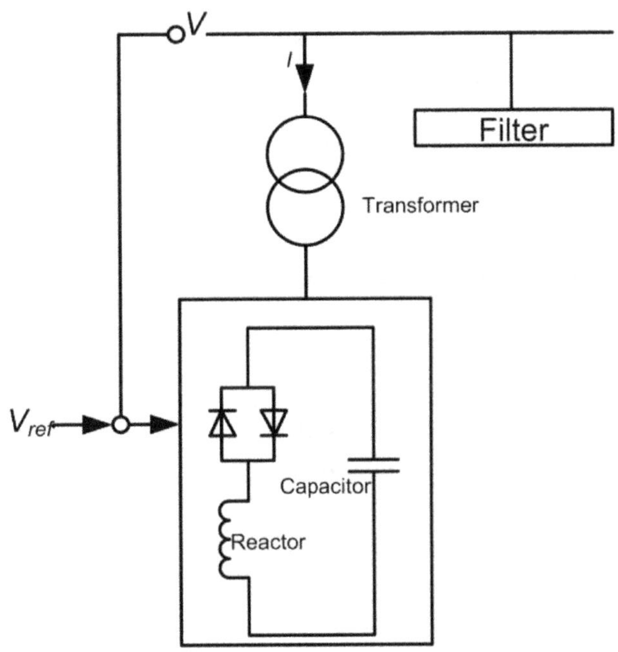

installed, the more voltage profile is enhanced, but it also takes lot of money. In addition, we also consider basic rights in operation power system, such as the balance power constraint, the limitation of bus voltage and the capacitor of SVC devices.

The Active Power Losses

The total power loss in a power system is given in literature as:

$$P_{loss} = \sum_{l=1}^{br} R_l I_l^2 = \sum_{i=1}^{b} \sum_{\substack{j=1 \\ i \neq j}}^{b} \left[V_i^2 + V_j^2 - 2V_i V_j \cos\left(\delta_i - \delta_j\right) \right] Y_{ii} \cos\varphi_{ij} \tag{15}$$

where br and b are the number of lines and buses, respectively; R_l is the resistance of line l^{th}; I_l is the current through line 1th; V_i and δ_i are the magnitude and angle of voltage at the ith bus, respectively; Y_{ij} and φ_{ij} are the magnitude and angle of the line admittance between bus i^{th} and bus j^{th}.

The Voltage Deviation

The voltage deviation is a sum of voltage deviations at all buses in the power system from reference values. The below formula define the voltage deviation objective:

$$\Delta V_{\Sigma} = \sum_{i=1}^{b} \left(\frac{V_{ref,i} - V_i}{V_{ref,i}} \right)^2 \tag{16}$$

where $V_{ref,i}$ is the reference voltage at the ith bus.

The Investment Cost

The investment cost of each SVC device is a quadratic function of reactive power (Cai et al., 2004). Thus, the total investment cost as below:

$$C_{SVC} = \sum_{k=1}^{n} 0.0003Q_k^2 - 0.3051Q_k + 127.38 \tag{17}$$

Where n is the number of installed SVC, Q_k is the reactive of the kth SVC.

Operational Constraints

Optimizing placement and sizing of SVC needs to satisfy all of operational constraints such as the power balance constraint, limitation of bus voltages and limitation of transmission lines.

Power Balance Constraint

As other problems for operation in a power system, the balance of generating and demand powers must be satisfied at each node. Two below equations describe the balance of active and reactive powers in a power system:

$$P_{G,i} - P_{D,i} = V_i \sum_{i=1}^{b} \left[V_j \left[G_{ij} \cos\left(\delta_i - \delta_j\right) + B_{ij} \sin\left(\delta_i - \delta_j\right) \right] \right] \tag{18}$$

$$Q_{G,i} - Q_{D,i} = V_i \sum_{i=1}^{b} \left[V_j \left[G_{ij} \sin\left(\delta_i - \delta_j\right) - B_{ij} \sin\left(\delta_i - \delta_j\right) \right] \right] \tag{19}$$

where $P_{G,i}$ and $Q_{G,i}$ are the active and reactive generating powers at the ith bus; $P_{D,i}$ and $Q_{D,i}$ are the active and reactive of demand powers at the ith bus. G_{ij} and B_{ij} represent the real and imaginary components of element Y'_{ij} of the admittance matrix

Limitation of SVC Devices

Each SVC device only works in a range of reactive power:

$$Q_{i,\min} \leq Q_i \leq Q_{i,\max} \tag{20}$$

Limitation of Bus Voltages

To keep the power system operate in stability and commit power quality, bus voltage at each bus must be maintained around a nominal value.

$$V_{i,\min} \leq V_i \leq V_{i,\max} \tag{21}$$

THE FITNESS FUNCTION AND IMPLEMENTATION

Solution Vector

A solution for this problem is a vector with $2n$ elements; where n is the number of candidate SVC devices. The first n elements are positions of SVC devices. Each element is a natural number that represents the bus number where a SVC device is connected. The other elements are continuing values that represent optimal installed reactive power of SVC devices. Figure 5 shows the structure of a solution vector.

With above structure of solution, it may lead the search engine to duplicated solutions. Table 1 shows a duplicated solution of populations. Two populations actually give the same solution that we need to install SVC at three buses {2, 4 and 7}. Hence, to prevent this case, we proposed another constraint for positions of SVC as $x_1 < x_2 < \ldots < x_n$.

Figure 5. Structure of solution vector

$$\{\underbrace{x_1, x_2 \ldots x_n}_{\text{Positions}}, \underbrace{Q_1^{SVC}, Q_2^{SVC} \ldots Q_N^{SVC}}_{\text{Reactive powers}}\}$$

Table 1. Example of duplicated solutions

Selected buses			Injected reactive powers		
2	4	7	44.95	40.69	23.76
4	2	7	40.69	44.95	23.76

Fitness Function

Because three objectives have various natures, in order to describe all of them in a same mathematical function, we normalize each objective in a comparative manner with the base case (the system without SVC) and connect them together by weights. The equation 22 is the fitness function for this problem. With opinion that technical objectives are more important than economic one, the corresponding weights are set as $\alpha = 0.4$, $\beta = 0.4$, $\eta = 0.2$.

In order to handle operational constraints, we use penalty factors to combine with objective functions. The element *balance_flag* is a factor that equals to 0 if the power balance constraint is not violated and 1 otherwise. With the limits of bus voltages, we use a limited function, $V_i^{\lim}(x)$. The equation 23 describes the limited function. With the constraint for positions, we use a *counter* to find out the number of positions are violated. Through all tested cases, all penalty factors are 100.

$$FF = \alpha \frac{P_{loss}}{P_{loss,base}} + \beta \frac{\Delta V}{\Delta V_{base}} + \eta \frac{C_{SVC}}{C_{\max}} + \xi.balance\_flag + \nu \sum_{i=1}^{b} \left[V_i - V_i^{\lim}(V_i) \right]^2 + \varsigma counter \quad (22)$$

$$V^{\lim}(x) = \begin{cases} x_{\max}, \text{if } x > x_{\max} \\ x, \text{if } x_{\min} \le x \le x_{\max} \\ x_{\min}, \text{if } x < x_{\min} \end{cases} \quad (23)$$

where:

- P_{loss}: active power loss
- ΔV: voltage deviation index
- C_{SVC}: total SVC cost

- $P_{loss,base}$, ΔV_{base} and C_{max} are the total base case active power loss in the network, the total base case voltage deviation and the maximum investment cost, respectively

Limitation of Solution Vector and Initialization

According to the structure of solution vector, the positions of candidate SVC devices cannot be over the number of buses in the power system. Thus, x_{max} is the number of buses and x_{min} is equal to 1. On other hand, the injected reactive power of SVC devices cannot be out of its capacitor in the constraint (20).

Similar to other population-based methods, in the Cuckoo search algorithm, the nests also lay randomly between upper and lower bounds. However, for this problem, the first n elements of nests are natural numbers. Hence, we use the round function *round*(*x*) to return the value *x* to the nearest natural number. Equation (24) and (25) describe the initialization of search space:

$$Nest_i = UpB + rand().\left(UpB - LowB\right) \tag{24}$$

$$Nest_i\left(1:n\right) = round\left(Nest_i(1:n)\right) \tag{25}$$

where:

- $Nest_i$ is the ith nest in populations
- *UpB* and *LowB* are the upper and lower bound vectors.

$$UpB = \left\{x_{\max}, x_{\max}, \ldots, x_{\max}, Q_{\max}, Q_{\max}, \ldots, Q_{\max}\right\}$$

$$LowB = \left\{x_{\min}, x_{\min}, \ldots, x_{\min}, Q_{\min}, Q_{\min}, \ldots, Q_{\min}\right\}$$

Overall Procedure

The overall procedure for the implementation of the Hybrid Cuckoo search algorithm to determine optimal placement and sizing of SVC devices is following. Figure 6 shows the flowchart of this procedure.

Step 1: Choose controlling parameters for the Cuckoo search algorithm. They include the probability of discovering Cuckoo eggs, $p_a = 0.4$, $p_h = 0.5$, the number of nests *NP* and the number of iterations It_{max}. Calculate base values $P_{loss,base}$, ΔV_{base} and C_{max} for the power system without SVC devices.

Step 2: Create randomly initial nests *currentNest*. Round the first *n* elements of nests, which are presented the installed positions of SVC devices.

Step 3: Evaluate value of the fitness function *FF* in equal (22), while using Newton-Raphson method for calculating the power flow.

Step 4: Determine the best value of the fitness function FF_{best} and the best nest $Nest_{best}$. Set the iteration counter *iter* = 1.

Figure 6. Flowchart of implementation of Hybrid Cuckoo search algorithm to determine optimal placement and sizing of SVC devices

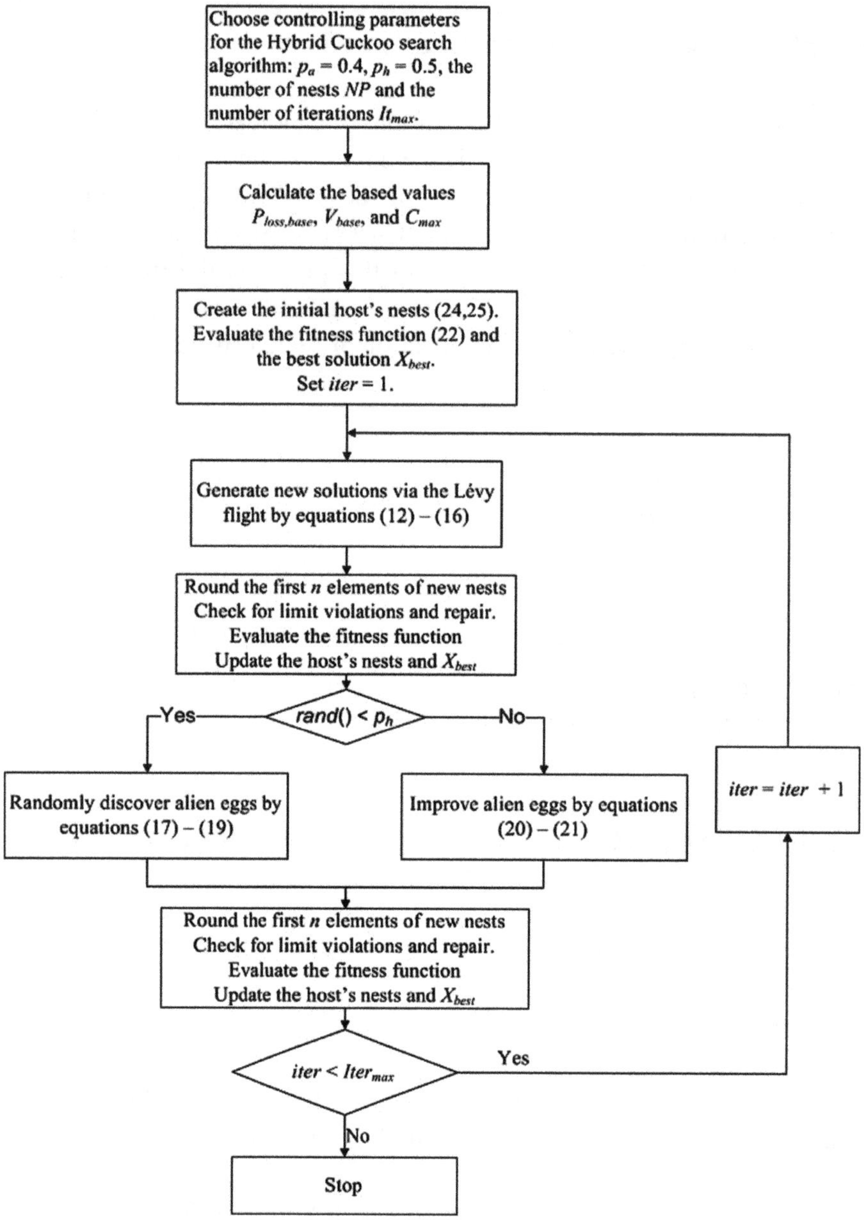

Step 5: Create Cuckoo eggs via Lévy flight and the new nests *newNest* (5) – (9).

Step 6: Round the first *n* elements of new nests *newNest* and modify the eggs that violate the limitations of SVC device constraints and the limitation of bus numbers.

Step 7: Evaluate the fitness function for new nests; we have new values of the fitness function FF_{new}

Step 8: Compare the new values FF_{new} to the current ones *FF* to pick up the better nests. Update the *currentNest*, the best value of fitness function FF_{best} and the best nest $Nest_{best}$.

Step 9: Randomly decide either discovering alien eggs or improving alien eggs, using equation (10) – (14)

Step 10: Round the first *n* elements of new nests *newNest* and modify the eggs that violate the limitations of SVC device constraints and the limitation of bus numbers.

Step 11: Once again, evaluate the fitness function FF_{new} for new nests *newNest*

Step 12: Compare the new values FF_{new} to the current ones *FF* to pick up the better nests. Update the *currentNest*, the best value of fitness function FF_{best} and the best nest $Nest_{best}$.

Step 13: Check if the iteration counter *iter* is lower than the maximum iteration $Iter_{max}$, increase *iter* and return step 5. Otherwise, stop.

NUMERICAL RESULTS

Problem Data

The Hybrid Cuckoo search algorithm has been applied to identify optimal placement and sizing of SVC devices in three various IEEE power systems. The first tested system is the modified IEEE 30-bus system. This system consists of six generators, 41 transmission lines and transformers. It supplies for 189.2 MW load power. Another larger system is also a standard IEEE system with 7 generators, 57 buses and 80 transmission lines-transformers. The last benchmark is the standard IEEE 118-bus system. This system has 54 generators, 118 buses and 186 transmission lines-transformers. All diagrams of case studies are shown in Appendix. The obtained numerical results are compared with the conventional Cuckoo search algorithm and Teaching-learning-based optimization. All applications are coded in Matlab and run in a personal computer with a 3GHz Core 2Duo processor and 4GB RAM. For each method, each benchmark is run 100 independent trials. In order to calculate power flow, we used the Newton-Raphson method by the Matpower toolbox (Zimmerman et al., 2011).

In a same tested case, both of methods have a same size of search space and the number of iterations. Table 2 shows the dimension, size of search space and number of iterations for each benchmark.

Case Study 1: IEEE 30-Bus System

In this tested system, the proposed Hybrid Cuckoo search algorithm gives the best solution and outperform than other methods. Bus data and branch data of this case were shown in Table A.1 and Table A.2 in Appendix. According to numerical results in Table 3, both of the proposed method and the Teaching-learning-based optimization are slightly better than Cuckoo search algorithm in finding the best solution. However, the Hybrid Cuckoo search algorithm is completely higher performance than the Teaching-learning-based optimization with lower average value and lower standard deviation. Table 4

Table 2. Size of search space and number of iterations

	30-Bus Case	57-Bus Case	118-Bus Case
Number of candidate SVC	5	6	10
Dimension of search space	10	12	20
Size of search space	30	50	30
Iteration	500	5000	1000

Table 3. Numerical results of CSA, TLBO and HCSA for IEEE 30-bus system

Fitness Function Value	CSA	TLBO	HCSA
Minimum	1.4562	1.4502	1.4502
Average	1.4801	1.4810	1.4619
Maximum	1.5115	1.5089	1.4985
Standard deviation	0.0112	0.0139	0.0099

Table 4. Best solutions of proposed methods

CSA		TLBO		HCSA	
Selected Bus	Reactive Power	Selected Bus	Reactive Power	Selected Bus	Reactive Power
8	47.6068	8	46.8054	8	46.8054
12	18.4960	12	29.1442	12	29.1442
19	14.5040	19	11.8746	19	11.8746
26	4.6094	26	4.6557	26	4.6557
30	7.3163	30	7.1452	30	7.1452

Figure 7. Voltage profiles of proposed methods

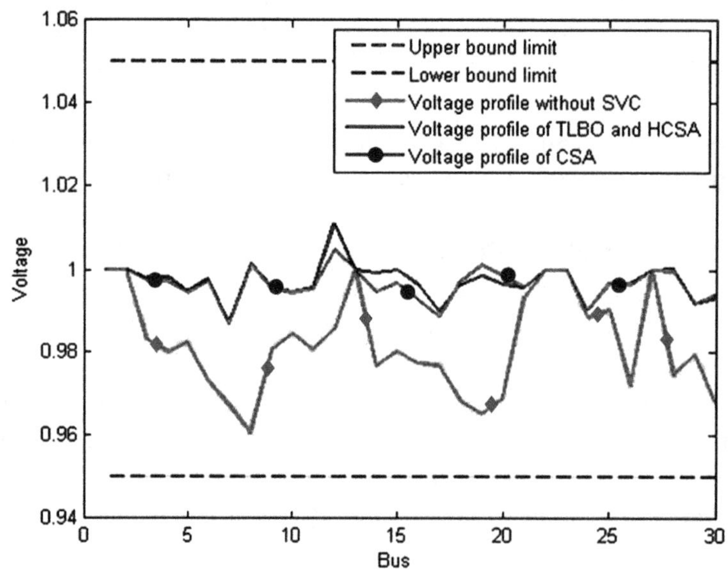

shows the best solutions of two methods. The proposed method and TLBO propose the same solution while the best solution of conventional Cuckoo search algorithm gives different reactive powers. Figure 6 shows the voltage profile of power system for each solution. All methods can handle the constraint of limited bus voltages.

Case Study 2: IEEE 57-Bus System

This benchmark is the IEEE 57-bus system, it includes 57 buses and 80 transmission lines. Bus data and branch data were shown in Table A.3 and Table A.4 in Appendix. Results of this benchmark show that the proposed Hybrid Cuckoo search algorithm is powerful in finding the global optimum solution with high performance. Table 5 shows the Monte Carlo numerical results. The Hybrid Cuckoo search algorithm is clearly better than others. The proposed method gives better solutions, even though its performance is slightly worse than the Cuckoo search algorithm. Detailed optimum solutions of proposed methods are shown in table 6. The installed SVC devices of HCSA are slightly different from given solutions by other methods. According to Figure 8, the Teaching-learning-based optimization reaches to a local optimum and is stuck there. On another hand, the Hybrid Cuckoo search algorithm converges more slowly at the first stage but it can find the global optimum at the end of search process. Figure 9 shows that both of three methods can handle that constraint by different ways. This benchmark is a difficult case study, because in the original network, there are four buses violated the constraint of limited voltage. All of proposed methods suggest to absorb reactive power at the 46th and 49th buses to reduce the over voltages.

Case Study 3: 118-Bus System

The tested system includes 118 buses and 186 transmission lines. Table A.5 and Table A.6 in Appendix show the bus data and branch data of this case. Once again, the proposed Hybrid Cuckoo search algorithm gives better solution than other methods. Detailed best solutions of compared methods are shown in Table 7. Both of the proposed method and the TLBO try to inject reactive power as much as possible while the standard Cuckoo search algorithm absorbs the reactive power at the ninth bus. However, the solution of the HCSA is slightly better than the one of TLBO.

Table 5. Numerical results of various methods for IEEE 57-bus system

Fitness Function Value	CSA	TLBO	HCSA
Minimum	62.5933	62.5924	60.9178
Average	68.1194	83.7811	67.7559
Maximum	73.1696	102.6498	85.7896
Standard deviation	3.1417	10.8205	5.3786

Table 6. Numerical results of various methods for IEEE 57-bus system

CSA		TLBO		HCSA	
Selected Bus	Reactive Power	Selected Bus	Reactive Power	Selected Bus	Reactive Power
11	48.71567	11	48.5833	11	50
20	7.7468	20	7.7580	20	7.4729
31	5.6677	31	5.6375	26	15.6464
36	29.4846	36	29.4992	34	26.1206
46	-50	46	-49.9887	46	-50
49	-45.7687	49	-45.6252	49	-49.0057

Figure 8. Comparison about convergences of proposed methods in the IEEE 57-bus case study

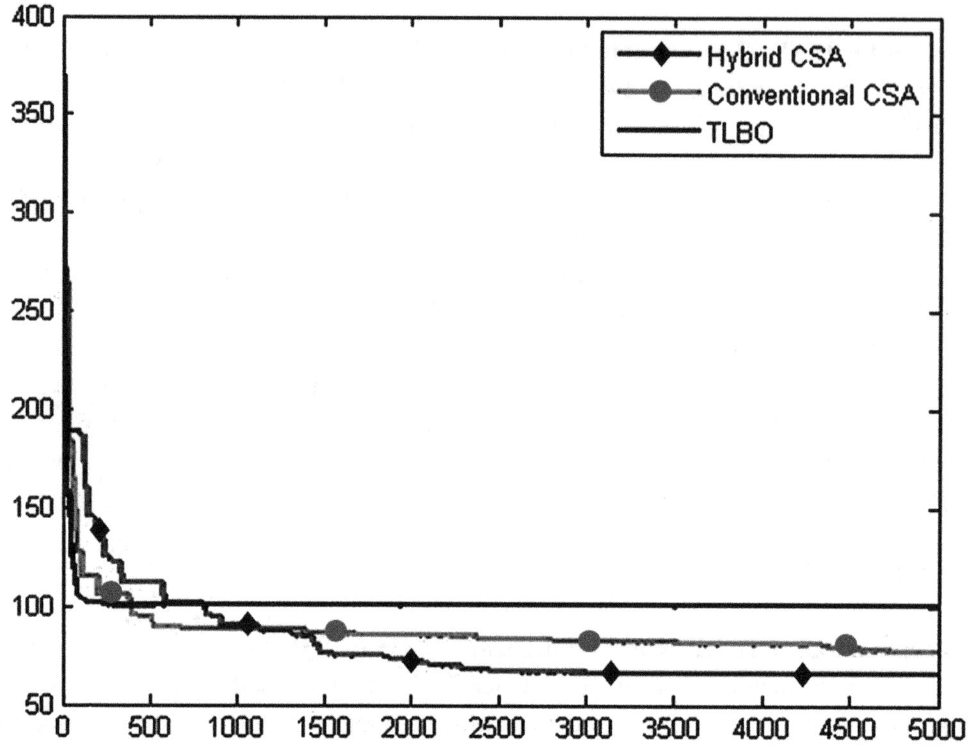

Figure 9. Voltage profile of proposed methods

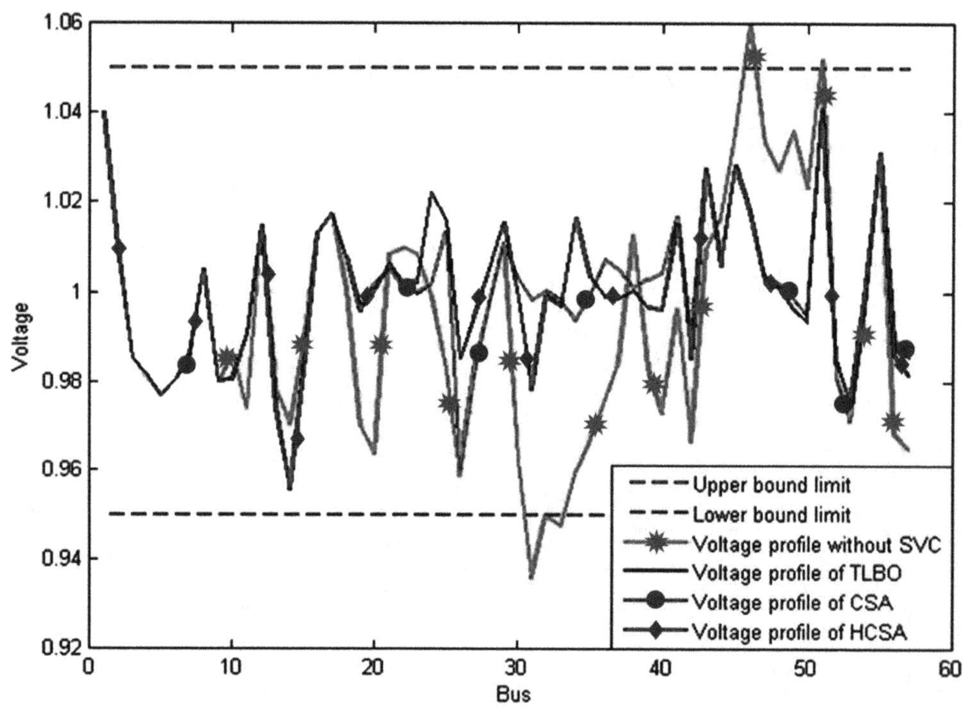

Table 7. Detailed optimum solutions of compared methods

Index	CSA		TLBO		HCSA	
	Selected Bus	Reactive Power	Selected Bus	Reactive Power	Selected Bus	Reactive Power
1	9	-6.7565	2	50	2	50
2	20	30.0106	13	50	20	50
3	29	44.7030	14	32.4255	28	50
4	33	43.0173	20	50	53	49.4629
5	38	12.0283	28	50	58	49.7824
6	57	2.0783	39	50	75	50
7	58	49.5730	52	50	84	41.3998
8	82	33.9050	109	50	95	49.3883
9	102	29.0896	115	50	108	50
10	114	24.9869	118	50	115	50
Minimum fitness value	31.7731		23.9944		23.9325	

FUTURE RESEARCH DIRECTIONS

The proposed Hybrid Cuckoo search algorithm is a simple combination of the conventional Cuckoo search algorithm and the Teaching-learning-based optimization. With the problem of optimizing placement and sizing of SVC devices in power systems, the proposed Hybrid Cuckoo search is more effective than the conventional CSA and TLBO. Therefore, this method can be implemented for other problems that combine discrete and continuous unknowns. Moreover, the proposed method should be continued employing to other non-convex problems to obtain its effects.

CONCLUSION

The Hybrid Cuckoo search algorithm is totally powerful and effective in finding global optimum solution for the problem of determining location and size of SVC devices. The proposed method is a simple combination of the conventional Cuckoo search algorithm and the Teaching-learning-based optimization. The learner phase of TLBO is randomly selected to improve the performance of alien eggs in Cuckoo search algorithm. According to three benchmarks, the proposed Hybrid Cuckoo search algorithm always gives the best solutions with high performance. Comparing with the conventional Cuckoo search algorithm, this method is more effective in finding global solutions. Comparing with the Teaching-learning-based optimization, this method outperform in all of case studies. Furthermore, This problem is a complex one which combines continuous and discrete numbers with many equal and unequal constraints. Thus, the proposed method is also favorable to optimize for discrete and combinational problems.

REFERENCES

Cai, L. J., Erlich, I., & Stamtsis, G. (2004, October). Optimal choice and allocation of FACTS devices in deregulated electricity market using genetic algorithms. In Power Systems Conference and Exposition, 2004. IEEE PES (pp. 201-207). IEEE. doi:10.1109/PSCE.2004.1397562

Civicioglu, P., & Besdok, E. (2013). A conceptual comparison of the Cuckoo-search, particle swarm optimization, differential evolution and artificial bee colony algorithms. *Artificial Intelligence Review*, *39*(4), 315–346. doi:10.1007/s10462-011-9276-0

Del Valle, Y., Hernandez, J. C., Venayagamoorthy, G. K., & Harley, R. G. (2006, August). Optimal STATCOM Sizing and Placement Using Particle Swarn Optimization. In Transmission & Distribution Conference and Exposition: Latin America, 2006. TDC'06. IEEE/PES (pp. 1-6). IEEE.

Kennedy, J., & Ebemart, R. C. (1995). Particle swarm optimization. In *Proceedings of IEEE International Conference*. IEEE.

Mandal, B., & Roy, P. K. (2013). Optimal reactive power dispatch using quasi-oppositional teaching learning based optimization. *International Journal of Electrical Power & Energy Systems*, *53*, 123–134. doi:10.1016/j.ijepes.2013.04.011

Mantegna, R. N. (1994). Fast, accurate algorithm for numerical simulation of Levy stable stochastic processes. *Physical Review E: Statistical Physics, Plasmas, Fluids, and Related Interdisciplinary Topics*, *49*(5), 4677–4683. doi:10.1103/PhysRevE.49.4677 PMID:9961762

Martín García, J. A., & Gil Mena, A. J. (2013). Optimal distributed generation location and size using a modified teaching–learning based optimization algorithm. *International Journal of Electrical Power & Energy Systems*, *50*, 65–75. doi:10.1016/j.ijepes.2013.02.023

Ouaarab, A., Ahiod, B., & Yang, X. S. (2014). Discrete cuckoo search algorithm for the travelling salesman problem. *Neural Computing & Applications*, *24*(7-8), 1659–1669. doi:10.1007/s00521-013-1402-2

Pisica, I., Bulac, C., Toma, L., & Eremia, M. (2009, June). Optimal SVC placement in electric power systems using a genetic algorithms based method. In PowerTech, 2009 IEEE Bucharest (pp. 1-6). IEEE. doi:10.1109/PTC.2009.5281841

Rao, R., & Patel, V. (2012). An elitist teaching-learning-based optimization algorithm for solving complex constrained optimization problems. *International Journal of Industrial Engineering Computations*, *3*(4), 535–560. doi:10.5267/j.ijiec.2012.03.007

Rao, R. V., Savsani, V. J., & Vakharia, D. P. (2011). Teaching–learning-based optimization: A novel method for constrained mechanical design optimization problems. *Computer Aided Design*, *43*(3), 303–315. doi:10.1016/j.cad.2010.12.015

Rao, R. V., Savsani, V. J., & Vakharia, D. P. (2012). Teaching–learning-based optimization: An optimization method for continuous non-linear large scale problems. *Information Sciences*, *183*(1), 1–15. doi:10.1016/j.ins.2011.08.006

Roy, P. K. (2013). Teaching learning based optimization for short-term hydrothermal scheduling problem considering valve point effect and prohibited discharge constraint. *International Journal of Electrical Power & Energy Systems*, *53*, 10–19. doi:10.1016/j.ijepes.2013.03.024

Roy, P. K., & Bhui, S. (2013). Multi-objective quasi-oppositional teaching learning based optimization for economic emission load dispatch problem. *International Journal of Electrical Power & Energy Systems*, *53*, 937–948. doi:10.1016/j.ijepes.2013.06.015

Shair, E. F., Khor, S. Y., Abdullah, A. R., Jaafar, H. I., Ali, N. M., & Abidin, A. Z. (2014). A Brief Review of Cuckoo Search Algorithm (CSA) Research Progression from 2010 to 2013. *Romania*, *1*, 1–4.

Sirjani, R., Mohamed, A., & Shareef, H. (2011). Optimal placement and sizing of Static Var Compensators in power systems using Improved Harmony Search Algorithm. *Przeglad Elektrotechniczny*, *87*(7), 214–218.

Sirjani, R., Mohamed, A., & Shareef, H. (2012). Optimal allocation of shunt Var compensators in power systems using a novel global harmony search algorithm. *International Journal of Electrical Power & Energy Systems*, *43*(1), 562–572. doi:10.1016/j.ijepes.2012.05.068

Valian, E., Mohanna, S., & Tavakoli, S. (2011). Improved cuckoo search algorithm for feedforward neural network training. *International Journal of Artificial Intelligence & Applications*, *2*(3), 36–43. doi:10.5121/ijaia.2011.2304

Walton, S., Hassan, O., Morgan, K., & Brown, M. R. (2011). Modified cuckoo search: A new gradient free optimisation algorithm. *Chaos, Solitons, and Fractals*, *44*(9), 710–718. doi:10.1016/j.chaos.2011.06.004

Yang, X. S. (2010a). Firefly algorithm. *Engineering Optimization*, 221–230.

Yang, X. S. (2010b). Firefly algorithm, Levy flights and global optimization. In *Research and Development in Intelligent Systems XXVI* (pp. 209–218). Springer London. doi:10.1007/978-1-84882-983-1_15

Yang, X. S., & Deb, S. (2009, December). Cuckoo search via Lévy flights. In *Nature & Biologically Inspired Computing, 2009. NaBIC 2009. World Congress on* (pp. 210-214). IEEE.

Yang, X. S., & Deb, S. (2010). Engineering optimisation by cuckoo search. *International Journal of Mathematical Modelling and Numerical Optimisation*, *1*(4), 330–343. doi:10.1504/IJMMNO.2010.035430

Zheng, H., & Zhou, Y. (2012). A novel cuckoo search optimization algorithm based on Gauss distribution. *Journal of Computer Information Systems*, *8*(10), 4193–4200.

Zimmerman, R. D., Murillo-Sánchez, C. E., & Thomas, R. J. (2011). MATPOWER: Steady-state operations, planning, and analysis tools for power systems research and education. *Power Systems. IEEE Transactions on*, *26*(1), 12–19.

ADDITIONAL READING

Abdel-Moamen, M. A., & Padhy, N. P. (2003, September). Optimal power flow incorporating FACTS devices-bibliography and survey. In Transmission and Distribution Conference and Exposition, 2003 IEEE PES (Vol. 2, pp. 669-676). IEEE. doi:10.1109/TDC.2003.1335357

Abido, M. A. (2009). Power system stability enhancement using FACTS controllers: A review. *The Arabian journal for science and engineering, 34*(1B), 153-172.

AlRashidi, M. R., & El-Hawary, M. E. (2009). Applications of computational intelligence techniques for solving the revived optimal power flow problem. *Electric Power Systems Research, 79*(4), 694–702. doi:10.1016/j.epsr.2008.10.004

Arabkhaburi, D., Kazemi, A., Yari, M., & Aghaei, J. (2006, December). Optimal placement of UPFC in power systems using genetic algorithm. In Industrial Technology, 2006. ICIT 2006. IEEE International Conference on (pp. 1694-1699). IEEE. doi:10.1109/ICIT.2006.372474

Atwa, Y. M., & El-Saadany, E. F. (2010). Optimal allocation of ESS in distribution systems with a high penetration of wind energy. *Power Systems. IEEE Transactions on, 25*(4), 1815–1822.

Atwa, Y. M., & El-Saadany, E. F. (2011). Probabilistic approach for optimal allocation of wind-based distributed generation in distribution systems. *IET Renewable Power Generation, 5*(1), 79–88. doi:10.1049/iet-rpg.2009.0011

Blum, C., Puchinger, J., Raidl, G. R., & Roli, A. (2011). Hybrid metaheuristics in combinatorial optimization: A survey. *Applied Soft Computing, 11*(6), 4135–4151. doi:10.1016/j.asoc.2011.02.032

Chen, C., Duan, S., Cai, T., Liu, B., & Hu, G. (2011). Optimal allocation and economic analysis of energy storage system in microgrids. *Power Electronics. IEEE Transactions on, 26*(10), 2762–2773.

Das, D., Prasai, A., Harley, R. G., & Divan, D. (2009, September). Optimal placement of distributed FACTS devices in power networks using particle swarm optimization. In Energy Conversion Congress and Exposition, 2009. ECCE 2009. IEEE (pp. 527-534). IEEE. doi:10.1109/ECCE.2009.5316292

Edris, A. A. (1997). Proposed terms and definitions for flexible AC transmission system (FACTS). *IEEE Transactions on Power Delivery, 12*(4).

Elbeltagi, E., Hegazy, T., & Grierson, D. (2005). Comparison among five evolutionary-based optimization algorithms. *Advanced Engineering Informatics, 19*(1), 43–53. doi:10.1016/j.aei.2005.01.004

Elhadidy, M. A., & Shaahid, S. M. (1999). Optimal sizing of battery storage for hybrid (wind+ diesel) power systems. *Renewable Energy, 18*(1), 77–86. doi:10.1016/S0960-1481(98)00796-4

Elmitwally, A. (2011). Optimal sizing and location of capacitor banks in distorted distribution networks with distributed wind generation. In *21st International Conference on Electricity Distribution* (pp. 1-4).

Farhoodnea, M., Mohamed, A., Shareef, H., & Zayandehroodi, H. (2013, June). Optimum D-STATCOM placement using firefly algorithm for power quality enhancement. In Power Engineering and Optimization Conference (PEOCO), 2013 IEEE 7th International (pp. 98-102). IEEE. doi:10.1109/PEOCO.2013.6564523

Gerbex, S., Cherkaoui, R., & Germond, A. J. (2001). Optimal location of multi-type FACTS devices in a power system by means of genetic algorithms. *Power Systems. IEEE Transactions on*, *16*(3), 537–544.

Grainger, J. J., & Stevenson, W. D. (1994). *Power system analysis* (Vol. 31). New York: McGraw-Hill.

Gyugyi, L., Schauder, C. D., Williams, S. L., Rietman, T. R., Torgerson, D. R., & Edris, A. (1995). The unified power flow controller: A new approach to power transmission control. *Power Delivery. IEEE Transactions on*, *10*(2), 1085–1097.

Idris, R. M., Khairuddin, A., & Mustafa, M. W. (2010). Optimal allocation of FACTS devices in deregulated electricity market using bees algorithm. *WSEAS Transactions on Power Systems*, *5*(2), 108–119.

Jordehi, A. R., & Jasni, J. (2012, June). Approaches for FACTS optimization problem in power systems. In *Power Engineering and Optimization Conference (PEDCO) Melaka, Malaysia, 2012 Ieee International* (pp. 355-360). IEEE. doi:10.1109/PEOCO.2012.6230889

Jordehi, A. R., Jasni, J., Wahab, N. A., & Kadir, M. Z. A. A. (2013, June). Particle swarm optimisation applications in FACTS optimisation problem. In Power Engineering and Optimization Conference (PEOCO), 2013 IEEE 7th International (pp. 193-198). IEEE.

Jumaat, S. A., Musirin, I., Othman, M. M., & Mokhlis, H. (2011, December). Optimal location and sizing of SVC using Particle Swarm Optimization technique. In Informatics and Computational Intelligence (ICI), 2011 First International Conference on (pp. 312-317). IEEE. doi:10.1109/ICI.2011.58

Kazemi, A., & Andami, H. (2004, April). FACTS devices in deregulated electric power systems: a review. In *Electric Utility Deregulation, Restructuring and Power Technologies, 2004.(DRPT 2004). Proceedings of the 2004 IEEE International Conference on* (Vol. 1, pp. 337-342). IEEE. doi:10.1109/DRPT.2004.1338518

Konishi, R., & Takahashi, M. (2014, June). Optimal allocation of photovoltaic systems and energy storages in power systems considering power shortage and surplus. In *Electric Power Quality and Supply Reliability Conference (PQ)* (pp. 127-132). IEEE. doi:10.1109/PQ.2014.6866796

Mínguez, R., Milano, F., Zarate-Miano, R., & Conejo, A. J. (2007). Optimal network placement of SVC devices. *Power Systems. IEEE Transactions on*, *22*(4), 1851–1860.

Mohandas, N., Balamurugan, R., & Lakshminarasimman, L. (2015). Optimal location and sizing of real power DG units to improve the voltage stability in the distribution system using ABC algorithm united with chaos. *International Journal of Electrical Power & Energy Systems*, *66*, 41–52. doi:10.1016/j.ijepes.2014.10.033

Ongsakul, W., & Jirapong, P. (2005, May). Optimal allocation of FACTS devices to enhance total transfer capability using evolutionary programming. In*Circuits and Systems, 2005. ISCAS 2005. IEEE International Symposium on*(pp. 4175-4178). IEEE. doi:10.1109/ISCAS.2005.1465551

Ouaarab, A., Ahiod, B., & Yang, X. S. (2014). Discrete cuckoo search algorithm for the travelling salesman problem. *Neural Computing & Applications*, *24*(7-8), 1659–1669. doi:10.1007/s00521-013-1402-2

Phadke, A. R., Fozdar, M., & Niazi, K. R. (2009, December). Multi-objective fuzzy-GA formulation for optimal placement and sizing of shunt FACTS controller. In Power Systems, 2009. ICPS'09. International Conference on (pp. 1-6). IEEE. doi:10.1109/ICPWS.2009.5442652

Phadke, A. R., Fozdar, M., & Niazi, K. R. (2012). A new multi-objective fuzzy-GA formulation for optimal placement and sizing of shunt FACTS controller. *International Journal of Electrical Power & Energy Systems*, *40*(1), 46–53. doi:10.1016/j.ijepes.2012.02.004

Rahimzadeh, S., & Bina, M. T. (2011). Looking for optimal number and placement of FACTS devices to manage the transmission congestion. *Energy Conversion and Management*, *52*(1), 437–446. doi:10.1016/j.enconman.2010.07.019

Saravanan, M., Slochanal, S. M. R., Venkatesh, P., & Abraham, J. P. S. (2007). Application of particle swarm optimization technique for optimal location of FACTS devices considering cost of installation and system loadability. *Electric Power Systems Research*, *77*(3), 276–283. doi:10.1016/j.epsr.2006.03.006

Sode-Yome, A., Mithulananthan, N., & Lee, K. Y. (2007, June). A comprehensive comparison of FACTS devices for enhancing static voltage stability. In *Power Engineering Society General Meeting*, 2007. IEEE (pp. 1-8). IEEE. doi:10.1109/PES.2007.385687

Tibin, J., Sini, X., Chitra, S., & Cherian, V. I. (2011). PSO Based Optimal Placement and Setting of FACTS Devices for Improving the Performance of Power Distribution System. Bonfring International Journal of Power Systems and Integrated Circuits, 1(Inaugural Special Issue), 60-64.

Varshney, S., Srivastava, L., & Pandit, M. (2011, December). Optimal location and sizing of STATCOM for voltage security enhancement using PSO-TVAC. In Power and Energy Systems (ICPS), 2011 International Conference on (pp. 1-6). IEEE. doi:10.1109/ICPES.2011.6156646

Vasant, P. (2015). *Handbook of Research on Artificial Intelligence Techniques and Algorithms* (pp. 1–873). Hershey, PA: IGI Global; doi:10.4018/978-1-4666-7258-1

Vasant, P., Barsoum, N., & Webb, J. (2012). *Innovation in Power, Control, and Optimization: Emerging Energy Technologies* (pp. 1–396). Hershey, PA: IGI Global; doi:10.4018/978-1-61350-138-2

Vasant, P. M. (2013). *Meta-Heuristics Optimization Algorithms in Engineering, Business, Economics, and Finance* (pp. 1–734). Hershey, PA: IGI Global; doi:10.4018/978-1-4666-2086-5

Vasant, P. M. (2014). *Handbook of Research on Novel Soft Computing Intelligent Algorithms: Theory and Practical Applications* (pp. 1–1018). Hershey, PA: IGI Global; doi:10.4018/978-1-4666-4450-2

Wetter, M., & Wright, J. (2004). A comparison of deterministic and probabilistic optimization algorithms for nonsmooth simulation-based optimization. *Building and Environment*, *39*(8), 989–999. doi:10.1016/j.buildenv.2004.01.022

Yang, H., Lu, L., & Zhou, W. (2007). A novel optimization sizing model for hybrid solar-wind power generation system. *Solar Energy*, *81*(1), 76–84. doi:10.1016/j.solener.2006.06.010

KEY TERMS AND DEFINITIONS

Cuckoo Search Algorithm: An optimization technique developed by Yang and Deb in 2009. This method inspires parasitic behaviors of Cuckoo species.

Flexible AC Transmission System (FACTS): A system concludes power electronic and other static elements to enhance quality and efficiency of the AC transmission system.

Power Flow: A numerical analysis of the electric power flow in steady-state operation. Power flow study determines voltage angles and magnitudes at buses in power system for specified load demand, real power and voltage magnitude of generators.

Power Loss: The amount of power lost in transmission lines. Resistive elements in transmission lines consume most of power loss in a power system.

Static VAR Compensator: A type of shunt connected FACTS. Static VAR Compensator includes many capacitors or reactors and it is often controlled by thyristors.

Teaching-Learning-Based Optimization: An evolutionary algorithm, which inspires behaviors of teachers and learners in class. Learners get knowledge from teachers and discuss together to improve their marks.

Voltage Deviation: A sum of voltage deviations at all buses in the power system from reference values. It is an important index in operating the electric power system.

ABBREVIATION

CSA: Cuckoo search algorithm
FACTS: Flexible AC transmission system
HCSA: Hybrid Cuckoo search algorithm
IEEE: Institute of Electrical and Electronics Engineering
IHS: Improved Harmony Search
PSO: Particle swarm optimization
STATCOM: Static synchronous compensator
SVC: Static VAR compensator
TLBO: Teaching-learning-based optimization

NOMENCLATURE

Γ(): the gamma function
ΔV: voltage deviation index
ΔV_{base}: the total voltage deviation in the base case
br, *b*: the number of lines and buses, respectively;
C_{SVC}: total SVC cost
C_{max}: the maximum investment cost of a SVC device
G_{ij}, B_{ij}: the real and imaginary components of element Y'_{ij} of the admittance matrix
I_l: the current through line l^{th};
n: the number of installed SVC,
$Nest_{best}$: the best solution
$Nest_i$: the ith nest in populations
p_h: the hybrid factor
$P_{G,i}$, $Q_{G,i}$: the active and reactive generating powers at the ith bus;
$P_{D,i}$, $Q_{D,i}$: the active and reactive of demand powers at the ith bus.
P_{loss}: active power loss
$P_{loss,base}$: the total base case active power loss in the base case
p_a: the probability of discovering Cuckoo eggs
Q_k: the reactive of the kth SVC
R_l: the resistance of line l^{th};
step: the step length for Lévy flight
UpB, *LowB*: the upper and lower bound vectors
V_i, δ_i: the magnitude and angle of voltage at the ith bus, respectively;
$V_{ref,i}$ is the reference voltage at the ith bus
Y_{ij}, φ_{ij}: the magnitude and angle of the line admittance between bus i^{th} and bus j^{th}

APPENDIX

Table 8. Bus data of the IEEE 30-bus system

Bus Number	Bus Type	Voltage (pu)	Angle	Real Power Load (MW)	Reactive Power Load (MVAr)	Real Power Generator (MW)	Reactive Power Generator (MVAr)	Max Q (MVAr)	Min Q (MVAr)	Bus Number
1	3	1.06	0	0	0	260.2	-16.1	132	0	0
2	2	1.043	-5.48	21.7	12.7	40	50	132	50	-40
3	0	1.021	-7.96	2.4	1.2	0	0	132	0	0
4	0	1.012	-9.62	7.6	1.6	0	0	132	0	0
5	2	1.01	-14.37	94.2	19	0	37	132	40	-40
6	0	1.01	-11.34	0	0	0	0	132	0	0
7	0	1.002	-13.12	22.8	10.9	0	0	132	0	0
8	2	1.01	-12.1	30	30	0	37.3	132	40	-10
9	0	1.051	-14.38	0	0	0	0	1	0	0
10	0	1.045	-15.97	5.8	2	0	0	33	0	0
11	2	1.082	-14.39	0	0	0	16.2	11	24	-6
12	0	1.057	-15.24	11.2	7.5	0	0	33	0	0
13	2	1.071	-15.24	0	0	0	10.6	11	24	-6
14	0	1.042	-16.13	6.2	1.6	0	0	33	0	0
15	0	1.038	-16.22	8.2	2.5	0	0	33	0	0
16	0	1.045	-15.83	3.5	1.8	0	0	33	0	0
17	0	1.04	-16.14	9	5.8	0	0	33	0	0
18	0	1.028	-16.82	3.2	0.9	0	0	33	0	0
19	0	1.026	-17	9.5	3.4	0	0	33	0	0
20	0	1.03	-16.8	2.2	0.7	0	0	33	0	0
21	0	1.033	-16.42	17.5	11.2	0	0	33	0	0
22	0	1.033	-16.41	0	0	0	0	33	0	0
23	0	1.027	-16.61	3.2	1.6	0	0	33	0	0
24	0	1.021	-16.78	8.7	6.7	0	0	33	0	0
25	0	1.017	-16.35	0	0	0	0	33	0	0
26	0	1	-16.77	3.5	2.3	0	0	33	0	0
27	1.023	-15.82	0	0	0	0	33	0	0	0
28	-11.97	0	0	0	0	132	0	0	0	0
29	0	1.003	-17.06	2.4	0.9	0	0	33	0	0
30	0	0.992	-17.94	10.6	1.9	0	0	33	0	0

Table 9. Branch data of the IEEE 30-bus system

No. Line	From Bus	To Bus	R (pu)	X (pu)	Charging B (pu)	Transformer's tap ratio
1	1	2	0.0192	0.0575	0.0528	0
2	1	3	0.0452	0.1652	0.0408	0
3	2	4	0.057	0.1737	0.0368	0
4	3	4	0.0132	0.0379	0.0084	0
5	2	5	0.0472	0.1983	0.0418	0
6	2	6	0.0581	0.1763	0.0374	0
7	4	6	0.0119	0.0414	0.009	0
8	5	7	0.046	0.116	0.0204	0
9	6	7	0.0267	0.082	0.017	0
10	6	8	0.012	0.042	0.009	0
11	6	9	0	0.208	0	0.978
12	6	10	0	0.556	0	0.969
13	9	11	0	0.208	0	0
14	9	10	0	0.11	0	0
15	4	12	0	0.256	0	0.932
16	12	13	0	0.14	0	0
17	12	14	0.1231	0.2559	0	0
18	12	15	0.0662	0.1304	0	0
19	12	16	0.0945	0.1987	0	0
20	14	15	0.221	0.1997	0	0
21	16	17	0.0524	0.1923	0	0
22	15	18	0.1073	0.2185	0	0
23	18	19	0.0639	0.1292	0	0
24	19	20	0.034	0.068	0	0
25	10	20	0.0936	0.209	0	0
26	10	17	0.0324	0.0845	0	0
27	10	21	0.0348	0.0749	0	0
28	10	22	0.0727	0.1499	0	0
29	21	22	0.0116	0.0236	0	0
30	15	23	0.1	0.202	0	0
31	22	24	0.115	0.179	0	0
32	23	24	0.132	0.27	0	0
33	24	25	0.1885	0.3292	0	0
34	25	26	0.2544	0.38	0	0
35	25	27	0.1093	0.2087	0	0
36	28	27	0	0.396	0	0.968
37	27	29	0.2198	0.4153	0	0
38	27	30	0.3202	0.6027	0	0
39	29	30	0.2399	0.4533	0	0
40	8	28	0.0636	0.2	0.0428	0
41	6	28	0.0169	0.0599	0.013	0

Table 10. Bus data of the IEEE 57-bus system

Bus Number	Bus Type	Voltage (pu)	Angle	Real Power Load (MW)	Reactive Power Load (MVAr)	Real Power Generator (MW)	Reactive Power Generator (MVAr)	Max Q (MVAr)	Min Q (MVAr)
1	3	1.04	0	55	17	128.9	-16.1	0	0
2	2	1.01	-1.18	3	88	0	-0.8	50	-17
3	2	0.985	-5.97	41	21	40	-1	60	-10
4	0	0.981	-7.32	0	0	0	0	0	0
5	0	0.976	-8.52	13	4	0	0	0	0
6	2	0.98	-8.65	75	2	0	0.8	25	-8
7	0	0.984	-7.58	0	0	0	0	0	0
8	2	1.005	-4.45	150	22	450	62.1	200	-140
9	2	0.98	-9.56	121	26	0	2.2	9	-3
10	0	0.986	-11.43	5	2	0	0	0	0
11	0	0.974	-10.17	0	0	0	0	0	0
12	2	1.015	-10.46	377	24	310	128.5	155	-150
13	0	0.979	-9.79	18	2.3	0	0	0	0
14	0	0.97	-9.33	10.5	5.3	0	0	0	0
15	0	0.988	-7.18	22	5	0	0	0	0
16	0	1.013	-8.85	43	3	0	0	0	0
17	0	1.017	-5.39	42	8	0	0	0	0
18	0	1.001	-11.71	27.2	9.8	0	0	0	0
19	0	0.97	-13.2	3.3	0.6	0	0	0	0
20	0	0.964	-13.41	2.3	1	0	0	0	0
21	0	1.008	-12.89	0	0	0	0	0	0
22	0	1.01	-12.84	0	0	0	0	0	0
23	0	1.008	-12.91	6.3	2.1	0	0	0	0
24	0	0.999	-13.25	0	0	0	0	0	0
25	0	0.982	-18.13	6.3	3.2	0	0	0	0
26	0	0.959	-12.95	0	0	0	0	0	0
27	0	0.982	-11.48	9.3	0.5	0	0	0	0
28	0	0.997	-10.45	4.6	2.3	0	0	0	0
29	0	1.01	-9.75	17	2.6	0	0	0	0
30	0	0.962	-18.68	3.6	1.8	0	0	0	0
31	0	0.936	-19.34	5.8	2.9	0	0	0	0
32	0	0.949	-18.46	1.6	0.8	0	0	0	0
33	0	0.947	-18.5	3.8	1.9	0	0	0	0
34	0	0.959	-14.1	0	0	0	0	0	0
35	0	0.966	-13.86	6	3	0	0	0	0
36	0	0.976	-13.59	0	0	0	0	0	0

continued on following page.

Table 10. Continued

Bus Number	Bus Type	Voltage (pu)	Angle	Real Power Load (MW)	Reactive Power Load (MVAr)	Real Power Generator (MW)	Reactive Power Generator (MVAr)	Max Q (MVAr)	Min Q (MVAr)
37	0	0.985	-13.41	0	0	0	0	0	0
38	0	1.013	-12.71	14	7	0	0	0	0
39	0	0.983	-13.46	0	0	0	0	0	0
40	0	0.973	-13.62	0	0	0	0	0	0
41	0	0.996	-14.05	6.3	3	0	0	0	0
42	0	0.966	-15.5	7.1	4.4	0	0	0	0
43	0	1.01	-11.33	2	1	0	0	0	0
44	0	1.017	-11.86	12	1.8	0	0	0	0
45	0	1.036	-9.25	0	0	0	0	0	0
46	0	1.05	-11.89	0	0	0	0	0	0
47	0	1.033	-12.49	29.7	11.6	0	0	0	0
48	0	1.027	-12.59	0	0	0	0	0	0
49	0	1.036	-12.92	18	8.5	0	0	0	0
50	0	1.023	-13.39	21	10.5	0	0	0	0
51	0	1.052	-12.52	18	5.3	0	0	0	0
52	0	0.98	-11.47	4.9	2.2	0	0	0	0
53	0	0.971	-12.23	20	10	0	0	0	0
54	0	0.996	-11.69	4.1	1.4	0	0	0	0
55	0	1.031	-10.78	6.8	3.4	0	0	0	0
56	0	0.968	-16.04	7.6	2.2	0	0	0	0
57	0	0.965	-16.56	6.7	2	0	0	0	0

Table 11. Branch data of the IEEE 57-bus system

No. Line	From Bus	To Bus	R (pu)	X (pu)	Charging B (pu)
1	2	0.0083	0.028	0.129	0
2	3	0.0298	0.085	0.0818	0
3	4	0.0112	0.0366	0.038	0
4	5	0.0625	0.132	0.0258	0
4	6	0.043	0.148	0.0348	0
6	7	0.02	0.102	0.0276	0
6	8	0.0339	0.173	0.047	0
8	9	0.0099	0.0505	0.0548	0
9	10	0.0369	0.1679	0.044	0
9	11	0.0258	0.0848	0.0218	0
9	12	0.0648	0.295	0.0772	0
9	13	0.0481	0.158	0.0406	0
13	14	0.0132	0.0434	0.011	0
13	15	0.0269	0.0869	0.023	0
1	15	0.0178	0.091	0.0988	0
1	16	0.0454	0.206	0.0546	0
1	17	0.0238	0.108	0.0286	0
3	15	0.0162	0.053	0.0544	0
4	18	0	0.555	0	0.97
4	18	0	0.43	0	0.978
5	6	0.0302	0.0641	0.0124	0
7	8	0.0139	0.0712	0.0194	0
10	12	0.0277	0.1262	0.0328	0
11	13	0.0223	0.0732	0.0188	0
12	13	0.0178	0.058	0.0604	0
12	16	0.018	0.0813	0.0216	0
12	17	0.0397	0.179	0.0476	0
14	15	0.0171	0.0547	0.0148	0
18	19	0.461	0.685	0	0
19	20	0.283	0.434	0	0
21	20	0	0.7767	0	1.043
21	22	0.0736	0.117	0	0
22	23	0.0099	0.0152	0	0
23	24	0.166	0.256	0.0084	0
24	25	0	1.182	0	1
24	25	0	1.23	0	1
24	26	0	0.0473	0	1.043
26	27	0.165	0.254	0	0
27	28	0.0618	0.0954	0	0
28	29	0.0418	0.0587	0	0

continued on following page.

Table 11. Continued

No. Line	From Bus	To Bus	R (pu)	X (pu)	Charging B (pu)
7	29	0	0.0648	0	0.967
25	30	0.135	0.202	0	0
30	31	0.326	0.497	0	0
31	32	0.507	0.755	0	0
32	33	0.0392	0.036	0	0
34	32	0	0.953	0	0.975
34	35	0.052	0.078	0.0032	0
35	36	0.043	0.0537	0.0016	0
36	37	0.029	0.0366	0	0
37	38	0.0651	0.1009	0.002	0
37	39	0.0239	0.0379	0	0
36	40	0.03	0.0466	0	0
22	38	0.0192	0.0295	0	0
11	41	0	0.749	0	0.955
41	42	0.207	0.352	0	0
41	43	0	0.412	0	0
38	44	0.0289	0.0585	0.002	0
15	45	0	0.1042	0	0.955
14	46	0	0.0735	0	0.9
46	47	0.023	0.068	0.0032	0
47	48	0.0182	0.0233	0	0
48	49	0.0834	0.129	0.0048	0
49	50	0.0801	0.128	0	0
50	51	0.1386	0.22	0	0
10	51	0	0.0712	0	0.93
13	49	0	0.191	0	0.895
29	52	0.1442	0.187	0	0
52	53	0.0762	0.0984	0	0
53	54	0.1878	0.232	0	0
54	55	0.1732	0.2265	0	0
11	43	0	0.153	0	0.958
44	45	0.0624	0.1242	0.004	0
40	56	0	1.195	0	0.958
56	41	0.553	0.549	0	0
56	42	0.2125	0.354	0	0
39	57	0	1.355	0	0.98
57	56	0.174	0.26	0	0
38	49	0.115	0.177	0.003	0
38	48	0.0312	0.0482	0	0
9	55	0	0.1205	0	0.94

Table 12. Bus data of the IEEE 118-bus system

Bus Number	Bus Type	Voltage (pu)	Angle	Real Power Load (MW)	Reactive Power Load (MVAr)	Real Power Generator (MW)	Reactive Power Generator (MVAr)	Max Q (MVAr)	Min Q (MVAr)
1	2	0.955	10.67	51	27	0	0	15	-5
2	0	0.971	11.22	20	9	0	0	0	0
3	0	0.968	11.56	39	10	0	0	0	0
4	2	0.998	15.28	30	12	-9	0	300	-300
5	0	1.002	15.73	0	0	0	0	0	0
6	2	0.99	13	52	22	0	0	50	-13
7	0	0.989	12.56	19	2	0	0	0	0
8	2	1.015	20.77	0	0	-28	0	300	-300
9	0	1.043	28.02	0	0	0	0	0	0
10	2	1.05	35.61	0	0	450	0	200	-147
11	0	0.985	12.72	70	23	0	0	0	0
12	2	0.99	12.2	47	10	85	0	120	-35
13	0	0.968	11.35	34	16	0	0	0	0
14	0	0.984	11.5	14	1	0	0	0	0
15	2	0.97	11.23	90	30	0	0	30	-10
16	0	0.984	11.91	25	10	0	0	0	0
17	0	0.995	13.74	11	3	0	0	0	0
18	2	0.973	11.53	60	34	0	0	50	-16
19	2	0.963	11.05	45	25	0	0	24	-8
20	0	0.958	11.93	18	3	0	0	0	0
21	0	0.959	13.52	14	8	0	0	0	0
22	0	0.97	16.08	10	5	0	0	0	0
23	0	1	21	7	3	0	0	0	0
24	2	0.992	20.89	0	0	-13	0	300	-300
25	2	1.05	27.93	0	0	220	0	140	-47
26	2	1.015	29.71	0	0	314	0	1000	-1000
27	2	0.968	15.35	62	13	-9	0	300	-300
28	0	0.962	13.62	17	7	0	0	0	0
29	0	0.963	12.63	24	4	0	0	0	0
30	0	0.968	18.79	0	0	0	0	0	0
31	2	0.967	12.75	43	27	7	0	300	-300
32	2	0.964	14.8	59	23	0	0	42	-14
33	0	0.972	10.63	23	9	0	0	0	0
34	2	0.986	11.3	59	26	0	0	24	-8
35	0	0.981	10.87	33	9	0	0	0	0
36	2	0.98	10.87	31	17	0	0	24	-8
37	0	0.992	11.77	0	0	0	0	0	0
38	0	0.962	16.91	0	0	0	0	0	0
39	0	0.97	8.41	27	11	0	0	0	0
40	2	0.97	7.35	20	23	-46	0	300	-300

continued on following page.

Table 12. Continued

Bus Number	Bus Type	Voltage (pu)	Angle	Real Power Load (MW)	Reactive Power Load (MVAr)	Real Power Generator (MW)	Reactive Power Generator (MVAr)	Max Q (MVAr)	Min Q (MVAr)
41	0	0.967	6.92	37	10	0	0	0	0
42	2	0.985	8.53	37	23	-59	0	300	-300
43	0	0.978	11.28	18	7	0	0	0	0
44	0	0.985	13.82	16	8	0	0	0	0
45	0	0.987	15.67	53	22	0	0	0	0
46	2	1.005	18.49	28	10	19	0	100	-100
47	0	1.017	20.73	34	0	0	0	0	0
48	0	1.021	19.93	20	11	0	0	0	0
49	2	1.025	20.94	87	30	204	0	210	-85
50	0	1.001	18.9	17	4	0	0	0	0
51	0	0.967	16.28	17	8	0	0	0	0
52	0	0.957	15.32	18	5	0	0	0	0
53	0	0.946	14.35	23	11	0	0	0	0
54	2	0.955	15.26	113	32	48	0	300	-300
55	2	0.952	14.97	63	22	0	0	23	-8
56	2	0.954	15.16	84	18	0	0	15	-8
57	0	0.971	16.36	12	3	0	0	0	0
58	0	0.959	15.51	12	3	0	0	0	0
59	2	0.985	19.37	277	113	155	0	180	-60
60	0	0.993	23.15	78	3	0	0	0	0
61	2	0.995	24.04	0	0	160	0	300	-100
62	2	0.998	23.43	77	14	0	0	20	-20
63	0	0.969	22.75	0	0	0	0	0	0
64	0	0.984	24.52	0	0	0	0	0	0
65	2	1.005	27.65	0	0	391	0	200	-67
66	2	1.05	27.48	39	18	392	0	200	-67
67	0	1.02	24.84	28	7	0	0	0	0
68	0	1.003	27.55	0	0	0	0	0	0
69	3	1.035	30	0	0	516.4	0	300	-300
70	2	0.984	22.58	66	20	0	0	32	-10
71	0	0.987	22.15	0	0	0	0	0	0
72	2	0.98	20.98	0	0	-12	0	100	-100
73	2	0.991	21.94	0	0	-6	0	100	-100
74	2	0.958	21.64	68	27	0	0	9	-6
75	0	0.967	22.91	47	11	0	0	0	0
76	2	0.943	21.77	68	36	0	0	23	-8
77	2	1.006	26.72	61	28	0	0	70	-20
78	0	1.003	26.42	71	26	0	0	0	0
79	0	1.009	26.72	39	32	0	0	0	0
80	2	1.04	28.96	130	26	477	0	280	-165

continued on following page.

Table 12. Continued

Bus Number	Bus Type	Voltage (pu)	Angle	Real Power Load (MW)	Reactive Power Load (MVAr)	Real Power Generator (MW)	Reactive Power Generator (MVAr)	Max Q (MVAr)	Min Q (MVAr)
81	0	0.997	28.1	0	0	0	0	0	0
82	0	0.989	27.24	54	27	0	0	0	0
83	0	0.985	28.42	20	10	0	0	0	0
84	0	0.98	30.95	11	7	0	0	0	0
85	2	0.985	32.51	24	15	0	0	23	-8
86	0	0.987	31.14	21	10	0	0	0	0
87	2	1.015	31.4	0	0	4	0	1000	-100
88	0	0.987	35.64	48	10	0	0	0	0
89	2	1.005	39.69	0	0	607	0	300	-210
90	2	0.985	33.29	78	42	-85	0	300	-300
91	2	0.98	33.31	0	0	-10	0	100	-100
92	2	0.993	33.8	65	10	0	0	9	-3
93	0	0.987	30.79	12	7	0	0	0	0
94	0	0.991	28.64	30	16	0	0	0	0
95	0	0.981	27.67	42	31	0	0	0	0
96	0	0.993	27.51	38	15	0	0	0	0
97	0	1.011	27.88	15	9	0	0	0	0
98	0	1.024	27.4	34	8	0	0	0	0
99	2	1.01	27.04	0	0	-42	0	100	-100
100	2	1.017	28.03	37	18	252	0	155	-50
101	0	0.993	29.61	22	15	0	0	0	0
102	0	0.991	32.3	5	3	0	0	0	0
103	2	1.001	24.44	23	16	40	0	40	-15
104	2	0.971	21.69	38	25	0	0	23	-8
105	2	0.965	20.57	31	26	0	0	23	-8
106	0	0.962	20.32	43	16	0	0	0	0
107	2	0.952	17.53	28	12	-22	0	200	-200
108	0	0.967	19.38	2	1	0	0	0	0
109	0	0.967	18.93	8	3	0	0	0	0
110	2	0.973	18.09	39	30	0	0	23	-8
111	2	0.98	19.74	0	0	36	0	1000	-100
112	2	0.975	14.99	25	13	-43	0	1000	-100
113	2	0.993	13.74	0	0	-6	0	200	-100
114	0	0.96	14.46	8	3	0	0	0	0
115	0	0.96	14.46	22	7	0	0	0	0
116	2	1.005	27.12	0	0	-184	0	1000	-1000
117	0	0.974	10.67	20	8	0	0	0	0
118	0	0.949	21.92	33	15	0	0	0	0

Table 13. Branch data of the IEEE 118-bus system

No. Line	From Bus	To Bus	R (pu)	X (pu)	Charging B (pu)
1	2	0.0303	0.0999	0.0254	0
1	3	0.0129	0.0424	0.01082	0
4	5	0.00176	0.00798	0.0021	0
3	5	0.0241	0.108	0.0284	0
5	6	0.0119	0.054	0.01426	0
6	7	0.00459	0.0208	0.0055	0
8	9	0.00244	0.0305	1.162	0
8	5	0	0.0267	0	0.985
9	10	0.00258	0.0322	1.23	0
4	11	0.0209	0.0688	0.01748	0
5	11	0.0203	0.0682	0.01738	0
11	12	0.00595	0.0196	0.00502	0
2	12	0.0187	0.0616	0.01572	0
3	12	0.0484	0.16	0.0406	0
7	12	0.00862	0.034	0.00874	0
11	13	0.02225	0.0731	0.01876	0
12	14	0.0215	0.0707	0.01816	0
13	15	0.0744	0.2444	0.06268	0
14	15	0.0595	0.195	0.0502	0
12	16	0.0212	0.0834	0.0214	0
15	17	0.0132	0.0437	0.0444	0
16	17	0.0454	0.1801	0.0466	0
17	18	0.0123	0.0505	0.01298	0
18	19	0.01119	0.0493	0.01142	0
19	20	0.0252	0.117	0.0298	0
15	19	0.012	0.0394	0.0101	0
20	21	0.0183	0.0849	0.0216	0
21	22	0.0209	0.097	0.0246	0
22	23	0.0342	0.159	0.0404	0
23	24	0.0135	0.0492	0.0498	0
23	25	0.0156	0.08	0.0864	0
26	25	0	0.0382	0	0.96
25	27	0.0318	0.163	0.1764	0
27	28	0.01913	0.0855	0.0216	0
28	29	0.0237	0.0943	0.0238	0
30	17	0	0.0388	0	0.96
8	30	0.00431	0.0504	0.514	0
26	30	0.00799	0.086	0.908	0
17	31	0.0474	0.1563	0.0399	0
29	31	0.0108	0.0331	0.0083	0
23	32	0.0317	0.1153	0.1173	0

continued on following page.

Table 13. Continued

No. Line	From Bus	To Bus	R (pu)	X (pu)	Charging B (pu)
31	32	0.0298	0.0985	0.0251	0
27	32	0.0229	0.0755	0.01926	0
15	33	0.038	0.1244	0.03194	0
19	34	0.0752	0.247	0.0632	0
35	36	0.00224	0.0102	0.00268	0
35	37	0.011	0.0497	0.01318	0
33	37	0.0415	0.142	0.0366	0
34	36	0.00871	0.0268	0.00568	0
34	37	0.00256	0.0094	0.00984	0
38	37	0	0.0375	0	0.935
37	39	0.0321	0.106	0.027	0
37	40	0.0593	0.168	0.042	0
30	38	0.00464	0.054	0.422	0
39	40	0.0184	0.0605	0.01552	0
40	41	0.0145	0.0487	0.01222	0
40	42	0.0555	0.183	0.0466	0
41	42	0.041	0.135	0.0344	0
43	44	0.0608	0.2454	0.06068	0
34	43	0.0413	0.1681	0.04226	0
44	45	0.0224	0.0901	0.0224	0
45	46	0.04	0.1356	0.0332	0
46	47	0.038	0.127	0.0316	0
46	48	0.0601	0.189	0.0472	0
47	49	0.0191	0.0625	0.01604	0
42	49	0.0715	0.323	0.086	0
42	49	0.0715	0.323	0.086	0
45	49	0.0684	0.186	0.0444	0
48	49	0.0179	0.0505	0.01258	0
49	50	0.0267	0.0752	0.01874	0
49	51	0.0486	0.137	0.0342	0
51	52	0.0203	0.0588	0.01396	0
52	53	0.0405	0.1635	0.04058	0
53	54	0.0263	0.122	0.031	0
49	54	0.073	0.289	0.0738	0
49	54	0.0869	0.291	0.073	0
54	55	0.0169	0.0707	0.0202	0
54	56	0.00275	0.00955	0.00732	0
55	56	0.00488	0.0151	0.00374	0
56	57	0.0343	0.0966	0.0242	0
50	57	0.0474	0.134	0.0332	0
56	58	0.0343	0.0966	0.0242	0

continued on following page.

Table 13. Continued

No. Line	From Bus	To Bus	R (pu)	X (pu)	Charging B (pu)
51	58	0.0255	0.0719	0.01788	0
54	59	0.0503	0.2293	0.0598	0
56	59	0.0825	0.251	0.0569	0
56	59	0.0803	0.239	0.0536	0
55	59	0.04739	0.2158	0.05646	0
59	60	0.0317	0.145	0.0376	0
59	61	0.0328	0.15	0.0388	0
60	61	0.00264	0.0135	0.01456	0
60	62	0.0123	0.0561	0.01468	0
61	62	0.00824	0.0376	0.0098	0
63	59	0	0.0386	0	0.96
63	64	0.00172	0.02	0.216	0
64	61	0	0.0268	0	0.985
38	65	0.00901	0.0986	1.046	0
64	65	0.00269	0.0302	0.38	0
49	66	0.018	0.0919	0.0248	0
49	66	0.018	0.0919	0.0248	0
62	66	0.0482	0.218	0.0578	0
62	67	0.0258	0.117	0.031	0
65	66	0	0.037	0	0.935
66	67	0.0224	0.1015	0.02682	0
65	68	0.00138	0.016	0.638	0
47	69	0.0844	0.2778	0.07092	0
49	69	0.0985	0.324	0.0828	0
68	69	0	0.037	0	0.935
69	70	0.03	0.127	0.122	0
24	70	0.00221	0.4115	0.10198	0
70	71	0.00882	0.0355	0.00878	0
24	72	0.0488	0.196	0.0488	0
71	72	0.0446	0.18	0.04444	0
71	73	0.00866	0.0454	0.01178	0
70	74	0.0401	0.1323	0.03368	0
70	75	0.0428	0.141	0.036	0
69	75	0.0405	0.122	0.124	0
74	75	0.0123	0.0406	0.01034	0
76	77	0.0444	0.148	0.0368	0
69	77	0.0309	0.101	0.1038	0
75	77	0.0601	0.1999	0.04978	0
77	78	0.00376	0.0124	0.01264	0
78	79	0.00546	0.0244	0.00648	0
77	80	0.017	0.0485	0.0472	0

continued on following page.

Table 13. Continued

No. Line	From Bus	To Bus	R (pu)	X (pu)	Charging B (pu)
77	80	0.0294	0.105	0.0228	0
79	80	0.0156	0.0704	0.0187	0
68	81	0.00175	0.0202	0.808	0
81	80	0	0.037	0	0.935
77	82	0.0298	0.0853	0.08174	0
82	83	0.0112	0.03665	0.03796	0
83	84	0.0625	0.132	0.0258	0
83	85	0.043	0.148	0.0348	0
84	85	0.0302	0.0641	0.01234	0
85	86	0.035	0.123	0.0276	0
86	87	0.02828	0.2074	0.0445	0
85	88	0.02	0.102	0.0276	0
85	89	0.0239	0.173	0.047	0
88	89	0.0139	0.0712	0.01934	0
89	90	0.0518	0.188	0.0528	0
89	90	0.0238	0.0997	0.106	0
90	91	0.0254	0.0836	0.0214	0
89	92	0.0099	0.0505	0.0548	0
89	92	0.0393	0.1581	0.0414	0
91	92	0.0387	0.1272	0.03268	0
92	93	0.0258	0.0848	0.0218	0
92	94	0.0481	0.158	0.0406	0
93	94	0.0223	0.0732	0.01876	0
94	95	0.0132	0.0434	0.0111	0
80	96	0.0356	0.182	0.0494	0
82	96	0.0162	0.053	0.0544	0
94	96	0.0269	0.0869	0.023	0
80	97	0.0183	0.0934	0.0254	0
80	98	0.0238	0.108	0.0286	0
80	99	0.0454	0.206	0.0546	0
92	100	0.0648	0.295	0.0472	0
94	100	0.0178	0.058	0.0604	0
95	96	0.0171	0.0547	0.01474	0
96	97	0.0173	0.0885	0.024	0
98	100	0.0397	0.179	0.0476	0
99	100	0.018	0.0813	0.0216	0
100	101	0.0277	0.1262	0.0328	0
92	102	0.0123	0.0559	0.01464	0
101	102	0.0246	0.112	0.0294	0
100	103	0.016	0.0525	0.0536	0
100	104	0.0451	0.204	0.0541	0

continued on following page.

Table 13. Continued

No. Line	From Bus	To Bus	R (pu)	X (pu)	Charging B (pu)
103	104	0.0466	0.1584	0.0407	0
103	105	0.0535	0.1625	0.0408	0
100	106	0.0605	0.229	0.062	0
104	105	0.00994	0.0378	0.00986	0
105	106	0.014	0.0547	0.01434	0
105	107	0.053	0.183	0.0472	0
105	108	0.0261	0.0703	0.01844	0
106	107	0.053	0.183	0.0472	0
108	109	0.0105	0.0288	0.0076	0
103	110	0.03906	0.1813	0.0461	0
109	110	0.0278	0.0762	0.0202	0
110	111	0.022	0.0755	0.02	0
110	112	0.0247	0.064	0.062	0
17	113	0.00913	0.0301	0.00768	0
32	113	0.0615	0.203	0.0518	0
32	114	0.0135	0.0612	0.01628	0
27	115	0.0164	0.0741	0.01972	0
114	115	0.0023	0.0104	0.00276	0
68	116	0.00034	0.00405	0.164	0
12	117	0.0329	0.14	0.0358	0
75	118	0.0145	0.0481	0.01198	0
76	118	0.0164	0.0544	0.01356	0
1	2	0.0303	0.0999	0.0254	0

Chapter 12
Modeling a Real Cable Production System as a Single Machine-Scheduling Problem:
Mathematical Model and Metaheuristic Approach

Sadegh Niroomand
Firouzabad Institute of Higher Education, Iran

Béla Vizvári
Eastern Mediterranean University, North Cyprus, Turkey

ABSTRACT

In cases where the size and colour of cable are changed, the cable industry is classified as a multi-product, mass production system. The paper provides a mixed integer linear programming model based on continuous time representation for a case study on the scheduling problem of the cable industry to minimize the total cost including setup cost, operating cost, and inventory holding cost. As the solution methodology, three grouping policies are proposed while Xpress solver could not give any feasible solution for the model. Cables of the same size and the same colour, respectively, of the different types of cable are grouped together. A metaheuristic based on a simulated annealing algorithm is applied to minimize the total cost of proposed solutions. Finally the solution with the smallest total cost is selected as the production schedule of the study case.

1. INTRODUCTION

In manufacturing organizations, goals like profitability, quality of products, on-time delivery, and so on play a critical role in keeping and increasing the market share of organizations. The scheduling of tasks in manufacturing helps to decrease production costs and to shorten the lead time of finished products. Classical scheduling problems are mostly time-based optimization problems, while the processing time

DOI: 10.4018/978-1-4666-9644-0.ch012

and setup time of tasks are assumed to be constant. A well-known type of scheduling problem is known as the single machine scheduling problem, which concerns the arrangement of jobs on a single machine/ system considering setup times and processing times of jobs. A good survey of scheduling problems with setup times is presented by Allahverdi *et al.* (2008).

Scheduling problems need efficient mathematical models in order to be solved effectively. Generally, mixed integer linear programming (MILP) models are used for such problems. Pinedo (2008) classifies these scheduling problems as complex optimization problems that belong to NP-hard problems. Floudas and Lin (2005) focused on the application of MILP in process scheduling while Barton and Lee (2004), Kallrath (1999), Lee and Grossman (2003) and Fahmy and ElMekkawy and Balakrishnan (2008) using mixed integer nonlinear programming (MINLP) to formulate scheduling problems. A mixed integer quadratic programming (MIQP) model was also applied by Dua and Morari (2004).

The mathematical models of scheduling problems are divided into two types, discrete and continuous formulations. In discrete formulation, the tasks must be started at the beginning or end of predetermined equal discrete time intervals. Floudas and Lin (2004) focused on the discrete formulation. These models need a large number of binary variables for representing the scheduling problem. They also represent an approximate time domain (Stefansson *et al.* (2011)). On the other hand, continuous formulation allows the tasks to be started at any time under the condition that there is no overlap with other tasks. Most of the researchers used continuous formulations to model scheduling problems, for example, Giannelos and Georgiadis (2002), Harjunkoski and Grossmann (2002), Janak and lin and Floudas (2005), Castro and Grossmann and Novais (2006), and others. Figure 1 shows the arrangement of tasks based on discrete and continuous models.

Different solution approaches, for example exact and non-exact, are used to solve the mathematical model of scheduling problems. A well-known exact approach is Branch and Bound's method, which can be applied by heuristic algorithms or optimization software, for example Cplex, Xpress, Lingo, and so on. This method was applied by Rios-Mercado *et al.* (1999), Rabadi *et al.* (2004), Dunstall *et al.* (2005), Ranjbar (2013), Neto *et al.* (2013), and so on. Different metaheuristic algorithms, for example simulated annealing (SA), genetic algorithm, tabu search, ant colony, and so on, were proposed as non-exact methods in order to obtain a good feasible solution where the degree of difficulty of the model is high and exact methods are not able to solve the mathematical model in reasonable time. Studies of Gilland (2002), Kim *et al.* (2002), Mendes *et al.* (2002), Bilge *et al.* (2004), Low (2005), Dupuy *et al.* (2005), Ruiz *et al.* (2006), Hendizadeh *et al.* (2007), Valente (2007), Damodaran *et al.* (2009), Wong *et al.* (2012) and Yang and Shen (2012) Yousefi and Yusuff (2013) can be mentioned as authors who have applied metaheuristics to scheduling problems.

Figure 1. Arrangement of a set of tasks based on discrete and continuous models

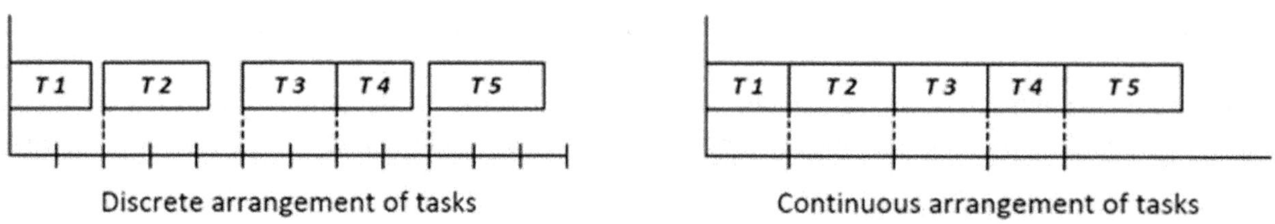

This paper introduces a continuous mathematical model to minimize the total production cost including inventory holding cost, setup cost, and operating cost of sequence-dependent setup time tasks. The model is introduced for a real-world multi-product scheduling problem of the cable industries.

The next section of the paper describes the study case. Section 3 introduces the mathematical model. Section 4 describes the solution methodology and SA metaheuristic that is applied to solve the model. Computational experiments are done in Section 5. The conclusions are drawn in Section 6 of the paper.

2. CASE STUDY

A real-world problem from a cable production company is considered in this paper. There is a single production system/machine in the company. The company produces cables in different sizes (diameters) and different cover colours. There are u different colours and v different sizes, making a total of $n = uv$ different cable types (all combinations of colour and size are demanded). The given data are as follows:

- Operating time and cost (machine cost per time unit) for a batch of cables. Different sizes of cables have different operating times, while changing the colour does not affect the operating time.
- Labour cost per time unit
- Sequence-dependent setup times and costs (e.g. if cable type B is produced after cable type A) as follows:
    - Case 1 (*S*1): If cable type B has a different colour from A (but the same size) and the setup time is s_1 seconds, there is a fixed amount of scrap cable cover for different production amounts.
    - Case 2 (*S*2): If cable type B has a different size from A (but the same colour) and the setup time is s_2 seconds, there is a fixed amount of scrap copper.
    - Case 3 (*S*3): If cable type B has a different size and different colour from A and the setup time is s_3 seconds, there is a fixed amount of scrap copper and cable cover.
    - $s_3 > s_2 > s_1$
    - The setup cost is calculated as the labour cost for the setup time plus the scrap cost.
    - $\text{Setup cost}_{case3} > \text{Setup cost}_{case1} > \text{Setup cost}_{case2}$
- All demands are due on the same due date
- The inventory holding unit cost is independent of the type of the cable and the produced cables are stored until the due date.

The problem is to determine the best sequence of cables to minimize the total cost including the setup cost, operating cost, and inventory holding cost.

3. EXACT MATHEMATICAL FORMULATION

A mathematical model is introduced in this part of the paper to minimize the total cost to the company. The model is constructed based on the assumptions below:

- The n types of cables are ordered and will be produced according to the demanded quantities.

- The whole demand of a cable type is produced once.
- The inventory time and cost of each type of cable are calculated from its completion time to its due date.
- The demands cannot be met before the due date.

The model is based on some variables and parameters, which are introduced as follows:

n: the number of product types to be produced (parameter),
i, j: the indices of products (index),
k, p: the indices of positions of products in their production order (index),
s_{ij} : the setup time of product j if it is produced after product i (parameter),
q_i : the production quantity (demand) of product i (parameter),
t_i : the unit operating time of product i (parameter),
c^P : the process cost for a unit of product per time unit (parameter),
c^l : the labour cost per time unit (parameter),
c^h : the inventory holding unit cost for a unit of product per time unit (parameter),
T: the due date for responding to the demands (parameter),
T_0 : starting time of the planning period (variable, $T_0 < T$),
x_{ip} : a binary variable indicating the assignment of product i to position p in the order of the products if its value is 1; otherwise 0 (variable).

3.1. Assignment Constraints

Each product is assigned to only one position in the product sequence and only one product can be assigned to each position of the sequence based on the following constraints.

$$\sum_{p=1}^{n} x_{ip} = 1 \qquad i = 1,2,\ldots,n \tag{1}$$

$$\sum_{i=1}^{n} x_{ip} = 1 \qquad p = 1,2,\ldots,n \tag{2}$$

$$x_{ip} \in \{0,1\} \qquad i,p = 1,2,\ldots,n \tag{3}$$

3.2. Operating Cost Calculations

The process cost of product i is obtained by $q_i t_i c^P$; therefore the total operating cost (C^P) of all products is:

$$C^P = \sum_{i=1}^{n} c^P q_i t_i$$

3.3. Setup Cost Calculations

The setup cost of products *i* and *j* is equal to $s_{ij}c^l$ if *j* is produced after *i*; therefore it is multiplied by $\sum_{p=1}^{n-1} x_{ip}x_{jp+1}$. This sum is 1 if *j* is produced after *i*. So, the total setup cost (C^S) of all products is:

$$C^S = \sum_{i=1}^{n}\sum_{\substack{j=1 \\ j\neq i}}^{n} s_{ij}c^l \sum_{p=1}^{n-1} x_{ip}x_{jp+1}$$

3.4. Inventory Holding Cost Calculations

According to the assumptions of the model, the storage time of each product starts from its completion time and ends at the due date. If the due date is later than the completion time of the last product type, a variable idle time (T_0) is considered at the beginning of production; therefore, the completion time of the last product type can be the same as the due date in order to decrease the inventory time.

Without considering setup times, the completion times of the first and p^{th} product types are calculated using the following formulas (Figure 2):

$$1^{st} \text{ completion time} = T_0 + \sum_{i=1}^{n} q_i t_i x_{i1}$$

$$p^{th} \text{ completion time} = T_0 + \sum_{k=1}^{p}\sum_{i=1}^{n} q_i t_i x_{ik}$$

$$T_0 \geq 0 \qquad (4)$$

Figure 2. Completion time of products without and with setup time

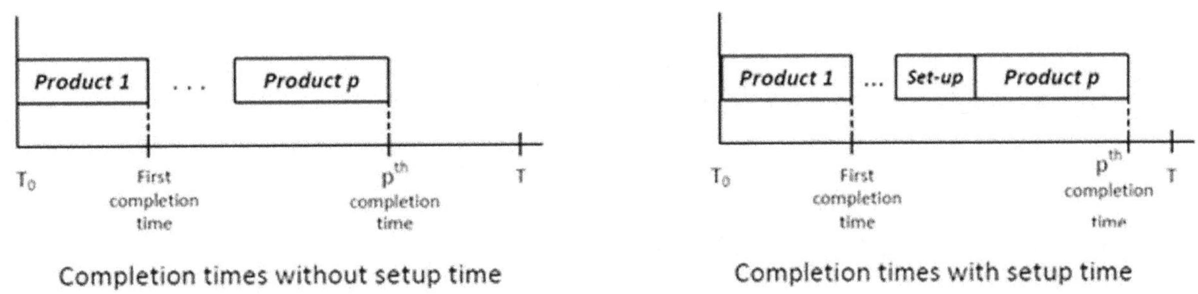

The first and k^{th} setup times are also calculated in the same way. The setup time between any pair of product types is considered if and only if they are assigned to two consecutive positions in the sequence of products.

$$1^{st} \text{ setup time} = \sum_{i=1}^{n}\sum_{\substack{j=1\\ j\neq i}}^{n} s_{ij}x_{i1}x_{j2}$$

$$k^{th} \text{ setup time} = \sum_{i=1}^{n}\sum_{\substack{j=1\\ j\neq i}}^{n} s_{ij}x_{ik}x_{jk+1}$$

Then, the real completion time of the pth product is the sum of idle time of the system (T_0) and the first p completion times without considering the setup time and the first $p-1$ setup times (Figure 2).

$$p^{th} \text{ completion time} = T_0 + \sum_{k=1}^{p}\sum_{i=1}^{n} q_i t_i x_{ik} + \sum_{k=1}^{p-1}\sum_{i=1}^{n}\sum_{\substack{j=1\\ j\neq i}}^{n} s_{ij}x_{ik}x_{jk+1}$$

The inventory holding time of the pth product is calculated by subtracting the pth completion time from the due dates of the demands. The storage performance of a product is the holding time multiplied by the quantity of that product. It is also called part parity. Then, by summing the part parity of all products the total storage performance is calculated, which is

$$\sum_{p=1}^{n}\left(\left(\sum_{l=1}^{n} q_l x_{lp}\right)\left(T - T_0 - \sum_{k=1}^{p}\sum_{i=1}^{n} q_i t_i x_{ik} - \sum_{k=1}^{p-1}\sum_{i=1}^{n}\sum_{\substack{j=1\\ j\neq i}}^{n} s_{ij}x_{ik}x_{j,k+1}\right)\right)$$

$$= \sum_{p=1}^{n}\sum_{l=1}^{n} q_l x_{lp}(T - T_0) - \sum_{p=1}^{n}\left(\sum_{l=1}^{n} q_l x_{lp}\right)\left(\sum_{k=1}^{p-1}\sum_{i=1}^{n} q_i t_i x_{ik}\right) - \sum_{p=1}^{n}\left(\sum_{l=1}^{n} q_l x_{lp}\right)\left(\sum_{k=1}^{p-1}\sum_{i=1}^{n}\sum_{\substack{j=1\\ j\neq i}}^{n} s_{ij}x_{ik}x_{j,k+1}\right)$$

$$= \left(T - T_0\right)\left(\sum_{l=1}^{n} q_l \sum_{p=1}^{n} x_{lp}\right) - \sum_{p=1}^{n}\left(\sum_{l=1}^{n} q_l x_{lp}\right)\left(\sum_{k=1}^{p-1}\sum_{i=1}^{n} q_i t_i x_{ik}\right) - \sum_{p=1}^{n}\left(\sum_{l=1}^{n} q_l x_{lp}\right)\left(\sum_{k=1}^{p-1}\sum_{i=1}^{n}\sum_{\substack{j=1\\ j\neq i}}^{n} s_{ij}x_{ik}x_{j,k+1}\right)$$

$$= \left(T - T_0\right)\left(\sum_{l=1}^{n} q_l\right) - \sum_{p=1}^{n}\left(\sum_{l=1}^{n} q_l x_{lp}\right)\left(\sum_{k=1}^{p-1}\sum_{i=1}^{n} q_i t_i x_{ik}\right) - \sum_{p=1}^{n}\left(\sum_{l=1}^{n} q_l x_{lp}\right)\left(\sum_{k=1}^{p-1}\sum_{i=1}^{n}\sum_{\substack{j=1\\ j\neq i}}^{n} s_{ij}x_{ik}x_{j,k+1}\right)$$

The total inventory holding cost (C^H) is calculated by multiplying the total storage time by the inventory holding cost per time unit (c^h).

$$C^H = \left(T - T_0\right)\left(\sum_{l=1}^{n} c^h q_l\right) - \sum_{p=1}^{n} c^h \left(\sum_{l=1}^{n} q_l x_{lp}\right)\left(\sum_{k=1}^{p-1}\sum_{i=1}^{n} q_i t_i x_{ik}\right) - \sum_{p=1}^{n} c^h \left(\sum_{l=1}^{n} q_l x_{lp}\right)\left(\sum_{k=1}^{p-1}\sum_{i=1}^{n}\sum_{\substack{j=1 \\ j\neq i}}^{n} s_{ij} x_{ik} x_{j,k+1}\right)$$

3.5. Objective Function of the Model

The total cost to the company is the sum of the total operating cost, the total setup cost, and the total inventory holding cost. The value of the total operating cost is constant but the other costs are variable and should be minimized. The total setup and total inventory holding costs are calculated from the same order of products; the best order that minimizes the total inventory holding cost does not minimize the total setup cost and the inverse situation also happens (Section 4 of the paper focuses on this issue). Therefore, the following objective function is used for minimization of the total cost.

$$\min C^P + C^S + C^H$$

The mathematical model (1)-(5) is solved to obtain the best order of cables to be produced.

4. SOLUTION METHODOLOGY

Most of the mathematical models of scheduling problems are classified as NP-complete and NP-hard problems. Therefore, metaheuristic methods are applied to obtain a good feasible solution (near to optimal) in a reasonable time.

The proposed formulation contains the quadratic and cubic terms in the objective function and is difficult to solve optimally by exact methods. Although there are some well-known and efficient linearization techniques for the quadratic assignment problem (see Chaovalitwongse *et al.* (2004) and He *et al.* (2012)), no efficient method has been introduced for linearizing the cubic assignment problem. Therefore, a metaheuristic method is proposed in this part of the paper to solve the model while Xpress solver could not give even a feasible solution.

As mentioned in Section 3.5, the objective function of the model consists of two independent parts and the optimality of either of them leads to a non-optimal solution of the other one. The proposed metaheuristic gives three different solutions, which are detailed in this part. At the end the solutions are compared to select the best one.

4.1. Metaheuristic Solution Based on Grouping Cables of the Same Colour (SA-1)

This method gives the optimal solution for setup costs and tries to decrease the inventory holding cost.

Each cable colour is considered as a set of cables that contains all sizes of that colour. So, there are u sets of cables and each set consists of v different sizes of cable of the same colour.

Step 1: Minimization of the Total Cost of a Random Sequence of Sets

A randomly ordered set of the cables should be reordered so as to minimize the total cost. The operating cost is always constant, so the minimization concerns the inventory holding cost (time) and the setup cost (time).

Step 1.1: Minimization of the Inventory Holding Cost (Phase 1)

The production starts with the first set and all sizes of the set are produced. Then the next set is started and so on (Figure 3). The order of sets is important for decreasing the inventory holding cost. For this purpose a good approximate order of sets is obtained by the following relation (in the following relation, the inventory time is approximate because the storage of the products in a set is assumed to start from the time of completion of the last product of that set). If two sets A and B are consecutive sets in the sequence (Figure 3), by interchanging them, the inventory holding cost is decreased if *Inventory time of interchanged order* ≤ *Inventory time of initial order.*

Therefore

$$\left(T-S-\sum_{i\in B} t_i q_i\right)\left(\sum_{i\in B} q_i\right)+\left(T-S-\sum_{i\in B} t_i q_i-S-\sum_{i\in A} t_i q_i\right)\left(\sum_{i\in A} q_i\right)\leq\left(T-S-\sum_{i\in A} t_i q_i\right)\left(\sum_{i\in A} q_i\right)+\left(T-S-\sum_{i\in A} t_i q_i-S-\sum_{i\in B} t_i q_i\right)\left(\sum_{i\in B} q_i\right)$$

S is the sum of setup times of products of each set ($S = vs_2$). This sum is the same in all sets except for the first set (the reason will be explained in the next steps). Finally,

$$\left(S+\sum_{i\in B} t_i q_i\right)\Big/\sum_{i\in B} q_i \geq \left(S+\sum_{i\in A} t_i q_i\right)\Big/\sum_{i\in A} q_i$$

So, for any pair of sets, for example A and B, set B is produced before set A if the relation obtained above is true.

Figure 3. A random order of products that gives optimal setup time

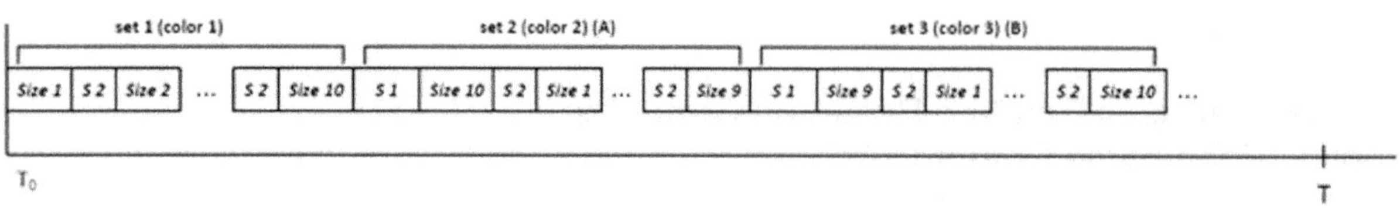

Step 1.2: Minimization of the Setup Cost

In order to minimize the setup cost, the necessary condition is that the first product of each set must have the same size as the last product of the previous set. Neither the order of sets nor the order of products of each set affects the optimality of the setup time (see Figure 3).

This ordering gives the optimal setup time because only Cases 1 and 2 of the setup times happen ($u - 1$ times s_1 and $n - u - 2$ times s_2). Case 3 ($S3$) of the setup times does not happen with this sequence of products.

Step 1.3: Minimization of the inventory holding cost (phase 2).

In the order of sets obtained from Step 1.2, the positions of the first and last products of each set are fixed but the positions of other products (including $v - 2$ products in each set except the first and last sets, which include $v - 1$ products) can be changed within that set to obtain a better inventory holding cost (see Figure 4).

Now, for any pair of products within a set (except the first and last products of the set), the interchange of Figure 4 decreases the inventory holding cost if *Inventory time of interchanged order* $\le$ *Inventory time of initial order.*

Therefore

$$\left(T-\left(s_2+t_jq_j\right)\right)q_j+\left(T-s_2-t_iq_i-s_2-t_jq_j\right)q_i \le \left(T-\left(s_2+t_iq_i\right)\right)q_i+\left(T-s_2-t_iq_i-s_2-t_jq_j\right)q_j$$

and finally

$$t_jq_iq_j + s_2q_i \le t_iq_iq_j + s_2q_j$$

Thus, for any pair of products i and j from a set of products (except the first and last products of the set), if the above inequality is true, product j is produced before i in order to decrease the inventory holding cost.

Figure 4. Interchanging a pair of products in a set

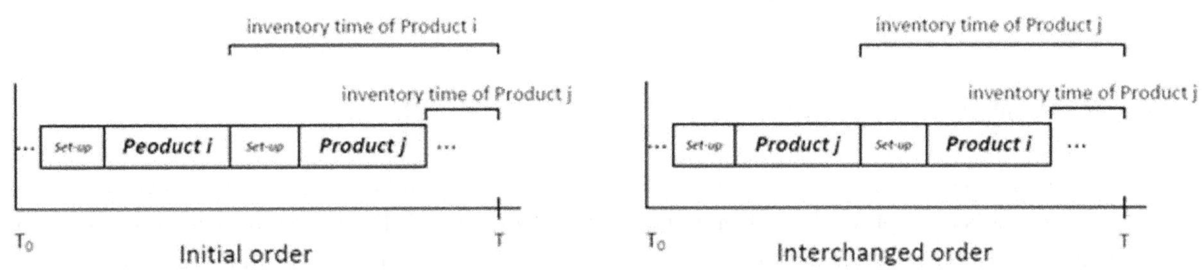

Therefore, the inequality obtained above is used to rearrange the products of all sets of the order of sets obtained from Step 1.2 to decrease the inventory holding cost.

Step 2: Simulated Annealing (SA) Algorithm for Improvement of the Solution Obtained

The SA method is used as a metaheuristic optimization method to solve combinatorial optimization problems. This method was introduced by Kirkpatrick, Gelatt, and Vecchi (1983). The SA algorithm is based on randomization techniques and works as a probabilistic local search method. Generally the SA applies an iterative improvement algorithm to a predetermined initial solution and its objective function. A generation mechanism is used to change the solution to another solution, which is called the neighbouring solution. The neighbouring solution is reached in a one-step transition. The method is also called Monte Carlo annealing. The objective function value of the solutions obtained by the SA algorithm is usually close to the optimal value.

The most important parameters of any SA algorithm are the following:

- Initial temperature (T)
- Final temperature (T_f)
- A cooling ratio or Boltzmann constant (r) such that $0 < r < 1$
- A random number (R, $0 < R < 1$)
- Number of iterations
- Initial solution
- The set of neighbouring solutions

To apply the SA method to this scheduling problem, the following steps should be followed.

Step 2.1

The solution obtained from Step 1 (rearranged order of sets) is considered as the initial solution of the SA algorithm. The initial solution is evaluated and its total cost (*TC*, including operating cost, setup cost, and inventory holding cost) is calculated. The initial solution is saved as the "current solution" and "best solution".

Step 2.2

The initial temperature is saved as $T_{current}$ and the following steps are carried out.

Step 2.3

Neighbour generation (see Figure 5)

1. Two random integers, for example a and b, are generated. Random number a selects one of the sets of products (except for the last one; therefore $1 \leq a \leq u - 1$). Random number b selects one of the products of the set selected by a (except the first and last product of the set; therefore $2 \leq b \leq v - 1$).

Figure 5. The method used to generate the neighbor solution

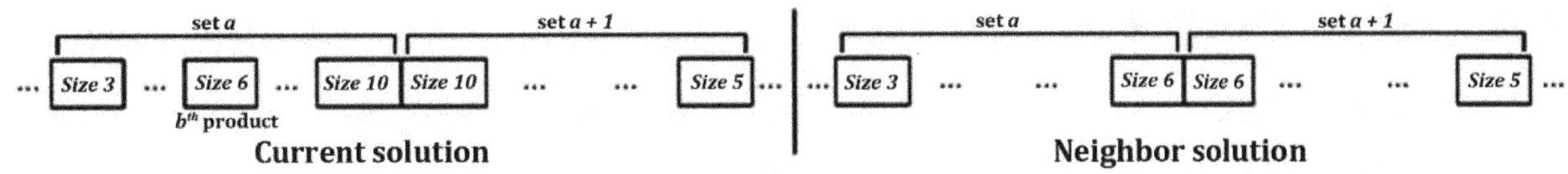

2. In set a, the bth product and the last product of the set are interchanged. Then Step 1.3 is repeated for this set to rearrange the products of the set (except the first and last products) to decrease the inventory holding cost.
3. In set $a + 1$, a product which has the same size as the last product of Set a (currently became the last one) is interchanged with the first product of the set. Then Step 2 is repeated for this set to rearrange the products of the set (except for the first and last products) to decrease the inventory holding cost.

In any neighbouring solution the setup cost remains optimal but the inventory holding cost may change.

Step 2.4

The neighbouring solution is evaluated and its total cost is obtained. Based on its total cost the following cases can happen.

- If $TC_{neighbor} < TC_{current}$ and $TC_{neighbor} < TC_{best}$, the best solution and current solution are updated by the neighbouring solution.
- If $TC_{neighbor} < TC_{current}$ and $TC_{neighbor} \geq TC_{best}$ the best solution is not changed but the current solution is updated by the neighbouring solution.
- If $TC_{neighbor} \geq TC_{current}$, but $\exp\left(-\left(TC_{neighbor} - TC_{current}\right) / T_{current}\right) \geq R$, the best solution is not changed but the current solution is updated by the neighbouring solution.
- If $TC_{neighbor} \geq TC_{current}$, but $\exp\left(-\left(TC_{neighbor} - TC_{current}\right) / T_{current}\right) < R$, the best solution and the current solution are not changed.

Step 2.5

Steps 2.3 and 2.4 are repeated at the current temperature to reach a limited number of iterations.

Step 2.6

The current temperature is updated by $T_{current} = r(T_{current})$. Then, Steps 2.3, 2.4, and 2.5 are repeated for the updated temperature.

The SA algorithm continues while $T_{current} \geq T_f$. Then the "best solution" of the algorithm is introduced as the best solution obtained by this proposed method.

4.2. Metaheuristic Solution Based on Grouping the Cables of the Same Size (SA-2)

This solution minimizes the inventory holding cost and setup cost dependently by considering each size of cable as a set. The following steps are considered in this solution.

Step 1: Minimization of the Total Cost

Each size of cable is considered as a set of cables. So, each set contains all colours of cable (Figure 6). Therefore there are v sets of products and each set contains cables of u colours of the same size. These sets should be ordered to decrease the total cost.

Step 1.1: Minimization of the Inventory Holding Cost (Phase 1)

A good approximate order of sets is obtained by the following relation (where the inventory time is approximate because the storage of the products of a set is assumed to be started from the completion time of all products of that set). If two sets A and B are consecutive sets in the order (Figure 6), by interchanging them, the inventory holding cost is decreased if *Inventory time of interchanged order ≤ Inventory time of initial order.*

Therefore

$$\left(T-S-\sum_{i\in B}t_iq_i\right)\left(\sum_{i\in B}q_i\right)+\left(T-S-\sum_{i\in B}t_iq_i-S-\sum_{i\in A}t_iq_i\right)\left(\sum_{i\in A}q_i\right)\le\left(T-S-\sum_{i\in A}t_iq_i\right)\left(\sum_{i\in A}q_i\right)+\left(T-S-\sum_{i\in B}t_iq_i-S-\sum_{i\in A}t_iq_i\right)\left(\sum_{i\in B}q_i\right)$$

S is the sum of setup times of products of each set ($S = us_1$). This sum is the same in all sets except the first set (the reason will be explained in the next steps). Finally,

$$\left(S+\sum_{i\in B}t_iq_i\right)\Big/\sum_{i\in B}q_i \ge \left(S+\sum_{i\in A}t_iq_i\right)\Big/\sum_{i\in A}q_i$$

So, for any pair of sets, for example A and B, set B is produced before set A if the relation obtained above is true.

Figure 6. The sets of products based on their size

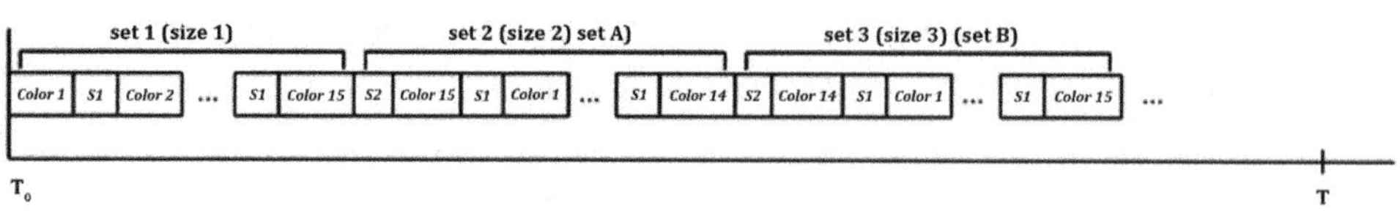

Step 1.2: Minimization of the Total Setup Cost

After reordering the sets, the last cable of each set should have the same colour as the first cable of the next set (Figure 6). This ordering logic for cables within the sets gives the best setup time (cost) based on the obtained order of sets, because only Cases 1 and 2 of the setup times happen ($v - 1$ times s_2 and $n - v - 2$ times s_1). Case 3 ($S3$) of the setup times does not happen with such a sequence of products within the sets. The total setup times of all sets are equal, except for the first set, whose setup time is less because there is no setup for its first product.

Step 1.3: Minimization of the Inventory Holding Cost (Phase 2)

In the order of sets obtained from Step 1.2, the positions of the first and last products of each set are fixed but the positions of other products (including $u - 2$ products in each set except for the first and last sets, which include $u - 1$ products) can be changed within that set to obtain a better inventory holding cost (see Figure 4). For any pair of products within a set (except the first and last products), the interchange shown in Figure 4 decreases the inventory holding cost if *Inventory time of interchanged order* $\leq$ *Inventory time of initial order*.

Therefore

$$\left(T - \left(s_1 + tq_j\right)\right)q_j + \left(T - s_1 - tq_i - s_1 - tq_j\right)q_i \leq \left(T - \left(s_1 + tq_i\right)\right)q_i + \left(T - s_1 - tq_i - s_1 - tq_j\right)q_j$$

The operating time and setup time of products of a set do not change (t and s); then, $q_j \leq q_i$

Thus, in order to decrease the inventory holding cost, for any pair of products i and j of a set of products (except for the first and last products), product j is produced before i if it has less demand.

Step 2: Simulated Annealing Algorithm for Improvement of the Obtained Solution

All steps of the SA algorithm for this method are the same as for the SA algorithm applied in Section 4.1 except for the generation of the neighbouring solution, which is detailed here.

The following steps are proposed to generate the neighbouring solution:

1. Two random numbers, for example a and b, are generated. Random number a selects one of the sets of products (except for the last one; therefore $1 \leq a \leq v - 1$). Random number b selects one of the products of the set which was selected by a (except for the first and last products of the set; therefore $2 \leq b \leq u - 1$).
2. In set a, the bth product and the last product of the set are interchanged. Then Step 1.3 is repeated for this set to rearrange the products of the set (except the first and last products) to decrease the inventory holding cost.
3. In set $a + 1$, a product which has the same colour as the last product of set a (currently became the last one) is interchanged with the first product of the set. Then Step 1.3 is repeated for this set to rearrange the products of the set (except for the first and last products) to decrease the inventory holding cost.

In any neighbouring solution the setup cost is the same but the inventory holding cost may change. The "best solution" of the SA algorithm is introduced as the best solution for this proposed method.

4.3. Metaheuristic Solution Based on Minimizing the Inventory Holding Cost (SA-3)

In order to compare the efficiency of grouping methods, a solution based on a non-grouping order of cells is introduced here.

Step 1: General order of all products

Regardless of the setup time of each product, all products (n types of cable) are ordered from the largest unit operating time to the smallest one in order to decrease the inventory time (cost). This order is used because for any pair of products of the order the interchange of Figure 4 (by ignoring setup times) decreases the inventory holding cost if *Inventory time of interchanged order* $\leq$ *Inventory time of initial order*.Therefore

$$\left(T - t_j q_j\right) q_j + \left(T - t_i q_i - t_j q_j\right) q_i \leq \left(T - t_i q_i\right) q_i + \left(T - t_i q_i - t_j q_j\right) q_j$$

Then

$$t_i \leq t_j .$$

Step 2: Simulated Annealing Algorithm for Improvement of the Obtained Solution

All steps of the SA algorithm for this method are the same as for the SA algorithm applied in Sections 4.1 and 4.2 except for the generation of the neighbouring solution, which is detailed here.

The following steps are used for generating the neighbouring solution:

1. Two different random numbers, for example a and b, are generated. Each random number selects one of the products of the general order (therefore $1 \leq a, b \leq n$).
2. By interchanging the selected products, a new general order (neighbour) is obtained.

In any neighbour solution the setup cost and the inventory holding cost may change.

The "best solution" of the SA algorithm is introduced as the best solution for this proposed method.

5. COMPUTATIONAL EXPERIMENTS

The combination of parameters is a very important issue in the SA algorithm. Three factors that should be determined are the initial temperature, cooling ratio and number of iterations. High value of the parameters causes longer run times. In the cases where the initial temperature and the number of iterations are too small, a good solution is not obtained. For the cooling ratio, any value from 0.8 to 0.99 can be used (Heragu, 1992). More details about parameter selection with the SA algorithm can be found in Laarhoven and Aarts (1987).

The SA algorithms proposed above were coded in Matlab. The programs were run on a computer with an Intel Core 2 Duo 2.53 GHz processor and 4.00 Gb Ram. Several combinations of parameters were tested to obtain a better solution using the SA algorithm. For each parameter of the SA algorithm (except final temperature which was considered to be constant as 5), 3 levels were considered as follow,

- Initial temperature: 50, 150, 300
- Cooling ratio: 0.85, 0.95, 0.99
- Number of iterations: 50, 250, 400

For totally 27 combinations obtained from the above-mentioned levels, each combination was run 5 times and the results are reported in Table 1.

The results of Table 1 are based on currency of money. The highlighted values in each column show the minimum value obtained for that column among all possible combinations of parameters. Although the SA algorithm is a random search method, the higher value of factors causes better solution. Based on the results of Table 1, the best minimum, average and maximum solution of each metaheuristic method is obtained in the combination that at least two parameters are at their highest level.

6. CONCLUSION

This study was done based on a real-world scheduling problem of the cable industry. An exact continuous mathematical model was developed for this scheduling problem. Because of computational complexity, metaheuristic methods based on the simulated annealing (SA) algorithm were proposed to solve the problem. The solution methodology of the paper can be summarized as grouping the cables of the same colour and the same size as the initial solution and then improving them using the SA algorithm in order to minimize the total cost of the system. Several combinations of factors were used to run the SA procedure. The results of the SA algorithm show that producing the products in sets of the same colour (first metaheuristic solution) has a lower total cost for the cable company.

ACKNOWLEDGMENT

We are grateful to our former student Eng. Halenur Arabacı for providing the data of this real-world problem.

Table 1. The results of the proposed SA algorithms

Run	*T*	*r*	Number of Iterations	Final Temperature	Cost obtained by SA-1			Cost obtained by SA-2			Cost obtained by SA-3		
					Min.	Average	Max.	Min.	Average	Max.	Min.	Average	Max.
1	50	0.85	50	5	187707.08	187717.132	187724.05	206780.34	206787.900	206791.50	208025.05	208283.290	208348.75
2	50	0.85	200	5	187701.97	187704.966	187710.97	206779.98	206780.196	206780.34	208348.75	208348.750	208348.75
3	50	0.85	400	5	187691.47	187696.522	187699.12	206779.98	206779.980	206779.98	208339.07	208346.814	208348.75
4	50	0.95	50	5	187704.26	187709.746	187715.71	206779.98	206782.140	206788.26	208280.72	208335.144	208348.75
5	50	0.95	200	5	187693.86	187698.142	187704.48	206779.98	206779.980	206779.98	208125.90	208304.180	208348.75
6	50	0.95	400	5	187691.56	187696.424	187702.83	206779.98	206779.980	206779.98	208196.62	208318.324	208348.75
7	50	0.99	50	5	187690.97	187698.026	187705.29	206779.98	206779.980	206779.98	208348.75	208348.750	208348.75
8	50	0.99	200	5	187690.17	187693.240	187695.58	206779.98	206779.980	206779.98	207786.14	208236.228	208348.75
9	50	0.99	400	5	187689.48	187692.982	187696.52	206779.98	206779.980	206779.98	208199.94	208296.248	208348.75
10	150	0.85	50	5	187712.32	187721.896	187733.36	206780.34	206792.796	206808.78	208099.11	208298.822	208348.75
11	150	0.85	200	5	187694.49	187708.986	187721.03	206779.98	206780.124	206780.34	208152.98	208309.596	208348.75
12	150	0.85	400	5	187697.62	187702.510	187708.54	206779.98	206779.980	206779.98	208348.75	208348.750	208348.75
13	150	0.95	50	5	187700.70	187706.488	187715.67	206779.98	206780.916	206782.14	207731.08	208225.216	208348.75
14	150	0.95	200	5	187689.66	187697.974	187706.61	206779.98	206779.980	206779.98	208348.75	208348.750	208348.75
15	150	0.95	400	5	187693.76	187696.510	187702.68	206779.98	206779.980	206779.98	207684.35	208124.968	208348.75
16	150	0.99	50	5	187698.82	187700.088	187701.52	206779.98	206779.980	206779.98	208348.75	208348.750	208348.75
17	150	0.99	200	5	187692.88	187694.444	187697.79	206779.98	206779.980	206779.98	208294.37	208337.874	208348.75
18	150	0.99	400	5	187691.00	187693.206	**187695.22**	206779.98	206779.980	206779.98	207594.76	208197.952	208348.75
19	300	0.85	50	5	187708.98	187717.840	187726.96	206779.98	206783.076	206794.74	208348.75	208348.750	208348.75
20	300	0.85	200	5	187691.19	187703.210	187710.67	206779.98	206780.124	206780.34	207916.92	208262.384	208348.75
21	300	0.85	400	5	187693.79	187701.948	187711.35	206779.98	206779.980	206779.98	208009.28	208219.444	208348.75
22	300	0.95	50	5	187695.85	187702.428	187708.56	206779.98	206782.212	206790.78	207109.11	208100.822	208348.75
23	300	0.95	200	5	187694.82	187701.636	187705.98	206779.98	206779.980	206779.98	208301.60	208339.320	208348.75
27	300	0.95	400	5	187693.93	187697.850	187706.51	206779.98	206779.980	206779.98	208218.09	208322.618	208348.75
25	300	0.99	50	5	187694.88	187698.664	187702.93	206779.98	206779.980	206779.98	208034.84	208255.886	208348.75
26	300	0.99	200	5	187689.13	**187691.628**	187695.39	206779.98	206779.980	206779.98	208125.36	208284.320	208348.75
27	300	0.99	400	5	**187690.06**	187693.284	187697.10	**206779.98**	**206779.980**	**206779.98**	**207075.42**	**208052.156**	**208348.75**

REFERENCES

Allahverdi, A., Ng, C. T., Cheng, T. C. E., & Kovalyov, M. Y. (2008). A survey of scheduling problems with setup times or costs. *European Journal of Operational Research, 187*(3), 985–1032. doi:10.1016/j.ejor.2006.06.060

Barton, P. I., & Lee, C. K. (2004). Design of process operations using hybrid dynamic optimization. *Computers & Chemical Engineering, 28*(6-7), 955–969. doi:10.1016/j.compchemeng.2003.09.015

Bilge, U., Kirac, F., Kurtulan, M., & Pekgun, P. A. (2004). Tabu search algorithm for parallel machine total tardiness problem. *Computers & Operations Research, 31*(3), 397–414. doi:10.1016/S0305-0548(02)00198-3

Castro, P. M., Grossmann, I. E., & Novais, A. Q. (2006). Two new continuous-time models for the scheduling of multistage batch plants with sequence dependent changeovers. *Industrial & Engineering Chemistry Research, 45*(18), 6210–6226. doi:10.1021/ie060449m

Chaovalitwongse, W., Pardalos, P. M., & Prokoyev, O. A. (2004). A new linearization technique foe multi-quadratic 0-1 programming problems. *Operations Research Letters, 32*(6), 517–522. doi:10.1016/j.orl.2004.03.005

Damodaran, P., Hirani, N. S., & Velez-Gallego, M. C. (2009). Scheduling identical parallel batch processing machines to minimise makespan using genetic algorithms. *European Journal of Industrial Engineering, 3*(2), 187–206. doi:10.1504/EJIE.2009.023605

Dhingra, A., & Chandna, P. (2010). Multi-objective flow shop scheduling using hybrid simulated annealing. *Measuring Business Excellence, 14*(3), 30–41. doi:10.1108/13683041011074191

Dua, V. P. E., & Morari, M. (2004). *Hybrid Systems Modeling, Parametric Programming, and Model Predictive Control-Impact on Process Operations*. Paper presented at the Sixth International Conference on Foundations of Computer-Aided Process Design (FOCAPD), Austin, TX.

Dunstall, S., & Wirth, A. (2005). A comparison of branch-and bound algorithms for a family scheduling problem with identical parallel machines. *European Journal of Operational Research, 167*(2), 283–296. doi:10.1016/j.ejor.2004.03.022

Dupuy, M., Lamothe, J., Gaborit, P., & Dupont, L. (2005). Efficient neighbors in simulated annealing algorithm to optimize the lead time in a parallel multipurpose machine scheduling problem with setup and calendar constraints. In *Proceedings of the International Conference on Industrial Engineering and Systems Management.*

Fahmy, S. A., ElMekkawy, T. Y., & Balakrishnan, S. (2008). Deadlock-free scheduling of flexible job shops with limited capacity buffers. *European Journal of Industrial Engineering, 2*(3), 231–252. doi:10.1504/EJIE.2008.017685

Floudas, C. A., & Lin, X. X. (2004). Continuous-time versus discrete-time approaches for scheduling of chemical processes: A review. *Computers & Chemical Engineering, 28*(11), 2109–2129. doi:10.1016/j.compchemeng.2004.05.002

Floudas, C. A., & Lin, X. X. (2005). Mixed integer linear programming in process scheduling: Modeling, algorithms, and applications. *Annals of Operations Research*, *139*(1), 131–162. doi:10.1007/s10479-005-3446-x

Giannelos, N. F., & Georgiadis, M. C. (2002). A simple new continuous-time formulation for short-term scheduling of multipurpose batch processes. *Industrial & Engineering Chemistry Research*, *41*(9), 2178–2184. doi:10.1021/ie010399f

Gilland, W. G. (2002). A simulation study comparing performance of CONWIP and bottleneck-based release rule. *Production Planning and Control*, *13*(2), 211–219. doi:10.1080/09537280110069784

Harjunkoski, I., & Grossmann, I. E. (2002). Decomposition techniques for multistage scheduling problems using mixed-integer and constraint programming methods. *Computers & Chemical Engineering*, *26*(11), 1533–1552. doi:10.1016/S0098-1354(02)00100-X

He, X., Chen, A., Chaovalitwongse, W., & Liu, H. (2012). An improved linearization technique for a class of quadratic 0-1 programming problems. *Optimization Letters*, *6*(1), 31–41. doi:10.1007/s11590-010-0249-z

Hendizadeh, S. H., ElMekkawy, T. Y., & Wang, G. G. (2007). 207. "Bi-criteria scheduling of a flowshop manufacturing cell with sequence dependent setup times. *European Journal of Industrial Engineering*, *1*(4), 391–413. doi:10.1504/EJIE.2007.015388

Heragu, S. S., & Alfa, A. S. (1992). Experimental analysis of simulated annealing based algorithms for the facility layout problem. *European Journal of Operational Research*, *57*(2), 190–202. doi:10.1016/0377-2217(92)90042-8

Janak, S. L., Lin, X. X., & Floudas, C. A. (2005). Enhanced continuous-time unit-specificevent-based formulation for short-term scheduling of multipurpose batch processes: Resource constraints and mixed storage policies (vol. 43, p. 2529, 2002). *Industrial & Engineering Chemistry Research*, *44*(2), 426. doi:10.1021/ie048866r

Kallrath, J. (1999). Mixed-Integer nonlinear Programming Applications. In T. A. Ciriani, S. Gliozzi, E. l. Johnson, & R. Tadei (Eds.), *Operational Research in Industry*. Basingstoke: Macmillan Press.

Kim, D. W., Kim, K. H., Jang, W., & Chen, F. F. (2002). Unrelated parallel machine scheduling with setup times using simulated annealing. *Robotics and Computer-integrated Manufacturing*, *18*(3-4), 223–231. doi:10.1016/S0736-5845(02)00013-3

Kirkpatrick, S., Gelatt, C. D. Jr, & Vecchi, M. P. (1983). Optimization by simulated annealing. *Science*, *220*(4598), 671–680. doi:10.1126/science.220.4598.671 PMID:17813860

Laarhoven, P. J. M., & Aarts, E. H. L. (1987). *Simulated Annealing: Theory and Applications*. Boston: Kluwer Academic. doi:10.1007/978-94-015-7744-1

Lee, S., & Grossmann, I. E. (2003). Global optimization of nonlinear generalized disjunctive programming with bilinear equality constraints: Applications to process networks. *Computers & Chemical Engineering*, *27*(11), 1557–1575. doi:10.1016/S0098-1354(03)00098-X

Low, C. (2005). Simulated annealing heuristic for flow shop scheduling problems with unrelated parallel machines. *Computers & Operations Research, 32*(8), 2013–2025. doi:10.1016/j.cor.2004.01.003

Mendes, A. S., Muller, F. M., Franca, P. M., & Moscato, P. (2002). Comparing meta-heuristic approach for parallel machine scheduling problems. *Production Planning and Control, 13*(2), 143–154. doi:10.1080/09537280110069649

Neto, T., Constantino, M., Martins, I., & Pedroso, J. P. (2013). A branch-and-bound procedure for forest harvest scheduling problems addressing aspects of habitat availability. *International Transactions in Operational Research, 20*(5), 689–709. doi:10.1111/itor.12003

Pinedo, M. L. (2008). *Scheduling: Theory, Algorithms, and Systems* (3rd ed.). New York: Springer.

Rabadi, G., Mollaghasemi, M., & Anagnostopoulos, G. C. (2004). A branch-and-bound algorithm for the early/tardy machine scheduling problem with a common due-date and sequence dependent setup time. *Computers & Operations Research, 31*(10), 1727–1751. doi:10.1016/S0305-0548(03)00117-5

Ranjbar, M. (2013). A branch-and-bound algorithm for scheduling of new product development projects. *International Transactions in Operational Research, 20*(2), 251–266. doi:10.1111/j.1475-3995.2012.00873.x

Rios-Mercado, R. Z., & Bard, J. F. (1999). A branch-and-bound algorithm for permutation flow shops with sequence-dependent setup times. *IIE Transactions, 31*(8), 721–731. doi:10.1080/07408179908969871

Ruiz, R., Sivrikaya-Serifoglu, F., & Urlings, T. (2006). *Evolutionary approach to realistic hybrid flexible flowshop schedulingproblems. Technical Report. Polytechnic University of Valencia.*

Stefansson, H., Sigmarsdottir, S., Jensson, P., & Shah, N. (2011). Discrete and continuous time representations and mathematical models for large production scheduling problems: A case study from the pharmaceutical industry. *European Journal of Operational Research, 215*(2), 383–392. doi:10.1016/j.ejor.2011.06.021

Valente, J. M. S. (2007). Heuristics for the single machine scheduling problem with early and quadratic tardy penalties. *European Journal of Industrial Engineering, 1*(4), 431–448. doi:10.1504/EJIE.2007.015391

Wong, T. N., Zhang, S., Wang, G., & Zhang, L. (2012). Integrated process planning and scheduling – multi-agent system with two-stage ant colony optimisation algorithm. *International Journal of Production Research, 50*(21), 6188–6201. doi:10.1080/00207543.2012.720393

Yang, T., & Shen, Y. (2012). Heuristic algorithms for a practical-size dynamic parallel-machine scheduling problem: Integrated-circuit wire bonding. *Production Planning and Control, 23*(1), 67–78. doi:10.1080/09537287.2011.572090

Yousefi, M., & Yusuff, R. M. (2013). Minimising earliness and tardiness penalties in single machine scheduling against common due date using imperialist competitive algorithm. *International Journal of Production Research, 51*(16), 4797–4804. doi:10.1080/00207543.2013.774475

Chapter 13
Hybrid Genetics Algorithms for Multiple Sequence Alignment

John Tsiligaridis
Heritage University, USA

ABSTRACT

The purpose of this chapter is to present a set of algorithms and their efficiency for the consistency based Multiple Sequence Alignment (MSA) problem. Based on the strength and adaptability of the Genetic Algorithm (GA) two approaches are developed depending on the MSA type. The first approach, for the non related sequences (no consistency), involves a Hybrid Genetic Algorithm (GA_TS) considering also Tabu Search (TS). The Traveling Salesman Problem (TSP) is also applied determining MSA orders. The second approach, for sequences with consistency, deals with a hybrid GA based on the Divide and Conquer principle (DCP) and it can save space. A consistent dot matrices (CDM) algorithm discovers consistency and creates MSA. The proposed GA (GA_TS_VS) also uses TS but it works with partitions. In conclusion, GAs are stochastic approaches that are proved very beneficial for MSA in terms of their performance.

INTRODUCTION

Sequence Alignment is the procedure of comparing two (pair-wise alignment) or more (multiple sequence alignment) DNA or protein sequences by searching for a series of individual characters or character patterns that are in the same order in the sequences. Sequence alignment is useful for discovering functional, structural and evolutionary information in DNA or protein sequences. It is important to obtain the best possible, so-called "optimal", alignment to discover this information (Mount 2004). Sequences that are very much similar in the parlance of sequence analysis probably have the same function. Additionally, if two sequences from different organisms are similar, there may have been a common ancestor sequence, and the sequences are then defined as homologous. An alignment indicates the changes that could have occurred between the two homologous sequences and a common ancestor sequence during the evolution as shown in Figure 1.

DOI: 10.4018/978-1-4666-9644-0.ch013

Figure 1. The evolutionary relationship between two similar sequences

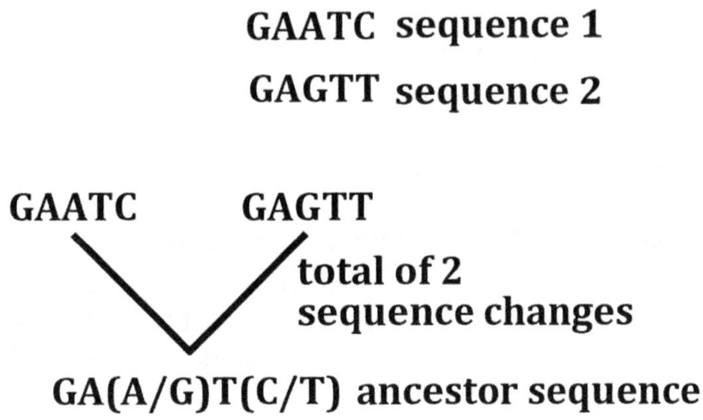

In Figure 1 there is an origin of similar sequences. Sequences 1 and 2 are each assumed to be derived from a common ancestor sequence. Some of the ancestor sequences can be inferred from conserved positions in the two sequences. For positions that vary there are two possible choices at these sites in the ancestor. In the phylogenetic analysis of three or more similar sequences, the separate distances from the ancestor can be estimated (Mount, 2004).

There are two principal methods of pair-wise alignment: the Dot Matrix analysis and the Dynamic Programming (DP) algorithm. Dot Matrix should be considered as a pair-wise (two sequences) alignment method that displays any possible sequence alignment in diagonal. The DP develops a process that generates a matrix of numbers that represents all possible alignments between two sequences (Jones, Pevzner, 2004). The highest set of sequential scores in the matrix defines an optimal alignment. For proteins, an amino acid substitution matrix, such as the Dayhoff percent accepted mutation matrix 250 (PAM250) is used for score matches and mismatches (Mount, 2004)

Multiple Sequence Alignment (MSA) is among the most important tasks in computational biology.

MSA is a very important extension of paiwise sequence alignment where there is a mutual alignment of three or more sequences. Usually we can find large families of similar sequences by identifying homologues in many different species (Lesk, 2012). In biological sequence comparison, emphasis is given to the simultaneous alignment of several sequences. Genetic Algorithms (GAs) are stochastic approaches for efficient and robust search that can play a significant role for MSA. GA is an evolutionary technique for large space search. It is based on principles inspired by the genetic and evolution mechanisms observed in natural systems (Goldberg,1989; Davis,1991). The near global optimal solution for objective function for Fuzzy Non-Linear Industrial Production Planning Problems is obtained by hybrid meta-heuristics optimization algorithms such as line search, genetic algorithms, and simulated annealing (Vasant, 2012).

Solutions for non linear programming problems that arise in industrial production planning in uncertain environments, using a methodology with GA and Hybrid GA are developed (Vasant, 2011). A very detailed discussion on the Fuzzy Rule-Based System is presented along with a procedure that evolves solutions using genetic operators and provide immediate medical decision (Mankand, 2014). A novel hybrid GA-GSA algorithm to permit the reliability analyst to increase the performance of the system by utilizing the uncertain data is presented. The optimal design parameters corresponding to each component of the system are evaluated by solving a nonlinear optimization problem and comparing their results with other methods (Garg, 2015). A hybrid unsupervised learning algorithm, which is termed as Evolutionary Rough Multi-Objective Optimization (ERMOO) algorithm is proposed. The performance is demonstrated on real-life breast cancer dataset for identification of Cancer Associated Fibroblasts (CAFs) within the tumor stroma (Sarkar, & Maulik, 2015). GA can contribute to small Enterpise default prediction modeling. To this end a multiple discriminant analysis and a logistic regression (the two main traditional techniques in default prediction modeling) were used to benchmarking GAs (Gordini 2014).

A solution for multi-objective Optimal Power Flow (OPF) problem via a genetic fuzzy formulation algorithm (GA-FMOPF) has been developed. The objectives and constraints are modelled as fuzzy mathematical programming problem. A GA is used in order to seek a feasible optimal solution to the environmental/economic dispatch problem (Derghal, & Golea 2014). A review of new hybrid methods that deal with the continuous local and global optimization problems for constrained industrial production planning problems with details about all types of optimization methods and approaches for the local and global optimization are highlighted. A GA is used in the hybridization process (Vasant, 2014).

An optimization algorithm is based on the Strength Pareto Evolutionary Algorithm (SPEA-II) and the fitness function is defined so as to minimize the electric vehicle cost (EVC), the electric vehicle weight (EVW) and the losses in the electric vehicle (EVL). (Hadj et all., 2014).

For solving global optimization problems, Algorithmic simplicity and effectiveness of swarm intelligence (SI) techniques have made it a powerful tool. The SI techniques can efficiently applied for clustering (Vora M., Mirnalinee T., 2015). Particle Swarm optimization (PSO) is used also for early detection of malignant melanoma (Popa, 2014).

Approximated solutions for the TSP can be found using a Monte Carlo algorithm (Cerny, 1985). The availability of an Linear Programming routine that can add constraints and reoptimize, makes it possible to adopt an integer programming approach to the TSP (Miliotis, 1975). An efficient technique is developed to solve TSP using GA with the onc point crossover operator (Sharadindu at al,, 2014).

The TSP with precedence constraints (TSPPC) using GA and topological sort is presented (Moon et al, 2002).

Tabu search is stochastic by nature. It is a meta-heuristic approach that is found to be useful in solving combinatorial optimization problems. It can not guarantee to always find an optimal solution even if one exists. Many hybrid Genetic Algorithms have been developed for various purposes. The (Thamilselvan & Balasurbramanie,2009) deals with the job shop and Yard Cranes scheduling problems respectively.

BACKGROUND

One of the most important contributions of biological sequences to evolutionary analysis is the discovery that sequences of different organisms are often related (Mount, 2004). Because the potential for learning about the structure and function of molecules by multiple sequence alignment (MSA) is so great, the necessary computational methods have received a great deal of attention. In MSA, sequences are aligned optimally by bringing the greatest number of similar characters into register in the same column of the alignment.

For the mathematical model of MSA we consider a set S of k strings $v_1,v_2,..v_k$ of length $n_1,n_2, ..n_k$ over an alphabet A. An extended alphabet $A \cup \{ - \}$, it can be also considered when the space is included. The MSA for the set S of strings is specified by a k x n matrix, M, where $n \geq \max_{1 \leq i \leq k} n_i$. The

Figure 2. MSA with consensus

```
            -20                                 1                      20
             |                                  |                       |
talA        CTTTTCAAGG  AGTATTTCCT  ATGAACGAGT  TAGACGGCAT
evgA        CATTGCAAAG  GGAATAATCT  ATGAACGCAA  TAATTATTGA
ypdI        CATTTTCAGG  ATAACTTTCT  ATGAAAGTAA  ACTTAATACT
nirB        GAAAAGAAAT  CGAGGCAAAA  ATGAGCAAAG  TCAGACTCGC
hmpA        TGCAAAAAAA  GGAAGACCAT  ATGCTTGACG  CTCAAACCAT
narQ        TTTTTGTGGA  GAAGACGCGT  GTGATTGTTA  AACGACCCGT
gltF        GTTATTAAGG  ATATGTTCAT  ATGTTTTTCA  AAAAGAACCT
intS        TACCCACCGG  ATTTTTACCC  ATGCTCACCG  TTAAGCAGAT
yfdF        AATCAAAATG  GAATAAAATC  ATGCTACCAT  CTATTTCAAT
dsdX        ATCACAGGGG  AAGGTGAGAT  ATGCACTCTC  AAATCTGGGT
suhB        ACATCCAGTG  AGAGAGACCG  ATGCATCCGA  TGCTGAACAT

Consensus   AATTTAAAGG  AGAATTACCT  ATGAACGCAA  TAATAAACAT
```

M matrix contains as rows the strings after their alignment. No column in a MSA contains only spaces. The score of the MSA is defined to be the sum of scores of columns. The optimal alignment can provide the maximum value of scores. This project deals with hybrid methods to achieve an optimum score of the sequence alignment.

MSA has a variety of applications e.g. finding motifs and conserved regions in protein families, analyzing evolutionary relations between proteins and secondary/tertiary structure prediction. It is a basic tool when grouping proteins into functionality related families. From a MSA can create a consensus, which is a pseudo-sequence representing the complete alignment. The consensus in turn can be used to identify conserved regions. MSA can also be used when deciding if a protein belongs to a family or not by aligning the qyery-sequence together with known proteins belonging to the family.

If the amount of sequence variation is great, it may be very difficult to find an optimal alignment of the sequences because so many combinations of substitutions, insertions, and deletions, each predicting a different alignment, are possible. In Figure 2 there is an example of MSA with the consensus.

In DNA, promoter sites or other DNA binding sites are highly conserved (see Figure 2). This is also the case for repressor sites as seen for the Cro repressor of bacteriophage λ. When aligning such sequences, regardless of whether they are highly variable or highly conserved at specific sites, it is very difficult to generate a consensus sequence which covers the actual variability of a given position. A consensus sequence is also given at the last line from the alignment of the previous sequences. The highly conserved area define the consistent area (consistent case) while the highly variable is considered as the inconsistent area (the inconsistent case). In addition MSA can also be used as a starting point for secondary or tertiary structure prediction. After the alignment of the related sequences, the conserved regions can give a hint of secondary or tertiary structure, since the structure is usually conserved during the evolution, but the actual sequence is likely to vary in certain parts. The tertiary structure is more evolutionary conserved than in primary structure (Berg, Tymoczko, Stryer, 2002). For example, the conservation is apparent in the tertiary structures of the globins for the human hemoglobin (a chain), human myoglobin, and lupine leghemoglobin (Figure 3).

The main purpose of this work is to solve the MSA problem by combining Genetic Algorithm and Tabu Search (TS) methods. The main reason of using GA is that it performs a global search rather than a local search. Tabu search is basically used for a local search. This heuristic will provide better searching possibilities since it combines algorithms that complete with each other on their operations. In the

Figure 3. Conservation of three-dimensional structure

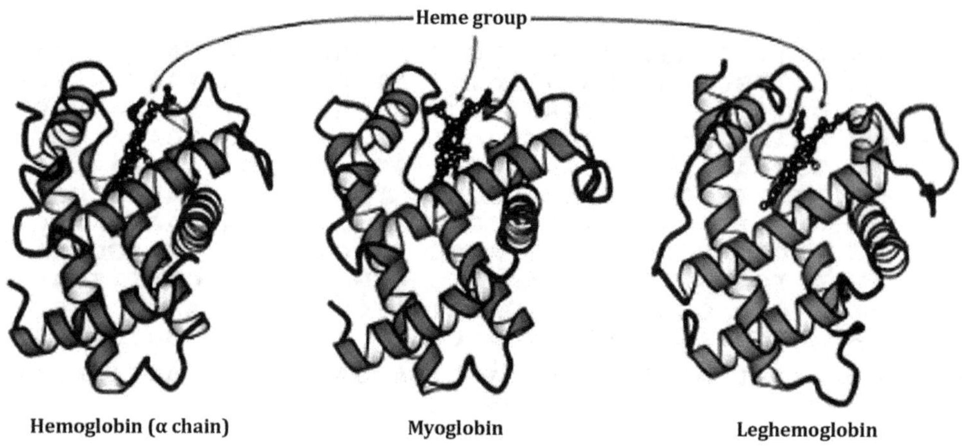

MSA the order of sequences' process plays significant role. Unfortunately it is not always possible to combine optimal pairwise alignments into a multiple alignment since some pairwise alignments may be incompatible (Jones, & Pevzner, 2004). As a result, some MSA algorithms attempt to combine some compatible subset of optimal pairwise alignments into a multiple alignment.

GA, with its global search, can select many features at a time while TS selects one feature at a time. Another approximation algorithm for finding MSA is developed using the Traveling Salesman Problem (TSP) based on recursion. The TSP provides the order at which the strings will be processed for the MSA in order to provide optimum multiple alignments. Although TSP is NP-hard, it has been studied extensively.

Sequences with similar regions are related sequences and have consistency while the no consistency ones have not similar regions. Two approaches are developed; the no consistency (no related sequences) and the consistency case.

For the no consistency case, a Hybrid Genetic Algorithm (GA_TS) has been developed for solving the MSA problem based on the TSP approach. The Traveling Salesman Problem (TSP) algorithm determines a circular tour in which the sequences are aligned. The TS can solve the TSP. The combination of GA, TS incorporates additional mechanisms for the alignment and provides better results than the classical GA solution for the MSA. This improvement is particularly useful for the global optimum search. GA improves the whole population while TS aims at producing one best solution.

For the consistency case, a hybrid GA based on Divide and Conquer Principle (DCP) is applied. The Divide and Conquer principle can ensure undisturbed consistency during vertical sequences' segmentations. The Consistent Dot Matrices (CDM) algorithm discovers subsequences with consistent areas and creates MSA using dot matrices that can provide faster solutions. CDM, with its internal phase, can minimize the number of pairwise comparisons in MSA thus giving a better performance. Based on the ability to keep the consistency with CDM, the hybrid GA, with vertical sections (GA_TS_VS), provides a solution with smaller size sequences giving a new dimension to the MSA problem.

The proposed DCP, having the property of using the predefined area of the cut points, can diminish the search space and provide better performance than the GA. Notation with meanings is shown in Table 1.

Table 1. Useful Acronyms

DCP	Divide and Conquer
CDM	Consistent Dot Matrix
GA_TS	Genetic Algorithm with Tabu Search
GA_TS_VS	Genetic Algorithm with Tabu Search and vertical segmentation
TSP	Traveling Salesman Problem
MSA	Multiple Sequence Alignment
TS	Tabu Search
DP	Dynamic Programming

The chapter is organized as follows. Section 2 contains the consistent case with DCP, CDM and GA_TS_VS. In Section 3, the inconsistent case with MSA, GA, GA_TS are described. Finally, simulation results are provided in Section 4.

THE CONSISTENT CASE

Divide and Conquer

If the size of the sequences, is greater than a bound then it is better to work with divide and conquere method with GA. Divide and conquer principle: each of the given k sequences is being cut at an appropriate chosen site somewhere near to its midpoint, this reduces the original alignment problem to two subproblems by aligning the two resulting groups of prefix and suffix subsequences, respectively. The main problem is to find a tuple of slicing sites $(c_1,c_2,c_3,..,c_k)$ so that the simple concatenation of the two optimal alignments of the prefixes $s_1(\leq c_1),s_2(\leq c_2)\ldots,s_k(\leq c_k)$ and the suffixes $s_1(>c_1)$, $s_2(>c_2)$, $\ldots,s_k(>c_k)$ forms an optimal alignment of the original sequences .

Consistent Dot Matrix

The consistency alignable sequence fragments for a set of sequences can be implemented by using dot matrices. The criterion of consistency can guarantee the creation of consistent regions. By applying this criterion CDM discovers common subsequences.

CDM has three phases: (1) the dot matrices preparation (2) the scan phase, and (3) the link phase. In the first, second and third phase the dot matrices, diagonal similarity, and vertical continuity, are examined, respectively.

Working with a set of dot matrices poses the problem of how to define *consistency.* Scanning of graphs for the series of 1s reveals similarity or a string of same characters. The steps of the algorithm:(1) prepare the dotplot (Davis, 1991) examine the horizontal continuity, preparation phase, (2) find gaps (horizontal), examine the diagonal location, (3) scan phase, (4) examine the vertical continuity, the link phase. The (1),(2) belong to the preparation phase. Create vectors that hold the locations of 1s (as it is shown in the example 1). In the link phase the locations are combined. The advantage of CDM is that with the link phase, only (n-1) comparisons are needed (instead of n), where n is the # of sequences.

```
  CDM /*input: the strings into 2 dim array in[][]  */
/*output:  MSA using pointers method, out[][] */
/*variable: ma[][], hold 1's in similar positions */
/*fin[][]: hold the similar locations of all the pairs (there are k-1 pairs
     for k sequences)  */
/*hor[][]: for horizontal/vertical gaps '-' in locations */
/*V arrays for similar items (k-1) arrays for k sequences*/
/*scan phase */
for each pair of strings
find the similar items, seed (m: location)    extend it
           (m+1,...)
/*update the positions of fin[][] with similar items  */
fin (i,j) =1;
/*gap */
if  a(j) = b(j) { there is no gap }
  if  a(j) = b(k)  (k≠j)
      /*there are k-j+1 gaps */
      /* horizontal (when k>j), vertical (when k<j)  */
      for j=1; j<=k;j++)
         hor(i,j)="-";
/*link phase */
/*   examine  the vertical continuity using V arrays */
   if (the two vectors are in sequence)
       { link the two vectors
        append the location in the final vector}
```

Example 1

Lets consider the sequences: A: a b c d e, B: b c x, C: a b c

Phase One: Preparation

The dotplot A-B:

	a	B	c	d	e
b	-	1			
c			1		
x					

There is a gap(horizontal) before the match b-b. Hold the horizontal gap into an array hor[i,j]: i: the number of the string, j: is the location of the gap.

The dotplot B-C:

	b	c	x
a	-		
b		1	
c			1

Phase Two: Scan

Read plot A-B: Find position 2(A) and 1(B) have similar item: b . Find position 3(A) and 2(B) have similar item: c

There is continuity (horizontal) since there are numbers the one next to the other (2(A), 3(A)), (1(B),2(B)). There is a gap horizontally.

Create the vectors with the vertical locations of the same item b: V_2: 2, 1. Create the vectors with the vertical locations of the same item c: V_3: 3, 2

Read plot B-C: Find position 1(B) and 2(C) have similar item: b. There is a vertical gap. Find position 2(B) and 3(C) have similar item: c

There is also continuity (horizontal), since there are numbers next to each other (1(B), 2(B)), (2(C),3(C))

Create the vectors with the vertical locations (a) of the same item b: V_1: 1, 2 and (b) of the same item c: V_2: 2, 3

Phase Three: Link

The vectors can be appended since they have something in common: for $V_2 - V_1$: 1 (last), $V_3 - V_2$: 2(last)

Hence, the final vectors are: FV_2: 2, 1, 2 for b and FV_3: 3, 2, 3 for c. After that, and by using the hor[] the MSA, there is vertical continuity for b and c as follows:

A:	a	b	c	d	e
B:	-	b	c	x	
C:	a	b	c		

CDM can cooperate with GA under DCP by using the Divide and Conquer method. The DCP framework that contains both CDM and GA provides better performance than the classical GA.

Divide and Conquer Genetic Algorithm

In the second approach, where there are sequences with consistency of a Hybrid GA, the DCP is developed. The Divide and Conquer technique (Stoye *et al.*, 1997; Chen & Lin 2007) can effectively reduce the space complexity for MSA so that the consistency achieved from the CDM is not broken. The cut points are defined so that they do not disturb the consistency. The subsequences, not covered by CDM, could be at any segment of the sequences (left, right, intermediate).

Example 2

A1= A A T T C C T

A2= A A T C T

A3= A A T T C G A

The sequences need to be cut according to the cut-points (the locations with no consistency) so that they do not disturb the consistency. The subsequences: set SA11= A A T SA12= T C C T

SA21= A A T SA22= C T

SA31= A A T SA32= T C G A

Scores for each column (pair work): match pair (2), mismatch pair (-1), space (-2). Then we calculate the score of multiple DNA by taking the summation of the score of each column of the sequences. We can merge columns in order to increase the score of alignment and decrease its length.

Genetic Algorithm with Tabu Search and Vertical Selection

The GA_TS_VS follows the DCP. GA_TS_VS works as GA_TS but with sequences of smaller size. Hence, the GA_TS_VS is applied first for the subsequences, after making the vertical partitions, and secondly for the merge operation. GA_TS_VS has superiority over the GA where the chromosome is used from DP without any cut. The pseudocode is as follows:

```
GA_TS_VS: input: the alignment of the sequences
output: new alignment
Step 1: prepare the vertical segmentation
Step 2: prepare the new alignment using GA for
        the not covered by CDM
        subsequences (or segment)
```

THE INCONSISTENT CASE

Multiple Sequence Alignment

Due to computational complexity, MSA is identified as one of the most challenging tasks in bioinformatics. This problem is proved to be a NP-complete (Wang & Jiang, 1994) with a complexity O(n, k) = $(n+1)^k(2^k-1)k^2$ when Dynamic Programming (DP) is used to optimize the sum of pairs of k sequences with maximum length n. Even today, MSA is solved through heuristics such as progressive alignment or clustering methods which offer a partial solution and operate in reasonable time for a large search space. In (Edgar, 2004), a multiple sequence alignment method with improved accuracy and speed is shown. The (Ergezer & Leblebicioglu, 2004) presents a multiple sequence alignment method using the hidden Markov model. In the (Hunt, et al.,2003) a linear programming based algorithm has been developed for multiple sequence alignment. The distance table is a two dimensional table that has scores between two sequences provided by the Needleman-Wunsh DP algorithm. For n sequences of length L, the time complexity for multidimensional dynamic programming is O($2^n L^n$) and the space complexity is $O(L^n)$. This approach is so computationally expensive that it is considered impractical for n > 4.

Genetic Algorithm

GA is an evolutionary technique for large space search. It is based on principles inspired by the genetic and evolution mechanisms observed in natural systems (Goldberg,1989; Davis,1991).

A GA begins with a set of solutions represented by a group of chromosomes called the population. A new population can be generated by borrowing solutions from the current population or by applying genetic operators such as selection, crossover, and mutation to current population. The new population must be better than the old one.

By imitating the evolution process of organisms in the complicated natural environment, an evolutionary system contains the operators of selection, crossover and mutation. In genetic algorithms, chromosomes form a population and a set of genes form a chromosome. There are three operations are: (a) Selection: The system selects chromosomes to perform crossover and mutation operations according to the fitness values of the chromosomes. The larger the fitness value of a chromosome, the higher the chance of the chromosome to be chosen. The roulette wheel selection method is used. (b) Crossover: The system randomly chooses a pair of chromosomes from a population, and a crossover point is randomly generated. The single-point crossover method is used. (c) Mutation: a selected chromosome will mutate or not.

GA includes three phases for searching: (1) creation of an initial population, (2) evaluation of fitness function, (3) production of a new generation. GAs perform a globalized search for solutions whereas most of the other clustering procedures perform just a localized search. A fitness value is associated with the distance between two sequences.

For the GA, the new population must be better than the old one. GA improves the whole population while TS aims at producing the best solution.

The pseudocode of GA is:

```
Generate the initial population P;
  for i = 1 to genSize do
   for j = 1 to (popSize/2) do
       Select two chromosomes X and Y
                  from P by Selection operation;
       Crossover(X, Y) and let X'and Y'   be the offspring;
        Mutate(X');          Mutate(Y');
        Put X'and Y'into the mating pool M
        /*M is the new  population*/
   end
end
```

Tabu Search

Tabu search is a meta-heuristic approach that is found to be useful in solving combinatorial optimization problems designed to guide the search operator in avoiding the traps of local optimality(Golver & Lagura, 1997). Tabu Search (TS) can be directly applied to virtually any kind of optimization problem. TS also allows to escape from sub-optimal solution by the use of a tabu list (TL).

The general idea of TS is to search the space by choosing a point and going to its neighbour that is not in the TL. The TL restricts the search to a subset of admissible moves. The goal is to permit "good" moves in each iteration with exchanges and with no re-visiting solutions. The TL is generated by adding the last solution to the beginning of the list and discarding the oldest solution from the list (Satapathy & Subhashini.,2008). The update of the list is based on some memory structure (short-term memory).

Basically, the TS algorithm employs a short-term memory to make a move by a controlled randomization and a long-term memory to diversify the search space using the frequency information of the moves. The short term memory is used for storing the components utilizing a list with information for how long an element should remain in memory. A move should remain in Tabu for 5 iterations. There are two memory based principles of TS, the intensification and diversification. Intensification strategies are based on modifying the choice rules that encourage the move combinations and solution features that have been historically found good (Thamilselvan & Balasurbramanie,2009). Diversification strategies on the other hand encourage the search process to examine unvisited regions and generate solutions that are significantly different from those already visited.

In addition, two switch operations are developed; (a) the insert, where two selected elements of an individual change positions and the later element takes earlier position. (b) the disorder, where elements of a sublist are rearranged around the mid-point. If f1={1,2,3,4,5,6,7,8} the disorder operation produces f1' = {1,2,3,6,7,5,4,8}.

Tabu search algorithm is used to generate new members in the new population (up to 15%) as neighbours to randomly selected solutions in the current population

Travelling Salesman Problem

The Traveling Salesman Problem is the following: given a set of n nodes and distances for each pair of nodes, find a round trip of minimal total length visiting each node only once. In the symmetric traveling salesman problem, the distance from node i to node j is the same as from node j to node i. Pairwise alignments are performed using the Needleman-Wunsch DP alignment. The SP scores are computed and converted to distances so that the TSP approach is used. In (Edgar, 2004) the MSA is solved using TSP. The combination of GA and TS is used to solve the MSA problem according to the TSP approach.

The aim of MSA problem is to find the order of alignment, so that the sum of pairs is optimal. In TSP the objective is similar since it seeks the cities' visitation order so that the total distance traveled is minimized. In this sense, the MSA problem can be treated like a TSP problem where a sequence s_i is viewed as the ith city and the distance $d(s_i; s_j)$ is related to the SP score of the s_i and s_j sequences pair. In the TSP algorithm sequences are aligned progressively by repeatedly merging the closest nodes along the TSP tour (Korostensky & Gonnet, 1999). For GA the fitness is the reverse of the distance.

Genetic Algorithm with Tabu Search

The pseudocode of GA_TS is as follows:

```
GA_TS: input: set of sequences, n_it: is the # of iteration, t: size of TL,
halting="no"
output: MSA
/*prepare the distance matrix */
D= {dij} between each pair of sequences
/* GA, initialize the variables of GA */
initialize Tabu List (TL) = empty
create the initial population (j=1)//randomly selection of
                                                      chromosomes
while (j≤ n_it)
  {evaluate each chromosome
   calculate the fitness function of each chromosome
   /*TS,apply operators of GA to generate new population */
   select the best solution with TS //randomly
```

```
  generate a single initial  solution S //new smaller size
                                                         population
  let S* =S be the best solution so far  (for alignment)
  do
   {select the next  points in neighborhood of S (N(S))
    main and various swap operation in the neighborhood
    create the alignments for N(S) (or get the alignments scores for
        the distance table)
    calculate fitness of each neighbour in the set
    sort neighbours (ascending order) according to their fitness
    //test for feasibility for the set of neighbours
    //let B is the alignment of the N(S) sequence (considering the
      order of the alignment)
    if (B is not Tabu: B ∉ TL)                /* Step A */
     {
        accept move B, S=B }                  /* Step B */
        update TL (TL= TL ∪B)

        /*TSP approach */
        if  (cost(B) < cost(S*)  //better solution
            { accept move: S=B
               update best: S*=B  }
        else (best move: S*   -remain the same-)
         if (tabu list (TL) > t)
             {remove the oldest solution from  it}
        /*test for finding different composition points */
        /*switch operations, different composition   /*step C*/
        /*test for  feasibility   (the same: step A, step B)
        /*continue after step B ..step C */
      if (B is in Tabu: B ∈ TL)
        {continue with the next move in a new point
            (the next point of the search space from
                neighborhood)
          }
      if (solution with the smaller cost has found)
          {halting = "yes"}
  }
   while (halting ="no")  //TS end
/*GA continue */
apply crossover to complete the # of the new population
current population = new population
}/*end GA */
```

The steps (A),(B) are relevant to the feasibility test for MSA.

With the main swap operation the next candidate solution is discovered from the neighbourhood (intensification). After finding a solution, the test for discovering a "different composition" solution is examined by (step C) switch operations (diversification). In the tabu list (TL or tlist1) if the length is greater than a predefined value t then the oldest solution is removed from it. As far as the combination of the two parts of the GA_TS are concerned, for solving the MSA problem, the GA may need more generations to get a success.

The acceptance criterion determines whether a solution from the heuristic can be used as a starting point for the next iteration. It provides a balance between intensification and diversification of the search space. In general, GA_TS follows procedures similar to the GA as outlined previously but for the selection process, GA_TS uses the TS process including the switch operations for providing diversifying search.

The flow chart of Figure 4 describes the main step of the GA_TS algorithm.

SIMULATION

a. *GA_TS_VS vs GA*. The GA_TS_VS and GA have been developed with five sequences. In Figure 5 GA_TS_VS (using the predefined area of cut points) has superiority over the GA. This is due to the fact that the vertical segmentation of the sequences at the cut points is defined by CDM, minimizes the search space efficiently. On the other hand, GA is more time consuming since it uses the DP for the alignment.
b. *GA_TS_VS for various sequence sizes*. Different size of sequences is compared with their running time using GA_TS_VS. In Figure 6 there is the case for two sets (5, 8) of sequences where the smaller size sequences have better performance.
c. *GA_TS_VS vs GA_TS*: Thc GA_TS_VS outperforms GA_TS because it works with subsequences which have smaller size than the sequences of GA_TS. In addition, Figure 7, GA_TS execution time is longer than GA_TS_VS since it uses DP. DP has exponential run time for large size sequences.
d. *GA_TS_VS vs GA_TS with longer similar subsequences:* It is assumed that there are sequences of the same size as the ones of the previous case. Unlike the previous case, we assume: (a) greater size of similar sebsequences than the previous ones and (b) lower number of subsequences. The GA_TS_VS outperforms again the GA_TS. In addition, in Figure 8, GA_TS_VA has lower value (11.8) than the one of the GA_TS_VS (12.45) of Figure 3. The greater the size of similar sebsequences, the lower the value of the execution time with the GA_TS_VS. This is due to the fact that sequences of the same size with more similar items result to smaller number of groups to connect (link phase).
e. Comparing: CRUSTALW and our methods: using the various data sets from the sets of (Zhang and Wong, 1997). Table 2 contains the results of the comparison CRUSTALW and our proposed algorithms

CLUSTALW is a program that produces sequence alignment and is based on pairwise alignment of all the sequences (Mount, 2004).

The scores for GA_TS are computed for each column of the alignment, the match pair score, the mismatch pair score and the pace pair score are +2,-1 and -2 respectively. Data set 1 contains 10 sequences which are very similar and the average size of data set 1 is 212 nucleotides

Figure 4. Flow chart of GA_TS algorithm

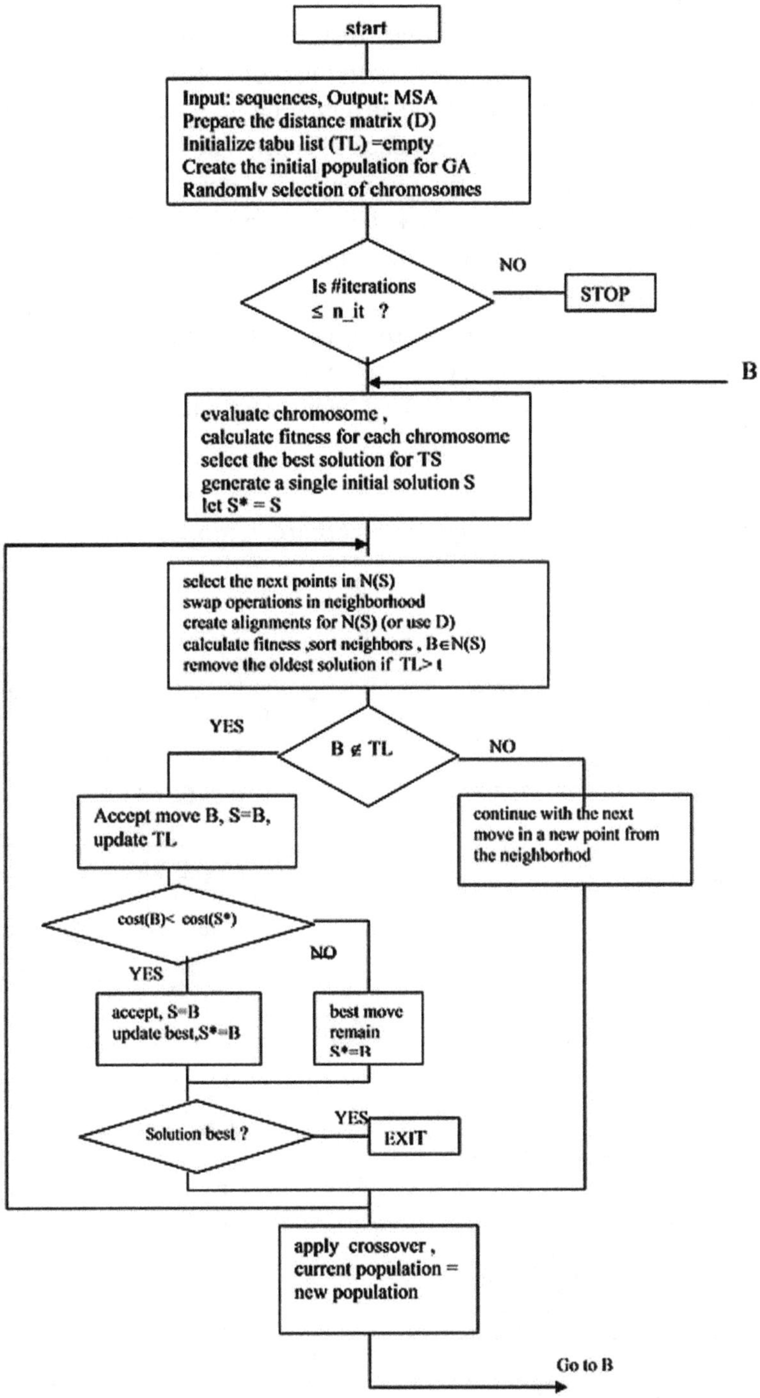

Data set2 contains 8 sequences (16S rRNA) from bacteria. This data set is similar to data set 1 but with higher average length.

Data set 3 contains sequences from the histone H3-II gene from 21 species (Tetrahymena). It has about the half the size of the data set 1 and twice the number of sequences relative to data set1.

The GA_TS and GA_TS_VS spend longer time on data set with smaller number of sequences. Hence CLUSTALW give better time for the shorter sequence data set. The time of GA_TS_VS is greater that the one of GA_TS because of the additional time to merge the subsequences.

For the algorithm programming, Java and C++ were preferred. For the GA the binary string representation for the chromosomes was used. This allows binary mutation and crossover. For d. experiment the long size sequences had 100-200 characters.

All data from experiments a to d were created by a set of programs that generate randomized DNA sequences using different parameters such as: the number of strings, the maximum and minimum length of sequences. Programs also made random DNA fragments from nucleotides and considered them into

Figure 5. GA_TS_VS and GA for various sizes of sequences

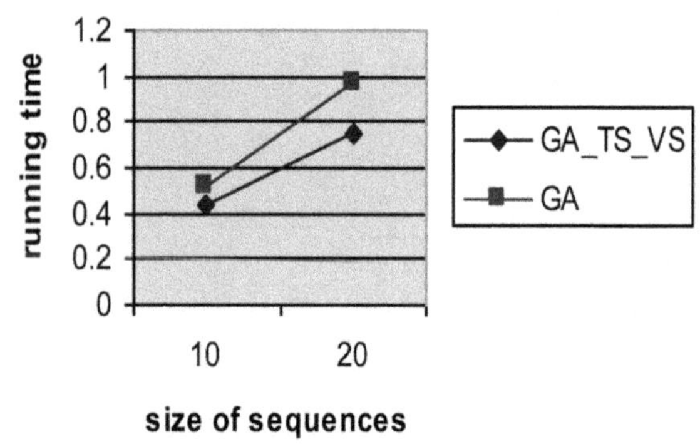

Figure 6. MSA for two sets (5, 8) of sequences

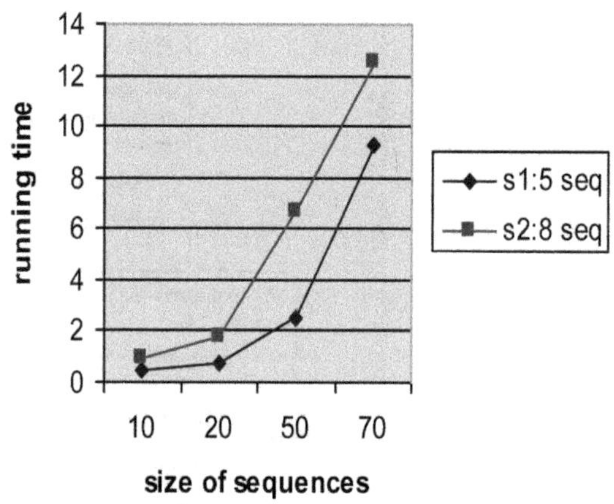

Figure 7. GA_TS_VS better than GA_TS

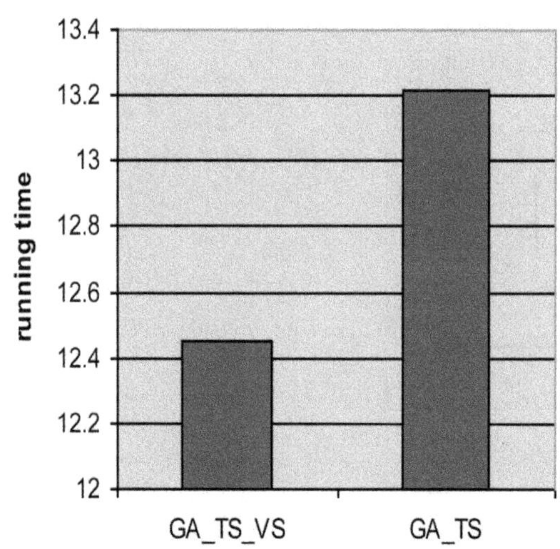

Figure 8. GA_TS_VS better than GA_TS for longer similar subsequences

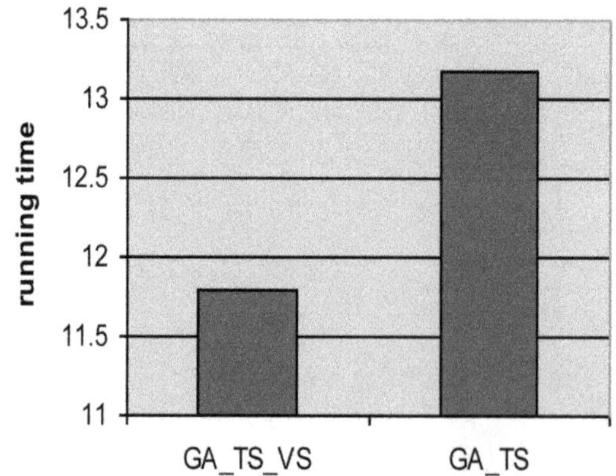

Table 2. CRUSTALW vs GA_TS and GA_TS_VS

Data Set	Number of Sequences	Average Size (Max,Min)	CRUSTALW No of Matches Time (tu)	GA_TS No of Matches Time (tu)	GA_TS_VS No of Matches Time (tu)
Set1	10	211(211,211)	198, 0.42	198, 0.72	198,0.75
Set2	8	457(457,457)	449, 0.68	449, 0.56	449,0.60
Set3	21	122(122,122)	109, 0.57	109,0.61	109,0.64

the sequences for making the matching. After that, the sequences were processed by the main programs of GA. For the experiment e the three real data sets (set1,set2,set3) are from the Zhang and Wong, 1997 paper.

For the GA_TS_VS implementation, the discovery of consistency areas is first needed. To this end the CDM is used. The data for c and d experiments has been created by using the additional parameter of the common size subsequences, in addition to the previous parameters.

DISCUSSION

In general the GA_TS follows procedures similar to the GA, outlined previously. For the selection process GA_TS uses the TS process including the switch operations for providing diversifying search. Before starting the MSA process it is better to know if the sequences have similarities in order to apply either one of the proposed solutions. Another point is the use of cut points for larger size sequences.

The Divide and Conquer Algorithm based on keeping the consistency can provide the opportunity to have better results because the cut points define smaller size of sequences which in turn speed up the process with GA and TS. The advantage of GA_TS and GA_TS_VS is that they provide better time for a small number of long size sequences. On the contrary, for any number of smaller size sequences these algorithms do not work so well. This is a limitation of this work. For GA_TS_VS the cut-point set of multiple DNA sequence set need to be discovered and then work to be continued with each subsequence. Generally a disadvantage of GA is that it can be very slow especially for long size sequences. The problem is that once a local maximum has been reached, it can often be a long time before a string is produced that escape from the local maximum and finds another higher, maximum. In addition, we can't see how well the GA is doing since we do not know anything about the fitness landscape.

FUTURE RESEARCH DIRECTIONS

Future work could include the MSA with Neural Networks for secondary structure prediction using consensus. The problem of over-fitting will also be examined. Measures for prediction accuracy will also be selected. Another option could be evolution study and phylogeny estimation from molecular sequences. Phylogenetic trees and MSA will also be examined. The problem will focus on the be construction of a more accurate tree when highly divergent sequences are examined. The possibility of inferring phylogenies as an alternative to MSA could be examined.

An alternative method for pattern searching is to produce trial alignments and then improve these alignments using statistical methods. To this end, the Expectation maximization (EM) algorithm to identify conserved domains in unaligned protein sequences will be used. Comparison with other statistical methods in sequence analysis will also be of our interest. All of these topics are very important to the Bioinformatics research area.

CONCLUSION

Based on Genetic Algorithm and Tabu Search a set of algorithms for finding MSA has been developed. The GA_TS with the use of TSP based on DP provides a solution that is applied basically for non related sequences. On the other hand the hybrid GA (GA_TS_VS), following the Divide and Conquer method (DCP) and using the CDM as internal part, can provide a solution for MSA utilizing appropriate cut points. It provides a faster MSA solution for the sequences with consistency. This last approach can facilitate the solving of the MSA problem and it is preferable for sequences that have common subsequences.

ACKNOWLEDGMENT

The author would like to thank all reviewers for their valuable and constructive comments that significantly contributed to the improvement of this paper. Their careful and detailed reviews led to a clarified presentation of this paper.

REFERENCES

Berg, M., Tymoczko, L., & Stryer, L. (2006). Biochemistry (6th ed.). New York, NY: Freeman W.

Cerny, V. (1985). Thermodynamical Approach to the Traveling Salesman Problem: An Efficient Simulation Algorithm. *Journal of Optimization Theory and Applications*, *45*(1), 41–52. doi:10.1007/BF00940812

Chen, S., & Lin, C. (2007). Multiple DNA sequence alignment based on genetic algorithms and divide and conquer techniques. *Information and Management Science Journal*, *18*(2), 76–111.

Davis, L. (1991). *Handbook of Genetic Algorithm*. New York, NY: Van Nostrand.

Derghal, A., & Golea, N. (2014). Multi-Objective Generation Scheduling Using Genetic-Based Fuzzy Mathematical Programming Technique. In P. Vasant (Ed.), Handbook of Research on Novel Soft Computing Intelligent Algorithms: Theory and Practical Applications (pp. 450-474). IGI Global. doi:10.4018/978-1-4666-4450-2.ch015

Edgar, R. C. (2004). MUSCLE: Multiple sequence alignment with improved accuracy and speed. In *Proceedings of the 2004 IEEE Computational Systems Bioinformatics Conference* (pp. 728-729). doi:10.1109/CSB.2004.1332560

Ergezer, H., & Leblebicioglu, K. (2004). Multiplesequence alignment using hidden Markov model. In *Proceedings of the 2004 IEEE 12th Signal Processing and Communications Applications Conference* (pp.434-437). doi:10.1109/SIU.2004.1338556

Garg, H. (2015). A Hybrid GA-GSA Algorithm for Optimizing the Performance of an Industrial System by Utilizing Uncertain Data. In P. Vasant (Ed.), Handbook of Research on Artificial Intelligence Techniques and Algorithms (pp. 620-654). IGI Global. doi:10.4018/978-1-4666-7258-1.ch020

Glover, F., & Laguna, M. (1997). *Tabu Search*. Boston, MA: Kluwer Academic Publishers. doi:10.1007/978-1-4615-6089-0

Golberg, D. (1989). *Genetic Algorithm in Search, Optimization and Machine Learning*. Cambridge, MA: Addison Wesley.

Gordini, N. (2014). Genetic Algorithms for Small Enterprises Default Prediction: Empirical Evidence from Italy. In P. Vasant (Ed.), Handbook of Research on Novel Soft Computing Intelligent Algorithms: Theory and Practical Applications (pp. 258-293). IGI Global. doi:10.4018/978-1-4666-4450-2.ch009

Hadj, N., Kammoun, J., & Neji, R. (2014). Application of Evolutionary Algorithm for Triobjective Objective: Electric Vehicle. *International Journal of Energy Optimization and Engineering*, *3*(3), 1–19. doi:10.4018/ijeoe.2014070101

Hunt, F. Y., Kearsley, A. J., & O'Gallagher, A. (2003). A linear programming based algorithm for multiple sequence alignments. In *Proceedings of the 2003 IEEE Bioinformatics Conference* (pp. 532-533). doi:10.1109/CSB.2003.1227398

Jones, N., & Pevzner, P. (2004). *An Introduction to Bioinformatics Algorithms*. Cambridge, MA: MIT Press.

Korostensky, C., & Gonnet, G. (1999). Near optimal multiple sequence alignments using a traveling salesman problem approach. In *Proceedings of String Processing and Information Retrieval Symposium, and International Workshop on Groupware* (pp. 105–114). doi:10.1109/SPIRE.1999.796584

Lesk, A. (2012). *Introduction to Genomics*. New York, NY: Oxford University Press.

Mankad, K. (2015). An Intelligent Process Development Using Fusion of Genetic Algorithm with Fuzzy Logic. In P. Vasant (Ed.), Handbook of Research on Artificial Intelligence Techniques and Algorithms (pp. 44-81). IGI Global doi:10.4018/978-1-4666-7258-1.ch002

Miliotis, P. (1976). Integer Programming approaches to the Traveling Salesman Problem. *Journal of Mathematical Programming*, *10*(1), 367–378. doi:10.1007/BF01580682

Moon, C., Kim, J., Choi, G., & Seo, Y. (2002). An efficient genetic algorithm for the traveling salesman problem with precedence constraints. *European Journal of Operational Research, EJOR*, *140*(3), 606–617. doi:10.1016/S0377-2217(01)00227-2

Mount, D. (2004). Bioinformatics: Sequence and Genome Analysis (2nd ed.). New York, NY: Cold Spring Harbor Laboratory Press.

Popa, R. (2014). Melanocytic Lesions Screening through Particle Swarm Optimization. In P. Vasant (Ed.), Handbook of Research on Novel Soft Computing Intelligent Algorithms: Theory and Practical Applications (pp. 355-384). IGI Global doi:10.4018/978-1-4666-4450-2.ch012

Sarkar, A., & Maulik, U. (2015). Cancer Biomarker Assessment Using Evolutionary Rough Multi-Objective Optimization Algorithm. In P. Vasant (Ed.), Handbook of Research on Artificial Intelligence Techniques and Algorithms (pp. 509-535). IGI Global doi:10.4018/978-1-4666-7258-1.ch016

Satapathy, J., & Subhashini, K. (2008). A tabu based Neural Network training algorithm for equalization of communication channels. In *Proceedings of International Conference on Artificial Intelligence* (pp.79-85).

Sharadindu, R., Uttam, P., & Nikhil, S. (2014). Efficient technique to solve Travelling Salesman Problem using Genetic Algorithm. *International Journal of Computer Engineering and Technology*, *5*(3), 132–137.

Stoye, J., Dress, A., & Perrey, S. (1997). Improving the Divide and conquer Approach to Sum-of-pairs Multiple Sequence Alignment. *Applied Mathematics Letters*, *10*(2), 67–73. doi:10.1016/S0893-9659(97)00013-X

Thamilselvan, R., & Balasubramanie, P. (2009). Integrating Genetic Algorithm, Tabu search Approach for Job Shop scheduling. *International Journal of Computer Science and Information Security*, *2*(1).

Vasant, P. (2011). Solving Fuzzy Optimization Problems of Uncertain Technological Coefficients with Genetic Algorithms and Hybrid Genetic. In P. Vasant (Ed.), Innovation in Power, Control, and Optimization: Emerging Energy Technologies (pp. 344-368). IGI Global.

Vasant, P. (2012). Hybrid Linear Search, Genetic Algorithms, and Simulated Annealing for Fuzzy Non-Linear Industrial Production Planning Problems. In P. Vasant (Ed.), Meta-Heuristics Optimization Algorithms in Engineering, Business, Economics and Finance (pp. 87-109). IGI Global.

Vasant, P. (2014). Hybrid Optimization Techniques for Industrial ProductionPlanning: A Review. In P. Vasant (Ed.), Handbook of Research on Novel Soft Computing Intelligent Algorithms: Theory and Practical Applications (pp.1-40), IGI Global.

Vora, M., & Mirnalinee, T. (2015). From Optimization to Clustering: A Swarm Intelligence Approach. In P. Vasant (Ed.), Handbook of Research on Artificial Intelligence Techniques and Algorithms (pp. 594-619). IGI Global doi:10.4018/978-1-4666-7258-1.ch019

Wang, L., & Jiang, T. (1994). On the complexity of multiple sequence alignment. *Journal of Computational Biology*, *1*(4), 337–348. doi:10.1089/cmb.1994.1.337 PMID:8790475

ADDITIONAL READING

Blum, C., & Roli, A. (2003). Metaheuristics in combinatorial optimization: Overview and conceptual comparison. *ACM Computing Surveys*, *35*(3), 268–308. doi:10.1145/937503.937505

Fitch, W. M., & Smith, T. (1983). Optimal sequence alignments. *Proceedings of the National Academy of Sciences of the United States of America*, *80*(5), 1382–1386. doi:10.1073/pnas.80.5.1382 PMID:16593289

Huang, X. (1994). On global sequence alignment. *Journal of Computer Applications in the Biosciences,* (10), 227-235

Huang, X., Hardison, R.C., & Miller, W. (1990). A space-efficient algorithm for local similarities. *Journal of Computer Applications in the Biosciences,* (6), 373-381

Huang, X., & Miller, W. (1991). A time–efficient, linear-space local similarity algorithm. *Advances in Applied Mathematics*, *12*(12), 337–357. doi:10.1016/0196-8858(91)90017-D

Kim, J., Pramanik, S., & Chung, M. J. (1994). Multiple Sequence alignment using simulated annealing. *Computer Applications in the Biosciences*, (4): 419–426. PMID:7804875

Needleman, S. B., & Wunsch, C. (1970). A general method applicable to the search for similarities in the amino acid sequence of two proteins. *Journal of Molecular Biology*, *48*(3), 443–453. doi:10.1016/0022-2836(70)90057-4 PMID:5420325

Pearson, W.R., & Miller, W. (1992). Dynamic programming algorithm for biological sequence comparison. *Journal of Methods in Enzymology*, (210), 575-601.

Pevzner, A. P. (2000). *Computational Molecular Biology: An Algorithmic Approach*. Cambridge, MA: MIT Press.

Thomson, J. D., Plewniak, F., Ripp, R., Thierry, J., & Poch, O. (2001). Towards a reliable objective functions for Multiple sequence Alignment. *Journal of Molecular Biology*, *314*(4), 937–951. doi:10.1006/jmbi.2001.5187

Weaver, F. R. (2002). Molecular Biology 2ed. New York, NY: McGraw-Hill.

Zhang, H., & Sun, G. (2002). Feature Selection using tabu search method. *Pattern Recognition*, *35*(3), 701–711. doi:10.1016/S0031-3203(01)00046-2

KEY TERMS AND DEFINITIONS

Consensus: The calculated order of most frequent residues, either nucleotide or amino ascid, found at each position in a sequence alignment.

Diversification: To allow the process to search other parts of the solution space, driving it into new regions.

DNA: A double-stranded, helical molecule made up of sequences of four nucleotides (A, G, C, T) in each strand and backbones of sugars and phosphates.

Dot Matrix: Diagrams that provide a graphical method for comparing two sequences. Dots are placed within the graph if the same letter appears at the corresponding positions in the sequences. A diagonal series of dots appearing as lines on the graph indicates an alignment of a series of positions in the sequences.

Dynamic Programming: Programming that solves the problem of finding the optimal alignment between two sequences by building a table.

Genome: The entire set of DNA sequence information for an organism that is passed along from one generation to the next. It includes coding sequences for macromoleculus and noncoding sequences.

Homologous Genes: Genes whose sequences are so similar that they almost certainly arose from a common ancestor gene.

Homology: A statement of common evolutionary origin.

Motif: A conserved pattern of amino acids that is found in two or more proteins.

Multiple Sequence Alignment (MSA): An alignment of three or more sequences such that each column of the alignment is an attempt to represent the evolutionary changes in one sequence position, including substitutions, insertions, and deletions.

Phylogenetic Analysis: An investigation of the evolutionary relationships among a group of related sequences by producing a tree representation of the relationships.

Progressive Alignment: A procedure for generating an MSA that reduces the construction of the MSA to a series of pair-wise alignments.

Sequence Alignment: The comparison of two or more sequences by searching for a series of individual characters or character patterns that are in the same order in the sequences.

Sequence Similarity: A measurement of the matching characters in the alignment.

Sum of Pairs: In the MSA of a set of sequences is the sum of the alignment scores for each pair of sequences in the MSA.

Chapter 14
Cuckoo Search Algorithm for Hydrothermal Scheduling Problem

Thang Trung Nguyen
Ton Duc Thang University, Vietnam

Dieu Ngoc Vo
Ho Chi Minh City University of Technology, Vietnam

ABSTRACT

This chapter proposes a Cuckoo Search Algorithm (CSA) and a Modified Cuckoo Search Algorithm (MCSA) for solving short-term hydrothermal scheduling (ST-HTS) problem. The CSA method is a new meta-heuristic algorithm inspired from the obligate brood parasitism of some cuckoo species by laying their eggs in the nests of other host birds of other species for solving optimization problems. In the MCSA method, the eggs are first classified into two groups in which ones with low fitness function are put in top group whereas others with higher fitness function are put in abandoned group. In addition, an updated step size in the MCSA changes and tends to decrease as the iteration increases leading to near global optimal solution. The robustness and effectiveness of the CSA and MCSA are tested on several systems with different objective functions of thermal units. The results obtained by the CSA and MCSA are analyzed and compared have shown that the two methods are favorable for solving short-term hydrothermal scheduling problems.

INTRODUCTION

Electrical energy is one of the most popular and economic energies due to it's easy production at high performance and it's reasonable cost. Development of a country's economy is partially dependent on the country's electrical energy improvement policy. In fact, most of developing and developed countries are focusing on industry such as producing car, machine, etc. consuming a large number of electricity among total generated energy. Therefore, a main task is to establish electrical power systems which have to works flexibly, continually and economically. Thermal plants using fossil fuel account for large rate

DOI: 10.4018/978-1-4666-9644-0.ch014

power plants among several kinds of electrical power plants including conventional plants like thermal plants and hydro plants, and renewable energy based ones like wind power plants, solar power plants, photovoltaic, etc.

Planning optimal operation of isolated or inter connected power systems show a large variety of challenging problems, the solution of which requires application of several mathematical techniques from various branches of it. In power system optimization operation, there is a set of many available power plants which are being waiting for operation from experts. The task is to determine how many Megawatts each plant should supply to power system gird and transfer to customer load. The problem has been formulated as an optimization problem containing an objective function and equality constraints and inequality constraints. As the objective function is obtained, it means that fuel cost for producing electricity at thermal plants is minimized and all equality constraints and inequality constraints are satisfied over considered schedule time. This problem is defined as an economic load dispatch problem.

For a larger power system, both all thermal plants and hydro plants are connected via transmission line. The optimal operation task of the system has become more complicated since the hydro plants had a significant contribution to total electrical energy supplying to load. In fact, there are several complex constraints in hydro plants including equality constraint such as available water volume and inequality constraints such as upper and lower limits on power output of generator, on water discharges and on reservoir volume. However, the objective function of the hydro plants can be ignored as their primary fuel for producing electricity, water, is considered negligible. Therefore, the objective function of hydrothermal scheduling problem is also to minimize fuel cost of thermal plants while satisfying all constraints of the hydrothermal system such as load balance constraint, available water constraint, maximum and minimum limits on thermal and hydro generations, on water discharge and on volume reservoir.

Several conventional methods have been used to solve the short-term hydrothermal scheduling problem such as Newton method since the end of twentieth century (Zaghlool & Trutt, 1988; Basu, 2003), method based on the combination of Lagrange function and Newton method (Rashid & Nor, 1991), lambda-gamma iteration method (Wood & Wollenberg, 1996), dynamic programming (DP) (Wood & Wollenberg, 1996), gradient search technique (GS) (Wood & Wollenberg, 1996), mixed integer programming (MIP) (Nilsson & Sjelvgren, 1996), decomposition and coordination method (Li, Svoboda, Chung-Li Tseng & Johnson, 1997), Lagrange relaxation (LR) (Salam Nor KM, Hamdan, 1998; Xiaohong, Luh & Zhang, 1995; Al-Agtash, 2001), network flow (Franco, Carvalho & Soares, 1994; Heredia & Nabona, 1995; Oliveira & Soares,1995), hydrothermal scheduling algorithm (HSA) (Cavalho & Soares, 1987), peak shaving (Simopoulos, Kavatza & Vournas, 2007), and progressive optimality algorithm (Nanda & Bijwe, 1981). The conventional methods have the same characteristic that the objective function and all constraints have to be represented as piecewise linear or differentiable functions. The manner means that applicability of the methods are restricted on problems with more complex objective function considering valve point loading effect and complex constraints.

In recent years, many meta heuristic algorithms have been proposed for solving the HTS problem such as simulated annealing approach (SA) (Wong & Wong, 1994), evolutionary programming (EP) (Yang, Yang & Huang, 1996; Hota, Chakrabarti & Chattopadhyay, 1999; Sinha, Chakrabarti & Chattopadhaya, 2003; Türkay, Mecitoğlu & Baran, 2011), genetic algorithm (GA) (Chan & Chang, 1996; Orero & Irving, 1998; Gil, Bustos & Rudnick, 2003; Sasikala J, Ramaswamy, 2010, Kumar & Naresh, 2007), Hopfield neural network (HNN) (Basu, 2003), differential evolution (DE) (Lakshminarasimman & Subramanian, 2006), particle swarm optimization (PSO) (Yu, Yuan & Wang, 2007, Mandal, Basu &

Chakraborty, 2008; Hota, Barisala & Chakrabarti, 2009), artificial immune system (AIS) (Basu, 2011), teaching learning based optimization (TLBO) (Roy, 2013), evolutionary optimization (Cau & Keyes, 2002), and cultural algorithm (Yuan & Yuan, 2006). Most of the meta-heuristic algorithms can obtain good solution; however, they must suffer the low convergence speed. Compared to conventional methods, the methods are slower for convergence but more feasible for complicated system with complex constraints and nonconvex fuel cost function. In addition, there have been also hybrid methods proposed for solving the HTS problem such as interactive fuzzy satisfying method based on evolutionary programming (Basu, 2004), hybrid differential evolution (Yuan et al., 2008), hybrid simulation-optimisation approach (Soares & Ohishi, 1995), simulated annealing-based goal-attainment method (Basu, 2005), hybrid enhanced Lagrangian relaxation and quadratic programming (Petcharaks & Ongsaul, 2007), improved merit order and augmented Lagrange Hopfield network (Dieu & Ongsakul, 2009), and hybrid DE–SQP algorithm (Sivasubramani & Shanti Swarup, 2011). The hybrid method can offer better solution quality than the single methods. However, they may suffer long computational time due to the combination of single methods.

BACKGROUND

Cuckoo search algorithm (CSA) is a meta-heuristic first developed by Yang and Deb in 2009 being inspired from Cuckoo bird's reproduction behavior. Cuckoo bird lays its egg in a random nest of other bird species. There are two cases happening to the egg including 1) The egg is certainly hatched and carried over to the next generation if the host birds do not identify it as a strange one in their nest; 2) the Cuckoo egg is thrown out of the nest by the host bird or both cuckoo egg and the host nest are forsaken by the host bird if the Cuckoo egg is identified as a strange one by the host birds. Since it was developed in 2009, the CSA method have gotten many attentions from huge number of researches and widely and successfully been applied to engineering optimization problems such as economic load dispatch (Basu & Chowdhury, 2013; Dieu &Ongsakul, 2013), photovoltaic system (Ahmed & Salam, 2014), micro grid power dispatch problem (Basu & Chowdhury, 2013). By testing on ten standard optimization Benchmark functions and several engineering optimization problems above, the applicability of CSA, GA, PSO and other meta-heuristics compared together has revealed that CSA is superior to the methods in terms of high quality solution, execution time and successful convergence rate. However, the CSA is judged slow convergence as compared with Modified Cuckoo Search (MCSA) developed by Walton et al in 2011 (Walton, Hassan, Morgan & Brown, 2011). In MCSA version, the egg group is divided into two groups including top group with better fitness function value and abandoned group with higher fitness function value. Via Lévy Flights the two groups are then generated new solutions by performing some changes on conventional CSA of Yang and Deb. The two new ones are integrated into one group only and then generated the second new solution time via discovery of alien eggs as described in the second case for a Cuckoo egg.

In this Chapter, both CSA and MCSA are applied for solving short-term hydrothermal scheduling problem where systems with convex or nonconvex fuel cost function are considered. The comparison of fuel cost and convergence characteristic has shown that the MCSA is more efficient than CSA as stated by Walton et al.

LIMITATION OF THE RESEARCH

In this chapter, the research is focused on the implementation of the CSA for solving the HTS problem. The considered problem in this chapter is a short-term fixed-head hydrothermal scheduling with water discharged limit. Moreover, power losses in transmission lines are also included in the research.

CUCKOO SEARCH ALGORITHM

Cuckoo Breeding Behavior in Nature

In nature, cuckoo bird is one of the bird species having an intelligent reproduction strategy by laying their eggs in the other species nest. In fact, once a Cuckoo chick is hatched in another specie's nest, the first instinct action of the chick is to kick the host baby bird or the host's eggs out of the nest in aim to increase food receiving from the host bird. Of course, the little cuckoos are carried over to the next generation without their parents' feed. On the other hand, there is a probability that the cuckoo egg is discovered and the host birds can simply throw it away or abandon their nest and built a new one.

Lévy Flights

Lévy flights are a special class of random walks whose increments are distributed according to a heavy-tailed probability distribution. The step lengths of Lévy flights are not constant but rather are chosen from a probability distribution with a power-law tail (Viswanathan, 2000).

In nature, species search for food in a either random or quasi-random manner. Generally, the foraging path of the species is actually a random walk since the next move is based on the current location and the transition probability to the next location. Therefore, the chosen direction depends implicitly on a probability which can be mathematically modeled. Recently, some studies (Brown, Liebovitch & Glendon, 2007; Yang & Deb, 2010) have shown that the flight behavior of many species has demonstrated the typical characteristics of Lévy flights. Furthermore, in some other researches (Shlesinger, 2006; Pavlyukevich, 2007) such behaviors have been applied on optimization and optimal search, and the obtained preliminary results have shown their potential capability.

Pseudo-Code Cuckoo Search Algorithm

Based on the existent strategy of Cuckoo bird in nature as described above, the CSA method are summarized by three idealized rules as follows:

- Each cuckoo lays and dumps its egg in a randomly chosen nest among the fixed and available number of host nests;
- The Cuckoo egg with better quality will be carried over to the next generation;
- The Cuckoo egg is intentionally discovered by the host bird with a probability of $p_a \in [0, 1]$. In this case, either the Cuckoo egg is evicted or both Cuckoo egg and the host bird's egg are abandoned.

Based on these three main principles, the CSA is simply summarized in the basic steps of pseudo-code as follows (Yang and Deb, 2009).

```
Algorithm 1: Cuckoo Search
Objective function f(X), C = (X_1, ..., X_d)^T;
Initialize population of Np host nests X_d  randomly (d = 1, 2, ..., Np);
while (current iteration <Maximum iteration)
        Generate new solution via Lévy flights, X_d^new1
        evaluate new solution quality
        compare X_d and X_d^new1 to keep a better one, called X_d^1
        A fraction (p_a) of worse nests
        are abandoned and new ones are built at the second time, X_d^new2
        evaluate new solution quality
        compare fitness function value of X_d^1 and X_d^new2 to keep a better one
        Find the current best solution
end while
```

When a new solution $x_i^{(t+1)}$ for cuckoo i is generated, the Lévy flight is performed:

$$x_i^{(t+1)} = x_i^{(t)} + \alpha \oplus L\acute{e}vy(t) \tag{1}$$

where $\alpha > 0$ is the step size for updating the new solution; $\oplus$ is the entry-wise multiplications; and *Lévy(t)* is the Lévy flights with their random steps drawn from a Lévy stable distribution via the power law of index β (Viswanathan et al, 2000)

$$L\acute{e}vy(t) \sim t^{-\beta}, \ \left(1 < \beta \le 3\right) \tag{2}$$

where t is flight length.

This Lévy step-length distribution has an infinite variance with an infinite mean.

MODIFIED CUCKOO SEARCH ALGORITHM

In the MCSA method, the eggs obtained by random initialization are first ranked in a descending order based on their corresponding fitness function value and then classified into two groups, top group with lower fitness function value and abandoned group with higher fitness function value. Two modifications are fulfilled on the two groups as below.

The First Modification on the Abandoned Group

In the first modification, the value of step size α in Equation (1) is a variable during the process of generating new solution via Lévy flights in aim to find an optimal solution within the local search zone. In the conventional CSA method, the step size is fixed at a predetermined value based on the experiment; however, α value in the MCSA method for the abandoned group is set to $1/\sqrt{G}$ where G is the current iteration. Clearly, the step size is reduced as the iteration is increased.

The Second Modification on the Top Group

In the second modification, exchanged information from two eggs including one chosen intentionally and one chosen randomly is used to generate new solution via Lévy flights. The action aims to speed up convergence and enhance solution quality. During the process of exchanging information between each two chosen eggs, there are four cases for the two ones consisting of 1) the same egg is picked twice; 2). The two eggs have the same fitness; 3) the randomly chosen egg has lower fitness value than the predetermined egg; 4) the randomly chosen egg has higher fitness function value than the predetermined egg. In case that an egg is selected twice, a new egg is produced via Lévy flights from the egg with the step size $\alpha=1/G^2$. A new egg is generated at the midpoint of the two egg for the case 2, two eggs with the same fitness. For cases 3 and 4, a new egg is produced at a location along the line connecting the two eggs by using inverse of the golden ratio $\phi = \left(1+\sqrt{5}\right)/2$. On the other hand, there is another parameter needs to be determined in MCSA method, ratio of top group and abandoned group. Walton et al have suggested that the best value for the ratio is 1:3 and the value was successfully applied on another study carried out by Bhargava et al in 2013. In spite of the statement from the two studies, by experiment with careful comparison the value is selected 1:1 for two test systems.

The steps involved in the modified cuckoo search are shown in detail in Algorithm 2.

```
Algorithm 2: Modified Cuckoo Search
Randomly Initialize Np nests, X_d (d =1,2, ..., N_p)
Calculate fitness function of all nests F_d = f(X_d)
Set current iteration G to 1
while Current iteration G < Maximum iteration G_max do
G ←G +1
Sort nests by order of fitness function and classify into two groups, top
group and abandoned group
for all abandoned eggs do
Current position X_d
Generate new solution via Lévy flights with step size α = 1/√G
Evaluate fitness function to keep the better abandoned eggs.
end for
for all of the top eggs do
Current position X_d
Randomly select one egg among the top eggs, X_r
if X_d = X_r (the same egg is picked twice) then
Generate new solution via Lévy ñights with step size α=1/G²
end if
if (F_d > F_r) do
```

```
dx = |X_d - X_r|/φ
```
$X_d^{new} = X_d + dx$
```
end if
if (F_d < F_r) do
dx = |X_d - X_r|/φ
```
$X_d^{new} = X_r + dx$
```
end if
if (F_d = F_r) do
dx = |X_d + X_r|/2
```
$X_d^{new} = X_d + dx$
```
end if
Evaluate the new top eggs
Compare fitness function value of the new and the old top egg
Keep better ones
end for
Integrate the two group into one group
A fraction (p_a) of worse nests
        are abandoned and new ones are built at the second time, X_d^new
        evaluate new solution quality
        compare fitness function value of new and old ones to keep a better
one
        Find the current best solution
end while
```

HYDROTHERMAL SCHEDULING PROBLEM

Problem Formulation

The main task of the short-term hydrothermal scheduling problem is to determine power output of each hydro plant and each thermal plant in aim to minimize electrical generation fuel cost of operating thermal plants since the operation cost of hydro plants is not considerable and negligible. Besides, the obtained set of hydrothermal power output must satisfy all equality and inequality constraints of each individual plant and whole system over the scheduled time period such as upper and lower bounds, available water volume and load power balance. The considered short-term HTS problem consisting of N_1 thermal units and N_2 hydro units scheduled in M sub-intervals with t hours for each is formulated as follows.

The objective functions for considering quadratic and nonconvex fuel cost curve of thermal units are respectively expressed in Equations. (3) and (4)

$$\text{Min } C_T = \sum_{m=1}^{M}\sum_{i=1}^{N_1} t_m \left(a_{si} + b_{si}P_{si,m} + c_{si}P_{si,m}^2\right), \tag{3}$$

$$\text{Min } C_T = \sum_{m=1}^{M}\sum_{i=1}^{N_1} t_m \left[a_{si} + b_{si}P_{si,m} + c_{si}P_{si,m}^2 + \left| d_{si} \times \sin\left(e_{si} \times \left(P_{si}^{\min} - P_{si,m} \right) \right) \right| \right]. \quad (4)$$

In addition to achieving the objective functions, a set of following constraints must be met exactly.

Power Balance Constraint

The total power generation from thermal and hydro plants must satisfy the total load demand and power loss in each subinterval:

$$\sum_{i=1}^{N_1} P_{si,m} + \sum_{j=1}^{N_2} P_{hj,m} - P_{L,m} - P_{D,m} = 0\,;\ m = 1,\ldots, M. \quad (5)$$

where the power losses in transmission lines are calculated using Kron's formula:

$$P_{L,m} = \sum_{i=1}^{N_1+N_2}\sum_{j=1}^{N_1+N_2} P_{i,m}B_{ij}P_{j,m} + \sum_{i=1}^{N_1+N_2} B_{0i}P_{i,m} + B_{00}. \quad (6)$$

Water Availability Constraint

The total available water discharge of each hydro plant for the whole scheduled time horizon is limited by:

$$\sum_{m=1}^{M} t_m q_{j,m} = W_j\,;\ j = 1,\ldots, N_2. \quad (7)$$

where the rate of water flow from hydro plant j in subinterval m is determined by:

$$q_{j,m} = a_{hj} + b_{hj}P_{hj,m} + c_{hj}P_{hj,m}^2. \quad (8)$$

Generator Operating Limits

Each thermal and hydro units have their upper and lower generation limits:

$$P_{si,\min} \le P_{si,m} \le P_{si,\max}\,;\ i = 1,\ \ldots,\ N_1;\ m = 1,\ \ldots,\ M, \quad (9)$$

$$P_{hj,\min} \le P_{hj,m} \le P_{hj,\max}\,;\ j = 1,\ldots,\ N_2;\ m = 1,\ \ldots,\ M. \quad (10)$$

Calculation of Power Output for Slack Thermal and Hydro Units

Meta-heuristic algorithms normally use a set of slack variables for handling equality constraints while other variables can be obtained by using strategies of new solution generation from the own algorithms. The slack technique is also employed during the CSA and MCSA iterative procedure. In the HTS problem, there are M system power balance constraint Equations (5) and N_2 available water volume constraint Equations (7) need to be satisfied. Two sets of power output including hydro power outputs from units 1 to N_2 for the first (M-1) subintervals and thermal power outputs from units 2 to N_1 for all M sub-intervals are generated at initialization step and at two phases of new solution generation.

By substituting the given N_2 hydro power outputs at the first (M-1) subintervals into Equation (8), water discharge at the first (M-1) subintervals of each hydro unit is obtained. The water discharge and power output at M^{th} subinterval of each hydro unit are respectively calculated as in following Equations (11) and (12)

$$q_{j,M} = (W_j - \sum_{m=1}^{M-1} t_m q_{j,m}) / t_M \,, \tag{11}$$

$$P_{hj,M} = \frac{-b_{hj} \pm \sqrt{b_{hj}^2 - 4c_{hj}(a_{hj} - q_{j,M})}}{2c_{hj}} \; ; j = 1, 2, \ldots, N_2. \tag{12}$$

where $b_{hj}^2 - 4 \times c_{hj} \times (a_{hj} - q_{j,M}) \geq 0$.

On the other hand, the known power output of all hydro units and (N-1) thermal units are then substituted into Equation (5) to obtain power output of the first thermal unit at M subintervals P_{slm} (m = 1, …, M) as below.

$$P_{s1,m} = P_{D,m} + P_{L,m} - \sum_{i=2}^{N_1} P_{si,m} - \sum_{j=1}^{N_2} P_{hj,m} \,. \tag{13}$$

The power loss in Equation (6) is rewritten with respect to the slack thermal unit 1 as follows:

$$P_{L,m} = B_{11} P_{s1,m}^2 + \left(2 \sum_{i=2}^{N_1+N_2} B_{si} P_{im} + B_{0s} \right) P_{s1,m} + \sum_{i=2}^{N_1+N_2} \sum_{j=2}^{N_1+N_2} P_{im} B_{ij} P_{jm} + \sum_{i=2}^{N_1+N_2} B_{0i} P_{im} + B_{00} \,. \tag{14}$$

Substituting (14) into (13), a quadratic equation with respect to the power output of the slack thermal unit 1 at subinterval m is obtained as follow:

$$A \times P_{s1,m}^2 + B \times P_{s1,m} + C = 0 \,. \tag{15}$$

where

$$A = B_{11}, \tag{16}$$

$$B = 2\sum_{i=2}^{N_1} B_{1i} P_{si,m} + 2\sum_{j=1}^{N_2} B_{1j} P_{hj,m} + B_{01} - 1, \tag{17}$$

$$\begin{aligned} C = &\sum_{i=2}^{N_1}\sum_{j=2}^{N_1} P_{si,m} B_{ij} P_{sj,m} + \sum_{i=1}^{N_2}\sum_{j=1}^{N_2} P_{hi,m} B_{ij} P_{hj,m} + 2\sum_{i=2}^{N_1}\sum_{j=1}^{N_2} P_{si,m} B_{ij} P_{hj,m} \\ &+\sum_{i=2}^{N_1} B_{0i} P_{si,m} + \sum_{j=1}^{N_2} B_{0j} P_{hj,m} + B_{00} + P_{D,m} - \sum_{i=2}^{N_1} P_{si,m} - \sum_{j=1}^{N_2} P_{hj,m} \end{aligned} \tag{18}$$

Finally, the solution for the second order Equation (15) is obtained by:

$$P_{s1,m} = \frac{-B \pm \sqrt{B^2 - 4AC}}{2A}. \tag{19}$$

where $B^2 - 4AC \geq 0$.

IMPLEMENTATION OF CSA FOR HYDROTHERMAL SCHEDULING PROBLEM

The main steps for the proposed CSA are described as follows:

Initialization

Similar to other meta-heuristic algorithm in the proposed CSA there is a population of N_p host nests included and represented by $X = [X_1, X_2, .., X_d, .., X_{Np}]^T$, in which each X_d is randomly initialized in Equation (20) and (21) including $P_{si,m,d}$ ($i = 2,..., N_1$; $m = 1,..., M$) and $P_{j,m,d}$ ($j = 1,.., N_2$; $m = 1,..., M$-1) satisfying $P_{si,min} \leq P_{si,m,d} \leq P_{si,max}$ and $P_{hj,min} \leq P_{hj,m,d} \leq P_{hj,max}$.

$$P_{si,m,d} = P_{si,\min} + rand_1 * (P_{si,\max} - P_{si,\min});\ i = 2,..., N_1;\ m = 1,..., M, \tag{20}$$

$$P_{hj,m,d} = P_{hj,\min} + rand_2 * (P_{hj,\max} - P_{hj,\min});\ j = 1,..., N_2;\ m = 1,..., M-1. \tag{21}$$

where $rand_1$ and $rand_2$ are uniformly distributed random numbers in [0,1].

Each nest corresponding to each solution needs to be evaluated in aim to improve obtained solution after each iteration. Therefore, the evaluation must be performed based on fitness function as below.

$$FT_d = \sum_{m=1}^{M}\sum_{i=1}^{N_1} F_i(P_{si,m,d}) + K_s \sum_{m=1}^{M} (P_{s1,m,d} - P_{s1}^{\lim})^2 + K_q \sum_{j=1}^{N_2} (q_{j,M,d} - q_j^{\lim})^2. \tag{22}$$

where K_s and K_q are penalty factors for the slack thermal unit 1 and available water at subinterval *M*, respectively; $P_{s1,m,d}$ is the power output of the slack thermal unit corresponding to nest d in the population; $q_{j,M}$ is the water discharge of all hydro plants at the subinterval *M* corresponding to nest *d* in the population. Both $P_{s1,m,d}$ and $q_{j,M}$ are obtained in section "Calculation of Power Output for Slack Thermal and Hydro Units".

The limits for the slack thermal unit 1 and the Mth subinterval water discharge in (22) have not been predetermined but in use to penalize the invalid nest for handling equality constraints including available water and load balance, and inequality constraints consisting of thermal generations and water discharge limits. These limits are obtained by.

$$P_{s1}^{\lim} = \begin{cases} P_{s1,\max} & if\ P_{s1,m,d} > P_{s1,\max} \\ P_{s1,\min} & if\ P_{s1,md} < P_{s1,\min} \\ P_{s1,m,d} & \text{otherwise} \end{cases}, \tag{22}$$

$$q_j^{\lim} = \begin{cases} q_{j,\max} & if\ q_{j,M,d} > q_{j,\max} \\ q_{j,\min} & if\ q_{j,M,d} < q_{j,\min} \\ q_{j,M,d} & \text{otherwise} \end{cases}. \tag{24}$$

where $P_{s1,max}$ and $P_{s1,min}$ are the maximum and minimum power outputs of slack thermal unit 1, respectively; $q_{j,max}$ and $q_{j,min}$ are the maximum and minimum water discharges of hydro plant *j*.

The *Np* initialized host nests is set to the best local nest $Xbest_d$ ($d = 1, \ldots, N_d$) and only one nest with the lowest fitness function in (22) is set to the best global nest *Gbest* in the population.

Generation of New Solution via Lévy Flights

As mentioned in three main principles of conventional CSA method, there are two phases of generating new solution including Lévy flight and replacement of a fraction of eggs. In this subsection, generation of new solution thank to Lévy Flights is described. Finally, each new nest X_d is created as follows:

$$X_d^{new} = Xbest_d + \alpha_1 \times rand_3 \times \Delta X_d^{new}. \tag{25}$$

where $\alpha_1 > 0$ is the updated step size; $rand_3$ is a normally distributed random number in [0, 1] and the increased value ΔX_d^{new} is determined by:

$$\Delta X_d^{new} = v \times \frac{\sigma_x(\beta)}{\sigma_y(\beta)} \times (Xbest_d - Gbest). \tag{26}$$

where

$$v = \frac{rand_x}{\left|rand_y\right|^{1/\beta}}. \tag{27}$$

where $rand_x$ and $rand_y$ are two normally distributed stochastic variables with standard deviation $\sigma_x(\beta)$ and $\sigma_y(\beta)$ given by:

$$\sigma_x(\beta) = \left[\frac{\Gamma(1+\beta)\times\sin\left(\frac{\pi\beta}{2}\right)}{\Gamma\left(\frac{1+\beta}{2}\right)\times\beta\times 2^{\left(\frac{\beta-1}{2}\right)}}\right]^{1/\beta}, \tag{28}$$

$$\sigma_y(\beta) = 1. \tag{29}$$

where β is the distribution factor ($0.3 \le \beta \le 1.99$) and $\Gamma(.)$ is the gamma distribution function.

For the newly obtained solution, its lower and upper limits should be satisfied according to the unit's limits:

$$P_{si,m,d} = \begin{cases} P_{si,\max} & \text{if } P_{si,m,d} > P_{si,\max} \\ P_{si,\min} & \text{if } P_{si,m,d} < P_{si,\min} \\ P_{si,m,d} & \text{otherwise} \end{cases} ; i = 2,..., N_1;\ m = 1,..., M, \tag{30}$$

$$P_{hj,m,d} = \begin{cases} P_{hj,\max} & \text{if } P_{hj,m,d} > P_{hj,\max} \\ P_{hj,\min} & \text{if } P_{hj,m,d} < P_{hj,\min} \\ P_{hj,m,d} & \text{otherwise} \end{cases} ; j = 1,..., N_2;\ m = 1,..., M-1. \tag{31}$$

The power outputs of the slack hydro unit j at subinterval M, P_{hjMd} and the slack thermal unit 1 at each subinterval, P_{s1md} are calculated from section "Calculation of Power Output for Slack Thermal and Hydro Units". The fitness function value of the new egg is calculated using (22) and then compared to that from the old egg. The egg with better fitness function value is considered as the new solution

Alien Egg Discovery and Randomization

In the subsection, the second time of new solution generation is performed in aim to improve quality of obtained solution. This different manner of the CSA methods with other meta-heuristic leads to better solution and convergence speed achieved. In Cuckoo bird behavior, there is a possibility that an alien egg is identified by host bird and it can be then thrown away out of the nest or it and the nest are forsaken by the host bird. Like the Lévy flights, the discovery action of alien eggs with the probability of P_a can also generate a new solution for an optimization problem. The new solution is created by.

$$X_d^{dis} = Xbest_d + \alpha_2 \times K \times \Delta X_d^{dis} . \tag{32}$$

where $\alpha_2 > 0$ is also a updated step size like α_1 and K is the updated coefficient determined based on the probability of a host bird to discover an alien egg in its nest:

$$K = \begin{cases} 1 & \text{if } rand_4 < p_a \\ 0 & \text{otherwise} \end{cases} . \tag{33}$$

and the increased value ΔX_d^{dis} is determined by:

$$\Delta X_d^{dis} = rand_5 \times \left[randp_1(Xbest_d) - randp_2(Xbest_d) \right] . \tag{34}$$

where $rand_4$ and $rand_5$ are the distributed random numbers in [0, 1] and $randp_1(Xbest_d)$ and $randp_2(Xbest_d)$ are the random perturbation for positions of the nests in $Xbest_d$. For the newly obtained solution, its lower and upper limits should be also satisfied constraints (30) and (31). The value of the fitness function is calculated using (22) and the nest corresponding to the best fitness function is set to the best nest *Gbest*.

Stopping Criterion

In the proposed CSA method, the stopping criterion for the algorithm is based on the maximum number of iterations. The algorithm is terminated as the maximum number of iterations reached.

Overall Procedure

The flowchart of the proposed CSA for solving the problem is given in Figure 1.

Figure 1. The flowchart of CSA for solving ST-HTS

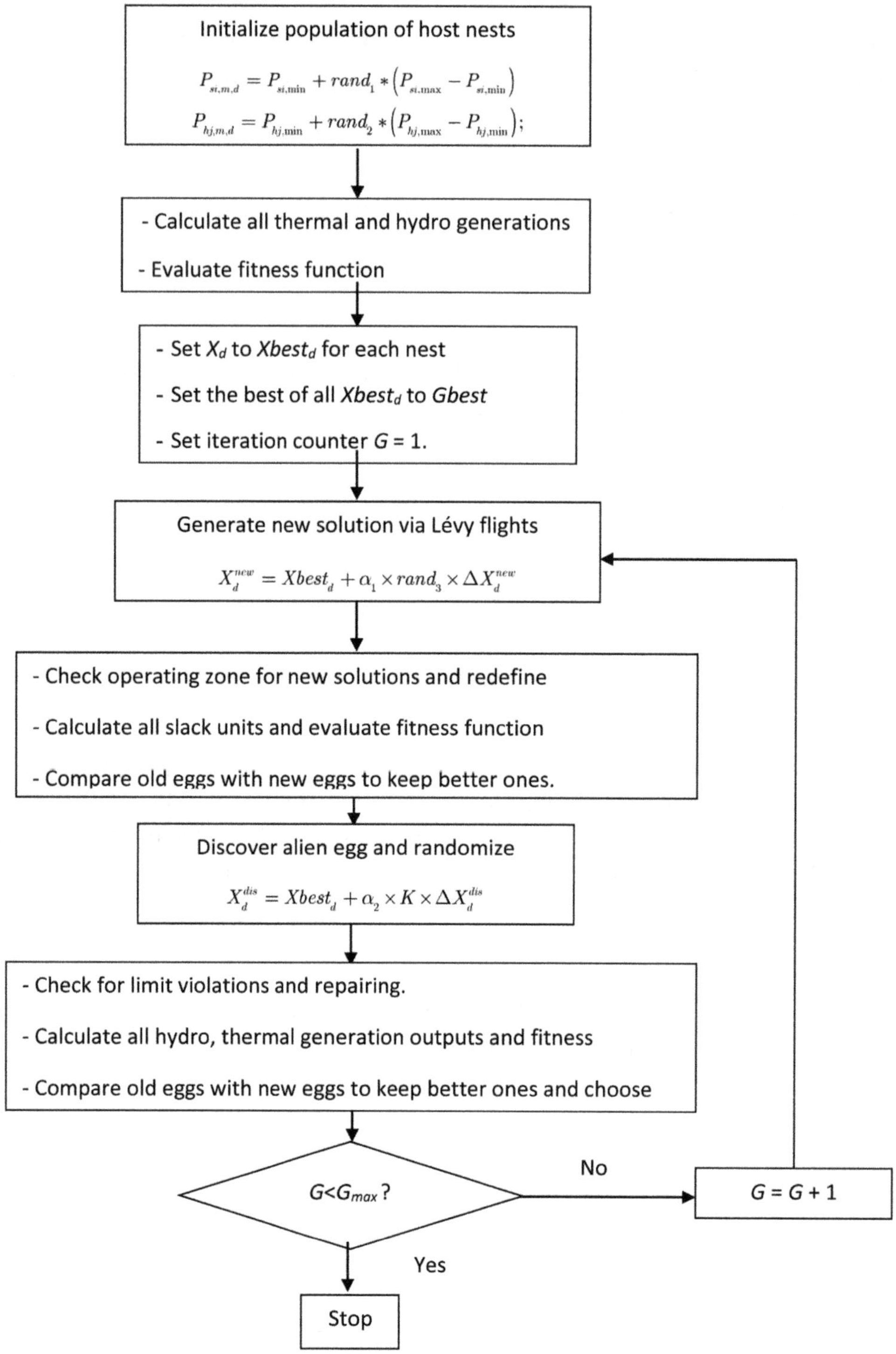

IMPLEMENTATION OF MCSA FOR HYDROTHERMAL SCHEDULING PROBLEM

Initialization

Similar to CSA, in the proposed MCSA there are also a population of N_p host nests generated including $X = [X_1, X_2, \ldots, X_{Np}]^T$, in which $X_d = [P_{si,m,d}\ P_{hj,m,d}]$. The rest of the initialization process is performed as in CSA method from Equations (20) to (24).

The nests are then sorted in the descending order based on their fitness function value and then classified into two groups. The nests with higher fitness function value are put in an abandoned group whereas the other ones are put in a top group, in which each nest in the abandoned group and in the top group are respectively named $Xbest\_discard_d$ and $Xbest\_nodiscard_d$. During new solution generation process of the top group, a nest randomly picked among the $Xbest\_nodiscard_d$ nests is named randomly picked nest $Xbest\_nodiscard_r$. Among all nests in the population one nest having the lowest fitness function value is set to the best global nest *Gbest*.

Generation of New Solution via Lévy Flights

Generation of New Solution for the Abandoned Group via Lévy Flights

Based on the modification applied to the abandoned eggs ($d = Notop+1, \ldots, N_d$), the optimal path for the Lévy flights is calculated using Mantegna's algorithm as follows:

$$Xdiscard_d^{new} = Xbest\_discard_d + \alpha_3 \times rand_6 \times \Delta Xdiscard_d^{new}. \tag{35}$$

where $rand_6$ is the distributed random number in [0, 1], the step size $\alpha_3 = 1/\sqrt{G}$ is determined and $\Delta Xdiscard_d^{new}$ is obtained by:

$$\Delta Xdiscard_d^{new} = \nu \times \frac{\sigma_x(\beta)}{\sigma_y(\beta)} \times (Xbest\_discard_d - Gbest). \tag{36}$$

where

$$\nu = \frac{rand_x}{\left|rand_y\right|^{1/\beta}}. \tag{37}$$

where $rand_x$ and $rand_y$ are two normally distributed stochastic variables with standard deviation $\sigma_x(\beta)$ and $\sigma_y(\beta)$ given by:

$$\sigma_x(\beta) = \left[\frac{\Gamma(1+\beta) \times \sin\left(\frac{\pi\beta}{2}\right)}{\Gamma\left(\frac{1+\beta}{2}\right) \times \beta \times 2^{\left(\frac{\beta-1}{2}\right)}}\right]^{1/\beta}, \tag{38}$$

$$\sigma_y(\beta) = 1. \tag{39}$$

where β is the distribution factor ($0.3 \le \beta \le 1.99$) and $\Gamma(.)$ is the gamma distribution function.

Generation of New Solution for the Top Egg Group via Lévy Flights

The modification applied to the eggs in the top group ($d = 1, \ldots, Notop$) is described above. There are three cases for the new generated eggs based on the information exchange among the top eggs. The optimal path for the Lévy flights is calculated using Mantegna's algorithm as follows:

$$Xnodiscard_d^{new} = Xbest\_nodiscard_d + \alpha_4 \times rand_7 \times \Delta Xnodiscard_d^{new}. \tag{40}$$

where $rand_7$ is the distributed random numbers in [0, 1]. The value of α_4 and $\Delta Xnodiscard_d^{new}$ will be selected depending on one of the considered cases below:

Case 1: The same egg is picked twice

$$\Delta Xnodiscard_d^{new} = v \times \frac{\sigma_x(\beta)}{\sigma_y(\beta)} \times (Xbest\_nodiscard_d - Gbest). \tag{41}$$

where $\alpha_4 = 1/G^2$, v and $\frac{\sigma_x(\beta)}{\sigma_y(\beta)}$ are calculated as in Section 3.4.2.1.

Case 2: Both eggs have the same fitness value function

$$\Delta Xnodiscard_d^{new} = (Xbest\_nodiscard_r - Xbest\_nodiscard_d) / 2. \tag{42}$$

where $\alpha_4 = 1$.

Case 3: The random egg has lower fitness than the egg d

$$\Delta Xnodiscard_d^{new} = (Xbest\_nodiscard_r - Xbest\_nodiscard_d) / \phi. \tag{43}$$

where $\alpha_4 = 1$ and $\phi = \left(1 + \sqrt{5}\right) / 2$.

Case 4: The random egg has higher fitness than the egg d

By modifying Equation (40), the new solution is obtained as below

$$Xnodiscard_d^{new} = Xbest\_nodiscard_r + \alpha_4 \times rand_8 \times \Delta Xnodiscard_d^{new}, \tag{44}$$

where

$$\Delta Xnodiscard_d^{new} = (Xbest\_nodiscard_d - Xbest\_nodiscard_r) / \phi. \tag{45}$$

and $\alpha_4 = 1$ and $\phi = \left(1+\sqrt{5}\right)/2$.

For the newly obtained solution, its lower and upper limits should be satisfied according to the unit's limits:

$$P_{si,m,d} = \begin{cases} P_{si,\max} & \text{if } P_{si,m,d} > P_{si,\max} \\ P_{si,\min} & \text{if } P_{si,m,d} < P_{si,\min} \\ P_{si,m,d} & \text{otherwise} \end{cases} ; i = 2,..., N_1;\ m = 1,..., M, \tag{46}$$

$$P_{hj,m,d} = \begin{cases} P_{hj,\max} & \text{if } P_{hj,m,d} > P_{hj,\max} \\ P_{hj,\min} & \text{if } P_{hj,m,d} < P_{hj,\min} \\ P_{hj,m,d} & \text{otherwise} \end{cases} ; j = 1,..., N_2;\ m = 1,..., M-1. \tag{47}$$

The power outputs of the slack hydro unit j at subinterval M, P_{hjMd} and the slack thermal unit 1 at each subinterval, P_{s1md} are calculated. The fitness function value of the new egg is calculated using (22) and then compared to that from the old egg. The egg with better fitness function value is considered as the new solution.

Alien Egg Discovery and Randomization

In the section, all top eggs and abandoned eggs are put in a group of eggs. New eggs are then generated as in the section of "alien egg discovery and randomization" in CSA method.

Stopping Criterion

In the proposed MCSA method, the stopping criterion for the algorithm is based on the maximum number of iterations. The algorithm is terminated as the maximum number of iterations reached.

Overall Procedure

The overall procedure of the proposed MCSA for solving the fixed head short-term HTS problem is described as follows.

Step 1: Select parameters for the MCSA including number of host nests N_p, probability of a host bird to discover an alien egg in its nest p_a, and maximum number of iterations N_{max}.

Step 2: Initialize a population of N_p host nests and calculate the power output for the slack thermal unit 1 and slack hydro units. Set the iteration counter $G = 1$.

Step 3: Evaluate fitness function to choose $Xbest\_discard_d$, $Xbest\_nodiscard_d$, $Xbest\_nodiscard_r$ and *Gbest* based on the value of their fitness function.
Step 4: Generate new solutions for abandoned eggs via Lévy flights
Step 5:

- Check for limit violations and repairing
- Calculate all hydro and thermal generation outputs
- Calculate the fitness function of new eggs and compare to that of old eggs. Keep better eggs with lower fitness function.

Step 6: Generate new solution for top eggs via Lévy flights
Step 7:

- Check for limit violations and repairing
- Calculate all hydro and thermal generation outputs.
- Calculate the fitness function of new eggs and compare to that of old eggs. Keep better eggs with lower fitness function.

Step 8: Put new eggs generated in step 4 and 6 in a group of egg
Step 9: Discover alien egg and randomize
Step 10:

- Check for limit violations and repairing
- Calculate all hydro and thermal generation outputs
- Evaluate fitness function to choose new *Gbest.*

Step 11: If $G \le G_{max}$, $G = G + 1$ and return to Step 3. Otherwise, stop.

NUMERICAL RESULTS

In order to demonstrate the performance of the CSA and MCSA, two test systems are employed where system 1 considers quadratic fuel cost function of thermal unit and system 2 considers nonconvex fuel cost function of thermal units.

Example 1: A test system consists of two thermal plants with quadratic fuel cost and two hydropower plants supplying a load for four scheduled subintervals. The load demand is given in Table 1. Thermal generator data and hydro system data are respectively given in Tables 2 and 3.

The transmission loss formula coefficients of the test system with quadratic fuel cost function

$$B = 10^{-5}\begin{bmatrix} 4.0 & 1.0 & 1.5 & 1.5 \\ 1.0 & 3.5 & 1.0 & 1.2 \\ 1.5 & 1.0 & 3.9 & 2.0 \\ 1.5 & 1.2 & 2.0 & 4.9 \end{bmatrix}$$

For implementation of CSA and MCSA the number of nests and the maximum number of iterations are respectively set to 20 and 4000 for CSA, and 6 and 4000 for MCSA in advance whereas the

Table 1. Load demand for scheduled time

Subinterval	1	2	3	4
Duration (hour)	6	6	6	6
PD (MW)	1200	1400	2000	1800

Table 2. Thermal generator data of the test system 1 with quadratic fuel cost function of thermal units

Thermal plant	a_{si} (\$/h)	b_{si} (\$/MWh)	c_{si} (\$/MW²h)	$P_{si,min}$ (MW)	$P_{si,max}$ (MW)
1	380	7.0	0.03	200	450
2	600	5.5	0.05	300	1000

Table 3. Hydro system data of the test system 5 with quadratic fuel cost function of thermal units

Hydro Plant	a_{hj} (acre-ft/h)	b_{hj} (acre-ft/MWh)	c_{hj} (acre-ft/MW²h)	W_j (acre-ft)	$P_{hj,min}$ (MW)	$P_{hj,max}$ (MW)
1	250	8.9	0.01	125000	0	450
2	250	9.4	0.015	286000	0	500

probability P_a is set to the range from 0.1 to 0.9 with a step of 0.1. Each test case is performed twenty independent trials for both CSA and MCSA. The obtained results including minimum total cost, average total cost, maximum total cost, standard deviation and computational time are indicated in Table 4 for CSA and Table 5 for MCSA. As observed from the Table, CSA obtains the best minimum total cost of \$ 493890.08 at Pa=0.7 whereas the best Pa that MCSA gets the best cost of \$ 493786.2 is 0.9. The minimum costs of MCSA is decreased as Pa is increased to Pa=0.9 while the minimum cost obtained by CSA does not follow value of Pa. The comparison of the best minimum cost, maximum cost, average cost and deviation cost from the two methods is given in Table 6.

The table shows that the best minimum cost, the best maximum cost, the best average cost and the best standard deviation cost obtained by CSA and MCSA are respectively \$ 493890.1 and \$493786.2, \$494361.7 and \$493795.3, \$495273.3 and \$493809.8, \$320.9526 and \$6.7321. The comparison reveals that MCSA is more efficient than CSA in terms of cost. The solution for the best minimum cost obtained

Table 4. Results by CSA for the first system with quadratic fuel cost of thermal units with different values of P_a

P_a	Min. Cost (\$)	Avg. Cost (\$)	Max. Cost (\$)	Std. Dev. (\$)	CPU (s)
0.1	493938.17	494361.7144	495457.98	363.1774	12.3
0.2	494072.92	494498.1515	495334.84	320.9526	12.6
0.3	493969.26	494429.3982	495480.49	355.5008	12.4
0.4	493905.59	494394.6898	495590.73	426.9667	11.8
0.5	493903.3	494391.0146	495451.05	400.6581	11.9
0.6	493933.93	494556.9506	495328.79	379.7789	12.3
0.7	**493890.08**	494572.2655	495763.63	613.4754	12.4
0.8	493896.16	494614.0145	496134.93	526.9662	11.9
0.9	494022.74	494514.5222	495273.26	384.5920	12.6

Table 5. Results by MCSA for the first system with quadratic fuel cost of thermal units with different values of P_a

P_a	Min. Cost ($)	Avg. Cost ($)	Max. Cost ($)	Std. Dev. ($)	CPU (s)
0.1	494157.8	495119.2	496691.4	671.7766	11.6
0.2	494002.8	494668.1	496062.2	453.1256	11.4
0.3	494046.4	494264.6	494522.3	118.1958	11.2
0.4	493912.9	494053.4	494157.9	71.6254	11.7
0.5	493843.3	493920	494059.3	65.0017	11.1
0.6	493817.5	493853.3	493883.9	18.9460	11.5
0.7	493796.4	493830.5	493897.5	21.8787	11.8
0.8	493791.8	493807.6	493832.9	9.3906	11.3
0.9	493786.2	493795.3	493809.8	6.7321	11.4

Table 6. Obtained Result comparison by CSA and MCSA

The Best Min. Cost ($)		The Best Aver. Cost ($)		The Best Max. Cost ($)		The Best Std. Cost ($)	
CSA	MCSA	CSA	MCSA	CSA	MCSA	CSA	MCSA
493890.1	493786.2	494361.7	493795.3	495273.3	493809.8	320.9526	6.7321

by CSA and MCSA are given in Tables 7 and 8. Figure 2 shows the cost convergence characteristic for the two methods.

Example 2: In the example, a test system with two thermal plants with nonconvex fuel cost function and two hydropower plants scheduled in three subintervals with eight for each is considered. All data for load, thermal plants and hydropower plants are given in Tables 9 to 11. The transmission loss formula coefficient is below.

$$B = 10^{-5}\begin{bmatrix} 14 & 1.0 & 1.5 & 1.5 \\ 1.0 & 6.0 & 1.0 & 1.3 \\ 1.5 & 1.0 & 6.8 & 6.5 \\ 1.5 & 1.3 & 6.5 & 7.0 \end{bmatrix}$$

For implementation of the CSA and MCSA to the second system, the maximum number of iterations is set to 2000 for both CSA and MCSA whereas the number of nests is respectively set to 20 and 6 for CSA and MCSA. The methods are run twenty independent trials for each value of Pa and the obtained result are given in Table 12 and 13 for CSA and MCSA, respectively. The best minimum cost, the best average cost, the best maximum cost and the best deviation cost obtained by the methods are compared and reported in Table 14. Obviously, MCSA obtains less all costs than CSA, especially, the best minimum

Table 7. The best solution for CSA for the system 1

P_{S1} (MW)	P_{S2} (MW)	P_{h1} (MW)	P_{h2} (MW)
438.8592	422.1166	110.484	259.5431
449.9947	514.7655	151.1961	326.1577
450	844.6172	319.7971	472.9452
450	734.2729	251.3044	434.7062

Table 8. The best solution for MCSA for the system 1

P_{S1} (MW)	P_{S2} (MW)	P_{h1} (MW)	P_{h2} (MW)
448.4278	429.4031	98.6842	254.6191
449.5741	527.5067	151.7621	313.1653
449.9939	828.8038	319.9746	488.723
449.8622	727.7219	259.8851	432.7018

Table 9. Load demand for scheduled time

Subinterval	1	2	3
Duration (hour)	8	8	8
PD (MW)	900	1200	1100

Table 10. Hydro system data of the test system with non-convex fuel cost of thermal units

Hydro Plant	a_{hj} (MCF/h)	b_{hj} (MCF/MWh)	c_{hj} (MCF /MW²h)	W_j (MCF)	$P_{hj,min}$ (MW)	$P_{hj,max}$ (MW)
1	1.980	0.306	0.000216	2600	0	400
2	0.936	0.612	0.000360	2200	0	300

Table 11. Thermal generator data of the test system with non-convex fuel cost of thermal units

Thermal plant	a_{si} (\$/h)	b_{si} (\$/MWh)	c_{si} (\$/MW²h)	d_{si} (\$/h)	e_{si} (1/MW)	$P_{si,min}$ (MW)	$P_{si,max}$ (MW)
1	25	3.2	0.0025	12	0.0550	50	300
2	30	3.4	0.0008	14	0.0450	50	700

cost and the best deviation cost. The cost convergence characteristic comparison is shown in Figure 3. The best solution obtained by CSA and MCSA for system 2 is given in Tables 15 and 16.

FUTURE RESEARCH DIRECTIONS

For the future research, the proposed ORCSA can be also implemented for solving other complex hydrothermal scheduling problems such as hydrothermal scheduling with reservoir constraint, hydrothermal scheduling with cascaded hydro power plants, multiobjective hydrothermal scheduling, etc. Moreover, the proposed ORCSA can be also implemented for solving other optimization problem in power systems such as economic dispatch problems, optimal power flow, optimal reactive power dispatch, optimal reconfiguration, etc. In addition, the proposed method can ben also improved or combined with other methods to enhance its search ability and reduce computational time.

Figure 2. Cost convergence characteristic of two methods for system 1

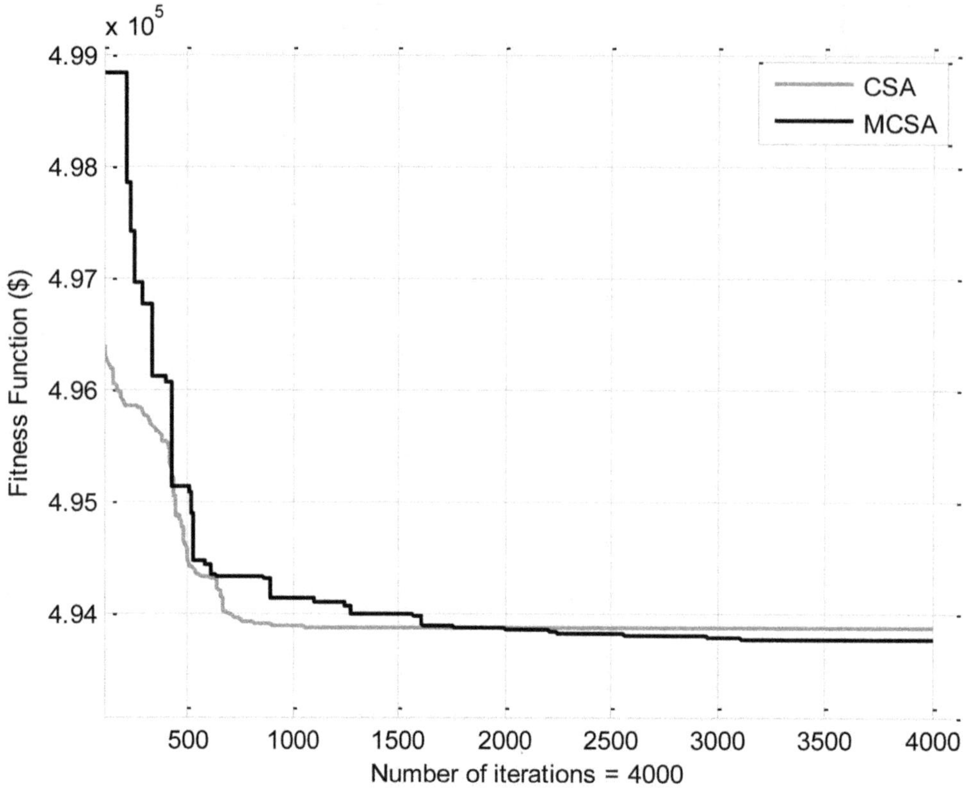

Table 12. Results by CSA for the first system with nonconvex fuel cost of thermal units with different values of P_a

P_a	Min. Cost ($)	Avg. Cost ($)	Max. Cost ($)	Std. Dev. ($)	CPU (s)
0.1	64571.976	64588.93936	64613.952	13.7094	4.35
0.2	64571.75	64586.30996	64618.667	16.6094	4.4
0.3	64572.176	64586.98624	64612.05	11.7601	4.4
0.4	64571.546	64593.15425	64618.962	15.7454	4.5
0.5	64571.94	64592.65454	64620.645	12.5657	4.32
0.6	64572.537	64591.12227	64643.343	15.8581	4.54
0.7	64572.112	64593.6007	64672.623	23.1083	4.4
0.8	64571.491	64591.81448	64702.207	28.7174	4.6
0.9	64572.334	64613.59386	64784.796	58.7183	4.5

Table 13. Results by MCSA for the system with nonconvex fuel cost of thermal units with different values of P_a

P_a	Min. Cost ($)	Avg. Cost ($)	Max. Cost ($)	Std. Dev. ($)	CPU (s)
0.1	64571.65	64587.77324	64615.9744	13.4931	4.5
0.2	64571.544	64577.98174	64594.9018	7.7099	4.56
0.3	64571.554	64573.19888	64580.2082	1.9442	4.4
0.4	64571.72	64572.5117	64576.4614	1.0061	4.4
0.5	64571.437	64572.18442	64572.9813	0.5221	4.6
0.6	64571.328	64571.95179	64572.7424	0.3609	4.5
0.7	64571.458	64571.931	64572.4829	0.3102	4.4
0.8	64571.41	64571.89365	64572.4854	0.3620	4.5
0.9	64571.258	64571.73413	64572.5063	0.3460	4.4

Table 14. Obtained result comparison by CSA and MCSA

The Best Min. Cost ($)		The Best Aver. Cost ($)		The Best Max. Cost ($)		The Best Std. Cost ($)	
CSA	MCSA	CSA	MCSA	CSA	MCSA	CSA	MCSA
64571.49	64571.26	64586.31	64571.73	64612.05	64572.48	11.7601	0.3102

Figure 3. Cost convergence characteristic of two methods for system 2

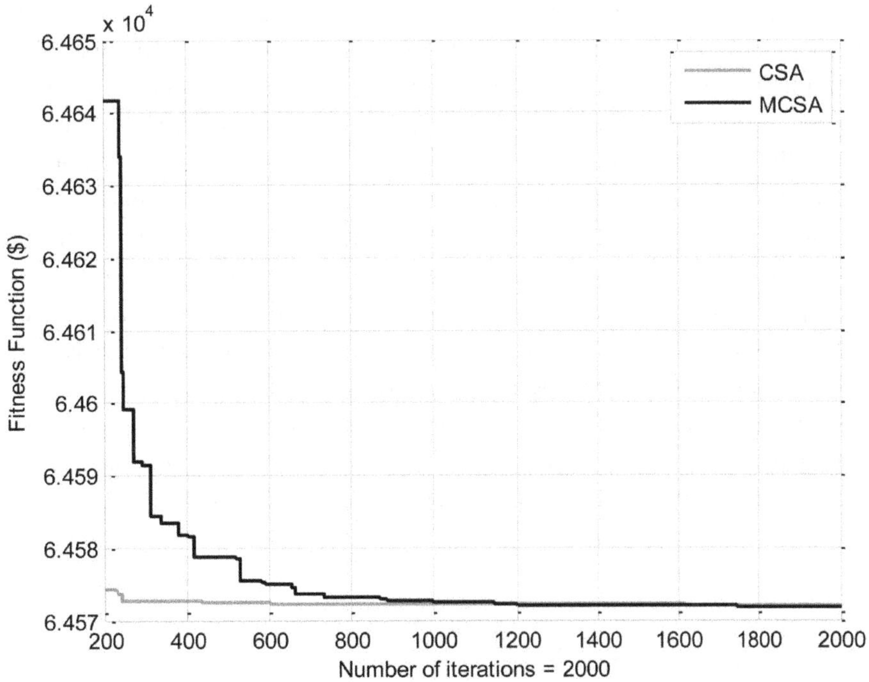

Table 15. The best solution for CSA for the system 2

Subinterval	P_{s1} (MW)	P_{s2} (MW)	P_{h1} (MW)	P_{h2} (MW)
1	165.3143	399.1777	264.7682	109.4243
2	221.4229	545.6849	324.0522	177.2164
3	221.3653	538.6968	276.3451	123.612

Table 16. The best solution for MCSA for the system 2

Subinterval	P_{s1} (MW)	P_{s2} (MW)	P_{h1} (MW)	P_{h2} (MW)
1	164.3881	399.3818	265.8184	109.1244
2	221.3598	546.4414	322.7086	177.8205
3	221.3616	538.6976	276.7143	123.2699

CONCLUSION

In this chapter, two methods, cuckoo search algorithm and modified Cuckoo search algorithm, are proposed for solving short-term hydrothermal scheduling problem. The objective function of the problem is electrical generation fuel cost of thermal plants over the scheduled period of time whereas its constraints needing to be exactly meet are composed of thermal plants constraint, hydro plants constraints and system constraint. For both thermal and hydro plants, each generator must operate within their capacity while load balance constraint of the system is an equation where total power output of thermal units and hydro units is equal to the sum of power losses and load demand. In addition, water discharge through hydro turbine and water volume of hydro reservoir must be retained within the upper and lower limits during operation process. Cuckoo search algorithm is inspired from natural phenomenon of Cuckoo specie's reproduction behavior. The Cuckoo lays its egg in a nest belonging to another specie and the egg may be hatched and becomes a Cuckoo chick with intelligent existing instinct that it kicks the host egg or the host chick out of the host and mimic the call of the host chick to increase food provided by the host parents. There are two phases of new solution generation via Lévy Flights and discovery of alien eggs. The new solutions just generated are evaluated and compared to old ones and which ones having better fitness function value is retained. This manner means it is not that new solution is always better than old with respect to quality. The CSA is known as a few parameters and high successful rate method. In modified CSA, the egg group in CSA is separated into top group with better fitness function value and abandoned group containing other eggs and two modifications are then performed on the two groups. The effectiveness and robustness of the methods are compared each other via two systems with two thermal and two hydro plants and quadratic fuel cost or nonconvex fuel cost function of thermal units. The obtained result has reported that MCSA is more efficient than CSA in terms of solution quality and convergence speed.

REFERENCES

Al-Agtash, S. (2001). Hydrothermal scheduling by augmented Lagrangian: Consideration of transmission constraints and pumped-storage units. *IEEE Transactions on Power Systems*, *16*(4), 750–756. doi:10.1109/59.962422

Basu, M. (2003). Hopfield neural networks for optimal scheduling of fixed head hydrothermal power systems. *Electric Power Systems Research*, *64*(1), 11–15. doi:10.1016/S0378-7796(02)00118-9

Basu, M. (2003). Hopfield neural networks for optimal scheduling of fixed head hydrothermal power systems. *Electric Power Systems Research*, *64*(1), 11–15. doi:10.1016/S0378-7796(02)00118-9

Basu, M. (2004). An interactive fuzzy satisfying method based on evolutionary programming technique for multiobjective short-term hydrothermal scheduling. *Electric Power Systems Research*, *69*(2-3), 277–285. doi:10.1016/j.epsr.2003.10.003

Basu, M. (2005). A simulated annealing-based goal-attainment method for economic emission load dispatch of fixed head hydrothermal power systems. *International Journal of Electrical Power & Energy Systems*, *27*(2), 147–153. doi:10.1016/j.ijepes.2004.09.004

Basu, M. (2011). Artificial immune system for fixed head hydrothermal power system. *Energy*, *36*(1), 606–612. doi:10.1016/j.energy.2010.09.057

Basu, M., & Chowdhury, A. (2013). Cuckoo search algorithm for economic dispatch. *Energy*, *60*, 99–108. doi:10.1016/j.energy.2013.07.011

Bhargava, V., Fateen, S. E. K., & Bonilla-Petriciolet, A. (2013). Cuckoo Search: A new nature-inspired optimization method for phase equilibrium calculations. *Fluid Phase Equilibria*, *337*, 191–200. doi:10.1016/j.fluid.2012.09.018

Binghui, Y., Xiaohui, Y., & Jinwen, W. (2007). Short-term hydro-thermal scheduling using particle swarm optimization method. *Energy Conversion and Management*, *48*(7), 1902–1908. doi:10.1016/j.enconman.2007.01.034

Brown, C., Liebovitch, S. L., & Glendon, R. (2007). Lévy ñights in Dobe Ju/'hoansi foraging patterns. *Human Ecology*, *35*(1), 129–138. doi:10.1007/s10745-006-9083-4

Carvalho, M. F., & Soares, S. (1987). An efficient hydrothermal scheduling algorithm. *IEEE Transactions on Power Systems*, *2*(3), 537–542. doi:10.1109/TPWRS.1987.4335164

Cau, T. D. H., & Keyes, R. J. (2002). Evolutionary optimisation method for multistorage hydrothermal scheduling. *IEE Proceedings. Generation, Transmission and Distribution*, *149*(2), 1359–7051. doi:10.1049/ip-gtd:20020120

Chan, P. H., & Chang, H. C. (1996). Genetic aided scheduling of hydraulically coupled plants in hydrothermal coordination. *IEEE Transactions on Power Systems*, *11*(2), 975–981. doi:10.1109/59.496183

Dieu, N. V., Peter, S., & Ongsakul, W. (2013). Cuckoo search algorithm for non-convex economic dispatch. *IET Generation. Transmission & Distribution*, *7*(6), 645–654. doi:10.1049/iet-gtd.2012.0142

Dieu, V. N., & Ongsakul, W. (2009). Improved merit order and augmented Lagrange Hopfield network for short term hydrothermal scheduling. *Energy Conversion and Management*, *50*(12), 3015–3023. doi:10.1016/j.enconman.2009.07.021

Franco, P. E. C., Carvalho, M. F., & Soares, S. (1994). A network flow model for short-term hydro-dominated hydrothermal scheduling problems. *IEEE Transactions on Power Systems*, *9*(2), 1016–1022. doi:10.1109/59.317642

Gil, E., Bustos, J., & Rudnick, H. (2003). Short-term hydrothermal generation scheduling model using a genetic algorithm. *IEEE Transactions on PWRS*, *18*, 1256–1264.

Guan, X. (1995). Nonlinear approximation method in Lagrangian relaxation-based algorithms for hydrothermal scheduling. *IEEE Transactions on Power Systems*, *10*(2), 772–778. doi:10.1109/59.387916

Heredia, F. J., & Nabona, N. (1995). Optimum short-term hydrothermal scheduling with spinning reserve through network flows. *IEEE Transactions on Power Systems*, *10*(3), 1642–1651. doi:10.1109/59.466476

Hota, P. K., Chakrabarti, R., & Chattopadhyay, P. K. (1999). Short-term hydrothermal scheduling through evolutionary programming technique. *Electric Power Systems Research*, *52*(2), 189–196. doi:10.1016/S0378-7796(99)00021-8

Hotaa, P. K., Barisala, A. K., & Chakrabarti, R. (2009). An improved PSO technique for short-term optimal hydrothermal scheduling. *Electric Power Systems Research*, *79*(7), 1047–1053. doi:10.1016/j.epsr.2009.01.001

Jubaer, A., & Zainal, S. A. (2014). Maximum Power Point Tracking (MPPT) for PV system using Cuckoo Search with partial shading capability. *Applied Energy*, *119*, 118–130. doi:10.1016/j.apenergy.2013.12.062

Kumar, S., & Naresh, R. (2007). Efficient real coded genetic algorithm to solve the non-convex hydrothermal scheduling problem. *International Journal of Electrical Power & Energy Systems*, *29*(10), 738–747. doi:10.1016/j.ijepes.2007.06.001

Lakshminarasimman, L., & Subramanian, S. (2006). Short-term scheduling of hydrothermal power system with cascaded reservoirs by using modified differential evolution. *IEE Proceedings. Generation, Transmission and Distribution*, *153*(6), 693–700. doi:10.1049/ip-gtd:20050407

Li, C., Svoboda, A. J., Tseng, C. V., & Johnson, R. B. (1997). Hydro unit commitment in hydro-thermal optimization. *IEEE Transactions on Power Systems*, *12*(2), 764–769. doi:10.1109/59.589675

Mandal, K. K., Basu, M., & Chakraborty, N. (2008). Particle swarm optimization technique based short-term hydrothermal scheduling. *Applied Soft Computing*, *8*(4), 1392–3399. doi:10.1016/j.asoc.2007.10.006

Nanda, J., & Bijwe, P. R. (1981). Optimal Hydrothermal Scheduling with cascaded plants using progressive optimality algorithm. *IEEE Transactions on Power Apparatus and Systems*, *PAS-100*(4), 2093–2099. doi:10.1109/TPAS.1981.316486

Narang, N., Dhillon, J. S., & Kothari, D. P. (2012). Multiobjective fixed head hydrothermal scheduling using integrated predator-prey optimization and Powell search method. *Energy*, *47*(1), 237–252. doi:10.1016/j.energy.2012.09.004

Nilsson, O., & Sjelvgren, D. (1996). Mixed-integer programming applied to short-term planning of a hydro-thermal system. *IEEE Transactions on Power Systems*, *11*(1), 281–286. doi:10.1109/59.486107

Oliveira, G. G., & Soares, S. (1995). A second order network flow algorithm for hydrothermal scheduling. *IEEE Transactions on Power Systems*, *10*(3), 1625–1641. doi:10.1109/59.466477

Orero, S. O., & Irving, M. R. A. (1998). Genetic algorithm modeling framework and solution technique for short term optimal hydrothermal scheduling. *IEEE Transactions on PWRS*, *13*, 501–518.

Pavlyukevich, I. (2007). Lévy flights, non-local search and simulated annealing. *Journal of Computational Physics*, *226*(2), 1830–1844. doi:10.1016/j.jcp.2007.06.008

Petcharaks, N., & Ongsakul, W. (2007). Hybrid enhanced Lagrangian relaxation and quadratic programming for hydrothermal scheduling. *Electric Power Components and Systems*, *35*(1), 19–42. doi:10.1080/15325000600815449

Rashid, A., & Nor, K. M. (1991). An efficient method for optimal scheduling of fixed head hydro and thermal plants. *IEEE Transactions on Power Systems*, *6*(2), 632–636. doi:10.1109/59.76706

Roy, P. K. (2013). Teaching learning based optimization for short-term hydrothermal scheduling problem considering valve point effect and prohibited discharge constraint. *International Journal of Electrical Power & Energy Systems*, *53*, 10–19. doi:10.1016/j.ijepes.2013.03.024

Salam, N. K. M., & Hamdan, A. R. (1998). Hydrothermal scheduling based Lagrangian relaxation approach to hydrothermal coordination. *IEEE Transactions on Power Systems*, *13*(1), 226–235. doi:10.1109/59.651640

Sasikala, J., & Ramaswamy, M. (2010). Optimal gamma based fixed head hydrothermal scheduling using genetic algorithm. *Expert Systems with Applications*, *37*(4), 3352–3357. doi:10.1016/j.eswa.2009.10.015

Shlesinger, M. F. (2006). Mathematical physics: Search research. *Nature*, *443*(7109), 281–282. doi:10.1038/443281a PMID:16988697

Simopoulos, D. N., Kavatza, S. D., & Vournas, C. D. (2007). An enhanced peak shaving method for short term hydrothermal scheduling. *Energy Conversion and Management*, *48*(11), 3018–3024. doi:10.1016/j.enconman.2007.07.001

Sinha, N., Chakrabarti, R., & Chattopadhaya, P. K. (2003). Fast evolutionary programming techniques for short-term hydrothermal scheduling. *Electric Power Systems Research*, *66*(2), 97–103. doi:10.1016/S0378-7796(03)00016-6

Sivasubramani, S., & Shanti Swarup, K. (2011). Hybrid DE–SQP algorithm for non-convex short term hydrothermal scheduling problem. *Energy Conversion and Management*, *52*(1), 757–761. doi:10.1016/j.enconman.2010.07.056

Soares, S., & Ohishi, T. (1995). Hydro-dominated short-term hydrothermal scheduling via a hybrid simulation-optimisation approach: A case study. *IEE Proceedings. Generation, Transmission and Distribution*, *142*(6), 569–575. doi:10.1049/ip-gtd:19952156

Türkay, B., Mecitoğlu, F., & Baran, S. (2011). Application of a Fast Evolutionary Algorithm to Short-term Hydro-thermal Generation Scheduling. *Energy Sources, Part B: Economics, Planning, and Policy*, *6*(4), 395–405. doi:10.1080/15567249.2010.489098

Viswanathan, G. M., Afanasyev, V., Buldyrev, S. V., Havlin, S., Luz, M., Raposo, E. P., & Stanley, H. E. (2000). Lévy flights in random searches. *Physica A: Statistical Mechanics and its Applications, 282*, 1-12.

Walton, S., Hassan, O., Morgan, K., & Brown, M. R. (2011). Modified cuckoo search: A new gradient free optimisation algorithm. *Chaos, Solitons, and Fractals*, *44*(9), 710–718. doi:10.1016/j.chaos.2011.06.004

Wong, K. P., & Wong, Y. W. (1994). Short-term hydrothermal scheduling, part-I: Simulated annealing approach. *IEE Proceedings. Generation, Transmission and Distribution*, *141*(5), 497–501. doi:10.1049/ip-gtd:19941350

Wood, A. J., & Wollenberg, B. F. (1996). *Power generation, operation and control* (2nd ed.). New York: John Wiley & Sons.

Xiaohui, Y., Bo, C., Bo, Y., & Yanbin, Y. (2008). Hydrothermal scheduling using chaotic hybrid differential evolution. *Energy Conversion and Management*, *49*(12), 3627–3633. doi:10.1016/j.enconman.2008.07.008

Yang, P. C., Yang, H. T., & Huang, C. L. (1996). Scheduling short-term hydrothermal generation using evolutionary programming technique. *IEE Proceedings. Generation, Transmission and Distribution*, *143*(4), 371–276. doi:10.1049/ip-gtd:19960463

Yang, X. S., & Deb, S. (2010). Engineering optimisation by cuckoo search. *International Journal of Mathematical Modelling and Numerical Optimisation*, *1*(4), 330–343. doi:10.1504/IJMMNO.2010.035430

Yuan, X., & Yuan, Y. (2006). Application of cultural algorithm to generation scheduling of hydrothermal systems. *Energy Conversion and Management*, *47*(15-16), 2192–2201. doi:10.1016/j.enconman.2005.12.006

Zaghlool, M. F., & Trutt, F. C. (1988). Efficient methods for optimal scheduling of ďxed head hydrothermal power systems. *IEEE Transactions on Power Systems*, *3*(1), 24–30. doi:10.1109/59.43176

ADDITIONAL READING

Basu, M. (2014). Improved differential evolution for short-term hydrothermal scheduling. *Electrical Power and Energy Systems*, *58*, 91–100. doi:10.1016/j.ijepes.2013.12.016

Cau, T. D. H., & Kaye, R. J. (2002). Evolutionary optimisation method for multi-storage hydrothermal scheduling. *IEE Proceedings. Generation, Transmission and Distribution*, *149*(2), 152–156. doi:10.1049/ip-gtd:20020120

Chen, P. H. (2008). Pumped-storage scheduling using evolutionary particle swarm optimization. *IEEE Transactions on Energy Conversion*, *23*(1), 294–301. doi:10.1109/TEC.2007.914312

Chiang, C. L. (2007). Optimal economic emission dispatch of hydrothermal power systems. *Electrical Power and Energy Systems*, *29*(6), 462–469. doi:10.1016/j.ijepes.2006.11.004

Dieu, N. V., & Ongsakul, W. (2012). Hopfield Lagrange Network for Economic Load Dispatch. Published. In P. Vasant (Ed.), *Innovation in Power, Control, and Optimization: Emerging Energy Technologies*. IGI. doi:10.4018/978-1-61350-138-2.ch002

Dieu, V. N., & Ongsakul, W. (2008). Enhanced merit order and augmented Lagrange Hopfeld network for hydrothermal scheduling. *International Journal of Electrical Power & Energy Systems*, *30*(2), 93–101. doi:10.1016/j.ijepes.2007.06.022

Ferreira, L. A. F. M. (1994). On the convergence of the classic hydro-thermal coordination algorithm. *IEEE Transactions on Power Systems*, *9*(2), 1002–1008. doi:10.1109/59.317644

Fogel, D. B. (1995). A comparison of evolutionary programming and genetic algorithms on selected constrained optimization problems. *Simulation*, *64*(6), 397–404. doi:10.1177/003754979506400605

Guan, X., Ni, E., Li, R., & Luh, P. B. (1997). An optimization-based algorithm for scheduling hydro-thermal power systems with cascaded reservoirs and discrete hydro constraints. *IEEE Transactions on Power Systems*, *12*(4), 1775–1780. doi:10.1109/59.627890

Hawary, E.I., & Ravindrannath, K.M. (1988). Optimal operation of variable head hydrothermal systems using the Glim-Kirchmayer model and the Newton-Raphson method, *electric power systems research, 14,* 11-22.

Hawary, E. I., & Tsang, D. H. (1986). The hydrothermal optimal load flow, a practical formulation and solution techniques using Newton's approach. *IEEE Transactions on Power Systems*, *PWRS-1*(3), 157–167. doi:10.1109/TPWRS.1986.4334975

Hinojosa, V. H., & Leyton, C. (2012). Short-term hydrothermal generation scheduling solved with a mixed-binary evolutionary particle swarm optimizer. *Electric Power Systems Research*, *92*, 162–170. doi:10.1016/j.epsr.2012.06.006

Huifeng, Z., Jianzhong, Z., Yongchuan, Z., Youlin, L., & Yongqiang, W. (2013). Culture belief based multi-objective hybrid differential evolutionary algorithm in short term hydrothermal scheduling. *Energy Conversion and Management*, *65*, 173–184. doi:10.1016/j.enconman.2012.04.006

Jingrui, Z., Shuang, L., & Weixia, Q. (2015). A modifed chaotic differential evolution algorithm for short-term optimal hydrothermal scheduling. *Electrical Power and Energy Systems*, *65*, 159–168. doi:10.1016/j.ijepes.2014.09.041

Khodier, M. (2013). Optimisation of antenna arrays using the cuckoo search algorithm. *IET Microwaves. Antennas & Propagation*, *7*, 458–464.

Kothari, D. P., & Dhillon, J. S. (2006). *Power system optimization*. New Delhi: Prentice Hall of India Private Limited.

Lakshminarasimman, L., & Subramanian, S. (2008). A modifed hybrid differential evolution for short-term scheduling of hydrothermal power systems with cascaded reservoirs. *Energy Conversion and Management*, *49*(10), 2513–2521. doi:10.1016/j.enconman.2008.05.021

Lee, T., & Chen, N. (1992). The effect of pumped storage and battery energy storage systems on hydrothermal generation coordination. *IEEE Transactions on Energy Conversion*, *7*(4), 631–637. doi:10.1109/60.182644

Lu, S., & Sun, C. (2011). Quadratic approximation based differential evolution with valuable trade off approach for bi-objective short-term hydrothermal scheduling. *Expert Systems with Applications*, *38*, 13950–13960.

Lu, Y., Zhou, J., Qin, H., Wang, Y., & Zhang, Y. (2010). An adaptive chaotic differential evolution for the short-term hydrothermal generation scheduling problem. *Energy Conversion and Management*, *51*(7), 1481–1490. doi:10.1016/j.enconman.2010.02.006

Mandal, K. K., & Chakraborty, N. (2008). Differential evolution technique-based short-term economic generation scheduling of hydrothermal systems. *Electric Power Systems Research*, *78*(11), 1972–1979. doi:10.1016/j.epsr.2008.04.006

Mandal, K. K., & Chakraborty, N. (2009). Short-term combined economic emission scheduling of hydrothermal power systems with cascaded reservoirs using differential evolution. *Energy Conversion and Management*, *50*(1), 97–104. doi:10.1016/j.enconman.2008.08.022

Mandal, K. K., & Chakraborty, N. (2011). Short-term combined economic emission scheduling of hydrothermal systems with cascaded reservoirs using particle swarm optimization technique. *Applied Soft Computing*, *11*(1), 1295–1302. doi:10.1016/j.asoc.2010.03.006

Mandal, K. K., & Chakraborty, N. (2013). Parameter study of differential evolution based optimal scheduling of hydrothermal systems. *Journal of Hydro-environment Research*, *7*(1), 72–80. doi:10.1016/j.jher.2012.04.001

Momoh, J. A., Dias, L. G., Guo, S. X., & Adapa, R. (1995). Economic operation and planning of multi-area interconnected power systems. *IEEE Transactions on Power Systems*, *10*(2), 1044–1053. doi:10.1109/59.387950

Nima, A., & Hassan, R. S. (2010). Daily Hydrothermal Generation Scheduling by a new Modiđed Adaptive Particle Swarm Optimization technique. *Electric Power Systems Research*, *80*(6), 723–732. doi:10.1016/j.epsr.2009.11.004

Parrilla, E., & García-González, J. (2006). Improving the B&B search for large-scale hydrothermalweekly scheduling problems. *International Journal of Electrical Power & Energy Systems*, *28*(5), 339–348. doi:10.1016/j.ijepes.2005.12.008

Saadat, H. (2002). *Power system analysis* (2nd ed.). Boston: McGraw-Hill.

Salam, S., Nor, K. M., & Hamdan, A. R. (1997). Comprehensive algorithm for hydrothermal coordination. *IEE Proceedings. Generation, Transmission and Distribution*, *144*(5), 482–488. doi:10.1049/ip-gtd:19970819

Santos, T. N., & Diniz, A. L. (2009). A new multiperiod stage deđnition for the multistage Benders decomposition approach applied to hydrothermal scheduling. *IEEE Transactions on Power Systems*, *24*(3), 1383–1392. doi:10.1109/TPWRS.2009.2023265

Sasikala, J., & Ramaswamy, M. (2012). PSO based economic emission dispatch for fixed head hydro-thermal systems. *Electrical Engineering*, *94*(4), 233–239. doi:10.1007/s00202-012-0234-x

Shlesinger, M. F. (2006). Mathematical physics: Search research. *Nature*, *443*(7109), 281–282. doi:10.1038/443281a PMID:16988697

Sifuentes, W., & Vargas, A. (2007). Short-term hydrothermal coordination considering an AC network modeling. *International Journal of Electrical Power & Energy Systems*, *29*(6), 488–496. doi:10.1016/j.ijepes.2006.11.008

Sifuentes, W. S., & Vargas, A. (2007). Hydrothermal scheduling using benders decomposition: Accelerating techniques. *IEEE Transactions on Power Systems*, *22*(3), 1351–1359. doi:10.1109/TPWRS.2007.901751

Songfeng, L., Chengfu, S., & Zhengding, L. (2010). An improved quantum-behaved particle swarm optimization method for short-term combined economic emission hydrothermal scheduling. *Energy Conversion and Management*, *51*(3), 561–571. doi:10.1016/j.enconman.2009.10.024

Swain, R. K., Barisal, A. K., Hota, P. K., & Chakrabarti, R. (2011). Short-term hydrothermal scheduling using clonal selection algorithm. *Electrical Power & Energy Systems*, *33*(3), 647–656. doi:10.1016/j.ijepes.2010.11.016

Thang, T. N., & Dieu, N. V. (2013). Augmented Lagrange Hopđeld Network Based Method for Multi-objective Hydrothermal Scheduling. In Lecture Notes in Electrical Engineering, 282:57-66

Thang, T. N., & Dieu, N. V. (2014). Multi-Objective Short-Term Fixed Head Hydrothermal Scheduling Using Augmented Lagrange Hopfield Network. *Journal of Electrical Engineering and Technology*, *9*(6), 1882–1890. doi:10.5370/JEET.2014.9.6.1882

Wadhwa, C. L. (2006). *Electrical power systems* (4th ed.). New Age International Publisher.

Wang, Y., Zhou, J., Mo, L., Ouyang, S., & Zhang, Y. (2012). A clonal real-coded quantum-inspired evolutionary algorithm with Cauchy mutation for short-term hydrothermal generation scheduling. *International Journal of Electrical Power & Energy Systems*, *43*(1), 1228–1240. doi:10.1016/j.ijepes.2012.06.036

Wang, Z., He, G.Y., Chen, X., & Zhou, S. (2004). A novel model of large-scale hydrothermal power system optimal scheduling. *Power Systems Conference and Exposition 2004 (IEEE/PES)*, *1*,518–523.

Wu, Y. G., Ho, C. Y., & Wang, D. Y. (2000). A diploid genetic approach to short-term scheduling of hydro-thermal system. *IEEE Transactions on Power Systems*, *15*(4), 1268–1274. doi:10.1109/59.898100

Xiangtao, L., & Minghao, Y. (2013). A hybrid cuckoo search via Lévy flights for the permutation flow shop scheduling problem. *International Journal of Production Research*, *15*, 4732–4754.

Yan, H., Luh, P. B., Guan, X., & Rogan, P. M. (1993). Scheduling of hydrothermal power systems. *IEEE Transactions on Power Systems*, *8*(3), 1358–1365. doi:10.1109/59.260857

Yan, H., Luh, P. B., & Zhang, L. (1994). Scheduling of hydrothermal power systems using the augmented Lagrangian decomposition and coordination technique. *American Control Conference*, *2*, 1558–1562.

Yang, J.-S., & Chen, N. (1989). Short term hydrothermal coordination using multi-pass dynamic programming. *IEEE Transactions on Power Systems*, *4*(3), 1050–1056. doi:10.1109/59.32598

Ying, W., Jianzhong, Z., Chao, Z., Yongqiang, W., Hui, Q., & Youlin, L. (2012). An improved self-adaptive PSO technique for short-term hydrothermal scheduling. *Expert Systems with Applications*, *39*(3), 2288–2295. doi:10.1016/j.eswa.2011.08.007

Yu, Z., Sparrow, F. T., Bowen, B., & Smardo, F. J. (2000). On convexity issues of short-term hydrothermal scheduling. *International Journal of Electrical Power & Energy Systems*, *22*(6), 451–457. doi:10.1016/S0142-0615(00)00004-1

Zhu, J. (2009). *Optimization of power system operation*. John Wiley & Sons, Inc., Publication. doi:.10.1002/9780470466971

Zoumas, C. E., Bakirtzis, A. G., Theocharis, J. B., & Petridis, V. (2004). A genetic algorithm solution approach to the hydrothermal coordination problem. *IEEE Transactions on Power Systems*, *19*(3), 1356–1364. doi:10.1109/TPWRS.2004.825896

KEY TERMS AND DEFINITIONS

Abandoned Group: A fraction of initial group of eggs with better fitness function value is first chosen for the top group. The other eggs with higher fitness value are put in the abandoned group.

Cuckoo Search Algorithm: The cuckoo search algorithm (CSA) is inspired from the obligate brood parasitism of some cuckoo species by laying their eggs in the nests of other bird species. In this behavior of reproduction, there are two possible cases for a cuckoo egg dumped into a host bird nest including: the host bird does not recognize the cuckoo egg and the cuckoo egg will hatch and carry over to the next generation or the host bird identifies the cuckoo egg and either throw it away or abandon its nest to build a new one. The two mentioned phenomena have been inspired in the CSA method for two phases of new solution generation including the exploration phase via Lévy flights (the first phenomenon) and the exploitation phase via replacement of a fraction of eggs (the second phenomenon).

Economic Load Dispatch: This is one of the most important optimization operation problems. In this problem, the main task is to determine power output of a set of available generating units so that total fuel cost for producing electricity is minimized while satisfying all constraints including power balance, power generation limits, ramp rate constraints, etc.

Hydrothermal Scheduling: This is also one of the most important optimization operation problems. However, this problem is more complicated than economic load dispatch since several hydro plants are added to the system supplying electricity to load. The main task of the problem is to calculate power output of both thermal plants and hydro plants so that only total fuel cost of thermal plants is minimized while satisfying power balance constraint, power generation limits on thermal and hydro units, water discharge limits and water volume limits.

Nonconvex Fuel Cost Function: This is not only a cost curve of thermal units but also a non-differentiable objective function of economic load dispatch problem and hydro scheduling problem where the valve point loading effect on thermal units is considered. As considering the valve point loading effect on thermal units the fuel cost curve of thermal units is more realistic to contain higher order nonlinearity and discontinuity.

Quadratic Fuel Cost Function: This is a differentiable objective function of economic load dispatch problem and hydrothermal scheduling problem where the valve point loading effect on thermal units is not considered.

Top Group: This is a group of eggs which is separated from the initial group of egg. The top group contains eggs with better fitness function value.

NOMENCLATURE

a_{hj}, b_{hj}, c_{hj} Water discharge coefficients of hydro plant *j*,
a_{si}, b_{si}, c_{si} Fuel cost coefficients of thermal plant *i*,
d_{si} e_{si} Fuel cost coefficients of thermal plant i reflecting valve-point effects,
B_{ij}, B_{0i}, B_{00} *B*-matrix coefficients for transmission power loss,
N_1 Total number of thermal units,
N_2 Total number of hydro units,
M Total number of subintervals,
$P_{D,m}$ Total system load demand at subinterval *m,*
$P_{hj,m}$ Power output of hydro plant *j* in subinterval *m,*
$P_{hj,max}$ Maximum power output of hydro plant *i,*
$P_{hj,min}$ Minimum power output of hydro plant *i,*
$P_{L,m}$ Total transmission loss at subinterval *m,*
$P_{si,m}$ Power output of thermal plant j in subinterval *m,*
$P_{si,max}$ Maximum power output of thermal plant *i,*
$P_{si,min}$ Minimum power output of thermal plant *i,*
$q_{j,m}$ Rate of water flow from hydro plant j in subinterval *m,*
t_m Duration for subinterval *m,*
W_j Volume of water available for generation by hydro unit j during the scheduling period.

Chapter 15
Imprecise Solutions of Ordinary Differential Equations for Boundary Value Problems Using Metaheuristic Algorithms

Ali Sadollah
Korea University, South Korea

Joong Hoon Kim
Korea University, South Korea

ABSTRACT

In this chapter, a general strategy is recommended to solve variety of linear and nonlinear ordinary differential equations (ODEs) with boundary value conditions. With the aid of certain fundamental concepts of mathematics, Fourier series expansion, and metaheuristic algorithms, ODEs can be represented as an optimization problem. The purpose is to reduce the weighted residual error (error function) of the ODEs. Boundary values of ODEs are considered as constraints for the optimization model. Inverted generational distance metric is utilized for evaluation and assessment of approximate solutions versus exact solutions. Four ODEs having different orders and features are approximately solved and compared with their exact solutions. The optimization task is carried out using different optimizers including the particle swarm optimization and the water cycle algorithm. The optimization results obtained show that the proposed method equipped with metaheuristic algorithms can be successfully applied for approximate solving of different types of ODEs.

INTRODUCTION

Mathematical formulation of most physical and engineering problems involves differential equations (DEs). DEs have applications in all areas of science and engineering. Hence, it is important for engineers and scientists to know how to set up DEs and solve them. With regard to real life problems, which are highly nonlinear, many problems in engineering often include one or more ordinary DEs (ODEs).

DOI: 10.4018/978-1-4666-9644-0.ch015

Analytical approaches are often incompetent in tackling ODEs (except the simple ODEs). Therefore, approximate analytical methods are applied to obtain solutions of ODEs. A number of analytical methods have been used to obtain approximate analytical solutions for engineering problems, such as the variational iteration method (Coşkun & Atay, 2008), the homotopy analysis method (Domairry & Fazeli, 2009), and differential transformation method (Zhou, 1986).

In particular, many studies have applied approximation methods for solving various types of integro-differential equations (linear/nonlinear) (Darania & Ebadian, 2007; Darania & Ivaz, 2008; Roul & Meyer, 2011). However, each of these numerical approximation techniques has its own operational limitations that strictly narrow its functioning domain. Hence, it is possible that these approximate techniques fail to overcome a specific problem (Torabi et al., 2012; Yaghoobi & Torabi, 2012).

Despite there being a wide range of approximate methods for solving ODEs, there is a lack of a general approach that meets all the engineering demands having unconventional and nonlinear ODEs. It should be very interesting to solve linear and nonlinear ODEs having arbitrary boundaries values.

Metaheuristic algorithms are usually devised by observing the phenomena occurring in nature. With the emergence of metaheuristic algorithms, complex problems having large dimensions may optimally solve. Metaheuristic optimization algorithms have demonstrated their efficiency in finding near-optimal solutions where exact and analytical approaches may not be able to produce superior solutions within reasonable computational time (Glover & Kochenberger, 2003; Yang, 2010).

Up to date, applications that use metaheuristic algorithms for finding estimated solution of ODEs have significantly increased. For instance, the genetic algorithms (Mastorakis, 2006), the particle swarm optimization (PSO) (Lee, 2006; Babaei, 2013), the genetic programming (Cao et al., 2000). However, the strategies applied in these methods are different with each other.

Recently, the concept of Fourier series expansion has been used as a base approximate function for finding the approximate solution of ODEs. Hence, the ODEs problem is modeled as an optimization problem and solved using the PSO (Babaei, 2013). For simplicity, the weight function considered in the literature was set to unit weight function (i.e., constant weigh function equal to one) (Babaei, 2013). However, this assumption may not help us in obtaining better results for different types of ODEs. Recently, Sadollah et al. (2015) has shown superiority of a new proposed weight function for several ODEs and mechanical problems and compared with the unit weight function.

In this paper, using the concept of Fourier series as the base approximate function, other optimizers, including the water cycle algorithm (WCA) (Eskandar et al., 2012) and the PSO (Kennedy & Eberhart, 1995) are applied for optimization purposes. Furthermore, new weight function called least square weight function is applied instead of unit weight function in this chapter. The WCA, recently used metaheuristic algorithms, is inspired by observation of water cycle process particularly surface runoff. Both the PSO and WCA have been implemented in optimization phase in this chapter.

The reminder of this chapter is given as follows: next section provides necessary information and steps of the proposed approximate approach along with its performance metric. Afterwards, the process, steps, and strategy of considered optimization methods namely, the PSO and WCA have been described. Later, four ODE test problems having different orders and features have been given with their exact and approximate solutions by the reported optimizers. Graphical comparisons between exact and estimated solutions are given in this section. Then, comparisons of statistical results related to the error function and performance metric are provided for both optimizers. Finally, this chapter ends by conclusions section.

PROPOSED ESTIMATED METHOD

In the following sections, the idea behind of the proposed approach has been provided. An evaluation performance metric has been used for evaluating the goodness of approximate solutions given in the following sections.

Approximate Base Function

Based on the suggestion in the literature (Babaei, 2013), Fourier series is an expansion of a periodic function $f(x)$ in terms of an infinite sum of sine and cosine terms. Fourier series employs the orthogonal relationships of the sine and cosine functions.

Generally, using Fourier series, one can estimate any periodic function having finite terms. In this chapter, Fourier series as our base approximate function is used. There are three main reasons for this selection; 1) A powerful theory backs its convergence for a wide variety of continuous functions (Kreyszig, 2009); 2) It contains sine and cosine terms, which are differentiable up to any order; 3) A unique form of the approximation function can be utilized for different kinds of ODEs.

In general, the suggested approximate base function, which is the partial sum of the Fourier series (finite simple terms of sine and cosine functions) with center x_0, given as follows (Kreyszig, 2009):

$$y(x) \approx Y_{appx}(x) = a_0 + \sum_{m=1}^{NT} \begin{bmatrix} a_m \cos\left(\dfrac{m\pi(x-x_0)}{L}\right) + \\ b_m \sin\left(\dfrac{m\pi(x-x_0)}{L}\right) \end{bmatrix}. \tag{1}$$

In order to create a weighted residual function, Equation (1) and its derivatives are replaced in the implicit form of ODE equation. In Equation (1), x_0 and x_n are the lower and upper bounds of the interval solution, respectively, L is the length of the interval solution, which is x_0 - x_n, and NT is the number of approximation terms of the sine and cosine functions.

Depending on the complexity of the ODEs and their exact solutions, different values of NT may be used. The unknown coefficients in Equation (1) (i.e., a_0, a_m, and b_m), which are the coefficient of Fourier series, should be calculated using metaheuristic methods. In fact, the coefficients of Fourier series are considered as design variables of ODE optimization model. The aim of the optimization task is to compute the best values for the design variables such that it has minimal error when compared to the exact solution.

Weighted Residual Function

An optimization model requires an objective function to estimate the error function of the ODEs. In order to obtain a numerical measure of error, the weighted residual function (WRF) is considered as an error function that needs to be minimized. Indeed, the WRF assumes that a solution can be approximated analytically or piecewise analytically. Therefore, general mathematical form of the WRF is given as follows (Bathe, 1996):

$$WRF = \int_D |W(x)| \times |R(\mathrm{x})| \, dx, \tag{2}$$

where $W(x)$ and $R(x)$ are called the weight and residual functions, respectively. $R(x)$ is obtained by substituting the approximate function $Y(x)$ and its derivatives in the implicit form of the ODE as follows:

$$R(x) = f(x, Y(x), Y'(x), ..., Y^{(n)}(x)). \tag{3}$$

The optimum value of the error function (i.e., the weighted residual function given in Equation (2)) is zero. In fact, the notion in the WRF is to force the residual to zero in some average sense over the domain. When the value of error function approaches zero, better approximate solutions and more accuracy can be obtained.

Talking about weight function, in this chapter, instead of assuming unit weight function (Babaei, 2013), a least square weight function is proposed for approximate solving of ODEs. If the continuous summation of all the squared residuals is minimized, the rationale behind the name can be seen. In other words, a minimum of the following equation needs to be computed as follows:

$$WRM = \int_D R(x) \times R(x) dx = \int_D R^2(x) dx. \tag{4}$$

Inverted Generational Distance Performance Metric

To perform a fair quantitative evaluation between the approximate and exact solutions, inverted generational distance (IGD) metric, considered as an assessment criterion widely used in the literature (Coello & Pulido, 2005; Gong et al., 2009), has been studied in this chapter. The IGD was first introduced by Zitzler et al. (2003).

The main objective of this criterion is to clarify the capability of the WRF and optimizers in finding an approximate solution having the least distance when compared with the exact solution. It is evident that the obtained approximate solution with the minimum value of IGD has the best convergence to the exact solution. The following equation shows the formulation of IGD performance metric Zitzler et al. (2003):

$$IGD = \frac{\sum_{i=1}^{n} d_i}{n}, \tag{5}$$

where n is the number of points in the approximate and exact solutions, and d is the Euclidean distance between point i in approximate solution and the nearest point in the exact solution. Lower values of IGD imply that obtained approximate solution must be very close to the exact solution. In this chapter, for comparison purposes, the chosen value for n is set to 100 for all considered ODE problems. Figure 1 demonstrates the application of the IGD in a clear way.

Figure 1. Schematic view of IGD performance metric as considered evaluator metric

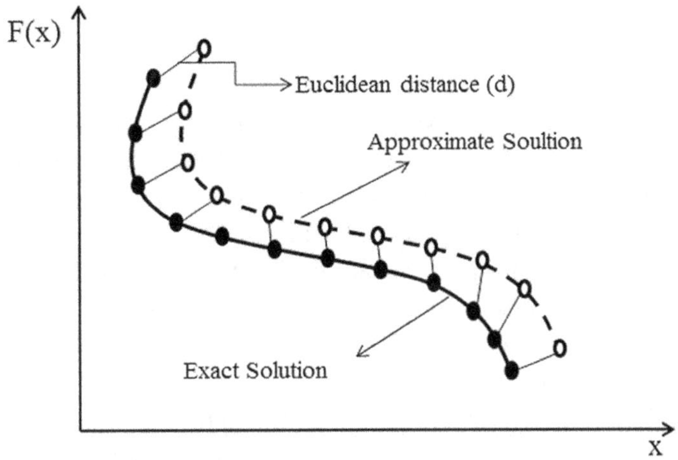

Steps of the Suggested Approximate Method

In summary, the objective function of the ODEs (minimization of the weighted residual function) and its constraints are as follows:

$$\text{Minimize} \quad WRF = \int_D |W(x)| \times |R(x)|\, dx, \tag{6}$$

subject to the boundary values (BVs) as constraints:

$$g_j^{BVs}(x_0 / x_n) = \begin{cases} 0 & j = 1, 2, \ldots, n_{BVs} \\ y_0 & j = 1, 2, \ldots, n_{BVs} \end{cases}. \tag{7}$$

Equation (7) is considered as the constraints for the boundary conditions. Finally, steps of the approximate method are given as follows:

Step 1: Convert the given ODE problem in its implicit form.
Step 2: Prepare the constraints (boundary conditions) in the proper format as seen in Equation (7).
Step 3: Choose an appropriate number of terms (*NT*) for the Fourier series, as used in Equation (1).
Step 4: Initiate user parameter of considered metaheuristic algorithm.
Step 5: Launch the optimizer in order to find the best coefficients for the approximate function (Fourier series). Error function and constraints are embedded in the optimizer for optimization purposes.
Step 6: Report the obtained coefficients of the Fourier series.
Step 7: Calculate the IGD metric to evaluate the goodness of the approximate solution against the exact solution.
Step 8: Plot the obtained approximate and exact solutions in the solution interval from x_0 to x_n.

METAHEURISTIC OPTIMIZATION ALGORITHMS

Particle Swarm Optimization

Particle swarm optimization (PSO) is an evolutionary computation technique for solving the global optimization problems developed by Kennedy and Eberhart (1995). This computation technique functions through individual improvement, population cooperation, and competition. The PSO is based on the simulation of simplified social models, such as bird flocking, fish schooling, and the swarm theory.

Researchers found that the synchrony of an animal's behavior was through maintaining optimal distances between individual members and its neighbors. The PSO exhibits common evolutionary computation attributes including initialization with a population of random solutions and searching for optima by updating generations (Dong et al., 2005).

Potential solutions, called birds or particles, are then flown through the problem space by following the current optimum particles. Each particle keeps track of the coordinates in the problem space that are associated with the best solution it has achieved so far. This value is called *pBest*. Another best value, which is tracked by the global version of the PSO, is the overall best value and this location obtained is the best so far by any particle in the population. This location is called *gBest*.

The PSO concept consists of, at each step, changing the velocity of each particle toward its *pBest* and *gBest* locations. The basic swarm position and velocity are updated using the following equations (Kennedy & Eberhart, 1995):

$$\vec{V}_{i+1} = w\vec{V}_i + c_1 r_1 (p\vec{Best}_i - \vec{X}_i) + c_2 r_2 (g\vec{Best}_i - \vec{X}_i), \tag{8}$$

$$\vec{X}_{i+1} = \vec{X}_i + \vec{V}_{i+1}, \tag{9}$$

where w is the inertia weight for previous velocity ($0 < w < 1$), X_i is the current location for the particle i, V_i is the updated velocity of particle i, $pBest_i$ is the best position found by the particle i, $gBest_i$ is the best location found by the swarm, r_1 and r_2 are uniform random numbers in the range of [0, 1], c_1 is the cognitive component, and c_2 is the social component. The role of w is crucial for the convergence of the PSO. It is employed to control the impact of the previous velocities on the current particle velocity (Dong et al., 2005).

Water Cycle Algorithm

In the water cycle algorithm (WCA), it is assumed that there are some rain or precipitation phenomena in the search space. Streams are created using the water from the rain. The WCA is a population-based algorithm; therefore, an initial population of designs variables (i.e., population of streams) is randomly generated between upper (UB) and lower (LB) bounds. The best individual, classified in terms of having the minimum cost function (or maximum fitness), is chosen as the sea.

Then, a number of good individuals, (i.e., cost function values close to the current best solution) are chosen as rivers, while the rest of individuals are called streams which flow to rivers and sea. In an *N* dimensional optimization problem, a stream is an array of 1×*N*. This array is defined as follows:

$$A\ \ Candidate\ \ Stream = [x_1, x_2, x_3, \ldots, x_N], \tag{10}$$

where *N* is the number of design variables (i.e., problem dimension). To initiate an optimization algorithm, an initial population demonstrating a matrix of individuals of size $N_{pop} \times N$ is created. Therefore, the matrix *X* which is randomly generated is given as follows:

$$Total\ \ Population = \begin{bmatrix} Sea \\ River_1 \\ River_2 \\ River_3 \\ \vdots \\ Stream_{Nsr+1} \\ Stream_{Nsr+2} \\ Stream_{Nsr+3} \\ \vdots \\ Stream_{N_{pop}} \end{bmatrix} = \begin{bmatrix} x_1^1 & x_2^1 & x_3^1 & \cdots & x_N^1 \\ x_1^2 & x_2^2 & x_3^2 & \cdots & x_N^2 \\ \vdots & \vdots & \vdots & \vdots & \vdots \\ x_1^{N_{pop}} & x_2^{N_{pop}} & x_3^{N_{pop}} & \cdots & x_N^{N_{pop}} \end{bmatrix}, \tag{11}$$

where N_{pop} is the population size. As the first step, N_{pop} individual are generated (see Equation (11)). A number of N_{sr} from the best individuals (minimum values) are chosen as a sea and rivers. The individual which has the minimum value among others is considered as the sea. In fact, N_{sr} is the summation of number of rivers (i.e., defined by the user) and a single sea as given in Equation (12). The rest of the population (i.e., streams which flow to the rivers or may directly flow to the sea) is calculated using Equation (13) as given follows:

$$N_{sr} = Number\ of\ Rivers + \underbrace{1}_{Sea}, \tag{12}$$

$$N_{Streams} = N_{pop} - N_{sr}. \tag{13}$$

Depending on the value of cost function (fitness function), the intensity of flow for each stream is calculated as follows for the sea and rivers:

$$C_i = Cost_i - Cost_{Nsr+1} \qquad n = 1, 2, 3, \ldots, N_{sr}, \tag{14}$$

$$NS_i = Round\{\left|\frac{C_i}{\sum_{i=1}^{N_{sr}} C_i}\right| \times N_{Streams}\} \quad , \quad i = 1,2,...,N_{sr}, \tag{15}$$

where NS_i is the number of streams which flow to the specific rivers or sea. Streams tend to move towards sea and rivers based on their magnitude of flow (intensity of flow). It means more streams flow into the sea than rivers. Hence, one of the best approaches to hand out the streams among sea and rivers in proportional way is to use the cost functions (fitness functions) of the sea and rivers.

Using Equations (14) and (15), the best solution (i.e., sea) will be able to control and possess more streams. It is worth pointing out that streams will be randomly selected among the population of streams. Each stream is controlled by one of the best individual (i.e., sea or a river). Hence, it is not possible to assign a stream to more than one best individual.

As it happens in nature, streams are created from the waters and join each other and flow to the rivers. Some streams may even flow directly to the sea. All rivers and streams end up in the sea that corresponds to the current best individual. Therefore, as the exploitation phase in the WCA, the new position for streams and rivers can be given in the following equations (Eskandar et al., 2012):

$$\vec{X}_{Stream}(t+1) = \vec{X}_{Stream}(t) + rand \times C \times (\vec{X}_{Sea}(t) - \vec{X}_{Stream}(t)), \tag{16}$$

$$\vec{X}_{Stream}(t+1) = \vec{X}_{Stream}(t) + rand \times C \times (\vec{X}_{River}(t) - \vec{X}_{Stream}(t)), \tag{17}$$

$$\vec{X}_{River}(t+1) = \vec{X}_{River}(t) + rand \times C \times (\vec{X}_{Sea}(t) - \vec{X}_{River}(t)), \tag{18}$$

where $1 < C < 2$ and the best value for C may be chosen as 2. Setting $C > 1$ allows streams to flow in different distances and directions towards rivers. *rand* is an uniformly distributed random number ([0,1]). Equations (16) and (17) are considered as updating equations for new positions of streams who flow to rivers and sea, respectively, while Equation (18) is the updated equation for the rivers who pour to the sea. Notations having vector sign corresponds vector values, otherwise the rest of notations and parameters consider as scalar values.

If the solution obtained by a stream is improved than its connecting river, the positions of river and stream are switched (i.e., the stream becomes a river and the river becomes a stream). Therefore, the previous river act as a stream and the better stream (in term of cost function value) assigns as a new river. Indeed, the new river controls all previous streams. A similar interchange can be carried out for a river and the sea and for the streams and sea as well.

In this condition (same as interchange between river and stream), the improved river will act as new sea and the previous sea will be a new river with its own streams which was directly moved toward it. Hence, the streams related to the previous river (which is currently new updated sea) will act as the streams moving directly toward the new sea.

Figure 2. A schematic view of Equations (16) to (18) used in the WCA

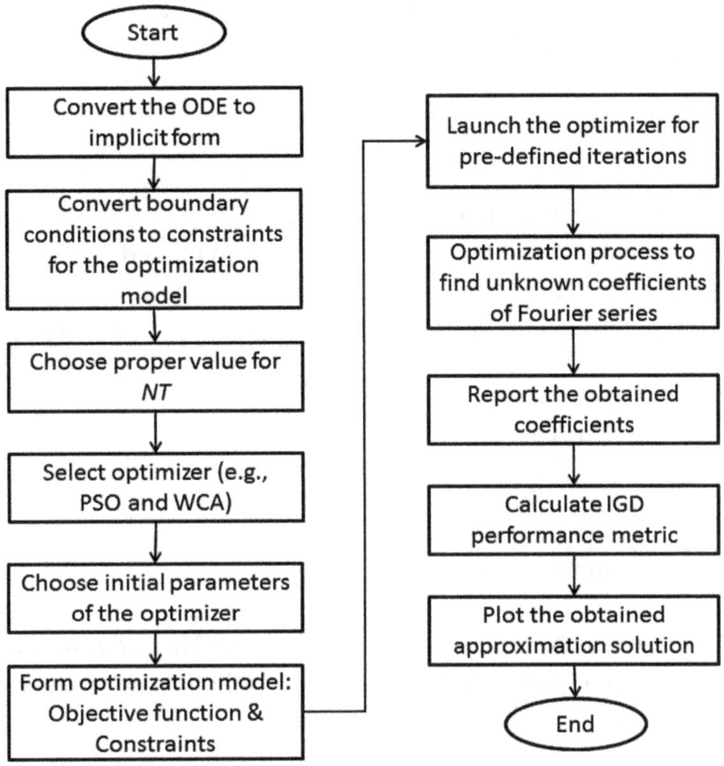

The development of the WCA optimization process is illustrated by Figure 2 where circles, stars, and the diamond correspond to streams, rivers, and sea, respectively. The white (empty) shapes denote the new positions taken by streams and rivers.

Also, evaporation and raining process are defined in the WCA to enhance its performance in global search. To be more precise, the evaporation concept operates similar to mutation in GAs and it causes the algorithm to escape from local optima. In fact, this concept enables the algorithm to avoid premature (immature) convergence to local optima (Eskandar et al., 2012). Basically, evaporation causes sea water to evaporate as rivers/streams flow into the sea. This leads to new precipitations. The evaporation operator is applied for both rivers and streams that flow into the sea. Therefore, rivers and streams should be checked whether they are close enough to the sea to make the evaporation process occur. For that purpose, the following criterion is utilized:

$$\begin{aligned} & if \left\| \vec{X}_{Sea} - \vec{X}^{i}_{River} \right\| < d_{\max} \quad or \quad rand < 0.1 \qquad i = 1, 2, \ldots, N_{sr} - 1 \\ S1: \quad & \quad Perform \ raining \ process \ by \ Equation \ (19) \qquad , \\ & end \end{aligned}$$

where d_{max} is a small number close to zero. After evaporation, the raining process is applied and new streams in the different locations are formed. For specifying the new locations of the newly formed streams, the following equation is used:

$$\vec{X}^{New}_{Stream}(t+1) = \vec{LB} + rand \times (\vec{UB} - \vec{LB}) . \tag{19}$$

This condition should also check for the streams belong to the sea and if the evaporation condition is satisfied for a specified stream, it will move to the new location based on Eq. (19).

By setting a large value for d_{max} the number of searches is limited, while small values leads to search near the sea. Therefore, d_{max} controls the search intensity near the sea. The value of d_{max} adaptively decreases as follows:

$$d_{\max}(t+1) = d_{\max}(t) - \frac{d_{\max}(t)}{Max\_Iteration} . \tag{20}$$

The steps of the WCA are summarized as follows:

Step 1: Set the initial parameters of the WCA: N_{sr}, d_{max}, N_{pop}, and *Max_Iteration.*
Step 2: Generate a random initial population and form the initial streams, rivers, and sea using Equations (11), (12), and (13).
Step 3: Calculate the cost function (fitness function) of streams, rivers, and sea.
Step 4: Determine the intensity of flow for rivers and sea using Equations (14) and (15).
Step 5: Streams flow directly to the sea by Eq. (16).
Step 6: Exchange positions of river with a stream which gives the best solution.
Step 7: Exchange positions of sea with a stream which gives the best solution.
Step 8: Streams flow to the rivers by Eq. (17).
Step 9: Check Steps 6 and 7 for updating rivers and sea.
Step 10: Rivers flow to the sea using Eq. (18).
Step 11: Exchange positions of river with a stream which gives the best solution.
Step 12: Check the evaporation condition with the reported pseudo-code for the rivers and streams.
Step 13: If the evaporation condition is satisfied, the raining process will occur using Equation (19).
Step 14: Reduce the user-defined parameter d_{max} via Equation (20).
Step 15: If the stopping criterion is not satisfied return to Step 5.

In the WCA, rivers (a number of best selected points except the best one (sea)) act as guidance points for guiding other individuals in the population toward better positions (as shown in Figure 2) in addition to minimize or prevent searching in regions containing local optima (see Equation (18)). Indeed, rivers are not fixed locations and move toward the sea (the best solution). This procedure (moving streams to the rivers and, then, moving rivers to the sea) leads to indirectly moving toward the optimum solution.

TESTED ODES

In this section, four different types of ODEs are considered for evaluating the approximate solutions using the proposed strategy. Note that this paper does not aim to investigate the ODE problems from the analytical point of view. Regarding the selection of optimization method, any optimization method such as heuristics and metaheuristics can be applied for the optimization phase and equipped with the proposed approximate approach. In the following sections, different types of ODEs are listed and their approximate solutions have been given.

ODE Problem 1

Assume the following ODE test problem in its implicit form:

$$\begin{aligned} &\frac{dy}{dx}=\frac{y^2+x^2}{yx} \\ &y(0)=0 \quad , \\ &y(3)=7.468 \end{aligned} \tag{21}$$

where the exact solution calculates using the following equation:

$$Y_{Exact}(x)=x\sqrt{2\ln x+4}\,. \tag{22}$$

Consider that the interval of the solution for this BVP varies from 0 to 3. Regarding the complexity of this problem, the chosen value for *NT* was set to three. Therefore, we have seven design variables in order to find the approximate solution. The lower and upper bounds of the design variables (i.e., a_0, a_m, and b_m) are considered between -10 and 10 for this test problem.

The best optimized solution reported by the WCA is given as follows:

$$\begin{aligned} Y_{appx} &= (4.0250)-(3.9221)\cos\left(\pi x/3\right)-(0.9624)\sin\left(\pi x/3\right)+(0.2910)\cos\left(2\pi x/3\right) \\ &+(0.9669)\sin\left(2\pi x/3\right)+(0.1881)\cos\left(\pi x\right)+(0.0766)\sin\left(\pi x\right) \end{aligned}\,. \tag{23}$$

Figure 3 depicts the existed error between the exact solution and the WRF method equipped with WCA, so called WRF-WCA. Looking at Figure 3, we can observe that the approximate solution (Equation (23)) provides acceptable solutions compared with its exact solutions.

Figure 3. Comparison of the best solution obtained by the WRF-WCA versus exact solution for the test ODE problem 1

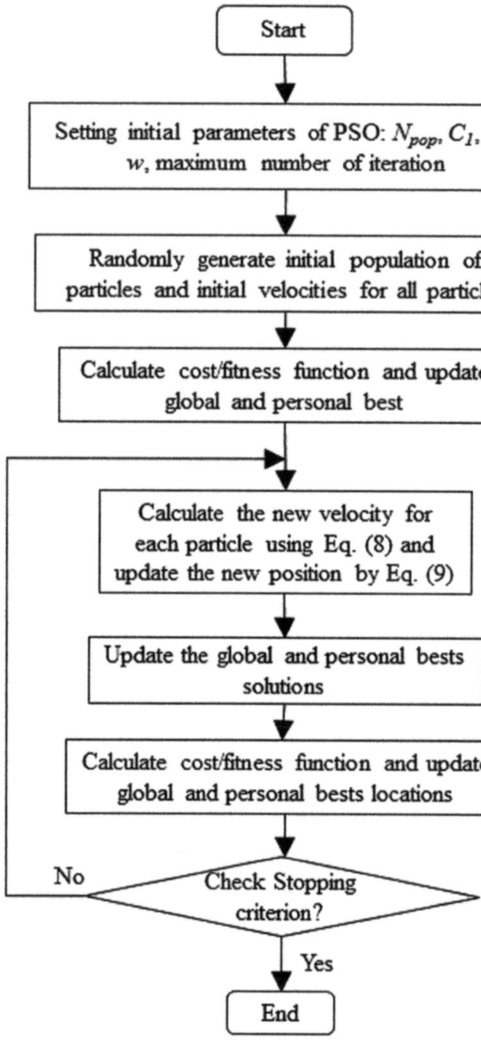

ODE Problem 2

Implicit form of the second tested ODE problem accompanied with its boundary condition and exact solution have been given in the following equations:

$$\begin{aligned}&\frac{dy}{dx} = 2\sin x + 5x^3\\&y(0) = 1 \qquad ,\\&y(\pi) = 5 + \left(5/4\right)\pi^4\end{aligned} \tag{24}$$

Figure 4. Comparison of estimated solution attained by the WRF-WCA contrast with exact solution for the ODE problem 2

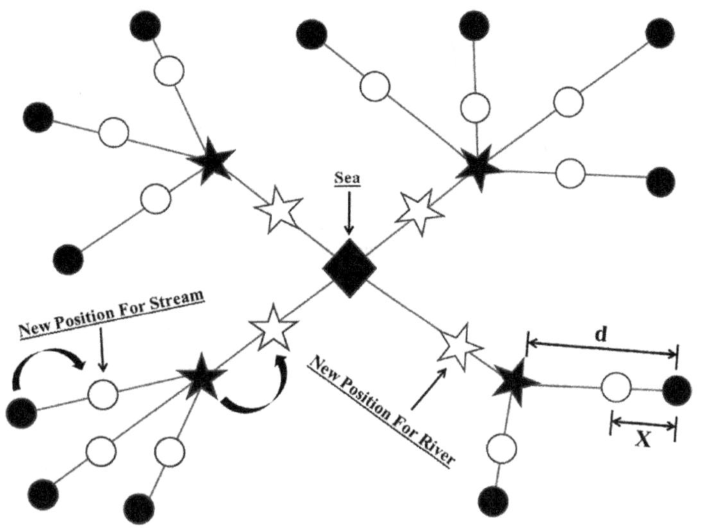

$$Y_{Exact}(x) = -2\cos x + \frac{5}{4}x^4 + 3. \tag{25}$$

We assumed solution interval spans from 0 to π. The number of cosine and sins terms, *NT*, was set to 3 for this problem. In fact, we need to find 7 design variables in order to reduce the error caused by weighted residual function. Lower and upper bounds of unidentified confidences were -100 and 100, respectively, for this problem. Coefficients of Equation (26) have been found using the proposed approximate method using the WCA:

$$\begin{aligned} Y_{appx} &= (76.9515) - (70.5549)\cos(x) - (78.2423)\sin(x) - (13.0708)\cos(2x) \\ &+ (35.1265)\sin(2x) + (7.6742)\cos(3x) + (1.1037)\sin(3x) \end{aligned}. \tag{26}$$

Figure 4 plots the approximate solution obtained (see Equation (26)) versus the exact solution (see Equation (25)) for the assumed solution interval ([0, π]). From Figure 4, suggested method shows its efficiency and performance in order to find estimated solution excellently covered the exact solution.

ODE Problem 3

The third test problem considered in this chapter, a second order ODE, in its implicit form is given as follows:

$$\begin{aligned} &\frac{d^2y}{dx^2} + 5\frac{dy}{dx} + 6y = 0 \\ &y(0) = 2 \\ &y'(0) = 3 \\ &y(5) = 9e^{-10}\left(1 - \frac{7}{9}e^{-5}\right) \end{aligned}. \tag{27}$$

Table 1. Optimum coefficients for the approximate solution using the proposed method for the ODE problem 3

Coefficient	Value	Coefficient	Value	Coefficient	Value
a_0	0.2978	b_4	0.3220	b_8	0.1120
a_1	0.2748	a_5	0.2832	a_9	0.1522
b_1	0.1017	b5	0.1960	b_9	0.0426
a_2	0.3063	a_6	0.0869	a_{10}	0.1025
b_2	0.4352	b_6	0.0062	b_{10}	-0.0732
a_3	0.2653	a_7	0.0315	a_{11}	-0.0073
b_3	0.2009	b_7	0.1316	b_{11}	-0.0523
a_4	0.0957	a_8	0.1106	-	-

Based on the calculus and mathematics rules, exact solution of ODE as seen in Equation (27) can be computed as:

$$Y_{Exact}(x) = 9e^{-2x}\left(1 - \tfrac{7}{9}e^{-x}\right). \tag{28}$$

For this ODE problem, the solutions span from 0 to 5. The number of assumed *NT* was taken as 11 which means that 23 design variables (coefficients) have to be optimally found. Lower and upper bounds of unknown confidences were -1 and 1, respectively, for this ODE problem. Table 1 tabulates all computed coefficients (decision variables) by the proposed method equipped with the WCA (WRF-WCA).

Figure 5 shows the comparison between exact and estimated solutions obtained by the WRF-WCA. Approximate solution found by the proposed method slightly has diverged from the exact solution as can be seen in Figure 5. However, overall estimation shows acceptable solution in spite of existence error.

ODE Problem 4

Assume the following ODE test problem having boundary conditions as it given in its implicit form:

$$\begin{aligned} &\frac{d^2y}{dx^2} - \frac{dy}{dx} = 0 \\ &y(0) = 1 \qquad , \\ &y'(0) = 2 \\ &y(1) = 3.893 \end{aligned} \tag{29}$$

where the exact solution can be found using the following equation:

$$Y_{Exact}(x) = \cosh x + 2\sinh x. \tag{30}$$

Figure 5. Approximate solution obtained by the WRF-WCA versus exact solution for the ODE problem 3

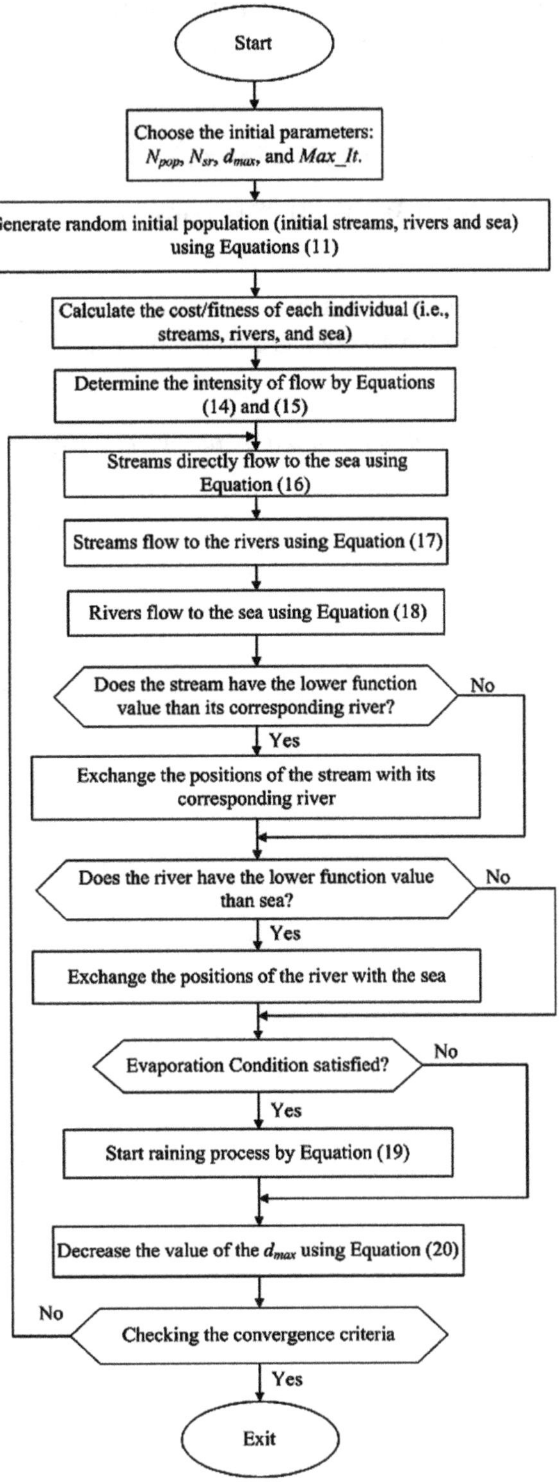

Consider that the interval of the solution for this BVP varies from 0 to 1. For this ODE problem NT was set to three (i.e., 7 design variables). Lower and upper bounds of unknown confidences were -5 and 5, respectively, for this problem. Hence, the best optimized solution obtained by the WCA is given in the following equation:

$$Y_{appx} = (2.4188) - (1.5499)\cos(\pi x) - (0.2373)\sin(\pi x) + (0.0276)\cos(2\pi x) \\ +(0.4691)\sin(2\pi x) + (0.1034)\cos(3\pi x) - (0.0214)\sin(3\pi x) \quad (31)$$

Figure 6 illustrates the graphical comparison between the exact and approximate for the ODE problem 4. By observing Figure, the WRF-WCA finely has covered the exact solution.

Figure 6. Comparison of the best solution obtained by the WRF-WCA against the exact solution for the ODE problem 4

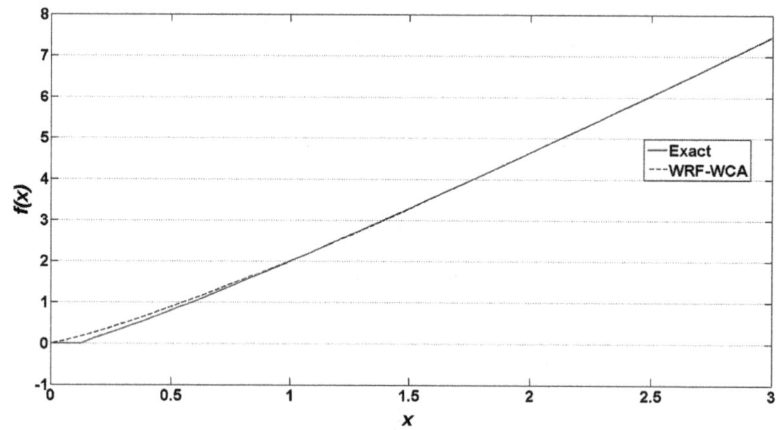

Figure 7.

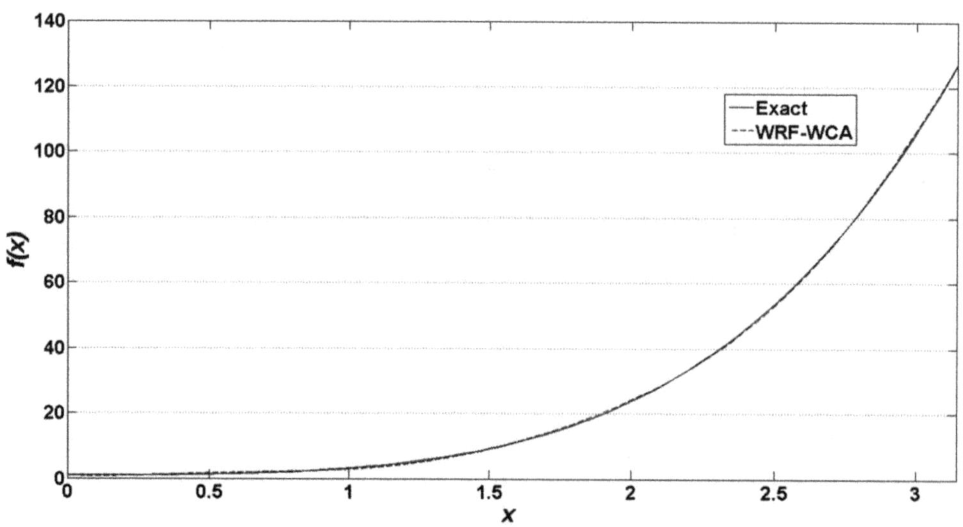

Figure 8.

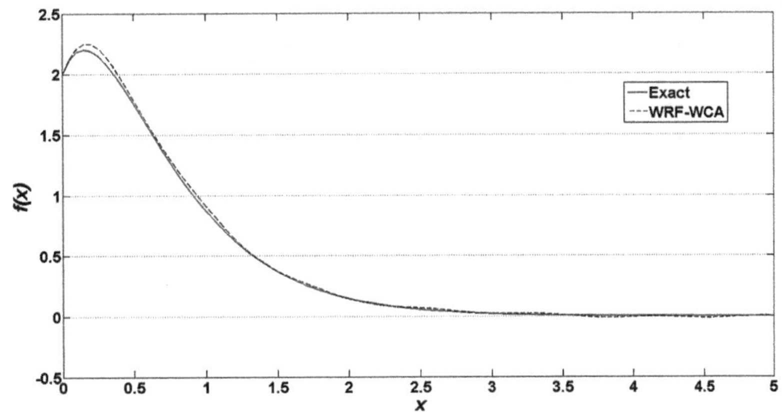

Figure 9.

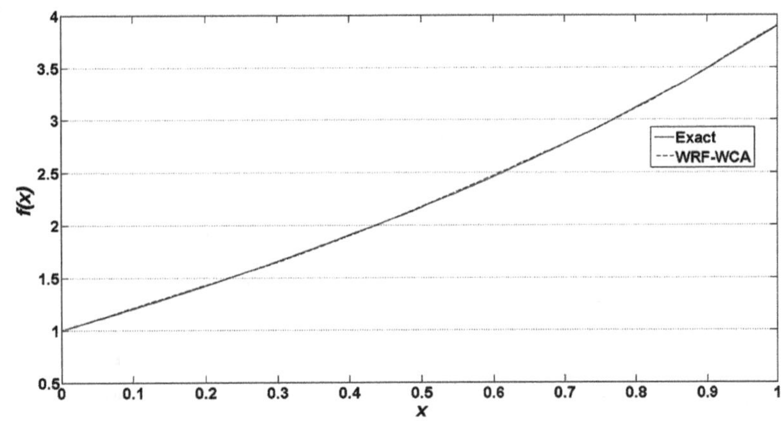

NUMERICAL OPTIMIZATION RESULTS AND DISCUSSIONS

In this section, statistical optimization results obtained by two metaheuristic algorithms including the WCA and PSO are provided. The task of optimization was carried out using 50 independent runs for all reported ODE problems. The chosen values for user parameters of the WCA are given as follows: population size (N_{pop}), number of sea and rivers (N_{sr}), and d_{max} were set to 50, 4, and 1e-16, respectively, for all test ODE problem in this chapter.

The initial parameters for the PSO were as follows; the inertia weight for velocities was 0.9, and the cognitive and social components (c_1 and c_2) were 1.5. Similar to the WCA, the population size of 50 was considered for the PSO. Maximum number of function evaluations (NFEs) was set to 300,000 as stopping condition for both considered optimizers.

Tables 2 and 3 represent the statistical optimization results of the error function and the IGD performance metric obtained by the reported optimizers, respectively. Note that the optimal value of the weighted residual function (i.e., error function) is zero.

Table 2. Comparison of obtained statistical optimization results using the two optimizers. "SD" stands for standard deviation

ODEs	Worst Error		Average Error		Best Error		Error SD	
	WCA	PSO	WCA	PSO	WCA	PSO	WCA	PSO
1	2.4505	11.2700	1.2425	2.7566	0.3316	0.0346	0.6596	3.7895
2	27.0125	31.9042	19.2245	14.0401	13.2774	12.0310	3.8677	6.2769
3	4.4308	43.9493	4.0460	22.0632	3.6121	6.3257	0.2875	10.1417
4	3.5489	3.7975	3.4884	3.5127	3.4327	3.4121	4.51e-02	0.1233

Table 3. Comparison of obtained statistical optimization results for the IGD metric using the applied optimizers

ODEs	Worst IGD		Average IGD		Best IGD		IGD SD	
	WCA	PSO	WCA	PSO	WCA	PSO	WCA	PSO
1	4.96e-02	8.81e-02	3.56e-02	4.48e-02	2.07e-02	1.07e-02	9.33e-03	2.95e-02
2	0.3121	0.3375	0.2553	0.2136	0.2102	0.1973	2.75e-02	4.36e-02
3	1.27e-02	0.2746	1.11e-02	0.1708	9.34e-03	3.97e-02	1.20e-03	6.30e-02
4	9.60e-03	1.01e-02	7.49e-03	8.84e-03	5.74e-03	7.30e-03	1.34e-03	9.17e-04

As can be seen from Tables 2 and 3, the WCA and PSO show close competitions, while providing more accurate solutions by the WCA for ODE problem 3. In terms of IGD metric, both WCA and PSO have found approximate solutions close enough to their exact solutions. Reported optimization results given in Tables 2 and 3 prove excellent performance of the proposed method accompanied with optimization methods for approximate solving of ODEs having boundary conditions.

CONCLUSION

Finding the exact solution of ordinary differential equations (ODEs) often requires substantial computational effort; however, it is not a reachable mission sometimes. In this situation, the need for approximate solutions arises. Metaheuristic optimization algorithms are famous approximate methods for solving optimization problems. Using the Fourier series as a base approximate function, ODEs can be modeled as an optimization problem. The aim is to minimize the error of the weighted residual function (i.e., error function), which is optimally zero. Boundary conditions are imposed as constraints. In order to compare the results obtained using the metaheuristic methods with the exact solutions, the inverted generational distance metric (IGD) was utilized. Four test ODEs were examined using the suggested approach. For optimizing purposes, two metaheuristic methods including the water cycle algorithm (WCA) and the particle swarm optimization (PSO) were applied and compared in terms of statistical results. Regarding the metaheuristic methods, the WCA offered a better optimized approximate solution (less error) as compared to the PSO. Being easily differentiable up to any order is considered as one of the most important aspects of the used base function (i.e., Fourier series). Based on the obtained results and the discussions given, it is observed that approximate solving of the ODE problems using Fourier series, equipped with metaheuristic optimization algorithms and the least square weight function (instead of the unit weight function) can lead us to the better simulation results. Also, the proficiency of metaheuristic algorithms in finding the optimum values for the unknown coefficients of the Fourier series is a proof of their skill in approximate solving of the ODEs.

ACKNOWLEDGMENT

This work was supported by the National Research Foundation of Korea (NRF) grant funded by the Korean government (MSIP) (NRF-2013R1A2A1A01013886).

REFERENCES

Babaei, M. (2013). A general approach to approximate solutions of nonlinear differential equations using particle swarm optimization. *Applied Soft Computing*, *13*(7), 3354–3365. doi:10.1016/j.asoc.2013.02.005

Bathe, K. J. (1996). *Finite Element Procedures* (2nd ed.). Prentice Hall.

Boyce, W. E., & Diprima, R. C. (1997). *Elementary Differential Equations and Boundary Value Problems* (6th ed.). New York: John Wiley & Sons.

Cao, H., Kang, L., Chen, Y., & Yu, J. (2000). Evolutionary modeling of systems of ordinary differential equations with genetic programming. *Genetic Programming and Evolvable Machines*, *1*(4), 309–337. doi:10.1023/A:1010013106294

Coello, C. A. C., & Pulido, G. T. (2005). Multiobjective structural optimization using a micro genetic algorithm. *Structtural and Multidisciplinary Optimization*, *30*(5), 388–403. doi:10.1007/s00158-005-0527-z

Coşkun, S. B., & Atay, M. T. (2008). Fin efficiency analysis of convective straight fins with temperature dependent thermal conductivity using variational iteration method. *Applied Thermal Engineering*, *28*(17-18), 2345–2352. doi:10.1016/j.applthermaleng.2008.01.012

Darania, P., & Ebadian, A. (2007). A method for the numerical solution of the integro-differential equations. *Applied Mathematics and Computation*, *188*(1), 657–668. doi:10.1016/j.amc.2006.10.046

Darania, P., & Ivaz, K. (2008). Numerical solution of nonlinear Volterra–Fredholm integro-differential equations. *Computers & Mathematics with Applications (Oxford, England)*, *56*(9), 2197–2209. doi:10.1016/j.camwa.2008.03.045

Domairry, G., & Fazeli, M. (2009). Homotopy analysis method to determine the fin efficiency of convective straight fins with temperature-dependent thermal conductivity. *Communications in Nonlinear Science and Numerical Simulation*, *14*(2), 489–499. doi:10.1016/j.cnsns.2007.09.007

Dong, Y., Tang, J., Xu, B., & Wang, D. (2005). An application of swarm optimization to nonlinear programming. *Computers & Mathematics with Applications (Oxford, England)*, *49*(11-12), 1655–1668. doi:10.1016/j.camwa.2005.02.006

Eskandar, H., Sadollah, A., Bahreininejad, A., & Hamdi, M. (2012). Water cycle algorithm – a novel metaheuristic optimization method for solving constrained engineering optimization problems. *Computers & Structures, 110-111*, 151–161. doi:10.1016/j.compstruc.2012.07.010

Glover, F., & Kochenberger, G. A. (2003). *Handbook of Metaheuristics*. Kluwer Academic Publishers. doi:10.1007/b101874

Gong, W., Cai, Z., & Zhu, L. (2009). An efficient multiobjective differential evolution algorithm for engineering design. *Structural and Multidisciplinary Optimization*, *38*(2), 137–157. doi:10.1007/s00158-008-0269-9

Kennedy, J., & Eberhart, R. (1995). Particle swarm optimization. *IEEE International Joint Conference on Neural Networks* (vol. 4, pp. 1942-1948). Perth, Australia: IEEE.

Kreyszig, E. (2009). *Advanced Engineering Mathematics* (9th ed.). New York: John Wiley & Sons.

Lee, Z. Y. (2006). Method of bilaterally bounded to solution Blasius equation using particle swarm optimization. *Applied Mathematics and Computation*, *179*(2), 779–786. doi:10.1016/j.amc.2005.11.118

Mastorakis, E. N. (2006). Unstable ordinary differential equations: solution via genetic algorithms and the method of Nelder-Mead. In *Proceeding of 6th WSEAS International Conference on Systems Theory & Scientific Computation* (vol. 119, pp. 1-6). Elounda, Greece.

Roul, P., & Meyer, P. (2011). Numerical solutions of systems of nonlinear integro-differential equations by Homotopy-perturbation method. *Applied Mathematical Modelling*, *35*(9), 4234–4242. doi:10.1016/j.apm.2011.02.043

Sadollah, A., Eskandar, H., Yoo, D. G., & Kim, J. H. (2015). Approximate solving of nonlinear ordinary differential equations using least square weight function and metaheuristic algorithms. *Engineering Applications of Artificial Intelligence*, *40*, 117–132. doi:10.1016/j.engappai.2015.01.014

Torabi, M., Yaghoobi, H., & Fereidoon, A. (2012). Application of differential transformation method and padé approximant for the Glauert-jet problem. *Recent Patents on Mechanical Engineering*, *5*, 150–155.

Veldhuizen, D. A. V., & Lamont, G. B. (1998). *Multiobjective evolutionary algorithm research: A history and analysis.* Technical Report TR-98-03. Department of Electrical and Computer Engineering, Graduate School of Engineering, Air Force Institute of Technology, Wright-Patterson AFB, Ohio.

Yaghoobi, H., & Torabi, M. (2012). Novel solution for acceleration motion of a vertically falling non-spherical particle by VIM–Padé approximant. *Powder Technology*, *215-216*, 206–209. doi:10.1016/j.powtec.2011.09.049

Yang, X. S. (2010). *Nature-Inspired Metaheuristic Algorithms* (2nd ed.). Luniver Press.

Zhou, J. K. (1986). *Differential transform and its applications for electrical circuits*. Wuhan: Huarjung University Press.

Zitzler, E., Thiele, L., Laumanns, M., Fonseca, C. M., & da Fonseca, V. G. (2003). Performance assessment of multiobjective optimizers: An analysis and review. *IEEE Transactions on Evolutionary Computation*, *7*(2), 117–132. doi:10.1109/TEVC.2003.810758

KEY TERMS AND DEFINITIONS

Approximate Solution: Sometimes it is difficult to solve an equation exactly. However, an approximate solution may be accurate enough for solving the considered equation.

Bundary Value Problems: A solution to a boundary value problem is a solution to the differential equation which also satisfies the boundary conditions.

Fourier Series: A Fourier series is an expansion of a periodic function f(x) in terms of an infinite sum of sines and cosines. The computation and study of Fourier series is known as harmonic analysis and is extremely useful as a way to break up an arbitrary periodic function into a set of simple terms.

Inverted Generational Distance: The IGD is an assessment criterion, used for calculating Euclidean distance between exact and apprxixmate solutions.

Metaheuristic Algorithms: A metaheuristic is defined as an iterative approach which guides a subordinate heuristic by combining intelligently different concepts for exploring and exploiting search space. They are inspired by observing the phenomena occurring in nature.

Ordinary Differential Equations: An ordinary differential equation (ODE) is an equality involving a function and its derivatives.

Particle Swarm Optimization: A population-based optimization algorithm inspired by the swarm movements of flocks (e.g., fish, birds).

Water Cycle Algorithm: A population-based optimization method inspired by the water cycle process and observation of how rivers and streams flow to sea.

Chapter 16
A Service Cost-Base Supply Balance of Sustainable Power Generation

Junzo Watada
Waseda University, Japan

Haydee Rocio Melo
Waseda University, Japan

Jaeseok Choi
Gyeongsang National University, South Korea

ABSTRACT

Recently, renewable power sources such as WTG and PV have become viable economic options for generating sustainable energy. However, WTG and PV have an inconstant power production problem. To solve this problem, multi-state models have been proposed. The electricity generated from these units varies with different time scales: hourly, daily and seasonally. Since planning an optimal size generates cost losses to the customer, three models should be built: a load model, generation model, and service cost model. Loss of energy expectation (LOEE) and loss of load expected (LOLE) are calculated for the load and generation model. The reliability value is calculated to determine the number of required renewable generators. As a result, the system is constructed to have sufficient capacity, and the utility cost became the main objective of the total service costs.

1. INTRODUCTION

The Kyoto Protocol was adopted by the UNFCCC (United Nations Framework Convention on Climate Change) on December 11, 1997 in Kyoto, Japan. However, Japan did not enforce the Protocol until February 16, 2005. The Kyoto Protocol created the foundation of political strategies for the utilization of renewable energy as a major goal on fighting global warming for all the industries (Yaakob, Watada, Tsuguhiro & Okamoto, 2010). Renewable energy, as stated, is the energy generated from natural sources

DOI: 10.4018/978-1-4666-9644-0.ch016

such as wind, sun- light, rain, geothermal heat and tides (Yaakob, Watada, & Liu, 2010). Not only does this help reduce carbon emissions, but also it becomes a sustainable energy for the future. In 2007 a break- through occurred, investing more than $100 billion in research and development, leading to the innovation of manufacturing plants and increasing renewable energy capacity. However, rapid changes in recent years had created a gap between public awareness and the reality of renewable energy.

1.1 Power Supply Systems with Renewable Energy Application

Renewable energy power supply system generates electricity in a safe and clean manner; the electricity utilization from a renewable source makes the system environmentally friendly. There is no pollution or emissions that can cause harm on the end user. Electricity production has a large impact and long-lasting consequences on not only on local and regional, but also in global environment.

Most of these consequences are generated from sulfur oxides (SOX) emissions, nitrogen oxides (NOX) emissions, and solid particles that can be diminished with investment in emission reduction technology and abatement facilities. The most harmful damage is caused by carbon dioxide (CO2). Carbon dioxide is the main cause of global warming, or greenhouse effect, and ocean acidification, since carbon dioxide can exist in the atmosphere and dissolves in water in the form of carbonic acid (Adelsberger, Collis & Pawlowski, 2002). Moreover its regulation and cost is complex and expensive. Since the power industry has been a major contributor on global CO2 emissions; renewable resources such as solar and wind energy generation have attracted the attention. In 2007, 240GW of renewable electricity was generated worldwide (which was a 50% increase since 2004), becoming 5% of power capacity and 3.4% of power generation in the world (El-Ashry & Chairman, 2007). Wind energy and solar energy became an alternative to fossil power generation with photovoltaic (PV) and wind turbine generators (WTG) that are economic and emissions free. In 2007, WTG generation capacity reached 95 GW, representing an increase of 28%; in comparison with 2006, the annual capacity additions levels increased to 40%. The grid-connected PV system is a fast growing technology with an annual installed capacity increase of 50% that represents a 7.7 GW between 2006 and 2007. In other words, the worldwide PV grid increased with 1.5 million home solar systems that provide the grid (Choi, Jeong, Bo, Park, El-Keib & Watada, 2007). One of the disadvantages of these renewable sources is the hourly, daily and seasonally inconstant power generation. The short-term prediction of wind plant prediction variability has a negative influence on the reliability of design/operation of PV and WTG systems.

1.2 Power System Reliability Evaluation

In an electrical power system, it is necessary to supply electricity to final users economically and with an acceptable reliability level. The definition of reliability is the ability of a system of component to perform its required functions under stated conditions for a specified period of time according to IEEE. The assurance of power system reliability provides confidence to the final users on the service quality and continuity. However, while assuring the service reliability and continuity, the finals users' needs and related costs must also be met. Higher-level system re- liability is proportional to a considerable investment. Power system designers and engineers have an unreal task to design a high reliable system with a reasonable budget. When designing a power system, the power system size is determined by an optimal reliability-cost value analysis. By assuring a considerable reliability level, the system ability to achieve its basic function is also assured. Power system reliability can be viewed as two interrelated components: system security and system adequacy (Billinton & Allan, 1996). There are three factors that are considered when power system reliability is evaluated: energy generation, transmission, and the distribution network. The power systems can be divided in three hierarchical levels (HL) according to the different functional zones. In this study the first level (HL-I), which includes the system generation and its ability to supply the load points, is considered.

The generation reliability assessment considers that the system can meet the load requirements. In this level, the influence of associated transmission systems and distribution facilities of the overall system are disregarded. The hierarchical level II (HL-II) evaluation is used to assess the adequacy of a system at both the generation and transmission utilities, as well as the complex system ability to supply to the bulk points. The evaluation refers to as composite or bulky system adequacy. The hierarchical level III (HL-III) includes not only the generation and transmission system but also the distribution facilities, by including all functional zones. Most of the analyses are focused on the distribution systems, as result of the complexity and enormity of the problem of including all the functional zones for evaluation (Billinton & Gan, 1991).

1.3 Reliability-Cost Balance Problem for a Renewable Power Generation System

In renewable systems (solar or wind energy), there are some possible approaches:

1. Solar/wind power system with a storage system,
2. Solar or wind power system without a storage system,
3. Solar or wind power system without a storage system, combined with a conventional power system (diesel engine), and
4. Solar or wind power system with a storage system, combined with a conventional power system (diesel engine). These four approaches are depicted in Figures 1-4.
   a. PV and WTG power systems without a storage system.
   b. PV and WTG power systems with a storage system.
   c. PV and WTG power systems without a storage system combined with conventional power system (diesel engine).
   d. PV and WTG power systems with a storage system combined with conventional power system (diesel engine).

The solar/wind power systems with or without storage systems are normally small, efficient and not network connected with generators. These systems are used in remote areas where it is challenging and costly to install an electrical power net- work due to the high cost of fuel and maintenance. These systems called isolated power systems (SIPS) may or may not contain an energy storage system depending on the site resources and the characteristic of the application. Wind and solar energy systems have no constant operation due to fluctuations of several factors, such as daily variations, cloud coverage and wind speed. The required schedule energy demand cannot meet the customer's needs with PV or WTG output. The conversion system's fluctuation can be smoothed with a storage system to meet customer demands. The use of diesel generators with renewable generators helps reduce the cost to maintain a reliability level in isolated communities.

Figure 1. Block diagram of WTG and PV based systems

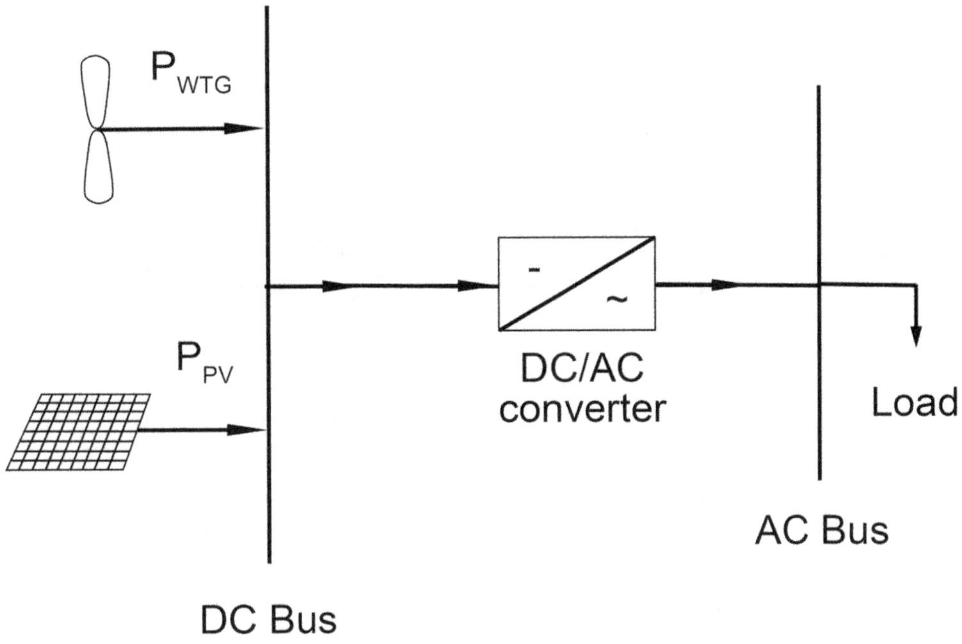

Figure 2. Block diagram of WTG and PV based without energy store

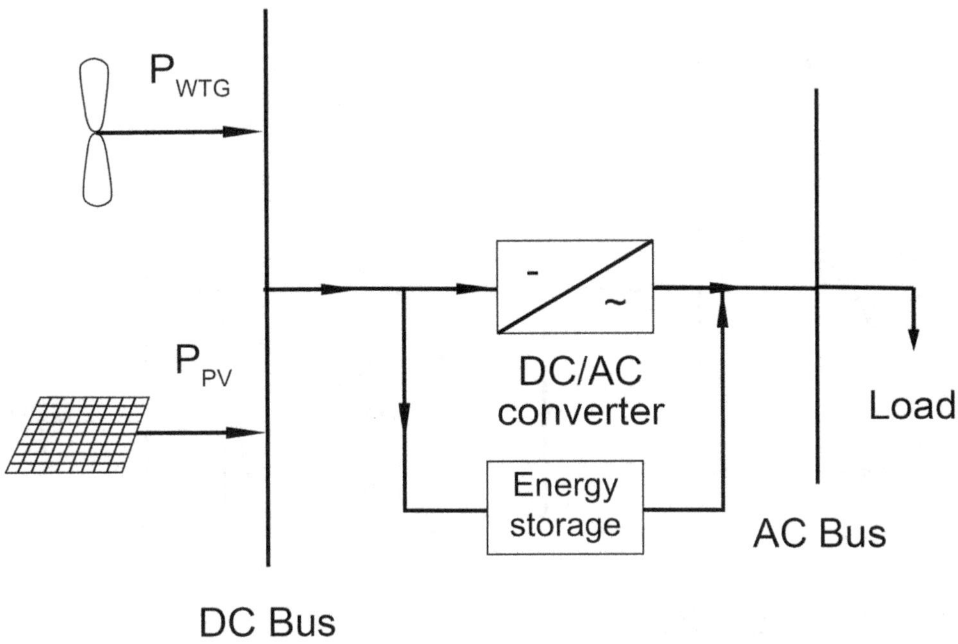

Figure 3. Wind or solar energy based systems

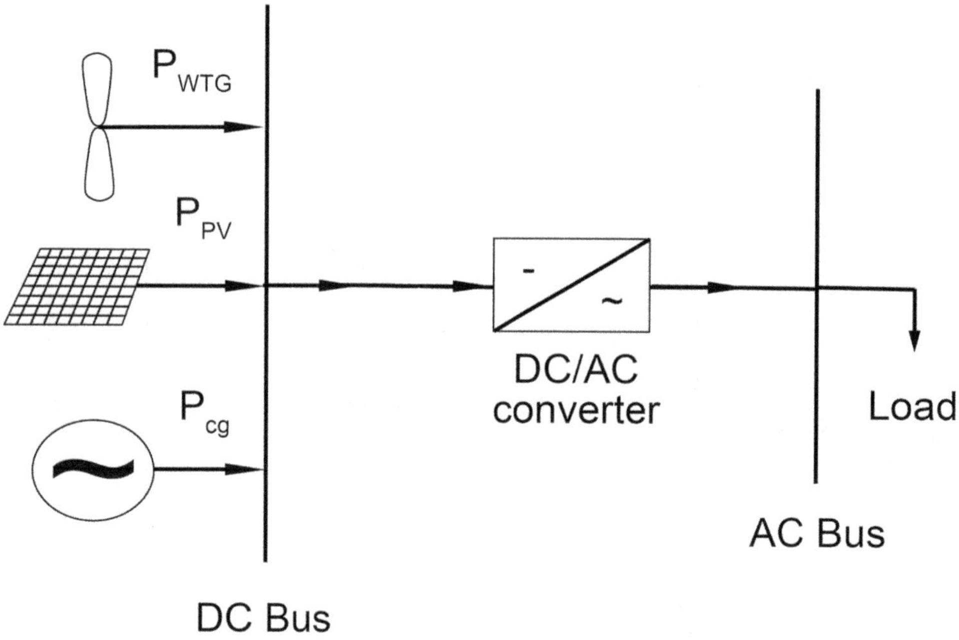

Figure 4. Wind or solar energy based systems without energy store combined with conventional generation

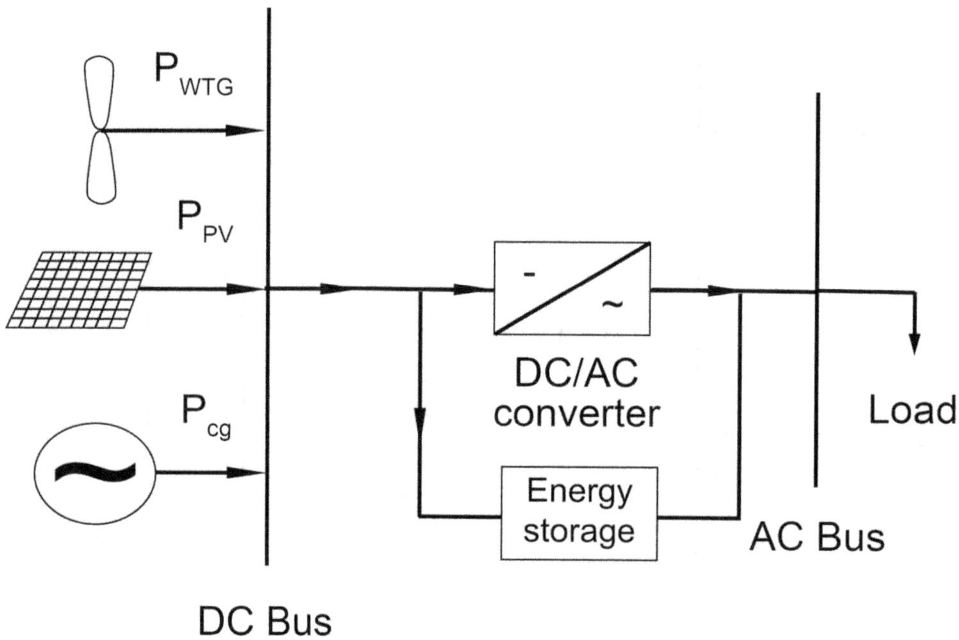

1.3.1 Reliability Evaluation of Renewable Power Generation

The second and fourth approach cannot be analyzed as due to absence of data about the storage system. Therefore, only the first and third approaches are considered in this paper. Conventional systems such as hydro or thermal generation systems are modeled at three capacities and states as the base for generating power. Advances on renewable energy generation technology (such as solar and wind), stimulate their utilization as non-conventional systems. Construct a model with acceptable cost-performance relationship, a respective evaluation method is needed to determine the wind/solar energy assessments, which determine the proper methods and techniques for the design/control of wind and solar power systems. Adequate techniques and models for designing renewable power systems will encourage the use of these systems. Significant work has been done to calculate the wind/solar energy-including power system's reliability. Researches have principally focused on power system reliability of a single wind farm that operates in parallel since 1988 (Allen, Billinton, Breipohl, & Grigg, 1994), (Allen, Billinton, Breipohl, & Grigg, 1999), (Billinton, Chen & Ghajar, 1996), (Billinton, Fotuhi-Firuzabad & Bertling, 2001). Wind power generation system influence on the system reliability can be calculated with a simple method presented in Spall, 2003. Electric power systems with non-conventional systems can be evaluated with chronological simulation methods as introduced in Billinton, Chen & Ghajar, 1996, Desrochers, Blanchard & Sud, 1986, Huynh, Nakamori, Lawry, Inuiguchi, 2010. Solar and/or wind power system reliability and economic estimation base on conventional power system has barely sufficient consideration. Small-scale stand-alone renewable energy system reliability evaluation is studied in the following references Abouzahr & Ramakumar, 1990, Abouzahr & Ramakumar, 1991, Gen (2005), Bakirtzis, 1992, Gavanidou, Bakirtzis & Dokopoulos, 1992, Karaki, Chedid & Ramadan, 1999a, Karaki, Chedid & Ramadan, 1999b. In Abouzahr & Ramakumar, 1990 a stand-alone system with a sole wind turbine supplying a load by a battery was introduced. Later, the study was applied to a stand-alone solar energy system (Abouzahr & Ramakumar, 1991). In this method, a uniform distributed customer load with a storage system is considered for analyzing. Loss of load expected (LOLE) and expected energy nor supplied (EENS) calculation for a wind power system in parallel with a storage system (battery) method is presented in Bakirtzis, 1992; this method is similar to the one presented in Abouzahr and Ramakumar (1990, 1991).

Wind, hybrid-wind and solar systems with storage evaluations are presented in Karaki, Chedid & Ramadan, 1999a and Karaki, Chedid & Ramadan, 1999b. Well distribution and β -distribution are assumed for the solar radiation and wind speed in these methods. Above-mentioned methods use approximate analytical techniques that do not characterize the chronological and variable nature of wind and solar energies. Ignoring the effects of random nature on renewable energy generation systems outputs cause to not fully recognize the load model and properly incorporate it. In addition, there has been little research on the development of techniques to calculate the reliability and asses the economic values of small-scaled power systems with renewable energy (such as wind and so- lar units) and storage system that operates in parallel with conventional systems (diesel generators) had been scarcely. Traditional techniques to calculate the mean values for accurate information to evaluate the system performance can be modeled. The lack of reliability index distribution data at hierarchical level HL-I especially on solar/wind energy and storage battery, made the deterministic models than probabilistic models for planning capacity reserves in small-scaled isolated systems (Gordon, 1988). Probabilistic methods are harder to comprehend to determine the risk index compared to deterministic methods. Deterministic methods can be combined with probabilistic methods to ease the interpretation of the risk index and provide relevant data to electrical power systems. The combination of these methods helps to overcome the complexities of understanding the risk index and providing data in a probabilistic method for system planning. As mentioned in Section 1.1, solar and wind energy applications are still operating and gaining new markets at the present time. Designing a system with reliable power supply and demand event at variable weather conditions (which influence solar radiation and wind speed), expected performance, and system components cost is challenging. Therefore, assessment of the renewable energy generation systems reliability is also challenging. Normally, in a conventional system, a unit can be considered available (up) or out of service (down); in non-conventional systems such as wind and solar energy generation, this cannot be applied. Furthermore, de-rated states (states when the units operate at reduced capacity) need to be considered. Different de-rated states are considered in a multi-state model other than full capacity or failed states. Generating unit de-rated states recognition has an important part in the evaluation of the adequate generating capacity. Multi-state model can be applied for conventional systems with thermal units and non-conventional systems such as PV and WTG units. Usually, solar radiation intensity and wind speed determines the number of capacity states at the generating units. Every generator with a multi-state model can generate a simulation of the power generation with small discrepancies with the real power generation. Additionally, an adequate reliability evaluation can be applied to the obtained data.

1.3.2 Reliability Evaluation of Renewable Power Generation

Many researches had been carried to design an ideal power system at adequate costs, considering both conventional systems and renewable energy systems (Choi, Jeong, Bo, Park, El-Keib & Watada, 2007) (Wang, Dai & Thomas, 1984). These researches considered real option (Noro, Dobashi, Wang, Takahashi, Okamoto, Tanaka, & Watada, 2006), (Yaakob, Takahashi, Okamoto, Tanaka, Minh, Watada & Zhu, 2008) mean variance (Watada, Yaakob, Takahashi & Okamoto, 2010) (Xue & Watada, 2010), rough sets (Wang, Watada, Yaakob & Choi, 2010) methods. Normally the applied models are more complex. Consequently the multi-state or bi-level models are considered in certain situations (Yaakob, Watada, Tsuguhiro & Okamoto, 2010)(Yaakob, Watada & Liu, 2010). An accurate forecast with past load date model is necessary (Ramli & Watada, 2010) (Singh, & Lago-Gonzalez, 1985) (Wei, Takahashi, Okamoto, Tanaka & Watada, 2006). Mathematical model as linear programming, non-linear programming, and dynamic program and mix-integer programming with some simplifications were primary used. Later, meta-heuristics algorithms such as simulating annealing (SA), Genetic Algorithm (GA), expert systems and particle swarm optimization (PSO) where developed with non-linearity and discrete points at the generation expansion planning. In the meta-heuristics proposed solution, a considerable solution space needs to be considered. However, a high-level reliability power system is hardly designed. One of the most used methods is the Optimal Utility Cost Method (OUCM). This method's objective is to find the minimum reliability-cost relationship and annual utility cost.

Capital investment, power generation or operation costs are considered as utility costs that are considered for conventional and renewable systems. Capital investment costs imply the necessary costs to build a power plant and generate electricity, or in other words the necessary cost to generate electricity. Initial capital investment in a power plant (fixed charges), fuel, operation and maintenance costs (O&M) are contemplated in the power generation costs. The power generation costs can be categorized in two subcategories: fixed and variable costs as shown in Figure 5. The fuel and O&M costs can be considered as a fixed and variable costs. If the in- vestment cost depend on capital investment, then a dash line will be indicated in the figure. The technology that is evaluated causes a considerable variation in the cost levels that are shown in Figure 2. In the OUCM method not only the optimal design for a power

Figure 5. Costs of power generation

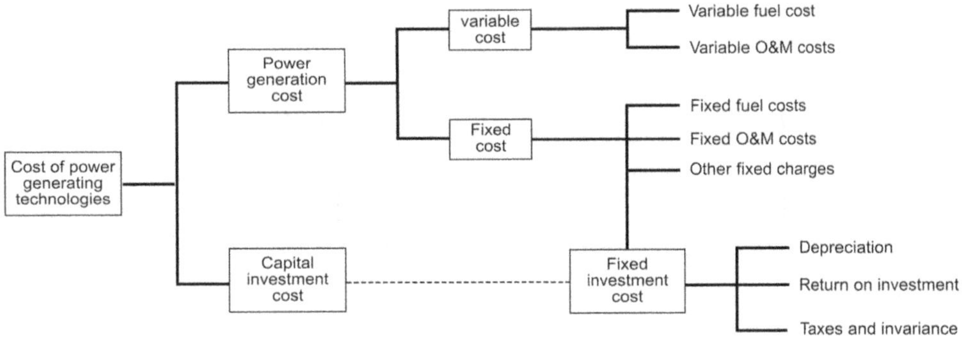

system under certain reliability level is determined, but also the previously mentioned costs are taken into account. However, the OUCM does not include the system components outage probabilities that substantially help to plan an optimal power generation system.

In other words, the leaser probability for these incidents to occur the higher the reliability is. However, the reliability value cannot be calculated because the nonexistence. The reliability level is proportional to the cost that the customer is willing to pay. Thus it can be considered that reliability is linked with customer interruption costs (CIC) that are originated from the power failures. A customer is willing to pay the increase on the reliability level with a favorable reaction if the interruption costs are decreased. Certain customers are willing to pay higher reliability while others are willing to pay for lower reliability. Due to completion, higher reliability level at lower cost is provided to the utilities.

There are many aspects that have to be considered in decision making such as social, economic, environmental and government making arduous task. To assist the decision making, system interrupt cost analysis includes analysis such as cost analysis and quantitative analysis to the usual framework. The method that reckons customer interruption and utilizes costs is called reliability cost/worth method (RCWM). RCWM is used for a complete evaluation of a power generation systems that includes WTG and PV that has not been considered in previous research. In table 1, a comparison between outage cost models and conventional model is shown. RCWM is a model that considers an objective function with outage cost (reliability level) associated with LOEE or EENS.

1.3.3 Problem Background

In the present environment, difficulty is placed in the cost of services and the reliability from the renewable energy sources. The non-conventional energy sources application has a substantial influence

Table 1. OUCM and RCWM comparison

	Model 1	Model 2
Objective function	Minimize J=	Minimize J=
	Investment cost +Operating cost	Investment cost +Operating cost- Outage cost
Main constrain	Reliability criterion (LOLE)	
Method for assessing outage output	-	LOEE Customer interruption cost

Figure 6. Connection of three models

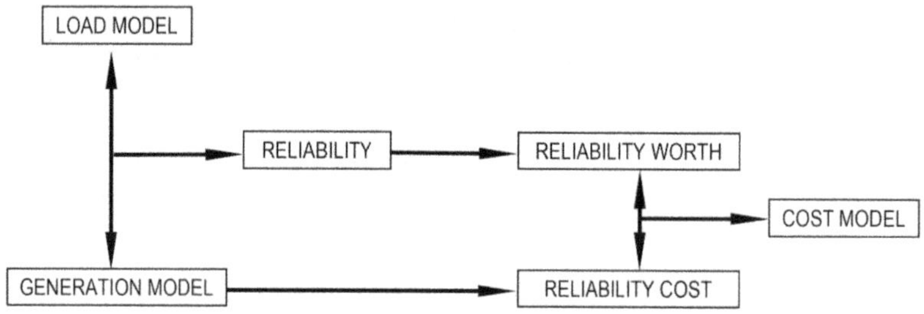

on service costs and power system reliability. The principal purpose in this research is to build a model that assesses the associated economic evaluation and reliability with PV and WTG as the base of a conventional generation system. The load model, generation model and the service cost model need to be referred to achieve this objective.

LOEE and LOLE are calculated with the load and generation model. Later the number of renewable generators is decided with the reliability evaluation of the existing generation units system. Then, by considering the reliability cost and value (Figure 6) a service cost model can be built. RCWM evaluation method can be considered as an evaluation of reliability value. The assessment of service costs that are generated from offering a reliable performance in electric power systems is acceptable and well founded.

1.3.4 Paper Summary

For designing an optimal power system that includes renewable energy sources such as solar and wind energies, a RCWM is applied in this research. Load prediction and generation planning is used to assess the reliability on renewable energy generation. Next, a feedback for economic evaluation is carried out. A case with data from IEEE Reliability Test System (IEEE RTS) and Canadian customer damage function data with analogous values from UK survey (Billinton & Allan, 1996) is presented. The paper is explained as a follows: Section 2 describes the evaluation. In this section, power systems re- liability basic concepts, analysis and assessment methods, and the different indices that are used to calculate the reliability are explained. In Section 3, the conventional generation systems are expanded by including renewable energy power generation to assess the reliability cost and reliability value with methods that evaluates the economic cost of power systems with WTG and PV. A stand-alone model is simulated to compare a system with a conventional generation system. In section 4, results are presented and future works such as evaluation that contains storage system. Finally, the use of renewable energy sources in a conventional power system is discussed.

2. EVALUATION OF RELIABILITY

Reliability evaluation usually considers the distribution system, energy transmission, and energy generation evaluation as mentioned in Section 1. The energy generation system reliability is based on the evaluation of its capacity to provide the sufficient load as depicted in Figure 1. Availability can be calculated based on the load and generation model, as depicted in Figure 3. A load and a generation model need

Table 2. Reliability criteria of electrical companies for selected countries

Country (company)	LOLE(d) (d/a)	LOLE(h) (h/a)	Country (company)	LOLE(d) (d/a)	LOLE(h) (h/a)
Belgium		5 ~ 7	Japan	0.3	
Belgium		15 ~ 6	Jordan	0.4	
Brazil	2.5		Sweden	0.4	
Canada	0.1	2.5	U.S.A	0.41	
CIS		35	Norway		3
U.K.	1.8		Spain	0.1	
Finland		9	Romania	0.1 ~ 1.2	
France	0.2	2	Slovenia		20
Ireland	1.5	9	South Africa	6	20

to be constructed while the reliability power system capacity is assessed due to the influence that it has on the operation of the generation system. The calculation of the generation capacity and load and the construction of both models determine the reliability level that meets the demand. Therefore, the load and the generation model create the availability risk model. Recently, there has been an advance on the calculation methods that are based on reliability. In section 2, a method that calculate the reliability and the indices to measure it, are introduced. Then the LOLEE and LOLE indices that are used in the cost evaluation model are explained. Since, a maintenance plan considerable affects the power system reliability; the supply capacity and maintenance plan has to be built based on the previous plan for assessing the power supply reliability. The system information considers a long-term collected data and experience for evaluating the reliability of the respective method or model. In worldwide companies, practical indices are suggested and applied. Table 2 shows a reference of the criteria of different countries or companies (Gordon, 1988).

In table 2, d/a stands for day per annual year and h/a stands for hour per annual year. Some countries usually use day as a unit while other use hours. Different countries have different reliability criteria. Furthermore, the criteria differ between day per annual and day per annual bases. The reliability base on h/a is the strictest.

2.1 Reliability Evaluation Methods

The absence of practical generation data for statistics influences the calculation of reliability, accurate failure rate estimates and force outage rate in a single generation making difficult. Every single component random failure affects the power system reliability. In traditional probabilistic analysis methods, the mean reliability indices of these components are used.

2.1.1 Reliability Indices

A reliability evaluation model is built based on the forced outage rate that combines the generation capacity model with a generator and load model. For combining the load and the generating model, there is wide-range of methods for a large electric power grid or a stand-alone system. Reliability characterizes the forecasting process for having a better understanding of the system or the components' reliability.

Therefore, the outage can be calculated to a complete extent with reliability indices. These indices are subdivided in: conventional risk and well-being indices.

2.1.1.1 Conventional Risk Indices

In power system reliability evaluation, particularly the generation capacity adequacy evaluation for conventional large electrical utility systems, the risk indices are the usually used indices. The most common indices used to plan the capacity adequacy are:

- Loss of load expectation (LOLE) h/ year.
- Loss of energy expectation (LOEE) kWh or MWh/ year Expected Loss of load duration (ELOLD) h/year Expected Loss of load frequency (ELOLF) occ/ year.

2.1.1.2 Well-Being Indices

As mentioned previously, deterministic criterion is included in the probabilistic evaluation on the well-being analysis of the system. The basic indices in this analysis are:

- Loss of health Expectation (LOHE) h/ year.
- Healthy state probability (HSP).
- Marginal state probability (MSP).
- Loss of load probability (LOLP).

LOLE and LOEE are the most used indices among them.

- **LOLE:** Expected time duration when the load exceeds the available capacity. This index is a criterion for comparing different system costs based on a suitable service cost model.
- **LOEE:** Expected energy not supplied by the generating system. This index is used in conjunction with customer cost functions to obtain a factor relating customer losses to the worth of electric service reliability.

For the reliability evaluation, LOLE and LOEE are used and their definition is given in this section. In Figure 7, the LOLE calculation is shown, by using a load- capacity relationship represented with DPLVC. If the capacity outage exceeds the reserve capacity it will cause the load loss for a time. The outage state can be de- fined as product of the probability by the time unit t_k, this value contributes to the system LOLE. The LOLE at a specified period, as presented in Equation (10), is the summation of all this products. If the capacity outage value is less than the reserve, then this value does not contribute to the system LOLE.

$$LOLE = \sum_{k=1}^{n} p_k X t_k \qquad (1)$$

Figure 7. Evaluation of LOLE

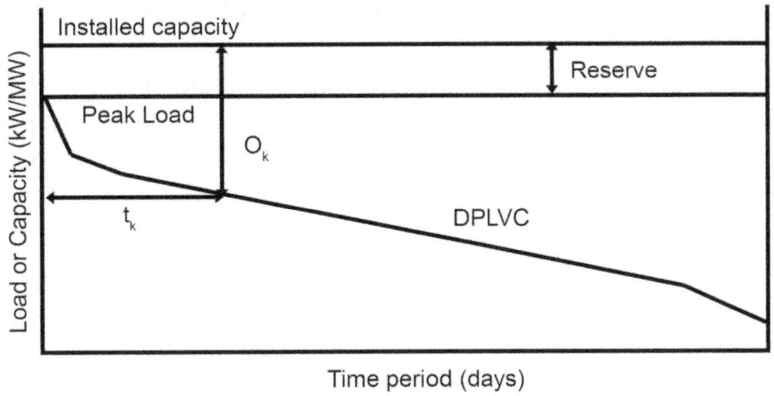

where n is the number of capacity outage states in excess of the reserve, p_k is the probability of the capacity outage, and t_k is the time for which load loss will occur. In this paper, the cost model reliability is evaluated with LOEE due to the intermittency that can cause a considerable power shortage in the WTG-PV power system. Expected energy not supplied (EENS) is more accurate than the expected time model to evaluate the reliability. However, in the reliability model, the LOLE is considered as an index to calculate the reliability. LOEE equation is modified to integrate cumulative probabilities as a follows.

$$\text{LOLEE} = \sum_{k=1}^{n} p_k X E_k \qquad (2)$$

where, n is the number of capacity outage state in excess of the reserve, p_k is the probability of the capacity outage, t_k is the time for which load loss will occur, E_k is the energy curtailment.

Energy limitation is calculated with the LDC and installed capacity as depicted in Figure 8. LOEE is calculated by multiplying the probability with the energy limitation, though the LDC time period is expressed in hours.

Figure 8. Evaluation of LOEE

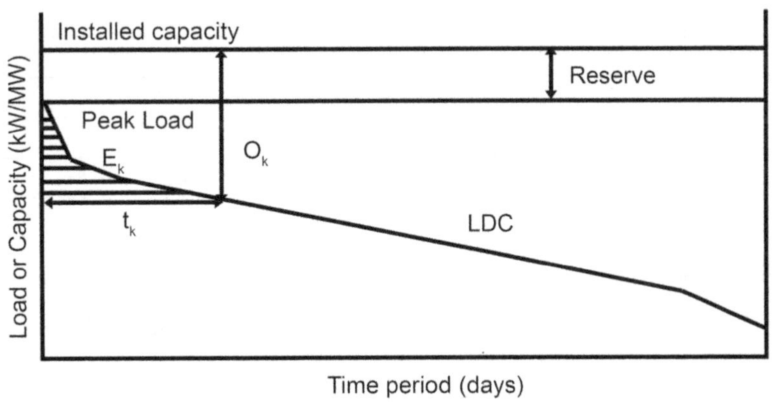

3. BUILDING A RELIABILITY EVALUATION MODEL FOR POWER GENERATION SYSTEMS WITH SOLAR AND WIND ENERGY

The power system reliability evaluation that includes solar and wind energy is presented in this section. The power generation model is expanded by included renew- able energy power generation by including probabilistic analysis method in the load model that was previously explained in Section 2. A comparison of reliability and costs between conventional generation systems and generation systems with WTG and PV is presented.

3.1 Load Model

In this paper, the load model has an annual peak load with 185 MW as the same in the previous section.

3.2 Generation Model

As explained in section 2, WTG and PV energy generation is de-rated because its intermittency. Therefore, numerous transitions states need to be considered between the down and up states in the generation mode. In this research, a five-state model is considered. In Table 3, the WTG and PV probabilities are shown. The capacity of WTG generators and PV generators are 10 MW and 10 MW, respectively.

Box 1.

Case: Total plan sizes are from one generator to 10 generators, respectively. So, nWTG and nPV are equal to 10 each when the system is a stand-alone system with conventional generating bases. The installed conventional generating capacity is 240 MW and the renewable generating capacity range will be 200 MW, which adds to the total range of planning capacity from 240 to yield 440 MW.

Table 3. Multi-state Models for a 10 MW WTG and 5MW PV

Unit Type	Capacity (MW)	Probability
WTG	10	0.2790
	7	0.0718
	5	0.0742
	3	0.1422
	0	0.4325
PV	10	0.6616
	7	0.1452
	5	0.0876
	3	0.0780
	0	0.0276

3.3 Energy Generation Systems Reliability Evaluation

The outage capacity probability for all the WTG and PV capacities is shown in Figure 9 and Figure 10. This energy generation systems has different characteristics that influences probability distribution creating a discrepant pattern. All possible capacity probability combinations for a renewable generation systems are depicted in Figure 11. According to the chart, if the constructed renewable energy generation is 200 MW, PV and WTG produce 10 MW, and the maximum generation is 130 MW. The capacity pattern and the probability function is similar if the number of PV and WTG are greater than 20. The simulated generation capacity is the combination of the curve for a whole year with a step of 8,736 hours. The outage energy and capacity probability can be calculated. If the number of PV and WTG units is increased from 20 to 30, the LOEE index is used for the calculation.

The case reliability is shown in Figure 12. In previous section, the WTG and PV constructed on conventional generators supply are analyzed. The WTG and PV number are from 0-10. The LOEE is 9.8897 in systems with no additional renewable generators and barely higher than the systems with only WTG and PV. Although as the numbers increase the energy decreases abruptly. In systems without conventional generation, the reliability demand level tends to be slower. The reliability level of WTG changes as the number of WTG increases as result of the WTG productive character.

3.4 Service Cost Models

The system has to be simulated for a whole year with 1-hour time steps and the reliability is calculated by using a cost/value method from the economic point of view. Furthermore, better specifications of all the systems components are required. Since the system operation time is 20 years is considered as the

Figure 9. Capacity probability of all possible capacity for PV

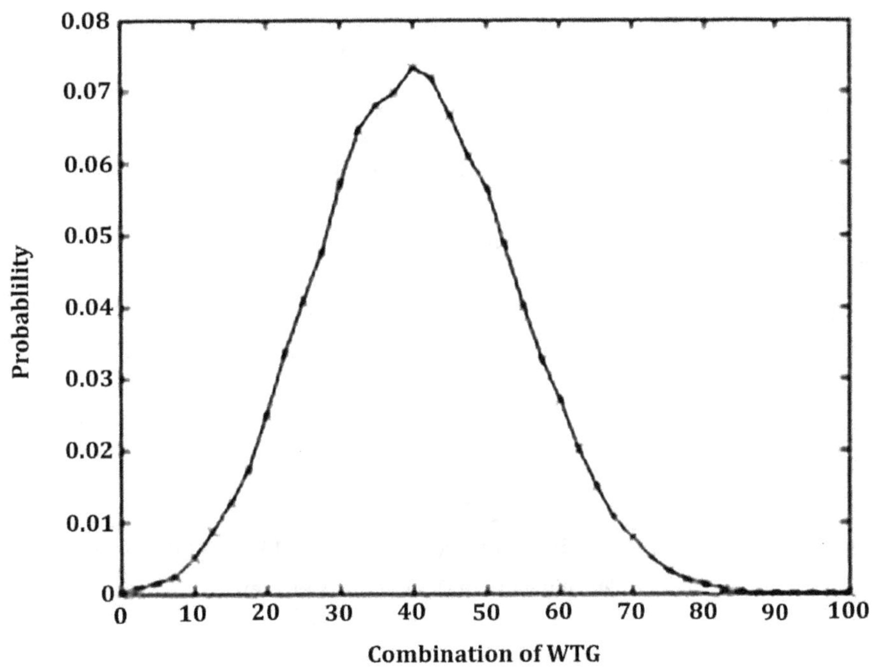

Figure 10.Capacity probability of all possible capacity for WTG

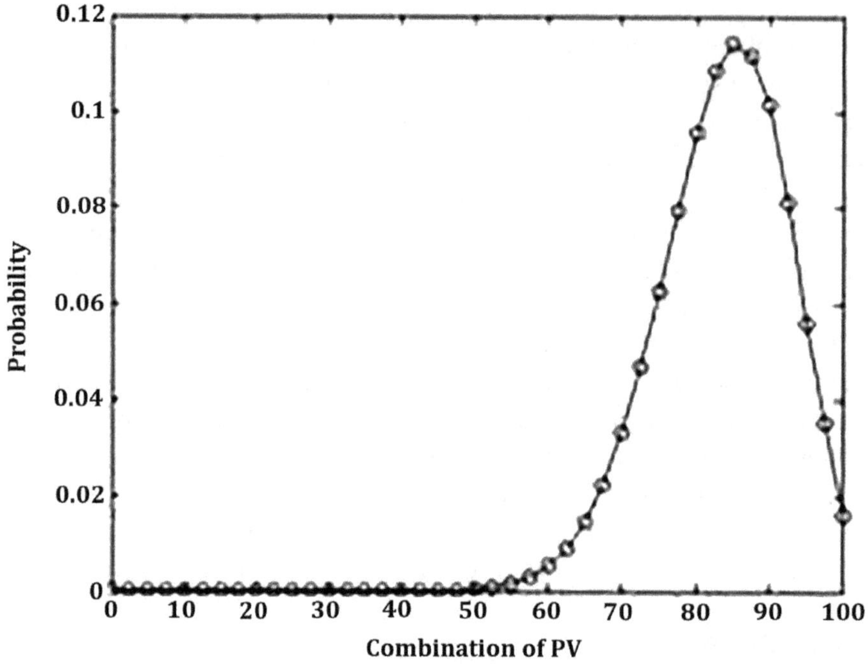

Figure 11. Total Probability of capacity combination for renewable generation

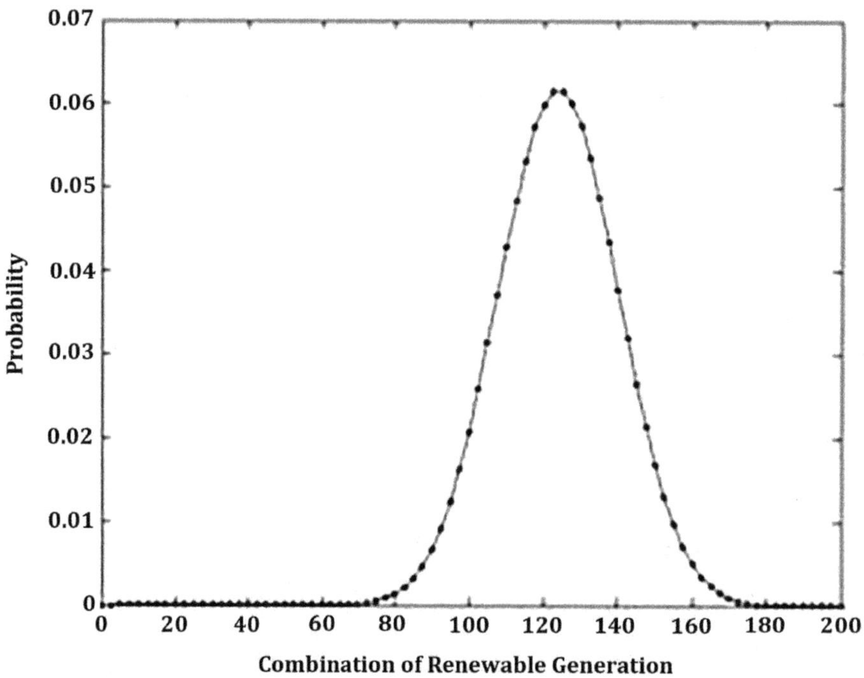

Figure 12. Numbers of WTG and PV to Reliability of the case

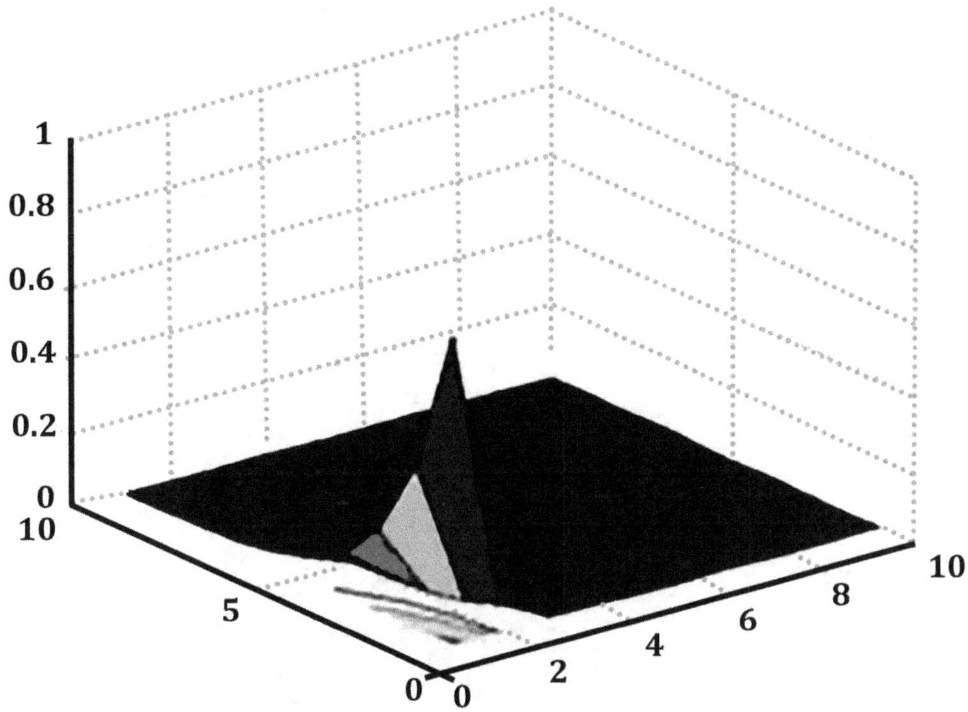

useful lifetime. The renewable generation cost data is shown in Table 4 (Gen, 2005). In comparison the conventional generation cost data that was acquired by IEEE RBTS is shown in Table 5. Moreover, the necessary data for the simulation were previously introduced in above sections.

3.4.1 Conventional Generators Based System Containing WTG and PV

In this section a generation system with conventional generator with PV and WTG are economically evaluated. Conventional generation system with its reliability value is depicted in section 3. The economic influence on total service cost that PV and WTG has when they are added to a conventional system. If a number of ten WTG or PV are included the range of the energy storage is 240 MW to 440MW. The annually customer interruption cost for a stand-alone system are summarized in the following table. Also the total service cost is shown in the final table and fig 10, 11, 12 and 13.

Table 4. The typical cost data for renewable generating units

Unit	Unit Cost	Other Constant Costs	Maintenance Cost
	($/kW)	($/kW)	(Percentage of Unit Costs)
WTG	1,200	450	2%
PV		11,000	0%

Table 5. The generating unit cost data of conventional generators

Unit Size (MW)	Type	Variable Cost ($/MW h)			Fixed Cost (k$/yr)	Capital Cost
		Fuel	Operat.	Total		
40	(Hydro)	0.45	0.05	0.50	100.00	160.00
20	(Hydro)	0.45	0.05	0.50	500.00	80.00
40	(Thermal)	9.50	2.50	12.00	790.00	80.00
20	(Thermal)	9.75	2.50	12.25	680.00	60.00
10	(Thermal)	0.00	2.50	12.50	600.00	40.00
20	(Hydro)	0.45	0.05	0.50	50.00	80.00
5	(Hydro)	0.45	0.05	0.50	12.50	40.00

Figure 13. Numbers of WTG and PV to annualized costs of the case

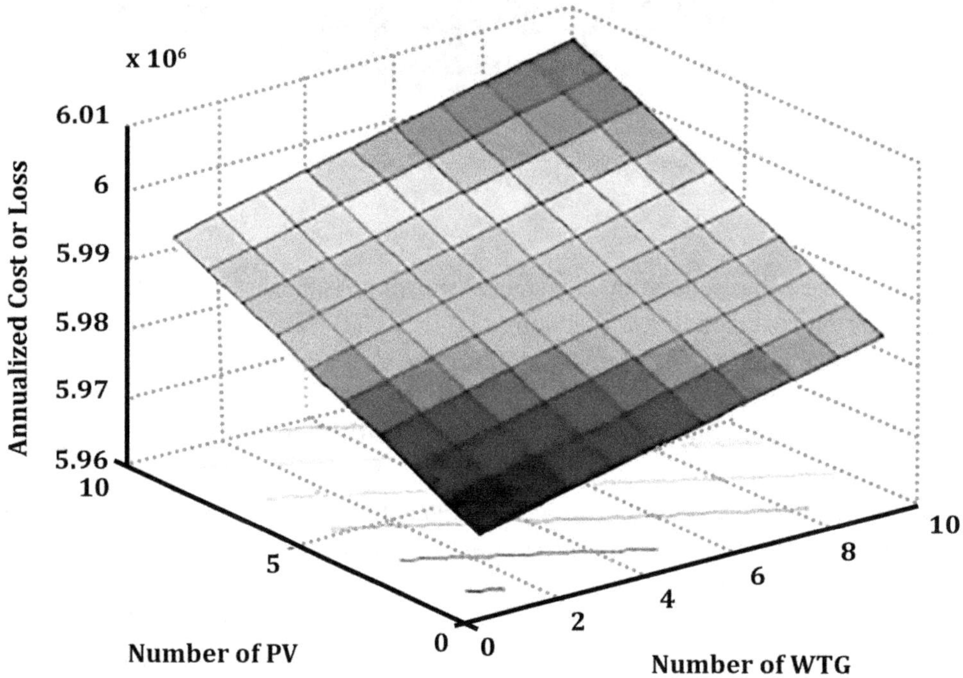

All the possible combinations are calculated and described in the previous charts. The calculations are carried as follows:

Minimumcost = 5.9737e + 006

nW T G = 4

nPV = 1

Although any constrains as reliability is not considered for calculating the optimal cost. For example, if the number of units for WTG and PV are 4 and 1, respectively; the demand cannot be met due to low reliability. Moreover, in order to ensure the supply level, a LOEE constraint value that is less than 0.001 MW is added. An optimal number of renewable generators are obtained, 5 units for WTG and 1 unit for PV. Even though the total service cost has a minor increase of $1,1310 per year, reliability increased significantly. The optimal values are:

Optimal n W T G = 5

Optimal n PV = 1

Total cost = 5.97501e + 006k/yr

From the result, it can be inferred that the fuel and operation costs have a great influence on the utility cost. It is observed in the total cost chart that if in a high level a forced outage occurred and the customer requirements are not met, the reliability has an impressive value. As result of the system is constructed to have sufficient capacity, the utility cost became the main objective of the total service costs where most of these costs are invested at conventional generators. In Figure 15, the previous conclusion is presented showing that it is stable. Furthermore, it can be concluded that investment and maintenance costs are not important compared with enormous variable costs, capital cost and fixed cost from the conventional generation systems. In case renewable energy generators substitute conventional generators in special thermal varieties, those utilities costs are not considered. Moreover, an assured reliability level for supplying the load is achieved.

3.5 Genetic Algorithm with Direct Search Toolbox Application

Since it is complicated and time consuming the calculation of the reliability of a power system, a simple model with an annual peak load of 18 MW for supplying the load model is simulated. Therefore, all combinations are calculated to obtain the model total service costs. A more complex system has a larger range of combinations that requires a more computation time and special environment to perform the simulation. GA and Direct Search Toolbox are used to lessen the problem. GA and Direct search toolbox that search for the optimal solution, determines the optimal value for the objective function by searching around the area around the current point. At each iteration, the search points are pulled closer to the optimal solution by calculating the objective values. The search stops when a point in the mesh that is the closer to the objective value is determined. The determined point is selected as the current point for the next iteration. Then, the current mesh size is multiplied by the Mesh Expansion factor 1. Furthermore, the pattern search algorithm performs an additional step. In this research, the GA is applied to the current point for the search step. As mentioned in previous study, the load condition and generation features are the same, but the range is change to 1-30. The matrix range is setting in two columns as [1 1] to [30 30]. This initial point influences the search of optimal cost.

Figure 15. Numbers of WTG and PV to total cost of the case

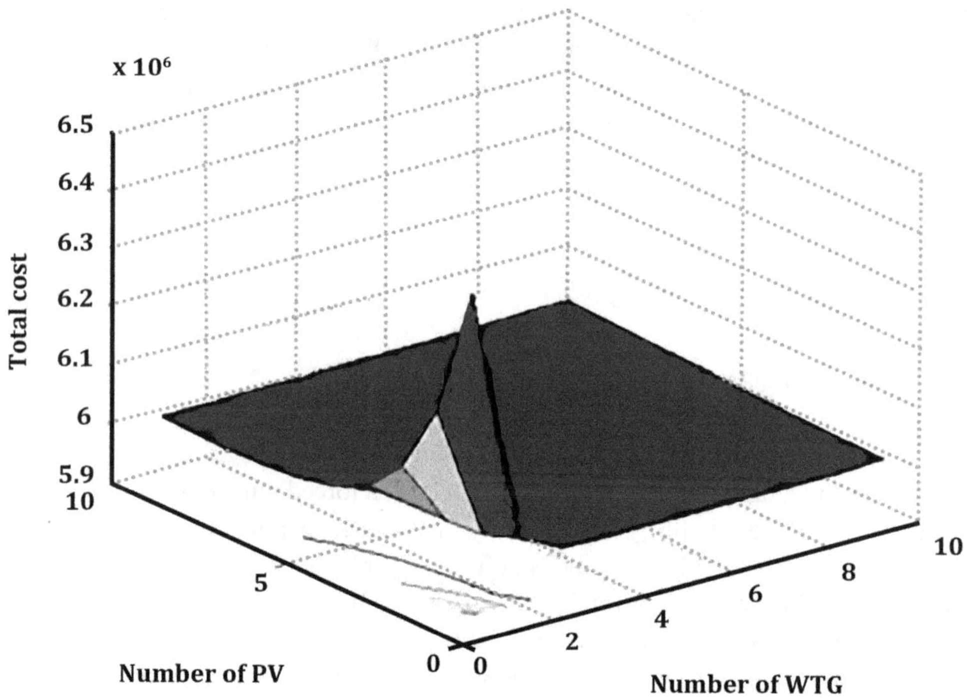

Figure 14. Numbers of WTG and PV to annualized costs of loss of the case

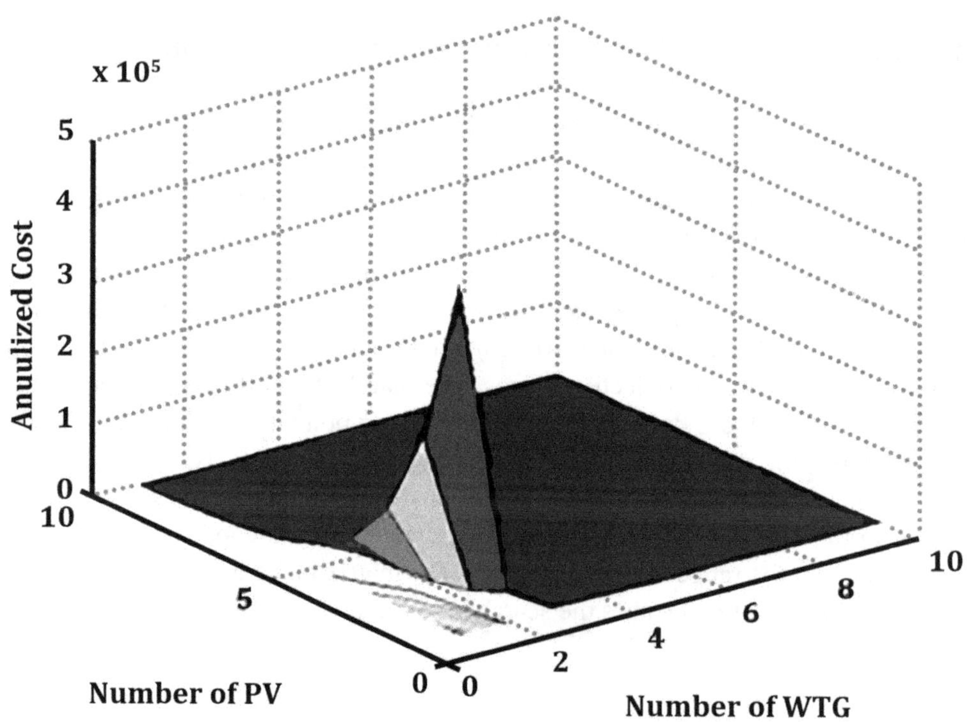

4. CONCLUSION

In this paper, the optimal power system supply design with renewable energy re- sources as solar and wind energy is discussed. Recently, renewable energy has get- ting more attention since global warming and environmental issues regarding conventional energy generation and shortages related to electricity over demand. Al- though renewable energy is unstable due to the variable nature of the solar radiation and wind speed compared with conventional energy generation. A hybrid system that combines conventional energy generation with renewable energy generation is proposed as a solution. The reliability assessment and the economic indices associated with these energy systems can be calculated with evaluation techniques for quantify its benefits. These techniques take into account the generation and load model development to calculate the reliability and economics indices. A reliability value-cost method that introduces the utility cost and customer interruption into the system is used for the cost evaluation. For calculating the optimal reliability value- cost relationship, the costs that are associated with the different system configurations and the associated differences values need to be taken into consideration. The main objective is to cautiously determine the costs related with the different combinations of WTG and PV at certain reliability value that is associated with these combinations. The optimal system design can be considered as a combinatorial problem. In this paper, a GA and reliability value-cost solution is presented. The GA is a meta-heuristic method that mimics the process of natural selection. It is considered as population-based algorithm which population is composed by chromosomes that are evolved with mutation and crossover operators. The main function and constrain function are input into the GA and Direct Search toolbox for the calculation. The GA and reliability value-cost combination balances the customer interruption and utility cost numbers. A series of reliability studies along with RBTS has been performed and the proposed method has been tested. In the test, systems that contained not only conventional generators, stand-alone systems that included only PV and WTG and conventional generation systems with WTG and PV are included. It can be concluded from the results that fuel and operation costs have a significant influence on the utility cost. When renewable generators are used instead of conventional generator especially the thermal varieties, the utility costs are not generated. MATLAB was used to depict the total service cost with the different systems combinations with renewable energy sources. Finally, GA and Direct Search Toolbox are considered as an alternative. The proposed method has demonstrated that reduces the necessary computation time and finds the global optimal solution.

REFERENCES

Abouzahr, I., & Ramakumar, R. (1990). Loss of power supply probability of stand-alone wind electric conversion systems: A closed form solution approach. *IEEE Transactions on Energy Conversion*, *5*(3), 445–451. doi:10.1109/60.105267

Abouzahr, I., & Ramakumar, R. (1991). Loss of power supply probability of stand-alone photovoltaic systems: A closed form solution approach. *IEEE Transactions on Energy Conversion*, *6*(1), 1–8. doi:10.1109/60.73783

Adelsberger, H. H., Collis, B., & Pawlowski, J. J. (2002). *Handbook on Information Technologies for Education and Training*. Springer-Verlag. doi:10.1007/978-3-662-07682-8

Alander, J. (1994). *An Indexed Bibliography of Genetic Algorithms: 1957-1993*. Espoo, Finland: Art of CAD Ltd.

Allen, R., Billinton, R., Breipohl, A. M., & Grigg, C. H. (1994). Bibliography on the application of probability methods in power system reliability evaluation 1987-1991. *IEEE Transactions on Power Systems*, *9*(1), 42–44.

Allen, R., Billinton, R., Breipohl, A. M., & Grigg, C. H. (1999). Bibliography on the application of probability methods in power system reliability evaluation. *IEEE Transactions on Power Systems*, *14*(1), 51–57. doi:10.1109/59.744483

Allen, R., Billinton, R., Shahidehpour, S. M., & Singh, C. (1988). Bibliography on the application of probability methods in power system reliability evaluation 1982-1987. *IEEE Transactions on Power Systems*, *3*(4), 1555–1564. doi:10.1109/59.192965

Bakirtzis, A. G. (1992). A probabilistic method for the evaluation of the reliability of stand-alone wind energy systems. *IEEE Transactions on Energy Conversion*, *7*(1), 99–107. doi:10.1109/60.124548

Billinton, R., & Allan, R. (1996). *Reliability evaluation of power systems* (2nd ed.). New York: Plenum Press. doi:10.1007/978-1-4899-1860-4

Billinton, R., Chen, H., & Ghajar, R. (1996). A sequential simulation technique for adequacy evaluation of generating systems including wind energy. *IEEE Transactions on Energy Conversion*, *11*(4), 728–734. doi:10.1109/60.556371

Billinton, R., Fotuhi-Firuzabad, M., & Bertling, L. (2001). Bibliography on the application of probability methods in power system reliability evaluation 22 1996-1999. *IEEE Transactions on Power Systems*, *16*(4), 595–602. doi:10.1109/59.962402

Billinton, R., & Gan, L. (1991). Use of Monte Carlo simulation in teaching generating capacity adequacy assessment. *IEEE Transactions on Power Systems*, *6*(4), 1571–1577. doi:10.1109/59.117004

Billinton, R., & Li, W. Y. (1994). *Reliability assessment of electrical power systems using Monte Carlo method*. New York: Plenum Press. doi:10.1007/978-1-4899-1346-3

Choi, J., Jeong, S., Bo, S., Park, J., El-Keib, A. A., & Watada, J. (2007). CO2, sox and NOx emissions constrained multi-criteria-best generation mix using fuzzy set theory. In *Proceedings, LESCOPE07-2007 Large Engineering Systems Conference on Power Engineering*.

Desrochers, G., Blanchard, M., & Sud, S. (1986). A Monte Carlo simulation method for the economic assessment of the contribution of wind energy to power systems. *IEEE Transactions on Energy Conversion*, *1*(4), 50–56. doi:10.1109/TEC.1986.4765774

El-Ashry, M. (2007). *REN21, Renewable energy policy network for the 21st century*. Renewables, Global Status Report, 51.

Gavanidou, E. S., Bakirtzis, A. G., & Dokopoulos, P. S. (1992). A probabilistic method for the evaluation of the performance of wind-diesel energy systems. *IEEE Transactions on Energy Conversion*, *7*(3), 418–425. doi:10.1109/60.148561

Gen, B. (2005). *Reliability and Cost/Worth Evaluation of Generating Systems Utilizing Wind and Solar Energy*. (Ph.D. Thesis). Department of Electrical Engineering, University of Saskatchewan Askatoon.

Gordon, G. (1988). *System simulation*. Englewood Cliffs, NJ: Prentice-Hall Inc.

Huynh, V.-N., Nakamori, Y., Lawry, J., & Inuiguchi, M. (Eds.). (2010). *Integrated Uncertainty Management and Applications*. Berlin: Springer-Verlag.

Karaki, S. H., Chedid, R. B., & Ramadan, R. (1999a). Probabilistic performance assessment of wind energy conversion systems. *IEEE Transactions on Energy Conversion*, *14*(2), 217–224. doi:10.1109/60.766986

Karaki, S. H., Chedid, R. B., & Ramadan, R. (1999b). Probabilistic performance assessment of autonomous solar-wind energy conversion systems. *IEEE Transactions on Energy Conversion*, *4*(3), 766–772. doi:10.1109/60.790949

Kaviani, A. K., Baghaee, H. R., & Riahy, G. H. (2007). Design and optimal sizing of a photovoltaic/ wind- generator system using particle swarm optimization. In *Proceedings of the 22nd Power System Conference*. Tehran, Iran: PSC.

Noro, R., Dobashi, Y., Wang, Z. Y., Takahashi, T., Okamoto, T., Tanaka, T., & Watada, J. (2006). Real Options Approach to Life Cycle Management of Power Equipment. In *Proceedings, CMD2006, International Conference on Condition Monitoring and Diagnosis Changwon.*

Ramli, A. A., & Watada, J. (2010). A novelty of real-time fuzzy switching regression analysis: a nuclear power plants case study. In *Proceedings of IUM2010*. Berlin: Springer-Verlag.

Singh, C., & Lago-Gonzalez, A. (1985). Reliability modeling of generation systems including un- conventional energy sources. *IEEE Transactions on Power Apparatus and Systems, 104*(5), 1049–1056. doi:10.1109/TPAS.1985.323455

Spall, J. C. (2003). *Introduction to Stochastic Search and Optimization*. Wiley. doi:10.1002/0471722138

The United Nations Framework Convention on Climate Change. (2005). Retrieved from http://unfccc.int/essentialbackground/convention/background/items/1353.php

Wang, B., Li, Y., & Watada, J. (2010). Fuzzy power system reliability model based on Value-at-Risk. In *Proceedings of 14th International Conference on Knowledge-Based and Intelligent Information and Engineering Systems*, (pp. 445-453). doi:10.1007/978-3-642-15390-7_46

Wang, X., Dai, H., & Thomas, R.J. (1984). Reliability modeling of large wind farms and associated electric utility interface system. *IEEE Transactions, 23*(3), 569-575.

Wang, Y. R., Watada, J., Yaakob, S. B., & Choi, J. (2010). Optimal Supply Balance of Power System Based on Sustainable Generators by Reliability Cost/Worth Method. In *Proceedings, Czeck-Japan Seminar.*

Watada, J., Yaakob, S. B., Takahashi, T., & Okamoto, T. (2010). The Diagnosis of Power Transformer Failures by Fuzzy Random Based Rough Sets Analysis. In *Proceedings, International Conference on Condition Monitoring and Diagnosis 2010 (CMD2010).*

Wei, J., Takahashi, T., Okamoto, T., Tanaka, T., & Watada, J. (2006). Mean-Variance Analysis for Replacing Unreliable Units in a Power Systems. In *Proceedings of CMD2006: International Conference on Condition Monitoring and Diagnosis*.

Xue, J. L., & Watada, J. (2010). Short-term Power Load Forecasting Method by Radial-basis-function Neural Net-work with Support Vector Machine Model. *ICIC Express Letters*, *5*(6B).

Yaakob, S. B., Takahashi, T., Okamoto, T., Tanaka, T., Minh, T. D., Watada, J., & Zhu, X. J. (2008). Structural learning approach to replacing unreliable units in a power system, In Proceedings of International Conference on Condition Monitoring and Diagnosis. *CMD. Current Medical Dialog*, *2008*, 570–575.

Yaakob, S. B., & Watada, J. (2010). Power System Equipment Investment Decision-Making under U- certainty: A Real Options Approach. *Advances in Intelligent Decision Technologies*, *14*, 699–708. doi:10.1007/978-3-642-14616-9_68

Yaakob, S. B., Watada, J., & Liu, S. W. (2010). A Hybrid Neural Network Approach for Solving Bi-level Programming Problems and its Application in Power System. *In Proceedings of ISME2010.*

Yaakob, S. B., Watada, J., Tsuguhiro, T., & Okamoto, T. (2010). A Hybrid Neural Network Approach for Solving Bi-level Programming Problems in a Power System Environment. In *Proceedings of CMD2010: International Conference on Condition Monitoring and Diagnosis 2010*.

Chapter 17
Software Module Clustering Using Bio-Inspired Algorithms

Kawal Jeet
D. A. V. College, India

Renu Dhir
Dr. B. R. Ambedkar National Institute of Technology, India

ABSTRACT

Nature has always been a source of inspiration for human beings. Large numbers of complex optimization problems have been solved by the techniques inspired by nature. Software modularization is one of such complex problems that have been encountered by software engineers. It is the process of organizing modules of a software system into optimal clusters. In this chapter, some bio-inspired algorithms such as bat, artificial bee colony, black hole and firefly algorithm have been proposed for the cause of software modularization. The hybrid of these algorithms with crossover and mutation operators of the genetic algorithm has also been proposed. All the algorithms along with their hybrids are tested on seven benchmark open source software systems. It has been evaluated from the results thus obtained that the hybrid of these algorithms proved to optimize better than the existing genetic and hill-climbing approaches.

INTRODUCTION

Human beings are always fascinated by nature. They have always found inspiration from nature in solving complex research and engineering problems. It could be easily depicted from various new optimization algorithms that are bio-inspired. Genetic Algorithms (GA)(Goldberg, 2006), Bat algorithm (Bat)(Yang, 2010b),Artificial Bee Colony algorithm (Mala & Mohan)(Yan, Zhu, Zou, & Wang, 2012), Black Hole algorithm (BH)(Hatamlou, 2013), Firefly algorithm (Kononoff, Lehman, & Heinrichs; Yang, 2010a) are the names of few of them. GA has been inspired by Darwin's theory of survival of the fittest bat algorithm has been inspired by the technique followed by bats to locate its prey in dark, ABC is an algorithm that has been inspired by swarm behavior of honey bees; BH has been inspired by the gravity of black holes in universe where gravitational force is so strong that even light can't get escape and FF has been inspired by the flashing characteristics of fireflies to attract other fireflies. These algorithms

DOI: 10.4018/978-1-4666-9644-0.ch017

are widely used for solving various complex optimization problems ranging from routing in wireless networks to image optimization.

In this paper, some of the bio-inspired algorithms have been used for modularization of software systems. Software modularization is the process of allocating modules to subsystems or clusters. This allocation should lead to efficient modularization of the software system concerned. It should be done in such a way that there should be minimum coupling and maximum cohesion among the subsystems and clusters thus generated. It is particularly useful to

- Recover the modularization of the system when source code is the only means available to get information about the system.
- Identify the best possible package to which classes of a java project should be allocated to its actual delivery.
- Combine the classes that could be downloaded together.

Software modularization is particularly useful during the maintenance phase of the software development life cycle. Maintenance is the largest and most expensive phase of the software development lifecycle (IEEE Standard for Software Maintenance). During this phase, the software system undergoes continuous change and is enhanced for a number of reasons. It is generally accompanied by the turnover of the software engineers involved in its development. It ultimately leads to the decay of the original clustering of the modules of the system concerned. It is called software erosion. For the effective and efficient maintenance of the software system, it is important to re-identify the modularization of this system. It might be a difficult task, especially in the absence of original design documents.

In this proposed work, bio-inspired algorithms such as bat, ABC, BH, and FF have been used for solving the problem of software modularization. These algorithms work to find optimized results but might get stuck at local optima. Experiments have been conducted to use the hybrid combination of these algorithms with crossover and mutation operators of genetic algorithm for software module clustering. Adding these operators has an inherent advantage of adding randomization which saves these algorithms to get stuck at local optima. Although we observed good quality modularization but we recommend the resulting modularization to be approved by the experts. In other words, it could be used as guidance to the experts but could not replace them.

The rest of the chapter is organized as discussed here. The literature of the field of software modularization, as well as applications of bio-inspired algorithms, has been discussed. Further, a concept of search based software engineering has been discussed along with the skeleton of bio-inspired algorithms. Software modularization using these bio-inspired algorithms has been elaborated. The experimental setup which includes the information about the samples, encoding scheme and algorithmic parameters used has been given. Results are discussed and analyzed followed by the discussion about threats to the validity of this research work, conclusion, and future work.

BACKGROUND

Wiggerts (Wiggerts, 1997) very well introduced clustering techniques that have been successfully applied to system modularization. Various other clustering techniques like Rigi (Müller, Orgun, Tilley, & Uhl, 1993) and Arch (Schwanke, 1991) work in a bottom-up fashion and produce the architecture of the software system by using its source code only. The main shortcoming of these tools is that they need the key involvement of the user.

In another technique called Algorithm for Comprehension Driven Clustering (ACDC) (Tzerpos, 2001) the authors used the pattern-based clustering technique that recovers subsystems using the incremental clustering. In this technique, a structure is created on the basis of patterns of subsystems and each newly introduced component is placed in the subsystem which seems most appropriate.

In another approach (Dietrich, Yakovlev, McCartin, Jenson, & Duchrow, 2008), the authors use Girvan- Newman clustering algorithm for the analysis of dependency graphs of object-oriented programs. This approach uses the concept of betweenness. The betweenness of an edge is defined as the number of shortest paths between all pairs of nodes in the graph passing through that edge. Since the edges that lie between clusters are expected to be those with highest betweenness, we can find a good separation of the graph into clusters by removing them recursively. The general structure of the Girvan-Newman algorithm is as follows:

1. Calculate the betweenness value for each of the edges.
2. Remove the edge(s) having the highest value.
3. Repeat the analysis on the resulting graph until a suitable separation of the graph into clusters has been achieved

Various search based optimization techniques have been successfully used for software modularization. Remarkable in this field include hill climbing, simulated annealing, GA which are available as the BUNCH tool (Mancoridis, Mitchell, Chen, & Gansner, 1999; Mancoridis, Mitchell, Rorres, Chen, & Gansner, 1998; Mitchell, 2002; Mitchell & Mancoridis, 2002, 2006). The traditional hill climbing algorithm implemented in BUNCH tool follows the following approach.

1. Take random initial modularizations.
2. Perform neighbourhood search for each initial modularizations. In search of better modularization, modules are moved between clusters.
3. Process iterates with improved modularization as basis.

Genetic Algorithm (GA) (Doval, Mancoridis, & Mitchell, 1999)has also been used for software modularization and has shown to produce good results in optimizing large problems. It contains a crossover operator and more defined mutation and crossover rates. Mutation and crossover operators are standard. A roulette wheel selection is used for the GA, complemented with elitism. These search based techniques work to optimize a fitness function called Modularization Quality (MQ). MQ is basically the ratio of cohesion and coupling.

Harman et al. (Harman, Hierons, & Proctor, 2002) experimented with fitness function (MQ) (Harman, 2007) for software modularization. They have presented a new encoding and crossover operator. This new representation allows only one representation per modularization. This GA with the novel crossover outperforms the traditional one. When properly tested, it has been observed that this algorithm quickly becomes trapped in local optima.

A Large number of bio-inspired optimization algorithms have been developed and used for solving complex engineering optimization problems. Load dispatch (Apostolopoulos & Vlachos, 2010), optimization of brushless DC motor (Bora, Coelho, & Lebensztajn, 2012), classification of microarray data

(Mishra, Shaw, & Mishra, 2012), estimation of full body human pose(Akhtar, Ahmad, & Abdel-Rahman, 2012), routing in wireless sensor networks (Karaboga, Okdem, & Ozturk, 2012), fault section estimation in power systems (Huang & Liu, 2013), optimal design of electromagnetic devices (Bouchekara, 2013) are few optimization problems that have found solution in bio-inspired algorithms. In case of software engineering, bio-inspired algorithms have been used to solve problems such as software effort estimation(Maleki & Gharehchopogh, 2014) and software testing (Dahiya, Chhabra, & Kumar, 2010; Jeya Mala, Mohan, & Kamalapriya, 2010). So the proposed work has been motivated to use bio-inspired algorithms for solving the problem of efficient modularization of software systems. It is observed that hybrid bio-inspired algorithms produced modularization better than existing approaches.

RESEARCH OBJECTIVES

The main objectives of the proposed work are listed below:

- To experimentally evaluate the behaviour of recent-inspired algorithms such as Bat, ABC, BH and FF algorithm on software modularization.
- To evaluate the impact of adding crossover and mutation operator to the bio-inspired algorithms mentioned above.
- Comparison of modularization obtained by applying these hybrid bio-inspired algorithms to that of existing search based approaches by taking MQ value as comparison criteria.
- Comparison of modularization obtained by applying these hybrid bio-inspired algorithms to that of existing search based approaches by taking number of iteration required to converge as comparison criteria.
- Using Wilcoxen test to show significant improvement in software modularization due to the addition of crossover and mutation operators to basic bio-inspired algorithms.

MAIN FOCUS OF THE CHAPTER

Search Based Software Engineering

A software system is composed of modules which could be a class, or variables which are related to each other due to procedure calls, inheritance relationships, variable references, etc. The syntactic structures of these systems could be represented as a graph called a Module Dependency Graph (MDG), where the nodes are the modules and edges, are the relations between the modules. These MDGs could be retrieved by parsing the source code to determine the modules of the software system and relationship between these modules. Large numbers of source code analysis tools such as JDepend (Clark, 2005), CIA, Acacia (Chen, Gansner, & Koutsofios, 1997), Chava (rey Korn & Chen)are available that could be used to retrieve these MDGs.

The problem of finding the best clustering for a given set of modules in an MDG is an NP-hard search problem (Mitchell, 2002). So, an automated assistance to partition MDG's is required that would help system maintainers to efficiently work in the absence of original design documentation.

Search-based software engineering (SBSE) techniques might be used to sort out this problem. SBSE starts with only two key ingredients (Harman, 2007):

1. The choice of the representation of the problem.
2. The definition of the fitness function.

This simplicity makes SBSE attractive. With just these two simple ingredients, the SBSE can implement search based optimization algorithms and get efficient results. Typically, a software engineer will have a suitable representation for their problem. Many problems in software engineering also have software metrics associated with them that naturally form good initial candidates for fitness functions. Further, infrastructural investment required to get started with SBSE is among the lowest of any approach one is likely to encounter.

Bio-inspired algorithms are inherently search based. So the implementation of these algorithms needs to have just these two ingredients.

Representation of the problem: In this proposed work, a technique has been used for representation of the problem (Doval, et al., 1999) in the form of encoded string. For encoding of this string, each node in the graph (MDG) has been assigned a unique numerical identifier. These unique identifiers define which position in the encoded string will be used to define that node's cluster. The value of each of these nodes depicts the cluster it belongs to.

Fitness Function: Modularization Quality (MQ) (Mitchell & Mancoridis, 2008) has been used as objective or fitness function for the proposed algorithms.MQ aims at finding a balance between cohesion and coupling. It means larger the MQ, better is the partition of the MDG and is closer to the desired system architecture. In other words, it measures the quality of the modularization of the system concerned.

$$MQ = \sum MF_k$$

where k is the number of clusters, MF_k is the modularization factor of cluster k.

MF_k=intra edges/(intra edges + half of inter edges)

Here intra edges depict cohesion (edges between modules in a sample cluster) and inter edges depict coupling (edges between modules in different clusters) of cluster k.

The purpose of these bio-inspired algorithms is to optimize the encoded string so as to maximize the fitness function MQ.

Skeleton of Bio-Inspired Algorithms

In this chapter, bio-inspired algorithms under consideration include Bat, ABC, BH, and FF. Let us discuss the brief introduction and inspiration to these algorithms.

Bio-Inspiration

Bat algorithm has been inspired by bat's capability of echolocation in order to detect prey while avoiding obstacles in the dark. Velocities of all the bats are initialized on the basis of the parameters to be optimized. Bats representing candidate modularization are viewed in space with dimensions equal to number of parameters to be optimized an i.e. number of modules of the software system to be clustered.

Pulse rate (P) and loudness (L) which are required to build the three-dimensional scenario of the surrounding of bats are initialized to random numbers between 0 and 1. At finding the prey, loudness usually decreases while the rate of pulse emission increases (Yang, 2010b, 2011).

ABC has been inspired by the intelligent foraging behaviour of honey bee swarms. According to this algorithm, the colony of bees contains three groups: employees, onlookers and scouts bees (Akbari, Hedayatzadeh, Ziarati, & Hassanizadeh, 2012; Yan, et al., 2012). So we also have three categories of candidate modularizations.

- **Employee Bee:** A bee that is associated to a particular food source or modularization under consideration and carries information about it.
- **Onlooker Bee:** A bee waiting in the dance area for making a decision to choose a food source or modularization on the basis of the information shared by employee bees.
- **Scout Bee:** This bee goes on a random search to discover new sources or modularization in case existing modularization could not be improved further.

Algorithm begins with employee bees.

Trial counter of each candidate modularization has been set to 1. This indicates number of times modularization has been improved.

BH has been inspired by the black hole theory of the universe. This theory describes a black hole as a place in space where gravitational force is so strong that even light can't escape (Hatamlou, 2013).

FF has been inspired by the flashing characteristics of fireflies. Fireflies are unisex creatures. So they attract each other on the basis of the intensity of light they emit; irrespective of their sex.

General Structure

The general structure of any bio-inspired evolutionary algorithm could be described as

Step 1: Generate initial population of candidate solutions.

The process of initialization is similar in all the bio-inspired algorithms under consideration.
Randomly generate candidate population of modularizations using Eq. (1)

$$y^j_i = y^j_{\min} + rand(0,1)(y^j_{\max} - y^j_{\min}) \tag{1}$$

where i=1, 2….Pop (Pop is the number of candidate solutions to be initially generated). By repeated execution of these algorithms and by manual tuning it has been observed that most appropriate value for

Table 1. Control factors for bio inspired algorithms and their hybrids for software modularization

Factor	Value	Comments
Initial population of candidate modularizations (Pop)	10*Number of modules to be clustered(Number of modules in MDG)	With increase in number of parameters to be optimized, initial size of population of candidate modularization increases proportionally.
Fitness/Objective function	Modularization Quality (MQ)	$MQ = \sum MF_k$ where MF_k=intra edges/(intra edges + half of inter edges) MF_k is modularization factor of cluster k.
Number of iterations	2500 or when value of Modularization Quality does not change for 500 iterations.	A stopping criteria
Number of variables to be optimized (par)	Number of nodes in MDG	Number of modules to be clustered

size of initial population (Pop) of candidate modularizations should be almost 10 times the number of parameters to be optimized i.e. 10 times the number of modules of the software system to be clustered.

j=1, 2…..par (par is the number of modules to be clustered).

Table 1 gives the detail of general control parameters to be used in the implementation of these algorithms.

y^j_{min} = Lower bound of the jth parameter (LB)=Least number of possible clusters=1.

y^j_{max} = Upper bound of jth parameter (UB)=Number of modules to be clustered.

y^j_i Specifies cluster to which j^{th} node (to be optimized) in i^{th} candidate modularization has been assigned.

Evaluate rank of each candidate in the population using fitness function MQ.

Step 2: Steps 3-6 are repeated until stopping criteria is met. The algorithm stops after 2500 iterations or when the modularization does not further improve for 500 consecutive iterations.

Step 3: By application of appropriate technique, identify neighbor of each current candidate solution. The technique is different for different algorithms.

Bat

Each bat is randomly assigned a frequency between η_{min} and η_{max} where

η_{min}=0 and

η_{max}=One-tenth of the number of parameters to be optimized

$\eta_i = \eta_{min} + (\eta_{max} - \eta_{min}) * \beta$

β is a random number between 0 and 1.

Empower the modularization with largest value of fitness function MQ as the best (y_{best}.)
Update velocity and location of each bat using Eq (2 and 3).

$$v_i(t+1) = v_i(t) + (y_i(t) - y_{best}) * \eta_i \tag{2}$$

$$y_i(t+1) = y_i(t) + v_i(t) \tag{3}$$

where vi(t+1) and vi(t) are velocities and yi(t+1) and yi(t) are locations of the ith bat at t+1 and t iterations respectively. This step identifies new velocity and location of each bat representing candidate modularization.

ABC

$$y^j{}_i(t+1) = y^j{}_i(t) + \Phi^j{}_i(y^j{}_i(t) - y^j{}_k(t)) \tag{4}$$

where $y_i^j(t)$ is value of jth parameter in i[th] modulariation in iteration t and $y_i^j(t+1)$ is its value in ietration (t+1), Φ is a random number between [-1, 1]. i and k lies between 1 and Pop (Pop is the number of candidate modularization), k is determined randomly and should be different from i.

BH

Empower the modularization with the largest value of objective function MQ (greater fitness) as the black hole $\left(y_{BH}\right)$ and rest as normal candidate star modularization.

Similar to the black hole, this modularization has the ability to absorb other normal modularizations. So, the candidate starts moving towards the black hole using Eq. (5).

$$y_i(t+1) = y_i(t) + rand * (y_{BH} - y_i(t)) \tag{5}$$

Where $y_i(t)$ and $y_i(t+1)$ are location of ith candidate modularization at iteration t and t+1 respectively, y_{BH} is the location of black hole, rand is a random number in interval [0,1].

FF

Calculate the light intensity of each fly using MQ. If light intensity of fly i is lesser than fly j, move i towards j using Eq. (6).

$$y_i = y_i + \beta e^{-\gamma r_{ij}^2}(y_j - y_i) + \alpha\varepsilon_i \tag{6}$$

where β=1, α=0.25, Υ=1, ε_i is drawn from Gaussian distribution.

Step 4: If the neighbor calculated in step 3 is better than the current candidate solution then replace this candidate solution with this new neighbor.

Bat

If this new bat location (modularization) is better than previous then generate a random number. If this random number generated on a fly is less than the loudness (L) of the candidate solution under consideration then replace this new solution as the candidate solution.

Update loudness (L) and pulse rate (P) of candidate modularizations using Eq. (7 and 8).

$$P = P * (1 - e^{-0.9*t}) \tag{7}$$

$$L = 0.9 * L \tag{8}$$

where t is the current iteration.

ABC

If the new neighbor of the bee (modularization) is better than the existing solution, keep this neighbor and ignore existing candidate modularization and reset trial counter to 1, else if the current solution can't be improved further increment its trial counter.

BH

While updating the location of stars using Eq. (5), a candidate modularization may reach a better location indicated by higher MQ than that of the black hole. In such a case, this star has the capability to act as a new black hole. So, exchange existing BH with that of this new candidate modularization.

FF

Evaluate new candidate solution using MQ and update their light intensity.

Step 5: Make a selection of the candidate on the basis of its fitness. More fit is the candidate more should be its probability of selection in next iteration.

Bat

For each bat, perform the local search for the better neighbor, a new neighbor is calculated using the random walk as shown in Eq. (9).

$$y_{new} = y_{old} + \varepsilon A_t \tag{9}$$

where y_{old} is candidate modularization,

y_{new} is modularization obtained after random walk
ε is a random number vector drawn from [−1, 1], A_t is the average loudness of all the bats at this iteration

ABC

Here starts the second phase of ABC algorithm. For selecting onlooker bees, calculate the probability of a modularization to be selected as shown in Eq. (10).

$$p_i = Fitness_i \,/\, Size\_of\_Population \tag{10}$$

where $Fitness_i$ is the fitness of the ith candidate modularization acting as employee bee calculated using MQ, size_of_population is the number of candidates in the population under consideration. This phase works similar to employee bee phase where new onlooker bee with better MQ is identified. The greedy selection process has been applied to keep the onlooker bees with a greater value of MQ (greater fitness) between neighbor newly created and already available onlooker bees.

BH

The black hole in the universe has an event horizon. This event horizon has a sphere-shaped region with a very high gravitational force. Similarly, the black hole candidate modularization has the event horizon. Any modularization that crosses this horizon gets absorbed by the black hole and vanishes. The radius of an artificial event horizon is calculated by using Eq. (11).

$$Radius\_of\_horizon = Fitness_{BlackHole} \,/\, \sum_{i=1}^{Pop} Fitness_i \tag{11}$$

$Fitness_{BlackHole}$ is the MQ value calculated from modularization empowered as Black Hole. $Fitness_i$ is the MQ of ith modularization.

FF

Rank the fireflies and find the current best

Step 6: If a candidate solution does not improve certain number of times, ignore it and generate a new random candidate.

Bat

No such mechanism is involved.

ABC

Here starts the scouting phase. Similar to scout bee, if a modularization can't be improved further during a pre-specified number of cycles called limit then that modularization (food source) is replaced by a new modularization (food source) created by using Eq. (1).

This newly created scout modularization is compared to existing solutions and best modularization achieved so far is memorized.

BH

If a black hole absorbs a candidate modularization in step 5, another candidate modularization is randomly generated using Eq. (1) and algorithm continues with this new modularization as candidate solution in the population of stars. This step has been done to keep the population of candidate modularization constant.

FF

No such mechanism is involved.

Software Modularization Using Bio-Inspired Algorithms

Software modularization is an important issue in a software development process. It is the process of allocating modules to appropriate clusters. As discussed above, the software could be represented in terms of an MDG. Clustering modules from MDG are an NP-hard problem. So some heuristic could be used to optimize their clustering. Bio-inspired algorithms have been used in wide variety of applications. So, they might be applied to the process of software modularization.

Further, it is observed that these algorithms may result in suboptimal results as compared to some traditional algorithms such as accent hill climbing algorithm. It is because these algorithms are more inclined towards local search and if crossover and mutation operator are added, the resulting modularization improves. This is because adding crossover and mutation operator leads to more diverse search capabilities. The proposed approaches are diagrammatically shown in Figure 1.

Addition of crossover and mutation phase takes following steps

- As an input use the candidates modularization modified by the algorithms under consideration. This will act as input population.

Figure 1. Proposed approaches

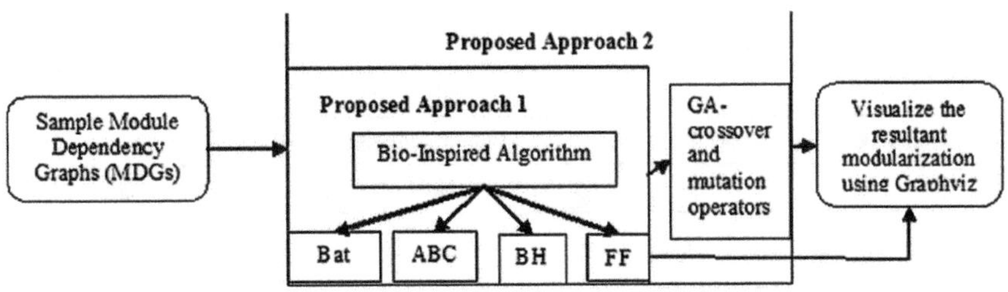

- Select the parent candidate modularizations from this population using the tournament selection technique. The quality of the candidate is calculated using MQ.

Generate new child modularizations by applying arithmetic crossover operator at the rate of 0.6. The equation for this purpose is defined as

Child=r1 * Parent1+ r2 * Parent2

where r1 and r2 are independent random numbers between 0 and 1

- To diversify the search and cover more search space apply mutation operator at the rate of 0.02.

Application of crossover and mutation operators enhances the exchange of information of the algorithm. The modularizations with higher quality are fully utilized too. The hybrids have been named Bat-GA, ABC-GA, BH-GA, and FF-GA.

The parameters that have been used to make hybrid nature inspired algorithms are discussed in Table 2.

EXPERIMENTAL SETUP

This section describes the experimental set up to carry out the comparison between the proposed approach using bio-inspired algorithms, their hybrids with crossover and mutation operators, traditional hill climbing and genetic algorithm for the modularization of software systems. The flowchart of the proposed bio-inspired algorithms bat, ABC, BH and FF after adding crossover and mutation phase has been shown in Figure 2.

Table 2. Control factors for crossover and mutation phase

Factors	Value	Comments
Parent Selection Algorithm	Tournament selection algorithm with size2	Parent modularizations for crossover; Parent1 and Parent2 are selected using this method.
Crossover function to generate children for next iteration population	Arithmetic crossover	Child=r1 * Parent1+ r2 * Parent2 Where r1 and r2 are independent random numbers between 0 and 1.
Mutation function	Uniform	Obtained by repeated execution of algorithms
Population	Candidate modularizations passed by bio inspired algorithm under consideration	Updated by all the phases of bio inspired algorithms.
Crossover fraction	0.6	Manually tuned to give best results.
Mutation fraction	0.02	Manually tuned to give best results.
Objective function	MQ	For backward compatibility and comparisons

Figure 2. Plot of MQ value for all the samples

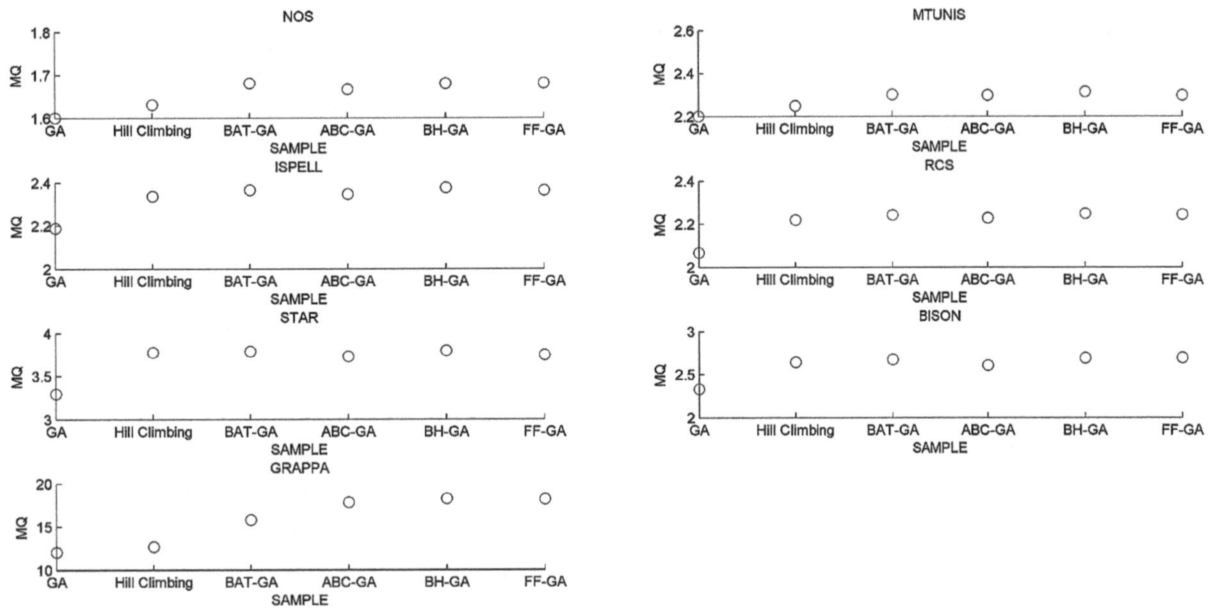

Sample Problems

The algorithms are applied to the MDGs of Table 3. These sample MDGs are some of the open source software. The dependencies between modules involved are taken as un-weighted arcs/edges. It means irrespective of the type and numbers of dependencies between modules the dependencies are considered. The main motivation for the usage of these MDGs is their usage by the researchers in the past. By using these sample MDGs, we can easily compare proposed approaches to that of existing search based approaches. The number of modules varies from 16 to 74.

Encoding

In order to represent the problem of software clustering, the encoding similar to Dovel et al. (Doval, et al., 1999) for Bunch tool has been used.

Algorithmic Parameters

The nature inspired algorithms and their hybrids proposed are implemented in MATLAB 8 (www.mathworks.in/products/matlab/). Table1 shows some common parameters and their values to be used in all the proposed

The other parameters which are used in these proposed algorithms are discussed in Tables 1, 2, 4-7.

Table 2 shows the algorithmic parameters along with their values to be used in crossover and mutation phase of all the algorithms. Table 4-7 shows the parameters used by Bat, ABC, BH and FF algorithms respectively.

Table 3. Description of the test software systems to be used in the experiment

Software System	Modules in MDG	Edges in MDG	System Description
Nos	16	52	C++ program that implements file system service
Mini-Tunis	20	57	A simple operating system
Ispell	24	103	An open source spell checker
Rcs	29	163	Open source version control tool
Star	36	89	Source code analysis tool to produce the MDG
Bison	37	179	Parser generator
Grappa	74	112	Graph visualization and drawing tool
Incl	174	360	Subsystem from a source code analysis system

Table 4. Control factors for Bat algorithm

Factor	Value	Comment
Qmin	0	Minimum frequency that has been assigned to each bat representing candidate modularization.
Qmax	Number of parameters to be optimized/10	Maximum frequency that has been assigned to each bat representing candidate modularization.
P	Random numbers between 0 and 1.	As suggested by the authors of Bat algorithm and has been tested manually.
L	Random numbers between 0 and 1.	As suggested by the authors of Bat algorithm and has been tested manually.

The reported MQ value is calculated from resulting modularization in dot format file of Graphviz (Ellson, Gansner, Koutsofios, North, & Woodhull, 2002).

Comparison Criteria

In order to compare software modularization using bio-inspired algorithms and their hybrids with crossover and mutation phase, MQ value and number of iterations required to converge the algorithm have been used.

MQ as Comparison Criteria

MQ has been selected as primary comparison criteria because of our requirement of comparison to earlier work in the field. The authors of some of the past research work in the field of search based software modularization (Harman, et al., 2002; Mahdavi, Harman, & Hierons, 2003; Mancoridis, et al., 1999; Mancoridis, et al., 1998; Mitchell, 2002; Mitchell & Mancoridis, 2002, 2006, 2008; Praditwong, Harman, & Yao, 2011) have used MQ to assess and evaluate the quality of modularization of the software systems concerned and hence comparison of techniques. According to these authors, higher the value of MQ, better is the quality of modularization.

Table 5. Control factors for ABC algorithm

Factor	Value	Comment
Size of colony	10*Number of modules to be clustered.	Employee bees and Onlooker bees together
Number of food sources (Pop)	Size of colony/2	Half of the colony size
Trial limit	100	Discard modularization which could not be further improved for 100trial limit.

Table 6. Control factors for BH algorithm

Factor	Value	Comment
Number of stars	10*Parameters to be optimized	Parameters is equal to number of modules in MDG
Radius of event horizon	$\frac{Fitness_{BlackHole}}{\sum_{i=1}^{Pop} Fitness_i}$	Ratio of Fitness of Black hole and sum of fitness of whole population

Table 7. Control factors for FF algorithm

Factor	Value	Comment
β_0	1	Attractiveness when distance between fireflies is 0.
α	[0,1]	Scaling parameter
γ	1	Light absorption parameter

Number of Iterations Required to Converge as Comparison Criteria

The number of iterations an algorithm takes to converge to a constant value is the characteristic of that algorithm. In this experiment, if the algorithm generates same output modularization (indicated by constant MQ) as output for 500 consecutive iterations, it has been accepted as converged. Lesser is this number better is the algorithm. So, this criterion could be used to compare the algorithms.

RESULT AND ANALYSIS

In this section, the result of the experiments carried out to compare the modularization (of sample software discussed in Table 3) obtained by applying the bio-inspired algorithms, their hybrids mentioned above and techniques already implemented have been discussed.

MQ as Comparison Criteria

Using encoding mentioned above and parameters mentioned in Tables 1, 2, 4-7, the bio-inspired algorithms under consideration are executed independently 30 times. The average of MQ value thus obtained on sample test software systems mentioned in Table 3 are shown in Table 8.If the values of Table 8 are analyzed, it is observed that traditional hill climbing algorithm implemented as part of Bunch tool produces modularization better than proposed bio-inspired algorithms. This is because MQ values obtained by executing these algorithms are lesser than the Hill-climbing algorithm.

In case of hybrid bio-inspired algorithms obtained by further addition of crossover and mutation operator, the average value MQ thus obtained after 30 consecutive executions are shown in Table 10. By analyzing this table, it could be seen that the value MQ is larger in case of these hybrids. In other words, better quality modularization has been observed. It means the resulting technique improves the quality of modularization of the software system. This is because, adding genetic operators have an inherent advantage of adding randomization which saves these algorithms to get stuck at local optima.

Figure 2 displays the value of fitness function MQ obtained by execution of all hybrid algorithms for all the samples. It is observed that ABC-GA outperforms the rest of the algorithms.

Number of Iterations Required to Converge as Comparison Criteria

Figures 3-8 shows the plot of number of iterations and MQ of the modularization for all the sample cases for all the hybrid bio-inspired algorithms under considerations along with traditional hill climbing implemented in BUNCH tool. By close evaluations of these figures, it could be seen that in most of the sample cases, ABC-GA algorithm converges in least number of iterations.

Standard Deviation

The standard deviation of the modularizations (depicted by MQ) is calculated for 30 consecutive runs of the algorithms. The resulting standard deviations (for the entire sample) without adding crossover and mutation operators are shown in Table 9 and after adding these operators are shown in Table 11. After

Table 8. The Average of the MQ of the Best Solutions found by Hill climbing, GA, Bat, ABC, BH and FF algorithms

Software System	Hill Climbing	GA	Bat	ABC	BH	FF
Nos	1.63038	1.60072	1.61211	1.60072	1.6292	1.64596
M-Tunis	2.249	2.20112	2.29345	2.28312	2.23091	2.11657
Ispell	2.337	2.18869	2.31265	2.28618	2.25667	2.12917
Rcs	2.218	2.06682	2.19723	2.16131	2.13123	2.06931
Star	3.777057	3.29476	3.4629	3.29476	3.71281	3.03418
Bison	2.639	2.33062	2.29011	2.39958	2.36557	2.17423
Grappa	12.676	12.0577	8.12334	9.71646	7.77892	8.48432

Table 9. The Standard deviation of the MQ of the Best Solutions found by Hill climbing, GA, Bat, ABC, BH and FF algorithms

Software System	Hill Climbing	GA	Bat	ABC	BH	FF
Nos	0.051536	0.048825	0.08667	0.054506	0.116782	0.039833
M-Tunis	0.054769	0.108065	0.11896	0.119806	0.122038	0.088972
Ispell	0.0255	0.076304	0.116756	0.095947	0.128503	0.109665
Rcs	0.020	0.109789	0.124321	0.103749	0.135543	0.117049
Star	0.057313	0.148447	0.223461	0.214032	0.140023	0.241789
Bison	0.041	0.104141	0.109877	0.10614	0.149844	0.199444
Grappa	0.017	1.173595	1.092234	0.787977	0.562231	1.03806

Table 10. The Average of the MQ of the Best Solutions found by Hill climbing, GA, Bat, ABC, BH and FF algorithms after adding crossover and mutation operator

Software System	Hill Climbing	GA	Bat-GA	ABC-GA	BH-GA	FF-GA
Nos	1.63038	1.60072	1.68059	1.666571	1.64973	1.66101
M-Tunis	2.249	2.20112	2.30112	2.31446	2.28781	2.29771
Ispell	2.337	2.18869	2.37575	2.34712	2.36734	2.36373
Rcs	2.218	2.06682	2.24237	2.24701	2.24868	2.24444
Star	3.777057	3.29476	3.72897	3.79903	3.73373	3.74392
Bison	2.639	2.33062	2.60112	2.68599	2.67694	2.69354
Grappa	12.676	12.0577	15.7833	17.7858	18.2335	18.1744

Figure 3. Plot number of generations and fitness value for 'Nos'

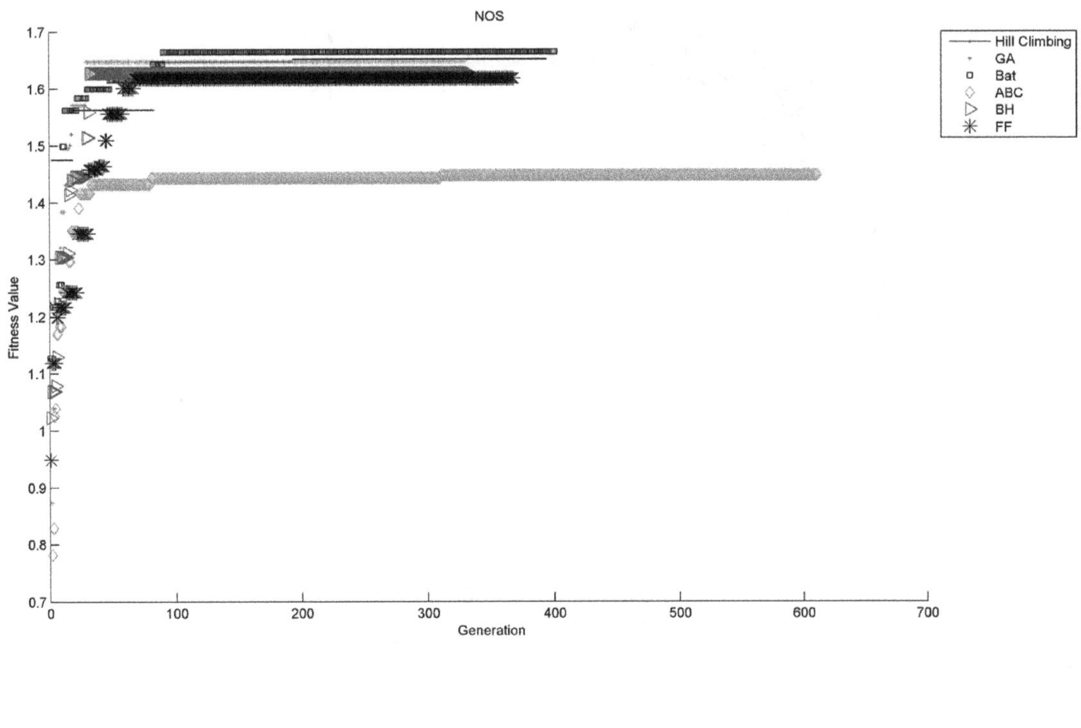

Figure 4. Plot number of generations and fitness value for 'Mtunis'

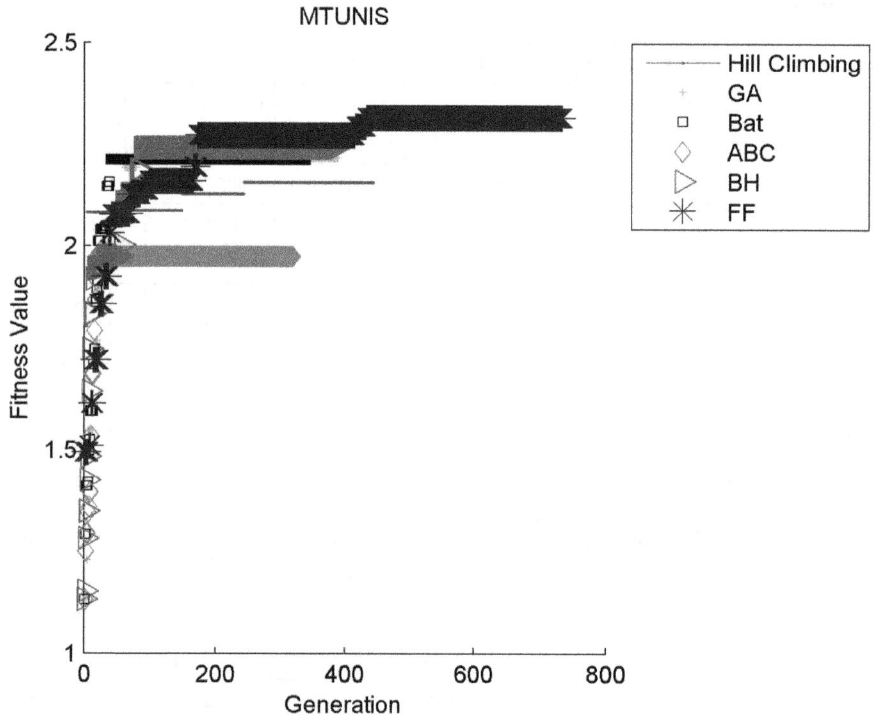

Figure 5. Plot number of generations and fitness value for 'Rcs

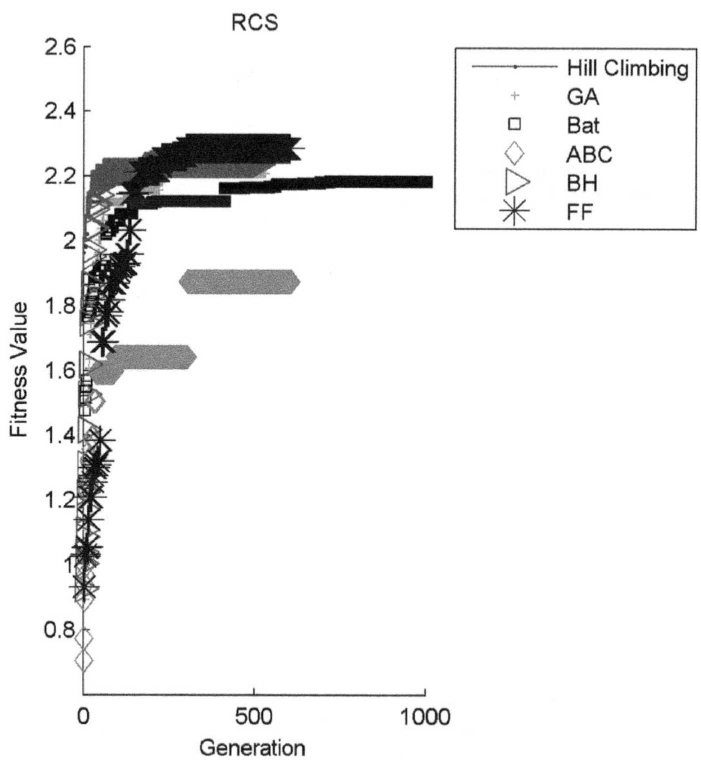

Figure 6. Plot number of generations and fitness value for 'Star

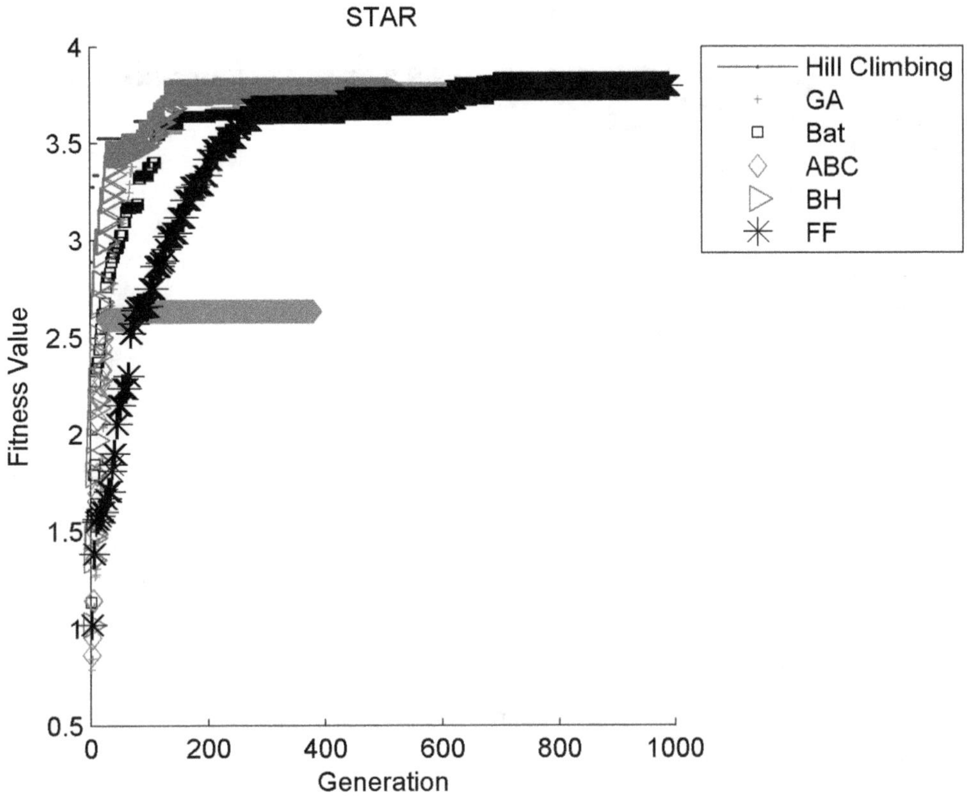

Figure 7. Plot number of generations and fitness value for 'Bison

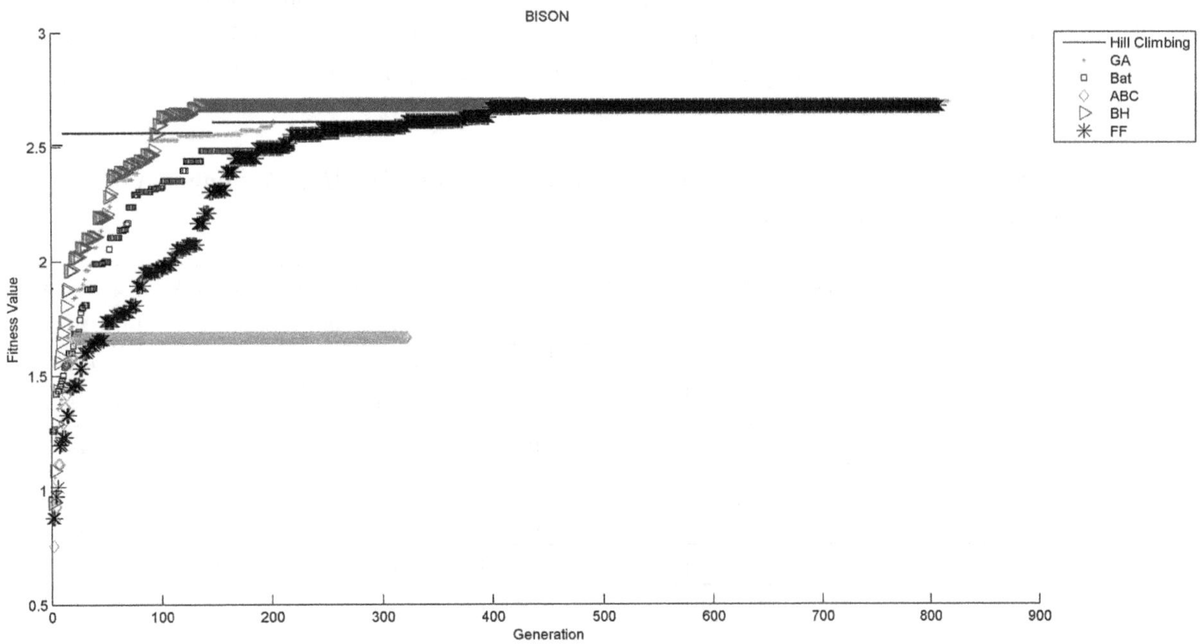

Figure 8. Plot number of generations and fitness value for 'Grappa

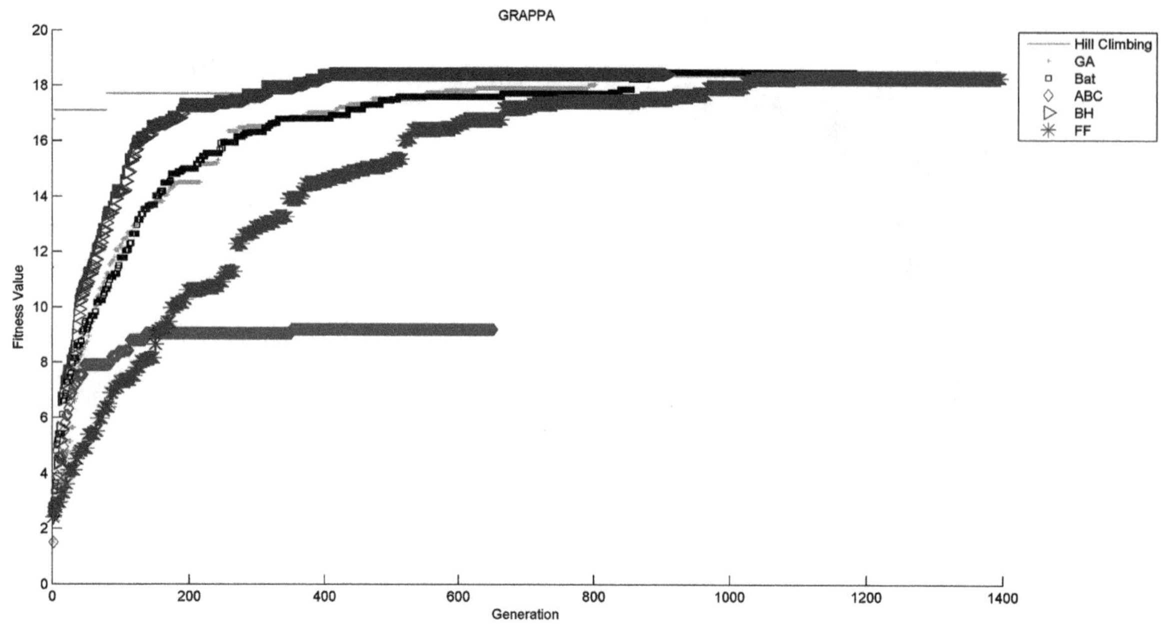

analyzing these tables, it is observed that a standard deviation of modularizations after adding genetic operators is lesser than without adding them.

Student's T-Test

It can be used to determine if two sets of data are significantly different from each other and is most commonly applied when the test statistic would follow a normal distribution (Sheskin, 2003). Even though, it is not sure that the given data is normal or not still t-test has been used for easy comparison with previous work. For the given sample, Student's t-test values are shown in Table 12. The value in bold

Table 11. The Standard deviation of the MQ of the Best Solutions found by Hill climbing, GA, Bat, ABC, BH and FF algorithms after adding crossover and mutation operator

Software System	Hill Climbing	GA	Bat-GA	ABC-GA	BH-GA	FF-GA
Nos	0.051536	0.08776	0.021173	0.026246	0.025843	0.021941
M-Tunis	0.054769	008912	0.051965	0.066064	0.040997	0.02332
Ispell	0.0255	003142	0.023752	0.04091	0.022981	0.02247
Rcs	0.020	0.03221	0.029043	0.028897	0.034613	0.024116
Star	0.057313	0.06082	0.094512	0.0012	0.087027	0.065992
Bison	0.041	0.053280	0.071006	0.046547	0.04895	0.022764
Grappa	0.78775	0.84714	0.469881	0.3231	0.154012	2.09415

shows the significant difference between the modularization obtained by traditional hill climbing and hybrid bio-inspired algorithms. It is observed that in most of the cases, hybrid bio-inspired algorithms produce modularization significantly better than traditional hill climbing algorithm.

Wilcoxon Test

The Wilcoxon signed-rank test is a non-parametric statistical hypothesis test that is used when two related samples are required to be compared (Sheskin, 2003). It has been used as an alternative to the paired Student's t-test as the population cannot be assumed to be normally distributed. The normality of data under consideration is not sure, so Wilcoxon test has been used to compare the modularization obtained by proposed hybrid Bat, ABC, BH and FF algorithms and traditional hill climbing implemented in BUNCH tool. The resulting value of Wilcoxon signed-rank test has been shown in Table 13.

The value in bold shows modularization obtained by applying these algorithms is significantly better than hill climbing. It has been observed by analyzing Table 13, that in most of the cases, the modularization is significantly better than traditional hill climbing after adding genetic operators.

THREAT TO VALIDITY

Any experimental not involving human subjects are vulnerable to only two potential threats to their validity. These are external and internal validity. As observed from experimental setup, human subjects are not involved in proposed approaches. So only these two threats are to be checked.

External validity agrees with the amount to which the results could be generalized to the wider class of software systems from which the sample has been drawn. Although, in this experiment test software are of different sizes even then there is threat to external validity because these samples are open source software which might affect the degree to which the results could be indiscriminately applied to other proprietary software.

Internal validity deals with the unsuitable statistical test or violation of statistical assumptions. In this experiment, we used the t-test for comparison of earlier studies even though we have not tested the normality of the data. But usage of the t-test is required for backward compatibility and comparison to earlier work. In order to avoid internal validation, so Wilcoxon signed-rank test has been used.

Table 12. The student's T-test at 58 degree of freedom

Software System	Bat-GA and Hill Climbing	ABC-GA and Hill Climbing	BH-GA and Hill Climbing	FF-GA and Hill Climbing
Nos	**2.5878**	0.9157	1.8387	**2.9225**
M-Tunis	1.8573	3.2059	-1.9871	0.1122
Ispell	**3.3398**	0.5325	0.9791	**3.7076**
Rcs	1.2670	**5.3202**	**5.1070**	**5.6807**
Star	**7.1312**	**2.2522**	**-2.2772**	**2.0768**
Bison	**6.2160**	**4.0396**	2.9999	**6.4960**
Grappa	**20.9152**	**86.4371**	**40.8902**	**103.4397**

Table 13. The Wilcoxon test

Software System	Bat-GA and Hill Climbing	ABC-GA and Hill Climbing	BH-GA and Hill Climbing	FF-GA and Hill Climbing
Nos	**2.0879**	0.4217	1.4913	**2.3553**
M-Tunis	**2.0261**	**2.6432**	**2.1700**	0.0103
Ispell	**3.1779**	0.8536	0.8742	**3.7076**
Rcs	1.2444	**4.0828**	**3.8154**	**4.3091**
Star	**4.5765**	**2.1905**	**2.5814**	**2.3962**
Bison	**3.8154**	**3.6097**	**2.6432**	**4.4119**
Grappa	**4.7821**	**4.7825**	**4.7825**	**4.7821**

CONCLUSION

In this proposed work, bio-inspired algorithms such as Bat, ABC, BH and FF and their hybrids with crossover and mutation have been applied to the problem of effective software modularization. These approaches have been applied to seven open source test problems. The results are compared to that of existing search based approaches such as hill climbing and GA implemented in Bunch tool.

The results thus obtained evaluates that the hybrid of this bio inspired algorithms with suitable crossover and mutation operator produces better modularization of software systems as compared to that of existing solutions.

FUTURE WORK

In future, this bio inspired algorithms could be hybridized with other evolutionary optimization approaches such as particle swarm optimization.

Impact of modules that do not fit into any particular cluster, but show belongingness to several clusters (omnipresent modules) and library modules on the problem of software modularization could be investigated in future(Wen & Tzerpos, 2005).

The impact of weights to the dependencies among the modules on the problem of software modularization has not been investigated yet. So, the quality of modularization by the usage of weighted relationships between the modules and omnipresent and library modules could be analyzed in future.

The problem of software clustering has been solved by using single objective nature inspired algorithms and using modularization quality (MQ) as an objective function. There are other objectives such as number of clusters, a number of isolated clusters, cyclic connectivity among modules etc. that can have an effect on efficient software modularization. The impact of these objectives by using multi-objective nature inspired algorithms is yet to be investigated. Further, more objectives having the impact on software modularization are yet to be inspected.

ACKNOWLEDGMENT

The authors are indebted to Dr. Spiros Mancoridis for providing both the Bunch tool and the MDGs used in this paper. We feel immense pleasure to thank Dr. Mark Harman for his valuable suggestions. We are also thankful to Mr. Harjit Singh, senior consultant, Tata Consultancy Services, New Delhi and Pavneet Kaur, component design engineer, Intel Corporation, USA, for their competent guidance and valuable suggestions.

REFERENCES

Akbari, R., Hedayatzadeh, R., Ziarati, K., & Hassanizadeh, B. (2012). A multi-objective artificial bee colony algorithm. *Swarm and Evolutionary Computation*, *2*, 39–52. doi:10.1016/j.swevo.2011.08.001

Akhtar, S., Ahmad, A., & Abdel-Rahman, E. (2012). *A metaheuristic bat-inspired algorithm for full body human pose estimation.* Paper presented at the Computer and Robot Vision (CRV), 2012 Ninth Conference on. doi:10.1109/CRV.2012.55

Apostolopoulos, T., & Vlachos, A. (2010). Application of the firefly algorithm for solving the economic emissions load dispatch problem. *International Journal of Combinatorics*, 2011.

Bora, T. C., Coelho, L. S., & Lebensztajn, L. (2012). Bat-inspired optimization approach for the brushless DC wheel motor problem. Magnetics. *IEEE Transactions on*, *48*(2), 947–950.

Bouchekara, H. (2013). *Optimal Design of Electromagnetic Devices Using a Black-Hole-Based Optimization Technique*. Academic Press.

Chen, Y.-F. R., Gansner, E. R., & Koutsofios, E. (1997). *A C++ data model supporting reachability analysis and dead code detection Software Engineering—ESEC/FSE'97*. Springer.

Clark, M. (2005). *JDepend*. Retrieved from http://www. clarkware. com/software/JDepend. html

Dahiya, S. S., Chhabra, J. K., & Kumar, S. (2010). *Application of artificial bee colony algorithm to software testing*. Paper presented at the Software Engineering Conference (ASWEC), 2010 21st Australian. doi:10.1109/ASWEC.2010.30

Dietrich, J., Yakovlev, V., McCartin, C., Jenson, G., & Duchrow, M. (2008). Cluster analysis of Java dependency graphs. In *Proceedings of the 4th ACM symposium on Software visualization*. doi:10.1145/1409720.1409735

Doval, D., Mancoridis, S., & Mitchell, B. S. (1999). *Automatic clustering of software systems using a genetic algorithm*. Paper presented at the Software Technology and Engineering Practice. doi:10.1109/STEP.1999.798481

Ellson, J., Gansner, E., Koutsofios, L., North, S. C., & Woodhull, G. (2002). *Graphviz—open source graph drawing tools*. Paper presented at the Graph Drawing.

Goldberg, D. E. (2006). *Genetic algorithms*. Pearson Education India.

Harman, M. (2007). *The current state and future of search based software engineering*. Paper presented at the 2007 Future of Software Engineering. doi:10.1109/FOSE.2007.29

Harman, M., Hierons, R. M., & Proctor, M. (2002). *A New Representation And Crossover Operator For Search-based Optimization Of Software Modularization*. Paper presented at the GECCO.

Hatamlou, A. (2013). Black hole: A new heuristic optimization approach for data clustering. *Information Sciences*, *222*, 175–184. doi:10.1016/j.ins.2012.08.023

Huang, S.-J., & Liu, X.-Z. (2013). Application of artificial bee colony-based optimization for fault section estimation in power systems. *International Journal of Electrical Power & Energy Systems*, *44*(1), 210–218. doi:10.1016/j.ijepes.2012.07.012

Jeya Mala, D., Mohan, V., & Kamalapriya, M. (2010). Automated software test optimisation framework-an artificial bee colony optimisation-based approach. *Software, IET*, *4*(5), 334–348. doi:10.1049/iet-sen.2009.0079

Karaboga, D., Okdem, S., & Ozturk, C. (2012). Cluster based wireless sensor network routing using artificial bee colony algorithm. *Wireless Networks*, *18*(7), 847–860. doi:10.1007/s11276-012-0438-z

Kononoff, P., Lehman, H., & Heinrichs, A. (2002). Technical note—a comparison of methods used to measure eating and ruminating activity in confined dairy cattle. *Journal of Dairy Science*, *85*(7), 1801–1803. doi:10.3168/jds.S0022-0302(02)74254-9 PMID:12201531

Mahdavi, K., Harman, M., & Hierons, R. M. (2003). *A multiple hill climbing approach to software module clustering*. Paper presented at the Software Maintenance, 2003. doi:10.1109/ICSM.2003.1235437

Mala, D. J., & Mohan, V. (2009). ABC tester—artificial bee colony based software test suite optimization approach. *International Journal of Software Engineering*, *2*(2), 15–43.

Maleki, I. E., & Gharehchopogh, L. (2014). Hybrid Approach of Firefly and Genetic Algorithms in Software Cost Estimation. *MAGNT Research Report*, *2*(6), 372–388.

Mancoridis, S., Mitchell, B. S., Chen, Y., & Gansner, E. R. (1999). *Bunch: A clustering tool for the recovery and maintenance of software system structures*. Paper presented at the Software Maintenance, 1999. doi:10.1109/ICSM.1999.792498

Mancoridis, S., Mitchell, B. S., Rorres, C., Chen, Y., & Gansner, E. R. (1998). *Using automatic clustering to produce high-level system organizations of source code*. Paper presented at the International Conference on Program Comprehension. doi:10.1109/WPC.1998.693283

Mishra, S., Shaw, K., & Mishra, D. (2012). A new meta-heuristic bat inspired classification approach for microarray data. *Procedia Technology*, *4*, 802–806. doi:10.1016/j.protcy.2012.05.131

Mitchell, B. S. (2002). *A heuristic search approach to solving the software clustering problem*. Drexel University.

Mitchell, B. S., & Mancoridis, S. (2002). *Using Heuristic Search Techniques To Extract Design Abstractions From Source Code*. Paper presented at the GECCO.

Mitchell, B. S., & Mancoridis, S. (2006). On the automatic modularization of software systems using the bunch tool. *Software Engineering. IEEE Transactions on*, *32*(3), 193–208.

Mitchell, B. S., & Mancoridis, S. (2008). On the evaluation of the Bunch search-based software modularization algorithm. *Soft Computing*, *12*(1), 77–93. doi:10.1007/s00500-007-0218-3

Müller, H. A., Orgun, M. A., Tilley, S. R., & Uhl, J. S. (1993). A reverse-engineering approach to subsystem structure identification. *Journal of Software Maintenance: Research and Practice*, *5*(4), 181–204. doi:10.1002/smr.4360050402

Praditwong, K., Harman, M., & Yao, X. (2011). Software module clustering as a multi-objective search problem. *Software Engineering. IEEE Transactions on*, *37*(2), 264–282.

Schwanke, R. W. (1991). *An intelligent tool for re-engineering software modularity*. Paper presented at the Software Engineering. doi:10.1109/ICSE.1991.130626

Sheskin, D. J. (2003). *Handbook of parametric and nonparametric statistical procedures. CRC Press*. doi:10.1201/9781420036268

Tzerpos, V. (2001). *Comprehension-driven software clustering*. University of Toronto.

Wen, Z., & Tzerpos, V. (2005). *Software clustering based on omnipresent object detection*. Paper presented at the Program Comprehension.

Wiggerts, T. A. (1997). *Using clustering algorithms in legacy systems remodularization*. Paper presented at the Reverse Engineering. doi:10.1109/WCRE.1997.624574

www.mathworks.in/products/matlab/

Yan, X., Zhu, Y., Zou, W., & Wang, L. (2012). A new approach for data clustering using hybrid artificial bee colony algorithm. *Neurocomputing*, *97*, 241–250. doi:10.1016/j.neucom.2012.04.025

Yang, X.-S. (2010a). Firefly algorithm, stochastic test functions and design optimisation. *International Journal of Bio-inspired Computation*, *2*(2), 78–84. doi:10.1504/IJBIC.2010.032124

Yang, X.-S. (2010b). *A new metaheuristic bat-inspired algorithm Nature inspired cooperative strategies for optimization (NICSO 2010)*. Springer. doi:10.1007/978-3-642-12538-6_6

Yang, X.-S. (2011). Bat algorithm for multi-objective optimisation. *International Journal of Bio-inspired Computation*, *3*(5), 267–274. doi:10.1504/IJBIC.2011.042259

KEY TERMS AND DEFINITIONS

Black Hole: A region in space with so much mass concentration that no object can escape its gravitational pull.

Firefly: A unisex creature which attracts other regardless of its sex.

Graphviz: Graph Visualization Software is open source software initiated by AT&T Labs which is used for drawing graphs.

Modularization Quality: Ratio of cohesion and coupling.

Module Dependency Graph: Graph representing the system modules as nodes and the module-level relationships as edges between these nodes.

Radius of Event Horizon: Sphere shaped boundary around the black hole where escape speed is equal to a speed of light. Nothing can escape once it enters this radius because nothing can have speed faster than the speed of light.

Search Based Software Engineering: Application of meta-heuristic search to a large body of software engineering where natural representation and fitness function suggest themselves.

Software Modularization: It is the process of organizing modules of a software system into optimal clusters.

Chapter 18
Hybrid Particle Swarm and Gravitational Search Optimization Techniques for Charging Plug-In Hybrid Electric Vehicles

Imran Rahman
Universiti Teknologi PETRONAS, Malaysia

Balbir Singh Mahinder Singh
Universiti Teknologi PETRONAS, Malaysia

Pandian Vasant
Universiti Teknologi PETRONAS, Malaysia

M. Abdullah-Al-Wadud
King Saud University, Saudi Arabia

ABSTRACT

Electrification of Transportation has undergone major modifications since the last decade. Success of combining smart grid technology and renewable energy exclusively depends upon the large-scale participation of Plug-in Hybrid Electric Vehicles (PHEVs) towards reach the desired pollution-free transportation industry. One of the key Performance pointers of hybrid electric vehicle is the State-of-Charge (SoC) which needs to be enhanced for the advancement of charging station using computational intelligence methods. In this Chapter, authors applied Hybrid Particle swarm and gravitational search Optimization (PSOGSA) technique for intelligently allocating energy to the PHEVs considering constraints such as energy price, remaining battery capacity, and remaining charging time. Computational experiment results attained for maximizing the highly non-linear fitness function estimates the performance measure of both the techniques in terms of best fitness value and computation time.

INTRODUCTION

Recent researches on green technologies for transportation sector are gaining popularity among the research communities from different areas. In this wake, Plug-in hybrid electric vehicles (PHEVs) have great future because of their charge storage system and charging facilities from traditional grid system. Several researchers have proved that a great amount of reductions in greenhouse gas emissions and the

DOI: 10.4018/978-1-4666-9644-0.ch018

increasing dependence on oil could be accomplished by electrification of transport sector (Caramanis & Foster 2009). Future transportation sector will depend much on the advancement of this emerging field of vehicle optimization. Indeed, the adoption of hybrid electric vehicles (HEVs) has brought significant market success over the past decade. Vehicles can be classified into three groups: internal combustion engine vehicles (ICEV), hybrid electric vehicles (HEV) and all- electric vehicles (AEV) (Tie & Tan, 2013). Plug-in hybrid electric vehicles (PHEVs) which is very recently introduced promise to boost up the overall fuel efficiency by holding a higher capacity battery system, which can be directly charged from conventional power grid system, that helps the vehicles to operate continuously in "all-electric-range" (AER). All-electric vehicles or AEVs is a kind of transport which use electric power as only sources to run the system. Plug-in hybrid electric vehicles with a connection to the smart grid can own all of these strategies. Hence, the widely extended adoption of PHEVs might play a significant role in the alternative energy integration into traditional grid systems (Lund & Kempton, 2008). There is a need of efficient mechanisms and algorithms for smart grid technologies in order to solve highly diverse problems like energy management, cost reduction, efficient charging station etc. with different objectives and system constraints (Hota, Juvvanapudi, & Bajpai, 2014).

According to Electric Power Research Institute (EPRI), about 62% of the whole United States (US) fleet will comprise of PHEVs within the year 2050 (Soares et al., 2013). Moreover, there is an increasing demand to implement this technology on the electric grid system. Large numbers of PHEVs have the capability to make threats to the stability of the power system. For example, in order to avoid disturbance when several thousand PHEVs are introduced into the system over a small period of time, the load on the power grid will need to be managed very carefully. One of the main targets is to facilitate the proper communication between the power grid and the PHEV. For the maximization of customer contentment and minimization of burdens on the grid, a complicated control appliance will need to be addressed in order to govern multiple battery loads from a numbers of PHEVs properly (Su & Chow, 2012a). The total demand pattern will also have an important impact on the electricity production due to differences in the needs of the PHEVs parked in the deck at certain time (Su & Chow, 2011). Proper management can ensure strain minimization of the grid and enhance the transmission and generation of electric power supply. The control of PHEV charging depending on the locations can be classified into two groups; household charging and public charging. The proposed optimization focuses on the public charging station for plug-in vehicles because most of PHEV charging is expected to take place in public charging location (Su & Chow, 2012). Wide penetration of PHEVs in the market depends on a well-organized charging infrastructure. The power demand from this new load will put extra stress on the traditional power grid (Morrow, Karner, & Francfort, 2008). As a result, a good number of PHEV charging stations with suitable facilities are essential to be built for recharging electric fleet, for this some strategies have been proposed by the researchers (Mayfield, Jul. 2012). Charging stations are needed to be built at workplaces, markets/shopping malls and home. Boyle (2007) proposed the necessity of building new smart charging station with effective communication among utilities along with sub-station control infrastructure in view of grid stability and proper energy utilization. Furthermore, sizeable energy storage, cost minimization; Quality of Services (QoS) and intelligent charging station for optimal power are underway (Hess et al., 2012). In this wake, numerous techniques and methods were proposed for deployment of PHEV charging stations (Z. Li, Sahinoglu, Tao, & Teo, 2010).

One of the main targets is to facilitate the proper interaction between the power grid and the PHEV. For the maximization of customer satisfaction and minimization of burdens on the grid, a complicated control mechanism will need to be addressed in order to govern multiple battery loads from a numbers of PHEVs appropriately (Su & Chow, 2012). Charging infrastructures are essential in order to facilitate the large-scale penetration of PHEVs. Different computational intelligence-based methods have been used by some researchers for charging station optimization of PHEV. Most of them applied traditional methods which are needed to be improved furthermore (Rahman, I., Vasant, P.M., Singh, B.S.M., & Abdullah-Al-Wadud, M., 2014).

Swarm intelligence came from the mimic of the living colony such as ant, bird, and fish in nature, which shows unparalleled excellence in swarm than in single in food seeking or nest building. Drawing inspiration from this, researches design many algorithms simulating colony living, such as ant colony algorithm, particle swarm optimization algorithm, artificial bee colony algorithm, and gravitational search algorithm, which shows excellent performance in dealing with complex optimization problems (Jia-zhao, C., Yu-xiang, Z., & Yin-sheng, L., 2012). The intrinsic characteristics of all the population-based meta-heuristic algorithms like Particle swarm optimization (PSO) and Gravitational search algorithm (GSA) are to maintain a good compromise between exploration and exploitation in order to solve the complex optimization problems (Rashedi, E. & Nezamabadi-Pour, 2009).

GSA is based on the law of gravity and mass interactions where the searcher agents are a collection of masses which interact with each other based on the Newtonian gravity and the laws of motion (Rashedi, E. & Nezamabadi-Pour, 2009). This method has also been used by the researchers for post-outage bus voltage magnitude calculations, solving economic dispatch with valve-point effects, optimal sizing and suitable placement for distributed generation (DG) in distribution system, Solving thermal unit commitment (UC) problem and finding out optimal solution for optimal power flow (OPF) problem in a power system (N M., Puteh, M., & Mahmood, M. R., 2013).

Moreover, PSOGSA-based optimization has already been used by the researchers for economic load dispatch (Dubey, Pandit et al., 2013), optimal static state estimation (Mallick, Ghoshal et al., 2013), dual channel speech enhance-ment (Kunche, Rao et al., 2014), training feed-forward neural networks (Mirjalili, Mohd Hashim et al., 2012) and multi-distributed generation planning (Tan, Hassan et al., 2013). Specifically, we are investigating the use of the Hybrid particle swarm optimization Gravitational Search Algorithm (PSOGSA) method for developing real-time and large-scale optimizations for allocating power.

The performance of PHEV depends upon proper utilization of electric power which is solely affected by the battery state-of-charge (SoC). In Plug-in hybrid electric vehicles (PHEVs), a key parameter is the state-of-charge (SoC) of the battery as it is a measure of the amount of electrical energy stored in it. It is analogous to fuel gauge on a conventional internal combustion (IC) car (Chiasson, J., & Vairamohan, B., 2005). State-of-charge determination becomes an increasingly vital issue in all the areas that include a battery. Previous operation policies made use of voltage limits only to guard the battery against deep discharge and overcharge. Currently, battery operation is changing to what could rather be called battery management than simply protection. For this improved battery control, the battery SoC is a key factor indeed (Piller, Perrin, & Jossen, 2001).

A charging station is one way that the operator of an electrical power grid can adapt energy production to energy consumption, both of which can vary randomly over time. Basically, PHEVs in a charging station are charged during times when production exceeds consumption and are discharged at times when consumption exceeds production (S. Li, Bao, Fu, & Zheng, 2014). There is a need of in-depth study on maximization of average SoC in order to facilitate intelligent energy allocation for PHEVs in a charging station.

The purpose of this chapter is to optimize state-of-charge, with respect to total cost, charging time, present SoC. Swarm intelligence-based method, Hybrid Particle swarm Optimization and Gravitational search algorithm (PSO-GSA) were applied for solving the optimization problem and then compared with Gravitational search algorithm (GSA) furthermore (Rahman, I., Vasant, P.M., Singh, B.S.M., & Abdullah-Al-Wadud, M., 2014) .

BACKGROUND

The vehicular network recently accounts for around 25% of CO_2 emissions and over 55% of oil consumption around the world. Carbon dioxide is the primary greenhouse gas emitted through human activities like combustion of fossil fuels (coal, natural gas, and oil) for energy and transportation. Several researchers have proved that a great amount of reductions in greenhouse gas emissions and the increasing dependence on oil could be accomplished by electrification of transport sector (Holtz-Eakin & Selden, 1995). Charging of PHEV/EV influences many parameters such as power rating, time of charging and location, cost, charging equipment, and effect on the power grid. Issues like charging time, distribution, standardization of demand policies for charging stations and proper regulatory procedures are needed to be addressed for the successful deployment of Electric vehicle charging station (Z. Li et al., 2010).

Most of the electric vehicles charging generally occur at charging area in one's house where the fleet can be connected to a garage outlet for Slow charging (Level-1). Level-2 charging is normally known as the primary technique for battery charging for both public and private utilities and needs an outlet of 240V. Future technologies focus on primary; fast charging and can be executed in most cases (Anegawa, 2009; Botsford & Szczepanek, 2009; Rawson & Kateley, 1999). Usually for Level-1 and 2 charging uses single-phase systems. Level-3(DC fast charging) is made for commercial and public applications and would operate just like a normal filling station. Off-board three-phase solutions are applied to Level-3 chargers and high power. Level-2 or 3 chargers installed in parking lots, shopping centers, hotels, theaters, restaurants, etc. are expected to use by the general public stations (Aggeler, Canales, Coccia, Butcher, & Apeldoorn, 2010).

Opportunity Charging (Level-1 Charging)

The slowest of all available methods is Level-1 charging. In the United States, Level-1 charging uses a standard 120V/15A single-phase outlet which is grounded, such as NEMA 5-15R. The connection may use a standard J1772 connector into the electric vehicle ac port. No additional infrastructure is required for home or business sites. At night, low off-peak rates for charging are likely to be available. The total cost of a residential Level-1 charging infrastructure has been estimated around $500 - $880 (De Sousa, Silvestre, & Bouchez, 2010; Morrow, Karner, & Francfort, 2008).

Primary Charging (Level-2 Charging)

Level-2 charging is the basic method for dedicated public and private facilities. At present, Level-2 equipment performs charging through 208V or 240V (at up to 80A, 19.2 kW). It may require dedicated equipment and a connection installation for home or public charging (Rawson & Kateley, 1999), although vehicles such as the Tesla have the power electronics on board. Most homes have 240 V service available, and Level-2 devices can charge a typical EV battery overnight. Owners seem likely to prefer Level-2 technology owing to its faster charging time and standardized vehicle-to-charger connection. A separate billing meter is typical. The cost of residential Level-2 infrastructure installation is around $2,150. For example, the Tesla Roadster charging system has imposed additional cost of $3,000 (Motors, 2009).

Fast Charging (Level-3 Charging)

Level-3 (DC fast charging) can be installed in highways and urban refueling points which is similar to petrol stations. It generally operates with a 480 V or higher three phase circuit and needs an off-board charger to provide regulated ac-dc conversion. Level-3 charging is very rear in the residential premises. Standards for dc plugs and hardware are in progress. CHAdeMO-a Japanese protocol is gaining worldwide recognition (Yilmaz & Krein, 2012). Installation cost is a vital issue. Level-3 charging infrastructure costs between $30,000 and $160,000 have been reported. An efficient energy management system is proposed (Dusmez, Cook, & Khaligh, 2011) which notably reduce total time of PHEVs charging in fast charging infrastructure by the use of additional super capacitors and flywheel. The simulations for two batteries between 10kWh and 15kWh show that the charging time on average is 15 min to charge from a minimum SOC 20% to maximum 95% in the latest configuration. Finally, Figure 1 summarizes the charging methods.

Charging Infrastructures

Maintenance of the charging infrastructures is another cost factor (Brown, Mikulin, Rhazi, Seel, & Zimring, 2010). There are increasing numbers of literatures on various aspects of the EV charging allocation strategies which includes the maintenance and scheduling of various chargers (Caramanis & Foster, 2009; Gan, Topcu, & Low, 2011; Kefayati & Caramanis, 2010; Ma, Callaway, & Hiskens, 2010; Pang, Dutta, Kim, Kezunovic, & Damnjanovic, 2010; Sojoudi & Low, 2011). Most of the works focus specially on residential charging schemes. Kulshrestha, Wang, Chow, and Lukic (2009) conducted studies based on simulation in energy management strategy (EMS) for PHEV/EV charging at parking areas where meta-heuristic algorithms for the purpose of efficient scheduling are applied. The electric vehicle charging for public garages is also considered (Su & Chow, 2011) where the objective is to maximize the throughput of service whereas the total cost of energy is not considered in the optimization. Subramanian et al. (2012) suggested a scheduling optimization using a combination of alternative energy and energy from the traditional grid.

The next section provides an overview of the charging infrastructure requirements for PHEVs/EVs in single-family household, multi-family household and commercial situations.

Figure 1. Changing infrastructure for PHEVs

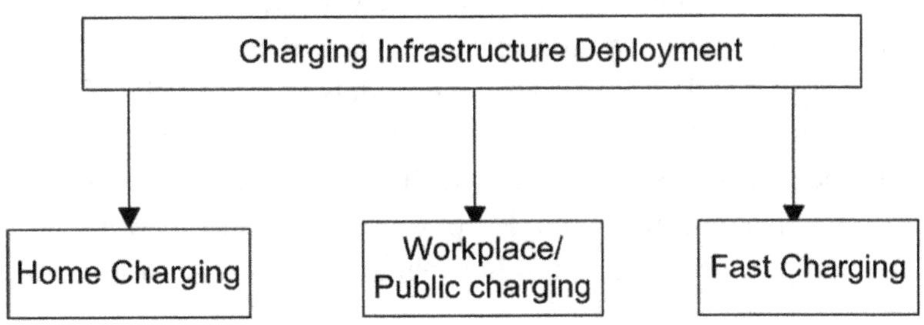

These scenarios include the following:

- Household garage charging
- Apartment complex charging
- Commercial complex charging
- Charging from renewable energy sources

Household Garage Charging

In order to install electric vehicle charging supply in a household garage, dedicated branch circuit from an existing house distribution panel to a convenience outlet or to a EVSE (Electric Vehicle Supply Equipment) is necessary (Morrow et al., 2008).

Apartment Complex Charging

Installation of the EV/PHEV charging supply in an apartment complex typically consists of installing new dedicated branch circuits from the central meter distribution panel to either a convenience outlet or to an EVSE (Morrow et al., 2008).

Commercial Complex Charging

Installation of the electric vehicle charging supply in a commercial complex parking lot typically consists of installing new dedicated branch circuits from the central meter distribution panel to an EVSE for Level-2 charging. Large parking lots provide an opportunity to control a fleet of PHEVs in an intelligent manner.

Effective use of PHEVs in parking areas to prevent the transmission lines getting overloaded and to act as shock observers when the wind power changes drastically is explored by Venayagamoorthy and Mitra (2011). A fuzzy logic controller was proposed (Mitra & Venayagamoorthy, 2010) which takes the total state of charge of a parking lot, instantaneous demand and wind power generated as inputs and gives control signals for charging/discharging of the PHEVs. Simulations on a12 bus system model show that when PHEVs charge and discharge according to the control signal, overloading of the transmission lines during high wind speeds can be prevented and the wind power supply fluctuations to the grid can be minimized.

Charging from Renewable Energy Sources

The ability of PHEVs/EVs to assist the integration of renewable energy sources into the existing power grid is potentially the most transformative impact on the electricity system. Deployment of large-scale photovoltaic (PV) charging equipment in a parking lot is explained by Neumann, Schär, and Baumgartner (2012). PV parking lot charging and different business models to charge PHEVs/EVs with solar energy are also studied by Letendre (2009). Economics and environmental impacts of PV based workplace charging station has also been discussed (Birnie, 2009; Tulpule, Marano, Yurkovich, & Rizzoni, 2013). The analysis shows the technical feasibility of a PV powered workplace parking lot with benefits to the owner of the vehicle as compared to facilities of household charging. Authors conclude that the owner will get the return of establishment and maintenance cost and profit within the lifespan of the photovoltaic panels. According to Birnie (2009), introducing a solar collector into a parking shade would result in a much more rapid pay-back-period, encouraging widespread installation of solar capacity. Zhang, Tezuka, Ishihara, and Mclellan (2012) explained smart control strategies for the integration of both EVs and PV together with the present electricity systems. Co-benefits of introducing large penetration of PHEVs and photovoltaic mechanisms have been analyzed by Denholm, Kuss, and Margolis (2013). The study came to a conclusion that PV has the capability of acting like a potential source of mid-day generation capacity for PHEVs as well as provide a dispatch able load during low demand periods (generally in the spring season). For this wake, a 2.1 kW PV charging station combined with the utility at Santa Monica is explained (Ingersoll & Perkins, 1996). Zhu, Yu, Ning, and Tang (2012) presented optimal charging control policy using stochastic semi-Markov decision (SMDP) process and later average reward was calculated using vehicle admission probability.

Smart grid has brought new opportunities and challenges for the development of electric vehicle Infrastructure facilities like charging station systems and parking lots. Recent advancement in renewable energy sector opens the option for a green infrastructure system which will minimize the burden of PHEVs in tradition grid-dependent charging stations.

Proper SoC will ensure smooth growth of PHEV in near future. Without proper maintenance and infrastructure facilities, large number of PHEVs charging will burden the existing grid hence instability of power grid. For this, researchers should apply proper techniques methods in order to solve various optimization problems related to this field. Energy allocation to PHEV charging station is subjected to various constraints such as charging time, SoC and price which will be highlighted in the problem formulation section. Different constraints make the entire search space limited to a particular suitable region. So, powerful optimization algorithms should be implemented in order to achieve high quality solutions with a stable convergence rate.

MAIN FOCUS OF THE CHAPTER

Problem Statement

One of the important constraints for accurate charging is State-of-Charge (SoC). Charging algorithm can precisely be managed by the precise state of charge evaluation (Shafiei & Williamson, 2010). An approximate graph of a typical Lithium-ion cell voltage versus SoC is shown in Figure 2 indicates that the slope of the curve below 20% and above 90% is high enough to result in a significant voltage difference to be depended on by measurement circuits and charge balancing control. There is a need of in-depth study on maximization of average SoC in order to facilitate intelligent energy allocation for PHEVs in a charging station.

The idea behind smart charging is to charge the vehicle when it is most favourable, which could be when electricity price, demand is lowest, when there is excess capacity (Su & Chow, 2012). When a vehicle is plugged in into a smart charging station a request for energy demand is sent to Substation Control Center (SCC), which decides based on the available energy from utility and either accepts the request or reject it. Performance of this kind of load management is measured in terms of delay, delivery ration and jitter. As a matter of fact EVs may be charged at any time of a day depending on requirement

Figure 2. Li-ion cell voltage vs. State-of-Charge

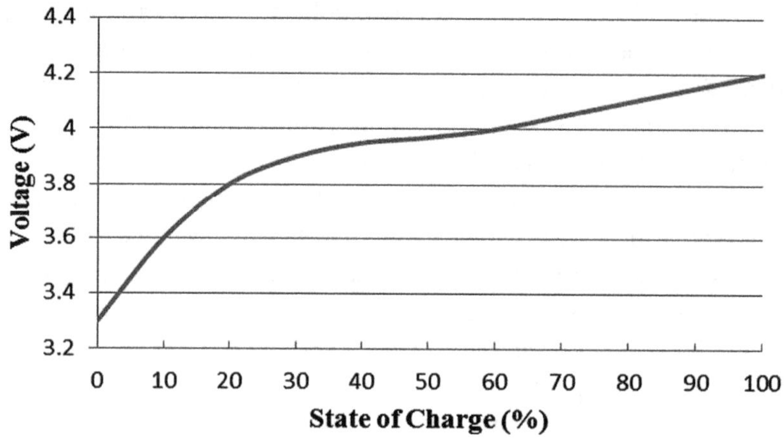

to top their batteries even during peak demand hours. Increasing load on the grid during peak hours may require extra power generation through any source which may increase the cost and greenhouse gases emission (Ul-Haq, Buccella, Cecati, & Khalid, 2013).

Suppose, there is a charging station with the capacity of total power P. Total N numbers of PHEVs need to be served in a day (24 hours). The proposed system should allow PHEVs to leave the charging station before their expected leaving time for making the system more effective. It is worth to mention that, each PHEV is regarded to be plugged-in to the charging station once. The main aim is to allocate power intelligently for each PHEV coming to the charging station. The State-of-Charge is the main parameter which needs to be maximized in order to allocate power efficiently. For this, the fitness function considered in this chapter is the maximization of average SoC and thus allocate energy for PHEVs at the next time step. The constraints considered are: charging time, present SoC and price of the energy (Rahman, I., Vasant, P.M., Singh, B.S.M., & Abdullah-Al-Wadud, M., 2014).

The fitness function is defined as:

$$\text{Max } J\left(k\right)=\sum_{i} w_{i}\left(k\right) SoC_{i}\left(k+1\right) \tag{1}$$

$$w_{i}\left(k\right)= f\left(C_{r,i}\left(k\right),\ T_{r,i}\left(k\right), D_{i}\left(k\right)\right) \tag{2}$$

$$C_{r,i}\left(k\right)=\left(1-SoC_{i}\left(k\right)\right)\cdot C_{i} \tag{3}$$

where $C_{r,i}\left(k\right)$ is the battery capacity (remaining) needed to be filled for i no. of PHEV at time step k; C_{i} is the battery capacity (rated) of the i no. of PHEV; remaining time for charging a particular PHEV at time step k is expressed as $T_{r,i}\left(k\right)$; the price difference between the real-time energy price and the price that a specific customer at the i no. of PHEV charger is willing to pay at time step k is presented by $D_{i}\left(k\right)$; $w_{i}\left(k\right)$ is the charging weighting term of the i no. of PHEV at time step k (a function of

Table 1. Parameter settings of the fitness function

Parameter	Values
Fixed Parameters	Maximum power, $P_{i,\max}$ = 6.7 kWh
	Charging station efficiency, η = 0.9
	Total charging time, Δt = 20 Minute
	Power allocation to each PHEV: 30 W
Variables	0.2 ≤ State-of-Charge (SoC) ≤ 0.8
	Waiting time ≤ 30 Minutes (1800 Seconds)
	16 kWh ≤ Battery Capacity (C_i) ≤ 40 kWh
Constraints	$\sum_i P_i(k) \leq P_{utility}(k) \times \eta$
	$0 \leq P_i(k) \leq P_{i,\max}(k)$
	$0 \leq SoC_i(k) \leq SoC_{i,\max}$
	$0 \leq SoC_i(k+1) - SoC_i(k) \leq \Delta SoC_{\max}$

charging time, present SoC and price of the energy); $SoC_i(k+1)$ is the state of charge of the i no. of PHEV at time step $k+1$. The illustration of the remaining battery capacity is given in Box 1.

Here, the weighting term indicates a bonus proportional to the attributes of a specific PHEV. For example, if a PHEV has a lower initial SoC and less charging time (remaining), but the driver is eager to pay a higher price, the system will provide more power to this particular PHEV battery charger:

$$w_i(k) \pm \left[C_{r,i}(k) + D_i(k) + 1/T_{r,i}(k)\right] \tag{4}$$

Box 1.

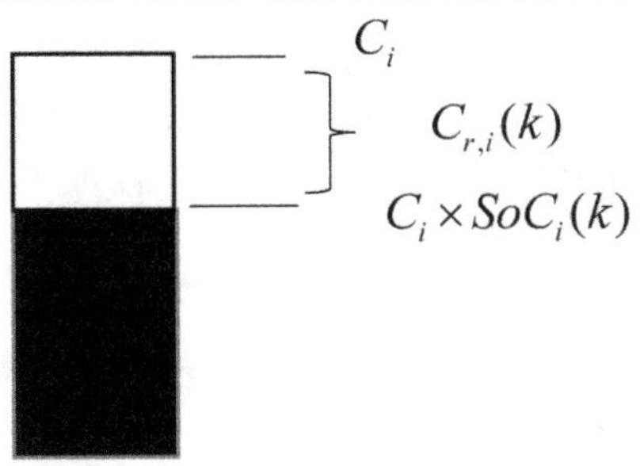

The charging current is also assumed to be constant over Δt.

$$\left[SoC_i\left(k+1\right) - SoC_i\left(k\right)\right] \cdot Cap_i = Q_i = I_i\left(k\right)\Delta t \tag{5}$$

$$SoC_i\left(k+1\right) = SoC_i\left(k\right) + \mathrm{I}_i\left(k\right)\Delta t \,/\, Cap_i \tag{6}$$

where the sample time Δt is defined by the charging station operators, and $\mathrm{I}_i\left(k\right)$ is the charging current over Δt.

The battery model is regarded as a capacitor circuit, where C_i is the capacitance of battery (Farad). The model is defined as

$$C_i \cdot \frac{dV_i}{dt} = I_i \tag{7}$$

Therefore, over a small time interval, one can assume the change of voltage to be linear,

$$C_i \cdot \left[V_i\left(k+1\right) - V_i\left(k\right)\right] / \Delta t = I_i \tag{8}$$

$$V_i\left(k+1\right) - V_i\left(k\right) = I_i\Delta t \,/\, C_i \tag{9}$$

Since the decision variable is the power allocated to the vehicles, replacing $I_i(k)$ with $P_i(k)$

$$\mathrm{I}_i\left(k\right) = P_i\left(k\right) / \, 0.5 \times [V_i\left(k+1\right) - V_i\left(k\right)] \tag{10}$$

$$V_i\left(k+1\right) = \sqrt{\frac{2P_i\left(k\right)\Delta t}{C_i} + V^2{}_i\left(\mathrm{k}\right)} \tag{11}$$

Substituting (10) into (6) yields

$$SoC_i\left(k+1\right) = SoC_i\left(k\right) + \frac{P_i\left(k\right)\Delta t}{0.5.\,C_i.\left[\sqrt{\frac{2P_i\left(k\right)\Delta t}{C_i} + V^2{}_i\left(\mathrm{k}\right)} + V_i\left(k\right)\right]} \tag{12}$$

Finally, the objective function becomes

$$J(k)=\sum w_i.\left[SoC_i(k)+\frac{P_i(k)\Delta t}{0.5.C_i.\left[\sqrt{\frac{2P_i(k)\Delta t}{C_i}+V^2_i(\mathrm{k})}+V_i(k)\right]}\right] \tag{13}$$

There are two kind of inequality constraints used here to optimize the fitness function- i) Power from the charging station operator and ii) individual PHEV's State-of-Charge (SoC). Power obtained from the utility ($P_{utility}$) and the maximum power ($P_{i,\max}$) absorbed by a specific PHEV are the primary energy constraints being considered in this chapter. The overall charging efficiency of a particular charging station is described by η. From the system point of view, charging efficiency is supposed to be constant at any given time step. Maximum battery SoC limit for the i no. of PHEV is $SoC_{i,\max}$. When SoC_i reaches the values close to $SoC_{i,\max}$, the i no. of battery charger shifts to a standby mode. The state of charge ramp rate is confined within limits by the constraint $\Delta SoC_{\max}$. The overall control system is changed the state when i) system utility data updates; ii) a new PHEV is plugged-in; iii) time period Δt has periodically passed. Obviously, SoC maximization method being considered in this chapter can provide a uniformly higher SoC for all PHEVs/PEVs at plug-out as compared with the alternative schemes. It also proves that the proposed function aims at ensuring some fairness in the SoC-distribution at each time step. This will help to ensure that a reasonable level of battery power is attained, even in the event of an early departure.

Table 1 shows all the fitness function parameters that were tuned for performing the optimization. There are total three (03) kinds of parameter: fixed, variables and constraints. Total charging time is fixed to 20 minutes and charging station efficiency assumed to be 0.9. The values are retrieved from various literatures (Hota, Juvvanapudi, & Bajpai, 2014; Su, 2012; Wencong & Mo-Yuen, 2011). Moreover, State-of-Charge is in the range of 0.2 to 0.8 (Chang, 2013).

The implementation of the proposed two optimization methods (GSA, PSOGSA) to our problem is given below in points:

1. Initializing the random and fixed values of both the algorithm and fitness function parameters.
2. Generation of a random solutions (masses) group in the feasible region. Since we normally have very little information about the global optima, these particles are scattered over the search space as uniformly as possible.
3. Evaluate the distance between the new solution and the desired solution based upon a fitness function.
4. Best fitness value as a solution.
5. Repeat steps 2-3 for another iteration until iteration=100.

Here, a single parking lot with the aggregation of distribution network-connected PHEVs is considered. We make use of historical data for office parking from the city of Livermore, CA (Downtown parking Study, Livermore, CA, 2006)

PROPOSED METHODS

Gravitational Search Algorithm

GSA is an optimization method which has been introduced by Rashedi et al. in the year of 2009. In GSA, the specifications of each mass (or agent) are total four, which is mass (inertial), position, mass (active gravitational) and mass (passive gravitational). The position of the mass presents a solution of a particular problem, and masses (gravitational and inertial) are obtained by using a fitness function. GSA can be considered as a collection of agents (candidate solutions), whose masses are proportional to their value of fitness function. During generations, all masses attract each other by the gravity forces between them. A heavier mass has the bigger attraction force. Therefore the heavier masses which are probably close to the global optimum attract the other masses proportional to their distances.

Law of gravity: The law states that particles attract each other and the force of gravitation between two particles is directly proportional to the product of their masses and inversely proportional to the distance between them.

Law of motion: The law states that the present velocity of any mass is the summation of the fraction of its previous velocity and the velocity variance. Variation in the velocity or acceleration of any mass is equal to the force acted on the system divided by inertia mass.

The gravitational force is expressed as follows:

$$F_{ij}^{d}(t) = G(t)\frac{M_{pi}(t) \times M_{aj}(t)}{R_{ij}(t) + \mu}\left(x_{j}^{d}(t) - x_{i}^{d}(t)\right) \quad (14)$$

where M_{aj} is the active gravitational mass related to agent j, M_{pi} is the passive gravitational mass related to agent i, G(t) is gravitational constant at time t, ε is a small constant and $R_{ij}(t)$ is the Euclidian distance between two agents i and j. The $G(t)$ is calculated as-

$$G(t) = G_{0} \times \exp\left(-\alpha \times \text{iter} / \text{maxiter}\right) \quad (15)$$

where α and G_0 are descending coefficient and primary value respectively, current iteration and maximum number of iterations are expressed as iter and maxiter. In a problem space with the dimension d, the overall force acting on agent i is estimated as following equation:

$$F_{i}^{d}(t) = \sum_{j=1, j\neq i}^{N} \text{rand}_{j} F_{ij}^{d}(t) \quad (16)$$

where rand_j is a random number with interval [0, 1]. From law of motion we know that, an agent's acceleration is directly proportional to the resultant force and inverse of its mass, so the acceleration of all agents should be calculated as follow:

$$\mathrm{ac}_i^d\left(t\right) = \frac{F_i^d\left(t\right)}{M_{ii}\left(t\right)} \tag{17}$$

where t is a specific time and M_{ii} is the mass of the object i. The velocity and position of agents are calculated as follow:

$$\mathrm{vel}_i^d\left(t+1\right) = \mathrm{rand}_i \times \mathrm{vel}_i^d\left(t\right) + \mathrm{ac}_i^d\left(t\right) \tag{18}$$

$$x_i^d\left(t+1\right) = x_i^d\left(t\right) + \mathrm{vel}_i^d\left(t+1\right) \tag{19}$$

where rand_i is a random number with interval [0, 1].

In Gravitational search algorithm, all agents are initialized first with random values. Each of the agents is a candidate solution. After initialization, velocities for all agents are defined using (18). Moreover, the gravitational constant, overall forces, and accelerations are determined by equations (15), (16) and (17) respectively. The positions of agents are calculated using (19). At the end, GSA will be terminated by meeting the stopping criterion of maximum 100 iterations. The parameter settings for GSA are demonstrated in Table 2. The GSA parameters were selected: Primary parameter, $G_0 = 100$, Acceleration coefficient, α =20 and No. of mass agents=100. Since each agent could observe the performance of the others, the gravitational force is an information-transferring tool.

Table 2. GSA parameter settings

Parameters	Values
Primary parameter, G_o	100
No. of mass agents, n	100
Acceleration coefficient, α	20
Constant parameter, ε	.01
Power of 'R'	1
Maximum iteration	100
Number of runs	50

The Algorithm Outline

The outline of gravitational search algorithm is given below:

```
1.  Initialization of  total N mass agents randomly
2.  Computation of  G(t), Fitness (Best and Worst)
3.  For each of the agent i, evaluate:
    3.1.  Finess_i
    3.2.  Mass_i
    3.3.  Force of Mass_i
    3.4.  Acceleration of Mass_i
    3.5.  Mass_i velocity update
    3.6. New position of Agent_i
            If (Probability_i > Threshold)
                          {
            If (Probability_i > Random_i)
          Then return Best Fitness solution so far
          Else
                Modification of solution
                      }
4.  Failed to meet stopping criteria,
          Go To Step 2, Else Stop
```

Moreover, the step involves in optimization using GSA is shown Figure 3. Here, we assume that the gravitational and the inertia masses are the same. However, for some applications different values for them can be used. A bigger inertia mass provides a slower motion of agents in the search space and hence a more precise search. Conversely, a bigger gravitational mass causes a higher attraction of agents. This permits a considerable convergence (Rashedi, Nezamabadi-Pour, & Saryazdi, 2009). When an algorithm finds an optimal solution to a given problem, one of the important factors is speed and rate of convergence to the optimal solution. For heuristics, the additional consideration of how close the heuristic solution comes to optimally is generally the primary concern of the researcher (Barr, Golden et al. 1995). In GSA, the stable convergence and better exploitation rate ensures good quality solution, which is expressed in terms of best fitness function.

The Hybrid PSOGSA Algorithm

Before starting the mechanism of hybrid PSOGSA algorithm, it is necessary to shade some light on standard Particle Swarm Optimization (PSO). The system initially has a population of random selective solutions. Each potential solution is called a particle. Each particle is given a random velocity and is flown through the problem space. The particles have memory and each particle keeps track of its previous best position (called the *pbest*) and its corresponding fitness. There exist a number of pbest for the respective particles in the swarm and the particle with greatest fitness is called the global best

Figure 3. Structural diagram of GSA

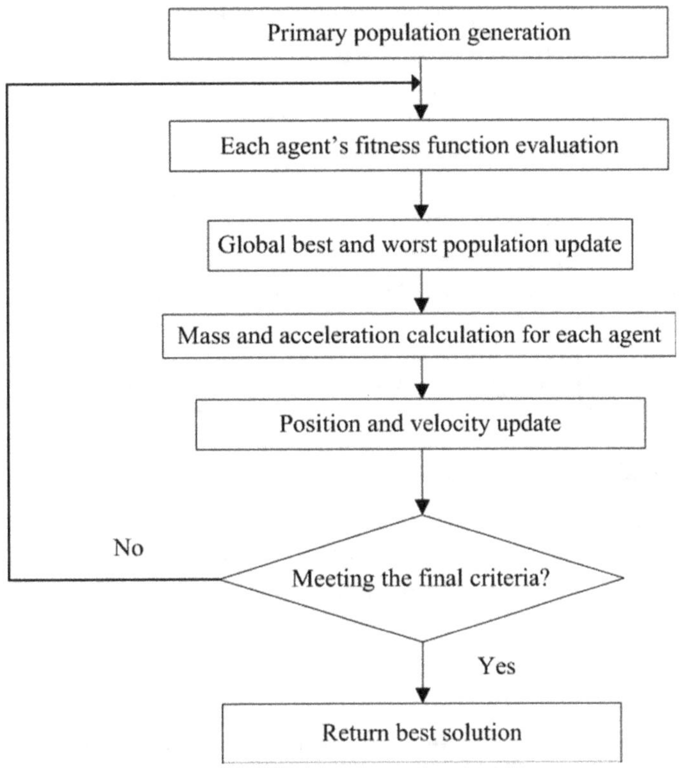

(*gbest*) of the swarm. The basic concept of the PSO technique lies in accelerating each particle towards its pbest and gbest locations, with a random weighted acceleration at each time step (Ganesan, Vasant, & Elamvazuthy, 2012).

In this chapter, a new hybrid population-based algorithm (PSOGSA) [14] is proposed with the combination of Particle Swarm Optimization (PSO) and Gravitational Search Algorithm (GSA). The basic idea is to fit in the exploitation ability in PSO with the exploration ability in GSA to synthesize both algorithms' strength. The basic idea of PSOGSA is to combine the ability of social thinking (gbest) in PSO with the local search capability of GSA. In order to combine these two algorithms, velocity update is proposed as

$$v_i(t+1) = w \times v_i(t) + \alpha' \times rand \times ac_i(t) + \beta' \times rand \times (gbest - x_i(t)) \tag{20}$$

where $v_i(t)$ is the velocity of agent i at iteration t, w is a weighting factor, *rand* is a random number between 0 and 1, $ac_i(t)$ is the acceleration of agent at iteration t, and gbest is the best solution so far. Here, α' and β' are the weighting factors. With adjusting α' and β', the abilities of global search and local search can be balanced. The position of the particle $x_i(t+1)$ in each iteration is updated using the equation

$$x_i(t+1) = x_i(t) + v_i(t+1) \tag{21}$$

Figure 4. Structural diagram of PSOGSA

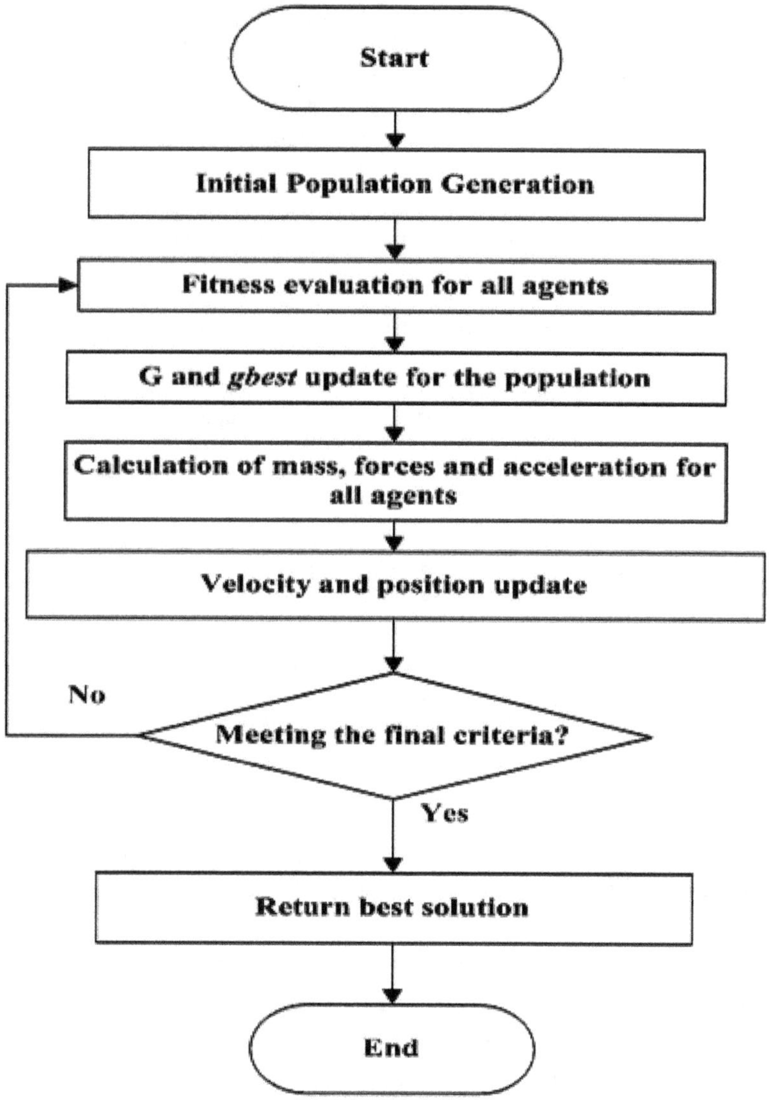

The flowchart of hybrid PSOGSA method is shown in figure 4.

The agents are adjusted randomly and each agent in the search space is attracted towards the agent having a good solution. The agents near the optimal solution moves more slowly and assures the exploitation step of algorithm. Here gbest is used to exploit the global best. The position and velocity are updated until it reaches to the stopping criterion (Dubey, Pandit et al., 2013).

The Algorithm Outline

The outline of gravitational search algorithm is given below:

```
1. Generate initial population
2. Evaluate fitness for all population
3. Update the G and gbest for the population
4. For each of the agent i, evaluate:
    4.1. Finess_i
    4.2. Mass_i
    4.3. Force of Mass_i
    4.4. Acceleration of Mass_i
5. Update velocity and position
6. Failed to meet stopping criteria,
        Go To Step 2, Else Stop
```

Solutions and Recommendations

The PSOGSA and GSA algorithm were applied to find out global best fitness of the fitness function. All the simulations were run on a Core™ i5-3470M CPU@ 3.20 GHz processor, 4.00 GB RAM and MATLAB R2013a.

Figure 5 show the convergence behavior of GSA. The result derived in this chapter reveals that each object of the standard GSA converges to a stable point. Here, the assumption was that the gravitational and inertia masses are the same. However, for some applications different values for them can be used. A heavier inertia mass provides a slower motion of agents in the search space and hence a more precise search (Rashedi et al., 2009). On the contrary, a heavier gravitational mass causes a higher attraction of agents. This allows a faster convergence. The analysis results confirm the convergence characteristics of GSA according to the given parameters ranges of the algorithm. The best fitness function convergences after 35 iterations for 100 numbers of PHEVs.

From figure 6, it can be apparently seen that although the PSOGSA algorithm has been set to run for maximum 100 iterations, the fitness value converges before 10 iterations and be-come stable. So, PSOGSA takes less iterations to converge than GSA method due to the weak local search ability. Moreover, there is an early convergence which may cause the fitness function to trap into local minima. This can be avoided by increasing the size of swarm hence the computational time will also be in-creased as well. As a result, a trade-off should be taken into consideration between the proper convergence and computational time.

PSOGSA with the parameter settings stated in Table 3 was also performed for the same fitness function and compared with the performance of gravitational search algorithm in terms of average best fitness. The swarm size and maximum iterations was set exactly same to that of GSA technique for the comparison purpose. The values of parameters c1, c2 and alpha were set as standard values, 0.5, 1.5 and 23 respectively.

Figure 5. Iteration vs. fitness value, J (k) for GSA [100 PHEVs]

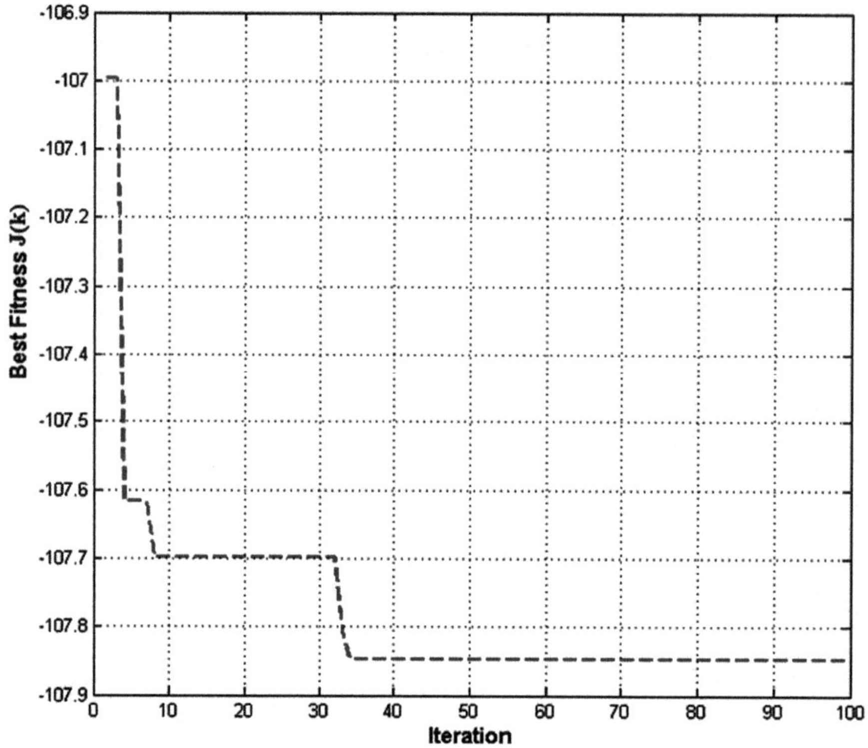

Figure 6. Iteration vs. fitness value, J(k) for PSOGSA [100 PHEVs]

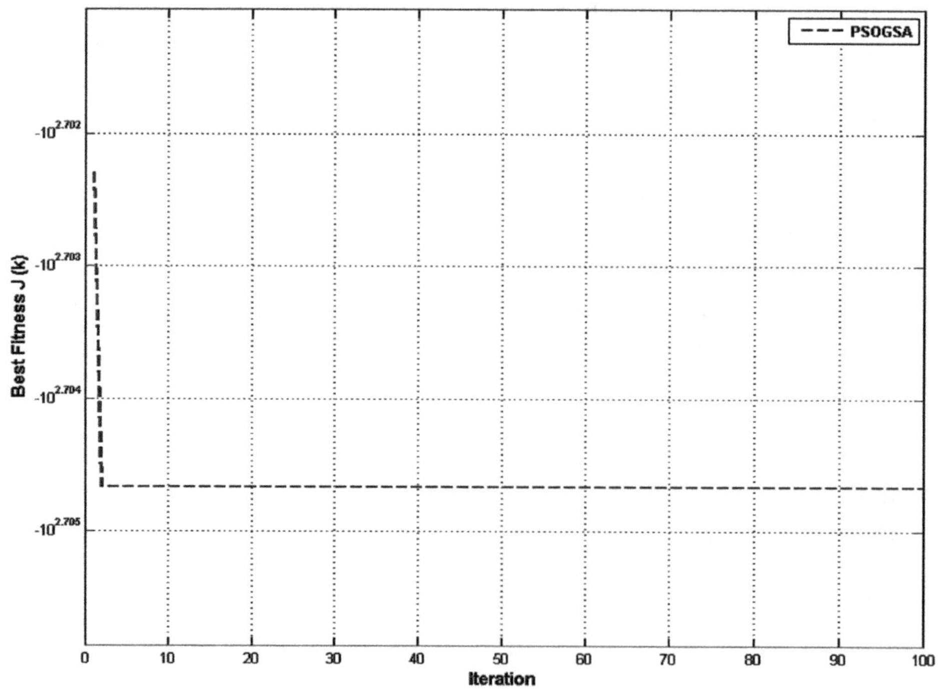

Table 3. Parameter settings of PSOGSA

Parameters	Values
Size of the swarm	100
Maximum iteration	100
PSO parameter, C1	0.5
PSO parameter, C2	1.5
Gravitational Constant, G_o	1
GSA Constant parameter, α	23
Number of runs	50

Comparison between PSOGSA and GSA

Table 4 summarizes the comparisons of GSA and PSO with PSOGSA algorithm in terms of average best fitness. Here, the average best fitness gives different values with the increment of PHEVs population. The convergence rate of mass agents in GSA is good through the fast information flowing among mass agents, so its diversity decreases very quickly in the successive iterations and lead to a suboptimal solution. In the case of PSOGSA, the algorithm cannot make full use of the feedback information in the system. There is also possibility of this algorithm to trap in the local optimal solution and lacks the searching capabilities within the whole search area.

Table 5 illustrates the advantages and disadvantages of both GSA and PSOGSA for solving different optimization problems. Energy scheduling at a PHEV charging station is subjected to different constraints that limit the search space to a certain feasible region. PSOGSA can easily handle the constraints separately, eliminating the need for additional parameters (Su & Chow, 2012). PSOGSA method is good for multi-objective optimization while GSA takes slightly more computational time with parameters tuning. The performance of both algorithms varies with the applications and different fitness functions.

Table 4. Average best fitness comparison between PSOGSA and GSA

Average Best Fitness for	PSOGSA	GSA	PSO (Rahman et al., 2014)
50 PHEVs	184.36	158.83	142.839
100 PHEVs	188.67	182.31	171.102
500 PHEVs	186.71	152.36	150.869
1000 PHEVs	185.82	161.52	156.802

Table 5. Advantages and disadvantages of PSOGSA and GSA

Optimization Method	Advantages	Disadvantages
PSOGSA	Easy constraint Good convergence rate	More parameters tuning Computational time
GSA	strong exploration ability among all evolutionary algorithms Local exploitation capability	Slow convergence rate More parameters tuning

Figure 7. Average best fitness vs. No. of PHEVs

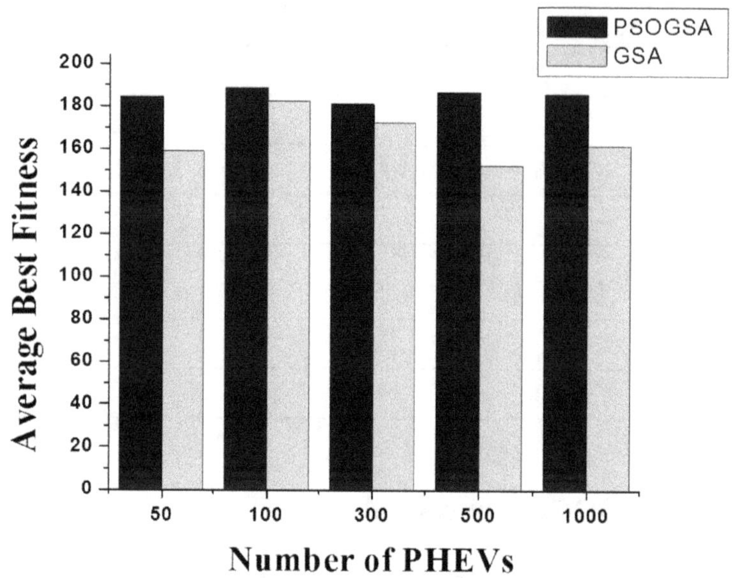

The average best fitness of both algorithms are represented with respect to number of vehicles (PHEVs) in Figure 7. From the figure it is clear that, Hybrid Particle Swarm Optimization and Gravitational Search Algorithm (PSOGSA) outperforms Gravitational Search Algorithm (GSA) in terms of Average best fitness. Here, the average best fitness gives different values with the increment of PHEVs population. The rate of convergence of mass agents in PSOGSA is good through the fast information flowing among mass agents, so its diversity decreases very quickly in the successive iterations and lead to a suboptimal solution. Starting from 50 numbers of PHEVs up to 1000 PHEVs, PSOGSA shows better fitness value than GSA.

Table 6 shows the computational time requirement for PSOGSA, GSA and PSO method. As the number of PHEVs increased from 100 to 500 and then 1000, standard PSO as well as GSA technique show better result than PSOGSA method in terms of computational time.

So, it can be concluded that, GSA obtains better result in terms of computational time while PSOGSA performs well for achieving the best fitness values compared to GSA.

Table 6. Computational time for PSOGSA and GSA

Computational Time (sec.)	PSOGSA	GSA	PSO (Rahman et al., 2014)
50 PHEVs	4.29	2.271	1.650
100PHEVs	7.90	4.439	1.686
500 PHEVs	36.82	18.165	1.990
1000 PHEVs	72.41	36.275	2.398

FUTURE RESEARCH DIRECTIONS

This paragraph summarizes the review results and suggests future directions of optimization techniques and procedures. The specific research field is relatively new and possible future perspectives have to be emphasized, so that new techniques can be realized.

Optimization Techniques

Possible characteristics of the future optimization tools are given below:

- The future optimization tools should be capable of performing parallel processing evaluations on the same computer by using modern multi-core processor technology or to distribute the calculations to a cluster of computers. Such ability will substantially improve the simulation runtime.
- Emerging mathematical methods like evolutionary algorithms, direct search methods, local search and other heuristic methods should be introduced in order to avoid the calculation of function derivatives. The experienced researcher should be able to choose the appropriate algorithm depending on the problem. Multi-objective capability should also be provided for multi-criteria optimization problems. Different swarm intelligence-based optimization methods like Ant Colony Optimization (ACO), Artificial Bee Colony (ABC) should also be explored by the researchers to solve problems related to PHEV charging.
- The future optimization tools should have the capability of stable convergence and thus provides good solution to the desired fitness functions.
- Exploration and exploitation of the search space is crucial in order to get desired solution within acceptable computation time.
- Advanced controlling mechanisms are necessary for allocating sufficient energy to a particular charging station in order to facilitate large-scale PHEV penetration in upcoming years.
- Finally, optimization of charging station needs proper selection of available resources as well as efficient available technique implementation.

Demand Side Management

Demand Side Management (DSM) is defined by Department of Energy (DOE) [69] as "Changes in electric usage by end-use customers from their normal consumption patterns in response to changes in the price of electricity over time, or incentive payments designed to induce lower electricity use at times of high wholesale market prices or when system reliability is jeopardized." Therefore, the demand side management programs should be incorporated into the existing Intelligent Energy Management System (iEMS) model in order to avoid voltage sag and blackout and to maximize the financial benefits. In addition, this under-utilized capacity could effectively power a national fleet of PHEVs with little need to increase the energy delivery capacity of the existing grid infrastructure (Gerkensmeyer, Kintner-Meyer, & DeSteese, 2010).

Trade-Off between Cost and Performance

Observing the effects of cost and performance on the marketability of PHEVs, the fitness function is defined to minimize drivetrain cost and driving performance requirements are selected as constraints to ensure that the vehicle performance is not sacrificed during the optimization. The battery is the key component within an electric vehicle (EV) which determines its overall capital cost and performance. Therefore, the task of determining the cost effectiveness of EVs is predominantly one of identifying the future trajectory of battery cost and performance. To meet power requirements: batteries have lower discharge power at low SoC and lower charge power at high SoC. To reduce safety risks, limiting the maximum SoC avoids overcharge situations.

CONCLUSION

Researchers are trying to design efficient controller for charging station and several literatures on optimization-based methods were published in this wake. These vehicles will help the government in its role of promoting energy security and environmental protection, when successfully marketed to consumers. Efforts are also to be taken for provision of affordable and accessible infrastructure for recharging. Hence, thrust in research and development on the aforementioned design considerations and technological challenges coupled with government support in terms of incentives to the automobile owners and to the manufacturers will go a long way in accelerating the deployment of large-scale PHEVs. In the future, more fitness functions (such as minimizing the overall charging time, etc.) should be considered in order to satisfy both client interests and the requirements of the power grid. However, conflicts can arise when multiple fitness functions are applied. The easiest solution to this problem involves combining all of the objectives into a single function. In this case, the weights assigned to each can be fixed or dynamically changed during the optimization process.

In this chapter, Hybrid Particle swarm and Gravitational search algorithm (PSOGSA)-based optimization were performed in order to optimally allocate power to each of the PHEVs entering into the charging station. A sophisticated controller will need to be designed in order to allocate power to PHEVs appropriately. For this wake, the applied algorithm is a step towards real-life implementation of such controller for PHEV charging stations. Here, four (04) different numbers of PHEVs were considered for MATLAB Simulation and then obtained results were compared with GSA in terms of average best fitness and computational time. The success of the electrification of transportation sector solely depends on charging infrastructure. Only proper charging control and infrastructure management can assure the larger penetration of PHEVs. The researchers should try to develop efficient control mechanism for charging infrastructure in order to facilitate upcoming PHEVs in highways. In future, more vehicles should be considered for intelligent power allocation strategy as well as other hybrid versions of computational techniques should be applied to ensure higher fitness value and low computational time.

ACKNOWLEDGMENT

The authors would like to thank Universiti Teknologi PETRONAS (UTP) for supporting the research under UTP Graduate Assistantship (GA) scheme.

REFERENCES

Aggeler, D., Canales, F., Coccia, A., Butcher, N., & Apeldoorn, O. (2010). Ultra-fast DC-charge infrastructures for EV-mobility and future smart grids. In Innovative Smart Grid Technologies Conference Europe (ISGT Europe), 2010 IEEE PES. doi:10.1109/ISGTEUROPE.2010.5638899

Anegawa, T. (2009). Desirable characteristics of public quick charger. In *Proceedings of PHEV09*.

Association of Bay Area Governments. (2010). *Bay area electrified vehicle charging infrastructure: Options for accelerating consumer*. Bay Area, CA: Renewable and Appropriate Energy Laboratory.

Barr, R. S., Golden, B. L., Kelly, J. P., Resende, M. G., & Stewart, W. R. Jr. (1995). Designing and reporting on computational experiments with heuristic methods. *Journal of Heuristics*, *1*(1), 9–32. doi:10.1007/BF02430363

Birnie, D. P. III. (2009). Solar-to-vehicle (S2V) systems for powering commuters of the future. *Journal of Power Sources*, *186*(2), 539–542. doi:10.1016/j.jpowsour.2008.09.118

Botsford, C., & Szczepanek, A. (2009). Fast charging vs. slow charging: Pros and cons for the new age of electric vehicles. In Proceedings of EVS24.

Boyle, G. (Ed.). (2007). *Renewable electricity and the grid: the challenge of variability*. Earthscan.

Caramanis, M., & Foster, J. M. (2009). Management of electric vehicle charging to mitigate renewable generation intermittency and distribution network congestion. In *Proceedings of the 48th IEEE Conference on Decision and Control*. Shanghai: IEEE. doi:10.1109/CDC.2009.5399955

Chang, W. Y. (2013). The state of charge estimating methods for battery: a review. *ISRN Applied Mathematics*.

Chiasson, J., & Vairamohan, B. (2005). Estimating the state of charge of a battery. *IEEE Transactions on Control Systems Technology*, *13*(3), 465–470. doi:10.1109/TCST.2004.839571

De Sousa, L., Silvestre, B., & Bouchez, B. (2010). A combined multiphase electric drive and fast battery charger for electric vehicles. In *Vehicle Power and Propulsion Conference (VPPC)*. Lille, France: IEEE. doi:10.1109/VPPC.2010.5729057

del Valle, Y., Venayagamoorthy, G. K., Mohagheghi, S., Hernandez, J.-C., & Harley, R. G. (2008). Particle swarm optimization: basic concepts, variants and applications in power systems. *Evolutionary Computation, IEEE Transactions on*, *12*(2), 171-195.

Denholm, P., Kuss, M., & Margolis, R. M. (2013). Co-benefits of large scale plug-in hybrid electric vehicle and solar PV deployment. *Journal of Power Sources*, *236*, 350–356. doi:10.1016/j.jpowsour.2012.10.007

Dubey, H. M., Pandit, M., Panigrahi, B. K., & Udgir, M. (2013). Economic Load Dispatch by Hybrid Swarm Intelligence Based Gravitational Search Algorithm. *International Journal of Intelligent Systems and Applications*, *5*(8), 21–32. doi:10.5815/ijisa.2013.08.03

Dusmez, S., Cook, A., & Khaligh, A. (2011). Comprehensive analysis of high quality power converters for level 3 off-board chargers. In *Vehicle Power and Propulsion Conference (VPPC)*. Chicago: IEEE. doi:10.1109/VPPC.2011.6043096

Eberhart, R. C., & Shi, Y. (2001). Particle swarm optimization: developments, applications and resources. In *Proceedings of the 2001 Congress on Evolutionary Computation*. Seoul, South Korea: IEEE. doi:10.1109/CEC.2001.934374

Gan, L., Topcu, U., & Low, S. (2011). Optimal decentralized protocol for electric vehicle charging. In *Decision and Control and European Control Conference (CDC-ECC)*. IEEE. doi:10.1109/CDC.2011.6161220

Ganesan, T., Vasant, P., & Elamvazuthy, I. (2012). A hybrid PSO approach for solving non-convex optimization problems. *Archives of Control Sciences*, *22*(1), 87–105. doi:10.2478/v10170-011-0014-2

Gerkensmeyer, C., Kintner-Meyer, M. C., & DeSteese, J. G. (2010). *Technical challenges of plug-in hybrid electric vehicles and impacts to the US power system: Distribution system analysis*. United States Department of Energy. doi:10.2172/974954

Hess, A., Malandrino, F., Reinhardt, M. B., Casetti, C., Hummel, K. A., & Barceló-Ordinas, J. M. (2012). Optimal deployment of charging stations for electric vehicular networks. In *Proceedings of the first workshop on urban networking*. doi:10.1145/2413236.2413238

Holtz-Eakin, D., & Selden, T. M. (1995). Stoking the fires? CO_2 emissions and economic growth. *Journal of Public Economics*, *57*(1), 85–101. doi:10.1016/0047-2727(94)01449-X

Hota, A. R., Juvvanapudi, M., & Bajpai, P. (2014). Issues and solution approaches in PHEV integration to the smart grid. *Renewable & Sustainable Energy Reviews*, *30*(0), 217–229. doi:10.1016/j.rser.2013.10.008

Ingersoll, J. G., & Perkins, C. A. (1996). The 2.1 kW photovoltaic electric vehicle charging station in the city of Santa Monica, California. In *Photovoltaic Specialists Conference, Conference Record of the Twenty Fifth IEEE*. doi:10.1109/PVSC.1996.564423

Jia-zhao, C., Yu-xiang, Z., & Yin-sheng, L. (2012). A Unified Frame of Swarm Intelligence Optimization Algorithm. *Knowledge Discovery and Data Mining*, 745-751.

Kefayati, M., & Caramanis, C. (2010). Efficient energy delivery management for PHEVs. In *Smart Grid Communications (SmartGridComm), 2010 First IEEE International Conference on*. doi:10.1109/SMARTGRID.2010.5621990

Kulshrestha, P., Wang, L., Chow, M. Y., & Lukic, S. (2009). Intelligent energy management system simulator for PHEVs at municipal parking deck in a smart grid environment. In Power & Energy Society General Meeting, 2009. PES'09. IEEE (pp. 1-6). IEEE. doi:10.1109/PES.2009.5275688

Kunche, P., Rao, G. S. B., Reddy, K., & Maheswari, R. U. (2014). A new approach to dual channel speech enhancement based on hybrid PSOGSA. *International Journal of Speech Technology*, 1–12.

Letendre, S. (2009). Solar Electricity as a Fuel for Light Vehicles. In *Proceedings of the 2006 American Solar Energy Society Annual Conference*.

Li, Z., Sahinoglu, Z., Tao, Z., & Teo, K. H. (2010). Electric vehicles network with nomadic portable charging stations. In Vehicular Technology Conference Fall (VTC 2010-Fall), 2010 IEEE 72nd (pp. 1-5). IEEE. doi:10.1109/VETECF.2010.5594437

Lund, H., & Kempton, W. (2008). Integration of renewable energy into the transport and electricity sectors through V2G. *Energy Policy*, *36*(9), 3578–3587. doi:10.1016/j.enpol.2008.06.007

Ma, Z., Callaway, D., & Hiskens, I. (2010). Decentralized charging control for large populations of plug-in electric vehicles. In *Decision and Control (CDC), 2010 49th IEEE Conference on* (pp. 206-212). IEEE.

Mallick, S., Ghoshal, S. P., Acharjee, P., & Thakur, S. S. (2013). Optimal static state estimation using improved particle swarm optimization and gravitational search algorithm. *International Journal of Electrical Power & Energy Systems*, *52*(0), 254–265. doi:10.1016/j.ijepes.2013.03.035

Mayfield, D. (2012). *Site Design For Electric Vehicle Charging Stations, ver. 1.0*. Sustainable Transportation Strategies.

Mirjalili, S., Mohd Hashim, S. Z., & Moradian Sardroudi, H. (2012). Training feedforward neural networks using hybrid particle swarm optimization and gravitational search algorithm. *Applied Mathematics and Computation*, *218*(22), 11125–11137. doi:10.1016/j.amc.2012.04.069

Mitra, P., & Venayagamoorthy, G. K. (2010). Intelligent coordinated control of a wind farm and distributed smartparks. In *Industry Applications Society Annual Meeting (IAS), 2010 IEEE* (pp. 1-8). IEEE. doi:10.1109/IAS.2010.5615930

Morrow, K., Karner, D., & Francfort, J. (2008). *Plug-in hybrid electric vehicle charging infrastructure review*. US Department of Energy-Vehicle Technologies Program.

Neumann, H. M., Schär, D., & Baumgartner, F. (2012). The potential of photovoltaic carports to cover the energy demand of road passenger transport. *Progress in Photovoltaics: Research and Applications*, *20*(6), 639–649. doi:10.1002/pip.1199

Pang, C., Dutta, P., Kim, S., Kezunovic, M., & Damnjanovic, I. (2010). PHEVs as dynamically configurable dispersed energy storage for V2B uses in the smart grid. *IET Conference Proceedings*. doi:10.1049/cp.2010.0903

Piller, S., Perrin, M., & Jossen, A. (2001). Methods for state-of-charge determination and their applications. *Journal of Power Sources*, *96*(1), 113–120. doi:10.1016/S0378-7753(01)00560-2

Rahman, I., Vasant, P., Singh, B., & Abdullah-Al-Wadud, M. (2015). Swarm intelligence based State-of-Charge optimization for charging Plug-in Hybrid Electric Vehicles. *Energy and Sustainability V. Special Contributions*, *206*, 261.

Rahman, I., Vasant, P. M., Mahinder Singh, B. S., & Abdullah-Al-Wadud, M. (2014). Swarm Intelligence-Based Smart Energy Allocation Strategy for Charging Stations of Plug-In Hybrid Electric Vehicles. *Mathematical Problems in Engineering*, *2014*, 620425.

Rahman, I., Vasant, P. M., Singh, B. S. M., & Abdullah-Al-Wadud, M. (2014). Intelligent energy allocation strategy for PHEV charging station using gravitational search algorithm. *AIP Conference Proceedings*, *1621*, 52–59. doi:10.1063/1.4898445

Rahman, I., Vasant, P. M., Singh, B. S. M., & Abdullah-Al-Wadud, M. (2014). Optimisation of PHEV/EV charging infrastructures: A review. *Int. J. Energy Technology and Policy*, *10*(3/4), 280–296. doi:10.1504/IJETP.2014.066878

Rashedi, E., Nezamabadi-Pour, H., & Saryazdi, S. (2009). GSA: A gravitational search algorithm. *Information Sciences*, *179*(13), 2232–2248. doi:10.1016/j.ins.2009.03.004

Rawson, M., & Kateley, S. (1999). *Electric vehicle charging equipment design and health and safety codes* (No. 1999-01-2941). SAE Technical Paper.

Sabri, N. M., Puteh, M., & Mahmood, M. R. (2013). A Review of Gravitational Search Algorithm. *Int. J. Advance. Soft Comput. Appl*, *5*(3).

Shafiei, A., & Williamson, S. S. (2010). Plug-in hybrid electric vehicle charging: Current issues and future challenges. In *Vehicle Power and Propulsion Conference (VPPC)* (pp. 1-8). IEEE. doi:10.1109/VPPC.2010.5729134

Soares, J., Sousa, T., Morais, H., Vale, Z., Canizes, B., & Silva, A. (2013). Application Specific Modified Particle Swarm Optimization for Energy Resource Scheduling Considering Vehicle-to-Grid. *Applied Soft Computing*, *13*(11), 4264–4280. doi:10.1016/j.asoc.2013.07.003

Sojoudi, S., & Low, S. H. (2011). Optimal charging of plug-in hybrid electric vehicles in smart grids. In *Power and Energy Society General Meeting, 2011 IEEE*. doi:10.1109/PES.2011.6039236

Su, W., & Chow, M. Y. (2010). An intelligent energy management system for PHEVs considering demand response. In *Proc. 2010 FREEDM Annual Conference*.

Su, W., & Chow, M. Y. (2011). Performance evaluation of a PHEV parking station using particle swarm optimization. In *Power and Energy Society General Meeting, 2011 IEEE* (pp. 1-6). IEEE. doi:10.1109/PES.2011.6038937

Su, W., & Chow, M.-Y. (2012a). Performance evaluation of an EDA-based large-scale plug-in hybrid electric vehicle charging algorithm. *Smart Grid. IEEE Transactions on*, *3*(1), 308–315.

Su, W., & Chow, M.-Y. (2012b). Computational intelligence-based energy management for a large-scale PHEV/PEV enabled municipal parking deck. *Applied Energy*, *96*, 171–182. doi:10.1016/j.apenergy.2011.11.088

Subramanian, A., Garcia, M., Dominguez-Garcia, A., Callaway, D., Poolla, K., & Varaiya, P. (2012, June). Real-time scheduling of deferrable electric loads. In *American Control Conference (ACC)*, (pp. 3643-3650). IEEE.

Tan, W. S., Hassan, M. Y., Rahman, H. A., Abdullah, M. P., & Hussin, F. (2013). Multi-distributed generation planning using hybrid particle swarm optimisation-gravitational search algorithm including voltage rise issue. *IET Generation. Transmission & Distribution*, *7*(9), 929–942. doi:10.1049/iet-gtd.2013.0050

Tesla Motors. (2009). *Tesla roadster spec sheet 2009*. Author.

Tie, S. F., & Tan, C. W. (2013). A review of energy sources and energy management system in electric vehicles. *Renewable & Sustainable Energy Reviews*, *20*, 82–102. doi:10.1016/j.rser.2012.11.077

Tulpule, P. J., Marano, V., Yurkovich, S., & Rizzoni, G. (2013). Economic and environmental impacts of a PV powered workplace parking garage charging station. *Applied Energy*, *108*, 323–332. doi:10.1016/j.apenergy.2013.02.068

Vasant, P., Ganesan, T., & Elamvazuthi, I. (2012). An improved PSO approach for solving non-convex optimization problems. In *ICT and Knowledge Engineering (ICT & Knowledge Engineering), 2011 9th International Conference on* (pp. 80-87). IEEE. doi:10.1109/ICTKE.2012.6152418

Venayagamoorthy, G. K., & Mitra, P. (2011). SmartPark shock absorbers for wind farms. *Energy Conversion. IEEE Transactions on*, *26*(3), 990–992.

Yilmaz, M., & Krein, P. T. (2012). Review of charging power levels and infrastructure for plug-in electric and hybrid vehicles. In *Electric Vehicle Conference (IEVC), 2012 IEEE International*. doi:10.1109/IEVC.2012.6183208

Zhang, Q., Tezuka, T., Ishihara, K. N., & Mclellan, B. C. (2012). Integration of PV power into future low-carbon smart electricity systems with EV and HP in Kansai Area, Japan. *Renewable Energy*, *44*, 99–108. doi:10.1016/j.renene.2012.01.003

Zhu, L., Yu, F. R., Ning, B., & Tang, T. (2012). Optimal charging control for electric vehicles in smart microgrids with renewable energy sources. In *Vehicular Technology Conference (VTC Spring), 2012 IEEE 75th* (pp. 1-5). IEEE. doi:10.1109/VETECS.2012.6240169

ADDITIONAL READING

Abraham, A., Guo, H., & Liu, H. (2006). Swarm intelligence: foundations, perspectives and applications. In *Swarm Intelligent Systems* (pp. 3–25). Springer Berlin Heidelberg. doi:10.1007/978-3-540-33869-7_1

Arumuggam, K., & Singh, B. S. M. (2013). Optimization of hybrid solar and wind power generation. *Journal of Applied Sciences*, *13*(6), 869–875. doi:10.3923/jas.2013.869.875

Axsen, J., & Kurani, K. S. (2013). Hybrid, plug-in hybrid, or electric—What do car buyers want? *Energy Policy*, *61*, 532–543. doi:10.1016/j.enpol.2013.05.122

Bauer, P., Zhou, Y., Doppler, J., & Stembridge, N. (2010). Charging of electric vehicles and impact on the grid. In *MECHATRONIKA, 13th International Symposium, Teplice* (pp. 121-127).

Bayram, I. S., Michailidis, G., Devetsikiotis, M., Granelli, F., & Bhattacharya, S. (2012). *Smart Vehicles in the Smart Grid: Challenges, Trends, and Application to the Design of Charging Stations Control and Optimization Methods for Electric Smart Grids* (pp. 133–145). Springer. doi:10.1007/978-1-4614-1605-0_6

Boschert, S. (20 Talatahari, S., Khalili, E., & Alavizadeh, S. (2013). Accelerated particle swarm for optimum design of frame structures. *Mathematical Problems in Engineering*, *2013*. 06). Plug-in hybrids: The cars that will recharge America: New Society Publishers.

Boulanger, A. G., Chu, A. C., Maxx, S., & Waltz, D. L. (2011). Vehicle electrification: Status and issues. *Proceedings of the IEEE*, *99*(6), 1116–1138. doi:10.1109/JPROC.2011.2112750

Ceylan, O., Ozdemir, A., & Dag, H. (2010, August). Gravitational search algorithm for post-outage bus voltage magnitude calculations. In *Universities Power Engineering Conference (UPEC), 2010 45th International* (pp. 1-6). IEEE.

Ceylan, O., Ozdemir, A., & Dag, H. (2012). Branch Outage Simulation Based Contingency Screening by Gravitational Search Algorithm. *International Review of Electrical Engineering*, *7*(1).

Chan, C. C. (2002). The state of the art of electric and hybrid vehicles. *Proceedings of the IEEE*, *90*(2), 247–275. doi:10.1109/5.989873

Contestabile, M., Offer, G. J., Slade, R., Jaeger, F., & Thoennes, M. (2011). Battery electric vehicles, hydrogen fuel cells and biofuels. Which will be the winner? *Energy & Environmental Science*, *4*(10), 3754–3772. doi:10.1039/c1ee01804c

Di Silvestre, M. L., Sanseverino, E. R., Zizzo, G., & Graditi, G. (2013). An optimization approach for efficient management of EV parking lots with batteries recharging facilities. *Journal of Ambient Intelligence and Humanized Computing*, *4*(6), 641–649. doi:10.1007/s12652-013-0174-y

Duman, S., Güvenç, U., Sönmez, Y., & Yörükeren, N. (2012). Optimal power flow using gravitational search algorithm. *Energy Conversion and Management*, *59*, 86–95. doi:10.1016/j.enconman.2012.02.024

Duman, S., Güvenç, U., & Yörükeren, N. (2010). Gravitational search algorithm for economic dispatch with valve-point effects. *International Review of Electrical Engineering*, *5*(6), 2890–2895.

Ehsani, M., Gao, Y., & Emadi, A. (2009). *Modern electric, hybrid electric, and fuel cell vehicles: fundamentals, theory, and design*. CRC press.

Eiben, A. E., & Smit, S. K. (2011). Parameter tuning for configuring and analyzing evolutionary algorithms. *Swarm and Evolutionary Computation*, *1*(1), 19–31. doi:10.1016/j.swevo.2011.02.001

Elgammal, A., & Sharaf, A. (2012). Self-regulating particle swarm optimised controller for (photovoltaic-fuel cell) battery charging of hybrid electric vehicles. *Electrical Systems in Transportation, IET*, *2*(2), 77–89. doi:10.1049/iet-est.2011.0021

Erickson, K., Schulke, A., Bodet, C., & Jabłonowski, R. (2012). Intelligent dynamic charging assignment for increasing efficiency in EV driving. In *Proc. 11tg IASTED European Conference on Power and Energy Systems*. doi:10.2316/P.2012.775-032

Fornarelli, G., & Mescia, L. (2013). *Swarm intelligence for electric and electronic engineering*. Engineering Science Reference. doi:10.4018/978-1-4666-2666-9

Ganesan, T., Elamvazuthi, I., Ku Shaari, K. Z., & Vasant, P. (2013). Swarm intelligence and gravitational search algorithm for multi-objective optimization of synthesis gas production. *Applied Energy*, *103*(0), 368–374. doi:10.1016/j.apenergy.2012.09.059

Ganesan, T., Vasant, P., & Elamvazuthi, I. (2014). Hopfield neural networks approach for design optimization of hybrid power systems with multiple renewable energy sources in a fuzzy environment. *Journal of Intelligent and Fuzzy Systems*, *26*(5), 2143–2154.

Hamadicharef, B. (2011). Bibliometric analysis of particle swarm optimization (PSO) research 2000-2010. In Artificial Intelligence and Computational Intelligence pp. 404-411.

Hannan, M. A., Azidin, F. A., & Mohamed, A. (2014). Hybrid electric vehicles and their challenges: A review. *Renewable & Sustainable Energy Reviews*, *29*(0), 135–150. doi:10.1016/j.rser.2013.08.097

Hendtlass, T. (2007, September). Fitness estimation and the particle swarm optimisation algorithm. In *Evolutionary Computation, 2007. CEC 2007. IEEE Congress on* (pp. 4266-4272). IEEE. doi:10.1109/CEC.2007.4425028

Inoa, E., Guo, F., Wang, J., & Choi, W. (2011). A full study of a PHEV charging facility based on global optimization and real-time simulation. In *Power Electronics and ECCE Asia (ICPE & ECCE), 2011 IEEE 8th International Conference on* (pp. 565-570). IEEE. doi:10.1109/ICPE.2011.5944611

Jiang, S., Ji, Z., & Shen, Y. (2014). A novel hybrid particle swarm optimization and gravitational search algorithm for solving economic emission load dispatch problems with various practical constraints. *International Journal of Electrical Power & Energy Systems*, *55*, 628–644. doi:10.1016/j.ijepes.2013.10.006

Karbowski, D., Rousseau, A., Pagerit, S., & Sharer, P. (2006, October). Plug-in vehicle control strategy: from global optimization to real time application. In *22nd Electric Vehicle Symposium, EVS22, Yokohama, Japan.*

Khajehzadeh, M., Taha, M. R., El-Shafie, A., & Eslami, M. (2012). A modified gravitational search algorithm for slope stability analysis. *Engineering Applications of Artificial Intelligence*, *25*(8), 1589–1597. doi:10.1016/j.engappai.2012.01.011

Krause, J., Cordeiro, J., Parpinelli, R. S., & Lopes, H. S. (2013). A survey of swarm algorithms applied to discrete optimization problems. *Swarm Intelligence and Bio-inspired Computation: Theory and Applications. Elsevier Science & Technology Books*, 169-191.

Markel, T. (2010). *Plug-in electric vehicle infrastructure: A foundation for electrified transportation.* National Renewable Energy Laboratory, US Department of Energy.

Martens, D., Baesens, B., & Fawcett, T. (2011). Editorial survey: Swarm intelligence for data mining. *Machine Learning*, *82*(1), 1–42. doi:10.1007/s10994-010-5216-5

Mirjalili, S., Mohd Hashim, S. Z., & Moradian Sardroudi, H. (2012). Training feedforward neural networks using hybrid particle swarm optimization and gravitational search algorithm. *Applied Mathematics and Computation*, *218*(22), 11125–11137. doi:10.1016/j.amc.2012.04.069

Mohamed, A. Z., Lee, S. H., Hsu, H. Y., & Nath, N. (2012). A faster path planner using accelerated particle swarm optimization. *Artificial Life and Robotics*, *17*(2), 233–240. doi:10.1007/s10015-012-0051-3

Mullan, J., Harries, D., Bräunl, T., & Whitely, S. (2012). The technical, economic and commercial viability of the vehicle-to-grid concept. *Energy Policy*, *48*(0), 394–406. doi:10.1016/j.enpol.2012.05.042

Pan, F., Bent, R., Berscheid, A., & Izraelevitz, D. (2010, October). Locating PHEV exchange stations in V2G. In *Smart Grid Communications (SmartGridComm), 2010 First IEEE International Conference on* (pp. 173-178). IEEE. doi:10.1109/SMARTGRID.2010.5622037

Panigrahi, B. K., Shi, Y., & Lim, M. H. (2011). *Handbook of Swarm Intelligence*. Berlin: Springer. doi:10.1007/978-3-642-17390-5

Preetham, G., & Shireen, W. (2012). Photovoltaic charging station for Plug-In Hybrid Electric Vehicles in a smart grid environment. In Innovative Smart Grid Technologies (ISGT), 2012 IEEE PES (pp. 1-8). IEEE. doi:10.1109/ISGT.2012.6175589

Quinn, C., Zimmerle, D., & Bradley, T. H. (2010). The effect of communication architecture on the availability, reliability, and economics of plug-in hybrid electric vehicle-to-grid ancillary services. *Journal of Power Sources*, *195*(5), 1500–1509. doi:10.1016/j.jpowsour.2009.08.075

Rahman, I., Vasant, P., Singh, B., & Abdullah-Al-Wadud, M. (2015). Hybrid Swarm Intelligence-Based Optimization for Charging Plug-in Hybrid Electric Vehicle. [Springer International Publishing.]. *Intelligent Information and Database Systems*, *9012*, 22–30.

Ramachandran, B., Srivastava, S. K., & Cartes, D. A. (2013). Intelligent power management in micro grids with EV penetration. *Expert Systems with Applications*, *40*(16), 6631–6640. doi:10.1016/j.eswa.2013.06.007

Razzaque, M. A., Hong, C. S., Abdullah-Al-Wadud, M., & Chae, O. (2008). A fast algorithm to calculate powers of a Boolean matrix for diameter computation of random graphs. In WALCOM: Algorithms and Computation (pp. 58-69). Springer Berlin Heidelberg. doi:10.1007/978-3-540-77891-2_6

Sabri, N. M., Puteh, M., & Mahmood, M. R. (2013). A Review of Gravitational Search Algorithm. *Int. J. Advance. Soft Comput. Appl*, *5*(3).

Sadrnia, A., Nezamabadi-Pour, H., Nikbakht, M., & Ismail, N. (2013). A Gravitational Search Algorithm Approach for Optimizing Closed-Loop Logistics Network. In P. Vasant (Ed.), *Meta-Heuristics Optimization Algorithms in Engineering, Business, Economics, and Finance* (pp. 616–638). Hershey, PA: Information Science Reference. doi:10.4018/978-1-4666-2086-5.ch020

Sandy Thomas, C. (2009). Transportation options in a carbon-constrained world: Hybrids, plug-in hybrids, biofuels, fuel cell electric vehicles, and battery electric vehicles. *International Journal of Hydrogen Energy*, *34*(23), 9279–9296. doi:10.1016/j.ijhydene.2009.09.058

Sarafrazi, S., Nezamabadi-Pour, H., & Saryazdi, S. (2011). Disruption: A new operator in gravitational search algorithm. *Scientia Iranica*, *18*(3), 539–548. doi:10.1016/j.scient.2011.04.003

Singh, J. D. (2012). Nonconvex Economic Load Dispatch Problem with Dynamic Constraint through Gravitational Search Algorithm. *Artificial Intelligent Systems and Machine Learning*, *4*(8), 494–501.

Strnad, I., Skrlec, D., & Tomisa, T. (2013). A model for the efficient use of electricity produced from renewable energy sources for electric vehicle charging. In *Energy (IYCE), 2013 4th International Youth Conference on* (pp. 1-8). IEEE. doi:10.1109/IYCE.2013.6604143

Tate, E. D., Harpster, M. O., & Savagian, P. J. (2008). *The electrification of the automobile: from conventional hybrid, to plug-in hybrids, to extended-range electric vehicles* (No. 2008-01-0458). SAE Technical Paper.

Tong, S. J., Same, A., Kootstra, M. A., & Park, J. W. (2013). Off-grid photovoltaic vehicle charge using second life lithium batteries: An experimental and numerical investigation. *Applied Energy*, *104*, 740–750. doi:10.1016/j.apenergy.2012.11.046

Tulpule, P., Marano, V., & Rizzoni, G. (2009). Effects of different PHEV control strategies on vehicle performance. In *American Control Conference, 2009. ACC'09.* (pp. 3950-3955). IEEE. doi:10.1109/ACC.2009.5160595

Vasant, P. (2014). Hybrid Line Search and Simulated Annealing For Production Planning System in Industrial Engineering. [IJMMME]. *International Journal of Manufacturing, Materials, and Mechanical Engineering*, *4*(2), 1–13. doi:10.4018/ijmmme.2014040101

Vasant, P., Ganesan, T., & Elamvazuthi, I. (2012). An improved PSO approach for solving non-convex optimization problems. In *ICT and Knowledge Engineering (ICT & Knowledge Engineering), 2011 9th International Conference on* (pp. 80-87). IEEE. doi:10.1109/ICTKE.2012.6152418

Vasant, P. M. (2013). *Handbook of Research on Novel Soft Computing Intelligent Algorithms: Theory and Practical Applications*. IGI Global.

Waraich, R. A., Galus, M. D., Dobler, C., Balmer, M., Andersson, G., & Axhausen, K. W. (2013). Plug-in hybrid electric vehicles and smart grids: Investigations based on a microsimulation. *Transportation Research Part C, Emerging Technologies*, *28*, 74–86. doi:10.1016/j.trc.2012.10.011

Wirasingha, S. G., Schofield, N., & Emadi, A. (2008). Plug-in hybrid electric vehicle developments in the US: Trends, barriers, and economic feasibility. In *Vehicle Power and Propulsion Conference, 2008. IEEE* (pp. 1-8). IEEE. doi:10.1109/VPPC.2008.4677702

Yang, J., He, L., & Fu, S. (2014). An improved PSO-based charging strategy of electric vehicles in electrical distribution grid. *Applied Energy*, *128*(0), 82–92. doi:10.1016/j.apenergy.2014.04.047

Zeng, J. C., & Cui, Z. H. (2004). A Guaranteed Global Convergence Particle Swarm Optimizer [J]. *Journal of computer research and development*, *8*, 1333-1338.

Zhang, L., Brown, T., & Samuelsen, S. (2013). Evaluation of charging infrastructure requirements and operating costs for plug-in electric vehicles. *Journal of Power Sources*, *240*, 515–524. doi:10.1016/j.jpowsour.2013.04.048

Zhou, F. Q., Lian, Z. W., Wang, X. L., Yang, X. H., & Xu, Y. S. (2010). Discussion on operation mode to the electric vehicle charging station. *Power System Protection and Control*, *38*(21), 63–67.

KEY TERMS AND DEFINITIONS

All Electric Range: All electric range is a mode of electric vehicle when it is only run by charged batteries in order to reduce the overall fuel consumption. Calculation of all electric range varies according to the designs of the hybrid electric vehicles. The "all electric range" (AER) test quantifies the electric-only miles possible with the battery for a particular configuration and vehicle class.

Charging Station: Charging station is an important component for the healthy growth of the electric vehicle industry. Charging station refers to an infrastructure similar to petrol station (for conventional vehicle) that provides electric energy for the charging of plug-in hybrid electric vehicles (PHEVs). Many charging stations are on-street facilities provided by electric utility companies, mobile charging stations have been recently introduced. From the grid standpoint, a charging station is one way that the operator of an electrical power grid can adapt energy production to energy consumption, both of which can vary randomly over time. Basically, EVs in a charging station are charged during times when production exceeds consumption and are discharged at times when consumption exceeds production. In this way, electricity production need is not drastically scaled up and down to meet momentary consumption, which would increase efficiency and lower the cost of energy production and facilitate the use of intermittent energy sources, such as photovoltaic and wind.

Energy Security: The interest in energy security is based on the notion that an uninterrupted supply of energy is critical for the functioning of an economy. However, an exact definition of energy security is hard to give as it has different meanings to different people at different moments in time. It has traditionally been associated with the securing of access to oil supplies and with impending fossil fuel depletion. With an increase in natural gas use, security concerns also arose for natural gas, widening the concept to cover other fuels. Because oil is nowadays a globally traded commodity, physical shortages show up in the price of oil on the world market, in the form of a long-term increase and of short-term fluctuations.

Fitness Function: A necessary characteristic of evolutionary structural testing is that the fitness function is constructed on the basis of the software under test. The fitness function itself is not of interest for the problem; however, a well-constructed fitness function may substantially increase the chance of finding a solution and reaching higher coverage. Better guidance of the search can result in optimizations with less iterations, therefore leading to savings in resource expenditure.

Gravitational Search Algorithm: Gravitational Search Algorithm (GSA) is a heuristic optimization algorithm which has been gaining much interest among the scientific community recently. GSA is a nature inspired algorithm based on the Newton's famous law of gravity and the law of motion. GSA is classified under population-based method and is reported to be more instinctive. In GSA, the agent has four parameters which are position, inertial mass, active gravitational mass, and passive gravitational mass. GSA is a memory-less algorithm. However, it works efficiently like the algorithms with memory.

Particle Swarm Optimization: Particle Swarm Optimization (PSO) algorithm was introduced by Kennedy and Eberhart in 1995, which is a heuristic global optimization method and a member of swarm intelligence family. PSO is a computational intelligence-based technique that is not largely affected by the size and nonlinearity of the problem, and can converge to the optimal solution in many problems where most analytical methods fail to converge.

Plug-in Hybrid Electric Vehicles: Plug-in Hybrid Electric Vehicles (PHEVs) are being made with relatively large sized batteries that can be charged during off-peak hours, and permit the vehicle owner to use exclusively electric made for 30 – 60 miles of driving as well as switching into traditional gasoline for longer trips. PHEVs offer customers the opportunity for fuel at gasoline-equivalent prices of less than $1.00 per gallon. For a given size battery bank, the range of a PHEV can be prolonged significantly before batteries need recharging by turning on the engine or fuel cell whenever the vehicle power demand exceeds some threshold.

Smart Charging: Smart charging refers to the intelligent control of electric vehicle charging by the assigned authority. Smart charging can be both direct and indirect depending upon the user demand and available infrastructure. The main concept of smart charging lies in the charging of vehicle when the price and demand are lowest as well as excess amount of available capacity.

Smart Grid: Smart grid is an intelligent bi-directional electrical power system. It ensures most advanced and efficient communication network between suppliers and consumers of electricity. Unlike traditional power grid, smart grid offers better system sustainability and network security. The "smart grid" includes advanced utility Supervisory Control and Data Acquisition (SCADA) systems that can keep track of thousands of data points of loads and resources, smart meters that can communicate to the utility SCADA center, and smart appliances that can respond instantaneously to economic or reliability imperatives.

State-of-Charge: State-of-Charge (SoC) of a PHEV battery is expressed as the ratio of its capacity of current $\mathrm{Q}(t)$ to the nominal capacity Q_n. The nominal capacity is known by the vehicle manufacturer and shows the maximum amount of charge that can be stored in the battery. SoC estimation is a fundamental challenge for battery use. The SoC of a battery, which is used to describe its remaining capacity, is a very important parameter for a control strategy. The SoC can be defined as follows: $SoC = {}^{Q(t)}\!/_{Q_n}$.

Vehicle-to-Grid: In vehicle-to-grid (V2G) concept, an electric vehicle acts both as a load and power source in smart grid environment. A V2G-capable vehicle offers reactive power support, active power regulation, tracking of variable renewable energy sources, load balancing, and current harmonic filtering. These technologies can enable ancillary services, such as voltage and frequency control and spinning reserve. Success of the V2G concept depends on standardization of requirements and infrastructure decisions, battery technology, and efficient and smart scheduling of limited fast-charge infrastructure.

APPENDIX: NOMENCLATURE

PHEVs: Plug-in hybrid electric vehicles.
EPRI: Electric power research institute.
V2G: Vehicle-to-grid.
SoC: State-of-charge.
ICEV: Internal combustion engine vehicles.
AEVs: All-electric vehicles.
HEVs: Hybrid electric vehicles.
AER: All-electric-range.
$\mathrm{I}_i(k)$**:** Charging current over Δt.
$V_i(k)$**:** Charging voltage over Δt.
$C_{r,i}(k)$**:** Remaining battery capacity required to be filled for i-th PHEV at time step k.
C_i**:** Rated battery capacity of the i-th PHEV (Farad).
$T_{r,i}(k)$**:** Remaining time for charging the i-th PHEV at time step k.
$D_i(k)$**:** Price difference.
$w_i(k)$**:** Charging weighting term of the i-th PHEV at time step k.
$SoC_i(k+1)$**:** State-of-charge of the i-th PHEV at time step k+1.
$SoC_{i,\max}$**:** User-defined maximum battery SoC limit for the i-th PHEV.
J (k): Fitness function.
$P_{utility}$**:** Power available from the utility.
$P_{i,\max}$**:** Maximum power that can be absorbed by a specific PHEV.
η**:** Overall charging efficiency of the charging station.
Δt**:** Total charging time.
PSO: Particle swarm optimization.
GSA: Gravitational search algorithm.
PSOGSA: (Hybrid) Particle swarm optimization and Gravitational search algorithm.
EMS: Energy management strategy.
EVSE: Electric vehicle supply equipment.
DG: Distributed generation.
DSM: Demand side management.
SMDP: Stochastic semi-Markov decision.

Chapter 19
Planning of a Project with Imprecise Activity Time

Sk. Md. Abu Nayeem
Aliah University, India

Madhumangal Pal
Vidyasagar University, India

ABSTRACT

A project where activity times are characterized by intervals or triangular fuzzy numbers is considered in this chapter. Classical project evaluation and review technique (PERT) cannot be directly applied to such projects. Comparison of two intervals or two triangular fuzzy numbers plays an important role in this problem. Ranking methods based on some fuzzy mathematical techniques are discussed and two algorithms for finding the critical path for such a project are given. An illustrative example is also provided.

INTRODUCTION

Project evaluation and review technique (PERT) is a method to analyze different tasks required in completing a given project, especially the time needed to complete each task, and to find out the minimum time needed to complete the total project. Any individual task which utilizes resources and which has a start and an end is called an activity. The activities are generally inter-related in such a way that some activities cannot be started before the completion of some others. Thus a combination of inter-related activities in a certain order constitutes the whole project. An event represents a point in time signifying the completion of some activities and the beginning of some new ones. The inter-dependencies and the precedence relationship among the activities of a project reveal efficiently when it is represented as a network diagram. Usually, the events of a project are represented by the set of nodes $V = \{v_1, v_2, \ldots, v_n\}$ of a network $N= (V, E)$, where E is the set of edges e_{ij} between two nodes i and j; e_{ij} determines the estimated time of completion of an activity within the events given by the nodes i and j.

In classical project management, the estimated completion time of each activity is taken as crisp real numbers. But, as time fluctuates with traffic conditions, payloads and so on, it is not practical to

DOI: 10.4018/978-1-4666-9644-0.ch019

consider each activity time as a deterministic value. For example, suppose the natural completion time for the construction of a building is 17 days provided the availability of a certain number of workers and instantaneous availability of the raw materials. But, if we take into account the possibility of bad weather, the allotted time to complete the work may be say, 20 days, being there is a chance of completing it earlier. So, it will be justified to consider the time as an interval number, say, [15, 20] days or a triangular fuzzy number as (15, 17, 20) days or some other way. Different kinds of uncertainties may occur for different problems.

Management of a project where the activity times are imprecise in nature requires specialized techniques for aggregation and comparison of quantities describing the imprecision. Dubois & Prade (1980) were the first to analyze the problem for fuzzy activity time by using the extended addition/ subtraction and fuzzy maximum/ minimum for the aggregation and comparison of fuzzy quantities. The main drawback of their proposal was that the path length determined may not correspond to the original one in the network. Also the double inclusion of the uncertainties of the fuzzy durations resulted in a more fuzzy quantity during the determination of the path. This is due to the backward traversal after the extended addition has been applied in the forward calculations. Another approach to fuzzy PERT by introducing the convolution law to aggregate the fuzzy duration times was analyzed by Mares & Horak (1983) and Mares (1989). They used enumeration to examine all possible paths individually to obtain the critical path. The overall fuzzy duration times were determined in terms of maximum/ minimum elements in the supports, the modes or the median elements depending upon the decision maker's attitude of pessimism or optimism so that the computational time can be tremendously high. Since then a lot of research in this direction have been carried out by a number of researchers by considering different kind of imprecision and different types of aggregation and comparison operations. Nayeem & Pal (2008) have studied the problem using ranking method based on 'acceptibility index' and deconvolution operation.

In fuzzy optimization theory, the most important fuzzy ranking methods are based on the possibility and necessity measures. The possibility theory was proposed by Zadeh (1978) and developed by many researchers such as Dubois & Prade (1988). Liu (2002) introduced the credibility measure of a fuzzy event as an average of the possibility and necessity measure. But, in reality, most decision-makers are neither absolutely optimistic/ optimistic, nor absolutely neutral. To balance between the optimism and pessimism, a convex combination of the possibility measure and the necessity measure is introduced by Yang & Iwamura (2008). It gives scope to the decision maker to set the degree of optimism/ pessimism. Thus m_λ measure is a generalization of the credibility measure. Liu (2004) shown that the expected value of a fuzzy variable can be defined in terms of the credibility measure. In a similar manner, the λ-expected value of a fuzzy variable can be defined in terms of the m_λ measure (Nayeem, 2012).

In this chapter, our aim is to review the problem of project management where activity times are imprecise using a combination of the above stated fuzzy mathematical tools.

ARITHMETIC AND RANKING OF IMPRECISE NUMBERS

In this section, we give the arithmetic and ranking methods of intervals and triangular fuzzy numbers.

An interval number is defined as $A = [a, b] = \{x: a \leq x \leq b\}$ where, a and b are real numbers called the left end point and the right end point of the interval A.

Or, alternatively, $A=\langle m(A),w(A)\rangle$, where $m(A)$ = midpoint of $A=\frac{a+b}{2}$ and $w(A)$ = half width of $A=\frac{a+b}{2}$.

Or, an interval $A=[a, b]$ is an equipossible fuzzy variable, whose membership function is given by

$$\mu_1(x)=\begin{cases}1; if \;\; a\le x\le b\\ 0; otherwise.\end{cases}$$

A crisp real number k may be considered as a degenerate interval $[k,k]=\langle 0,0\rangle$.

The sum of two interval numbers $A=[a_1,b_1]$ and $B=[a_2,b_2]$ is given by $A\oplus B=\left[a_1+a_2,a_2+b_2\right]$.

Alternatively, in mean-width notations, if $A=\langle m_1, w_1\rangle$ and $B=\langle m_2, w_2\rangle$ then, $A\oplus B=\langle m_1+m_2, w_1+w_2\rangle$.

In the following, we give the definition of *acceptability index* in connection with the ranking of two intervals, due to Sengupta & Pal (2000).

Definition 1: *The acceptability index (A-index) of the proposition '$A=\langle m_1, w_1\rangle$ is preferred to $B=\langle m_2, w_2\rangle$' is given by*

Using this A-index Nayeem & Pal (2005) have defined the following ranking orders.

Definition 2: *If $A(A<B)\geq 1$ then A is said to be totally dominating over B in case of minimization and converse in case of maximization and this is denoted by $A<B$.*

Definition 3: *If $0<A(A<B)<1$ then A is said to be 'partially dominating' over B in the sense of minimization and B is said to be 'partially dominating' over A in the sense of maximization. This is denoted by $A\prec_P B$.*

A triangular fuzzy number is given by a triplet $A=(a, b, c)$ with the membership function

$$\mu_2(x)=\begin{cases}\frac{x-a}{b-a}; if \;\; a\le x\le b\\ \frac{x-c}{b-c}; if \;\; b\le x\le c\\ 0; otherwise.\end{cases}$$

i.e., b is the point whose membership value is 1 and $b-a$ and $c-b$ are the left hand and right hand spreads respectively.

Another way of representation of a triangular fuzzy number is $(\underline{a}, a, \overline{a})$. Here, a has the membership value 1 and left hand spread and right hand spread are $a-\underline{a}$ and $\overline{a}-a$ respectively.

Let $a==\langle m_1, \alpha_1, \beta_1\rangle$ and $b=\left\langle m_2,\alpha_2,\beta_2\right\rangle$ be two triangular fuzzy numbers. Then the fuzzy sum of these two numbers is given by

$$a\oplus b=\left\langle m_1+m_2,\alpha_1+\alpha_2,\beta_1+\beta_2\right\rangle \quad (1)$$

In fuzzy optimization theory, the most important fuzzy ranking methods are based on the possibility and necessity measures. The possibility theory was proposed by Zadeh (1978) and developed by many researchers such as Dubois and Prade (1988). Let ξ be a fuzzy variable with membership function $\mu_\xi(x)$ and *B* be an arbitrary subset of *R*. Then the possibility measure of fuzzy event {ξ∈*B*} is defined as Pos $\{\xi \in B\} = \sup_{x\in B} \mu_\xi(x)$.

The necessity of this fuzzy event is defined as

$$\text{Nec}\{\xi\in B\}=1-\text{Pos}\{\xi \in B^C\} = 1 - \sup_{x\in B^C} \mu_\xi(x).$$

Liu (2002) introduced the credibility measure of a fuzzy event as an average of the possibility and necessity measure as below.

Definition 4: *The credibility measure for the chance of a fuzzy event* {ξ∈*B*} *is defined as,*

$$Cr\{\xi \in B\} = \frac{1}{2}(Pos\{\xi \in B\} + Nec\{\xi \in B\}).$$

But, in reality, most decision-makers are neither absolutely optimistic/ optimistic, nor absolutely neutral. To balance between the optimism and pessimism, a convex combination of the possibility measure and the necessity measure is introduced by Yang & Iwamura (2008). It gives scope to the decision maker to set the degree of optimism/ pessimism. Thus m_λ measure is a generalization of the credibility measure.

Definition 5: *Formally, the* m_λ *measure for the chance of a fuzzy event is defined as,*

$$m_\lambda\{\xi \in B\} = \lambda Pos\{\xi \in B\} + (1-\lambda)Nec\{\xi \in B\},$$

where, the parameter $\lambda\in[0,1]$ *is pre-determined by the decision-maker according to the degree of optimism or pessimism.*

Clearly, when $\lambda = \frac{1}{2}$, the m_λ measure reduces to the credibility measure.

Liu & Liu (2004) have shown that the expected value of a fuzzy variable can be defined in terms of the credibility measure as follows.

Definition 6: *Let ξ be a fuzzy variable. The expected value of ξ is defined as,*

$$E[\xi] = \int_0^\infty Cr\{\xi \geq r\}dr - \int_{-\infty}^0 Cr\{\xi \leq r\}dr,$$

provided that at least one of the two integrals is finite.

Based on the m_λ measure, the λ-expected value of a fuzzy variable is defined as follows (Nayeem, 2012).

Definition 7: *Let ξ be a fuzzy variable. The λ-expected value of ξ is defined as,*

$$E_\lambda[\xi] = \int_0^\infty m_\lambda\{\xi \geq r\}dr - \int_{-\infty}^0 m_\lambda\{\xi \leq r\}dr,$$

provided that at least one of the two integrals is finite and the parameter $\lambda \in [0,1]$ is pre-determined by the decision-maker according to the degree of optimism or pessimism.

Example 1: *Let ξ be a non-negative equipossible fuzzy variable (a,b), i.e., the interval [a,b] with a≥0. Then the λ-expected value of ξ is given by $E_\lambda[\xi]=\lambda b+(1-\lambda)a$.*

Example 2: *Let η be a non-negative triangular fuzzy variable (a,b,c). Then the λ-expected value of η is given by*

$$E_\lambda[\eta] = \frac{a+b}{2} + \frac{\lambda}{2}(c-a).$$

Deconvolution

The extended substraction for two L-R fuzzy numbers $A=(a_L, a_m, a_R)_{LR}$ and $B=(b_L, b_m, b_R)_{LR}$ is given by $A \Theta B = (a_L+b_R, a_m - b_m, a_R+b_L)$.

If this extended substraction is used then double inclusion of uncertainties will be resulted. In order to avoid this double inclusion, the deconvolution approach (Kaufman & Gupta, 1985) is used in the backward pass computations of PERT problem.

The deconvolution or backward extended substraction for two L-R fuzzy numbers $A=(a_L, a_m, a_R)_{LR}$ and $B=(b_L, b_m, b_R)_{LR}$ is given by

$$\mathrm{A}[-]B = (a_L - b_L, a_m - b_m, a_R - b_R)_{LR}.$$

Then consider the following two cases:

If $a_L - b_L < 0$ and $a_R - b_R \geq 0$, then

The mode, i.e., the point where the fuzzy number attains the membership value 1, is transformed to

$$l_m = \int_w [\mu_{A_L}^{-1}(w) - \mu_{B_L}^{-1}(w)]\,\mathrm{dw} \text{ and}$$

The left and right spreads are transformed to $l_L = l_m - (a_m - b_m)$ and $l_R = (a_R - b_R) - l_L$ respectively.

If $a_L - b_L \geq 0$ and $a_R - b_R < 0$, then

The mode is transformed to $r_m = \int_w [\mu_{A_R}^{-1}(w) - \mu_{B_R}^{-1}(w)]\,\mathrm{dw}$ and

The left and right spreads are transformed to $r_R = (a_m - b_m) - r_m$ and $r_L = (a_L - b_L) - r_R$ respectively.

For triangular fuzzy numbers l_m and r_m are given by

$$l_m = (a_m - b_m) - \frac{1}{2}(a_L - b_L)$$

and

$$r_m = (a_m - b_m) - \frac{1}{2}(a_R - b_R)$$

so that

$$l_L = -\frac{1}{2}(a_L - b_L), l_R = (a_R - b_R) + \frac{1}{2}(a_L - b_L)$$

and

$$r_L = (a_L - b_L) + \frac{1}{2}(a_R - b_R), r_R = -\frac{1}{2}(a_R - b_R).$$

If $(a_L - b_L)$ and $(a_R - b_R)$ are both negative, then the mode remains unchanged and the left spread is changed to $-(a_R - b_R)$ and the right spread is changed to $-(a_L - b_L)$.

Similarly, we define the deconvolution for two interval numbers A=,α⟩ and B=,β⟩ as $A[-]B=\langle a-b,|\alpha-\beta|\rangle$.

THE ALGORITHMS

A general algorithm to determine the critical path for a network with interval number or triangular fuzzy number edge weights is given by Nayeem and Pal (2008). Acceptibility index based ranking of intervals and triangular fuzzy numbers along with deconvolution of fuzzy numbers were used there. Here we give two algorithms: one is similar to that in Nayeem and Pal (2008) except for the ranking method and the other employing classical PERT method by fully defuzzifying the given network.

Let $G=(V,E)$ be a network representing a project, where $V = \{v_1, v_2, \ldots v_n$ is the set of nodes and E is the set of arcs a_{ij} such that $a_{ij} \in E$ if there is the edge between the nodes v_i and v_j and its weight e_{ij} being an interval number or a triangular fuzzy number, is the time duration of the activity represented by it. Also let, E_i and L_i be the earliest occurrence time and latest allowable occurrence time of the event given by the node i respectively and F_{ij} and B_{ij} be the earliest finishing time and latest beginning time of the activity given by the arc a_{ij} respectively.

Algorithm 1.Algorithm FuzzyPERT_1(G,n,λ,s,t,Π)

Input: A network $G=(V,E)$ with n nodes, a real number λ in [0,1] reflecting the decision makers level of optimism/ pessimism, source node s and sink node t, and the adjacency matrix $[e_{ij}]_{n\times n}$.

Output: The critical path Π between s and t.

Step 1: //forward pass computation//

Step 1.1: Set $j = s$, $E_j=0$, where, $0=\langle 0,0,0\rangle$ for triangular fuzzy number and $0=\langle 0,0\rangle$ for interval number and $V^*=V-\{s\}$.

Step 1.2: while $V^* \neq \phi$ do

Calculate $F_{ij} = E_j \oplus e_{ji}$ for all j such that $a_{ji} \in E$.

Find $F_{ij^*} = \max_{j\in P(i)} E_\lambda[F_{ij}]$, where $P(i)$ represents the set of events immediately preceding the node i.

Set $E_i = F_{ij^*}$ and $V^* = V^* - \{i\}$.

endwhile;

Step 2: //backward pass calculation//

Step 2.1: Set $j = t$ and $L_j = E_j$, $V^* = \{t\}$.

Step 2.2: while $V^* \neq V$ do

Calculate $B_{ij} = L_j[-]e_{ij}$ for all j such that $a_{ij} \in E$.

Find $B_{ij^*} = \min_{j\in S(i)} E_\lambda[B_{ij}]$, where $S(i)$ represents the set of events immediately following the node i.

Set $L_i = B_{ij^*}$ and $V^* = V^* \cup \{i\}$.

endwhile;

Step 3: //slack computation//

for all $i\in V$ do

$s_i = L_i[-]E_i$.

endfor;

Step 4: //finding the critical path//

Step 4.1: $i = s$, $\Pi = \phi$.

Step 4.2: while $i \neq t$ do

Find $s_{j^*} = \min_{j\in N(i)} E_\lambda[s_j]$, where $N(i)$ is the neighbour of i.

Set $\Pi \leftarrow \Pi \cup a_{ij^*}$ and $i = j^*$.

endwhile;

End FuzzyPERT_1

It may be noted that mixed type of activity times can be handled by the second algorithm.

An Illustrative Example

Here, the example considered by Nayeem and Pal (2008) is chosen. A planning network for the control of a project with imprecise activity times is shown in Figure 1.

Algorithm 2. Algorithm FuzzyPERT_2(G,n,λ,s,t,Π)

Input: A network $G=(V,E)$ with n nodes, a real number λ in [0,1] reflecting the decision makers level of optimism/ pessimism, source node s and sink node t, and the adjacency matrix $\left[e_{ij}\right]_{n\times m}$.

Output: The critical path Π between s and t.

Step 0: //Defuzzification//

for all $a_{ij} \in E$ do

$\widehat{e_{ij}} = E_{\lambda}\left[e_{ij}\right]$

endfor;

Step 1: //forward pass computation//

Step 1.1: Set $j = s$, $E_j=0$, and $V^* = V - \{s\}$.

Step 1.2: while $V^* \neq \phi$ do

Calculate $F_{ij} = E_j + \widehat{e_{ij}}$ for all j such that $a_{ji} \in E$.

Find $F_{ij^*} = \max_{j\in P(i)} F_{ij}$, where $P(i)$ represents the set of events immediately preceding the node i. Set $E_i = F_{ij^*}$ and $V^* = V^* - \{i\}$.

endwhile;

Step 2: //backward pass calculation//

Step 2.1: Set j and $L_{jj}, V^* = \{t\}$.

Step 2.2: while $V^* = V$ do

Calculate $B_{ij} = L_j - \widehat{e_{ij}}$ for all j such that $a_{ij} \in E$.

Find $B_{ij^*} = \min_{j\in S(i)} B_{ij}$, where $S(i)$ represents the set of events immediately following the node i. Set $L_i = B_{ij^*}$ and $V^* = V^* \cup \{i\}$.

endwhile;

Step 3: //slack computation//

for all $i \in V$ do

$s_i = L_i - E_i$.

endfor;

Step 4: //finding the critical path//

Step 4.1: $i = s$, $\Pi = \phi$.

Step 4.2: while $i=t$ do

Find $s_{j^*} = \min_{j\in N(i)} s_j$, where $N(i)$ is the neighbour of i.

Set $\Pi \leftarrow \Pi \cup a_{ij^*}$ and $i = j^*$.

endwhile;

End FuzzyPERT_2

Figure 1. A project network

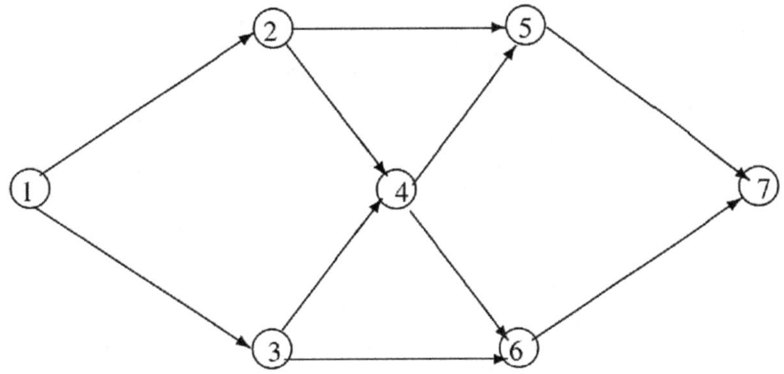

Activity times, i.e. the arc weights of the network and their λ-expected values for different values of λ are shown in Table 1 when those are given as intervals and in Table 2 when those are triangular fuzzy numbers. Significance of λ in a linguistic scale is shown in Table 3.

Both of the algorithms given in Section 3 produce the critical path **1→3→4→5→7** for all values of λ for triangular fuzzy activity times.

The critical paths obtained by both of the algorithms for interval activity times are also same. Those are **1→2→5→7** for λ>0.5 and **1→2→4→6→7** for λ≤0.5. The results obtained thus are in good agreement with those obtained by Nayeem & Pal (2008).

CONCLUSION

In this chapter, the problem of finding the critical path in planning of a project with imprecise activity times is studied. Two algorithms based on a new kind of expected values of the imprecise numbers are

Table 1. Interval activity times and their λ-expected values

Edge	Weight	λ-Expected Values										
		λ=0	λ=0.1	λ=0.2	λ=0.3	λ=0.4	λ=0.5	λ=0.6	λ=0.7	λ=0.8	λ=0.9	λ=1
e_{12}	[49,61]	49	50.2	51.4	52.6	53.8	55	56.2	57.4	58.6	59.8	61
e_{13}	[50,60]	50	51.0	52.0	53.0	54.0	55	56.0	57.0	58.0	59.0	60
e_{24}	[60,80]	60	62.0	64.0	66.0	68.0	70	72.0	74.0	76.0	78.0	80
e_{25}	[90,110]	90	92.0	94.0	96.0	98.0	100	102.0	104.0	106.0	108.0	110
e_{34}	[60,70]	60	61.0	62.0	63.0	64.0	65	66.0	67.0	68.0	69.0	70
e_{36}	[86,104]	86	87.8	89.6	91.4	93.2	95	96.8	98.6	100.4	102.2	104
e_{45}	[87,103]	87	88.6	90.2	91.8	93.4	95	96.6	98.2	99.8	101.4	103
e_{46}	[109,121]	109	110.2	111.4	112.6	113.8	115	116.2	117.4	118.6	119.8	121
e_{57}	[108,132]	108	110.4	112.8	115.2	117.6	120	122.4	124.8	127.2	129.6	132
e_{67}	[95,105]	95	96.0	97.0	98.0	99.0	100	101.0	102.0	103.0	104.0	105

Table 2. Triangular fuzzy activity times and their λ-expected values

Edge	Weight	λ-Expected Values										
		λ=0	λ=0.1	λ=0.2	λ=0.3	λ=0.4	λ=0.5	λ=0.6	λ=0.7	λ=0.8	λ=0.9	λ=1
e_{12}	(45,55,58)	50.0	50.7	51.3	52.0	52.6	53.25	53.9	54.55	55.2	55.85	56.5
e_{13}	(52,55,65)	53.5	54.2	54.8	55.5	56.1	56.75	57.4	58.05	58.7	59.35	60
e_{24}	(55,70,75)	62.5	63.5	64.5	65.5	66.5	67.5	68.5	69.5	70.5	71.5	72.5
e_{25}	(90,100,112)	95.0	96.1	97.2	98.3	99.4	100.5	101.6	102.7	103.8	104.9	106
e_{34}	(55,65,68)	60.0	60.7	61.3	62.0	62.6	63.25	63.9	64.55	65.2	65.85	66.5
e_{36}	(85,95,103)	90.0	90.9	91.8	92.7	93.6	94.5	95.4	96.3	97.2	98.1	99
e_{45}	(85,95,102)	90.0	90.9	91.7	92.6	93.4	94.25	95.1	95.95	96.8	97.65	98.5
e_{46}	(107,115,120)	111.0	111.7	112.3	113.0	113.6	114.25	114.9	115.55	116.2	116.85	117.5
e_{57}	(113,120,140)	116.5	117.9	119.2	120.6	121.9	123.25	124.6	125.95	127.3	128.65	130
e_{67}	(95, 100,105)	97.5	98.0	98.5	99.0	99.5	100	100.5	101	101.5	102	102.5

Table 3. Significance of λ in a linguistic scale

λ	Linguistic Scale
0	Extremely optimistic
0.1	Very highly optimistic
0.2	Highly optimistic
0.3	Moderately optimistic
0.4	Optimistic
0.5	Indifferent
0.6	Pessimistic
0.7	Moderately pessimistic
0.8	Highly pessimistic
0.9	Very highly pessimistic
1	Extremely pessimistic

given. The result obtained by these algorithms are same as obtained in some previous works. The advantage of the algorithms are that the decision maker's view is taken into consideration in a very robust scale in comparison to the earlier methods. Also one of the algorithms is enabled to handle mixed type of imprecise data.

REFERENCES

Dubois, D., & Prade, H. (1980). *Fuzzy Sets and Systems: Theory and Applications*. New York: Academic Press.

Dubois, D., & Prade, H. (1988). *Possibility Theory*. New York: Plenum. doi:10.1007/978-1-4684-5287-7

Kaufmann, A., & Gupta, M. M. (1985). *Introduction to Fuzzy Arithmetic: Theory and Applications*. New York: Van Nostrand Reinhold.

Liu, B. (2002). *Theory and Practice of Uncertain Programming*. Heidelberg, Germany: Physica-Verlag. doi:10.1007/978-3-7908-1781-2

Liu, B. (2004). *Uncertainty Theory: An Introduction to its Axiomatic Foundations*. Berlin: Springer-Verlag. doi:10.1007/978-3-540-39987-2

Mares, M. (1989). Network anaysis of fuzzy technologies. In G. W. Evans, W. Karwowski, & M. R. Wilhelm (Eds.), *Applications of Fuzzy Set Methodologies in Industrial Engineering*. Elsevier.

Mares, M., & Horak, J. (1983). Fuzzy quantities in networks. *Fuzzy Sets and Systems*, *10*(1-3), 123–134. doi:10.1016/S0165-0114(83)80110-9

Nayeem, S. M. A. (2012). An expected value model of quadratic clique problem on a graph with fuzzy parameters. In *Proceedings of 12th International Conference on Intelligent Systems Design and Applications* (pp. 592 – 597). IEEE, DOI: doi:10.1109/ISDA.2012.6416604

Nayeem, S. M. A., & Pal, M. (2005). Shortest path problem on a network with imprecise edge weight. *Fuzzy Optimization and Decision Making*, *4*(4), 293–312. doi:10.1007/s10700-005-3665-2

Nayeem, S. M. A., & Pal, M. (2008). PERT on a network with imprecise edge weight. *Journal of Fuzzy Mathematics*, *16*(4), 853–874.

Sengupta, A., & Pal, T. K. (2000). On comparing interval numbers. *European Journal of Operational Research*, *127*(1), 28–43. doi:10.1016/S0377-2217(99)00319-7

Yang, L., & Iwamura, K. (2008). Fuzzy chance-constrained programming with linear combination of possibility measure and necessity measure. *Applied Mathematical Sciences*, *2*, 2271–2278.

Zadeh, L. A. (1978). Fuzzy sets as a basis for a theory of possibility. *Fuzzy Sets and Systems*, *1*(1), 3–28. doi:10.1016/0165-0114(78)90029-5

Chapter 20
Swarm Intelligence for Multiobjective Optimization of Extraction Process

T. Ganesan
The University of Adelaide, Australia

I. Elamvazuthi
Universiti Technologi Petronas, Malaysia

P. Vasant
Universiti Technologi Petronas, Malaysia

ABSTRACT

Multi objective (MO) optimization is an emerging field which is increasingly being implemented in many industries globally. In this work, the MO optimization of the extraction process of bioactive compounds from the Gardenia Jasminoides Ellis fruit was solved. Three swarm-based algorithms have been applied in conjunction with normal-boundary intersection (NBI) method to solve this MO problem. The gravitational search algorithm (GSA) and the particle swarm optimization (PSO) technique were implemented in this work. In addition, a novel Hopfield-enhanced particle swarm optimization was developed and applied to the extraction problem. By measuring the levels of dominance, the optimality of the approximate Pareto frontiers produced by all the algorithms were gauged and compared. Besides, by measuring the levels of convergence of the frontier, some understanding regarding the structure of the objective space in terms of its relation to the level of frontier dominance is uncovered. Detail comparative studies were conducted on all the algorithms employed and developed in this work.

INTRODUCTION

Multi-criteria or multi-objective (MO) scenarios have become increasingly prevalent in industrial engineering environments (Statnikov & Matusov, 1995; Zhang and Li, 2007; Li and Zhou, 2011). MO optimization problems are commonly tackled using the concept of Pareto-optimality to trace-out the

DOI: 10.4018/978-1-4666-9644-0.ch020

non-dominated solution options at the Pareto curve (Zitzler & Thiele, 1998; Deb *et al.*, 2002). Other methods include the weighted techniques which involve objective function aggregation resulting in a master weighted function. This master weighted function is then solved for various weight values (which are usually fractional) (Fishburn, 1967; Triantaphyllou, 2000; Luyben. & Floudas, 1994; Das & Dennis, 1998). Using these techniques, the weights are used to consign relative importance or priority to the objectives in the master aggregate function. Hence, alternative near-optimal solution options are generated for various values of the scalars. In this chapter, the Normal Boundary Intersection (NBI) scheme (Das & Dennis, 1998) was used as a scalarization tool to construct the Pareto frontier. In Sandgren (1994) and Statnikov & Matusov (1995), detail examples and analyses on MO techniques for problems in engineering optimization are presented.

Many optimization techniques have been implemented for solving the extraction process problem (e.g. Hismath *et al.*, 2011; Jie and Wei, 2008). In addition, evolutionary techniques such as DE have also been employed for extraction process optimization (Ubaidullah *et al.*, 2012). The MO problem considered in this work was formulated by Shashi *et al*, (2010). This problem involves the optimization of the yields of certain chemical products which are extracted from the Gardenia *Jasminoides Ellis* fruit. The MO optimization model was developed in Shashi *et al*, (2010) to maximize the extraction yields which are the three bioactive compounds; crocin, geniposide and total phenolic compounds. The optimal extraction parameters which construct the most dominant Pareto frontier are then identified such that the process constraints remain unviolated. In Shashi *et al.*, (2010), the MO problem was tackled using the real-coded Genetic Algorithm (GA) approach to obtain a single individual optima and not a Pareto frontier. In that work, measurement metrics were not employed to evaluate the solution quality in detail. In addition, the work done in Shashi *et al.*, (2010) focused on modeling the system rather than optimizing it. The authors of that work employed only one optimization technique and did not carry out extensive comparative analysis on the optimization capabilities. Due to this setbacks, these factors are systematically addressed in this chapter to provide some insights on the optimization of the extraction process.

Over the past years, swarm intelligence-based meta-heuristic techniques have been applied with increasing frequency to industrial MO scenarios. Some of the most effective swarm approaches have been devised using ideas from Newtonian gravitational theory (Rashedi *et al.*, 2009), dynamics of fish movement (Neshat *et al.*, 2012) and birds flocking behaviors (Kennedy & Eberhart, 1995). In this work, three swarm-based techniques; gravitational search algorithm (GSA) (Rashedi *et al.*, 2009), particle swarm optimization (PSO) (Kennedy & Eberhart, 1995) and the novel Hopfield-Enhanced PSO (HoPSO) were employed to the extraction problem (Shashi *et al*, 2010). The measurement techniques; the convergence metric (Deb & Jain, 2002) and the Hypervolume Indicator (HVI) (Zitzler & Thiele, 1998) were used to analyze the solution spread produced by these algorithms.

The HVI is a set measure reflecting the volume enclosed by a Pareto front approximation and a reference set (Emmerich *et al.*, 2005). The convergence metric on the other hand measures the degree at which the solutions conglomerate towards optimal regions of the objective space. Using the values obtained by the measurement metrics, the correlation between the convergence and the degree of dominance (measured by the HVI) of the solution sets is obtained and discussed. The solutions constructing the Pareto frontier obtained using the developed HoPSO algorithm is also subjected to the analyses mentioned above. In this work, all computational procedures were developed using the Visual C++ Programming Language on a PC with an Intel i5-3470 (3.2 GHz) Processor.

This chapter is organized as follows: Section 2 presents an overview on industrial MO optimization while Section 3 discusses some fundamentals on pareto dominance. Section 4 gives the problem description followed by the NBI approach in Section 5. Section 6 provides some details regarding swarm intelligence techniques employed in this work. Section 7 provides information related to the measurement indicators while in Section 8 the computational results are discussed. Finally, this chapter ends with some conclusions and recommendations for future work.

INDUSTRIAL MULTIOBJECTIVE OPTIMIZATION

Over the past years, MO optimization has been introduced and applied into many engineering problems. Some of these developments shall be briefly discussed in the following. In Aguirre *et al.*, (2004), a MO evolutionary algorithm with an enhanced constraint handling mechanism was used to optimize the circuit design of a Field Programmable Transistor Array (FPTA). The authors used the Inverted Shrinkable Pareto Archived Evolution Strategy (ISPAES) for the MO optimization of the circuit design. Another MO problem involving engineering design was solved by Reddy and Kumar, (2007). In that work, a MO swarm intelligence algorithm was developed by incorporating the Pareto dominance relation into the standard particle swarm optimization (PSO) algorithm. Three engineering design problems, the 'two bar truss design (Palli *et al.*, 1999)', 'I-beam design (Yang *et al.*, 2002)' and the 'welded beam design (Deb *et al.*, 2000)' problems were successfully solved in Reddy and Kumar, (2007). In the area of thermal system design, the MO optimization of an HVAC (Heating, Ventilating, Air-Conditioning and Cooling) system was carried out by Kusiak *et al.*, (2010). In that work, a neural network was used to derive the MO optimization model. This model was then optimized using a multi-objective PSO algorithm (MOPSO). Using this algorithm, the authors identified the optimum control settings for the supply air temperature and static pressure to minimize the air handling unit energy consumption while maintaining air quality. On the other hand, in materials engineering, the MO optimization of the surface grinding process was carried out by Pai *et al.*, (2011). In that work, the enhanced version of the Non-dominated Sorting Genetic Algorithm (NSGA-II) was employed to optimize the machining parameters of the grinding process to increase the utility for machining economics and to increase the product quality (Azouzi & Guillot, 1998). Another application of the NSGA-II algorithm to engineering system design was done in Nain *et al.*, (2010). In Nain *et al.*, (2010), the authors optimized the structural parameters (area and length of the TEC elements) of the Thermoelectric Cooler (TEC). The Coefficient of Performance (COP) and the Rate of Refrigeration (ROR) was successfully maximized in that work.

Recently, MO optimization methods have also penetrated the power and energy industries. For instance, in Van Sickel *et al.*, (2008), the MO optimization of a fossil fuel power plant was done using multi-objective evolutionary programming (MOEP) and the multi-objective particle swarm optimization (MOPSO) algorithms. The MO techniques in that work were done to develop reference governors for power plant control systems. Another work on the MO optimization of reference governor design for power plants was done by Heo and Lee, (2006). In that work, PSO variant algorithms were used to find the optimal mapping between unit load demands and pressure set point of a fossil fuel power plant. By this approach, the optimal set points of the controllers under a large variety of operation scenarios were achieved. Similarly, in the works of Song and Kusiak (2010), temporal processes in power plants were optimized using MO techniques. In that work, the central theme was to maximize the boiler efficiency while minimizing the limestone consumption. Two approaches; the Data Mining (DM) and evolutionary strategy algorithms were combined to solve the optimization model. In Song and Kusiak, (2010) the MO optimization of temporal dependent processes were successfully completed by identifying the optimum control parameters. One other area at which MO optimization has been applied with considerable success is in the field of economic/environmental dispatch for power systems. For instance, in the works of Gunda and Acharjee, (2011), an MO economic/environmental dispatch problem was solved using the Pareto Frontier Differential Evolution (PFDE) approach. By using this technique, the authors managed to minimize the fuel consumption and emissions with minimal energy loss. This triple-objective problem was successfully solved without the violation of the system's security constraints. Another similar problem was tackled in King *et al.*, (2005). In that work, power generation optimization was done to minimize the total fuel costs as well as the amount of emission.

PARETO FRONTIER AND DOMINANCE LEVELS

To get a clearer picture of the idea the Pareto-compliance, the concepts of Pareto dominance needs to be defined. Pareto dominance can be categorized into three types which are; strictly dominates ($\succ$), weakly dominates ($\succeq$) and indifferent (~). Let two solution vectors be **a** and **b**. Then if the solution vector **a** dominates the vector **b** in all the objectives then **a** strictly dominates **b** (**a** $\succ$ **b**). If the solution vector **a** dominates the vector **b** in some of the objectives but not all, then **a** weakly dominates **b** (**a** $\succeq$ b). Finally, if the solution vector **a** does not dominate the vector **b** and the solution vector **b** does not dominate **a** as well in all the objectives, then **a** is indifferent to **b** (**a** ~ **b**). Strictly Pareto-compliant can be defined as the following: Let there be two solution sets say; *X* and *Y* for a specific MO problem. If the hypervolume coverage for *X* is greater than *Y*, then the solution set *X* $\succ$ Y or *X* $\succeq$ Y. The hypervolume measures the volume of the dominated section of the objective space and can be applied for multi-dimensional scenarios. Implementation of the hypervolume requires a reference point or a 'nadir point'. The nadir point is a point which is dominated by all the solutions from the approximate Pareto frontier. Relative to this point, the volume of the space of all dominated solutions can be computed. A bi-objective depiction of the hypervolume is given in Figure 1.

The larger the value of the hypervolume, the more dominant the solution is in the objective space. The hypervolume is strictly monotonic. Its computational effort is exponential to the amount of solution vectors however requires a bounding vector (nadir point).

PROBLEM DESCRIPTION

The model for MO problem considered in this work was developed by Shashi *et al.*, (2010). This problem involves the optimization of the yields of certain chemical products which are extracted from the Gardenia *Jasminoides Ellis* fruit. The phenolic compounds in Gardenia *Jasminoides Ellis* have high antioxidant capabilities which make this fruit valuable for medicinal uses (Li *et al.*, 2008). Compared to other natural food pigments, the colouring constituents of the fruit of Gardenia are non-toxic and chemically stable (Van Calsteren *et al.*, 1997) The constituents present in the Gardenia fruit (Osima *et al.*, 1988) are iridoid glycosides (for instance; gardenoside, geniposide, gardoside and scandoside methyl ester). These constituents could be converted into blue colorants under aerobic condition by enzymes or some microorganism. The Gardenia fruit extract in its rudimentary form also contains phenolic compounds high antioxidant capacity in abundance (Li *et al.*, 2008). The MO optimization model in Shashi *et al.*, (2010) was for the extraction process of bioactive compounds from gardenia with respect to the

Figure 1. Hypervolume construction for a two objective maximization problem

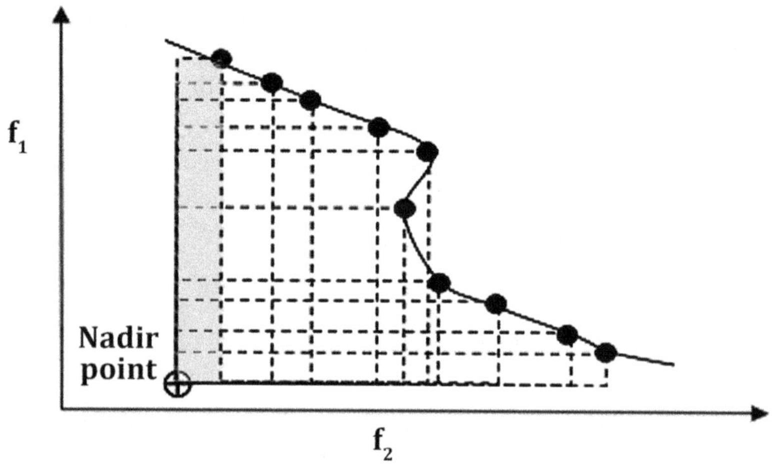

constraints. The MO optimization model was developed to maximize the yield consists of three bioactive compound; crocin (f_1), geniposide (f_2) and total phenolic compounds (f_3). This model is presented as the following:

Maximize → Crocin, f1

Maximize → Geniposide, f2

Maximize → Total phenolic compounds, f3

subject to *process constraints.* (1)

The objective functions represent the yields of each of the bioactive compound in the units of mg/g of dry powder as in Shashi *et al.*, (2010). The objective function modeled with respect to the decision variables are as the following:

$$\begin{aligned} f_1 = {} & 3.8384907903 + 0.0679672610X_1 + 0.0217802311X_2 + 0.0376755412X_3 \\ & -0.0012103181X_1^2 + 0.0000953785X_2^2 - 0.0002819634X_3^2 + 0.0005496524X_1X_2 \\ & - 0.0009032316X_2X_3 + 0.0008033811X_1X_3 \end{aligned} \quad (2)$$

$$\begin{aligned} f_2 = {} & 46.6564201287 + 0.6726057655X_1 + 0.4208752507X_2 + 0.9999909858X_3 \\ & -0.0161053654X_1^2 - 0.0034210643X_2^2 - 0.0116458859X_3^2 + 0.0122000907X_1X_2 \\ & -0.0095644212X_2X_3 + 0.0089464814X_1X_3 \end{aligned} \quad (3)$$

$$f_3 = -6.3629169281 + 0.4060552042X_1 + 0.3277005337X_2 + 0.3411029105X_3 \\ -0.0053585731X_1^2 - 0.0020487593X_2^2 - 0.0042291040X_3^2 + 0.0017226318X_1X_2 \\ -0.0011990977X_2X_3 + 0.0007814998X_1X_3 \quad (4)$$

The decision variables (which the process extraction parameters) are constrained as per the experimental setup described in Shashi *et al.*, (2010). The constraints are as the following:

$$X_1 \in [19.5, 80.5]$$

$$X_2 \in [27.1, 72.9]$$

$$X_3 \in [7.1, 52.9] \quad (5)$$

where X_1 is the concentration of ethanol in %, X_2 is the extraction temperature in ^oC and X_3 is the extraction time in minutes.

NORMAL BOUNDARY INTERSECTION METHOD

The NBI method (Das and Dennis, 1998) is a geometrically-inspired scalarization technique for solving MO optimization problems. As opposed to the Weighted Sum Method, the NBI method searches for a near-uniform spread of Pareto-optimal solutions in the frontier. This makes the NBI approach a more interesting alternative as compared to the Weighted Sum Method when solving non-convex MO problem. Given a generic triple-objective MO optimization problem:

$$Min_{x \in X} \ F(x) \qquad s.t.$$

$$X = \left\{ x : g(x) = 0; h(x) \leq 0, x \in [1,3] \right\}$$

$$F^* = \left(f_1^* \quad f_2^* \right)^T \quad (6)$$

where F^* is the utopia point for this MO problem. The individual minimum denoted as x^*_i is obtained for $i \in [1,3]$. This way the the simplex for the convex hull of the individual minima is generated:

$$\Gamma = \left\{ \phi \cdot Y : \phi = F(x_i^*); Z = \beta_i : i \in [1,3] \right\} \quad (7)$$

where ϕ forms a 3 x 3 matrix and $\sum_{i=1}^{3} \beta_i = 1$. Thus, the β-sub problem is then constructed as follows:

$$Max_{(X,t)}\ t \quad s.t.$$

$$\phi \cdot Z + tn = F(x) \quad and \quad x \in X \tag{8}$$

where t is the distance parameter, and n is the normal vector at the point towards the utopia point. Thus, the NBI approach aims to search for the maximum distance, t in the direction of the normal vector, n. This distance, t is thus between a point on the simplex and the utopia point. Then the scalar values, Z are then varied systematically to develop a near-uniform spread of the Pareto frontier.

SWARM INTELLIGENCE

In the field of optimization, an artificial swarm is a group of virtual organisms or agents that behave interactively to achieve some pre-defined goal. This form of interaction gives the individuals higher capabilities as well as better efficiency in achieving the goal as compared to a single individual. In this work, two conventional swarm-based techniques (GSA and PSO) are employed. In addition, an enhanced version of the conventional PSO approach (Ho-PSO) is developed and implemented on the extraction process problem.

1. Particle Swarm Optimization

The PSO algorithm was originally proposed by Kennedy and Eberhart (1995). This technique springs from two distinct frames of notions. The first notion was based on the investigation of swarming or flocking behaviors among some species of organisms (such as; birds, fishes, etc.). The second idea was inspired by the field of evolutionary computing. The PSO algorithm operates by exploring the search space for optimal candidate solutions and then evaluates these solutions with respect to some user defined fitness condition. The candidate optimal solutions obtained by this algorithm are achieved as a result of particles which are swarming through the fitness landscape. The velocity and position updating rule is critical to the optimization capabilities of this technique. The velocity and the position of each particle are updated using the following equations:

$$v_i(t+1) = wv_i(t) + c_1r_1[\hat{x}_i(t) - x_i(t)] + c_2r_2[g(t) - x_i(t)] \tag{9}$$

$$x_i(t+1) = x_i(t) + v_i(t+1) \tag{10}$$

where each particle is identified by the index i, $v_i(t)$ is the particle velocity and $x_i(t)$ is the particle position with respect to iteration (t). The terms, $c_1r_1[\hat{x}_i(t) - x_i(t)]\frac{\Delta y}{\Delta x}$ and $c_2r_2[g(t) - x_i(t)]$ represent the personal and social influence respectively. These terms control the level of exploration and exploitation of the PSO algorithm during the search process in the objective space. The parameters w, c_1 and c_2 are defined by the user while r_1 and r_2 are random variables. These parameters are typically constrained with the following ranges:

Table 1. PSO parameter settings

Parameters	Values
Initial parameter (c_1, c_2, w)	(1, 1.2, 0.8)
Number of particles	6
initial social influence (s_1, s_2, s_3, s_4, s_5, s_6)	(1.1, 1.05,1.033, 1.025, 1.02, 1.017)
initial personal influence (p_1, p_2, p_3, p_4, p_5, p_6)	(3, 4, 5, 6, 7,8)

$w \in [0,1.2], c_1 \in [0,2], c_2 \in [0,2], r_1 \in [0,1], r_2 \in [0,1]$. (11)

The stopping criterion is satisfied when all particles/candidate solutions have reached their fittest positions during the iterations. The execution procedures of the PSO technique and the initial parameters used in this work are shown in Algorithm 1 and Table 1 respectively.

In the event the position of all the particles converges during the iterative process, the solutions are considered feasible if:

- The specified ranges are respected and no constraint violation exists.
- All the decision variables are non-negative.
- No further optimization of the objective function occurs.

Therefore, based on these criteria it can be said that the fitness requirements have been met. The candidate solutions are hence at their fittest and in effect the program is stopped and the solutions are printed.

2. Gravitational Search Algorithm

The GSA algorithm is a meta-heuristic algorithm first developed by Rashedi *et al.*, (2009). This technique was developed by the using the gravitational laws and the idea of masses interactions as the basis. This algorithm uses the Newtonian gravitational laws where the search agents act as masses. Therefore, the

Algorithm 1. Particle Swarm Optimization (PSO) Algorithm

Step 1: Initialize no of particles, i and the algorithm parameters w, c_1, c_2, r_1, r_2, n_o

Step 2: Randomly set initial position x_i (t) and velocity, $v_i(t)$

Step 3: Compute individual and social influence

Step 4: Compute position $x_i(t+1)$ and velocity $v_i(t+1)$ at next iteration

Step 5: If the swarm evolution time, $t > t_o+T_{max}$, update position x_i and velocity v_i and go to Step 3, else proceed to Step 6

Step 6: Evaluate fitness swarm.

Step 7: If fitness criterion satisfied, halt and print solutions, else go to step 3.

gravitational forces influence the motion of these masses, where lighter masses gravitate towards the heavier masses during these interactions which signify the algorithm's progression towards the optima. The GSA algorithm randomly generates a distribution of masses, $m_i(t)$ and also sets the position for these masses, x_i^d at the initial stages. For a minimization problem, the least fit mass, $m_i^{worst}(t)$ and the fittest mass, $m_i^{best}(t)$ at time t with N number of masses are computed as follows:

$$m^{best}(t) = \min_{j\in[1,N]} m_j(t) \tag{12}$$

$$m^{worst}(t) = \max_{j\in[1,N]} m_j(t) \tag{13}$$

For a maximization problem, the operators are swapped for the best and worst masses respectively. The inertial mass, $m_i'(t)$ and gravitational masses, $M_i(t)$ are then computed based on the fitness map developed previously.

$$m_i'(t) = \frac{m_i(t) - m^{worst}(t)}{m^{best}(t) - m^{worst}(t)} \tag{14}$$

$$M_i(t) = \frac{m_i(t)}{\sum_{j=1}^{N} m_j(t)} \tag{15}$$

such that,

$$M_{ai} = M_{pi} = M_{ii} = M_i : i \in [1, N] \tag{16}$$

Then the gravitational constant, $G(t+1)$ and the Euclidean distance $R_{ij}(t)$ is computed as the following:

$$G(t+1) = G(t)\exp\left(\frac{-\alpha t}{T_{\max}}\right) \tag{17}$$

$$R_{ij}(t) = \sqrt{\left(\Delta x_i(t)\right)^2 - \left(\Delta x_j(t)\right)^2} \tag{18}$$

where α is some arbitrary constant and T_{max} is the maximum number of iterations, $x_i(t)$ and $x_j(t)$ are the positions of particle i and j at time t . The interaction forces at time t, $F_{ij}^d(t)$ for each of the masses are then computed:

$$F_{ij}^d(t) = G(t)\left(\frac{M_{pi}(t) \times M_{aj}(t)}{R_{ij}(t) + \varepsilon}\right) \times \left(x_j^d(t) - x_i^d(t)\right) \tag{19}$$

where ε is some small parameter. The total force acting on each mass *i* is given in a stochastic form as the following:

$$F_i^d(t) = \sum_{\substack{j=1 \\ i \neq j}}^{N} rand(w_j) F_{ij}^d(t) : rand(w_j) \in [0,1] \tag{20}$$

where *rand*(w_j) is a randomly assigned weight. Consequently, the acceleration of each of the masses, $a_i^d(t)$ is then as follows:

$$a_i^d(t) = \left(\frac{F_i^d(t)}{M_{ii}(t)} \right) \tag{21}$$

After the computation of the particle acceleration, the particle position and velocity is then calculated:

$$v_i^d(t+1) = rand(w_j) + v_i^d(t) + a_i^d(t) \tag{22}$$

$$x_i^d(t+1) = x_i^d(t(t) + v_i^d(t(t+1) \tag{23}$$

where *rand*(w_j) is a randomly assigned weight. The iterations are then continued until the all mass agents are at their fittest positions in the fitness landscape and some stopping criterion which is set by the user is met. The GSA algorithm is presented in Algorithm 2 and the parameter settings are given in Table 3:

3. Hopfield-Enhanced Particle Swarm Optimization

The HoPSO employed in this work merges the ideas from the Ising spin models (Amit *et al.*, 1986), the Hopfield Neural Networks (HNN) (Hopfield, 1982; Hopfield, 1984) and the PSO technique. The Ising model is a ferromagnetism model in statistical physics developed by Ernst Ising in 1925 (Dyson, (1969)). This model is constructed based on the concept that atomic configurations can be represented in terms

Algorithm 2. Gravitational Search Algorithm (GSA)

Step 1: Initialize no of particles, m_i and initial positions, $x_i(0)$
Step 2: Initialize parameters: $G(0)$, α.
Step 3: Compute gravitational & inertial masses based on the fitness map
Step 4: Compute the gravitational constant, $G(t)$
Step 5: Compute distance between agents, $R_{ij}(t)$
Step 6: Compute total force, $F_i^d(t)$ and the acceleration $a_i^d(t)$ of each agent.
Step 7: Compute new velocity $v_i(t)$ and position $x_i(t)$ for each agent
Step 8: If the fitness criterion is satisfied and $t = T_{max}$, halt and print solutions
else proceed to step 3

Table 3. HoPSO parameter settings

Parameters	Values
Initial parameter (c_1, c_2, r_1, r_2, w)	(1, 1.2, 0.5, 0.5, 0.8)
Number of particles	6
Initial social influence ($s_1, s_2, s_3, s_4, s_5, s_6$)	(1.1, 1.05,1.033, 1.025, 1.02, 1.017)
Initial personal influence ($p_1, p_2, p_3, p_4, p_5, p_6$)	(3, 4, 5, 6, 7,8)
Constant Parameter, U	100
Constant Parameter, θ	0.02

Table 2. GSA parameter settings

Parameters	Values
Initial parameter (G_o)	100
Number of mass agents, n	6
Constant parameter, α	20
Constant parameter, ε	0.01

of magnetic dipole moments (atomic spins – the quantization of magnetism) which are in either in state +1 or -1. These spin interactions are localized to their closest neighbour and are usually in a lattice arrangement. The central idea of the Ising model is to detect phase transitions in real substances. Hence, repetitive magnetization would under certain circumstances and after a period of time would cause the total energy of the magnetized material to converge into local minima.

The statistical physics models developed in the Ising model (Dyson, 1969) then inspired the development of a new type of neural net called the Hopfield Recurrent Artificial Neural Network (HNN) with improved convergence properties. The HNN was developed in 1982 by Hopfield, (1982) and Hopfield, (1984). These neural nets observed to have applications in optimization problems (for instance in Lee, Sode-Yome *et al.*, (1998) and Tank & Hopfield (1986)).

One of the key features of the HNN is that there is a decrease in the energy by a finite amount whenever there is a change in the network's state. This essential property confirms or accentuates the convergence of the output whenever the network state is changed. HNNs are usually constructed by a finite number of interlinked neurons. These neurons update their weights or their activation values (outputs from threshold neurons) independently relative to other neurons in the network. It is important to take note that the neurons in these sorts of networks are not directly connected to themselves and each neurons function as an input as well as output. In HNNs, the activation values are usually binary (+1 or -1) and all the weights of neurons are symmetric ($w_{ij} = w_{ji}$).

In this work, the ideas of discreet magnetic spin as well symmetric weight assumptions (from the Ising spin model) that were used in the HNN are applied to the PSO algorithm to improve its convergence capabilities. Similarly, a set of random weights w_{ij} was initialized and the symmetric property was imposed. Then, the modification of the particle position update equation in the PSO algorithm was done:

$$x_i(t+1) = x_i(t) + v_i(t) \tag{24}$$

where $x_i(t)$ is the particle position and $v_i(t)$ is the particle velocity and t is the iteration counter. The weight previously defined is then introduced as a coefficient that dampens the position of the previous iteration such as the following:

$$s_j(t+1) = w_{ij}(t)x_i(t) + v_i(t) \tag{25}$$

The particle position at the next iteration is computed in the following piece-wise form:

$$x_j(t+1) = \begin{cases} +1 & iff \quad s_j(t+1) > U \\ -1 & iff \quad s_j(t+1) < U \\ x_j(t) & iff \quad otherwise \end{cases} \tag{26}$$

where U is the threshold which user-defined. Then the energy function is defined as follows:

$$E = -\frac{1}{2}\sum_{j=1}\sum_{i=1} x_j(t+1)x_j(t)w_{ij} - \sum_{i=1}\theta x_j(t) \text{ such that } j \neq i \tag{27}$$

where θ is the coefficient which user-defined. This way the PSO algorithm is iterated until the total energy of the system reaches its local minima. To detect the instance when the HoPSO has reached the local minima, the difference between the energy levels at two consequent states is computed as follows:

$$dE = E_{n+1} - E_n \tag{28}$$

where n is the index that denotes the states. Thus, a new variant for the PSO algorithm is developed which in this work shall be termed as Hopfield PSO (HoPSO) algorithm. The algorithm and the flowchart for this HoPSO method are given in Algorithm 3 and Figure 2 respectively:

The parameter settings specified in the HoPSO algorithm in this work is as in Table 4.

MEASUREMENT INDICATORS

1. Convergence Metric

The convergence metric (Deb & Jain, 2002) was developed to measure the degree of convergence of a solution set relative to a reference/basis set. Since the Pareto optimal frontier was not known, a target

Figure 2. Flowchart for the HoPSO technique

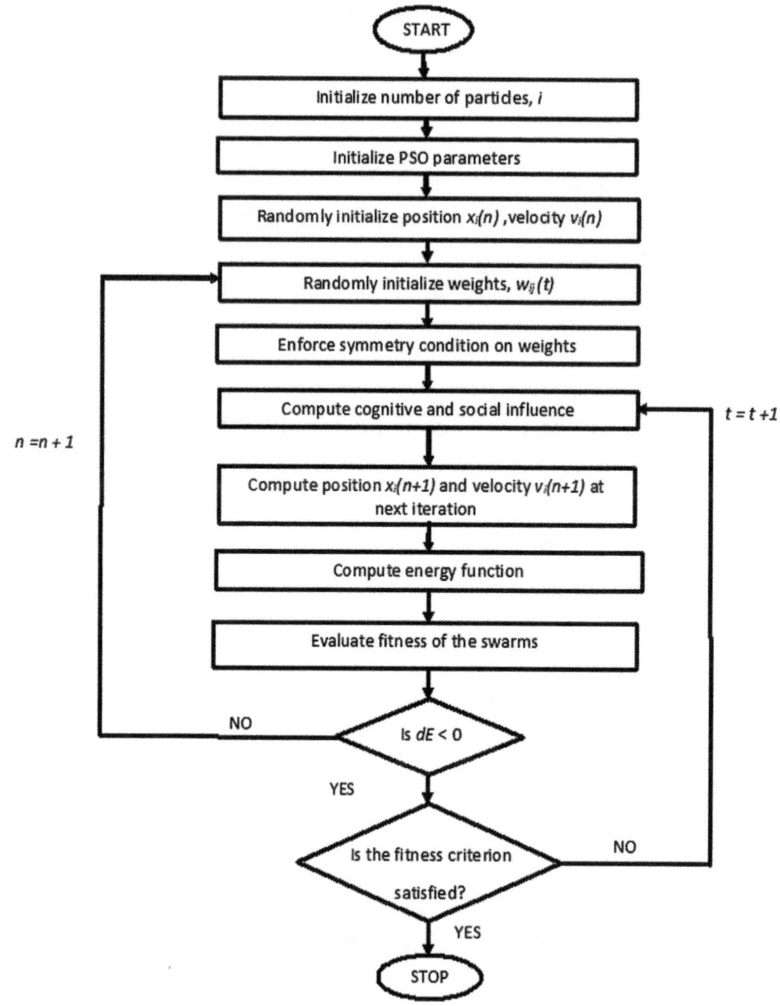

vector P^* which is the most dominant vector was employed as the reference set. For a set of solutions, the formulation to compute the convergence metric for a single run of the program is as follows:

$$d_i = \min_{j=1}^{|P^*|} \sqrt{\sum_{k=1}^{M} \left(\frac{f_k^i - f_k^j}{f_k^{\max} - f_k^{\min}} \right)} \tag{29}$$

where i and j denotes the subsequent objective function values, k is the index which denotes the objective function, f_k^{max} is the maximum objective function value, f_k^{min} is the minimum objective function value and M denotes the overall number of objectives. For this convergence metric, low the metric values indicate high convergence characteristics among the solution vectors.

Algorithm 3. Hopfield Particle Swarm Optimization (HoPSO)

```
Step 1: Set no of particles, i and the initialize parameter settings w, c1, c2,
r1, r2,no
Step 2: Randomly initialize particles' position xi (t) and velocity vi(t)
Step 3: Randomly initialize weights, wij (t)
Step 4: Enforce symmetry condition on weights
Step 5: Calculate cognitive and social components of the particles
Step 6: Compute position xi(t+1) and velocity vi(t+1) of the particles at next
iteration
Step 7: Compute energy function
Step 8:  Proceed with the evaluation of the fitness of each particle in the
swarm.
Step 8:  If the energy difference between states are < 0, proceed to Step 9
         else go to Step3.
Step 9:  If the fitness conditions are satisfied and t < Tmax, stop program and
print solutions,
 else go to Step 3.
```

Table 4. Ranked individual solutions generated by the GSA algorithm

Description		Best	Median	Worst
Objective Function	f_1	6.13334	5.90007	5.88888
	f_2	81.1329	75.1058	74.8027
	f_3	11.9987	9.8936	9.78725
Decision Variable	x_1	19.5	19.5	19.5
	x_2	27.1	27.1	27.1
	x_3	17.0987	7.5167	7.11013
HVI		5970.741	4384.15	4311.324

2. Hypervolume Indicator

One approach that has been effective in measuring the quality of the solution set that constructs the Pareto-frontier in cases where the Pareto frontier is unknown is the Hypervolume Indicator (HVI) (Zitzler & Thiele, 1998). The HVI is a Pareto-compliant gauge that is employed for the measurement of the quality of solution sets in MO optimization problems (Zitzler & Thiele, 1998; Beume *et al.*, 2007). Pareto-compliance is defined as follows:

Let two solution sets exists such that these sets are produced by some arbitrary algorithm for a particular MO problem. Then the property of Pareto compliance implies that the indicator would consistently produce a higher value for one of the solution set only if that solution set strictly dominates the other.

In the past couple of years, many investigations were conducted on methods to apply evolutionary algorithms to MO problems. Besides, many ideas have also been proposed on methods to attain a good solution spread over the Pareto frontier. Unfortunately, in all these works, the optimization goal remains fuzzy since there is no established technique to measure the quality of a solution set produced by any MO algorithm (especially in cases where there are more than two objectives). The impact of this issue magnifies especially in real-world problems where it is often that the Pareto frontier is unknown (unlike theoretical test functions where the Pareto frontier is known and can be used to benchmark the solution quality).

Recently, this indicator has been frequently applied in many works involving MO problems. The HVI is the only indicator which is strictly Pareto-compliant that can be used to measure the quality of solution sets in MO optimization problems. To get a clearer picture of the idea the Pareto-compliance, the concepts of Pareto dominance needs to be defined.

Pareto dominance can be categorised into three types which are; strictly dominates ($\succ$), weakly dominates ($\succeq$) and indifferent (~). Let two solution vectors be **a** and **b**. Then if the solution vector **a** dominates the vector **b** in all the objectives then **a** strictly dominates **b** (**a** $\succ$ b). If the solution vector **a** dominates the vector **b** in some of the objectives but not all, then **a** weakly dominates **b** (**a** $\succeq$ b). Finally, if the solution vector **a** does not dominate the vector **b** and the solution vector **b** does not dominate **a** as well in all the objectives, then **a** is indifferent to **b** (**a** ~ **b**).

Strictly Pareto-compliant can be defined as the following. Let there be two solution sets say; *X* and *Y* for a particular MO problem. If the HVI value for *X* is greater than *Y*, then the solution set $X \succ$ Y or $X \succeq Y$. The HVI measures the volume of the dominated section of the objective space and can be applied for multi-dimensional scenarios. When using the HVI, a reference point needs to be defined. In this work, we define a 'nadir point'. The nadir point is a point which is dominated by all the solutions from the approximate Pareto frontier. Relative to this point, the volume of the space of all dominated solutions can be computed. The HVI of a solution set $x_d \in X$ can be defined as follows:

$$HVI(X) = vol\left(\bigcup_{(x_1,\ldots x_d)\in X} [r_1, x_1]\times\ldots\times[r_d, x_d]\right) \tag{30}$$

where $r_{1,\ldots,}r_d$ is the reference point and *vol*(.) being the usual Lebesgue measure. The larger the value of the HVI, the more dominant the solution is in the objective space. The characteristics of the HVI are as follows:

1. Strictly monotonic indicator.
2. Its computational effort is exponential to the amount of solution vectors.
3. It requires an upper-bounding vector (nadir point).

In this work the HVI is used to measure the quality of the approximation of the Pareto front by the solution sets produced by the algorithms when used in conjunction with the NBI approach. In this work the HVI is used to measure the quality of the approximation of the Pareto front by the swarm algorithms when used in conjunction with the NBI framework.

COMPUTATIONAL RESULTS AND ANALYSIS

In this work, the Pareto frontier is constructed using the solution set produced by the swarm algorithms. In this work, 27 solution points were produced by all the algorithms for the construction of the Pareto frontier. The algorithms presented in this work were executed on a 2GHz Intel dual-core processor. The dominance levels of these solutions were measured using the HVI. The nadir point selected in this work is $(r_1, r_2, r_3) = (0, 0, 0)$. The individual solutions with their designated weights produced by the GSA and the PSO algorithms were gauged with the HVI and the ranked solutions were determined. These individual solutions for both the algorithms and their respective dominance levels are shown in Table 4 and 5 respectively.

The associated weights for the best, median and worst solution for the GSA and PSO algorithms are shown in Table 6.

The depiction of the solution spread of the approximate Pareto frontier obtained using the GSA and PSO algorithms are shown in Figures 3 and 4.

Similar to the GSA and PSO algorithms, the individual solutions for the HoPSO technique and the respective dominance levels are given in Table 7.

The associated weights (w_1, w_2, w_3) for the best, median and worst solutions produced by the HoPSO technique are (0.5, 0.4, 0.1), (0.1, 0.2, 0.7) and (0.4, 0.4, 0.2). The approximate Pareto frontiers obtained using the HoPSO algorithm is given in Figure 5.

For comparing the percentage of dominance among the techniques, a simple error metric (%) is utilized. The following equation provides the difference (%) of the performance of Algorithm A when compared against that of Algorithm B using the HVI:

$$\% \quad of \quad Difference = \left| \frac{HVI_A - HVI_B}{HVI_B} \right| \times 100\% \tag{28}$$

It can be observed using the HVI that the best solution obtained by HoPSO algorithm dominates the best solution produced by the PSO and GSA algorithms by 1.367% and 291.192%. The comparison of the best candidate solutions obtained by the methods employed in this work against the PSO method in Shashi *et al.*, (2010) is shown in Table 8.

Table 5. Ranked individual solutions generated by the PSO algorithm

Description		Best	Median	Worst
Objective Function	f_1	8.53054	8.5228	8.52069
	f_2	108.941	108.861	108.842
	f_3	24.7943	24.7803	24.7694
Decision Variable	x_1	49.6226	49.4068	49.4391
	x_2	72.8999	72.7734	72.6776
	x_3	30.6499	30.6636	30.5293
HVI		23041.98	22991.18	22971.36

Table 6. Ranked individual solutions generated by the GSA and PSO algorithms

Weights		Best	Median	Worst
GSA	w_1	0.4	0.4	0.1
	w_2	0.2	0.3	0.4
	w_3	0.4	0.3	0.5
PSO	w_1	0.5	0.3	0.2
	w_2	0.2	0.4	0.6
	w_3	0.3	0.3	0.2

Figure 3. Pareto frontier of the objectives obtained by the GSA method

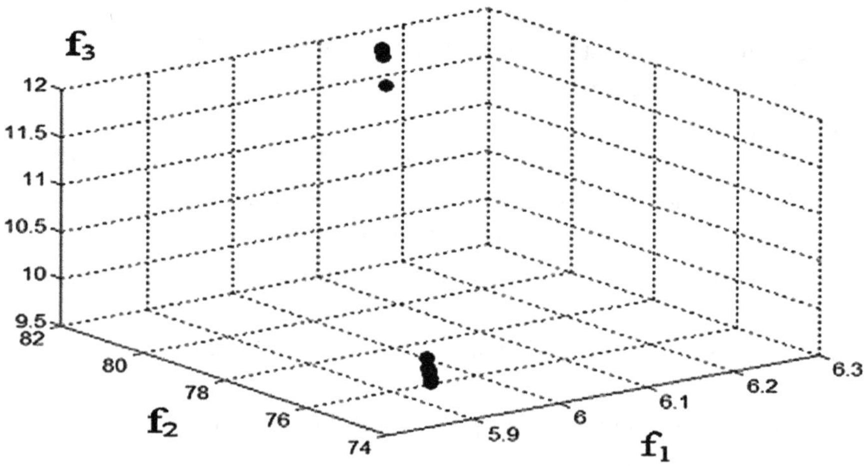

Figure 4. Pareto frontier of the objectives obtained by the PSO technique

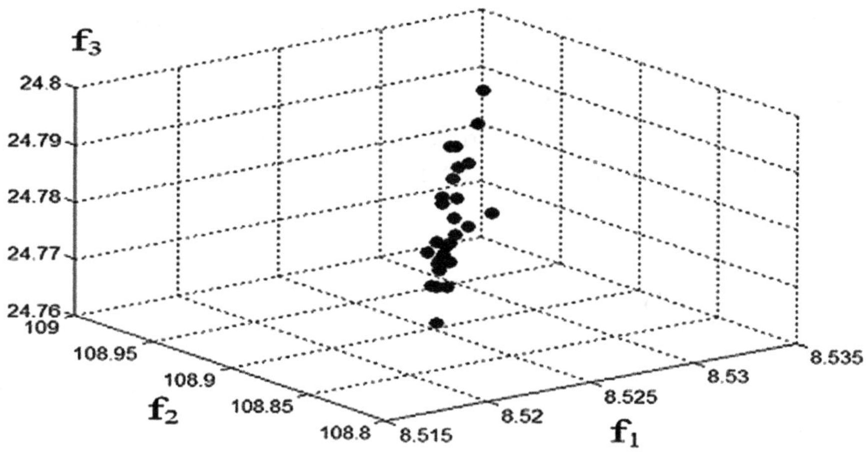

Figure 5. Pareto frontier of the objectives obtained by the HoPSO technique

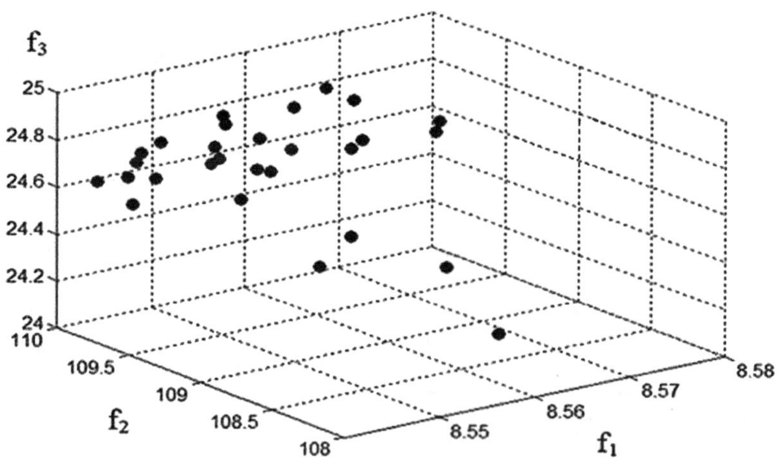

Table 7. Ranked individual solutions generated by the HoPSO algorithm

Description		Best	Median	Worst
Objective Function	f_1	8.5675	8.54683	8.56125
	f_2	109.929	109.903	108.326
	f_3	24.8	24.6649	24.1797
Decision Variable	x_1	55.6416	57.699	55.3454
	x_2	72.8689	72.5172	72.81
	x_3	33.5713	36.9105	22.7159
HVI		23357.05	23168.29	22424.4

In Table 8, it can be observed that the best solutions produces by HoPSO and PSO algorithms used in this work are more dominant than the real-coded GA approach (Shashi *et al.*, 2010) by 1.5005% and 0.1314% respectively. The HVI computed for the entire frontier of each solution set produced by an algorithm gives the true measure of dominance when compared with another algorithm. In this work, the HVI for the entire frontier was computed for each of the algorithm. The execution time for the GSA, PSO and the HoPSO algorithms to generate the entire frontier are 9.588, 13.113 and 13.268 seconds respectively. The HVI for the entire frontier for the solution sets produced by all the algorithms in this work is shown in Figure 6:

The Pareto frontier produced by the HoPSO algorithm is more dominant than the one produced by the GSA and PSO algorithms by 364.1% and 0.497% respectively. The level of convergence of the Pareto frontier for all the algorithms employed in this work is measured using the convergence metric (refer to equation 26). The level of convergence of the Pareto frontiers generated by the GSA, PSO and the HoPSO technique is shown in Figure 7.

As seen in Table 7, the individual best solution of the HoPSO technique maximizes all the objectives f_1,f_2 and f_3 very effectively as compared to the solutions obtained by using the GSA and PSO algorithms. Thus, it can be said that the HoPSO method in this work outweighs the overall optimization capabilities of PSO, GSA and the real-coded GA (Shashi *et al.*, 2010). The HVI values are show in Table 7. The HoPSO technique takes the longest computational time to produce the entire Pareto frontier followed by the PSO and GSA methods. The HoPSO algorithm produces Pareto frontier with the highest level

Table 8. The comparison of the best solutions obtained by the algorithms

Description		Real-coded GA (Shashi *et al.*, 2010)	PSO	GSA	HoPSO
Objective Function	f_1	8.43	8.53054	6.13334	8.5675
	f_2	110.026	108.941	81.1329	109.929
	f_3	24.81	24.7943	11.9987	24.8
Decision Variable	x_1	56.62	49.6226	19.5	55.6416
	x_2	72.9	72.8999	27.1	72.8689
	x_3	34.76	30.6499	17.0987	33.5713
HVI		23011.75	23041.98	5970.741	23357.05

Figure 6. Dominance comparison between Pareto frontiers produced by the GSA, PSO and HoPSO techniques

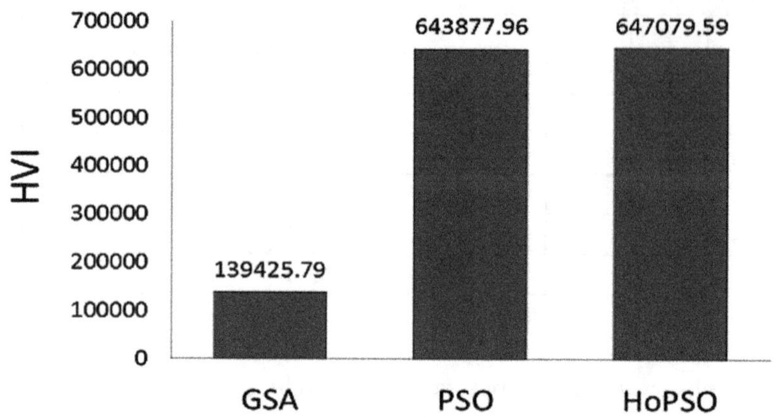

Figure 7. Comparison of the convergence level between Pareto frontiers produced by the PSO and HoPSO techniques

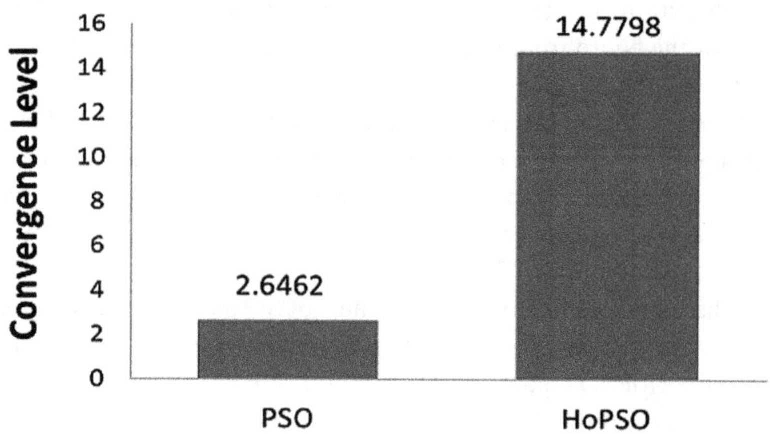

of dominance. However, due to high algorithmic complexity it sacrifices computational time the most as compared to the GSA and PSO methods. As can be seen in Figures 4 and 5, the frontiers produced by the PSO and HoPSO algorithms are more uniform and diversely spaced in the objective space as compared to the GSA algorithm in Figure 3. The frontier produced by the GSA algorithm seems to be confined to certain regions in the objective space. Thus, the solution spacing seems to heavily influence the ability of the algorithm to approximate the Pareto frontier dominantly. Localized solutions on the frontier such as the ones produced by the GSA algorithm omit some of the solutions in the objective space. This in effect causes the GSA algorithm to have a lower level of dominance as compared to the PSO and HoPSO algorithms.

In Figure 7, it can be observed that the Pareto frontier produced by the HoPSO method has a higher level of convergence as compared to the PSO technique. In this case, it could be inferred that the level of convergence as well as the level of dominance of the HoPSO algorithm is higher than that of the PSO technique. Therefore, in a post-analytical sense, it can be said that the level of convergence may be one of the properties of the objective space which is interlinked to the level of dominance.

CONCLUSION

In this work, a new optimal solution as well as a dominant Pareto frontier was achieved by the HoPSO method (refer to Table 7 and Figure 6). The HoPSO technique outperforms the GSA, PSO and the real-coded GA (Shashi *et al*, 2010) methods. Thus, it can be said that the HoPSO method in this work outweighs the overall optimization capabilities of GSA and PSO techniques (refer to levels of dominance in Figure 6). In this work, it can be seen that the level of convergence is interconnected to the level of Pareto dominance. Due to this, the Hopfield enhancement in the HoPSO technique improves the generation of convergent solution which in effect increases the dominance level of the solutions. In this work, the swarm intelligence algorithms performed stable computations during the program executions. All Pareto-efficient solutions produced by the algorithms developed in this work were feasible and no constraints were compromised. One of the advantages of using the swarm-based algorithms as compared to the other algorithms used in this work is that it produces highly effective results in terms of approximating the Pareto frontier and is computationally efficient.

FUTURE RESEARCH DIRECTIONS

In future works, other measurement metrics such as spacing or diversity metrics could be employed and the relations between the levels of dominance (using the HVI (Ganesan *et al.*, 2013) could be obtained. As mentioned in Ganesan *et al.*, (2013), understanding the structure of the objective space is critical when developing algorithmic enhancements. Keeping this purpose in mind, testing MO with a wide range of algorithms and measurement metrics would improve our understanding on the problem morphology which would in effect provide insight on approaches to enhance our algorithms to achieve the global or near-global optima.

In addition, hybridization procedures (Elamvazuthi *et al.*, 2011) could also be utilized and compared with algorithmic enhancement approaches. This way, critical differences in terms of performance and solution quality may be identified. Besides, the approach employed in this work could also be extended towards problems containing various uncertainties (Vasant, *et al.*, 2010; Ganesan *et al.*, 2014).

ACKNOWLEDGMENT

The authors are very grateful to the reviewers for their efforts in providing useful and constructive insights that has been overlooked by the author during the preparation of the manuscript. The authors would also like to thank Universiti Teknologi Petronas for their support during progress of this work. Special thanks to our friends, colleagues and students who continuously inspire and support us throughout the phases of our research.

REFERENCES

Aguirre, A. H., Zebulum, R. S., & Coello Coello, C. (2004). Evolutionary multiobjective design targeting a Field Programmable Transistor Array. In *Proceedings of the Conference on Evolvable Hardware*. doi:10.1109/EH.2004.1310831

Amit, D. J., Gutfreund, H., & Sompolinsky, H. (1986). Spin-glass models of neural networks. *Physical Review A., 32*(2), 1007–1018. doi:10.1103/PhysRevA.32.1007 PMID:9896156

Beume, N., Naujoks, B., & Emmerich, M. (2007). SMS-EMOA: Multiobjective selection based on dominated hypervolume. *European Journal of Operational Research, 181*(3), 1653–1669. doi:10.1016/j.ejor.2006.08.008

Das, I., & Dennis, J. E. (1998). Normal-boundary intersection: A new method for generating the Pareto surface in nonlinear multicriteria optimization problems. *SIAM Journal on Optimization, 8*(3), 631–657. doi:10.1137/S1052623496307510

Deb, K., & Jain, S. (2002). *Running performance metrics for Evolutionary Multiobjective Optimization*. KanGAL Report No. 2002004. Kanpur Genetic Algorithms Laboratory, Indian Institute of Technology Kanpur.

Deb, K., Pratap, A., Agarwal, S., & Meyarivan, T. (2002). A Fast and Elitist Multiobjective Genetic Algorithm: NSGA-II. *IEEE Transactions on Evolutionary Computation, 6*(2), 182–197. doi:10.1109/4235.996017

Deb, K., Pratap, A., & Moitra, S. (2000), Mechanical component design for multiple objectives using elitist non-dominatedsorting GA. In *Proceedings of the Parallel Problem Solving from Nature VI Conference*. doi:10.1007/3-540-45356-3_84

Dyson, F. J. (1969). 'Existence of a phase-transition in a one-dimensional Ising ferromagnet'. *Communications in Mathematical Physics, 12*(2), 91–107. doi:10.1007/BF01645907

Elamvazuthi, I., Ganesan, T., & Vasant, P. (2011). A comparative study of HNN and Hybrid HNN-PSO techniques in the optimization of distributed generation (DG) power systems. *International Conference on Advanced Computer Science and Information System (ICACSIS)*.

Emmerich, M., Beume, N., & Naujoks, B. (2005). An EMO Algorithm Using the Hypervolume Measure as Selection Criterion. In *Conference on Evolutionary Multi-Criterion Optimization (EMO 2005)*. doi:10.1007/978-3-540-31880-4_5

Fishburn, P. C. (1967). *Additive Utilities with Incomplete Product Set: Applications to Priorities and Assignments*. Baltimore, MD: ORSA.

Ganesan, T., Elamvazuthi, I., Shaari, K. Z. K., & Vasant, P. (2013). An Algorithmic Framework for Multiobjective Optimization. The Scientific World Journal.

Ganesan, T., Vasant, P., & Elamvazuthi, I. (2014). Hopfield neural networks approach for design optimization of hybrid power systems with multiple renewable energy sources in a fuzzy environment. *Journal of Intelligent and Fuzzy Systems, 2*(5), 2143–2154.

Heo, J. S., Lee, K. Y., & Garduno-Ramirez, R. (2006). Multiobjective Control of Power Plants Using Particle Swarm Optimization Techniques. *IEEE Transactions on Energy Conversion*, *21*(2), 552–561. doi:10.1109/TEC.2005.858078

Hismath, I., Wan Aida, W. M., & Ho, C. W. (2011). Optimization of extraction conditions for phenolic compounds from neem (Azadirachta indica) leaves. *International Food Research Journal*, *18*(3), 931–939.

Hopfield, J. J. (1982). Neural Networks and Physical Systems with Emergent Collective Computational Abilities. *Proceedings of the National Academy of Sciences of the United States of America*, *29*(8), 2554–2558. doi:10.1073/pnas.79.8.2554 PMID:6953413

Hopfield, J. J. (1984). Neurons with Graded Response have Collective Computational Properties like those of Two-State Neurons. *Proceedings of the National Academy of Sciences of the United States of America*, *81*(10), 3088–3092. doi:10.1073/pnas.81.10.3088 PMID:6587342

Jie, H. X., & Wei, C. (2008). Optimization of extraction process of crude polysaccharides from wild edible BaChu mushroom by response surface methodology. *Carbohydrate Polymers*, *72*(1), 67–74. doi:10.1016/j.carbpol.2007.07.034

Kennedy, J., & Eberhart, R. (1995). Particle Swarm Optimization. In *Proceedings of the International Conference on Neural Networks*.

Kusiak, A., Xu, G., & Tang, F. (2010). Optimization of an HVAC System With A Strength Multi-Objective Particle-Swarm Algorithm. *Energy*, *36*(5), 184–192.

Lee, K. Y., Sode-Yome, A., & Park, J. H. (1998). Adaptive Hopfield Neural Networks for Economic Load Dispatch. *IEEE Transactions on Power Systems*, *13*(2), 519–526. doi:10.1109/59.667377

Li, C., & Zhou, J. (2011). Parameters identification of hydraulic turbine governing system using improved gravitational search algorithm. *Energy Conversion and Management*, *52*(1), 374–381. doi:10.1016/j.enconman.2010.07.012

Li, H. B., Wong, C. C., Chen, K. W., & Chen, F. (2008). Antioxidant properties in vitro and total phenolic contents in methanol extracts from medicinal plants. *LWT Food Science and Technology*, *41*(3), 385–390. doi:10.1016/j.lwt.2007.03.011

Luyben, M. L., & Floudas, C. A. (1994). Analyzing the interaction of design and control. 1. A multiobjective framework and application to binary distillation synthesis. *Computers & Chemical Engineering*, *18*(10), 933–969. doi:10.1016/0098-1354(94)E0013-D

Nain, P.K.S., Giri, J.M., Sharma, S. & Deb, K., (2010). Multi-objective Performance Optimization of Thermo-Electric Coolers Using Dimensional Structural Parameters. In *Proceedings of SEMCCO 2010* (LNCS), (vol. 6466, pp. 607-614). Springer-Verlag Berlin Heidelberg.

Neshat, M.A., Sepidnam, G.A., Sargolzaei, M.A. & Toosi, A.A. (2012). Artificial fish swarm algorithm: a survey of the state-of-the-art, hybridization, combinatorial and indicative applications. *Journal of Artificial Intelligence Review*, 1-33.

Oshima, T., Sagara, K., Yoshida, T., Tong, Y. Y., Zhang, G., & Chen, Y. H. (1988). Determination of geniposide, gardeniside, geniposidic acid and genipin-1-β-gentiobioside in Gardenia jasminoides by high-performance liquid chromatography. *Journal of Chromatography. A*, *455*, 410–414. doi:10.1016/S0021-9673(01)82148-8 PMID:3235620

Palli, N., Azram, S., McCluskey, P., & Sundararajan, R. (1999). An interactive multistage ε-inequality constraint method for multiple objectives decision making. *ASME Journal of Mechanical Design.*, *120*(4), 678–686. doi:10.1115/1.2829331

Rashedi, E., Nezamabadi-pour, H., & Saryazdi, S. (2009). GSA: A Gravitational Search Algorithm. *Information Sciences*, *179*(13), 2232–2248. doi:10.1016/j.ins.2009.03.004

Reddy, M. J., & Kumar, D. N. (2007). An efficient multi-objective optimization algorithm based on swarm intelligence for engineering design. In Engineering Optimization. Taylor & Francis.

Sandgren, E. (1994). Multicriteria design optimization by goal programming. In Advances in Design Optimization. Chapman & Hall.

Shashi, D. K., & Katiyar, V. K. (2010). Multiobjective Extraction Optimization of Bioactive Compounds from Gardenia Using Real Coded Genetic Algorithm. In *Proceedings of 6th World Congress of Biomechanics* (WCB 2010).

Statnikov, R. B., & Matusov, J. B. (1995). *Multicriteria Optimization and Engineering*. New York: Chapman and Hall. doi:10.1007/978-1-4615-2089-4

Tank, D. W., & Hopfield, J. J. (1986). Simple Neural Optimization Network: An A\D Converter, Signal Decision Circuit and a Linear Programming Circuit. *IEEE Transactions on Circuits and Systems*, *CAS-33*(5), 533–541. doi:10.1109/TCS.1986.1085953

Triantaphyllou, E., (2000). *Multi-Criteria Decision Making: A Comparative Study*. Dordrecht, The Netherlands: Kluwer Academic Publishers.

Ubaidullah, S., Upadhyay, R., Raut, S. A. & Rahman, I. (2012). Optimization of Continuous Extraction Column and Solvent Selection Using Differential Evolution Technique. *Journal of The Institution of Engineers (India): Series E*, *93*(1), 49-54.

Van Calsteren, M. R., Bissonnette, M. C., Cormier, F., Dufresne, C., Ichi, T., LeBlanc, J. C. Y., & Roewer, I. et al. (1997). Spectroscopic characterization of crocetin derivatives from Crocus sativus and Gardenia jasminoides. *Journal of Agricultural and Food Chemistry*, *45*(4), 55–1061. doi:10.1021/jf9603487

Vasant, P., Elamvazuthi, I., Ganesan, T., & Webb, J. F. (2010). Iterative fuzzy optimization approach for crude oil refinery industry. *Scientific Annals of Computer Science*, *8*(2), 261–280.

Yang, B. S., Yeun, Y.-S., & Ruy, W.-S. (2002). Managing approximation models in multiobjective optimization. *Structural Multidiscsiplinary Optimization*, *24*(2), 141–156. doi:10.1007/s00158-002-0224-0

Zhang, Q., & Li, H. (2007). MOEA/D: A multiobjective evolutionary algorithm based on decomposition. *IEEE Transactions on Evolutionary Computation*, *11*(6), 712–731. doi:10.1109/TEVC.2007.892759

Zitzler, E., & Thiele, L. (1998). Multiobjective Optimization Using Evolutionary Algorithms - A Comparative Case Study. *In Conference on Parallel Problem Solving from Nature (PPSN V)*. doi:10.1007/BFb0056872

ADDITIONAL READING

Abraham, A., Guo, H., & Liu, H. (2006). Swarm intelligence: foundations, perspectives and applications. In *Swarm Intelligent Systems* (pp. 3–25). Springer Berlin Heidelberg. doi:10.1007/978-3-540-33869-7_1

Arumuggam, K., & Singh, B. S. M. (2013). Optimization of hybrid solar and wind power generation. *Journal of Applied Sciences*, *13*(6), 869–875. doi:10.3923/jas.2013.869.875

Ceylan, O., Ozdemir, A., & Dag, H. (2012). Branch Outage Simulation Based Contingency Screening by Gravitational Search Algorithm. *International Review of Electrical Engineering*, *7*(1).

Ganesan, T., Elamvazuthi, I., Ku Shaari, K. Z., & Vasant, P. (2013). Swarm intelligence and gravitational search algorithm for multi-objective optimization of synthesis gas production. *Applied Energy*, *103*(0), 368–374. doi:10.1016/j.apenergy.2012.09.059

Ganesan, T., Elamvazuthi, I., & Vasant, P. (2015). Multiobjective Design Optimization of a Nano-CMOS Voltage-Controlled Oscillator Using Game Theoretic-Differential Evolution. *Applied Soft Computing*, *32*, 293–299. doi:10.1016/j.asoc.2015.03.016

Koza, J. R. (1992). *Genetic Programming: On the Programming of Computers by means of Natural Selection*. USA: MIT Press.

Krause, J., Cordeiro, J., Parpinelli, R. S., & Lopes, H. S. (2013). A survey of swarm algorithms applied to discrete optimization problems. Swarm Intelligence and Bio-inspired Computation: Theory and Applications. Elsevier Science & Technology Books, 169-191.

Martens, D., Baesens, B., & Fawcett, T. (2011). Editorial survey: Swarm intelligence for data mining. *Machine Learning*, *82*(1), 1–42. doi:10.1007/s10994-010-5216-5

Mirjalili, S., Mohd Hashim, S. Z., & Moradian Sardroudi, H. (2012). Training feedforward neural networks using hybrid particle swarm optimization and gravitational search algorithm. *Applied Mathematics and Computation*, *218*(22), 11125–11137. doi:10.1016/j.amc.2012.04.069

Mistree, F., Hughes, O. F., & Bras, B. A. (1993). *'The Compromise Decision Support Problem and the Adaptive Linear Programming Algorithm', Structural Optimization: Status and Promise* (pp. 247–286). Washington, D.C.: AIAA.

Mohamed, A. Z., Lee, S. H., Hsu, H. Y., & Nath, N. (2012). A faster path planner using accelerated particle swarm optimization. *Artificial Life and Robotics*, *17*(2), 233–240. doi:10.1007/s10015-012-0051-3

Mohanty, S. (2006). Multiobjective optimization of synthesis gas production using non-dominated sorting genetic algorithm. *Computers & Chemical Engineering*, *30*(6-7), 1019–1025. doi:10.1016/j.compchemeng.2006.01.002

Oplatkova, Z., & Zelinka, I. (2007). Creating evolutionary algorithms by means of analytic programming - design of new cost function. In *ECMS 2007* (pp. 271–276). European Council for Modeling and Simulation. doi:10.7148/2007-0271

Rahimpour, M. R., Parvasi, P., & Setoodeh, P. (2009). Dynamic optimization of a novel radial-flow, spherical-bed methanol synthesis reactor in the presence of catalyst deactivation using Differential Evolution (DE) algorithm. *International Journal of Hydrogen Energy*, *34*(15), 6221–6230. doi:10.1016/j.ijhydene.2009.05.068

Ramachandran, B., Srivastava, S. K., & Cartes, D. A. (2013). Intelligent power management in micro grids with EV penetration. *Expert Systems with Applications*, *40*(16), 6631–6640. doi:10.1016/j.eswa.2013.06.007

Razzaque, M. A., Hong, C. S., Abdullah-Al-Wadud, M., & Chae, O. (2008). A fast algorithm to calculate powers of a Boolean matrix for diameter computation of random graphs. In WALCOM: Algorithms and Computation (pp. 58-69). Springer Berlin Heidelberg. doi:10.1007/978-3-540-77891-2_6

Sabri, N. M., Puteh, M., & Mahmood, M. R. (2013). A Review of Gravitational Search Algorithm. Int. J. Advance. *Soft Comput. Appl, 5*(3).

KEY TERMS AND DEFINITIONS

Gardenia *Jasminoides Ellis*: A fruit that produces chemical products such as crocins, geniposide and the phenolic compounds which are widely used in the food industry as a natural food colorants (dyes). These compounds also have high antioxidant capabilities which make this fruit valuable for medicinal uses.

Gravitational Search Algorithms (GSA): A type of metaheuristic algorithm integrates ideas from Newtonian gravitational laws to search for optimal solutions in the objective space.

Hopfield-Enhanced Particle Swarm Optimization (HoPSO): A novel swarm intelligence technique which integrates concepts from the Hopfield Neural Nets and swarming behavior of organisms to improve the solution dominance of the standard PSO when solving multiobjective problems.

Measurement Metrics: Mathematical metrics used to identify and measure solution characteristics such as; degree of convergence, diversity and dominance.

Metaheuristics: A framework consisting of a class of algorithms employed to find good solutions to optimization problems by iterative improvement of solution quality.

Multiobjective Optimization: Optimization problems which are represented with more than one objective functions.

Normal Boundary Intersection: A solution method where a multiobjective optimization problem is geometrically reduced to a beta-sub problem and then solved as a weighted single objective optimization problem.

Particle Swarm Optimization (PSO): A type of metaheuristic algorithm that uses concepts from swarming behavior of organisms to search for optimal solutions.

APPENDIX: MISCELLANEOUS DATA

Table 9. Values of the objective function for the GSA algorithm for various scalarization

β_1	β_2	β_3	f_1	f_2	f_3	X_1	X_2	X_3	Iterations
0.1	0.2	0.7	5.89547	74.9813	9.84993	19.5	27.1	7.3491	46
0.1	0.3	0.6	6.13174	81.0941	11.9852	19.5	27.1	17.0239	183
0.1	0.4	0.5	5.88888	74.8027	9.78725	19.5	27.1	7.11013	22
0.1	0.5	0.4	6.13333	81.1325	11.9985	19.5	27.1	17.0981	183
0.1	0.6	0.3	6.13133	81.084	11.9817	19.5	27.1	17.0044	182
0.1	0.7	0.2	6.13216	81.1043	11.9887	19.5	27.1	17.0436	183
0.1	0.8	0.1	6.10442	80.4237	11.7523	19.5	27.1	15.7678	159
0.2	0.2	0.6	6.13106	81.0774	11.9794	19.5	27.1	16.9918	182
0.2	0.3	0.5	6.13101	81.0762	11.979	19.5	27.1	16.9895	182
0.2	0.4	0.4	5.88935	74.8154	9.79169	19.5	27.1	7.127	33
0.2	0.5	0.3	5.89061	74.8496	9.80371	19.5	27.1	7.17269	74
0.2	0.6	0.2	6.13329	81.1317	11.9983	19.5	27.1	17.0965	183
0.2	0.7	0.1	6.12652	80.9668	11.941	19.5	27.1	16.7801	179
0.3	0.2	0.5	5.89441	74.9525	9.83981	19.5	27.1	7.31041	52
0.3	0.3	0.4	5.89274	74.9072	9.82393	19.5	27.1	7.24972	56
0.3	0.4	0.3	5.89446	74.954	9.84033	19.5	27.1	7.3124	28
0.3	0.5	0.2	5.89877	75.0706	9.88126	19.5	27.1	7.46924	40
0.3	0.6	0.1	5.89774	75.0427	9.87148	19.5	27.1	7.43166	131
0.4	0.2	0.4	6.13334	81.1329	11.9987	19.5	27.1	17.0987	183
0.4	0.3	0.3	5.90007	75.1058	9.8936	19.5	27.1	7.5167	36
0.4	0.4	0.2	5.89856	75.065	9.87931	19.5	27.1	7.46167	181
0.4	0.5	0.1	6.13207	81.1019	11.9879	19.5	27.1	17.039	183
0.5	0.2	0.3	5.89744	75.0346	9.86861	19.5	27.1	7.42071	17
0.5	0.3	0.2	5.89236	74.8971	9.82036	19.5	27.1	7.23612	50
0.5	0.4	0.1	5.90698	75.2925	9.95911	19.5	27.1	7.76982	76
0.6	0.2	0.2	5.89499	74.9683	9.84537	19.5	27.1	7.33162	107
0.6	0.3	0.1	5.90615	75.27	9.95123	19.5	27.1	7.73926	31
0.7	0.2	0.1	5.89164	74.8774	9.81347	19.5	27.1	7.20983	73

Table 10. Values of the objective function for the PSO algorithm for various scalarization

β_1	β_2	β_3	f_1	f_2	f_3	X_1	X_2	X_3	*Iterations*
0.1	0.3	0.6	8.52396	108.876	24.795	49.3379	72.8828	30.932	309
0.1	0.4	0.5	8.51943	108.829	24.7781	49.2946	72.7417	30.7565	283
0.1	0.5	0.4	8.52782	108.909	24.7934	49.484	72.895	30.7332	240
0.1	0.6	0.3	8.52153	108.853	24.782	49.3628	72.7718	30.7571	269
0.1	0.7	0.2	8.52156	108.845	24.7794	49.3387	72.7722	30.6909	255
0.1	0.8	0.1	8.5208	108.842	24.7786	49.3429	72.752	30.7115	261
0.2	0.2	0.6	8.52055	108.836	24.7806	49.2918	72.7752	30.7579	276
0.2	0.3	0.5	8.52368	108.87	24.7832	49.4167	72.7983	30.6892	245
0.2	0.4	0.4	8.52432	108.879	24.7888	49.407	72.8369	30.7838	266
0.2	0.5	0.3	8.52756	108.896	24.779	49.549	72.812	30.4389	164
0.2	0.6	0.2	8.52069	108.842	24.7694	49.4391	72.6776	30.5293	203
0.2	0.7	0.1	8.52265	108.862	24.7798	49.4206	72.7631	30.663	236
0.3	0.2	0.5	8.52181	108.85	24.7743	49.4223	72.7248	30.58	217
0.3	0.3	0.4	8.52444	108.878	24.7854	49.4322	72.8161	30.7055	244
0.3	0.4	0.3	8.5228	108.861	24.7803	49.4068	72.7734	30.6636	238
0.3	0.5	0.2	8.52059	108.841	24.7759	49.3695	72.7267	30.6658	245
0.3	0.6	0.1	8.52269	108.864	24.7781	49.4506	72.7456	30.6276	220
0.4	0.2	0.4	8.5201	108.834	24.781	49.2812	72.7724	30.7884	285
0.4	0.3	0.3	8.52577	108.888	24.7898	49.4257	72.8627	30.739	250
0.4	0.4	0.2	8.52578	108.895	24.7885	49.4847	72.8344	30.7215	233
0.4	0.5	0.1	8.522	108.856	24.7882	49.3125	72.8262	30.8662	301
0.5	0.2	0.3	8.53054	108.941	24.7943	49.6226	72.8999	30.6499	202
0.5	0.3	0.2	8.5202	108.842	24.7822	49.3185	72.7635	30.8167	290
0.5	0.4	0.1	8.52491	108.885	24.7936	49.386	72.8764	30.8617	286
0.6	0.2	0.2	8.5251	108.888	24.7777	49.565	72.7498	30.5274	183
0.6	0.3	0.1	8.52539	108.883	24.7794	49.5095	72.7831	30.5408	189
0.7	0.2	0.1	8.52065	108.84	24.7806	49.3153	72.7677	30.7582	275

Table 11. Values of the objective function for the HPSO algorithm for various scalarization

β_1	β_2	β_3	f_1	f_2	f_3	X_1	X_2	X_3	Iterations
0.1	0.2	0.7	8.54683	109.903	24.6649	57.699	72.5172	36.9105	76
0.1	0.3	0.6	8.56443	109.547	24.6564	55.4598	72.616	29.3023	55
0.1	0.4	0.5	8.56161	109.771	24.8042	54.6208	72.7013	32.4585	67
0.1	0.5	0.4	8.54641	109.35	24.7936	52.1086	72.5603	30.8118	84
0.1	0.6	0.3	8.56848	109.741	24.6144	57.4313	72.8477	30.215	88
0.1	0.7	0.2	8.5439	109.74	24.5534	58.3004	72.891	40.5242	71
0.1	0.8	0.1	8.55699	109.725	24.59	57.8938	72.6076	30.9019	59
0.2	0.2	0.6	8.55445	109.542	24.7869	53.4146	72.5551	31.1148	57
0.2	0.3	0.5	8.57352	109.556	24.6483	55.5045	72.843	28.7859	42
0.2	0.4	0.4	8.54243	109.881	24.6273	58.1635	72.3826	36.6553	77
0.2	0.5	0.3	8.55299	109.679	24.8298	53.7879	72.6255	32.9866	71
0.2	0.6	0.2	8.54623	109.923	24.6032	58.693	72.5267	36.6219	58
0.2	0.7	0.1	8.56719	109.715	24.8006	54.1888	72.8216	31.4303	56
0.3	0.2	0.5	8.55619	109.907	24.7863	55.9271	72.7838	36.1053	74
0.3	0.3	0.4	8.54763	109.783	24.7766	55.363	72.379	34.124	70
0.3	0.4	0.3	8.57379	109.544	24.6937	54.6109	72.8529	28.9895	63
0.3	0.5	0.2	8.5531	109.028	24.3727	57.2096	72.6254	26.2125	60
0.3	0.6	0.1	8.54398	108.982	24.748	50.1817	72.8661	29.2429	68
0.4	0.2	0.4	8.55019	109.536	24.7396	54.2021	72.3286	30.8219	70
0.4	0.3	0.3	8.54072	109.474	24.8598	52.2739	72.6808	33.3906	58
0.4	0.4	0.2	8.56125	108.326	24.1797	55.3454	72.81	22.7159	59
0.4	0.5	0.1	8.55578	109.406	24.7608	52.8167	72.6043	30.0057	51
0.5	0.2	0.3	8.55857	109.167	24.4196	57.2167	72.7171	26.7412	71
0.5	0.3	0.2	8.54968	109.951	24.5644	59.4341	72.6607	35.4821	59
0.5	0.4	0.1	8.5675	109.929	24.8	55.6416	72.8689	33.5713	84
0.6	0.2	0.2	8.5503	109.192	24.7651	51.3036	72.7476	29.6299	55
0.6	0.3	0.1	8.55	109.56	24.7846	53.6643	72.4305	31.5473	56
0.7	0.2	0.1	8.5607	108.642	24.3953	53.428	72.5883	24.4832	36

Table 12. List of symbols

Symbols	Description
$G(t)$	Gravitational constant (GSA)
n	Number of mass agents (GSA)
α, ε	Constant parameter (GSA)
$R_{ij}(t)$	Distance between agents (GSA)
$F_i^d(t)$	Total force of each agent (GSA)
$a_i^d(t)$	Acceleration of each agent (GSA)
$v_i(t)$	Velocity of each mass agent (GSA)
$x_i(t)$	Position of each mass agent (GSA)
T	Function Evaluations (GSA)
w, c_1, c_2, r_1, r_2	PSO Parameters (PSO)
$v_i(t)$	Velocity of each particle (PSO)
$x_i(t)$	Position of each particle (PSO)
T_{max}	Maximum limit of function evaluations
s_i	Initial Social Influence (PSO)
p_i	Initial Individual Influence (PSO)
$\varphi = F*$	Utopia point (NBI)
M	Maximum number of objectives (NBI)
$X*_i$	Individual Minimum (NBI)
$g(x)$	Equality Constraints (NBI)
$h(x)$	Inequality Constraints (NBI)
$\beta \in Y$	Scalar constants (NBI)
t'	Distance Parameter (NBI)
$\vec{n}$	Normal Vector (NBI)
w_{ij}	Weights of neurons
$y_j(t)$	Activation values from the output
θ	Relaxation Constant
U	Threshold value
E	Energy of the system
dE	Energy Differential
$A_i(t)$	Transfer Function
$s_i(t)$	Piece-wise threshold function
P^j	Pareto frontier
c^j, d^j	Real-valued scalar
$vol(.)$	Lebesgue measure
$HVI(.)$	Hypervolume indicator

Chapter 21
Verification of Iterative Methods for the Linear Complementarity Problem:
Verification of Iterative Methods for LCPs

H. Saberi Najafi
Islamic Azad University, Lahijan Branch, Iran

S. A. Edalatpanah
Islamic Azad University, Lahijan Branch, Iran

ABSTRACT

In the present chapter, we give an overview of iterative methods for linear complementarity problems (abbreviated as LCPs). We also introduce these iterative methods for the problems based on fixed-point principle. Next, we present some new properties of preconditioned iterative methods for solving the LCPs. Convergence results of the sequence generated by these methods and also the comparison analysis between classic Gauss-Seidel method and preconditioned Gauss-Seidel (PGS) method for LCPs are established under certain conditions. Finally, the efficiency of these methods is demonstrated by numerical experiments. These results show that the mentioned models are effective in actual implementation and competitive with each other.

INTRODUCTION

For a given real vector $q \in R^n$ and a given matrix $A \in R^{n \times n}$ the linear complementarity problem abbreviated as *LCP (A, q)*, consists in finding vectors $z \in R^n$ such that

$$\begin{cases} w = Az + q \\ z \geq 0, w \geq 0 \\ z^T w = 0 \end{cases} \tag{1.1}$$

DOI: 10.4018/978-1-4666-9644-0.ch021

where, z^T denotes the transpose of the vector z . Note that the components of a solution (w, z) are *complementary* in the sense that if $w_i > 0$, then $z_i = 0$, and if $z_i > 0$, then $w_i = 0$. It may be the case that $w_i = z_i = 0$ however.

The Linear Complementarity Problems (*LCPs*) is one of the fundamental problems in optimization and mathematical programming (Murty & Yu 1988; Cottle, Pang et al., 1992). The Eq.(1.1) is a general problem which unifies bounteous practical problems in various scientific computing, economics and engineering areas. For example, linear and quadratic programming (Murty & Yu, 1988), economics and Nash equilibrium point of a bi-matrix game (Gale 1960; Lemke & Howson, 1964; Lemke, 1965), invariant capital stock (Cottle, Pang et al., 1992), optimal stopping in Markov chains(Cottle, Pang et al., 1992), contact and structural mechanics (Pfeiffer & Glocker, 2000), free boundary problem for journal bearings (Lin & Cryer 1985), traffic equilibriums(Ferris & Pang, 1997), circuit simulation (Van Eijndhoven, 1986), geometry (Du Val, 1940), etc. Therefore, it deserves to pay much more attention to the researches on both theoretical properties and numerical methods about the solution of this problem. So, many direct and iterative methods have been developed for its solution; see (Murty & Yu, 1988; Cottle, Pang et al, 1992) and the references therein.

The early motivation for studying the LCPs was because the KKT optimality conditions for linear and quadratic programs constitute an LCP of the form Eq.(1.1). However, the study of LCPs really came into prominence since the 1960s and in several decades, many methods for solving the LCP (A, q) have been introduced. Most of these methods originate from those for the system of the linear equations where may be classified into two principal kinds. The book by Cottle et al. (Cottle, Pang et al., 1992) is a good reference for direct methods developed to solve LCP. Another important class of methods used to tackle LCPs is the iterative methods. A more general iterative method, attributed to Christopherson (Christopherson, 1941), has been analyzed and clarified by Cryer (Cryer, 1971). In contrast to real attractiveness of iterative methods, lack of strength in these methods over direct methods and slow rate of convergence where unexpected, will make their real applications doubtful. The key of this problem is using preconditioning techniques. A preconditioner is defined as an auxiliary approximate solver which will be combined with an iterative method. According to critical importance of spectral radius, in preconditioning, we finding a more desired spectral radius. In other words, for the iterative solution of a linear system *Ax=b*, in order to accelerate the convergence of iterative solvers, preconditioning transforms the system to *PAx=Pb*, where *P* is a linear operator, called the preconditioner (Saberi Najafi & Edalatpanah, 2013, 2014) . Recently, Duan *et.al (2012)*, present a new preconditioned Gauss-Seidel iterative method for solving the linear complementarity problem, whose the certain elementary operations row are performed on system matrix *A* before applying the Gauss-Seidel iterative method. Moreover the sufficient conditions for guaranteeing the convergence of the new preconditioned Gauss-Seidel iterative method are presented. In this chapter, we study some models of iterative methods for linear complementarity problems (LCPs). Another main purpose in this chapter is further study on Duan *et.al* (2012) preconditioned method for LCP. To accomplish this purpose, we use these preconditioners and show that the preconditioned Gauss-Seidel (PGS) methods are superior to the basic Gauss-Seidel iterative method point of view rate of convergence and computing efficiency. Finally, numerical results are also given to illustrate the efficiency of these algorithms.

BACKGROUND

Numerous methods exist for solving the linear complementarity problems (LCPs). Numerical algorithms to solve Eq.(1.1) fall in two main classes, direct and iterative. Direct methods are those based on the process of pivoting, that is, exchanging the roles of dependent and independent variables (similar to basic and non-basic variables in a system of equations), while iterative methods are those which produce a (possibly infinite) sequence of iterates (trial solutions) which converge to a solution (Murty & Yu, 1988; Cottle, Pang et al., 1992); see also, Bard-type algorithms (Bard 1974; Stickney & Watson, 1978), the *n-cycle algorithms* (Watson, 1978; Doverspike, 1982), the block principal pivotal algorithm(Takriti & Murty, 1997).

Iterative methods, however, are considerably simpler than direct methods and require storage capacity of the order of n, whereas direct methods require storage capacity of the order of n^2. Moreover, for large scale sparse problems these direct methods may destroy sparsity and suffer from round off error. Thus, iterative methods are well suited and might prove important in the practical solution of large scale linear complementarity problems.

One of the initial attempts at devising iterative methods called "*cyclic coordinate descent method*" for solving problems that led to the linear complementarity problem is due to Hildreth (1954), who designed the procedure to solve a strictly convex quadratic program and proves its convergence. Hildreth stated its Kuhn-Tucker conditions and used the non-singularity of the Hessian matrix of the objective function to eliminate the primal variables. What remains after this operation is a linear complementarity problem in which the variables are Lagrange multipliers and the matrix A is symmetric and positive semi-definite. This model is simple, uses little storage, can exploit problem sparsity, and is practical for solving large scale problems.

A more general iterative method called "method of relaxation'', attributed to Christopherson (1941), has been analyzed and clarified by Cryer (1971), and it is often cited as Cryer's method (Benzi 1997). With the symmetric positive definite case, Cryer (1971) applied successive over-relaxation(SOR) to solve this special linear complementarity problem. According to the author, the method of Christopherson (Christopherson, 1941; Cryer, 1971) for solving free boundary problems for journal bearings leads to the consideration of such problems, in which moreover A is a finite-difference matrix. The author contends that the proposed method has advantages over conventional pivoting algorithms (Lemke, 1967; Cottle & Dantzig, 1968; Chandrasekaran, 1970; Saigal, 1970) for this problem because of the potentially very large size of A, its sparsity, and the density of A^{-1}.

Diamond (1974) makes repeated reference to a paper by Cryer (1971), in which the well-known equivalence between the LCPs and the Kuhn-Tucker conditions for the quadratic program was pointed out by Cryer .The main ideas of Diamond is to exhibit an alternative to solution of the problem via SOR. He gives an algorithm that in essence amounts to solving a finite number of systems of linear equations.

Cottle and Sacher (Cottle & Sacher, 1976) have discussed three algorithms for a class of large-scale LCPs. The three algorithms may be briefly described. The first algorithm is a modification of the principal pivoting method (Cottle & Dantzig, 1968). The second algorithm is a specialization of a method proposed by Chandrasekaran (Chandrasekaran, 1970) and employs LU factorizations. Main drawback of these algorithms is that pivoting methods generally do not preserve sparsity of the system. The third algorithm is a modification of the Cryer's point successive SOR technique for the solution of large linear complementarity problems in which the matrix has a block tridiagonal structure. The efficiency of the method is highly dependent on the ability to readily solve sub-problems which are LCPs. In 1977 (Cottle, Golub et al., 1977) partitioned the matrix A in Eq.(1.1) into m^2 blocks and proposed the block SOR method in which the sub-problems are linear complementarity problems.

An extension of Cryer's method along a different direction was proposed by Mangasarian (Mangasarian, 1977), which is also closely related to a gradient projection algorithm of Aganagic (Aganagic, 1978). Mangasarian (Mangasarian, 1977) presented a general iterative algorithm for solving the symmetric linear complementarity problem. This algorithm is in the following manner:

Let $z^0 \geq 0$ be an initial solution, then,

$$z^{i+1} = \lambda(z^i - wE^i[Mz^i + q + K^i(z^{i+1} - z^i)))_+ + (1-\lambda)z^i,$$

i=0,1,2,..., where $0 < \lambda \leq 10 <, w > 0$. $\{E^i\}$ and $\{K^i\}$ are bounded sequences of matrices in $R^{n\times n}$ with each E^i being a positive diagonal matrix satisfying $E^i > \alpha I$ for some $\alpha > 0$. Some of its realizations include a Jacobi-like algorithm, a generalization of the SOR method, and a generalization of the symmetric SOR method (SSOR method). It is shown that any accumulation point of the iterates generated by the above algorithm solves the Eq. (1.1), where A is symmetric matrix. Similar to Mangasarian's work in its algorithmic aspect, Ahn (1981) presented some convergence properties of the method for the non-symmetric cases. Under the blanket assumption that the matrix defining the linear complementarity problem is symmetric or non-symmetric, two above papers gave some general sufficient conditions for the convergence of the mentioned class of iterative scheme. However, necessary conditions are also important, because they depict the underlying limitations of the methods. Pang (1984) presented necessary and sufficient conditions for convergence for above class of iteration methods. These conditions are similar to those of the Ostrowski-Reich theorem(Varga, 2009). The results of the paper unified and extended a large number of known results. Pang (1986) also established necessary and sufficient conditions for the convergence of iterative methods for solving each individual linear complementarity problem with a fixed constant vector. These results improve the author's previous results (Pang, 1984). However, unlike the earlier paper the mentioned paper used only to the symmetric linear complementarity problem.

Han and Mangasarian (Han & Mangasarian, 1983) applied a new penalty function combined with SOR method to solve the quadratic programming. Following their work, Subramanian (1988) considered a dual exact penalty formulation for the LCP(A, q) when A is positive-definite, thus reducing the resulting problem to a maximization problem over R^n_+. If A is positive-semi-definite, then weak assumptions ensure that, for any pre-assigned tolerance δ, one can find approximate solutions by solving LCP(A+εI, q) for a specific ε>0 depending on δ. In this case, the solution of LCP(A, q) reduces to the solution of a sequence of positive-definite, symmetric quadratic programs.

Based on approximately solving an equivalent convex quadratic programming problem by SOR method (Agmon, 1954) which needs no matrix inversion and is sparsity preserving, Shiau (1988) presented a sparsity-preserving iterative algorithm for solving the Eq.(1.1), where A is only (not necessarily symmetric) positive semi-definite matrix.

In (De Pierro 1990), the author presents a family of iterative algorithms for solving a linear complementarity problem where, A is symmetric matrix with nonnegative elements and strictly positive diagonal. These algorithms are based on nonlinear relaxation methods (Gold, 1963, 1964; Chahine, 1970). The mentioned methods can be written in the form of the fundamental Mangasarian's iterative algorithm (Mangasarian, 1977). De Pierro and Iusem (De Pierro & Iusem, 1991), modified and extended the well-known SOR and Jacobi over-relaxation (JOR) iterative methods for solving a linear system to obtain the solutions of LCPs. Furthermore, it is shown that the modified SOR and JOR methods are strongly convergent with linear rate if the underlying matrix is symmetric and semi-definite; see also (De Pierro & Lopes, 1994).

Luo and Tseng (Luo and Tseng, 1991) develop a matrix splitting algorithm for the symmetric positive semi-definite LCPs. It is shown that if the splitting is regular (the matrix A write as: A=B+C. In the terminology of numerical analysis, such a pair (B, C) is called a splitting of A. If in addition B-C is positive definite, then (B, C) is called a regular splitting of A.), then the iterates generated by the algorithm are well defined and converge to a solution (see also (Mangasarian, 1991; Pang, 1993)). This result resolves in the affirmative a long standing question about the convergence of the point SOR method for solving this problem.

Mandel (Mandel, 1984) presented a multilevel iterative scheme for the large, sparse, symmetric and positive definite linear complementarity problems, which, for non-degenerate problems, behaves like linear iterates. Global convergence is guaranteed by a various type argument; see also (Brandt & Cryer, 1983; Hackbusch & Mittelmann, 1983; Oosterlee, 2003; Toivanen & Oosterlee, 2011) .

Schmidt is concerned with numerical methods for solving (a slightly generalized form of) the LCPs, where A is symmetric, positive definite and tridiagonal (Schmidt, 1987). When the problem size is large and when A has special structures, he proposed a special-purpose algorithm to solve LCPs. The method proposed is based on an equivalent formulation of the LCP in terms of an unconstrained optimization problem via construction of a weakly coupled program and its dualization.

In (He, 1992), the author, presented a new iterative method called "a projection and contraction method" for solving the linear complementarity problem, where A is a positive semi-definite (not necessarily symmetric) matrix. The method presented seems to be simple and robust, and is shown to be globally convergent with any starting point; see also (He, 1994, 1996, 1997; Cui & He, 1999; Liao & Wang, 2003) .

Using a column-wise procedure, Niethammer (1993) discussed projected SOR method for LCPs, which has advantages for the implementation on vector and parallel computers. In addition this procedure saves computational effort whenever a component of the iteration vector is set to zero. He also considered projected SOR method as follows:

For A = D + L + U, where D is diagonal and nonsingular, L lower and U upper triangular, the (projected) SOR method is given by:

$$z^{i+1} = [z^i - \omega D^{-1}(q + Lz^{i+1} + Dz^i + Uz^i)])_+, \qquad (i = 0,1,2,\ldots)$$

where, $\omega > 0, [a]_+ = \max(a,0)$.

The principal advantage to using an SOR based method is for solving very large sparse problems since it is a sparsity preserving method. It is simple to implement and there is no need to use complex sparsity preserving methods since the original data is never modified. The disadvantage to an SOR method is that it is generally characterized by fast movement into a neighborhood of the solution followed by slow progress toward the solution. Therefore, an SOR based algorithm called "a two-stage successive over-relaxation method (TSOR)method" for solving the symmetric LCPs was proposed in (Medhi, 1994) . After the first SOR preprocessing stage this model concentrates on updating a certain prescribed subset of variables which is determined by exploiting the complementarity property, see (Bertsekas, 1982). The author demonstrated that this method successfully solves problems with up to ten thousand variables; see also the original TSOR algorithm in (Thompson, 1987). Another two-step iterative methods was proposed in (Kocvara & Zowe, 1994). The authors established an algorithm which each step of the model combines an application of the projected SOR (to determine an approximation of the optimal active set) with the preconditioned conjugate gradient method (to solve the reduced residual systems of linear equations).

Fischer (1995) presented a damped and perturbed Newton-type method for the solution of the LCPs, with a given positive-semi-definite, but not necessarily symmetric, matrix *A*. Furthermore, the author used a special mapping to transform the linear complementarity problem into a nonlinear system of equations. By using the properties of this mapping and a perturbation technique it is observed that no additional assumption has to be made to ensure that all occurring Newton sub-problems are solvable; see also(Kanzow, 1996; Bai & Dong, 2006).

Another class of methods used to tackle LCPs is the interior point methods (IPMs). Most IPMs for LCPs can be viewed as natural extensions of the IPMs for linear programming. The most successful interior point method is the primal-dual method. Ye (1997) extended a homogeneous and self-dual linear programming algorithm (Xu, Hung et al., 1996) to the LCPs and proposed an iterative interior point method. The author considered the LCP(A,q) as: Minimize $z^T w$ subject to $w = Az + q, (z, w) \geq 0$, where *A* is a monotone matrix, *i.e.*, $A + A^T$ is positively semi-definite. This assumption is obviously equivalent to the assumption that *A* itself is positively semi-definite. Embedding the problem LCP(A,q) into two auxiliary problems, one-phase and two-phase algorithms for determining a solution of this problem are developed on the basis of algorithms developed by Kojima, Mizuno and Yoshisa (Kojima, Mizuno et al., 1991) under the assumption that the feasible set of LCP(A,q) has a nonempty interior. If the interior of this set is empty then the algorithms are modified in the paper in such a way that they yield a solution point with maximal number of positive components in (z, w); for other interior point methods, see (Liu, Shang et al. , Ai & Zhang, 2005, Liu & Potra, 2006; Yu, Huang et al., 2006; Illés, Nagy et al., 2010; Andreani, Júdice et al., 2011; Gurtuna, Petra et al., 2011; Liu, Liu et al., 2012; Wang & Bai, 2012; Wang, Yu et al., 2014).

Chen and Harker (Chen & Harker, 1993) presented a smoothing method. Kojima, Mizuno, and Yoshise first developed in 1989 a polynomial-time algorithm for a class of LCPs by following a "path of centers''(Kojima, Mizuno et al., 1989). In their approach, the complementarity condition is relaxed while the feasibility condition is maintained. Starting from an initial interior point, the main effort is spent on tracking the feasible path, or "the path of centers'', and preventing each iterate from going out of the feasible region. In Chen and Harker's works, the authors extend their "continuation method'', which was originally designed for solving monotone variational inequalities, to solve linear complementarity problems based on a new smooth equation formulation. Kanzow (Kanzow, 1996) considered a family of path-following methods for the linear complementarity problem having the distinction that they are not interior point methods. Burke and Xu (Burke & Xu, 1998), developed a noninterior path-following method for the LCPs. The method is based on Kanzow's algorithm(Kanzow, 1996) and is also closely related to the algorithm proposed by Chen and Harker (1993); for other smoothing methods, see(Qinggang, Jinling et al. 2010; Lu & Huang, 2014).

Gao (1999) proposed a neural network for solving large scale LCPs. Experimental and theoretical results of the mentioned method indicate that the trajectory of the proposed LCP neural network always asymptotically and stably converges to the solution of the primal and dual problems simultaneously, regardless of the location of the initial point; see also(Gao & Wang, 2011).

Li et al.(Li, Nie et al., 2008) presented a conjugate gradient method for solving the linear complementarity problem that involves a symmetrical M-matrix(see; Definition 1).Each step of this algorithm, solve a lower-dimensional system of linear equations by conjugate gradient method. The method also terminates at the exact solution of the problem after a finite number of iterations.

Yang et al.(Yang, Li et al., 2009) used a multiplicative Schwarz Iteration Method correspond to domain decomposition methods (Quarteroni & Valli, 1999) to solve the LCPs when A is an H-matrix(see; Definition 1). The scheme is an extension of the multiplicative Schwarz iteration scheme for the solution of linear systems, which was proposed by Bru et al.(Bru, Pedroche et al. 2004); see also (Yang & Li 2012; Xie, Sun et al. 2013).

All above models are in sequential computational environments. To solve Eq.(1.1) in the parallel computational environments, Mangasarian and De Leone() proposed a parallel successive over-relaxation (SOR) algorithm for solving the symmetric LCPs. It extends Mangasarian's earlier development of a serial SOR algorithm (). The parallel SOR is achieved by decomposing the matrix A into k blocks of disjoint row matrices and then applying one sweep of Mangasarian's serial SOR algorithm to each block simultaneously as one iteration. After this sweep, the information is shared among all k blocks for the next iteration. Under appropriate assumptions and a particular step-size condition, this parallel SOR method converges to an accumulation point that solves the problem Eq.(1.1). Thus, in this synchronous parallel SOR method, typically each processor was assigned a fixed portion of the problem data, and once a processor completed its computation it awaited all other processors to finish their computations as well, upon which all results were shared among them all and the process was repeated. Therefore, in () the authors proposed asynchronous parallel SOR version for the symmetric LCPs. In this model, the problem data may be assigned initially to each processor (static asynchronous SOR) or dynamically selected by each processor (dynamic asynchronous SOR) and, just as importantly, the results of each processor are broadcast instantly to all other processors as soon as they are obtained. The asynchronous SOR approach consists of letting each of the k processors solve simultaneously a linear complementarity problem in one variable, with each processor broadcasting its newly computed optimal variable value as soon as it is computed and proceeding to the solution of another one-dimensional linear complementarity problem without waiting for any of the other processors to finish but using the latest values for all variables (). These works are really much more motivating for the deepened research on the parallel solution of the LCPs (Mangasarian & De Leone, 1988;). In Machida, Fukushima et al. (1995), the authors extended the idea of multi-splitting for a system of linear equations (O'Leary & White, 1985) to the symmetric LCPs. They Also established a convergence theorem for the multi-splitting method to the parallel SOR method (Mangasarian & De Leone, 1987); see also (). With the consideration of the concrete characteristic of the LCPs as well as the intrinsic superiority of the parallel multiprocessor system, Benassi and White

(1994) considered some algorithms which utilize multiprocessing computers for the numerical solution of variational inequalities. By making use of the projecting technique described in (Cryer, 1971), they studied parallel versions of the SOR method with projection to the constraint set. in accordance with the fact that the Eq.(1.1) is equivalent to a fixed point system of equations (Mangasarian, 1977; Ahn, 1981), Bai (Bai, 1996) extended the parallel SOR method of (Benassi & White, 1994) to two-parameter cases called accelerated over-relaxation method (AOR) (Hadjidimos, 1978) and sets up a class of parallel matrix multi-splitting AOR algorithms suitable for multiprocessor systems for LCPs. The convergence of the new algorithm is established when the problem's matrix is either an H-matrix (see, Definition1) with positive diagonal elements or a symmetric positive definite matrix. (Bai & Evans, 1997) set up a class of matrix multi-splitting AOR type methods suitable to the SIMD systems for solving Eq.(1.1) and then established their asynchronous variants suitable to the MIMD systems, so that the available information is used as efficiently as possible and the synchronous delays between different processors are mostly avoided; see also (). In () the authors again by using the matrix multi-splitting methodology and the AOR technique, proposed a class of multi-parameter relaxed parallel multi-splitting methods for solving the LCPs on the SIMD multiprocessor systems. This class of methods is made up of two sweeps of relaxations within each iteration, and each sweep possibly includes its own pair of relaxation parameters. Thus, this class of methods can cover all the known and a lot of practical parallel multi-splitting relaxation methods such as the multi-splitting of Jacobi, symmetric Jacobi, Gauss-Seidel, symmetric Gauss-Seidel, SOR and the symmetric SOR methods. Bru et al.(Bru, Elsner et al., 1988) proposed two models of parallel multi-splitting chaotic iterations for solving large nonsingular systems of equations Ax=b. In the first model each processor can carry out an arbitrary number of local iterations before the next global approximation to the solution is formed and in the second model any processor can update the global approximation which resides in the central processor at any time. For the solution of E.q (1.1),quite similar to (Bru, Elsner et al., 1988), Bai and Evans (Bai & Evans, 1998) established a class of parallel multi-splitting chaotic relaxation methods, for which each processor can carry out a varying number of local iterations until a mutual phase time is reached when all processors are ready to contribute towards the global iteration; see also (Dong, 2009; Dehghan & Hajarian, 2011; Zhang, Zuo et al., 2014).

Much attention has recently been paid on two classes of iterative methods. One class of iterative methods, called *modulus algorithms*, was introduced by van Bokhoven (Van Bokhoven, 1980) and was extended by Kappel and Watson (Kappel & Watson, 1986) to *block modulus algorithm*. In particular, a new iterative methods, called *modulus-based matrix splitting iterative methods*, was introduced by Bai (Bai, 2010) and was extended in some works that followed; see, e.g.,(Hadjidimos, Lapidakis et al. 2012; Bai & Zhang, 2013; Zhang & Ren, 2013; Zheng & Yin 2013, Cvetković & Kostić 2014, Ke & Ma 2014, Li & Zheng 2014, Xu & Liu, 2014; Zhang, 2014; Zhang & Li, 2014; Zheng & Yin, 2014; Liu, Zheng et al. 2015; Xu, 2015; Zeng & Zhang, 2015; Zhang, Zhang et al. 2015). Another class is based on splitting methods called *projective iterative algorithms* and was extended in many works; see(Yuan & Song, 2003; Cvetković & Rapajić, 2005; Jiang & Dong, 2005; Li & Dai, 2007; Xu & Luan, 2007; Salmi & Toivanen, 2011; Duan, Zeng et al. 2012; Saberi Najafi & Edalatpanah, 2012; Saberi Najafi & Edalatpanah, 2013; Zheng & Yin, 2013; Hadjidimos & Tzoumas, 2014; Zhang & Li, 2014) .

Now, we study some models of projective iterative methods for linear complementarity problems (LCPs).

PREREQUISITE

In order to study iteration methods for solving *LCP (A, q)*, let us introduce some definitions and lemmas.

Definition 1: (Berman & Plemmons, 1979; Frommer & Szyld, 1992: Varga, 2009).

a. A matrix $A = (a_{ij})_{n\times n}$ is nonnegative (positive) if $a_{ij} \geq 0$ ($a_{ij} > 0$). In this case we write A $\geq$ 0(A>0) . Similarly, for n-dimensional vectors *x*, by identifying them with $n\times 1$ matrices, we can also define x $\geq$ 0 (x>0).
b. A matrix $A = (a_{ij})_{n\times n}$ is called a *Z-matrix* if for any $i \neq j, a_{ij} \leq 0$;
c. *Z-matrix* is an *M-matrix*, if *A* is nonsingular, and $A^{-1} \geq 0$;
d. A matrix $A = (a_{ij})_{n\times n}$ is called *M-matrix* if $A = \alpha I - B$; $B \geq 0$ and $\alpha > \rho(B)$; (we denote the spectral radius of *B* by ρ *(B)*) .
e. For any matrix $A = (a_{ij})_{n\times n}$ the comparison matrix $\langle A\rangle = (m_{ij}) \in R^{n\times n}$ is defined by:

$$m_{ii} = |a_{ii}|, \quad m_{ij} = -|a_{ij}|, \quad i \neq j \quad 1 \leq i, j \leq n.$$

f. $A = (a_{ij})_{n\times n}$ is an *H-matrix* if and only if $\langle A\rangle$ is *M-matrix*.

Definition 2: (Bai and Evans 1997). For $x \in R^n$, vector x_+ is defined such that $(x_+)_j = \max(x_j, 0)$, j = 1,2,. . ., n. Then, for any $x, y \in R^n$, the following facts hold:

1. $(x + y)_+ \leq x_+ + y_+$;
2. $x_+ - y_+ \leq (x - y)_+$;
3. $|x| = x_+ + (-x)_+$;
4. $x \leq y$ implies $x_+ \leq y_+$.

Definition 3: (Berman and Plemmons 1979, Frommer and Szyld 1992, Varga 2009). Let *A* be a real matrix. The splitting *A=M- N* is:

a. Convergent if $\rho(M^{-1}N) < 1$;
b. Regular if $M^{-1} \geq 0$ and $N \geq 0$;
c. Weak regular if $M^{-1} \geq 0$ and $M^{-1}N \geq 0$;
d. *M-splitting* if *M* is *M-matrix* and $N \geq 0$.

Clearly an M-splitting is regular and a regular splitting is weak regular.

Lemma 1: (Berman and Plemmons 1979). Let *A* be a *Z-matrix*. Then *A* is *M-matrix* if and only if there is a positive vector *x* such that *Ax >0* .

Lemma 2: (Varga 2009). Let *A =M − N* be an *M-splitting* of A. Then $\rho(M^{-1}N) < 1$ if and only if *A* is *M-matrix.*

Lemma 3: (Frommer and Szyld 1992). Let *A, B* are *Z-matrix* and *A* is an *M-matrix*, if A$\leq$ B then *B* is also an *M-matrix* .

Lemma 4: (Varga 2009). If A ≥ 0, then;

a. A has a nonnegative real eigenvalue equal to its spectral radius;
b. For $\rho(A) > 0$, there corresponds an eigenvector x ≥ 0,
c. $\rho(A)$ does not decrease when any entry of A is increased.

Lemma 5: (Berman and Plemmons 1979). Let T ≥ 0. If there exist x > 0 and a scalar $\alpha > 0$ such that:

a. $Tx \leq \alpha x$,then $\rho(T) \leq \alpha$. Moreover, if $Tx < \alpha x$, then $\rho(T) < \alpha$.
b. $Tx \geq \alpha x$,then $\rho(T) \geq \alpha$. Moreover, if $Tx > \alpha x$, then $\rho(T) > \alpha$.

Lemma 6: (Bai and Evans 1997). Let $A \in R^{n\times n}$ be an *H-matrix* with positive diagonal elements. Then the $LCP(A,q)$ has a unique solution $z^* \in R^n$.

Lemma 7: (Ahn 1981). *LCP(A, q)* can be equivalently transformed to a fixed-point system of equations $z = (z - \alpha E(Az + q))_+$, where α is some positive constant and *E* is a diagonal matrix with positive diagonal elements.

PROJECTIVE ITERATIVE METHODS

Let us to consider Eq.(1.1) . Here, we study some important models of projective iterative methods.

GAOR Projective Methods for *LCP(A, q)*

Let the matrix *A* be as

$$A=D+L+U , \tag{3.1}$$

where, *D* diagonal, *L* and *U* are strictly lower and upper triangular matrices of *A*, respectively.

Then by choice of $\alpha E = D^{-1}$ and Lemma 7 we get,

$$z = (z - D^{-1}(Az + q))_+ . \tag{3.2}$$

So, in order to solve *LCP (A,q)*, Generalized Accelerated Over-relaxation (GAOR) iterative methods defined in (Li and Dai 2007) is:

$$z^{k+1} = (z^k - D^{-1}[\alpha\Omega L z^{k+1} + (\Omega A - \alpha\Omega L)z^k + \Omega q])_+ , \tag{3.3}$$

where, α is a real parameter and $\Omega = (w_1, w_2, \ldots, w_n)$ is a real diagonal relaxation matrix.

The operator $f : R^n \longrightarrow R^n$, is defined such that $f(z) = \xi$, where ξ is the fixed point of the system

$$\xi = (z - D^{-1}[\alpha\Omega L\xi + (\Omega A - \alpha\Omega L)z + \Omega q])_+ . \tag{3.4}$$

In next Theorem, we have the convergence theorem, proposed in (Li and Dai 2007) for the GAOR methods.

Theorem1: Let $A \in R^{n \times n}$ be an *H-matrix* with positive diagonal elements. Moreover, let

$$G = I - \alpha\Omega D^{-1}\left|L\right| \quad , \qquad F = \left|I - D^{-1}(\Omega M - \alpha\Omega L)\right|,$$

then, for any initial vector $z^0 \in R^n$, the iterative sequence $\{ z^k \}$ generated by the GAOR method converges to the unique solution z^* of the *LCP (A, q)* and

$$\rho(G^{-1}F) \le \underset{1 \le i \le n}{Max}\{\left|1 - w_i\right| + w_i\rho(\left|J\right|)\} < 1,$$

if

$$0 < w_i < \frac{2}{1 + \rho(\left|J\right|)}, \qquad 0 \le \alpha \le 1,$$

where $\rho(\left|J\right|)$ is the spectral radius Jacobi iteration matrix ($J = D^{-1}(L + U)$). Some wider convergence area for the GAOR method presented in (Saberi Najafi and Edalatpanah 2013, Hadjidimos and Tzoumas 2014).

Corollary 1: By choosing special parameters in GAOR methods, it can be obtained the similar results for other well known iterative methods. For example,

1. GSOR (generalized SOR) methods (Li and Dai 2007) for $\alpha = 1$.
2. AOR (accelerated Overrelaxation) methods (Hadjidimos 1978) for $\alpha = r / w$ & $\Omega = wI$.
3. EAOR (extrapolated AOR) methods (Evans and Martins 1992) for $\alpha = r^2 / w^2$ & $\Omega = (w^2 / r)I$.
4. SOR methods for $\alpha = 1$ & $\Omega = wI$.
5. JOR (Jacobi Overrelaxation) methods (Truyen and Cornelis 1995) for $\alpha = 0$ & $\Omega = wI$.
6. Gauss-Seidel method for $\alpha = 1$ & $\Omega = I$.
7. Jacobi method for for $\alpha = 0$ & $\Omega = I$.

MAOR Projective methods for *LCP(A,q)*

Another model of projective iterative algorithms is for class of 2×2 block matrices. Let the matrix *A* be the following form:

$$A = \begin{bmatrix} D_1 & H \\ K & D_2 \end{bmatrix},$$

where, D_1, D_2 are nonsingular diagonal matrices of orders n_1 and n_2 respectively and $H \in R^{n_1 \times n_2}$, $K \in R^{n_2 \times n_1}$ and

$$D = \begin{bmatrix} D_1 & 0 \\ 0 & D_2 \end{bmatrix}, L = \begin{bmatrix} 0 & 0 \\ K & 0 \end{bmatrix}, U = \begin{bmatrix} 0 & H \\ 0 & 0 \end{bmatrix}.$$

Then, the Modified Accelerated Over-relaxation (MAOR) iterative methods is defined in(Yuan and Song 2003) as follows:

$$z^{(k+1)} = (z^{(k)} - D^{-1}[\gamma L z^{(k+1)} + (\Omega A - \gamma L) z^{(k)} + \Omega q])_+ \tag{3.5}$$

where

$$\Omega = diag(w_1 I_1, w_2 I_2), \Gamma = diag(\gamma_1 I_1, \gamma_2 I_2)$$

with $w_1, w_2 \neq 0$ and $I_1 \in R^{n_1 x n_1}$ .Let,

$$Q = I - \gamma D^{-1} \left|L\right| \quad \& R = \left|I - D^{-1}(\Omega M - \gamma L)\right|$$

Then in next Theorem, we have the convergence theorem, proposed in (Yuan and Song 2003) for the MAOR method.

Theorem 2. Let $A \in R^{n \times n}$ be an *H-matrix* with positive diagonal elements. Then, for any initial vector $z^0 \in R^n$, the iterative sequence $\{ z^{(k)} \}$ generated by the MAOR method converges to the unique solution z^* of the *LCP (A, q)* and

$$\rho(Q^{-1}R) \leq \underset{I=1,2}{Max}\{\left|1 - w_i\right| + w_i \rho(\left|J\right|)\} < 1$$

Whenever

$$0 < w_i < \frac{2}{1 + \rho(\left|J\right|)}, \qquad 0 \leq \gamma \leq w_2$$

where

$$J = D^{-1}(L + U).$$

SSOR-Like Method for Non-Hermitian Positive Definite LCP(A, q)

The above projective iterative algorithms for LCPs exploit particular features of matrices such as the sparsity and the block structure and in these methods. However, most convergence results have been established for the case that the system matrix A is symmetric positive definite, M-matrix, H-matrix or diagonally dominant. The case that A is Non-Hermitian is much more difficult and, as we have known, only a few results were reported about this case of matrix for *LCP (A, q)*. Saberi Najađ and Edalatpanah(Saberi Najafi and Edalatpanah 2012) developed a new model for this drawback. Let the matrix A is split as Eq.(3.1).Then the iterative method successive over-relaxation like method (SSOR-Like method) for LCP(A, q) by the following (Algorithm 1).

Algorithm 1. SSOR-Like method for LCPs

Step1.choos an initial vector $z^0 \in R^n$ parameter w and set k=0
Step 2. For k=0,1,2,… do

$$z^{k+1} = (z^k - D^{-1}[w(L-U^*)z^{k+1} + [w(2-w)A - w(L-U^*)]z^k + w(2-w)q])_+.$$

Step 3. If $z^{k+1} = z^k$, then stop; otherwise, set k=k+1 and go to step 2.

Note: U^* denotes the conjugate transpose of the matrix U.

In next Theorem, we have the existence and uniqueness of the solution of SSOR-Like method proposed in (Saberi Najafi and Edalatpanah 2012), when the coefficient matrix is non-Hermitian positive definite.

Theorem 3: Let $A \in C^{n\times n}$ be non-Hermitian positive definite with $H = (A + A^*)/2$ its Hermitian part, $\eta = \lambda_{\min}(B)$ and $\mu = \lambda_{\min}(C)$ be the smallest eigenvalue of $B = H - 2D^{-1}(U + U^*)$ and $C = H - 2D^{-1}(L + L^*)$ respectively. Then for any initial vector z^0, Algorithm1 convergence to the unique solution of LCP(A,q), if the following conditions are true:

$$\begin{cases} w \in (0,1), & if \quad \eta \geq 0, \mu \geq 0; \\ w \in (0, \dfrac{2}{2-\eta}), & if \quad \eta < 0 \leq \mu; or; \eta \leq \mu < 0; \\ w \in (0, \dfrac{2}{2-\mu}), & if \quad \mu < 0 \leq \eta; or; \mu \leq \eta < 0. \end{cases}$$

KRYLOV SUBSPACE METHODS FOR LCP(A, q)

The Krylov subspace methods are based on refinement residuals algorithms. For example, if we consider the MAOR for LCP based on refinement methods we have the following Algorithm 2 (Saberi Najafi and Edalatpanah 2013).

Therefore, based on the concept of refinement methods and Algorithm 2, we can explain the Krylov subspace methods for LCP. Now, we shortly describe a Krylov subspace method called Conjugate Gradient Squared Method (CGS) for *LCP(A, q)*.

The conjugate gradient squared method is a related algorithm that attempts to improvement the some problems of Bi-Conjugate Gradient method (BiCG). Additionally, often one observes a speed of convergence for CGS that is about twice as fast as for the biconjugate gradient method; see (Barrett, Berry et al. 1994).

Remark: All of these techniques and their results are also applicable for parallel computing such as *multi-splitting* methods (Mangasarian & De Leone 1987, Machida, Fukushima et al. 1995, Bai & Evans 1997, Bai, Evans et al. 1997b, Bai & Evans 1998, Zhang, Zuo et al. 2014)[23, 35, 46-49].

Algorithm 2. MAOR (A, q)

Step 1. Choose an initial vector $z^{(0)} \in R^n$, parameter w .

Step 2. For k=0,1,2,… do

$$r^{(k+1)} = -q - [\gamma\Omega^{-1}Lz^{(k+1)} + (A - \gamma\Omega^{-1}L)z^{(k)}]$$

$$z_i^{(k+1)} = z_i^{(k+1)} + \Omega D^{-1} r_i^{(k+1)}$$

$$z_i^{(k+1)} = \max\{0, z_i^{(k+1)}\}$$

Step 3. If $z^{(k+1)} = z^{(k)}$, then stop; otherwise, set *k=k+1* and go to Step 2.

Algorithm 3. CGS for LCP (A, q)

1.Compute $r^{(0)} = -q - Az^{(0)}$,

2.Set $\tilde{r} = r^{(0)}$,

3. for k=1,…, until convergence do

$$\rho_{k-1} = \tilde{r}^T r^{(k-1)},$$

if $\rho_{k-1} = 0$, then method fails,

if $k = 1$

$$u^{(1)} = r^{(1)},$$

$$l^{(1)} = u^{(1)},$$

else

$$\beta_{k-1} = \rho_{k-1} / \rho_{k-2}$$

$$u^{(k)} = r^{(k-1)} + \beta_{k-1} y_{k-1},$$

$$l^{(k)} = u^{(k)} + \beta_{k-1}(y_{k-1} + \beta_{k-1} l^{(k-1)}),$$

endif

Solve $\hat{l} = l^{(k)}$,

$$\hat{v} = A\hat{l},$$

$$\alpha_k = \rho_{k-1} / \tilde{r}^T \hat{v}$$

$$y_k = u_k - \alpha_k \hat{v},$$

Solve $\hat{u} = u^{(k)} + y_k$,

$$\hat{l} = A\hat{u},$$

$$z^{(k)} = z^{(k-1)} + \alpha_k \hat{u},$$

$$z^{(k)} = \max\{0, z^{(k)}\},$$

$$r^{(k)} = -q - Az^{(k)},$$

4.end.

PRECONDITIONED GAUSS-SEIDEL(PGS)ITERATIVE METHOD FOR LCP (A, q)

The development of efficient and authentic preconditioned iterative methods is the key for the successful application of scientific computation to the solution of many large scale problems. Since, the convergence rate of iterative methods depends on spectral properties of the coefficient matrix, in preconditioning schemes attempt to transform the original system into another one that has the same solution but more desirable properties for iterative solution.

In this section, PGS methods for LCP and the effect of preconditioning for these methods, is investigated. In literature, various authors have suggested different model of *(I+S)-type* preconditioner for linear systems $A=I\text{-}L\text{-}U$; where I is the identity matrix and L,U are strictly lower and strictly upper triangular matrices of A, respectively; see(Milaszewicz 1987, Gunawardena, Jain et al. 1991, Evans, Martins et al. 2001, Hirano 2001, Saberi Najafi & Edalatpanah 2013, Saberi Najafi, Edalatpanah et al. 2014, Saberi Najafi & Edalatpanah 2014) and the references therein.

These preconditioners have reasonable effectiveness and low construction cost. For example In 1987 Milaszewicz presented the preconditioner of (I+ S)-type, where the elements of the first column below the diagonal of A eliminate (Milaszewicz, 1987). Gunawardena, Jain and Snyder(Gunawardena, Jain et al. 1991) considered the alternative preconditioner, which eliminates the elements of the first upper diagonal. Saberi Najafi and Edalatpanah (2014) established a new family of preconditioned modeling, which the convergence rates of the proposed methods are superior to the other similar existing methods. In the present section, we consider some preconditioners of (I+S)-type model for A, introduced in(Duan, Zeng et al. 2012).

Without lost generality, let $D=I$ in Eq.(3.1), where I is the identity matrix; *i.e*:

$$A=I\text{-}L\text{-}U\,. \tag{4.1}$$

Then preconditioning in A is;

$$\bar{A} = PA = \bar{D} + \bar{L} + \bar{U}, \tag{4.2}$$

where $\bar{D}, \bar{L}, \bar{U}$ are diagonal, strictly lower and strictly upper triangular parts of $\bar{A}$ and

$$\bar{q} = Pq. \tag{4.3}$$

Now, consider the following preconditioner for LCP;

$$P = (I + S_{\alpha}^{\beta}) = (I + S_{\alpha} + S^{\beta}),$$

where

$$S_{\alpha} = \begin{bmatrix} 0 & 0 & \cdots & 0 \\ -\alpha_2 a_{21} & 0 & \cdots & 0 \\ -\alpha_3 a_{31} & 0 & \cdots & 0 \\ \vdots & \vdots & \cdots & \vdots \\ -\alpha_n a_{n1} & 0 & \cdots & 0 \end{bmatrix}, \quad s^{\beta} = \begin{bmatrix} 0 & -\beta_1 a_{12} & 0 & \cdots & 0 \\ 0 & 0 & -\beta_2 a_{23} & \ddots & \vdots \\ \vdots & \vdots & \vdots & \ddots & \vdots \\ 0 & 0 & 0 & \ddots & -\beta_{n-1} a_{n-1,n} \\ 0 & 0 & 0 & \cdots & 0 \end{bmatrix}.$$

and, we have,

$$\begin{cases} A_{\alpha}^{\beta} = (I + S_{\alpha}^{\beta})A = I - L - U + S_{\alpha}^{\beta} - S_{\alpha}^{\beta}(L+U), \\ q_{\alpha}^{\beta} = (I + S_{\alpha}^{\beta})q, \\ S_{\alpha}^{\beta}(L+U) = \hat{D} + \hat{L} + \hat{U}, \\ A_{\alpha}^{\beta} = M_{\alpha}^{\beta} - N_{\alpha}^{\beta} = \underbrace{(I - \hat{D} - L + S_{\alpha} - \hat{L})}_{M_{\alpha}^{\beta}} - \underbrace{(U - S^{\beta} + \hat{U})}_{N_{\alpha}^{\beta}}. \end{cases} \tag{4.4}$$

where $\hat{D}, \hat{L}$ and $\hat{U}$ are diagonal, strictly lower and strictly upper triangular parts of $S_{\alpha}^{\beta}(L+U)$.

Then the PGS method for the LCP(A,q) proposed in (Duan, Zeng et al. 2012) is as the following (Algorithm 4).

Furthermore, these authors proposed the following convergence theorems:

Theorem 4: Let $A \in R^{n\times n}$ be an *H-matrix* with its diagonal elements are all equal one,

$$< A_{\alpha}^{\beta} > = < M_{\alpha}^{\beta} > - < N_{\alpha}^{\beta} >.$$

If $a_{i,i+1} \neq 0 \quad (i = 1, 2, \ldots, n-1), a_{n1} \neq 0$, suppose that

For β_1, α_n, there hold

Algorithm 4. Preconditioned Gauss-Seidel method for LCP

Step1. Choose an initial point $z^0 \in R^n$, set k=0.

Step 2. Let z^{k+1} be an arbitrary solution of the following LCP: find z^{k+1}, such that

$$\begin{cases} z^{k+1} \geq 0, \\ (I + S_{\alpha}^{\beta})q - N_{\alpha}^{\beta} z^k + M_{\alpha}^{\beta} z^{k+1} \geq 0, \\ (z^{k+1})^t((I + S_{\alpha}^{\beta})q - N_{\alpha}^{\beta} z^k + M_{\alpha}^{\beta} z^{k+1}) = 0 \end{cases} \tag{4.5}$$

Step3. If $z^{k+1} = z^k$, then stop; otherwise, set k=k+1 and go to step2.

$$0 \le \beta_1 < 1 + \frac{|a_{12}| + 1}{|a_{12}|(2\| < A^{-1} > \|_\infty - 1)}$$

and

$$0 \le \alpha_n < 1 + \frac{|a_{n1}| + 1}{|a_{n1}|(2\| < A^{-1} > \|_\infty - 1)}.$$

For any $\beta_i, \alpha_i \quad (i = 2, 3, \ldots, n-1)$, there hold $0 \le \alpha_i \le 1$ and:

$$0 \le \beta_i < 1 + \frac{\alpha_i |a_{i1}| + 1}{|a_{i,i+1}|(2\| < A^{-1} > \|_\infty - 1)},$$

or $\alpha_i > 1$ and

$$\alpha_i |a_{i,1}| + \beta_i |a_{i,i+1}| < \frac{2\| < A^{-1} > \|_\infty |a_{i,1}| + 1}{2\| < A^{-1} > \|_\infty - 1}.$$

Then, for any initial vector z^0, the iterative sequence $\{z^k\}$, generated by the PGS iterative method converges to the unique solution z^* of the *LCP(A,q)*.

Theorem 5: Let $A \in R^{nxn}$ be an *H-matrix* with its diagonal elements are all equal one,

$$< A_\alpha^\beta > = < M_\alpha^\beta > - < N_\alpha^\beta >.$$

If

$$a_{i,1} \ne 0 \quad (i = 2, \ldots, n-1), a_{12} \ne 0,$$

suppose that

For β_1, α_n, there hold

$$0 \le \beta_1 < 1 + \frac{|a_{12}|+1}{|a_{12}|(2\| < A^{-1} > \|_\infty - 1)}$$

and

$$0 \le \alpha_n < 1 + \frac{|a_{n1}|+1}{|a_{n1}|(2\| < A^{-1} > \|_\infty - 1)}.$$

For any $\beta_i, \alpha_i \quad (i = 2, 3, \ldots, n-1)$,there hold $0 \le \beta_i \le 1$ and

$$0 \le \alpha_i < 1 + \frac{\beta_i |a_{i1}|+1}{|a_{i,1}|(2\| < A^{-1} > \|_\infty - 1)},$$

or $\beta_i > 1$ and

$$\alpha_i |a_{i,1}| + \beta_i |a_{i,i+1}| < \frac{2\| < A^{-1} > \|_\infty |a_{i,i+1}| + 1}{2\| < A^{-1} > \|_\infty - 1}.$$

Then, for any initial vector z^0, the iterative sequence $\{z^k\}$,generated by the PGS iterative method converges to the unique solution z^* of the *LCP(A,q)* .

Furthermore, it can to show that the basic Gauss –Seidel converges to the unique solution of the *LCP(A,q)* when *A* be *H-matrix* with positive diagonal elements ; see (Li and Dai 2007).

In this section, we show that PGS method is superior to the Basic Gauss-Seidel method.

We know that since $LCP(q_\alpha^\beta, A_\alpha^\beta)$ is equivalent to the following zero-finding formulation;

$$\min(z, (A_\alpha^\beta z + q_\alpha^\beta)) = 0. \tag{4.6}$$

And the zero-finding formulation is equivalent to the following fixed-point formulation;

$$\max(0, z - (A_\alpha^\beta z + q_\alpha^\beta)) = z. \tag{4.7}$$

Then for any iteration we have;

$$\max(0, z^{k+1} - (M_\alpha^\beta z^{k+1} - N_\alpha^\beta z^k + q_\alpha^\beta)) = z^{k+1} \tag{4.8}$$

By (4.8), we know that, if $(z^{k+1} - (M_\alpha^\beta z^{k+1} - N_\alpha^\beta z^k + q_\alpha^\beta))_i < 0$ then $z_i^{k+1} = 0$, otherwise

$$\begin{aligned}
&(z^{k+1} - (M_\alpha^\beta z^{k+1} - N_\alpha^\beta z^k + q_\alpha^\beta))_i = z_i^{k+1},\\
&\Rightarrow z_i^{k+1} - (M_\alpha^\beta z^{k+1} - N_\alpha^\beta z^k + q_\alpha^\beta)_i = z_i^{k+1},\\
&\Rightarrow M_\alpha^\beta z_i^{k+1} = N_\alpha^\beta z_i^k - (q_\alpha^\beta)_i = -((q_\alpha^\beta)_i + (A_\alpha^\beta - M_\alpha^\beta) z_i^k),\\
&\Rightarrow M_\alpha^\beta z_i^{k+1} = (M_\alpha^\beta z^k - (q_\alpha^\beta + \bar{A} z^k))_i.
\end{aligned}$$

Now, since

$$M_\alpha^\beta = (I - \hat{D}) - (L - S_\alpha + \hat{L}) = D_\alpha^\beta - L_\alpha^\beta,$$

where $D_\alpha^\beta = (I - \hat{D}), L_\alpha^\beta = (L - S_\alpha + \hat{L})$ are diagonal, strictly lower triangular parts of M_α^β, in order to solve $LCP(q_\alpha^\beta, A_\alpha^\beta)$, we have,

$$\begin{aligned}
&(D_\alpha^\beta - L_\alpha^\beta) z_i^{k+1} = ((D_\alpha^\beta - L_\alpha^\beta) z^k - (q_\alpha^\beta + A_\alpha^\beta z^k))_i.\\
&\Rightarrow D_\alpha^\beta z_i^{k+1} = (D_\alpha^\beta - L_\alpha^\beta) z_i^k - (q_\alpha^\beta + A_\alpha^\beta z^k)_i + L_\alpha^\beta z_i^{k+1}.\\
&\Rightarrow z_i^{k+1} = z_i^k - (D_\alpha^\beta)^{-1} L_\alpha^\beta z_i^k - (D_\alpha^\beta)^{-1} (q_\alpha^\beta + A_\alpha^\beta z^k)_i + (D_\alpha^\beta)^{-1} L_\alpha^\beta z_i^{k+1}.
\end{aligned}$$

Therefore we get following formula of PGS method

$$z^{k+1} = \max(0, z^k - (D_\alpha^\beta)^{-1}(q_\alpha^\beta + (A_\alpha^\beta + L_\alpha^\beta) z^k - L_\alpha^\beta z^{k+1})).$$

Now by Definition 2 we have the following Algorithm (Algorithm 5).

Similar to the proving process of [(Sun 2006); Theorem 3.4] Duan *et.al*, presented the following lemma(Duan, Zeng et al. 2012):

Algorithm 5.

Step1. Initialization. Choose an initial point $z^0 \in R^n$, set k=0

Step 2. General iteration. Let z^{k+1} be an arbitrary solution of the following LCP:

find z^{k+1}, such that

$$z^{k+1} = [z^k - (D_\alpha^\beta)^{-1}(q_\alpha^\beta + (A_\alpha^\beta + L_\alpha^\beta) z^k - L_\alpha^\beta z^{k+1})]_+.$$

Step3.If $z^{k+1} = z^k$, then stop; otherwise, set k=k+1 and go to step2.

Lemma 8: Let A be an *H-matrix*, then the preconditioned A_{α}^{β} also is *H-matrix*.

The operator $f: R^n \longrightarrow R^n$, is defined such that $f(z) = \zeta_{\alpha}^{\beta}$, where ζ_{α}^{β} is the fixed point of the system:

$$\zeta_{\alpha}^{\beta} = [z - (D_{\alpha}^{\beta})^{-1}(q_{\alpha}^{\beta} + (A_{\alpha}^{\beta} + L_{\alpha}^{\beta})z - L_{\alpha}^{\beta}\zeta_{\alpha}^{\beta})]_{+}. \tag{4.9}$$

Let,

$$Q_{\alpha}^{\beta} = I - (D_{\alpha}^{\beta})^{-1}\left|L_{\alpha}^{\beta}\right|, \quad \& R_{\alpha}^{\beta} = \left|I - (D_{\alpha}^{\beta})^{-1}(A_{\alpha}^{\beta} + L_{\alpha}^{\beta})\right|. \tag{4.10}$$

Then in next theorems, we have the comparison analysis for the mentioned iterative methods.

Theorem 6: Let $A \in R^{n\times n}$ with positive diagonal elements be an *H-matrix*, $A_{\alpha}^{\beta} = (I + S_{\alpha}^{\beta})A$ is preconditioned form of A and $Q = I - \left|L\right|, R = \left|I - (A + L)\right|$ be non-preconditioned forms of (3.10). Then we have;

$$\rho((Q_{\alpha}^{\beta})^{-1}R_{\alpha}^{\beta}) \le \rho(Q^{-1}R) < 1.$$

Proof: By Lemma 8 A_{α}^{β} is an *H-matrix* .Hence $< A_{\alpha}^{\beta} >= Q_{\alpha}^{\beta} - R_{\alpha}^{\beta}$ is *M-matrix* and by Lemma 2 $\rho((Q_{\alpha}^{\beta})^{-1}R_{\alpha}^{\beta}) < 1$. Since $(< A_{\alpha}^{\beta} >) < Q_{\alpha}^{\beta}$, by Lemma 3 Q_{α}^{β} is *M-matrix*. As same demonstration Q is *M-matrix* too.

Thus ;

$$\begin{cases} (Q_{\alpha}^{\beta})^{-1} \ge 0, (Q_{\alpha}^{\beta})^{-1}R_{\alpha}^{\beta} \ge 0, \\ Q^{-1} \ge 0, Q^{-1}R \ge 0. \end{cases}$$

Then by Lemma 4 (Perron-Frobenius Theorem) there exist a positive vector X such that:

$$(Q^{-1}R)X = \rho(Q^{-1}R)X.$$

$$\Rightarrow (Q - R)X = < A > X = Q(I - Q^{-1}R)X = \frac{1 - \rho(Q^{-1}R)}{\rho(Q^{-1}R)}RX \ge 0$$

Furthermore, we have:

$$< A_\alpha^\beta >=< (I + S_\alpha^\beta)A >=< (I + S_\alpha^\beta)(D - L - U) >=$$
$$< D - L - U + SD - \underbrace{SL}_{D_1+L_1+U_1} - SU >=$$
$$< (D - D_1) - (L + L_1) - (U + U_1 + SU - SD) > = \bar{D} - |\bar{L}| - |\bar{U}|.$$

where,

$$\bar{D} = D - D_1,$$
$$|\bar{L}| = |-(L + L_1)| = |(L + L_1)| \geq |L|,$$
$$|\bar{U}| = |(U + U_1 + SU - SD)|.$$

Thus, $Q_\alpha^\beta \leq Q$ and in view of the fact that both Q_α^β, Q are *M-matrices* we have;

$$(Q_\alpha^\beta)^{-1}(I + S_\alpha^\beta) \geq (Q_\alpha^\beta)^{-1} \geq Q^{-1}.$$

Therefore,

$$0 \leq [(Q_\alpha^\beta)^{-1}(I + S_\alpha^\beta) - (Q_\alpha^\beta)^{-1}](Q - R)X$$
$$= (I - (Q_\alpha^\beta)^{-1} R_\alpha^\beta)X - (I - Q^{-1}R)X$$
$$= Q^{-1}RX - (Q_\alpha^\beta)^{-1} R_\alpha^\beta X = \rho(Q^{-1}R)X - (Q_\alpha^\beta)^{-1} R_\alpha^\beta X.$$

And by Lemma 5 we have; $\rho((Q_\alpha^\beta)^{-1}R) \leq \rho(Q^{-1}R)$.

And the proof is completed. ∎

Now, we show that in LCPs the convergence rate of PGS methods is faster than of the Gauss-Seidel method.

Theorem 7: Let $A \in R^{n \times n}$ with positive diagonal elements be an *H-matrix* and $A_\alpha^\beta = (I + S_\alpha^\beta)A$ is preconditioned form of A. Then convergence rate of PGS methods is faster than of the Gauss-Seidel method.

Proof: Let, iterative sequence $\{z^k\}$,k=0,1,…,generated by (4.8). From the assumption that *A* is an *H-matrix*, it follows, by Lemma 8 A_α^β is an *H-matrix* and therefore by Lemma 6,the vector sequence $\{z^k\}$, is uniquely defined and the *LCP(M, q)* has a unique solution $z^\bullet \in R^n$. By Eq.(4.9), consider the following equation

$$\psi_\alpha^\beta = f(x) = [x - (D_\alpha^\beta)^{-1}(q_\alpha^\beta + (A_\alpha^\beta + L_\alpha^\beta)x - L_\alpha^\beta \psi_\alpha^\beta)]_+. \tag{4.11}$$

By subtracting Eq.(4.9) and Wq.(4.11), we get:

$$\zeta_\alpha^\beta - \psi_\alpha^\beta = [z - (D_\alpha^\beta)^{-1}(q_\alpha^\beta + (A_\alpha^\beta + L_\alpha^\beta)z - L_\alpha^\beta \zeta_\alpha^\beta)]_+ - [x - (D_\alpha^\beta)^{-1}(q_\alpha^\beta + (A_\alpha^\beta + L_\alpha^\beta)x - L_\alpha^\beta \psi_\alpha^\beta)]_+$$
$$\leq [(z-x) - (D_\alpha^\beta)^{-1}((A_\alpha^\beta + L_\alpha^\beta)(z-x) - L_\alpha^\beta(\zeta_\alpha^\beta - \psi_\alpha^\beta))]_+$$
$$= [(I - (D_\alpha^\beta)^{-1}(A_\alpha^\beta + L_\alpha^\beta))(z-x) + (D_\alpha^\beta)^{-1} L_\alpha^\beta(\zeta_\alpha^\beta - \psi_\alpha^\beta)]_+.$$

Furthermore, we can obtain:

$$\psi_\alpha^\beta - \zeta_\alpha^\beta \leq [(I - (D_\alpha^\beta)^{-1}(A_\alpha^\beta + L_\alpha^\beta))(x-z) + (D_\alpha^\beta)^{-1} L_\alpha^\beta(\psi_\alpha^\beta - \zeta_\alpha^\beta)]_+.$$

Therefore, by Definition 2 and above relations we have:

$$\left|\zeta_\alpha^\beta - \psi_\alpha^\beta\right| = (\zeta_\alpha^\beta - \psi_\alpha^\beta)_+ + (\psi_\alpha^\beta - \zeta_\alpha^\beta)_+ \leq [(I - (D_\alpha^\beta)^{-1}(A_\alpha^\beta + L_\alpha^\beta))(z-x)]_+ +$$
$$[(I - (D_\alpha^\beta)^{-1}(A_\alpha^\beta + L_\alpha^\beta))(x-z)]_+ + (D_\alpha^\beta)^{-1} L_\alpha^\beta(\zeta_\alpha^\beta - \psi_\alpha^\beta)]_+ + (D_\alpha^\beta)^{-1} L_\alpha^\beta(\psi_\alpha^\beta - \zeta_\alpha^\beta)]_+$$
$$= \left|(I - (D_\alpha^\beta)^{-1}(A_\alpha^\beta + L_\alpha^\beta))(z-x)\right| + \left|(D_\alpha^\beta)^{-1} L_\alpha^\beta(\zeta_\alpha^\beta - \psi_\alpha^\beta)\right|$$
$$\leq \left|(I - (D_\alpha^\beta)^{-1}(A_\alpha^\beta + L_\alpha^\beta))\right|\left|(z-x)\right| + (D_\alpha^\beta)^{-1}\left|L_\alpha^\beta\right|\left|(\zeta_\alpha^\beta - \psi_\alpha^\beta)\right|.$$

It follows that:

$$\left|\zeta_\alpha^\beta - \psi_\alpha^\beta\right| = (\zeta_\alpha^\beta - \psi_\alpha^\beta)_+ + (\psi_\alpha^\beta - \zeta_\alpha^\beta)_+ \leq (Q_\alpha^\beta)^{-1} R_\alpha^\beta \left|z - x\right|, \tag{4.12}$$

Thus from the definition of the PGS method and Eq.(4.12) we can write:

$$\left|z^{k+1} - z^*\right| = \left|f(z^k) - f(z^*)\right| \leq (Q_\alpha^\beta)^{-1} R_\alpha^\beta \left|z^k - z^*\right|,$$

Hence, the iterative sequence $\{z^k\}$,k=0,1,…, converges to z^* if $\rho((Q_\alpha^\beta)^{-1} R_\alpha^\beta) \leq 1$ and since by Theorem 6, $\rho((Q_\alpha^\beta)^{-1} R) \leq \rho(Q^{-1}R)$, we conclude that for solving LCP, the preconditioned Gauss-Seidel iterative methods are better than of the Gauss-Seidel method form point of view of the convergence speed and the proof is completed. ■

NUMERICAL RESULTS

Here we give some examples, to illustrate the results obtained in previous section. In these experiment, the initial approximation of z^0 is $z^0 = (1,1,\ldots,1)^T$ and as a stopping criterion we choose:

$$\begin{cases} \left\|\min(Mz^k + q, z^k)\right\|_\infty \leq 10^{-6}, \\ \left\|\min(\bar{M}z^k + \bar{q}, z^k)\right\|_\infty \leq 10^{-6}. \end{cases}$$

Furthermore, we report the CPU time and number of iterations for the corresponding Gauss-Seidel and preconditioned Gauss-Seidel methods by *CPU* and *Iter*, respectively. All the numerical experiments presented in this section were computed with MATLAB 7 on a PC with a 1.86GHz 32- bit processor and 1GB memory.

Example 1: Consider LCP (A, q) with following system $A \in R^{n\times n}$ and $q \in R^n$.

Matrix A is a banded matrix select from (Lam 1999), which use in Quantum Chemistry, *i.e.*:

$$A = \begin{bmatrix} 1 & 0.21 & 1.2 & 0 & 0.13 & 1.42 & & & & & \\ 0.11 & 2 & 0.21 & 1.2 & 0 & 0.13 & 1.42 & & & & \\ 0.12 & 0.11 & 3 & 0.21 & 1.2 & 0 & \ddots & \ddots & & & \\ 0 & 0.12 & 0.11 & 4 & \ddots & \ddots & \ddots & \ddots & \ddots & & \\ 0.34 & 0 & 0.12 & \ddots & \ddots & \ddots & \ddots & \ddots & 0.13 & 1.42 \\ 0.45 & 0.34 & 0 & \ddots & \ddots & \ddots & 0.21 & 1.2 & 0 & 0.13 \\ & 0.45 & \ddots & \ddots & \ddots & 0.11 & 997 & 0.21 & 1.2 & 0 \\ & & \ddots & \ddots & \ddots & 0.12 & 0.11 & 998 & 0.21 & 1.2 \\ & & & \ddots & 0.34 & 0 & 0.12 & 0.11 & 999 & 0.21 \\ & & & & 0.45 & 0.34 & 0 & 0.12 & 0.11 & 1000 \end{bmatrix}_{1000\times1000}$$

and

$$q = (-1, 1, \dots, (-1)^n)^T \in R^n$$

Evidently, A is Non-Hermitian positive definite. Then LCP (M, q) has a unique solution and we solved the 1000 × 1000 matrix yielded by the SSOR-like iterative methods.In this experiment, the sign of $\eta = \lambda_{\min}(B)$, $\mu = \lambda_{\min}(C)$ are positive.

In Table 1, with several values we report the CPU time (CPU) and number of iterations (Iter) for the corresponding SSOR-like methods(Saberi Najafi and Edalatpanah 2012) (when n=1000). From the table, we can see that our SSOR iterative methods are useful for solving the LCP.

Example 2. This test problem comes from the finite difference discretization of the one side obstacle problem (Ciarlet 1978),

$$\langle -\Delta u - b, v - u \rangle \geq 0, \qquad \forall v \in K,$$

where,

$$K = \{v \in H_0^1(\Omega) : v \geq 0\}, b = 4\sin(4xy), \Omega = (0,1) \times (0,1).$$

Table 1. Shows the results of example 1 for SSOR methods

Method	SSOR	
w	Iter	CPU
0.3	29	0.328000
0.4	22	0.250000
0.5	17	0.203000
0.6	14	0.156000
0.7	11	0.125000
0.8	9	0.109000
0.9	11	0.125000
1.0	21	0.125000
1.5	∞	-

By discretization, we obtain the problem as *LCP(M,q)*,where,

$$h = 1/m, \quad n = m^2, \quad q = (4h^2 \sin(4ij/m^2))_{ij}, \quad i, j = 1,\dots,m,$$

$$M = \begin{bmatrix} A & -I & & & \\ -I & A & & & \\ & \ddots & \ddots & \ddots & \\ & & \ddots & \ddots & -I \\ & & & -I & A \end{bmatrix} \in R^{n\times n},$$

I is the identity matrix of *m -dimension*, and

$$A = \begin{bmatrix} 4 & -1 & & & \\ -1 & 4 & & & \\ & \ddots & \ddots & \ddots & \\ & & \ddots & \ddots & -1 \\ & & & -1 & 4 \end{bmatrix} \in R^{m\times m}.$$

Evidently, *M* is an *H-matrix* with positive diagonal elements. Then *LCP (M, q)* has a unique solution. Then, we solved the *H-matrix* yielded by the iterative methods, and the preconditioned forms. In this experiment, we choose Duan et al.'s model (Duan, Zeng et al. 2012)[38], as preconditioner. In Table 2, we report the CPU time and number of iterations for the corresponding Gauss- Seidel and preconditioned Gauss-Seidel methods. Here, we choose $\alpha = \beta = 1$.

In Table 3, we report the CPU time and number of iterations for the corresponding GAOR and preconditioned GAOR methods. Furthermore, the *N* parameters w_i , are taken from the *N* equal-partitioned

Table 2. Shows the results of example 2 for Gauss-Seidel method

Method	Gauss-Seidel		PGS	
n	Iter	CPU	Iter	CPU
100	109	0.003063	69	0.001262
225	237	0.078000	147	0.037000
400	450	0.391000	287	0.250000
625	657	1.640000	420	0.987000
900	965	4.360000	614	2.735000
1600	1695	21.265000	1065	15.422000

points of the interval [1, 1.02] and alpha is one . Here, the preconditioned GAOR methods is denoted by PREC(GAOR).

From the tables, we can see that the preconditioned iterative method is superior to the basic iterative method.

CONCLUSION

In this chapter, we have studied the iterative methods for linear complementarity problems and established the convergence for these methods under certain conditions. Furthermore, we have used the Krylov subspace methods such as conjugate gradient squared method for LCPs. Moreover, we have considered the preconditioned iterative method for linear complementarity problem and analyzed the convergence for these methods under certain conditions. The results show that these mentioned models are effective in actual implementation and competitive with each other.

Table 3. Shows the results of example 2 for GAOR method

Method	GAOR		PREC(GAOR)	
n	Iter	CPU	Iter	CPU
100	102	0.004054	82	0.003262
225	198	0.064214	172	0.063753
400	345	0.289446	307	0.248204
625	472	0.839852	436	0.726877
900	608	1.960319	600	1.628224
1225	744	3.688457	740	3.625643
1600	907	8.138793	859	7.415670

REFERENCES

Aganagic, M. (1978). *Iterative methods for linear complementarity problems.* Systems Optimization Laboratory, Department of Operations Research, Univ.

Agmon, S. (1954). The relaxation method for linear inequalities. *Canadian Journal of Mathematics, 6*(3), 382–392. doi:10.4153/CJM-1954-037-2

Ahn, B. (1981). Solution of nonsymmetric linear complementarity problems by iterative methods. *Journal of Optimization Theory and Applications, 33*(2), 175–185. doi:10.1007/BF00935545

Ai, W., & Zhang, S. (2005). An O (\sqrtnL) Iteration Primal-dual Path-following Method, Based on Wide Neighborhoods and Large Updates, for Monotone LCP. *SIAM Journal on Optimization, 16*(2), 400–417. doi:10.1137/040604492

Andreani, R., Júdice, J. J., Martínez, J. M., & Patrício, J. (2011). A projected–gradient interior–point algorithm for complementarity problems. *Numerical Algorithms, 57*(4), 457–485. doi:10.1007/s11075-010-9439-0

Bai, Z. (1998). On the monotone convergence of matrix multisplitting relaxation methods for the linear complementarity problem. *IMA Journal of Numerical Analysis, 18*(4), 509–518. doi:10.1093/imanum/18.4.509

Bai, Z.-Z. (1996). The convergence of parallel iteration algorithms for linear complementarity problems. *Computers & Mathematics with Applications (Oxford, England), 32*(9), 1–17. doi:10.1016/0898-1221(96)00172-1

Bai, Z.-Z. (1999). On the convergence of the multisplitting methods for the linear complementarity problem. *SIAM Journal on Matrix Analysis and Applications, 21*(1), 67–78. doi:10.1137/S0895479897324032

Bai, Z. Z. (2010). Modulus-based matrix splitting iteration methods for linear complementarity problems. *Numerical Linear Algebra with Applications, 17*(6), 917–933. doi:10.1002/nla.680

Bai, Z.-Z., & Dong, J.-L. (2006). A modified damped Newton method for linear complementarity problems. *Numerical Algorithms, 42*(3-4), 207–228. doi:10.1007/s11075-006-9028-4

Bai, Z.-Z., & Evans, D. (1997a). Matrix multisplitting relaxation methods for linear complementarity problems. *International Journal of Computer Mathematics, 63*(3-4), 309–326. doi:10.1080/00207169708804569

Bai, Z.-Z., & Evans, D. (1998). Chaotic iterative methods for the linear complementarity problems. *Journal of Computational and Applied Mathematics, 96*(2), 127–138.

Bai, Z.-Z., Evans, D. J., & Wang, D. E.-R. E. N. (1997b). A class of multi-parameter relaxed parallel multisplitting methods for large sparse linear complementarity problems. *Parallel Algorithms and Applications, 11*(1-2), 113–127. doi:10.1080/10637199708915590

Bai, Z.-Z., & Zhang, L.-L. (2013). Modulus-based synchronous two-stage multisplitting iteration methods for linear complementarity problems. *Numerical Algorithms, 62*(1), 59–77. doi:10.1007/s11075-012-9566-x

Bard, Y. (1974). *Nonlinear parameter estimation.* Academic Press New York.

Barrett, R. (1994). *Templates for the solution of linear systems: building blocks for iterative methods.* Siam. doi:10.1137/1.9781611971538

Benassi, M., & White, R. (1994). Parallel numerical solution of variational inequalities. *SIAM Journal on Numerical Analysis, 31*(3), 813–830. doi:10.1137/0731044

Benzi, M. (1997). Remarks on the numerical solution of certain linear complementarity problems. *Journal of Computational and Applied Mathematics, 83*(1), 137–143. doi:10.1016/S0377-0427(97)00103-9

Berman, A. & R. J. Plemmons (1979). *Nonnegative matrices.* The Mathematical Sciences, Classics in Applied Mathematics.

Bertsekas, D. P. (1982). Projected Newton methods for optimization problems with simple constraints. *SIAM Journal on Control and Optimization, 20*(2), 221–246. doi:10.1137/0320018

Brandt, A. & C. W. Cryer (1983). Multigrid algorithms for the solution of linear complementarity problems arising from free boundary problems. *SIAM Journal on Scientific and Statistical Computing, 4*(4), 655-684.

Bru, R., Elsner, L., & Neumann, M. (1988). Models of parallel chaotic iteration methods. *Linear Algebra and Its Applications, 103*, 175–192. doi:10.1016/0024-3795(88)90227-3

Bru, R., Pedroche, F., & Szyld, D. B. (2004). Overlapping additive and multiplicative Schwarz iterations for H-matrices. *Linear Algebra and Its Applications, 393*, 91–105. doi:10.1016/j.laa.2003.10.022

Burke, J. V., & Xu, S. (1998). The global linear convergence of a noninterior path-following algorithm for linear complementarity problems. *Mathematics of Operations Research, 23*(3), 719–734. doi:10.1287/moor.23.3.719

Chahine, M. T. (1970). Inverse problems in radiative transfer: Determination of atmospheric parameters. *Journal of the Atmospheric Sciences, 27*(6), 960–967. doi:10.1175/1520-0469(1970)027<0960:IPIRTD>2.0.CO;2

Chandrasekaran, R. (1970). A special case of the complementary pivot problem. *Opsearch, 7*, 263–268.

Chen, B., & Harker, P. T. (1993). A non-interior-point continuation method for linear complementarity problems. *SIAM Journal on Matrix Analysis and Applications, 14*(4), 1168–1190. doi:10.1137/0614081

Christopherson, D. G. (1941). A new mathematical method for the solution of film lubrication problems. *Proceedings - Institution of Mechanical Engineers, 146*(1), 126–135. doi:10.1243/PIME_PROC_1941_146_027_02

Ciarlet, P. G. (1978). *The finite element method for elliptic problems.* Elsevier.

Cottle, R. W. (1992). *The linear complementarity problem.* Siam.

Cottle, R. W., & Dantzig, G. B. (1968). Complementary pivot theory of mathematical programming. *Linear Algebra and Its Applications, 1*(1), 103–125. doi:10.1016/0024-3795(68)90052-9

Cottle, R. W., Golub, G. H., & Sacher, R. S. (1977). On the solution of large, structured linear complementarity problems: The block partitioned case. *Applied Mathematics & Optimization*, *4*(1), 347–363. doi:10.1007/BF01442149

Cottle, R. W., & Sacher, R. S. (1976). On the solution of large, structured linear complementarity problems: The tridiagonal case. *Applied Mathematics & Optimization*, *3*(4), 321–340. doi:10.1007/BF01448184

Cryer, C. W. (1971). The method of Christopherson for solving free boundary problems for infinite journal bearings by means of finite differences. *Mathematics of Computation*, *25*(115), 435–443. doi:10.1090/S0025-5718-1971-0298961-7

Cryer, C. W. (1971). The solution of a quadratic programming problem using systematic overrelaxation. *SIAM Journal on Control*, *9*(3), 385–392. doi:10.1137/0309028

Cui, Y., & He, B. (1999). A class of projection and contraction methods for asymmetric linear variational inequalities and their relations to Fukushima's descent method. *Computers & Mathematics with Applications (Oxford, England)*, *38*(11), 151–164. doi:10.1016/S0898-1221(99)00294-1

Cvetković, L., & Kostić, V. (2014). A note on the convergence of the MSMAOR method for linear complementarity problems. *Numerical Linear Algebra with Applications*, *21*(4), 534–539. doi:10.1002/nla.1896

Cvetković, L., & Rapajić, S. (2005). How to improve MAOR method convergence area for linear complementarity problems. *Applied Mathematics and Computation*, *162*(2), 577–584. doi:10.1016/j.amc.2003.12.108

De Leone, R., & Mangasarian, O. L. (1988). Asynchronous parallel successive overrelaxation for the symmetric linear complementarity problem. *Mathematical Programming*, *42*(1-3), 347–361. doi:10.1007/BF01589410

De Leone, R., Mangasarian, O. L., & Shiau, T.-H. (1990). Multi-sweep asynchronous parallel successive overrelaxation for the nonsymmetric linear complementarity problem. *Annals of Operations Research*, *22*(1), 43–54. doi:10.1007/BF02023047

De Pierro, A., & Lopes, J. (1994). Accelerating iterative algorithms for symmetric linear complementarity problems. *International Journal of Computer Mathematics*, *50*(1-2), 35–44. doi:10.1080/00207169408804240

De Pierro, A. R. (1990). Nonlinear relaxation methods for solving symmetric linear complementarity problems. *Journal of Optimization Theory and Applications*, *64*(1), 87–99. doi:10.1007/BF00940024

De Pierro, A. R., & Iusem, A. N. (1991). On the convergence of SOR-and JOR-type methods for convex linear complementarity problems. *Linear Algebra and Its Applications*, *154*, 601–614. doi:10.1016/0024-3795(91)90396-E

Dehghan, M., & Hajarian, M. (2011). Two class of synchronous matrix multisplitting schemes for solving linear complementarity problems. *Journal of Computational and Applied Mathematics*, *235*(15), 4325–4336. doi:10.1016/j.cam.2011.03.021

Diamond, M. A. (1974). The solution of a quadratic programming problem using fast methods to solve systems of linear equations. *International Journal of Systems Science*, *5*(2), 131–136. doi:10.1080/00207727408920083

Dong, J.-L. (2009). Inexact multisplitting methods for linear complementarity problems. *Journal of Computational and Applied Mathematics*, *223*(2), 714–724. doi:10.1016/j.cam.2008.02.013

Doverspike, R. D. (1982). Some perturbation results for the linear complementarity problem. *Mathematical Programming*, *23*(1), 181–192. doi:10.1007/BF01583787

Du Val, P. (1940). The unloading problem for plane curves. *American Journal of Mathematics*, *62*(1/4), 307–311. doi:10.2307/2371454

Duan, B. X., Zeng, W. Y., & Zhu, X. P. (2012). A Preconditioned Gauss-Seidel Iterative Method for Linear Complementarity Problem in Intelligent Materials System. *Advanced Materials Research*, *340*, 3–8. doi:10.4028/www.scientific.net/AMR.340.3

Evans, D., & Martins, M. (1992). On the convergence of the extrapolated AOR method. *International Journal of Computer Mathematics*, *43*(3-4), 161–171. doi:10.1080/00207169208804083

Evans, D., Martins, M. M., & Trigo, M. E. (2001). The AOR iterative method for new preconditioned linear systems. *Journal of Computational and Applied Mathematics*, *132*(2), 461–466. doi:10.1016/S0377-0427(00)00447-7

Ferris, M. C., & Pang, J.-S. (1997). Engineering and economic applications of complementarity problems. *SIAM Review*, *39*(4), 669–713. doi:10.1137/S0036144595285963

Fischer, A. (1995). A Newton-type method for positive-semidefinite linear complementarity problems. *Journal of Optimization Theory and Applications*, *86*(3), 585–608. doi:10.1007/BF02192160

Frommer, A., & Szyld, D. B. (1992). H-splittings and two-stage iterative methods. *Numerische Mathematik*, *63*(1), 345–356. doi:10.1007/BF01385865

Gale, D. (1960). *The theory of linear economic models*. University of Chicago press.

Gao, X. (1999). Neural network for solving monotone linear complementary problems. *Communications in Nonlinear Science and Numerical Simulation*, *4*(2), 128–132. doi:10.1016/S1007-5704(99)90025-4

Gao, X., & Wang, J. (2011). *A Neural Network for a Class of Horizontal Linear Complementary Problems*. Computational Intelligence and Security (CIS), 2011 Seventh International Conference on. doi:10.1109/CIS.2011.85

Gold, R. (1963). *An iterative solution of the matrix representation of detection systems*. Argonne National Lab.

Gold, R. (1964). *An iterative unfolding method for response matrices*. Argonne National Lab. doi:10.2172/4634295

Gunawardena, A. D., Jain, S. K., & Snyder, L. (1991). Modified iterative methods for consistent linear systems. *Linear Algebra and Its Applications*, *154*, 123–143. doi:10.1016/0024-3795(91)90376-8

Gurtuna, F., Petra, C., Potra, F. A., Shevchenko, O., & Vancea, A. (2011). Corrector-predictor methods for sufficient linear complementarity problems. *Computational Optimization and Applications, 48*(3), 453–485. doi:10.1007/s10589-009-9263-4

Hackbusch, W., & Mittelmann, H. D. (1983). On multi-grid methods for variational inequalities. *Numerische Mathematik, 42*(1), 65–76. doi:10.1007/BF01400918

Hadjidimos, A. (1978). Accelerated overrelaxation method. *Mathematics of Computation, 32*(141), 149–157. doi:10.1090/S0025-5718-1978-0483340-6

Hadjidimos, A., Lapidakis, M., & Tzoumas, M. (2012). On Iterative Solution for Linear Complementarity Problem with an H_+-Matrix. *SIAM Journal on Matrix Analysis and Applications, 33*(1), 97–110. doi:10.1137/100811222

Hadjidimos, A., & Tzoumas, M. (2014). On the Solution of the Linear Complementarity Problem by the Generalized Accelerated Overrelaxation Iterative Method. *Journal of Optimization Theory and Applications*, 1–18.

Han, S.-P., & Mangasarian, O. L. (1983). A dual differentiable exact penalty function. *Mathematical Programming, 25*(3), 293–306. doi:10.1007/BF02594781

He, B. (1992). A projection and contraction method for a class of linear complementarity problems and its application in convex quadratic programming. *Applied Mathematics & Optimization, 25*(3), 247–262. doi:10.1007/BF01182323

He, B. (1994). A new method for a class of linear variational inequalities. *Mathematical Programming, 66*(1-3), 137–144. doi:10.1007/BF01581141

He, B. (1996). A modified projection and contraction method for a class of linear complementarity problems. *Journal of Computational Mathematics-International Edition, 14*, 54–63.

He, B. (1997). A class of projection and contraction methods for monotone variational inequalities. *Applied Mathematics & Optimization, 35*(1), 69–76. doi:10.1007/BF02683320

Hildreth, C. (1954). Point estimates of ordinates of concave functions. *Journal of the American Statistical Association, 49*(267), 598–619. doi:10.1080/01621459.1954.10483523

Hirano, H., & Niki, H. (2001). Application of a preconditioning iterative method to the computation of fluid flow. *Numerical Functional Analysis and Optimization, 22*(3), 405–418. doi:10.1081/NFA-100105110

Illés, T., Nagy, M., & Terlaky, T. (2010). A polynomial path-following interior point algorithm for general linear complementarity problems. *Journal of Global Optimization, 47*(3), 329–342. doi:10.1007/s10898-008-9348-0

Jiang, M., & Dong, J. (2005). On convergence of two-stage splitting methods for linear complementarity problems. *Journal of Computational and Applied Mathematics, 181*(1), 58–69. doi:10.1016/j.cam.2004.11.018

Kanzow, C. (1996). Global convergence properties of some iterative methods for linear complementarity problems. *SIAM Journal on Optimization, 6*(2), 326–341. doi:10.1137/0806019

Kanzow, C. (1996). Some noninterior continuation methods for linear complementarity problems. *SIAM Journal on Matrix Analysis and Applications*, *17*(4), 851–868. doi:10.1137/S0895479894273134

Kappel, N. W., & Watson, L. T. (1986). Iterative algorithms for the linear complementarity problem. *International Journal of Computer Mathematics*, *19*(3-4), 273–297. doi:10.1080/00207168608803522

Ke, Y.-F., & Ma, C.-F. (2014). On the convergence analysis of two-step modulus-based matrix splitting iteration method for linear complementarity problems. *Applied Mathematics and Computation*, *243*, 413–418. doi:10.1016/j.amc.2014.05.119

Kocvara, M., & Zowe, J. (1994). An iterative two-step algorithm for linear complementarity problems. *Numerische Mathematik*, *68*(1), 95–106. doi:10.1007/s002110050050

Kojima, M., Mizuno, S., & Yoshise, A. (1989). A polynomial-time algorithm for a class of linear complementarity problems. *Mathematical Programming*, *44*(1-3), 1–26. doi:10.1007/BF01587074

Kojima, M., Mizuno, S., & Yoshise, A. (1991). AnO(\sqrt n L) iteration potential reduction algorithm for linear complementarity problems. *Mathematical Programming*, *50*(1-3), 331–342. doi:10.1007/BF01594942

Lam, S. (1999). *The role of CSP in reduced chemistry modeling*. IMA Symposium, Minnesota University.

Lemke, C. E. (1965). Bimatrix equilibrium points and mathematical programming. *Management Science*, *11*(7), 681–689. doi:10.1287/mnsc.11.7.681

Lemke, C. E. (1967). *On complementary pivot theory*. Department of Mathematics, Rensselaer Polytechnic Institute.

Lemke, C. E., Howson, J. Jr, & Joseph, T. (1964). Equilibrium points of bimatrix games. *Journal of the Society for Industrial and Applied Mathematics*, *12*(2), 413–423. doi:10.1137/0112033

Li, D.-H., Nie, Y.-Y., Zeng, J.-P., & Li, Q.-N. (2008). Conjugate gradient method for the linear complementarity problem with S-matrix. *Mathematical and Computer Modelling*, *48*(5), 918–928. doi:10.1016/j.mcm.2007.10.017

Li, W., & Zheng, H. (2014). Some new error bounds for linear complementarity problems of H-matrices. *Numerical Algorithms*, *67*(2), 257–269. doi:10.1007/s11075-013-9786-8

Li, Y., & Dai, P. (2007). Generalized AOR methods for linear complementarity problem. *Applied Mathematics and Computation*, *188*(1), 7–18. doi:10.1016/j.amc.2006.09.067

Liao, L.-Z., & Wang, S. (2003). A self-adaptive projection and contraction method for linear complementarity problems. *Applied Mathematics & Optimization*, *48*(3), 169–180. doi:10.1007/s00245-003-0767-5

Lin, Y., & Cryer, C. (1985). An alternating direction implicit algorithm for the solution of linear complementarity problems arising from free boundary problems. *Applied Mathematics & Optimization*, *13*(1), 1–17. doi:10.1007/BF01442196

Liu, C., Shang, Y., & Liu, H. (2015). An O (\ sqrt {n} L) iteration Mehrotra-type predictor-corrector algorithm for monotone linear complementarity problem. *Optimization Letters*, 1–16.

Liu, H. Liu, X. & Liu, Changhe. (2012). Mehrotra-type predictor–corrector algorithms for sufficient linear complementarity problem. *Applied Numerical Mathematics*, *62*(12), 1685–1700.

Liu, S., Zheng, H., & Li, W. (2015). A general accelerated modulus-based matrix splitting iteration method for solving linear complementarity problems. *Calcolo*, 1–11.

Liu, X., & Potra, F. A. (2006). Corrector-predictor methods for sufficient linear complementarity problems in a wide neighborhood of the central path. *SIAM Journal on Optimization*, *17*(3), 871–890. doi:10.1137/050623723

Lu, N., & Huang, Z.-H. (2014). A smoothing Newton algorithm for a class of non-monotonic symmetric cone linear complementarity problems. *Journal of Optimization Theory and Applications*, *161*(2), 446–464. doi:10.1007/s10957-013-0436-z

Luo, Z.-Q., & Tseng, P. (1991). On the convergence of a matrix splitting algorithm for the symmetric monotone linear complementarity problem. *SIAM Journal on Control and Optimization*, *29*(5), 1037–1060. doi:10.1137/0329057

Machida, N., Fukushima, M., & Ibaraki, T. (1995). A multisplitting method for symmetric linear complementarity problems. *Journal of Computational and Applied Mathematics*, *62*(2), 217–227. doi:10.1016/0377-0427(94)00103-2

Mandel, J. (1984). A multilevel iterative method for symmetric, positive definite linear complementarity problems. *Applied Mathematics & Optimization*, *11*(1), 77–95. doi:10.1007/BF01442171

Mangasarian, O., & De Leone, R. (1988). Parallel gradient projection successive overrelaxation for symmetric linear complementarity problems and linear programs. *Annals of Operations Research*, *14*(1), 41–59. doi:10.1007/BF02186473

Mangasarian, O. L. (1977). Solution of symmetric linear complementarity problems by iterative methods. *Journal of Optimization Theory and Applications*, *22*(4), 465–485. doi:10.1007/BF01268170

Mangasarian, O. L. (1991). Convergence of iterates of an inexact matrix splitting algorithm for the symmetric monotone linear complementarity problem. *SIAM Journal on Optimization*, *1*(1), 114–122. doi:10.1137/0801009

Mangasarian, O. L., & De Leone, R. (1987). Parallel successive overrelaxation methods for symmetric linear complementarity problems and linear programs. *Journal of Optimization Theory and Applications*, *54*(3), 437–446. doi:10.1007/BF00940194

Medhi, K. T. (1994). A two-stage successive overrelaxation algorithm for solving the symmetric linear complementarity problem. *Mathematical Programming*, *65*(1-3), 365–380. doi:10.1007/BF01581703

Milaszewicz, J. (1987). Improving jacobi and gauss-seidel iterations. *Linear Algebra and Its Applications*, *93*, 161–170. doi:10.1016/S0024-3795(87)90321-1

Murty, K. G., & Yu, F.-T. (1988). *Linear complementarity, linear and nonlinear programming*. Berlin: HeldermannVerlag.

Niethammer, W. (1993). A note on the implementation of the successive overrelaxation method for linear complementarity problems. *Numerical Algorithms*, *4*(1), 197–200. doi:10.1007/BF02142746

O'Leary, D. P., & White, R. E. (1985). Multi-splittings of matrices and parallel solution of linear systems. *SIAM Journal on Algebraic Discrete Methods*, *6*(4), 630–640. doi:10.1137/0606062

Oosterlee, C. W. (2003). On multigrid for linear complementarity problems with application to American-style options. *Electronic Transactions on Numerical Analysis, 15*(165-185), 2-7.

Pang, J. (1984). Necessary and sufficient conditions for the convergence of iterative methods for the linear complementarity problem. *Journal of Optimization Theory and Applications*, *42*(1), 1–17. doi:10.1007/BF00934130

Pang, J.-S. (1986). More results on the convergence of iterative methods for the symmetric linear complementarity problem. *Journal of Optimization Theory and Applications*, *49*(1), 107–134. doi:10.1007/BF00939250

Pang, J.-S. (1993). Convergence of splitting and Newton methods for complementarity problems: An application of some sensitivity results. *Mathematical Programming*, *58*(1-3), 149–160. doi:10.1007/BF01581264

Pfeiffer, F., & Glocker, C. (2000). *Multibody dynamics with unilateral contacts*. Springer Science & Business Media.

Qinggang, W., Jinling, Z., & Qingzhi, Y. (2010). Some non-interior path-following methods based on a scaled central path for linear complementarity problems. *Computational Optimization and Applications*, *46*(1), 31–49. doi:10.1007/s10589-008-9188-3

Quarteroni, A., & Valli, A. (1999). *Domain decomposition methods for partial differential equations numerical mathematics and scientific computation*. Oxford University Press New York.

Saberi Najafi, H., & Edalatpanah, S. (2012). A kind of symmetrical iterative methods to solve special class of LCP (M, q). *International Journal of Applied Mathematics and Applications*, *4*(2), 183–189.

Saberi Najafi, H., & Edalatpanah, S. (2013a). Comparison analysis for improving preconditioned SOR-type iterative method. *Numerical Analysis and Applications*, *6*(1), 62–70. doi:10.1134/S1995423913010084

Saberi Najafi, H., & Edalatpanah, S. (2013b). Modification of iterative methods for solving linear complementarity problems. *Engineering Computations*, *30*(7), 910–923. doi:10.1108/EC-10-2011-0131

Saberi Najafi, H., & Edalatpanah, S. (2013c). On the convergence regions of generalized accelerated overrelaxation method for linear complementarity problems. *Journal of Optimization Theory and Applications*, *156*(3), 859–866. doi:10.1007/s10957-012-0135-1

Saberi Najafi, H., & Edalatpanah, S. A. (2014b). A new family of (I+S)-type preconditioner with some applications. *Computational & Applied Mathematics*, 1–15.

Saberi Najafi, H., Edalatpanah, S. A., & Sheikhani, A. H. (2014a). Convergence analysis of modified iterative methods to solve linear systems. *Mediterranean Journal of Mathematics*, *11*(3), 1019–1032. doi:10.1007/s00009-014-0412-3

Saigal, R. (1970). *Lemke's algorithm and a special linear complementarity problem*. Berkeley, CA: Center for Research in Management Science, University of California.

Salmi, S., & Toivanen, J. (2011). An iterative method for pricing American options under jump-diffusion models. *Applied Numerical Mathematics*, *61*(7), 821–831. doi:10.1016/j.apnum.2011.02.002

Schmidt, J. W. (1987). On tridiagonal linear complementarity problems. *Numerische Mathematik*, *51*(1), 11–21. doi:10.1007/BF01399692

Shiau, T. (1988). An LP-based successive overrelaxation method for linear complementarity problems. *Journal of Optimization Theory and Applications*, *59*(2), 247–259.

Stickney, A., & Watson, L. (1978). Digraph models of Bard-type algorithms for the linear complementarity problem. *Mathematics of Operations Research*, *3*(4), 322–333. doi:10.1287/moor.3.4.322

Subramanian, P. (1988). A dual exact penalty formulation for the linear complementarity problem. *Journal of Optimization Theory and Applications*, *58*(3), 525–538. doi:10.1007/BF00939395

Sun, L. Y. (2006). Some extensions of the improved modified Gauss–Seidel iterative method for H-matrices. *Numerical Linear Algebra with Applications*, *13*(10), 869–876. doi:10.1002/nla.498

Takriti, S., & Murty, K. G. (1997). On the convergence of the block principal pivotal algorithm for the LCP. *European Journal of Operational Research*, *102*(3), 657–666. doi:10.1016/S0377-2217(96)00243-3

Thompson, K. M. (1987). *A two-stage successive overrelaxation algorithm for solving the linear complementarity problem*. University of Wisconsin-Madison, Computer Sciences Department.

Toivanen, J. & C. W. Oosterlee (2011). *A projected algebraic multigrid method for linear complementarity problems*. Available at SSRN 1739902.

Truyen, B., & Cornelis, J. (1995). Adiabatic layering: A new concept of hierarchical multi-scale optimization. *Neural Networks*, *8*(9), 1373–1378. doi:10.1016/0893-6080(95)00076-3

Van Bokhoven, W. (1980). A class of linear complementarity problems is solvable in polynomial time. Unpublished Paper, Dept. of electrical engineering, university of technology, the Netherlands.

Van Eijndhoven, J. (1986). Solving the linear complementarity problem in circuit simulation. *SIAM Journal on Control and Optimization*, *24*(5), 1050–1062. doi:10.1137/0324062

Varga, R. S. (2009). *Matrix iterative analysis*. Springer Science & Business Media.

Wang, G., Yu, C., & Teo, K. (2014). A full-Newton step feasible interior-point algorithm for P_*(\kappa)-linear complementarity problems. *Journal of Global Optimization*, *59*(1), 81–99. doi:10.1007/s10898-013-0090-x

Wang, G.-Q., & Bai, Y.-Q. (2012). A class of polynomial interior point algorithms for the Cartesian P-matrix linear complementarity problem over symmetric cones. *Journal of Optimization Theory and Applications*, *152*(3), 739–772. doi:10.1007/s10957-011-9938-8

Watson, L. T. (1978). An algorithm for the linear complementarity problem. *International Journal of Computer Mathematics*, *6*(4), 319–325. doi:10.1080/00207167808803148

Xie, S., Sun, Z., & Zeng, Y. (2013). A two-level domain decomposition algorithm for linear complementarity problem. *Journal of Inequalities and Applications*, (1): 1–9.

Xu, M., & Luan, G. (2007). A rapid algorithm for a class of linear complementarity problems. *Applied Mathematics and Computation*, *188*(2), 1647–1655. doi:10.1016/j.amc.2006.11.184

Xu, W. (2015). Modified modulus-based matrix splitting iteration methods for linear complementarity problems. *Numerical Linear Algebra with Applications*, *22*(4), 748–760. doi:10.1002/nla.1985

Xu, W.-, & Liu, H. (2014). A modified general modulus-based matrix splitting method for linear complementarity problems of H-matrices. *Linear Algebra and Its Applications*, *458*, 626–637. doi:10.1016/j.laa.2014.06.022

Xu, X., Hung, P.-F., & Ye, Y. (1996). A simplified homogeneous and self-dual linear programming algorithm and its implementation. *Annals of Operations Research*, *62*(1), 151–171. doi:10.1007/BF02206815

Yang, H., & Li, Q. (2012). Overlapping restricted additive Schwarz method applied to the linear complementarity problem with an H-matrix. *Computational Optimization and Applications*, *51*(1), 223–239. doi:10.1007/s10589-010-9334-6

Yang, H., Li, Q., & Xu, H. (2009). A multiplicative Schwarz iteration scheme for solving the linear complementarity problem with an H-matrix. *Linear Algebra and Its Applications*, *430*(4), 1085–1098. doi:10.1016/j.laa.2008.10.005

Ye, Y. (1997). On homogeneous and self-dual algorithms for LCP. *Mathematical Programming*, *76*(1), 211–221. doi:10.1007/BF02614384

Yu, Q., Huang, C., & Wang, X. (2006). A combined homotopy interior point method for the linear complementarity problem. *Applied Mathematics and Computation*, *179*(2), 696–701. doi:10.1016/j.amc.2005.12.010

Yuan, D., & Song, Y. (2003). Modified AOR methods for linear complementarity problem. *Applied Mathematics and Computation*, *140*(1), 53–67. doi:10.1016/S0096-3003(02)00194-7

Zeng, M.-L., & Zhang, G.-F. (2015). Modulus-based GSTS Iteration Method for Linear Complementarity Problems. *J. Math. Study*, *48*(1), 1–17. doi:10.4208/jms.v48n1.15.01

Zhang, L.-L. (2014). Two-stage multisplitting iteration methods using modulus-based matrix splitting as inner iteration for linear complementarity problems. *Journal of Optimization Theory and Applications*, *160*(1), 189–203. doi:10.1007/s10957-013-0362-0

Zhang, L.-L., & Ren, Z.-R. (2013). Improved convergence theorems of modulus-based matrix splitting iteration methods for linear complementarity problems. *Applied Mathematics Letters*, *26*(6), 638–642. doi:10.1016/j.aml.2013.01.001

Zhang, L.-T., & Li, J.-L. (2014). The weaker convergence of modulus-based synchronous multisplitting multi-parameters methods for linear complementarity problems. *Computers & Mathematics with Applications (Oxford, England)*, *67*(10), 1954–1959. doi:10.1016/j.camwa.2014.04.018

Zhang, L.-T., Zhang, Y.-X., Gu, T.-X., Liu, X.-P., & Zhang, L.-W. (2015). New convergence of modulus-based synchronous block multisplitting multi-parameter methods for linear complementarity problems. *Computational & Applied Mathematics*, 1–12.

Zhang, L.-T., Zuo, X.-Y., Gu, T.-X., & Liu, X.-P. (2014). Improved convergence theorems of multisplitting methods for the linear complementarity problem. *Applied Mathematics and Computation*, *243*, 982–987. doi:10.1016/j.amc.2014.06.038

Zheng, N., & Yin, J. (2013). Modulus-based successive overrelaxation method for pricing American options. *Appl Math Informatics*, *31*(5), 769–784. doi:10.14317/jami.2013.769

Zheng, N., & Yin, J.-F. (2013). Accelerated modulus-based matrix splitting iteration methods for linear complementarity problem. *Numerical Algorithms*, *64*(2), 245–262. doi:10.1007/s11075-012-9664-9

Zheng, N., & Yin, J.-F. (2014). Convergence of accelerated modulus-based matrix splitting iteration methods for linear complementarity problem with an H+-matrix. *Journal of Computational and Applied Mathematics*, *260*, 281–293.

Chapter 22
Parameter Optimization of Photovoltaic Solar Cell and Panel Using Genetic Algorithms Strategy

Benmessaoud Mohammed Tarik
University of Science and Technology of Oran, Algeria

Amine Boudghene Stambouli
University of Science and Technology of Oran, Algeria

Fatima Zohra Zerhouni
University of Science and Technology of Oran, Algeria

Mustapha Tioursi
University of Science and Technology of Oran, Algeria

Aouad M'harer
University of Tissemsilt, Algeria

ABSTRACT

In this chapter, we propose to perform a numerical technique based on genetic algorithms (GAs) to identify the electrical parameters (Is, Iph, Rs, Rsh, and n) of photovoltaic (PV) solar cells and modules. The one diode type approach is used to model the I–V characteristic of the solar cell. To extract electrical parameters, the approach is formulated as optimization problem. The GAs approach was used as a numerical technique in order to overcome problems involved in the local minima in the case optimization criteria. Compared to other methods, we find that the GAs is a very efficient technique to estimate the electrical parameters of photovoltaic solar cells and modules. Compared with other parameter extraction techniques, based on statistical study, results indicate the consistency and uniformity of method in terms of the quality of final solutions. In parallel, the simulated data with the extracted parameters of method base with GAs are in very good agreement with the experimental data in all cases.

INTRODUCTION

Metaheuristic algorithms have become high powerful tools for modeling and optimization. Metaheuristic optimization algorithms are used extensively for solving complex optimization problems. Compared

DOI: 10.4018/978-1-4666-9644-0.ch022

to conventional methods based on formal logics or mathematical programming, these metaheuristic algorithms are generally more powerful (Sahab, Toropov & Gandomi, 2013).

Optimization techniques can be divided in two groups, mathematical programming and metaheuristic algorithms. In general, the existing metaheuristic algorithms may be divided into two main categories as follows, Evolutionary algorithm and Evolutionary algorithms (Gandomi, 2014).

An evolutionary algorithm (EA) is a population-based meta-heuristic optimization algorithm inspired by the concepts of biological evolution such as reproduction, mutation, recombination and selection. They are randomized and stochastic techniques which simulate the evolutionary pressure of selection to select high fit individuals and use the evolutionary operators "crossover" and "mutation" to evolve better individuals. EAs generally start with a set (population) of random candidate solutions called individuals or chromosomes. In case of de novo drug design, chromosomes are represented by candidate chemical compounds. Evolutionary algorithms use the scoring function or the fitness function over the current population to determine the quality of the candidate solutions. Various selection methods such as "Roulette wheel", "Tournament", and "Rank" selection are used to populate the right candidates into the mating pool, where they are subjected to crossover and mutation. The crossover operator combines the genetic traits of two individuals (parents) and produces a new off-spring. A mutation event introduces new piece of information into an existing population. Thus, the crossover and mutation operators provide variations among the current population and evolve better and better solutions. The evolution process continues tilluser-specified termination criterion (Surbey, Devia, Sathyaa & Coumarb, 2015). Evolutionary strategy (ES) (Fogell, Owens & Walsh, 1966), genetic algorithm (GA) (Holland, 1975), and differential evolution (DE) (Storn & Price, 1997) are the most well-known paradigms of the evolutionary algorithms which use biological mechanisms such as crossover and mutation. These algorithms have been improved a lot and new variants of them have been introduced (e.g. cellular GA (Canyurt & Hajela, 2010)). Genetic programming (GP) is known as an extension of GAs (Koza, 1992). Several attempts have been made to improve the performance of GP (e.g.multi-stage GP (Gandomi & Alavi, 2011)). The mentioned evolutionary algorithms have been widely used to solve engineering optimization problems (e.g. Yüzgeç, Türker and Hocalar (2009)). Harmony search (HS) algorithm is another evolutionary algorithm proposed by Geemetal (Hedar & Fukushima, 2006). This algorithm simulates the musicians' behavior when they are searching for better harmonies.

Genetic algorithms (GAs) are adaptive heuristic search algorithm based on the evolutionary ideas of natural selection and genetics. As such they represent an intelligent exploitation of a random search used to solve optimization problems (Singh, Agrawal, Tiwari, Al-Helal & Avasthi, 2015). The performance of genetic algorithms depends to a large degree on the parameters which are under the control of the researcher, requiring adjustments to deal with the specific problem at hand. These parameters and namely the fitness function there- fore have to be carefully selected to match the specifics of credit scoring (Kozeny, 2015).

The algorithms that we use to determine parameters of PV generators (solar cells, modules and arrays) must be efficient and sufficiently accurate for process optimization and photovoltaic systems design tasks. These algorithms are of two types: those that use selected parts of the current–voltage (I–V) characteristic (Charles, Abdelkrim, Muoy & Mialhe, 1981; Charles, Ismail & Bordure, 1985; Laplaze & Youm, 1985; Chan & Phang, 1987). Accurate extraction and optimization of solar cells and solar panel parameters are very important in improving the device quality during fabrication and in device modeling and simulation (Gottschalg, Rommel, Infield & Kearney, 1999; Ortiz-Conde, Ma, Thomson, Santos, Garcia, Lei, Finol & Layman, 2000). To simulate and characterized a photovoltaic cell, and therefore the various parameters that affect these characteristics, several methods have been proposed for the identification of these parameters, to simulate its behavior, and optimize different characteristics.

The sensitivities of the parameters is directly related to the current-voltage characteristic of the solar cell, fives parameters were substituted into equation described in the following paragraph to draw an I–V curve. The shape of the I–V curve in the voltage source region is depressed horizontally with a gradual increase in the value of resistance series from zero. As shunt resistance decreases, the current diverted through the shunt resistor increases for a given level of junction voltage. The ideality factor and the reverse saturation current have a large influence on the open circuit voltage of the solar cell.

In practice, two main equivalent circuit models have been widely used: single and double diode models (Martil & Diaz, 1992; AlRashidi, AlHajri, El-Naggar & Al-Othman, 2011; AlHajri, El-Naggar, AlRashidi & Al-Othman, 2011). A relative of the photovoltaic generator empirical model, it is currently the most widely used due to its quality of results: the model with a one diode. Generally, in the literature, there are two types of approaches for the purpose of parameter extraction of solar cell models: analytical and numerical approaches.

In this chapter, we propose a technique based on genetic algorithms. The objective consisted to extract by genetic algorithm, parameters from experimental data using genetic algorithms (GAs) via an explicit sum equation of the photovoltaic cell. We decided to work with a solar cell and a solar panel, make our choice is based on the parameters in existing literature, in order to properly evaluate our proposed method.

BACKGROUND

Genetic Algorithms (GAs): Basic Concepts

The first step in the implementation of any genetic algorithm is first step in the implementation to generate an initial population. In the canonical genetic algorithm each member of this population will be a binary string of length L which corresponds to the problem encoding. Each string is sometimes referred to as a *genotype* (Charles, Ismail & Bordure, 1985) or, alternatively, a *chromosome* (Holland, 1985). In most cases the initial population is generated randomly. After creating an initial population, each string is then evaluated and assigned a *fitness* value (Schaffer, 1987).

Genetic algorithms (GAs) originated from the studies of cellular automata conducted by John Holland and his colleagues. Holland's book in 1975 is generally acknowledged as the beginning of GA research and only recently has their potential for solving combinatorial optimization problems been explored. Genetic algorithms owe their name to an early emphasis on representing and manipulating individuals in terms of their genetic make-up rather than using a phenotypic representation. The basic idea is to maintain a population of candidate solutions which evolves under a selective pressure that favours the survival of the fittest. Hence they can be viewed as a class of local search methods that employ a *solution-generation* mechanism which operates on attributes of a set of solutions rather than attributes of a single solution using a *move-generation* mechanism (Whitley).

The fundamental idea of GA is that it operates on a finite population of N chromosomes (solutions). The chromosomes are fixed strings with binary values (alleles) at each position (or locus). An allele is the 0 or 1 value in the bit string, and the "locus" is the position at which the 0 or 1 value is present in each location of the chromosome. Each chromosome of the population is evaluated according to some fitness function. Members of the population are selectively interbred in pairs to produce offspring. The fitter a member of the population the more likely it is to produce offspring. Genetic operators are used to facilitate the breeding process that results in offspring inheriting properties from their parents. The offspring are evaluated and placed in the population, possibly replacing the weaker members of the last generation. Thus, the search mechanism consists of three phases: evaluation of the fitness of each chromosome, selection of the parent chromosomes, and applications of *mutation* and *re-combination (crossover)* operators to the parent chromosomes. The new chromosomes resulting from these operations form the population for the next generation and the process is repeated until the system ceases to improve. The survival of the fittest principle ensures that the overall quality of solutions increases as the algorithm progresses from one generation to the next (Ibrahim, 1996).

A genetic algorithm starts with a population of randomly generated individuals, although it is also possible to use a previously saved population or a population of individuals encoding for solutions provided by a human expert or by another heuristic algorithm (Figure 1). Once an initial population has been created, the genetic algorithm enters a loop. At the end of each iteration a new population will have been created by applying a certain number of stochastic operators to the previous population (Koehn, 1994; Tomassini, 2015).

One such iteration is referred to as a generation.

The first operator to be applied is *selection*: in order to create a new intermediate population of n "parents", n independent extractions of an individual from the old population are performed, where the probability for each individual of being extracted is linearly proportional to its fitness.

Once the population of parents, that is of individuals that have been selected for reproduction, has been extracted, the individuals for the next generation will be produced through the application of a number of reproduction operators, which can involve just one parent, in which case we speak of *mutation*, or more parents, in which case we speak of recombination. In genetic algorithms two reproduction operators are used: *crossover* and *mutation*.

To apply *crossover*, couples are formed with all parent individuals; then, with a certain probability, called crossover rate p_{cross}, each couple actually undergoes crossover: the two bit strings are cut at the same random position and their second halves are swapped between the two individuals, thus yielding two novel individuals, each containing characters from both parents (Koehn, 1994; Tomassini, 2015).

After *crossover*, all individuals undergo mutation. The purpose of mutation is to simulate the effect of transcription errors that can happen with a very low probability (p_{mut}) when a chromosome is dupli-

Figure 1. The principle structure of a genetic algorithm (Ibrahim, 1996)

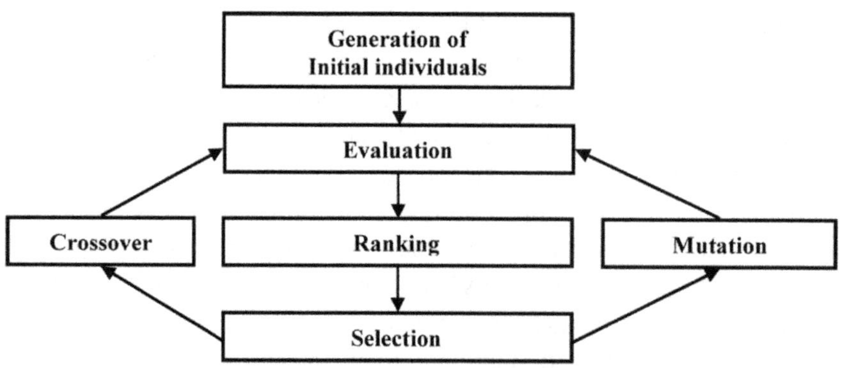

Box 1. Genetic algorithm

```
{
   Generate initial population P(t)
   Evaluate population  P(t)
    while stopping criteria not satisfied REPEAT
{
        Select some elements from P(t) to put into P(t+1)
        Crossover some elements of  P(t) and put them into P(t+1)
        Mutate some elements of  P(t) and put them into P(t+1)
        Ealaute new population  P(t+1)
        P(t)=P(t+1)
}
}
```

cated. This is accomplished by flipping each bit in every individual with a very small probability, called mutation rate. In other words, each "0" has a small probability of being turned into a "1" and vice versa.

In principle, the above-described loop is infinite, but it can be stopped when a given termination condition specified by the user is met. Possible termination conditions are: a pre-determined number of generations or time has elapsed or a satisfactory solution has been found or no improvement in solution quality has been taking place for a pre-determined number of generations. At this point, it is useful to recall that evolutionary algorithms are stochastic iterative techniques that are not guaranteed to converge although they usually succeed in finding good enough solutions (Tomassini, 2015; Langdon et al, 2008).

The evolutionary cycle can be summarized by the following pseudo-code (Langdon et al, 2008) (shown in Box 1).

OPTIMIZATION PARAMETERS SOLAR CELL: THEORY AND FORMULATION

In general, most semiconductor solar cells are based on single p-n junctions. Description of cell behavior can therefore be given in terms of a model based on the diode equation (AlRashidi, AlHajri, El-Naggar & Al-Othman, 2011). There are many possible mathematical relationships of varying complexity which can be used to describe PV current-voltage behavior. In this model, the solar cell is modeled as a current source connected in parallel with a diode. So a shunt resistance and a series resistance component are added to the model. The model of the solar cells for which the superposition principle is applicable can be represented by the equivalent circuit in Figure 2 and expressed as Equation (1) which includes light-generating current source (I_L), series resistance (R_s) and shunt resistance (R_p) of diode with the diode equation shown by Equation (1) (Van Peteghem & Vanhoucke, 2010).

Substituting these into the first equation produces the characteristic equation of a solar cell, which relates solar cell parameters to the output current and voltage:

$$I = I_l - I_0\left[\mathrm{Exp}\left(\frac{q\left(V + IR_s\right)}{nKT}\right) - 1\right] - \frac{V + IR_s}{R_p} \tag{1}$$

Figure 2. Simplified conceptual model, with one diode, for a photovoltaic solar cell

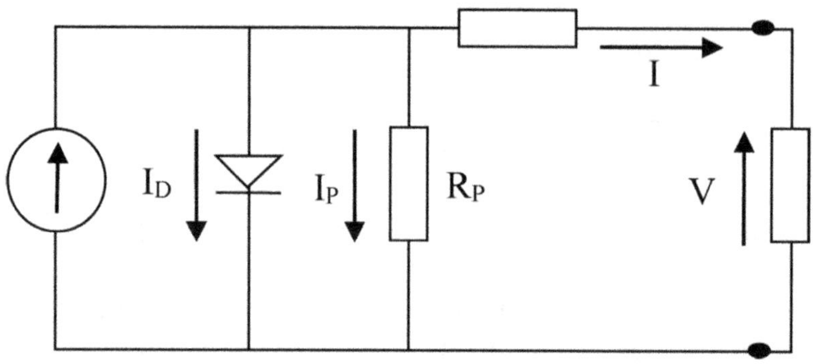

I_0 is the reverse saturation current, n the quality factor, k the Boltzmann's constant, T the absolute temperature, q the electron charge.

We noticed that the implicit nonlinear equation transcendent above (1) cannot be solved explicitly in general I or V using common elementary functions. To solve the equation and extract its parameters, there are explicit analytical solutions based on the use of the Lambert function, a special function which is not expressible in terms of two basic analytic functions. This approach has also been used recently to study the bright solar cells (Townsend, 1989; Ghani, Duke & Carson, 2013; Celik & Acikgoz, 2007).

In this context, we exploited the offering of circuit (Figure 1) and solving equation 2 in terms of the Lambert function. The five parameter model has been shown to offer improved accuracy over the more basic four parameter model which neglects the effects of shunt resistance (Jung & Guziewicz, 2009) while avoiding the additional complexity of the double diode model (Chen & Phang, 1987).

The current–voltage relation for a solar cell modified by the diode ideality factor is given by:

$$I = -\frac{V + IR_s}{R_p} - I_0\left[\mathrm{Exp}\left(\frac{V + IR_s}{nV_{th}}\right) - 1\right] + I_{ph} \qquad (2)$$

where n is the diode ideality factor, V_{th} is thermal voltage.

SOLAR CELL PARAMETERS EXTRACTION

Genetic Algorithms (GA) are direct, parallel, stochastic method for global search and optimization, which imitates the evolution of the living beings, described by Charles Darwin. GA is part of the group of Evolutionary Algorithms (EA). A canonical genetic algorithm (also called simple or standard GA) (Goldberg, 1989) is the one which uses binary representation, one-point crossover and bit-flipping mutation. A canonical genetic algorithm can be implemented as shown in Figure 2.

The organization follows the steps (Figure 3):

Step 1: The measured data are located, the voltages and currents are calculated by bais of Equation 1 the parameter values are estimated. A first generation of chromosomes is generated.

Figure 3. Flow diagram for the GA crossing-mates algorithm

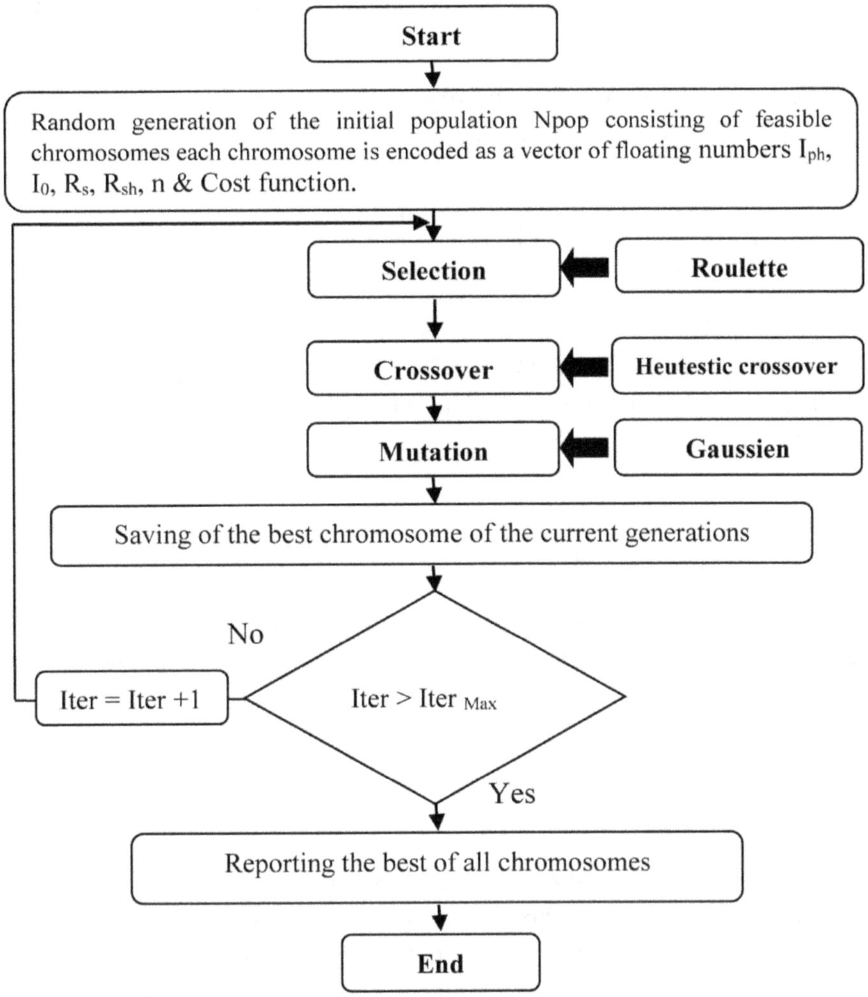

Step 2: A feature (I,V) of each individual in the current generation is calculated.
Step 3: The fitness values of all individuals of the current generation are calculated using equation (2).
Step 4: Evaluation of the number of iterations, if N is not equal to the number N_{Max}, then:
Step 5: We create a new generation reserving the best individuals of the current generation.
Step 6: The growth of the new generation continues with the operation croissement. In this case based on the roulette operation.
Step 7: The new generation is performed using the Gaussian mutation operation.

By adjusting the curve to the measured data, the solar cells parameters extraction involves minimizing the cost function f with respect to the five parameters (I_{ph}, I_0, R_s, n):

$$fitness = \sum_{i=1}^{N} \left(I_m - I_e(V_i, \theta)\right)^{1/2} \tag{3}$$

where, N is the number of experimental data. In equation (5), for the single diode model, (I_m, V_m) are respectively the measured current and voltage, I_e is a calculated value by substitution V_m in the equation (2). $\theta = \left(I_{ph}, I_S, R_S, n\right)$.

Equation (2) is explicit in I; one way of simplifying the computation of $I\left(V_i, \theta\right)$ is to substitute I_i and V_i in equation (2). Hence, we obtain equation (6).

$$I\left(V_i, \theta\right) = I_{ph} - I_s\left[\exp\left(\frac{q\left(V_i + R_s I\right)}{nKt}\right) - 1\right] - \frac{\left(V_i + R_s I\right)}{R_{sh}} \tag{4}$$

In Figure 3, we give the flow chart of the GAs. The chromosome here is the vector , containing the five parameters Is, I_{ph}, n, and R_s. The initial population (N_{POP}) of chromosomes is a matrix given by Eq. 2. (Zagrouba et al. (2010).

$$Initial\,Pop = \left(h_i - I_0\right).random\left[N_{ipop}, N_{par}\right] + I_0 \tag{5}$$

where N_{ipop} is the initial number of chromosomes in *Initial Pop*, Npar is the number of parameters in the chromosome (N_{par} = 5 in our case), I_o and h_i are respectively the lowest and the highest values of parameters I_s, I_{ph}, R_s, R_{sh} and n.

RESULTS AND DISCUSSION

The formulate optimization problem of the system is solved, in this work, by using the GA under MATLAB software, which may provide a number of potential solutions to the given problem. Real values measured of the characterization current-tension data for solar cell is used to implementation search its performance (Diego, Cuevas & Pajares, 2014). In this study, the single diode model is used to examine the proposed methodology in estimating the solar cell and panel solar parameters. The proposed estimation method is tested in this section for the solar cell single model case. The objective function as illustrated in equation (5) is minimized in order to reach an optimal set of parameters that reflects the solar cell characteristics. In order to test the quality of the fit to the experimental data, statistical analysis of the results is performed. The root mean squared error (RMSE), the mean bias error (MBE) and the mean absolute error (MAE) are the fundamental measures of accuracy (Gong & Cai, 2013; Oliva, Cuevas & Pajares, 2014).

The RMSE, MBE and MAE are given by:

$$RMSE = \sqrt{1/N\sum_{i=1}^{N} e_i^2} \tag{6}$$

$$MBE = \sum_{i=1}^{N} e_i \tag{7}$$

$$MAE = 1/N\sum_{i=1}^{N}|e_i| \tag{8}$$

where

e: is the relative error, i.e.

$$e_i = \frac{I_{mesured} - I_{calculed}}{I_{mesured}}$$

N: is the number of measurements.

In the first case, the solar cell is tested to validate the technique of genetic algorithms.

The objective function is present by equation (4) is minimized in order to be to reach an optimal set of parameters That Reflects the solar cell characteristics. A value of zero for the objective function would yield an optimal solution. Experimental data of a silicon solar cell are taken from the bibliography. Voltage and current data measurements were taken using a 57 mm diameter commercial (R.T.C France) silicon solar cell under 1 sun (1000 W/m^2) at 33°C. The obtained results are compared with different sets of results reported from references and results obtained using GA.

The extracted parameters obtained using the method proposed here for the silicon solar cell and module is given in Table 1 and Table 2. These methods represent different estimation methodologies used in parameter estimation, namely direct based techniques, indirect methods (metaheuristics optimization algorithms). Table 1 and Table 2 shows that most parameters extracted using the method study (Genetic Algorithms) are very close to those reported in other references.

Satisfactory agreement is obtained for most of the extracted parameters. A comparison with different methods is also given, and good agreement is reported. Statistical indicators of accuracy for the method of this work are shown in Table 3.

Table 1. Extracted parameters by the current method and the previous work

	Solar cell RTC Si SC						
Parameter	**AlHajri et al. (2012)**	**Askarzadeh & Rezazadeh, (2012)**	**Askarzadeh & Rezazadeh, (2013)**	**AlHajri et al. (2012)**	**El-Naggar et al. (2012)**	**Huang et al. (2011)**	**This work**
I_{ph} (A)	0.760776	0.7608	0.7608	0.7617	0.7620	0.7607	0.7608
I_0(μA)	0.323021	0.3435	0.3062	0.9980	0.4798	0.4000	0.1459
a	1.481184	1.4874	1.4758	1.6000	1.5172	1.5033	1.4022
R_s (Ω)	0.036377	0.0361	0.0366	0.0313	0.0345	0.0354	0.0404
R_{sh} (kΩ)	53.718526	53.2845	52.2845	64.1026	43.2845	64.1026	50.1424

Table 2. Extracted parameters by the current method and the previous work

Parameter	MSX-60 Panel solar				
	Bonkoungou, Koalaga, & Njomo (n.d.).	Shu-xian (2013).	El Tayyan (2011)	Zhang et al. (2011)	This work
I_{ph} (A)	3.8119	3.82006	3.812	3.812	3.8119
$I_0(\mu A)$	0.1859	0.2446	0.1866	0.1802	0.1845
a	1.360	0.9001	1.358	1.277	1.34
R_s (Ω)	0.180	0.3862	0.178	0.171	0.1774
R_{sh} ($k\Omega$)	360.00	146.2	3376.71	3585.69	334.85

Table 3. Statistical indicators of accuracy for the method of this work

	RMSE (%)	MBE (%)	MAE (%)
Solar cell RTC Si SC	0.0155	-0.0502	0.0502
MSX-60 Panel solar	0.0261	-0.0505	0.05388

The computed values of the statistical factors show high accuracy in both cases. It has been found that RMSE, MBE, and MAE values are respectively 0.0155,-0.0502, and 0.0502 for the solar cell case. The corresponding values for the PV module are 0.0261, -0.0505, and 0.0538.

Figures 4 –5 show the plots of I–V experimental characteristics and the fitted curves derived from Eq. (4) with the parameters shown in Tables 1 and 2 for the silicon solar cell and module. A curve fitting is performed next to evaluate the goodness of fit of the obtained solution as shown in Tables 1 and 2 for the solar cell and PV module, respectively.

Unlike other methods that have been developed for this purpose, the genetic algorithms, does not require prior in-depth knowledge of complex parameters. Interesting targets with Genetic Algorithm procedure is here that it has been successfully applied to experimental IV characteristics of different types of solar cells with solar cells.

The results of the analysis presented in white figures 4 and 5 confirm the superior quality of results. We can say that the method presented, tested for selected cases, is more reliable to obtain meaningful physical parameters and is simple and easy to use.

CONCLUSION

We present in this chapter an application based in Genetic Algorithm. Genetic algorithms belong to the family of evolutionary algorithms on low natural mechanisms: selection, crossover and mutations. Their goal is to obtain an approximate solution to an optimization problem, when there is no live classical method to solve the adequacy with time. Genetic algorithms use the concept of natural selection and apply it to a population of potential solutions to the given problem. In this chapter, the use of genetic algorithm to accurately estimate the parameter of solar cells has been presented. In our vision, the estimation process

Figure 4. Solar cell Characteristic I(V) obtained experimentally (experimental data) and the calculated characteristic I(V) –solar cell RTC

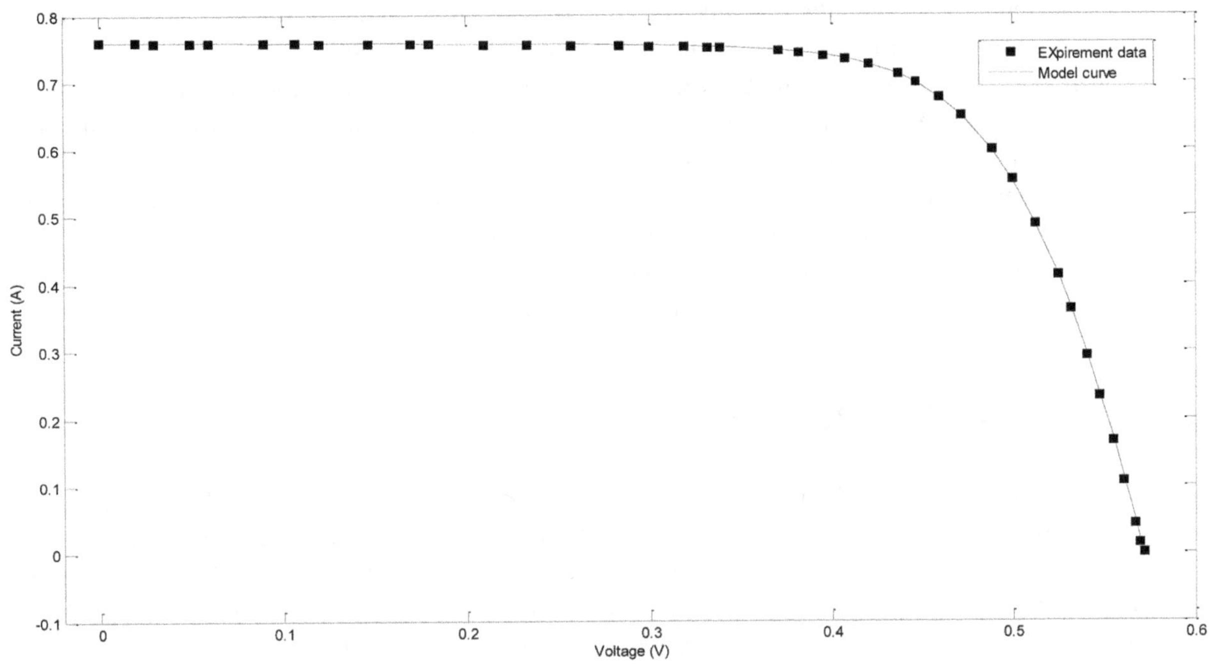

Figure 5. Solar cell Characteristic I(V) obtained experimentally (experimental data) and the calculated characteristic I(V) – panel solar 60W

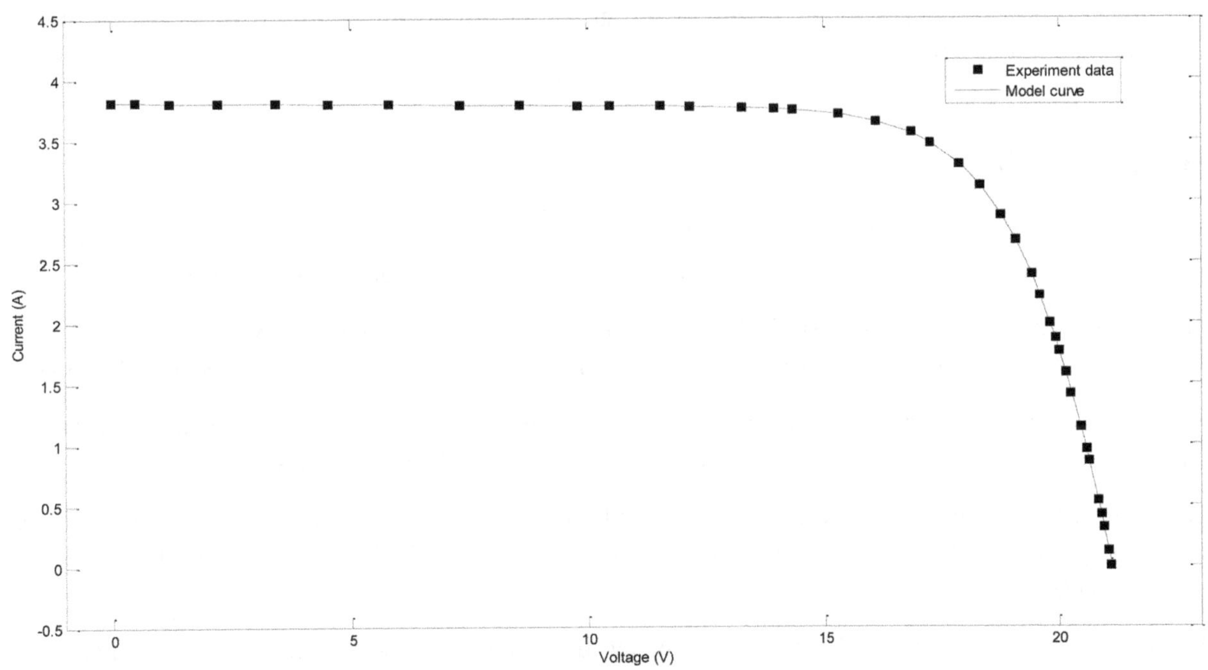

is considered as an optimization problem. A simple method for extracting the solar cell parameters based on measured current–voltage data is considered. The method has been successfully applied to a silicon solar cell, and module. The results obtained are in good agreement with those published previously. The proposed approach encodes the parameters of the solar cell as a candidate solution. An objective function evaluates the matching quality between a candidate solution and the experimental data. Guided by the values of this objective function, the set of encoded candidate solutions is evolved by using the operators defined by genetic algorithm so that the parameters that produce the best possible approximation to the I-V measurements obtained by the true solar cell can be found. The proposed approach has been compared with other similar techniques proposed in the literature. The efficiency of the algorithm has been evaluated in terms of accuracy and robustness. The method is very simple to use in physical models, optimized to their theoretical or empirical parameters.

FUTURE RESEARCH DIRECTIONS

The work presented is registered, in the context of renewable energy development and research, in collaboration of two laboratories, Electrical Engineering Laboratory of Oran, and Laboratory of Materials in Electrical Engineering, University of Science and Technology of Oran (USTO-MB) BP 1505 El M'naouer, 31000 Oran, Algeria. The work presented in this chapter is to provide parameters optimization methodologies for mathematical model or physical discussed. We believe that the analysis and the results presented here can motivate further study of phenomena represented by a parametric model. My search below in this area has focused on modeling, optimizing a system with advanced modern methods.

Prospects in the near future are:

1. Research the methodologies of global optimization algorithms, and simple to implement;
2. Integration of the algorithms in the appropriate models to obtain an approximate solution to an optimization problem.

REFERENCES

AlHajri, M., El-Naggar, K., AlRashidi, M., & Al-Othman, A. (2012). Optimal extraction of solar cell parameters using pattern search. *Renewable Energy*, *44*(0), 238–245. doi:10.1016/j.renene.2012.01.082

AlHajri, M., El-Naggar, K., AlRashidi, M., & Al-Othman, A. (2012). Optimal extraction of solar cell parameters using pattern search. *Renewable Energy*, *44*(0), 238–245. doi:10.1016/j.renene.2012.01.082

AlHajri, M. F., El-Naggar, K. M., AlRashidi, M. R., & Al-Othman, A. K. (2012). Optimal extraction of solar cell parameters using pattern search. *Renewable Energy*, *44*, 238–245. doi:10.1016/j.renene.2012.01.082

AlRashidi, M., AlHajri, M., El-Naggar, K., & Al-Othman, A. (2011). A new estimation approach for determining the I–V characteristics of solar cells. *Solar Energy*, *85*(7), 1543–1550. doi:10.1016/j.solener.2011.04.013

Askarzadeh, A., & Rezazadeh, A. (2012). Parameter identification for solar cell models using harmony search-based algorithms. *Solar Energy*, *86*(11), 3241–3249. doi:10.1016/j.solener.2012.08.018

Askarzadeh, A., & Rezazadeh, A. (2013). Artificial bee swarm optimization algorithm for parameters identification of solar cell models. *Applied Energy*, *102*(0), 943–949. doi:10.1016/j.apenergy.2012.09.052

Bonkoungou, Koalaga, & Njomo. (n.d.). Modelling and simulation of photovoltaic module considering signale-diode equivalent circuit model in MATLAB. *International Journal of Emerging Technology and Advanced Engineering*.

Canyurt, O. E., & Hajela, P. (2010). Cellular genetic algorithm technique for the multicriterion design optimization. *Struct Multidiscip Optim*, *40*(1–6), 201–214. doi:10.1007/s00158-008-0351-3

Cartwright, N. (1983). *How the Laws of Physics Lie*. Oxford University Press. doi:10.1093/0198247044.001.0001

Celik, A. N., & Acikgoz, N. (2007). Modelling and experimental verification of the operating current of mono-crystalline photovoltaic modules using four-and five-parameter models. *Applied Energy*, *84*(1), 1–15. doi:10.1016/j.apenergy.2006.04.007

Chan, D. S., & Phang, J. C. H. (1987). Analytical methods for the extraction of solar-cell singleand double-diode model parameters from I-V characteristics. *IEEE Transactions on Electron Devices*, *34*(2), 286–293. doi:10.1109/T-ED.1987.22920

Chan, D. S. H., & Phang, J. C. H. (1987). Analytical methods for the extraction of solar-cell single- and double-diode model parameters from I–V characteristics. *IEEE Transactions on Electron Devices*, *ED-34*(2), 286–293. doi:10.1109/T-ED.1987.22920

Charles, J. P., Abdelkrim, M., Muoy, Y. H., & Mialhe, P. (1981). A practical method of analysis of the current–voltage characteristics of solar cells. *Solar Cells*, *4*(2), 169–178. doi:10.1016/0379-6787(81)90067-3

Charles, J. P., Ismail, M. A., & Bordure, G. (1985). A critical study of the effectiveness of the single and double exponential models for I–V characterization of solar cells. *Solid-State Electronics*, *28*(8), 807–820. doi:10.1016/0038-1101(85)90068-1

Charles, J. P., Ismail, M. A., & Bordure, G. (1985). A critical study of the effectiveness of the single and double exponential models for I–V characterization of solar cells. *Solid-State Electronics*, *28*(8), 807–820. doi:10.1016/0038-1101(85)90068-1

El-Naggar, K., AlRashidi, M., AlHajri, M., & Al-Othman, A. (2012). Simulated annealing algorithm for photovoltaic parameters identification. *Solar Energy*, *86*(1), 266–274. doi:10.1016/j.solener.2011.09.032

El Tayyan, A. A. (2011). Article. *Journal of Electron Devices*, *9*, 335.

Fogel, L. J., Owens, A. J., & Walsh, M. J. (1966). *Artificial intelligence thorough simulated evolution*. Chichester, UK: John Wiley&Sons, Ltd.

Gandomi, A. H. (2014). Interior search algorithm (ISA): A novel approach for global optimization, ISA Transactions. *ISA Transactions*, *53*(4), 1168–1183. doi:10.1016/j.isatra.2014.03.018 PMID:24785823

Gandomi, A. H., & Alavi, A. H. (2011). Multi-stage genetic programming: A new strategy to nonlinear system modeling. *InfSci*, *181*, 5227–5239.

Genetic Algorithm. (n.d.). Retrieved from http://en.wikipedia.org

Ghani, F., Duke, M., & Carson, J. (2013). Numerical calculation of series and shunt resistance of a photovoltaic cell using the Lambert W-function: Experimental evaluation. *Solar Energy*, *87*, 246–253. doi:10.1016/j.solener.2012.11.002

Goldberg, D. E. (1989). *Genetic Algorithms in Search, Optimization, and Machine Learning*. Boston: Addison-Wesley.

Gong, W., & Cai, Z. (2013). Parameter extraction of solar cell models using repaired adaptive differential evolution. *Solar Energy*, *94*, 209–220. doi:10.1016/j.solener.2013.05.007

Gottschalg, R., Rommel, M., Infield, D. G., & Kearney, M. J. (1999). The influence of the measurement environment on the accuracy of the extraction of physical parameters of solar cells. *Meas. Sci. Technol.*, *10*, 796–804.

Hacking, I. (1983). *Representing and Intervening. Introductory Topics in the Philosophy of Natural Science*. Cambridge University Press. doi:10.1017/CBO9780511814563

Hedar, A., & Fukushima, M. (2006). Derivative-Free Filter Simulated Annealing Method for Constrained Continuous Global Optimization. *JGlobalOptim*, *35*, 521–549.

Holland, J. (1975). *A aptation in natural and artificial systems*. University of Michigan Press.

Holland, J. (1985). *Adaptation in Natural and Artificial Systems*. University of Michigan Press.

Huang, W., Jiang, C., Xue, L., & Song, D. (2011). Extracting solar cell model parameters based on chaos particle swarm algorithm. In *2011 International Conference on Electric Information and Control Engineering (ICEICE)*. doi:10.1109/ICEICE.2011.5777246

Ibrahim, H. (1996). Metaheuristics: A bibliography, Institute of Mathematics and Statistics. *Annals of Operations Research*, *63*, 513–623.

Jung, W., & Guziewicz, M. (2009). Schottky diode parameters extraction using LambertWfunction. *Materials Science and Engineering B*, *165*(1-2), 57–59. doi:10.1016/j.mseb.2009.02.013

Koehn. (1994). *Combining Genetic Algorithms and Neural Networks: The Encoding Problem*. (Master of Science Degree Thesis). The University of Tennessee, Knoxville, TN.

Koza, J. R. (1992). *Genetic programming: on the programming of computers by natural selection*. Cambridge, MA: MIT Press.

Kozeny, V. (2015). Genetic algorithms for credit scoring: Alternative fitness function performance comparison, Expert Systems with Applications. *Expert Systems with Applications*, *42*(6), 2998–3004. doi:10.1016/j.eswa.2014.11.028

Langdon, W. B. (2008). Genetic Programming: An Introduction and Tutorial, with a Survey of Techniques and Applications. *Studies in Computational Intelligence, 115*, 927–1028. Retrieved from www.springerlink.com

Laplaze, D., & Youm, I. (1985). Modélisation dune cellule photovoltaïque: Détermination des paramètres à partir de la caractéristique courant-tension sous e'clairement. *Solar Cells*, *14*(2), 167–177. doi:10.1016/0379-6787(85)90039-0

Martil, I., & Diaz, G. G. (1992). Determination of the dark and illuminated characteristic parameters of a solar cell for IV characteristicsEur. *Journal of Physics*, 13193–13197.

Mathematical Optimization. (n.d.). Retrieved from http://en.wikipedia.org

Metaheuristic. (n.d.). Retrieved from http://en.wikipedia.org

Mitchell, M. (1996). *An Introduction to Genetic Algorithms*. Cambridge, MA: MIT Press.

Oliva, D., Cuevas, E., & Pajares, G. (2014). Parameter identification of solar cells using artificial bee colony optimization. *Energy*, *72*, 93–102. doi:10.1016/j.energy.2014.05.011

Ortiz-Conde, A., Ma, Y., Thomson, J., Santos, E., Liou, J. J., Garcia, F. J., & Layman, P. et al. (2000). Parameter extraction using lateral and vertical optimisation. In *Proc. 22nd Int. Conf. on Microelectronics, MIEL 2000*.

Pearce, J. (2002). Open access "Photovoltaics – A Path to Sustainable Futures". *Futures*, *34*(7), 663–674. doi:10.1016/S0016-3287(02)00008-3

Peteghem, V. V., & Vanhoucke, M. (2010). A genetic algorithm for the preemptive and non-preemptive multi-mode resource-constrained project scheduling problem. *European Journal of Operational Research*, *201*(2), 409–418. doi:10.1016/j.ejor.2009.03.034

Photovoltaics. (n.d.). Retrieved from http://en.wikipedia.org

Sahab, M. G., Toropov, V. V., & Gandomi, A. H. (2014). Traditional and Modern Structural Optimization Techniques – Theory and Application. In Metaheuristic Applications in Structures and Infrastructures. Elsevier.

Schaffer, J. D. (1987) Some Effects of Selection Procedures on Hyperplane Sampling by Genetic Algorithms. In Genetic Algorithms and Simulated Annealing. Colorado State University.

Scientific Modelling. (n.d.). Retrieved from http://en.wikipedia.org

Shu-xian, L. (2013). An explicit approximate I–V characteristic model of a solar cell based on pade´ approximants. *Solar Energy*, *92*, 147–159.

Singh, S., Agrawal, S., Tiwari, A., Al-Helal, I. M., & Avasthi, D. V. (2015). Modeling and parameter optimization of hybrid single channel photovoltaic thermal module using genetic algorithms. *Solar Energy*, *113*, 78–87. doi:10.1016/j.solener.2014.12.031

Solar Cell. (n.d.). Retrieved from www.wikipedia.org

Solar Cells. (n.d.). Retrieved from www.chemistryexplained.com

Solar Panel. (n.d.). Retrieved from wikipedia.org

Storn, R., & Price, K. V. (1997). Differential evolution, a simple and efficient heuristic for global optimization over continuous spaces. *Journal of Global Optimization*, *11*(4), 341–359. doi:10.1023/A:1008202821328

Survey, R. (2015). Evolutionary algorithms for de novo drug design, Applied Soft Computing. *Applied Soft Computing*, *27*, 543–552.

The Nature of Mathematical Programming. (n.d.). Mathematical Programming Glossary, INFORMS Computing Society.

Tomassini. (n.d.). *Parallel and Distributed Evolutionary Algorithms: A Review*. Institute of Computer Science, University of Lausanne.

Towsend, T. U. A. (1989). Method for estimating the long-term performance of direct-coupled photovoltaic systems. (M.S. Thesis). Solar Energy Laboratory, University of Wisconsin.

Yüzgeç, U., Türker, M., & Hocalar, A. (2009). On-line evolutionary optimization of an industrial fed-batch yeast fermentation process. *ISA Transactions*, *48*(1), 79–92. doi:10.1016/j.isatra.2008.09.001 PMID:18849027

Zagrouba, M., Sellami, A., Bouaıcha, M., & Ksouri, M. (2010). Identification of PV solar cells and modules parameters using the genetic algorithms: Application to maximum power extraction. *Solar Energy*, *84*(5), 860–866. doi:10.1016/j.solener.2010.02.012

Zhang, C., Zhang, J., Hao, Y., Lin, Z., & Zhu, C. (2011). A simple and efficient solar cell parameter extraction method from a single current-voltage curve. *Journal of Applied Physics*, *110*(6), 064504. doi:10.1063/1.3632971

ADDITIONAL READING

Araujo F, Sanchez E and Marti M 1982 Determination of the two-exponential solar cell equation parameters from empirical data *Solar Cells* 5 199–204.

Aubray, V., & Meyer, F. (1994). Schottky diodes with high series resistance: Limitations of forward *I–V* methods. *Journal of Applied Physics*, *76*(12), 7973–7984. doi:10.1063/1.357909

Cai, T. Application of Soft Computing Techniques for Renewable Energy Network Design and Optimization, http://www.igi-global.com/chapter/application-of-soft-computing-techniques-for-renewable-energy-network-design-and-optimization

Chan, D. S. H., & Phang, J. C. H. (1987). Analytical methods for the extraction of solar-cell single- and double-diode model parameters from I–V characteristics. *Electron Dev IEEE Trans*, *34*(2), 286–293. doi:10.1109/T-ED.1987.22920

Charles, J.-P., Bordure, G., Khoury, A., & Mialhe, P. (1985). Consistency of the double exponential model with physical mechanisms of conduction for a solar cell under illumination. *Journal of Physics. D, Applied Physics*, *18*(11), 2261–2268. doi:10.1088/0022-3727/18/11/015

Chegaar, M., Azzouzi, G., & Mialhe, P. (2006). Simple parameter extraction method for illuminated solar cells. *Solid-State Electronics*, *50*(7-8), 1234–1237. doi:10.1016/j.sse.2006.05.020

Chegaar, M., Ouennoughi, Z., & Guechi, F. (2004). Extracting dc parameters of solar cells under illumination. *Vacuum*, *75*(4), 367–372. doi:10.1016/j.vacuum.2004.05.001

De Soto, W., Klein, S. A., & Beckman, W. A. (2006). Improvement and validation of a model for photovoltaic array performance. *Solar Energy*, *80*(1), 78–88. doi:10.1016/j.solener.2005.06.010

E. A. Barakhtenko et al., New Results in Development of Methods for Optimization of Heat Supply System Parameters and Their Software Implementation, http://www.igi-global.com/article/new-results-in-development-of-methods-for-optimization-of-heat-supply-system-parameters-and-their-software-implementation

Gottschalg, R., Rommel, M., Infield, D. G., & Kearney, M. J. (1999). The influence of the measurement environment on the accuracy of the extraction of the physical parameters of solar cells. *Measurement Science & Technology*, *10*(9), 796–804. doi:10.1088/0957-0233/10/9/306

Han, X., Wang, Y., & Zhu, L. (2011). Electrical and thermal performance of silicon concentrator solar cells immersed in dielectric liquids. *Applied Energy*, *88*(12), 4481–4489. doi:10.1016/j.apenergy.2011.05.037

Haouari-Merbah, M., Belhamel, M., Tobías, I., & Ruiz, J. M. (2005). Extraction and analysis of solar cell parameters from the illuminated current-voltage curve. *Solar Energy Materials and Solar Cells*, *87*(1-4), 225–3. doi:10.1016/j.solmat.2004.07.019

Ishaque, K., & Salam, Z. (2011). An improved modeling method to determine the model parameters of photovoltaic (PV) modules using differential evolution (DE). *Solar Energy*, *85*(9), 2349–2359. doi:10.1016/j.solener.2011.06.025

Ishaque, K., Salam, Z., & Syafaruddin, . (2011). Syafaruddin. A comprehensive MATLAB Simulink PV system simulator with partial shading capability based on two-diode model. *Solar Energy*, *85*(9), 2217–2227. doi:10.1016/j.solener.2011.06.008

Ishaque, K., Salam, Z., Taheri, H., & Shamsudin, A. (2011). A critical evaluation of EA computational methods for Photovoltaic cell parameter extraction based on two diode model. *Solar Energy*, *85*(9), 1768–1779. doi:10.1016/j.solener.2011.04.015

Ishaque, K., Salam, Z., Taheri, H., & Syafaruddin, . (2011). Syafaruddin. Modeling and simulation of photovoltaic (PV) system during partial shading based on a two-diode model. *Simulation Modelling Practice and Theory*, *19*(7), 1613–1626. doi:10.1016/j.simpat.2011.04.005

Kaminski, A., Marchand, J. J., Fave, A., & Laugier, A. 1997 New method of parameters extraction from dark *I–V* curve *Proc. 26th PVSC (Anaheim, CA, 1997)* pp 203–5.

Kuznetsov, I. A., Greenfield, M. J., Mehta, Y. U., Merchan-Merchan, W., Salkar, G., & Saveliev, A. V. (2011). Increasing the solar cell power output by coating with transition metal-oxide nanorods. *Applied Energy*, *88*(11), 4218–4221. doi:10.1016/j.apenergy.2011.04.033

Lee, J. I., Brini, J., & Dimitriadis, C. A. (1998). Simple parameter extraction method for non-ideal Schottky barrier diodes. *Electronics Letters*, *34*(12), 1268–1269. doi:10.1049/el:19980831

Li, M., Ji, X., Li, G., Wei, S., Li, Y., & Shi, F. (2011). Performance study of solar cell arrays based on a trough concentrating photovoltaic/thermal system. *Applied Energy*, *88*(9), 3218–3227. doi:10.1016/j.apenergy.2011.03.030

Li, M., Liu, Y., Wang, H., & Shen, H. (2011). Synthesis of TiO2 submicro-rings and their application in dye-sensitized solar cell. *Applied Energy*, *88*(3), 825–830. doi:10.1016/j.apenergy.2010.09.017

Liu, C.-C., Chen, C.-Y., Weng, C.-Y., Wang, C.-C., Jenq, F.-L., Cheng, P.-J., & Houng, M.-P. et al. (2008). Physical parameters extraction from current–voltage characteristic for diodes using multiple nonlinear regression analysis. *Solid-State Electronics*, *52*(6), 839–843. doi:10.1016/j.sse.2007.12.010

Liu, C.-C., Chen, C.-Y., Weng, C.-Y., Wang, C.-C., Jenq, F.-L., Cheng, P.-J., & Houng, M.-P. et al. (2008). Physical parameters extraction from current–voltage characteristic for diodes using multiple nonlinear regression analysis. *Solid-State Electronics*, *52*(6), 839–843. doi:10.1016/j.sse.2007.12.010

Lyakas, M., Zaharia, R., & Eizenberg, M. (1995). Analysis of non-ideal Schottky and pn junction diodes, extraction of parameters from *I–V* plots. *Journal of Applied Physics*, *78*(9), 5481–5489. doi:10.1063/1.359664

Mankad, K. B. An Intelligent Process Development Using Fusion of Genetic Algorithm with Fuzzy Logic, http://www.igi-global.com/chapter/an-intelligent-process-development-using-fusion-of-genetic-algorithm-with-fuzzy-logic

Martil, I., & Diaz, G. (1992, July 01). Determination of the dark and illuminated characteristic parameters of a solar cell for *I–V* characteristics. *European Journal of Physics*, *13*(4), 193–197. doi:10.1088/0143-0807/13/4/009

Nakanishi, F., Ikegami, T., Ebihara, K., Kuriyama, S., & Shiota, Y. Modeling and operation of a 10kW photovoltaic power generator using equivalent electric circuit method. In: Photovoltaic specialists conference, 2000. Conference record of the twenty-eighth IEEE; 2000. p. 1703–6.

Ortiz-Conde, A., García Sánchez, F. J., & Muci, J. (2006). New method to extract the model parameters of solar cells from the explicit analytic solutions of their illuminated I–V characteristics. *Solar Energy Materials and Solar Cells*, *90*(3), 352–361. doi:10.1016/j.solmat.2005.04.023

Ortiz-Conde, A., Ma, Y., Thomson, J., Santos, E., Liou, J. J., Garcia, F. J., & Layman, P. et al. 2000 Parameter extraction using lateral and vertical optimisation *Proc. 22nd Int. Conf. on Microelectronics, MIEL 2000 (Nis*, 2000*)* vol 1, pp 165–8.

Ouennoughi, Z., & Cheggar, M. (1999). A simpler method for extracting solar cell parameters using the conductance method. *Solid-State Electronics*, *43*(11), 1985–1988. doi:10.1016/S0038-1101(99)00174-4

Phang J and Chan D 1986 A comparative study of extraction methods for solar cells *I–V* characteristics *Solar Cells* 18 1–12.

Qiao, L., Wang, D., Zuo, L., Ye, Y., Qian, J., Chen, H., & He, S. (2011). Localized surface Plasmon resonance enhanced organic solar cell with gold nanospheres. *Applied Energy*, *88*(3), 848–852. doi:10.1016/j.apenergy.2010.09.021

Sah, C. T., Noyce, R. N., & Shockley, W. Carrier generation and recombination in P–N junctions and P–N junction characteristics. *Proc IRE* 1957;45:1228–43.

Walker, G. (2001). Evaluating MPPT converter topologies using a MATLAB PV model. *Journal of Electrical and Electronics Engineering, Australia, 21*, 8.

Wang, Y., Fang, Z., Zhu, L., Huang, Q., Zhang, Y., & Zhang, Z. (2009). liquids The performance of silicon solar cells operated in. *Applied Energy*, *86*(7-8), 1037–1042. doi:10.1016/j.apenergy.2008.08.020

Werner, J. H. (1988). Schottky barrier and pn-junction *I–V* plots, small signal evaluation. *Applied Physics. A, Materials Science & Processing*, *47*(3), 291–300. doi:10.1007/BF00615935

Wolf, M., Noel, G. T., & Stirn, R. J. (1997). Investigation of double exponential in the current–voltage characteristics of silicon solar cells. *IEEE Transactions on Electron Devices*, *24*(4), 419–428. doi:10.1109/T-ED.1977.18750

Xu, H. B., Hong, R. J., Ai, B., Zhuang, L., & Shen, H. (2010). Application of phosphorus diffusion gettering process on upgraded metallurgical grade Si wafers and solar cells. *Applied Energy*, *87*(11), 3425–3430. doi:10.1016/j.apenergy.2010.03.026

Zhang, W., Zhu, R., Liu, B., & Ramakrishna, S. (2012). High-performance hybrid solar cells employing metal-free organic dye modified TiO2 as photoelectrode. *Applied Energy*, *90*(1), 305–308. doi:10.1016/j.apenergy.2011.03.037

KEY TERMS AND DEFINITIONS

Genetic Algorithm: In the field of artificial intelligence, a genetic algorithm (GA) is a search heuristic that mimics the process of natural selection. This heuristic (also sometimes called a metaheuristic) is routinely used to generate useful solutions to optimization and search problems. Genetic algorithms belong to the larger class of evolutionary algorithms (EA), which generate solutions to optimization problems using techniques inspired by natural evolution, such as inheritance, mutation, selection, and crossover (Wikipedia, 2015; Mitchell, 1996).

Metaheuristic Algorithms: In computer science, artificial intelligence, and mathematical optimization, a *heuristic* is a technique designed for solving a problem more quickly when classic methods are too slow, or for finding an approximate solution when classic methods fail to find any exact solution (Wikipedia, 2015).

Modeling: Scientific modelling is a scientific activity, the aim of which is to make a particular part or feature of the world easier to understand, define, quantify, visualize, or simulate by referencing it to existing and usually commonly accepted knowledge. It requires selecting and identifying relevant aspects of a situation in the real world and then using different types of models for different aims, such as conceptual models to better understand, operational models to operationalize, mathematical models to quantify, and graphical models to visualize the subject. Modelling is an essential and inseparable part of scientific activity, and many scientific disciplines have their own ideas about specific types of modelling (Mathematical Programming; Wikipedia, 2015; Cartwright, 2983; Hacking, 1983).

Optimization: In mathematics, computer science, economics, or management science, mathematical optimization (alternatively, optimization or mathematical programming) is the selection of a best element (with regard to some criteria) from some set of available alternatives. In the simplest case, an optimization problem consists of maximizing or minimizing a real function by systematically choosing input values from within an allowed set and computing the value of the function. The generalization of optimization theory and techniques to other formulations comprises a large area of applied mathematics.

More generally, optimization includes finding "best available" values of some objective function given a defined domain (or a set of constraints), including a variety of different types of objective functions and different types of domains (Wikipedia, 2015).

Photovoltaic: Photovoltaics (PV) is a method of converting solar energy into direct current electricity using semiconducting materials that exhibit the photovoltaic effect. A photovoltaic system employs solar panels composed of a number of solar cells to supply usable solar power. Power generation from solar PV has long been seen as a clean sustainable (Wikipedia, 2015; Pearce, 2002).

Solar Cell: A solar cell, or photovoltaic cell, is an electrical device that converts the energy of light directly into electricity by the photovoltaic effect, which is a physical and chemical phenomenon (Chemistry explained, 2015). It is a form of photoelectric cell, defined as a device whose electrical characteristics, such as current, voltage, or resistance, vary when exposed to light. Solar cells are the building blocks of photovoltaic modules, otherwise known as solar panels.

Solar Panel: Solar panel refers either to a photovoltaics (PV) module, a solar hot water panel, or to a set of solar photovoltaics modules electrically connected and mounted on a supporting structure. A PV module is a packaged, connected assembly of solar cells. Solar panels can be used as a component of a larger photovoltaic system to generate and supply electricity in commercial and residential applications (Wikipedia, 2015).

The Five-Parameter Model: The five-parameter model used in this work evaluates individually each parameter $(I_s, I_{ph}, R_s, R_{sh}, \text{and } n)$ of the nonlinear equation of I-V characteristic at STC and any other operating condition from analytical calculations.

Chapter 23
Dantzig-Wolfe Decomposition and Lagrangean Relaxation-Based Heuristics for an Integrated Production and Maintenance Planning with Time Windows

Najib Mohamed Najid
Université de Nantes, France

Marouane Alaoui-Selsouli
Ortems, France

Abdemoula Mohafid
Université de Nantes, France

ABSTRACT

In this chapter, two approaches are developed to solve the integrated production planning and maintenance problem. Moreover, Some Propositions and mathematical properties were suggested and applied in the proposed heuristics to solve the problem. The first heuristic developed is based on Dantzig-Wolfe decomposition. The Dantzig-Wolfe Decomposition principle reformulates the original model and Column generation is then used to deal with the huge number of variables of the reformulated model. A simple rounding heuristic and a smoothing procedure are finally carried out in order to obtain integer solutions. The second heuristic is based on Lagrangean relaxation of the capacity constraints and sub-gradient optimization. At every step of sub-gradient method, feasibility and improvement procedures are applied to the solution of the Lagrangean problem. Computational experiments are carried out to show the results obtained by our approaches and compared to those of commercial solver.

DOI: 10.4018/978-1-4666-9644-0.ch023

I. INTRODUCTION

Maintenance is a task closely related to production scheduling in industrial settings. It is the operations or techniques that allow maintaining or restoring equipment to a specific state and guaranteeing a given service. Both activities conflict since, as it is known, maintenance is generally a secondary process in companies that have production as their core business. Indeed, production management often views maintenance in the context of hours or days out of service and fails to realize the strategic importance of incorporating maintenance in the manufacturing planning. On the other hand, management for the maintenance function attempts to impose constraints on production that it deems necessary to achieve complete equipment reliability. As we can see, the production planning and maintenance are dealt separately in the literature and also in the industry. Therefore, maintenance planning should be an integral part of the overall business strategy and should be coordinated and scheduled with manufacturing activities. So, we should consider maintenance as integral parts of the production plan rather than as interruptions to that plan. Any violation of the maintenance schedule is treated as a violation of the production plan integrity.

As a remedy to this problem, we develop an integrated model of production and maintenance planning at the tactical level. In the core of production planning, we find the single stage multi item capacitated lot sizing problem with demand shortages in which the schedules and lot sizes are planned for the production of multiple items that share capacity constrained resource. We deal with problems with tight capacities, and when we are in lack of capacity to produce the total demand, we try to spread the capacity among the items by minimizing the total amount of demand shortages. Maintenance planning problem is to determine the dates of preventive maintenance in time windows according to reliability of production equipment. When preventive maintenance actions are carried out the production line is restored to an 'as-good-as-new *(AGAN)*' status, and when a production line fails, a minimal repair is performed to restore it to an 'as-bad-as-old *(ABAO)*' status. The resulting problem is modelled as a linear mixed-integer program to minimize production, inventory, setup, demand shortage, preventive and corrective maintenance costs.

We develop two approaches to solve the integrated production planning and maintenance problem. The first heuristic is based on Dantzig-Wolfe decomposition (Dantzig and wolfe, 1960). The Dantzig-Wolfe Decomposition principle reformulates the original model. Column generation is then used to deal with the huge number of variables of the reformulated model. Finally, we apply a simple rounding heuristic and smoothing procedure in order to obtain integer solutions. The second heuristic is based on Lagrangean relaxation (Ficher, 1981) of the capacity constraints and sub-gradient optimization. At every step of sub-gradient method, feasibility and improvement procedures are applied to the solution of the Lagrangean problem.

To our knowledge, there are only few works dealing with this issue. (Weinstein and Chung, 1999) presented a three part-model to resolve the conflicting objectives of system reliability and profit maximization. An aggregate production plan is first generated, and then a master production schedules is developed to minimize the weighted deviations from the specified aggregate production goals. Finally, work-center loading requirements, determined through rough cut capacity planning, are used to simulate equipment failures during the aggregate planning horizon. (Aghezzaf et al. 2007) studied a production system that subjects to random failures. They assume that any maintenance action carried out on the system, in a period, reduces the system's available production capacity during that period. The objective is to find an integrated lot-sizing and preventive maintenance strategy of the system that satisfies the demand for all items over the entire horizon without backlogging, and which minimizes the expected sum of production and maintenance costs. An extension of the above work is treated by (Aghezzaf and Najid, 2008). They developed models which take into consideration the parameters of reliability of the system and consider parallel production lines. They solved the problem using a lagrangian-based heuristic procedure. Recently, (Najid et al. 2011) tackle the problem of integrating production and maintenance. They study a comparison between integrated and separate models and show the effectiveness of the integrated model. They used a commercial solver for solving the integrated problem, but the solver was not able to provide an optimal solution in reasonable computation time when instances are large. Nevertheless, we can say that the integrated production planning and maintenance problem is close to the multi items capacitated lot sizing problems denoted MCLSP. (Nourelfath et al., 2010) integrated preventive maintenance with tactical production planning in multi-state systems with binary-states, and *s*-independent component. The system is subject to cyclical PM with minimal repair at failure. The proposed model coordinates the production with the maintenance decisions, so that the total expected cost is minimized. (Alaoui et al, 2012) solved the model proposed by (Najid et al,2011) using a lagrangean-based heuristic. The computational results show that the integration can reduce the total maintenance and production costs, and the removal of periodicity constraint is directly affected by the demand fluctuation and can also reduce the total cost. (Fitouhi & Nourelfath, 2014) extended the model in Aghezzaf and Najid.(2008) into an integrated problem with non-cyclical PM with minimal repair at failure and tactical production planning for a single machine. (Nourelfath & Chatelet, 2012) extended the model in (Aghezzaf and Najid.2008) by taking into account the presence of economic dependence and common cause failures in parallel systems. (Yalaoui et al. 2014) developed an optimal relaxation technique based on the polyhedral theory a and a ''Fix and Relax heuristic'' to resolve the Aghezzaf and Najid Model.

The integrated production planning and maintenance problem is close to the multi items capacitated lot sizing problems denoted MCLSP. In the literature, there are several studies that address the single stage capacitated lot sizing problem. Since the papers published by (Wagner & Whitin, 1958) and (Manne, 1958), a lot of research has been done on lot-sizing problems. The single-item problem has been given special interest for its relative simplicity and for its importance as a sub-problem of some more complex lot-sizing problems. For a complete review, the reader can refer to (Brahimi et al. 2006a). When several items are involved, additional complexity is introduced by resource interactions. (Chen & Thizy, 1989) have proved that multi-item capacitated lot-sizing problem with set up time is strongly NP-hard. (Maes et al. 1991) shows that even the feasibility problem is NP complete. So, obtaining optimal and sometimes even feasible solution remains challenging. (Trigeiro et al. 1989) were among the first to try to solve such problem. They include setup costs and solve the problem by Lagrangean relaxation and subgradient optimization. A heuristic procedure generates feasible schedules from the solution of the Lagrangean problem. (Diaby et al. 1992) and (Ozdamar & Bozyel 2000) assume negligible setup costs, but include the cost of overtime available as a second source of capacity. (Brahimi and al. 2006b) treat a new family of capacitated multi-item lot-sizing problems, namely, lot-sizing problems with time windows. Two classes of the problem are analyzed and solved using different Lagrangian heuristics. For problem dealing with shortage costs, there are few papers. To our knowledge, (Absi & Kedad-Sidhoum, 2007) proposed some mixed integer programming (MIP) based heuristics to solve the multi capacitated lot sizing problem with set up time and shortage cost with additional constraints such as setup groups, operating sets, and lower and upper bounds on production. The same authors (Absi & Kedad-Sidhoum, 2008) developed also a branch-and-cut algorithm for the same problem. Recently, (Absi & Kedad-Sidhoum, 2009) consider safety stock in their above model, and solve the problem by using Lagrangean heuristic with several smoothing algorithms. For MCLSP solved by using column generation, we find the work of (Vanderbeck, 1998) who developed a Branch-and-Price method that combines column generation and cutting method for resolving the MCLSP with start-up time. Recently, (Degreavs & Jans, 2007) develop a new Dantzig-Wolf decomposition for MCLSP with set-up time and solve it by using a Branch and Price method. (Lasdon & Terjung, 1971) proposed a reformulation of the linear program to provide an approach based on a column generation technique. (Chen & Thizy, 2002) proposed a method of column generation for the MCLSP without setup time. (Bahl, 1983) and (Catrysse et al. 1990) have also used the column generation method for resolving the MCLSP. For a survey on multi-item capacitated lot-sizing problems, the reader can refer to (Karimi et al. 2003) and recently (Jans & Degreaves, 2007, 2008).

The main contribution of this chapter is twofold. First, we develop heuristics based on Lagrangean relaxation and Dantzig-Wolf decomposition to obtain a good integer solution for the integrated production planning and maintenance problem. Second, we propose a good feasible solution in a reasonable execution time.

The remainder of the paper is organized as follows. In second section, the description and mathematical formulation of the problem are presented. The heuristics used to solve the integrated problem are presented in the third section. Some computational results are showed in the fourth section, and we end up with conclusion and outlooks in the last section.

II. PROBLEM DESCRIPTION AND MATHEMATICAL FORMULATION

1. Production Planning Problem

Consider a planning horizon H of length $T = N \times \hat{o}$ covering N periods of fixed length $\hat{o}$, and a set of item $i \in P$ to be produced on a capacitated production line. During each period $\forall t \in H$, a demand d_{it} of item $i \in P$ should be satisfied (figure 1). Items are produced on a production line with known capacities given in unit time, and processing time expressed in unit time per item. Furthermore, we allow demand shortage unfulfilled due to insufficient capacity, and we will assign a high unit cost for each item lost in each period. Let define the parameters and the mathematical formulation of the production planning problem with demand shortage, denoted MCLSP-SC.

The notations are given below:

Index:

i: Items.

t: Periods.

Figure 1. Production planning

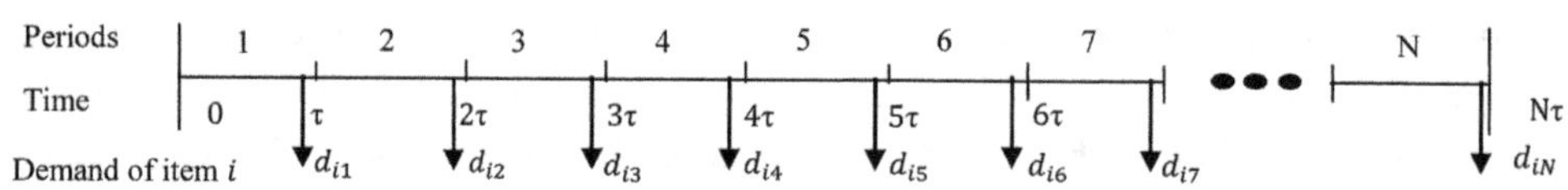

Model parameters:

d_{it} : Demand of item *i* to satisfy during period *t*.
K (t) : Available capacity in period *t*.
f_{it} : Set-up cost of producing one unit of item i in period t.
p_{it} : Fixed cost of producing one unit of item i period t.
h_{it} : Variable cost of holding one unit of item i by the end of period t.
$ö_{it}$: Unit cost for requirements not met regarding demand for item i in period t.
$ñ_i$: Processing time for each item i.

$$d^i_{tN} = \sum_{k=t}^{N} d_{ik}$$

Decision variables:

y_{it} : Binary set up variable of item i in period t.
x_{it} : Quantity of item i produced in period t.
I_{it} : Inventory of item i at the end of period t.
r_{it} : Demand shortage for item i in period t.

The aggregate formulation of the MCLS-SC model is stated as follows:
MCLSP-SC:

$$\text{Minimize } Z_1 = \sum_{i \in P} \sum_{t \in H} f_{it} y_{it} + p_{it} x_{it} + h_{it} I_{it} + ö_{it} r_{it} \quad (1)$$

Subject to:

$$x_{it} + r_{it} + I_{i,t-1} - I_{it} = d_{it} \qquad \forall i \in P, \ \forall t \in H \quad (2)$$

$$\sum_{I \in P} ñ_i x_{it} \le K(t) \qquad \forall i \in P, \ \forall t \in H \quad (3)$$

$$x_{it} - d^i_{tN} y_{it} \le 0 \qquad \forall i \in P, \ \forall t \in H \quad (4)$$

$$r_{it} \le d_{it} \qquad \forall i \in P, \ \forall t \in H \quad (5)$$

$$x_{it}, I_{it}, r_{it} \geq 0 \qquad \forall i \in P, \ \forall t \in H \qquad (6)$$

$$y_{it} \in \{0,1\} \qquad \forall i \in P, \ \forall t \in H$$
$$I_0 = 0 \qquad (7)$$

The objective function (1) minimizes the sum of the set up, holding, production, demand shortage costs over the whole N-period horizon. Constraints (2) are the inventory balance equations. Constraints (3) are the capacity constraints. Constraints (4) relate the continuous production variables to the binary setup variables. Constraints (5) express that quantity lost of item i in period t must be less than demand of item $i \in P$ in period $t \in H$. The constraints (6) and (7) express non-negativity and integrality constraints.

2. Maintenance Planning Problem

We denote by maintenance planning the determination of the execution dates of maintenance activities, in accordance with other plans, the work preparation and the determination of required maintenance capacity. For that, although thousands of maintenance models have been published, there are a limited number of maintenance policies which all maintenance models can be based on. For example, models deal with age replacement policy, other with periodic preventive maintenance policy (periodic MP policy) or failure limit policy etc. For more details on maintenance models and different policies, the readers can refer to review of (Wang, 2002).

In our model, we plan preventive maintenance actions in time windows. The idea is based on the periodic MP policy, see e.g. Barlow and Hunter (1960), Nakagawa (1981a, b), Wang and Pham (1999). Indeed, in classical periodic MP policy, equipment is preventively maintained at fixed time intervals $k \times T^*_{PM}$ (k=1, 2...) and the optimal length of preventive maintenance period is $T^*_{PM} = n^* \times \hat{o}$ ($\hat{o}$ is the length of period $t \in H$). This means that preventive maintenance tasks will be performed periodically in the beginning of period's

$$t = 1, \ n^* + 1, \ 2n^* + 1, \ 3n^* + 1, \ 4n^* + 1, \ \text{etc.}.$$

In our study, the purpose is to plan preventive maintenance actions in time windows

$$\left[\left(pn^* - k\right) \times \tau, \left(pn^* + k\right) \times \tau\right].$$

In other words, a preventive maintenance task will be carried out at the earliest in the beginning of the period $t = pn^* + 1 - k$ or at the latest in the beginning of the period $t = pn^* + 1 + k$, and will complete within the period in which it started, with

$$p = 1, 2, \ldots, \frac{T - (k+1)^* \hat{o}}{T^*_{PM}}$$

($\frac{T-(k+1)*\tau}{T_{PM}^{*}}$ is the number of preventive maintenance actions), and k is chosen to avoid the overlaps between time windows:

$$k = \begin{cases} \frac{n^{*}-1}{2} & \textit{if } n^{*} \textit{ is an even number} \\ \frac{n^{*}-1}{2} & \textit{else} \end{cases} \tag{8}$$

Moreover, we assume that each preventive or corrective maintenance action carried out on the production line consumes capacity units, at the beginning of the planning horizon the production line is in "as good as new" (*AGAN*) status, the production line is considered here as a complex system and the failure rate, denoted r(u), is an overall rate of the whole line. When a preventive maintenance is carried out, the production line is restored to *AGAN* status and when a production line fails, a minimal repair is carried out to restore it to "as bad as old" (*ABAO*) status. Finally we assume that the expected failures increase with elapsed time since the last preventive maintenance.

To determine time windows, we have to compute the length of preventive maintenance period $T_{PM}^{*} = n^{*} \times \tau$ when maintenance is independent of production. The algorithm to compute T_{PM}^{*} is stated in Algorithm 1.

Algorithm 1.

```
Initialization:
Parameters of the used failure distribution;
Preventive maintenance cost (C_p);
Corrective maintenance cost (C_r);
Main program:
    for        t = 1: to Number_period do
          Computation of expected number of failures in each period t:
```

$$NB(t) = \int_{0}^{t} r(u)\,du .$$

```
          Computation of expected maintenance (preventive and corrective) cost
in each period t:
```

$$C_{exp} = C_p + C_r \times NB(t) .$$

```
     Computation of expected maintenance cost per unit time in each period t:
```

$$E(t) = \frac{C_{exp}(t)}{t}$$

```
End for
Determination of the optimal length of preventive maintenance period T*_PM which
minimizes the expected maintenance cost per unit time E(T*_MP).
```

Based on the value of the optimal length of preventive maintenance period T^*_{PM} in classical case, we can determine n^*. Then, we compute value of k (eq. (8)) and finally, we obtain time windows $\left[\left(pn^* - k\right) \times \hat{o}, \left(pn^* + k\right) \times \hat{o}\right]$.

Example: If we consider a horizon of 9 periods and an optimal length of preventive maintenance period $T^*_{PM} = 3\tau$ $(n^* = 3)$, the maintenance planning, if we don't take into account production, is shown in figure 2. By using equation (8), we obtain k =1 and then we can define time windows in the whole of horizon as shown in figure 3.

3. Mathematical Formulation of Integrated Production and Maintenance Problem

The integrated problem is a combination of production and maintenance planning problems described below, denoted (MCLSP-SC-TW). The aim is to find an optimal plan of production and maintenance to minimize production, Set up, inventory, demand shortage, preventive and corrective maintenance costs.

The mathematical formulation of MCLSP-SC-TW problem is an extension of MCLSP-SC problem. We joint additional constraints and decision variables modeling preventive and corrective maintenance to MCLSP-SC model. We keep the same parameters as in Section a) and we add new parameters and decision variables defined below:

Parameters:

$\theta_p(t)$: Capacity consumed by preventive maintenance in period t.

$\theta_r(t)$: Capacity consumed by corrective maintenance in period t.

Figure 2. Maintenance plan without taking account production when $T^*_{PM} = 3\tau$

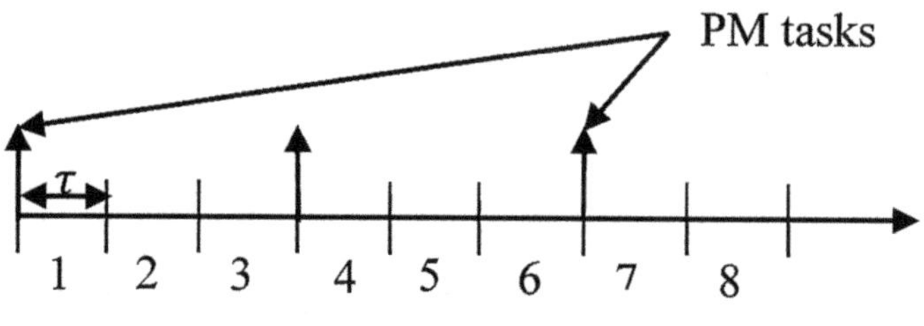

Figure 3. Time windows of preventive maintenance in integrated case

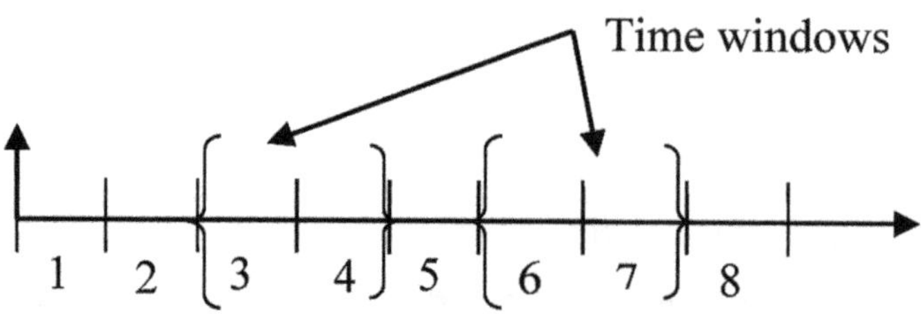

NB : Vector of N elements contains the expected number of failures in each period t nb(t) [nb(1), nb(2), nb(3)… nb(T)]

nb(t) : Expected number of failure in period t, computed as follows:

$$nb(t) = \int_{t-1}^{t} r(u)\,du$$

Let now define how we calculate maintenance cost $C_{pr}(t)$ and capacity $\grave{e}_{pr}(t)$ consumed when a preventive maintenance action is planned. We observe that if a preventive maintenance is performed at a given period t, the maintenance cost in this period t is the sum of preventive and corrective maintenance costs. The corrective maintenance cost in period t is the expected number of failures in that period t multiplied by the cost of one corrective maintenance action. Notice that if no preventive maintenance action was performed, the maintenance cost in period t is the corrective maintenance cost. The same reasoning can be applied for the capacity consumed by maintenance actions in a given period t when preventive maintenance was carried out. Equations (9) and (10) express respectively maintenance cost and capacity consumed by maintenance action when a preventive maintenance was planned in a given period t:

$$C_{pr}(t) = C_p + C_r nb(1) \tag{9}$$

$$0_{pr}(t) = 0_p(t) + 0_r(t) * nb(1) \tag{10}$$

Let recall that if preventive maintenance is planned at any period t, the failure rate is renewed. Then, the expected number of failures in this period t is the first element of vector NB which is nb(1). This justifies equations (9), and (10).

Decision variables:

δ_t : Binary preventive maintenance variable ($\delta_t = 1$ if preventive maintenance is carried out in the beginning of period t, 0 otherwise).

z_{jt} : Binary variable ($z_{jt} = 1$ if in period t the last preventive maintenance ended in period j, 0 otherwise).

MCLSP-TW-SC:

$$\text{Minimize } Z = \sum_{i\in P}\sum_{t\in H} f_{it} y_{it} + p_{it} x_{it} + h_{it} I_{it} + \ddot{o}_{it} r_{it} +$$

$$\sum_{t=2}^{N} (C_{pr}(t)\delta_t + C_r \sum_{j=1}^{t-1} NB(t-j+1) z_{jt}) + C_{pr}(1)\delta_1 \tag{11}$$

Subject to:

$$x_{it} + r_{it} + I_{i,t-1} - I_{it} = d_{it} \qquad \forall i \in P,\ \forall\ t \in H \qquad (12)$$

$$\sum_{I \in P} \rho_i x_{it} + \theta_{pr}(t)\delta_t \leq K(t) - \theta_r(t)\sum_{j=1}^{t-1} NB(t-j) z_{jt} \qquad t = 2..N \qquad (13.1)$$

$$\sum_{i \in P} \rho_i x_{it} \quad + \theta_{pr}(t)\delta_t \leq K(t) \qquad t = 1 \qquad (13.2)$$

$$x_{it} - d^{i}_{tN} y_{it} \leq 0 \qquad \forall i \in P,\ \forall\ t \in H \qquad (14)$$

$$r_{it} \leq d_{it} \qquad \forall i \in P,\ \forall\ t \in H \qquad (15)$$

$$\sum_{t=pn+1-k}^{pn+1+k} \delta_t = 1 \qquad p = 1, 2, \ldots, \frac{T-(k+1)^{*}\tau}{T^{*}_{PM}} \qquad (16)$$

$$\delta_t + \delta_{t-1} \leq 1 \qquad t = 1, 2, \ldots, N-1 \qquad (17)$$

$$z_{jt} \leq 1 - \delta_k \qquad \forall t \in H,\ j = 1, \ldots t-1,\ k = j+1, \ldots,\ t \qquad (18)$$

$$z_{jt} \leq \delta_j \qquad \forall t \in H, j = 1, \ldots, t-1 \qquad (19)$$

$$z_{jt} \geq \sum_{k=j+1}^{t} (1-\delta_k) + \delta_j - (t-j) \qquad \forall t \in H,\ j = 1, \ldots, t-1 \qquad (20)$$

$$x_{it}, I_{it}, r_{it} \geq 0 \qquad \forall i \in P,\ \forall\ t \in H \qquad (21)$$

$$y_{it}, z_{jt}, \delta_t \in \{0,1\} \qquad \forall i \in P,\ \forall t \in H,\ j = 1,\dots,\ t-1$$
$$I_0 = 0,\ \delta_1 = 1 \tag{22}$$

The objective function (11) minimizes the sum of the set up, holding, production, demand shortage, and maintenance (preventive and corrective) costs over the whole N-period horizon. Constraints (12) are the inventory balance equations. Constraints (13.1) and (13.2) are the capacity constraints that consider preventive and corrective maintenance. Constraints (14) relate the continuous production variables to the binary setup variables. Constraints (15) express that quantity lost of item i in period t must be less than demand of item $i \in P$ in period t. The constraints (16) ensure that one maintenance is carried out in the interval $\left[(pn+1-k),(pn+1+k)\right]$. The constraints (17) ensure that two preventive maintenance actions cannot be carried out successively. The variable z_{it} must equal one when the last preventive maintenance was occurred in period j, and we are now in period t. To ensure this, we define z_{it} as: $z_{it} = (1-\delta_t)(1-\delta_{t-1})\dots(1-\delta_{j+1})\delta_j$. This is a nonlinear expression, but can be solved by introducing the following three linear constraints (18) to (20). The constraints (21) and (22) express non-negativity and integrality constraints.

III. HEURISTIC FOR ULSP-TW-SC

When we use a decomposition method, we decompose our integrated production and maintenance problem to a set of sub-problems. Each sub-problem is a single item uncapacitated lot sizing problem with time windows and shortage cost called ULSP-TW-SC. This sub-problem is combined from the single item capacitated lot sizing problem with shortage cost (ULSP-SC) treated by (Aksen et al, 2003) solved in $O(T^2)$ and a maintenance problem where preventive maintenance tasks are planned in time windows. Also, this sub-problem is an exponential problem. Therefore, we develop a heuristic to solve ULSP-SC which will be incorporated in our heuristics based on Dantzig-Wolfe decomposition and Lagrangean relaxation in the following section.

Proof: To find an optimal solution, if we solve the problem using a Branch and Bound method, the set of possible solution is equal to *m* possibilities where each possibility is a production and maintenance planning in the whole horizon and m is given by the formula below:

$$m = (C^1_{2*k+1})^{\frac{T-(k+1)\tau}{n^* \times \tau}}$$

The number of possibility *m* is a combination of one preventive maintenance activity among the possible number of preventive maintenance activities in a time window ($2*k+1$) to the power of the number of time window $\dfrac{T-(k+1)\tau}{n^* \times \tau}$.

The algorithm of the heuristic is recalled below:

Step 1: Solve the single item uncapacitated lot Sizing problem with Shortage cost (ULSP-SC) to optimality using the dynamic algorithm addressed by [Aksen et al, 03] based on the structural characteristics stated in lemmas 1 to 3.

Step 2: Select the first time window, plan a preventive maintenance task in the period when the total cost of production and Maintenance is minimal.

Step 3: Update the total cost of production and maintenance, select the next time window and repeat step 2.

Lemma 1:

$$x_t^* * r_t^* = 0$$

The first lemma suggests that there is an optimal solution such that demand in a given period will be fully satisfied if procurement is made in that period. A proof for this lemma easily follows from our assumption that $\varphi_t > p_t$.

Lemma 2:

$$x_t^* * I_{t-1}^* = 0$$

The second lemma suggests that there is an optimal solution such that we will procure in a given period only if the inventory level at the end of the preceding period drops to zero. This principle is also known as the zero-inventory ordering policy of the *Wagner–Whitin* solution, according to which beginning inventory in a period of procurement activity is always zero. In our lost demand model, it is slightly altered such that we might have both $I_{t-1} = 0$ and $x_t = 0$ if the demand d_t is determined to be lost.

Lemma 3:

$$r_t^* * (d_t - r_t^*) = 0$$

The third lemma suggests that there is an optimal solution such that if we lose any demand in a given period, then we should lose the entire demand in that period. In other words, it prohibits partial loss of demand. If $r_t^* > 0$, then r_t^* must equal d_t .

IV. HEURISTICS FOR INTEGRATED PROBLEM

We develop two approaches to solve the integrated production planning and maintenance problem. The first heuristic is based on Dantzig-Wolfe decomposition (Dantzig and wolfe, 1960). The Dantzig-Wolfe Decomposition principle reformulates the original model. Column generation is then used to deal with the huge number of variables of the reformulated model. In order to obtain integer solutions, we apply a simple rounding heuristic and smoothing procedure. The second heuristic is based on Lagrangean relaxation (Ficher, 1981) of the capacity constraints and sub-gradient optimization. This heuristque is the same as that applied by Alaoui and al. (Alaoui and al. 2012) for this problem. At Every step of sub-gradient method, feasibility and improvement procedures are applied to the solution of the Lagrangean problem.

1. Heuristic Based on Dantzig-Wolfe Decomposition (DWH)

1.1 Dantzig-Wolfe Decomposition

Dantzig-Wolfe decomposition is standard way to decompose an integer-programming model into a master problem and one or several sub-problems. The key idea of Dantzig-Wolfe decomposition is to reformulate the problem by substituting the original variables with a convex combination of the extreme points and a linear combination of extreme rays of the polyhedron corresponding to a substructure of the formulation. Moreover, it is alternative method for obtaining tighter lower bound from LP relaxation of the master problem. Decomposition approach for this problem hinge on the observation that when we leave out the capacity constraints, the problem is decomposed into uncapacitated lot sizing problem for each item *i* with time windows and shortage cost solved in section II.

In Dantzig-Wolfe decomposition, we keep the capacity constraints in the master problem and add a convexity constraint for each item (Manne, 1958; Dzielinsky & Gomory, 1965). The new columns represent a production and maintenance plan for a specific item over the full time horizon. Let Q_i be the set of all production and maintenance plans for item *i*. Let $\psi_{i,j}$ the new decision variable representing production and maintenance plan *j* for item *i*. Let $C_{i,j}$ the total cost of set up, production, inventory, and preventive and corrective maintenance according to plan *j* for item *i* and $r_{it,j}$ the capacity usage of the production and capacity consumed by maintenance during period *t* according to plan *j* for item *i*. Finally, let each point from

$$X = \left\{ \left(\left(\mathrm{y}_{\mathrm{it}},\ \mathrm{x}_{\mathrm{it}},\ \mathrm{r}_{\mathrm{it}}, \mathrm{I}_{\mathrm{it}},\ \delta_{\mathrm{t}}, \mathrm{z}_{\mathrm{jt}} \right) \mid [11]-[22] \right) \right\}$$

denoted as:

$$\mathrm{y}_{\mathrm{it}} = \sum_{j\in Q_i} y_{it}^{j} * \psi_{i,j}\ ,\ \mathrm{x}_{\mathrm{it}} = \sum_{j\in Q_i} x_{it}^{j} * \psi_{i,j}\ ,\ \mathrm{r}_{\mathrm{it}} = \sum_{j\in Q_i} r_{it}^{j} * \psi_{i,j}\ ,\ \mathrm{I}_{\mathrm{it}} = \sum_{j\in Q_i} I_{it}^{j} * \psi_{i,j}\ ,$$

$$\ddot{\mathrm{a}}_{\mathrm{t}} = \sum_{j\in Q_i} \delta_t^{j} * \psi_{i,j}\ ,\ \mathrm{z}_{\mathrm{jt}} = \sum_{j\in Q_i} z_{mt}^{j} * \psi_{i,j}\ ,\ \text{such as} \sum_{j\in Q_i} \psi_{i,j} = 1 \quad \forall\ \mathrm{i} \in \mathrm{P},$$

Then, the formula of different parameters explained above are stated as follows:

$$C_{i,j} = \sum_{\mathrm{t}\in\mathrm{H}} \mathrm{f}_{\mathrm{it}} y_{it}^{j} + \mathrm{p}_{\mathrm{it}} x_{it}^{j} + \mathrm{h}_{\mathrm{it}} I_{it}^{j} + \ddot{\mathrm{o}}_{\mathrm{it}} r_{it}^{j} + \left(\mathrm{C}_{\mathrm{p}} + \mathrm{C}_{\mathrm{r}} * \mathrm{nb}(1) \right) * \delta_1^{j} +$$

$$\sum_{t\in H\setminus\{1\}}\left[\left(C_p + C_r{*}nb\left(1\right)\right)*\delta_t^j + \sum_{m=1}^{t-1} C_r{*}NB\left(t-m+1\right)*z_{mt}^j\right]$$

$$r_{it,j} = \rho_i x_{it}^j + \left(\theta_p\left(t\right)+\theta_r\left(t\right)*nb\left(1\right)\right)*\delta_t^j + \sum_{m=1}^{t-1}\theta_r\left(t\right)*NB\left(t-m+1\right)*z_{mt}^j$$

(If t =1 then $r_{it,j} = \rho_i x_{it}^j + \left(\theta_p\left(t\right)+\theta_r\left(t\right)*nb\left(1\right)\right)$)

The LP relaxation of the master problem then looks as follows:

$$\sum_{i\in P}\sum_{j\in Q_i} C_{i,j} * \psi_{i,j} \tag{23}$$

Sc:

$$\sum_{i\in P}\sum_{j\in Q_i} r_{it,j} * \psi_{i,j} \le K\left(t\right) \qquad \mu_t \qquad \forall\ t\in H\setminus\{1\} \tag{24.1}$$

$$\sum_{i\in P}\sum_{j\in Q_i} r_{it,j} * \psi_{i,j} \le K\left(t\right) \qquad \mu_1 \qquad t=1 \tag{24.2}$$

$$\sum_{j\in Q_i}\psi_{i,j} = 1 \qquad \pi_i \qquad \forall\ i\in P \tag{25}$$

$$\psi_{i,j} \ge 0 \qquad \forall\ i\in P, \forall j\in Q_i \tag{26}$$

The objective function (23) minimizes the total cost. Constraints (24.1) and (24.2] are the capacity constraints. Constraints (25) are the convexity constraints: choose a convex combination of schedules for each item. μ_t,π_i are the dual prices on the capacity and convexity constraint, respectively. The constraints (26) express non-negativity and integrality constraints.

We notice that this formulation has a large number of columns, The existence of such a huge number of variables in the various formulations makes the problem well suited for column generation. Column generation starts with a feasible restricted master with only a few columns and we add new columns iteratively as they are needed. At each iteration of the columns generation procedure, we check for each item *i* if we can generate a new column by solving the following single item uncapacitated lot sizing sub-problem with time windows and shortage cost UCLSP-TW-SC:

$$Min\sum_{t\in H}[(f_{it}y_{it}+p_{it}x_{it}+h_{it}I_{it}+\varphi_{it}r_{it})+C_{pr}(t)\delta_t+C_r\sum_{j=1}^{t-1}NB(t-j+1)z_{jt}]+C_{pr}(1)\delta_1-$$

$$\pi_i+\sum_{t\in H\setminus\{1\}}\mu_t(\rho_i x_{it}+\theta_{pr}(t)\delta_t-\theta_r(t)\sum_{j=1}^{t-1}NB(t-j)z_{jt})+\mu_1\left(\sum_{i\in P}\rho_i x_{it}+\theta_{pr}(t)\delta_t\right)$$

If such a column with negative reduced cost is found, we add it to the restricted master problem, reoptimize this problem and perform another pricing iteration, otherwise, we have found the optimal Dantzig-Wolfe bound. To be sure that we will obtain an optimal Dantzig-Wolfe solution, theorem 1 and proposition1 should be applied to the master model.

Theorem 1: An optimal Dantzig-Wolfe solution is guaranteed if and only if, at most one set Q_i whose extreme points of the maintenance plan are nonzero.

Proof: Each preventative or corrective maintenance activity is defined by the period when it should be planned. From this observation, we can say that corrective or preventive maintenance activity does not depend on the production of a single item but the period where all items can be produced. If the extreme points of maintenance associated to each production and maintenance plan *j* are non-zero in more than one set Q_i representing item *i*, then the available capacity in each period will not be fully exploited and therefore we will not get an optimal solution.

Proposition 1: The set Q_i where the extreme points of maintenance associated to each production and maintenance plan *j* will be nonzero is chosen randomly $\forall i \in P$.

Proof: As we mention it in proof of theorem 1, maintenance activities do not depend on the items but on the periods. So, if we integrate the non-zero extreme points of maintenance in a set Q_n or in a set Q_m such as $m \neq n$, the final solution will be the same or may not change considerably.

1.2 Rounding and Smoothing Procedures

After the first step of the heuristic by solving Dantzig-Wolfe decomposition with columns generation and obtaining lower bound for the original problem, we apply two following procedures to construct a feasible solution for the MCLSP-TW-SC problem.

1.2.1 Rounding Procedure

This procedure consists in rounding the linear solution provided by Dantzig-Wolfe decomposition. For each point from X, the rounding procedure is done according to the algorithm stated as follows:

Rounding procedure:

1. $\boldsymbol{x}_{it}=\boldsymbol{x}_{it}$, $\boldsymbol{I}_{it}=\boldsymbol{I}_{it}$, $\boldsymbol{r}_{it}=\boldsymbol{r}_{it}$
2. $y_{it}=\begin{cases}1 & if\ \ x_{it}\geq 0\\ 0 & otherwise\end{cases}$

δ_t is rounded to the nearest integral value.

1.2.2 Smoothing Procedure

When we apply the rounding procedure to the solution, we should verify that the capacity constraints are not violated by the rounding procedure and also, try to improve the upper bound. Therefore, we established a procedure that will be able to smooth the production and construct an improved solution. This smoothing procedure is three steps that can be described below:

Step 1: It is a backward step where we will shift a quantity of one or several items produced in period t, where the available capacity is exceeded, at a previous period t-1 where the available capacity is not fully used. We can be based on proposition 2 to choose the items that will be shifted.

Step 2: From the solution of the previous step, we can transform the quantity lost of one or several items in period t+1 to a quantity produced in period t if the capacity is available, and then we will consider this quantity as stock of the period t. The proposition 3 is an answer on which item to transform from a quantity lost to a quantity produced. Otherwise, we will look for a period t where the available capacity is not fully used and we will transform the quantity lost (demand shortage) of one or several items in the same period to a quantity produced according to proposition 3.

Step 3: After the step 2, the smoothing of production is normally achieved, but there is a final check to do. We must verify that in each period the available capacity is not exceeded. Otherwise, we select according to proposition 4 one or several items to make in shortage in the same period.

Proposition 2: The item to be shifted is selected randomly.

Proof: In computational result, we will assume that production, inventory and setup costs are constant $\forall i \in P,\ \forall\ t \in H$, so if we shift from period *t* to period *t-1* a quantity *x* of item *m* or item *n* such as $m \neq n$, it will be the same in terms of total cost.

Proposition 3: Let $r \leq d_{it}$ the quantity lost and that will be transformed to a quantity produced, the item to select is the item which has a total shortage cost equal to $\max_{i \in P}(\varphi_i * r)$.

Proof: Our aim is to minimize the total cost of production and maintenance. Then among all items, the item to select and to plan in production is the item which costs very expensive when it is lost.

Proposition 4: Let $x \leq d_{it}$ the quantity produced and that will be transformed to a quantity lost, the item to select is the item which has a total shortage cost equal to $\min_{i \in P}(\varphi_i * x)$.

Proof: Our aim is to minimize the total cost of production and maintenance. Then among all items, the item to select and to consider as shortage is the item which costs less when it will be lost.

2. Heuristic Based on Lagrangean Relaxation (LH)

2.1 Lagrangean Relaxation

The general idea behind the Lagrangean relaxation heuristic is to decompose our integrated production planning and maintenance problem to N sub-problems easy to solve. This is possible by relaxing the resource capacity constraints (10) while using a set of Lagrange multipliers $u_t \geq 0$ in the objective function of the MCLSP-TW-SC model. We derive then a Lagrangean function denoted by $\mathcal{L}(x, y, I, r, \delta, z, u)$. The mathematical formulation of relaxed problem MULSP-SC-TW is stated below:

$$Min\,\mathcal{L}(x,y,I,r,\delta,z,u) = \sum_{i \in P}\sum_{t \in H} f_{it}y_{it} + p_{it}x_{it} + h_{it}I_{it} + \varphi_{it}r_{it} +$$

$$\sum_{t=2}^{N} C_{pr}(t)\delta_t + C_r \sum_{j=1}^{t-1} NB(t-j+1)z_{jt} +$$

$$\sum_{t=2}^{N} u_t (\sum_{i \in P} \rho_i x_{it} + \theta_{pr}(t)\delta_t + \theta_r(t)\sum_{j=1}^{t-1} NB(t-j)z_{jt} - K(t)) +$$

$$u_1 \sum_{i \in P} \rho_i x_{i1} + \theta_{pr}(1) - k(1) \tag{28}$$

Subject to: (12) and [14] – (22)

The Lagrangean relaxation of the capacity constraints of the MCLSP-TW-SC problem decomposes the model into N single-item uncapacitated lot-sizing problems with shortage cost and time windows, denoted ULSP-TW-SC solved in section II.

From Lagrangean relaxation theory (Ficher, (1981)), we know that $\mathcal{L}(x, y, I, r, ä, z, u)$. is a lower bound of the optimal solution of MCLSP-SC-TW problem. We also know that the greatest lower bound attainable with the Lagrangean relaxation is provided by multipliers obtained by solving the following Lagrangean dual problem (LD) which can be solved efficiently by a sub-gradient optimization procedure (see Ficher, (1981)).

$$L_D(u) = max_u\{w(u) = \min \mathcal{L}(x, y, I, r, \delta, z, u)\}$$

Subject to:

$$u \geq 0$$

The main advantage of using a Lagrangean relaxation is that it, usually, preserves most of the original problem structure. This makes it easier to use the relaxed problem solution to generate a good feasible solution for the original problem. This suggests that a very efficient heuristic method to solve MCLSP-SC-TW problem could be obtained by applying the sub-gradient optimization procedure and checking, at each iteration, if the solution provided by the primal sub problem is a feasible solution of MCLSP-SC-TW problem, i.e. .If

$$\sum_{i\in P}\rho_i x_{it} + \theta_{pr}(t)\delta_t + \theta_r(t)\sum_{j=1}^{t-1} NB(t-j)z_{jt} - K(t) = 0,$$

then this solution is optimal. Otherwise, this solution can be modified by using a perturbation procedure to generate a feasible solution for MCLSP-SC-TW problem. A detailed heuristic based on this idea will be presented in the following subsection.

2.2 Lagrangean Heuristic Algorithm

Our overall solution method to solve MCLSP-CS-TW problem is a modified sub-gradient optimization procedure. At each iteration of this procedure, a lower bound and a feasible solution of MCLSP-SC-TW problem are generated. The procedure works as follows. At a given iteration, if the Lagrangean solution is not feasible for MCLSP-SC-TW, this solution is modified using the heuristic described in following sub-section b) to find a new feasible solution for MCLSP-SC-TW. If the value of this new feasible solution is better than the current upper bound, then the value of obtained feasible solution becomes the new upper bound. The Lagrangean multipliers are initially set to zero and updated in each iteration to maximize the objective function of dual relaxed problem (LD) according to the formula:

$$u_t^{k+1} = \max(0;\ u_t^k + t^k s_t^k / \|s^j\|)$$

where s_t^k is the sub-gradient of $w(u^k)$ given by:

$$s_t^k := \sum_{i\in P}\rho_i x_{it}^k + \theta_{pr}(t)\delta_t^k + \theta_r(t)\sum_{j=1}^{t-1} NB(t-j)z_{jt}^k - K(t).$$

and t^k is the sub-gradient step size, see below:

$t^k := \lambda_k(UB_k - w(u^k)) / \|s^j\|$; $\|s^j\|$ (is the sub-gradient vector norm)

$\lambda_k \in [0,2]$, we start with $\lambda_0 = 2$ and divide λ_{k-1} by 2 if any improvement of $w(u^k)$ is seen after some iterations. Finally, the stopping criterion is based on maximum number of iteration or when the Gap between upper and lower bounds is smaller than a value $å_1$. A detailed description of the Lagrangean heuristic is found in Algorithm 2.

2.3 Smoothing Heuristic

In order to find a feasible solution at each step of the Lagrangean relaxation algorithm, we propose a Lagrangean based upper bound denoted as AMN. It is based on the Lagrangean solution obtained at each step of the Lagrangean heuristic algorithm. Since the capacity constraints (3) are relaxed, the Lagrangean solution violates them. AMN heuristic is mainly based on a smoothing procedure to lower the shortages

Algorithm 2.

1. Initialization:

$u_t^0 = 0$ t= 1, 2…T. (Lagrange multipliers)

k =1 (Iteration counter)

$\lambda_0 = 2$ (Multipliers)

$LB_0 = -\infty$ (Lower bound value)

$UB_0 = M$ (Upper bound value where M is a large number)

2. For a given iteration k:

(a) Solving the Lagrangean problem $\mathcal{L}(x, y, I, r, \delta, z, u)$ with u_t^{k-1}.

$$w(u^k) = \min_{X^k \in S} \mathcal{L}(x, y, I, r, \delta, z, u)$$

If Lagrangean solution is feasible then

$UB_k = LB_k$

Stop the algorithm.

(b) Compute the new lower bound:

If $LB_k > LB_{k-1}$ then $LB_k := LB_{k-1}$

(c) Perturbation procedure: a heuristic is used to find a feasible solution using a smoothing procedure as described in the following subsection b).

If $UB_k < UB_{k-1}$ then $UB_k := UB_{k-1}$

(d) Compute sub-gradient s_t^k of $w(u^k)$.

(e) Compute sub-gradient step size t^k.

(f) Updating Lagrange multipliers u_t^{k+1}.

If (no improvement after more than K iterations) then

$\lambda_{k+1} := \lambda_k / 2$

Else $\lambda_{k+1} := \lambda_k$

(f) Stopping criteria:

- Maximum number of iteration is reached.

Or when Gap is less than a value ε_1 ($\varepsilon_1 > 0$).

by reusing missing resource capacities. The heuristic is based on the work of (Trigeiro et al., 1989) who proposed an efficient Lagrangean relaxation heuristic for the classical multi-item capacitated lot sizing problem with setup times. Recently, (Brahimi et al. 2006b) proposed a generalization of (Trigeiro et al, 1989) smoothing heuristic to solve the multi item capacitated lot sizing problem with time windows. We need formula below, it computes the overtimes and the available hours in each period, and propositions 5 and 6 to develop the procedure to find feasible solution and to improve it.

$$s_t = \left(\sum_{i \in P} \rho_i x_{it} + \theta_{pr}(t)\delta_t + \theta_r(t)\sum_{j=1}^{t-1} NB(t-j) z_{jt}\right) - K(t)$$

Proposition 5: At a given period, $s_t > 0$ and $s_{t+l} < 0, \forall l \in (t+1 \ldots s-1)$ and s is the period where we had a new set up of production, it means that in period *t* were produced amounts to meet demand for the current period and one or several next periods.

Proof: At each iteration, we solve Lagrangean problem where capacity constraint are violated. If $s_t > 0$ and $s_{t+l} < 0 \ \forall l \in (t+1 \ldots s-1)$ then according to Lemma 2 $(x_{it}^* * I_{it-1}^* = 0 \ \forall i \in P)$ we have

$$x_{it}^* = \sum_{l=t}^{s-1} d_{il} \ , \ x_{il}^* = 0 \, and \, I_{il} = I_{il-1} - d_{il-1} \, \forall l \in (t+1 \ldots s-1) \, .$$

Proposition 6: At a given period, $s_t < 0$ and $s_{t+l} \geq 0 \ \forall l \in (1 \ldots t-1, \, t+1 \ldots s)$, it means that in period *t* were not produced amounts to meet demand for the current period but the demand was lost.

Proof: If

$$s_t < 0 \text{ and } s_{t+l} \geq 0 \ \forall l \in (1 \ldots t-1, \, t+1 \ldots s) \text{ and } s \leq T$$

and, then according to Lemma 1 $(x_{it}^* * r_{it}^* = 0 \ \forall i \in P)$ we have $x_{it}^* = 0 \forall i \in P$ and according to Lemma 3 $(r_t^* * (d_t - r_t^*) = 0)$ we have $r_{it}^* = d_{it} \forall i \in P$.

The perturbation procedure is described as follows:

Step 1: After solving the relaxed problem, several cases are arisen. We find the case according to proposition 5, and then we shift the surplus amounts produced from period t to period t+1 using proposition 2.

Step 2: Another case can be presented after solving the relaxed problem, it is one of proposition 6. Therefore, we should transform the quantity lost (demand shortage) of one or several items according with proposition 3 to a quantity produced in the current period.

Step 3: From the solution obtained from steps 1 and 2, if at a given period $s_t < 0$ and $s_{t+1} > 0$ then we can shift a quantity of one or several items produced in period *t+1*, where the available capacity is exceeded, at a previous period *t* where the available capacity is not fully used according to proposition 2.

Step4*:* After the step 3, the smoothing of production is normally achieved, but there is a final check to do. First, we must verify that, in each period, the available capacity is not exceeded. Otherwise, if at a given period, $s_t > 0$ and $s_{t+l} = 0 \ \forall l \in (1 \ldots t-1, \, t+1 \ldots s)$ and $s \leq T$ then we select according to proposition 4 one or several items to make in shortage in the same period. Else if $s_t < 0$ and $s_{t+l} = 0 \ \forall l \in (1 \ldots t-1, \, t+1 \ldots s)$ and $s \leq T$ then, we can transform the quantity lost of one or several items in period *t+1* to a quantity produced in period *t* according to proposition 4, and then we will consider this quantity as inventory for the items selected in the period *t*.

V. COMPUTATIONAL RESULTS

In this section, we present different results resulting from the application of the heuristics of the capacity constraints of the MCLP-TW-SC. Our algorithms were implemented in the Java programming language. The computations were performed on an Intel Core 2 CPU 2.2GHz PC with 4GB RAM.

We have performed computational tests on a series of extended instances from the lot-sizing library LOTSIZELIB, initially described inTrigeiro et al. (1989). These instances are denoted by trN−T, where N = 6, 12, 24 is the number of items and T = 15, 30 is the number of periods. These instances are characterized by variable resource consumption equal to one, and enough capacity to satisfy all demands over the planning horizon. They are also characterized by important setup costs, small setup times. Since these instances have enough capacity to satisfy all demands over the planning horizon, we make some modifications to induce shortages. Also, we consider instances with N= 36 and 48 as number of items.

We consider a following planning horizon composed of T production periods of fixed length ô = 1 to produce a set of items on production line with an available capacity. The production, set-up, and holding costs are respectively 10, 30, and 5. Four parameters are considered for the analysis:

- **Problem Dimension**: The problem dimensions represented by the number of items N $\in [6, 48]$ and the number of periods T = 15 and 30.
- **Production Capacity**: The capacity required, in each period, is then computed as if lot-for-lot solutions were implemented. The capacity is then obtained by dividing the later result by the target average utilization of capacity ñ . The factor ñ is set to 0.95 and 1.1 corresponding respectively to situations with tight and too tight capacity constraints.
    - **Demand Pattern**: The demand of each item at each period is generated randomly from the interval [20,100].
    - **Shortage Cost**: The shortage cost is considered as penalty cost and its value for each item is generated from the follows intervals [I1], [I2] and [I3].
        - [I1]: [0.5*(Production cost+ setup cost), 1.5*(Production cost+ setup cost)]
        - [I2]: [0.5*(Production cost+ setup cost), 2.5*(Production cost+ setup cost)]
        - [I3]: [0.5*(Production cost+ setup cost), 3.5*(Production cost+ setup cost)]

Six classes of instances are created:

- Class A is a class of instance when capacity is too tight and shortage cost for each item is generated from [I1].
- Class B is a class of instance when capacity is too tight and shortage cost for each item is generated from [I2].
- Class C is a class of instance when capacity is too tight and shortage cost for each item is generated from [I3].
- Class D is a class of instance when capacity is tight and shortage cost for each item is generated from [I1].
- Class E is a class of instance when capacity is tight and shortage cost for each item is generated from [I2].

- Class F is a class of instance when capacity is tight and shortage cost for each item is generated from [I3].

All problem tests are generated with Weibull distribution of production line. The shape and scale parameters are respectively $\beta = 3$, and $\lambda = 4$. The cost of preventive maintenance action is set to $C_p = 28$, and the cost of minimal repair action is given by $C_r = 35$. The capacity lost when a preventive maintenance task and minimal repair action are carried out, is respectively $\theta_p = 0.067K(t)$, and $\theta_r = 0.33K(t)$. Table 1 shows the expected number of failures in each period nb(t) as a function of system's age.

We assume that the system lifetime is distributed according to weibull distribution with Failure probability density function f(t) and Failure rate function r(t).

$$f(t) = \left(\frac{\beta}{\lambda}\right)\left(\frac{t}{\lambda}\right)^{\beta-1} \exp\left(-\frac{t}{\lambda}\right)^{\beta}$$

$$r(t) = \left(\frac{\beta}{\lambda}\right)\left(\frac{t}{\lambda}\right)^{\beta-1}$$

The shape and scale parameters are respectively $\beta = 3$, and $\lambda = 4$. The cost of preventive maintenance action is set to $C_p = 28$, and the cost of minimal repair action is given by $C_r = 35$. The capacity lost, when a preventive maintenance task and minimal repair action are carried out, is respectively $\theta_p = 0.067K(t)$, and $\theta_r = 0.33K(t)$. Table 1 shows the expected number of failures in each period nb(t) as a function of system's age.

To have a meaningful comparison, we compare the results of heuristics to those obtained by XpressMP solver. The computational results of the heuristics (DWH and LH) and the solver are shown in the following tables 2 and 3. We compute the gaps between the upper bounds of heuristics and the solver with the best lower bounds obtained or optimal solution. The different gaps (eq. (27), eq. (28) and eq. (29)) are calculated by the given formula suggested by Millar and Yang (1994):

$$\text{Gap1}(\%) = 200 \times \frac{UB(\text{XpressMP}) - LB(\text{XpressMP})}{UB(\text{XpressMP}) + LB(\text{XpressMP})} \quad (27)$$

$$\text{Gap2}(\%) = 200 \times \frac{UB(\text{DWH}) - LB(\text{XpressMP})}{UB(\text{DWH}) + LB(\text{XpressMP})} \quad (28)$$

Table 1. Expected number of failures

Periods	Expected Number of Failures	Periods	Expected Number of Failures
1	0.0157	16	11.2657
2	0.1095	17	12.7657
3	0.2970	18	14.3595
4	0.5782	19	16.0470
5	0.9532	20	17.8282
6	1.4220	21	19.7032
7	1.9845	22	21.6720
8	2.6407	23	23.7345
9	3.3907	24	25.8907
10	4.2345	25	28.1407
11	5.1720	26	30.4845
12	6.2032	27	32.9220
13	7.3282	28	35.4532
14	8.5470	29	38.0782
15	9.8595	30	40.7970

$$\text{Gap3}\left(\%\right) = 200 \times \frac{\text{UB}\left(\text{LH}\right) - \text{LB}\left(\text{XpressMP}\right)}{\text{UB}\left(\text{LH}\right) + \text{LB}\left(\text{XpressMP}\right)} \quad (29)$$

The stopping criterion of the XpressMP computation is time limit which is equal to 3600 seconds, for Lagrangean heuristic is when a number of sub-gradient iteration reaches a maximum of iteration $Q = 50$ and for both of them if the gap is small than $\varepsilon = 0.05$.

Tables 2 and 3 summarize the computational behavior of the Lagrangean heuristic, of the Dantzig-Wolfe based heuristic and the computational results given by the solver. The gaps and CPU time are computed for each instance characterized by the following parameters: the number of items (N), the number of periods (T), shortage cost for each item and capacity tightness.

The results obtained with solver are very interesting, we notice that almost of instances are solved to optimality or very close to optimal solution, but also require a significant amount of CPU time. Therefore, we developed heuristics to reduce computation time and obtain a high quality solution. First, we implement a heuristic based on lagrangean relaxation (Lagrangean Heuristic (LH)). The results provided by this heuristic (LH) are shown in table 2 when capacity is too tight and in table 3 when capacity is more large (tight capacity). We notice from tables 2 and 3 that the heuristic (LH) can solve some instances to optimality. Others instances which is not solved to optimality have very small gaps and the upper bounds of the Lagrangean heuristic are very close to the upper bounds obtained by the solver, the deviation from the solver do not exceed 0.97%. Also, we observe that the CPU time of the Lagrangean heuristic increase partially when we increase the number of items and considerably when we increase the number of periods. Finally, we can say that the computation time of the heuristic is much smaller than the computation time of the solver for the same result or a result very close. Second, we develop a heuristic

Table 2. Computation results: HR vs. DWH vs. XpressMP when capacity is too tight

		Solver		DWH		LH	
Class A							
Items	Periods	Time (s)	Gap1(%)	Time (s)	Gap2(%)	Gap3(%)	Time (s)
6	15	49.52	**0.00**	7.582	0.05	0.07	9
12	15	226.3	**0.00**	10.109	0.01	0.09	11
24	15	2677.44	**0.00**	20.311	**0.00**	**0.00**	16
36	15	2693.89	**0.00**	22.043	**0.00**	**0.00**	25
48	15	2641.43	**0.00**	22.714	**0.00**	0.01	30
6	30	2723.36	0.01	117.531	**0.00**	0.10	35
12	30	2804.70	0.00	209.218	**0.00**	0.05	60
24	30	2170.73	0.03	466.94	0.02	0.01	70
36	30	2009.33	0.03	628.322	0.14	0.13	96
48	30	1512.41	0.04	658.149	0.21	0.08	130
Class B							
6	15	2677.87	0.02	6.645	0.05	0.18	10
12	15	2644.63	0.02	9.438	0.03	0.05	11
24	15	2657.81	**0.00**	15.366	0.01	**0.00**	17
36	15	2629.07	**0.00**	17.55	**0.00**	**0.00**	22
48	15	2629.07	0.01	34.055	0.01	0.10	28
6	30	191.123	**0.00**	158.246	0.02	0.10	27
12	30	84.194	**0.00**	233.907	0.00	0.06	43
24	30	2666.68	**0.00**	491.495	0.03	0.04	69
36	30	2678.21	0.02	655.794	0.22	0.25	92
48	30	2562.49	0.03	693.931	0.02	0.03	119
Class C							
6	15	0.718	**0.00**	5.179	**0.00**	**0.00**	8
12	15	2632.24	0.05	6.817	0.15	0.34	13
24	15	2652.22	0.03	18.626	0.02	0.39	19
36	15	2650.65	0.01	19.079	0.01	0.09	24
48	15	2667.17	0.01	26.895	0.00	0.03	33
6	30	2657.43	0.01	123.007	0.04	0.03	31
12	30	2660.38	0.01	200.601	0.91	0.98	41
24	30	2672.74	0.03	588.604	0.02	0.09	68
36	30	2685.30	0.02	432.417	0.04	0.14	88
48	30	2640.75	0.03	658.477	0.02	0.06	123

based on Dantzig-Wolfe decomposition (DWH). We notice the same behavior of the heuristic DWH. The gaps are very close to the gaps obtained by the lagrangean heuristic and the solver. The computation time of the heuristic DWH is much smaller than computation time required by the solver, and it is better than computation time of the lagrangean heristic when the number of period is small (15 periods), but

Table 3. Computation results: HR vs. DWH vs. XpressMP when capacity is tight

		Solver		DWH		LH	
Class D							
Items	**Periods**	**Time (s)**	**Gap1(%)**	**Time (s)**	**Gap2(%)**	**Gap3(%)**	**Time (s)**
6	15	2666.58	0.01	5.304	0.04	0.16	11
12	15	2642.53	0.01	8.924	0.03	0.02	13
24	15	2679.54	**0.00**	13.853	0.01	**0.00**	18
36	15	2658.76	**0.00**	23.759	0.06	0.09	27
48	15	2677.53	0.01	9.484	0.13	0.10	36
6	30	1662.35	**0.00**	156.422	0.04	0.10	27
12	30	7.036	**0.00**	289.177	0.04	0.09	46
24	30	2676.89	0.01	550.057	0.03	0.06	69
36	30	1663.35	0.02	652.67	0.07	0.06	102
48	30	2532.2	0.03	872.65	0.11	0.08	150
Class E							
6	15	2656.7	0.05	7.098	0.02	0.28	8
12	15	2651.08	0.02	8.924	**0.00**	0.03	13
24	15	2649.21	**0.00**	16.661	0.03	**0.00**	17
36	15	2641.19	0.01	28.548	0.02	**0.00**	23
48	15	2687.65	**0.00**	24.975	0.03	0.04	35
6	30	16.708	**0.00**	185.859	0.17	0.36	36
12	30	2635.11	0.01	282.111	0.04	0.03	43
24	30	2703.74	**0.00**	591.819	0.01	**0.00**	63
36	30	2661.22	0.01	642.799	0.18	0.10	99
48	30	2677.51	0.01	675.526	0.34	0.16	141
Class F							
6	15	1.263	**0.00**	6.224	0.03	0.16	12
12	15	2624.67	**0.00**	9.189	0.03	0.01	13
24	15	4.134	**0.00**	15.163	0.05	0.01	19
36	15	2653.64	**0.00**	21.029	**0.00**	0.02	25
48	15	2682.07	**0.00**	20.374	**0.00**	**0.00**	35
6	30	2771.33	**0.00**	170.758	**0.00**	0.07	27
12	30	2670.3	**0.00**	315.073	0.02	0.03	41
24	30	2657.09	**0.00**	449.952	0.17	0.02	68
36	30	2659.6	**0.00**	565.907	0.27	0.03	91
48	30	2680.87	0.01	787.131	0.27	0.33	120

once the number of period becomes more important (30 periods) we notice that computation time of the heuristic DWH increase and exceeds computation time of the Lagrangean heuristic.

To conclude, the Lagrangean heuristic (LH) and Dantzig-Wolfe decomposition based heuristic (DWH) seem a good trade-off between the solution quality and time execution. Therefore, for a decision maker who is interested in a good solution quality and a short execution time, our Lagrangean heuristic can be an appropriate approach to resolution when the number of period is important and our Dantzig-Wolfe decomposition based heuristic can be applied when tne number of period is small.

VI. CONCLUSION AND PERSPECTIVE

We have formulated a mixed-integer linear programming model to plan jointly production and maintenance activities. The model takes into account reliability of production equipment and its capacity to develop an integrated production planning and maintenance. Preventive maintenance is carried out in time windows, and corrective maintenance is performed to restore into the operating state without changing its failure rate function. Propositions and mathematical properties were suggested and applied in two heuristics to solve the problem. First heuristic developed is based on Lagrangean relaxation (Lagrangean heuristic (LH)) and second one is based on Dantzig-Wolfe decomposition (Dantzig-Wolfe based heuristic (DWH)). Computation results show that LH and DWH represent a good trade-off between quality solution and execution time. Also, we notice that DWH is faster than LH when the number of period is small, but when this number increase LH becomes faster.

It will be interesting to develop a meta-heuristic to solve this problem in order to get the same result with better execution time.

REFERENCES

Absi, N., & Kedad-Sidhoum, S. (2007). MIP-based heuristics for multi-item capacitated lot-sizing problem with setup times and shortage costs. RAIRO—. *Operations Research*, *41*(2), 171–192. doi:10.1051/ro:2007014

Absi, N., & Kedad-Sidhoum, S. (2008). The multi capacitated lot sizing problem with set up time and shortage costs. *European Journal of Operational Research*, *185*(3), 1351–1374. doi:10.1016/j.ejor.2006.01.053

Absi, N., & Kedad-Sidhoum, S. (2009). The multi-item capacitated lot-sizing problem with setup times, safety stock and demand shortages. *European Journal of Operational Research*, *36*, 2926–2936.

Aghezzaf, E. H., Jamali, M. A., & Ait-Kadi, D. (2007). An integrated production and preventive maintenance planning model. *European Journal of Operational Research*, *181*(2), 676–685. doi:10.1016/j.ejor.2006.06.032

Aghezzaf, E. H., & Najid, N. M. (2008). Integrated production and preventive maintenance in production systems subject to random failures. *Information science*, *178*, 3382–3392. doi:10.1016/j.ins.2008.05.007

Alaoui-Selsouli, M., Mohafid, A., & Najid, N. M. (2012). Lagrangian relaxation heuristic for integrated production and maintenance planning problem. *International Journal of Production Research*, *50*(13), 3630–3642. doi:10.1080/00207543.2012.671586

Bahl, H. C. (1983). Column generation based algorithm for multi-item scheduling. *IIE Transactions*, *15*(2), 136–141. doi:10.1080/05695558308974624

Barlow, R. E., & Hunter, L. C. (1960). Optimum preventive maintenance policies. *Operations Research*, *8*(1), 90–100. doi:10.1287/opre.8.1.90

Brahimi, N., Dauzère-Pérès, S., & Najid, M. N. (2006). Capacitated multi-item lot-sizing problems with time windows. *Operations Research*, *54*(5), 951–967. doi:10.1287/opre.1060.0325

Brahimi, N., Dauzere-Peres, S., Najid, N. M., & Nordli, A. (2006). Single items lot sizing problems. *European Journal of Operational Research*, *168*(1), 1–16. doi:10.1016/j.ejor.2004.01.054

Cattrysse, D., Maes, J., & Van Wassenhove, L. N. (1990). Set partitioning and column generation heuristics for capacitated dynamic lotsizing. *European Journal of Operational Research*, *46*(1), 38–47. doi:10.1016/0377-2217(90)90296-N

Chen, W. H., & Thizy, J. M. (1990). Analysis of relaxations for the multi-item capacitated lot- sizing problem. *Annals of Operations Research*, *26*(1), 29–72. doi:10.1007/BF02248584

Chen & Thizy. (2002). *An efficient column generation algorithm for the multi-item capacitated lot-sizing problem.* Technical Report 02-52, University of Ottawa.

Dantzig, G. B., & Wolfe, P. (1960). Decomposition principle for linear programs. *Operations Research*, *8*(1), 101–111. doi:10.1287/opre.8.1.101

Degraeve, Z., & Jans, R. (2007). A New Dantzig- Wolfe Reformulation and Branch-and-Price Algorithm for the Capacitated Lot-Sizing Problem with Setup Times. *Operations Research*, *55*(5), 909–920. doi:10.1287/opre.1070.0404

Diaby, M., Bahl, H. C., Karwan, M. H., & Zionts, S. (1992). A Lagrangean relaxation approach for very large scale capacitated lot-sizing. *Management Science*, *38*(9), 1329–1340. doi:10.1287/mnsc.38.9.1329

Dzielinski, B. P., & Gomory, R. E. (1965). Optimal programming of lot sizes, inventory, and labor allocations. *Management Science*, *11*(9), 874–890. doi:10.1287/mnsc.11.9.874

Fisher, M. L. (1981). The Lagrangian relaxation method for solving integer programming problems. *Management Science*, *27*(1), 1–18. doi:10.1287/mnsc.27.1.1

Fitouhi, M., & Nourelfath, M. (2014). Integrating non cyclical preventive maintenance scheduling and production planning for multi-state systems. *Reliability Engineering & System Safety*, *121*, 175–186. doi:10.1016/j.ress.2013.07.009

Jans, R., & Degraeve, Z. (2007). Meta-heuristics for dynamic lot sizing: A review and comparison of solution approaches. *European Journal of Operational Research*, *177*(3), 1855–1875. doi:10.1016/j.ejor.2005.12.008

Jans, R., & Degraeve, Z. (2008). Modeling industrial lot sizing problems: A review. *International Journal of Production Research*, *46*(6), 1619–1643. doi:10.1080/00207540600902262

Karimi, B., Fatemi Ghomi, S. M. T., & Wilson, J. M. (2003). The capacitated lot sizing problem: A review of models and algorithms. *Omega*, *31*(5), 365–378. doi:10.1016/S0305-0483(03)00059-8

Lasdon, L. S., & Terjung, R. C. (1971). An efficient algorithm for multi-item scheduling. *Operations Research*, *19*(4), 946–969. doi:10.1287/opre.19.4.946

Maes, J., & Van Wassenhove, L. (1988). Multi-item single-level capacitated dynamic lot- sizing heuristics: A general review. *The Journal of the Operational Research Society*, *39*(11), 991–1004. doi:10.1057/jors.1988.169

Manne, A. S. (1958). Programming of economic lot-sizes. *Management Science*, *4*(2), 115–135. doi:10.1287/mnsc.4.2.115

Millar, H. H., & Yang, M. (1994). Lagrangian heuristics for the capacitated multi-item lot-sizing problem with backordering. *International Journal of Production Economics*, *34*(1), 1–15. doi:10.1016/0925-5273(94)90042-6

Najid, N. M., Alaoui-Selsouli, M., & Mohafid, A. (2011). An integrated production and maintenance planning model with time windows and shortage cost. *International Journal of Production Research*, *49*(8), 2265–2283. doi:10.1080/00207541003620386

Nakagawa, T. (1981a). A summary of periodic replacement with minimal repair at failure. *Journal of the Operations Research Society of Japan*, *24*, 213–228.

Nakagawa, T. (1981b). Modified periodic replacement with minimal repair at failure. *IEEE Transactions on Reliability*, *30*(2), 165–168. doi:10.1109/TR.1981.5221018

Nourelfath, M., & Châtelet, E. (2012). Integrating production, inventory and maintenance planning for a parallel system with dependent components. *Reliability Engineering & System Safety*, *101*, 59–66. doi:10.1016/j.ress.2012.02.001

Nourelfath, M., Fitouhi, M., & Machani, M. (2010). A genetic algorithm for integrated production and preventive maintenance planning in multi state systems. *IEEE Transactions on Reliability*, *59*(3), 496–506. doi:10.1109/TR.2010.2056412

Ozdamar, L., & Bozyel, M. A. (2000). The capacitated lot sizing problem with overtime decisions and setup times. *IIE Transactions*, *32*(11), 1043–1057. doi:10.1080/07408170008967460

Trigeiro, W., Thomas, L. J., & McLain, J. O. (1989). Capacitated lot sizing with setup time. *Management Science*, *35*(3), 353–366. doi:10.1287/mnsc.35.3.353

Vanderbeck, F. (1998). Lot-sizing with start-up times. *Management Science*, *44*(10), 1409–1425. doi:10.1287/mnsc.44.10.1409

Wagner, H. M., & Whitin, T. M. (1985). A dynamic version of the economic lot size model. *Management Science*, *5*(1), 89–96. doi:10.1287/mnsc.5.1.89

Wang, H. (2002). A survey of maintenance policies of deteriorating systems. *European Journal of Operational Research*, *139*(3), 469–489. doi:10.1016/S0377-2217(01)00197-7

Wang, H., & Pham, H. (1999). Some maintenance models and availability with imperfect maintenance in production systems. *Annals of Operations Research*, *91*, 305–318. doi:10.1023/A:1018910109348

Weinstein, L., & Chung, C. (1999). Integrating maintenance and production decisions in a hierarchical production planning environment. *Computers & Operations Research*, *26*(10-11), 1059–1074. doi:10.1016/S0305-0548(99)00022-2

Yalaoui, A., Chaabi, K., & Yalaoui, F. (2014). Integrated production planning and preventive maintenance in deteriorating production systems. *Information Sciences*, *278*, 841–861. doi:10.1016/j.ins.2014.03.097

Chapter 24
TOPSIS in Generalized Intuitionistic Fuzzy Environment

Amal Kumar Adak
Jafuly Deshpran High School (H.S.), India

Monoranjan Bhowmik
Asam Central University (Silchar), India

Debashree Manna
Damda Jr. High School, India

Madhumangal Pal
Vidyasagar University, India

ABSTRACT

The aim of this chapter is to investigate the multiple attribute decision making problems to a selected project with generalized intuitionistic fuzzy information in which the information about weights is completely known and the attributes values are taken from the generalized intuitionistic fuzzy environment. Here, we extend the technique for order performance by similarity to ideal solution (TOPSIS) for the generalized intuitionistic fuzzy data. In addition, obtained the concept of possibility degree of generalized intuitionistic fuzzy numbers and used to solve ranking alternative in multi-attribute decision making problems.

INTRODUCTION

A lot of multiple criteria decision making (MCDM) approaches have been developed and applied to diverse fields, like engineering, management, economics etc. As one of the known classical MCDM approaches, TOPSIS (technique for the order preference by similarity to ideal solution) was first developed by Hwang and Yoon (1981). The primary concept of TOPSIS approach is that the most preferred alternative should not only have the shortest distance from the positive ideal solution but also have the farthest distance from the negative ideal solution. The advantages for the TOPSIS include (a) simple, rationally, comprehensive concept, (b) good computational efficient, (c) ability to measure the relative performance for each alternative in a simple mathematical form.

Zadeh (1965) introduced first the theory of fuzzy sets. Later on, many researchers have been working on the process of dealing with fuzzy decision making problems by applying fuzzy set theory. Atanassov (1986) introduced the concept of intuitionistic fuzzy sets (IFSs). IFSs are proposed using two characteristic functions expressing the degree of membership and the degree of non-membership of elements

DOI: 10.4018/978-1-4666-9644-0.ch024

of the universal set to the IFS. Adak (2011b, 2012b) *et al.* investigated some interesting properties of intuitionistic fuzzy sets and matrices. Mondal and Samanta (2002) defined generalized intuitionistic fuzzy sets (GIFSs). Motivated by the definition of GIFSs, Bhowmik (2008) investigated generalized intuitionistic fuzzy matrices (GIFMs). Generalized intuitionistic fuzzy matrices has been applied in multi-criteria decision making problem including conflicting criteria such as political environment, proximity to markets and customers, quality of life issues, culture issues, etc.

In this chapter, we introduce the concept of generalized intuitionistic fuzzy preference relation and develop the TOPSIS method in multi-criteria decision making problem with generalized intuitionistic fuzzy information.

BACKGROUND

Decision-making problems are the process of finding the best option from all of the feasible alternatives. In almost all such problems the multiplicity of criteria for judging the alternatives is pervasive. That is for many such problems, the decision maker wants to solve a multiple criteria decision making (MCDM) problem. Technique for order performance by similarity to ideal solution (TOPSIS), one of the known classical MCDM method, was first developed by Hwang and Yoon [16] for solving a MCDM problem. It bases upon the concept that the chosen alternative should have the shortest distance from the positive ideal solution (PIS) and the farthest from the negative ideal solution (NIS). In the process of TOPSIS, the performance ratings and the weights of the criteria are given as crisp values.

Under many conditions, crisp data are inadequate to model real-life situations. Since human judgments including preferences are often vague and cannot estimate his preference with an exact numerical value. A more realistic approach may be to use linguistic assessments instead of numerical values, that is, to suppose that the ratings and weights of the criteria in the problem are assessed by means of linguistic variables. In this paper, we further extended the concept of TOPSIS to develop a methodology for solving multi-person multi-criteria decision-making problems in generalized intuitionistic fuzzy environment. Considering the generalized intuitionistic fuzziness in the decision data and group decision-making process, linguistic variables are used to assess the weights of all criteria and the ratings of each alternative with respect to each criterion. We can convert the decision matrix into a generalized intuitionistic fuzzy decision matrix and construct a weighted normalized generalized intuitionistic fuzzy decision matrix once the decision makers' generalized intuitionistic fuzzy ratings have been pooled. According to the concept of TOPSIS, we define the generalized intuitionistic fuzzy positive ideal solution (GIFPIS) and the generalized intuitionistic fuzzy negative ideal solution (GIFNIS). Using the mathematical formula, we calculate the distance of each alternative from GIFPIS and GIFNIS, respectively. Finally, a closeness co-efficient of each alternative is defined to determine the ranking order of all alternatives. The higher value of closeness co-efficient indicates that an alternative is closer to GIFPIS and farther from GIFNIS simultaneously.

PRELIMINARIES

Here we recall some preliminaries, definitions of IFSs and GIFSs. Also, we define Hamming distance between two GIFSs.

2.1 Fuzzy Set and Instuitionistic Fuzzy Set

Fuzzy set and Instuitionistic fuzzy set .

Definition 1: (Fuzzy set (FS)) A fuzzy set A in a universal set X is defined as $A = \{\langle x, \mu_A(x)\rangle \mid x \in X\}$, where $\mu_A : X \rightarrow [0,1]$ is a mapping called the membership function of the fuzzy set A.

Definition 2: (Instuitionistic fuzzy set (IFS)) An instuitionistic fuzzy set (IFS) A over X is an object having the form $A = \langle x, \mu(x), v(x) | x \in X\rangle$, where $\mu(x): X \rightarrow [0,1]$ and $v(x): X \rightarrow [0,1]$, Where $\mu(x)$ and $v(x)$ are called the membership and non-membership value of x in A satisfying the condition $0 \le \mu(x) + v(x) \le 1$.

An element x of X is called significant with respect to an intuitionistic fuzzy subset A of X if the of membership $\mu_A(x) > 0.5$, otherwise insignificant . If member ship is significant then non-membership $v_A(x) = 1 - \mu_A(x)$ cannot be significant. Further, for an IFS $A = \langle x, \mu_A(x), v_A(x) | x \in X\rangle$, it is observe that $0 \le \mu(x) + v(x) \le 1$, for all $x \in X$ and hence $\mu_A(x) \wedge v_A(x) \le 0.5$, for all $x \in X$.

Definition 3: (Generalized instuitionistic fuzzy set (GIFS)) A generalized intuitionistic fuzzy set A of X is an object having the form $A = \{\langle x, \mu_A(x), v_A(x)\rangle \mid x \in X\}$, where the function $\mu_A : X \rightarrow [0,1]$ and $v_A : X \rightarrow [0,1]$ define respectively the degree of membership and degree of non-membership of the element $x \in X$ to the set A, which is a subset of X and for every $x \in X$ satisfy the conditions $0 \le \mu_A(x) \wedge v_A(x) \le 0.5$, for all $x \in X$.

This condition is called generalized intuitionistic fuzzy condition (GIFC).

Since, $0 \le \mu_A(x) \le 1$, $0 \le v_A(x) \le 1$ and $0 \le \mu_A(x) \wedge v_A(x) \le 0.5$, therefore from the definition we can write $0 \le \mu_A(x) + v_A(x) \le 1.5$, for all $x \in X$.

Now, we define another parameter $\pi_A(x)$, where

$$\pi_A(x) = 1.5 - \mu_A(x) - v_A(x). \text{ Obviously, } 0 \le \pi_A(x) \le 1.5.$$

In a GIFSs A, the parameter $\pi_A(x)$ is known as generalized intuitionistic fuzzy index or hesitation degree of whether x belongs to A or not.

If the value of $\pi_A(x)$ is very small, then knowledge about x is certain; if $\pi_A(x)$ is large, then the knowledge about x is more uncertain.

Definition 4: (Generalized intuitionistic fuzzy preference relation) A generalized intuitionistic fuzzy preference relation B on the set X is represented by a GIFM

$$B = (b_{ij}) \text{ with } b_{ij} = \langle (x_i, x_j), \mu(x_i, x_j), v(x_i, x_j)\rangle \text{ for all } i, j = 1, 2, \ldots, n.$$

For convenience, we let $b_{ij} = (\mu_{ij}, v_{ij})$, for all $i, j = 1, 2, \ldots, n$, where b_{ij} be a generalized intuitionistic fuzzy value, composed by the certainty degree μ_{ij} to which x_i is positively preferred to x_j and certainty degree v_{ij} to which x_i is negatively preferred to x_j. Moreover, μ_{ij}, v_{ij} satisfy GIFC and $0 \le \mu_{ij} + v_{ij} \le 1.5$, $\mu_{ij} = v_{ji}$, $v_{ij} = \mu_{ji}$ an $\mu_{ii} = v_{ii} = 0.5$.

Definition 5: Hamming distance between GIFSs Let A and B be two GIFSs on X, where

$A=\{x,\langle\mu_A(x),\nu_A(x)\rangle : x\in X\}$ *and* $B=\{x,\langle\mu_B(x),\nu_B(x)\rangle : x\in X\}$. Then,

Hamming distance between GIFSs A and B is given as follows:

$$d(A,B)=\frac{1}{2}[|\mu_A(x)-\mu_B(x)|+|\nu_A(x)-\nu_B(x)|+|\pi_A(x)-\pi_B(x)|].$$

Definition 6: (Generalized intuitionistic fuzzy matrix (GIFM)) Generalized intuitionistic fuzzy matrix (GIFM) of order $m\times n$ is defined as $A=\langle a_{ij\mu},a_{ij\nu}\rangle$, where $a_{ij\mu}$ and $a_{ij\nu}$ are the membership value and non-membership value of the ij-th element in A satisfying the GIFC and the maximum value of $\mu_A(x)$ and $\nu_A(x)$ is 1.0, for all i, j.

AN INTEGRATED GENERALIZED INTUITIONISTIC FUZZY MULTI-CRITERIA DECISION MAKING METHOD

In this section, the technique for order preference by similarity to ideal solution (TOPSIS) method is extended to generalized intuitionistic fuzzy environment, which is a very suitable for solving decision making problem.

Let $A=\{A_1,A_2,\ldots,A_m\}$ be a set of alternatives and $C=\{C_1,C_2,\ldots,C_m\}$ be a set of criteria. Generalized intuitionistic fuzzy TOPSIS method consists of the following steps which are given, as follows:

1. Construct a generalized intuitionistic fuzzy preference relation matrix.

Let $B=(b_{ij})_{n\times n}$ be a generalized intuitionistic preference matrix of criteria as follows

$$B=\begin{bmatrix}\tilde{b}_{11} & \tilde{b}_{12} & \cdots & \tilde{b}_{1n}\\ \tilde{b}_{21} & \tilde{b}_{22} & \cdots & \tilde{b}_{2n}\\ \vdots & \vdots & \ddots & \vdots\\ \tilde{b}_{n1} & \tilde{b}_{n2} & \cdots & \tilde{b}_{nn}\end{bmatrix},$$

where $\tilde{b}_{ij}=\langle b_{ij\mu},b_{ij\nu}\rangle, i,j=1,2,\ldots,n,$ and satisfies the generalized intuitionistic fuzzy preference relation.

Now, calculate

$$(\mu_{ij})^*=\max\left\{\mu_{ij},\max_p\left\{\frac{\mu_{ip}\mu_{pj}}{\mu_{ip}\mu_{pj}+(1.5-\mu_{ip})(1.5-\mu_{pj})}\right\}\right\} \quad (1)$$

$$(v_{ij})^* = \min\left\{v_{ij}, \min_p \left\{\frac{v_{ip}v_{pj}}{v_{ip}v_{pj} + (1.5 - v_{ip})(1.5 - v_{pj})}\right\}\right\}, \tag{2}$$

where $(\mu_{ij})^*$ and $(v_{ij})^*$, the element of aggregated generalized intuitionistic fuzzy preference matrix $(B)^*$. If $0 \le (\mu_{ij})^* + (v_{ij})^* \le 1.5$ for all i, j, $k = 1, 2, \cdots, n$, then we call $(B)^*$ a multiplicative consistent generalized intuitionistic fuzzy preference relation. If $(B)^*$ does not satisfy the condition of $0 \le (\mu_{ij})^* + (v_{ij})^* \le 1.5$ for all i, j, $k = 1, 2, \cdots, n$, then we call $(B)^*$ an inconsistent multiplicative generalized intuitionistic fuzzy preference relation.

2. Obtain the priority vector of criteria.

After construction of aggregated generalized intuitionistic fuzzy preference matrix, the priority vector of $\mathrm{w} = (\mathrm{w}_1, \mathrm{w}_2, \ldots, \mathrm{w}_n)^T$ can be estimated with the following equation

$$\mathrm{w}_j = \left[\mathrm{w}_j^L, \mathrm{w}_j^U\right] = \left(\frac{1}{\sum_{j=1}^{n}\left(\frac{1.5 - \tilde{\mu}_{ij}^*}{\tilde{\mu}_{ij}^*}\right)}, \frac{1}{\sum_{j=1}^{n}\left(\frac{1.5 - \tilde{v}_{ij}^*}{\tilde{v}_{ij}^*}\right)}\right). \tag{3}$$

3. Construct a generalized intuitionistic fuzzy decision matrix.

Let $\tilde{\mathrm{R}} = (\tilde{\mathrm{r}}_{ij})_{n \times n}$ be a generalized intuitionistic fuzzy decision matrix such that

$$R = \begin{bmatrix} \tilde{r}_{11} & \tilde{r}_{12} & \cdots & \tilde{r}_{1n} \\ \tilde{r}_{21} & \tilde{r}_{22} & \cdots & \tilde{r}_{2n} \\ \vdots & \vdots & \ddots & \vdots \\ \tilde{r}_{n1} & \tilde{r}_{n2} & \cdots & \tilde{r}_{nn} \end{bmatrix},$$

where $\tilde{\mathrm{r}}_{ij} = (\mu_{ij}, v_{ij}, \pi_{ij})$, i, j = 1,2,…,n, which are contained in a generalized intuitionistic fuzzy decision matrix.

4. Determine the generalized intuitionistic fuzzy positive ideal solution and the generalized intuitionistic fuzzy negative ideal solution.

Let J_1 be the set of benefit criteria, J_2 be the set of cost criteria, A^* be the generalized intuitionistic fuzzy positive ideal solution and A^- be the generalized intuitionistic fuzzy negative ideal solution, then A^* and A^- can be determined respectively as

$$A^* = (r_1^*, r_2^*, \ldots, r_j^*),\ \mathrm{r}_j^* = (\mu_j^*, v_j^*, \pi_j^*),\ \mathrm{j} = 1,2,\ldots,\mathrm{n} \tag{4}$$

$$A^- = (r_1^-, r_2^-, \ldots, r_j^-),\ \mathrm{r}_j^- = (\mu_j^-, v_j^-, \pi_j^-),\ \mathrm{j} = 1,2,\ldots,\mathrm{n} \tag{5}$$

$$(\mu_j)^* = \left\{\left(\max_i\{\mu_{ij}\} \mid j \in J_1\right), \left(\min_i\{\mu_{ij}\} \mid j \in J_2\right)\right\} \tag{6}$$

$$(v_j)^* = \left\{\left(\min_i\{v_{ij}\} \mid j \in J_1\right), \left(\max_i\{v_{ij}\} \mid j \in J_2\right)\right\} \tag{7}$$

$$(\mu_j)^- = \left\{\left(\min_i\{\mu_{ij}\} \mid j \in J_1\right), \left(\max_i\{\mu_{ij}\} \mid j \in J_2\right)\right\} \tag{8}$$

$$(v_j)^- = \left\{\left(\max_i\{v_{ij}\} \mid j \in J_1\right), \left(\min_i\{v_{ij}\} \mid j \in J_2\right)\right\} \tag{9}$$

$$(\pi_j^*) = \left\{\left(1.5 - \max_i\{\mu_{ij}\} - \min_i\{v_{ij}\} \mid j \in J_1\right), \left(1.5 - \min_i\{\mu_{ij}\} - \max_i\{v_{ij}\} \mid j \in J_2\right)\right\} \tag{10}$$

$$(\pi_j^-) = \left\{\left(1.5 - \min_i\{\mu_{ij}\} - \max_i\{v_{ij}\} \mid j \in J_1\right), \left(1.5 - \max_i\{\mu_{ij}\} - \min_i\{v_{ij}\} \mid j \in J_2\right)\right\} \tag{11}$$

5. Calculate the weighted separation measure.

The weighted Hammining distance is used to obtain separation measures. The weighted lower and upper separation measure $(S_i^*)^L$, $(S_i^*)^U$ and $(S_i^-)^L$, $(S_i^-)^U$ of each alternative from the generalized intuitionistic fuzzy positive ideal solution and the generalized intuitionistic fuzzy negative ideal solution are respectively calculated as follows

$$(S_i^*)^L = \frac{1}{2}\sum_{j=1}^{n} w_j^L \left[| \mu_{ij} - \mu_j^* | + | \nu_{ij} - \nu_j^* | + | \pi_{ij} - \pi_j^* | \right] \quad (12)$$

$$(S_i^*)^U = \frac{1}{2}\sum_{j=1}^{n} w_j^U \left[| \mu_{ij} - \mu_j^* | + | \nu_{ij} - \nu_j^* | + | \pi_{ij} - \pi_j^* | \right] \quad (13)$$

$$(S_i^-)^L = \frac{1}{2}\sum_{j=1}^{n} w_j^L \left[| \mu_{ij} - \mu_j^- | + | \nu_{ij} - \nu_j^{*-} | + | \pi_{ij} - \pi_j^- | \right] \quad (14)$$

$$(S_i^-)^U = \frac{1}{2}\sum_{j=1}^{n} w_j^U \left[| \mu_{ij} - \mu_j^- | + | \nu_{ij} - \nu_j^{*-} | + | \pi_{ij} - \pi_j^- | \right] \quad (15)$$

6. Calculate the relative closeness co-efficient of each alternative to the generalized intuitionistic fuzzy positive and negative ideal solutions.

The relative closeness coefficient of an alternative A_i with respect to the generalized intuitionistic fuzzy positive ideal solution A^* and the generalized intuitionistic fuzzy negative ideal solution A^- is defined as follows

$$\left((C_i^*)^L, (C_i^*)^U\right) = \left\{ \left(\frac{(S_i^-)^L}{(S_i^-)^U + (S_i^-)^U} \right), \left(\frac{(S_i^-)^U}{(S_i^-)^L + (S_i^-)^L} \right) \right\} \quad (16)$$

7. Rank the alternative according to the descending order of the relative closeness co-efficient

$$C_i^* = \left((C_i^*)^L, (C_i^*)^U\right). \quad (17)$$

In order to rank alternatives the possibility degree formula is used.

Definition7: Let $a = [a^U, a^L]$ and $b = [b^U, b^L]$ be two interval numbers then the possibility degree ofa ≥ b is defined as:

$$p(a \geq b) = \max\left\{ 1 - \max\left(\frac{b^U - a^L}{b^U - b^L + a^U - a^L}, 0 \right), 0 \right\} \quad (18)$$

That is a superior to b degree of $a \overset{p(a \geq b)}{\succ} b$.

Let $P = p_{ij} = p(a_i \geq a_j)$ be the complementary generalized intuitionistic fuzzy matrix and given as follows:

$$P = \begin{bmatrix} p_{11} & p_{12} & \cdots & p_{1n} \\ p_{21} & p_{22} & \cdots & p_{2n} \\ \vdots & \vdots & \cdots & \vdots \\ p_{n1} & p_{n2} & \cdots & p_{nn} \end{bmatrix},$$

where $p_{ij} \geq 0,\ p_{ij} + p_{ji} = 1,\ p_{ii} = 0.5,\ i, j = 12, \ldots, n.$

Summing all elements in each row of matrix P, then

$$p_i = \sum_{j=1}^{n} p_{ij},\ i = 1, 2, \ldots, n. \tag{19}$$

Alternatives are ranked according to descending order of p_i.

ILLUSTRATE EXAMPLE

Suppose a manufacturing company finds a location for building a new plant. Assume that there are four candidates places A_1, A_2, A_3 and A_4 are chosen for further evaluation. In order to select final location, expansion possibility (C_1), availability of material (C_2), distance to market (C_3) and labour cost (C_4) are considered as evaluation factor.

1. Construct an generalized intuitionistic fuzzy preference matrix.

Let $B = (b_{ij})_{4\times 4}$ generalized intuitionistic fuzzy preference relation matrix

$$B = \begin{bmatrix} \langle 0.50, 0.50 \rangle & \langle 0.60, 0.20 \rangle & \langle 0.80, 0.30 \rangle & \langle 0.60, 0.40 \rangle \\ \langle 0.20, 0.60 \rangle & \langle 0.50, 0.50 \rangle & \langle 0.50, 0.20 \rangle & \langle 0.40, 0.70 \rangle \\ \langle 0.30, 0.80 \rangle & \langle 0.20, 0.50 \rangle & \langle 0.50, 0.50 \rangle & \langle 0.50, 0.60 \rangle \\ \langle 0.40, 0.60 \rangle & \langle 0.70, 0.40 \rangle & \langle 0.60, 0.50 \rangle & \langle 0.50, 0.50 \rangle \end{bmatrix}$$

Here $B = (b_{ij})_{4\times 4}$ has been consistent generalized intuitionistic fuzzy preference relation matrix due to satisfying the condition (1) and (2) and the matrix existing as follows:

$$B^* = \begin{bmatrix} \langle 0.50, 0.50 \rangle & \langle 0.60, 0.20 \rangle & \langle 0.80, 0.30 \rangle & \langle 0.60, 0.40 \rangle \\ \langle 0.20, 0.60 \rangle & \langle 0.50, 0.50 \rangle & \langle 0.50, 0.20 \rangle & \langle 0.40, 0.70 \rangle \\ \langle 0.30, 0.80 \rangle & \langle 0.20, 0.50 \rangle & \langle 0.50, 0.50 \rangle & \langle 0.50, 0.60 \rangle \\ \langle 0.40, 0.60 \rangle & \langle 0.70, 0.40 \rangle & \langle 0.60, 0.50 \rangle & \langle 0.50, 0.50 \rangle \end{bmatrix}$$

2. Obtain the priority vector of criteria.

The priority vector of criteria has been estimated by utilizing Eq.3 as follows:

$w_1 = \langle 0.170, 0.788 \rangle$, $w_2 = \langle 0.075, 0.436 \rangle$, $w_3 = \langle 0.083, 0.355 \rangle$ and $w_4 = \langle 0.135, 0.492 \rangle$.

3. Construct the generalized intuitionistic fuzzy decision matrix. The generalized intuitionistic fuzzy decision matrix has been constructed as follows in Table 1.
4. Determine the generalized intuitionistic fuzzy positive ideal solution and the generalized intuitionistic fuzzy negative ideal solution.

Considering that expansion possibility, availability of acquirement material are the benefit criteria $J_1 = \{C_1, C_2\}$ and distance to the market and labour cost are the cost criterion $J_2 = \{C_3, C_4\}$. Then the generalized intuitionistic fuzzy positive ideal solutions and generalized intuitionistic fuzzy negative ideal solutions have been obtained by employing Eq.(4)-Eq.(11) as follows:

$$A^* = \begin{cases} \langle 0.85, 0.23, 0.42 \rangle \langle 0.78, 0.38, 0.34 \rangle \\ \langle 0.48, 0.65, 0.37 \rangle \langle 0.69, 0.45, 0.36 \rangle \end{cases}$$

$$A^- = \begin{cases} \langle 0.74, 0.46, 0.30 \rangle \langle 0.66, 0.45, 0.39 \rangle \\ \langle 0.71, 0.36, 0.43 \rangle \langle 0.82, 0.35, 0.33 \rangle \end{cases}$$

5. Calculate the weighted separation measures.

Generalized intuitionistic fuzzy negative and positive separation measures based on the weighted lower and upper Hamming distance for each candidate have been calculated by utilizing Eq.(12)-Eq.(15) and given in Table 2.

Table 1.

Candidates	Criteria			
	C_1	C_2	C_3	C_4
A_1	$\langle 0.74,0.44,0.32 \rangle$	$\langle 0.72,0.38,0.40 \rangle$	$\langle 0.64,0.44,0.62 \rangle$	$\langle 0.69,0.45,0.36 \rangle$
A_2	$\langle 0.85,0.42,0.23 \rangle$	$\langle 0.78,0.45,0.27 \rangle$	$\langle 0.71,0.36,0.43 \rangle$	$\langle 0.82, 0.42,0.26 \rangle$
A_3	$\langle 0.76,0.36,0.38 \rangle$	$\langle 0.66,0.38,0.46 \rangle$	$\langle 0.48,0.65,0.37 \rangle$	$\langle 0.70,0.40,0.40 \rangle$
A_4	$\langle 0.78,0.46,0.26 \rangle$	$\langle 0.75,0.42,0.30 \rangle$	$\langle 0.68,0.48,0.34 \rangle$	$\langle 0.75,0.35,0.40 \rangle$

Table 2.

Candidates	$(S^*)^U$	$(S^*)^L$	$(S^-)^L$	$(S^-)^U$
A_1	0.134	0.625	0.101	0.464
A_2	0.199	0.901	0.182	0.797
A_3	0.171	0.805	0.162	0.735
A_4	0.120	0.536	0.116	0.496

6. Calculate the relative closeness co-efficient of each candidate to the generalized intuitionistic fuzzy positive ideal solutions and the generalized intuitionistic fuzzy negative ideal solutions. The relative closeness co-efficient of each candidate to the generalized intuitionistic fuzzy positive and negative ideal solutions have been calculated by using Eq.(16) as follows:

$$\left((C_1^*)^L,(C_1^*)^U\right)=\left\{\left(\frac{0.101}{0.625+0.464}\right),\left(\frac{0.464}{0.134+0.101}\right)\right\}=\left(0.092,1.794\right).$$

Similarly, $\left((C_1^*)^L,(C_1^*)^U\right)=\langle 0.107,2.091\rangle,\left((C_1^*)^L,(C_1^*)^U\right)=\langle 0.105,2.207\rangle,\left((C_1^*)^L,(C_1^*)^U\right)=\langle 0.112,2.101\rangle.$

7. Rank of the candidates according to the descending order of the relative closeness co-efficient. Four candidate locations have been ranked according to the descending order of relative closeness co-efficient. The candidates have been ranked by using the possibility degree formula and the following matrix has been constructed as follows:

$$P=\begin{pmatrix} 0.50 & 0.52 & 0.53 & 0.52 \\ 0.48 & 0.50 & 0.51 & 0.50 \\ 0.47 & 0.49 & 0.50 & 0.49 \\ 0.48 & 0.50 & 0.51 & 0.50 \end{pmatrix}.$$

The candidates have been ranked as A_1, A_2, A_4 and A_3 according to the descending order p_i, $i = 1,2,3,4$. Thus A_1 has been selected as the most desirable facility location among the candidates.

CONCLUSION

The success of companies depends on their capability on making right strategic decisions. Facility location selection is one of these strategic decisions, which it is a costly and difficult to reverse activity for companies. Therefore, this has presented the generalized intuitionistic fuzzy preference relation and generalized intuitionistic fuzzy TOPSIS method for selecting the most desirable facility location. The generalized intuitionistic fuzzy preference relation has been applied to derive the weights of criteria and generalized intuitionistic fuzzy TOPSIS method has been used to rank alternative. Thus, this method has an enormous chances of success for multi-criteria decision making problems due to great superiority on dealing with uncertainty in-formation. In future, the proposed method can be used for dealing with uncertainty in a variety of multi-criteria decision making problems.

REFERENCES

Adak, A. K., & Bhowmik, M. (2011a). Interval cut-set of interval-valued intuitionistic fuzzy sets. *African Journal of Mathematics and Computer Sciences*, *4*(4), 192–200.

Adak, A. K., Bhowmik, M., & Pal, M. (2011b). Semi-ring of interval-valued intuitionistic fuzzy matrices. *Global Journal of Computer Application and Technology*, *1*(3), 340–347.

Adak, A. K., Bhowmik, M., & Pal, M. (2012a). Interval cut-set of generalized interval-valued in-tuitionistic fuzzy sets. *International Journal of Fuzzy System Applications*, *2*(3), 35–50. doi:10.4018/ijfsa.2012070103

Adak, A. K., Bhowmik, M., & Pal, M. (2012b). Intuitionistic fuzzy block matrix and its some properties. *Annals of Pure and Applied Mathematics*, *1*(1), 13–31.

Adak, A. K., Bhowmik, M., & Pal, M. (2012c). Some properties of generalized intuitionistic fuzzy nil-potent matrices over distributive lattice. *International Journal of Fuzzy Information and Engineering*, *4*(4), 371–387. doi:10.1007/s12543-012-0121-1

Adak, A. K., Bhowmik, M., & Pal, M. (2013). Distributive Lattice over Intuitionistic Fuzzy Matrices. *The Journal of Fuzzy Mathematics*, *21*(2), 401–416.

Atanassov, K. (1986). Intuitionistic fuzzy sets. *Fuzzy Sets and Systems*, *20*(1), 87–96. doi:10.1016/S0165-0114(86)80034-3

Atanassov, K. (1994). New operations defined over the intuitionistic fuzzy sets. *Fuzzy Sets and Systems*, *61*(2), 137–142. doi:10.1016/0165-0114(94)90229-1

Bhowmik, M., & Pal, M. (2008). Generalized intuitionistic fuzzy matrices. *Far East Journal of Mathematical Sciences*, *29*(3), 533–554.

Boran, F. E. (2011). An integrated intuitionistic fuzzy multi criteria decision making method for facility location selection. *Mathematical and Computational Applications*, *16*(2), 487–496.

Cook, W. D., & Kress, M. (1991). A multiple criteria decision model with ordinal preference data. *European Journal of Operational Research*, *54*(2), 191–198. doi:10.1016/0377-2217(91)90297-9

Dai, Y. Q., & Xu, Z. S. (2007). Approach to multiple attribute decision making based on different intuitionistic preference structures. *J Southeast University*, *37*(4), 706–710.

Fan, Z. P., & Liu, Y. (2010). An approach to solve group decision making problems with ordinal interval Numbers. *IEEE Trans Syst Man Cybern Part B*, *40*(5), 1413–1423. doi:10.1109/TSMCB.2009.2039477 PMID:20172834

Herrera, F., Herrera-Viedma, E., & Chiclana, F. (2001). Multiperson decision-making based on multiplicative preference relations. *European Journal of Operational Research*, *129*(2), 372–385. doi:10.1016/S0377-2217(99)00197-6

Hong, D. H., & Choi, C. H. (2000). Multicriteria fuzzy decision making problems based on vague set theory. *Fuzzy Sets and Systems*, *114*(1), 103–113. doi:10.1016/S0165-0114(98)00271-1

Hwang, C. L., & Yoon, K. (1981). *Multiple attribute decision methods and applications: A state of the art survey*. New York: Springer Verlag. doi:10.1007/978-3-642-48318-9

Khamseh, A. A., & Mahmoodi, M. (2014). A new fuzzy topsis-todim hybrid method for green supplier selection using fuzzy time function. *Advances in Fuzzy Systems, Article ID 841405.*

Li, X., & Chen, X. (2014). Extension of the TOPSIS method based on prospect theory and trapezoidal intuitionistic fuzzy numbers for group decision making. *Journal of Systems Science and Systems Engineering*, *23*(2), 231–247. doi:10.1007/s11518-014-5244-y

Mammadova, M., & Jabrayilova, Z. (2014). Application of Fuzzy Optimization Method in Decision-Making for Personnel Selection. *Computer Science & Communications*, *5*(4), 190–204.

Mondal, T. K., & Samanta, S. K. (2002). Generalized intuitionistic fuzzy sets. *The Journal of Fuzzy Mathematics*, *10*(4), 839–862.

Nayagama, V. L. G., Muralikrishnan, S., & Sivaraman, G. (2010). Multi-criteria decision-making method based on interval-valued intuitionistic fuzzy sets. *Expert Systems with Applications*, *38*, 1464–1467.

Pal, M., Khan, S. K., & Shyamal, A. K. (2002). Intuitionistic fuzzy matrices. *Notes on IntuitionisticFuzzy Sets*, *8*(2), 51–62.

Rouyendegh, B. D., & Saputro, T. E. (2014). Supplier Selection Using Integrated Fuzzy TOPSIS and MCGP: A Case Study. *Procedia: Social and Behavioral Sciences*, *116*(21), 3957–3970. doi:10.1016/j.sbspro.2014.01.874

Shyamal, A. K., & Pal, M. (2005). Distance between fuzzy matrices and its applications-I, *J. Nature. Physical Science (London)*, *19*(1), 39–58.

Suder, A., & Kahraman, C. (2015). Minimizing Environmental Risks Using Fuzzy TOPSIS: Location Selection for the ITU Faculty of Management, *Human and Ecological Risk Assessment. International Journal (Toronto, Ont.)*, *21*(5), 1326–1340.

Xu, Z. S. (2007). Some similarity measure of intuitionistic fuzzy sets and their applications to multi attribute decision making. *Fuzzy Optimization and Decision Making*, *6*(2), 109–121. doi:10.1007/s10700-007-9004-z

Yeh, C. H. (2002). A problem based selection of multi attribute decision making methods. *International Transactions in Operational Research*, *9*(2), 169–181. doi:10.1111/1475-3995.00348

Zadeh, L. A. (1965). Fuzzy sets. *Information and Control*, *8*(3), 338–353. doi:10.1016/S0019-9958(65)90241-X

Zhang, X., Yue, G., & Teng, S. Z. (2009). Possibility degree of interval-valued fuzzy numbers and its application. *International Symposium on Information Processing*.

Chapter 25
Bi-Criteria Optimization for Finding the Optimal Replacement Interval for Maintaining the Performance of the Process Industries

Harish Garg
Thapar University, India

ABSTRACT

The optimization of the maintenance decision making can be defined as an attempt to resolve the conflicts of decision situation in such a way that variable under the control of the decision maker take their best possible value. One of the most important controllable parameters is the time interval between maintenance. Most of the researchers have kept the fact that whenever the suitable maintenance interval is reached, the system is replaced with the original one. However the improvement of a system life not only depends on the replacement of deteriorated components, but also on the effectiveness of the maintenance. Taking care about this fact, the effects of maintenance of a multi-component system by combining the three main different PM actions, namely (1a), (1b) and (2p)-maintenance actions. Thus, the main purpose of an effective maintenance program is to present a technique for finding the optimal maintenance interval for the system by considering the multiple goals of the organization viz. maximum availability, minimum maintenance cost.

1. INTRODUCTION

With the advance in technology, a designer always wants to manufacture the equipment and systems of greater capital cost, complexity and capacity which results in increasing the reliability of the system. Also at the same time the unfortunate penalty of low availability and high maintenance cost need to be improved for their survival. To achieve this end, availability and reliability of equipment in the process must be maintained at the higher order. Thus, reliability and maintainability concepts are mainly applicable at the design stage of a machinery or plant layout, while the availability concept is mostly applicable after

DOI: 10.4018/978-1-4666-9644-0.ch025

commissioning the plant or after a steady state of production is reached. Modern technology has developed a tendency to design and manufacture equipment and systems of greater capital cost, sophistication, complexity, and capacity. The disastrous consequences of low availability and high maintenance cost of such systems led to the desire for high reliability, high maintainability, and low mean time to support. Due to design problems and poor product support, manufacturer equipment and systems are not able to meet these requirements. However, to improve the quality and quantity of a manufactured associated prospectus, there is a need to accentuate more on operational management and the proper consideration of reliability, availability and maintainability (RAM) in the design, manufacturing and installation phase will could be reduced the number of failures and their consequences minimized (Saraswat and Yadava, 2008). Thus, for this reason and to reduce the number of likelihood failures, there is a great interest in dealing with the main feature of the reliability parameters which affects the system performance directly i.e. reliability, availability and maintainability. In that direction, various researchers have addressed the issue of RAM analysis by adapting the suitable maintenance strategies for increasing the performance of the system (Saraswat & Yadava, 2008; DuJulio & Leet, 1988; Wood, 1989; Madu & Kuci, 1994; Hwang, 1996; Zerwick, 1996). To improve the system reliability and availability, implementation of appropriate maintenance strategies play an important role. The high performance of these units can be achieved with highly reliable subunits and perfect maintenance. To this effect the knowledge of behavior of system, their component(s) is customary in order to plan and adapt suitable maintenance strategies. Thus, maintainability is also to be a key index to enhance the performance of these systems.

In a system where a certain amount of failures is allowed, the efficient repair and/or replacement of these failures is critical to the continued usefulness of the system. This repair and replacement of failures are called maintenance. Maintenance has a definite influence on operating costs, either through its own (maintenance) labor or through its effect of system downtime and efficiency. In reliability, maintainability can also be used to increase the probability that a system will continue to operate efficiently, given that it is allowed a certain amount of downtime of repairs. The purpose of maintainability is to return a failed or deteriorating system to a satisfactory operating state. To do this, there are two extreme maintenance policies that can be applied. The first is to unplanned (corrective) maintenance while the second one is planned (preventive) maintenance. In unplanned, corrective strategy, no maintenance action is carried out until the component or structure breaks down or when its cost of operation becomes creeping or wear-out failures. This is called corrective maintenance (CM) or emergency repair. A study on the effect of CM in maintenance policies was done by Samrout et al. (2009). CM activities are categorized into minimal repair (1C). 1C-maintenance makes no change in system time and restores the system reliability to it when it had failed. Upon failure, the associated repair time is large and thus leading to large downtimes and high costs. In this approach, efforts are undertaken to achieve small mean times to repair (MTTRs). To avoid failures at occasions that have high cost consequences preventive maintenance is normally chosen. This allows that inspections and upgrading can be planned for periods, which have the lowest impact on production or availability of the systems. The main function of planned maintenance is to restore equipment to the "as good as new" condition; periodical inspections must control equipment condition and both actions will ensure equipment availability. PM, by its nature, can be scheduled and controlled for a minimum cost; corrective action cannot. PM includes such things as inspecting, tuning and major overhauls. In some types of system (especially mechanical systems such as automobiles), PM can also include painting, waxing, chemical solution monitoring, and water treatment. PM strategy involves preventive tasks such as repair, replacement or inspection at a pre-determined interval

(Gertsbakh, 1977). This interval is determined by using a scientific approach such as numerical analysis, optimization technique, etc., which are summarized in the following sections.

2. RELATED WORKS AND BACKGROUND

In this section, a brief literature review regarding availability/maintainability evaluation and optimization is given. The gaps found from literature review are also addressed in it.

2.1. Performance Analysis of the System

Woodman (1964) gave a model for replacing the components when its age will exceed a control limit by using dynamic programming. DuJulio and Leet (1988) have presented space station synergetic RAM-Logistics analysis, this study emphasizes to analyze the maintenance activities and processes that can be accomplished on-orbit within the known design and support constraints of the space station. Wood (1989) has developed equations for the availability of a system with exhaustible spares. Cockerill (1990) has presented a RAM analysis of a turbine - generator system. McFadden (1990) has proposed the techniques for developing the database for a RAM improvement program for an industrial plant or commercial building. Prince and Haire (1991) presented some Markov models to evaluate the impact of maintenance system availability on the overall plant. Hansen et al. (1992) developed a RAM expert system to assist a weapon system manager to conduct weapon system RAM performance analysis. Zerwick (1996) pointed out in the context of pressure vessels that a systematic strategy based on RAM principles helps to evaluate changes in inspection frequency, maintenance actions or condition monitoring strategies leading to decrease in frequency of planned shut downs, increase of the time period between statutory inspections, and reduction in maintenance cost. Hwang (1996) has developed a general design and performance evaluation model for the system designers in the initial design phase of the integrated manufacturing system based on RAM and logistic cost circuit (LCC). Dhudshia (1997) has described SEMI E10 (standard for definition and measurement of equipment reliability, availability and maintainability), which contains key definitions and formulas as well as confidence limits to characterize semiconductor manufacturing equipment utilization. Singh and Swaminathan (2002) have presented a methodology for determining sample sizes for system availability under the assumption that the time between failures and the times to repair are independent and exponentially distributed. Markeset and Kumar (2003) have presented an approach for the design and development of product support and maintenance needs for industrial systems in a multinational environment. Barabady and Kumar (2005, 2008) had presented a methodology for improving the availability of a repairable system by using the concept of important measures. The empirical data of two crushing plants at the Jajarm bauxite mine of Iran are used as a case study for reliability and availability analysis. The reliability and maintainability analysis of strudel, peach, bread production line at machine, workstation and entire line level was developed in Tsarouhas et al. (2009); Tsarouhas and Arvanitoyannis (2010,2011). The descriptive statistics of the failure and repair data was carried out, and the best index of fit among the common theoretical distribution and their corresponding parameters was determined. Rao and Naikan (2006) proposed a condition based preventive maintenance (CBPM) policy for Markov deteriorating systems by considering deterioration and random failures with minimal and major preventive maintenance. Nepal et al. (2007) presented an integrated and analytical approach to modularise the product architecture by focusing on reliability and maintainability (R & M) issues upfront at a conceptual stage so that the redesign problems at the later stage can be minimized. In addition, they also considered cost as a criterion for modularisation and fuzzy logic was used to evaluate the candidate modules with respect to R & M and cost metrics. Yao et al. (2001) present a two-layer hierarchical model that optimizes the preventive maintenance scheduling in semiconductor manufacturing operations. They develop a Markov decision process and optimize this model via a mixed-integer linear programming model. Later, Yao et al. (2004) extend their previous model to be more general, apply this extended model in a production line of a semiconductor manufacturing system, and show the application of it via numerical examples. Jayakumar and Asagarpoor (2004) present a linear programming model in order to optimize the maintenance policy for a component with deterioration and random failure rate. They determine optimal mean times of minor and major preventive maintenance actions based on maximizing the availability of the component by formulating and solving a linear programming model of Markov decision process. Canto (2006) presents an optimization model to schedule a preventive maintenance of a power plant. He considers the total cost of various operations as the objective function and uses the Bender's decomposition method to solve a mixed-integer linear programming model. Tam et al. (2006) developed three non-linear optimization models: one that minimizes total cost subject to satisfying a required reliability, one that maximizes reliability at a given budget, and one that minimizes the expected total cost including expected breakdown outages cost and maintenance cost. They utilize MS-Excel Solver as the optimization software that uses a generalized reduced gradient (GRG) algorithm to solve the non-linear optimization models. Kuo and Chang (2007) develop an integrated maintenance scheduling and production planning, optimization model for a single machine based on preventive maintenance strategies on production schedules in order to minimize total tardiness. Sharma and Kumar (2008) presented the application of RAM analysis in a process industry by using a Markovian approach as a tool to model the system behavior. In their approach, crisp historical data are utilized in the analysis without quantification of involved uncertainties. Rajpal et al. (2006) explored the application of artificial neural networks to model the behavior of a complex, repairable system. A composite measure of RAM parameters called as RAM-Index has been proposed for measuring the system performance by simultaneously considering all the three key indices which influence the system performance directly. Barabadi et al. (2011) studied the effect of time-dependent covariates on the analysis of maintainability performance. Garg and Sharma (2012) have discussed a two-phase approach for analyzing the reliability and maintainability analysis of the industrial system. The crankcase unit of the two wheeler manufacturing industry has been taken as an illustrative example and gave a recommendation to the system analyst.

2.2. Maintenance Optimization

Optimization model is a mathematical model that refers to choosing the best solution from all feasible solutions and it has been widely developed and used to find optimal preventive maintenance schedules for a variety of systems. For instance, Woodman (1964) gave a model for replacing the components when its age will exceed a control limit by using dynamic programming. Toshio and Shnji (1976) developed a method for evolving PM policies to maximize availability for 2-unit standby redundant system. Kay (1976) concluded that when the component operated in the increasing failure rate region the preventive replacement is one of the most appropriate maintenance strategies. He addressed the problem of PM schedules based on maximizing availability, minimizing maintenance cost and maximizing profit. Nagakava (1977) developed a model for imperfect maintenance in which the effective age of the system was reduced by a fraction of time due to each PM action. Frank (2001) discussed a policy for optimal scheduling replacement intervals of technical systems only on the basis of cost parameters. It was based on the assumption that a system is replaced by a new one as soon as the maintenance cost within a replacement cycle reaches or exceeds a given level. Nakamura (2001) discussed the application of maintenance scheduling for pump systems in thermal power stations in order to reduce the maintenance cost during the whole period of operation, while keeping the current reliability level of pump systems. Wang (2002) carried out an extensive survey in maintenance policies on deteriorating systems and gave an account of various maintenance and replacement problems of deteriorating systems have been studied in the literature. Bris et al. (2003) attempted to optimize the maintenance policy, for each component of the system, minimizing the cost function, with respect to the availability constraints using genetic constraints. Mariappan (2004) extended the model of Kay (1976) using 2 parameter Weibull distribution for determining the PM schedules. Samrout et al. (2005) made the claim of improving the results of Bris et al. (2003) by developing a new method based on Ant colony optimization. Lapa et al. (2006) presented a methodology for PM policy evaluation based upon a cost-reliability model using a genetic algorithm. Leou (2006) proposed a formulation considering both reliability and cost reduction for maintenance scheduling. The genetic algorithm combined with the simulated annealing method was adopted as a solution method.

Preventive maintenance scheduling has been proved as an NP hard combinatorial optimization problem in many situations and systems. Because of complexity of the optimization models developed for the preventive maintenance problem, meta-heuristic algorithms have been widely used in several research papers as a major optimization approach. Usher et al. (1998) present an optimization, maintenance and the replacement model for a single-component system. They determine an optimal preventive maintenance schedule for a new system subject to deterioration, by considering the time value of money in all future costs, increasing rate of occurrence of failure over time and the use of the improvement factor to provide for the case of imperfect maintenance actions. Levitin and Lisnianski (2000) present a research in which an optimization model was developed in order to determine the optimal replacement scheduling in multi-state series-parallel systems. They consider an increasing failure rate based on the expected number of failures during time intervals and define summation of maintenance activities cost along with cost of unsupplied demand due to failures of components as the objective function. Suresh and Kumarappan (2006) present a mathematical programming model to determine preventive maintenance schedules for power systems, using a combination of genetic algorithms and simulated annealing solution approaches. Verma and Ramesh (2007) apply a multi-objective initial preventive maintenance scheduling for large engineering plants. They model this problem as a constrained non-linear multi-objective mathematical program and use a genetic algorithm to solve the model. However, a number of papers (Dekker and Scarf, 1998; Hipkin and Lockett, 1995; Pierskalla and Voelkar, 1976; Valdez-Flores and Feldman, 1989) present overviews of the research work in the field of maintenance optimization models, assessing their impact and applications. The optimization of maintenance decision making can be defined as an attempt to resolve the conflicts of a decision situation in such a way that the variables under the control of the decision maker take their best possible value. Saraswat and Yadav (2008) reviewed the literature from 1988 to 2005 on RAMS engineering. One of the main controllable variable in their optimization is the time interval between maintenance. The preventive maintenance schedule is determined with mathematical formulation by setting one or more parameters like cost, reliability, availability, life cycle cost as the constraints (Chareonsuk et al., 1997; Gopalakrishnan et al., 2001; Ntuen,1991; Sachdeva et al., 2008; Schabe, 1995).

Advances on the field of study continue to be an important area of research and provides the analyst with new and improved methods for handling system evaluations. Under a repairable series-parallel system framework, availability is one of the most important measures than reliability for measuring the effectiveness of maintaining systems, because it includes reliability as well as maintainability. Generally, system availability can be improved either by incremental improvements of component availability or by provision of redundant components in parallel; both methods result in an increase in system cost. Various mathematical programming methods such as linear programming, homogenous linear programming, integer programming, dynamic programming, and nonlinear programming have been applied for solving such types of problems (Prasad and Kuo, 2000; Gen and Yun, 2006). These methods use gradient information to search the solution space near an initial starting point. In general, gradient-based methods converge faster and can obtain solutions with higher accuracy compared to stochastic approaches in fulfilling the local search task. However, for effective implementation of these methods, the variables and cost function of the generators need to be continuous. Furthermore, a good starting point is vital for these methods to be executed successfully. In many optimization problems, prohibited zones, side limits, and non-smooth or non-convex cost functions need to be considered. As a result, these non-convex optimization problems cannot be solved by the traditional mathematical programming methods. As an alternative to the conventional mathematical approaches, the meta-heuristic optimization techniques have been used to obtain global or near-global optimum solutions. Among them, genetic algorithms (GAs) (Goldberg, 1989; Holland, 1975), particle swarm optimization (PSO) (Kennedy & Eberhart, 1995; Eberhart & Kennedy, 1995), bibliography based optimization (BBO) (Simon, 2008) etc., are attractive because they do not apply mathematical assumptions to the optimization problems and have better global search abilities over conventional optimization algorithms. A lot of researchers have investigated the theoretical problems of availability modeling (Cao, 2002; Lee, 2000; Sherwin, 2000; Wood, 1994) and many others (Das et al. 1993; Garg & Sharma, 2012a; Grover, 1999; Lai et al. 2002; Ma et al. 2001; Robino & Sericola, 1995) have their studies on the specific applications. Based on these algorithms, various researchers have investigated the various tasks in different disciples. For instance, Garg and Sharma (2012) had discussed the two-phase approach for analyzing the reliability and maintainability analysis of the industrial system. The crankcase unit of the two wheeler manufacturing industry has been taken as an illustrative example and gave a recommendation to the system analyst. Garg and Sharma (2013) have investigated the multi-objective reliability-redundancy allocation problem of a repairable industrial system with PSO and GA. Garg et al. (2012) have solved the reliability optimization problem with ABC algorithm and compared their performance with other evolutionary algorithm. For more details about the application and methodology of evolutionary algorithm in the field of various optimization problems, we may refer (Wu et al., 2011; Herder et al., 2008; Racioppi et al., 2007; Zio et al., 2006; Garg & Rani, 2013; Garg et al., 2014a; Garg, 2013; Garg et al., 2013, Garg et al., 2013a; Gen & Yun, 2006; Ganesan et al., 2012a,c,b; Tsoulos & Vasant, 2009; Coelho et al., 2005; Garg & Sharma, 2013; Ganesan et al., 2012d; Garg, 2013a; Vasant et al., 2010; Garg & Sharma, 2012a).

2.3. Shortcoming and Overcoming of the Literature

Almost all the researchers kept in the view of the fact that whenever suitable or optimal preventive maintenance interval is reached, then the system is as good as the original one with the replaced new components. However the improvement of a system life of a system which consists of several sub-components not only depends on the replacement of deteriorated components, but also on the effectiveness of the maintenance that depends on both the improved levels and maintenance-costs of the sub-components. Taking care about this fact, the effects of maintenance of a multi-component system by combining the following three main different PM actions.

- **(1a)-maintenance (mechanical service):** This type of action emphasizes on maintaining a system on normal operating condition. It just improves the deteriorated environment so that it can tune the sub-components to a better condition by lubricating, tightening the loose parts, cleaning the dust, jam and rust, etc.
- **(1b)-maintenance (repair):** This type of action is mainly adopted for some sub - components which are expensive and/or acquiring them is not easy. It generally includes the activities of (1a) and repairing/replacing for some simple parts such as springs, seals, belts etc.
- **(2p)-maintenance (replacement):** This type of actions is to replace the sub-component/ subsystem with a new one. It is frequently adopted for the key sub-components to avoid serious damage occurred.

Thus, in the present chapter, we have proposed an algorithm to find the optimal maintenance scheduling for all components of the system which exhibit linearly increasing hazard rate and constant repair rate. For this, a bi-criteria optimization model has been formulated by considering the maintenance cost and availability as an objective function for finding the periodic interval of time between preventive maintenance actions for each component of the system. The model has optimized the performance with a natural restriction that corrective/establishment costs are greater than the maintenance cost and corrective availability is less than the preventive availability. The so obtained optimization model has been solved with the nature-inspired algorithm namely Biogeography based algorithm. Based on the optimal maintenance interval of each sub-component, the system interval is determined and the corresponding PM action (as defined by 1a, 1b, 2p) for individual sub-component is taken. The proposed approach has been demonstrated through the case study of the industrial system.

The rest of the chapter is summarized as follows. Section 2 presented the basic notations and assumptions that have used in the entire chapter. The procedure for maintenance planning and scheduling by considering the preventive cost and availability is discussed in section 3 along with the formulation and the procedure for taking maintenance-benefit analysis. Section 5 discuss an illustrative example to demonstrate the approach by taking the system from the repairable process industry. Finally, some concrete conclusion has been drawn in section 6.

3. NOTATIONS AND ASSUMPTION

3.1. Notations

The following notations have been used in the entire paper.

$f(\cdot)$ probability distribution function.
$F(\cdot)$ Commutative distribution function.
$R(\cdot)$ Reliability function.

$h(\cdot)$ Hazard rate function per hours.
M Mean Time Between Failure in hours.
T Scheduled period in hours.
m_c mean repair time for CM in hours.
m_p mean repair time for PM in hours.
c_c maintenance cost per unit time for CM in dollors/hours.
c_p maintenance cost per unit time for PM in dollors/hours.
C_c average maintenance cost rate in case of CM in dollors/hours.
C_p average maintenance cost rate in case of PM in dollors/hours.
$\bar{T} = \int_0^T R(x)\,dx$ is mean time between replacement(MTBR).
θ scale parameter for Weibull distribution in hours.
β shape parameter for Weibull distribution.

3.2. Assumptions

The following assumptions are made during the computation:

- The time-dependent failure, exponential repair time distributions and the degraded behaviors of the subsystems can be acquired.
- The dynamic behaviors of reliability about the PM components are given.
- The unpredictable failures of the system can be recovered by 1C-maintenance as well as by minimal repair.
- CM time is larger than the PM time.
- The improvement factors, the maintenance times and the related costs of the subsystems in maintenance can be identified.

4. MAINTENANCE PLANNING AND SCHEDULING

In real-life modeling, increasing the reliability or availability of the system will a great challenge to the system analyst, plant personnel or decision makers for increasing the performance and productivity of the system. For this, either increasing the reliability of each component of the system or to add a redundant components in parallel will increase the system reliability. But added a redundant component will increase the system cost, weight and volume also. Thus, there is a need of proper technique which will increase the system performance under some given resources. For this, preventive maintenance (PM) scheduling is one of the ways to achieve the system reliability target. These scheduling not only reduce the unexpected failures of systems, but also have a significant impact on the system life. But the main task in the PM scheduling is to determine the optimal PM interval which helps the DM to take the necessary action related to repair or replace the subsequent component of the system. On the other hand, plant personnel will always look for a solution of PM interval in such a way that their goals are reached up to a desired degree of accuracy, i.e., he wants to maximize the availability of the system and simultaneously minimize the maintenance cost of the system. Thus, it is necessary to construct an availability-cost optimization model based on PM scheduling which gives an idea to the system analyst or plant personnel for maintaining the system components in terms of repair, replacement, etc. The mathematical expression of these availability and cost under the effect of the PM scheduling is obtained as below.

4.1. PM Optimization Model

The availability of the system, in case of corrective maintenance, is given by

$$A_c = \frac{M}{M + m_c} \tag{1}$$

while, in case of preventive maintenance, it is given by

$$A_p = \frac{\bar{T}}{\bar{T} + m_p R(T) + m_c[1 - R(T)]} \tag{2}$$

If $\frac{\bar{T}}{M} = \alpha$, $\frac{m_c}{M} = \mu$, $\frac{m_p}{m_c} = \gamma$ then Eqs. (1) and (2) becomes

$$A_c = \frac{1}{1+\mu} \quad ; \quad A_p = \frac{\alpha}{\alpha + \mu[1 - k_1 R(T)]} \tag{3}$$

where $k_1 = 1 - \gamma$

The average maintenance cost in terms of corrective maintenance is (Kay, 1976)

$$C_c = \frac{c_c m_c}{M + m_c}, \text{ i.e. } C_c = \frac{c_c \mu}{1+\mu} \tag{4}$$

while, in case of preventive maintenance it is

$$\begin{aligned} C_p &= \frac{m_c c_c[1 - R(T)] + m_p c_p R(T)}{\bar{T} + m_p R(T) + m_c[1 - R(T)]} \\ &= \frac{\mu c_c[1 - (1 - \delta\gamma)R(T)]}{\alpha + \mu[1 - (1 - \gamma)R(T)]} \end{aligned} \tag{5}$$

where $\alpha = \frac{\bar{T}}{M} = \frac{1}{M}\int_0^T R(x)\,dx$, $\mu = \frac{m_c}{M}$, $\delta = \frac{c_p}{c_c}$ and $\gamma = \frac{m_p}{m_c}$.

Now, for PM action,

$$C_p - C_c < 0;\ A_p - A_c > 0 \tag{6}$$

Using the value of C_c and C_p from (4) and (5), the Eq.(6) becomes

$$\frac{\mu \cdot c_c[1-(1-\delta\gamma)R(T)]}{\alpha+\mu[1-(1-\gamma)R(T)]} < \frac{c_c \cdot \mu}{1+\mu} \text{ and } \frac{\alpha}{\alpha+\mu[1-k_1R(T)]} - \frac{1}{1+\mu} > 0$$

$$\Rightarrow \alpha > 1-k_2R(T) \text{ and } \alpha > 1-k_1R(T)$$

where $k_2' = (1-\delta\gamma)+\mu\gamma(1-\delta)$ and $\delta = \frac{c_p}{c_c}$ represents the maintenance cost ratio under PM and CM. Since the factor $\mu\gamma(1-\delta)$ is very less as compared to $(1-\delta\gamma)$, therefore k_2 can be approximated by

$$k_2 \cong (1-\delta\gamma)\,. \tag{7}$$

For the preference of PM over CM, the ratios δ and γ both need to be less than 1 and consequently

$$k_2 \geq 1-\gamma = k_1(say) \tag{8}$$

Thus the average maintenance cost in case of preventive maintenance, in view of (7) and (8) takes the form

$$C_p = c_c\left[\frac{1-k_2R(T)}{\frac{\alpha}{\mu}+(1-k_1R(T))}\right] \tag{9}$$

and hence PM schedule is optimized together with minimizing the maintenance cost per unit time i.e we minimize C_0 defined as below

$$C_0 = \frac{C_p}{c_c} = \frac{1-k_2R(T)}{\alpha/\mu + 1 - k_1R(T)} \tag{10}$$

Chitra (2003) discussed that life cycle cost (LCC) is another constraint, including in the model which is given as

$$LCC = n_p c_p + n_c c_c \tag{11}$$

where n_p and n_c are respectively the number of components replaced through preventive maintenance and corrective maintenance during a sufficiently long operating time T_0, then the total number of the component replaced is given by

$$N = n_p + n_c \tag{12}$$

Let MTBR represent Mean Time Between Replacements. Since the total operating time .. is very large compared to MTBR, we have

$$N = \frac{T_0}{MTBR} \tag{13}$$

Let $R(x)$ represent the reliability of the original component which is automatically replaced after a time T. Then

$$MTBR = \int_0^{\infty} \tilde{R}(x)\ dx \tag{14}$$

where $\tilde{R}(x) = R(x),\ \ for\ \ x \le T$ and zero otherwise.

The fraction of the components that survive until the next preventive maintenance is given by $R(T)$ then the number of surviving and failed component that are replaced through preventive and corrective maintenance are given by

$$n_p = R(T) \cdot N, \tag{15}$$

and

$$n_c = [1 - R(T)] \cdot N \tag{16}$$

Consequently, Eq. (11) takes the form

$$LCC = c_p \frac{R(T) \cdot T_0}{\int_0^T R(x)\ dx} + c_c \frac{[1 - R(T)] \cdot N}{\int_0^T R(x)\ dx} \tag{17}$$

The value of T corresponding to minimum cost is obtained by,

$$\frac{d}{dt}(LCC) = 0$$

$$\Rightarrow \frac{T_0\left[\int_0^T R(x)\,dxR'(T)\,(c_p - c_c) - R^2(T)(c_p - c_c) - R(T)\,c_c\right]}{\left[\int_0^T R(x)\,dx\right]^2} = 0$$

$$\Rightarrow R(T) + h(T)\int_0^T R(x)\,dx = \frac{c_c}{c_c - c_p}$$

$$\Rightarrow R(T) + h(T)\int_0^T R(x)\,dx = \frac{1}{1-\delta} \qquad (18)$$

The equation (18) gives the optimum value of T_0 at which the value of life cycle cost become minimum. So for using this criteria as a constraint for obtaining PM schedule along with the objective it should be ensured that the value of T (PM schedule) does not exceed the minimum life cycle cost (i.e. $T \le T_0$). For this constraint, equation (18) will convert into an inequality

$$R(T) + h(T)\int_0^T R(x)\,dx \le \frac{1}{1-\delta} \qquad (19)$$

As during the maintenance scheduling, the important factor is to find the optimal maintenance interval. So, the PM schedule need to optimized based on the minimizing the availability and cost of the maintenance. For this an optimization model has been constructed by taking both the preventive maintenance cost and availability as an objective function with the natural restriction that corrective/establishment costs are greater than the maintenance cost and corrective availability is less than the preventive availability i.e.,

$$Max \ \frac{A_p}{C_0}$$

$$\text{where} \quad C_0 = \frac{1-k_2R(T)}{\alpha/\mu + 1 - k_1R(T)} \ ; \ A_p = \frac{\alpha}{\alpha + \mu[1-k_1R(T)]}$$

$$\begin{aligned} \text{subject to} \quad & \alpha > 1 - k_2R(T) \\ & \alpha > 1 - k_1R(T) \\ & 0 \le \alpha \le 1 \\ & A_p \ge \min A_{pi} \\ & R(T) + h(T)\int_0^T R(x)\,dx \le \frac{1}{1-\delta} \end{aligned} \qquad (20)$$

Since the problem is nonlinear in nature, so it requires an efficient technique to solve this problem. A variety of methods and algorithms exist for optimization of such problems and applied in various technological fields during the last decades. Evolutionary algorithms (EAs) are found to be very promising global optimizers because it does not require the derivative of the nonlinear constraints. Among the existing algorithm, EAs, BBO probabilistically share information between candidate solutions. Biogeography not only gives a description of species distributions, but also a geographical explanation. Biogeography is modeled in terms of such factors as habitat area and immigration rate and emigration rate, and describes the evolution, extinction and migration of species. After tests on many benchmarks, and comparisons with many other widely used heuristic algorithms like GAs, PSO and others, BBO outperformed most of the other algorithms on most of the benchmarks (Gong et al., 2010; Simon et al., 2011; Ma & Simon, 2011). This shows that BBO is an algorithm that has much promise and merits further development and investigation. BBO has certain features in common with other population-based optimization methods like GAs and PSO, BBO can share information between solutions. This makes BBO applicable to many of the same types of problems that GAs and PSO are used for, including unimodal, multimodal and deceptive functions. In BBO and PSO, each solution stays survive to the end of optimization procedure, but in most of evolutionary based algorithms, solutions die at the end of each generation. However, BBO also has some unique features that clearly differ from other population-based optimization methods. In some of evolutionary due to crossover step, good solutions lose their efficiency, but in BBO do not have crossover step. One more characteristic of BBO is that its set of solutions is maintained and improved from one generation to the next by migrating. Also, for each generation, BBO uses the fitness of each solution to determine its emigration and immigration rate. Motivated by this, the present study uses BBO as a tool for finding the optimal PM interval. The description of the BBO algorithm is the next section.

4.2. Reliability Under PM

In order to improve the performance and hence reliability of the system, then the necessary PM actions is to be identified before scheduling the PM program. For this an improvement factor has been introduced in the maintenance scheduling program. The advantage of this factor is to provide the maintenance action in the system other than replacing the system with the new one and hence the corresponding reliability of the system on the jth PM stage, is based on the initial reliability $(R_{0,j})$ of the jth stage along with the percent of survival parts of the system $(R_{v,j}(t))$ when it is maintained, and is defined as

$$R_j(t) = R_{0,j} R_{v,j}(t) \tag{21}$$

where

$$R_{o,j} = \begin{cases} R_{o,j-1} \cdot R(T^*) & ; \text{if (1a)} - \text{maintenace} \\ R_{f,j-1} + m_2(R_0 - R_{f,j-1}) & ; \text{if (1b)} - \text{maintenace} \\ R_0 & ; \text{if (2p)} - \text{maintenace} \end{cases} \tag{22}$$

where R_0 denotes the initial reliability of the new system, $R_{0,j-1}$ and $R_{f,j-1}$ indicate the initial and final reliabilities of systems on the $(j-1)th$ stage; and

$$R_{v,j}(t) = R\left(\frac{t-(j-1)T^*}{m_1}\right); \quad (j-1)T^* \le t \le jT^* \tag{23}$$

where T^* is the periodical preventive maintenance interval, m_1 $(0 \leq m_1 \leq 1)$ is the improvement factor of (1a)-maintenance which can be defined as the ratio of the life of the surviving part to the original life. In particular, $m_1 = 1$ if the survival parts are restored to as good as its original condition.

As the optimal replacement time (T^*) of each component of a system have been derived individually after solving the Eq.(20) and hence based on that, if sub-components of the system have been replaced then due to frequent system-down occurrence, the availability of the system would decrease rapidly. To avoid this, the system replacement time has been computed by taking minimum among them i.e. system PM interval, $T = Min(T^*)$. On the other hand, simultaneously (1a) and (1b) PM actions have been taken for those components whose maintenance time is greater than T i.e. $T^* > T$. Finally the subcomponent of the system has been maintained at the stage when its reliability at the next stage i is less than the already set minimum reliability requirement, i.e. $R(2T) < R_{\min}$ and their corresponding PM actions to be taken are decided on the basis of maintenance-benefit of any component on the jth stage and is defined in Tsai et al. (2004) is given by

$$B_{i,k} = \frac{\int_{t_j}^{\infty} R_{i,j+1}(x)\ dx - \int_{t_j}^{\infty} R(x)\ dx}{C_{i,k}} \tag{24}$$

where the subscripts i and $k = (1, 2, 3)$ denote the ith component and the three actions, respectively. The numerator indicates the extended life of ith component by action k whereas the denominator is the corresponding maintenance cost. The action which leads to the maximum maintenance benefit, i.e. $B_i^* = \max(B_{i,k})$, would be selected for the component. Normally, the system availability will gradually decline following the PM times. The PM scheduling would be terminated if the system life reaches to the expected life.

5. BIOGEOGRAPHY BASED OPTIMIZATION

Biogeography-based optimization (BBO), proposed by Simon (2008), is a stochastic optimization technique for solving multimodal optimization problems. It is based on the concept of biogeography, which deals with the distribution of species that depend on different factors such as rainfall, diversity of topographic features, temperature, land area etc. Biogeography describes how species migrate one island or habitat to another, how new species arise and how species become extinct. A habitat is an island which is geographically isolated from other habitats. The habitat having high suitability index (HSI) means that the habitat is geographically well suited for species living. The species have more possibility to migrate into the neighboring habitat when (i) the high HSI habitats have more species (ii) habitat has low HSI. This process is named as emigration. In immigration, the species move towards the high HSI habitat having few species. The immigration and emigration of species in a habitat are known as migration. The migration of species in a single habitat is explained in Fig. 1. From this figure, it has been seen that when there is no species in a single habitat, then the immigration rate (λ) of the habitat will be maximized (I). On the other hand, when the number of species increases, then only a few can successfully survive. Therefore, the immigration rate is decreased and become zero at $S = S_{\max}$. Similarly, considering the emigration curve in the figure, the emigration rate (μ) increases with the increases of the species and hence the emigration rate is higher for the more crowded habitat than the less populated

Figure 1. Species model of an island

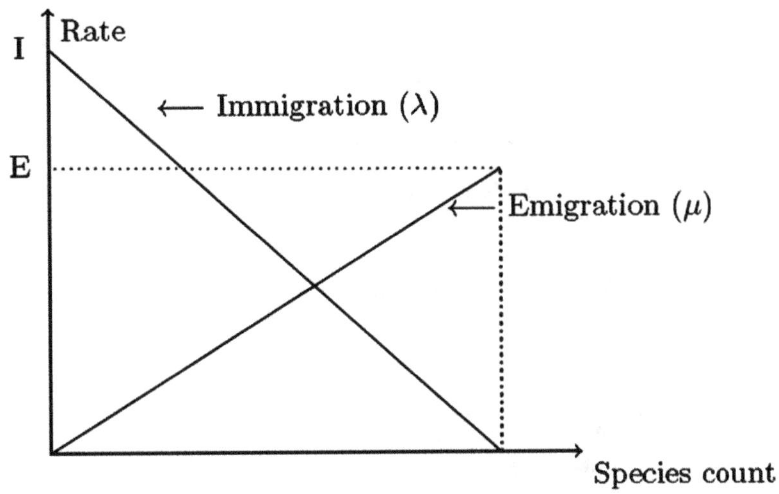

habitat. The optimization steps of the BBO algorithm are described as follows while the flow chart of the proposed approach is described completely in Fig. 2.

5.1. Initializing the BBO Parameters and Habitats

The optimization problem, in general, is expressed as

$$\begin{aligned} &Max\,/\,Min \quad f(x) \\ &\quad s.t. \quad g(x) \le 0 \\ &\quad\quad x_i^{\min} \le x_i \le x_i^{\max} \quad ; \; i = 1, 2, \ldots, n \end{aligned}$$

where $f(x)$ is the objective function, $g(x)$ is the set of all constraints, $x = (x_1, x_2, \ldots, x_n)$ be the decision variables; $x_i^{\min}$ and $x_i^{\max}$ be the lower and upper limit of the variable x_i for $i = 1, 2, \ldots, n$. The initialization phase of the BBO algorithm is to set the parameter of BBO algorithm. The parameters are maximum species count $(S_{\max})$, maximum migration rate (E), immigration rate (I), the maximum mutation rate $(m_{\max})$, habitat modification probability (p^{mod}), the number of habitat and the number of iterations. In BBO algorithm, each individual or decision variable is considered as a "habitat" with a habitat suitability index (HSI) to measure the individual where length of the habitat is depending on the number of decision variables. This HSI measures the goodness of the solution and is same as that of fitness value in population-based optimization algorithms. The corresponding variables of individuals that characterize habitability feature is called as suitability index variables (SIVs). These initialized randomly within the maximum and minimum limits as $x_i^{\min} + Ud(x_i^{\max} - x_i^{\min})$ where Ud is the uniformly distributed random number lies between (0, 1).

Figure 2. Flow chart of proposed approach

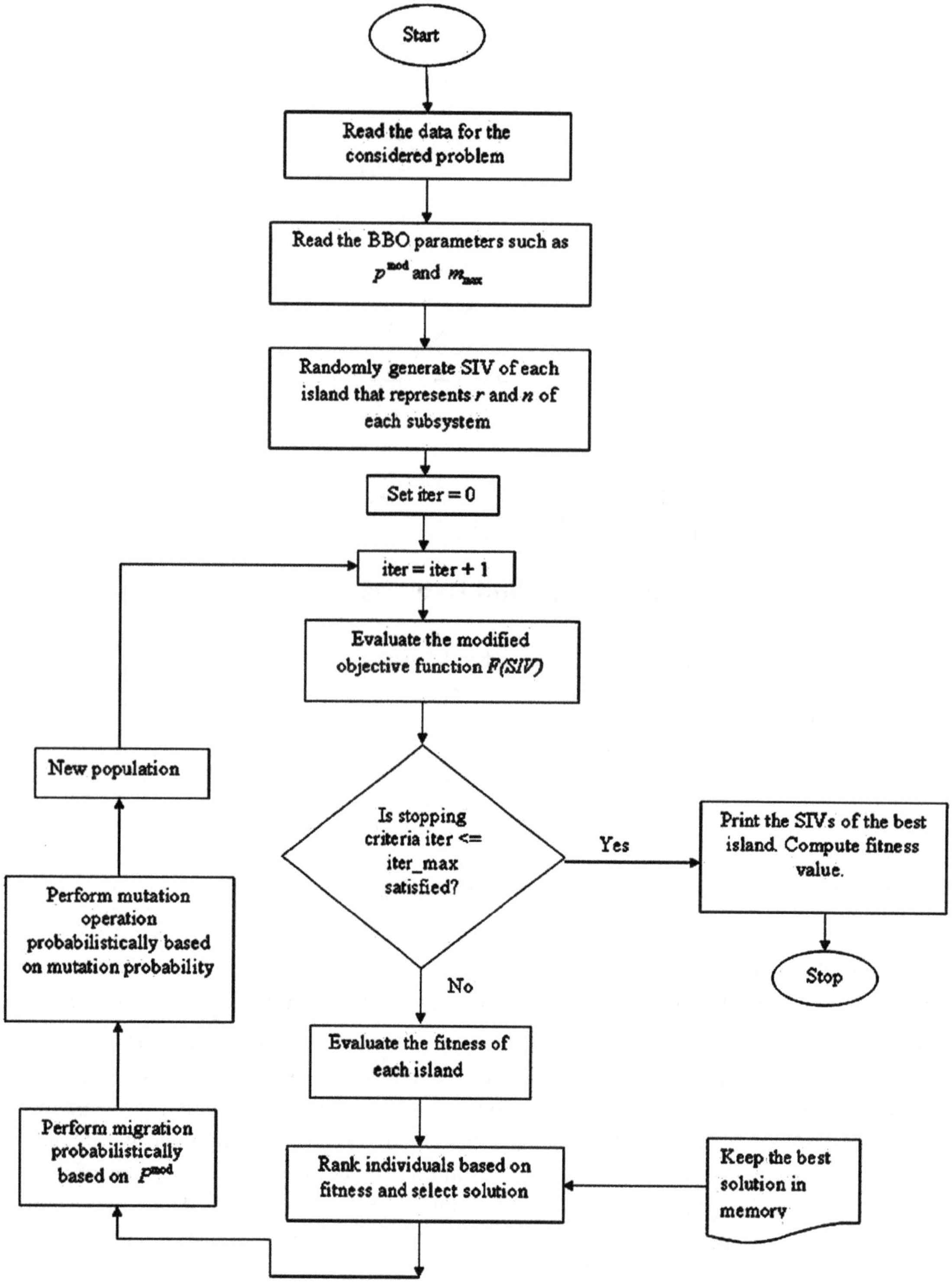

5.2. Perform BBO Operators

There are mainly two main operators: migration (which include both emigration and immigration) and mutation. Migration is the adaptive process which is used to change the existing habitat's SIV. Each individual has its own immigration rate λ and emigration rate μ for improving the solution and thus evolve a solution to the optimization problem and is modified based on its probability p^{mod}, that is user defined parameter. Based on this probability, its immigration rate (λ) of each habitat is deciding whether to modify each SIV's or not. Another habitat is selected based on emigration rate μ, its SIV will migrate randomly to the selected habitat's SIV. A good solution is a habitat with relatively high μ and low λ, while the converse is true for a poor solution. High HSI habitat resists change more the low HSI habitat. Poor solutions accept more useful information from good solution, which improve the exploitation ability of the algorithm. At initial, the immigration and emigration rates are the functions of the number of species in the habitat and they are evaluated as

$$\lambda_k = I\left(1 - \frac{k}{S_{\max}}\right); \ \mu_k = E\left(\frac{k}{S_{\max}}\right)$$

where I is the maximum possible immigration rate; E is the maximum possible emigration rate; k is the number of the species of the kth individual; and $S_{\max}$ is the maximum number of species. I and E are often set equal to 1, or slightly less than 1.

On the other hand, in BBO, mutation is used to increase the diversity of the population to get the best solution. The low and high HSI valued habitats have less possibility to mutate while average HSI valued habitats have more possibility to mutate. For instance, consider there are a total of $S_{\max}$ habitat and the Sth habitat, there are S species. For the Sth island, the immigration rate is λ_s and emigration rate is μ_s. By considering P_s as the probability that there are exactly S species live in a habitat so P_s changes from time t to $t+\Delta t$ is given as

$$P_S(t+\Delta t) = P_s(t)(1-\lambda_s\Delta t - \mu_s\Delta t) + P_{S-1}(t)\lambda_s\Delta t + P_{S+1}(t)\mu_{k+1}\Delta t \tag{25}$$

where P_{S-1}, P_S and P_{S+1} are the species count probability; λ_{S-1}, λ_S, λ_{S+1} are immigration rate and μ_{S-1}, μ_S, μ_{S+1} are the emigration rate of the habitat with $S-1$, S and $S+1$ species respectively, $S_{\max}$ is the maximum species count in the habitat. The steady state probability of the quantity of each species is given as below

$$P_0 = \left[1 + \sum_{j=1}^{S_{\max}} \prod_{i=1}^{l} \frac{\lambda_{i-1}}{\mu_i}\right]^{-1} \tag{26}$$

$$P_k = P_0 \prod_{i=1}^{k} \frac{\lambda_{i-1}}{\mu_i} = \frac{1}{1 + \sum_{j=1}^{S_{\max}} \prod_{i=j}^{l} \frac{\mu_i}{\lambda_{i-1}} + \sum_{l=k+1}^{S_{\max}} \prod_{i=k+1}^{l} \frac{\lambda_{i-1}}{\mu_i}}, \; k \in \{1, 2, \ldots, S_{\max}\} \tag{27}$$

Thus, based on that the randomly selected SIV habitat modified it based on mutation rate m, in the case of $E = I$, and is expressed as

$$m(S) = m_{\max} \times \left(1 - \frac{P_S}{p_{\max}}\right)$$

where $m_{\max} \in [0,1]$ is the predefined parameter and $p_{\max} = \max(P_1, P_2, \ldots, P_n)$, n is the number of habitats, P_S is the probability of habitat S which can be calculated by Eq. (26) and (27). This mutation scheme tends to enhance the diversity among the population, which helps to decrease the chance of getting trapped in local optima.

5.3. Update the Habitat

The generation process is repeated until a number of generations has been met, as a termination criteria.

6. AN ILLUSTRATIVE EXAMPLE

The above approach has been illustrated through a case study of the process industry, namely as a paper mill situated in the northern part of India and producing approximately 200 tons of paper per day. The paper mills are large capital oriented engineering systems, comprising of units/subsystems, namely, feeding, pulping, washing, screening, bleaching, forming, dryer and press, arranged in a predefined configuration (Garg et al., 2013; Garg, 2013). The present analysis is based on the study of one of the important unit, i.e., washing unit. The brief description of this unit is as follows.

6.1. System Description

The Washing of prepared pulp is done in three to four stages, to get it free from blackness and to prepare the fine fibers of the pulp. The washing system consists of four main subsystems, defined as (Garg, 2013):

- **Filter (A):** It consists of single unit which is used to drain black liqueur from the cooked pulp.
- **Cleaners (B):** In this subsystem three units of cleaners are arranged in parallel configuration. Each unit may be used to clean the pulp by centrifugal action. Failure of anyone will reduce the efficiency of the system as well as quality of paper.

- **Screeners (C):** Herein two units of screeners are arranged in series. These are used to remove oversized, uncooked and odd shaped fibers from pulp through straining action. Failure of any one will cause the complete failure of the system.
- **Deckers (D):** Two units of deckers are arranged in parallel configuration. The function of deckers is to reduce the blackness of pulp. Complete failure of decker occurs when both the components will fail.

The system diagram for representing the various components of the system is given in Fig. 3. As all the units of the system are arranged in series and hence system reliability is simplified as

$$R_s(t) = R_C(t)\prod_{u=1}^{4} R_u(t)$$

Here, $u(=1,2,3,4)$ corresponding to the filter, cleaners, screeners and decker units and $R_C(t)$ is the reliability of the surplus part of the system excluding the subcomponents. The surplus part is main-

Figure 3. Systematic diagram of the system

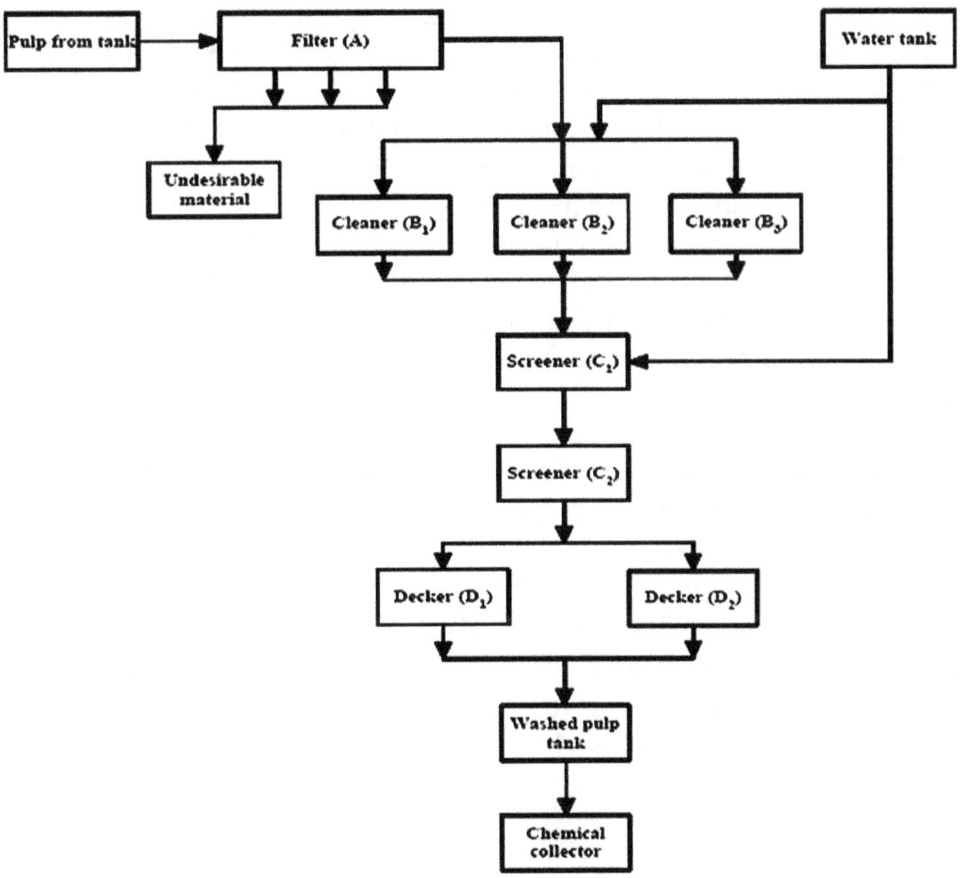

Table 1. Input data for the Washing System

Components	Failure Data		Repair Time
	Weibull Distribution		
	Scale (θ) hrs	Shape (β)	m_c hrs
Filter $(i = 1)$	337	1.33	3.5
Cleaners $(i = 2, 3, 4)$	170	1.88	2.0
Screeners $(i = 5, 6)$	426	1.46	4.0
Deckers $(i = 7, 8)$	252	1.76	2.5

tained with 1C and is modeled with an exponential function which possesses constant failure rate i.e. $R_C(t) = \exp(-\lambda t)$. Here, it is set to $\lambda = 0.0002$.

6.2. Information Collection Phase

As the system goes failure after a certain interval of time and hence the corresponding failure-repair data of the major components of the system are obtained from the historical records/logbook etc. In this study, Weibull distribution and exponential distribution has been used for representing the equipment failure and repair time respectively. It is an established fact that constant failure data i.e. which follow an exponential distribution is not useful for establishing the PM actions and hence the components having shape factor sufficiently large than one (parameter of the Weibull distribution) have been taken for the further analysis i.e. which have more deterministic deteriorating characteristics. Hence the input data related to shape (β) and scale (θ) parameters of the Weibull distribution and mean time to repair (m_c) of exponential distribution for the entire major components of the system are given in Table 1.

6.3. Analysis of the Results

In order to analyze the performance of the system in terms of its PM interval, the optimization model (20) has been formulated for the considered system and hence the optimal design PM interval corresponding to each component of the system are obtained by solving the problem with Biogeography based optimization (BBO) technique. The following BBO parameters have been used for evaluating the performance of the system: habitat modification probability $p^{\text{mod}} = 0.8$, maximum immigration and migration rates of each island is 1 and mutation probability $m_{\max}$ is 0.2. Habitat size is taken as 50. Based on these parameters, the results' corresponding to it has been summarized in Table 2 associated with their component availability and cost. From this table it has been observed that

Table 2. Decision parameters and optimal PM schedule for Washing unit

Units	Decision Parameters	MTTF (hrs)	Cost		Availability	
			Without PM	With PM	Without PM	With PM
Filter	$\theta = 337;\ \beta = 1.33;$ $\gamma = 0.3; \mu = 0.011;$ $\delta = 0.45$	309.86	0.010880	BBO: 0.010087 PSO: 0.010090	0.989119	BBO: 0.989120 PSO: 0.989123
Cleaner	$\theta = 170; \beta = 1.88;$ $\gamma = 0.35; \mu = 0.013;$ $\delta = 0.4$	223.72	0.012833	BBO: 0.004723 PSO: 0.004725	0.987166	BBO: 0.989931 PSO: 0.989845
Screener	$\theta = 426; \beta = 1.46;$ $\gamma = 0.45; \mu = 0.01;$ $\delta = 0.6$	240.04	0.009901	BBO: 0.009615 PSO: 0.009618	0.990099	BBO: 0.990099 PSO: 0.990100
Decker	$\theta = 252; \beta = 1.79;$ $\gamma = 0.4; \mu = 0.016;$ $\delta = 0.4$	297.40	0.015748	BBO: 0.008043 PSO: 0.008049	0.984251	BBO: 0.984481 PSO: 0.984122

- The PM interval for subcomponents components, namely filter, cleaner, screener and decker are $T^* = \{358.57,\ 109.44,\ 413.44,\ 139.53\}$ respectively by BBO while $\{360.31,\ 107.54,\ 415.65,\ 134.71\}$ respectively by PSO algorithm.
- The cost of the each component under PM and without PM action have been computed and observed that the cost of the system components are decreases after PM scheduling and decrease in cost (in %) from CM to PM are 7.288%, 63.17%, 2.97% and 48.92% respectively by BBO while 7.253%, 63.182%, 2.856% and 48..884% by PSO respectively to the filter, cleaner, screener and decker units. On the other hand, the overall availability of the system and each component availability is increased after PM scheduling.
- Thus, it will suggest the system analysts and show that simultaneously considering (1a), (1b) and (2p) can efficiently reduce the maintenance cost and increase system availability so that it is more advantageous than only replacement (2p-maintenance) considered.

6.4. PM Scheduling Actions

In order to find the maintenance action for improving the performance of the system, maintenance-benefit analysis has been conducted by using Eq. (24). For this, the cost of (1a)-maintenance is set to be 30%, while the cost of (1b)- maintenance is set to be 60% of (2p)-maintenance. On the other hand, the expected life of the system is set to be 700 hrs and the initial reliability of all the subcomponents are set to be 0.9999. Therefore, for judging whether the maintenance action is taken or not on the component of the system, the predetermined reliability i.e., minimum reliability corresponding to each component of the system are set to be {0.80, 0.80, 0.65, 0.75} respectively, while the improvement factor of each component on (1a)- and (1b)- maintenance are set to be $m_1 = m_2 = \{0.82, 0.7, 0.9, 0.8\}$. If each sub-component of the system has been replaced directly on the PM intervals then due to frequent system-down occurrence, the availability of the system would decrease rapidly. To avoid this, the system replacement time has been computed by taking minimum among them i.e. PM interval for the whole system is 109 hrs.

Figure 4. Effect of (1a), (1b) and (2p)- maintenance of each component and entire system

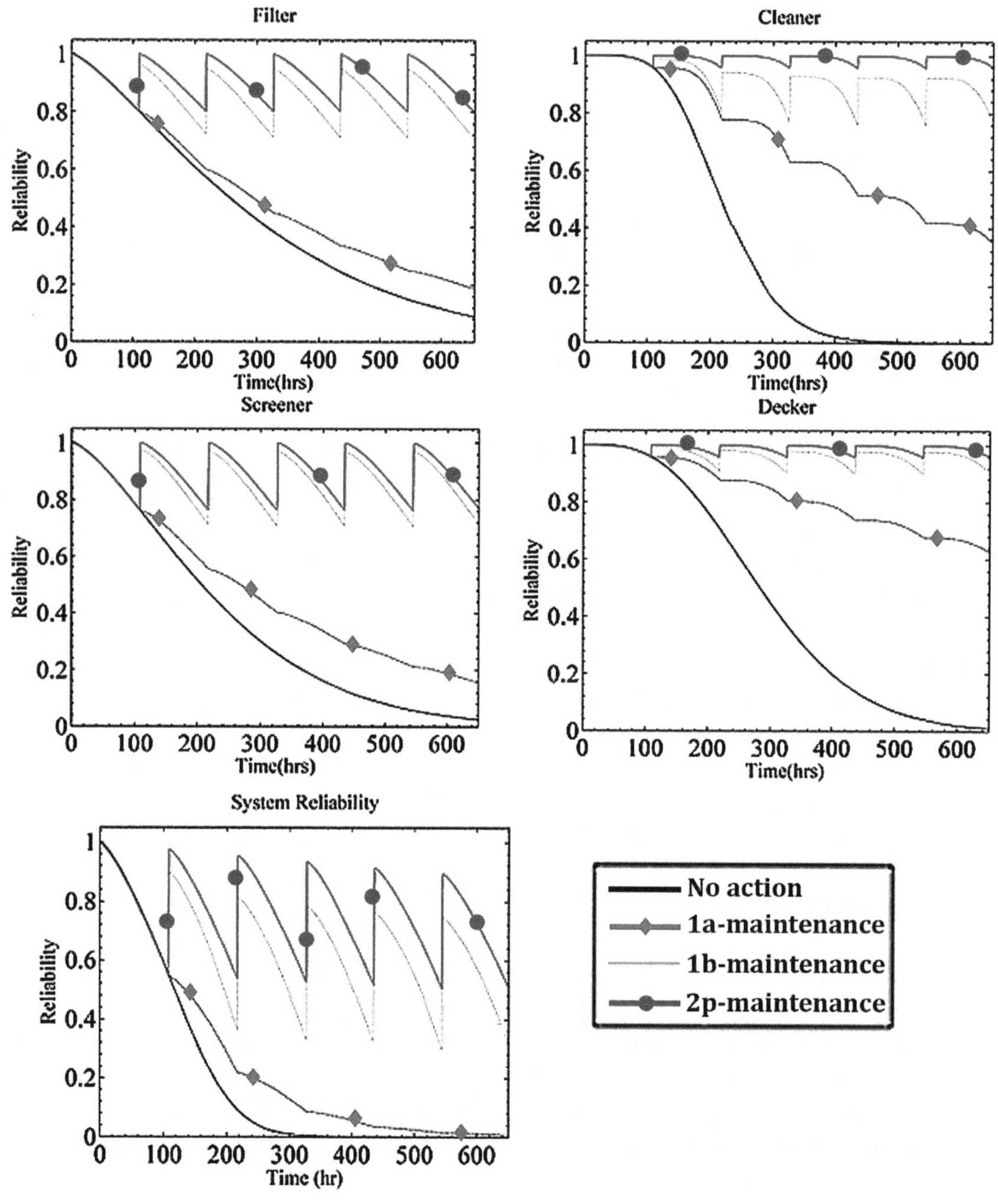

The effect of each maintenance action on the performance of the entire system as well as on each component of the system is shown in Fig. 4. From this figure, it has been clearly seen that the reliability of the system is gradually decreased when no PM action has been taken while the variations in the form of (1a), (1b) and (2p)-maintenance actions play a significant effect on the system performance. Thus the maintenance should be done at a regular interval of time instead of replacement of the system i.e. (2p)-maintenance. This complete analysis will provide the performance of each component to the system analyst so that the decision maker or plant personnel may depict these effects and then adopt the suitable maintenance action in a regular interval of time for increasing the performance and productivity of the system.

On the other hand, the behavior of the each component after taking the effect of (1a)-, (1b) and (2p)-maintenance along with the system performance are depicted graphically in Fig. 5. From this figure, it has been concluded that the maintenance actions for improving the performance of system should be undergo (2p)-maintenance for the first component, i.e, filter and (1a)-maintenance for cleaner, no-maintenance action for screener and (1a) for decker. Similarly, at the second stage of the system, i.e. when $j = 2$ or after $T = 218$ hrs, system analysts would perform (3, 3, 0, 1) maintenance action respectively on the subsequent components before reaching the expecting life of the system as compared to the CM maintenance. The behavior of the component reliability along with the system reliability at different PM interval is depicted graphically in Fig. 3 corresponding to the entire system as well as the each component. The complete maintenance action schedule corresponding to each component at different stages of the interval is computed and summarized in Table 3. Here in this table, 0 represents for the no action while 1, 2 and 3 respectively, represent for the (1a)-, (1b)- and (2p)- maintenance type actions. Complete information will not only increase the performance of the system, but also decreases the maintenance cost in a regular interval of time and thus beneficial for the system analyst.

Table 3. PM schedule of the system

Stage	Time (hrs)	PM actions			
$j = 1$	109	3	1	0	1
$j = 2$	218	3	3	0	1
$j = 3$	327	3	1	0	1
$j = 4$	436	3	3	0	3
$j = 5$	545	3	1	0	1
$j = 6$	654	3	3	0	1

Figure 5. Behavior of each component after taking the effect of (1a), (1b) and (2p)- maintenance

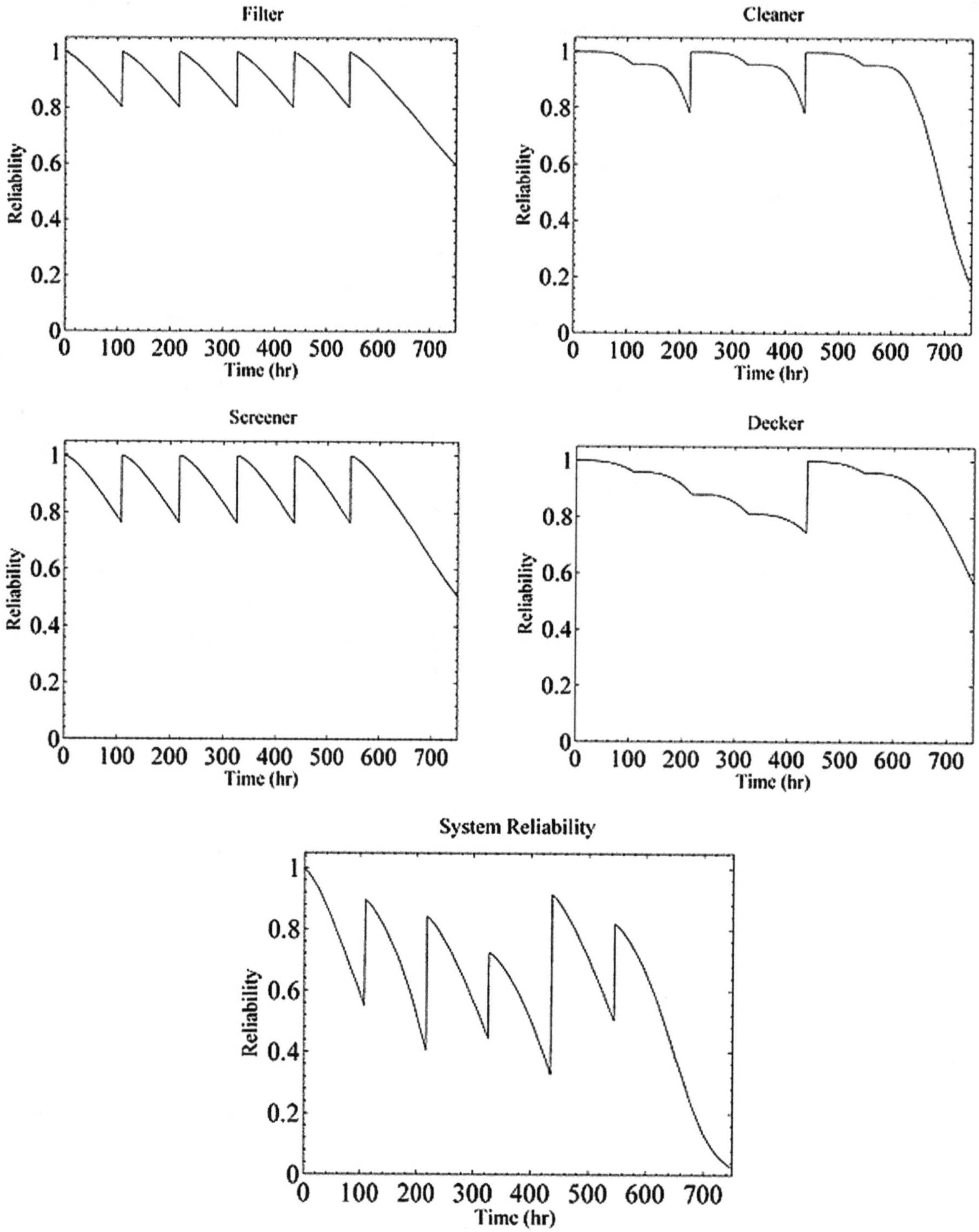

7. CONCLUSION

The present chapter studies the preventive maintenance in simultaneously considering three actions, mechanical service (1a), repair (1b) and replacement (2p) for a complex repairable industrial system. During the formulation of the model, bi-criteria optimization model has been considered by taking the preventive maintenance costs as well as preventive availability is taken as a objective function. Since these functions are highly nonlinear, so it is sufficient to require a suitable algorithm which will optimize the performance of these systems in terms of its objective functions. However, the traditional or heuristic algorithms require derivative for all the non-linear constraint functions to evaluate the system performance. But in real-life problem, it is difficult to derive it due to the highly computational complexity of the system. Thus to overcome this, nature inspired algorithm are the latest state of art algorithms \& works well with optimization problems as well as other problems than the classical methods because classical methods are inflexible in nature. Therefore, Biogeography based optimization algorithm has been used for obtaining the optimal preventive interval for maintaining the performance of the system by solving the corresponding optimization problem. The minimum among the intervals is chosen as the PM interval of the system for determining the periodical PM policy. The degradation of components is analyzed from its failure mechanisms and the improvements of various actions to it in reliability are measured by using two improved factors. While scheduling the PM program, the components, whether maintained or not was settled depending on reliability check, and the action option for them were decided by maintenance-benefit analyzing until the extended life of the system meets the expected system life. Based on the PM scheduling action recommended, the plant personnel may be able to plan the future maintenance of the system accordingly. On the other hand, the simultaneously considering the (1a), (1b) and (2p)-maintenance actions will be more beneficial than only considering the replacement action, i.e. (2p) in terms of reducing the maintenance cost and increasing the availability of the system. Thus, it will facilitate the management in reallocating the resources, making maintenance decisions, achieving long run availability of the system, and enhancing the overall productivity of the system.

8. FUTURE RESEARCH DIRECTIONS

Although, we have focused on the series–parallel systems in this chapter, the general idea presented here could also be applicable to many other systems like parallel-system, k-out-of- n system, multi-state system, circular structure systems and so on. Moreover, the approach has been extended for the repairable multi-state system by considering the other parameters which affect the system performance directly. The presented methodology will be further extended and improved using other optimization tools/algorithms and artificial neural network will be used to handle the complex nature of the systems. Moreover, we can extend this one by taking different types of fuzzy membership functions. The work can also be extended to devise a suitable methodology for

1. Conducting cost analysis.
2. Developing inventory and spare parts, maintenance management system.
3. Redundancy allocation problem.
4. suitable maintenance index after understanding the behavior dynamics associated with the functioning of the system.

Also, authors are engaged in the approach for further analysis such as resource allocation, facility planning and management, inventory control, network analysis, and job shop scheduling for an arbitrary failure rate model instead of constant failure rate model.

REFERENCES

Barabadi, A., Barabady, J., & Markeset, T. (2011). Maintainability analysis considering time-dependent and time-independent covariates. *Reliability Engineering & System Safety*, *96*(1), 210–217. doi:10.1016/j.ress.2010.08.007

Barabady, J., & Kumar, U. (2005). *Maintenance schedule by using reliability analysis: A case study at jajarm bauxite mine of Iran. 20th World Mining Congress and EXPO*, Tehran, Iran.

Barabady, J., & Kumar, U. (2008). Reliability analysis of mining equipment: A case study of a crushing plant at jajarm bauxite mine in iran. *Reliability Engineering & System Safety*, *93*(4), 647–653. doi:10.1016/j.ress.2007.10.006

Bris, R., Chatelet, E., & Yalaoui, F. (2003). New method to minimize the preventive maintenance cost of series-parallel systems. *Reliability Engineering & System Safety*, *82*(3), 247–255. doi:10.1016/S0951-8320(03)00166-2

Chitra, T. (2003). Life based maintenance policy for minimum cost. In *Proceedings of the annual reliability and maintainability symposium*, (pp. 470 – 474). doi:10.1109/RAMS.2003.1182034

Dhudshia, V. H. (1997). SEMI E10 – equipment reliability, availability and maintainability. *Semiconductor International*, *20*(6), 167–174.

DuJulio, E. T., & Leet, J. H. (1988). Space station synergetic RAM - logistic analysis. In *Proceedings of Annual Reliability and Maintainability Symposium*. IEEE.

Ganesan, T., Vasant, P., & Elamvazuthi, I. (2012c). A hybrid PSO approach for solving non-convex optimization problems. *Archives of Control Sciences*, *22*(1), 87–105. doi:10.2478/v10170-011-0014-2

Garg, H. (2013). Reliability analysis of repairable systems using Petri nets and Vague Lambda-Tau methodology. *ISA Transactions*, *52*(1), 6–18. doi:10.1016/j.isatra.2012.06.009 PMID:22789401

Garg, H., & Rani, M. (2013). An approach for reliability analysis of industrial systems using PSO and IFS technique. *ISA Transactions*, *52*(6), 701–710. doi:10.1016/j.isatra.2013.06.010 PMID:23867122

Garg, H., Rani, M., & Sharma, S. P. (2013). Preventive maintenance scheduling of the pulping unit in a paper plant. *Japan Journal of Industrial and Applied Mathematics*, *30*(2), 397–414. doi:10.1007/s13160-012-0099-4

Garg, H., & Sharma, S. P. (2012). A two-phase approach for reliability and maintainability analysis of an industrial system. *International Journal of Reliability Quality and Safety Engineering*, *19*(3), 1250013. doi:10.1142/S0218539312500131

Garg, H., & Sharma, S. P. (2013). Multi-objective reliability-redundancy allocation problem using particle swarm optimization. *Computers & Industrial Engineering*, *64*(1), 247–255. doi:10.1016/j.cie.2012.09.015

Garg, H., Sharma, S. P., & Rani, M. (2012). Cost minimization of washing unit in a paper mill using artificial bee colony technique. *International Journal of System Assurance Engineering and Management*, *3*(4), 371–381. doi:10.1007/s13198-012-0128-3

Gertsbakh, I. B. (1977). *Models of Preventive Maintenance*. New York: North-Holland Publishing Company.

Gong, W., Cai, Z., Ling, C., & Li, H. (2010). A real-coded biogeography-based optimization with mutation. *Applied Mathematics and Computation*, *216*(9), 2749–2758. doi:10.1016/j.amc.2010.03.123

Gopalakrishnan, M., Mohan, S., & He, Z. (2001). A tabu search heuristic for preventive maintenance scheduling. *Computers & Industrial Engineering*, *40*(1-2), 149–160. doi:10.1016/S0360-8352(01)00014-6

Grover, W. D. (1999). High availability path design in ring-based optical networks. *IEEE/ACM Transactions on Networking*, *7*(4), 558–574. doi:10.1109/90.793028

Hansen, W. A., Edson, B. N., & Larter, P. C. (1992). Reliability, availability and maintainability expert system. In *Proceedings of Annual Reliability and Maintainability Symposium*. IEEE.

Jayakumar, A., & Asagarpoor, S. (2004). Maintenance optimization of equipment by linear programming. In *International conference on probabilistic methods applied to power systems*, (pp. 145 – 149).

Kay, E. (1976). The effectiveness of preventive maintenance. *International Journal of Production Research*, *14*(3), 329–344. doi:10.1080/00207547608956606

Lapa, C. M. F., Pereira, C. M., & Barros, M. P. D. (2006). A model for preventive maintenance planning by genetic algorithms based on cost and reliability. *Reliability Engineering & System Safety*, *91*(2), 233–240. doi:10.1016/j.ress.2005.01.004

Lee, K. W. (2000). Stochastic models for random-request availability. *IEEE Transactions on Reliability*, *49*(1), 80–84. doi:10.1109/24.855539

Leou, R. (2006). A method for unit maintenance scheduling considering reliability and operation expense. *Electrical Power and Energy Systems*, *28*(7), 471–481. doi:10.1016/j.ijepes.2006.02.009

Levitin, G., & Lisnianski, A. (2000). Optimal replacement scheduling in multi – state series-parallel systems. *Quality and Reliability Engineering International*, *16*(2), 157–162. doi:10.1002/(SICI)1099-1638(200003/04)16:2<157::AID-QRE311>3.0.CO;2-J

Ma, H., & Simon, D. (2011). Blended biogeography-based optimization for constrained optimization. *Engineering Applications of Artificial Intelligence*, *24*(3), 517–525. doi:10.1016/j.engappai.2010.08.005

Ma, Y., Han, J. J., & Trivedi, K. S. (2001). Composite performance and availability analysis of wireless communication networks. *IEEE Transactions on Vehicular Technology*, *50*(5), 1216–1223. doi:10.1109/25.950322

Madu, N. C., & Kuci, C. H. (1994). A group decision support system (GDSS) framework for adjusting system availability levels. *International Journal of Quality & Reliability Management*, *11*(9), 90–100. doi:10.1108/02656719410074314

Mariappan, V. (2004). *Some studies on reliability centered maintenance*. (Ph.D. thesis). IIT Bombay.

McFadden, R. H. (1990). Developing a database for a reliability availability and maintainability improvement program for an industrial plant or commercial building. *IEEE Transactions on Industry Applications*, *26*(4), 735–740. doi:10.1109/28.56001

Nagakava, T. (1977). Optimum preventive maintenance policies for repairable system. *IEEE Transactions on Reliability*, *26*(3), 168–173. doi:10.1109/TR.1977.5220105

Ntuen, C. (1991). An economic maintenance scheduling model with truncated gamma function. *Reliability Engineering & System Safety*, *31*(1), 31–38. doi:10.1016/0951-8320(91)90034-5

Pierskalla, W. P., & Voelker, J. A. (1976). A survey of maintenance models: The control and surveillance of deteriorating systems. *Naval Research Logistics Quarterly*, *23*(2), 353–389. doi:10.1002/nav.3800230302

Prince, B. E., & Haire, M. J. (1991). The influence of maintenance system reliability and maintainability characteristics on plant availability. In *Proceedings of Annual Reliability and Maintainability Symposium*. IEEE. doi:10.1109/ARMS.1991.154508

Racioppi, G., Monaci, G., Michelassi, C., Saccardi, D., Borgia, O., & Carlo, F. D. (2007). A methodology to assess the availability of a sour gas injection plant for the production of oil. In *Proceedings of the European safety and reliability conference 2007 – risk, reliability and societal safety*. Taylor and Francis.

Rajpal, P. S., Shishodia, K. S., & Sekhon, G. S. (2006). An artificial neural network for modeling reliability, availability and maintainability of a repairable system. *Reliability Engineering & System Safety*, *91*(7), 809–819. doi:10.1016/j.ress.2005.08.004

Robino, G., & Sericola, S. (1995). Interval availability analysis using denumerable markov process: Application to multiprocessor subject to break-downs and repair. *IEEE Transactions on Computers*, *44*(2), 286–291. doi:10.1109/12.364539

Sachdeva, A., Kumar, D., & Kumar, P. (2008). Planning and optimizing the maintenance of paper production systems in a paper plant. *Computers & Industrial Engineering*, *55*(4), 817–829. doi:10.1016/j.cie.2008.03.004

Samrout, M., Chatelet, E., Kouta, R., & Chebbo, N. (2009). Optimization of maintenance policy using the proportional hazard model. *Reliability Engineering & System Safety*, *94*(1), 44–52. doi:10.1016/j.ress.2007.12.006

Saraswat, S., & Yadava, G. (2008). An overview on reliability, availability, maintainability and supportability (RAMS) engineering. *International Journal of Quality & Reliability Management*, *25*(3), 330–344. doi:10.1108/02656710810854313

Schabe, H. (1995). A new approach to optimal replacement times for complex systems. *Microelectronics and Reliability*, *35*(8), 1125–1130. doi:10.1016/0026-2714(94)00166-L

Sharma, R. K., & Kumar, S. (2008). Performance modeling in critical engineering systems using RAM analysis. *Reliability Engineering & System Safety, 93*(6), 913–919. doi:10.1016/j.ress.2007.03.039

Sherwin, D. J. (2000). Steady-state series availability. *IEEE Transactions on Reliability, 49*(2), 131–132. doi:10.1109/24.877327

Simon, D. (2008). Biogeography - based optimization. *IEEE Transactions on Evolutionary Computation, 12*(6), 702–713. doi:10.1109/TEVC.2008.919004

Simon, D., Ergezer, M., Du, D., & Rarick, R. (2011). Markov models for biogeography-based optimization. *IEEE Transactions on Systems, Man, and Cybernetics. Part B, Cybernetics, 41*(1), 299–306. doi:10.1109/TSMCB.2010.2051149 PMID:20595090

Tam, A. S. B., Chan, W. M., & Price, J. W. H. (2006). Optimal maintenance intervals for multi - component system. *Production Planning and Control, 17*(8), 769–779. doi:10.1080/09537280600834452

Toshio, N., & Shnji, O. (1976). Reliability analysis of a one unit system with unrepairable spare units and its optimization applications. *Operations Research, 27*(1), 101–110. doi:10.1057/jors.1976.10

Tsai, Y., Wang, K., & Tsai, L. (2004). A study of availability - centered preventive maintenance for multi-component systems. *Reliability Engineering & System Safety, 84*(3), 261–270. doi:10.1016/j.ress.2003.11.011

Tsarouhas, P. H., & Arvanitoyannis, I. S. (2010). Reliability and maintainability analysis of bread production line. *Critical Reviews in Food Science and Nutrition, 50*(4), 327–343. doi:10.1080/10408390802544488 PMID:20301018

Tsoulos, I. G., & Vasant, P. (2009). Product mix selection using an evolutionary technique. *Proceedings of AIP Conference* (Vol. 1159, pp. 240 – 247). doi:10.1063/1.3223936

Usher, J. S., Kamal, A. H., & Hashmi, S. W. (1998). Cost optimal preventive maintenance and replacement scheduling. *IIE Transactions, 30*(12), 1121–1128. doi:10.1080/07408179808966568

Wang, H. (2002). A survey of maintenance policies of deteriorating systems. *European Journal of Operational Research, 139*(3), 469–489. doi:10.1016/S0377-2217(01)00197-7

Wood, A. (1989). Availability calculations with exhaustible spares. *IEEE Transactions on Reliability, 38*(3), 388–391. doi:10.1109/24.44189

Yao, X., Fernandez-Gaucherand, E., Fu, M. C., & Marcus, S. I. (2004). Optimal preventive maintenance scheduling in semiconductor manufacturing, Semiconductor Manufacturing. *IEEE Transactions on, 17*(3), 345–356.

Zio, E., Baraldi, P., & Patelli, E. (2006). Assessment of the availability of an offshore installation by Monte Carlo simulation. *International Journal of Pressure Vessels and Piping, 83*(4), 312–320. doi:10.1016/j.ijpvp.2006.02.010

ADDITIONAL READING

Canto, S. V. (2006). Application of benders' decomposition to power plant preventive maintenance scheduling. *European Journal of Operational Research*, *184*(2), 759–777. doi:10.1016/j.ejor.2006.11.018

Cao, Y., Sun, H., Trivedi, K. S., & Han, J. J. (2002). System availability with non-exponentially distributed outages. *IEEE Transactions on Reliability*, *51*(2), 193–198. doi:10.1109/TR.2002.1011525

Chareonsuk, C., Nagarur, N., & Tabucanon, M. (1997). A multicriteria approach to the selection of preventive maintenance intervals. *International Journal of Production Economics*, *49*(1), 55–64. doi:10.1016/S0925-5273(96)00113-2

Cockerill, A. W. (1990). RAM analysis helps cut turbine - generator systems costs. *Power Engineering*, *94*(7), 27–29.

Coelho, J. P., Oliveira, P. N. M., & Cunha, J. B. (2005). Greenhouse air temperature predictive control using the particle swarm optimization algorithm. *Computers and Electronics in Agriculture*, *49*(3), 330–344. doi:10.1016/j.compag.2005.08.003

Das, C. R., Mohapatra, P., Tien, L., & Bhuyan, L. N. (1993). An availability model for min-based multiprocessors. *IEEE Transactions on Parallel and Distributed Systems*, *4*(10), 1118–1129. doi:10.1109/71.246073

Dekker, R., & Scarf, P. A. (1998). On the impact of optimization models in maintenance decision making: The state of the art. *Reliability Engineering & System Safety*, *60*(2), 111–119. doi:10.1016/S0951-8320(98)83004-4

Eberhart, R., & Kennedy, J. (1995). A new optimizer using particle swarm theory. *Proceedings of the Sixth International Symposium on Micro Machine and Human Science*. pp. 39 – 43. doi:10.1109/MHS.1995.494215

Frank, B. (2001). A replacement policy based on limiting the cumulative maintenance cost. *International Journal of Quality & Reliability Management*, *18*(1), 76–83. doi:10.1108/02656710110364459

Ganesan, T., Elamvazuthi, I., Shaari, K. Z. K., & Vasant, P. (2012a). Swarm intelligence and gravitational search algorithm for multi-objective optimization of synthesis gas production. *Applied Energy*. doi:10.1016/j.apenergy.2012.09.059

Ganesan, T., Vasant, P., & Elamvazuthi, I. (2012b). Hybrid neuro-swarm optimization approach for design of distributed generation power system. *Neural Computing & Applications*. doi:10.1007/s00521-012-0976-4

Ganesan, T., Vasant, P., & Elamvazuthi, I. (2012d). Normal-boundary intersection based parametric multi-objective optimization of green sand mould system. *Journal of Manufacturing Systems*. doi:10.1016/j.jmsy.2012.10.004

Garg, H. (2013). An approach for analyzing fuzzy system reliability using particle swarm optimization and intuitionistic fuzzy set theory, Journal of Multiple-Valued Logic and Soft Computing, 21(3 – 4), 335 – 354.

Garg, H. (2013a). Performance analysis of complex repairable industrial systems using PSO and fuzzy confidence interval based methodology. *ISA Transactions*, *52*(2), 171–183. doi:10.1016/j.isatra.2012.09.010 PMID:23098922

Garg, H. (2014). Performance and behavior analysis of repairable industrial systems using Vague Lambda-Tau methodology. *Applied Soft Computing*, *22*, 323–338. doi:10.1016/j.asoc.2014.05.027

Garg, H. (2014a). Reliability, availability and maintainability analysis of industrial systems using PSO and fuzzy approach, *MAPAN – Journal of Metrology Society of India*, 29(2), 115 – 129.

Garg, H. (2014b). A hybrid GA-GSA algorithm for optimizing the performance of an industrial system by utilizing uncertain data. In P. Vasant (Ed.), *Handbook of Research on Artificial Intelligence Techniques and Algorithms* (pp. 625–659). USA: IGI Global.

Garg, H. (2015). Predicting uncertain behavior in critical engineering systems under vague environment [Old City Publishers.]. *Journal of Multiple-Valued Logic and Soft Computing*, *25*(1), 1–20.

Garg, H., Rani, M., & Sharma, S. P. (2013a). An efficient two phase approach for solving reliability-redundancy allocation problem using artificial bee colony technique. *Computers & Operations Research*, *40*(12), 2961–2969. doi:10.1016/j.cor.2013.07.014

Garg, H., Rani, M., & Sharma, S. P. (2014a). An approach for analyzing the reliability of industrial systems using soft computing based technique. *Expert Systems with Applications*, *41*(2), 489–501. doi:10.1016/j.eswa.2013.07.075

Garg, H., & Sharma, S. P. (2011). *Behavior and system performance optimization for an industrial system by using particle swarm optimization*, proceeding in "2011 IEEE International Conference on Intelligent Computing and Intelligent Systems (ICIS 2011)" held at Guangzhou, China, pp. 237-241.

Garg, H., & Sharma, S. P. (2012a). Stochastic behavior analysis of complex repairable industrial systems utilizing uncertain data. *ISA Transactions*, *51*(6), 752–762. doi:10.1016/j.isatra.2012.06.012 PMID:22795801

Gen, M., & Yun, Y. S. (2006). Soft computing approach for reliability optimization: State-of- the-art survey. *Reliability Engineering & System Safety*, *91*(9), 1008–1026. doi:10.1016/j.ress.2005.11.053

Goldberg, D. E. (1989). *Genetic Algorithm in Search, Optimization and Machine Learning*. MA: Addison-Wesley.

Herder, P., Van Luijk, J., & Bruijnooge, J. (2008). Industrial application of RAM modeling: Development and implementation of a RAM simulation model for the lexan plant at GE industrial, plastics. *Reliability Engineering & System Safety*, *93*(4), 501–508. doi:10.1016/j.ress.2006.10.019

Hipkin, I., Lockett, A. (1995). A study of maintenance technology implementation, Omega - International Journal of Management Science 23(1), 79 – 88.

Holland, J. H. (1975). *Adaptation in Natural and Artificial Systems*. Ann Arbor, MI: The University of Michigan Press.

Hwang, H. (1996). A performance evaluation model for fms based on RAM and LCC using FACTOR/AIM. *Computers & Industrial Engineering*, *31*(3/4), 593–598. doi:10.1016/S0360-8352(96)00292-6

Kennedy, J., & Eberhart, R. C. (1995). Particle swarm optimization. *IEEE International Conference on Neural Networks*. Vol.~IV. Piscataway, NJ, Seoul, Korea, pp. 1942 – 1948.

Kuo, Y., & Chang, Z. A. (2007). Integrated production scheduling and preventive maintenance planning for a single machine under a cumulative damage failure process. *Naval Research Logistics*, *54*(6), 602–614. doi:10.1002/nav.20232

Lai, C. D., Xie, M., Poh, K. L., Dai, Y. S., & Yang, P. (2002). A model for availability analysis of distributed software/hardware systems. *Information and Software Technology*, *44*(6), 343–350. doi:10.1016/S0950-5849(02)00007-1

Markeset, T., & Kumar, U. (2003). Design and development of product support and maintenance concepts for industrial systems. *Journal of Quality in Maintenance Engineering*, *9*(4), 376–392. doi:10.1108/13552510310503231

Nakamura, M., Katafuchi, T., & Hatazaki, H. (2001). Decisions for maintenance-intervals of equipment in thermal power stations based on few data. *IEEE Transactions on Reliability*, *50*(4), 360–364. doi:10.1109/24.983395

Nepal, B., Monplaisir, L., & Singh, N. (2007). A framework to integrate design for reliability and maintainability in modular product design. *International Journal of Product Development*, *4*(5), 459–484. doi:10.1504/IJPD.2007.013043

Prasad, V. R., & Kuo, W. (2000). Reliability optimization of coherent systems. *IEEE Transactions on Reliability*, *49*(3), 323–330. doi:10.1109/24.914551

Rao, P. N. S., & Naikan, V. N. A. (2006). An optimization methodology for condition based minimal and major preventive maintenance. *International Journal of Economic Quality Control*, *21*(1), 127–141.

Samrout, M., Yalaoui, F., Chatelet, E., & Chebbo, N. (2005). New methods to minimize the preventive maintenance cost of series-parallel systems using ant colony optimization. *Reliability Engineering & System Safety*, *89*(3), 346–354. doi:10.1016/j.ress.2004.09.005

Singh, V. P., & Swaminathan, S. (2002). Sample sizes for system availability, in: *Proceedings of Annual Reliability and Maintainability Symposium*, IEEE, New York, NY. pp. 51 -- 55.

Suresh, K., & Kumarappan, N. (2006). Combined genetic algorithm and simulated annealing for preventive unit maintenance scheduling in power system. In *Power Engineering Society General Meeting* (p. 5). IEEE. doi:10.1109/PES.2006.1709254

Tsarouhas, P., & Arvanitoyannis, I. S. (2011). Quantitative analysis for peach production line management. *Journal of Food Engineering*, *105*(1), 28–35. doi:10.1016/j.jfoodeng.2010.11.029

Tsarouhas, P., Varzakas, T., & Arvanitoyannis, I. (2009). Reliability and maintainability analysis of strudel production line with experimental data; a case study. *Journal of Food Engineering*, *91*(2), 250–259. doi:10.1016/j.jfoodeng.2008.09.002

Valdez-Flores, C., & Feldman, R. M. (1989). A survey of preventive maintenance models for stochastically deteriorating dingle unit systems. *Naval Research Logistics*, *36*(4), 419–446. doi:10.1002/1520-6750(198908)36:4<419::AID-NAV3220360407>3.0.CO;2-5

Vasant, P., Elamvazuthi, I., & Webb, J. F. (2010). Fuzzy technique for optimization of objective function with uncertain resource variables and technological coefficients. *International Journal of Modeling, Simulation, and Scientific Computing*, *1*(3), 349–367. doi:10.1142/S1793962310000225

Verma, A. K., & Ramesh, P. G. (2007). Multi-objective initial preventive maintenance scheduling for large engineering plants. *International Journal of Reliability Quality and Safety Engineering*, *14*(3), 241–250. doi:10.1142/S0218539307002611

Wood, A. (1994). Availability modeling. *IEEE Circuits and Devices Magazine*, *10*(3), 22–27. doi:10.1109/101.283651

Woodman, R. C. (1964). Replacement policies for components that deteriorate. *Operations Research*, *18*(3), 267–280. doi:10.1057/jors.1967.43

Wu, P., Gao, L., Zou, D., & Li, S. (2011). An improved particle swarm optimization algorithm for reliability problems. *ISA Transactions*, *50*(1), 71–81. doi:10.1016/j.isatra.2010.08.005 PMID:20850737

Yao, X., Fu, M., Marcus, S. I., & Fernandez-Gaucherand, E. (2001). Optimization of preventive maintenance scheduling for semiconductor manufacturing systems: models and implementation, in: Control Applications, 2001. (CCA '01). Proceedings of the 2001 IEEE International Conference on, pp. 407 - 411.

Zerwick, A. Y. (1996). A focused approach to reliability, availability and maintainability for critical pressure vessels and piping. *International Journal of Pressure Vessels and Piping*, *66*(1-3), 155–160. doi:10.1016/0308-0161(95)00092-5

KEYTERMS AND DEFINITIONS

Evolutionary Algorithm (EA): A collective term for all variants of (probabilistic) optimization and approximation algorithms that are inspired by Darwinian evolution. Optimal states are approximated by successive improvements based on the variation-selection paradigm. Thereby, the variation operators produce genetic diversity and the selection directs the evolutionary search.

Exploitation: Exploitation is the process using information gathered from previously visited points in the search space to determine which places might be profitable to visit next.

Exploration: The process of visiting entirely new regions of a search space, to see if anything promising may be found there.

Preventive Maintenance (PM): It is planned maintenance of plant and equipment that is designed to improve equipment life and avoid any unplanned maintenance activity. It includes painting, lubrication, cleaning, adjusting and minor component replacement to extend the life of equipment and facilities. Its purpose is to minimize breakdowns and excessive depreciation. Neither equipment nor facilities should be allowed to go to the breaking point. In its simplest form, preventive maintenance can be compared to the service schedule for an automobile.

Reliability: It is a characteristic of an item (component or system), expressed by the probability that the item (component/system) will perform its required function under given conditions for a stated time interval.

Chapter 26
Performances of Adaptive Cuckoo Search Algorithm in Engineering Optimization

Pauline Ong
Universiti Tun Hussein Onn Malaysia, Malaysia

S. Kohshelan
Universiti Tun Hussein Onn Malaysia, Malaysia

ABSTRACT

A new optimization algorithm, specifically, the cuckoo search algorithm (CSA), which inspired by the unique breeding strategy of cuckoos, has been developed recently. Preliminary studies demonstrated the comparative performances of the CSA as opposed to genetic algorithm and particle swarm optimization, however, with the competitive advantage of employing fewer control parameters. Given enough computation, the CSA is guaranteed to converge to the optimal solutions, albeit the search process associated to the random-walk behavior might be time-consuming. Moreover, the drawback from the fixed step size searching strategy in the inner computation of CSA still remain unsolved. The adaptive cuckoo search algorithm (ACSA), with the effort in the aspect of integrating an adaptive search strategy, was attached in this study. Its beneficial potential are analyzed in the benchmark test function optimization, as well as engineering optimization problem. Results showed that the proposed ACSA improved over the classical CSA.

INTRODUCTION

Nature inspired metaheuristic algorithms such as genetic algorithms, particle swarm optimization and artificial bee colony algorithm are widely used for solving optimization problems (Holland, 1975; Karaboga & Basturk, 2008; Kennedy & Eberhart, 1995). A metaheuristic algorithm is an iterative generation process which combines different concepts for exploring the search space and finding near-optimal solutions (Gomathi & Sharmila, 2014). In short, metaheuristic algorithms attend to find an optimal solution and obtain a good solution in a specific period of time. Two crucial characteristics of metaheuristic algorithms

DOI: 10.4018/978-1-4666-9644-0.ch026

are: intensification and diversification, or in other words, exploitation and exploration. Intensification or exploitation means to search around the current best solution in the local region and select the best solution which will eventually converge to the optimality. Meanwhile, diversification or exploration means to explore the search space on the global scale in order to avoid the solution from getting trapped at local optimal, and at the same time, increase the diversity of the solution. It is important to balance up between these two characteristics to ensure that the metaheuristic algorithms achieve the global optimal successfully. Metaheuristic algorithms are usually developed to deal with complex optimization problems where other optimization methods have failed to be either effective or efficient (Senvar, Turanoglu & Kahraman, 2013). These methods are known as one of the most practical approaches for solving many complex problems, which is especially true for real-world problems. Consequently, the field of metaheuristic in engineering optimization problems is a rapidly growing field of research. This is due to the importance of solving the mentioned optimization problems for the scientific as well as the industrial world. The practical advantage of metaheuristic lies in both their effectiveness and general applicability (Senvar, Turanoglu & Kahraman, 2013; Mukherjee & Ray, 2008).

Cuckoo search algorithm (CSA) is one of the nature inspired metaheuristic algorithms which is based on the brood parasitism of some cuckoo species (X.-S. Yang & Deb, 2010; X. S. Yang & Deb, 2009). CSA is a simple and effective algorithm that has been applied to solve a range of real-world optimization problem (Bhandari, Singh, Kumar, & Singh, 2014; Durgun & Yildiz, 2012; Malik, Ahsan, & Mohsin, 2014; Yildiz, 2013). CSA is enhanced by using Lévy flights, rather that simple isotropic random walk which has an infinite variance with an infinite mean (Ong, 2014; Ouaarab, Ahiod, & Yang, 2014; X.-S. Yang & Deb, 2010; X. S. Yang & Deb, 2009). The purpose of utilization of Lévy flight is to speed up the local search whereby CSA can achieve global optimal solutions rapidly. Nevertheless, due to the behavior of random walk in the standard CSA, it is difficult to achieve the optimal solution in a specified given time (Walton, Hassan, Morgan, & Brown, 2011). In addition, it is important for CSA to achieve the optimal solution more efficiently without getting trapped in a local optimum.

To serve for this intention, this chapter has proposed adaptive cuckoo search algorithm (ACSA) by integrating CSA with an accelerated searching strategy in its computation. The main thrust of this study is thus to improve the performance of the current CSA in achieving the optimal solution faster at a specific given time. The objectives of this study can be summarized as follows:

- To improve the optimization capability of the standard CSA
- To develop and apply a novel technique, i.e., the ACSA in function optimization for the purpose of accelerating its convergence characteristic
- To assess the potential benefits of the proposed ACSA in the context of function optimization, through empirical approaches with simulated as well as real world engineering problem, i.e., optimization of pulp and paper properties.

In the next section, the related works in CSA are provided. A detail description of the CSA and ACSA also are presented. The benchmark optimization functions used are discussed briefly. Meanwhile, the performances of CSA and ACSA in optimizing the benchmark functions and real world engineering problem are evaluated. Finally, some conclusions are drawn.

RELATED WORKS

The CSA has contributed immensely in solving optimization problems. It has been proven to give a better performance in solving the optimization problems as compared to other algorithms such as particle swarm optimization (PSO) and genetic algorithm (GA) (Jianzhou et al, 2014; Seif-Eddeen & Adrián, 2014; Yang, 2014). Previous study by Jianzhou et al (2014) on the power load prediction showed that the CSA outperformed the PSO in terms of accuracy. However, achieving the optimum solution effectively in the CSA still become a crucial issue. Previously, Yang (2014) has highlighted the importance of adaptation of search moves in CSA to balance the exploration and exploitation. Furthermore, the author also emphasize that the imbalance in the CSA search moves will required infinite amount of time to achieve the optimum solution. Moreover, the current behavior of random walk may contribute to an issue in achieving the optimal solution at a specific time due to the search process (Walton et al., 2011). Therefore, wesuggest the improvement of CSA to overcome the issuein achieving the optimal solution at a specific time.

BACKGROUND

Introduction to Cuckoo Search Algorithm

In recent years, the metaheuristic optimization algorithms have been a noticeable part for solving real mathematics and engineering problems, especially the Cuckoo search algorithm (CSA) (Jianzhou et al, 2014). In this section, the CSA is utilized and describe briefly. CSAis inspired by the obligate interspecific brood parasitism of some cuckoo species that lay their eggs in the nests of other host birds (X. S. Yang & Deb, 2009). The specialty of cuckoo eggs is it closely mimics the eggs of the host. Of course, in some cases, the host birds discover the alien eggs in their nests. Therefore, if a host bird discovers the eggs, which are not their own, it will either throw the alien eggs or abandon its nest and build a new nest elsewhere. This surprising behavior is the metaphor of the cuckoo search metaheuristic approach for solving optimization problems. Each egg in the nest represents the solution, and cuckoo egg represents new solution. The ultimate goal is to use the new or better solutions to replace low quality solution in the nests. This replacement, is carried out iteratively, which eventually will lead to near-optimal solution of the problem. In the simple form of cuckoo search, each nest contains only one egg while new solution is generated by Levy flight. Basically, the searching process of CSA is based on three idealized rules which statethat:

1. Each cuckoo lays one egg at a time and dumps it in a randomly chosen nest.
2. The best nests with high quality of eggs will be carried over to the next generation.
3. The number of available host nests is fixed, and a host can discover an alien egg with a probability $pa_{\in}$ [0, 1]. In this case, the host bird can either throwthe egg away or abandon the nest to build a new nest in a new location.

Based on these three idealized rules, pseudo code for CSA is (Yang & Deb, 2009):

beginGenerate initial population of q host nest $\mathbf{x}_i$, $i = 1,2,\ldots,q$
for all $\mathbf{x}_i$ **do**
Evaluate the fitness function $F_i = f(\mathbf{x}_i)$
end for
while (*iter<MaxGeneration*) or (*stopping criterion*)
Generate a cuckoo egg $\mathbf{x}_j$ from random host nest by using Lévy flight
Calculate the fitness function $F_j = f(\mathbf{x}_j)$
Get a random nest i among q host nest
if $(F_j > F_i)$ **then**
Replace $\mathbf{x}_i$ with $\mathbf{x}_j$
Replace F_i with F_j
end if
Abandon a fraction p_a of the worst nests
Build new nests randomly to replace nests lost
Evaluate the fitness of new nests
end while
end

In CSA, while generating new solution $(t+1)$ for a cuckoo i, a Levy flight is performed using the following:

$$xi_{(t}+1) = xi\,(t)_{+}\,\alpha\oplus \text{Levy}\,(\lambda)$$

where α>0 represents the step size. This parameter should be related to the scales of problem the algorithm is trying to solve. In most cases, the value α is 1. The symbol ⊕ is an entry-wise multiplication. Basically, Levy flights provide a random walk while their random steps are drawn from a Levy distribution for large steps as follows:

$$\text{Levy} \sim u = t{-}\lambda$$

wher$^{e\,\lambda}$ $\in (0,3]$.

This has an infinite variance with an infinite mean. Here, the steps essentially form a random walk process with a power-law step-length distribution with a heavy tail.

THE PROPOSED ADAPTIVE CUCKOO SEARCH ALGORITHM

Basically, the CSA involved the random walk process where the current behavior of random walk may contribute to an issue in achieving the optimal solution at a specific time due to the search process (Walton et al., 2011). In addressing this issue, Walton et al. enumerated considerations, which, the addition of information exchange between the best solutions was included (Walton et al., 2011). Li et al. approached

Figure 1. CSA Flowchart (based on Yang & Deb (2009)

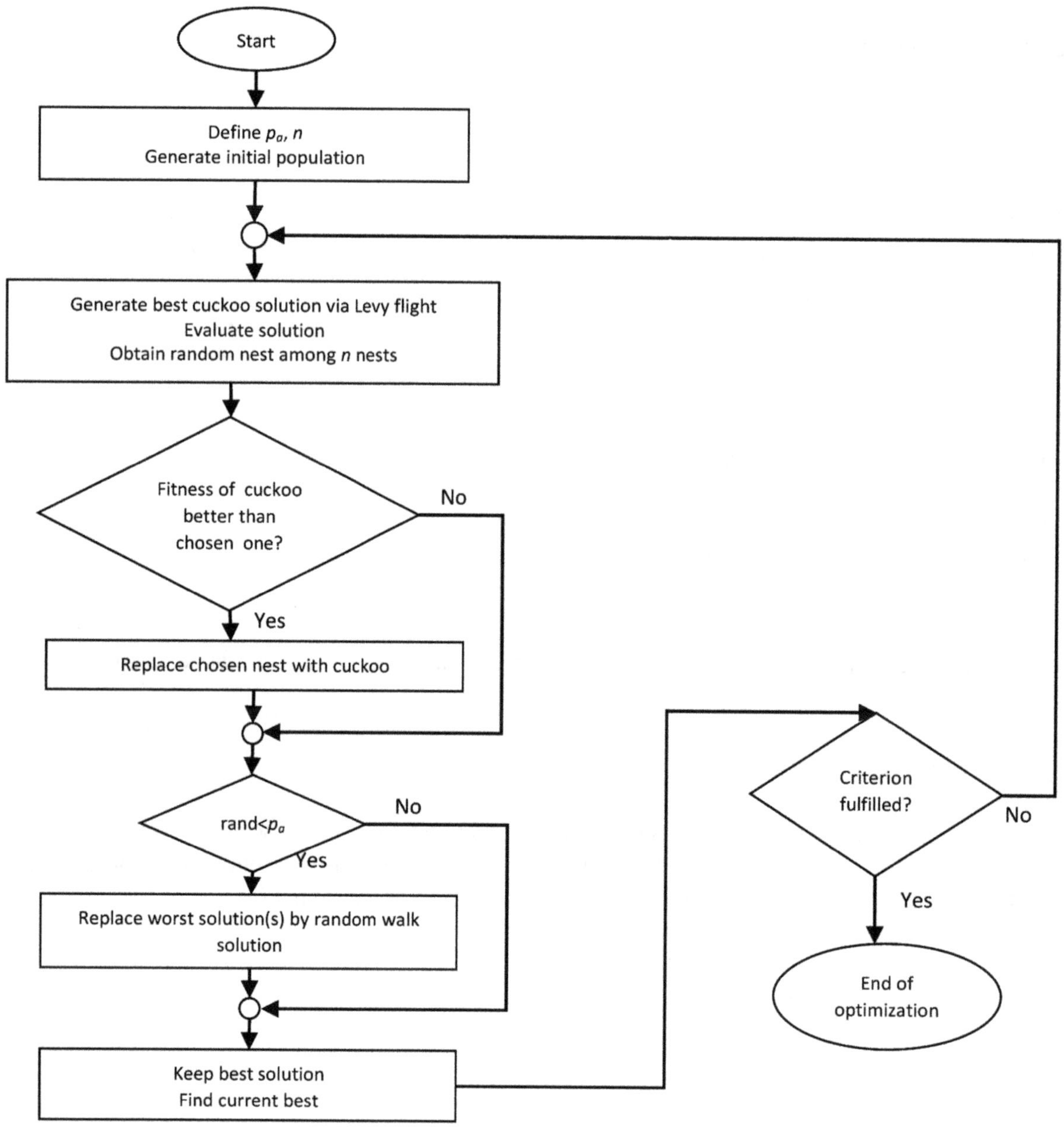

the CSA through the integration of the orthogonal learning method (Li, Wang, & Yin, 2014). In order to add new variations to existing search, the orthogonal design was applied to generate the new solution in each run. Marichelvam et al. provided insights into the requirements of generating some of the initial solutions by using some constructive heuristics into the CSA(Marichelvam, Prabaharan, & Yang, 2014).

While prior research make great contribution to sufficiently excavating some efficient means in expediting the convergence of CSA, this chapter has proposed the improved CSA, specifically, the ACSAby integrating the original CSA with an accelerated searching strategy. This is due to the fact that the success rate in achieving the optimal solution is highly dependent on the search process.Compared to the original CSA, ACSA adjusts the step size, α adaptively. By default, to balance the local and global searching, α is chosen as 1 in standard CSA. In this work, the step size, α in ACSA is determined adaptively as follows:

$$\alpha = \begin{cases} \alpha_L + (\alpha_U - \alpha_L)\dfrac{F_j - F_{\min}}{F_{avg} - F_{\min}}, & F_j \le F_{avg} \\ \dfrac{\alpha_U}{\sqrt{t}}, & F_j > F_{avg} \end{cases},$$

where αL is the predefined minimum step size, αU is the predefined maximum step size, Fj is the fitness value of the *j*th cuckoo egg, and *F*min and *F*avg denote the minimum and the average fitness values of all host nests, respectively.

The fitness function in the Equation (3) is responsible to perform the desired evaluation to fulfill the condition by returning a positive integer number (fitness value). This fitness value, Fj reflects how optimal the solution is: the higher the number, the better the solution (Yang & Deb, 2015). On the other hand, the adaptive step size strategy is based on the assumption that the cuckoos tend to seek for environment with a higher egg survival rate. It can be noticed from Equation that if the current environment shows a higher survival rate, or in other words, this region, probably, may contain the optimal solution, cuckoos will search more intensely in this region. On the contrary, if the current environment is considered no longer suitable to live in due to high mortality rate, cuckoos will search more rigorously in far-field area for the new host nest. The step size α is defined such that the performance of ACSA in terms of exploitation and exploration is more effective in order to achieve the best optimal solution at the specific time.

The pseudo code for ACSA is given as follow:

begin
Generate initial population of q host nest xi, i = 1, 2,…,q
Define minimum step size value αL
Define maximum step size value αU
for all xi **do**
Evaluate the fitness function Fi= f(xi)
end for
while (iter<*MaxGeneration*) or (*stopping criterion*)
Calculate the average fitness value Favg of all the host nests
Find the minimum fitness value Fmin among all the host nests
for all host nests **do**
Current position $\mathbf{x}_i$
Calculate the step size α for Lévy flight using Equation (3)
Generate a cuckoo egg $\mathbf{x}_j$ from host nest $\mathbf{x}_i$ by using Lévy flight
if ($\mathbf{x}_j$ falls outside the bounds) **then**
Replace $\mathbf{x}_j$ with $\mathbf{x}_i$
end if
Calculate the fitness function $F_j = f(\mathbf{x}_j)$
if ($F_j > F_i$) **then**
Replace $\mathbf{x}_i$ with $\mathbf{x}_j$

Replace F_i with F_j
end if
end for
Abandon a fraction p_a of the worst nests
Build new nests randomly to replace nests lost
Evaluate the fitness of new nests
end while
end

OPTIMIZATION OF BENCHMARK FUNCTION

Next, as compared to the conventional CSA, the effectiveness of the proposed ACSA is proven by simulation on five well-known benchmark functions, in term of finding the optimal solution after a predefined tolerance error. These benchmarks are widely used in evaluating the performance of nature inspired metaheuristic algorithms. Some of the benchmark functions are unimodal, while others are multimodal. Basically, a benchmark function is called unimodal if it consists of one optimum position, meanwhile multimodal benchmark functions consists of two or more local optima. The parameter setting is designed such that the initial population of 20 host nest (n) is generated randomly for each test function. The standard value of CSA for step size (a) and the fraction (p_a) is 1 and 0.25, respectively (Yang & Deb, 2010). The average value of the Euclidean distance between the known global minimum and the location of the best host nest is measured for each loop from all the 30 trials.

To test the performance of ACSA, the following set of benchmark test functions are used:

- Ackley
- De Jong
- Griewank
- Rastrigin
- Rosenbrock

Ackley function is a multimodal, continuous function obtained by modulating an exponential function with a cosine wave of moderate amplitude. Formulation:

$$f(x) = -20\exp\left(-0.2\sqrt{\frac{1}{d}\sum_{i=1}^{d} x_i^2}\right) - \exp\left(\frac{1}{d}\sum_{i=1}^{d}\cos(2\pi x_i)\right) + 20 + e,$$

where $x_i \in [-32.768, 32.768]$ and data dimension, $d = 50$, has a global minima of $f_* = 0$, at $x_* = (0, 0, \ldots, 0)$. The α_L and α_M are chosen as 0.2∗domain x_i and 0.8∗domain x_i, respectively.

De Jong function is the simplest test function, known as sphere model. It is continuous, convex and unimodal. Formulation:

$$f(x) = \sum_{i=1}^{d} x_i^2,$$

where $x_i \in [-5.12, 5.12]$ and data dimension, $d = 50$, has a global minima of $f_* = 0$, at $x_* = (0, 0, ..., 0)$. The α_L and α_M are chosen as 0.2∗ domain xi and 0.8∗domain xi, respectively.

Griewank function has many widespread local minima. However, the locations of the minima are regularly distributed. Formulation:

$$f(x) = \frac{1}{4000}\sum_{i=1}^{d} x_i^2 - \prod_{i=1}^{d} \cos\left(\frac{x_i}{\sqrt{i}}\right) + 1,$$

where $xi \in [-600, 600]$ and data dimension, $d = 100$. This is a high dimensional multimodal function, with the known global minima of $f_* = 0$, at $x_* = (0, 0, ..., 0)$. The α_L and α_M are chosen as 0.2∗domain x_i and 0.8∗domain x_i, respectively.

Rastrigin function is based on sphere function with the addition of cosine modulation to produce many local minima. Formulation:

$$f(x) = 10d + \sum_{i=1}^{d} [x_i^2 - 10\cos(2\pi x_i)],$$

where $x_i \in [-5.12, 5.12]$ an$_d$ data dimension, $d = 100$, has a global minima of $f* = 0$, at $x* = (0, 0, ..., 0)$. The α_L and α_M are chosen as 0.2∗domain x_i and 0.8∗domain x_i, respectively.

Rosenbrock function, also referred to as the Valley or Banana function, is unimodal, and the global minimum lies in a narrow, parabolic valley. Formulation:

$$f(x) = \sum_{i=1}^{d-1} [(1 - x_i)^2 + 100(x_{i+1} - x_i^2)^2],$$

where $x_i \in [-100, 100]$ and data dime$_{ns}$ion, $d = 10$. This is a high dimensional multimodal function, with the known global minima of $f* = 0$, at $x* = (0, 0, ..., 0)$. The α_L and α_M are chosen as 0.2∗domain x_i and 0.8∗domain x_i, respectively.

The performances comparison of the proposed ACSA with standard CSA for each benchmark functions are presented in terms of iteration number and Euclidean distance from known global optima.

Case 1: Ackley Function

Figure 2 shows the performance comparison of the original CSA and ACSA for the Ackley function. In Figure 2, ACSA outperformed CSA by achieving optimal solution in less number of iterations. This shows that the searching process of ACSA to achieve the global optimum in this multimodal function is faster than the original CSA.

Case 2: De Jong Function

Figure 3 shows the performance comparison of the original CSA and ACSA for the De Jong function. Similar to Ackley function, ACSA shows a good performance meanwhile, CSA requires a longer process-

Figure 2. Performance Comparison of CSA and ACSA for the Ackley function

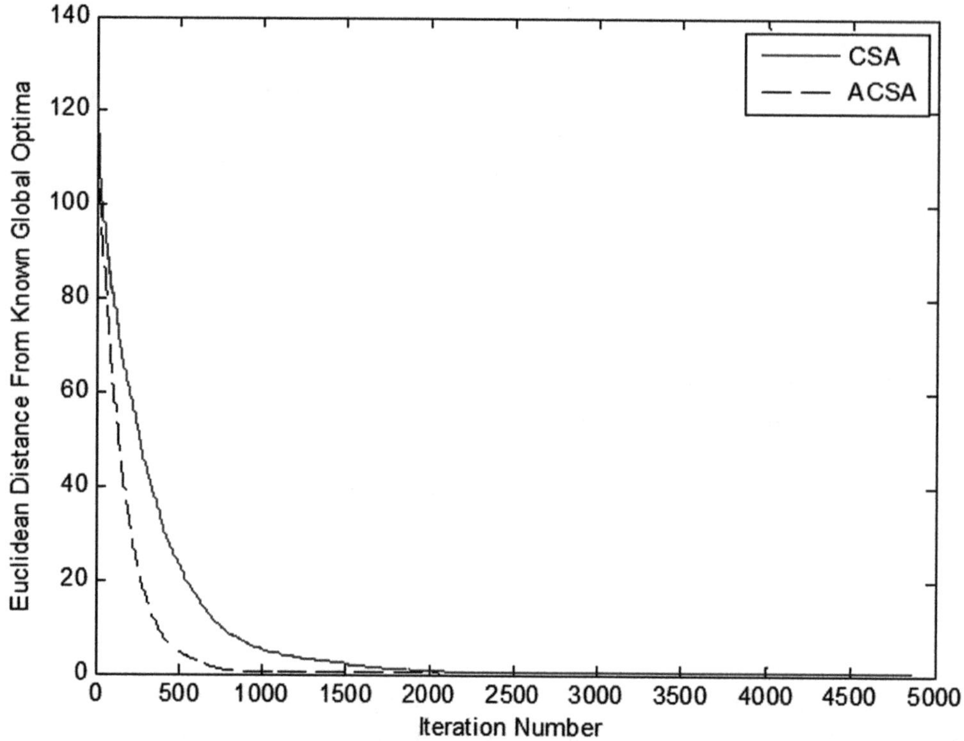

Figure 3. Performance comparison of CSA and ACSA for the De Jong function

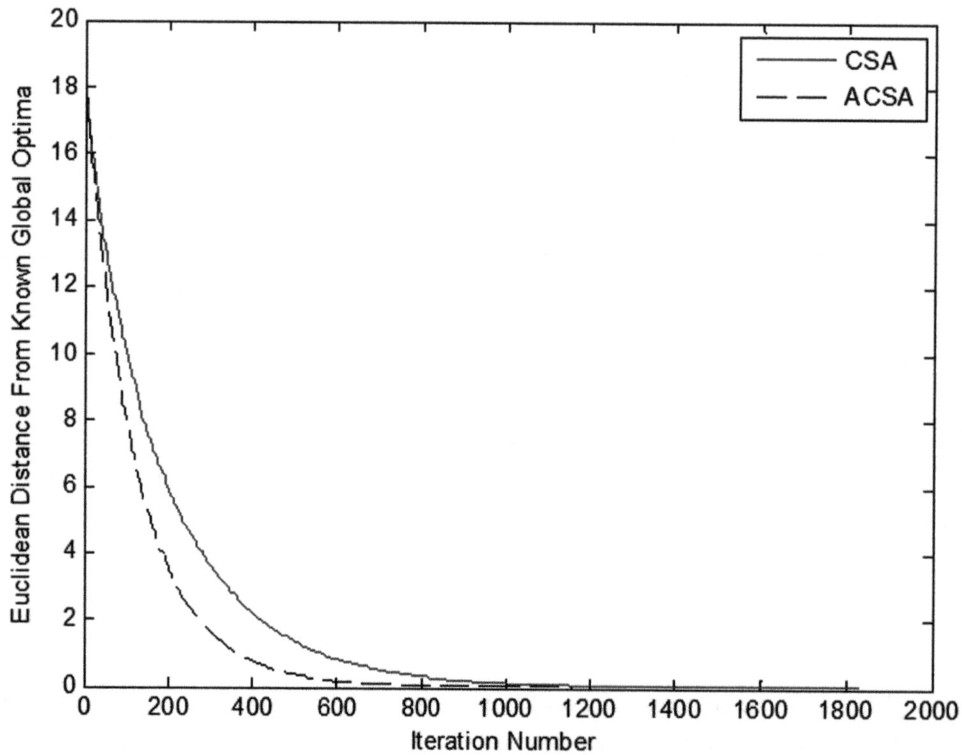

Figure 4. Performance comparison of CSA and ACSA for the Griewank function

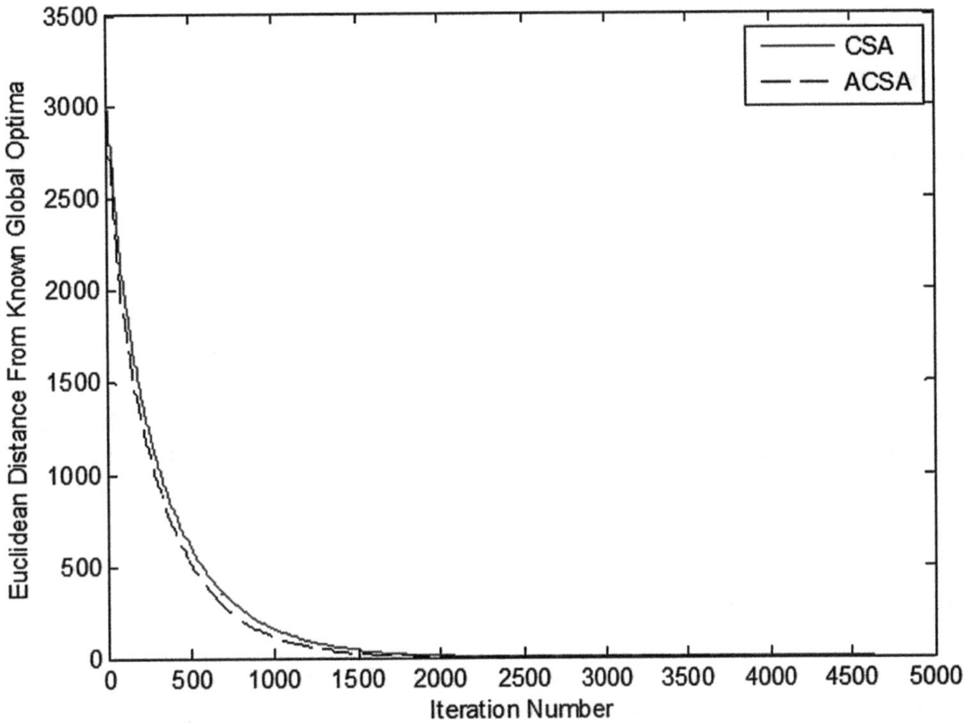

Figure 5. Performance comparison of CSA and ACSA for the Rastrigin function

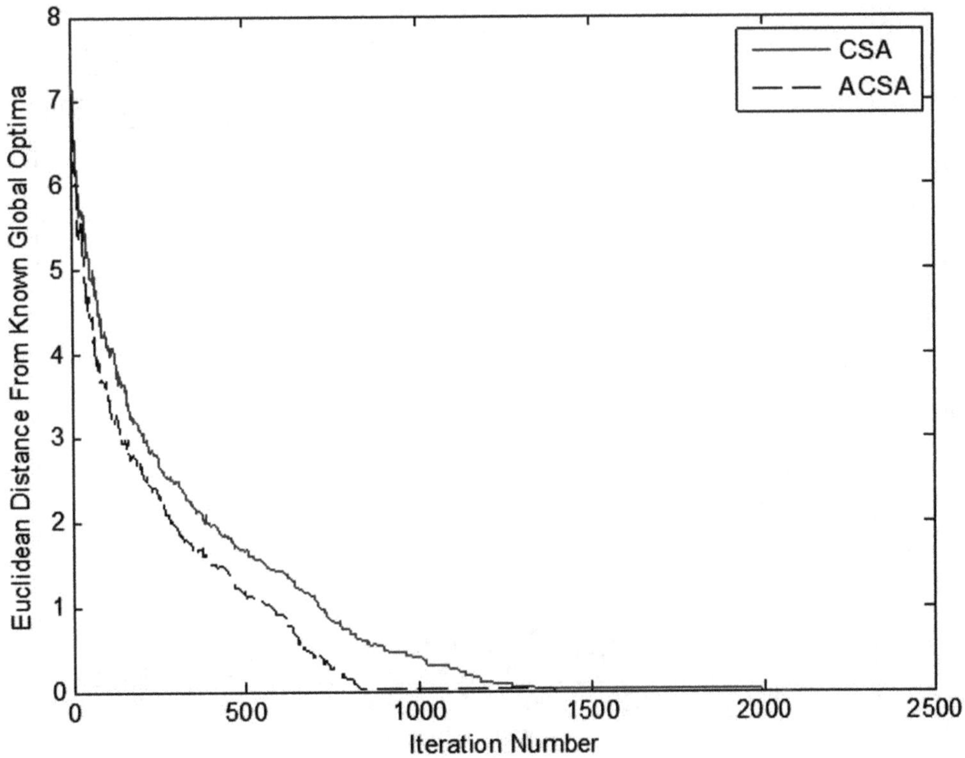

ing time to reaches the known global optimum. In this simplest function, ACSA has shownthe capability to achieve the optimum solution effectively with a small number of iterations.

Case 3: Griewank Function

Figure 4 shows the performance comparison of the original CSA and ACSA for the Griewank function. In Figure 4, ACSA has shown a slightly improvement in terms of number of iterations compared to CSA. Furthermore, in this benchmark function, ACSA has successfully demonstrated that it can provide a better performance in achieving the optimum solution.

Case 4: Rastrigin Function

Figure 5 shows the performance comparison of the original CSA and ACSA for the Rastrigin function. It is obvious that ACSA has shown a good performance than CSA. This show that, in this sphere function, ACSA has succeed to achieve the optimum solution rapidly with a small number of iterations compared to the original CSA.

Case 5: Rosenbrock Function

Figure 6 shows the performance comparison of the original CSA and ACSA for the Rosenbrock function. It can be noted that the average number of iteration needed by the CSA is much longer than ACSA.

Figure 6. Performance comparison of CSA and ACSA for the Rosenbrock function

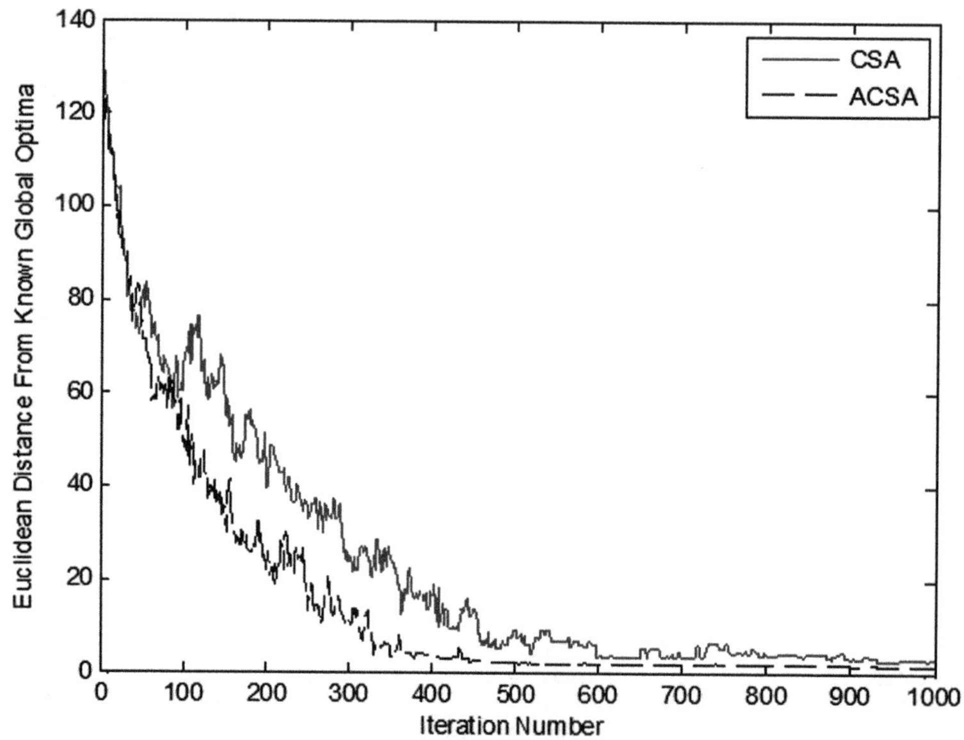

Table 1. Performance comparison of ACSA with CSA

Function		CSA	ASCA	Statistically Significant?
Ackley	Best Mean Worst SD	3884 4862 4222 250.74	1912 2068 1989 41.03	Yes
De Jong	Best Mean Worst SD	1690 1833 1763 37.55	1087 1158 1121 19.81	Yes
Griewank	Best Mean Worst SD	4287 4632 4499 75.63	3617 3880 3725 63.77	Yes
Rastrigin	Best Mean Worst SD	1582 1994 1734 118.73	1190 1389 1286 50.69	Yes
Rosenbrock	Best Mean Worst SD	17519 33929 25402 4592.24	9131 28803 22164 5416.74	Yes

Therefore, the result indicated that the searching process of ACSA is faster than the original CSA in terms of achieving the optimum solution with a small number of iterations.

Overall performances in terms of the number of iterations needed to achieve the global optima are tabulated in Table 1. Values for best, mean, worst and standard deviation (SD) are shown. Interestingly, ACSA has outperformed CSA with less number of iterations in all cases.This shows that the ACSA can achieve the optimum solution faster that may lead to a small number of iterations. In addition, two-tailed t-test is used to valid*at*e the significance of the obtained results.

Discussion

Considering Table 1 which summarizes the average iterations needed for both CSA and ACSA to meet the stopping criterion, it can be concluded that the CSA takes more iteration to converge. For instance, the ACSA needs approximately 2500 iterations in converging to global optimality; however, the CSA only reaches the global solution after 4000 iterations, which shown approximately 38% of improvement in terms of convergence rate.

The analysis on De Jong function indicates that a noteworthy decrease in iteration needed can be observed, where the ACSA takes a mean of 1000 iterations in order to reach the known global optimum, while the CSA requires 1700 iterations on the other hand, which is about 41% slower than the proposed ACSA.Moreover, it can be inferred from Table 1 that the ACSA outperformed the CSA in terms of the best, worst, average and standard deviation, for the number of iterations needed in order to reach the known global optimum. In average, a 16% of improvement in convergence rate can be observed.

Considering Figure 4, for Rastrigin function, the CSA shows faster convergence than the ACSA initially. However, as the evolution continues, the CSA, presumably, is getting trapped in local optima, most probably due to it makes too small and cautious steps while exploring the search space, i.e., the α which controls t*he* scale of the search patterns is specified as 1. On the other hand, the ACSA is able to find the global optima in the average of 20000 iterations, albeit there are many local optima exist in this function. This suggests that a more rigorous search through the solution space in ACSA might accelerate the convergence characteristics and help to escape from the local optima.

Comparing the best obtained results in among all the 30 trials for Rosenbrock function, it can be observed that the proposed ACSA speeds up the computation more than one time faster, as compared to the standard CSA. The optimization capability of the ACSA can be said is encouragingly, since it can be noted that the worst result achieved by the ACSA, i.e., 22164 itera*tions*, is much superior than the average number of iterations needed by the CSA, i.e., 339929 iter*atio*ns.

REAL WORLD APPLICATION

Cuckoo search algorithm (CSA) has been used in various real world applications. For instance, Chandrasekaran & Simon (2014)have implemented CSA to solve the unit commitment problem (UCP) for hybrid power system. Moreover, they also show the effectiveness of the CSA by comparing the performance with other techniques. On the other hand, Sallem et al. (2015) has used CSA to solve the conflict problem in the analog and RF circuitdesign. Therefore, in their studies, they integrated CSA with other metaheuristic algorithms to improve the performance of techniques in solving the conflict issues. Furthermore, Goel, Sharma & Panchal (2014) has implemented CSA in anticipatory system to identify the best probable location for deployment of enemy forces. Their results successfully showed the accurate predictions in anticipating the enemy's movements. In the next section, we have applied CSA in the real world engineering application in finding the optimal experimental conditions to achieve the desired properties of pulp and paper.

REAL WORLD ENGINEERING APPLICATION: OPTIMIZATION OF PULP AND PAPER PROPERTIES

Throughout the pulping process in pulp and paper industry, different experimental conditions, for instance, the temperature, pressure, concentration of different chemicals, will produce significantly different pulp and paper properties, such as tensile strength, and kappa number. Interpreting the relationship between experimental conditions and the properties of the pulp and paper is complex. Moreover, the optimal experimental conditions for each response are often conflicting with each other. For instance, medium experimental milieu might be needed in order to maximize the tensile index due to fiber degradation while a severe experimental condition might be needed in getting a high tensile index on the other hand. As such, finding the optimal experimental conditions in order to achieve the desired properties of optimal material is important, which can be done by using the proposed ACSA.

Table 2 presents the experimental results of the obtained pulp and paper properties under different experimental conditions from 30 experimental trials(Zainuddin, Wan Daud, Pauline, & Shafie, 2011). The effects of four types of pulping variables (sodium hydroxide (NaOH), ethanol (EtOH) concentration, temperature and time) on the properties of pulp and paper (screened yield, kappa number, tensile index and tear index) are studied. Finding the optimal experimental conditions in which it maximizes the screened yield, tensile and tear indexes while minimizes the kappa number at the same time is performed using the ACSA. The quadratic models which correlate the pulping variables to the response variables are formed initially, which are given as(Zainuddin et al., 2011):

$$Y_{ScreenedYield} = 28.51 - 4.66X_A - 0.30X_{Et} + 0.44X_T - 0.19X_t - 2.69X_AX_T - 1.27X_A^2$$

$$Y_{KappaNumber} = 32.2667 - 36.63X_A - 7.60X_{Et} - 7.67X_T - 3.72X_t + 4.20X_AX_{Et} + 3.40X_AX_T$$

Table 2. Values of pulping variables and experimental values of the pulp and paper properties

No	Experimental Conditions				Experimental Values for the Paper Qualities Response			
	NaOH (%)	EtOH (%)	Temperature (°C)	Time (min)	Screened Yield (%)	Kappa Number	Tensile Index (N m/g)	Tear Index (mN N²/g)
1	10	45	170	120	28.25	102.0	28.59	2.83
2	20	45	170	120	28.12	32.6	62.99	4.13
3	25	60	180	90	24.51	23.3	46.5	3.84
4	25	60	160	150	22.52	24.9	64.74	4.18
5	15	30	180	150	32.92	59.6	46.06	3.55
6	15	60	180	90	31.71	57.5	44.88	3.38
7	30	45	170	120	25.91	25.4	51.26	3.64
8	20	45	170	180	29.45	30.9	51.29	3.95
9	25	30	180	150	24.07	25.4	56.88	3.56
10	20	45	170	60	28.75	39.5	56.12	4.13
11	15	60	160	150	28.98	58.7	39.52	3.35
12	15	30	160	150	31.50	72.7	33.53	2.96
13	25	30	180	90	25.60	28.1	60.59	4.65
14	20	45	170	120	29.54	31.7	57.1	4.23
15	25	30	160	90	25.65	34.6	59.82	4.45
16	15	60	180	150	32.10	53.0	49.6	3.56
17	15	30	160	90	29.03	75.1	33.14	3.01
18	20	45	190	120	26.84	31.6	59.1	4.23
19	25	60	160	90	25.99	28.5	47.06	3.73
20	15	60	160	90	30.54	62.5	35.28	3.06
21	25	60	180	150	23.24	20.2	48.4	3.45
22	20	15	170	120	26.57	47.0	54.27	3.65
23	20	75	170	120	28.01	32.0	50.52	3.5
24	20	45	170	120	27.86	32.5	58.4	3.95
25	20	45	150	120	27.15	48.8	50.32	3.9
26	25	30	160	150	25.67	31.0	64.69	4.68
27	15	30	180	90	31.66	63.3	41.83	3.56

$$+4.10X_{Et}X_T + 31.60X_A^2 + 7.40X_{Et}^2 + 8.10X_T^2 + 3.10X_t^2$$

$$Y_{TensileIndex} = 59.50 + 14.18X_A - 2.34X_{Et} + 2.88X_T + 2.06X_t - 12.50X_AX_{Et} - 16.21X_AX_T$$

$$-20.84X_A^2 - 8.36X_{Et}^2 - 6.05X_T^2 - 7.06X_t^2$$

$$Y_{TearIndex} = 4.10 + 0.64X_A - 0.18X_{Et} + 0.07X_T - 0.06X_t - 0.60X_AX_{Et} - 0.80X_AX_T$$

$$-0.56X_TX_t - 0.90X_A^2 - 0.56X_{Et}^2$$

where X_A, X_{Et}, X_T and X_t are the wt% of NaOH, EtOH, temperature and time, respectively.

Although the underlying problem is a multi-objectives optimization problem in which maximizing the screened yield, tensile index and tear index while minimizing the kappa number form the main concern, it is transformed into a single-objective maximization problem, where the objective function is formed as:

$$Objective\_Function = Y_{ScreenedYield} + Y_{TensileIndex} + Y_{TearIndex} - Y_{KappaNumber}$$

In Table 2, it can be concluded that the optimal experimental conditions are found when the screened yield, tensile index and the tear index are maximized while the kappa number is minimized simultaneously. Moreover, the optimization process of ACSA continues unless the absolute difference between the best fitness values of consecutive iteration is less than 1e-10. The obtained optimal pulping conditions and its best fitness value by the proposed ACSA are summarized in Table 3. In addition, the obtained optimal experimental conditions by wavelet neural network (WNN) and response surface methodology (RSM) on the same dataset are used for comparison purposes(Zainuddin et al., 2011). It should be noted that the method with higher fitness value indicates better optimization results, since the underlying problem is modeled as a maximization problem. It can be inferred from Table 3 that the proposed ACSA outperformed the others in this regards. Moreover, the optimal experiment condition achieved for ACSA required ahighoptimal experimental condition of sodium hydroxide (NaOH) with low ethanol (EtOH) concentration compared to RSM and WNN. Furthermore, the ACSA required only a small amount of time compared to other optimization method.

Table 3. The obtained optimal experimental conditions

Optimization Method	Optimal Experimental Conditions				Fitness Value
	NaOH	EtOH	T	t	
ACSA	26.60	19.87	187.18	153.09	356.2663
RSM	18.15	38.62	165	170	295.3275
WNN	18.15	38.12	165.23	167.71	294.7659

FUTURE RESEARCH DIRECTIONS

This work provides another insight for the enhancement on the CSA, from the aspect of integration of adaptive step size strategy. However, much effort needs to be done for further improvement. To increase the optimization ability of CSA, besides by updating the Lévy flight step size adaptively, including the information exchange among the good solutions can be integrated. Utilization of quantum computing and chaotic map will be an interesting topic to study. In addition, the CSA can be further improve to enhance the performance of the algorithms. Furthermore, in the future, the application of CSA can be used in different type of optimization problems such as medical or modelling optimization, beside engineering optimization. CSA also can be applied to the audio optimization problems such as to solve the problem in the amplitude of the audio signal by optimizing the audio signal (Kohshelan & Wahid, 2014). On the other hand, the metaheuristic algorithm also can be integrated with another algorithm to improve the performance of CSA in achieving the optimum solution with a specific time in a different kind of issue or problems.

CONCLUSION

The proposed ACSA, accelerates the convergence characteristic of the standard CSA, with the utilization of the adaptive step size searching strategy. To evaluate the performance of ACSA, this chapter has used five benchmark optimization functions, namely Ackley, De Jong, Griewank, Rastrigin and Rosenbrock. Overall, the experiment results are encouraging, illustrating that the accelerated searching strategy leads to a better performance in terms of less number of iterations to achieve the optimal solution compared to CS. In addition, the ACSA has been used in finding the optimal experimental conditions for pulping process. The obtained optimal solutions are compared to the results available in literature, in which its superiority is shown. Overall, ACSA has shown a better performance than the original CSA in terms of achieving the optimal solution with a small number of iterations. Moreover, ACSA also have achieved a better fitness value compared to RSM and WNN. This proved the capability of ACSA to perform well in the benchmark functions as well as in real world optimization problem. The future work will emphasize on enhancing the convergence characteristic of CSA with integration with quantum computing and chaotic map.

REFERENCES

Bhandari, A. K., Singh, V. K., Kumar, A., & Singh, G. K. (2014). Cuckoo search algorithm and wind driven optimization based study of satellite image segmentation for multilevel thresholding using Kapur's entropy. *Expert Systems with Applications*, *41*(7), 3538–3560. doi:10.1016/j.eswa.2013.10.059

Chandrasekaran, K., & Simon, S. P. (2014). Wind-Thermal Integrated Power System Scheduling Problem Using Cuckoo Search Algorithm. *International Journal of Operations Research and Information Systems*, *5*(3), 81–109. doi:10.4018/ijoris.2014070104

Durgun, İ., & Yildiz, A. R. (2012). Structural Design Optimization of Vehicle Components Using Cuckoo Search Algorithm. *Materials Testing*, *54*(3), 185–188. doi:10.3139/120.110317

Fateen, S.-E. K., & Bonilla-Petriciolet, A. (2014). Gradient-Based Cuckoo Search for Global Optimization. *Mathematical Problems in Engineering, Article ID 493740*. doi:10.1155/2014/493740

Goel, S., Sharma, A., & Panchal, V. K. (2014). Multiobjective Cuckoo Search for Anticipating the Enemy's Movements in the Battleground. *International Journal of Applied Metaheuristic Computing*, *5*(4), 26–46. doi:10.4018/ijamc.2014100102

Gomathi, R., & Sharmila, D. (2014). A Novel Adaptive Cuckoo Search for Optimal Query Plan Generation. *TheScientificWorldJournal*, *7*. doi:10.1155/2014/727658 PMID:25215330

Holland, J. H. (1975). *Adaptation in Natural and Artificial Systems*. University of Michigan Press.

Karaboga, D., & Basturk, B. (2008). On the performance of artificial bee colony (ABC) algorithm. *Applied Soft Computing*, *8*(1), 687–697. doi:10.1016/j.asoc.2007.05.007

Kennedy, J., & Eberhart, R. (1995, Nov/Dec). *Particle swarm optimization*. In *Proceedings of IEEE International Conference on Neural Networks*. doi:10.1109/ICNN.1995.488968

Kohshelan & Wahid, N. (2014). Improvement of Audio Feature Extraction Techniques in Traditional Indian Musical Instrument. Recent Advances on Soft Computing and Data Mining, 507-516

Li, X., Wang, J., & Yin, M. (2014). Enhancing the performance of cuckoo search algorithm using orthogonal learning method. *Neural Computing & Applications*, *24*(6), 1233–1247. doi:10.1007/s00521-013-1354-6

Malik, M., Ahsan, F., & Mohsin, S. (2014). Adaptive image denoising using cuckoo algorithm. *Soft Computing*. doi:10.1007/s00500-014-1552-x

Marichelvam, M. K., Prabaharan, T., & Yang, X. S. (2014). Improved cuckoo search algorithm for hybrid flow shop scheduling problems to minimize makespan. *Applied Soft Computing*, *19*, 93–101. doi:10.1016/j.asoc.2014.02.005

Mukherjee, I., & Ray, P. K. (2008). Artificial Neural Network and Metaheuristic Strategies: Emerging Tools for Metal Cutting Process Optimization. In D. Laha & P. Mandal (Eds.), *Handbook of Computational Intelligence in Manufacturing and Production Management* (pp. 366–397). Hershey, PA: Information Science Reference; doi:10.4018/978-1-59904-582-5.ch019

Ong, P. (2014). Adaptive Cuckoo Search Algorithm for Unconstrained Optimization. *TheScientific-WorldJournal*, *8*. doi:10.1155/2014/943403 PMID:25298971

Ouaarab, A., Ahiod, B., & Yang, X. S. (2014). (in press). Random-key cuckoo search for the travelling salesman problem. *Soft Computing*.

Sallem, A., Kotti, M., Fakhfakh, M., Tlelo-Cuautle, E., & Siarry, P. (2015). Combination of Metaheuristics to the Optimal Design of Analog and RF Circuits. In M. Fakhfakh, E. Tlelo-Cuautle, & M. Fino (Eds.), *Performance Optimization Techniques in Analog, Mixed-Signal, and Radio-Frequency Circuit Design* (pp. 210–234). Hershey, PA: Engineering Science Reference; doi:10.4018/978-1-4666-6627-6.ch009

Senvar, O., Turanoglu, E., & Kahraman, C. (2013). Usage of Metaheuristics in Engineering: A Literature Review. In P. Vasant (Ed.), *Meta-Heuristics Optimization Algorithms in Engineering, Business, Economics, and Finance* (pp. 484–528). Hershey, PA: Information Science Reference; doi:10.4018/978-1-4666-2086-5.ch016

Walton, S., Hassan, O., Morgan, K., & Brown, M. R. (2011). Modified cuckoo search: A new gradient free optimisation algorithm. *Chaos, Solitons, and Fractals*, *44*(9), 710–718. doi:10.1016/j.chaos.2011.06.004

Wang, J., Jin, S., Qin, S., & Jiang, H. (2014). Swarm Intelligence-Based Hybrid Models for Short-Term Power Load Prediction. *Mathematical Problems in Engineering*, *Article ID 712417*. doi:10.1155/2014/712417

Yang, X. (2014). Diversity and Mechanisms in Swarm Intelligence. [IJSIR]. *International Journal of Swarm Intelligence Research*, *5*(2), 1–12. doi:10.4018/ijsir.2014040101

Yang, X., & Deb, S. (2015). Cuckoo Search for Optimization and Computational Intelligence. In M. Khosrow-Pour (Ed.), *Encyclopedia of Information Science and Technology* (3rd ed.; pp. 133–142). Hershey, PA: Information Science Reference; doi:10.4018/978-1-4666-5888-2.ch014

Yang, X. S., & Deb, S. (2009). *Cuckoo Search via Levy flights.* Paper presented at the World Congress on Nature & Biologically Inspired Computing (NaBIC 2009) doi:10.1109/NABIC.2009.5393690

Yang, X.-S., & Deb, S. (2010). Engineering optimisation by cuckoo search. *International Journal of Mathematical Modelling and Numerical Optimisation*, *1*(4), 330–343. doi:10.1504/IJMMNO.2010.035430

Yildiz, A. (2013). Cuckoo search algorithm for the selection of optimal machining parameters in milling operations. *International Journal of Advanced Manufacturing Technology*, *64*(1-4), 55–61. doi:10.1007/s00170-012-4013-7

Zainuddin, Z., Wan Daud, W. R., Pauline, O., & Shafie, A. (2011). Wavelet neural networks applied to pulping of oil palm fronds. *Bioresource Technology*, *102*(23), 10978–10986. doi:10.1016/j.biortech.2011.09.080 PMID:21996481

ADDITIONAL READING

Adamidis, P., & Kynigopoulos, G. (2015). EvoWebReg: Web-Based Course Registration and Optimization of Student Personal Schedules with Evolutionary Algorithms. In Research Methods: Concepts, Methodologies, Tools, and Applications (pp. 1523-1539). Hershey, PA: Information Science Reference. doi:10.4018/978-1-4666-7456-1.ch065

Alaa, T., Saoussen, K., & Adel, G. (2015). Genetic Algorithm for Solving the Resource Constrained Project Scheduling Problem. [IJAMC]. *International Journal of Applied Metaheuristic Computing*, *6*(2), 45–60. doi:10.4018/IJAMC.2015040104

Anbuudayasankar, S. P., Ganesh, K., & Lee, T. (2011). Meta-heuristic Approach to Solve Mixed Vehicle Routing Problem with Backhauls in Enterprise Information System of Service Industry. In Enterprise Information Systems: Concepts, Methodologies, Tools and Applications (pp. 1537-1552). Hershey, PA: Business Science Reference. doi:10.4018/978-1-61692-852-0.ch608

Banerjee, A. (2012). Multi-Objective Genetic Algorithm for Robust Clustering with Unknown Number of Clusters. [IJAEC]. *International Journal of Applied Evolutionary Computation*, *3*(1), 1–20. doi:10.4018/jaec.2012010101

Bensouyad, M., Guidoum, N., & Saïdouni, D. (2015). An Efficient Evolutionary Algorithm for Strict Strong Graph Coloring Problem. In Research Methods: Concepts, Methodologies, Tools, and Applications (pp. 2002-2015). Hershey, PA: Information Science Reference. doi:10.4018/978-1-4666-7456-1.ch086

Cadenas, J. M., Carmen Garrido, M., Muñoz, E., Cruz-Corona, C., Pelta, D. A., & Verdegay, J. L. (2009). Different Approaches for Cooperation with Metaheuristics. In J. Rabuñal Dopico, J. Dorado, & A. Pazos (Eds.), *Encyclopedia of Artificial Intelligence* (pp. 480–487). Hershey, PA: Information Science Reference; doi:10.4018/978-1-59904-849-9.ch073

Castillo-Villar, K. K., & Smith, N. R. (2014). Supply Chain Design Including Quality Considerations: Modeling and Solution Approaches based on Metaheuristics. In P. Vasant (Ed.), *Handbook of Research on Novel Soft Computing Intelligent Algorithms: Theory and Practical Applications* (pp. 102–140). Hershey, PA: Information Science Reference; doi:10.4018/978-1-4666-4450-2.ch004

Chiong, R., Neri, F., & McKay, R. I. (2012). Nature That Breeds Solutions. [IJSSS]. *International Journal of Signs and Semiotic Systems*, *2*(2), 23–44. doi:10.4018/ijsss.2012070102

Cuevas, E., & Reyna-Orta, A. (2014). A Cuckoo Search Algorithm for Multimodal Optimization. The Scientific World Journal, vol. 2014. *Article ID*, *497514*, 1–20. doi:10.1155/2014/497514

Das, S., & Konar, A. (2008). Swarm Intelligence in Production Management and Engineering. In D. Laha & P. Mandal (Eds.), *Handbook of Computational Intelligence in Manufacturing and Production Management* (pp. 345–365). Hershey, PA: Information Science Reference; doi:10.4018/978-1-59904-582-5.ch018

Deroussi, L., & Lemoine, D. (2013). Discrete Particle Swarm Optimization for the Multi-Level Lot-Sizing Problem. In P. Yin (Ed.), *Trends in Developing Metaheuristics, Algorithms, and Optimization Approaches* (pp. 99–113). Hershey, PA: Information Science Reference; doi:10.4018/978-1-4666-2145-9.ch007

Freschi, F., Coello Coello, C. A., & Repetto, M. (2009). Multiobjective Optimization and Artificial Immune Systems: A Review. In H. Mo (Ed.), *Handbook of Research on Artificial Immune Systems and Natural Computing: Applying Complex Adaptive Technologies* (pp. 1–21). Hershey, PA: Medical Information Science Reference; doi:10.4018/978-1-60566-310-4.ch001

Gaitonde, V. N., Karnik, S. R., & Davim, J. P. (2012). Application of Particle Swarm Optimization for Achieving Desired Surface Roughness in Tungsten-Copper Alloy Machining. In J. Davim (Ed.), *Computational Methods for Optimizing Manufacturing Technology: Models and Techniques* (pp. 144–161). Hershey, PA: Engineering Science Reference; doi:10.4018/978-1-4666-0128-4.ch006

Ganesan, T., Elamvazuthi, I., Shaari, K. Z. K., & Vasant, P. (2013). Swarm intelligence and gravitational search algorithm for multi-objective optimization of synthesis gas production. *Applied Energy*, *103*, 368–374. doi:10.1016/j.apenergy.2012.09.059

Gaspar-Cunha, A., Mendes, F., Duarte, J., Vieira, A., Ribeiro, B., Ribeiro, A., & Neves, J. (2010). Feature Selection for Bankruptcy Prediction: A Multi-Objective Optimization Approach. [IJNCR]. *International Journal of Natural Computing Research*, *1*(2), 71–91. doi:10.4018/jncr.2010040105

Gonsalves, T. (2015). Hybrid Swarm Intelligence. In M. Khosrow-Pour (Ed.), *Encyclopedia of Information Science and Technology* (3rd ed., pp. 175–186). Hershey, PA: Information Science Reference; doi:10.4018/978-1-4666-5888-2.ch018

Gordini, N. (2014). Genetic Algorithms for Small Enterprises Default Prediction: Empirical Evidence from Italy. In P. Vasant (Ed.), *Handbook of Research on Novel Soft Computing Intelligent Algorithms: Theory and Practical Applications* (pp. 258–293). Hershey, PA: Information Science Reference; doi:10.4018/978-1-4666-4450-2.ch009

Jiang, P., Zhou, Q., Jiang, H., & Dong, Y. (2014). An Optimized Forecasting Approach Based on Grey Theory and Cuckoo Search Algorithm: A Case Study for Electricity Consumption in New South Wales. Abstract and Applied Analysis, vol. 2014. *Article ID*, *183095*, 1–13. doi:10.1155/2014/183095

Kanthababu, M. (2013). Multi-Objective Optimization of Manufacturing Processes Using Evolutionary Algorithms. In Industrial Engineering: Concepts, Methodologies, Tools, and Applications (pp. 352-376). Hershey, PA: Engineering Science Reference. doi:10.4018/978-1-4666-1945-6.ch021

Kessentini, S., Barchiesi, D., Grosges, T., Giraud-Moreau, L., & Lamy de la Chapelle, M. (2013). Adaptive Non-Uniform Particle Swarm Application to Plasmonic Design. In P. Yin (Ed.), *Trends in Developing Metaheuristics, Algorithms, and Optimization Approaches* (pp. 114–124). Hershey, PA: Information Science Reference; doi:10.4018/978-1-4666-2145-9.ch008

Köse, U. (2013). Newly Developed Nature-Inspired Algorithms and their Applications to Recommendation Systems. In S. Dehuri, M. Patra, B. Misra, & A. Jagadev (Eds.), *Intelligent Techniques in Recommendation Systems: Contextual Advancements and New Methods* (pp. 214–229). Hershey, PA: Information Science Reference; doi:10.4018/978-1-4666-2542-6.ch011

Kose, U. (2015). Present State of Swarm Intelligence and Future Directions. In M. Khosrow-Pour (Ed.), *Encyclopedia of Information Science and Technology* (3rd ed., pp. 239–252). Hershey, PA: Information Science Reference; doi:10.4018/978-1-4666-5888-2.ch023

Lienland, B., & Zeng, L. (2015). A Review and Comparison of Genetic Algorithms for the 0-1 Multidimensional Knapsack Problem. [IJORIS]. *International Journal of Operations Research and Information Systems*, *6*(2), 21–31. doi:10.4018/ijoris.2015040102

Madureira, A. M. (2012). Hybrid Meta-Heuristics Based System for Dynamic Scheduling. In Machine Learning: Concepts, Methodologies, Tools and Applications (pp. 428-435). Hershey, PA: Information Science Reference. doi:10.4018/978-1-60960-818-7.ch305

Mahmoudzadeh, H., & Eshghi, K. (2012). A Metaheuristic Approach to the Graceful Labeling Problem. In P. Yin (Ed.), *Modeling, Analysis, and Applications in Metaheuristic Computing: Advancements and Trends* (pp. 217–232). Hershey, PA: Information Science Reference; doi:10.4018/978-1-4666-0270-0.ch013

Mankad, K. B. (2015). An Intelligent Process Development Using Fusion of Genetic Algorithm with Fuzzy Logic. In P. Vasant (Ed.), *Handbook of Research on Artificial Intelligence Techniques and Algorithms* (pp. 44–81). Hershey, PA: Information Science Reference; doi:10.4018/978-1-4666-7258-1.ch002

Mishra, B. S., Dehuri, S., Mall, R., & Ghosh, A. (2013). Parallel Single and Multiple Objectives Genetic Algorithms: A Survey. In W. Samuelson Hong (Ed.), *Modeling Applications and Theoretical Innovations in Interdisciplinary Evolutionary Computation* (pp. 73–110). Hershey, PA: Information Science Reference; doi:10.4018/978-1-4666-3628-6.ch006

Mitropoulou, Chara Ch., Vagelis Plevris and Nikos D. Lagaros. "Metaheuristic Optimization in Seismic Structural Design and Inspection Scheduling of Buildings." *Structural Seismic Design Optimization and Earthquake Engineering: Formulations and Applications.* IGI Global, 2012. 152-173. Web. 31 Mar. 2015. doi:10.4018/978-1-4666-1640-0.ch007

Montoya-Torres, J. R., Gómez-Vizcaíno, L. S., Solano-Charris, E. L., & Paternina-Arboleda, C. D. (2012). Global Bacteria Optimization Meta-Heuristic Algorithm for Jobshop Scheduling. In J. Wang (Ed.), *Innovations in Information Systems for Business Functionality and Operations Management* (pp. 182–193). Hershey, PA: Business Science Reference; doi:10.4018/978-1-4666-0933-4.ch011

Morais, R. M., & Pinto, A. N. (2013). Topological Design Using Genetic Algorithms. In Y. Kavian & Z. Ghassemlooy (Eds.), *Intelligent Systems for Optical Networks Design: Advancing Techniques* (pp. 153–173). Hershey, PA: Information Science Reference; doi:10.4018/978-1-4666-3652-1.ch007

Mukherjee, I., & Ray, P. K. (2008). Artificial Neural Network and Metaheuristic Strategies: Emerging Tools for Metal Cutting Process Optimization. In D. Laha & P. Mandal (Eds.), *Handbook of Computational Intelligence in Manufacturing and Production Management* (pp. 366–397). Hershey, PA: Information Science Reference; doi:10.4018/978-1-59904-582-5.ch019

Narula, P., Misra, S., & Kumar Dhurandher, S. (2009). Swarm Intelligence Approach for Ad-Hoc Networks. In J. Rabuñal Dopico, J. Dorado, & A. Pazos (Eds.), *Encyclopedia of Artificial Intelligence* (pp. 1530–1536). Hershey, PA: Information Science Reference; doi:10.4018/978-1-59904-849-9.ch224

Nesmachnow, S. (2015). Using Metaheuristics as Soft Computing Techniques for Efficient Optimization. In M. Khosrow-Pour (Ed.), *Encyclopedia of Information Science and Technology* (3rd ed., pp. 7390–7399). Hershey, PA: Information Science Reference; doi:10.4018/978-1-4666-5888-2.ch727

Nesmachnow, S., Cancela, H., & Alba, E. (2010). Nature-Inspired Informatics for Telecommunication Network Design. In R. Chiong (Ed.), *Nature-Inspired Informatics for Intelligent Applications and Knowledge Discovery: Implications in Business, Science, and Engineering* (pp. 323–371). Hershey, PA: Information Science Reference; doi:10.4018/978-1-60566-705-8.ch014

Nguyen, S., & Kachitvichyanukul, V. (2010). Movement Strategies for Multi-Objective Particle Swarm Optimization. [IJAMC]. *International Journal of Applied Metaheuristic Computing*, *1*(3), 59–79. doi:10.4018/jamc.2010070105

Nguyen, S., & Kachitvichyanukul, V. (2012). Movement Strategies for Multi-Objective Particle Swarm Optimization. In P. Yin (Ed.), *Modeling, Analysis, and Applications in Metaheuristic Computing: Advancements and Trends* (pp. 109–130). Hershey, PA: Information Science Reference; doi:10.4018/978-1-4666-0270-0.ch008

Nobre, T., Vergilio, S. R., & Pozo, A. (2012). Reducing Interface Mutation Costs with Multiobjective Optimization Algorithms. [IJNCR]. *International Journal of Natural Computing Research*, *3*(3), 21–40. doi:10.4018/jncr.2012070102

Ohki, M. (2014). Nurse Scheduling by Cooperative GA with Effective Virus Operator. [IJAEC]. *International Journal of Applied Evolutionary Computation*, *5*(1), 19–29. doi:10.4018/ijaec.2014010102

Pham, D., & Castellani, M. (2015). The Bees Algorithm as a Biologically Inspired Optimisation Method. In M. Khosrow-Pour (Ed.), *Encyclopedia of Information Science and Technology* (3rd ed., pp. 285–294). Hershey, PA: Information Science Reference; doi:10.4018/978-1-4666-5888-2.ch027

Preve, N. (2012). Balanced Job Scheduling Based on Ant Algorithm for Grid Network. In Grid and Cloud Computing: Concepts, Methodologies, Tools and Applications (pp. 1114-1131). Hershey, PA: Information Science Reference. doi:10.4018/978-1-4666-0879-5.ch506

Ries, J., Beullens, P., & Wang, Y. (2015). Instance-Specific Parameter Tuning for Meta-Heuristics. In Research Methods: Concepts, Methodologies, Tools, and Applications (pp. 1660-1693). Hershey, PA: Information Science Reference. doi:10.4018/978-1-4666-7456-1.ch072

Roeva, O., Slavov, T., & Fidanova, S. (2014). Population-Based vs. Single Point Search Meta-Heuristics for a PID Controller Tuning. In P. Vasant (Ed.), *Handbook of Research on Novel Soft Computing Intelligent Algorithms: Theory and Practical Applications* (pp. 200–233). Hershey, PA: Information Science Reference; doi:10.4018/978-1-4666-4450-2.ch007

Roy, R., Dehuri, S., & Cho, S. B. (2013). A Novel Particle Swarm Optimization Algorithm for Multi-Objective Combinatorial Optimization Problem. In P. Yin (Ed.), *Trends in Developing Metaheuristics, Algorithms, and Optimization Approaches* (pp. 1–16). Hershey, PA: Information Science Reference; doi:10.4018/978-1-4666-2145-9.ch001

Salcedo-Sanz, S., Camps-Valls, G., & Bousoño-Calzón, C. (2009). Hybrid Genetic Algorithms in Data Mining Applications. In J. Wang (Ed.), *Encyclopedia of Data Warehousing and Mining* (2nd ed., pp. 993–998). Hershey, PA: Information Science Reference; doi:10.4018/978-1-60566-010-3.ch154

Stanimirović, Z., & Mišković, S. (2014). Efficient Metaheuristic Approaches for Exploration of Online Social Networks. In W. Hu & N. Kaabouch (Eds.), *Big Data Management, Technologies, and Applications* (pp. 222–269). Hershey, PA: Information Science Reference; doi:10.4018/978-1-4666-4699-5.ch010

Taleizadeh, A. A., & Cárdenas-Barrón, L. E. (2013). Hybrid Metaheuristics Algorithms for Inventory Management Problems. In P. Vasant (Ed.), *Meta-Heuristics Optimization Algorithms in Engineering, Business, Economics, and Finance* (pp. 312–356). Hershey, PA: Information Science Reference; doi:10.4018/978-1-4666-2086-5.ch011

Torres, S. P., Castro, C. A., & Rider, M. J. (2013). Transmission Expansion Planning by using DC and AC Models and Particle Swarm Optimization. In G. Fornarelli & L. Mescia (Eds.), *Swarm Intelligence for Electric and Electronic Engineering* (pp. 260–284). Hershey, PA: Engineering Science Reference; doi:10.4018/978-1-4666-2666-9.ch013

Toscano, R. (2010). Stochastic Methods for Hard Optimization: Application to Robust Control and Fault Diagnosis of Industrial Systems. In G. Rigatos (Ed.), *Intelligent Industrial Systems: Modeling, Automation and Adaptive Behavior* (pp. 182–220). Hershey, PA: Information Science Reference; doi:10.4018/978-1-61520-849-4.ch007

Tuzkaya, G., Gülsün, B., & Bildik, E. (2011). Reverse Logistics Network Design Using a Hybrid Genetic Algorithm and Simulated Annealing Methodology. In I. Mahdavi, S. Mohebbi, & N. Cho (Eds.), *Electronic Supply Network Coordination in Intelligent and Dynamic Environments: Modeling and Implementation* (pp. 168–186). Hershey, PA: Business Science Reference; doi:10.4018/978-1-60566-808-6.ch007

Unay, A. T., & Guzey, M. (2013). A Swarm Intelligence Heuristic Approach to Longest Common Subsequence Problem for Arbitrary Number of Sequences. *Metabolomics, 3*(2), 1–6. doi:10.4172/2153-0769.1000120

Vasant, P. (2014). Hybrid Evolutionary Optimization Algorithms: A Case Study in Manufacturing Industry. In Z. Luo (Ed.), *Smart Manufacturing Innovation and Transformation: Interconnection and Intelligence* (pp. 59–95). Hershey, PA: Business Science Reference; doi:10.4018/978-1-4666-5836-3.ch004

Vora, M., & Mirnalinee, T. T. (2015). From Optimization to Clustering: A Swarm Intelligence Approach. In P. Vasant (Ed.), *Handbook of Research on Artificial Intelligence Techniques and Algorithms* (pp. 594–619). Hershey, PA: Information Science Reference; doi:10.4018/978-1-4666-7258-1.ch019

Wang, Y. (2012). A Sociopsychological Perspective on Collective Intelligence in Metaheuristic Computing. In P. Yin (Ed.), *Modeling, Analysis, and Applications in Metaheuristic Computing: Advancements and Trends* (pp. 267–285). Hershey, PA: Information Science Reference; doi:10.4018/978-1-4666-0270-0.ch015

Weise, T., & Chiong, R. (2010). Evolutionary Approaches and Their Applications to Distributed Systems. In R. Chiong (Ed.), *Intelligent Systems for Automated Learning and Adaptation: Emerging Trends and Applications* (pp. 114–149). Hershey, PA: Information Science Reference; doi:10.4018/978-1-60566-798-0.ch006

Xing, B., & Gao, W. (2015). Applications of Swarm Intelligence in Remanufacturing. In M. Khosrow-Pour (Ed.), *Encyclopedia of Information Science and Technology* (3rd ed., pp. 66–74). Hershey, PA: Information Science Reference; doi:10.4018/978-1-4666-5888-2.ch007

Yang, B. (2009). Artificial Life Optimization Algorithm and Applications. In H. Mo (Ed.), *Handbook of Research on Artificial Immune Systems and Natural Computing: Applying Complex Adaptive Technologies* (pp. 423–449). Hershey, PA: Medical Information Science Reference; doi:10.4018/978-1-60566-310-4.ch020

Yang, N., Wang, S., & Schonfeld, P. (2015). Simulation-Based Scheduling of Waterway Projects Using a Parallel Genetic Algorithm. [IJORIS]. *International Journal of Operations Research and Information Systems*, *6*(1), 49–63. doi:10.4018/ijoris.2015010104

Yuce, B., Mastrocinque, E., Packianather, M. S., Lambiase, A., & Pham, D. T. (2015). The Bees Algorithm and Its Applications. In P. Vasant (Ed.), *Handbook of Research on Artificial Intelligence Techniques and Algorithms* (pp. 122–151). Hershey, PA: Information Science Reference; doi:10.4018/978-1-4666-7258-1.ch004

KEY TERMS AND DEFINITIONS

Cuckoo: The birds which employ unique reproduction strategy in which the female cuckoos lay their fertilized eggs in the nest of other bird species.

Cuckoo Search Algorithm: An optimization algorithm inspired by the obligate interspecific brood parasitism of some cuckoo species that lay their eggs in the nests of other host birds.

Exploitation: The searching process around the current best solution in the local region and select the best solution which will converge to the optimality.

Exploration: The searching process around the search space on the global scale in order to avoid the solution from getting trapped at local optimal and improve diversity of solution.

Lévy Flight: A random walk in which the step-lengths have a probability distribution that is heavy-tailed.

Nature Inspired Algorithm: The optimization algorithms which inspired by how nature adapts to challenging circumstances.

Optimization: The action to identify the best available solutions which either maximize or minimize a given objective function, while conforming to a set of prescribed constraints.

Test Function: The functions used to validate the performances of the developed optimization algorithm.

Chapter 27
Gravitational Search Algorithm: Concepts, Variants, and Operators

Hossein Nezamabadi-pour
Shahid Bahonar University of Kerman, Iran

Fatemeh Barani
Higher Education Complex of Bam, Iran

ABSTRACT

During the last decades, several metaheuristics have been developed to solve complex engineering optimization problems which most of them have been inspired by natural phenomena and swarm behaviors. Metaheuristics are the most selected techniques to find optimal solution intelligently in many areas of scheduling, space allocation, decision making, pattern recognition, document clustering, control objectives, image processing, system and filter modeling, etc. These algorithms have promised better solutions in single and multi-objective optimization. Gravitational search algorithm (GSA) is one of the recent created metaheuristic search algorithms, which is inspired by the Newtonian laws of gravity and motion. GSA was first proposed by Rashedi et al. and in the short time it became popular among the scientific community and researchers resulting in a lot of variants of the basic algorithm with improved performance. This chapter book presents a detailed review of the basic concepts of GSA and a comprehensive survey of its advanced versions. We propose a number of suggestions to the GSA community that can be undertaken to help move the area forward.

1 INTRODUCTION

With the growth of computer technology, storage devices, and soft computing (SC) techniques, it is now very accessible to solve more difficult and complex real-world problems in the fields of system modelling and optimization. For years, the researchers have been looking into nature to find heuristic approaches for inspiration to handle complicated computational problems. Optimization is at the heart of various natural processes like Darwinian evolution. Over millions of years, each species had to adapt itself to fit to the environment it was in (Das et al., 2011). The evolutionary computation (EC) techniques as an

DOI: 10.4018/978-1-4666-9644-0.ch027

important paradigm of computational intelligence (CI) have been developed by mimicking biological evolution to accomplish complex real-world optimization. EC is a form of stochastic optimization search.

Optimization tasks are unavoidable in many disciplines ranging from arts and design, business and finance to science and engineering (Ong et al., 2009). Classical optimization algorithms cannot provide a suitable solution in so many complex fields due to the increase of search space with the problem size, dependency of these algorithms on initial solutions, etc. Therefore, solving these problems using classical techniques is impractical and this causes a growth interest in metaheuristic search algorithms (Talbi et al., 2002; Sarafrazi et al., 2012).

There are two distinct forms of nature-inspired EC techniques which are evolutionary algorithms (EA), and swarm intelligence (SI)-based algorithms (Karaboga et al., 2014; Davarynejad et al., 2014). EAs are metaheuristic global search methods and optimization algorithms modeled from natural genetic principles such as natural selection. The basic idea of natural selection is "Select the best, discard the rest". It means that better individuals get higher chance to survive. The important methods in the field of EAs are Genetic Algorithms (GA) proposed by Holland (1975), Evolutionary Programming (EP) proposed by Fogel et al. (1966), Evolutionary Strategies (ES) proposed by Rechenberg (1973) and Schwefel (1975), Genetic Programming proposed by Koza (1992) and Differential Evolution (DE) proposed by Storn and Price (1995).

Swarm Intelligence (SI) refers to a newly developed group of population-based algorithms for multi-agent search and optimization. SI studies the collective behavior of systems made up of a population of simple agents interacting locally with each other and with their environment. In the SI systems the agents follow very simple rules, although there is no centralized control structure dictating how individual agents should behave. Social interactions (locally shared knowledge) provide the basis for unguided problem solving. In recent years, the swarm intelligence metaheuristics have received tremendous attention in research, mainly as Ant Colony Optimization (ACO) and Particle Swarm Optimization (PSO). ACO is the first family of SI-based search algorithms which was proposed by Dorigo et al. (1991), where the behavior of ants is modeled in finding the shortest path from nest to food source. PSO, which was proposed by Kennedy and Eberhart (1995), mimics the flocking behavior of birds and fish. Some other successful instances of swarm intelligence metaheuristics are bees algorithm (Jung, 2003; Karaboga, 2005), bacterial foraging optimization (BFO) (Passino, 2002), monkey algorithms (Soleimanpour, 2013), and gravitational search algorithm (GSA) (Rashedi et al., 2009).The interested reader could follow refs. (Vasant, 2012; Vasant, 2014).

In the late-2000s, gravitational search algorithm was introduced by E. Rashedi and her colleagues (Rashedi, 2007a, 2007b, 2007c, 2007d, 2009) as a novel nature inspired population based metaheuristic for solving continuous (real-valued) optimization problems. GSA belongs to the class of metaheuristics which are stochastic global optimization methods. It is inspired by Newtonian laws of gravity and motion and notion of mass interactions. The GSA uses the theory of Newtonian physics and its searcher agents are the collection of objects. In GSA, there is an isolated system of objects. Using the gravitational force, every object in the system can see the situation of other objects. The gravitational force is therefore a way of transferring information between different objects (Rashedi et al., 2009). In other words, in GSA, searcher agents are considered as objects which their masses are considered corresponding to their performance. All these objects attract each other by a gravity force, and this force causes a movement of all objects globally towards the objects with heavier masses. The heavy objects correspond to good solutions of the problem. The position of the agent corresponds to a solution of the problem, and its mass is determined using a fitness function. By lapse of time, objects are attracted by the heaviest object, which would ideally present the optimum solution in the search space.

Since GSA has been introduced, various advances and enhances of it have been developed, such as some versions for continuous, binary, discrete, multimodal, single-objective and multi-objective optimization. Furthermore, In order to improve the performance of GSA, it has been combined with some traditional and evolutionary optimization algorithms, such as PSO, Neural network, Support vector machine and so on. Also, GSA due to its good exploration and exploitation capabilities has been applied in various applications in a wide range of areas.

To the best of our knowledge, so far there is no survey article reviewing recent advances and new variants of GSA for solving different optimization problems as well as new operators and hybrid algorithms to empower GSA in solving complex problems. In this book chapter, we will try to give a detailed analysis of the process of GSA and a review of its recent advances that have been developed so far. This book chapter may be useful for a broad audience including students, scholars, academicians, researchers who study environmental services and work in the different fields of engineering, such as computer sciences, civil, electronic, industrial, and so on.

The rest of the chapter is organized as follows. Section 2 introduces the concepts and the basic GSA with original formulation in real-valued space. Section 3 provides an overview of several prominent operators used in GSA. Section 4 describes the GSA's variants for solving different types of optimization problems including continuous (real-valued), binary-valued, triple-valued, discrete, multimodal, multi-objective optimization problems. Section 5 introduces a survey on the hybridization of GSA with other heuristic algorithms. Section 6 provides the parameter adaptation and control schemes for GSA. Finally, the paper is concluded in the last Section. In this section drawbacks of GSA are pointed out and a number of future research issues related to GSA are highlighted.

2 GRAVITATIONAL SEARCH ALGORITHM

Rashedi et al. introduced a new optimization algorithm for solving real-valued (continuous) problems, called Gravitational Search Algorithm (GSA), based on the law of gravity and mass interactions (Rashedi and Nezamabadi-pour, 2007a;Rashedi et al.,2007b;Rashedi et al., 2007c;Rashedi 2007d;Rashedi et al.2009). The searcher agents are a collection of objects which interact with each other by simulation of the Newtonian gravity and the laws of motion to find the optimum solution. The mass of each agent is suggested by its performance and the gravity force causes a global movement of all agents towards the agents with heavier masses. In the algorithm, heavy masses correspond to good solutions that move slowly than lighter masses. This case guarantees the exploitation ability of the algorithm.

Authors were considered four features for each object in their proposed algorithm:

1. Position
2. Inertial mass
3. Active gravitational mass
4. Passive gravitational mass

The position of the object corresponds to a candidate solution of the problem, and its gravitational and inertial masses are determined using a fitness function. The heaviest mass attracts other masses and presents an (sub-) optimum solution in the search space. Objects obey the following modified laws (Rashedi et al., 2009):

- **Modified Law of Gravity:** Each object attracts every other object and the gravitational force between two objects is directly proportional to the product of their masses and inversely proportional to the distance between them, R. In GSA R is used instead of R^2, because according to the authors experience, R provides better results than R^2 in all experimental cases.
- **Modified Law of Motion:** The current velocity of any object is equal to the sum of the fraction of its previous velocity and the variation in the velocity. Variation in the velocity or acceleration of any object is equal to the force acted on the system divided by its mass of inertia.

In GSA, an n-dimensional solution space with s searcher agents are considered in which the position of ith agent is represented by the vector X_i:

$$X_i = \left(x_i^1, \ldots, x_i^d, \ldots, x_i^n\right);\ i=1,2,\ldots,s \tag{1}$$

where x_i^d presents the position of the ith object in the dth dimension. Authors defined the force acting on object i from object j at time t as follows.

$$F_{ij}^d(t) = G(t)\frac{M_{pi}(t)\times M_{aj}(t)}{R_{ij}(t)+\varepsilon}\left(x_j^d(t) - x_i^d(t)\right) \tag{2}$$

where M_{aj} and M_{pi} are the active gravitational mass related to object j and the passive gravitational mass related to object i, respectively. The term ε is a very small value used in order to avoid from division by zero error whenever the Euclidean distance between two objects i and j is equal to zero. $R_{ij}(t)$ is the Euclidean distance between two objects i and j. $G(t)$ is gravitational constant at time t and it is initialized at the beginning by G_0 and will be reduced with time to control the search accuracy.

Overall force that acts on the ith object in the dimension d from other agents is calculated using following equation:

$$F_i^d(t) = \sum_{j=1, j\neq i}^{S} rand_j F_{ij}^d(t), \tag{3}$$

where $rand_j$ is a random number in the interval [0,1]. Based on the law of motion, the acceleration of the agent i at time t, in dimension d, is given by Equation (4):

$$a_i^d(t) = \frac{F_i^d(t)}{M_{ii}(t)}, \tag{4}$$

where M_{ii} is the inertial mass of ith object. Assuming the equality of the gravitational and inertia mass, the above equation is rewritten as follows.

$$a_i^d(t) = \sum_{j=1, j\neq i}^{s} rand_j G(t)\frac{M_j(t)}{R_{ij}(t)+\varepsilon}\left(x_j^d(t) - x_i^d(t)\right), \tag{5}$$

The gravitational and inertial masses are updated using the map of fitness by following equations:

$$m_i(t) = \frac{fit_i(t) - worst(t)}{best(t) - worst(t)}, \tag{6}$$

$$M_i(t) = M_{pi}(t) = M_{ai}(t) = M_{ii}(t) = \frac{m_i(t)}{\sum_{j=1}^{s} m_j(t)}, \tag{7}$$

where $fit_i(t)$ is the fitness value of the object i at time t, and, $worst(t)$ and $best(t)$ for maximization problems are defined as follows:

$$best(t) = \max_{j \in \{1,\dots,s\}} fit_j(t), \tag{8}$$

$$worst(t) = \min_{j \in \{1,\dots,s\}} fit_j(t), \tag{9}$$

Furthermore, the next velocity of the ith object in dth dimensional is calculated as a fraction of its current velocity added to its acceleration by Equation (10). Then, the next position of the ith object in dimension d can be calculated using Equation (11):

$$v_i^d(t+1) = rand_i \times v_i^d(t) + a_i^d(t), \tag{10}$$

$$x_i^d(t+1) = x_i^d(t) + v_i^d(t+1), \tag{11}$$

where $rand_i$ is a uniform random variable in the interval [0,1]. Authors used random numbers $rand_i$ and $rand_j$ to give a randomized characteristic to the search and to enhance the exploration ability of GSA.

One way to perform a good compromise between exploration and exploitation is to reduce the number of agents with lapse of time in Equation (3). Hence, the authors suggested only a set of objects with bigger mass apply their force to the other. Hence, Equation (3) is modified as follows:

$$F_i^d(t) = \sum_{j \in Kbest, j \neq i}^{S} rand_j F_{ij}^d(t), \tag{12}$$

where *Kbest* is the set of first *K* objects with the best fitness value and biggest mass. *Kbest* is a function of time, with the initial value K_0 at the beginning and decreasing with time. The different steps of GSA are as follows:

```
Generate the initial population;
Evaluate the fitness for each agent;
```

```
While stopping criteria is not satisfied Do
   Update Kbest, worst, best;
   Calculate the mass for each agent by Equation (7);
   Calculate the force of each agent by Equation (12);
   Calculate the acceleration and velocity of each agent by Equations (5) and
(10);
   Update the position of each agent by Equation (11);
   Evaluate the fitness for each agent;
Endwhile
Output: Best solution found.
```

Some of characteristics of GSA, stated in (Rashedi et al., 2009), are reported in the following.

- The gravitational force is playing the role of an information-transferring tool that by using it each agent (object) could observe the performance of the others in the search space.
- Each object, due to the force that acts on it from its neighborhood objects, can see and percept space around itself.
- In the theory of gravity, a heavy mass has a large effective attraction radius and hence a great intensity of attraction. Therefore, in GSA the objects with a higher performance have a greater gravitational mass. As a result, the objects tend to move toward the best object.
- The inertia mass is against the motion and make the mass movement slow. Therefore, agents with heavy inertia mass move slowly and hence search the space more locally. So, it can be considered as an adaptive learning rate.
- The accuracy of the search can be adjusted by gravitational constant, so by controlling it with time, it is possible to balance the exploration and exploitation through different steps of the algorithm.
- Although GSA is a memory-less algorithm; it works effectively like the algorithms with memory. The experimental results show the good convergence rate of the GSA. According to the complexity of problems and conditions the problem optimized, one could use memorized GSA to get higher performance.
- The authors assume that the gravitational and the inertia masses are the same. However, for some applications different values for them can be used. A bigger inertia mass provides a slower motion of agents in the search space and hence a more precise search. Conversely, a bigger gravitational mass causes a higher attraction of agents. This permits a faster convergence.

3 OPERATORS IN GRAVITATIONAL SEARCH ALGORITHM

The metaheuristic algorithms must have a good balance between exploration and exploitation to achieve both an efficient global and local search. In this way, they can effectively solve the optimization problems. The exploration is the ability to investigate the search space for finding new and better solutions and the exploitation is the ability to look for the optimal solution near a good one. The abilities of exploration and exploitation in every metaheuristic algorithm are applied with the specific operators. Since each operator has its own abilities of exploration and exploitation, the operators should be artfully hybridized together for a good trade-off between exploitation and exploration. Hence, new operators are designed or

Table 1. Summarization of main operators of GSA

Operator	Scholars (years)	Contributions
Disruption	Sarafrazi et al. (2011) Liu et al. (2012)	They proposed a new operator inspired by the disruption phenomenon in astronomy to improve GSA
Mutation	Nobahari et al. (2012)	They modeled mutation phenomenon inspired by natural genetic in original GSA using two mutation operators, called sign and reordering, to improve the diversity within swarm.
	Tabatabaei (2014)	The author presented four new mutation operators for GSA.
Chaotic	Han and Chang(2012)	They equipped GSA with a novel chaotic operator. The chaotic operator is used in the updating phase of the position of agents into GSA to enhance the escaping ability.
Crossover	Shang (2013) Liu et al. (2014)	Author applied the crossover operator to enhance GSA for the local search capacity.
Escape	Rashedi and Nezamabadi-pour (2013)	The operator was inspired by the concept of escape velocity in physics.
Discrete Local Search (DLS)	Doraghinejad et al. (2014)	DLS operator was proposed to increase the exploitation ability in a discrete integer-valued space.
Black hole	Doraghinejad and Nezamabadi-pour (2014)	Black hole operator is inspired by some characteristics of the black hole phenomenon in astronomy.
Kepler	Sarafrazi et al. (2015)	Kepler operator is based on the first Kepler law and inspired by the astrophysics.

the available operators are redesigned in order to add specific capabilities to metaheuristic algorithms for solving some problems (Sarafrazi et al., 2011). The main operators defined to enhance GSA in solving different optimization problems are summarized in Table 1.

3.1 Disruption Operator

Disruption phenomenon in astronomy is the sudden inward fall of a swarm of gravitationally particles under the influence of the force of gravity. Sarafrazi et al. (2011) proposed a new operator inspired by the disruption phenomenon in astronomy to improve GSA. To implement disruption, the best agent is considered as the star of the system whereas the other agents may be disrupted under the gravity force of the star. If the ratio of the distance between agent i and its nearest neighbor to its distance from star $(\frac{R_{ij}}{R_{i,best}})$ is smaller than a given threshold, $C = c_0(1-\frac{t}{T})$. The disruption operator is applied on position of agent i as follows:

$$X_i(t)=X_i(t).D$$

$$D = \begin{cases} R_{i,j}.U\left(-0.5,0.5\right) & \text{if } R_{i,best} \geq 1 \\ 1+\rho.U\left(-0.5,0.5\right) & otherwise \end{cases} \tag{13}$$

where ρ is a small number and U(–0.5,0.5) is a uniformly distributed pseudo random number in interval [-0.5,0.5]. If $Ri_{,best\geq}1$, the algorithm explores the search space and otherwise, it exploits around the best agent.

Liu et al. (2012) introduced an improved disruption operator to decrease the computational complexity of original disruption operator. Also, the authors used the elitism updating strategy to improve the coverage rate of GSA. Based on this strategy only the agents whose fitness value has improved are disrupted.

3.2 Mutation Operators

The close objects may impact to each other and so the direction of their movement may change. To simulate the irregularities in real object, Nobahari et al. (2012) modeled mutation phenomenon in original GSA using two mutation operators, called sign and reordering, to improve the diversity within swarm and decrease the probability of trapping in local optimum. In sign mutation, the sign of velocity of each agent is changed with a predefined probability as follows.

$$v_i'^d\left(t+1\right) = sign_i^d v_i^d\left(t+1\right) d = 1,\ldots,n, i = 1,\ldots,s \tag{14}$$

$$sign_i^d = \begin{cases} -1 & rand < p_s \\ 1 & otherwise \end{cases} \tag{15}$$

$$x_i^d\left(t+1\right) = x_i^d\left(t\right) + v_i'^d\left(t+1\right) \tag{16}$$

where $v_i'^d\left(t+1\right)$ is the mutated velocity by the sign mutation operator and *rand* is a uniform random number in the interval [0,1]. p_s is the probability of sign mutation. In reordering mutation, velocity of some agents is rearranged randomly with a predefined probability, $R_{ij}(t)$, as shown in Figure1.

Figure 1. Mutation operators

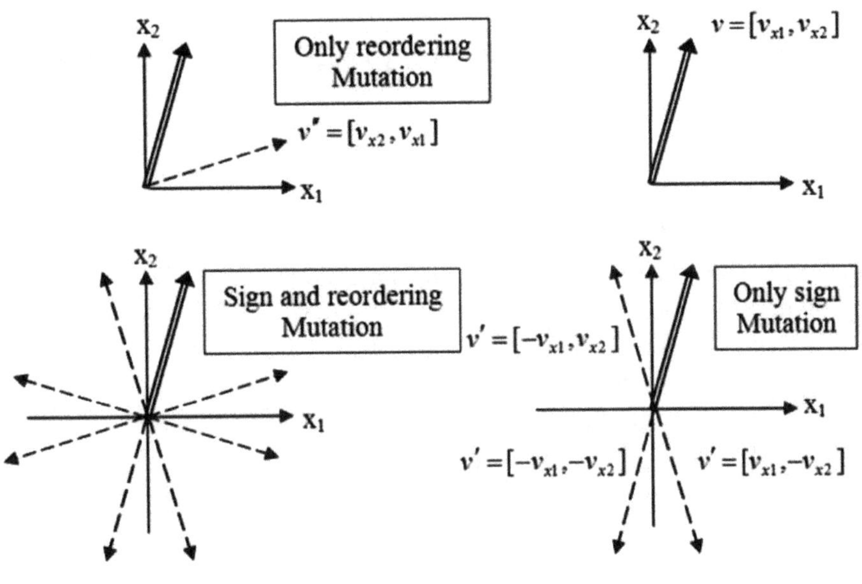

In general, updating position of each agent based on mutation operators is done as follows:

$$v_i(t+1) = \omega(t) \times v_i(t) + a_i(t)$$

$$v_i^{'}(t+1) = Sign\_Mutate(v_i(t+1))$$

$$v_i^{''}(t+1) = Reordering\_Mutate(v_i^{'}(t+1))$$

$$x_i(t+1) = x_i(t) + v_i^{''}(t+1) \quad (17)$$

where $\omega(t) = w_0 - (w_0 - w_1)\frac{t}{T}$ is a time varying coefficient to improve exploration and exploitation abilities in the beginning and the ending iterations, respectively.

Tabatabaei (2014) presented a modified gravitational search algorithm based on four novel mutation methods to overcome some problems of standard GSA such as stagnation, weak exploration, and trapping in local optima.

3.3 Chaotic Operator

Han et al. (2012) applied an equipped GSA with a novel chaotic operator to propose a chaotic secure communication scheme. In the updating phase of the position of agents, the chaotic operator (which is a kind of nonlinear dynamic) is employed in GSA to enhance the escape ability from premature convergence and local optimum. A n-dimensional random vector c_i in the interval [0,1] is generated and then new chaotic vectors are produced at each iteration as follows.

$$c_i^d = 4c_{i-1}^d\left(1 - c_{i-1}^d\right),\ i=2,\ldots,s \quad (18)$$

where s is the number of agents in system. The velocity updating of each agent in GSA is modified as follows.

$$v_i^{d'}(t+1) = \left[rand_i \times v_i^d(t) + \xi\left(c_i^d - 0.5\right)\right] + a_i^d(t),\ i=2,\ldots,s \quad (19)$$

where ξ is a factor to control the chaos scope and c_i is a chaotic vector. The position updating is performed as the following equation:

$$x_i^{d'}(t+1) = x_i^d(t) + v_i^{d'}(t+1),\ i=2,\ldots,s \quad (20)$$

3.4 Crossover Operator

Shang (2013) applied the neighborhood crossover operator (NCO) to enhance GSA for the local search capacity. A scale factor, w, was used to balance between GSA and NCO to obtain best results. Let r is a random number in the interval [0,1]. If $r<w$, the gravitational search algorithm is used to search the solution space. Otherwise, the neighborhood crossover operator is used to generate new position of agents as follows:

$$X_i(t+1) = rand_i \times X_i(t) + U(-1,1) \times \left(rand_i \times X_i(t) - X_i(t)\right),\ i=1,\ldots,s \tag{21}$$

where $U(-1,1)$ and $rand_i$ are random numbers in the interval [-1,1] and [0,1], respectively. Authors introduced scale factor as follows:

$$w = w_{max} - (w_{max} - w_{min}) \times t/T \tag{22}$$

where w_{max} and w_{min} are the maximum and the minimum of the scale factor, respectively. t is the current number of iteration and T is the maximum number of iterations.

Liu et al. (2014) merged the crossover operator of the genetic algorithm and the niche technology with GSA and proposed an improved GSA which enhances the search speed and the solution quality of the standard GSA.

3.5 Escape Operator

Rashedi et al. (2013) presented an image segmentation algorithm based on GSA (SGISA) that equipped with a new operator, called escape. This operator is inspired by the concept of escape velocity in physics. The term escape velocity refers to a scalar speed which is independent of direction. For a given gravitational potential energy at a given position, the escape velocity is the minimum speed that an object needs to be able to ‘‘escape’’ from the gravity. In SGISA, each agent escapes from the cluster that it belonging to with a probability and will be absorbed by its nearest cluster in the feature space. The escape velocity of agent i to become free from its corresponding cluster k with mass M_k is calculated as follows.

$$Ve_i^k = \sqrt{\frac{2GM_k}{r_i}} \tag{23}$$

where G is constant gravity and r_i is the Euclidean distance between agent i and cluster center k (agent k). Escape probability of agent i is calculated as follows:

$$Pe_i^k = \begin{cases} 1 & Ve_i^k \leq Ve_{min} \\ \dfrac{Ve_{max} - Ve_i^k}{Ve_{max} - Ve_{min}} & Ve_{min} < Ve_i^k < Ve_{max} \\ 0 & Ve_i^k \geq Ve_{max} \end{cases} \tag{24}$$

where $Ve_{\max} = \sqrt{\frac{2G(t)}{d_{\min}}}$ is the escape velocity of an agent close to the cluster center less than a given value $d_{\min}$ and $Ve_{\min} = \sqrt{\frac{2G(t)}{d_{\max}}}$ is the escape velocity of an agent far from the cluster center more than a given value $d_{\max}$.

3.6 Black Hole Operator

Doraghinejad and Nezamabadi-pour (2014) introduced a new operator for GSA, called black hole, inspired by some characteristics of the black hole phenomenon in astronomy. To simulate the black hole phenomenon, the best agent is considered as star (black hole) that attracts other agents. Each swarm's agent is categorized into heavy or light agent based on the amount of its mass. Position updating of the heavy and light agents is done in two different ways as follows.

$$x_i^d(t+1) = \begin{cases} x_i^d(t) + rand.\left(x_{BH}^d(t) - x_i^d(t)\right) & \text{if agent } i \text{ is heavy and } r < R_s \\ rand.\left(x_i^d(t).\dfrac{r'}{R_s'}\right) & \text{if agent } i \text{ is light and } r' < R_s' \end{cases} \quad (25)$$

where r and r' denote the distance of each heavy agent and light agent from the black hole, respectively. *rand* is an uniform distributed random number in the interval [0,1] and x_{BH}^d is the position of black hole in dimension d. R_s and R_s' are the Schwarzschild radiuses for heavy and light agents, respectively that are defined as follows.

$$\begin{aligned} R_s &= G\,M\frac{\nu^2}{t} \\ R_s' &= M\log\left(t\right) \end{aligned} \quad (26)$$

where M is the mass of black hole and G is the gravity constant, t and v denote the counter of iteration and the velocity of the agent, respectively. New operator was hybridized with GSA (BH-GSA) to improve exploration and exploitation ability and avoid the premature converge of GSA. In BH-GSA, after obtaining of position agents by Equation (11), new operator is applied to update the position of agents by Equation (25).

3.7 Discrete Local Search Operator

Doraghinejad et al. (2014) equipped GSA with a discrete local search operator to solve Channel Assignment (CA) problem in mesh networks. The discrete local search (DLS) operator was proposed to increase the exploitation ability in the around of the best solution (heaviest mass). It is applied to the best solution with probability $P(t) = \frac{t}{2T}$. A few nodes of the best solutions are chosen randomly and then the DLS

Figure 2. The elliptical motion of planets around the Sun

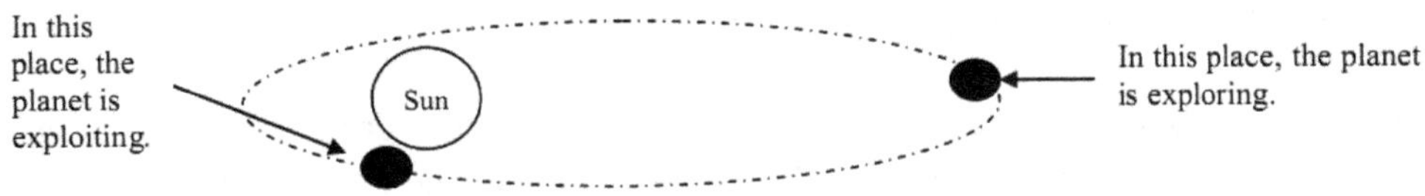

operator changes randomly a few channels allocated with each node. The DLS operator is suitable to solve problems with integer-valued variables.

3.8 Kepler Operator

Sarafrazi et al. (2015) introduced a new operator, called Kepler algorithm, and combined it with GSA to enhance the performance of GSA. Kepler operator is based on the first Kepler law and inspired by the astrophysics. The first Kepler law says: *the planets move around the Sun on an elliptical orbit; the Sun stays in one focus of the ellipse*. It means that the planets have different distances to the sun in different times. Inspired by the law, authors in (Sarafrazi et al., 2015) introduced the Kepler operator in which the best solution is the sun and the planets are the other solutions. The elliptical motion of planets causes to exploit and explore the search space effectively (Figure 2).

In the Kepler algorithm, the new position of selected planet was calculated by Equation (27). Also, to escape the best solution (sun) from the local optima, inspired by the movement of the sun in the space, its position is changed by Equation (28).

$$X_{i,new}\left(t+1\right)=X_{best}\left(t\right)+R_{i,best}\times U\left(-2,2\right),i=2,\ldots,s \tag{27}$$

$$X_{best,new}\left(t+1\right)=X_{best}\left(t\right)\times U\left(-2,2\right) \tag{28}$$

where $U(-2,2)$ is a uniformly distributed pseudo random number at the interval [-2,2]. If $U(-2,2)$ returns a number near 1, the Kepler operator will perform exploitation; otherwise it will perform exploration. The new position of the planets and the sun was determined by the elitism strategy using Equation (29).

$$X_i\left(t+1\right)=\begin{cases}X_{i,new}\left(t\right) & \text{if } fit\left(X_{i,new}\right)>fit\left(X_i\right)\\ X_i\left(t\right) & \text{otherwise}\end{cases} \tag{29}$$

The pseudo code of the Kepler algorithm is shown as follows:

1. Select K agents from $X_i(t)$ for i=1,2,…,s
2. Calculate the Euclidean distance between two agents i and the best agent
3. Calculate the new positions of k agents $X_{j,new}$ for j=1,2,…,k and best agent by Equations 27, 28
4. Update the position agents $X_j(t+1)$ for j=1,2,…,K on elitism (Equation 29)

5. $t=t+1$
6. Repeat steps 1 to 5 until the stopping criterion is reached

In the GSA-Kepler algorithm, the GSA and the Kepler algorithm are performed one after another and used the output of one as input to another. The pseudo code of the GSA-Kepler algorithm is shown as follows:

1. Search space identification, $t=0$
2. Initialize Population randomly $X_i(t)$ for $i=1,2,\ldots,s$
3. Evaluate fitness of agents
4. Update $G(t)$, $worst(t)$ and $M_i(t)$ for $i=1,2,\ldots,s$
5. Calculate the total force in different directions
6. Calculate acceleration and velocity of each agent
7. Update agents' position $X_i(t+1)$ for $i=1,2,\ldots,s$, $t=t+1$
8. Implement the Kepler operator
9. Repeat steps 3 to 8 until the stopping criterion is reached

4 OVERVIEW OF GSA'S VARIANTS

In this section, we describe the GSA's variants for solving different types of optimization problems including continuous (real-valued), binary-valued, triple-valued, discrete, multimodal, multi-objective optimization problems. The improvements of GSA in solving different optimization problems are summarized in Table 2.

4.1 Improved Real-Valued GSA

There are many improvements of GSA variants. In this section, some basic ones are introduced.

4.1.1 Opposition-Based GSA

The basic concept of opposition-based learning (OBL) was proposed by Tizhoosh (2005). The main idea behind applying OBL in metaheuristic optimization algorithms is the concurrent consideration of a candidate solution and its opposite candidate solution to find a better approximation for candidate solution which is closer to the global optimum. Shaw et al. (2012) introduced an opposition-based GSA, called OGSA, based on OBL to improve the coverage rate of GSA. The OBL strategy was applied on GSA in two steps: Opposition-based population initialization and Opposition-based generation jumping. Let $X_i = \left(x_i^1,\ldots,x_i^d,\ldots,x_i^n\right)$ with $x_i^d \in [a_i^d, b_i^d](X_i \in P)$ be a candidate solution (object/mass). The opposition of X_i is $\widehat{X}_i = \left(\widehat{x_i^1},\ldots,\widehat{x_i^d},\ldots,\widehat{x_i^n}\right)$ ($\widehat{X}_i \in \widehat{P}$) where $\widehat{x_i^d} = a_i^d + b_i^d - x_i^d$. Then s fittest objects are selected from set of $\left\{X,\widehat{X}\right\}$ as initial population P. In the process of agents' position updating,

Table 2. Summarization of variants of GSA

Variants	Scholars (years)	Contributions
Real-valued GSA	Shaw et al. (2012)	An opposition-based GSA, based on the concept of opposition-based learning is proposed to improve the coverage rate of GSA
	Liu et al. (2012)	Opposition-based disruption operator
	Khajezadeh et al. (2012)	A modification of GSA with an adaptive maximum velocity constraint is proposed to control the exploration ability of the standard GSA.
	Li et al. (2012) Duman et al. (2015) Jayaprakasam et al. (2015) Mallick et al. (2013) Mirjalili and Hashim (2010) Mirjalili et al. (2012) Tsa et al. (2013) Zhu et al. (2013) Gu and Pan (2013)	Authors applied the idea of memory of PSO into the improved GSA and redesigned the velocity updating formula.
	Aziz et al. (2013)	The authors proposed the *asynchronous* updating procedure to increase the potential of parallelization of GSA
Binary-valued GSA	Rashedi et al. (2010) Rashedi and Nezamabadi-pour (2014)	Reformulation of GSA for binary-valued optimization problems.
Triple-valued GSA	Gao et al. (2014)	Reformulation of GSA for triple-valued optimization problems.
Discrete GSA	Shamsudin et al. (2012) Dowlatshahi et al. (2014)	Reformulation of GSA for discrete-valued optimization problems.
Multimodal GSA	Yazdani et al. (2014)	Reformulation of GSA to support niching in multimodal problems
Multi-objective GSA	Hassanzadeh et al. (2010) Nobahari et al. (2012) Ibrahim et al. (2012) Tabatabaei (2014) Abbasian et al. (2015)	The authors proposed different types of multi-objective GSA.
Quantum GSA	Soleimanpour-moghadam et al. (2011) Soleimanpour and Nezamabadi-pour (2012) Soleimanpour-moghadam et al. (2014)	The authors proposed three versions of real observation quantum behaved GSA.
	Ibrahim et al. (2012) Nezamabadi-pour (2015)	Suggestion of quantum inspired GSA for binary-valued optimization problems.
Mixed GSA	Sarafrazi and Nezamabadi-pour (2013) Rashedi et al. (2013)	The authors presents a new object representation scheme to encode continuous and binary variables and optimize by GSA.

opposition-based generation jumping is performed with probability J_r on each dimension of candidate solutions as follows:

$$\widehat{x_i^d} = \min_p^d + \max_p^d - x_i^d \tag{30}$$

where $\min_p^d$ and $\max_p^d$ are minimum and maximum values of the dth dimension in the current population, respectively. It is mentioned that an improved disruption operator using opposition based learning has been introduced by Liu et al. (2012).

4.1.2 Velocity Constraint GSA

Khajezadeh et al. (2012) proposed a modification of GSA with an adaptive maximum velocity constraint, called MGSA, to control the exploration ability of the standard GSA and present faster convergence. Based on Equation (10) in the standard GSA, the velocity of each agent is a stochastic variable and its movement in various dimensions has an uncontrolled trajectory. In order to control these undulations, authors imposed the following restrictions on the velocity of each agent:

$$-V_{\max} \le v_i^d \le V_{\max} \tag{31}$$

where V_{max} is a time varying maximum velocity and a problem-dependent parameter. If its value is too large, the agents move unsteadily and may be away from the optimal solution. On the other hand, if it is too small, the movement of agent is limited and the optimal solution may not be reached. V_{max} is calculated as follows:

$$V_{\max} = \left[1 - \left(\frac{t}{t_m}\right)^h\right] \times V_{\max 0} \tag{32}$$

where h is a positive constant selected using trial and error. V_{max0} is the initial velocity constraint and defined as follows:

$$V_{\max 0} = \alpha \times \left(x_{\max} - x_{\min}\right), 0 \le \alpha \le 1 \tag{33}$$

After calculating the acceleration and velocity for each agent using Equation (5) and (10), the MGSA determines the maximum velocity V_{max} to prevent the violation of constraints. If the velocity of an agent violated the velocity constraint, the velocity is replaced by its respective constraints. Then, the position of agents is updated using Equation (11). Authors applied MGSA to search for the minimum factor of safety and minimum reliability index in both deterministic and probabilistic slope stability analysis.

4.1.3 GSA with Memory

Li et al. (2012) introduced an improved GSA to solve the path planning of uninhabited aerial vehicle (UAV) problem. The standard GSA is memory-less and the position updating of each agent is based on the total force by other agents. Authors applied the idea of memory of PSO into the improved GSA and defined the velocity updating as follows:

$$v_i^d(t+1) = rand_i v_i^d(t) + a_i^d(t) + c_1 rand_j\left(p_{best}^d - x_i^d(t)\right) + c_2 rand_k\left(g_{best}^d - x_i^d(t)\right) \tag{34}$$

where c_1 and c_2 are parameters in the interval [0,1] which balance the effectiveness of the gravity law and the memory. $rand_\alpha$ is a random variables in the interval [0,1]. The survival of fittest strategy with reference to the differential evolution (DE) algorithm is used in the proposed GSA during selection process as follows:

$$x_i(t+1) = \begin{cases} x_i^{new}(t+1) & f\left(x_i^{new}(t+1)\right) < f\left(x_i(t)\right) \\ x_i(t) & \text{otherwise} \end{cases} \tag{35}$$

They utilized the weight-based principle after the mass updating of agents to have the faster convergence and to enhance the global search ability. The weight value $K_i(t)$ is added to mass of agent i at the iteration t as follows:

$$K_i(t) = \frac{C_{\max} M_{\max} - C_{\min} M_{\min}}{M_{\max} - M_{\min}} - M_i \tag{36}$$

where $C_{\max}$ and $C_{\min}$ are minimum and maximum weight value and $M_{\max}$ and $M_{\min}$ are minimum and maximum mass of each agent, respectively. According to the weight-based principle, agents with heavy mass become heavier and agents with small mass become smaller.

GSA applies only the current position of the agents for updating procedures and does not use any memory for this purpose. In other words, GSA is a memory-less optimization algorithm. PSO due to use of local optimum solution ($pbest_i$) and global optimum solution ($gbest$) is an optimization algorithm with memory. Gu and Pan (Gu, 2013) inspired by PSO and adopted the idea of $pbest_i$ and $gbest$ for updating the velocity into GSA and developed a modified GSA (called as MGSA) to enhance agent memory ability and search accuracy of GSA. They modified Equation (10) as follows:

$$v_i^d(t+1) = rand_0 \times v_i^d(t) + c_1 rand_1 \times \left(pbest_i^d(t) - x_i^d(t)\right) + c_2 rand_2 \times \left(gbest^d(t) - x_i^d(t)\right) + a_i^d(t) \tag{37}$$

$$pbest_i = \left\{pbest_i^1, pbest_i^2, \ldots, pbest_i^n\right\} \tag{38}$$

$$gbest = \left\{gbest^1, gbest^2, \ldots, gbest^n\right\} \tag{39}$$

where c_1 and c_2 are positive constants that adjust the movement length of agent toward local optimum position and global optimum position, respectively. $pbest_i$ denotes the best previous position of the ith agent and $gbest$ denotes the best previous position among all the agents in the population. $rand_1$ and $rand_2$

are two random variables in the interval [0,1]. Authors presented the convergence property analysis of MGSA and applied it to solve high-dimensional benchmark optimization functions. MGSA is suitable for high-dimension function optimization no matter it is unimodal or multimodal.

Many researchers hybridized GSA with PSO to gain from advantage of PSO memory. The idea behind all is approximately the same and there is a little difference in the formulation. Some versions of GSA-PSO variants are Duman et al. (2015), Jayaprakasam et al. (2015), Mallick et al. (2013), Mirjalili and Hashim (2010), Mirjalili et al. (2012), Tsa et al. (2013), Zhu et al. (2013).

Also, some researchers have used the elitism to save the best solution found. For example, Kazak et al. (2012) presented a modified version of GSA, called Member-Satellite algorithm (MSGSA). Each agent in the search space is named as a member and the certain number of satellites is considered the around of members. The members and their satellites are used to find optimal solution in search space. If the fitness value of a member is better than its satellites, the member maintains its soubriquet. Otherwise, the satellite with the best fitness value is replaced as new member. MSGSA in the other steps is similar to standard GSA.

Zhang et al. (2012) presented an improved version of GSA, called immunity-based GSA (IGSA), to overcome the disadvantages of the convergence speed in later search steps and diversity of agents. IGSA is inspired by the biological immune system (BIS) that used the mechanisms of immunity memory and antibodies diversity to improve performance of GSA. The immunity memory and the antibody diversity are used to enhance precision of solution and the solution space, respectively.

4.1.4 Asynchronous GSA

Asynchronous GSA (A-GSA) was proposed in (Aziz et al, 2013& 2014). The authors believed that asynchronous update method increases the potential of parallelization of the algorithms. In A-GSA, the position and velocity of each agent are updated once that the performance of agent is evaluated without waiting for the entire population to be evaluated. Therefore the best and worst agents are identified using mix of information from current and previous iterations. Information from current iteration is from the agents that had been updated before the current agent, while information from previous iteration is from the agents that have not been updated yet.

Aziz et al. (Aziz, 2014) investigated the performance of the synchronous GSA and the asynchronous GSA by statistical analysis. Based on the obtained results in that research, both implementations without parallelization have similar performance but A-GSA has a higher parallelization potential. Due to lack of synchronization of information, asynchronous metaheuristic algorithms have been introduced to improve exploration ability. The original GSA is synchronous GSA (S-GSA) and each agent updates its mass via the best and worst masses after calculating the performance of whole population. On the other hand, before identifying the best and worst masses, the performance of whole population is calculated and then each agent updates its position and velocity synchronously.

4.2 Binary Gravitational Search Algorithm (BGSA)

The original version of GSA was designed for solving continuous optimization problems and search spaces of real valued vectors. However, many optimization problems are set in binary space, such as feature selection and dimensionality reduction, unit commitment, and cell formation. The binary search space is considered as a hypercube in which an agent may move to nearer and farther corners of the hypercube by flipping various numbers of the bits.

Rashedi et al. (2010) has introduced a binary version of the algorithm. In the original version of GSA, the algorithmic "gravitational forces" lead directly to changes in the position of agents in a many-dimensional continuous search space. In the binary version of GSA (BGSA), the outcome of these forces is converted into a probability value for each element of the binary vector, which guides whether that element will take on the value 0 or 1. The main difference between continuous and binary GSA is that in the binary algorithm, the position updating means a switching between "0" and "1" values.

In BGSA, velocity of agent is updated by Equation (10) that is similar to RGSA. But new position of agent is considered to be either 0 or 1 with a proper probability function. This function maps the velocity to the probability of position updating, so that should be bounded within interval [0,1] and increases with increasing velocity. Authors have defined the function $Tf\left(v_i^d\right)$ as follows:

$$Tf\left(v_i^d(t)\right) = \left|\tanh v_i^d(t)\right| \tag{40}$$

and have updated agent's position according to Equation (41):

$$\text{if } rand < Tf\left(v_i^d(t+1)\right) \text{ then } x_i^d(t+1) = \text{complement } \left(x_i^d(t)\right) \text{ else } x_i^d(t+1) = x_i^d(t) \tag{41}$$

One of the negative situations that may occur in swarm intelligence optimization techniques is stagnation. Stagnation is a situation that swarm agents fall in a local optimum. If the stagnation situation is occurred in BGSA, agent velocities become zero and based on Equation (40) and (41), relocating probability of agents turns to zero. Saha et al. (2015) used the BGSA algorithm with a novel modification to improve the prediction accuracy of classifier with reduced feature set. Rashedi et al. (2014) introduced an improved version of BGSA, called IBGSA, to overcome the stagnation situation by replacing the transfer function (Equation (40) is replacing by Equation (42)). This function reinforces the exploration ability and avoids fall in local optimum.

$$Tfn\left(v_i^d(t)\right) = A + (1-A) \times \left|\tanh v_i^d(t)\right| \tag{42}$$

where A is a parameter that prevents the stagnation situation. In introduced function, if the velocity of an agent becomes zero, the relocating probability becomes A. This modification is similar to the mutation operator in genetic algorithms. A is a state-dependent parameter that is calculated by the following equation:

$$A = k_1\left(1 - \exp\left(\frac{F_c}{k_2}\right)\right) \tag{43}$$

where k_1 and k_2 are constant parameter and time constant defined based on the dimension of algorithm. F_c denotes failure counter. A failure happens when the best-so-far solution does not change after one iteration. Above equation shows that if F_c=0, the value of *A* turns into 0 and the mutation does not occurred. In this paper, authors have used elitist strategy that the position of an agent is updated if the new position provides a better fitness value than the previous position. Elitism strategy ensures that the good results are not eliminated through the movement operator. The following equation donates using elitism strategy in IBGSA for maximization problems:

$$X_i(t+1)=\begin{cases} X_i\,(t+1) & \text{if } fit\left(X_i\,(t+1)\right)\geq fit\left(X_i(t)\right) \\ X_i(t) & \text{otherwise} \end{cases} \tag{44}$$

Ibrahim et al. (2014) proposed an approach based on BGSA to solve an assembly sequence planning (ASP) problem with the aim of generating the optimal feasible assembly sequence to minimize production time, thereby decreasing the assembly costs. In proposed approach, authors represented each agent by a feasible assembly sequence. An assembly sequence is feasible, if it satisfies all precedence constraints. The determination of feasibility of an assembly sequence is done based on the precedence matrix (PM). For this purpose, a few modifications in the initial population and evolving the updated assembly sequence of each agent to feasible assembly sequence is applied to BGSA. Also, a swap mutation operator was used to update position of each agent.

4.3 Triple-Valued Gravitational Search Algorithm (TGSA)

Gao et al. (2014) proposed a triple-valued gravitational search algorithm, called TGSA, to solve the graph planarization problem (GPP). TGSA considers the search space as a triangular hypercube and uses triple-valued encoding scheme to define the position of each agent in population. For this case, authors considered the triple-valued set $\{U,L,E\}$ to denote the encoding scheme.

In TGSA, the total force acted on an agent is calculated using Equation (12). Also the acceleration and velocity of an agent are computed in the same way based on Equation (4) and Equation (10), respectively. The position updating of *i*th agent is done based on two indices: *velocity index* and *population index*.

$$x_i^d(t+1)=\begin{cases} U & \text{if } r_1<\tanh(v_i^d(t+1)) \text{ and } r_2<\dfrac{\left|x_i^d(t)\right|_U}{\left|x_i^d(t)\right|_U+\left|x_i^d(t)\right|_L+\left|x_i^d(t)\right|_E} \\ L & \text{if } r_1<\tanh(v_i^d(t+1)) \text{ and } r_2<\dfrac{\left|x_i^d(t)\right|_U+\left|x_i^d(t)\right|_L}{\left|x_i^d(t)\right|_U+\left|x_i^d(t)\right|_L+\left|x_i^d(t)\right|_E} \\ x_i^d(t) & \text{otherwise} \end{cases} \tag{45}$$

where r_1 and r_2 are two uniform random variables in the interval [0,1]. $\left|x_i^d(t)\right|_z$ denotes the number of dimensions *d* in current population with the value of *Z*. the velocity index is a function of the current velocity to determine the occurrence or non-occurrence of updating progress. The population index is a cumulative index which is calculated based on the collected information of the whole population.

TGSA uses a distance metric $|.|_T$ to calculate distance between agents which is defined by the following equations:

$$|U-L| = |U-E| = |L-E| = 1,\ |U-U| = |L-L| = |E-E| = 0 \tag{46}$$

$$R_{ij}(t) = \left|X_i(t), X_j(t)\right|_T = \sum_{d=1}^{n}\left|x_i^d - x_j^d\right| \tag{47}$$

According to Equation (45), an agent with a large value of velocity has a high probability of position updating and it will move to the position where most other agents stand. Also,

4.4 Discrete Gravitational Search Algorithm (DGSA)

BGSA is a special type of discrete gravitational search algorithm (DGSA). In BGSA a discrete optimization problem was modeled by a sequence of binary bits. Shamsudin et al. (2012) introduced a Fast Discrete GSA (FDGSA) that models a discrete optimization problem by a vector of integer values. In FDGSA the velocity update is still the same as standard GSA, but position of an agent is updated based on direction and velocity of it. Authors calculated the next position of the agent as follows.

$$x_i^d(t+1) = \begin{cases} \text{random}\left\{x_i^d(t), x_i^d(t)+1, \ldots, x_i^d(t)+v_i^d(t+1)\right\} & \text{if } v_i^d(t+1) > 0 \\ \text{random}\left\{x_i^d(t), x_i^d(t)-1, \ldots, x_i^d(t)-v_i^d(t+1)\right\} & \text{if } v_i^d(t+1) < 0 \end{cases} \tag{48}$$

They have compared their algorithm with BGSA and claimed that FDGSA is able to find better solutions and converges faster than BGSA.

Dowlatshahi et al. (2014) proposed a discrete gravitational search algorithm and introduced the movement operator as the main component of GSA in accomplishing the search in the solution space. Movement length of each agent is its velocity. They classified the movement length into two classes: Independent Movement Length (IML) and Dependent Movement Length (DML). IML indicates the movement length obtained for each agent without knowing position of any other agent. In other words, IML depends only on its previous movement length and in the standard GSA, it is a fraction of the previous movement length. DML indicates the movement length obtained for each agent by considering the position of all members of *Kbest* set. DML is the same as the acceleration of agent under influence of other agents.

A modification version of local search algorithm was used as independent movement operator of DGSA and an IML was added as the input parameter of algorithm to determine the movement length of local search. The value of the input IML is calculated by multiplication of a random number in interval [0,1] to the previous movement length of agent. Independent movement operator similar to the original local search algorithms is strongly dependent on the starting solution.

Authors presented the concept of neighborhood space by an undirected graph $H=(N,E)$ to define the dependent movement operator of DGSA. Nodes in N correspond to candidate solutions and E corresponds to moves in the neighborhood space. A small move operator $\oplus$ (as example a swap operator) was applied to generate a new neighbor for solution $y \in N$.

The neighborhood space H is a connected graph. Based on the idea of Path Re-linking (PR) strategy, they introduced an algorithm, called Simple Movement in Neighborhood Space (SMNS) algorithm, to simulate the DMO concept in discrete spaces. The SMNS algorithm generates the trajectory of a starting solution to a target solution by the small move operator such that the distance between the attributes of the starting and target solutions should decrease. The algorithm, unlike PR algorithm, returns the lth constructed solution in the trajectory between starting and target solutions. l is movement length that is equaled with the acceleration that target solution applies to the starting solution.

In the proposed DGSA, the Multi Target Movement in Neighborhood Space (MTMNS) algorithm is used to accomplish the function of DMO.

Dependent movement length of the ith agent toward the jth agent (l_{ij}) is calculated as follows:

$$F_{ij}\left(t\right) = rand \times G\left(t\right) \times \frac{M_j\left(t\right) M_i\left(t\right)}{NR_{ij}\left(t\right)} \times R_{ij}\left(t\right) \tag{49}$$

$$l_{ij}\left(t\right) = a_{ij}\left(t\right) = \frac{F_{ij}\left(t\right)}{M_i\left(t\right)} = rand \times G\left(t\right) \times \frac{M_j\left(t\right)}{NR_{ij}\left(t\right)} \times R_{ij}\left(t\right) \tag{50}$$

where $R_{ij}(t)$ is the distance between two agents i and j, defined as the number of applying the small move operator $\oplus$ that converts agent i to agent j. $NR_{ij}(t)$ is a normalized value of $R_{ij}(t)$ in interval [0.5,1] which is obtained by expression $0.5 + \frac{R_{ij}(t)}{2 + Max_{dist}}$, where Max_{dist} is the maximum possible distance between each arbitrary pair of solutions in the neighborhood space.

4.5 Multimodal Gravitational Search Algorithm

Population based algorithms are suitable alternatives for solving hard optimization problems. The algorithms typically are able to converge to a single optimal solution. Many real-world problems compose of objective functions with multimodal behavior. A multimodal problem has several global optimal solutions. When dealing with multimodal problems some modifications to population based algorithms are needed to find all optimal solutions, may it be local or global ones. Many theoretical and empirical studies have been realized to find multiple solutions to multimodal problems. Niching methods are able to find the location of optimal solutions in multimodal problems. Niching Methods partition the population in such a way that each group of the main population focuses on a different possible solution in a single run of the search.

In GSA the searcher objects tend to converge towards each other, which in turn means convergence to the global optimum of the problem at hand. Therefore, a simple GSA similar to other EA algorithms, is unable to provide multiple solutions in multimodal problems in a single run of the search. The authors in (Yazdani & Nezamabadi-pour, 2011; Yazdani et al., 2014) applied some simple modifications to standard GSA to support niching in multimodal problems and they called new algorithm as Niche GSA (NGSA). In NGSA, K-nearest neighbors (K-NN) strategy is used to divide the swarm into sub-swarms and only K-nearest neighbors of each agent are allowed to apply the gravity force to attract it. Also, active gravitational mass of each agent dose not remain constant for all objects. Authors considered the active gravitational mass of each agent as a vector containing s elements such that each element corresponds to an object of the system.

$$M_{aj} = [m_{aj1}, m_{aj2}, \ldots, m_{ajN}] \forall j \in \{1, 2, \ldots, s\} \tag{51}$$

where s is the number of agents in the swarm and m_{aij} denotes the active gravitational mass of agent j that acts on agent i to attract it which is calculated by following equation:

$$m_{aji}(t) = \begin{cases} \dfrac{fit_j(t) - worst_i(t)}{best_i(t) - worst_i(t)} & if\ agent\ j \in KNN_i \\ 0 & otherwise \end{cases} \tag{52}$$

where $fit_j(t)$ denotes the fitness value of the agent j at time t. KNN_i is the set of K-nearest neighbors of agent i. For a maximization problem $worst_i(t)$ and $best_i(t)$ are defined on the neighborhood of agent i by following equation:

$$worst_i(t) = \min_{l \in KNN_i} fit_l(t) \tag{53}$$

$$best_i(t) = \max_{l \in KNN_i} fit_l(t) \tag{54}$$

Here, selecting the number of neighbors, K, is a critical problem. Authors calculated K based on two constant parameter K_i and K_f that determine the number of neighbors at the beginning and the end of the search process, respectively ($0<K_i<K_f<1$):

$$K(t) = Round\left(\left[K_i - (K_i - K_f) * \frac{t}{T}\right] N\right) \tag{55}$$

A small value of K leads the algorithm to generate virtual niches and a large value of K makes the algorithm to lose some actual niches. According to above equation, K is initialized with a small value at the beginning to form niches and then it is increased by time. It causes that the near virtual niches merge to a real niche. The force acting on object i from object j is defined as:

$$F_{ij}^d(t) = G(t) \frac{M_{pi}(t) \times M_{aij}(t)}{R_{ij}(t) + \varepsilon} \left(x_j^d(t) - x_i^d(t)\right) \tag{56}$$

The total force that acts on agent i in the dimension d is computed as follows:

$$F_i^d(t) = \sum_{j=1, j \neq i}^{S} rand_j F_{ij}^d(t), \tag{57}$$

The acceleration of each agent and its next velocity are calculated like the standard GSA. They used elitism strategy for updating the position of each agent that ensures the new position is a solution with higher quality.

4.6 Multi-Objective Gravitational Search Algorithm

Many real-world problems deal with concurrent optimization of two or more conflicting objectives rather than single objective. These problems are called as multi-objective problems. In such problems, there is a set of non-dominated solutions called Pareto optimal solutions. The original version of GSA proposed to optimize single real-value problems. Hassanzadeh et al. (2010) presented a multi-objective version of GSA, called MOGSA and applied same idea of Multi-objective PSO (MOPSO) (Cagnina et al., 2005) for this purpose. To apply the multi-objective concepts, they used a "Uniform Mutation Operator" and an "Elitist Policy" to store non-dominated solutions in an archive which has a grid structure.

To create the grid structure in the objective space, each dimension was divided into $2^{\sum_{i=1}^{k} n}$ equal segments where i and k denote dimension index and the number of optimization objectives, respectively. New non-dominated solutions are added to the archive and once the number of non-dominated solutions in it reaches to maximum length, a random solution (object) from the most crowded hypercube should be selected to remove. Also, those solutions that become dominated by new solutions should be removed from the archive.

The mass of searching objects is set to one and the mass of archived objects is updated based on the distance to the nearest neighbor among other archived objects. Authors considered only $k$best non-dominated solutions in archive as the source of gravitational force to attract the others and after each updating position; the uniform mutation is applied to the new positions.

Nobahari et al. (2012) proposed another multi-objective version of GSA, called Non-dominated Sorting Search Algorithm (NSGSA). Like the algorithm presented by Hassanzadeh et al. (2010), the NSGSA algorithm has an external archive to store the Pareto optimal solutions and to provide some elitism. Objects are classified into the archived objects and the moving objects. NSGSA uses the non-dominated sorting algorithm (Srinivas, 1994) for ranking the moving objects in different layers. The rank of each object was considered as its fitness.

In NSGSA, to update the moving objects list and the updating of their masses, the m objects imported from the extreme regions of the external archive and the m members imported from the least crowded area of the archive are ranked as 1 and the $P_{elitism}$ percent of the archive members chosen randomly and inserted to the moving objects list are ranked as 2. The members remained from the last iteration get ranks 3 and more, regarding the rank of the layers they belong to. To update the external archive, new non-dominated objects are added to the external archive. If the number of archived objects to be more than the maximum archive length ($n_{archive}$), NSGSA by an indicator determines the most crowded area to remove a random member from it. The spread indicator (δ) denotes the spread of the particles in the Pareto front. Authors to calculate δ used distance measure ($d_{c,i}$) that is similar to crowding distance measure proposed by Deb et al. (2002).

$$d_{c,i} = \sqrt{\sum_{j=1}^{m} \left(d_{c,i}^{j}\right)^{2}} \left(i = 1, \ldots, n_{archive}\right)$$

$$d_{c,i}^{j} = fit_{i+1}^{j} - fit_{i-1}^{j}, \left(i = 1, \ldots, n_{archive}, j = 1, \ldots, m\right) \tag{58}$$

where $n_{archive}$ is the length of archive and fit_{i+1}^{j} is the *j*th fitness of object *i+1* when the members of the archive are sorted based on the *i*th objective function. Crowding distances ($d_{c,i}$) have to be nearly equal. The deviation of these distances from their average ($\bar{d}_{c}$) should be minimized. Finally the spread indicator (δ) was calculated as follows:

$$\delta = \frac{\sum_{i=1, i \notin E}^{n_{archive}} \left| d_{c,i} - \bar{d}_{c} \right|}{\left[n_{archive} - m\right] \bar{d}_{c}} \tag{59}$$

Because the new objects at first should be moved slowly to search around the Pareto front, the velocity of each new object was initialized to zero. Authors used parameter G_0 in the calculation of the gravitational constant ($G\left(t\right) = G_{0}\left(1 - \frac{t}{t_{\max}}\right)$) as follows:

$$G_{0} = \beta \max_{d \in \{1, \ldots, n\}} \left(\left| x_{u}^{d} - x_{l}^{d} \right| \right) \tag{60}$$

The NSGSA algorithm has eight control parameters. β is one of its parameters. In reality, the close objects may impact to each other and consequently the direction of their movement may change. To simulate the irregularities, authors proposed two new mutation operators, called sign and reordering mutation. The combination of these mutation operators was used to update the position of particles.

Ibrahim et al. (2012) presented a new algorithm, called vector evaluated GSA (VEGSA), for multi-objective optimization. VEGSA was inspired by some optimization algorithms such as vector evaluated genetic algorithm (VEGA) and vector evaluated particle swarm optimization (VEPSO). VEGSA uses *M* populations with *N* objects. Each population optimizes an objective function using objects with maximum and minimum fitnesses in its neighbor population. Non-dominated solutions are stored in an archive and are updated at each iteration. Authors introduced two version of VEGSA.

In VEGSA I, the *best*(*t*) and *worst*(*t*) for a maximum problem are updated as follows:

$$best_{m}\left(t\right) = \max\left(\max_{j \in \{1, \ldots, N\}} fit_{j}^{m}\left(t\right), \max_{m-1} \right), \quad m = \left\{1, 2, \ldots, M\right\} \tag{61}$$

$$worst_m(t) = \min\left(\min_{j \in \{1,\ldots,N\}} fit_j^m(t), \min_{m-1} \right), \quad m = \{1, 2, \ldots, M\} \tag{62}$$

where fit_j^m denotes the fitness of *j*th object in *m*th population. $\min_{m-1}$ and $\max_{m-1}$ are the minimum and maximum fitnesses of neighbor population. In VEGSA II, direction of each object in population *m* depends on force acted from all objects on it with the addition of force acted from the best object in its neighbor population *m-1*. The total force that acts on *i*th object was calculated as the following.

$$F_i^{d,m}(t) = rand \times F_{i,m-1}^d(t) + \sum_{j=1, j \neq i}^{N} rand_j \times f_{ij}^{d,m}(t) \tag{63}$$

$$F_{i,m-1}^d(t) = G(t) \times \frac{M_{pi}^m \times M_{a,m-1}^m}{R_{i,m-1}^k + \varepsilon} \left(x_{m-1}^d(t) - x_i^d(t) \right) \tag{64}$$

where $F_i^{d,m}$ is the total force acted on *d*th dimension of *i*th object of population *m*. $F_{i,m-1}^d$ denotes a force acted on *i*th particle by the best particle in its neighbor population. Muhammad et al. (2014) evaluated the performance of the VEGSAII algorithm. The obtained results based on ZDT test problems showed that it is still immature and needs further enhancements to employ it in any application.

Tabatabaei (Tabatabaei, 2014) proposed a multi-objective GSA based on a fuzzy function method. The suggested fuzzy function maps a value $\mu_j(X_j)$ to each object X_j based on it fitness $f_i(X_j)$ in objective function *i*. Then, author introduced a new parameter $\mu_j(X_j)$ for each object X_j to make a single value from all objective functions.

$$\mu_i(X_j) = \frac{f_{max,i} - f_i(X_j)}{f_{\max,i} - f_{\min,i}} \tag{65}$$

where $f_i(X_j)$ is the value of *i*th objective function of *j*th agent. $f_{\min,i}$ and $f_{\max,i}$ are the minimum and maximum values of *i*th objective function.

$$\mu(X_j) = \sum_{i=1}^{N_{of}} \omega_i \mu_i(X_j) \tag{66}$$

where $\mu(X_j)$ is sum of the fitness values of all objective functions. ω_i and N_{of} denote the weight factor of the *i*th objective function and the number of objective functions, respectively. The normalized value of $\mu(X_j)$ was calculated as follows:

$$N\mu\left(X_j\right)=\frac{\mu\left(X_j\right)}{\sum_{i=1}^{s}\mu\left(X_i\right)} \tag{67}$$

In this algorithm, at first non-dominated solutions are added to the first repository and then at each step the repository is updated. New objects which are not dominated with previous objects are added to new-repository. Then, the objects of repository and new-repository are compared to exploit non-dominated solutions and update the repository. At each step of the algorithm, the sorted solutions in new-repository are ranked into three groups: first-repository, second-repository and third-repository. The n-repository consists of objects which are dominated by n other objects. The speed of the algorithm depends on the size of repository. In this paper, the unnecessary and near objects are eliminated of the repository to reduce the size repository.

Abbasian et al. (2015) proposed a novel multi-objective GSA, called clustering based archive multi-objective GSA (CA-MOGSA). The CA-MOGSA is based on the Pareto principles and non-dominated solutions are stored in the external archive. The archive is clustered and for each agent a random cluster is selected to apply the gravitational force to attract it. Solutions with less crowding distance and extra solutions are removed from archive. Hence, the size of archive was considered variable.

At each iteration of the CA-MOGSA algorithm, the population and the archive members are combined and non-dominated solutions are selected for adding to the next archive. Then for each agent in the population a k best set is selected from archive to update its position and acceleration. To do this, the archive members are clustered in the objective space and a cluster is selected as the k best set for an agent with a probability based on the Euclidean distance of all cluster representatives to the population members. Clustering is done using the agglomerative single-link (or nearest-neighbors) method. In the single-link method, the closest two clusters are merged to form a new cluster, where the distance between two cluster is the distance between their closest members. Authors to have a best trade-off between exploration and exploitation used two different strategies to select clusters for agents at different iterations based on Equation (68). At the beginning the clusters are selected based on their Euclidian distances to the agents and with the increase of the iteration number the selection probability of the closest cluster to the agent is increased.

if *rand*<0.4 tanh $\left(0.987\frac{t}{T}\right)$ then select the closest cluster to the agent i as its k best set else

$$P_{ij}=\frac{\left(d_{ij}\right)^{-1}}{\sum_{u=1}^{Cl}\left(d_{iu}\right)^{-1}}\quad i=1,\ldots,s \text{ and } j=1,\ldots,Cl \tag{68}$$

where T, s and Cl are the maximum iterations, the population size and the clusters numbers, respectively. P_{ij} denotes the selection probability of cluster j as the k best set of agent i. After determining the k best set for each agent, the masses of the archive's members are calculated using Equation (69).

$$M_i(t)=CD(X_i),\ X_i\in A(t) \tag{69}$$

where $A(t)$ denotes the archive at iteration t. $CD(X_i)$ is the crowding distance of solution X_i in the archive. The crowding distance of a problem with K objective functions, where $F_1,\ldots,F_R$ are its non-dominated fronts, is calculated as follows:

$$CD_k\left(X[i,k]\right)=\frac{f_k(X(i+1,k]-f_k\left(X[i-1,k]\right)}{f_k^{\max}-f_k^{\min}} \quad i=2,\ldots,L-1 \tag{70}$$

where L is the length of the archive and $X[i,k]$ denotes the solution i in the archive with respect to the objective function k. $CD_k(X[1,k]) = CD_k(X[L,k]) = B$ and $B>\max_{2<i<L-1}\{CD_k(X[i,K]\}$.

$$CD\left(X\right)=\sum_{i=1}^{K}CD_i\left(X\right) \tag{71}$$

where $f_k(X)$, $f_k^{\max}$ and $f_k^{\min}$ are the objective values of solution X, the maximum and minimum of the objective values regarding the kth objective function, respectively. The updating of acceleration, velocity and position of each agent are done by Equations (5), (10), (11), respectively.

There are many versions of multi-objective GSA which were employed in optimization of different problems. Some versions of multi-objective GSA include but are not limited to: (Ganesan et al., 2013; Baniassadi et al. 2011; Abbasian et al. 2012; Rubio-Largo et al. 2011; Ajami and Armaghan 2013; Rubio-Largo and Vega-Rodríguez 2013)

4.7 Quantum Gravitational Search Algorithm

The word "Quantum Computing (QC)" refers to creating a new generation of computers, called quantum mechanical computers, which work with a focus on the role of quantum mechanics in the theory of computation. Research on integrating Evolutionary Computing (EC) and QC has been started since late 1990s. The existing algorithms in this field are classified by Zhang (Zhang, 2011) in three categories as follows:

- Evolutionary-designed quantum algorithms (EDQAs) which try to automate the synthesis of new quantum algorithms using EAs.
- Quantum evolutionary algorithms (QEAs) which focus on implementation of EC algorithms in a quantum computation environment.
- Quantum-inspired evolutionary algorithms (QIEAs) which concentrate on generating new EC algorithms using some concepts and principles of QC.

So far, different variants of QIEAs have been proposed that can be categorized into three types (Zhang, 2011): Binary observation QIEA (BQIEA), Real observation QIEA (RQIEA) and QIEA-like algorithms. A comprehensive survey on quantum-inspired evolutionary algorithms along with a limited comparison of important ones was given by Zhang. The interested readers could refer to it for more details. Recently, binary observation and real observation versions of QIEA for GSA have been introduced. In the succeeding subsections, we will describe these algorithms.

4.7.1 Real Observation Quantum GSA

In Newtonian mechanics, agent's state is described by vectors of position, velocity, acceleration and force and the agent's motion is done under Newtonian mechanics laws. In quantum mechanics, an agent is described by wave function (Ψ) instead of position that each agent moves under Quantum mechanics. In quantum space, an agent in GSA system has quantum behavior and GSA algorithm is bounded to work in a different fashion. The time-dependent Schrodinger equation is governing equation in quantum mechanics and is the equivalent of Newton's law. The general form of this equation is as follows:

$$j\hbar\frac{\partial}{\partial t}\psi(X,t) = H(X)\psi(X,t) \tag{72}$$

where $H(X)$ is a time-independent Hamiltonian operator given by:

$$H(X) = -\frac{\hbar^2}{2m}\nabla^2 + V(X) \tag{73}$$

where $\hbar$ is Planck's constant, ∇^2 and m are Laplacian operator and the mass of the object, respectively. $V(X)$ is the potential energy distribution. $Q=|\Psi|^2$ is the probability measure for the agent's position at the time t that depicts the probability of appearing it in a specific region. In quantum mechanics, an agent can exist in many positions at once. The probability of appearing an agent at a specific position in the quantum search space is mapped into its actual position in the solution space by a technique called ''collapsing'' (Neumann, 1955). $|\Psi|^2$ as the probability density function should satisfy the condition $\int_{-\infty}^{+\infty}|\Psi|^2 dx = \int_{-\infty}^{+\infty} Q dx = 1$. The wave function is considered as follows:

$$\Psi(X,t) = \Psi(x)f(t) \tag{74}$$

Equation (72) is rewritten with substituting Equation (74) in it:

$$\frac{j\hbar}{f(t)}\cdot\frac{d}{dt}f(t) = \frac{1}{\psi(X)}\cdot\left(-\frac{\hbar^2}{2m}\nabla^2 + V(X)\right)\psi(X) \tag{75}$$

The two sides of Equation (75) must be equal to a constant (E) that represents the energy of the agent. Therefore, we have:

$$\frac{j\hbar}{f(t)}\cdot\frac{d}{dt}f(t) = E \rightarrow f(t) = e^{\frac{-jEt}{\hbar}} \tag{76}$$

and

$$\frac{1}{\psi(X)}\cdot\left(-\frac{\hbar^2}{2m}\nabla^2+V(X)\right)\psi(X)=E\rightarrow H\psi(X)=E\psi(X) \tag{77}$$

and Equation (74) becomes:

$$\Psi(X,t)=\psi(x)e^{\frac{-jEt}{\hbar}} \tag{78}$$

Therefore, $|\Psi(x,t)|^2 = |\psi(x)|^2$. By solving Equation (77) and calculating the wave function $\psi(x)$, we can achieve the probability density function $|\Psi(x,t)|^2$.

To simplify the formulation, without loss of generality, consider an object in a quantum system with the 1-dimensional search space. In this case the position vector X is reduced to scaler x. Let $y=x-c$, where c is the position which is defined by the position of potential well (one promising solution). To avoid divergence of QIGSA, y should approach zero. Therefore, we need to apply an attractive potential field centered at zero. Any type of potential well can be used but based on researchers' experiences the simplest and most effective one is the delta potential well (Equation 79) (Soleimanpour, 2011), (Sun, 2004), (Soleimanpour, 2014). Therefore, assume that we have one 1-dimensional Delta potential well.

$$V(y)=-\gamma\delta(y) \tag{79}$$

where $\gamma\geq 0$ is the depth of the potential well. Therefore, for this model the Schrodinger equation becomes:

$$E\psi(y)=\left(-\frac{\hbar^2}{2m}\nabla^2-\gamma\delta(y)\right)\psi(y) \tag{80}$$

For $y\neq 0$ the equation can be written as,

$$\frac{d^2\psi}{dr^2}+\frac{2m}{\hbar^2}E\psi=0 \tag{81}$$

Equation (81) is calculated as follows:

$$\psi(y)=e^{\frac{\sqrt{-2mE}}{\hbar}|y|}\rightarrow Q(y)=|\psi(y)|^2=e^{\frac{-2\sqrt{-2mE}}{\hbar}|y|} \tag{82}$$

The precise information from agent's position is needed to evaluate its fitness value. Therefore, to collapse the quantum state to the classical state, Monte Carlo method was used. To this aim, a random variable (*rand*) is generated uniformly distributed in the interval (0,1) and then Equation (82) is substituted with it.

$$\left|\frac{-2\sqrt{-2mE}}{\hbar}\right| y = \ln(rand) \rightarrow y = x - c = \pm\frac{\hbar}{2\sqrt{-2mE}}\ln\left(\frac{1}{rand}\right) \tag{83}$$

Thus, the agent's position accurately is measured as follows.

$$x = c \pm \frac{\hbar}{2\sqrt{-2mE}}\ln\left(\frac{1}{rand}\right) \tag{84}$$

It is assumed that $\frac{\hbar}{2\sqrt{-2mE}} = g \mid c - x \mid$, where g is a constant and $|c{-}x|$ is used to scale the variation of the new position of object around potential well. Therefore, for an n-dimensional search space system one can summarize that:

$$\begin{cases} x_i^d(t+1) = c_i^d + g\left|c_i^d - x_i^d(t)\right|.\ln\left(\frac{1}{rand}\right) & if\ r \geq 0.5 \\ x_i^d(t+1) = c_i^d - g\left|c_i^d - x_i^d(t)\right|.\ln\left(\frac{1}{rand}\right) & otherwise \end{cases} \tag{85}$$

where *rand* and *r* are two random numbers generated by uniform probability density function in the interval (0,1), *g* is a parameter which determines the balance between exploration and exploitation and c^d is the dimension *d* of the position (center) of a potential.

Soleimanpour et al. (Soleimanpour, 2011) proposed a quantum-behaved gravitational search algorithm (QGSA) that each agent moves toward a Delta potential well (c_i) in search space with a center which is weighted average of all members of *k* best set ($Mbest_i$) and the best position found by that agent ($Pbest_i$). They represented the Delta potential well for agent *i* as follows.

$$Mbest_i = \frac{\sum_{j \in kbest} \frac{X_j}{d_{i,j}}.}{\sum_{j \in kbest} \frac{1}{d_{i,j}}.} \tag{86}$$

where

$$d_{i,j} = ||X_i - X_j|| \tag{87}$$

and

$$c_i^d = \frac{r_1 Mbest_i^d + r_2 Pbest_i^d}{r_1 + r_2} \tag{88}$$

where r_1 and r_2 are two random numbers generated by uniform probability distribution in the interval [0,1]. $Pbest_i$ is the best position that object i have ever seen.

QGSA is an optimization algorithm with outstanding result for unimodal functions. But like some other optimization algorithm, QGSA, duo to the fast flow of information between objects suffers from the premature convergence. Authors in (Soleimanpour, 2012) presented an improved Quantum-behaved gravitational search algorithm, called IQGSA, for both unimodal and multi-modal optimization. IQGSA enhances diversity and exploration capability to escape from local optima within constructing a new $Mbest_i$ as follows:

$$Mbest_i = \frac{\sum_{j \in kbest} \frac{rnd_j M_i X_j}{d_{i,j}}.}{\sum_{j \in kbest} \frac{rnd_j M_i}{d_{i,j}}.} \tag{89}$$

where *rnd* is a random vector with size of equal to *k*best set in [0,1] that is added to keep the stochastic characteristic of algorithm. M_i is the mass of agent *i*. In finding the optimum results, IQGSA is much better than QGSA and GSA for unimodal and multi-modal functions.

Soleimanpour et al. (Soleimanpour, 2014) introduced another version of quantum-inspired GSA (QIGSA) by increasing the exploration ability of the algorithm at the beginning iterations. Updating agent's position in QIGSA is performed using Equation (85), where c_i^d is the dimension *d* of the position of a potential well selected among from *k* best set randomly, *g* is a parameter to balance exploration and exploitation. QIGSA has only one control parameter and is faster and more precise than GSA itself. It should be noted that the way of selecting one of the *k* best members is so similar to the roulette wheel selection in genetic algorithms. In this kind of selection, individuals are given a probability of being selected that is directly proportionate to their fitness. The pseudo code of QIGSA is given in the following.

```
Identification of search space;
Initialization of agent positions;
While stopping criteria is not satisfied do
   Evaluate the fitness for each agent;
   Update kbest set;
   Choose center of well for each agent by a probabilistic procedure;
   Update the position of each agent by Equation (85);
Endwhile
Output: Best solution found.
```

4.7.2 Binary Observation Quantum GSA

The metaheuristic search algorithm inspired by quantum mechanics, QIEAs, use quantum-inspired bits (Q-bits), quantum-inspired gates (Q-gates) and observation process to form their structure. The smallest unit of information that can be stored in a two-state quantum computer (Benioff, 1980) is called quantum bit (Q-bit) (Han, 2002). Quantum computers work by manipulating Q-bits and applying a sequence of quantum operators to them. A Q-bit is able to present states "1", "0", or any superposition of these two. The state of a Q-bit, $|ø\rangle$, is presented by

$$|ø\rangle = \acute{a}\,|0\rangle + \hat{a}\,|1\rangle \tag{90}$$

where α and β are complex numbers that indicate probability amplitudes of the states "0" and "1", respectively. The values $|\alpha|^2$ and $|\beta|^2$ are the probabilities that the Q-bit is found in the states "0" or "1", respectively. Normalization requires that

$$|\alpha|^2 + |\beta|^2 = 1 \tag{91}$$

Therefore, a Q-bit is defined as a pair of number (α,β) as

$$q = \begin{bmatrix} \alpha \\ \beta \end{bmatrix} \tag{92}$$

So, in a typical quantum system, an agent (object) is presented as a string with *n* Q-bits as follows (Han, 2002):

$$q = \left[q^1, q^2, \ldots, q^n\right] = \left[\begin{array}{c|c|c|c|c} \acute{a}^1 & \acute{a}^2 & \acute{a}^3 & \cdots & \acute{a}^n \\ \hat{a}^1 & \hat{a}^2 & \hat{a}^3 & & \hat{a}^n \end{array}\right] \tag{93}$$

where $|\alpha^d|^2 + |\beta^d|^2 = 1$, $d=1,2,\ldots,n$. In general, a quantum system, $|ø_n\rangle$, with *n* Q-bits can be in an arbitrary superposition of states simultaneously up to2^n different states, as (Zhang, 2011):

$$|\psi_n\rangle = \sum_{i=1}^{2^n} C_i\,|S_i\rangle \tag{94}$$

where C_i is the probability amplitude of the state *i*, and S_i is described by the binary string $(x^1, x^2, \ldots, x^n)$ where $x^d \in \{0,1\}$, $d=1,2,\ldots,n$. In contrast, a normal computer can only be in one of these 2^n states at any one time. However, in the act of observing a quantum state, it collapses to a single state. The pseudo code of observation process for *dth* Q-bit is given by Box 1. In this box, *rand* [0, 1] is a random number produced by a uniformly distributed probability function in the range [0, 1].

Box 1. The pseudo code of observation process in quantum-inspired algorithms

If $rand(0,1)<(\alpha^d)^2$ ***Then*** $x^d=0$ ***Else*** $x^d=1$

Another essential concept is Q-gate which refers to a variation operator in QIEAs. A Q-gate gets a Q-bit's state, (α,β), as input and changes the state such that the updated Q-bit should be satisfied the normalization $\alpha'^2 + \beta'^2 = 1$, where a' and β' are the probability amplitudes of updated Q-bit (Han, 2002). The Rotation Q-gate (RQ-gate) due to its capability and effectiveness is used more than others in the QIEAs. The RQ-gate was firstly proposed by Han and Kim (Han, 2002). It is defined as a variation operator to drive individuals toward better solutions and eventually find global optimum. To describe it, consider two binary strings $B(t) = (b^1,\ldots,b^n)$ and $X(t) = (x^1,\ldots,x^n)$ which are the best solution, and current solution observed by collapsing Q-string $q(t)$ at time t, respectively. In QIEA the RQ-gate is defined as:

$$\begin{bmatrix} \alpha^d(t+1) \\ \beta^d(t+1) \end{bmatrix} = \begin{bmatrix} Cos(\Delta\theta^d) & -Sin(\Delta\theta^d) \\ Sin(\Delta\theta^d) & Cos(\Delta\theta^d) \end{bmatrix} \begin{bmatrix} \alpha^d(t) \\ \beta^d(t) \end{bmatrix} \text{for } d=1,2,\ldots,n \quad (95)$$

where $\Delta\theta^d = \Delta \times S(\alpha^d,\beta^d)$ is a rotation angle of dth Q-bit of $q(t)$ toward either 0 or 1 state depending on its sign, in which Δ and S are the value and the sign of $\Delta\theta^d$, respectively. The parameter Δ should be designed in compliance with the application problem. Han and Kim specified $\Delta\theta^d$ for maximization of the knapsack problem in (Han, 2002)

Ibrahim et al. (2012b) presented a quantum-inspired binary gravitational search algorithm, called QBGSA, by combining the BGSA and the concept of quantum computation to improve the exploration ability.

In QBGSA, the concept of acceleration updating process of BGSA is modified by replacing the gravitational mass with the magnitude of the rotation factor (θ). Decision parameter, γ_i, of agent i is utilized instead of both mass of other agents and distance between agent i to other agents. The rotation factor is expressed as follows:

$$\Delta\theta_i^d(t) = \sum_{j\in Kbest,\, j\neq i} \left[\theta \times \gamma_{ij} \times (x_i^d - x_j^d(t)\right] \quad (96)$$

where θ is the magnitude of rotation factor which monotonously decreases from θ_{max} to θ_{min} by lapse of time. Decision parameter γ_{ij} is calculated as the follows:

$$\lambda_{ij} = \begin{cases} 1 & if\, M_j > M_i\, and\, R_{ij} \leq \tau \\ 0 & otherwise \end{cases} \quad (97)$$

$$\gamma_{ij} = \begin{cases} \lambda_{ij} + 1 & if\ f(X_k) > f(X_{best}) \\ \lambda_{ij} & otherwise \end{cases} \tag{98}$$

where, τ is a maximum of different number of bits between agent *i* and agent *j*. According to this equation, the force acting on the agent *i* from far agents is ignored. Ibrahim et al. (2012a & 2012b) used QBGSA to solve the optimal power quality monitor (PQM) placement problem in power systems for voltage sag assessment. Ji et al. (2014) applied a combination of QBGSA and constrained programming to solve the thermal unit commitment problem with wind power integration (TUCPW). They used a local mutation adjustment strategy and a priority list of thermal units to reduce searching time and avoid the premature convergence.

Nezamabadi-pour (2015) presented a binary search algorithm by combination of GSA and quantum computing, called Binary Quantum-Inspired Gravitational Search Algorithm (BQIGSA). A modified rotation Q-gates strategy was used with the structure of GSA to solve binary encoded problems. In BQIGSA, the set *SB(t)* is used to maintain the best solution found by each agent through iterations. At *t=0* all solutions are transferred into *SB(0)*. In binary encoded problems, current solutions $SW(t) = \{X_1(t),\ldots,X_N(t)\}$ are made by doing the observing process on Q-bit objects. In a quantum system, the angular velocity determines the value of movement toward 0 or 1 and it is used to perform the RQ-gate on Q-bits. The angular velocity is related with the linear velocity by equation $\omega = \frac{v}{r}$ that *r* is the radius of the circular system. In a system with *r=1*, the angular velocity of each Q-bit is calculated as follows:

$$a_i^d(t) = \sum_{j \in kbest} G(t) \frac{M_j(t)}{R_{ij}(t) + \varepsilon} \left(b_j^d - x_i^d(t)\right) \tag{99}$$

$$\omega_i^d(t+1) = rand_i \times \omega_i^d(t) + a_i^d(t) \tag{100}$$

where *k*best indicates the set of first *k* solution in the *SB(t)*. In the circular system, the amount of displacement is $\Delta\theta = \omega.\Delta t$ that in a discrete system Δt is considered to be 1. Therefore, $\Delta\theta_i^d$ is calculated as follows.

$$\Delta\theta_i^d = \begin{cases} \omega_i^d(t+1) & if\left(\alpha_i^d(t+1)\beta_i^d(t+1)\right) \geq 0 \\ -\omega_i^d(t+1) & if\left(\alpha_i^d(t+1)\beta_i^d(t+1)\right) < 0 \end{cases} \tag{101}$$

Based on calculated $\Delta\theta_i^d$ the updating process for each Q-bit object is done by Equation (95).

4.8 Mixed Gravitational Search Algorithm

Sarafrazi and Nezamabadi-pour (2013) presented a GSA-SVM hybrid system to improve classification accuracy with a suitable feature subset. For this purpose, authors combined a binary GSA with a real-valued

GSA and introduced a mixed version of GSA with objects including continuous and discrete variables, called "super mass". Each super mass consists of two real-valued variables for SVM parameters (C and γ) and p binary variables for feature selection. p is the number features of dataset. The feature mask is Boolean that ''1'' represents the feature is selected, and ''0'' indicates the feature is not selected. For this problem, fitness function with the two criteria of classification accuracy and the number of selected features was calculated as follows:

$$fit_i = acc_i \times \omega_A + \left[1 - \frac{\sum_{j=1}^{p} f_j}{p}\right] \times \omega_F \tag{102}$$

where ω_A and ω_B are the weight of the SVM classification accuracy and the weight of the number of selected features, respectively. acc_i is the SVM classification accuracy. f_j denotes the value of feature mask. In the updating procedure, the Euclidean distance and the Hamming distance were used for the continuous part and discrete part of super mass, respectively.

5 HYBRID GRAVITATIONAL SEARCH ALGORITHM

In this section, we describe the various combinations of GSA with other algorithms and methods. The hybridization of GSA in solving different optimization problems are summarized in Table 3.

Sarafrazi et al. (2010) and Sarafrazi and Nezamabadi-pour (2013) introduced a novel GSA-SVM hybrid system to improve the classification accuracy of SVM. As mentioned in the sub-section 4.8, the proposed hybrid system has been used as a mixed GSA (Sarafrazi and Nezamabadi-pour, 2013) to improve classification accuracy with a suitable feature subset. Mirjalili et al. (2010) proposed a new hybrid GSA with PSO, called PSOGSA, that integrate the exploitation ability of PSO with the exploration ability of GSA to escape from local optima and obtain the better solutions.

David et al. (2012) used the PSO, the GSA and the PSO-GSA for the optimal tuning of Takagi-Sugeno-Kang PI-fuzzy controllers (T-S-K PI-FCs). They presented new T-S-K PI-FCs with a reduced parametric sensitivity with respect to the process small time constant, and a tuning method and then discussed implementation aspects related to the optimization algorithms. Mirjalili et al. (2012) used PSOGSA as new training method for Feed forward Neural Networks (FNNs) and investigated its efficiency in comparison with GSA and PSO. Also, the number of authors (Mallick, 2013; Tsai, 2013; Zhu, 2013; Duman, 2015; Jayaprakasam, 2015) introduced other versions of hybrid GSA with PSO.

Ghalambaz et al. (2011) presented a hybrid GSA and neural network method. The GSA was used to train a multi-layer perceptron neural network to solve the well-known Wessinger's equation approximately. GSA was applied to gain optimum adjustable parameters (weights and biases) for the corresponding multi-layered perceptron regarding to minimize error function.

Yin et al. (Yin, 2011) proposed an improved GSA and then integrated IGSA into K-harmonic means (KHM) algorithm to introduce a hybrid clustering method, called IGSAKHM. The KHM clustering uses the harmonic mean of the distance from a data point to all cluster centers rather than the minimum

Table 3. Summarization of hybrid GSAs

Hybridization	Scholars (years)	Contributions
GSA-SVM	Sarafrazi et al. (2010)	The hybrid system was used to improve the classification accuracy in binary problems.
	Sarafrazi et al. (2013)	Mixed-GSA was used to improve classification accuracy with a suitable feature subset.
GSA-PSO	Mirjalili et al. (2010) David et al. (2012) Mirjalili et al. (2012) Mallick et al. (2013) Tsai et al.(2013) Zhu et al.(2013) Duman et al. (2015) Jayaprakasam et al.(2015)	Authors applied the idea of memory of PSO into GSA and they the modified velocity by combining with PSO velocity term.
GSA-NN	Ghalambaz et al.(2011) Ojugo et al. (2013)	Authors proposed a hybrid system based on GSA and neural network method
GSA-KHM	Yin et al. (2011)	Authors proposed a hybrid clustering method based on IGSA into K-harmonic means algorithm, called IGSAKHM.
GSA-DE	Li et al. (2011)	DE-GSA is a hybrid GSA with the differential evolution
GSA-SA	Chen et al. (2011)	Authors proposed a hybrid GSA with simulated annealing and applied it to traveling salesman problem
GSA-HS	Hatamlou et al. (2011)	The hybrid system integratedGSA with a heuristic search algorithm for data clustering.
GSA-KM	Hatamlou et al. (2012)	Authors proposed a hybrid GSA with K-means for data clustering
GSA-ABC	Guo et al. (2012)	The hybrid system is based on combining the GSA and the artificial bee colony algorithm to balance the exploitation and exploration
GSA-GA	Sheikhpour et al., (2013)	Authors proposed a training method based on a hybrid GSA with genetic algorithmto optimize parameters of neural networks
GSA-K-NN	Han et al. (2013)	Authors presented a hybrid system based on the quantum-inspired binary GSA and the K-nearest neighbor method to improve classification accuracy with a suitable feature subset

distance of a data point to the cluster centers in KM. The objective function of KHM clustering is computed as follows:

$$KHM\left(X,C\right)=\sum_{i=1}^{n}\frac{k}{\sum_{j=1}^{k}\frac{1}{x_i-c_j^{\;p}}} \tag{103}$$

where p is an input parameter ($p\geq 2$), x_i and c_j are data point i and cluster centroids j, respectively. k and n denote the numbers of clusters and the data to be clustered, respectively. The membership of each data x_i in the class j is calculated based on the below equation:

$$m\left(c_j|x_i\right) = -\frac{\left\|x_i - c_j\right\|^{-p-2}}{\sum_{j=1}^{k}\left\|x_i - c_j\right\|^{-p-2}} \tag{104}$$

The improved GSA enhances the coverage rate and object's diversity by two proposed strategies. In the standard GSA, if the position of an object goes out of range of its values, it is returned to the boundary. If there are too many objects on the boundary, and also there exist local optimum solutions on the boundary, the standard GSA will be trapped in the local optimum solution. In the first strategy, the object position back to the upper bound or lower bound and its velocity is recalculated as follows:

$$\begin{aligned} &\text{if } x_i^d > up_d \Rightarrow x_i^d = up_d \\ &v_i^d = -rand \times v_i^d \end{aligned} \tag{105}$$

where *rand* is the random number in the interval [0,1]. up_d and dw_d are the maximum and minimum values of *d*th dimension in the data set. In the second strategy, authors redefined the gravity constant *G(t)* to increase the coverage rate of GSA as follows:

$$G(t) = G_0 \exp\left(\alpha\left(\frac{t}{t_{max}}\right)^2\right) \tag{106}$$

In the hybrid clustering algorithm (IGSAKHM), the position of each object is represented by a vector of dimension $k \times d$, where d is the number of dimension and k is the number of clusters. The fitness of each object is calculated based on the objective function of the KHM algorithm. Of course, it should be noted that IGSAKHM requires more runtime compared to KHM and when time is critical, IGSAKHM is not appropriate.

Li et al. (2011) proposed a hybrid GSA with the differential evolution (DE) algorithm, called DE-GSA. The exploration ability of the DE algorithm is good and it can be used as a complement of the GSA algorithm. In DE-GSA, the searching process is started with DE and then GSA is used to improve the solution quality. Also, authors introduced a new method in updating strategy of GSA to restrain the least and most bounds of the individual. In the new updating strategy, the speed of agents is restricted and when an agent moves outside the boundary, it is scattered in a feasible region away from the boundary. The pseudo code of the DE-GSA is shown as follows:

```
Input
T (Maximum number of iterations)
NP (Population size)
D (The dimension of the agent)
Output
X_best (The best optimal solution)
Step 1: Initialization
  Generate an initial population P with NP agents.
```

```
  Evaluate the fitness for each agent. Set the best position to X_best, t=0.
Step 2: Stopping criterion
  If (t=T) then
    Return the optimal solutions X_best and stop
  End if
Step 3: Mutation and Crossover
  Select a, b, c randomly and a≠b≠c
  Select j_rand∈D
  For i=1 to D do
    If (rand()<CR or i=j_rand) then U_i(t) = X_a(t) + F*(X_b(t) - X_c(t))
    End if
 End for
Step 4: Selection
  Evaluate the fitness for each solution
  For j=1 to NP do
    If (fit(U_j)<fit(X_j)) then X_j(t) = U_j(t)
    Else
      The GSA algorithm to be active to find a new solution GX_j
      Evaluate the fitness of GX
      If (fit(GX_j)<fit(X_j)) then X_j(t)= GX_j
      Else
         X_j(t) = X_j+1(t)
      End if
 End if
End for
Step 5: Let the best optimal solution to X_best
  If (fit(X_best)) keeps fixed at consecutive L steps) then
    For j=1 to NP do
      U_i(t) = X_best(t) + F*(X_a(t) + X_b(t) - X_c(t))
    Endfor
 Endif
Step6: t=t+1; goto Step2
```

where $CR\in(0,1)$ is the crossover rate and $F\in(0,2)$ is the scaling factor. L is a parameter to balance the effect of GSA and DE algorithm. CR, F and L are set to 0.1, 0.5 and 15 by trial and error, respectively.

Chen et al. (2011) proposed a novel hybrid GSA with simulated annealing (SA), called Hr-GSA, and applied it to traveling salesman problem (TSP). A random-key encoding technique is used to the representation of the problem. Authors integrated GSA with a multi-type local search scheme based on SA to get better solutions and enhance the search ability of agents. The position of each agent is initialized randomly in the interval [0,1) and then the position is decoded from the random-key space into the real solution space. The proposed multi-type local search scheme is composed of four operations: swapping, insertion, inversion and long-distance movement. In Hr-GSA, one agent is selected with a given probability to enhance by the operations. Each of operations is performed on the position of agent with a given probability. If fitness of new position is worse than that of old position, the worst position is accepted to replace with a probability. Otherwise, the best position can be accepted according to a threshold that is generated by SA.

Hatamlou et al. (2011a) extended the gravitational search algorithm to perform data clustering. Each agent represents a candidate solution with *k* cluster centroids in a *n*-dimensional problem space. At the beginning, the candidate solutions are randomly generated. The mass of each agent was calculated by an objective function to determine the goodness of a clustering.

The Newtonian gravity law introduced in the standard GSA is used to move the agents in the problem space to find good solutions for data clustering. Also, Hatamlou et al. (2011b) proposed a hybrid GSA with a new heuristic search algorithm, called GSA-HS, for data clustering. Initially, the gravitational search algorithm is utilized to find a near optimum solution for clustering problem. Then the proposed heuristic search algorithm is applied on the output of GSA to improve it.

In another work, Hatamlou et al. (2012) proposed a novel hybrid GSA with K-means for data clustering, called GSA-KM. A good initial population is very efficient to increase performance of GSA. Authors used the k-means algorithm to generate a near optimal solution as one of candidate solutions in GSA. Solutions were represented using one-dimensional arrays that length of arrays is equal to $d \times k$ array; d is the number of features and k is the number of clusters.

The GSA-KM algorithm consists of three main steps. 1) Applying k-means algorithm on selected dataset to generate near optimal centroids for k clusters, 2) Producing other candidate solutions of initial population in three ways: minimum of dataset, maximum dataset, mean of data in each dataset and selecting at random, 3) Determining an optimal solution for the clustering problem by employing GSA on the candidate solutions. The GSA-KM helps the k-means algorithm to escape from local optima and increase the convergence rate of GSA.

Guo (2012) proposed a hybrid algorithm based on combining the GSA and the artificial bee colony (ABC) algorithm to balance the exploitation and exploration. The selection mechanism of ABC that carried out by three types of bees and the updating method of GSA were used. After calculating the masses of agents $M_i(t)$ and $G(t)$, the total forces $F_i^d(t)$ and the acceleration $a_i^d(t)$ of each agent in all dimensions, the employed bees update the agents' positions with the same way in standard GSA. The onlooker bees select their agents based on a selected probability. The scout bee replaces an agent if it is not updated after a given limit time. The best solution found so far is memorized and then the previous steps are repeated until the stopping conditions are met.

Ojugo et al. (2013) applied a hybrid GSA with the artificial neural network (ANN) to develop GSANN model. The neural network learns from experience by generalizing from previous examples to new ones. The main control parameters are its weights and biases which are optimized by GSA.

Sheikhpour et al. (2013) proposed a novel training method based on a hybrid GSA with genetic algorithm (GA) to optimize parameters of neural networks. In the initial steps, the GSA was used to perform the global search and then the GA was applied to do the local search around the global optimum solution. Authors considered a network with only one hidden layer that consists of n neurons in the input layer, m neurons in the hidden layer and p neurons in the output layer. Hence, each individual/object was represented by the weights and biases in the layers. The fitness function of each chromosome is calculated as follows:

$$fitness = \min\left(Error\right) = \frac{1}{2}\sum_{i=1}^{p}\left(t(i) - O_o(t)\right)^2 \qquad (107)$$

where $O_o(i)$ and $t(i)$ are the actual and desired output of neuron i in the output layer, respectively.

Han et al. (2013) presented a new hybrid system which hybridizes the quantum-inspired binary GSA (QBGSA) with the K-nearest neighbor (K-NN) method to enhance classification accuracy with a suitable feature subset in binary problems. In the hybrid system, the position of each agent is represented by a binary vector of length equal to the number of all features in a given dataset. In other words, feature mask is selected as the position of each agent. If the value of the ith bit equals 1, the ith feature is selected. Otherwise, it is ignored. In this system, two terms of classification accuracy and the number of selected features are used to calculate the fitness function.

Su et al. (2015) proposed a novel hybrid algorithm which combines GSA and gauss pseudo-spectral method (GPM) and applied it for solving the reusable launch vehicle approach and landing trajectory optimization problem. To enhance the coverage speed and improve the exploration ability of GSA, they improved GSA by adopting the elite memory reservation strategy and an adaptive gravitational constant adaption with individual optimum fitness feedback.

6 PARAMETER ADAPTATION AND CONTROL SCHEMES FOR GSA

Making a good balance between exploration and exploitation is a difficult issue in applying a certain metaheuristic algorithm to solve different problems (Nezamabadi-pour et al., 2011). When the exploration increases, the algorithm tends to search new points in the search space. While this capability is vital for a metaheuristic search, the algorithm also needs to exploit around the good solutions for finding the optima. Furthermore, the exploration and exploitation cannot be occurred simultaneously. In other words, once a metaheuristic algorithm has a high exploration, it cannot converge to a near-optimum solution with a high probability. Besides, a high exploitation may guide the algorithm to stagnation situation and premature convergence to a local optimum.

One of the major steps in applying any metaheuristic algorithm to a particular problem is how to choose its components such as different operators and their parameters. The value of these parameters significantly concludes exploration and exploitation abilities; determine whether the algorithm will find a near-optimum solution and whether it will find such a solution efficiently. Eiben et al. (1999) distinguished two major forms of setting parameter values: parameter tuning, and parameter control (Eiben et al., 1999). By parameter tuning, one tries to find good value for the parameters before running the algorithm. Afterward, the algorithm will be run using these values which remain fixed during the run.

Parameter control techniques could be classified into three categories: deterministic, self-adaptive, and adaptive. Deterministic parameter control occurs when the value of a parameter is changed by a set of deterministic rule. Here, the rules determine the parameter value without any feedback from the search and performance indices. This method usually uses a time-varying function. The self-adaptive parameter control terminology uses the idea of "evolution of evolution". In this policy, the parameters to be adapted are encoded as added genes into the chromosomes to adjust during the reproduction.

The adaptive terminology of control uses some suitable feedback from the search to perceive the direction of search, performance indices, convergence situation, etc. to change the parameters. The feedback signals can be selected from genotype, phenotype or both of them to monitor the search conditions before changing the parameters and then based on these signals the parameters are set such that to conduct the search to a desired state. In this approach, based on the existing conditions of search and using a set of heuristic rule, one can control the parameter dynamically to earn desired abilities of exploration and exploitation. Generally, adaptive terminology of control uses current value of the parameters beside the performance feedback signals to produce the new value of the parameters (Nezamabadi-pour, 2011).

GSA has a few parameters to be adjusted (e.g. gravity constant, G, and the size of $k$best set). Generally, $G(t)$ is adjusted by a deterministic manner. In the RGSA a nonlinear exponential decreasing function is used to control the gravity constant (Equation 108) whereas a linear decreasing function is used to adjust the gravity constant in the BGSA (Equation 109. The size of $k$best set usually is controlled by a time varying linear decreasing function (Rashedi et al., 2009; Rashedi et al., 2010). However, in some literature we can rarely find other techniques for controlling these two parameters.

$$G(t) = G_0 e^{-\alpha \frac{t}{T}} \tag{108}$$

$$G(t) = G_0 \left(1 - \frac{t}{T}\right) \tag{109}$$

where in these equations G_0 is the gravity constant at time $t=0$, and α is the damping factor.

However, controlling parameters did not restricted to these two parameters in GSA. For example, as above-mentioned, Khajezadeh et al. (Khajehzadeh, 2012) proposed a modification of GSA with an adaptive maximum velocity constraint to control the exploration ability of the standard GSA and present faster convergence (See sub-section 4.1.2).

An important approach for adaptive control of parameters is adaptive parameter control technique using Fuzzy Logic Controllers (FLC). The GSA that uses a FLC to dynamically adjust its parameters is known as Fuzzy Adaptive Gravitational search Algorithm (FAGSA).

A few researchers try to address the issue of balancing between exploration and exploitation by FAGSA (Saeidi-Khabisi and Rashedi, 2012), (Askariand Zahiri 2012). Saeidi-khabisi et al. (2012) utilized FLC to control one of the parameters of GSA. Their proposed method gets better results with less iterations of algorithm in solving optimization problems and increases convergence rate. The parameter of α in Equation (110) is effective in increasing or decreasing the exploration and exploitation capability. They used two parameters of ED and CM to measure the population diversity and the population progress as follows:

$$ED = \frac{R^{ave} - R^{min}}{R^{max} - R^{min}} \tag{109}$$

$$CM = \frac{fit^{ave}(t) - fit^{ave}(t-1)}{fit^{ave}(t)} \tag{110}$$

where R^{ave}, R^{min} and R^{max} are the average, minimum and maximum distance between agents and best result, respectively. For four inputs IT (number of current iteration), CM, ED and $\alpha(t-1)$ and one output $\alpha(t)$, membership functions are defined. The set of linguistic labels associated with IT, $\alpha(t-1)$, and $\alpha(t)$

are Low, Medium, and High and those related to CM and ED are Low and High. The proposed fuzzy system have 36 rules that designed by knowledge of the expert.

Askari and Zahiri (2012) proposed new classifier based on GSA for decision function estimation without requirement to prior knowledge, called Intelligent GSA based classifier (IGSA-classifier). In IGSA, fuzzy rule base is used to update two effective parameters of GSA intelligently. Those are gravitational coefficient and the number of effective objects (*k*best). Then the intelligent GSA is employed to construct a novel decision function estimation algorithm from feature space. Authors constructed the fuzzy block with three inputs and two outputs. The outputs are the number of effective masses and the gravitational coefficient. The maximum fitness value among the all objects in iteration t (f_{best}), the number of iterations, and the variance of the obtained fitnesses in iteration t ($VAR\_fit(t)$) are considered as inputs of the fuzzy block designed. The eight fuzzy rules are extracted based on the linguistic descriptions of effective parameters of GSA to control search process intelligently. The fitness of each object is defined based on the number of misclassified training points by *i*th mass. Gonzalez et al. (2015) used the fuzzy GSA to find the optimal number of modules, layers and nodes of the neural network for pattern recognition applications.

7 CONCLUSION AND FUTURE RESEARCH DIRECTION

Nowadays, we encounter with the increasing complexity of many real world optimization problems. In fact, this causes to increase the demand for fast and accurate metaheuristic search algorithms as good optimizers. So, the interest of many researchers from various fields has been attracted to work on problem-solvers of this kind. Gravitational search algorithm is one of the recent created metaheuristic stochastic population-based search algorithms, which is inspired by the Newtonian laws of gravity and motion. In the short time it could become popular among the scientific community and researchers and could attract many scholars to itself. The references presented show that the last few years were committed by a great amount of interest and work in the GSA.

GSA developed as a simple and effective technique for global optimization over continuous spaces. Despite the fact that GSA is just about 6 years old, various advances and enhances of it have been developed, such as some versions for continuous, binary, discrete, multimodal, single-objective and multi-objective optimization. Furthermore, In order to improve the performance of GSA, it has been combined with some traditional and evolutionary optimization algorithms, such as PSO, neural network, Support Vector Machine and so on.

However, current document clearly indicates that GSA exhibits remarkable performance in optimizing a wide variety of multi-dimensional, multimodal and multi-objective optimization problems including real, binary, triple as well as discrete valued variables. This research tried to provide an overall picture of the state-of-the-art research on GSA. Starting with a brief overview to the basic steps of GSA, it discussed the different operators used in GSA, then overviewed several promising variants and modifications of GSA for tackling binary-valued, triple-valued, discrete-valued, multi-modal, and multi-objective optimization problems and finally introduced the schemes of parameter control and adaptation for GSA.

Although over the past years, research on GSA has reached an impressive state, there are still many open problems for the algorithm. In sequel, some important future directions of research in the area of GSA are disclosed. Only a few operators have been designed for GSA and employed by it. Among them Kepler, Black hole, Disruption and escape were inspired by astronomy and other were originated from other EAs like mutation and crossover. It is possible to think of some new operators for GSA in different application fields. Of course, many aspects of existing operators have not been studied yet. So, the effectiveness of these operators in solving practical optimization problems should be investigated in future. It is clear that each operator may has merit in the specific tasks, however more work should be done in the application domain to reveal the ability of the operators.

It is obvious that different optimization problems, depending on the nature of problem, require different exploration and exploitation abilities which provided by different parameter values. As reported in the section 6, only few works have been investigated the problem of parameter adaptation and control schemes for GSA. Therefore, much work should be devoted to the presentation of new schemes for adapting the GSA's parameters in different applications. Also, it is necessary to theoretically investigate the effects of each parameter on the convergence rate and getting trapped in local optima.

So far, theoretical analysis of GSA family has not been challenged considerably whilst it is necessary for good understanding of the algorithm. Especially a study of the probabilistic convergence properties of GSA remains an open problem. Besides, working on some variants of GSA for solving problems including constraint handling, dynamic environments, and large scales optimization problems is still open for further considerations.

In this chapter, we have focused on algorithmic aspects including operators, variants, hybridization and parameter adaptation of GSA. However, many researchers have contributed to employ GSA and apply it to many applications in various fields of science and engineering. The outcomes of this research indicates the fact that GSA will continue to remain an active field of multi-disciplinary research in the future years, and the growth of the GSA family will be exceeded and expected because there is still much more to be done in this area. We hope that this chapter will be useful for readers interested in GSA.

ACKNOWLEDGMENT

The authors would like to thank the anonymous reviewers for their very helpful suggestions.

REFERENCES

Abbasian, M. A., & Nezamabadi-pour, H. (2012). Multi Objective Gravitational Search Algorithm Using Non-dominated Fronts. *Journal of Electrical Engineering*, *41*(1), 67–80.

Abbasian, M. A., Nezamabadi-pour, H., & Amoozegar, M. (2015). A Clustering based Archive Multi Objective Gravitational Search Algorithm. *Journal of Fundamenta Informaticae*, *138*(4), 387–409. doi:10.3233/FI-2015-1218

Ajami A., &Armaghan M., (2013). A multi-objective gravitational search algorithm based approach of power system stability enhancement with UPFC. *Journal of Central South University, 20*(6), 1536-1544.

Askari, H., & Zahiri, S. H. (2012). Decision Function Estimation Using Intelligent Gravitational Search Algorithm. *International Journal of Mach. Learn. and Cyber.*, *3*(2), 163–172. doi:10.1007/s13042-011-0052-x

Aziz, N. A. A. B., Ibrahim, Z., Ibrahim, I., Zaidi, M., & Tumari, M. (2013). Synchronous vs Asynchronous Gravitational Search Algorithm. In *Proceedings of First International Conference on Artificial Intelligence, Modeling & Simulation.*

Aziz, N.A. AB., Ibrahim, Z., Nawawi, S.W., Sudin, Sh., Mubin, M. & Aziz, K. A. B. (2014). *Synchronous Gravitational Search Algorithm vs Asynchronous Gravitational Search Algorithm: A Statistical Analysis, Frontiers in Artificial Intelligence and Applications.* IOS Press. Doi: 10.3233/978-1-61499-434-3-160

Baniassadi, Z., Nezamabadi-pour, H., & Farsangi, M. M. (2011). A multi-objective Solution of Gravitational Search Algorithm for Benchmark Functions and Placement of SVC. *Intelligent Systems in Electrical Engineering*, *1*(1), 59–78.

Benioff, P. (1980). The computer as a physical system: A microscopic quantum mechanical Hamiltonian model of computers as represented by Turing machines. *Journal of Statistical Physics*, *22*(5), 563–591. doi:10.1007/BF01011339

Cagnina, L., Esquivel, S., & Coello Coello, C. A. (2005). A Particle Swarm Optimizer for Multi-Objective Optimization. *JCST*, *4*(5), 204–210.

Chen, H., Li, S., & Tang, Z. (2011). Hybrid GSA with Random-Key Encoding Scheme Combined with Simulated Annealing. *Journal of Computer Science and Network Security*, *11*, 208–217.

Das, S., & Suganthan, P. N. (2011). Differential Evolution: A Survey of the State-of-the-Art. *IEEE Transactions on Evolutionary Computation*, *15*(1), 4–31. doi:10.1109/TEVC.2010.2059031

Davarynejad, M., Berg, J., & Rezaei, J. (2014). Evaluating center-seeking and initialization bias: The Case of Particle Swarm and Gravitational Search Algorithms. *Journal of Information Science*, *278*, 802–821. doi:10.1016/j.ins.2014.03.094

David, R. C., Precup, R., Petriu, E. M., Purcaru, C., & Preitl, S. (2012). PSO and GSA algorithms for fuzzy controller tuning with reduced process small time constant sensitivity. In *Proceedings of 16th International Conference on System Theory, Control and Computing (ICSTCC).*

De, M. H., Neto, G. F., Silva, A. M., Mariani, V. C., & Coelho, L. S. (2014). A Modified Gravitational Search Algorithm for Continuous Optimization. In *Proceedings of the conference companion on Genetic and evolutionary computation companion* (GECCO). ACM.

Deb, K., Pratap, A., Agarwal, S., & Meyarivan, T. (2002). A Fast and Elitist Multi-Objective Genetic Algorithm: NSGA-II. *IEEE Transactions on Evolutionary Computation*, *6*(2), 182–197. doi:10.1109/4235.996017

Doraghinejad, M. & Nezamabadi-pour, H. (2014). Black Hole: A New Operator for Gravitational Search Algorithm. *Journal of Computational Intelligence Systems, 7*(5), 809-826.

Doraghinejad, M., Nezamabadi-pour, H., & Mahani, A. (2014). Channel assignment in multi-radio wireless mesh networks using an improved gravitational search algorithm. *Journal of Network and Computer Applications*, *38*, 163–171. doi:10.1016/j.jnca.2013.04.007

Dorigo, M., Maniezzo, V., & Colorni, A. (1991). *Positive Feedback as a Search Strategy.* Technical Report no.91016. Politecnico di Milano.

Dowlatshahi, M. B., Nezamabadi-pour, H., & Mashinchi, M. (2014). A discrete gravitational search algorithm for solving combinatorial optimization problems. *Information Sciences*, *258*, 94–107. doi:10.1016/j.ins.2013.09.034

Duman, S., Yorukeren, N., & Altas, I. H. (2015). A novel modified hybrid PSOGSA based on fuzzy logic for non-convex economic dispatch problem with valve-point effect. *Journal of Electrical Power & Energy Systems*, *64*, 121–135. doi:10.1016/j.ijepes.2014.07.031

Eiben, A. E., Hinterding, R., & Michlewicz, Z. (1999). Parameter control in evolutionary algorithms. *IEEE Transactions on Evolutionary Computation*, *3*(2), 124–141. doi:10.1109/4235.771166

Fogel, L., Owens, A., & Walsh, M. (1966). *Artificial Intelligence Through Simulated Evolution*. New York: Wiley.

Ganesan, T., Elamvazuthi, I., Shaari, K. Z. K., & Vasan, P. (2013). Swarm Intelligence and Gravitational Search Algorithm for Multi-objective Optimization of Synthesis Gas Production. *Journal of Applied Energy*, *103*, 368–374. doi:10.1016/j.apenergy.2012.09.059

Gao, S., Todo, Y., Gong, T., Yang, G., & Tang, Z. (2014). Graph Planarization Problem Optimization Based on Triple-Valued Gravitational Search Algorithm. *Journal of IEEJ Transaction on Electrical and Electrical Engineering*, *9*(1), 39–48. doi:10.1002/tee.21934

Ghalambaz, M., Noghrehabadi, A. R., Behrang, M. A., Assareh, E., Ghanbarzadeh, A., & Hedayat, N. (2011). A Hybrid Neural Network and Gravitational Search Algorithm (HNNGSA) method to solve well known Wessinger's equation, Journal of World Academy of Science. *Engineering and Technology*, *73*, 803–807.

Gonzalez, B., Valdez, F., Melin, P., & Prado-Arechiga, G. (2015). Fuzzy logic in the gravitational search algorithm for the optimization of modular neural networks in pattern recognition. *Journal of Expert Systems with Application*, *42*(14), 5839–5847. doi:10.1016/j.eswa.2015.03.034

Gu, B., & Pan, F. (2013). Modified Gravitational Search Algorithm with Particle Memory Ability and its Application. *International Journal of Innovative Computing, Information, & Control*, *9*(11), 4531–4544.

Guo, Z., (2012). A Hybrid Optimization Algorithm Based on Artificial Bee Colony and Gravitational Search Algorithm. *International Journal of Digital Content Technology and its Applications, 6*(17), 620-626.

Han, K. H., & Kim, J. H. (2002). Quantum-inspired evolutionary algorithm for a class of combinatorial optimization. *IEEE Transactions on Evolutionary Computation*, *6*(6), 580–593. doi:10.1109/TEVC.2002.804320

Han, X., & Chang, X. (2012). A Chaotic Digital Secure Communication based on a Modified Gravitational Search Algorithm Filter. *Journal of Information Science*, *208*, 14–27. doi:10.1016/j.ins.2012.04.039

Han, X. H., Quan, L., Xiong, X. Y., & Wu, B. (2013). Facing the Classification of Binary Problems with a Hybrid System based on Quantum-Inspired Binary Gravitational Search Algorithm and K-NN Method. *Journal of Engineering Applications of Artificial Intelligence*, *26*(10), 2424–2430. doi:10.1016/j.engappai.2013.05.011

Hassanzadeh, H. R., & Rohani, M. (2010). A multi-objective gravitational search algorithm. In *Proceedings of the Second International Conference on Computational Intelligence, Communication Systems and Networks*.

Hatamlou, A., Abdullah, S., & Nezamabadi-pour, H. (2011a). Application of Gravitational Search Algorithm on Data Clustering. In *Proceedings of Rough Sets and Knowledge Technology* (pp. 337–346). Berlin: Springer. doi:10.1007/978-3-642-24425-4_44

Hatamlou, A., Abdullah, S., & Nezamabadi-pour, H. (2012). A Combined Approach for Clustering based on K-means and Gravitational Search Algorithms. *Journal of Swarm and Evolutionary Computation*, *6*, 47–52. doi:10.1016/j.swevo.2012.02.003

Hatamlou, A., Abdullah, S., & Othman, Z. (2011b). *Gravitational Search Algorithm with Heuristic Search for Clustering Problems. In Proceeding of 3rd IEEE on Data Mining and Optimization* (pp. 190–193). Selangor, Malaysia: DMO.

Holland, J. H. (1975). *Adaptation in Natural and Artificial Systems*. Ann Arbor, MI: University of Michigan Press.

Ibrahim, A., Mohamed, A., & Shareef, H. (2012). Application Of Quantum-Inspired Binary GSA For Optimal Power Quality Monitor Placement. In *Proceedings of The 11th International Conference On Artificial Intelligence, Knowledge Engineering And Data Bases*. World Scientific and Engineering Academy and Society (WSEAS).

Ibrahim, A., Mohamed, A., & Shareef, H. (2012). A novel quantum-inspired binary gravitational search algorithm in obtaining optimal power quality monitor placement. *Journal of Applied Science*, *12*(9), 822–830. doi:10.3923/jas.2012.822.830

Ibrahim, I., Ibrahim, Z., Ahmad, H., Yusof, Z. MD., Shapiai, M.I., Nawawi, S.W. & Mubin, M., (2014). An Assembly Sequence Planning Approach with BGSA. *Journal of New Trends in Software Methodologies, Tools and Techniques*, 179-193.

Ibrahim, Z., Muhammad, B., Ghazali, K. H., Lim, K. S., Nawawi, S. W., & Yusof, Z. (2012). Vector Evaluated Gravitational Search Algorithm (VEGSA) for Multi-Objective Optimization Problems. In *Proceedings of the Fourth International Conference on Computational Intelligence, Modeling and Simulation*. doi:10.1109/CIMSim.2012.29

Jayaprakasam, S., Rahim, S. K. A., & Leow, C. Y. (2015). PSOGSA-Explore: A new hybrid metaheuristic approach for beam-pattern optimization in collaborative beam-forming. *Journal of Applied Soft Computing*, *30*, 229–237. doi:10.1016/j.asoc.2015.01.024

Ji, B., Yuan, X., Li, X., Huang, Y., & Li, W. (2014). Application of Quantum-Inspired Binary Gravitational Search Algorithm for Thermal Unit Commitment with Wind Power Integration. *Journal of Energy Conversion and Management*, *87*, 589–598. doi:10.1016/j.enconman.2014.07.060

Jung, S. H. (2003). Queen-bee Evolution for Genetic Algorithms. *Electronics Letters*, *39*(6), 575–576. doi:10.1049/el:20030383

Karaboga, D. (2005). *An Idea based on Honey Bee Swarm for Numerical Optimization*. Technical Report TR06. Computer Engineering Department, Engineering Faculty, Erciyes University.

Karaboga, D., Gorkemli, B., Ozturk, C., & Karaboga, N. (2014). A Comprehensive Survey: Artificial Bee Colony (ABC) Algorithm and Applications. *Journal of Artificial Intelligence Research*, *42*(1), 21–57. doi:10.1007/s10462-012-9328-0

Kazak, N., & Duysak, A. (2012). Modified gravitational search algorithm. *International Symposium on Innovations in Intelligent Systems and Applications*.

Kennedy, J., & Eberhart, R. C. (1995). In Particle swarm optimization. In *Proceedings of IEEE International Conference on Neural Networks*. doi:10.1109/ICNN.1995.488968

Khajehzadeh, M., Taha, M. R., El-Shafie, A., & Eslami, M. (2012). A Modified Gravitational Search Algorithm for Slope Stability Analysis. *Journal of Engineering Applications of Artificial Intelligence*, *25*(8), 1589–1597. doi:10.1016/j.engappai.2012.01.011

Koza, J. R. (1992). *Genetic Programming: On the Programming of Computers by Means of Natural Selection*. MIT Press.

Li12Li, P., & Duan, H. B. (2012). Path Planning of Unmanned Aerial Vehicle based on Improved Gravitational Search Algorithm. *Science China*, *55*(10), 2712–2719. doi:10.1007/s11431-012-4890-x PMID:23108865

Li, X., Yin, M., & Ma, Z. (2011). Hybrid Differential Evolution and Gravitation Search Algorithm for Unconstrained Optimization. *International Journal of the Physical Sciences*, *6*(25), 5961–5981.

Liu, H., Ding, G., & Sun, H. (2012). An Improved Opposition-based Disruption Operation Gravitational Search Algorithm. In *Processing Fifth International Symposium on Computational Intelligence and Design*.

Liu, N., Qiu, Y., & Zhang, B. (2014). An Improved Gravitation Search Algorithm forVehicle Routing Problem. *Journal of Modern Transportation*, *3*(2), 39–48.

Mallick, S., Ghoshal, S. P., Acharjee, P., & Thakur, S. S. (2013). Optimal Static State Estimation using Improved Particle Swarm Optimization and Gravitational Search Algorithm. *International Journal of Electrical Power & Energy Systems*, *52*, 254–265. doi:10.1016/j.ijepes.2013.03.035

Mirjalili, S., & Mohd Hashim, S. Z. (2010). A New Hybrid PSOGSA Algorithm for Function Optimization. In *Proceeding of International Conference on Computer and Information Application* (ICCIA 2010).

Mirjalili, S., Mohd Hashim, S. Z., & Moradian Sardroudi, H. (2012). Training Feed forward Neural Networks using Hybrid Particle Swarm Optimization and Gravitational Search Algorithm. *Journal of Applied Mathematics and Computation*, *218*(22), 11125–11137. doi:10.1016/j.amc.2012.04.069

Muhammad, B., Ibrahim, Z., Ghazali, K. H., Ghazali, M. R., Saealal, M. S., Lim, K. Sh., . . . Mokhtar, N. (2014). Performance Evaluation of Vector Evaluated Gravitational Search Algorithm II. In New Trends in Software Methodologies, Tools and Techniques. IOS Press.

Neumann, J. V. (1955). *Mathematical Foundations of Quantum Mechanics*. Princeton University Press.

Nezamabadi-pour, H., Yazdani, S., Farsangi, M. M., & Neyestani, M. (2011). A Solution to Economic Dispatch Problem by Fuzzy Adaptive Genetic Algorithm. *Iranian Journal of Fuzzy System*, *8*(3), 1–21.

Nobahari, H., Nikusokhan, M., & Siarry, P. (2012). A Multi-Objective Gravitational Search Algorithm Based on Non-Dominated Sorting. *International Journal of Swarm Intelligence Research*, *3*(3), 32–49. doi:10.4018/jsir.2012070103

Ojugo, A. A., Emudianughe, J., Yoro, R. E., Okonta, E. O., & Eboka, A. O. (2013). A Hybrid Artificial Neural Network Gravitational Search Algorithm for Rainfall Runoff Modeling and Simulation in Hydrology. *Journal of Intelligent Computing and Applications*, *2*(1), 22–33.

Ong, Y. S., Lim, M. H., Neri, F., & Ishibichi, H. (2009). Emerging Trends in Soft Computing: Memetic Algorithms. *Journal of Soft Computing*, *13*(8-9), 739–740. doi:10.1007/s00500-008-0353-5

Passino, K. M. (2002). Bio mimicry of Bacterial Foraging for Distributed Optimization and Control. *IEEE Control Systems Magazine*, *22*(3), 52–67. doi:10.1109/MCS.2002.1004010

Rashedi, E. (2007d). *Gravitational Search Algorithm*. (M.Sc. Thesis). Shahid Bahonar University of Kerman, Kerman, Iran. (in Farsi)

Rashedi, E., & Nezamabadi-pour, H. (2007a). Gravitational Search Algorithm. In *Proceeding of Fifth Iranian Conference Electrical Engineering*. (in Farsi)

Rashedi, E., & Nezamabadi-pour, H. (2013). A Stochastic Gravitational Approach to Feature based Color Image Segmentation. *Journal of Engineering Applications of Artificial Intelligence*, *26*(4), 1322–1332. doi:10.1016/j.engappai.2012.10.002

Rashedi, E., & Nezamabadi-pour, H. (2014). Feature Subset Selection using Improved Binary Gravitational Search Algorithm. *Journal of Intelligent and Fuzzy Systems*, *26*(3), 1211–1221.

Rashedi, E., Nezamabadi-pour, H., & Saryazdi, S. (2007c). Binary Gravitational Search Algorithm, In *Proceeding of First Joint Conference on Fuzzy and Intelligent Systems*. (in Farsi)

Rashedi, E., Nezamabadi-pour, H., & Saryazdi, S. (2009). GSA: A Gravitational Search Algorithm. *Journal of Information Science*, *179*(13), 2232–2248. doi:10.1016/j.ins.2009.03.004

Rashedi, E., Nezamabadi-pour, H., & Saryazdi, S. (2010). BGSA: Binary Gravitational Search Algorithm. *Journal of Natural Computing*, *9*(3), 727–745. doi:10.1007/s11047-009-9175-3

Rashedi, E., Nezamabadi-pour, H., & Saryazdi, S. (2013). A simultaneous feature adaptation and feature selection method for content-based image retrieval systems. *Knowledge-Based Systems*, *39*, 85–94. doi:10.1016/j.knosys.2012.10.011

Rashedi, E., Nezamabadi-pour, H., Saryazdi, S., & Farsangi, M. M. (2007b). Allocation of Static Var Compensator Using Gravitational Search Algorithm. In *Proceeding of First Joint Conference on Fuzzy and Intelligent Systems*.

Rechenberg, I. (1973). *Evolutions Strategie: Optimierung Technischer Systemenach Prinzipien der biologischen Evolution*. Stuttgart: Frommann-Holzboog.

Rubio-Largo, Á., & Vega-Rodríguez, M. A. (2013). A Multi-objective Approach Based on the Law of Gravity and Mass Interactions for Optimizing Networks, Evolutionary Computation in Combinatorial Optimization. *Lecture Notes in Computer Science*, *7832*, 13–24. doi:10.1007/978-3-642-37198-1_2

Rubio-Largo, Á., Vega-Rodríguez, M. A., Gómez-Pulido, J. A., & Sánchez-Pérez, J. M. (2011). A Multi-objective Gravitational Search Algorithm Applied to the Static Routing and Wavelength Assignment Problem", Applications of Evolutionary Computation. *Lecture Notes in Computer Science*, *6625*, 41–50. doi:10.1007/978-3-642-20520-0_5

Saeidi-Khabisi, F.S. & Rashedi, E., (2012). Fuzzy Gravitational Search Algorithm. In *Proceeding of Computer and Knowledge Engineering* (ICCKE).

Saha, S., & Chakraborty, D. (2015). Improved Prediction Accuracy with Reduced Feature Set Using Novel Binary Gravitational Search Optimization, Computational Advancement in Communication Circuits and Systems. *Lecture Notes in Electrical Engineering*, *335*, 177–183. doi:10.1007/978-81-322-2274-3_22

Sarafrazi, S., & Nezamabadi-pour, H. (2012). A Hybrid Method of Gravitational Search Algorithm and an Adaptive Stochastic Local Search with Application to Function Optimization. *Journal of Soft Computing and Information Technology*, *1*(4), 13–28.

Sarafrazi, S., & Nezamabadi-pour, H. (2013). Facing the Classification of Binary Problems with a GSA-SVM Hybrid System. *Journal of Mathematical and Computer Modeling*, *57*(1-2), 270–278. doi:10.1016/j.mcm.2011.06.048

Sarafrazi, S., Nezamabadi-pour, H., & Brahman, M. (2010). A GSA-SVM Hybrid System for Classification of Binary Problems. In *Proceeding of 4th Global Conference on Power Control and Optimization* (PCO'2010).

Sarafrazi, S., Nezamabadi-pour, H., & Saryazdi, S. (2011). Disruption: A new operator in gravitational search algorithm. *Journal of Scientia Iranica-D*, *18*(3), 539–548. doi:10.1016/j.scient.2011.04.003

Sarafrazi, S., Nezamabadi-pour, H. & Seydnejad, S.R., (2015). A Novel Hybrid Algorithm of GSA with Kepler Algorithm for Numerical Optimization. *Journal of King Saud University - Computer and Information Sciences, 27*(3), 288-296.

Schwefel, H. P. (1975). *Evolutionsstrategie und Numerische Optimierung*. (PhD dissertation). Technische Berlin, Germany.

Shamsudin, H. C., Irawan, A., Ibrahim, Z., Abidin, A. F. Z., Wahyudi, S., Rahim, M. A. A., & Khalil, K. (2012). A Fast Discrete Gravitational Search Algorithm. In *Proceeding of Fourth International Conference on Computational Intelligence, Modeling and Simulation.*

Shang, Z. (2013). Neighborhood Crossover Operator: A new Operator in Gravitational Search Algorithm. *International Journal of Computer Science Issues*, *10*(2).

Shaw, B., Mukherjee, V., & Ghoshal, S. P. (2012). A Novel Opposition-based Gravitational Search Algorithm for Combined Economic and Emission Dispatch Problems of Power Systems. *Journal of Electrical Power and Energy Systems*, *35*(1), 21–33. doi:10.1016/j.ijepes.2011.08.012

Sheikhpour, S., Sabouri, M., & Zahiri, S.-H. (2013). A Hybrid Gravitational Search Algorithm-Genetic Algorithm for Neural Network Training. In *Proceeding of 21st Iranian Conference of Electrical Engineering* (ICEE).

Soleimanpour, M., & Nezamabadi-pour, H. (2013). A Modified Monkey Algorithm for Real-Parameter Optimization. *Journal of Multiple-Valued Logic and Soft Computing*, *21*, 453–477.

Soleimanpour-moghadam, M., & Nezamabadi-pour, H. (2012). An Improved Quantum Behaved Gravitational Search Algorithm. In *Proceeding of 20th Iranian Conference on Electrical Engineering* (ICEE2012).

Soleimanpour-moghadam, M., Nezamabadi-pour, H., & Farsangi, M. M. (2011). A quantum behaved gravitational search algorithm. *Journal of Intelligent Information Management*, *4*(6), 390–395. doi:10.4236/iim.2012.46043

Soleimanpour-moghadam, M., Nezamabadi-pour, H., & Farsangi, M. M. (2014). A Quantum Inspired Gravitational Search Algorithm for Numerical Function Optimization. *Journal of Information Science*, *267*, 83–100. doi:10.1016/j.ins.2013.09.006

Srinivas, N., & Deb, K. (1994). Multi-objective Optimization using Non-Dominated Sorting in Genetic Algorithms. *Journal of Evolutionary Computation*, *2*(3), 221–248. doi:10.1162/evco.1994.2.3.221

Storn, R., & Price, K. (1995). Differential Evolution – A Simple and Efficient Heuristic for Global Optimization over Continuous Spaces. International Computer Science Institute.

Su, Z., & Wang, H. (2015). A Novel Robust Hybrid Gravitational Search Algorithm for Reusable Launch Vehicle Approach and Landing Trajectory Optimization. *Journal of Neurocomputing*, *162*, 116–127. doi:10.1016/j.neucom.2015.03.063

Sun, J., Feng, B., & Xu, W. (2004). Particle swarm optimization with particles having quantum behavior. In Proceeding of Congress on Evolutionary Computation.

Tabatabaei, S. (2014). A new Gravitational Search Optimization Algorithm to Solve Single and Multi-objective Optimization Problems. *Journal of Intelligent and Fuzzy Systems*, *26*(2), 993–1006.

Talbi, E. G. (2002). A Taxonomy of Hybrid Metaheuristic. *Journal of Heuristics*, *8*(5), 541–546. doi:10.1023/A:1016540724870

Tizhoosh, H. R. (2005). Opposition-Based Learning: A New Scheme for Machine Intelligence. In *Proceedings of the International Conference on Computational Intelligence for Modeling, Control and Automation, and International Conference on Intelligent Agents, Web Technologies and Internet Commerce*. doi:10.1109/CIMCA.2005.1631345

Tsai, H. C., Tyan, Y. Y., Wu, Y. W., & Lin, Y. H. (2013). Gravitational Particle Swarm. *Journal of Applied Mathematics and Computation*, *219*(17), 9106–9117. doi:10.1016/j.amc.2013.03.098

Vasant, P. (2012). *Meta-Heuristics Optimization Algorithms in Engineering, Business, Economics, and Finance*. IGI Global Publishing.

Vasant, P. (2014). *Handbook of Research on Artificial Intelligence Techniques and Algorithms* (2 vols.). IGI Global Publishing.

Yazdani, S., & Nezamabadi-pour, H. (2011). A new Gravitational Solution for Multimodal Optimization. In *Proceeding of the 18th Iranian Conference on Electrical Engineering* (ICEE20111).

Yazdani, S., Nezamabadi-pour, H., & Kamyab, S. (2014). A Gravitational Search Algorithm for Multimodal Optimization. *Journal of Swarm and Evolutionary Computation*, *14*, 1–14. doi:10.1016/j.swevo.2013.08.001

Yin, M., Hu, Y., Yang, F., Li, X., & Gu, W. (2011). A Novel Hybrid K-harmonic Means and Gravitational Search Algorithm Approach for Clustering. *Journal of Expert Systems with Applications*, *38*(8), 9319–9324. doi:10.1016/j.eswa.2011.01.018

Zhang, G. (2011). Quantum-inspired evolutionary algorithms: A survey and empirical study. *Journal of Heuristics*, *17*(3), 303–351. doi:10.1007/s10732-010-9136-0

Zhang, Z., Li, Y., Xia, Y., & Luo, F. (2012). Immunity-based Gravitational Search Algorithm. In *Proceeding of 3rd International Conference on Information Computing and Applications.*

Zhu, T., Xue-ming, D., & Can, L. (2013). T-S Model Identification Based on a New Hybrid GSAPSO Algorithm. *Journal of University of Shanghai for Science and Technology*, *4*, 351–354.

KEY TERMS AND DEFINITIONS

Gravitational Search Algorithm: A swarm intelligence algorithm based on the Newtonian gravity and the laws of motion.

Swarm Algorithms: The algorithms mainly inspired by social behavior patterns of organisms which locate the global optimum without any central control.

Chapter 28
Single Batch-Processing Machine Scheduling Problem with Fuzzy Due-Dates:
Mathematical Model and Metaheuristic Approaches

Sadegh Niroomand
Firouzabad Institute of Higher Education, Iran

Ali Mahmoodirad
Islamic Azad University (Masjed-Soleiman Branch), Iran

Saber Molla-Alizadeh-Zavardehi
Islamic Azad University (Masjed-Soleiman Branch), Iran

ABSTRACT

This paper focuses on a problem of minimizing total weighted tardiness of jobs in a real-world single batch-processing machine (SBPM) scheduling in existence of fuzzy due date. In this paper, first a fuzzy mixed integer linear programming model is developed. Then, due to the complexity of the problem, which is NP-hard, we design two hybrid metaheuristics called GA-VNS and VNS-SA applying the advantages of genetic algorithm (GA), variable neighborhood search (VNS) and simulated annealing (SA) frameworks. Besides, we propose three fuzzy earliest due date heuristics to solve the given problem. Through computational experiments with several random test problems, a robust calibration is applied on the parameters. Finally, computational results on different-scale test problems are presented to compare the proposed algorithms.

1. INTRODUCTION

A batch processing machine (BPM) is a special variant of a scheduling problem, in which several jobs can be simultaneously processed in such a way that all the jobs in a batch start and complete their processing at the same time. The main advantage is to reduce setups and/or facilitation of material handling. The problem of BPM scheduling is often encountered in real industries. The industrial application of these

DOI: 10.4018/978-1-4666-9644-0.ch028

machines can be found in semiconductor burn-in operations, environmental stress-screening (ESS) chambers, chemical, food and mineral processing, pharmaceutical and construction materials industries, etc.

The BPM scheduling problem is important because the scheduling of batching operations has a significant economic impact. It is mainly motivated by an industrial application, namely, the burn-in operation found in the final testing phase in semiconductor manufacturing (Uzsoy *et al.*, 1992; Uzsoy *et al.*, 1994). In the semiconductor manufacturing, the jobs have different processing times and sizes that are both required by the customers. The jobs are grouped in batches where a batch means a subset of jobs. The BPM can process a batch of jobs as long as the sum of all the job sizes in the batch does not violate the capacity of the machine. The processing time of a batch is equal to the longest processing time of all the jobs in that batch.

Ikura and Gimple (1986) were the first researcher who studied the BPM problem and Lee *et al.* (1992) first presented a detailed description for burn-in operation. As reported in the studies, the exact algorithms have a slow convergence rate and they can solve only small instances to optimality.

As this study addresses SBPM with fuzzy due dates using metaheuristics, the review on SBPM scheduling under a fuzzy environment and the application of metaheuristics to these problems is carried out. For an extensive review on BPM scheduling problems, we refer to Potts and Kovalyov (2000) and Mathirajan and Sivakumar (2006).

In BPM scheduling problems, Wang and Uzsoy (2002) firstly proposed a metaheuristic algorithm. Considering dynamic job arrivals, they combined a dynamic programming algorithm with a random key genetic algorithm (GA) to minimize the maximum lateness. Melouk *et al.* (2004) used a simulated annealing (SA) to minimize the makespan. Koh *et al.* (2005) proposed a random key representation-based GA for the problems of minimizing the makespan and total weighted completion time. Sevaux and Peres (2003), Husseinzadeh Kashan *et al.* (2006) and Damodaran *et al.* (2006) used a GA and redesigned the coding and decoding methods.

Mönch et al. (2006) presented a GA combined with dominance properties to minimize the earliness–tardiness of the jobs. Chou *et al.* (2006) and Wang *et al.* (2007) presented a hybrid GA and a hybrid forward/backward approach to minimize the makespan. Husseinzadeh Kashan and Karimi (2008) developed two versions of an ant colony optimization (ACO) framework under the situation considered in Koh *et al.* (2005). Chou and Wang (2008), Mathirajan *et al.* (2010) and Wang (2011) proposed a hybrid GA, SA and iterated heuristic for the objective of the total weighted tardiness, respectively. Husseinzadeh Kashan *et al.* (2010) considered bi-criteria scheduling for the simultaneous minimization of the makespan and maximum tardiness.

In the classic scheduling problems, it is usually assumed that the aspects of the problem in hand are certain. Most existing models neglect the presence of uncertainty within a scheduling environment. In many real-world scheduling problems; however, uncertainty and vagueness in due date often do exist that make the models more complex. This uncertainty may come about because of production problems (e.g., defect in raw material, machine malfunctioning) or problems with delivery itself (e.g., transportation delay, traffic jam). Although classic BPM scheduling models are extensively studied in the literature, there are only three studies on fuzzy-based BPM models.

Ishii *et al.* (1992) introduced the concept of fuzzy due dates to scheduling problems, fuzzy due dates scheduling problems have been investigated by many researchers. Harikrishnan and Ishii (2005) presented a polynomial time algorithm for bi-criteria scheduling of serial-batching problem with fuzzy due dates to minimize the total weighted resource consumption and maximize the minimal satisfaction degree. Yimer & Demirli (2009) considered a fuzzy goal programming problem for batch scheduling of jobs in a two-stage flow shop to minimize the total weighted flow time of jobs. Cheng *et al.* (2010) proposed ACO to minimize the fuzzy makespan on an SBPM with triangular fuzzy processing times.

Till now, none has considered the objective of minimizing the fuzzy total weighted fuzzy tardiness penalties. So, a new approach to solving a fuzzy SBPM (FSBPM) is proposed and a related fuzzy number are considered for due dates and modeled by fuzzy sets, in which the corresponding membership functions represent satisfaction degree with respect to jobs' completion times. Hence, for the first time, we present a new programming approach. Since the problem is NP-hard for solving the addressed problem, two hybrid metaheuristics (GA-VNS and VNS-SA) are developed to obtain better results.

The remainder of this paper is as follows. Section 2 describes the problem in detail and presents the fuzzy mathematical model. Section 3 explains the proposed algorithms. Section 4 describes the experimental design and compares the computational results. Finally, conclusions are provided and some areas of further research are then suggested in section 5.

2. FUZZY MATHEMATICAL MODEL AND PROBLEM DESCRIPTIONS

2.1. Deterministic Model

The objective of this problem is to minimize the total weighted tardiness penalties. Suppose there are n jobs to be processed and each job $j \in J$ has a processing time p_j and a corresponding size s_j. The total size of all the jobs in a batch does not exceed machine capacity S. The processing time of a batch b is given by the longest job in the batch (i.e., $P^b = \max\{p_j \mid j \in \text{batch}B\}$). The formulation is as follows:

Notations

Sets:
J Jobs, $j \in J$
B Batches, $b \in B$
Parameters:
p_j Processing time of job j
s_j Size of job j
S Machine capacity
β_j Tardiness penalty (/unit/h) of job j
d_j Due date of job j
Decision variables:
X_{jb} A binary variable indicates the assignment of job j to batch b
P^b Processing time of batch b
c_j Completion time of job j
C^b Completion time of batch b
T_j Tardiness of job j

According to the mentioned sets, parameters and decision variables, the mathematical formulation of the total weighted tardiness penalties can be written below:

$$Min\ Z = \sum_{j \in J} \beta_j T_j s_j \tag{1}$$

s.t:

$$\sum_{b \in B} X_{jb} = 1 ; \forall j \in J \tag{2}$$

$$\sum_{j \in j} s_j X_{jb} \leq S ; \forall b \in B \tag{3}$$

$$P^b \geq p_j X_{jb}; \forall j \in J, \forall b \in B \tag{4}$$

$$C^b = \sum_{i=1}^{b} P^i ; \forall b \in B \tag{5}$$

$$c_j \geq C^b - M(1 - M_{jb}); \forall j \in J, \forall b \in B \tag{6}$$

$$T_j \geq c_j - d_j; \forall j \in J \tag{7}$$

$$X_{jb} \in \{0.1\}; \forall j \in J, \forall b \in B \tag{8}$$

The objective function is to minimize the total weighted tardiness penalties of jobs. Constraint set (2) ensures that each job can be processed in only one batch. Constraint set (3) ensures that the machine capacity is not exceeded when jobs are assigned to a batch. Constraint set (4) states that the processing time of a batch is the longest processing time among all the jobs in that batch. Constraint set (5) determines the completion time of each batch. Constraint set (6) defines the completion time of each job as the completion time of the batch that it is processed in. Constraint set (7) defines the tardiness of a job as the difference between the due date of a job and its completion time or 0 if it is negative. Constraint set (8) specifies the type of decision variableX_{jb}.

Due to minimization of just only tardiness or total weighted tardiness penalties in the objective function, the model chooses the minimum P^b in the constraint sets (4) to reach the longest processing time among all the jobs in that batch. The smaller the completion time of jobs, the more desirable the objective function. Similarly, the model finds the minimum c_j and T_j in the constraint sets (6) and (7).

2.2. Fuzzy Model

We briefly introduce some basic concepts and results about fuzzy measure theory.

Definition 1: If X is a collection of objects denoted generically by x, then a fuzzy set in X is a set of the ordered pairs:

$$\tilde{d} = \left\{ x, \tilde{d}(x) \mid x \in X \right\},$$

where $\tilde{d}(x)$ is called the membership function that associates with each $x \in X$ a number in [0,1] indicating to what degree x is a number.

Definition 2: $\tilde{d}_j = \left\{ d_{j,l}, d_{j,u} \right\}$ denote a fuzzy number as shown in Figure1.

As mentioned in the literature, the concept of fuzzy due dates has been used in scheduling problems. Here, this concept is being firstly utilized in the BPM scheduling problem. In a fuzzy due date, the membership function assigned to each job represents the customer satisfaction degree for the delivery or completion time of that job. The membership function of a fuzzy due date of a job is represented below.

$$\mu_j(c_j) = \begin{cases} 1 & \text{if } c_j \le d_{j,l} \\ \dfrac{d_{j,u} - c_j}{d_{j,u} - d_{j,l}} & \text{if } d_{j,l} < c_j < d_{j,u} \\ 0 & \text{if } c_j \ge d_{j,u} \end{cases} \tag{9}$$

Figure 1. Membership function

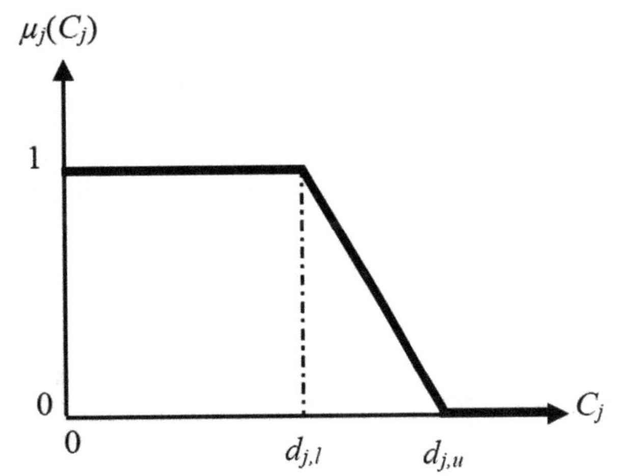

From Figure 1, we can see that the full satisfaction (i.e., $\mu_j(c_j)=1$) is attained if $c_j \le d_{j,l}$, and the satisfaction grade is positive if $d_{j,l}<c_j<d_{j,u}$ in the membership function (9). If $d_{j,l}$=dj,u, the fuzzy due date is transformed to interval due date or due window.

According to the mentioned fuzzy due date, the studied problem can be formulated as a maximization problem of the total degree of satisfaction over given jobs or equivalently, a minimization problem of the total degree of dissatisfaction. For the fuzzy mathematical formulation $\sum_{j\in J}(^{2}{}_{j}T_{j}+\pm_{j}E_{j})s_{j}$, the objective function (10) and constraint sets (11) and (12) are replaced instead of objective function (1) and constraint set (7) to calculate the total degree of satisfaction.

$$Max\ Z=\sum_{j\in J} w_j\,\mu_j\,s_j \tag{10}$$

$$\mu_j=1 \text{ if } c_j\le d_{j,l} \tag{11}$$

$$\mu_j=\frac{d_{j,u}-c_j}{d_{j,u}-d_{j,l}} \text{ if } d_{j,l}<c_j<d_{j,u} \tag{12}$$

We can also use the objective function (13) to calculate the total degree of satisfaction instead of expressions (10-12).

$$\text{Max Z} =\sum_{j\in J} w_j\,s_j\left(\left(\frac{\max(0,d_{j,l}-c_j)}{d_{j,l}-c_j}\right)+\left(\frac{\max(0,\ c_j-d_{j,l})}{c_j-d_{j,l}}\right)\left(\frac{\max(0,d_{j,u}-c_j)}{d_{j,u}-c_j}\right)\left(\frac{d_{j,u}-c_j}{d_{j,u}-d_{j,l}}\right)\right) \tag{13}$$

It is clear that $\frac{\max(0,d_{j,l}-c_j)}{d_{j,l}-c_j}=1$ results in $c_j\le d_{j,l}$ while $\frac{\max(0,\ c_j-d_{j,l})}{c_j-d_{j,l}}$ and $\frac{\max(0,d_{j,u}-c_j)}{d_{j,u}-c_j}$ = 1 result in $d_{j,l}<c_j<d_{j,u}$. For more explanation of

$$\left(\frac{\max(0,d_{j,l}-c_j)}{d_{j,l}-c_j}\right)+\left(\frac{\max(0,\ c_j-d_{j,l})}{c_j-d_{j,l}}\right)\left(\frac{\max(0,d_{j,u}-c_j)}{d_{j,u}-c_j}\right)\left(\frac{d_{j,u}-c_j}{d_{j,u}-d_{j,l}}\right),$$

consider a simple example of $\tilde{d}_j=\{d_{j,l},d_{j,u}\}=\{4,7\}$ with different completion times:

1. $c_j=3\Rightarrow\left(\frac{\max(0,4-3)}{4-3}\right)+\left(\frac{\max(0,\ 3-4)}{3-4}\right)\left(\frac{\max(0,7-3)}{7-3}\right)\left(\frac{7-3}{7-4}\right)=1+0\times1\times\frac{4}{3}=1$
2. $c_j=5\Rightarrow\left(\frac{\max(0,4-5)}{4-5}\right)+\left(\frac{\max(0,\ 5-4)}{5-4}\right)\left(\frac{\max(0,7-5)}{7-5}\right)\left(\frac{7-5}{7-4}\right)=0+1\times1\times\frac{2}{3}=\frac{2}{3}$

3. $c_j = 8 \Rightarrow \left(\frac{\max(0, 4-8)}{4-8}\right) + \left(\frac{\max(0, 8-4)}{8-4}\right)\left(\frac{\max(0, 7-8)}{7-8}\right)\left(\frac{7-8}{7-4}\right) = 0 + 1 \times 0 \times \frac{-1}{3} = 0$

As mentioned above, similarly to the expressions (10-13), expressions (1) and (14-16) can be used for the equivalent fuzzy mathematical formulation of the total degree of dissatisfaction as follows:

$$T_j = \frac{c_j - d_{j,l}}{d_{j,u} - d_{j,l}} \text{ if } d_{j,l} < c_j < d_{j,u} \quad (14)$$

$$T_j = 1 \text{ if } c_j \geq d_{j,u} \quad (15)$$

$$\text{Min Z} = \sum_{j \in J} w_j s_j \left(\left(\frac{\max(0, c_j - d_{j,l})}{c_j - d_{j,l}}\right)\left(\frac{\max(0, d_{j,u} - c_j)}{d_{j,u} - c_j}\right)\left(\frac{c_j - d_{j,l}}{d_{j,u} - d_{j,l}}\right) + \left(\frac{\max(0, c_j - d_{j,u})}{c_j - d_{j,u}}\right)\right) \quad (16)$$

Linearization

Obviously, the proposed fuzzy model is a nonlinear mathematical model because of the conditional expressions in the constraint sets (11), (12), (14) and (15). Also, multiplication of variables and max function in the objective functions (13) and (16) are used. An attempt is made in this part to linearize the fuzzy model via introducing binary variable. Hence, the following constraints should be used instead of nonlinear constraint sets (11) and (12).

$$d_{j,u} - c_j \geq M(y_j - 1);\ \forall j \in J \quad (17)$$

$$\mu_j \leq \frac{d_{j,u} - c_j}{d_{j,u} - d_{j,l}} + M(1 - y_j);\ \forall j \in J \quad (18)$$

$$\mu_j \leq y_j i;\ \forall j \in J \quad (19)$$

$$y_j \in \{0,1\};\ \forall j \in J \quad (20)$$

Similarly, the constraint (20), following objective function and constraints should be used instead of objective function (1) and nonlinear constraint sets (14) and (15).

$$Min\ Z = \sum_{j \in J} w_j (T_{j,1} + T_{j,2}) s_j \quad (21)$$

$$c_j - d_{j,u} < MT_{j,1};\ \forall j \in J \tag{22}$$

$$d_{j,u} - c_j \leq My_j;\ \forall j \in J \tag{23}$$

$$T_{j,2} \geq \frac{c_j - d_{j,l}}{d_{j,u} - d_{j,l}} - M(1 - y_j);\ \forall j \in J \tag{24}$$

3. SOLUTION APPROACH

The evolutionary computation community has shown for many years significant interest in optimization problems, in particular in the global optimization of real valued problems, for which exact and analytical methods are not productive. These techniques have shown great promise in several real-world applications (Zhang et al., 2013). Hence, these methods are often utilized in order to solve the problem in a shorter run time.

3.1. Proposed Earliest Due Date Heuristics

In this subsection, we propose three constructive greedy heuristics based on EDD as a well-known heuristic method related to the due date. The details of these proposed heuristics are as follows:

- Calculate the index of jobs to be scheduled.
- Sort jobs in increasing order of their index.
- Apply the first-first (FF) heuristic to group jobs into batches.

Accordingly, the details of these three variants of EDDs are as follows:

- **EDD Algorithm:** In this variant, the indexes are equal to the EDD of the respective jobs. The centroid-based distance method is used for ranking fuzzy numbers as follows.

$$\text{crisp due date } (d) = \frac{\int_0^{d_l} x\, dx + \int_0^{d_l} x \frac{d_u - x}{d_u - d_l} dx}{\int_0^{d_l} dx + \int_0^{d_l} x \frac{d_u - x}{d_u - d_l} dx} = \frac{1}{3}\left(d_l + d_u - \frac{d_l d_u}{d_l + d_u}\right) \tag{25}$$

The jobs are sorted in increasing order of their crisp due dates. So, the job that has the earliest due date will be allotted first.

- **EDDL Algorithm:** Sort jobs in increasing order of their d_l.

- **EDDU Algorithm:** Sort jobs in increasing order of their d_u.

3.2. Encoding Scheme and Initialization

As mentioned earlier in the literature, the random key (RK) method is used for solving BPM scheduling problems. To generate a sequence by this method, random real numbers between zero and one are generated for each job. By ascending sorting of the value corresponding to each job, the sequence of job is obtained and then the FF heuristic is applied to group the jobs into the batches. After having a permutation and forming the batches, we can use it to compute the objective function value of this solution.

3.3. Hybrid Metaheuristics

Over the last years, considerable research has been conducted in hybrid metaheuristics in the field of optimization. The trade-off between intensification and diversification mechanisms is the main aspect of these algorithms. Generally, metaheuristics can be categorized into two main classes; local search methods and population based methods. Population-based methods deal with a set of solutions in every iteration of the algorithm while local search heuristics only deal with a single solution.

Although local search heuristics only deal with a single solution, it has shown its potential in both exploring and exploiting the promising regions in the search space with high quality solutions such as VNS. On the other hand, the basic scheme of VNS and its extensions requires few and sometimes no parameters. However, it is still prone to inferior solutions due to the limited exploration and exploitation ability.

There are two major approaches to hybridize the VNS with other metaheuristics to improve its performance: hybridizing with a local base metaheuristic and hybridizing with a population based metaheuristic. The first idea is to embedding SA into VNS, so that it is replaced with local search, whereas SA in hybrid VNS addresses how to get out of large valleys. Beside, SA acts as the local search method, because it is good at searching the neighborhood of a solution. The three neighborhoods employed are swap, insertion and inversion.

As one of the most well-known population based methods, genetic algorithm (GA) shows robust performance with various problems. Usually, GA has been proven to be very good at shuffling the solution space or global exploration ability, but fail to intensify the search towards promising regions. Nevertheless, GA usually takes more computing efforts to locate the optimal in the region of convergence (Wu et al., 2004), owing to the lack of local search ability. Therefore, hybridization with local search methods may overcome this weakness and lead to powerful search schemes. So, the second idea is to embedding VNS as a local search into GA and may be a likely choice to consider the hybridization of them. In GA, VNS is applied as a local search to a subset of offspring generated by one-point crossover and swap mutation operator to search for better solutions.

4. COMPUTATIONAL EXPERIMENTS

4.1. Instances

To compare the proposed algorithms, some test problems are needed. In this regard, we generate the required data that can affect the performance of the algorithms including the number of jobs (n), range of processing time of jobs (p_j), size of jobs (s_j), tardiness costs (β_i) and due date of jobs (d_j). The crisp due dates in Tavakkoli-Moghaddam *et al.* (2007) test problems are generated from a uniform distribution. We use such procedure with some modifications to adapt the procedure for our problem as follows:

$$\bar{P} = \frac{\sum_{j=1}^{n} P_{ij}}{n} \tag{26}$$

$$\bar{B} = \frac{\sum_{j=1}^{n} s_j}{0.8 \times Cap} \tag{27}$$

$$BP = \bar{B} \times \bar{P} \tag{28}$$

After generating the BP, the $d_{j,l}$ and $d_{j,u}$ are generated as explained in Table 1.

4.2. Parameter Setting

Because of the dependency of metaheuristic algorithms on the correct selection of parameters and operators, we study the behavior of different parameters of proposed algorithms. The parameters of proposed algorithms are as follows: initial temperature (T_0), number of neighborhood search (n_{max}), reduction ratio of temperature (α) population size (*popsize*), crossover percentage (p_c), and mutation probability (p_m). Levels of these factors are illustrated in Table 2.

Table 1. Test problems characteristics

Parameters	Levels	Count
Number of jobs (N)	10, 20, 30, 50, 75, 100, 125, 150, 175 and 200	10
Processing time of jobs (P)	Uniform distributions [1, 10], [1, 20]	2
Size of jobs (S)	Uniform distributions [1, 10], [2, 4], [4, 8]	3
Tardiness cost (T)	[5, 8]	1
$d_{j,u}$	Uniform distributions ($0.7 \times BP$, BP)	1
$d_{j,l}$	Uniform distributions ($0.6 \times d_{j,u}$, $0.8 \times d_{j,u}$)	1
Cap	10	1
Total number of problem instances		60

Table 2. Factors and their levels

SA, VNS and VNS-SA				GA and GA-VNS		
Parameters	SA Levels	VNS Levels	VNS-SA Levels	Parameters	GA Levels	GA-VNS Levels
T_0	A(1)- 300	-	A(1)- 200	*popsize*	A(1)- 45	A(1)- 30
	A(2)- 350	-	A(2)- 250		A(2)- 50	A(2)- 35
	A(3)- 400	-	A(3)- 300		A(3)- 60	A(3)- 40
n_{max}	B(1)- 600	A(1)- 400	B(1)- 400	p_c	B(1)- 80%	B(1)- 80%
	B(2)- 650	A(2)- 450	B(2)- 450		B(2)- 85%	B(2)- 85%
	B(3)- 700	A(3)- 500	B(3)- 500		B(3)- 90%	B(3)- 90%
α	C(1)- 0.91	-	C(1)- 0.89	p_m	C(1)- 0.1	C(1)- 0.1
	C(2)- 0.92	-	C(2)- 0.9		C(2)- 0.15	C(2)- 0.15
	C(3)- 0.93	-	C(3)- 0.91		C(3)- 0.2	C(3)- 0.2
				n_{max}		D(1)- 300
						D(2)- 350
						D(3)- 400

In order to be fair, the stopping criterion for all algorithms is equal to 6×*n* milliseconds. This criterion is sensitive to the problem size. Using this stopping criterion, searching time increases according to the rise in number of jobs. To yield more reliable information and due to having stochastic nature of algorithms, we tackle each test problem ten times. Because the scale of objective functions in each instance is different, they cannot be used directly. To solve this problem, the relative percentage deviation (RPD) is used for each instance. The RPD is obtained by the following formula:

$$RPD = \frac{A\lg_{spl} - Min_{sol}}{Min_{sol}} \times 100$$

where Alg_{sol} and Min_{sol} are the obtained objective value and minimum objective value found from both proposed algorithms for each instance, respectively. So, we use the RPD measure in the proposed algorithms.

After obtaining the results of the test problems, the results are transformed into RPD measures. The RPD measures are averaged and its value is depicted in Figures 2 through 6. In SA, better robustness is happened when parameters T_0, $n_{\max}$ and α are 350, 650 and 0.92, respectively, as depicted in Figure 2. In Figure 3, the RPD measure for the single parameter of VNS ($n_{\max}$) is depicted and the second level or 450 is the best. Also, for hybrid VNS, as illustrated in Figure 4, T_0, $n_{\max}$ and α are defined as 250, 450 and 0.9. In conformity with Figure 5, best magnitude for *popsize*, p_c and p_m in GA are 50, 85% and 15. Besides, in accordance with Figure 6, in the proposed GA-VNS, best quantity for *popsize*, p_c, p_m and n_{max} are 35, 85%, 15, and 350, respectively.

Figure 2. Mean RPD plot for each level of the factors in SA

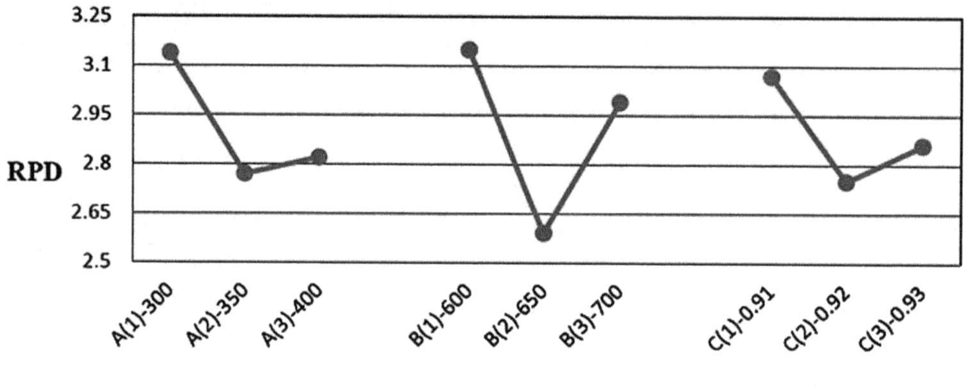

Figure 3. Mean RPD plot for each level of the factors in the VNS

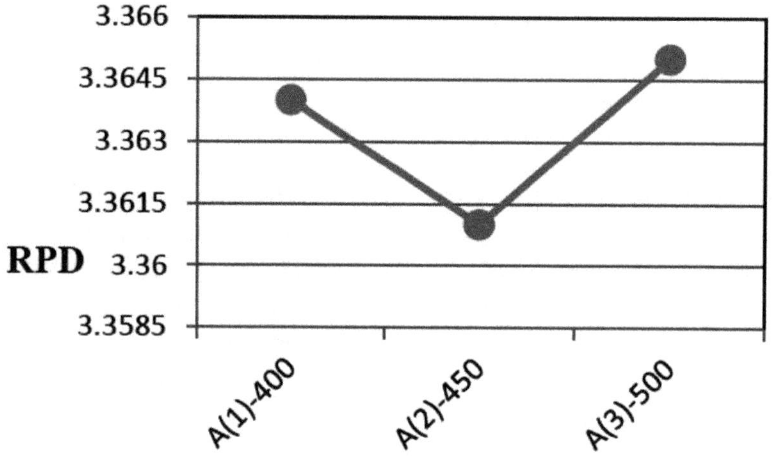

Figure 4. Mean RPD ratio plot for each level of the factors in VNS-SA

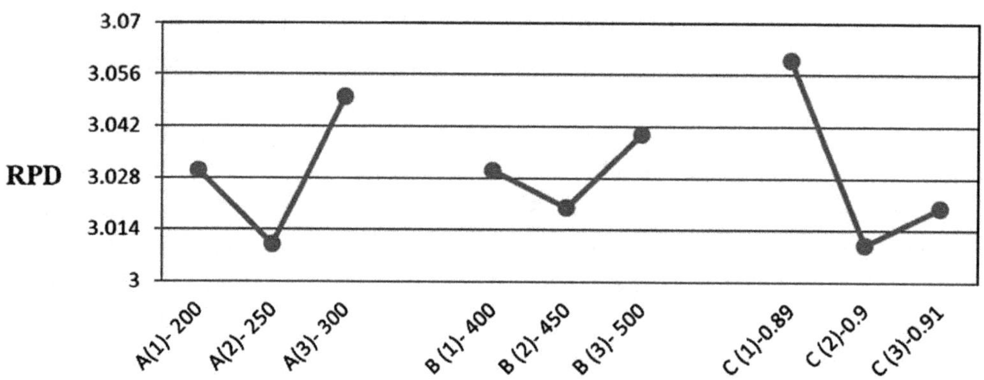

Figure 5. Mean RPD ratio plot for each level of the factors in GA

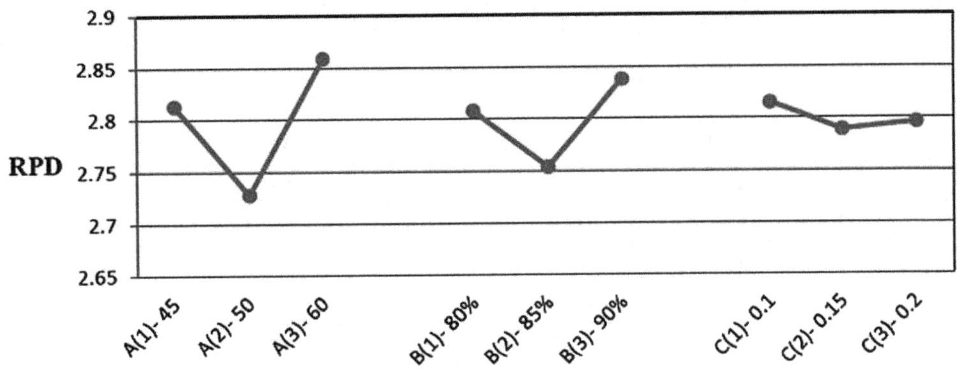

Figure 6. Mean RPD ratio plot for each level of the factors in GA-VNS

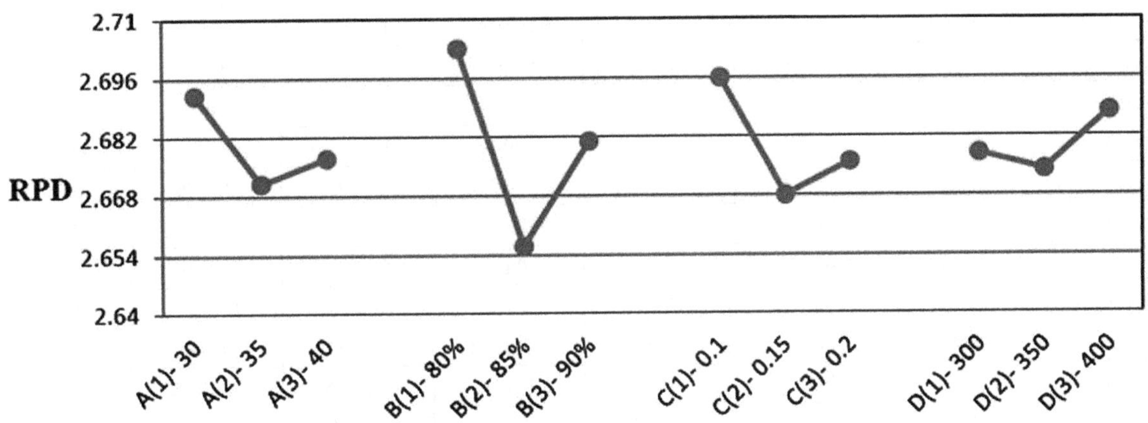

4.3. Experimental Results

In this section, we present and compare the results of EDDL, EDDU, SA, VNS, GA, VNS-SA and GA-VNS with the EDD dispatching rule as a well-known heuristic algorithm related to the due date. As mentioned above, we have 60 problem instances, in which each one includes 10 performed replications to achieve the more reliable results. Table 3 demonstrated the results obtained from EDD, EDDL and EDDU, in which the first and fourth columns represent the data sets characteristics and the remaining columns show the results on instances.

According to Table 3, among heuristics, EDDL has worst results, and it can be concluded that EDDU is better than EDD. In order to analyze the interaction between quality of the algorithms and different problem sizes more concisely, the RPD results are calculated for test problems and averaged for each problem size. The average RPDs obtained by each algorithm are shown in Figures 7 and 8. In these figures, each point represents the average results obtained from six test problems considered in each size of problems with ten replications in each algorithm.

Table 3. Results of EDD, EDDL and EDDU on test problems

Problem	EDD	EDDL	EDDU	Problem	EDD	EDDL	EDDU
10jp1s1	113.9	**101.15**	113.9	100jp1s1	**573.12**	631.37	655.84
10jp1s2	**140.49**	189.19	**140.49**	100jp1s2	874.41	932.05	**864.1**
10jp1s3	123.16	**91.03**	132.06	100jp1s3	**1015.52**	1032.15	1052.11
10jp2s1	**119.33**	130.42	185.23	100jp2s1	**991.97**	1047	946.9
10jp2s2	**113.42**	114.74	**113.42**	100jp2s2	810.87	**716.69**	636.58
10jp2s3	**155.56**	166.44	185.35	100jp2s3	1365.03	1402.12	**1342.32**
20jp1s1	203.7	208.63	**189.57**	125jp1s1	1263.82	1602.84	**1236.04**
20jp1s2	196.16	190.16	**175.72**	125jp1s2	978.41	**910.7**	989.08
20jp1s3	256.24	204.22	**203.58**	125jp1s3	1090.05	1125.03	**1088.53**
20jp2s1	**120.96**	134.36	122.45	125jp2s1	**968.01**	1303.24	1148.23
20jp2s2	214.69	193.68	**162.22**	125jp2s2	962.34	**942.8**	947.46
20jp2s3	153.54	172.27	**153.37**	125jp2s3	1261.83	**1250.44**	1254.45
30jp1s1	195.61	**193.39**	198.37	150jp1s1	1523.47	1473.88	**1199.04**
30jp1s2	252.3	267.01	**252.26**	150jp1s2	1185.04	**1079**	1417.27
30jp1s3	**488.24**	772.72	569.51	150jp1s3	1654.88	1730.38	**1579.02**
30jp2s1	**191.91**	234.04	197.27	150jp2s1	992.93	**992.75**	897.83
30jp2s2	**351.17**	362.43	379.27	150jp2s2	1167.75	1279.56	**1143.29**
30jp2s3	**396.51**	400.85	410.41	150jp2s3	1727.18	1684.63	**1676.66**
50jp1s1	**539.52**	686.09	476.37	175jp1s1	1150.47	1261.93	**1025.12**
50jp1s2	424.3	443.42	**398.52**	175jp1s2	**1303.52**	1414.09	1335.53
50jp1s3	708.06	682.57	**701.39**	175jp1s3	1888.85	**1805.08**	2013.34
50jp2s1	**559.46**	591.87	665.65	175jp2s1	**1623.61**	1761.51	1876.28
50jp2s2	**410.91**	423.84	412.4	175jp2s2	1408.78	1437.47	**1406.21**
50jp2s3	527.77	**390.2**	565.44	175jp2s3	1651.22	1699.95	**1520.91**
75jp1s1	761.27	707.47	**701.15**	200jp1s1	1198.46	**1142.2**	1160.17
75jp1s2	594.23	635.03	**515.88**	200jp1s2	**1556.72**	1828.91	1724.02
75jp1s3	**682.1**	774.99	716.68	200jp1s3	2266.57	2261.79	**1935.25**
75jp2s1	**896.54**	1085.56	1117.63	200jp2s1	**1486.81**	1935.52	1912.32
75jp2s2	517.24	**507.16**	507.64	200jp2s2	**1708.57**	1774.06	1731.21
75jp2s3	872.84	**852.19**	917.32	200jp2s3	1892.47	2006.77	**1885.73**

It is noticeable that with increasing the problem size, gradually the RPDs of the proposed EDD, EDDL and EDDU decrease. In spite of decreasing the RPDs, they are not capable to be completive even in the four last sizes.

As it can be seen, the GA-VNS keeps its robust performance in all ranges of problem sizes. On the contrary to SA and VNS, both VNS-SA and GA have good results. VNS-SA has better performance in 10j and 30j; however, in the last six problem sizes, GA outperforms it. In the first problem size, SA

Figure 7. Means plot for the interaction among heuristic algorithms

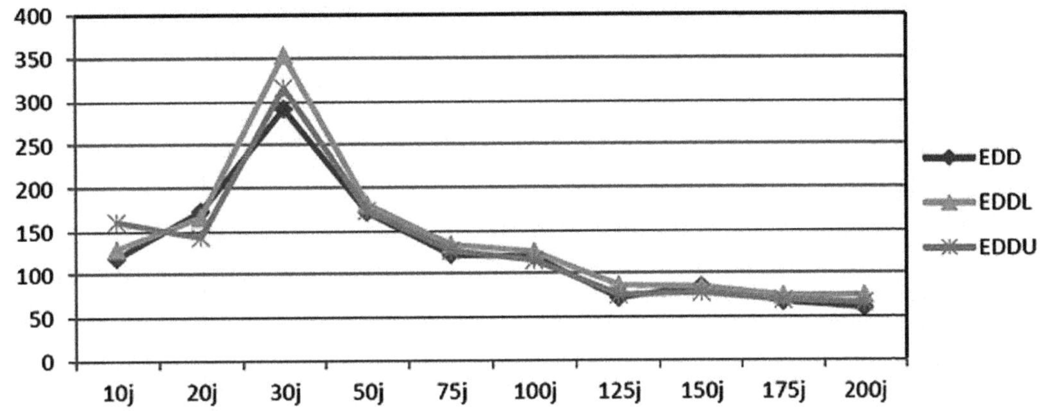

Figure 8. Means plot for the interaction among metaheuristic algorithms

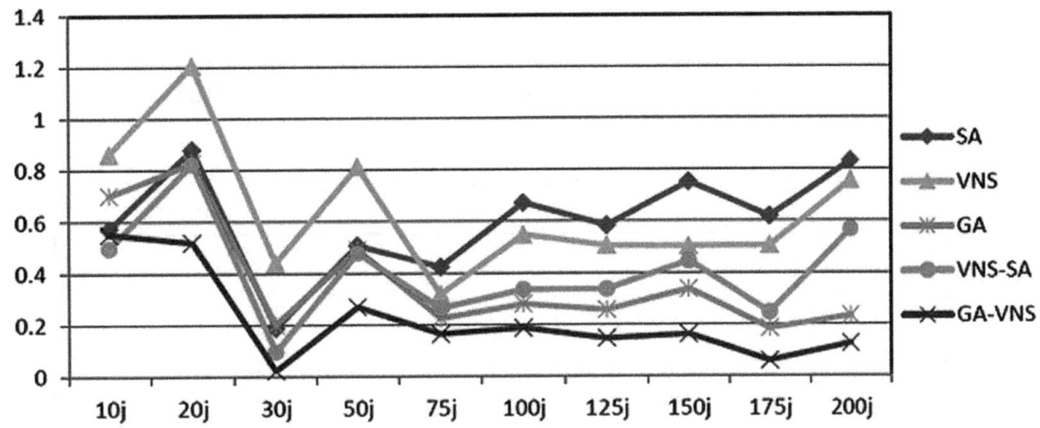

yields the best result, but with increasing the problem size, gradually its RPDs increase. In the first four Problem sizes, VNS Does not show a good performance, but with increasing the problem size its RPDs decrease while in latter sizes it outperforms SA.

From Figures 7 and 8, there is no significant difference between proposed EDD and EDDU or SA and VNS. So, we perform an analysis of variance (ANOVA) to accurately analyze the results among them. The means plot and LSD intervals (at the 95% confidence level) for the presented algorithms are shown in Figures 9 and 10. According to the results, the average RPD obtained by the proposed EDD, EDDL and EDDU are 128.04, 141 and 132.33 respectively. So, EDD is better than EDDU and EDDL.

Also, the average RPD obtained by the proposed GA-VNS is 0.22, while for GA, VNS-SA, SA and VNS are 0.37, 0.41, 0.6, and 0.65, respectively. As is evident, GA-VNS has outperformed other algorithm. As it can be seen, between GA and VNS-SA, and also between SA and VNS, there is not a significant difference. However, they failed to statistically overcome each other. However, based on on the results, we conclude that the proposed GA-VNS can be used to effectively solve the problem.

Figure 9. Means plot and LSD intervals for proposed heuristics

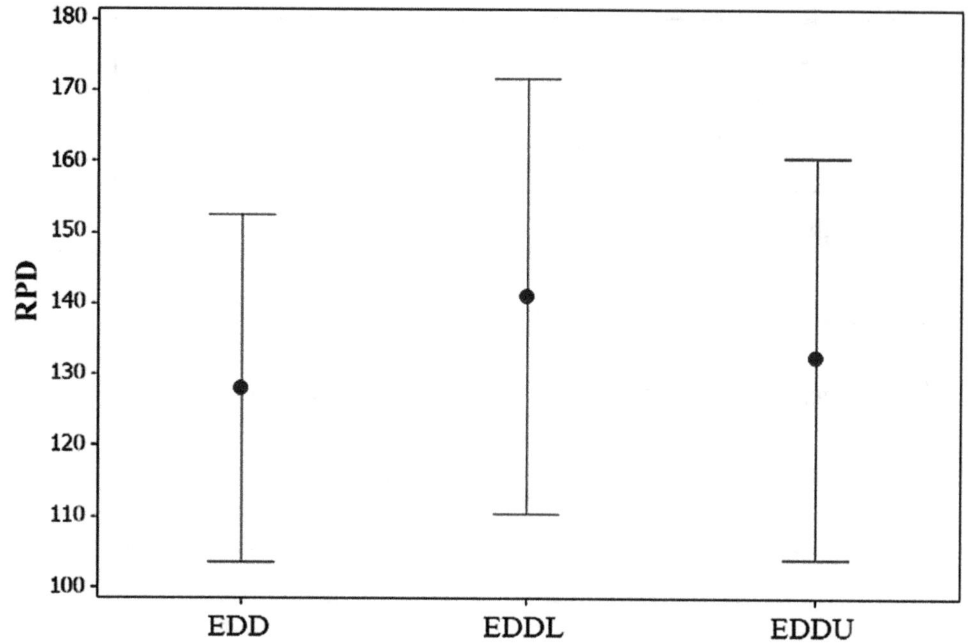

Figure 10. Means plot and LSD intervals for proposed metaheuristics

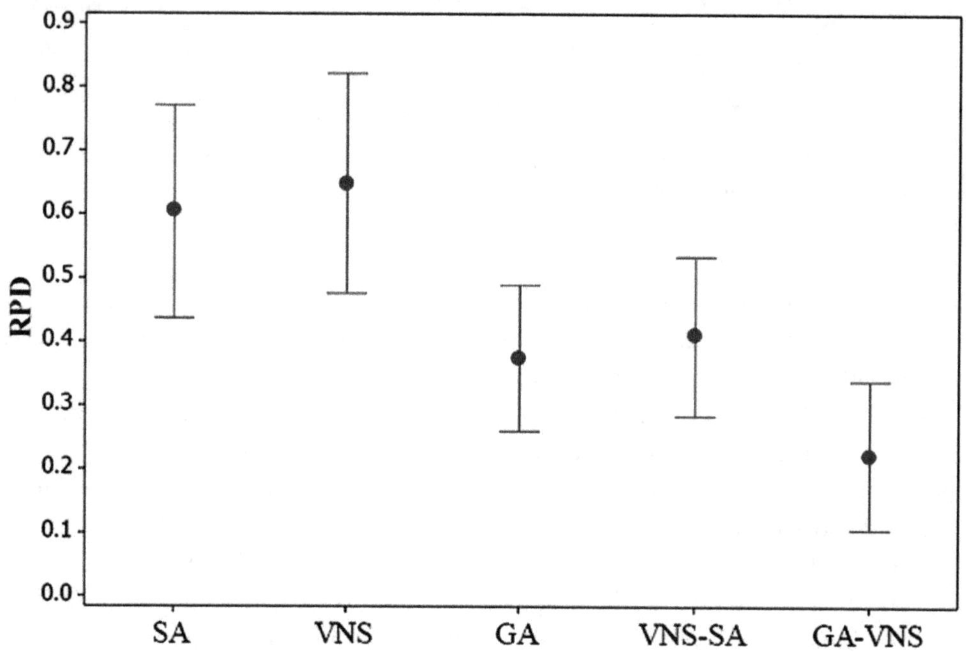

5. CONCLUSION AND FUTURE RESEARCH DIRECTIONS

In this paper we discussed the single batch-processing machine (SBPM) scheduling problem in the presence of fuzzy due date to minimize he total weighted tardiness. We developed a mixed integer linear programming model with the objective functions of the total satisfaction or dissatisfaction degree. To solve this model, three heuristics (EDD, EDDL, EDDU), three metaheuristics (GA, VNS, SA) and two hybrid metaheuristics (GA-VNS and VNS-SA) are developed. Also, a plan was developed and utilized to generate test problems in a fuzzy environment. To enhance the performance of the proposed method, the experimental design method was used by setting their parameters. The computational results showed that the hybrid GA-VNS were robust and superior to other proposed algorithms. As a future work, total weighted earliness tardiness can be considered as the objective function and the same proposed algorithms can be developed for it. Another direction is to work on other algorithm, such as Cuckoo Optimization Algorithm (Balochian and Ebrahimi, 2013), Honey Bees Optimization (Celik, and Ulker, 2013), Differential Evolution (Choi *et al.*, 2013), Cuckoo Search (Zhou and Zheng, 2013) and Firefly Algorithm (Zhang *et al.*, 2013).

REFERENCES

Alebachew, D., & Yimer, K. D. (2009). Fuzzy scheduling of job orders in a two-stage flowshop with batch-processing machines. *International Journal of Approximate Reasoning*, *50*(1), 117–137. doi:10.1016/j.ijar.2007.08.013

Balochian, S. & Ebrahimi, E. (2013). Parameter Optimization via Cuckoo Optimization Algorithm of Fuzzy Controller for Liquid Level Control. *Journal of Engineering*. doi:.10.1155/2013/982354

Celik, Y., & Ulker, E. (2013). An Improved Marriage in Honey Bees Optimization Algorithm for Single Objective Unconstrained Optimization. *TheScientificWorldJournal*, *2013*, 1–11. doi:10.1155/2013/370172 PMID:23935416

Cerny, V. (1985). Thermodynamical approach to the traveling salesman problem: An efficient simulation algorithm. *Journal of Optimization Theory and Applications*, *45*(1), 41–51. doi:10.1007/BF00940812

Cheng, B., Li, K., & Chen, B. (2010). Scheduling a single batch-processing machine with non-identical job sizes in fuzzy environment using an improved ant colony optimization. *Journal of Manufacturing Systems*, *29*(1), 29–34. doi:10.1016/j.jmsy.2010.06.007

Choi, T. J., Ahn, C. W., & An, J. (2013). An Adaptive Cauchy Differential Evolution Algorithm for Global Numerical Optimization. *TheScientificWorldJournal*, *2013*, 1–12. doi:10.1155/2013/969734 PMID:23935445

Chou, F. D., Chang, P. C., & Wang, H. M. (2006). A hybrid genetic algorithm to minimize makespan for the single batch machine dynamic scheduling problem. *International Journal of Advanced Manufacturing Technology*, *31*(3-4), 350–359. doi:10.1007/s00170-005-0194-7

Chou, F. D., & Wang, H. M. (2008). A scheduling for a single semiconductor batch-processing machine to minimize total weighted tardiness. *Journal of the Chinese Institute of Industrial Engineers*, *25*(2), 136–147. doi:10.1080/10170660809509079

Damodaran, P., Manjeshwar, P. K., & Srihari, K. (2006). Minimizing makespan on a batch processing machine with non-identical job sizes using genetic algorithms. *International Journal of Production Economics*, *103*(2), 882–891. doi:10.1016/j.ijpe.2006.02.010

Harikrishnan, K. K., & Ishii, H. (2005). Single machine batch scheduling problem with resource dependent setup and processing time in the presence of fuzzy due date. *Fuzzy Optimization and Decision Making*, *4*(2), 141–147. doi:10.1007/s10700-004-5870-9

Husseinzadeh Kashan, A., & Karimi, B. (2008). Scheduling a single batch-processing machine with arbitrary job sizes and incompatible job families: An ant colony framework. *The Journal of the Operational Research Society*, *59*(9), 1269–1280. doi:10.1057/palgrave.jors.2602448

Husseinzadeh Kashan, A., Karimi, B., & Jolai, F. (2006). Effective hybrid genetic algorithm for minimizing makespan on a single-batch-processing machine with nonidentical job sizes. *International Journal of Production Research*, *44*(12), 2337–2360. doi:10.1080/00207540500525254

Husseinzadeh Kashan, A., Karimi, B., & Jolai, F. (2010). An effective hybrid multi-objective genetic algorithm for bi-criteria scheduling on a single batch processing machine with non-identical job sizes. *Engineering Applications of Artificial Intelligence*, *23*(6), 911–922. doi:10.1016/j.engappai.2010.01.031

Ikura, Y., & Gimple, M. (1986). Efficient scheduling algorithms for a single batch processing machine. *Operations Research Letters*, *5*(2), 61–65. doi:10.1016/0167-6377(86)90104-5

Ishii, H., Tada, M., & Masuda, T. (1992). Two scheduling problems with fuzzy due-dates. *Fuzzy Sets and Systems*, *46*(3), 339–347. doi:10.1016/0165-0114(92)90372-B

Kirkpatrick, S., Gelatt, C. D. Jr, & Vecchi, M. P. (1983). Optimization by simulated annealing. *Science*, *220*(4598), 671–680. doi:10.1126/science.220.4598.671 PMID:17813860

Koh, S. G., Koo, P. H., Kim, D. C., & Hur, W. S. (2005). Scheduling a single batch processing machine with arbitrary job sizes and incompatible job families. *International Journal of Production Economics*, *98*(1), 81–96. doi:10.1016/j.ijpe.2004.10.001

Mathirajan, M., Bhargav, V., & Ramachandran, V. (2010). Minimizing total weighted tardiness on a batch-processing machine with non-agreeable release times and due dates. *International Journal of Advanced Manufacturing Technology*, *48*(9-12), 1133–1148. doi:10.1007/s00170-009-2342-y

Mathirajan, M., & Sivakumar, A. I. (2006). A literature review, classification and simple meta-analysis on scheduling of batch processors in semiconductor. *International Journal of Advanced Manufacturing Technology*, *29*(9-10), 990–1001. doi:10.1007/s00170-005-2585-1

Melouk, S., Damodaran, P., & Chang, P. Y. (2004). Minimizing makespan for single batchprocessing machine with non-identical job sizes using simulated annealing. *International Journal of Production Economics*, *87*(2), 141–147. doi:10.1016/S0925-5273(03)00092-6

Mladenovic, N., (1995), A variable neighborhood algorithm: a new metaheuristic for combinatorial optimization. In *Abstracts of papers presented at Optimization Days: Montre ́al.*

Mönch, L., Unbehaun, R., & Choung, Y. I. (2006). Minimizing earliness–tardiness on a single burn-in oven with a common due date and maximum allowable tardiness constraint. *OR-Spektrum*, *28*(2), 177–198. doi:10.1007/s00291-005-0013-4

Potts, C. N., & Kovalyov, M. Y. (2000). Scheduling with batching: A review. *European Journal of Operational Research*, *120*(2), 228–249. doi:10.1016/S0377-2217(99)00153-8

Sevaux, M., & Peres, S. D. (2003). Genetic algorithms to minimize the weighted number of late jobs on a single machine. *European Journal of Operational Research*, *151*(2), 296–306. doi:10.1016/S0377-2217(02)00827-5

Tavakkoli-Moghaddam, R., Rahimi-Vahed, A., & Mirzaei, A. H. (2007). A hybrid multi-objective immune algorithm for a flow shop scheduling problem with bi-objectives: Weighted mean completion time and weighted mean tardiness. *Information Sciences*, *177*(22), 5072–5090. doi:10.1016/j.ins.2007.06.001

Uzsoy, R., Lee, C., & Martin-Vega, L. (1992). A review of production planning and scheduling models in the semiconductor industry, Part I: System characteristics, performance evaluation and production planning. *IIE Transactions*, *24*(4), 47–60. doi:10.1080/07408179208964233

Uzsoy, R., Lee, C., & Martin-Vega, L. (1994). A review of prod planning and scheduling models in the semiconductor industry, Part II: Shop floor control. *IIE Transactions*, *26*(5), 44–55. doi:10.1080/07408179408966627

Wang, C. S., & Uzsoy, R. (2002). A genetic algorithm to minimize maximum lateness on a batch processing machine. *Computers & Operations Research*, *29*(12), 1621–1640. doi:10.1016/S0305-0548(01)00031-4

Wang, H. M. (2011). Solving single batch-processing machine problems using an iterated heuristic. *International Journal of Production Research*, *49*(14), 4245–4261. doi:10.1080/00207543.2010.518995

Wang, H. M., Chang, P. C., & Chou, F. D. (2007). A hybrid forward/backward approach for single batch scheduling problems with non-identical job sizes. *Journal of the Chinese Institute of Industrial Engineers*, *24*(3), 191–199. doi:10.1080/10170660709509033

Wu, A., Yu, H., Jin, S., Lin, K., & Schiavone, G. (2004). An incremental genetic algorithm approach to multiprocessor scheduling. *IEEE Transactions on Parallel and Distributed Systems*, *15*(9), 824–834. doi:10.1109/TPDS.2004.38

Zhang, Y., Wang, S., Ji, G., & Dong, Z. (2013). Genetic Pattern Search and Its Application to Brain Image Classification. *Mathematical Problems in Engineering*. doi:10.1155/2013/580876

Zhang, Y., Wu, L., & Wang, S. (2013). Solving Two-Dimensional HP Model by Firefly Algorithm and Simplified Energy Function. *TheScientificWorldJournal*. doi:10.1155/2013/398141

Zhou, Y., & Zheng, H. (2013). A Novel Complex Valued Cuckoo Search Algorithm. *TheScientificWorldJournal*, *2013*, 1–6. doi:10.1155/2013/597803 PMID:23766699

Chapter 29
Innovative Hierarchical Fuzzy Logic for Modelling Using Evolutionary Algorithms

M. Mohammadian
University of Canberra, Australia

R. J. Stonier
University of Canberra, Australia

ABSTRACT

This paper considers issues in the design and construction of a fuzzy logic system to model complex (nonlinear) systems. Several important applications are considered and methods for the decomposition of complex systems into hierarchical and multi-layered fuzzy logic sub-systems are proposed. The learning of fuzzy rules and internal parameters is performed using evolutionary computing. The proposed method using decomposition and conversion of systems into hierarchical and multi-layered fuzzy logic sub-systems reduces greatly the number of fuzzy rules to be defined and improves the learning speed for such systems. However such decomposition is not unique and may give rise to variables with no physical significance. This can raise then major difficulties in obtaining a complete class of rules from experts even when the number of variables is small. Application areas considered are: the prediction of interest rate, hierarchical control of the inverted pendulum, robot control, feedback boundary control for a distributed optimal control system and image processing.

1. INTRODUCTION

Conventional modelling approaches have been to construct a mathematical model and examine the system's evolution or its control. Such approaches fail are not completely successful for many of the very large non-linear and complex systems such as for example robotic systems with large input-output data. These systems have typically high dimensionality and have degrees of uncertainty in many parameters. Progress in solving these problems has been with the aid of new advanced high-speed computers and the application of artificial intelligence paradigms, particularly neural networks and fuzzy logic for

DOI: 10.4018/978-1-4666-9644-0.ch029

nonlinear modelling and evolutionary algorithms for optimisation methods. Application of fuzzy logic in particular, to model and solve industrial problems is now wide spread and has universal acceptance.

Fuzzy logic systems typically have a knowledge base consisting of a set of rules of the form:

If (x_1 is A_1^ℓ and x_2 is A_2^ℓ and $\cdots$ and x_n is A_n^ℓ)
Then (z_1 is B_1^ℓ else z_2 is B_2^ℓ else $\cdots$ else z_m is B_m^ℓ)

where A_k^ℓ, k=1,…,n are normalised fuzzy sets for n input variables x_k, k=1,…,n, and where B_k^ℓ, k=1,…,m are normalised fuzzy sets for m output variables z_k, k=1,…,m. The heart of the fuzzy logic system is the inference engine that applies principles of intelligent human reasoning to interpret the rules to output an action from inputs. There are many types of inference engines in the literature, including the popular Mamdani inference engine (Wang, 1997).

Given a fuzzy rule base with M rules and n antecedent variables, a fuzzy controller as given in Equation 1 uses a singleton fuzzifier, Mamdani product inference engine and centre average defuzzifier to determine output variables, has the general form for a single output variable, say z_k:

$$z_k = \frac{\sum_{\ell=1}^{M} \bar{y}_k^l \left(\prod_{i=1}^{n} \mu_{A_i^\ell}(x_i)\right)}{\sum_{\ell=1}^{M} \left(\prod_{i=1}^{n} \mu_{A_i^\ell}(x_i)\right)} \tag{1}$$

where $\bar{y}_k^\ell$ are centres of the output sets B_k^ℓ and membership function μ defines for each fuzzy set A_i^ℓ the value of x_i in the fuzzy set, namely, $\mu_{A_i^\ell}(x_i)$. Common shapes of the membership function are typically, triangular, trapezoidal and Gaussian. A first step in the construction of a fuzzy logic system is to determine which variables are fundamentally important. Any number of these decision variables may appear, but the more that are used, the larger the rule set that must be found. It is known (Raju, Zhou & Kisner, 1990), that the total number of rules in a system is an exponential function of the number of system variables. This curse of dimensionality can be handled in a variety of ways:

1. Fusing variables before input into the inference engine, thereby reducing the number of rules in the knowledge base,
2. Grouping the rules into prioritised levels to design hierarchical or multi-layered structures,
3. Reducing the size of the inference engine directly using notions of passive decomposition of fuzzy relations, and
4. Decomposing the system into a finite number of reduced-order subsystems, eliminating the need for a large-sized inference engine.

In this paper we shall concentrate our discussion on (2). In a hierarchical fuzzy logic structure (HFLS), typically the most influential parameters are chosen as the system variables in the first level, the next most important parameters are chosen as the system variables in the second level, and so on, (Raju & Zhou, 1993). In this hierarchy, the first level gives an approximate output which is then modified by the second level rule set. This procedure can be repeated in succeeding levels of hierarchy. The number of rules in a complete rule set is so reduced to a linear function of the number of variables, but this number may still be high.

But which variables in a given system are the most influential and in what order should variables be chosen? For a general multi-layered structure it is not necessary that the output be improved through the layers, but these questions are still valid. These questions are yet unanswered and an understanding of fundamentals is required to determine the level of interdependence of input variables. Further, given that different hierarchical and multi-layered structures can exist, how can the fuzzy knowledge base and associated parameters in each layer be effectively learnt? We will discuss here evolutionary learning approaches for the determination of these knowledge bases and associated parameters.

2. VARIABLE SELECTION AND RULE BASE DECOMPOSITION

2.1 Interest Rate Prediction

In Kingham, Mohammadian and Stonier (1998), the authors used hierarchical fuzzy logic structures and multi-layered neural network structures for modelling and prediction of the Australian interest rate with 14 input variables, on actual data of key economic indicators that was a limited data set. Using expert knowledge from an economist the input variables were placed into 5 different groupings, namely,

1. Employment (Job Vacancies, Unemployment Rate)
2. Country (Gross Domestic Product, Consumer Price Index)
3. Savings (Household Saving Ratio, Home Loans, Average Weekly Earnings)
4. Foreign (Current Account, RBA Index, Trade Weighted Index)
5. Company (All Industrial Index, Company Profit, New Motor Vehicles)

which then were formed into a two layered fuzzy system, see Figure 1.

The current interest rate was input into each of the five fuzzy systems in the first layer and the final output of the second layer was the updated interest rate. It is assumed that the first layer gives a first iteration of the new interest rate and they are input into the second layer. But the output variables from the first layer do not necessarily have to be identified with the interest rate. Assuming there are five membership sets for all variables, including those entering the second layer, there are 5250 fuzzy rules

Figure 1. Interest rate prediction

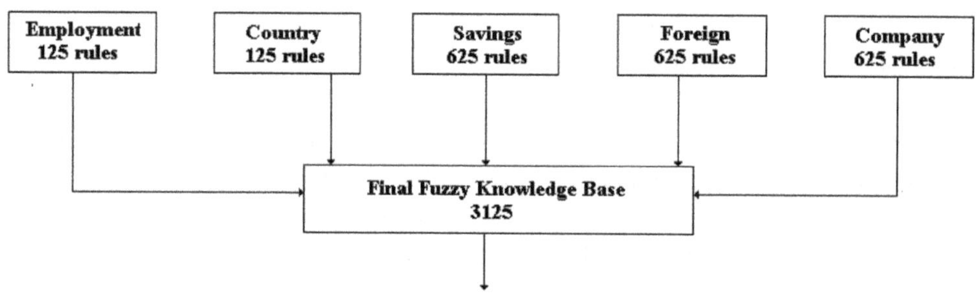

in this structure. If all fourteen variables were input into a single layer structure there would be some 6 million rules (5^{16}). Hence there is a considerable reduction in the number of rules for this simple two layered structure. But it is clear that this in not the only decomposition that could have been formed in grouping the variables, or in number of levels of the multi-layered structure. A genetic algorithm (Goldberg, 1989) was used to learn the rules in this fuzzy system, and it was found that hierarchical fuzzy structure was more accurate in the prediction of quarterly interest rate that the neural network structure. Further research on this problem discussing different hierarchical fuzzy structures of three, four and five layers, and the learning of the fuzzy rule bases, can be found in Mohammadian and Kingham (2004).

2.2 Inverted Pendulum Control

The design of the hierarchical and multi-layered structures is not unique and can be undertaken in such a way that variables output from a given layer, have some physical meaning. Consider the control of the simple inverted pendulum which can be seen in Figure 2 (Stonier, Stacey & Messom, 1998).

Two structures for the fuzzy controller were considered. The first was a single rule base case with all four of the variables x_1 (position of the cart in [–1,1]), x_2 (velocity of the cart), x_3 (the current angle of the pole) and x_4 (the angular velocity of the pole) are used as inputs to the fuzzy rule set to produce a control output u. There were five linguistic sets for each input variable and seven for the output control. All fuzzy membership functions are assumed to be triangular. The fuzzy rule base had therefore 625 rules. The second structure was a two layer fuzzy control system. In the two fuzzy knowledge base case the first knowledge base has the two inputs, x_1 and x_2 to produce as output an offset angle ω. The offset angle was then added to x_3. This updated value of x_3 and x_4 were then used as input in the second knowledge base to produce the control output u. The structure was based on an actual fuzzy control structure used in the annual robot competition held in Singapore.

The output from the first layer was identified with physical meaning in this research. The same problem was considered in Magdalena (1998), but this time no physical was given to the output of the first layer. The latter method of using the output of one layer to act as an input modifier to the next layer (Madgalena, 1998) was seen also to be very effective in reducing the number of rules in the rule base without degrading performance. As pointed out by the author of Hierarchical Fuzzy Control of a Complex System using Metaknowledge, more theoretical study of the testing of the method with more complex systems other than the control of the inverted pendulum is required. The simple inverted pendulum has only four input variables and is a much smaller problem than the prediction of the Australian interest rate. Further research on this problem examining different two layered hierarchical control structures and the learning of the fuzzy rules in each layer can be found in a recent paper (Stonier & Zajaczkowski, 2003).

2.3 Collision-Avoidance in a Robot System

Multi-layered structures can also be used without prioritisation. As illustration, consider the following collision-avoidance problem in a simulated, point mass, two robot system. A three-level hierarchical, fuzzy logic system was proposed to solve the problem, full details can be found in Mohammadian and Stonier (1998), see also Mohammadian and Stonier (1995). In the first layer, two knowledge bases, one for each robot, are developed to find the steering angle to control each robot to its target. In the second layer two new knowledge bases are developed using the knowledge in the first layer to control the speed of each robot so that each robot approaches its target with near zero speed. Finally in the third layer, a

Figure 2. Inverted pendulum

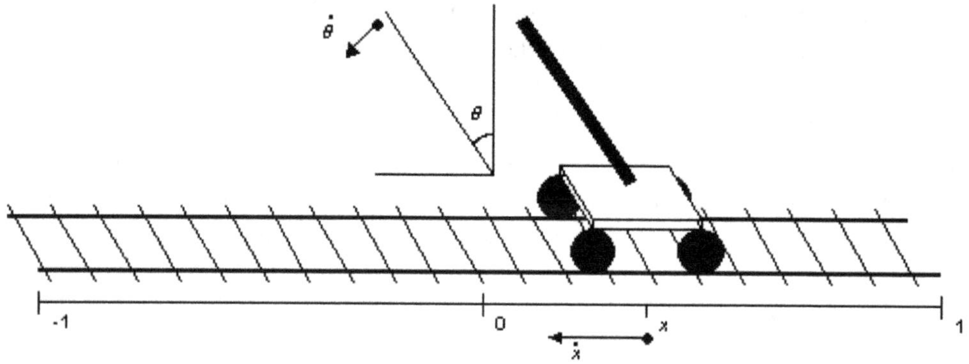

single knowledge base is developed to modify the controls of each robot to avoid collision in a restricted common workspace, see Figure 3.

In Figure 3, x, y gives the physical position of the robot on the plane, ϕ is the directional heading of the robot, θ is the steering angle, D_1 and D_2 are the distances of the two robots from their respective targets, S_1 and S_2 are the speeds of the two robots, $\hat{D}$ is the distance between the two robots and $\theta_1^{'}$, $\theta_2^{'}$, $\theta_1^{''}$, $\theta_2^{''}$, $S_1^{'}$ and $S_2^{'}$ are the updates of variable outputs for the last two layers. An important issue in this example is that of learning knowledge in a given layer sufficient for use in higher layers. Consider the knowledge base of a single robot in layer one. It is not sufficient to learn a fuzzy knowledge base from an initial configuration and use this knowledge base for information on the steering angle of the robot to learn fuzzy controllers in the second layer. Quite clearly this knowledge base is only guaranteed to be effective from this initial configuration as not all the fuzzy rules will have fired in taking the robot to its target. We have to find a knowledge base that is effective to some acceptable measure, in controlling the

Figure 3. Hierarchical structure for collision-avoidance

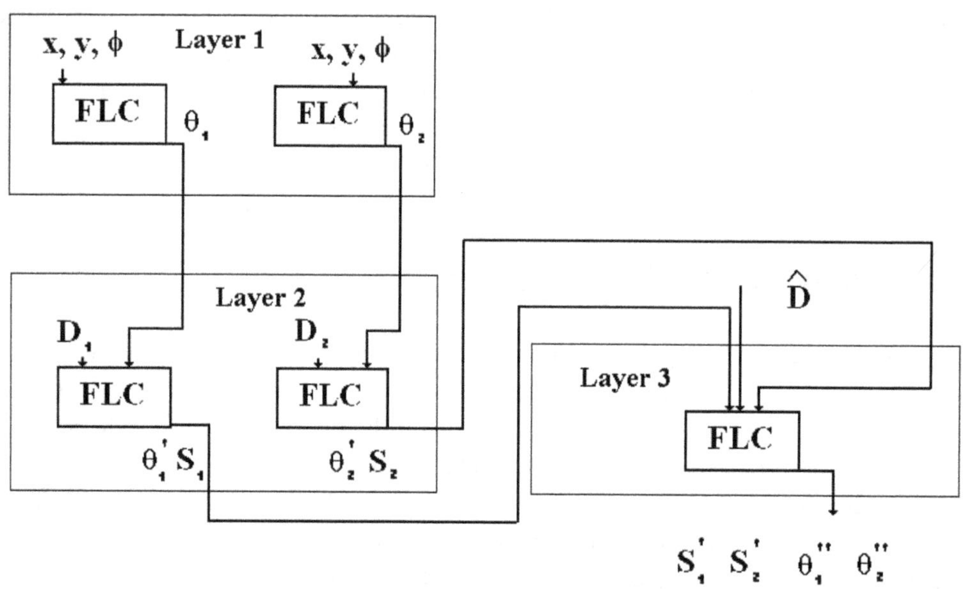

Figure 4. Relative co-ordinate parameters

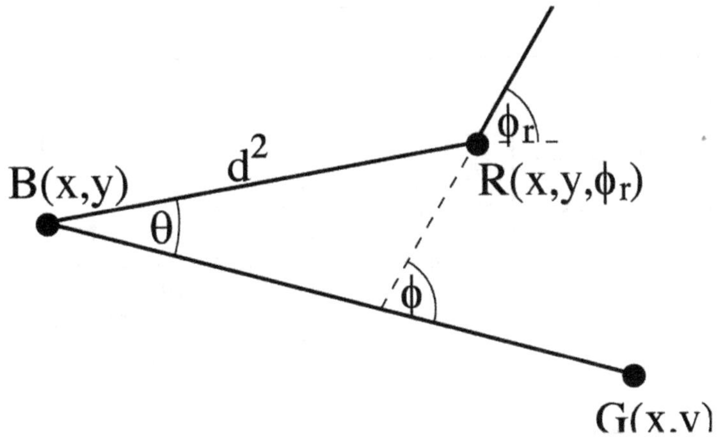

robot to its target from 'any' initial configuration. One way is to first learn a set of local fuzzy controllers, each knowledge base learnt by an evolutionary algorithm from a given initial configuration within a set of initial configurations spread uniformly over the configuration space. These knowledge bases can then be *fused* through a *fuzzy amalgamation* process (Stonier & Mohammadian, 1995) into the global (final), fuzzy control knowledge base. An alternative approach (Mohammadian & Stonier, 1996, 1998), is to develop an evolutionary algorithm to learn directly the 'final' knowledge base by itself over the region of initial configurations. For more details, we refer the reader to the cited papers.

2.4 Micro-Robot Control

In Thomas and Stonier (2003), the authors developed a three layer hierarchical fuzzy knowledge base for a fuzzy controller to control a micro-robot from any configuration belonging to a grid of initial configurations to hit the ball along the ball to goal line of sight. The inputs to the fuzzy controller are variables from a relative coordinate system and output change in two wheel velocities. Forward and reverse facing impact with the ball and physical dimensions were included.

A single layer fuzzy control system using five input variables $x_1=\phi$, $x_2=d^2$, $x_3=\theta$, $x_4=v_L$ and $x_5=v_R$ using seven membership sets per input results in $7^5=16807$ rules. Here v_L is the instantaneous speed at the left wheel of the robot, v_R is the instantaneous speed at the right wheel of the robot, the angle ϕ describes the orientation of the robot relative to the line from the ball to the goal, d is the distance from the centre of the robot to the centre of the ball, and θ is the angle between robot to ball line and the ball to goal line (see Figure 4).

By using a three layer hierarchical fuzzy system the number of rules is reduced to $3\times7^3=1029$. The importance of the input variables was unknown and an arbitrary selection of importance chosen as: ϕ, v_L and v_R for the first fuzzy layer, d; Δv_L^1 and Δv_R^1 for the second and θ; Δv_L^2 and Δv_R^2 for the third. Where the change in wheel speeds Δv_L^k and Δv_R^k, for k=1,2, are input from the output of the previous fuzzy layer. Figure 5 shows the hierarchical configuration used. For further discussion on this research and the development of a 5 input 2 output controller in robot soccer, see Thomas and Stonier (2003).

Figure 5. Hierarchical controller for micro-robot control

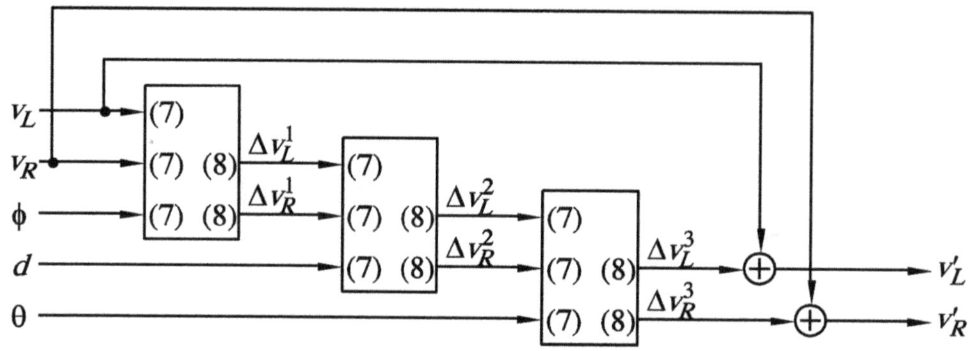

2.5 Multi Traffic Light Control

In Nainar, Mohammadian, Stonier and Millar (1996) and Nainar, Mohammadian and Millar (1996) the supervised and unsupervised fuzzy concept learning using evolutionary Algorithms is considered. The paper explores the design and development of supervisory hierarchical fuzzy logic systems using an evolutionary algorithm. The development of hierarchical fuzzy logic systems is considered by a new method which allows communication among fuzzy logic controller using a supervisory fuzzy logic controller. The advantages and disadvantages of using hierarchical fuzzy logic systems for modelling is also considered. Finally evolutionary algorithm is then used to design a fuzzy logic system from a set of data in an unsupervised learning manner. Specifically it's application to urban traffic control is considered.

3. RULE BASE IDENTIFICATION

Added to this problem of non-unique decomposition, there is the problem of identifying the rules in the knowledge base. For a system with a small number of rules we can use expert knowledge of the system and intuition to possibly define a complete set of rules but how do we find the complete set 5150 rules for the interest rate prediction problem, or a smaller rule set which yields an 'effective' predictor of interest rate? Further, how would do we give understanding for a rule which has an output which has no physical significance in the system?

One way to overcome this problem is to develop an intelligent learning algorithm to learn the fuzzy rules of the knowledge base using just input-output data. Research into this area has been described as genetic fuzzy systems using the classical genetic algorithm has been surveyed Cordon (2002), see also Cord'on (2001). We shall refer to the research area as evolutionary fuzzy systems(EFS) for it is more common to use evolutionary algorithms (EAs) rather than genetic algorithms to learn in the fuzzy rule bases in our research. The most used EFS involves an evolutionary algorithm(EA) being employed to learn or tune different components of a fuzzy logic system such as the fuzzy knowledge base and the membership functions in the inference process. This is usually accomplished by evolving an evolutionary algorithm to produce the "best" fuzzy rules and membership functions/parameters with respect to an optimisation criterion. There are three main approaches in the literature for learning the rules in a fuzzy knowledge base. They are, the Pittsburgh approach, the Michigan approach and the iterative rule-learning approach, cf (Cord'on, 2001). The Pittsburgh and Michigan approaches are the most commonly used methods in the area of evolutionary fuzzy systems.

Michigan Approach: In the Michigan approach individuals of the population are encoded as separate rules and the knowledge base is represented by the entire population. Crossover serves to provide a new combination of rules and mutation yield new rules. In some cases variable length rule bases are used employing modified genetic operators for dealing with variable length individuals.

Pittsburgh Approach: In this approach the entire fuzzy rule set is encoded as an individual in the population that is then evolved maintaining a population of candidate rule sets and using genetic operators such as selection, crossover and mutation to produce new rule sets.

Iterative Approach: In the iterative approach, individuals code separate rules, and a new rule is adapted and added to the rule set, in an iterative fashion, at every generation of the evolutionary algorithm. The difference between the Michigan approach and this approach is that, only the best individual is considered to form part of the solution discarding the rest of the individuals in the population.

Research by the authors, colleagues and postgraduate students has predominately used the Pittsburgh approach with success in learning the fuzzy rules in complex systems, across hierarchical and multi-layered structures in problems, such as the classical cart-balancing problem (Stonier, Stavey & Messom, 1998; Stonier & Zajaczkowski, 2003; Young & Stonier, 2003), interest rate prediction (Kingham, Mohammadian & Stonier, 1998; Mohammadian & Kingham, 2004), modelling and prediction (Mohammadian, 2002), supervisory traffic flow (Nainar, Mohammadian, Stonier & Millar, 1996), multi-robot collision avoidance and tracking problems (Mohammadian & Stonier, 1998; Stonier & Mohammadian, 1995,1996), robot soccer (Thomas & Stonier, 2003, 2003, 2003), image enhancement (Anver & Stonier, 2002, 2002, 2003, 2003, 2003) and closed loop controllers in boundary optimal control (Stonier, Stacey, Mohammadian & Smith, 1999)(a five layer, hierarchical fuzzy controller).

A discussion on aspects of evolutionary learning in fuzzy logic including encoding the representation of the knowledge base as a string in an evolutionary algorithm and the development of various evolutionary operators has been given for the collision-avoidance problem in Stonier, Stacey, Mohammdian and Smith (1999), and for the evolutionary learning of a fuzzy logic controller over a region of initial states in Stonier (1999). The reader is referred to these papers for further details. Some recent results in evolutionary learning concerning the decomposition a fuzzy logic system and the learning of rules in layers is presented here. We begin with the control of the inverted pendulum in some detail to show the process of encoding of the fuzzy knowledge bases and the details of the evolutionary learning.

3.1 Control of the Inverted Pendulum

The control of the inverted pendulum or simulated pole-cart system has been undertaken using linear and nonlinear dynamics and includes both classical and fuzzy logic control techniques. It has already been demonstrated that this system can be controlled by a fuzzy logic controller, the rules for which can be learnt by evolutionary algorithms. The objective in Stonier and Zajczkowski (2003) was to examine the different topologies of a hierarchical fuzzy control structure with single layer and two layers, with different variable input in the two layered structure. The results obtained were used to examine the efficiencies in learning by the evolutionary algorithms and the control of the pendulum.

3.1.1 Inverted Pendulum System

The nonlinear system to be controlled consists of the cart and a rigid pole hinged to the top of the cart. The cart is free to move left or right on a straight bounded track and the pole can swing in the vertical plane determined by the track. It is modelled by the formula:

$$\begin{aligned}
\dot{x}_1 &= x_2 \\
\dot{x}_2 &= u + m\ell(\sin(x_3)x_4^2 - \dot{x}_4\cos(x_3))/(M+m) \\
\dot{x}_3 &= x_4 \\
\dot{x}_4 &= \frac{g\sin(x_3) + \cos(x_3)(u - m\ell x_4^2\sin(x_3))/(M+m)}{\ell(4/3 - m\cos^2(x_3)/(M+m))}
\end{aligned}$$

where x_1 is the position of the cart, x_2 is the velocity of the cart, x_3 is the angle of the pole, x_4 is the angular velocity of the pole, u is the control force on the cart, m is the mass of the pole, M is the mass of the cart, ℓ is the length of the pole, and g is gravitational acceleration. The control force is applied to the cart to prevent the pole from falling while keeping the cart within the specified bounds on the track. We take m=0.1kg, M=1kg, ℓ=0.5m, g=9.81ms^{-2}, with state limits: $-1.0 \le x_1 \le 1.0$, & $^{-\pi}/_6 \le x_3 \le {}^{\pi}/_6$. The goal is to determine fuzzy controllers necessary to stabilise the system about the unstable reference position $\underset{\sim}{x} = \underset{\sim}{0}$ as quickly as possible, whilst maintaining the system within the specified bounds given above.

3.1.2 Single Layer Fuzzy System

An initial analysis of the learning of fuzzy rules in a single layered system was given in Stonier, Stacey and Messom (1998). In that paper a min-max inference engine was used and all variables were normalised to lie in [–1,1]. This was changed to the Mamdani product inference engine with no normalisation in order to compare with the two layered hierarchical fuzzy system discussed next. Each domain region for x_i was divided into 5 overlapping intervals and assigned membership sets A_i^k , k=1,…,5, which we encode numerically as integers from 1 to 5 respectively. Membership sets for x_1 and x_2 are assumed the same. The output variable u is divided into 7 overlapping regions with 7 membership sets B^k, k=1,…,7, with integer encoding 1 to 7. All fuzzy membership functions are assumed to be triangular, see Figures 6, 7, and 8. There are 625 rules in the rule base. The output variable u was calculated using the Mamdani product inference engine:

$$u = \frac{\sum_{\ell=1}^{M} \bar{u}^{\ell}(\prod_{i=1}^{n}\mu_{A_i^{\ell}}(x_i))}{\sum_{\ell=1}^{M}(\prod_{i=1}^{n}\mu_{A_i^{\ell}}(x_i))} \qquad (2)$$

where $\bar{u}^{\ell}$ are centres of the output sets B^{ℓ}.

Figure 6. Membership Functions x_1 and x_2

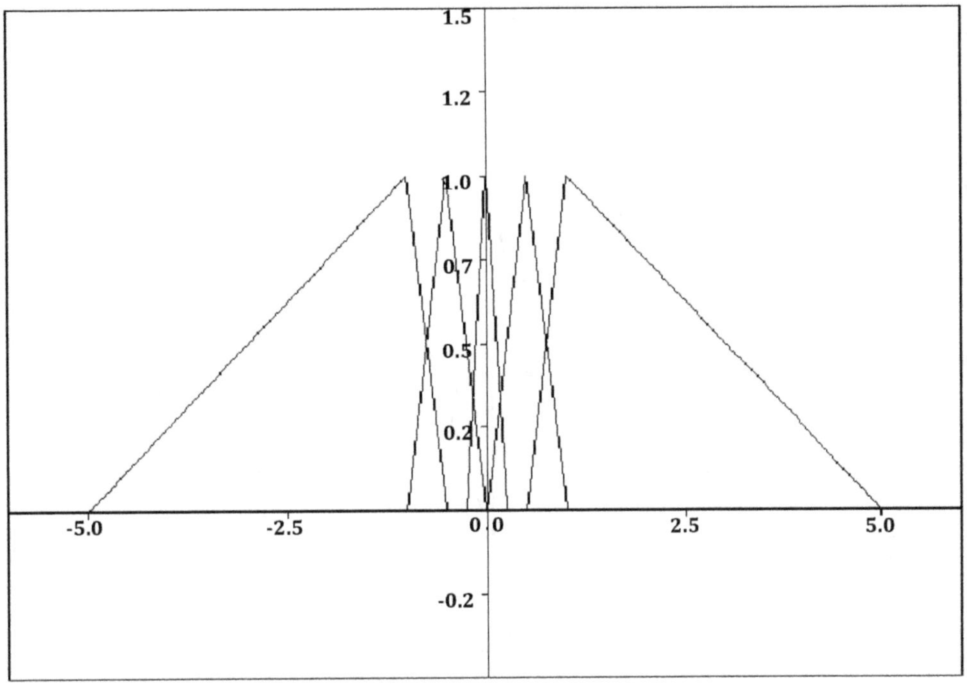

Figure 7. Membership functions x_3

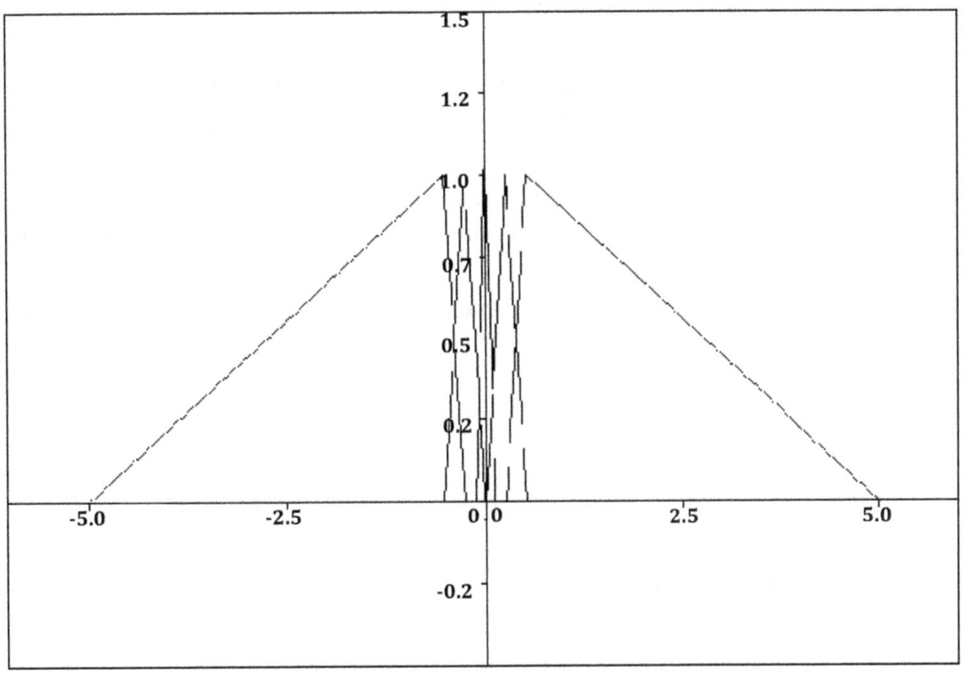

Figure 8. Membership Functions x_4

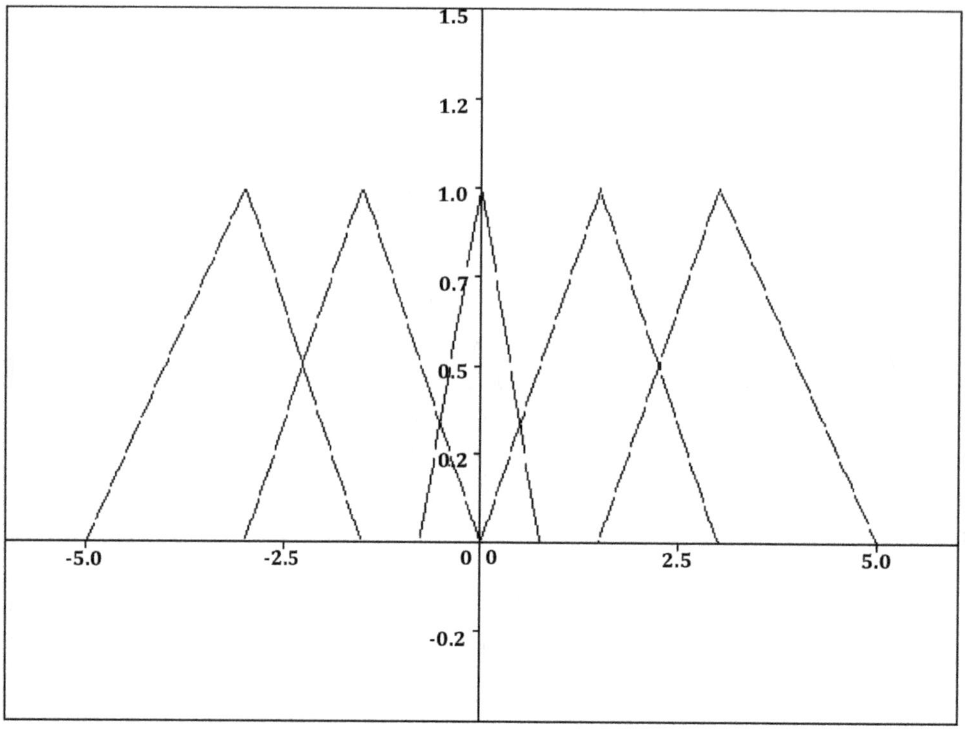

3.1.3 Two Layered Hierarchical System

There are a number of different architectures or topologies of the two layered hierarchical fuzzy system. The first architecture we consider is shown in Figure 9. For this system the first knowledge base KB_1 has the two inputs, x_1 (position of the cart) and x_2 (velocity of the cart) to produce as output a first approximation to the control u'. This u' together with x_3 (the current angle of the pole) and x_4 (the angular velocity of the pole) are used as input in the second knowledge base KB_2 to produce the final control output u.

In the first layer there are $25=(5)^2$ rules in the knowledge base. In general, we may write the ℓth fuzzy rule has the form:

If (x_1 is A_1^ℓ) and (x_2 is A_2^ℓ) Then (u' is B^ℓ)

where A_k^ℓ, k=1,2 are normalised fuzzy sets for input variables x_k, k=1,2, respectively, and where B^ℓ are normalised fuzzy sets for output variable u'. For the second layer there are $175=7(5)^2$ rules in the knowledge base and we may write the ℓth fuzzy rule has the form:

If (u' is C^ℓ) and (x_3 is A_3^ℓ) and (x_4 is A_4^ℓ) Then (u is B^ℓ).

where C^ℓ are normalised fuzzy sets for the input control variable u. There are a total of 200 fuzzy rules in this hierarchical structure with the output variable of each layer calculated using the appropriate

Figure 9. Hierarchical topology 1

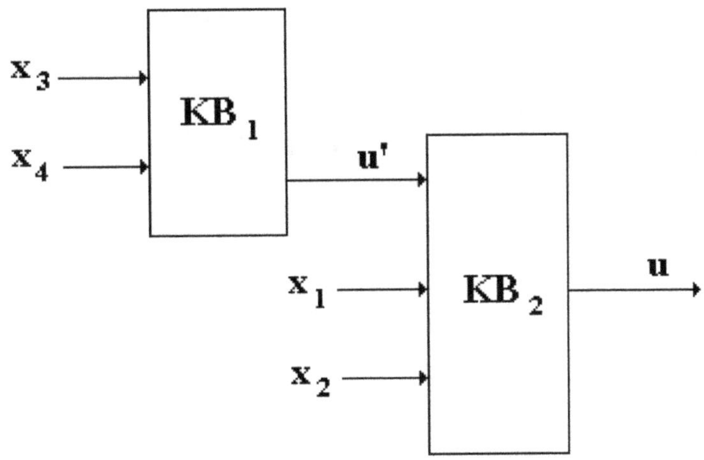

Equation 2. The seven (7) centres associated with the output sets B^{ℓ} are -10.0, -4.5, -2.5, 0.0, 2.5, 4.5, 10.0. These values were obtained by examining the values of u' and u obtained as output from the integration of the state equations. This hierarchical system is different to that described in Stonier, Stacey, and Messon (1998) where the output variable was an offset angle added to x_3 before input into the second layer. Furthermore this new fuzzy system uses a product inference engine rather than a min-max inference engine and results in a different range of values for u' and control u, requiring adjustments to membership functions and centres. The output from the first layer is identified as u', in reality this variable may have not have any physical representation. It is an artificial variable connecting the two layers, see Magdalena (1998). What is important is the final value of control from layer two. There are other possible topologies for the two layered hierarchical structure. The three possibilities are shown in Figures 10, 11 and 12. In the first the variable input is reversed to that in the initial configuration of Figure 10. In Figure 11 the state variables are input into the first layer, and in Figure 12, the speed variables are input into the first layer, reversing the order in Figure 9.

Calculation of the output for each system is obtained using the Mamdani inference engine as described for the first topology, with the appropriate change of variable and associated membership functions for that variable.

3.1.4 Evolutionary Learning

In this section we apply an evolutionary algorithm (Michalewicz, 1994), to learn the fuzzy rules in the two knowledge bases. As illustration, the analysis will refer to the initial topology shown in Figure 9, but is clearly general for all four structures. Each individual string in the evolutionary population is to uniquely represent the hierarchical structure and this is achieved as follows. In the knowledge base of any layer each fuzzy rule is uniquely defined by the consequent part, it being represented by an integer in the range [1,7] defining the output linguistic set. In a similar manner each fuzzy rule in the second knowledge base is defined by its consequent. The two fuzzy rule bases can therefore be represented as a linear individual string of 25+175=200 consequents, $\underset{\sim}{p}_k = [a_1, \cdots, a_{200}]$ where a_j is an integer in [1,7] for j=1,…,200. An initial population $P(0)=\{P_k: k=1,\ldots,M\}$, where M is the number of strings (size

Figure 10. Hierarchical topology 2

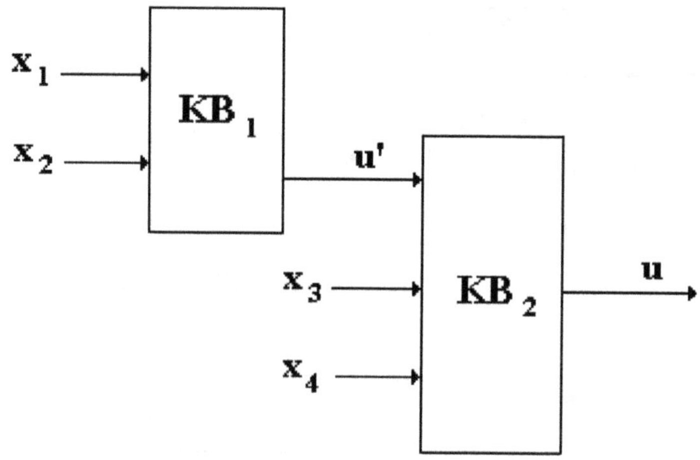

Figure 11. Hierarchical topology 3

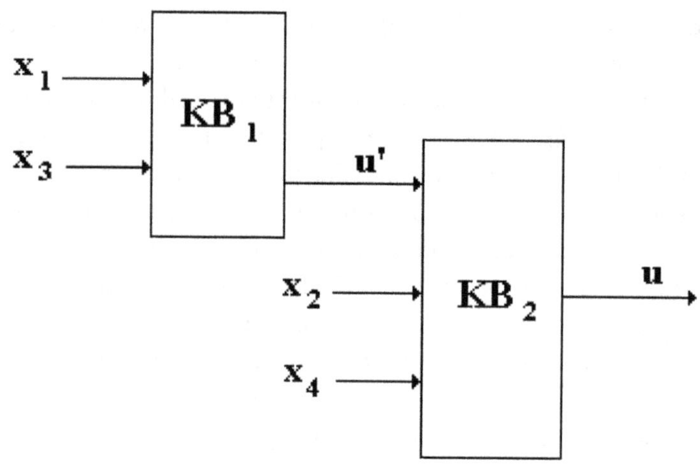

Figure 12. Hierarchical topology 4

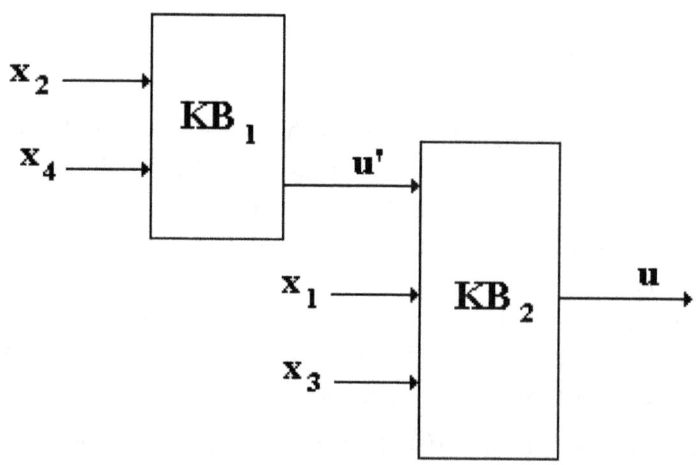

of the EA), was determined by choosing the a_j as a random integer in {1,...,7}. Given an initial state of the system we decode each string $\underset{\sim}{p}_k$ into the two components defining the fuzzy knowledge base for each layer, then use the Mamdani formula to evaluate u' and u to find the final control to be applied at each value of the state $\underset{\sim}{x}$. The system state equations are integrated by the Runge-Kutta algorithm (RK4) with step size 0.02 over a sufficiently long time interval [0,*T*]. A string's fitness f_i, to be minimised, is calculated based on some measures of the behaviour of the system over the time interval. Such measures might include, the accumulated sum of normalised absolute deviations of x_1 and x_3 from zero, the average deviation from vertical, the average deviation from the origin or $T–T_S$ where T_S, the survival time, is taken to mean the total time before the pole and cart break some bounds. The current objective function has the form:

$$f_i = w_1 f_1 + w_2 f_2 + w_3 f_3 + w_4 f_4 + w_5 f_5,$$

with

$$f_1 = \frac{1}{N}\sum_1^N \frac{|x_1|}{x_{\max}},\quad f_2 = \frac{1}{N}\sum_1^N \frac{|x_2|}{\dot{x}_{\max}},\quad f_3 = \frac{1}{N}\sum_1^N \frac{|x_3|}{\theta_{\max}},\quad f_4 = \frac{1}{N}\sum_1^N \frac{|x_4|}{\dot{\theta}_{\max}},\quad f_5 = \frac{1}{T}(T - T_S)$$

where x_{max}=1.0, θ_{max}=π/6, $\dot{x}_{\max} = 1.0$, $\dot{\theta}_{\max} = 3.0$, *N* is the number of iteration steps, survival time T_S=0.02**N*, *T*=0.002*N_{max} with the maximum number of iterations N_{max}=1200, and ω_k are selected positive weights. A penalty of 1000 was added to the objective if the final state breaks the following bounds $|x_1| \le 0.1$, $|x_2| \le 0.1$, $|x_3| \le \pi/24$. $|x_4| \le 3.0$.

In determining successive populations full replacement policy was used and tournament selection with size n_T=4 and a modified mutation operator. An elitism policy was also used with copies of the best string from a given generation passed to the next generation. The number was dependent upon the size of the population. Typically, for a population of 100, two or four copies of the best individual were passed to the next generation. For a population of 500, four copies of the top five individuals were passed to the next generation. Mutation was undertaken with probability $p_{m,}$ whose value was determined by a mutation schedule that decreases from 0.7 to 0.001.

```
if (mutate){
if (ak = 7) ak = ak - rnd(1,3);
  else if (ak = 1) ak = ak + rnd(1,3);
  else if (flip(0.5)) ak= ak + rnd(1,3);
           else ak = ak - rnd(1,3);
     if (ak > 7) ak = 7;
     if (ak < 1) ak = 1;}
```

3.1.5 Simulation Results

The initial state chosen for illustration was (0.5, 0.0, 0.1, 0.0). Parameters ω_1=3000.0, ω_2=2000.0, ω_3=0.0, ω_4=0.0, ω_5=5000.0, were chosen for the first three simulations and ω_1 increased to 5000 for the last simulation. Population size was usually set at 500. The algorithm was terminated usually at 700 generations. Typical results of evolutionary learning of the fuzzy controller, state and control graphs, for

Figure 13. Topology 1 State variables x_k

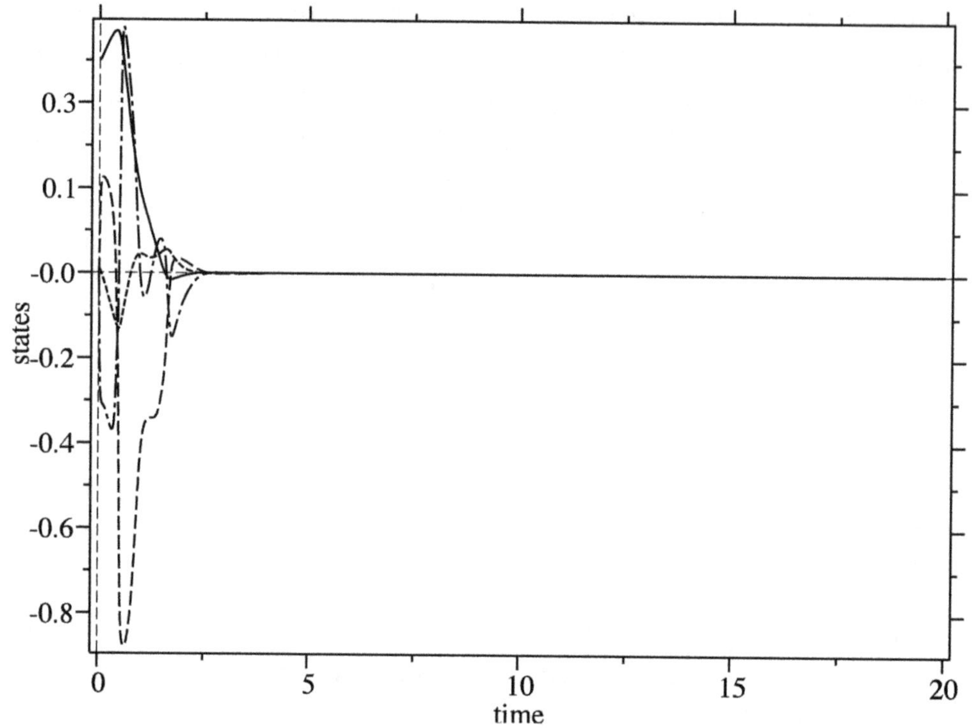

each of the four different topologies are shown in Figures 13, 14, 15, and 16. The evolutionary algorithm was run with many different initial random populations and convergence of the state variables to within the specific bounds was usually found to occur within 300-500 generations and typically within the 4-6 time unit range. Differing parameters ω_k, yielded substantially different convergence of the algorithm and convergence time for the state variables to lie within the specified bounds.

The evolutionary algorithm can be run with a much lower population size, for example 100, quite successfully and this clearly decreases computation time substantially.

Of the four hierarchical structures the easiest to learn using the evolutionary algorithm was topology 2. With state variables x_3 and x_4 being the key variables in the first layer, convergence was usually very quick for a wide range of parameters ω_k, and it was typically to 5 decimal place accuracy on all variables. Furthermore a smaller control effort was required with control values in [−1.5,1]. The fuzzy control for topology 3 also performed well in terms of accuracy but required a higher magnitude of control. The fuzzy controller for topology 4 was the most difficult to learn, not necessarily in finding an appropriate set of parameters, but finding a control which would ensure strong convergence to the final desired state. This can be seen by examining the graphs in Figure 17 with Figures 18, 19, and 20.

3.1.6 Comments

It appears from the results obtained that it is important to control the inverted pendulum, by examining first its angular position and angular speed, then the position displacement and speed. There are of

Figure 14. Topology 1 State control u

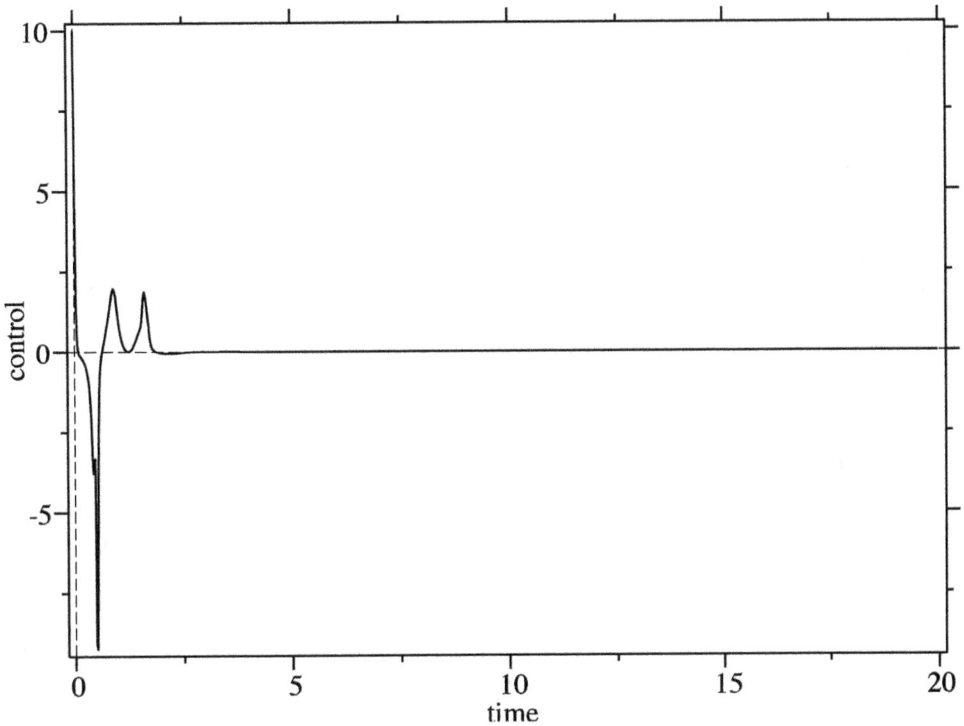

Figure 15. Topology 2 State variables x_k

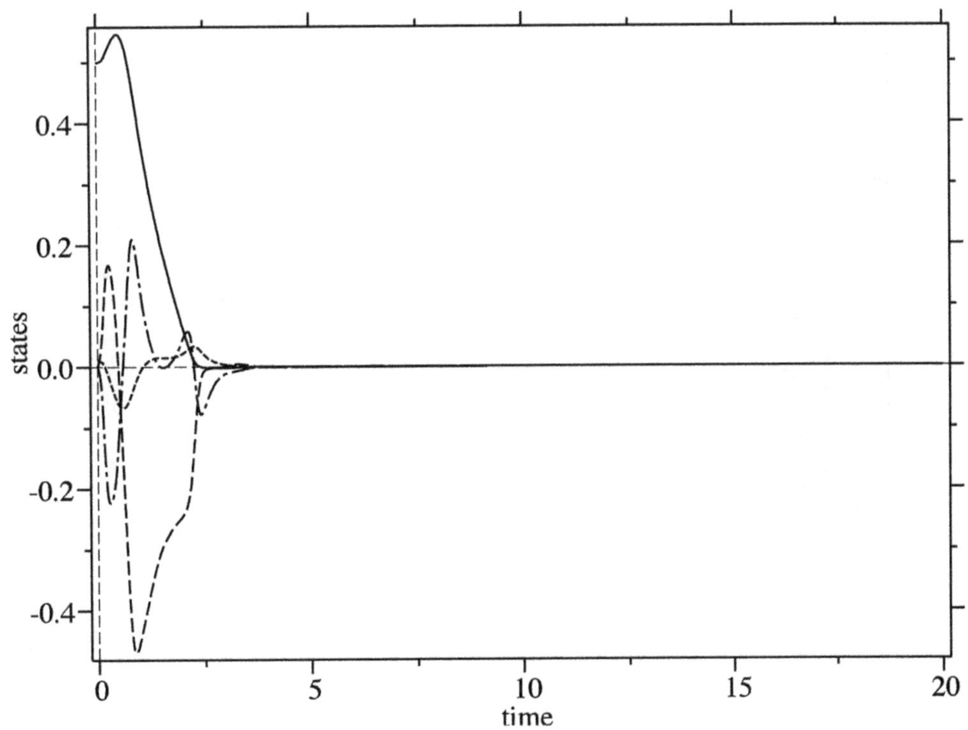

Figure 16. Topology 2 Control u

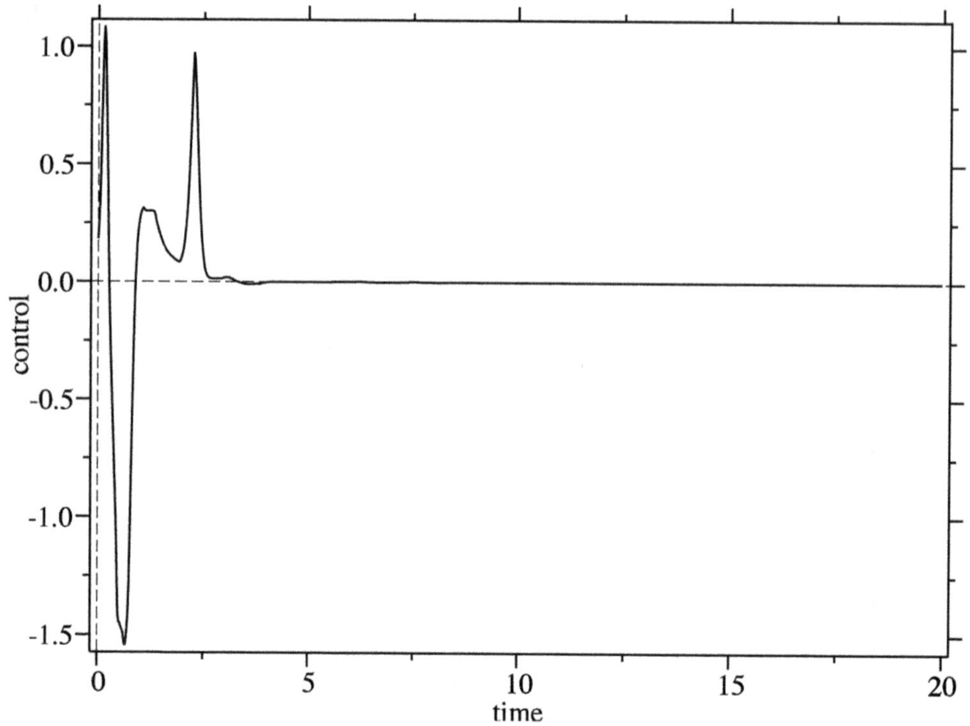

Figure 17. Topology 3 State variables x_k

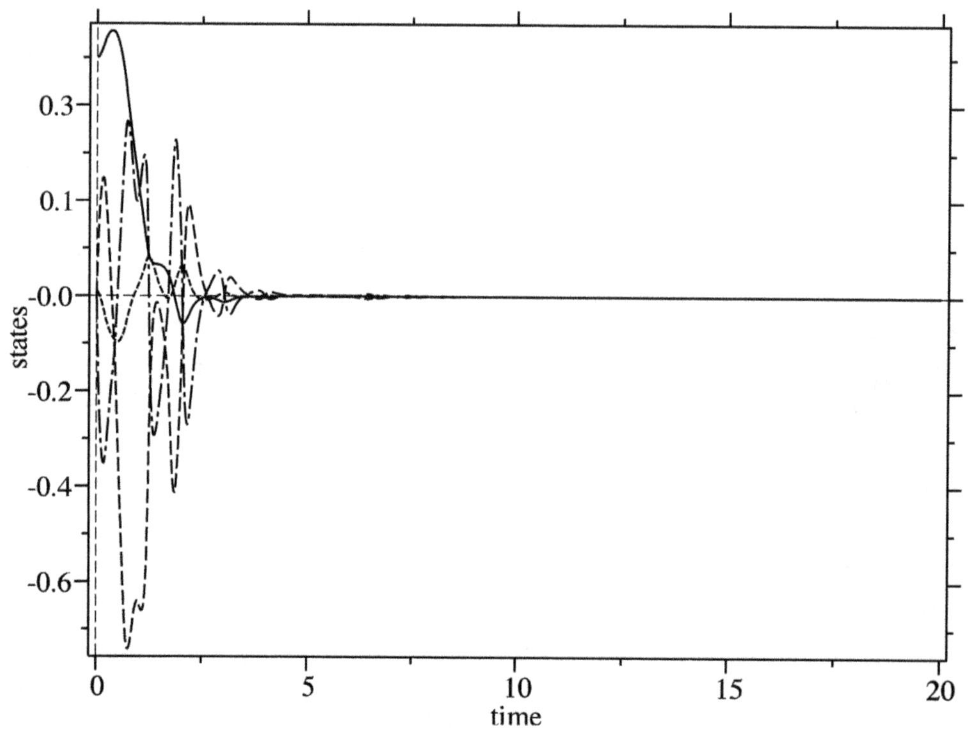

Figure 18. Topology 3 Control u

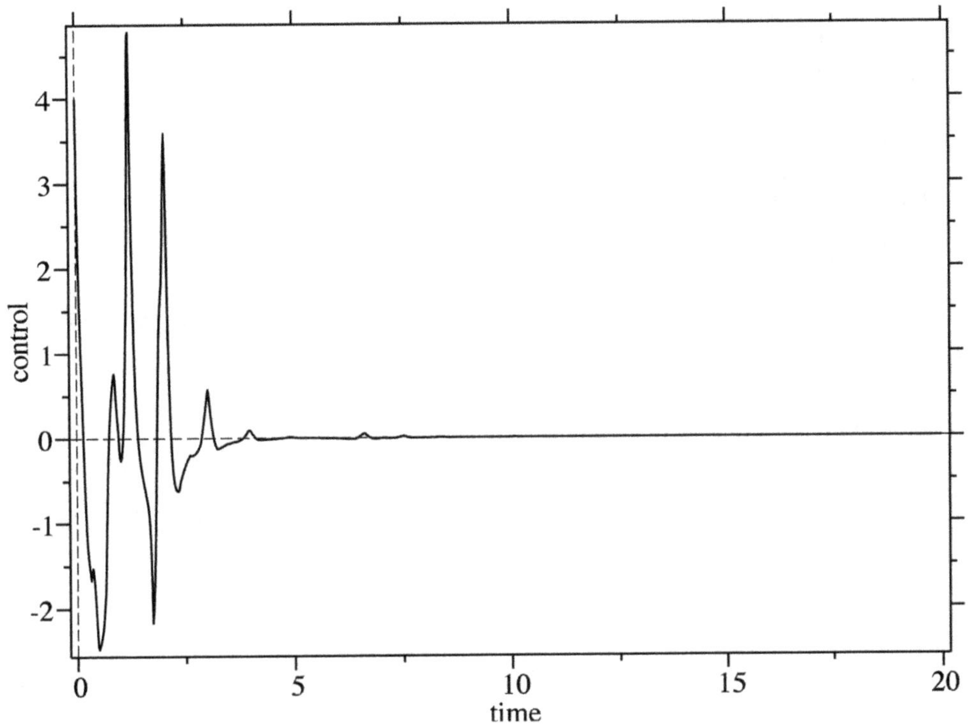

Figure 19. Topology 4 State variables x_k

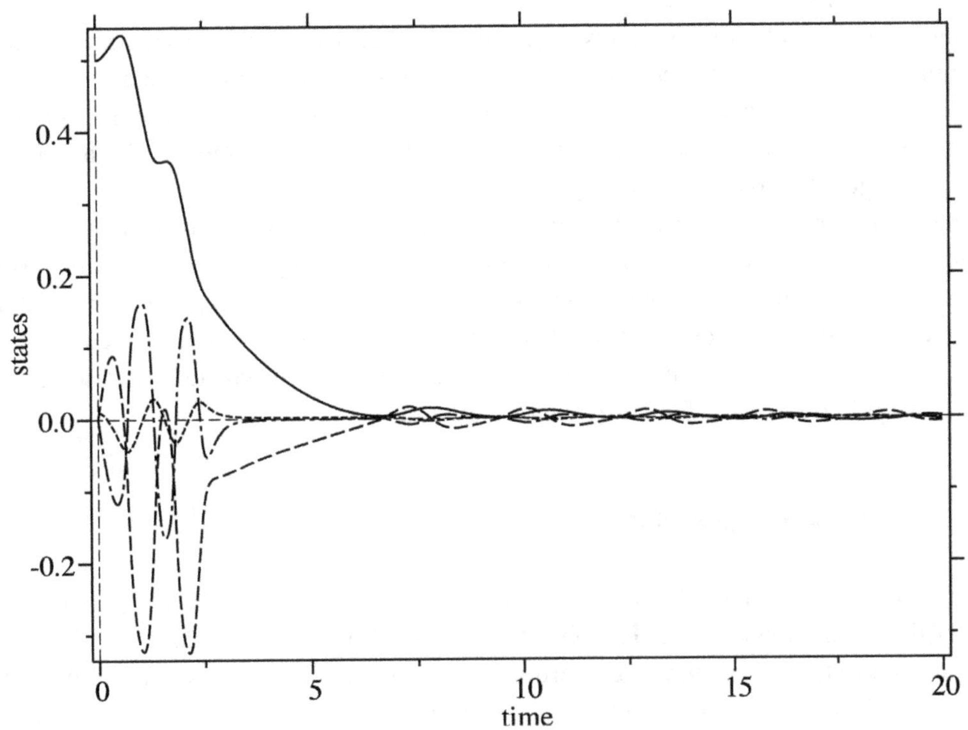

Figure 20. Topology 4 Control u

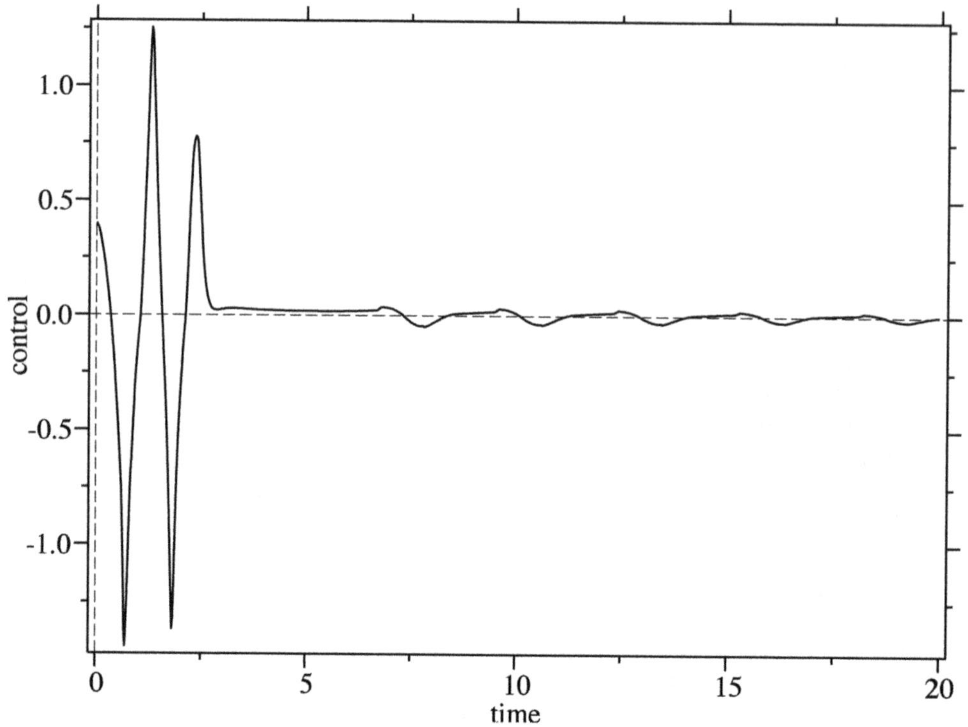

course other topologies or structures for this control problem with a three layered hierarchical structure and these are currently being researched.

Increasing the number of output sets, say from 7 to 31, increases the granularity of the output control u' and u. Typically the associated centres of these sets are set uniformly in the range [–10,10]. It comes with a computational cost and a need to carefully select the amount of mutation in the mutation operator. For problems such as this, it is seen necessary to include both a large and small random mutation in the evolutionary algorithm to ensure convergence.

Increasing the number of input membership functions for each variable is necessary if we wish to consider control from a range of initial states. In this case the string length is increased and it is necessary to consider what affects this has on mutation and crossover. It was found that a variation of arithmetic crossover which incorporates knowledge of when a rule fires is necessary to obtain convergence of the evolutionary algorithm over a region of initial states. This region of initial states is also constrained by the maximum magnitude allowed for the evaluation of control.

4 CO-EVOLUTIONARY LEARNING

In [19] the authors also examined the problem of learning a two layer hierarchical fuzzy controller for the control of the inverted pendulum. The objective was to show the fuzzy rules can be learnt by co-evolving two evolutionary algorithms, one defining the first layer and the second defining the second layer. The results of this co-evolutionary learning were compared with the learning of the rules in both layers by a classical evolutionary algorithm. Layered fuzzy logic systems exploit the modularity present in many systems. Specifically, in the context of fuzzy logic knowledge bases, the output influenced by one closely related group of input variables may be largely independent of the values of other variables. Therefore, a layered fuzzy logic system can decompose the rule base along these lines of weak interdependence and still maintain a high level of accuracy. The curse of dimensionality afflicts, among others, fuzzy logic and evolutionary algorithms. Its effects can be mitigated in fuzzy logic by using a layered structure and so too in evolutionary algorithms by using co-evolutionary algorithms Co-evolutionary algorithms (CEAs) were first proposed by Hills in 1990 (Hills, 1994), using predator-prey co-evolution.

These algorithms model competitive evolving of species one against another which can be seen in nature. Instead of using a single population to define the problem domain, two(2) competing populations are used having an inverse fitness interaction. This often results in an arms race in which the complexity (suitability) of both predator and prey increase stepwise. The above mentioned variety of co-evolutionary algorithms are classified as competitive co-evolutionary algorithms.

Cooperative co-evolutionary genetic algorithms were introduced by Potter and De Jong [34] as a way of managing the increasing complexity in evolutionary problems. The search space that an evolutionary algorithm must cover increases exponentially with each additional element in the string representing an individual; the search space can be dramatically reduced by using multiple populations that each encode only part of the complete solution.

Cooperative co-evolution takes a string of many elements that completely specifies a solution and breaks it into several smaller strings that each encode only part of the complete solution. The partial strings are kept in genetically isolated populations; individuals in the same population are strings that encode the same part of the complete solution. Each population is evolved by its own evolutionary algorithm instance. It is impossible to evaluate the fitness of an individual from a single population in isolation, because each individual encodes only part of the complete solution. To form a complete solution a representative is selected from each population (except the population being evaluated), and these representatives are recombined with the individual being evaluated. This complete solution can then be evaluated by the original fitness function. The fitness calculated for this complete solution is assigned to the individual being evaluated. The fitness of a representative from another population is not changed. There are two ways of selecting representatives from the other populations: select the fittest string as the representative, or select a random string as the representative. The former is greedy; the latter is explorative and performs twice as many fitness evaluations than the greedy algorithm.

As long as there is little interdependence, or epistasis, between the partial strings in different populations the narrower evolutionary focus does not adversely affect cooperative co-evolution, and it benefits from a reduced search space. A faster convergence result is expected and was shown by simulated experiments.

For further details the reader is referred to the cited paper. It is typical to also tune or learn many of the internal parameters of the fuzzy logic system, for example, the parameters defining the membership functions on the antecedent side of the knowledge base, see Cord'on, Herrera, Hoffman, and Magdalena (2001).

This can also be achieved using co-evolutionary algorithms, including the type of membership function to select that best suits the problem. In a recently published paper (Anver & Stonier, 2003), the authors present a co-evolutionary approach for optimisation of image enhancement filters.

In this paper it was demonstrated how co-evolutionary algorithms could be employed for optimisation of image enhancement filters. Specifically, an example of an impulse noise filter constructed using the fuzzy paradigm was taken, and it was shown how a co-evolutionary algorithm could be effectively used for its optimisation. From research into fuzzy impulse filters it was seen that the shape and the corresponding parameters of the membership functions used in the fuzzy inference process, play a major role in the quality of the enhanced image, apart from the proper selection of the fuzzy rule base. This is true, both in terms of objective and subjective evaluations of the processed image.

A co-evolutionary algorithm having a blend of cooperativeness and competitiveness, was employed to optimise the rule base and to select the best shape of the membership function used in the fuzzy inference process. Application was made for several real images corrupted by noise from 30% to 60% to show the effectiveness of the proposed approach.

Using the Pittsburgh approach, the fuzzy rule base defining the fuzzy filter was encoded as a string the population $POP_{FuzzyRule}$. Three other populations were formed defining membership functions and their parameters. They were POP_{Gauss}, POP_{Trap} and POP_{Exp} for Gaussian, Trapezoidal, and the Exponential functions whose strings encoded the parameters (λ_1,λ_2), (c_1,c_2) and (β_1,β_2), respectively, see Figures 21, 22, and 23.

Figure 24 shows a schematic representation of the co-evolutionary algorithm which was used that incorporated a blend of cooperativeness and competitiveness. Here, three (3) membership functions (represented by three (3) membership populations) competed with each other to cooperate with the best fuzzy rule base that was evolving in parallel. For specific details of the co-evolutionary algorithm which included greedy and tournament selection combination strategies for the evaluation of fitness, the reader is referred to the cited paper and its contents. The best membership function turned out to be the Gaussian function with parameters, λ_1=0.0040 and λ_2=0.0071. The best membership function was decided on the basis that the best individual of POP_{Guass} had the best (minimum) Mean Square Error (MSE) compared to that of POP_{Trap} and POP_{Exp}. The image that was used for training was the 'Lena' image of size 512 × 512 with 30% random-valued impulse noise corruption.

Figure 21. Gaussians with parameters (λ_1,λ_2)

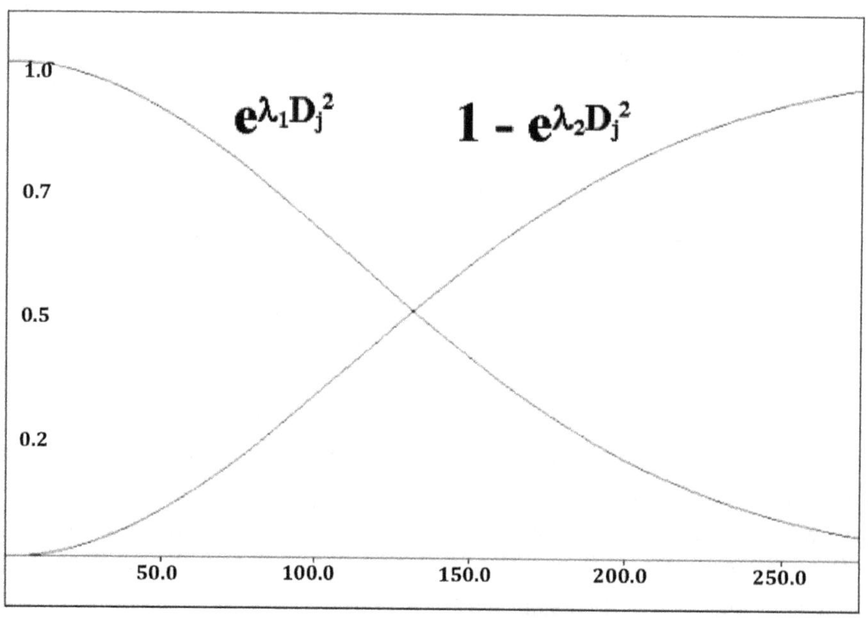

Figure 22. Trapezoidals with parameters (c_1,c_2)

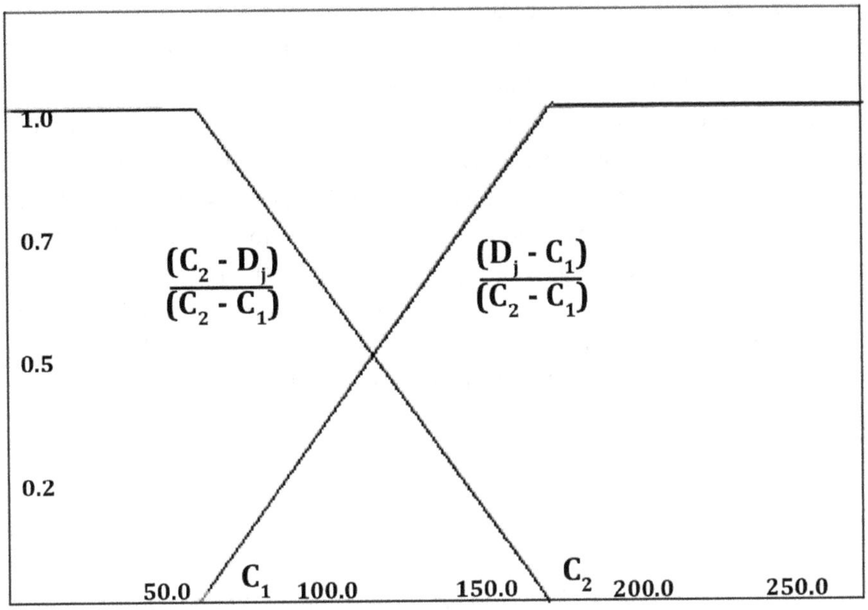

Figure 23. Exponentials with parameters (β_1,β_2)

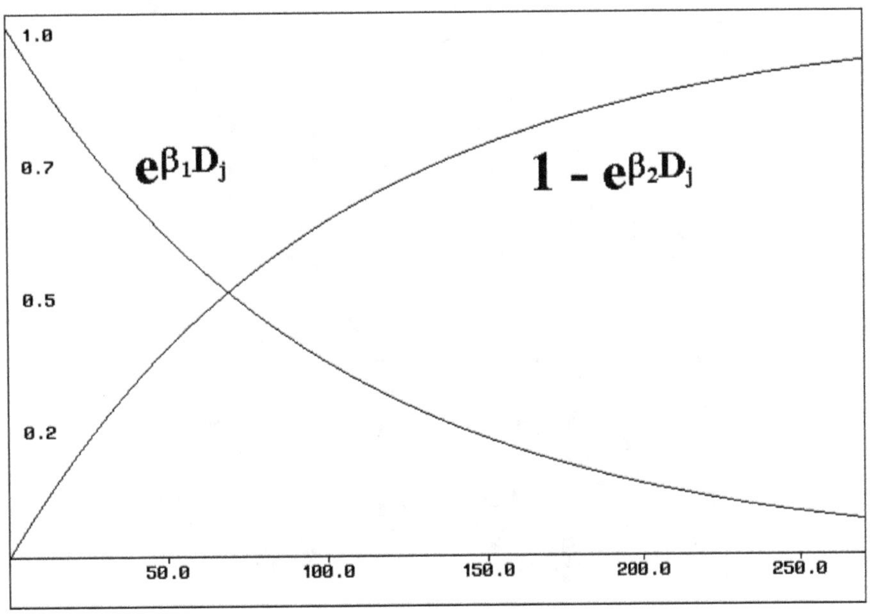

The minimum MSE obtained for the above mentioned 'Lena' image was 22.08 with the Gaussian membership function (λ_1=0.0040, λ_2=0.0071) and the best fuzzy rule base. Trapezoidal (c_1=5.0, c_2=180.0) and Exponential (β_1=1.0, β_2=1.0) membership functions resulted in higher MSE values of 24.67 and 30.13 respectively showing the superiority of the Gaussian membership function in this case. The algorithm was re-run with different random initialising seeds achieving minimum MSE around the same value.

Although this filter was implemented on a 'Lena' test image having 30% corruption, it was found that it performs very well even at noise levels as high as 50%. The results for the 512×512 image are shown in Figure 25 for the 'Lena', 'Pepper' and the 'Hill' image having a corruption level of 30%. The MSE values obtained for corruption levels ranging from 10% to 60% for the 'Lena' image are given in Table 1 along with MSE values obtained using some other existing filters detailed in the literature. It is seen that the proposed system which uses a CEA to determine the fuzzy knowledge base as well as to find the best shape of the membership function (together with its parameters) outperforms many of the existing filters.

Figure 25. (a) Corrupted 'Lena' Image (30%); (b) Restored 'Lena' Image (30%) MSE: 22.08; (c) Corrupted 'Pepper' Image (30%); (d) Restored 'Pepper' image (30%) MSE: 27.69; (e) Corrupted 'Hill' Image (30%); (f) Restored 'Hill' Image (30%) MSE: 52.36

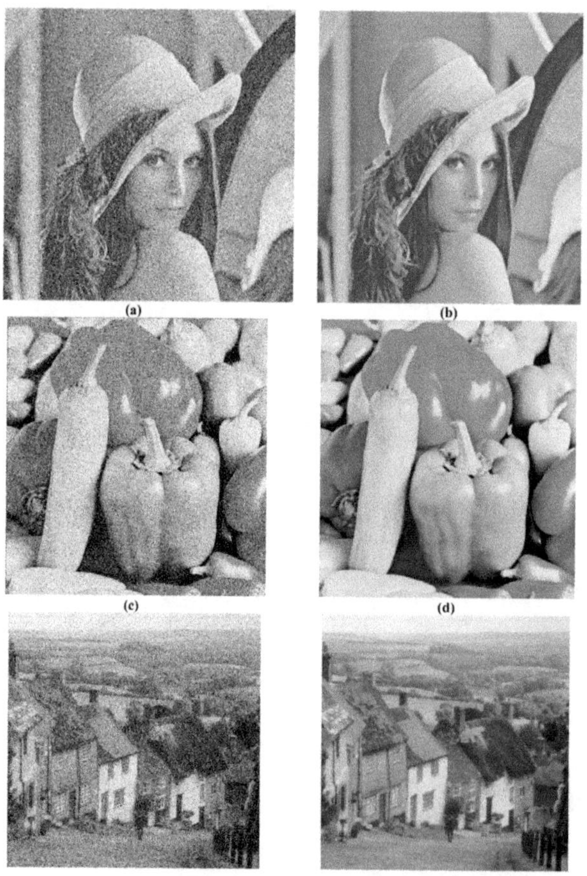

Table 1. Noise % vs MSE for the 512 'Lena' image

Noise %	Median	WFBF [35]	MLFL [25]	Proposed
10%	23.94	15.04	14.33	11.29
20%	30.64	24.49	23.81	16.51
30%	39.87	36.68	36.41	22.08
40%	50.65	54.06	52.11	33.98
50%	67.90	80.25	73.76	56.51
60%	109.57	135.90	194.97	97.95

The approach taken here to determine parameters and membership function selection, as well as determine the underlying set of rules was seen to be fast (results within 100 generations) and effective (as can be seen from the image restoration).

5. CONCLUSION

In this paper we have examined some issues in the construction of a fuzzy logic system to model a complex (nonlinear) system, namely the decomposition into hierarchical/multilayered fuzzy logic sub-systems and the learning of fuzzy rules and internal parameters. Whilst the decomposition into hierarchical/multi-layered fuzzy logic sub-systems reduces greatly the number of fuzzy rules to be defined and to be learnt, other issues arise such as the decomposition is not unique and that it may give rise to variables with no physical significance. This can raise then major difficulties in obtaining a complete class of rules from experts even when the number of variables is small. For a problem with a large number of input variables, for example, the problem of interest rate prediction, the non-uniqueness of the decomposition

Figure 24. Structure of the proposed CEA

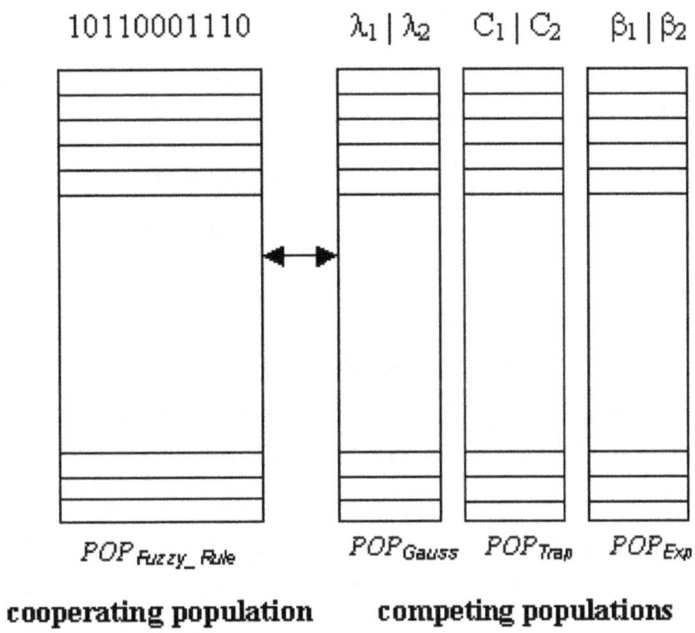

yields numerous different structures to examine in order to find one which in some sense, is the 'best' structure. We have examined the learning of fuzzy rules in such systems using evolutionary algorithms and seen that it is feasible in problems of small dimension to come to grips with some of the resulting problems, selection of operators for evolutionary convergence, determination of parameters in the fuzzy rule base, etc.

In using the Pittsburgh approach the coding of the fuzzy rule base as a linear string in an evolutionary algorithm has its drawbacks other than the string may even be relatively large in length under decomposition into multi-layer and hierarchical structures. One is that this is a specific linear encoding of a nonlinear structure and typical one-point crossover when implemented introduces bias when reversing the coding to obtain the fuzzy logic rule base. Using co-evolutionary algorithms to evolve each component rule base of the structure is one way in keeping the encoding small and therefore effective to evolutionary learning. Currently research is being undertaken to examine a more direct encoding of the fuzzy rule base to enable a variant of multi-dimensional crossover to be effective in the learning of the fuzzy rule base.

Typically, the problems now being examined, are more complex and non-linear and especially large input-output data sets need to be modelled effectively. These problems may have solutions realised using a fuzzy paradigm by finding the best decomposition into a multi-layer or hierarchical fuzzy structure.

ACKNOWLEDGMENT

The authors wish to thank those colleagues and students who have helped in this research and associated publications.

REFERENCES

Cord'on, O., Herrera, F., Hoffmann, F., & Magdalena, L. (2001). Genetic Fuzzy Systems: Evolutionary Tuning and Learning of Fuzzy Knowledge Bases. World Scientific Publishing.

Cordon, O., Herrera, F., & Zwir, I. (2002). Francisco Herrera & Igor Zwir, Linguistic Modeling of Hierarchical Systems of Linguistic Rules. *IEEE Transactions on Fuzzy Systems*, *10*(1), 2–20. doi:10.1109/91.983275

Goldberg, D. (1989). *Genetic Algorithms in Search, Optimisation and Machine Learning*. Addison–Wesley.

Hills, D. W. (1990). Co-evolving parasites improve simulated evolution as an optimization procedure. In Artificial Life II. Addison Wesley.

Kingham, M., Mohammadian, M., & Stonier, R. J. (1998). Prediction of Interest Rate using Neural Networks and Fuzzy Logic. In *Proceedings of ISCA 7th International Conference on Intelligent Systems*.

Magdalena, L. (1998). Hierarchical Fuzzy Control of a Complex System using Metaknowledge. In *Proceedings of the 7th International conference on Information Processing and Management of Uncertainty in Knowledge-based Systems*.

Mao, X., Stonier, R. J., & Thomas, P. (2001). An Effective Fuzzy Image Filter. In *Proceedings of the International Conference on Computational Intelligence for Modelling Control and Automation (CIMCA'01)*.

Michalewicz, Z. (1994). *Genetic Algorithms + Data Structures = Evolution Programs*. Springer Verlag. doi:10.1007/978-3-662-07418-3

Mohammadian, M. (2002). Designing customised hierarchical fuzzy systems for modelling and prediction. In *Proceedings of the International Conference on sinulated Evolution and Learning (SEAL'02)*.

Mohammadian, M., & Kingham, M. (2004, January-March). An adaptive hierarchical fuzzy logic system for modelling of financial systems. *Journal of Intelligent Systems in Accounting, Finance and Management, Wiley Interscience, 12*(1), 61–82. doi:10.1002/isaf.241

Mohammadian, M., & Stonier, R. J. (1998). Hierarchical Fuzzy Control. In *Proceedings of the 7th International conference on Information Processing and Management of Uncertainty in Knowledge-based Systems*.

Mohammadian, M., & Stonier, R. J. (1995). Adaptive Two Layer Fuzzy Logic Control of a Mobile Robot System. In *Proceedings of IEEE Conference on Evolutionary Computing*.

Mohammadian, M., & Stonier, R. J. (1996). Fuzzy Rule Generation by Genetic Learning for Target Tracking. In *Proceedings of the 5th International Intelligent Systems Conference*.

Mohideen Anver, M., & Russel, J. (2002). Stonier, Design of an Hierarchical Fuzzy Filter for Removal of Heavy Impulse Noise from Corrupted Images. In *Proceedings of the Seventh International Conference on Control, Automation, Robotics and Vision (ICARCV2002)*.

Mohideen Anver, M., & Russel, J. (2003). Parameter Optimization and Rule Base Selection for Fuzzy Impulse Filters using Evolutionary Algorithms. In *Proceedings of IEEE Tencon 2003*, Allied Publishers Private Ltd.

Mohideen Anver, M., & Stonier, R. J. (2003). A Coevolutionary Approach for Optimization of Image Enhancement Filters. In *Proceedings of the International Congress on Evolutionary Computation (CEC2003)*.

Nainar, I., & Mohammadian, M. (1996). *Adaptive Control of Traffic Signals using Fuzzy Logic and Genetic Algorithms. Fourth International Conference on Control, Automation, Robotics and Vision*, Westin Stamford, Singapore.

Nainar, I., Mohammadian, M., Stonier, R. J., & Millar, J. (1996). An adaptive fuzzy logic controller for control fo traffic signals. In *Proceedings of the 4th International Conference on Control, Automation, Robotics and Computer Vision (ICARCV'96)*.

Nainar, I., Mohammadian, M., Stonier, R. J., & Millar, J. W. L. (1996). *An adaptive fuzzy logic controller for control of traffic signals. Fourth International Conference on Control, Automation, Robotics and Vision*, Westin Stamford, Singapore.

Peter, J. (2003). Evolutionary Learning of a 5I2O Fuzzy Controller Including Wheel Lift Constraint. In *Proceedings of the 2nd International Conference on Computational Intelligence, Robotics and Autonomous Systems (CIRAS2003)*.

Potter, M. A., & De Jong, K. (1994). *A Cooperative Coevolutionary Approach to Function Optimization, Parallel Problem Solving from Nature - PPSN III*. Berlin: Springer.

Raju, G. V. S., & Zhou, J. (1993). Adaptive Hierarchical Fuzzy Controller. *IEEE Transactions on Systems, Man, and Cybernetics*, *23*(4), 973–980. doi:10.1109/21.247882

Raju, S., Zhou, J., & Kisner, R. A. (1990). Fuzzy logic control for steam generator feedwater control. In *Proceedings of American Control Conference*.

Stonier & Anver. (2003). A new multi-layered fuzzy image filter for removing impulse noise. *Journal of Systemics, Cybernetics and Informatics*, *1*(4).

Stonier, R. J. (1999). Evolutionary learning of fuzzy logic controllers over a region of initial states. In *Proceedings of the 1999 Congress on Evolutionary Computation*. doi:10.1109/CEC.1999.785538

Stonier, R. J., & Anver, M. (2002). *Multi-layered Fuzzy Image Filters for removing Impulse Noise*. ISCA 11th International Conference on Intelligent Systems on Emerging Technologies (ICIS-2002), Boston, MA.

Stonier, R. J., Drumm, M. J., & Bell, J. (2003). Optimal boundary control of a tracking problem for a parabolic distributed system using hierarchical fuzzy control and evolutionary algorithms. In *Proceedings of the 18th International Conference on Computers and their Applications*.

Stonier, R. J., & Mohammadian, M. (1995). Self Learning Hierarchical Fuzzy Logic Controller in Multi-Robot Systems. In *Proceedings of the IEA Conference Control95*.

Stonier, R. J., & Mohammadian, M. (1998). Knowledge Acquisition for Target Capture. In *Proceedings of the International Conference on Evolutionary Computing ICEC'98*.

Stonier, R. J., & Mohammadian, M. (1996). Intelligent Hierarchical Control for Obstacle- Avoidance. In Computational Techniques and Applications: CTAC95. World Scientific.

Stonier, R. J., Stacey, A., Mohammadian, M., & Smith, S. F. (1999). Application of evolutionary learning in fuzzy logic and optimal control. In *Proceedings of the International Conference on Computational Intelligence for Modelling, Control and Automation*.

Stonier, R. J., Stacey, A. J., & Messom, C. (1998). Learning fuzzy controls for the inverted pendulum. In *Proceedings of ISCA 7th International Conference on Intelligent Systems*.

Stonier, R. J., & Zajaczkowski, J. (2003). Hierarchical fuzzy controllers for the inverted pendulum. In *Proceedings of the 2nd International Conference on Computational Intelligence, Robotics and Autonomous Systems (CIRAS 2003)*.

Thomas & Stonier. (2003). Fuzzy control in robot-soccer, evolutionary learning in the first layer of control. *Journal of Systemics, Cybernetics and Informatics*, *1*(1).

Thomas, P. J., & Stonier, R. J. (2003). Hierarchical Fuzzy Control in Robot Soccer using Evolving Algorithms. In *Proceedings of the International Congress on Evolutionary Computation (CEC2003)*. doi:10.1109/CEC.2003.1299393

Wang. (1997). *A Course in Fuzzy Systems and Control*. Prentice-Hall Inc.

Young, N., & Stonier, R. J. (2003). Co-evolutionary learning and hierarchical fuzzy control for the inverted pendulum. In *Proceedings of the International Congress on Evolutionary Computation (CEC2003)*.

Chapter 30
Solving Non-Linear Bi-Level Programming Problem Using Taylor Algorithm

Eghbal Hosseini
Payamenur University of Tehran, Iran

Isa Nakhai Kamalabadi
University of Kurdistan, Iran

Fatemah Daneshfar
University of Kurdistan, Iran

ABSTRACT

In recent years the bi-level programming problem (BLPP) is interested by many researchers and it is known as an appropriate tool to solve the real problems in several areas such as economic, traffic, finance, management and so on. Also it has been proved that the general BLPP is an NP-hard problem. The literature shows a few attempts for using approximate methods. In this chapter we attempt to develop an effective approach based on Taylor theorem to obtain an approximate solution for the non-linear BLPP. In this approach using the Karush-Kuhn–Tucker, the BLPP has been converted to a non-smooth single problem, and then it is smoothed by the Fischer – Burmeister function. Finally the smoothed problem is solved using an approach based on Taylor theorem. The presented approach achieves an efficient and feasible solution in an appropriate time which is evaluated by comparing to references and test problems.

1. INTRODUCTION

The bi-level programming problem (BLPP) is a nested optimization problem, which has two levels in hierarchy. The first level is called leader and the second one is called follower. They have their own objective functions and constraints. The leader actions first, and the follower reacts to the leader decision. The follower should optimize its objective function according to the leader decision and delivered answers of the leader. In fact, the leader inflicts his decision on and obtains reaction of the follower.

DOI: 10.4018/978-1-4666-9644-0.ch030

It has been proved that the BLPP is an NP- Hard problem even to seek for the locally optimal solutions (Bard, 1991; Vicente et.al, 1994). Nonetheless the BLPP is an applicable problem and a practical tool to solve decision making problems. It is used in several areas such as transportation, finance and so on. Therefore finding the optimal solution has a special importance to researchers.

Several algorithms have been presented for solving the BLPP (Lv, et. al., 2007; Allende, et. al, 2013; Mathieu, 1994; Wang, et. al., 2003; Weng, et. Al, 2000; Bard, 1991; Jiang, et. al. 2015). These algorithms are divided into the following classes: Transformation methods (Lv, et. al. 2007; Allende, et. al, 2013; Brotcorne, et. al, 2013; Dempe, et. al, 2012), Fuzzy methods (Sakawa, et. al. 1998; Shih, et. al. 1996; Pramanik, et. al. 2007; Sakawa, et. al. 2001; sakawa, et. al, 2012), Global techniques (Nocedal, 2006; Khayyal, 1985; Mathieu, 1994; Wang, et. al. 2003), Primal–dual interior methods (Weng, et. Al, 2000), Enumeration methods, (Thoai, et. al, 2004), Meta heuristic approaches (Hejazi, et. al, 2002; Wang, et. al, 2008; Hu, et. al, 2010; Pal, et. al. 2010; Wan, et. al. 2013, 25]. In the remaining of section, two classes of methods have been explained.

1.1 Enumeration Methods

Branch and bound is an optimization algorithm that uses the basic enumeration. But in these methods we employ clever techniques for calculating upper bounds and lower bounds on the objective function by reducing the number of search steps. In these methods, the main idea is that the vertex points of achievable domain for BLPP are basic feasible solutions of the problem and optimal solution is among them.

1.2 Transformation Methods

An important class of methods for constrained optimization seeks the solution by replacing the original constrained problem with a sequence of unconstrained sub-problems or a problem with simple constraints. These methods are interested by some researchers for solving BLPP, so that they transform the follower problem by methods such as penalty functions, barrier functions, Lagrangian relaxation method or KKT conditions. In fact these techniques convert the BLPP into a single problem and then it is solved by other methods.

In this chapter, we propose a procedure based on Taylor theorem and consider the linear bi-level programming problem. In the remaining of pages, in section 2 basic concepts and definitions of the BLPP are proposed. We provide main theoretical results in section 3. Section 4 describes the steps of the presented algorithm. Computational results are proposed in Section 5 and finally, the chapter is finished in section 6.

2. THE NON-LINEAR BLPP AND SMOOTHING METHOD

The BLPP is used frequently by problems with decentralized planning structure. It is defined as (Bard, 1999):

$$\begin{aligned}&\min_{x} F\left(x,y\right)\\ &\text{s.t}\ \ \min_{y} f\left(x,y\right)\\ &\text{s.t}\ \ g\left(x,y\right)\leq 0,\quad x,y\geq 0.\end{aligned}$$

where

$$F : R^{n\times m} \rightarrow R^1, \quad f : R^{n\times m} \rightarrow R^1, \\ g : R^{n\times m} \rightarrow R^q, \quad x \in R^n, \quad y \in R^m. \tag{1}$$

Also F and f are objective functions of the leader and follower respectively.
The feasible region of the non-linear BLP problem is

$$S = \{(x, y) \mid g(x, y) \leq 0, x, y, \geq 0\} \tag{2}$$

Using KKT conditions problem (1) can be converted into the following problem:

$$\begin{aligned} &\min_{x,y\mu} F(x, y, \mu) \\ &\text{s.t } \nabla_y L(x, y, \mu) = 0, \\ &\quad \mu g(x, y) = 0, \\ &\quad g(x, y) \leq 0, \quad \mu \geq 0. \end{aligned} \tag{3}$$

where L is the Lagrange function and $L(x, y, \mu) = f(x, y) + \mu g(x, y)$.

Because problem (3) has a complementary constraint, it is not convex and it is not differentiable. Fortunately Facchinei et al, 1999 proposed smooth method for solving problem with complementary constraints and we use this method to smooth problem (3).

In general the BLPP is a non-convex optimization problem therefore there is no general algorithm to solve it. This problem can be non-convex even when all functions and constraints are bounded and continuous.

Definition 2.1: Fischer – Burmeister is the following function,

$$\phi: \mathrm{R}^2 \rightarrow \mathrm{R}, \ \phi(\mathrm{a}, \mathrm{b}) = \mathrm{a} + \mathrm{b} - \sqrt{\mathrm{a}^2 + \mathrm{b}^2}$$

or ϕ: $\mathrm{R}^3 \rightarrow \mathrm{R}$, $\phi(\mathrm{a}, \mathrm{b}, \blacklozenge) = \mathrm{a} + \mathrm{b} - \sqrt{\mathrm{a}^2 + \mathrm{b}^2 + \blacklozenge}$.

Using Fischer – Burmeister function $\phi(\mathrm{a}, \mathrm{b}, \blacklozenge) = \mathrm{a} + \mathrm{b} - \sqrt{\mathrm{a}^2 + \mathrm{b}^2 + \blacklozenge}$ in problem (3) we obtain the following problem:

$$\begin{aligned} &\min \ F(x, y, \mu) \\ &\text{s.t } \nabla_y L(x, y, \mu) = 0, \\ &\quad \mu_i - g_i(x, y) - \sqrt{\mu_i^2 + g_i^2(x, y) + \varepsilon} = 0, \ i = 1, 2, \ldots, m, \\ &\quad x, y, \mu_i \geq 0, \ i = 1, \ldots, m. \end{aligned} \tag{4}$$

which $g_i(x, y) = a^i{}_{x+b}{}^i y - r$, and a^i, b^i are i-th row of A, B respectively.

Let:

$$G(x, y, \mu)=\begin{bmatrix} \mu_1 - g_1(x, y) - \sqrt{\mu_1^2 + g_1^2(x, y) + \blacklozenge} \\ \mu_2 - g_2(x, y) - \sqrt{\mu_2^2 + g_2^2(x, y) + \blacklozenge} \\ \vdots \\ \mu_m - g_m(x, y) - \sqrt{\mu_m^2 + g_m^2(x, y) + \blacklozenge} \end{bmatrix}, H(x, y, \mu) = \nabla_y L(x, y, \mu). \tag{5}$$

Problem (4) can be written as follows,

$$\begin{aligned} \min\ & F(x, y, \mu) \\ \text{s.t}\ & H(x, y, \mu) = 0, \\ & G(x, y, \mu) = 0, \\ & x, y, \mu \geq 0. \end{aligned} \tag{6}$$

where $t = (x, y, \mu)$

By applying Taylor theorem at a feasible point such as t^k for function G, H, F and take only two linear part of them, the following linear functions is constructed:

$$\begin{aligned} & G_i(t^k) + \nabla G_i(t^k)(t - t^k) = 0, \quad i = 1, 2, \ldots m \\ & H_i(t^k) + \nabla H_i(t^k)(t - t^k) = 0, \quad i = 1, 2, \ldots m \\ & F_i(t^k) + \nabla F_i(t^k)(t - t^k) = 0, \quad i = 1, 2, \ldots m \end{aligned} \tag{7}$$

Because the obtained problem by using Taylor theorem is linear programming, it can be solved using simplex methods.

A summary of important properties for convex problem as follows, which $f: S \dot{\rightarrow} R^n$ and S is a nonempty convex set in R^n.

1. The convex function f is continuous on the interior of S.
2. Every local optimal solution of f over a convex set $X \subseteq S$ is the unique global optimal solution.
3. If $\nabla f(\bar{x}) = 0$, then $\bar{x}$ is unique global optimal solution of f over S.

Since in problem (3), most of the equality constraints are not linear then it concerns that the above problem is a non-convex programming problem, which indicates there are local optimal solutions that are not global solutions. Therefore solving the problem (3) will be complicated and we use the following method for solving this problem.

3. MAIN THEORETICAL RESULTS

Definition 3.1: A function f is a continuous function at x=a when f(a) is defined,

$\lim_{x\to a} f(x)$ exists, $\lim_{x\to a} f(x) = f(a)$.

Theorem 3.1: All polynomials continuous everywhere. Additionally $x^n, \sqrt[n]{x}$ are continuous for all x., when n is odd and for x>0, when n is even.

Proof: Suppose that P(x) is a polynomial of degree n≥0,

$$P(x) = c_n x^n + c_{n-1} x^{n-1} + \ldots + c_1 x^1 + c_0.$$

Then,

$$\lim_{x\to a}[P(x)]$$

$$= \lim_{x\to a}\left[c_n x^n + c_{n-1} x^{n-1} + \ldots + c_1 x^1 + c_0\right] = c_n \lim_{x\to a} x^n + c_{n-1} \lim_{x\to a} x^{n-1} + \ldots + c_1 \lim_{x\to a} x^1 + \lim_{x\to a} c_0 .$$

This finished the proof. The rest of the theorem follows in a similar way.

Theorem 3.2: Suppose that f and g are continuous at x=a. Then f+g, f–g are continuous at x=a.

Proof: Since f and g are continuous at x=a, then:

$$\lim_{x\to a}\left[f(x) \pm g(x)\right] = \lim_{x\to a} f(x) \pm \lim_{x\to a} g(x) = f(a) \pm g(a) = (f \pm g)(a).$$

Thus $\lim_{x\to a}\left[f(x) \pm g(x)\right] = (f \pm g)(a)$.

This shows that f+g, f–g are continuous at x=a.

Theorem 3.3: Suppose that $\lim_{x\to a} g(x) = L$ and f is continuous at L. Then,

$$\lim_{x\to a} f(g(x)) = f\left(\lim_{x\to a} g(x)\right) = f(L)$$

By our continuity assumptions, we know that

$$\forall_{\varepsilon_1}\exists_{\delta_1} s.t\, |x-L|<\delta_1 \rightarrow |f(x)-f(L)|<\varepsilon_1$$

$$\forall_{\varepsilon_2}\exists_{\delta_2} s.t\, |x-a|<\delta_2 \rightarrow |g(x)-L|<\varepsilon_2$$

So for this ε, choose $\varepsilon_1=\varepsilon$, which gives a $\delta_1>0$ so that

$|x-L| < \delta_1 \rightarrow | f(x) - f(L) | < \varepsilon$

Next set $\varepsilon_2=\delta_2>0$. This gives a $\delta_2>0$ so that

$|x-a| < \delta_2 \rightarrow | g(x)-L| < \varepsilon_2$

Finally, let $\varepsilon_2=\delta_2$, corresponding to in the continuity requirement for g. therefore we can write:

$$\begin{aligned}&|x-a|<\delta \leftrightarrow |x-a|<\delta_2\\ &\rightarrow |g(x)-g(a)|<\varepsilon_2 \leftrightarrow |g(x)-L|<\delta_1\\ &\rightarrow |f(g(x))-f(L)|<\varepsilon_1 \leftrightarrow |f(g(x))-f(g(a))|<\varepsilon\end{aligned}$$

Corollary 1: Suppose that g is continuous at a and f is continuous at g(a). Then, the composition *fog* is continuous at a.

Proof: From above theorem, we have:

$$\lim_{x\to a}(fog)(x)=\lim_{x\to a} f(g(x)) = f(\lim_{x\to a} g(x)) = f(g(a)) = (fog)(a)$$

since g is continuous at a. This finished the proof.

Because functions G, H in (6) is always continuous everywhere and it is possible to use Theorems and corollary, Taylor Theorem for it in (6) and F should be continuous too.

Theorem 3.4 (Taylor Theorem) (Protter, 1998): Suppose f has n +1 continuous derivatives on an open interval containing a. Then for each x in the interval,

$$f(x)=\left[\sum_{k=0}^{n}\frac{f^k(a)}{k!}(x-a)^k\right]+R_{n+1}(x) \tag{8}$$

where the error term $R_{n+1}(x)$, for some c between a and x, satisfies

$$R_{n+1}(x)=\frac{f^{(n+1)}(c)}{(n+1)!}(x-a)^{n+1} \tag{9}$$

This form for the error $R_{n+1}(x)$, is called the Lagrange formula for the reminder.
The infinite Taylor series converge to f,

$$f(x)=\left[\sum_{k=0}^{\infty}\frac{f^{k}(a)}{k!}(x-a)^{k}\right] \tag{10}$$

If only if $\lim_{n\to\infty} R_{n+1}(x)=0$.

Proof: The proof of this theorem was given by (Facchinei, et. al. 1999).

In mathematics, an approximation of a k-times differentiable function near to a point is given by Taylor's theorem. Taylor's theorem is one of the fundamental tools in pure mathematics and it is the starting point of advanced asymptotic analysis, also it is usually used in applied fields of mathematics. If a real-valued function f is differentiable at the point a then it has a linear approximation at the point a. This means that there exists a function g such that

$$f(x)=f(a)+f'(a)(x-a)+g(x)(x-a),\quad \lim_{x\to a} g(x)=0. \tag{11}$$

Here

$$P_1(x)=f(a)+f'(a)(x-a) \tag{12}$$

is the linear approximation of f at the point a.
Steps of the proposed algorithm as follows:

Step 1: Initialization

The feasible point t^1 is created randomly, error $�_1$ is given and suppose k=1.
$�_1$ is a small and appropriate given error and finishing the algorithm depends to $�_1$ such that it is finished whenever difference between produced solutions by the algorithm in two consecutive iterations is less than $�_1$.

Step 2: Finding Solution

According to the step 1, k=1 and feasible solution t^1 has been defined. Using these assumptions and Taylor theorem for G(t), H(t), and F(t) at t^k, we obtain following problem:

$$\begin{aligned}\min\ & F_i\left(t^k\right)+\nabla F_i\left(t^k\right)\left(t-t^k\right)\\ \text{s.t}\ & H_i\left(t^k\right)+\nabla H_i\left(t^k\right)\left(t-t^k\right)=0,\ \ i=1,2,\ldots m\\ & G_i\left(t^k\right)+\nabla G_i\left(t^k\right)\left(t-t^k\right)=0,\ \ i=1,2,\ldots m\\ & x,y,\mu_i\geq 0,\ \ i=1,\ldots,m.\end{aligned} \tag{13}$$

Solve the problem (13) using simplex method (by MATLAB 7.1). By solving this problem, an optimal solution such as t^{k+1} is obtained for (13).

Step 3: Keeping the Present Best Solution

Because (13) is an approximation for (6) by Taylor theorem, therefore optimal solution for (13) is an approximation of optimal solution for (6). Thus t^{k+1} can be a good approximation of problem (6) optimal solution. Therefore let $t^*=t^{k+1}$ and go to next step.

Step 4: Termination

If $t^{k+1}\left|F\left(t^{k+1}\right)-F\left(t^k\right)\right|<\blacklozenge_1$ then the algorithm is finished and t* is the best solution by the proposed algorithm. Otherwise, let k=k+1 and go to the step 2.

4. COMPUTATIONAL RESULTS

Example 1(Lv, et. al. 2008): Consider the following non-linear bi-level programming problem:

$$\begin{aligned}\min_{x}\ & x^2+(y-10)^2\\ \text{s.t}\ & \min_{y\geq 0}\ \ (x+2y-30)^2\\ \text{s.t}\ & x-y^2\geq 0,\\ & 20-x-y^2\geq 0\\ & 0\leq x\leq 15.\end{aligned}$$

Using KKT conditions the following problem is obtained:

$$\begin{aligned}
\min_{x}\ & x^2+(y-10)^2 \\
\text{s.t}\ & 4(x+2y-30)=0, \\
& 2y(\lambda_1+\lambda_2)=0, \\
& \lambda_1(y^2-x)=0, \\
& \lambda_2(y^2+x-20)=0, \\
& \lambda_3(x-15)=0 \\
& y^2-x\le 0, \\
& y^2+x-20\le 0, \\
& x-15\le 0, \\
& \lambda_1,\lambda_2,\lambda_3\ge 0.
\end{aligned}$$

By the Fischer – Burmeister function, the above problem as follows:

$$\begin{aligned}
\min\ & x^2+(y-10)^2 \\
\text{s.t}\ & 4(x+2y-30)=0, \\
& (\lambda_1+\lambda_2)-2y-\sqrt{(\lambda_1+\lambda_2)^2+(2y)^2+\varepsilon}=0, \\
& \lambda_1-(y^2-x)-\sqrt{\lambda_1^2+(y^2-x)^2+\varepsilon}=0, \\
& \lambda_2-(y^2+x-20)-\sqrt{\lambda_2^2+(x+y^2-20)^2+\varepsilon}=0, \\
& \lambda_3-(x-15)-\sqrt{\lambda_3^2+(x-15)^2+\varepsilon}=0,
\end{aligned}$$

We solve this problem using the proposed line search algorithm and we present the optimal solution in Table 1. By solving this problem the best solutions are found according to Table 1. It declares that the best solutions by the proposed algorithm are better than the best solution by the references in less time.

Behavior of the variables in Example 1 has been show in Figure 1 that variables x and y will be stable after 5000 and 4850 iterations respectively.

Example (Allende, et. al. 2013) according to the following theorem, consider the following linear bi-level programming problem.

Table 1. Comparison optimal solutions: Example 1

Best Solution by Our Method ε=0.001		Best Solution According to Reference (Thomas, et. al. 1988)		Optimal Solution	
(x*,y*)	**z***	**(x*,y*)**	**z***	**(x*,y*)**	**z***
(2.600,1.612)	-77.11	(2.600,1.613)	-77.10	(2.600,1.612)	-77.11

Figure 1. The transient behavior of the variables in Example 1

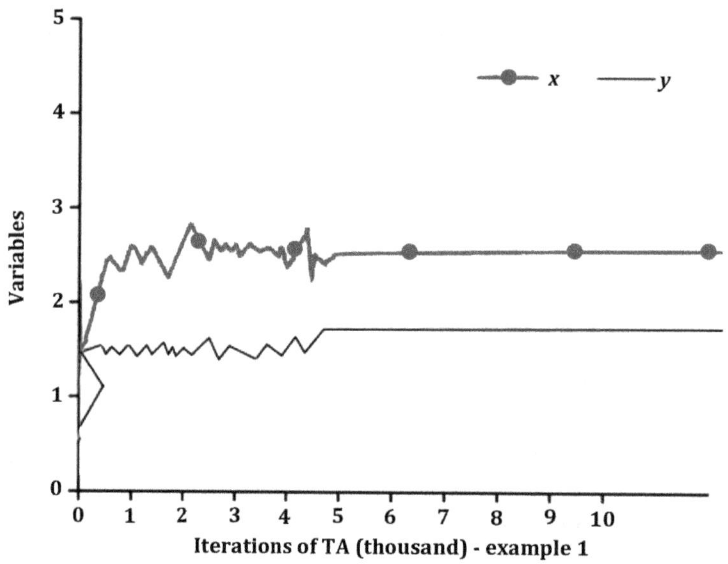

$$\begin{aligned}
\min_{x} \quad & -x_1^2 - 2x_1 + x_2^2 - 2x_2 + y_1^2 + y_2^2 \\
\text{s.t} \quad & \min \ y_1^2 - 2x_1y_1 + y_2^2 - 2x_2y_2 \\
\text{s.t} \quad & .25 - (y_1 - 1)^2 \geq 0, \\
& .25 - (y_2 - 1)^2 \geq 0.
\end{aligned}$$

After applying KKT conditions and smoothing method, and then proposed penalty function above problem will be transformed to the following problem:

$$\begin{aligned}
\min_{x} \quad & -x_1^2 - 2x_1 + x_2^2 - 2x_2 + y_1^2 + y_2^2 \\
\text{s.t} \quad & +\mu_1(2y_1 - 2x_1 + 2y_2 - 2x_2) \\
& +\mu_2(\lambda_1 - (y_1 - 1)^2 + .25 - \sqrt{\lambda_1^2 + ((y_1 - 1)^2 + .25)^2 + \varepsilon}\ )^2 \\
& +\mu_3(\lambda_2 + .25 - (y_2 - 1)^2 - \sqrt{\lambda_2^2 + ((y_2 - 1)^2 + .25)^2 + \varepsilon}\ )^2
\end{aligned}$$

The optimal solution is obtained using our method according to Table 2.

Table 2. Comparison optimal solution: Example 2

Best Solution by Our Method **ε=0.001**		**Best Solution According to Reference** **(Allende, et. al. 2013)**		**Optimal Solution**	
(x^*, y_1^*, y_2^*)	z^*	(x^*, y_1^*, y_2^*)	z^*	(x^*, y_1^*, y_2^*)	z^*
(0.51,0.51,0.51,0.51)	-1.598	(0.5,0.5,0.5,0.5)	-1.5	(0.51,0.51,0.51,0.51)	-1.598

Behavior of the variables in Example 2 has been show in Figure 2 that variables will be stable after 6 thousand iterations respectively.

More problems with deferent sizes have been solved by our approach and computation results have been proposed in Table 3. According to this Table, the best solutions by our algorithm are better than the best solution by the references. The algorithm is feasible and efficient according to the Tables.

We make program with MATLAB 7.1 and use a personal computer (CPU: Intel (R) Celeron(R) 1000 M @ 1.8 GHz, RAM: 4 GB) to execute the program. References of the examples in Table 3 as follows:

Example 3: (Lv, et. al. 2008),
Example 4: (Allende, et. al. 2013),
Example 5: (Dempe, 2012),
Example 6: (Zhongping, et. al, 2011).

Example 3 is minimization and examples 4, 5, 6 are maximization problems.

Figure 2. The transient behavior of the variables in Example 2

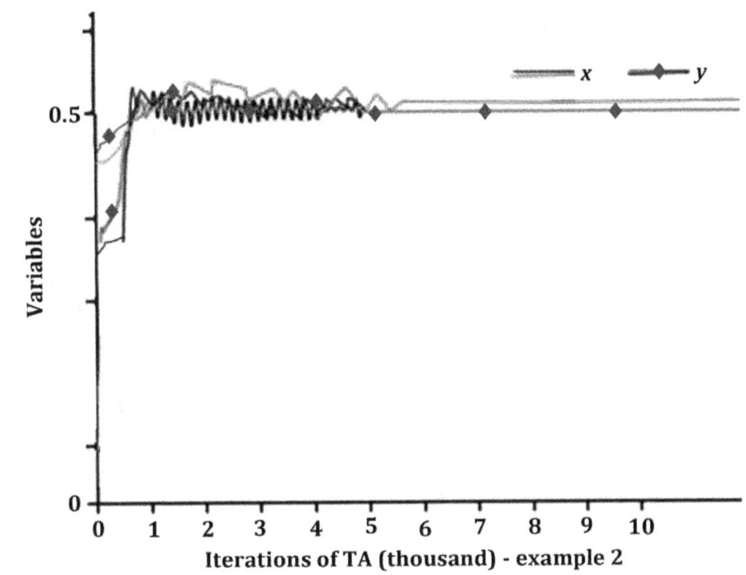

Table 3. Comparison optimal solutions and elapsed time with deferent Examples 3-6 of BLPP

	Best Solution by Our Method ε=0.001	Best Solution According to References (Allende, et. al. 2013, Lv, et. al. 2008- Zhongping, et. al, 2011)	Optimal Solution
Example 3	(1.889,0.888,0)	(1.883,0.891,0.003)	$(^{17}/_{9}, ^{8}/_{9}, 0)$
Example 4	(0,0)	(0,0)	(0,0)
Example 5	(1,0)	(1,0)	(1,0)
Example 6	(0001,0.73,0,0.54,0)	(0,0.75,0,0.5,0)	(0,0.75,0,0.5,0)

5. CONCLUSION AND FUTURE WORK

In this chapter we used the KKT conditions to convert the problem into the single level problem. Then using the Fischer – Burmeister function the problem is made simpler and convert to smooth programming problem. Finally using proposed algorithm based on Taylor theorem the smoothed problem was solved. Comparing with the results of previous methods, our algorithm has better numerical results and presents better solution in less time. Also the best solution produced by proposed algorithm is feasible unlike the previous best solution by references. In the future works, the following should be researched:

1. Examples in larger sizes can be supplied to illustrate the efficiency of the proposed algorithm.
2. Show the efficiency of the proposed algorithm for solving other kind of the BLP such as quadratic and non-linear BLP.

REFERENCES

Allende, G. B., & Still. (2013). Solving bilevel programs with the KKT-approach. *Mathematical Programming, 138*(1-2), 309-332.

Bard, J. F. (1991). Some properties of the bilevel programming problem. *Journal of Optimization Theory and Applications*, *68*(2), 371–378. doi:10.1007/BF00941574

Bard, J. F. (1999). *Practical Bilevel Optimization: Algorithms and Applications*. Academic Press.

Bazaraa, M. S., Sherali, H. D., & Shetty, C. M. (2013). *Nonlinear programming: theory and algorithms*. John Wiley & Sons.

Brotcorne, L., Hanafi, S., & Mansi, R. (2013). One-level reformulation of the bilevel knapsack problem using dynamic programming. *Discrete Optimization*, *10*(1), 1–10. doi:10.1016/j.disopt.2012.09.001

Dempe, S., & Dutta, J. (2012). Is bilevel programming a special case of a mathematical program with complementarity constraints? *Mathematical Programming*, *131*(1-2), 37–48. doi:10.1007/s10107-010-0342-1

Dempe, S., & Zemkoho, A. B. (2012). On the Karush–Kuhn–Tucker reformulation of the bilevel optimization problem. *Nonlinear Analysis: Theory. Methods & Applications*, *75*(3), 1202–1218.

Facchinei, F., Jiang, H., & Qi, L. (1999). A smoothing method for mathematical programs with equilibrium constraints. *Mathematical Programming*, *85*(1), 107–134. doi:10.1007/s101070050048

Hejazi, S. R., Memariani, A., Jahanshahloo, G., & Sepehri, M. M. (2002). Linear bilevel programming solution by genetic algorithm. *Computers & Operations Research*, *29*(13), 1913–1925. doi:10.1016/S0305-0548(01)00066-1

Hu, T., Guo, X., Fu, X., & Lv, Y. (2010). A neural network approach for solving linear bilevel programming problem. *Knowledge-Based Systems*, *23*(3), 239–242. doi:10.1016/j.knosys.2010.01.001

Jiang, Y., Li, X., Huang, C., & Wu, X. (2013). Application of particle swarm optimization based on CHKS smoothing function for solving nonlinear bilevel programming problem. *Applied Mathematics and Computation*, *219*(9), 4332–4339. doi:10.1016/j.amc.2012.10.010

Khayyal, A. (1985). AL Minimizing a Quasi-concave Function Over a Convex Set: A Case Solvable by Lagrangian Duality. In *Proceedings IEEE International Conference on Systems, Man, and Cybemeties*, (pp. 661-663).

Lan, K. M., Wen, U. P., Shih, H. S., & Lee, E. S. (2007). A hybrid neural network approach to bilevel programming problems. *Applied Mathematics Letters*, *20*(8), 880–884. doi:10.1016/j.aml.2006.07.013

Lv, Y., Hu, T., Wang, G., & Wan, Z. (2007). A penalty function method based on Kuhn–Tucker condition for solving linear bilevel programming. *Applied Mathematics and Computation*, *188*(1), 808–813. doi:10.1016/j.amc.2006.10.045

Lv, Y., Hu, T., Wang, G., & Wan, Z. (2008). A neural network approach for solving nonlinear bilevel programming problem. *Computers & Mathematics with Applications (Oxford, England)*, *55*(12), 2823–2829. doi:10.1016/j.camwa.2007.09.010

Mathieu, R., Pittard, L., & Anandalingam, G. (1994). Genetic algorithm based approach to bi-level linear programming. *Revue française d'automatique, d'informatique et de recherche opérationnelle. Recherche opérationnelle, 28*(1), 1-21.

Nocedal, J., & Wright, S. J. (2006). Conjugate gradient methods. *Numerical Optimization*, 101-134.

Pal, B. B., Chakraborti, D., & Biswas, P. (2010, July). A genetic algorithm approach to fuzzy quadratic bilevel programming. In *Computing Communication and Networking Technologies (ICCCNT), 2010 International Conference on* (pp. 1-7). IEEE. doi:10.1109/ICCCNT.2010.5591820

Pramanik, S., & Roy, T. K. (2007). Fuzzy goal programming approach to multilevel programming problems. *European Journal of Operational Research*, *176*(2), 1151–1166. doi:10.1016/j.ejor.2005.08.024

Protter, M. H., & Protter, P. E. (1988). *Calculus with analytic geometry*. Jones & Bartlett Learning.

Sakawa, M., & Matsui, T. (2012). Stackelberg solutions for random fuzzy two-level linear programming through possibility-based probability model. *Expert Systems with Applications*, *39*(12), 10898–10903. doi:10.1016/j.eswa.2012.03.001

Sakawa, M., & Nishizaki, I. (2001). Interactive fuzzy programming for two-level linear fractional programming problems. *Fuzzy Sets and Systems*, *119*(1), 31–40. doi:10.1016/S0165-0114(99)00066-4

Sakawa, M., Nishizaki, I., & Uemura, Y. (1998). Interactive fuzzy programming for multilevel linear programming problems. *Computers & Mathematics with Applications (Oxford, England)*, *36*(2), 71–86. doi:10.1016/S0898-1221(98)00118-7

Shih, H. S., Lai, Y. J., & Lee, E. S. (1996). Fuzzy approach for multi-level programming problems. *Computers & Operations Research*, *23*(1), 73–91. doi:10.1016/0305-0548(95)00007-9

Thoai, N. V., Yamamoto, Y., & Yoshise, A. (2005). Global optimization method for solving mathematical programs with linear complementarity constraints. *Journal of Optimization Theory and Applications, 124*(2), 467–490. doi:10.1007/s10957-004-0946-9

Vicente, L., Savard, G., & Júdice, J. (1994). Descent approaches for quadratic bilevel programming. *Journal of Optimization Theory and Applications, 81*(2), 379–399. doi:10.1007/BF02191670

Wan, Z., Wang, G., & Hou, K. (2008, October). An interactive fuzzy decision making method for a class of bilevel programming. In *Fuzzy Systems and Knowledge Discovery, 2008. FSKD'08. Fifth International Conference on* (Vol. 1, pp. 559-564). IEEE. doi:10.1109/FSKD.2008.537

Wan, Z., Wang, G., & Sun, B. (2013). A hybrid intelligent algorithm by combining particle swarm optimization with chaos searching technique for solving nonlinear bilevel programming problems. *Swarm and Evolutionary Computation, 8*, 26–32. doi:10.1016/j.swevo.2012.08.001

Wang, G., Jianq, B., Zhu, K., & Wan, Z. (2010). Global convergent algorithm for the bilevel linear fractional-linear programming based on modified convex simplex method. *Systems Engineering and Electronics. Journalism, 21*(2), 239–243.

Wang, G., Wan, Z., Wang, X., & Lv, Y. (2008). Genetic algorithm based on simplex method for solving linear-quadratic bilevel programming problem. *Computers & Mathematics with Applications (Oxford, England), 56*(10), 2550–2555. doi:10.1016/j.camwa.2008.05.006

Weng, W. T., & Wen, U. P. (2000). A primal-dual interior point algorithm for solving bilevel programming problem. *Asia-Pacific Journal of Operational Research, 17*(2), 213.

Zhongping, W., Guangmin, W., & Yibing, L. (2011). A dual-relax penalty function approach for solving nonlinear bilevel programming with linear lower level problem. *Acta Mathematica Scientia, 31*(2), 652–660. doi:10.1016/S0252-9602(11)60265-8

Chapter 31
Some Aspects of Energy Efficiency for Refrigeration and Air Conditioning

Abdeen Mustafa Omer
Energy Research Institute (ERI), UK

ABSTRACT

Over the years, all parts of a commercial refrigerator, such as the compressor, heat exchangers, refrigerant, and packaging, have been improved considerably due to the extensive research and development efforts carried out by academia and industry. However, the achieved and anticipated improvement in conventional refrigeration technology are incremental since this technology is already nearing its fundamentals limit of energy efficiency is described is 'magnetic refrigeration' which is an evolving cooling technology. The word 'green' designates more than a colour. It is a way of life, one that is becoming more and more common throughout the world. An interesting topic on 'sustainable technologies for a greener world' details about what each technology is and how it achieves green goals. Recently, conventional chillers using absorption technology consume energy for hot water generator but absorption chillers carry no energy saving. With the aim of providing a single point solution for this dual purpose application, a product is launched but can provide simultaneous chilling and heating using its vapour absorption technology with 40% saving in heating energy. Using energy efficiency and managing customer energy use has become an integral and valuable exercise. The reason for this is green technology helps to sustain life on earth. This not only applies to humans but to plants, animals and the rest of the ecosystem. Energy prices and consumption will always be on an upward trajectory. In fact, energy costs have steadily risen over last decade and are expected to carry on doing so as consumption grows. This article discusses the potential for such integrated systems in the stationary and portable power market in response to the critical need for a cleaner energy technology for communities. Throughout the theme several issues relating to renewable energies, environment and sustainable development are examined from both current and future perspectives.

DOI: 10.4018/978-1-4666-9644-0.ch031

INTRODUCTION

This section describes the different methods and techniques for providing energy for heating and cooling systems. It also, covers the optimisation and improvement of the operation conditions of the heat cycles and the performance of the ground source heat pump systems (GSHPs).

With the improvement of people's living standards and the development of economies, heat pumps have become widely used for air conditioning. The driver to this was that environmental problems associated with the use of refrigeration equipment, the ozone layer depletion and global warming are increasingly becoming the main concerns in developed and developing countries alike. With development and enlargement of the cities in cold regions, the conventional heating methods can severely pollute the environment. In order to clean the cities, the governments drew many measures to restrict citizen heating by burning coal and oil and encourage them to use electric or gas-burning heating. New approaches are being studied and solar-assisted reversible absorption heat pump for small power applications using water-ammonia is under development (Luo, Tondeur, Le Gall, & Corbel, 2007).

An air-source heat pump is convenient to use and so it is a better method for electric heating. The ambient temperature in winter is comparatively high in most regions, so heat pumps with high efficiency can satisfy their heating requirement. On the other hand, a conventional heat pump is unable to meet the heating requirement in severely cold regions anyway, because its heating capacity decreases rapidly when ambient temperature is below -10°C. According to the weather data in cold regions, the air-source heat pump for heating applications must operate for long times with high efficiency and reliability when ambient temperature is as low as -15°C. Hence, much researches and developments have been conducted to enable heat pumps to operate steadily with high efficiency and reliability in low temperature environments (Luo, Fan, & Tondeur, 2007). For example, the burner of a room air conditioner, which uses kerosene, was developed to improve the performance in low outside temperature (Philappacopoulus, & Berndt, 2001). Similarly, the packaged heat pump with variable frequency scroll compressor was developed to realise high temperature air supply and high capacity even under the low ambient temperature of –10 to –20°C (Jo, Katsumi, Benson, & Edil, 2001). Such a heat pump systems can be conveniently used for heating in cold regions. However, the importance of targeting the low capacity range is clear if one has in mind that the air conditioning units below 10 kW cooling account for more than 90% of the total number of units installed in the EU (Anandarajah, 2003).

EARTH-ENERGY SYSTEMS (EESs)

The earth-energy systems, EESs, have two parts; a circuit of underground piping outside the house, and a heat pump unit inside the house. And unlike the air-source heat pump, where one heat exchanger (and frequently the compressor) is located outside, the entire GSHP unit for the EES is located inside the house.

The outdoor piping system can be either an open system or closed loop. An open system takes advantage of the heat retained in an underground body of water. The water is drawn up through a well directly to the heat exchanger, where its heat is extracted. The water is discharged either to an aboveground body of water, such as a stream or pond, or back to the underground water body through a separate well. Closed-loop systems, on the other hand, collect heat from the ground by means of a continuous loop of piping buried underground. An antifreeze solution (or refrigerant in the case of a direct expansion 'DX' earth-energy system), which has been chilled by the heat pump's refrigeration system to several degrees colder than the outside soil, circulates through the piping, absorbing heat from the surrounding soil.

In some EESs, a heat exchanger, sometimes called a "desuperheater", takes heat from the hot refrigerant after it leaves the compressor. Water from the home's water heater is pumped through a coil ahead of the condenser coil, in order that some of the heat that would have been dissipated at the condenser is used to heat water. Excess heat is always available in the cooling mode, and is also available in the heating mode during mild weather when the heat pump is above the balance point and not working to full capacity. Other EESs heat domestic hot water (DHW) on demand: the whole machine switches to heating DHW when it is required.

Hot water heating is easy with EESs because the compressor is located inside. Because EESs have relatively constant heating capacity, they generally have many more hours of surplus heating capacity than required for space heating. In fact, there are sources of energy all around in the form of stored solar energy, which even if they have a low temperature, can provide the surroundings with enough energy to heat the soil, bedrock and ground water as a heat source for domestic dwellings as shown in Figure 1, for example. Some emphasis has recently been put on the utilisation of the ambient energy from ground source and other renewable energy sources in order to stimulate alternative energy sources for heating and cooling of buildings. Exploitation of renewable energy sources and particularly ground heat in buildings can significantly contribute towards reducing dependency on fossil fuels.

The Cooling Cycle

The cooling cycle is basically the reverse of the heating cycle. The reversing valve changes the direction of the refrigerant flow. The refrigerant picks up heat from the house air and transfers it directly in DX systems or to the ground water or antifreeze mixture. The heat is then pumped outside, into a water body or return well (in the case of an open system), or into the underground piping (in the case of a closed-loop system). Once again, some of this excess heat can be used to preheat domestic hot water.

Unlike air-source heat pumps, EESs do not require a defrost cycle. Underground temperatures are much more stable than air temperature, and the heat pump unit itself is located inside; therefore, problems with frost do not arise.

Figure 1. Using the soil, bedrock or groundwater as the heat source

Function of the GSHP Circuit

The collector liquid (cooling medium) is pumped up from the borehole in tubing and passed to the heat pump. Another fluid, a refrigerant, circulates in the heat pump in a closed system with the most important characteristic of having a low boiling point. When the refrigerant reaches the evaporator, which has received energy from the borehole, and the refrigerant evaporates. The vapour is fed to a compressor where it is compressed. This results in a high increase in temperature. The warm refrigerant is fed to the condenser, which is positioned in the boiler water. Here the refrigerant gives off its energy to the boiler water, so that its temperature drops and the refrigerant changes state from gas to liquid. The refrigerant then goes via filters to an expansion valve, where the pressure and temperature are further reduced. The refrigerant has now completed its circuit and is once more fed into the evaporator where it is evaporated yet again due to the effect of the energy that the collector has carried from the energy source (Figure 2).

Efficiencies of the GSHP systems are much greater than conventional air-source heat pump systems. A higher COP (coefficient of performance) can be achieved by a GSHP because the source/sink earth temperature is relatively constant compared to air temperatures. Additionally, heat is absorbed and rejected through water, which is a more desirable heat transfer medium because of its relatively high heat capacity. The GSHP systems rely on the fact that, under normal geothermal gradients of about 0.5°F/100 ft (30°C/km), the earth temperature is roughly constant in a zone extending from about 20 ft (6.1 m) deep to about 150 ft (45.7 m) deep. This constant temperature interval within the earth is the

Figure 2. Detail of the GSHP circuit

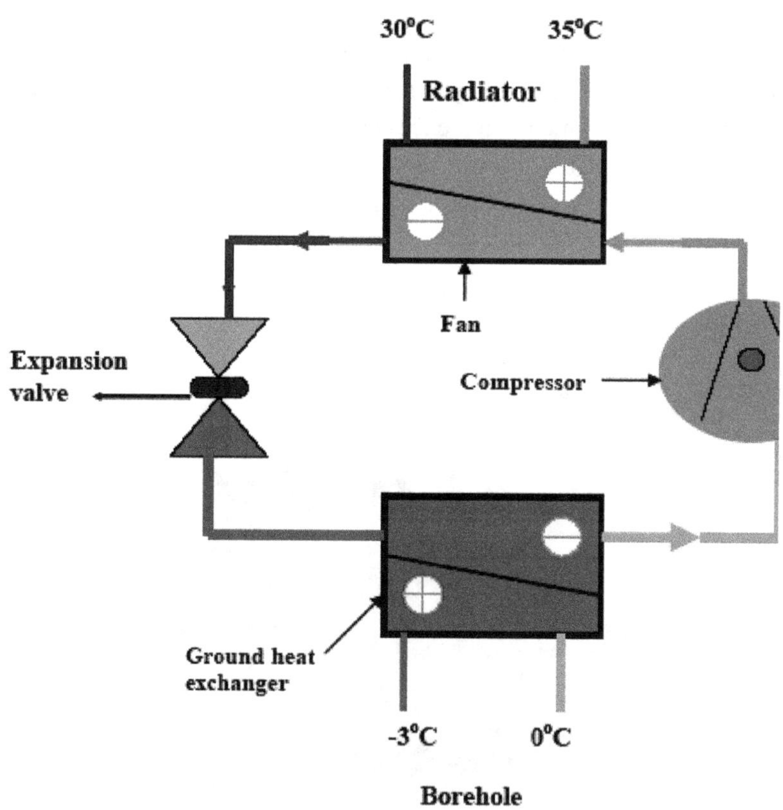

result of a complex interaction of heat fluxes from above (the sun and the atmosphere) and from below (the earth interior). As a result, the temperature of this interval within the earth is approximately equal to the average annual air temperature (Malin, & Alex, 2000). Above this zone (less than about 20 feet (6.1 m) deep), the earth temperature is a damped version of the air temperature at the earth's surface. Below this zone (greater than about 150 ft (45.7 m) deep), the earth temperature begins to rise according to the natural geothermal gradient. The storage concept is based on a modular design that will facilitate active control and optimisation of thermal input/output, and it can be adapted for simultaneous heating and cooling often needed in large service and institutional buildings (Omer, 2008). Loading of the core is done by diverting warm and cold air from the heat pump through the core during periods with excess capacity compared to the current need of the building (Brain, and Mark, 2007; Omer, 2009a; and Omer, 2009b). The cool section of the core can also be loaded directly with air during the night, especially in spring and fall when nights are cold and days may be warm.

Free Cooling

The installation can additionally be fitted with fan convectors, for example, in order to allow connections for free cooling (Figure 3). To avoid condensation, pipes and other cold surfaces must be insulated with diffusion proof material. Where the cooling demand is high, fan convectors with drip tray and drain connection are needed.

Figure 3. Diagram of cooling system

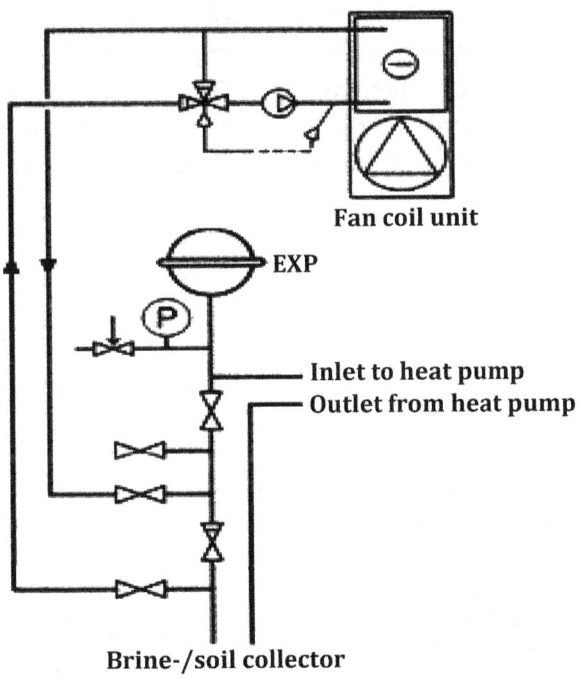

Refrigeration and Heat Pumps

The pressure (ps) is a function of how rapidly vapour can be removed through suction or formed through pressure. At equilibrium, the rate at which vapour is formed (determined by Q) equals the rate at which it is removed. Therefore, both the heat transfer rate into the liquid (Q) and the vapour removal rate (suction pump capacity) determines the pressure and hence Tsat(s) (Figure 4). This is governed by the following set of equations.

$$Q = m\, h_{fg} \tag{1}$$

$$m = \rho\, g\, V \tag{2}$$

$$Q = \rho\, g\, V\, h_{fg} \tag{3}$$

$$Q = V\, h_{fg}/v_g \tag{4}$$

Both h_{fg} and v_g depend on the saturation temperature (or pressure) as assumed in Figure 5, which describes the relationship represented by Equation 4.

The RHS of the Figure 6 is the 'converse' of the LHS, and constitutes a heat pump. Heat is 'pumped' from the LHS to the RHS. The main difference is that the vapour, after compression, will almost certainly be superheated and must cool to Tsat(c) before condensing will occur. The same reasoning (in converse) applies to the RHS as previously applied to LHS. Obviously, with the above system, the entire refrigerant would eventually end up on the RHS, and the heat pumping (& refrigeration) effect would cease.

Clearly, to ensure that the system can operate continuously liquid refrigerant needs to be fed from the RHS back to the LHS. This can be achieved by simply allowing it to flow back under its natural pressure difference. In this way a continuous closed circuit refrigeration (Or heat pump) system is obtained (Figure 7).

Figure 4. Refrigeration Cycle

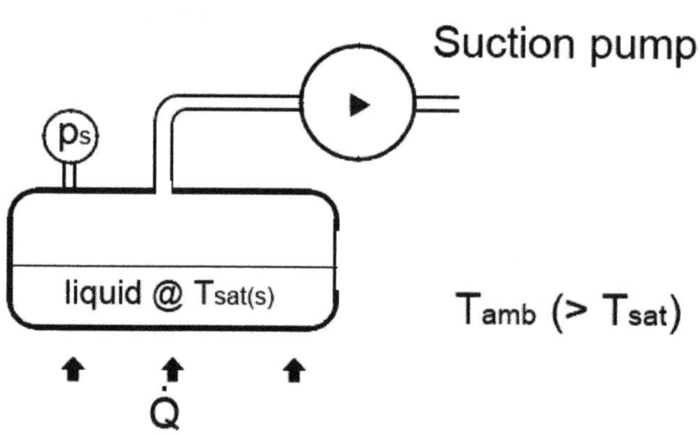

Figure 5. Heat transfer rate versus volume rate
pc = Condenser or 'high side' pressure.
ps = Evaporator, 'low side', or suction pressure.

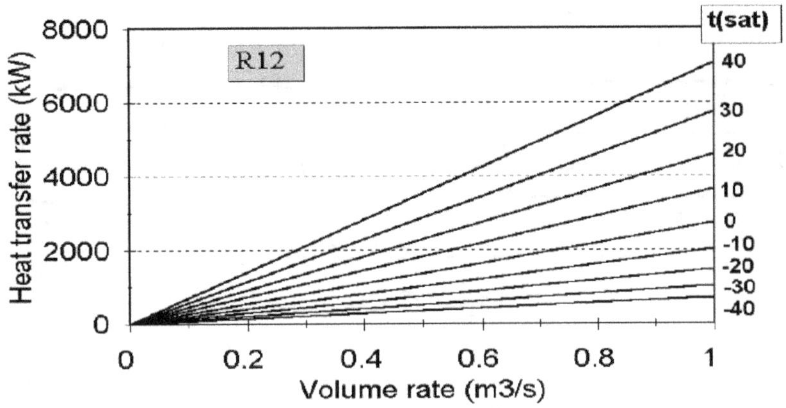

Figure 6. Heat pumps

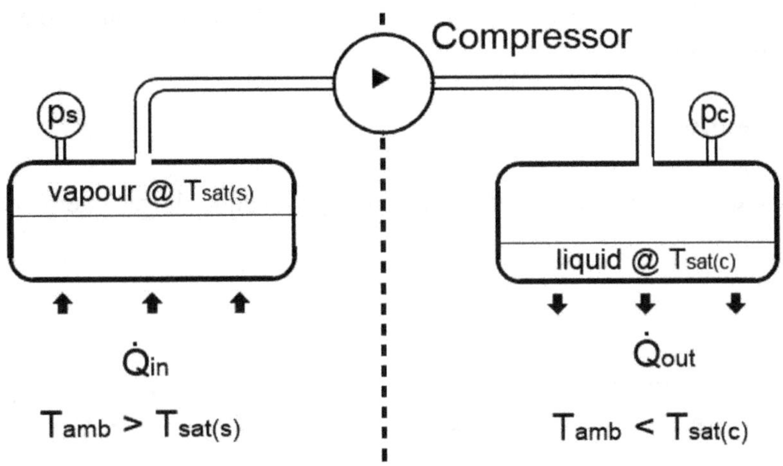

Figure 7. Simplified refrigeration system diagram

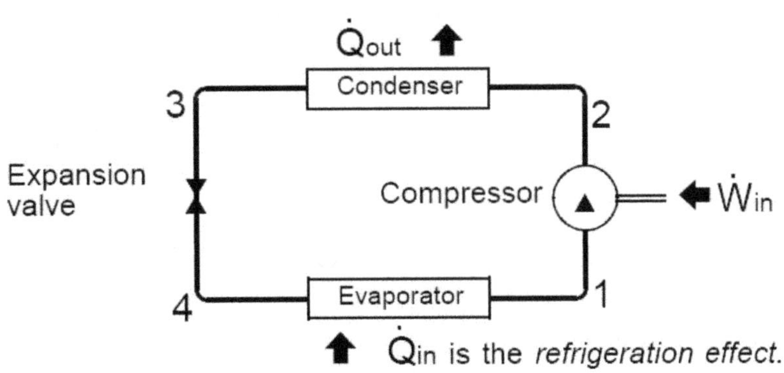

Figure 8. Heat pump refrigeration cycle

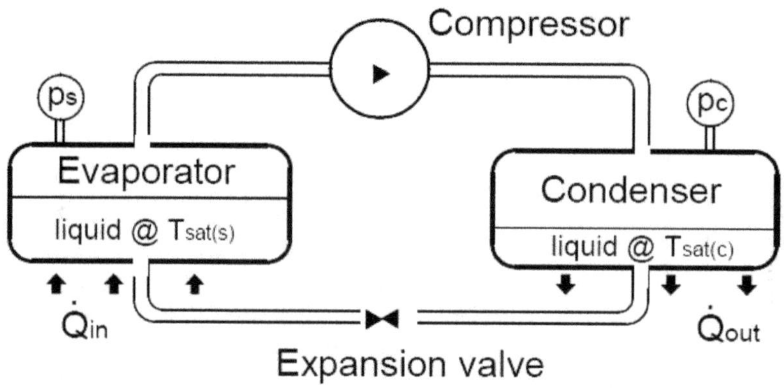

Control of the liquid flow rate is needed to ensure that it equals the vapour formation rate, and an appropriate balance of liquid quantities in the evaporator and condenser is maintained. When the liquid passes through the expansion valve it experiences a sudden drop in pressure, which causes instantaneous boiling (known as flashing). Vapour is formed using the liquid's sensible heat, which causes the liquid to drop in temperature to Tsat(s). A saturated liquid/vapour mixture will enter the evaporator. Figure 8 explains this cycle in practice.

System Performance

The system balance requires the overall work done to be equivalent to the net energy used by the system. Hence,

$$Q_{out} - Q_{in} = W_{in} \quad (5)$$

For operation as a refrigerator, a measure of system performance is the amount of heat absorbed per unit work supplied to drive the system. This is known as the Coefficient of Performance (Omer, 2009c).

$$COP_{ref} = Q_{in} / W_{in} \quad (6)$$

For operation as a heat pump, a measure of system performance is the amount of heat delivered per unit work supplied to drive the system. This is known as the Coefficient of Performance (Omer, 2009d).

$$COP_{hp} = Q_{out} / W_{in} \quad (7)$$

It follows that (for the same system):

$$COP_{hp} = COP_{ref} + 1 \quad (8)$$

Vapour Compression Refrigeration

The term "vapour compression refrigeration" is somewhat of a misnomer, it would be more accurately described as 'vapour suction refrigeration'. Vapour compression is used to reclaim the refrigerant and is more aptly applied to heat pumps. Vapour compression refrigeration exploits the fact that the boiling temperature of a liquid is intimately tied to its pressure. Generally, when the pressure on a liquid is raised its boiling (and condensing) temperature rises, and vice-versa. This is known as the saturation pressure-temperature relationship.

Refrigerant Properties

In practice, the choice of a refrigerant is a compromise, e.g., Ammonia is good but toxic and flammable while R12 is very good but detrimental to the Ozone layer. Figure 9 shows some commonly used refrigerants and their typical ranges of usability.

Ideally, a refrigerant will have the following characteristics.

- **Non-toxic:** For health and safety reasons.
- **Non-flammable:** To avoid risks of fire or explosion.
- **Operate at modest positive pressures:** To minimise pipe and component weights (for strength) and avoid air leakage into the system.
- **Have a high vapour density:** To keep the compressor capacity to a minimum and pipe diameters relatively small.
- **Easily transportable:** Because refrigerants are normally gases at SSL conditions they are stored in pressurised containers.
- **Environmentally friendly:** Non-polluting & non-detrimental to the atmosphere, water or ground.
- Easily re-cycleable, and relatively inexpensive to produce.
- Compatible with the materials of the refrigeration system - non-corrosive, miscible with oil, and chemically benign.

Figure 9. Refrigerant chart

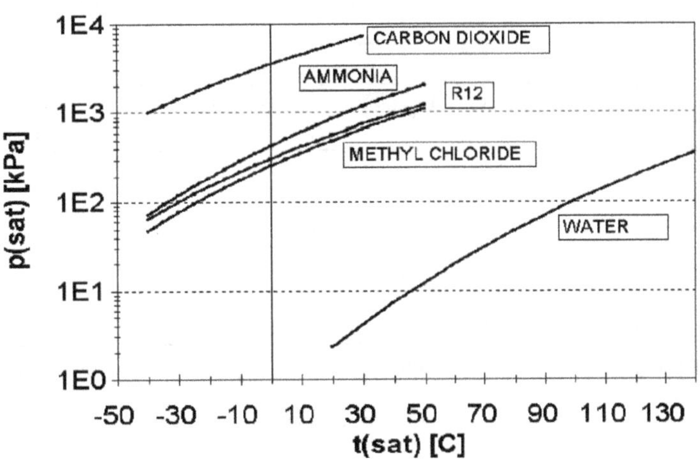

Cooling Mode

In the cooling mode, cool vapour arrives at the compressor after absorbing heat from the air in the building. The compressor compresses the cool vapour into a smaller volume, increasing its heat density. The refrigerant exits the compressor as a hot vapour, which then goes into the earth loop field. The loops act as a condenser condensing the vapour until it is virtually all liquid. The refrigerant leaves the earth loops as a warm liquid. The flow control regulates the flow from the condenser such that only liquid refrigerant passes through the control. The refrigerant expands as it exits the flow control unit and becomes a cold liquid.

Heat Pump Antifreeze

A potential negative effect of all geothermal heat pumps is the release of antifreeze solutions to the environment. Antifreeze solutions are required in colder climates to prevent the circulating fluid from freezing. Antifreeze chemicals include methanol, ethanol, potassium acetate, propylene glycol, calcium magnesium acetate (CMA), and urea. These chemicals are generally mixed with water when used as a heat exchange fluid. These chemicals can be released to the environment via spills or corrosion of system components. Approved antifreezes include methanol, ethanol, propylene glycol, calcium chloride, or ethylene glycol. These antifreezes must be mixed with water, at concentrations of 20% or less. Geothermal heat pumps for a single-family residence and the antifreezes for these units were evaluated by Heinonen et al., (1996) (Heinonen, Tapscott, Wildin, & Beall, 1996). These authors evaluated total energy consumption, corrosion due to the antifreeze, and the operational and environmental effects of six antifreeze solutions, namely methanol, ethanol, potassium acetate, propylene glycol, CMA, and urea. However, the excluded salt solutions, such as sodium and calcium chloride, from their study because they pose serious potential corrosion problems. The differences in total energy consumption for the studied antifreezes were considered minimal. Nevertheless, Heinonen et al., recommended that propylene glycol was a good choice based on its low health, fire, and environmental risks (Table 1). Unfortunately, these authors did not assess the leak potential of these antifreezes in the plastic pipe (e.g., HDPE & CPVC SDR-11) commonly used for the ground loop (Sims, 2007).

Table 1. Cost and risk factors for heat pump antifreeze (Heinonen et al., 1996)

Factor	Antifreeze					
	Methanol	Ethanol	Propylene Glycol	Potassium Acetate	CMA	Urea
Life cycle cost	3	3	2	2	2	3
Corrosion risk	2	2	3[a]	2	2	1
Leakage risk	3	2	2[a]	1[b]	1	1
Health risk	1	2	3	3	3	3
Fire risk	1[a]	1[c]	3	3	3	3
Environmental risk	2	2	3	2	2	3
Risk of future use	1	2	3	2	2	2

Notes: Ratings- 1 means potential problems and caution required, 2 means minor potential for problems, 3 means little or no potential problems.

a) DOWFROST HD; b) GS-4; c) Pure fluid only. Diluted antifreeze (25% solution) is rated 3.

However, the bond between the grout and borehole can be compromised by desiccation of the geologic materials near the borehole, as the heat from the borehole lowers the moisture content of the geologic materials and these materials contract. In areas with thick unsaturated zones, the bentonite grout may dry out over time, compromising the seal. To improve heat exchange, some advocate the use of spacers, which moves the heat conductor pipe to the side of the borehole, putting it in contact with the geologic materials. However, the use of spacers appears to increase the environmental risk of antifreeze leaking into groundwater, by reducing or removing the bentonite between the heat conductor pipe and geologic materials.

AIR DISTRIBUTION

The air distribution system can make a big difference in both the cost and the effectiveness of geothermal heating and cooling. It also has an important effect on personal comfort and health. The air-handling component is either a separate cabinet or is part of the cabinet that houses the geothermal heat pump, and includes the blower assembly that forces air through the ductwork. The supply ductwork carries air from the air handles to the rooms. Typically, each room has at least one supply duct and larger rooms may have several. The return ductwork moves air from the room back to the air handler. Most buildings have one or more main return ducts located in a central area. The cold liquid refrigerant is circulated through the air handler where it absorbs and removes the unwanted heat from the air and vaporises the refrigerant to a gas. The gas is compressed to increase its temperature and then the underground/underwater coils act as a condenser rather than an evaporator (as in the heating cycle) (Figure 10). The heat in the refrigerant is transferred to the ground/water as the refrigerant condenses.

Figure 10. Heating and cooling operations

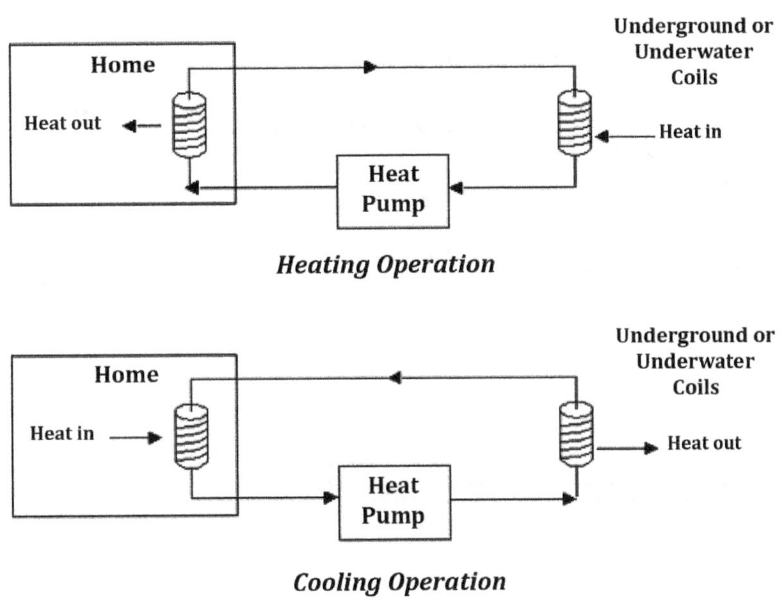

Refrigerants are present in the GSHP systems and so present the threat of the HCFCs and toxicity. However, new types and blends of refrigerant (some using CO_2) with minimal negative impacts are approaching the market as shown in Table 2. Because the GSHPs raise the temperature to around 40°C they are most suitable for underfloor heating systems or low-temperature radiators, which require temperatures of between 30° and 35°C. Higher outputs, such as to conventional radiators requiring higher temperatures of around 60° to 80°C can be obtained through use of the GSHP in combination with a conventional boiler or immersion heater.

The GSHPs come in 15 models from 4 kW up to 30 kW (even up to 300 kW when connected in parallel). At least 65% of the heating and hot water energy consumption of a house can be saved (65-75% of heating costs with a heat pump) as a result of using such a system. However, sizing of the heat pump and the ground loops is essential for the efficient operation of the system. If sized correctly, a GSHP can be designed to meet 100% of space heating requirements. The sizing of the system is very sensitive to heat loads and should therefore be installed into properties with high-energy efficiency standards, particularly new build (Trevor, 2007; and Steele, 1997). It is a good and practical idea to explore ways of minimising space heating and hot water demand by incorporating energy efficiency measures (Figure 11). This is known as the saturation pressure-temperature relationship (Figure 12). The refrigerant exits the compressor as a hot vapour, which then goes into the earth loop field (Figure 13).

Table 2. CO_2 emissions

System	Primary Energy Efficiency (%)	CO_2 emissions (kg CO_2/kWh heat)
Oil fired boiler	60 – 65	0.45 – 0.48
Gas fired boiler	70 – 80	0.26 – 0.31
Condensing gas boiler + low temperature system	100	0.21
Electrical heating	36	0.9
Conventional electricity + GHSP	120-160	0.20-027
Green electricity + GHSP	300-400	0.00

Figure 11. Schematic of heat transfer through a circular tube heat exchanger

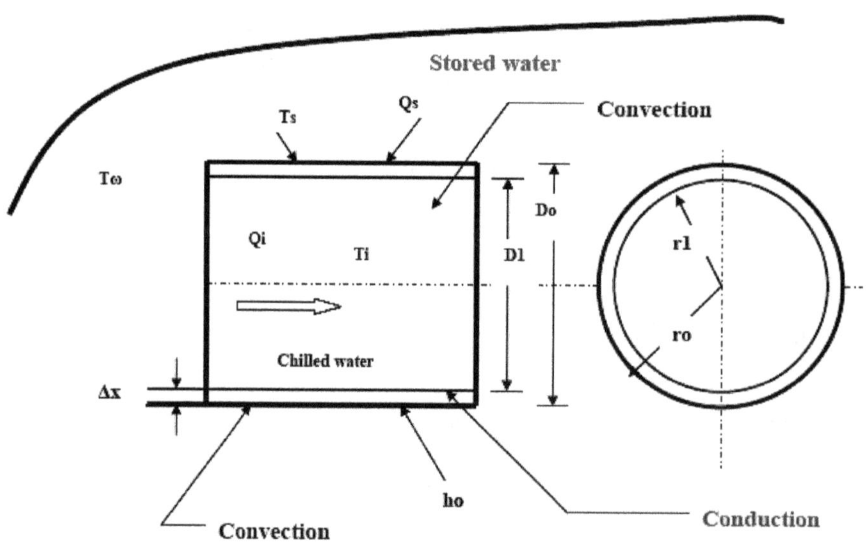

Figure 12. The ideal cycle on Pressure-Enthalpy diagram

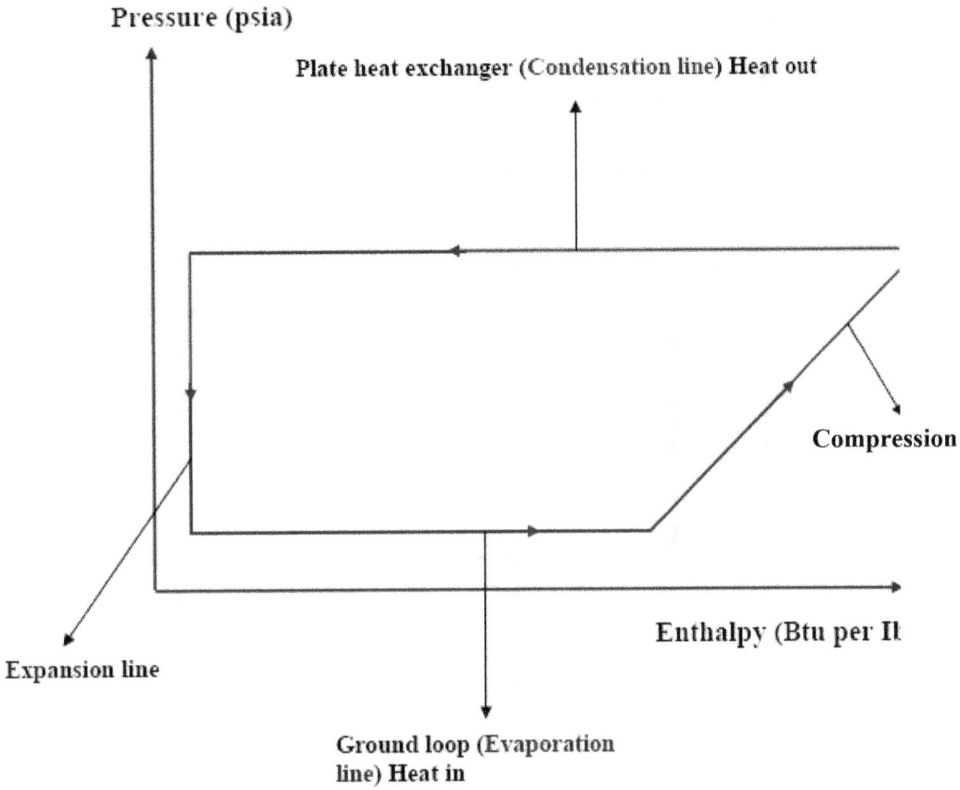

Some Definitions

1. The word "Efficiency" is defined as the ratio of useful heat output to energy input, e.g., if an open fireplace loses half its energy up the chimney it is said to be 50% efficient.
2. The COP or "Coefficient of performance" is found by dividing the useful heat output by the energy input, e.g., a heat pump that produces 3 kWatts of heat for 1 kWatt of input power has a COP of 3. The open fireplace example with 50% efficiency would have a COP of 0.5 (1/2).
3. The heat "Source" is the outside air, river or ground, wherever the heat is being extracted from. Sometimes is referred to as an ambient source.
4. The "Sink" is the name given to the part where the heat is usefully dissipated, such as radiators in the room, underfloor heating, hot water cylinder, etc.

Horizontal Collector

This can be either coiled 'Slinky' or straight pipes that are buried 1.5 m to 2 m deep in open ground (in gardens). The pipe is usually plastic and contains a Glycol antifreeze solution.

Figure 13. Water undergoing a charge of state

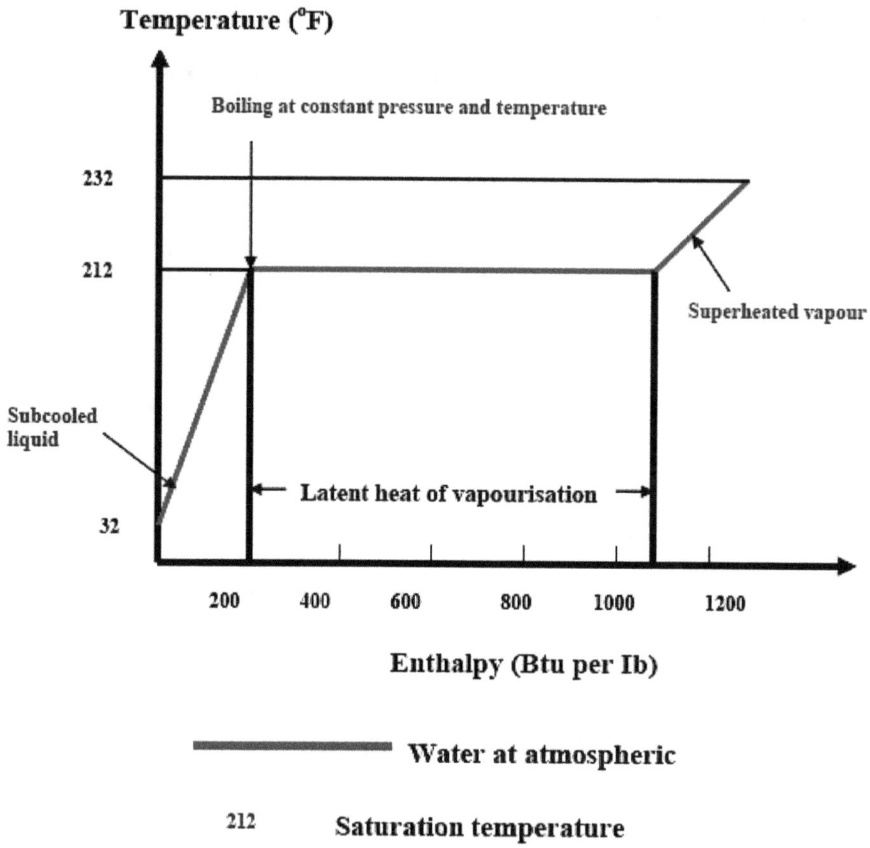

Antifreeze

This is simply an additive to water that makes its freezing point lower. Common salt does the same thing, but Ethylene or Propylene Glycol is more practical for heat pump systems.

Refrigerant

This is the working fluid within the heat pump. It evaporates in one part and condenses in another. By doing so, heat is transferred from cold to hot. This fluid is sealed in and will not degrade within the heat pumps life.

Heat Exchanger

This is a simple component that transfers heat from one fluid to another. It could be liquid-to-liquid, or liquid-to-air, or air-to-air. Two heat exchangers are housed within the heat pump, one for the hot side (the condenser), and one for the cold side (the evaporator).

Slinky

The name is given to the way that ground collector pipes can be coiled before buying in a trench.

Passive Heat Exchange

When waste hot water preheats cold input water, it is said to be 'passive'. This costs nothing to run. A heat pump is said to be 'active' it can extract heat from cold waste water but requires a relatively small power input.

Some Refrigeration Characteristics

The seasonal energy efficiency ratio (SFEE) may be applied to each of the components.

Assuming that KE & PE effects are negligible, i.e., the SSFEE is applicable; vis

$$Q + W = m\ \Delta h \tag{9}$$

Compressor

Compression assumed adiabatic:

$$\therefore Q = 0 \tag{10}$$

$$W12 = m\ (h2 - h1) \tag{11}$$

Or

$$W_{in} = m\ (h2 - h1) \tag{12}$$

Condenser:

$$W23 = 0 \tag{13}$$

$$\therefore Qout = m\ (h2 - h3) \tag{14}$$

Expansion valve:

$$W34 = 0\ \&\ Q34 = 0 \tag{15}$$

$$\therefore h3 = h4 \tag{16}$$

Evaporator:

$$W41 = 0 \tag{17}$$

$$\therefore Q_{in} = m\,(h1\text{-}h4) \tag{18}$$

Refrigeration Effect

It follows that:

$$COP_{ref} = (h1\text{-}h3)/(h2\text{-}h1) \tag{19}$$

$$COP_{hp} = (h2\text{-}h3)/(h2\text{-}h1) \tag{20}$$

In order to determine the above equations, the specific enthalpy values will be needed. Because refrigerants work in the liquid/vapour phases appropriate property charts or tables must be used.

The Ideal Refrigeration Cycle

- Isentropic compression ($1 \rightarrow 2$)
- Constant pressure cooling/condensation ($2 \rightarrow 3$)
- Throttling ($3 \rightarrow 4$)
- Constant pressure vaporisation/heating ($4 \rightarrow 1$)

The ideal refrigeration cycle plotted on the p-h chart as shown in Figure 14.

Figure 14. Refrigeration cycle

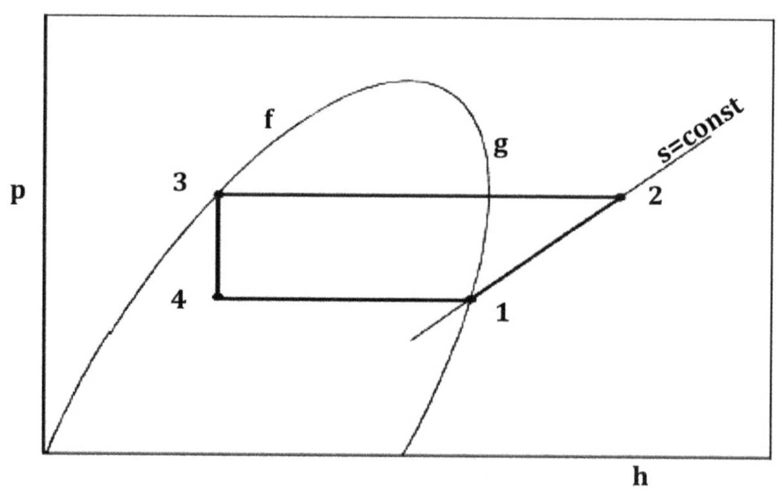

Real Refrigeration Systems

Evaporator superheat

g → 1 given in K above Tsat(s)

Isentropic Compressor Efficiency

η_{isen} = h2'-h1/h2-h1 (21)

Condenser Sub-Cooling

f → 3 given in K below Tsat(c)

NB: s1 = s2'

Refrigerant Properties (Charts and Tables)

Because refrigeration systems basically work between two pressures, and specific enthalpy is one of the most useful properties we need, refrigerant thermodynamic properties are normally presented in the form of a pressure - specific enthalpy (or p-h) chart.

This is done for convenience, and is simply an alternative way of presenting property data, instead of, e.g., p-V, or T-s, or h-s (Figures 15, 16, and 17).

Other useful properties are also shown on the chart, vis: specific entropy, specific volume, temperature and quality. Regard these properties as 'contours'.

The pressure axis (y-axis) is typically logarithmic.

Figure 15. Evaporator superheat

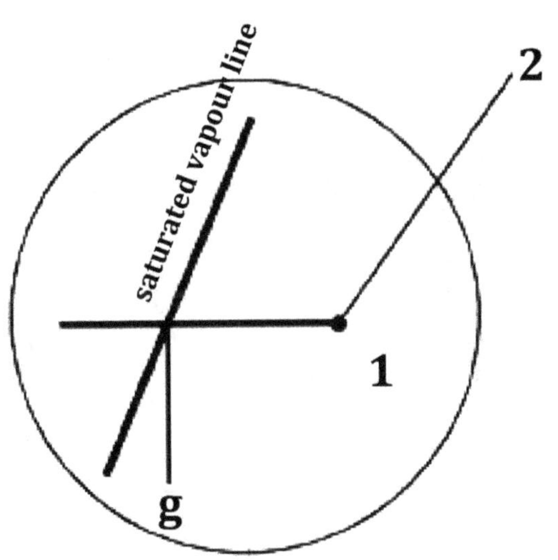

Figure 16. Isentropic compressor

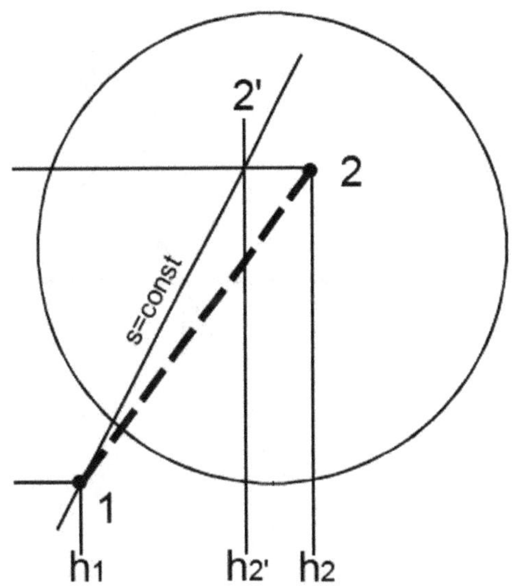

3.6. Pressure Drops in Evaporator and Condenser

Clearly, any or all of the above effects can be present, but the pressure drops are often small enough to be neglected (Figure 18).

Refrigeration System Performance Improvement

Liquid-Suction heat exchanger can be seen in Figures 19 and 20.

Assuming no losses:

$$H_{1b}\text{-}h_{1a} = h_{3a}\text{-}h_{3b} \quad (22)$$

Multiple Compression Using Flash Chambers

Figure 21 shows an illustration of the cycle.

Diagram of Multiple Compressions Using Flash Chamber

At point 3a, have a mixture of vapour and liquid, which is separated, in the flash chamber (Figure 21). The proportion of the total mass flow that is liquid (and proceeds to the evaporator) is given by:

$$x_f = h_{3\,(g)}\text{-}h_3/h_{3\,(g)}\text{-}h_{3\,(f)} \quad (23)$$

Figure 17. Condenser sub-cooling

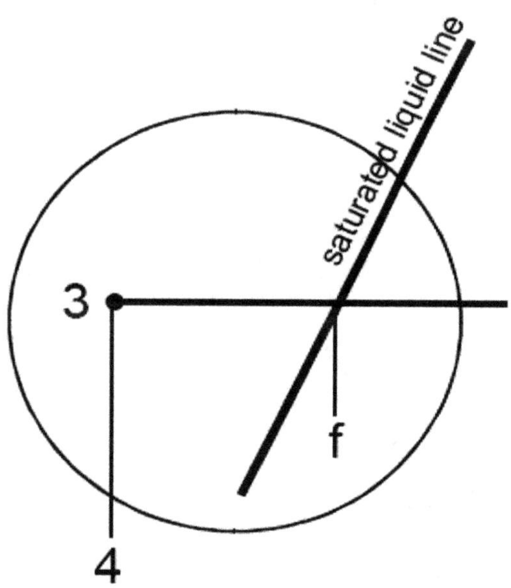

Figure 18. Pressure drops in evaporator and condenser

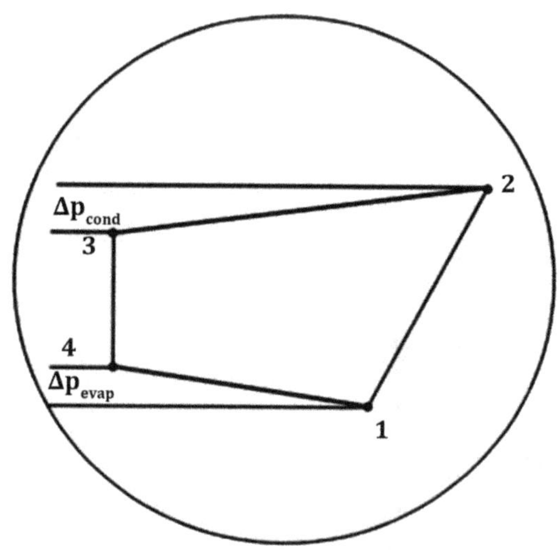

Figure 19. Diagram of liquid-suction heat exchanger

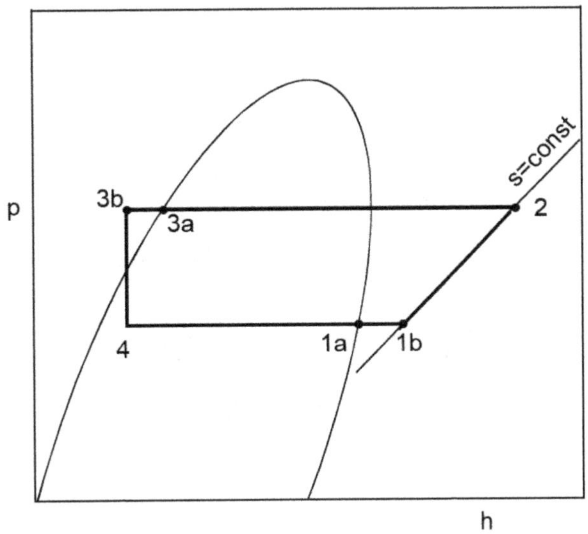

Figure 20. Liquid-suction heat exchanger cycle

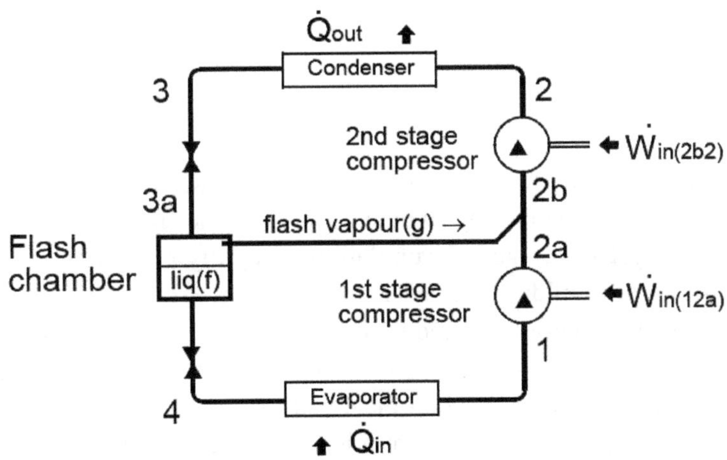

The remaining vapour mixes with the discharge from the first stage compressor to give different inlet conditions to the second stage.

Assuming adiabatic mixing:

$$1^* h_{2b} = x_f h_{2a} + (1-x_f)\, h3_{(g)} \tag{24}$$

A similar equation can be used to find s_{2b}. Finally the COP is given by:

$$COP = (x_f (h_1-h_4))/ (x_f (h_{2a}-h_1)+(h_2-h_{2b})) \tag{25}$$

Figure 21. Cycle of multiple compressions using flash chamber

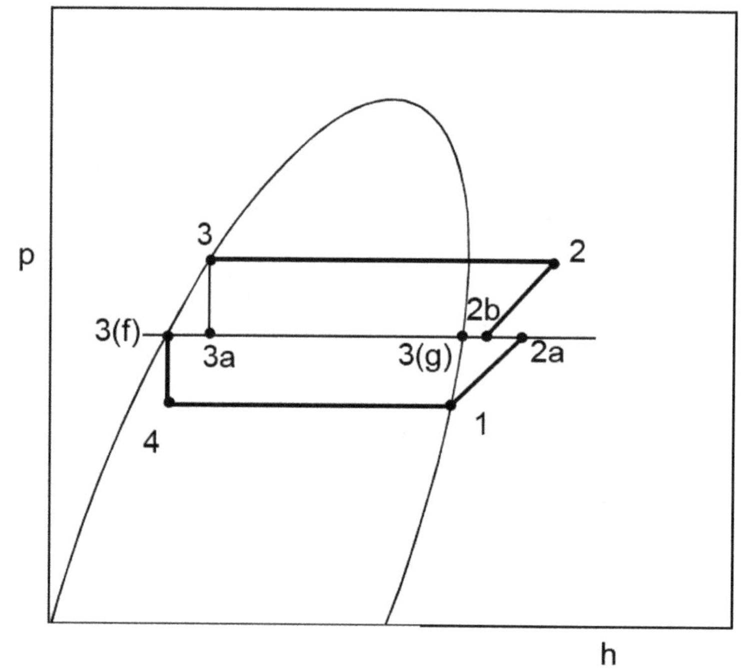

DISCUSSIONS

Thermal comfort is an important aspect of human life. Buildings where people work require more light than buildings where people live. In buildings where people live the energy is used for maintaining both the temperature and lighting. Hence, natural ventilation is rapidly becoming a significant part in the design strategy for non-domestic buildings because of its potential to reduce the environmental impact of building operation, due to lower energy demand for cooling. A traditional, naturally ventilated building can readily provide a high ventilation rate. On the other hand, the mechanical ventilation systems are very expensive. However, a comprehensive ecological concept can be developed to achieve a reduction of electrical and heating energy consumption, optimise natural air condition and ventilation, improve the use of daylight and choose environmentally adequate building materials. Energy efficiency brings health, productivity, safety, comfort and savings to homeowner, as well as local and global environmental benefits. The use of renewable energy resources could play an important role in this context, especially with regard to responsible and sustainable development. It represents an excellent opportunity to offer a higher standard of living to local people and will save local and regional resources. Implementations of GSHPs offer a chance for maintenance and repair services. It is expected that the pace of implementation will increase and the quality of work to improve in addition to building the capacity of the private and district staff in contracting procedures. The financial accountability is important and more transparent. Various passive techniques have been put in perspective, and energy saving passive strategies can be seen to reduce interior temperature and increase thermal comfort, and reducing air conditioning loads. The scheme can also be employed to analyse the marginal contribution of each specific passive measure working under realistic conditions in combination with the other housing elements. In regions where heating is important during winter months, the use of top-light solar passive strategies for spaces without an equator-facing façade can efficiently reduce energy consumption for heating, lighting and ventilation. The use of renewable energy resources could play an important role in this context, especially with regard to responsible and sustainable development. It represents an excellent opportunity to offer a higher standard of living to local people and will save local and regional resources. Implementation of greenhouses offers a chance for maintenance and repair services. Various passive techniques have been put in perspective, and energy saving passive strategies can be seen to reduce interior temperature and increase thermal comfort, and reducing air conditioning loads.

Renewable energy is the term to describe a wide range of naturally occurring, and replenishing energy sources. The use of renewable energy sources and the rational use of energy are the fundamental inputs for a responsible energy policy. The energy sector is encountering difficulties because increased production and consumption levels entail higher levels of pollution and eventually climate changes, with possibly disastrous consequences. Moreover, it is important to secure energy at acceptable cost to avoid negative impacts on economic growth. On the technological side, renewables have an obvious role to play. In general, there is no problem in terms of the technical potential of renewables to deliver energy and there are very good opportunities for renewable energy technologies to play an important role in reducing emissions of greenhouse gases into the atmosphere-certainly far more than have been exploited so far. But there are still technical issues to be addressed to cope with the intermittency of some renewables, particularly wind and solar. However, the biggest problem with replying on renewables to deliver the necessary cuts in greenhouse gas emissions is more to do with politics and policy issues than with technical ones. The single most important step governments could take to promote and increase the use of renewables would be to improve access for renewables to the energy market. That access to the market would need to be under favourable conditions and possibly under favourable economic rates. One move that could help-or at least justify-better market access would be to acknowledge that there are environmental costs associated with other energy supply options, and that these costs are not currently internalised within the market price of electricity or fuels. It could make significant difference, particularly if, appropriate subsidies were applied to renewable energy in recognition of environmental benefits it offers. Cutting energy consumption through end-use efficiency is absolutely essential. And this suggests that issues of end-use consumption of energy will have to come onto the table in the foreseeable future.

The scientific consensus is clear-climate change is occurring. Existing renewable energy technologies could play a significant mitigating role, but the economic and political climate will have to change first. Climate change is real, it is happening now, and greenhouse gases produced by human activities are significantly contributing to it. The predicted global temperature changes of between 1.5 and 4.5 degrees C could lead to potentially catastrophic environmental impacts-including sea level rise, increased frequency of extreme weather events, floods, droughts, disease migration from various places and possible stalling of the Gulf stream. This is why scientists argue that climate change issues are not ones that politicians can afford to ignore. And policy makers tend to agree, but reaching international agreements on climate change policies is no trivial task.

The most favourable orientation, which is due north, results in diminished excessive solar gains through the windows. However, most buildings cannot be oriented at will. If the only possible orientation is due south, and no external shade is used, the index reveals extra heat gains of some 0.26 over the value of totally shaded window. Application of the model results from exploring the relative importance of the thermal inertia of walls, floor and ceiling. Heat stored in building materials, as proven in old, massive buildings, can be compensated during high insolation hours with thermal losses at night and early morning hours, when ambient temperatures are below 25°C. Temperature variation will be lower for higher thermal capacities of building materials. However, it is known while thermal capacity increases the relative importance of individual heat flows change. For example, for lower wall temperatures, the contribution of radiative heat transfer will be reduced, and the relative importance of convective processes will increase, and thus the difficulty to calculate accurately the overall heat flows. The relevance of certain passive techniques is variable with prevailing weather.

Finally, the required temperature dependent air transport properties were evaluated by the following expression, which are valid between 2°C and 77°C with temperature expressed in k:

- **Thermal diffusivity:** $\alpha = 1.534 \times 10^{-3}\, T - 0.2386\ (10^{-4}\ m^2 s^{-1})$
- **Kinematics viscosity:** $v = 0.1016\, T - 14.8\ (10^{-6}\ m^2 s^{-1})$
- **Thermal conductivity:** $k = 7.58 \times 10^{-5}\, T + 3.5 \times 10^{-3}\ (Wm^{-1}K^{-1})$, and
- **Thermal expansion coefficient:** $\beta = T^{-1}\ (K^{-1})$

In order to depict the relative contribution of each of these techniques to inside temperature, a dimensionless index is defined as follows. When interior temperature exceeds 25°C, it will be considered as a temperature discomfort condition. This reference temperature is widely elements. Then the following expression:

$$F\,(t) = \max\,(T_t - 25.25) \tag{26}$$

I_s a time function of truncated temperature and it will be able to estimate the overall discomfort by means of the integration along the day for each different scenarios S:

$$A\,(S) = \int_S F(t)\, dt \tag{27}$$

Then, for each passive technique, let:

$$A_{max} = \max\,[A(S)\text{: for all scenarios } S] \tag{28}$$

Finally, the normalised temperature index for each scenario S is:

$$I(S) = A(S)/A_{max} \tag{29}$$

Naturally, it would be preferred, for comfort reasons that this index would be small, preferably nil. It may be seen that the variable is directly related to temperature discomfort: the larger the value of the index, the farthest will inside conditions be from expected wellbeing. Also, the use of electricity operated air conditioning systems will be more expensive the higher this variable is. Hence, energy expenditure to offset discomfort will be higher when comparing two index values; the ratio of them is proportional to the expected energy savings. When the external shade blocks the windowpane completely, the excessive heat gains belong to the lowest values in the set, and the dimensionless index will be constant with orientation. For the climate conditions of the locality, it can be seen that a naked window can produce undesirable heat gains if the orientation is especially unfavourable, when the index can have an increase of up to 0.3 with respect to the totally shaded window.

CONCLUSION

With increasing worldwide awareness of the serious environmental problems due to fossil fuel consumption, efforts are being made to develop energy efficient and environmentally friendly systems by utilisation of non-polluting renewable energy sources, such as solar energy, industrial waste heat or geothermal water. The GSHPs are suitable for heating and cooling of buildings and so could play a significant role in reducing CO_2 emissions. Ground source or geothermal heat pumps are a highly efficient, renewable energy technology for space heating and cooling. This technology relies on the fact that, at depth, the Earth has a relatively constant temperature, warmer than the air in winter and cooler than the air in summer.

A geothermal heat pump can transfer heat stored in the earth into a building during the winter, and transfer heat out of the building during the summer. Furthermore, special geological conditions, such as hot springs, are not needed for successful application of geothermal heat pumps. The GSHPs are receiving increasing interest because of their potential to reduce primary energy consumption and thus reduce emissions of the GHGs. The GSHP is generally recognised to be one of the most outstanding technologies of heating and cooling in both residential and commercial buildings, because it provides high coefficient of performance (COP), up to 3-4 for an indirect heating system and 3.5-5 for a direct heating system.

The main benefit of using the GSHPs is that the temperature of the subsurface is not subject to large variations experienced by air. It is currently the most common thermal energy source for the heat pumps, and so would allow construction of more efficient systems with superior performance. The GSHPs do not need large cooling towers and their running costs are lower than conventional heating and air conditioning systems. As a result, the GSHPs have increasingly been used for building heating and cooling with annual rate of increase of 10% in recent years.

REFERENCES

Anandarajah, A. (2003). Mechanism controlling permeability changes in clays due to changes in pore fluids. *Journal of Geotechnical and Geoenvironmental Engineering*, *129*(2), 163–172. doi:10.1061/(ASCE)1090-0241(2003)129:2(163)

Brain, G., & Mark, S. (2007). Garbage in, energy out: Landfill gas opportunities for CHP projects. *Cogeneration and On-Site Power*, *8*(5), 37–45.

Heinonen, E.W., Tapscott, R.E., Wildin, M.W., & Beall, A.N. (1996). *Assessment of antifreeze solutions for ground-source heat pumps systems*. New Mexico Engineering Research Institute NMERI 96/15/32580.

Jo, H. Y., Katsumi, T., Benson, C. H., & Edil, T. B. (2001). Hydraulic conductivity and swelling of non-prehydrated GCLs permeated with single- species salt solutions. *Journal of Geotechnical and Geoenvironmental Engineering*, *127*(7), 557–567. doi:10.1061/(ASCE)1090-0241(2001)127:7(557)

Luo, L., Fan, Y., & Tondeur, D. (2007). Heat exchanger: From micro to multi- scale design optimisation. *International Journal of Energy Research*, *31*(13), 1266–1274. doi:10.1002/er.1298

Luo, L., Tondeur, D., Le Gall, H., & Corbel, S. (2007). Constructal approach and multi- scale components. *Applied Thermal Engineering*, *27*(10), 1708–1714. doi:10.1016/j.applthermaleng.2006.07.018

Malin, N., & Alex, W. (2000). Ground-Source Heat Pumps: Are They Green? Environmental Building News, 9(7-8).

Omer, A. M. (2008). Development of integrated bioenergy for improvement of quality of life of poor people in developing countries. In F. L. Magnusson & O. W. Bengtsson (Eds.), *Energy in Europe: Economics, Policy and Strategy- IB* (pp. 341–373). New York, USA: NOVA Science Publishers, Inc.

Omer, A. M. (2009a). Environmental and socio-economic aspect of possible development in renewable energy use. In *Proceedings of the 4th International Symposium on Environment.*

Omer, A. M. (2009b). Energy use, environment and sustainable development. In *Proceedings of the 3rd International Conference on Sustainable Energy and Environmental Protection (SEEP 2009).*

Omer, A. M. (2009c). Energy use and environmental impacts: a general review. *Journal of Renewable and Sustainable Energy, 1.*

Omer, A. M. (2009d). Energy use, environment and sustainable development. In R. T. Mancuso (Ed.), *Environmental Cost Management* (pp. 129–166). New York: NOVA Science Publishers, Inc.

Philappacopoulus, A. J., & Berndt, M. L. (2001). Influence of de-bonding in ground heat exchangers used with geothermal heat pumps. *Geothermics, 30*(5), 527–545. doi:10.1016/S0375-6505(01)00011-6

Sims, R. H. (2007). Not too late: IPCC identifies renewable energy as a key measure to limit climate change. *Renewable Energy World, 10*(4), 31–39.

Steele, J. (1997). *Sustainable architecture: principles, paradigms, and case studies.* New York: McGraw-Hill Inc.

Trevor, T. (2007). Fridge recycling: Bringing agents in from the cold. *Waste Management World, 5*, 43–47.

Chapter 32
Modeling and Monitoring of Chemical System: CSTR Model

Majdi Mansouri
Texas A&M University at Qatar, Qatar

Hazem Numan Nounou
Texas A&M University at Qatar, Qatar

Mohamed Numan Nounou
Texas A&M University at Qatar, Qatar

ABSTRACT

This chapter addresses the problem of time-varying nonlinear modeling and monitoring of a continuously stirred tank reactor (CSTR) process using state estimation techniques. These techniques include the extended Kalman filter (EKF), particle filter (PF), and the more recently the variational Bayesian filter (VBF). The objectives of this chapter are threefold. The first objective is to use the variational Bayesian filter with better proposal distribution for nonlinear states and parameters estimation. The second objective is to extend the state and parameter estimation techniques to better handle nonlinear and non-Gaussian processes without a priori state information, by utilizing a time-varying assumption of statistical parameters. The third objective is to apply the state estimation techniques EKF, PF and VBF for time-varying nonlinear modeling and monitoring of CSTR process. The estimation performance is evaluated on a synthetic example in terms of estimation accuracy, root mean square error and execution times.

INTRODUCTION

Many process operations, such as modeling, monitoring, and control, require the availability of key process variables or parameters that are essential for the effective implementation of these operations. In some cases, however, measuring or estimating these variables or parameters might be practically difficult or costly. For example, it is challenging to accurately measure the concentrations inside a reactor or the compositions in a distillation column online, which are needed to control and/or monitor these processes.

DOI: 10.4018/978-1-4666-9644-0.ch032

In such cases, state estimation techniques are usually used to estimate the sought variables (called state variables) online from measurements of other variables(s). State estimation techniques can also be used to model various processes. For example, deriving dynamic models of chemical reactors can be achieved using material and energy balances. However, such models contain several parameters (such as the reaction rate constant, reaction order, heat transfer coefficient, heat of reaction, and others) that may not be known. Determining these parameters usually requires several experimental setups that can be challenging and expensive in practice. Estimating these parameters using state estimation techniques can be of a great value. In this chapter, state estimation will be utilized to model and monitor a continuously stirred tank reactor (CSTR) by estimating its model parameters and state variables (concentration and temperature inside the reactor), respectively. A comparative analysis involving various conventional as well as state-of the-art state estimation techniques will be performed to achieve this objective.

The contributions of this chapter are four-fold. The first contribution is to develop a variational Bayesian filter algorithm for nonlinear and non-Gaussian estimation. In case of standard PF, the latest observation is not considered for the evaluation of the weights of the particles as the importance function is taken to be equal to the prior density function. The second contribution is to extend the state and parameter estimation techniques (i.e., EKF, PF, and VBF) to better handle nonlinear and non-Gaussian processes without a priori state information, by utilizing a time-varying assumption of statistical parameters. The third contribution of this chapter is to investigate the effects of practical challenges on the performances of the EKF, PF, and VBF algorithms. The comparative analysis is conducted to study the effect of the number of states and parameters to be estimated on the estimation performances of various estimation techniques: EKF, PF, and VBF. Similarly, to investigate the effect of measurement noise on the estimation performances, several measurement noise contributions (e.g., different signal-to-noise ratios) will be considered. Then, the estimation performances of EKF, PF, and VBF will be compared for different noise levels. Also, to study the effect of the number of states and parameters to be estimated on the estimation performances of EKF, PF, and VBF, the estimation performance will be analyzed for different numbers of estimated states and parameters. The fourth contribution of this chapter is to apply the estimation techniques (EKF, PF, and VBF) to estimate the states and parameters of a chemical model representing the CSTR process.

The rest of the chapter is organized as follows. We first discuss related works and motivate the need for our proposed work in Section II. In Section III, a statement of the problem addressed in this chapter is presented. In Section IV, a description of variational Bayesian filter is presented. Then, in Section V, the performances of the various state estimation techniques are compared through their application to estimate the state variables and model parameters of a CSTR process. Finally, some concluding remarks are presented in Section VI.

RELATED WORKS

Several state estimation techniques have been developed and used to achieve this objective. These techniques include the extended Kalman filter (EKF) (Lillacci & Khammash, 2010), the unscented Kalman filter (UKF) (Gustafsson & Hendeby, 2012), the particle filter (PF) (Arulampalam et al., 2005), and more recently the variational Bayesian filter (VBF) (Mansouri et al., 2014). The classical Kalman Filter (KF) was developed in the 1960s (Kalman, 1960), and has been widely applied in various engineering and science areas, including communications, control, machine learning, neuroscience, and many others. In the case where the model describing the system is assumed to be linear and Gaussian, the KF provides an optimal solution (Simon, 2006; Grewal & Andrews, 2008). KF has also been formulated in the context of Takagi-Sugeno fuzzy systems, which can be described by a convex set of multiple linear models (Chen et al., 1998; Simon, 2003). It is known that KF is computationally efficient; however, it is limited by the non-universal linear and Gaussian modeling assumptions. To relax such assumptions, the Extended Kalman Filter (Simon, 2006; Grewal & Andrews, 2008; Julier et al., 1997; Ljung et al., 1979; Kim et al., 1994) and the Unscented Kalman Filter (Simon, 2006; Grewal & Andrews, 2008; Wan et al., 2000; Wan et al., 2001; Sarkka et al., 2001; Sarkka et al., 2007) have been developed. In extended Kalman filtering, the model describing the system is linearized at every time sample (which means that the model is assumed to be differentiable). Therefore, for highly nonlinear models, EKF does not usually provide a satisfactory performance. The UKF, on the other hand, instead of linearizing the model to approximate the mean and covariance matrix of the state vector, uses the unscented transformation to approximate these moments. In the unscented transformation, a set of samples (called sigma points) are selected and propagated through the nonlinear model to improve the approximation of these moments and thus the accuracy of state estimation. Other state estimation techniques use a Bayesian framework to estimate the state and/or parameter vector (Beal et al. 2003).

Other state estimation techniques use a Bayesian framework to estimate the state and/or parameter vector. The Bayesian framework relies on computing the probability distribution of the unobserved state given a sequence of the observed data in addition to the state evolution model.

Unfortunately, for most nonlinear systems and non-Gaussian noise observations, derivation of the closed form analytic expression of the posterior distribution of the state is untractable. To overcome this drawback, a non-parametric Monte Carlo sampling based method called particle filtering (Arulampalam et al., 2005; Mansouri et al., 2009; Lindenmeyer et al., 2002) has gained popularity. The Particle filter approximates the posterior probability distribution by a set of weighted samples, called particles. Since real-world problems usually involve high-dimensional random variables with complex uncertainty, the nonparametric and sample-based estimation of uncertainty (provided by the PF) has thus become quite popular to capture and represent the complex distribution P(z|y) for nonlinear and non-Gaussian process models (Arulampalam & Maskell, 2002). The PF has the ability to accommodate nonlinear and multi-modal dynamics, but at the cost of more computational complexity and storage requirements. Also, taking into account the stringent calculus and storage constraints, the propagation of a huge amount of particles has impeded the implementation of the PF in very challenging parameter estimation problems. A common problem in the particle filtering is the particle degeneracy, where almost all importance weights tend to zero after some iteration. To overcome this drawback, some modifications such as extended particle filter (Aggarwal & Nassar, 2007), Gaussian Particle Filter (GPF) (Liang et al., 2012), and unscented particle filter (Van Der Merwe et al., 2000, Van Der Merwe et al., 2001) have been proposed and their performance has been shown satisfactory.

Each of the above estimation methods has its advantages and disadvantages. The particle filtering can be applied to large parameter spaces, has better convergence properties and easier to implement than the UKF, and both of them can provide improved accuracy over the EKF (Mansouri et al., 2014). In addition, the variational Bayesian filter algorithm provides a significant improvement over the particle filtering because, unlike the PF which depends on the choice of sampling distribution used to estimate the posterior distribution, which also accounts for the observed data. The proposal sampling distribution is obtained by minimizing the Kullback-Leibler distance. But till now, the variational Bayesian filter algorithm has been used in (Mansouri et al., 2014, Mansouri, et al., 2013) to resolve the problem of state and parameter estimation, only where the statistical parameters are assumed to be constants.

PROBLEM STATEMENT

Proper operation of chemical reactors requires estimating or filtering some of the key process variables, such as the concentrations and temperature inside the reactor, given a dynamic model of the reacting system. The reactor's model structure can be usually derived from the basic conservation laws, namely material and energy balances. In many cases, however, the model parameters (such as the reaction rate constant, reaction order, heat transfer coefficient, heat of reaction, and others) are not known, and therefore, need to be estimated. Estimating these parameters, to completely define the reactor model, usually requires several experimental setups, which can be expensive and challenging in practice. In this chapter, we propose to use state estimation to simultaneously estimate the state variables (concentration and temperature) for process monitoring purposes as well as the reactor model parameters, which can simplify the task of modeling the chemical reactor compared to the conventional experimentally intensive methods. The idea here is that, if a dynamic model structure of the reactor is available, the model parameters as well as the state variables can be estimated using the process model and measurements of some of the process variables. As an example, consider the following dynamic model of a continuously stirred tank reactor (Mansouri et al., 2014):

$$\begin{cases} \frac{dC_A}{dt} = f_1\left(C_A, C_{Ai}, T, F, \theta\right) \\ \frac{dT}{dt} = f_2\left(T_i, T, C_A, F, \theta\right) \end{cases}$$

where, f_1 and f_2, are some nonlinear functions, C_{Ai} and C_A are the inlet and exit concentrations of a certain component i in the reactor, T_i and T are the temperatures of the inlet and exit streams, F is the volumetric flow rate into and out of the reactor, and θ is the unknown model parameter vector. If an experiment is designed in which the volumetric flow rate, F, is varied with time and then noisy dynamic measurements of the state variables (concentration and temperature) are collected, these measurements along with the dynamic process model can be used to estimate the unknown model parameters as well as to filter the measured state variables using a state estimation technique. Specifically, in this chapter, various state estimation techniques (which include EKF, PF, and VBF) will be used and compared when solving the following two problems:

1. Estimate the state variables (concentration and temperature inside a CSTR) given noisy measurements of these variables and a completely defined process model, including the model parameters.
2. Estimate the concentration and temperature inside the CSTR in addition to the model parameters, given noisy measurements of the state variables.

In problem 2) above, the effect of the number of estimated model parameters on the performances of the various state estimation techniques will also be assessed.

Problem Formulation

Here, the estimation problem of interest is formulated for a general system model. Let a nonlinear state space model be described as follows:

$$\begin{cases} \dot{x} = g\left(x, u, \theta, w\right) \\ y = l\left(x, u, \theta, v\right) \end{cases} \quad (1)$$

where, $x \in R^n$ is a vector of the state variables, $u \in R^p$ is a vector of the input variables (which can be changed as desired), $\theta \in R^q$ is an unknown parameter vector, $y \in R^m$ is a vector of the measured variables, g and l are nonlinear differentiable functions, and $w \in R^n$ and $v \in R^m$ are process and measurement noise, which quantify randomness in the process and errors in the measurements, respectively.

Discretizing the state space model (1), the discrete model can be written as follows:

$$\begin{cases} x_k = f\left(x_{k-1}, u_{k-1}, \theta_{k-1}, w_{k-1}\right) \\ y_k = h\left(x_k, u_k, \theta_k, v_k\right) \end{cases}, \quad (2)$$

which describes the state variables at some time step (k) in terms of their values at a previous time step (k–1). Note that in Equation (2). The process and measurement noise vectors have the following properties:

$$E[w_k]=0,\ E\left[w_k w_k^T\right] = Q_k,\ E[v_k]=0,\ E\left[v_k v_k^T\right] = R_k.$$

The function f is used to predict the value of the state vector at some time step (k) given its value at the previous time step (k–1), and the function h relates the measured vector (y_k) to the state vector (x_k) at the same time step. Also, defining the augmented vector, u_k is the vector of input variables, θ_k is a parameter vector (assumed to be known), y_k is the vector of the measured variables, w_k and v_k are respectively model and measurement noise vectors, and the matrices, Q_k and R_k, represent the covariance matrices of the process and measurement noise vectors, respectively. We assume that the error terms w_k and v_k have normal distributions with zero expectation, and that they are mutually independent.

Since we are interested to estimate the state vector x_k, as well as the parameter vector θ_k, let's assume that the parameter vector is described by the following model:

$$\theta_k = \theta_{k-1} + \gamma_{k-1} \quad (3)$$

where γ_{k-1} is white noise. In other words, the parameter vector model (3) corresponds to a stationary process, with an identity transition matrix, driven by white noise. We can define a new state vector that augments the two vectors together as follows:

$$z_k = \begin{bmatrix} x_k \\ \theta_k \end{bmatrix} = \begin{bmatrix} f\left(x_{k-1}, u_{k-1}, \theta_{k-1}, w_{k-1}\right) \\ \theta_{k-1} + \gamma_{k-1} \end{bmatrix} \tag{4}$$

where z_k is assumed to follow a Gaussian model as $z_k \sim N(\mu_k, \lambda_k)$, and where at any time k the expectation μ_k and the covariance matrix λ_k are both random. Also, defining the augmented vector,

$$\varepsilon_{k-1} = \begin{bmatrix} w_{k-1} \\ \gamma_{k-1} \end{bmatrix} \tag{5}$$

the model (2) can be written as:

$$\begin{aligned} z_k &= \Im\left(z_{k-1}, u_{k-1}, \varepsilon_{k-1}\right) \\ y_k &= \Re\left(z_k, u_k, v_k\right) \end{aligned} \tag{6}$$

Time-Variant Evolution Systems

Instead of the kinematic parametric model (Mansouri et al., 2014, Beal et al. 2003) which is usually used in estimation problems, we employ a time-variant evolution systems (TVES) (Beal et al. 2003). This model is more appropriate to practical non- linear and non-Gaussian situations where no a priori information on the state value is available. The state variable z_k at instant k is assumed to follow a Gaussian model, where the expectation μ_k and the precision matrix λ_k are both random. Gaussian distribution for the expectation and Whishart distribution for the precision matrix form a practical choice for the filtering implementation. The hidden state z_k is extended to an augmented state $\alpha_k = (z_k, \mu_k, \lambda_k)$, yielding a hierarchical model as follows,

$$\begin{cases} \mu_k \sim N\left(\mu_k \middle| \mu_{k-1}, \bar{\lambda}\right) \\ \lambda_k \sim W_n\left(\bar{S} \middle| \bar{n}\right) \\ z_k \sim N\left(\mu_k, \lambda_k\right) \end{cases} \tag{7}$$

where the fixed hyperparameters $\bar{\lambda}, \bar{S}$, and $\bar{n}$ are respectively the random walk precision matrix, the degrees of freedom and the precision of the Wishart distribution. Note that assuming random mean and covariance for the state z_k leads to a prior probability distribution covering a wide range of tail behaviors allowing discrete jumps in the state variable.

In fact, the marginal state distribution is obtained by integrating over the mean and precision matrix:

$$p(z_k \mid z_{k-1}) = \int p(z_k \mid \mu_k, \lambda_k) p(\mu_k, \lambda_k \mid x_{k-1}) d\mu_k \lambda_k \tag{8}$$

where the integration with respect to the precision matrix leads to the known class of scale mixture distributions introduced by Barndorff-Nielsen. Low values of the degrees of freedom $\bar{n}$ reflects the heavy tails of the marginal distribution $p(z_k \mid z_{k-1})$.

Particle filtering methods offer a number of significant advantages over other conventional methods. However, since they use the prior distribution as the importance distribution (Rui & Chen, 2001), the latest data observation is not considered and not taken into account when evaluating the weights of the particles. Even this choice of the importance sampling distribution has computational advantages, it can cause filtering divergence. In cases where the likelihood distribution is too small compared to the prior distribution, very few particles will have significant weights. Hence, a better proposal distribution that takes the latest observation data into account is needed. The VF yields an optimal choice of the sampling distribution over the state variable by minimizing the Kullback-Leibler (KL) divergence. In addition, compared to the particle filtering method, the computational cost and the memory requirements are dramatically reduced by the variational approximation in the prediction phase. In fact, the expectations involved in the computation of the predictive distribution have closed forms, avoiding the use of Monte Carlo integration. Next, we present the variational Bayesian filter algorithm.

DESCRIPTION OF VARIATIONAL BAYESIAN FILTERING TECHNIQUE

Variational Bayesian Filter

The variational approach consists in approximating $p(\alpha_k \mid z_{1:k})$ by a separable distribution $q(\alpha_k) = q(x_k)$ $q(\mu_k)q(\lambda_k)$ that minimizes the Kullback-Leibler divergence (KLD) between the true filtering distribution and the approximate distribution,

$$D_{KL}\left(q \| p\right) = \int q\left(\alpha_k\right) \log \frac{q\left(\alpha_k\right)}{p\left(\alpha_k \middle| \alpha_{1:k}\right)} d\alpha_k \tag{9}$$

The minimization is subject to constraint $\int q(\alpha_k) d\alpha_k = 1$. The Lagrange multiplier method used in [31, 32] shows that the updated separable approximating distribution $q(\alpha_k)$ has the following form:

$$\begin{aligned} &q(z_k) \propto p(y_k \mid z_k) N(z_k \mid \langle \mu_k \rangle, \langle \lambda_k \rangle) \\ &q(\mu_k) \propto N(\mu_k \mid \mu_k^*, \lambda_k^*) \\ &q(\lambda_k) \propto W_n(\lambda_k \mid S_k^*) \\ &q(\mu_k \mid \mu_{k-1}) \propto N(\langle \mu_k^p \rangle, \langle \lambda_k^p \rangle) \end{aligned} \tag{10}$$

where $\langle . \rangle$ denotes the expectation operator relative to the distribution q. The parameters are iteratively updated according to the following scheme:

$$\begin{aligned}
&\mu_k^* = \lambda_k^{*-1}(\langle\lambda_k\rangle\langle z_k\rangle + \mu_k^p\lambda_k^p)\\
&\lambda_k^* = \langle\lambda_k\rangle + \lambda_k^p\\
&n^* = \bar{n} + 1\\
&S_k^* = (\langle z_k z_k^T\rangle - \langle z_k\rangle\langle\mu_k\rangle^T + \langle\mu_k\rangle\langle z_k\rangle^T + \langle\mu_k\mu_k^T\rangle + \bar{S}^{-1})^{-1}\\
&\mu_k^p = \mu_{k-1}^*\\
&\lambda_k^p = (\lambda_{k-1}^{*\,-1} + \bar{\lambda}^{-1})
\end{aligned} \tag{11}$$

In fact, taking into account the separable approximate distribution at time k–1, the predictive distribution is written,

$$\begin{aligned}
p(\alpha_{k-1} \mid z_{1:k-1}) &\propto \int p(\alpha_k \mid \alpha_{k-1})q(\alpha_{k-1})d\alpha_{k-1}\\
&\propto p(z_k, \lambda_k \mid \mu_k)q_p(\mu_k)
\end{aligned} \tag{12}$$

The exponential form solution, which minimizes the Kullback-Leibler divergence between the predictive distribution $p(\alpha_k \mid z_{1:k-1})$ and the separable approximate distribution $q_{k|k-1}(\alpha)$, yields Gaussian distributions for the state and its mean and Wishart distribution for the precision matrix:

$$\begin{aligned}
&q_{k|k-1}(z_k) \propto N(\langle\mu_k\rangle_{q_{k|k-1}}, \langle\lambda_k\rangle_{q_{k|k-1}})\\
&q_{k|k-1}(\mu_k) \propto N(\mu^*_{k|k-1}, \lambda^*_{k|k-1})\\
&q_{k|k-1}(\lambda_k) \propto N_n(V^*_{k|k-1}, n^*_{k|k-1})
\end{aligned} \tag{13}$$

where the parameters are updated according to the same iterative scheme as and the expectations are exactly computed as follows:

$$\begin{aligned}
&\mu_k^p = \mu_{k-1}^*\\
&\lambda_k^p = (\lambda_{k-1}^{*\,-1} + \bar{\lambda}^{-1})\\
&\mu^*_{k|k-1} = \lambda^{*\,-1}_{k|k-1}(\langle\lambda_k\rangle_{k|k-1}\langle z_k\rangle_{k|k-1} + \mu_k^p\lambda_k^p)\\
&\lambda^*_{k|k-1} = \langle\lambda_k\rangle_{k|k-1} + \lambda_k^p\\
&n^*_{k|k-1} = \bar{n} + 1\\
&S_k^* = (\langle z_k z_k^T\rangle_{q_{k|k-1}} - \langle z_k\rangle_{q_{k|k-1}}\langle\mu_k\rangle^T_{q_{k|k-1}} + \langle\mu_k\rangle_{q_{k|k-1}}\langle z_k\rangle^T_{q_{k|k-1}} + \langle\mu_k\mu_k^T\rangle_{q_{k|k-1}} + \bar{S}^{-1})^{-1}
\end{aligned} \tag{14}$$

and the predictive expectations of the target state are now evaluated by the following expressions:

$$\begin{aligned} \langle z_k \rangle_{q_{k|k-1}} &= \langle \mu_k \rangle_{q_{k|k-1}} \\ \langle z_k z_k^T \rangle_{q_{k|k-1}} &= \langle \lambda_k \rangle_{q_{k|k-1}}^{-1} + \langle \mu_k \rangle_{q_{k|k-1}} \langle \mu_k \rangle_{q_{k|k-1}}^T \end{aligned} \tag{15}$$

In the next Section, these state estimation techniques (EKF, PF, and VBF) are used to estimate the state variables (concentration and temperature) as well as the model parameters of a CSTR process.

STATE AND PARAMETER ESTIMATION OF A CSTR PROCESS

In this section, the state estimation techniques (i.e., EKF, PF, and VBF) are compared through their utilization to estimate the states variables and parameters of a continuously stirred tank reactor. First, a description of the CSTR process model is presented, and then two comparative studies are conducted to assess the performances of these state estimation techniques. In the first comparative study, the three state estimation techniques are used to estimate the two state variables (concentration and temperature inside the reactor) from noisy measurements of these variables. In the second comparative study, the various state estimation techniques are compared when used to simultaneously estimate the state variables as well as the model parameters of the CSTR. In the second comparative study, the effect of the number estimated parameters on the performances of these state estimation techniques is also investigated.

CSTR Process Description

The process used in this example is a non-isothermal CSTR, in which the following exothermic irreversible reaction occurs (Vejtasa & Schmitz, 1970 from Hahn's paper):

$$2Na_2S_2O_3 + 4H_2O_2 \rightarrow Na_2S_3O_6 + Na_2SO_4 + 4H_2O \tag{16}$$

Denoting the compounds $Na_2S_2O_3$ and H_2O_2 by the letters "A" and "B", the reaction rate can be written as:

$$-r_A = k_o e^{-E/RT} C_A C_B \tag{17}$$

where, k_o is the reaction rate constant, E is the activation energy, T is the temperature inside the reactor, and C_A and C_B are the concentrations of "A" and "B", respectively. Applying material and energy balances on the reactor (assuming a constant reactor volume and constant properties, i.e., density and heat capacities in all streams), the following model is obtained:

$$\begin{aligned} \frac{dC_A}{dt} &= \frac{F}{V}\left(C_{Ai} - C_A\right) - k_o e^{-E/RT} C_A C_B \\ \frac{dT}{dt} &= \frac{F}{V}\left(T_i - T\right) - \frac{\Delta H}{\rho c_p} k_o e^{-E/RT} C_A C_B - \frac{UA}{V \rho c_p}\left(T - T_j\right) \end{aligned} \tag{18}$$

where, F is the flow rate in and out of the reactor, V is the reactor volume, C_{Ai} and C_A are the concentrations of "A" in the inlet and exit streams, C_B is the concentration of "B" in the exit stream, T_i and T are the temperatures of the inlet and exit streams, T_j is the temperature of the cooling fluid in the jacket, ΔH is the heat of reaction, U is the overall heat transfer coefficient, A is the area over through which heat transfer from the reactor to the jacket fluid, and ρ and c_p are the density and heat capacity of the reactor contents and in all streams.

Assuming a stoichiometric proportion of compounds "A" and "B" in the feed, one can assume that $C_B(t) = 2C_A(t)$. Thus, the model (18) becomes Equation (19). Discretizing the model using a sampling interval of Δt, it can be written as Equation (20).

$$\begin{aligned}\frac{dC_A}{dt} &= \frac{F}{V}(C_{Ai} - C_A) - 2k_o e^{-E/RT} C_A^2 \\ \frac{dT}{dt} &= \frac{F}{V}(T_i - T) - 2\frac{\Delta H}{\rho c_p} k_o e^{-E/RT} C_A^2 - \frac{UA}{V\rho c_p}(T - T_j)\end{aligned} \quad (19)$$

$$\begin{aligned}C_{A,k} &= \left[\frac{F_{k-1}}{V}(C_{Ai,k-1} - C_{A,k-1}) - 2k_o e^{-E/RT_{k-1}} C_{A,k-1}^2\right]\Delta t + C_{A,k-1} \\ T_k &= \left[\frac{F_{k-1}}{V}(T_i - T_{k-1}) - 2\frac{\Delta H}{\rho c_p} k_o e^{-E/RT_{k-1}} C_{A,k-1}^2 - \frac{UA}{V\rho c_p}(T_{k-1} - T_j)\right]\Delta t + T_{k-1}\end{aligned} \quad (20)$$

The model (20) assumes that the parameters, k_o, E, and ΔH are constants. If we are interested to estimate some (or all) of these parameters, the following equations that describe their evolution, are also needed:

$$\begin{aligned}E_k &= E_{k-1} + \gamma_{k-1}^1 \\ k_{o,k} &= k_{o,k-1} + \gamma_{k-1}^2 \\ \Delta H_k &= \Delta H_{k-1} + \gamma_{k-1}^3\end{aligned} \quad (21)$$

and the model (21) needs to incorporate the evolution of these parameters as:

$$\begin{aligned}\Im_1 \equiv C_{A,k} &= \left[\frac{F_k}{V}\left(C_{Ai,k-1} - C_{A,k-1}\right) - 2k_{o,k-1} e^{-E_{k-1}/RT_{k-1}} C_{A,k-1}^2\right]\Delta t + C_{A,k-1} \\ \Im_2 \equiv T_k &= \left[\frac{F_k}{V}\left(T_i - T_{k-1}\right) - 2\frac{\Delta H_{k-1}}{\rho c_p} k_{o,k-1} e^{-E/RT_{k-1}} C_{A,k-1}^2 - \frac{UA}{V\rho c_p}\left(T_{k-1} - T_j\right)\right]\Delta t + T_{k-1}\end{aligned}$$

Generation of Dynamic Data

In a practical setting, the data would be collected by changing the feed flow rate (which is chosen in this example to be the model input, i.e., $u_k = F_k$), and then measuring the state variables, i.e., the concentration and temperature as functions of time. However, since this is a simulated example, the data are generated from the discretized model (25), given some pre-defined model parameters. The CSTR model parameters as well as other physical properties are shown in Table 1. To generate dynamic data from the CSTR model, the flow rate (model input), F_k, is varied around its nominal value in a step-wise fashion, and the model is used to simulate the responses of the concentration and temperature as functions of time. These simulated states, which are assumed to be noise free, are then contaminated with zero mean Gaussian errors, i.e., the measurement noise $v_k \sim N(0, \sigma_v^2)$. Figure 1 shows the changes in the feed flow rate and the resulting changes in the state variables (concentration and temperature). In all simulations, the following parameters are used. The variance of the process noise (for states and parameters) is 0.01, i.e., $\sigma_v^2 = \sigma_w^2 = 0.01$, the measurement noise we used (which represent the experimental error) has variances of 0.02, and 0.08 in the two measured variables ($C_{A,k}$ and T_k), the number of data samples equals 200, and the number of particles used in PF and VBF equals 200. Finally, the sampling time is defined to be 0.012s. All the simulations performed in this chapter are implemented using MATLAB version 7.1, using an Intel Pentium CPU 3.4 GHz, 1.0 G of RAM PC.

Figure 1. Simulated CSTR data used in estimation: Input (flow rate) and state variables (concentration and temperature)

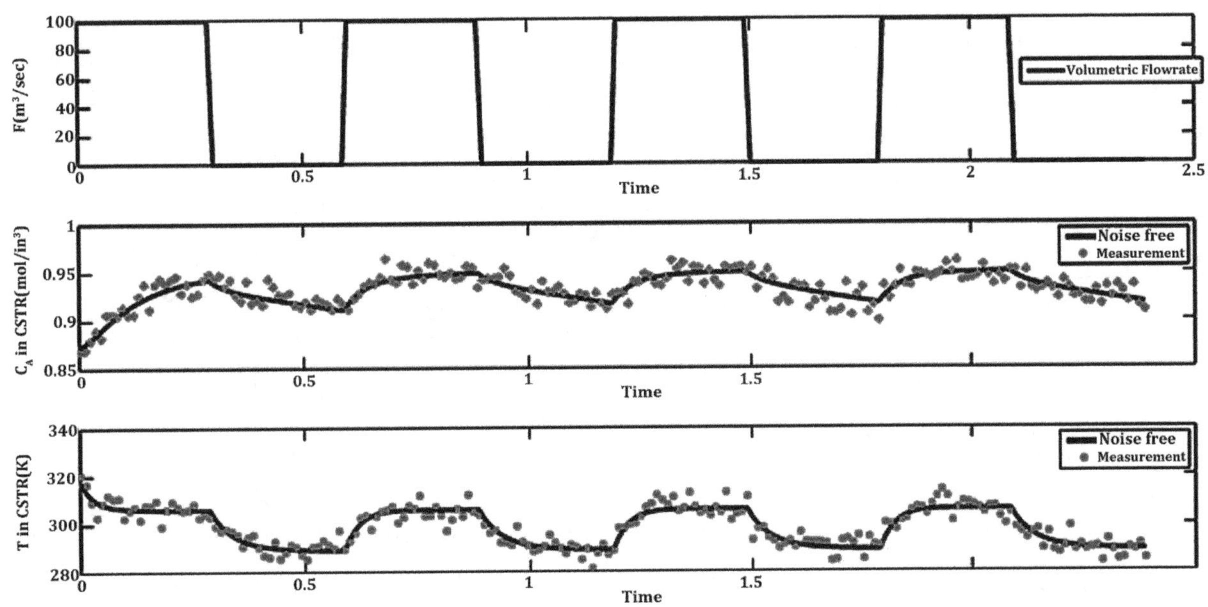

Table 1. CSTR model parameters and physical properties

Parameter	Value	Parameter	Value
E(J/mol)	76534	CAi(mol/l)	1
ΔH (J/mol)	596619	Ti (K)	350
k_o (l/min.mol)	4.11 E +13	R (J/mol.K)	8.31451
V (l)	100	Tj (K)	1
ρ (g/l)	1000	UA (W.K)	5 E +05
cp(J/g.K)	4.2		

Comparative Study 1: Estimation of State Variables from Noisy Measurements

In this comparative study, the objective is to compare the estimation accuracy of EKF, PF and VBF when they are used to estimate the two state variables of the CSTR process, i.e., concentration, $C_{A,k}$, and temperature, T_k. Hence, we consider the state vector that we wish to estimate, $z_k = x_k = [C_{A,k}, T_k]^T$, and model parameters, E, ΔH and k_o are assumed to be known. The various estimation techniques are compared by comparing the mean square errors (MSE) of the estimated states with respect to their noise-free values.

The simulation results of estimating the two states $C_{A,k}$ and T_k using EKF, PF and VBF are shown in Figures 2(a,b,c) and Figures 2(d,e,f), respectively. Also, the estimation mean square errors for the estimated states are shown in Table 2. It can be observed from Figure 2 as well as Table 2 that EKF resulted in the worst performance of all estimation techniques, which is expected due to the limited abil-

Figure 2. Estimation of state variables using various state estimation techniques

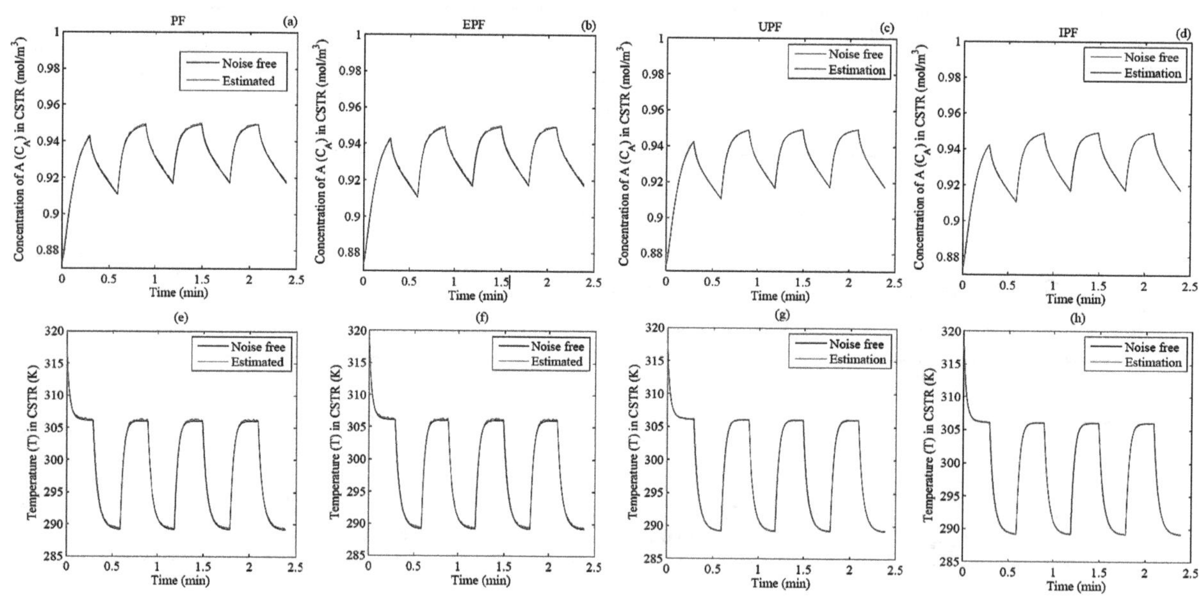

Table 2. Mean square errors of estimated state variables using EKF, PF and VBF

Technique	C_A	T
EKF	0.0021	0.7246
PF	4.524E – 04	0.1347
VBF	8.247E – 05	0.0289

ity of EKF to accurately estimate the covariance matrix of the estimated states through lineralization of the nonlinear process model. The results also show that the VBF provides a significant improvement over the PF, which is due to the fact that the VBF yields an optimal choice of the sampling distribution over the estimated states by minimizing a KL divergence criterion that also utilizes the observed data.

Comparative Study 2: Simultaneous Estimation of State Variables and Model Parameters

The main objective of this study are to compare the performances of the various estimation techniques when they are used to estimate the state variables as well as the model parameters of the CSTR process. Therefore, the state vector that we wish to estimate, zk, includes the state variables $C_{A,k}$ and T_k, as well as some (or all) of the model parameters (i.e., E, ΔH and k_o) that need to be estimated.

Defining θ_k to be a vector of the parameters to be estimated (as described by equation (4)), we can write the augmented CSTR model (that combines the states and unknown parameters) as shown in Equation (5).

To investigate the effect of the number of estimated model parameters on the performances of the different state estimators, this comparative study will be conducted through the following three cases, which are summarized below. In all cases, it is assumed that both state variables (concentration and temperature) are measured.

Case 1: The two state variables ($C_{A,k}$ and T_k) and one parameter (say E) will be estimated.
Case 2: The two state variables ($C_{A,k}$ and T_k) and two parameters (say E and ΔH) will be estimated.
Case 3: The two state variables ($C_{A,k}$ and T_k) and three parameters (say E, ΔH and k_o) will be estimated.

The estimation of the state variables and parameter(s) for these three cases were performed using the three state estimation techniques, EKF, PF, and VBF. The estimation results for the model parameters using the three estimation techniques (EKF, PF, and VBF) are shown in Figures 3, 4, and 5, respectively. For example, Figure 3(a) shows the estimation of the parameter E in case 1, Figures 3(b,c) show the estimation of the parameters E and ΔH in case 2, and Figures 3(d,e,f) show the estimation of all three parameters E, ΔH and k_o in case 3. Also, Tables 3, 4 and 5 compare the performances of the three estimation techniques for the three cases. For example, for case 1, Table 3 compares the estimation mean square errors for the two state variables $C_{A,k}$ and T_k (with respect to the noise-free data) and the mean of the estimated parameter E at steady state (i.e., after convergence of parameter(s)). Tables 4 and 5 present similar comparisons for cases 2 and 3, respectively. Moreover, Figures 6(a) and 6(b) present histograms comparing the estimation MSE for the two state variables ($C_{A,k}$ and T_k), respectively, using the various state estimation techniques. It can be seen from the results presented in Tables 3, 4, and 5 and illustrated in Figures 6(a) and 6(b) that in all cases, the PF outperforms the EKF (i.e., provides smaller MSE for the state variables), and that the VBF shows significant improvement over all other techniques. These results confirm the results obtained in the first comparative study, where only the state variables are estimated. The advantages of the VBF over the PF (and the PF over the EKF) can also be seen through their abilities to estimate the model parameters.

For example, EKF could perfectly estimate one parameter in case 1 (see Figure 3(a)), but it took longer to estimate a second parameter in case 2 (see Figures 3(b,c)), and it could not converge for the third parameter in case 3 (see Figure 3(d,e,f)), where it is used to estimate all three parameters. The PF, on the other hand, could estimate all parameters in all cases 1-3, even though it took longer to converge in case 3, where all three parameters are estimated (see Figure 4). The VBF, however, could estimate all parameters in all three cases, and converged faster than all other techniques.

The results also show that the number of estimated parameters affect the estimation accuracy of the estimated state variables. In other words, for all estimation techniques, the estimation MSE of CA and T increases from the first comparative study (where only the state variables are estimated) to case 1 (where only one parameter, E, is estimated) to case 3 (where all three parameters, E, ΔH and k_o, are estimated). For example, the MSEs obtained using EKF for $C_{A,k}$ in the first comparative study and cases 1-3 of the second comparative study are 0.0021, 0.0023, 0.0033, and 0.0046, respectively, which increase as the number of estimated parameters increases (refer to Tables 3, 4 and 5). This observation is valid for the other state variable T_k and for all other estimation techniques, PF and VBF.

Figure 3. Estimation of the CSTR model parameters using EKF for all cases: (a) case 1; (b) case 2; (c),(d),(e),(f) case 3

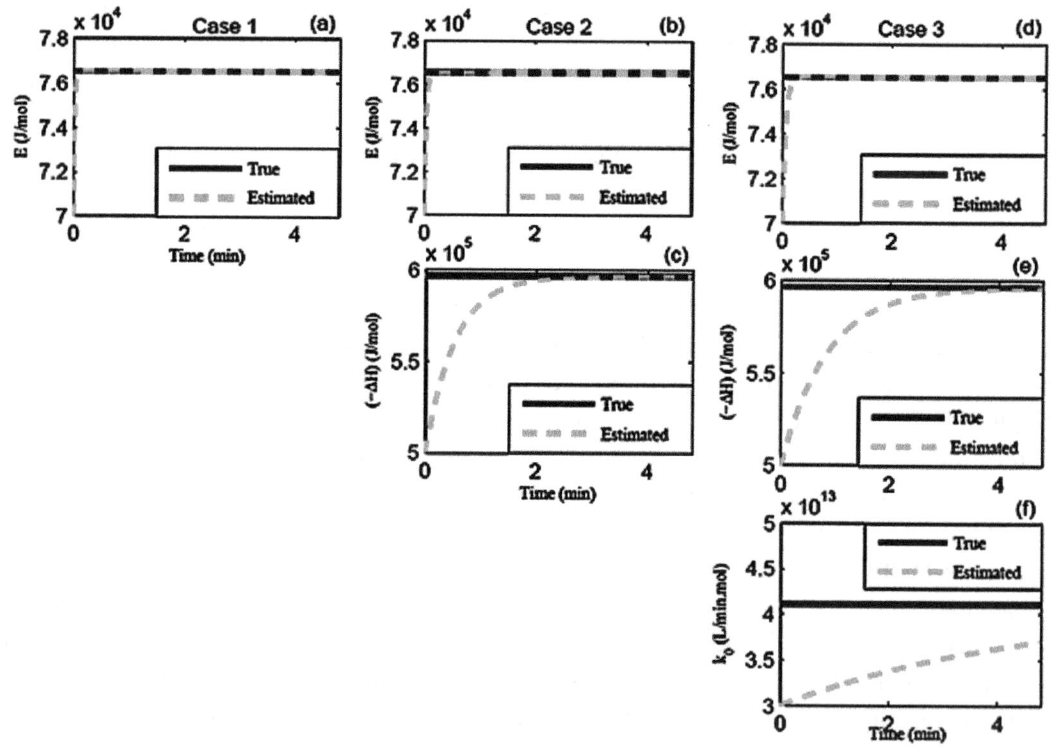

Figure 4. Estimation of the CSTR model parameters using PF for all cases: (a) case 1; (b),(c) case 2; (d),(e),(f) case 3

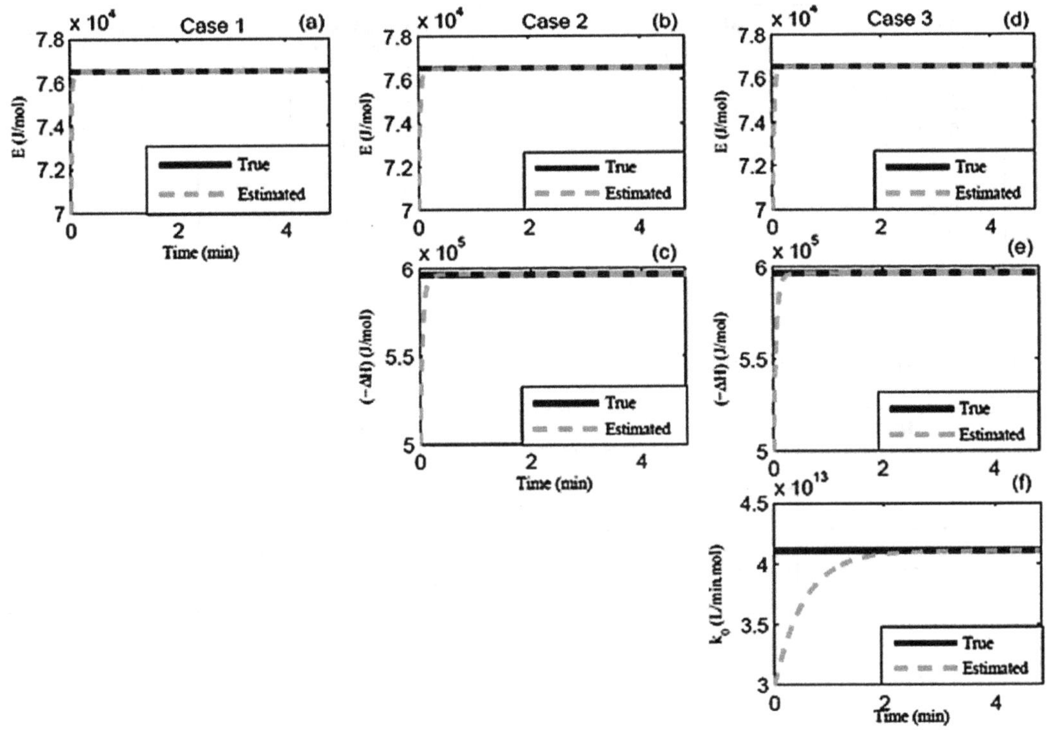

Figure 5. Estimation of the CSTR model parameters using VBF for all cases: (a) case 1; (b),(c) case 2; (d),(e),(f) case 3

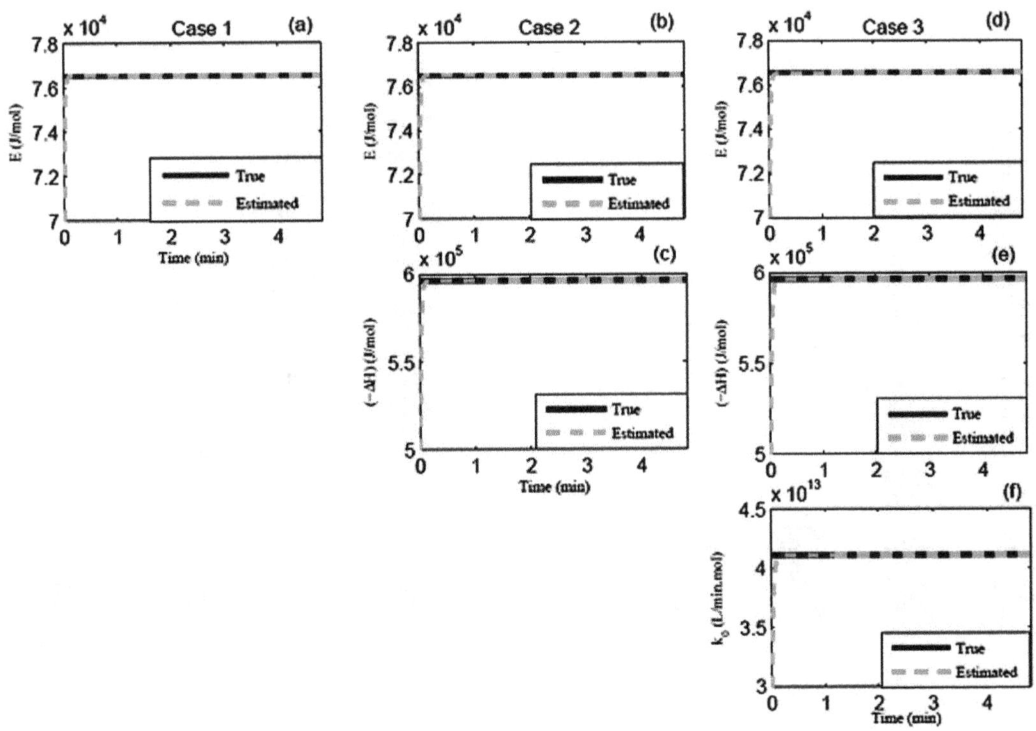

Table 3. Mean square errors of estimated state variables and mean of estimated parameter - case 1

RMSE			Mean at Steady State
Technique	C_A	T	E
EKF	0.0023	0. 7626	76534
PF	4.72E – 04	0.158	76534
VBF	8.61E – 05	0.03082	76534

Table 4. Mean square errors of estimated state variables and mean of estimated parameters- case 2

RMSE			Mean at Steady State	
Technique	C_A	T	E	ΔH
EKF	0.0033	0. 7923	76534	596619
PF	6.197E – 04	0.161	76534	596619
VBF	1.01E – 04	0. 0343	76534	596619

Table 5. Mean square errors of estimated state variables and mean of estimated parameters - case 3

RMSE			Mean at Steady State		
Technique	C_A	T	E	ΔH	k_0
EKF	0.0033	0. 7923	76534	596619	did not converge
PF	6.197E – 04	0.161	76534	596619	4.11E +13
VBF	1.01E – 04	0. 0343	76534	596619	4.11E +13

Figure 6. Histograms showing the MSE of estimated concentration and temperature (CA and T) for the three cases using EKF, PF and VBF

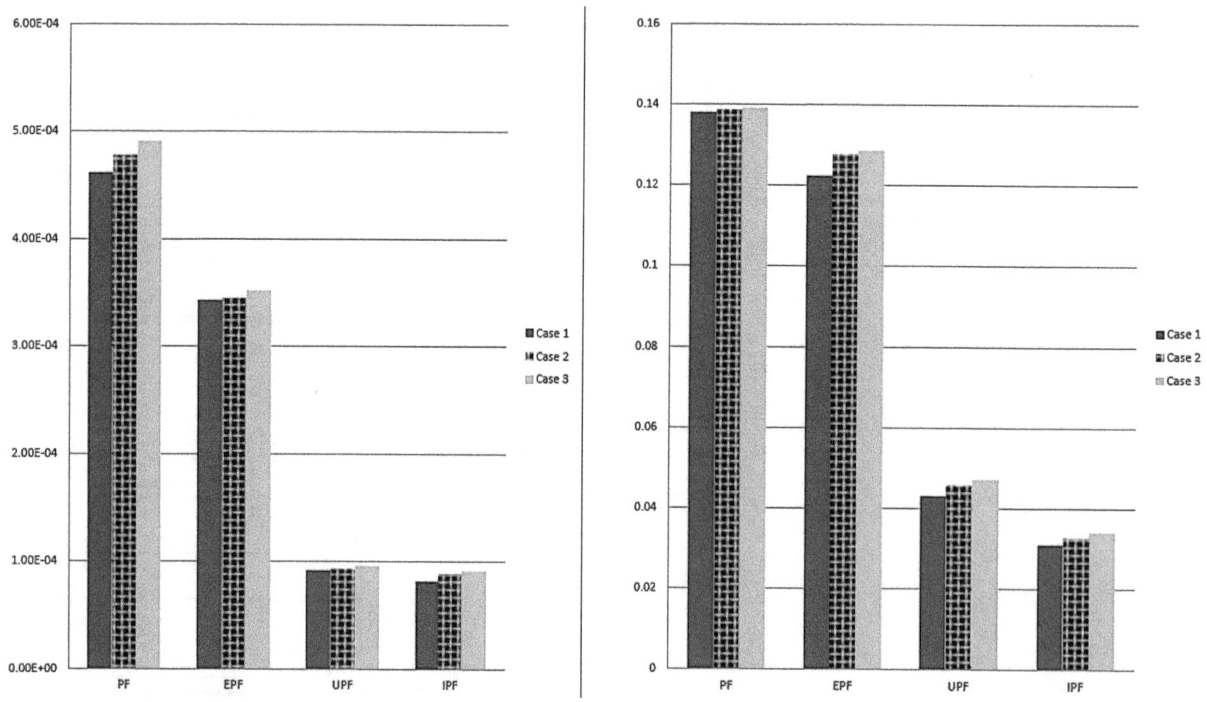

FUTURE RESEARCH DIRECTIONS

In future work, we will consider investigating the effects of practical challenges, such as the presence of measurement noise in the data and the availability of small number of measured data samples, on the estimation performances of the these estimators. The computational complexity of the state estimation techniques will also be studied to investigate their applicability in practice.

CONCLUSION

In this chapter, state estimation techniques are used to simultaneously estimate the state variables and model parameters of a continuously stirred tank reactor. Various state estimation techniques, which include the extended Kalman filter, particle filter, and variational Bayesian filter, are compared as they are used to achieve this objective. Two comparative studies have been conducted to compare the estimation performances of these three estimation techniques. In the first comparative study, EKF, PF and VBF are used to estimate the two state variables (the concentration and temperature) of the CSTR process. In the second comparative study, the state variables and model parameters are simultaneously estimated, and the effect of the number of estimated parameters on the performances of the three estimation techniques is investigated. The simulation results of both comparative studies show that the PF provides a higher accuracy than the EKF due to the limited ability of the EKF to deal with highly nonlinear process models. The results also show that the VBF provides a significant improvement over the PF. This is because, unlike the PF which depends on the choice of sampling distribution used to estimate the posterior, the VBF yields an optimum choice of the sampling distribution, which also utilizes the observed data. The results of the second comparative study show that, for all techniques, estimating more model parameters affects the estimation accuracy as well as the convergence of the estimated states and parameters. The VBF also provides advantages over the other methods with respect to the estimation accuracy and convergence.

ACKNOWLEDGMENT

This work was made possible by NPRP grant NPRP7-1172-2-439 from the Qatar National Research Fund (a member of Qatar Foundation). The statements made herein are solely the responsibility of the authors.

REFERENCES

Aggarwal, P., Gu, D., Nassar, S., Syed, Z., & El-Sheimy, N. (2007, September). Extended particle filter (EPF) for INS/GPS land vehicle navigation applications. In *Proceedings of the 20th international technical meeting of the satellite division of the institute of navigation (ION GNSS 2007),* (pp. 2619-2626).

Arulampalam, M., Maskell, S., Gordon, N. & Clapp, T. (2002). A tutorial on particle filters for online nonlinear/non-gaussian bayesian tracking. *Signal Processing, IEEE Transactions on*, 174–188.

Arulampalam, M. S., Maskell, S., Gordon, N., & Clapp, T. (2002). A tutorial on particle filters for online nonlinear/non-Gaussian Bayesian tracking. *Signal Processing. IEEE Transactions on, 50*(2), 174–188. doi:10.1109/78.978374

Barndorff-Nielsen, O. (1977). *Exponentially decreasing distributions for the logarithm of particle size.* Proc. Roy. Soc.

Barndorff-Nielsen, O. (1977). Exponentially decreasing distributions for the logarithm of particle size. *Proceedings of the Royal Society of London. A. Mathematical and Physical Sciences, 353*(1674), 401-419. doi:10.1098/rspa.1977.0041

Beal, M. (2003). *Variational algorithms for approximate bayesian inference.* (Unpublished doctoral dissertation). Gatsby Computational Neuroscience Unit, University College London.

Ben-Brahim, L., & Tadakuma, S. (1998). Speed control of induction motor without rotational transducers. In *Industry Applications Conference.* doi:10.1109/IAS.1998.732393

Chen, G., Xie, Q., & Shieh, L. (1998). Fuzzy kalman filtering. *Journal of Information Science, 109*(1-4), 197–209. doi:10.1016/S0020-0255(98)10002-6

Doucet, A., & Johansen, A. (2009). A tutorial on particle filtering and smoothing: Fifteen years later. In Handbook of Nonlinear Filtering, (pp. 656–704). Academic Press.

Doucet, A., & Tadic, V. (2003). Parameter estimation in general state-space models using particle methods. *Annals of the Institute of Statistical Mathematics, 55*(2), 409–422. doi:10.1007/BF02530508

Grewal, M., & Andrews, A. (2008). *Kalman Filtering: Theory and Practice using MATLAB.* John Wiley and Sons. doi:10.1002/9780470377819

Gustafsson, F., Gunnarsson, F., Bergman, N., Forssell, U., Jansson, J., Karlsson, R., & Nordlund, P. (2002). Particle filters for positioning, navigation, and tracking. *Signal Processing, IEEE Transactions on*, 425–437.

Gustafsson, F., & Hendeby, G. (2012). Some relations between extended and unscented Kalman filters. *Signal Processing. IEEE Transactions on, 60*(2), 545–555. doi:10.1109/TSP.2011.2172431

Guzinski, J., Abu-Rub, H., Diguet, M., Krzeminski, Z. & Lewicki, A. (2010). Speed and load torque observer application in high-speed train electric drive. *Industrial Electronics, IEEE Transactions on*, 565–574.

Ho, S., Hsieh, C., Yu, F., & Huang, H. (2007). An intelligent two-stage evolutionary algorithm for dynamic pathway identification from gene expression profiles. *Computational Biology and Bioinformatics, IEEE/ACM Transactions on*, 648–704.

Holtz, J. (1995). The representation of ac machine dynamics by complex signal flow graphs. *Industrial Electronics, IEEE Transactions on*, 263–271.

Holtz, J. (2002). Sensorless control of induction motor drives. *Proceedings of the IEEE*, 1359–1394.

Jia, J. (2009). Parameter estimation for nonlinear biological system model based on global sensitivity analysis. In *3rd International Conference on Bioinformatics and Biomedical Engineering*. doi:10.1109/ICBBE.2009.5163168

Julier, S., & Uhlmann, J. (1997). New extension of the kalman filter to nonlinear systems. *Proceedings of the Society for Photo-Instrumentation Engineers*, 182–193. doi:10.1117/12.280797

Kalman, R. E. (1960). A new approach to linear filtering and prediction problem, *Trans. ASME, Ser. D. Journal of Basic Engineering*, 34–45.

Kim, Y., Sul, S., & Park, M. (1994). Speed sensorless vector control of induction motor using extended kalman filter. *IEEE Transactions on Industry Applications*, *30*(5), 1225–1233. doi:10.1109/28.315233

Kotecha, J., & Djuric, P. (2003). Gaussian particle filtering. *IEEE Transactions on Signal Processing*, *51*(10), 2592–2601. doi:10.1109/TSP.2003.816758

Krzeminski, Z. (2008). Observer of induction motor speed based on exact disturbance model. In *13th IEEE on Power Electronics and Motion Control Conference*, (pp. 2294–2299). doi:10.1109/EPEPEMC.2008.4635605

Kubota, H., Matsuse, K. & Nakano, T. (1993). Dsp-based speed adaptive flux observer of induction motor. *Industry Applications, IEEE Transactions on*, 344–348.

Lee, J., & Ricker, N. (1994). Extended kalman filter based nonlinear model predictive control. *Industrial & Engineering Chemistry Research*, *33*(6), 1530–1541. doi:10.1021/ie00030a013

Lee, K., Lee, G., Yoon, T., Choy, I., & Song, J. (2005). Speed and flux estimation for an induction motor using a parameter estimation technique. *International Journal of Control, Automation and Systems*, 79–86.

Levi, E. & Wang, M. (2002). A speed estimator for high performance sensorless control of induction motors in the field weakening region. *Power Electronics, IEEE Transactions on*, 365–378.

Li, J., Xu, L., & Zhang, Z. (2005), An adaptive sliding-mode observer for induction motor sensorless speed control. *Industry Applications, IEEE Transactions on*, 1039–1046.

Liang, K., Lingfu, K., & Peiliang, W. (2012, July). Adaptive Gaussian particle filter for nonlinear state estimation. In *Control Conference (CCC), 2012 31st Chinese* (pp. 2146-2150). IEEE.

Lillacci, G., & Khammash, M. (2010). Parameter estimation and model selection in computational biology. *PLoS Computational Biology*, *6*(3), e1000696. doi:10.1371/journal.pcbi.1000696 PMID:20221262

Lindenmeyer, D., Dommel, H., Moshref, A., & Kundur, P. (2001). An induction motor parameter estimation method. *International Journal of Electrical Power & Energy Systems*, *23*(4), 251–262. doi:10.1016/S0142-0615(00)00060-0

Liu, J., & Chen, R. (1998). Sequential monte carlo methods for dynamic systems. *Journal of the American Statistical Association*, *93*(443), 1032–1044. doi:10.1080/01621459.1998.10473765

Ljung, L. (1979). Asymptotic behavior of the extended kalman filter as a parameter estimator for linear systems. IEEE Trans. on Automatic Control, pp. 36–50.

Mansouri, M., Ilham, O., Snoussi, H., & Richard, C. (2011). Adaptive quantized target tracking in wireless sensor networks. *Wireless Networks*, *17*(7), 1625–1639. doi:10.1007/s11276-011-0368-1

Mansouri, M., Nounou, H., & Nounou, M. (2013, March). State estimation of a chemical reactor process model-A comparative study. In Systems, Signals & Devices (SSD), 2013 10th International Multi-Conference on (pp. 1-6). IEEE. doi:10.1109/SSD.2013.6563998

Mansouri, M., Nounou, H., & Nounou, M. (2014, February). Improved particle filtering for state and parameter estimation-CSTR model. In Multi-Conference on Systems, Signals & Devices (SSD), 2014 11th International (pp. 1-6). IEEE. doi:10.1109/SSD.2014.6808794

Mansouri, M. M., Mohamed-Seghir, M. M., Nounou, H. N., Nounou, M. N., & Abu-Rub, H. A. (2014). Bayesian methods for time-varying state and parameter estimation in induction machines. *International Journal of Adaptive Control and Signal Processing*.

Mansouri, M. M., Nounou, H. N., Nounou, M. N., & Datta, A. A. (2014). State and parameter estimation for nonlinear biological phenomena modeled by S-systems. *Digital Signal Processing*, *28*, 1–17. doi:10.1016/j.dsp.2014.01.012

Mansouri, M. M., Nounou, H. N., Nounou, M. N., & Datta, A. A. (2014). Modeling of nonlinear biological phenomena modeled by S-systems. *Mathematical Biosciences*, *249*, 75–91. doi:10.1016/j.mbs.2014.01.011 PMID:24524881

Rui, Y., & Chen, Y. (2001). Better proposal distributions: Object tracking using unscented particle filter. In *Computer Vision and Pattern Recognition, 2001. CVPR 2001. Proceedings of the 2001 IEEE Computer Society Conference on* (*Vol. 2*, pp. II-786). IEEE.

Van Der Merwe, R., Doucet, A., De Freitas, N., & Wan, E. (2000, August). *The unscented particle filter*. NIPS.

Vermaak, J., Lawrence, N. D., & Pérez, P. (2003, June). Variational inference for visual tracking. In *Computer Vision and Pattern Recognition, 2003. Proceedings. 2003 IEEE Computer Society Conference on* (Vol. 1, pp. I-773). IEEE. doi:10.1109/CVPR.2003.1211431

Zuo, J., Jia, Y., Chen, Y., & Zhang, Y. (2012, April). A New Particle Filter Guided by Iterated EKF. In *Proceedings of the 2012 Second International Conference on Electric Information and Control Engineering* (pp. 124-127). IEEE Computer Society.

ADDITIONAL READING

Andrews, B. W., Yi, T. M., & Iglesias, P. A. (2006). Optimal noise filtering in the chemotactic response of escherichia coli. *PLoS Computational Biology*, *2*(11), e154. doi:10.1371/journal.pcbi.0020154 PMID:17112312

Arulampalam, M. S., Maskell, S., Gordon, N., & Clapp, T. (2002). A tutorial on particle filters for online nonlinear/non-Gaussian Bayesian tracking. *Signal Processing. IEEE Transactions on*, *50*(2), 174–188. doi:10.1109/78.978374

Arulampalam, M. S., Maskell, S., Gordon, N., & Clapp, T. (2005). A tutorial on particle filters for online nonlinear/non-gaussian bayesian tracking. *Signal Processing. IEEE Transactions on*, *50*(2), 174–188. doi:10.1109/78.978374

Burbea, J., & Rao, C. (1982). On the convexity of some divergence measures based on entropy functions. Information Theory. *IEEE Transactions on*, *28*(3), 489–495.

Cox, H. (1964). On the estimation of state variables and parameters for noisy dynamic systems. *Automatic Control. IEEE Transactions on*, *9*(1), 5–12.

Doucet, A., Gordon, N. J., & Krishnamurthy, V. (2001). Particle filters for state estimation of jump Markov linear systems. *Signal Processing. IEEE Transactions on*, *49*(3), 613–624. doi:10.1109/78.905890

Dumont, B., Leemans, V., Mansouri, M., Bodson, B., Destain, J. P., & Destain, M. F. (2014). Parameter identification of the STICS crop model, using an accelerated formal MCMC approach. *Environmental Modelling & Software*, *52*, 121–135. doi:10.1016/j.envsoft.2013.10.022

Ervadi-Radhakrishnan, A., & Voit, E. O. (2005). Controllabilty of non-linear biochemical systems. *Mathematical Biosciences*, *196*(1), 99–123. doi:10.1016/j.mbs.2005.03.012 PMID:15982674

Fritz, G., Koller, C., Burdack, K., Tetsch, L., Haneburger, I., Jung, K., & Gerland, U. (2009). Induction kinetics of a conditional ph stress response system in escherichia coli. *Journal of Molecular Biology*, *393*(2), 272–286. doi:10.1016/j.jmb.2009.08.037 PMID:19703467

Gordon, N. J., Salmond, D. J., & Smith, A. F. (1993, April). Novel approach to nonlinear/non-Gaussian Bayesian state estimation. *IEE Proceedings. Part F. Radar and Signal Processing*, *140*(2), 107–113. doi:10.1049/ip-f-2.1993.0015

Grewal, M., & Andrews, A. (2008). *Kalman Filtering: Theory and Practice using MATLAB*. John Wiley and Sons. doi:10.1002/9780470377819

Gustafsson, F., Gunnarsson, F., Bergman, N., Forssell, U., Jansson, J., Karlsson, R., & Nordlund, P. J. (2002). Particle filters for positioning, navigation, and tracking. *Signal Processing. IEEE Transactions on*, *50*(2), 425–437. doi:10.1109/78.978396

Han, S. P. (1977). A globally convergent method for nonlinear programming. *Journal of Optimization Theory and Applications*, *22*(3), 297–309. doi:10.1007/BF00932858

Haug, A. J. (2005). *A tutorial on bayesian estimation and tracking techniques applicable to nonlinear and non-gaussian processes*. McLean: MITRE Corporation.

Jeffreys, H. (1946). An invariant form for the prior probability in estimation problems. *Proceedings of the Royal Society of London. Series A, Mathematical and Physical Sciences*, *186*(1007), 453–461. doi:10.1098/rspa.1946.0056 PMID:20998741

Juang, B. H., & Rabiner, L. (1985). A probabilistic distance measure for hidden Markov models. *AT & T Bell Laboratories Technical Journal*, *64*(2), 391–408.

Kuper, C., & Jung, K. (2005). Cadc-mediated activation of the cadba promoter in escherichia coli. *Journal of Molecular Microbiology and Biotechnology*, *10*(1), 26–29. doi:10.1159/000090346 PMID:16491024

Lee, J. H., & Ricker, N. L. (1994). Extended kalman filter based nonlinear model predictive control. *Industrial & Engineering Chemistry Research, 33*(6), 1530–1541. doi:10.1021/ie00030a013

Li, Z. F., Osborne, M. R., & Prvan, T. (2003). Numerical algorithms for constrained maximum likelihood estimation. *Anziam Journal, 45*(1), 91–114. doi:10.1017/S1446181100013171

Liu, J., & West, M. (2001). Combined parameter and state estimation in simulation-based filtering. In *Sequential Monte Carlo methods in practice* (pp. 197–223). Springer New York. doi:10.1007/978-1-4757-3437-9_10

Mansouri, M., Dumont, B., & Destain, M. F. (2012). Bayesian methods for predicting LAI and soil moisture. In *11th International Conference on Precision Agriculture.*

Mansouri, M., Dumont, B., & Destain, M. F. (2013). Modeling and prediction of time-varying environmental data using advanced bayesian methods. *Exploring Innovative and Successful Applications of Soft Computing*, 112-137.

Mansouri, M., Dumont, B., & Destain, M. F. (2013). Modeling and prediction of nonlinear environmental system using Bayesian methods. *Computers and Electronics in Agriculture, 92*, 16–31. doi:10.1016/j.compag.2012.12.013

Mansouri, M., Dumont, B., & Destain, M.-F. Modeling and Prediction of Nonlinear Environmental System Using Bayesian Methods. In Computers and Electronics in Agriculture. 16-31, 2013. doi:10.1016/j.compag.2012.12.013

Mansouri, M., Dumont, B., & Destain, M. F. (2014, April). Predicting Grain Protein Content of Winter Wheat. In *22nd European Symposium on Artificial Networks, Computational Intelligence and Machine Learning.*

Mansouri, M., Dumont, B., Leemans, V., & Destain, M. F. (2014). Bayesian methods for predicting LAI and soil water content. *Precision Agriculture, 15*(2), 184–201. doi:10.1007/s11119-013-9332-7

Mansouri, M., Mohamed-Seghir, M., Nounou, H., Nounou, M., & Abu-Rub, H. (2014). Non linear and non Gaussian states and parameters estimation using Bayesian methods–comparatives studies. Handbook of Research on Novel Soft Computing Intelligent Algorithms: Theory and Practical Applications. IGI global. doi, 10, 978-1.

Mansouri, M. M., Mohamed-Seghir, M. M., Nounou, H. N., Nounou, M. N., & Abu-Rub, H. A. (2014). Bayesian methods for time-varying state and parameter estimation in induction machines. *International Journal of Adaptive Control and Signal Processing.*

Mansouri, M. M., Nounou, H. N., Nounou, M. N., & Datta, A. A. (2014). State and parameter estimation for nonlinear biological phenomena modeled by S-systems. *Digital Signal Processing, 28*, 1–17. doi:10.1016/j.dsp.2014.01.012

Meskin, N., Nounou, H., Nounou, M., Dattta, A., & Dougherty, E. R. (2011). Intervention in biological phenomena modeled by s-systems. *IEEE Transactions on Bio-Medical Engineering, 58*(5), 1260–1267. doi:10.1109/TBME.2010.2099658 PMID:21172748

Nounou, H., & Nounou, M. (2006). Multiscale fuzzy kalman filtering. *Engineering Applications of Artificial Intelligence*, *19*(5), 439–450. doi:10.1016/j.engappai.2005.11.001

Nounou, M. N., Nounou, H. N., & Mansouri, M. (2013). Model-based and model-free filtering of genomic data. *Network Modeling Analysis in Health Informatics and Bioinformatics*, *2*(3), 109–121. doi:10.1007/s13721-013-0030-1

Rife, D., & Boorstyn, R. (1974). Single tone parameter estimation from discrete-time observations, *Information Theory. IEEE Transactions on*, *20*(5), 591–598.

Savkin, A. V., & Petersen, I. R. (1995). Recursive state estimation for uncertain systems with an integral quadratic constraint. *IEEE Transactions on Automatic Control*, *40*(6), 1080–1083. doi:10.1109/9.388688

Schmidt, R. O. (1986). Multiple emitter location and signal parameter estimation. *Antennas and Propagation. IEEE Transactions on*, *34*(3), 276–280. doi:10.1109/TAP.1986.1143830

Schweppe, F. C. (1970). Power system static-state estimation, Part III: Implementation. *Power Apparatus and Systems, IEEE Transactions on*, (1), 130-135.

Storvik, G. (2002). Particle filters for state-space models with the presence of unknown static parameters. *IEEE Transactions on Signal Processing*, *50*(2), 281–289. doi:10.1109/78.978383

Tarantola, A. (2005). *Inverse problem theory and methods for model parameter estimation*. Society for Industrial Mathematics. doi:10.1137/1.9780898717921

Terejanu, G. A. (2009). Unscented kalman filter tutorial. In *Workshop on Large-Scale Quantification of Uncertainty, Sandia National Laboratories*, 1–6.

Torres, N. V. (1994). Modelization and experimental studies on the control of the glycolytic-glucogenolytic pathway in rat liver. *Molecular and Cellular Biochemistry*, *132*(2), 117–126. doi:10.1007/BF00926920 PMID:7969094

Vajesta, A., & Schmitz, R. (1970). An experimental study of steady-state multiplicity and stability in an adiabatic stirred reactor. *AIChE Journal. American Institute of Chemical Engineers*, *3*, 410–419.

Voit, E. O., & Almeida, J. (2004). Decoupling dynamical systems for pathway identification from metabolic profiles. *Bioinformatics (Oxford, England)*, *20*(11), 1670–1681. doi:10.1093/bioinformatics/bth140 PMID:14988125

Zhang, Z. (1997). Parameter estimation techniques: A tutorial with application to conic fitting. *Image and Vision Computing*, *15*(1), 59–76. doi:10.1016/S0262-8856(96)01112-2

KEY TERMS AND DEFINITIONS

Chemical Reaction: Chemical reaction engineering is a specialty in chemical engineering dealing with chemical reactors.

CSTR Process: The continuous flow stirred-tank reactor (CSTR) is a reactor type in chemical systems. A CSTR refers to a system used to estimate the key state variables from given measurements to reach a specified output.

Extended Kalman Filter (EKF): EKF is an estimation technique that is applicable to nonlinear and non-Gaussian models. As the Kalman filter is applicable to linear systems, the EKF can be viewed as an extension of the Kalman filter that is applied to a linearized version of the nonlinear model.

Particle Filter (PF): PF, which is based on Bayesian estimation, is a sequential Monte Carlo state estimation method for nonlinear and non-Gaussian systems. The EKF algorithm does not always provide a satisfactory performance, especially for highly nonlinear processes as model linearization does not necessarily provide good estimates of the mean of the state vector and the covariance matrix of the estimation error, which are used in state estimation. These issues are addressed by the PF.

States and Parameters Estimation: State and parameter estimation is the process of obtaining the estimators of state variables and model parameters based on measurements.

Variational Bayesian Filter (VBF): VBF algorithm provides a significant improvement over the particle filtering because, unlike the PF which depends on the choice of sampling distribution used to estimate the posterior distribution, which also accounts for the observed data.

Compilation of References

Aarts, E., & Korst, J. (1990). *Simulated Annealing and Boltzmann Machines. A Stochastic Approach to Combinatorial Optimization and Neural Computing*. Chichester, UK: John Wiley & Sons.

Abbasian, M. A., & Nezamabadi-pour, H. (2012). Multi Objective Gravitational Search Algorithm Using Non-dominated Fronts. *Journal of Electrical Engineering, 41*(1), 67–80.

Abbasian, M. A., Nezamabadi-pour, H., & Amoozegar, M. (2015). A Clustering based Archive Multi Objective Gravitational Search Algorithm. *Journal of Fundamenta Informaticae, 138*(4), 387–409. doi:10.3233/FI-2015-1218

Abdulhadi, A. E., & Abhari, R. (2011, 3-8 July 2011). *Dual printed meander monopole antennas for passive UHF RFID tags*. Paper presented at the Antennas and Propagation (APSURSI), 2011 IEEE International Symposium on.

Abouzahr, I., & Ramakumar, R. (1990). Loss of power supply probability of stand-alone wind electric conversion systems: A closed form solution approach. *IEEE Transactions on Energy Conversion, 5*(3), 445–451. doi:10.1109/60.105267

Abouzahr, I., & Ramakumar, R. (1991). Loss of power supply probability of stand-alone photovoltaic systems: A closed form solution approach. *IEEE Transactions on Energy Conversion, 6*(1), 1–8. doi:10.1109/60.73783

Absi, N., & Kedad-Sidhoum, S. (2007). MIP-based heuristics for multi-item capacitated lot-sizing problem with setup times and shortage costs. RAIRO—. *Operations Research, 41*(2), 171–192. doi:10.1051/ro:2007014

Absi, N., & Kedad-Sidhoum, S. (2008). The multi capacitated lot sizing problem with set up time and shortage costs. *European Journal of Operational Research, 185*(3), 1351–1374. doi:10.1016/j.ejor.2006.01.053

Absi, N., & Kedad-Sidhoum, S. (2009). The multi-item capacitated lot-sizing problem with setup times, safety stock and demand shortages. *European Journal of Operational Research, 36*, 2926–2936.

Abulleif, T., & Al Dossary, A. (2008). *Location Based Services (LBS)*. Report from Surveying Services Division, Saudi Aramco Dhahran, Saudi Arabia. Retrieved from: www.saudigis.org/FCKFiles/File/.../79_E_ThamerAbulleif_KSA.pdf

Adak, A. K., & Bhowmik, M. (2011a). Interval cut-set of interval-valued intuitionistic fuzzy sets. *African Journal of Mathematics and Computer Sciences, 4*(4), 192–200.

Adak, A. K., Bhowmik, M., & Pal, M. (2011b). Semi-ring of interval-valued intuitionistic fuzzy matrices. *Global Journal of Computer Application and Technology, 1*(3), 340–347.

Adak, A. K., Bhowmik, M., & Pal, M. (2012a). Interval cut-set of generalized interval-valued in-tuitionistic fuzzy sets. *International Journal of Fuzzy System Applications, 2*(3), 35–50. doi:10.4018/ijfsa.2012070103

Adak, A. K., Bhowmik, M., & Pal, M. (2012b). Intuitionistic fuzzy block matrix and its some properties. *Annals of Pure and Applied Mathematics, 1*(1), 13–31.

Adak, A. K., Bhowmik, M., & Pal, M. (2012c). Some properties of generalized intuitionistic fuzzy nilpotent matrices over distributive lattice. *International Journal of Fuzzy Information and Engineering*, *4*(4), 371–387. doi:10.1007/s12543-012-0121-1

Adak, A. K., Bhowmik, M., & Pal, M. (2013). Distributive Lattice over Intuitionistic Fuzzy Matrices. *The Journal of Fuzzy Mathematics*, *21*(2), 401–416.

Adelsberger, H. H., Collis, B., & Pawlowski, J. J. (2002). *Handbook on Information Technologies for Education and Training*. Springer-Verlag. doi:10.1007/978-3-662-07682-8

Afshinmanesh, F., Marandi, A., & Shahabadi, M. (2008). Design of a Single-Feed Dual-Band Dual-Polarized Printed Microstrip Antenna Using a Boolean Particle Swarm Optimization. *IEEE Transactions on Antennas and Propagation*, *56*(7), 1845–1852. doi:10.1109/TAP.2008.924684

Aganagic, M. (1978). *Iterative methods for linear complementarity problems*. Systems Optimization Laboratory, Department of Operations Research, Univ.

Aggarwal, P., Gu, D., Nassar, S., Syed, Z., & El-Sheimy, N. (2007, September). Extended particle filter (EPF) for INS/GPS land vehicle navigation applications. In *Proceedings of the 20th international technical meeting of the satellite division of the institute of navigation (ION GNSS 2007)*,(pp. 2619-2626).

Aggeler, D., Canales, F., Coccia, A., Butcher, N., & Apeldoorn, O. (2010). Ultra-fast DC-charge infrastructures for EV-mobility and future smart grids. In Innovative Smart Grid Technologies Conference Europe (ISGT Europe), 2010 IEEE PES. doi:10.1109/ISGTEUROPE.2010.5638899

Aghezzaf, E. H., Jamali, M. A., & Ait-Kadi, D. (2007). An integrated production and preventive maintenance planning model. *European Journal of Operational Research*, *181*(2), 676–685. doi:10.1016/j.ejor.2006.06.032

Aghezzaf, E. H., & Najid, N. M. (2008). Integrated production and preventive maintenance in production systems subject to random failures. *Information science*, *178*, 3382–3392. doi:10.1016/j.ins.2008.05.007

Agmon, S. (1954). The relaxation method for linear inequalities. *Canadian Journal of Mathematics*, *6*(3), 382–392. doi:10.4153/CJM-1954-037-2

Aguirre, A. H., Zebulum, R. S., & Coello Coello, C. (2004). Evolutionary multiobjective design targeting a Field Programmable Transistor Array. In *Proceedings of the Conference on Evolvable Hardware*. doi:10.1109/EH.2004.1310831

Ahmad, S. A. (2012). *A Study of Search Neighborhood in the Bees Algorithm*. (PhD Thesis). Cardiff University.

Ahn, B. (1981). Solution of nonsymmetric linear complementarity problems by iterative methods. *Journal of Optimization Theory and Applications*, *33*(2), 175–185. doi:10.1007/BF00935545

Ai, W., & Zhang, S. (2005). An O (\sqrtnL) Iteration Primal-dual Path-following Method, Based on Wide Neighborhoods and Large Updates, for Monotone LCP. *SIAM Journal on Optimization*, *16*(2), 400–417. doi:10.1137/040604492

Ajami A., &Armaghan M., (2013). A multi-objective gravitational search algorithm based approach of power system stability enhancement with UPFC. *Journal of Central South University, 20*(6), 1536-1544.

Akay, B. (2013). A study on particle swarm optimization and artificial bee colony algorithms for multilevel thresholding. *Applied Soft Computing*, *13*(6), 3066–3091. doi:10.1016/j.asoc.2012.03.072

Akbari, R., Hedayatzadeh, R., Ziarati, K., & Hassanizadeh, B. (2012). A multi-objective artificial bee colony algorithm. *Swarm and Evolutionary Computation*, *2*, 39–52. doi:10.1016/j.swevo.2011.08.001

Akhtar, S., Ahmad, A., & Abdel-Rahman, E. (2012). *A metaheuristic bat-inspired algorithm for full body human pose estimation.* Paper presented at the Computer and Robot Vision (CRV), 2012 Ninth Conference on. doi:10.1109/CRV.2012.55

Al Nuaimi, K., & Kamel, H. (2011). A survey of indoor positioning systems and algorithms. In *Proceedings of the IEEE International Conference on Innovation in Information Technology (IIT).* doi:10.1109/INNOVATIONS.2011.5893813

Al-Agtash, S. (2001). Hydrothermal scheduling by augmented Lagrangian: Consideration of transmission constraints and pumped-storage units. *IEEE Transactions on Power Systems*, *16*(4), 750–756. doi:10.1109/59.962422

Alander, J. (1994). *An Indexed Bibliography of Genetic Algorithms: 1957-1993.* Espoo, Finland: Art of CAD Ltd.

Alaoui-Selsouli, M., Mohafid, A., & Najid, N. M. (2012). Lagrangian relaxation heuristic for integrated production and maintenance planning problem. *International Journal of Production Research*, *50*(13), 3630–3642. doi:10.1080/00207543.2012.671586

Alarco, X. N,.J., Deleruyelle, T., Pannier, P., & Egels, M. (2010). *A new spiral antenna for passive UHF RFID tag on different substrates.* Paper presented at the Antennas and Propagation (EuCAP).

Alba, E., & Dorronsoro, B. (2008). A hybrid cellular genetic algorithm for the capacitated vehicle routing problem. *Studies in Computational Intelligence*, *82*, 379–422. doi:10.1007/978-3-540-75396-4_14

Alebachew, D., & Yimer, K. D. (2009). Fuzzy scheduling of job orders in a two-stage flowshop with batch-processing machines. *International Journal of Approximate Reasoning*, *50*(1), 117–137. doi:10.1016/j.ijar.2007.08.013

Aleman, R., & Hill, R. (2010). A tabu search with vocabulary building approach for the vehicle routing problem with split demands. *International Journal of Meta-heuristics*, *1*, 55–80.

Aleman, R., Zhang, X., & Hill, R. (2010). An adaptive memory algorithm for the split delivery vehicle routing problem. *Journal of Heuristics*, *16*(3), 441–473. doi:10.1007/s10732-008-9101-3

Alguacil, N., & Conejo, A. (2000). Multiperiod optimal power flow using benders decomposition. *IEEE Transactions on Power Systems*, *15*(1), 196–201. doi:10.1109/59.852121

AlHajri, M., El-Naggar, K., AlRashidi, M., & Al-Othman, A. (2012). Optimal extraction of solar cell parameters using pattern search. *Renewable Energy*, *44*(0), 238–245. doi:10.1016/j.renene.2012.01.082

Alhawari, A. R. H., Ismail, A., Jalal, A. S. A., Raja Abdullah, R. S. A., & Rasid, M. F. A. (2013). *U-shaped inductively coupled feed RFID tag antenna for gain enhancement.* Paper presented at the RFID-Technologies and Applications (RFID-TA), 2013 IEEE International Conference on.

Ali, A. F., Hassanien, A. E., & Snasel, V. (2014). Memetic artificial bee colony for integer programming. *Communications in Computer and Information Science*, *488*, 268–277. doi:10.1007/978-3-319-13461-1_26

Allahverdi, A., Ng, C. T., Cheng, T. C. E., & Kovalyov, M. Y. (2008). A survey of scheduling problems with setup times or costs. *European Journal of Operational Research*, *187*(3), 985–1032. doi:10.1016/j.ejor.2006.06.060

Allende, G. B., & Still. (2013). Solving bilevel programs with the KKT-approach. *Mathematical Programming, 138*(1-2), 309-332.

Allen, R., Billinton, R., Breipohl, A. M., & Grigg, C. H. (1994). Bibliography on the application of probability methods in power system reliability evaluation 1987-1991. *IEEE Transactions on Power Systems*, *9*(1), 42–44.

Allen, R., Billinton, R., Breipohl, A. M., & Grigg, C. H. (1999). Bibliography on the application of probability methods in power system reliability evaluation. *IEEE Transactions on Power Systems*, *14*(1), 51–57. doi:10.1109/59.744483

Allen, R., Billinton, R., Shahidehpour, S. M., & Singh, C. (1988). Bibliography on the application of probability methods in power system reliability evaluation 1982-1987. *IEEE Transactions on Power Systems*, *3*(4), 1555–1564. doi:10.1109/59.192965

AlRashidi, M., AlHajri, M., El-Naggar, K., & Al-Othman, A. (2011). A new estimation approach for determining the I–V characteristics of solar cells. *Solar Energy*, *85*(7), 1543–1550. doi:10.1016/j.solener.2011.04.013

Altiparmak, F., Gen, M., Lin, L., & Paksoy, T. (2006). A genetic algorithm approach for multi-objective optimization of supply chain networks. *Computers & Industrial Engineering*, *51*(1), 196–215. doi:10.1016/j.cie.2006.07.011

Ambrosino, D., & Sciomachen, A. (2007). A food distribution network problem: A case study. *IMA Journal of Management Mathematics*, *18*, 35–53.

Amit, D. J., Gutfreund, H., & Sompolinsky, H. (1986). Spin-glass models of neural networks. *Physical Review A.*, *32*(2), 1007–1018. doi:10.1103/PhysRevA.32.1007 PMID:9896156

Anandarajah, A. (2003). Mechanism controlling permeability changes in clays due to changes in pore fluids. *Journal of Geotechnical and Geoenvironmental Engineering*, *129*(2), 163–172. doi:10.1061/(ASCE)1090-0241(2003)129:2(163)

Anderson, N. J. (2001). Challenges of Total Asset Visibility. *Army Logistician, 33*(1). Retrieved from http://www.alu.army.mil/alog/issues/JanFeb01/MS602.htm

Andreani, R., Júdice, J. J., Martínez, J. M., & Patrício, J. (2011). A projected–gradient interior–point algorithm for complementarity problems. *Numerical Algorithms*, *57*(4), 457–485. doi:10.1007/s11075-010-9439-0

Anegawa, T. (2009). Desirable characteristics of public quick charger. In *Proceedings of PHEV09.*

Ang, M. C., Ng, K. W., Pham, D. T., & Soroka, A. (2013a). Simulations of PCB Assembly Optimization Based on the Bees Algorithm with TRIZ-Inspired Operators. In Advances in Visual Informatics (pp. 335-346).

Anglania, A., Griecoa, A., Pacellaa, M., & Toliob, T. (1996). Object-oriented modeling and simulation of flexible manufacturing systems: A rule based procedure. *Simulation Modeling Practice and Theory, 10*(3-4), 209-234.

Ang, M. C., Ng, K. W., & Pham, D. T. (2013b). Combining the Bees Algorithm and shape grammar to generate branded product concepts. *Proceedings of the Institution of Mechanical Engineers. Part B, Journal of Engineering Manufacture*, *227*(12), 1860–1873. doi:10.1177/0954405413494922

Apostolopoulos, T., & Vlachos, A. (2010). Application of the firefly algorithm for solving the economic emissions load dispatch problem. *International Journal of Combinatorics*, 2011.

Army, U. S. (1996). *Navstar GPS - user equipment introduction.* Technical Report released by the U.S. Government. Retrieved from www.navcen.uscg.gov/pubs/gps/gpsuser/gpsuser.pdf

Arulampalam, M., Maskell, S., Gordon, N. & Clapp, T. (2002). A tutorial on particle filters for online nonlinear/non-gaussian bayesian tracking. *Signal Processing, IEEE Transactions on*, 174–188.

Arulampalam, M. S., Maskell, S., Gordon, N., & Clapp, T. (2002). A tutorial on particle filters for online nonlinear/non-Gaussian Bayesian tracking. *Signal Processing. IEEE Transactions on*, *50*(2), 174–188. doi:10.1109/78.978374

Askari, H., & Zahiri, S. H. (2012). Decision Function Estimation Using Intelligent Gravitational Search Algorithm. *International Journal of Mach. Learn. and Cyber.*, *3*(2), 163–172. doi:10.1007/s13042-011-0052-x

Askarzadeh, A., & Rezazadeh, A. (2012). Parameter identification for solar cell models using harmony search-based algorithms. *Solar Energy*, *86*(11), 3241–3249. doi:10.1016/j.solener.2012.08.018

Askarzadeh, A., & Rezazadeh, A. (2013). Artificial bee swarm optimization algorithm for parameters identification of solar cell models. *Applied Energy*, *102*(0), 943–949. doi:10.1016/j.apenergy.2012.09.052

Association of Bay Area Governments. (2010). *Bay area electrified vehicle charging infrastructure: Options for accelerating consumer*. Bay Area, CA: Renewable and Appropriate Energy Laboratory.

Atamtürk, A., & Zhang, M. (2007). Two-stage robust network flow and design under demand uncertainty. *Operations Research*, *55*(4), 662–667. doi:10.1287/opre.1070.0428

Atanassov, K. (1986). Intuitionistic fuzzy sets. *Fuzzy Sets and Systems*, *20*(1), 87–96. doi:10.1016/S0165-0114(86)80034-3

Atanassov, K. (1994). New operations defined over the intuitionistic fuzzy sets. *Fuzzy Sets and Systems*, *61*(2), 137–142. doi:10.1016/0165-0114(94)90229-1

Azfanizam, A. S., Pham, D. T., & Faieza, A. A. (2014). Combination of Adaptive Enlargement and Reduction in the Search Neighborhood in the Bees Algorithm. *In Applied. Mechanics of Materials*, *564*, 614–618. doi:10.4028/www.scientific.net/AMM.564.614

Aziz, N.A. AB., Ibrahim, Z., Nawawi, S.W., Sudin, Sh., Mubin, M. & Aziz, K. A. B. (2014). *Synchronous Gravitational Search Algorithm vs Asynchronous Gravitational Search Algorithm: A Statistical Analysis, Frontiers in Artificial Intelligence and Applications*. IOS Press. Doi: 10.3233/978-1-61499-434-3-160

Aziz, N. A. A. B., Ibrahim, Z., Ibrahim, I., Zaidi, M., & Tumari, M. (2013). Synchronous vs Asynchronous Gravitational Search Algorithm. In *Proceedings of First International Conference on Artificial Intelligence, Modeling & Simulation.*

Babaei, M. (2013). A general approach to approximate solutions of nonlinear differential equations using particle swarm optimization. *Applied Soft Computing*, *13*(7), 3354–3365. doi:10.1016/j.asoc.2013.02.005

Bahl, H. C. (1983). Column generation based algorithm for multi-item scheduling. *IIE Transactions*, *15*(2), 136–141. doi:10.1080/05695558308974624

Bahl, P., & Padmanabhan, V. N. (2000). RADAR: An in-building RF-based user location and tracking system. In *Proceedings of the nineteenth Annual Joint Conference of the IEEE Computer and Communications Societies (INFOCOM 00)* (vol. 2, pp. 775-784). doi:10.1109/INFCOM.2000.832252

Bai, Z. (1998). On the monotone convergence of matrix multisplitting relaxation methods for the linear complementarity problem. *IMA Journal of Numerical Analysis*, *18*(4), 509–518. doi:10.1093/imanum/18.4.509

Bai, Z. Z. (2010). Modulus-based matrix splitting iteration methods for linear complementarity problems. *Numerical Linear Algebra with Applications*, *17*(6), 917–933. doi:10.1002/nla.680

Bai, Z.-Z. (1996). The convergence of parallel iteration algorithms for linear complementarity problems. *Computers & Mathematics with Applications (Oxford, England)*, *32*(9), 1–17. doi:10.1016/0898-1221(96)00172-1

Bai, Z.-Z. (1999). On the convergence of the multisplitting methods for the linear complementarity problem. *SIAM Journal on Matrix Analysis and Applications*, *21*(1), 67–78. doi:10.1137/S0895479897324032

Bai, Z.-Z., & Dong, J.-L. (2006). A modified damped Newton method for linear complementarity problems. *Numerical Algorithms*, *42*(3-4), 207–228. doi:10.1007/s11075-006-9028-4

Bai, Z.-Z., & Evans, D. (1997a). Matrix multisplitting relaxation methods for linear complementarity problems. *International Journal of Computer Mathematics*, *63*(3-4), 309–326. doi:10.1080/00207169708804569

Bai, Z.-Z., & Evans, D. (1998). Chaotic iterative methods for the linear complementarity problems. *Journal of Computational and Applied Mathematics*, *96*(2), 127–138.

Bai, Z.-Z., Evans, D. J., & Wang, D. E.-R. E. N. (1997b). A class of multi-parameter relaxed parallel multisplitting methods for large sparse linear complementarity problems. *Parallel Algorithms and Applications, 11*(1-2), 113–127. doi:10.1080/10637199708915590

Bai, Z.-Z., & Zhang, L.-L. (2013). Modulus-based synchronous two-stage multisplitting iteration methods for linear complementarity problems. *Numerical Algorithms, 62*(1), 59–77. doi:10.1007/s11075-012-9566-x

Bakirtzis, A. G. (1992). A probabilistic method for the evaluation of the reliability of stand-alone wind energy systems. *IEEE Transactions on Energy Conversion, 7*(1), 99–107. doi:10.1109/60.124548

Balanis, C. A. (Ed.). (2012). Theory, analysis and design (3rd ed.). John Wiley & Sons.

Balduzi, F., Giua, A., & Seatzu, C. (2001). Modelling and simulation of manufacturing systems with first order hybrid Petri nets. *International Journal of Production Research, 39*(2).

Balochian, S. & Ebrahimi, E. (2013). Parameter Optimization via Cuckoo Optimization Algorithm of Fuzzy Controller for Liquid Level Control. *Journal of Engineering*. doi:.10.1155/2013/982354

Baniassadi, Z., Nezamabadi-pour, H., & Farsangi, M. M. (2011). A multi-objective Solution of Gravitational Search Algorithm for Benchmark Functions and Placement of SVC. *Intelligent Systems in Electrical Engineering, 1*(1), 59–78.

Bansal, J. C., Sharma, H., Arya, K. V., & Nagar, A. (2013). Memetic search in artificial bee colony algorithm. *Soft Computing, 17*(10), 1911–1928. doi:10.1007/s00500-013-1032-8

Baquela, E. G., & Olivera, A. C. (2014, October). *Combining Genetic Algorithms with Traffic Simulations for restructuring traffic networks subject to increases in population.* Paper presented in the International Conference on Metaheuristics and Nature Inspired Computing. Marrakech, Morocco.

Barabadi, A., Barabady, J., & Markeset, T. (2011). Maintainability analysis considering time-dependent and time-independent covariates. *Reliability Engineering & System Safety, 96*(1), 210–217. doi:10.1016/j.ress.2010.08.007

Barabady, J., & Kumar, U. (2005). *Maintenance schedule by using reliability analysis: A case study at jajarm bauxite mine of Iran. 20th World Mining Congress and EXPO*, Tehran, Iran.

Barabady, J., & Kumar, U. (2008). Reliability analysis of mining equipment: A case study of a crushing plant at jajarm bauxite mine in iran. *Reliability Engineering & System Safety, 93*(4), 647–653. doi:10.1016/j.ress.2007.10.006

Bard, J. F. (1999). *Practical Bilevel Optimization: Algorithms and Applications*. Academic Press.

Bard, J. F. (1991). Some properties of the bilevel programming problem. *Journal of Optimization Theory and Applications, 68*(2), 371–378. doi:10.1007/BF00941574

Bard, Y. (1974). *Nonlinear parameter estimation*. Academic Press New York.

Barlow, R. E., & Hunter, L. C. (1960). Optimum preventive maintenance policies. *Operations Research, 8*(1), 90–100. doi:10.1287/opre.8.1.90

Barndorff-Nielsen, O. (1977). Exponentially decreasing distributions for the logarithm of particle size. *Proceedings of the Royal Society of London. A. Mathematical and Physical Sciences, 353*(1674), 401-419. doi:10.1098/rspa.1977.0041

Barndorff-Nielsen, O. (1977). *Exponentially decreasing distributions for the logarithm of particle size*. Proc. Roy. Soc.

Barnes, J. B., Rizos, C., Pahwa, A., Politi, N., & Van Cranenbroeck, J. (2007). The potential of Locata technology for structural monitoring applications. *Journal of Global Positioning Systems, 6*(2), 166–172. doi:10.5081/jgps.6.2.166

Barnes, J. B., Rizos, C., Wang, J., Small, D., Voigt, G., & Gambale, N. (2003). LOCATA: A new positioning technology for high precision indoor and outdoor positioning.*Proceedings of the 16th International Technical Meeting of the Satellite Division of The Institute of Navigation (ION GPS/GNSS 2003).*Portland, OR.

Barnes, J. B., Wang, J., Rizos, C., & Tsujii, T. (2002). The performance of a Pseudolite based positioning system for deformation monitoring. In *Proceedings of the 2nd Symposium on Geodesy for Geotechnical and Structural Engineering*. Berlin, Germany.

Barrett, R. (1994). *Templates for the solution of linear systems: building blocks for iterative methods*. Siam. doi:10.1137/1.9781611971538

Barr, R. S., Golden, B. L., Kelly, J. P., Resende, M. G., & Stewart, W. R. Jr. (1995). Designing and reporting on computational experiments with heuristic methods. *Journal of Heuristics*, *1*(1), 9–32. doi:10.1007/BF02430363

Bartolini, E., Cordeau, J. F., & Laporte, G. (2013). Improved lower bounds and exact algorithm for the capacitated arc routing problem. *Mathematical Programming*, *137*(1-2), 409–452. doi:10.1007/s10107-011-0497-4

Barton, P. I., & Lee, C. K. (2004). Design of process operations using hybrid dynamic optimization. *Computers & Chemical Engineering*, *28*(6-7), 955–969. doi:10.1016/j.compchemeng.2003.09.015

Basu, M. (2003). Hopfield neural networks for optimal scheduling of fixed head hydrothermal power systems. *Electric Power Systems Research*, *64*(1), 11–15. doi:10.1016/S0378-7796(02)00118-9

Basu, M. (2004). An interactive fuzzy satisfying method based on evolutionary programming technique for multiobjective short-term hydrothermal scheduling. *Electric Power Systems Research*, *69*(2-3), 277–285. doi:10.1016/j.epsr.2003.10.003

Basu, M. (2005). A simulated annealing-based goal-attainment method for economic emission load dispatch of fixed head hydrothermal power systems. *International Journal of Electrical Power & Energy Systems*, *27*(2), 147–153. doi:10.1016/j.ijepes.2004.09.004

Basu, M. (2011). Artificial immune system for fixed head hydrothermal power system. *Energy*, *36*(1), 606–612. doi:10.1016/j.energy.2010.09.057

Basu, M., & Chowdhury, A. (2013). Cuckoo search algorithm for economic dispatch. *Energy*, *60*, 99–108. doi:10.1016/j.energy.2013.07.011

Basu, S. (2012). Tabu search implementation on the traveling salesman problem and its variations: A literature survey. *American Journal of Operations Research*, *2*(02), 163–173. doi:10.4236/ajor.2012.22019

Bathe, K. J. (1996). *Finite Element Procedures* (2nd ed.). Prentice Hall.

Bayazit, O. (2005). Use of AHP in decision-making for flexible manufacturing systems. *Journal of Manufacturing Technology Management*, *16*(7), 808–819. doi:10.1108/17410380510626204

Bazaraa, M. S., Sherali, H. D., & Shetty, C. M. (2013). *Nonlinear programming: theory and algorithms*. John Wiley & Sons.

Beal, M. (2003). *Variational algorithms for approximate bayesian inference*. (Unpublished doctoral dissertation). Gatsby Computational Neuroscience Unit, University College London.

Behrisch, M., Bieker, L., Erdmann, J., & Krajzewicz, D. (2011). Sumo - simulation of urban mobility: An overview. In *SIMUL 2011,The Third International Conference on Advances in System Simulation* (pp. 63-68), Barcelona, Spain.

Belenguer, J., Benavent, E., Labadi, N., Prins, C., & Reghioui, M. (2010). Split delivery capacitated arc routing problem: Lower bound and metaheuristic. *Transportation Science*, *44*(2), 206–220. doi:10.1287/trsc.1090.0305

Belfiore, P., & Yoshizaki, H. (2009). Scatter search for a real-life heterogeneous fleet vehicle routing problem with time windows and split deliveries in Brazil. *European Journal of Operational Research, 199*(3), 750–758. doi:10.1016/j.ejor.2008.08.003

Belov, I. A. (2005). On the computation of the probability density function of α-stable distributions. Mathematical Modelling and Analysis.*Proceedings of the 10th International Conference MMA 2005.*

Belov, I., Kabasinskas, A., & Sakalauskas, L. (2006). A study of stable models of stock markets. *Information Technology and Control., 35*(1), 34–56.

Benassi, M., & White, R. (1994). Parallel numerical solution of variational inequalities. *SIAM Journal on Numerical Analysis, 31*(3), 813–830. doi:10.1137/0731044

Ben-Brahim, L., & Tadakuma, S. (1998). Speed control of induction motor without rotational transducers. In *Industry Applications Conference*. doi:10.1109/IAS.1998.732393

Beni, G., & Wang, J. (1989). Swarm Intelligence in Cellular Robotic Systems. *Advanced Workshop on Robots and Biological Systems*, Tuscany, Italy.

Benioff, P. (1980). The computer as a physical system: A microscopic quantum mechanical Hamiltonian model of computers as represented by Turing machines. *Journal of Statistical Physics, 22*(5), 563–591. doi:10.1007/BF01011339

Benjaar, S., & Elhafs, M. (2006). Production and inventory control of a single product assemble to order system with multiple customer classes. *Management Science, 52*(12), 1896–1912. doi:10.1287/mnsc.1060.0588

Ben-Tal, A., Goryashko, A., Guslitzer, E., & Nemirovski, A. (2004). Adjustable robust solutions of uncertain linear programs. *Mathematical Programming, 99*(2), 351–376. doi:10.1007/s10107-003-0454-y

Ben-Tal, A., & Nemirovski, A. (1999). Robust solutions of uncertain linear programs. *Operations Research Letters, 25*(1), 1–13. doi:10.1016/S0167-6377(99)00016-4

Ben-Tal, A., & Nemirovski, A. (2000). Robust solution of linear programming problems contaminated with uncertain data. *Mathematical Programming, 88*(3), 411–424. doi:10.1007/PL00011380

Ben-Tal, A., & Nemirovski, A. (2002). Robust Optimization-Methodology and applications. *Mathematical Programming, 92*(3), 453–480. doi:10.1007/s101070100286

Benzi, M. (1997). Remarks on the numerical solution of certain linear complementarity problems. *Journal of Computational and Applied Mathematics, 83*(1), 137–143. doi:10.1016/S0377-0427(97)00103-9

Berg, M., Tymoczko, L., & Stryer, L. (2006). Biochemistry (6th ed.). New York, NY: Freeman W.

Berman, A. & R. J. Plemmons (1979). *Nonnegative matrices*. The Mathematical Sciences, Classics in Applied Mathematics.

Bertsekas, D. P. (1982). Projected Newton methods for optimization problems with simple constraints. *SIAM Journal on Control and Optimization, 20*(2), 221–246. doi:10.1137/0320018

Bertsekas, D. P. (2005). *Dynamic programming and optimal control* (2 vols.). Athena Scientific Massachussets.

Bertsimas, D. P., & Sim, M. (2003). Robust discrete optimization and network flows. *Math. Prog. B, 98*(1-3), 49–71. doi:10.1007/s10107-003-0396-4

Bertsimas, D. P., & Sim, M. (2004). The price of robustness. *Operations Research, 52*(1), 35–53. doi:10.1287/opre.1030.0065

Beume, N., Naujoks, B., & Emmerich, M. (2007). SMS-EMOA: Multiobjective selection based on dominated hypervolume. *European Journal of Operational Research, 181*(3), 1653–1669. doi:10.1016/j.ejor.2006.08.008

Bhandari, A. K., Singh, V. K., Kumar, A., & Singh, G. K. (2014). Cuckoo search algorithm and wind driven optimization based study of satellite image segmentation for multilevel thresholding using Kapur's entropy. *Expert Systems with Applications*, *41*(7), 3538–3560. doi:10.1016/j.eswa.2013.10.059

Bhargava, V., Fateen, S. E. K., & Bonilla-Petriciolet, A. (2013). Cuckoo Search: A new nature-inspired optimization method for phase equilibrium calculations. *Fluid Phase Equilibria*, *337*, 191–200. doi:10.1016/j.fluid.2012.09.018

Bhatta, B., Saraswati, S., & Bandyopadhyay, D. (2010). Urban sprawl measurement from remote sensing data. *Applied Geography (Sevenoaks, England)*, *30*(4), 731–740. doi:10.1016/j.apgeog.2010.02.002

Bhowmik, M., & Pal, M. (2008). Generalized intuitionistic fuzzy matrices. *Far East Journal of Mathematical Sciences*, *29*(3), 533–554.

Bilge, U., Kirac, F., Kurtulan, M., & Pekgun, P. A. (2004). Tabu search algorithm for parallel machine total tardiness problem. *Computers & Operations Research*, *31*(3), 397–414. doi:10.1016/S0305-0548(02)00198-3

Billingsley, P. (1961). Statistical methods in Markov chains. *Annals of Mathematical Statistics*, *32*(1), 12–40. doi:10.1214/aoms/1177705136

Billinton, R., & Allan, R. (1996). *Reliability evaluation of power systems* (2nd ed.). New York: Plenum Press. doi:10.1007/978-1-4899-1860-4

Billinton, R., Chen, H., & Ghajar, R. (1996). A sequential simulation technique for adequacy evaluation of generating systems including wind energy. *IEEE Transactions on Energy Conversion*, *11*(4), 728–734. doi:10.1109/60.556371

Billinton, R., Fotuhi-Firuzabad, M., & Bertling, L. (2001). Bibliography on the application of probability methods in power system reliability evaluation 22 1996-1999. *IEEE Transactions on Power Systems*, *16*(4), 595–602. doi:10.1109/59.962402

Billinton, R., & Gan, L. (1991). Use of Monte Carlo simulation in teaching generating capacity adequacy assessment. *IEEE Transactions on Power Systems*, *6*(4), 1571–1577. doi:10.1109/59.117004

Billinton, R., & Li, W. Y. (1994). *Reliability assessment of electrical power systems using Monte Carlo method.* New York: Plenum Press. doi:10.1007/978-1-4899-1346-3

Bin, L., Jianhua, Z., Baiqiang, Y., & Weiming, X. (2008). *A dual-frequency ceramic spiral antenna with rectangle PBG structure array used for modern RFID system.* Paper presented at the Wireless, Mobile and Multimedia Networks (IC-WMMN 2008), IET 2nd International Conference on.

Binghui, Y., Xiaohui, Y., & Jinwen, W. (2007). Short-term hydro-thermal scheduling using particle swarm optimization method. *Energy Conversion and Management*, *48*(7), 1902–1908. doi:10.1016/j.enconman.2007.01.034

Birge, J. R. (1985). Decomposition and partitioning methods for multistage stochastic linear programs. *Operations Research*, *33*(5), 989–1007. doi:10.1287/opre.33.5.989

Birnie, D. P. III. (2009). Solar-to-vehicle (S2V) systems for powering commuters of the future. *Journal of Power Sources*, *186*(2), 539–542. doi:10.1016/j.jpowsour.2008.09.118

Bjorninen, T., Nikkari, M., Ukkonen, L., Fan, Y., Elsherbeni, A., Sydanheimo, L., & Kivikoski, M. (2008). *Design and RFID signal analysis of a meander line UHF RFID tag antenna.* Paper presented at the Antennas and Propagation Society International Symposium.

Bodin, L., & Golden, B. (1981). Classification in vehicle routing and scheduling. *Networks*, *11*(2), 97–108. doi:10.1002/net.3230110204

Bolduc, M., Laporte, G., Renaud, J., & Boctor, F. (2010). A tabu search heuristic for the split delivery vehicle routing problem with production and demand calendars. *European Journal of Operational Research, 202*(1), 122–130. doi:10.1016/j.ejor.2009.05.008

Bonenberg, L. K. (2014). *Closely-coupled integration of Locata and GPS for engineering applications.* (doctoral dissertation). The University of Nottingham.

Bonenberg, L. K., Roberts, G. W., & Hancock, C. M. (2012). Using Locata to augment GNSS in a kinematic urban environment. *Archives of Photogrammetry. Cartography and Remote Sensing, 12*, 63–74.

Bonkoungou, Koalaga, & Njomo. (n.d.). Modelling and simulation of photovoltaic module considering signale-diode equivalent circuit model in MATLAB. *International Journal of Emerging Technology and Advanced Engineering.*

Boran, F. E. (2011). An integrated intuitionistic fuzzy multi criteria decision making method for facility location selection. *Mathematical and Computational Applications, 16*(2), 487–496.

Bora, T. C., Coelho, L. S., & Lebensztajn, L. (2012). Bat-inspired optimization approach for the brushless DC wheel motor problem. Magnetics. *IEEE Transactions on, 48*(2), 947–950.

Botsford, C., & Szczepanek, A. (2009). Fast charging vs. slow charging: Pros and cons for the new age of electric vehicles. In Proceedings of EVS24.

Bouchekara, H. (2013). *Optimal Design of Electromagnetic Devices Using a Black-Hole-Based Optimization Technique.* Academic Press.

Boute, R., Disney, S., Lambrecht, M., & Houdt, B. (2007). An integrated production and inventory model to dampen upstream demand variability in the supply chain. *European Journal of Operational Research, 178*(1), 121–142. doi:10.1016/j.ejor.2006.01.023

Boyce, W. E., & Diprima, R. C. (1997). *Elementary Differential Equations and Boundary Value Problems* (6th ed.). New York: John Wiley & Sons.

Boyle, G. (Ed.). (2007). *Renewable electricity and the grid: the challenge of variability.* Earthscan.

Bozdogan, H. (2004). Intelligent statistical data mining with information complexity and genetic algorithms. In H. Bozdogan (Ed.), *Statistical Data Mining & Knowledge Discovery* (pp. 15–56). Boca Raton, FL: Chapman & Hall/CRC.

Brahimi, N., Dauzère-Pérès, S., & Najid, M. N. (2006). Capacitated multi-item lot-sizing problems with time windows. *Operations Research, 54*(5), 951–967. doi:10.1287/opre.1060.0325

Brahimi, N., Dauzere-Peres, S., Najid, N. M., & Nordli, A. (2006). Single items lot sizing problems. *European Journal of Operational Research, 168*(1), 1–16. doi:10.1016/j.ejor.2004.01.054

Brain, G., & Mark, S. (2007). Garbage in, energy out: Landfill gas opportunities for CHP projects. *Cogeneration and On-Site Power, 8*(5), 37–45.

Brandt, A. & C. W. Cryer (1983). Multigrid algorithms for the solution of linear complementarity problems arising from free boundary problems. *SIAM Journal on Scientific and Statistical Computing, 4*(4), 655-684.

Breiman, L., Friedman, J. H., Ohlsen, R. A., & Stone, C. J. (1984). *Classification and Regression Trees.* Belmont: Wadsworht.

Brisbourne, A. (2013, February 25). The Current State Of The Internet Of Things. *Field Technologies Online Magazine.*

Bris, R., Chatelet, E., & Yalaoui, F. (2003). New method to minimize the preventive maintenance cost of series-parallel systems. *Reliability Engineering & System Safety*, *82*(3), 247–255. doi:10.1016/S0951-8320(03)00166-2

Brooke, A., Kendrick, D., & Meeraus, A. (1998). *GAMS: A User's Guide*. The Scientific Press.

Brooke, A., Kendrick, D., & Meeraus, A. (2010a). *GAMS / Cplex 12. User Notes*. GAMS Development Corporation.

Brooke, A., Kendrick, D., & Meeraus, A. (2010b). *GAMS / DICOPT. User Notes*. GAMS Development Corporation.

Brotcorne, L., Hanafi, S., & Mansi, R. (2013). One-level reformulation of the bilevel knapsack problem using dynamic programming. *Discrete Optimization*, *10*(1), 1–10. doi:10.1016/j.disopt.2012.09.001

Brown, C., Liebovitch, S. L., & Glendon, R. (2007). Lévy ñights in Dobe Ju/'hoansi foraging patterns. *Human Ecology*, *35*(1), 129–138. doi:10.1007/s10745-006-9083-4

Bru, R., Elsner, L., & Neumann, M. (1988). Models of parallel chaotic iteration methods. *Linear Algebra and Its Applications*, *103*, 175–192. doi:10.1016/0024-3795(88)90227-3

Bru, R., Pedroche, F., & Szyld, D. B. (2004). Overlapping additive and multiplicative Schwarz iterations for H-matrices. *Linear Algebra and Its Applications*, *393*, 91–105. doi:10.1016/j.laa.2003.10.022

Burke, J. V., & Xu, S. (1998). The global linear convergence of a noninterior path-following algorithm for linear complementarity problems. *Mathematics of Operations Research*, *23*(3), 719–734. doi:10.1287/moor.23.3.719

Buzacot, J. A., & Shatikumar, J. (1993). *Stochastic Models of Manufacturing Systems*. Prentice Hall.

Caffery, J. J. (2000). *Wireless location in CDMA cellular radio systems*. Springer.

Caggiani, L., & Ottomanelli, M. (2011). Traffic equilibrium network design problem under uncertain constraints. *Procedia: Social and Behavioral Sciences*, *20*, 372–380. doi:10.1016/j.sbspro.2011.08.043

Cagnina, L., Esquivel, S., & Coello Coello, C. A. (2005). A Particle Swarm Optimizer for Multi-Objective Optimization. *JCST*, *4*(5), 204–210.

Cai, L. J., Erlich, I., & Stamtsis, G. (2004, October). Optimal choice and allocation of FACTS devices in deregulated electricity market using genetic algorithms. In Power Systems Conference and Exposition, 2004. IEEE PES (pp. 201-207). IEEE. doi:10.1109/PSCE.2004.1397562

Calabrese, C., & Marrocco, G. (2008). Meandered-Slot Antennas for Sensor-RFID Tags. *IEEE Antennas and Wireless Propagation Letters*, *7*, 5–8. doi:10.1109/LAWP.2007.914123

Canyurt, O. E., & Hajela, P. (2010). Cellular genetic algorithm technique for the multicriterion design optimization. *Struct Multidiscip Optim*, *40*(1–6), 201–214. doi:10.1007/s00158-008-0351-3

Cao, H., Kang, L., Chen, Y., & Yu, J. (2000). Evolutionary modeling of systems of ordinary differential equations with genetic programming. *Genetic Programming and Evolvable Machines*, *1*(4), 309–337. doi:10.1023/A:1010013106294

Caramanis, M., & Foster, J. M. (2009). Management of electric vehicle charging to mitigate renewable generation intermittency and distribution network congestion. In *Proceedings of the 48th IEEE Conference on Decision and Control*. Shanghai: IEEE. doi:10.1109/CDC.2009.5399955

Carpentier, P., Cohen, G., Culioli, J. C., & Renaud, A. (2012). Stochastic optimization of unit commitment: A new decomposition framework. *IEEE Transactions on Power Systems*, *11*(2), 1067–1073. doi:10.1109/59.496196

Cartwright, N. (1983). *How the Laws of Physics Lie*. Oxford University Press. doi:10.1093/0198247044.001.0001

Carvalho, M. F., & Soares, S. (1987). An efficient hydrothermal scheduling algorithm. *IEEE Transactions on Power Systems*, *2*(3), 537–542. doi:10.1109/TPWRS.1987.4335164

Castellani, M., Pham, Q. T., & Pham, D. T. (2012). Dynamic optimization by a modified bees algorithm. *Proceedings of the Institution of Mechanical Engineers. Part I, Journal of Systems and Control Engineering*, *226*(7), 956–971. doi:10.1177/0959651812443462

Castillo, W., & Trejos, J. (2002). Two-mode partitioning: review of methods and application of tabu search. In K. Jajuga, A. Sokolowski, & H.-H. Bock (Eds.), *Classification, Clustering and Data Analysis* (pp. 43–51). Berlin: Springer. doi:10.1007/978-3-642-56181-8_4

Castro, P. M., Grossmann, I. E., & Novais, A. Q. (2006). Two new continuous-time models for the scheduling of multistage batch plants with sequence dependent changeovers. *Industrial & Engineering Chemistry Research*, *45*(18), 6210–6226. doi:10.1021/ie060449m

Çatay, B. (2009). Ant colony optimization and its application to the vehicle routing problem with pickups and deliveries. In R. Chiong & S. Dhakal (Eds.), *Natural Intelligence for Scheduling, Planning and Packing Problems, Studies in Computational Intelligence* (Vol. 250, pp. 219–244). Heidelberg: Springer. doi:10.1007/978-3-642-04039-9_9

Cattrysse, D., Maes, J., & Van Wassenhove, L. N. (1990). Set partitioning and column generation heuristics for capacitated dynamic lotsizing. *European Journal of Operational Research*, *46*(1), 38–47. doi:10.1016/0377-2217(90)90296-N

Cau, T. D. H., & Keyes, R. J. (2002). Evolutionary optimisation method for multistorage hydrothermal scheduling. *IEE Proceedings. Generation, Transmission and Distribution*, *149*(2), 1359–7051. doi:10.1049/ip-gtd:20020120

Celik, A. N., & Acikgoz, N. (2007). Modelling and experimental verification of the operating current of mono-crystalline photovoltaic modules using four-and five-parameter models. *Applied Energy*, *84*(1), 1–15. doi:10.1016/j.apenergy.2006.04.007

Celik, Y., & Ulker, E. (2013). An Improved Marriage in Honey Bees Optimization Algorithm for Single Objective Unconstrained Optimization. *TheScientificWorldJournal*, *2013*, 1–11. doi:10.1155/2013/370172 PMID:23935416

Cerny, V. (1985). Thermodynamical Approach to the Traveling Salesman Problem: An Efficient Simulation Algorithm. *Journal of Optimization Theory and Applications*, *45*(1), 41–52. doi:10.1007/BF00940812

Ceylan, H., & Bell, M. (2005). Genetic algorithm solution for the stochastic equilibrium transportation networks under congestion. *Transportation Research Part B: Methodological*, *39*(2), 169–185. doi:10.1016/j.trb.2004.04.001

Chahine, M. T. (1970). Inverse problems in radiative transfer: Determination of atmospheric parameters. *Journal of the Atmospheric Sciences*, *27*(6), 960–967. doi:10.1175/1520-0469(1970)027<0960:IPIRTD>2.0.CO;2

Chambers, J. M., Mallows, C. L., & Stuck, B. W. (1976). A Method for Simulating Stable Random Variables. *JASA*, *71*(354), 340–344. doi:10.1080/01621459.1976.10480344

Chan, D. S., & Phang, J. C. H. (1987). Analytical methods for the extraction of solar-cell singleand double-diode model parameters from I-V characteristics. *IEEE Transactions on Electron Devices*, *34*(2), 286–293. doi:10.1109/T-ED.1987.22920

Chandra-Satapathy, S., Murthym, J. V. R., Prasad-Reddy, P. V. G. D., Misra, B. B., Dash, P. K., & Panda, G. (2009). Particle swarm optimized multiple regression linear model for data classification. *Applied Soft Computing*, *9*(2), 470–476. doi:10.1016/j.asoc.2008.05.007

Chandrasekaran, K., & Simon, S. P. (2014). Wind-Thermal Integrated Power System Scheduling Problem Using Cuckoo Search Algorithm. *International Journal of Operations Research and Information Systems*, *5*(3), 81–109. doi:10.4018/ijoris.2014070104

Chandrasekaran, R. (1970). A special case of the complementary pivot problem. *Opsearch*, *7*, 263–268.

Chang, W. Y. (2013). The state of charge estimating methods for battery: a review. *ISRN Applied Mathematics*.

Chan, P. H., & Chang, H. C. (1996). Genetic aided scheduling of hydraulically coupled plants in hydro-thermal coordination. *IEEE Transactions on Power Systems*, *11*(2), 975–981. doi:10.1109/59.496183

Chao, G. (2013). Production and availability policies through the Markov Decision Process and myopic methods for contractual and selective orders. *European Journal of Operational Research*, *225*(3), 383–392. doi:10.1016/j.ejor.2012.10.005

Chaovalitwongse, W., Pardalos, P. M., & Prokoyev, O. A. (2004). A new linearization technique foe multi-quadratic 0-1 programming problems. *Operations Research Letters*, *32*(6), 517–522. doi:10.1016/j.orl.2004.03.005

Charles, J. P., Abdelkrim, M., Muoy, Y. H., & Mialhe, P. (1981). A practical method of analysis of the current–voltage characteristics of solar cells. *Solar Cells*, *4*(2), 169–178. doi:10.1016/0379-6787(81)90067-3

Charles, J. P., Ismail, M. A., & Bordure, G. (1985). A critical study of the effectiveness of the single and double exponential models for I–V characterization of solar cells. *Solid-State Electronics*, *28*(8), 807–820. doi:10.1016/0038-1101(85)90068-1

Charman, P., Bhavaraju, P., & Billington, R. (1979). IEEE reliability test system. *IEEE Transactions on Power Apparatus and Systems*, *98*(6), 2047–2054.

Charnes, A., & Cooper, W. W. (1959). Chance Constrained Programming. *Management Science*, *6*(1), 73–79. doi:10.1287/mnsc.6.1.73

Chen & Thizy. (2002). *An efficient column generation algorithm for the multi-item capacitated lot-sizing problem*. Technical Report 02-52, University of Ottawa.

Chen, B., & Harker, P. T. (1993). A non-interior-point continuation method for linear complementarity problems. *SIAM Journal on Matrix Analysis and Applications*, *14*(4), 1168–1190. doi:10.1137/0614081

Chen, C. L., & Lee, W. C. (2004). Multi-objective optimization of multi-echelon supply chain networks with uncertain product demands and prices. *Computers & Chemical Engineering*, *28*(6), 1131–1144. doi:10.1016/j.compchemeng.2003.09.014

Chen, G., Xie, Q., & Shieh, L. (1998). Fuzzy kalman filtering. *Journal of Information Science*, *109*(1-4), 197–209. doi:10.1016/S0020-0255(98)10002-6

Cheng, B., Li, K., & Chen, B. (2010). Scheduling a single batch-processing machine with non-identical job sizes in fuzzy environment using an improved ant colony optimization. *Journal of Manufacturing Systems*, *29*(1), 29–34. doi:10.1016/j.jmsy.2010.06.007

Chen, H., Li, S., & Tang, Z. (2011). Hybrid GSA with Random-Key Encoding Scheme Combined with Simulated Annealing. *Journal of Computer Science and Network Security*, *11*, 208–217.

Chen, H., Zhu, Y., Ma, L., & Niu, B. (2014). Multiobjective RFID network optimization using multiobjective evolutionary and swarm intelligence approaches. *Mathematical Problems in Engineering*, 2014.

Chen, K., & Ji, P. (2007). A mixed integer programming model for advanced planning and scheduling (APS). *European Journal of Operational Research*, *181*(1), 515–522. doi:10.1016/j.ejor.2006.06.018

Chen, S. Y., & Hsu, P. (2004). CPW-fed folded-slot antenna for 5.8 GHz RFID tags. *Electronics Letters*, *40*(24), 1516–1517. doi:10.1049/el:20046780

Chen, S., & Lin, C. (2007). Multiple DNA sequence alignment based on genetic algorithms and divide and conquer techniques. *Information and Management Science Journal*, *18*(2), 76–111.

Chen, W. H., & Thizy, J. M. (1990). Analysis of relaxations for the multi-item capacitated lot- sizing problem. *Annals of Operations Research, 26*(1), 29–72. doi:10.1007/BF02248584

Chen, Y.-F. R., Gansner, E. R., & Koutsofios, E. (1997). *A C++ data model supporting reachability analysis and dead code detection Software Engineering—ESEC/FSE'97*. Springer.

Che, Z. H. (2012). A particle swarm optimization algorithm for solving unbalanced supply chain planning problems. *Applied Soft Computing, 12*(4), 1279–1287. doi:10.1016/j.asoc.2011.12.006

Che, Z. H., & Chiang, C. J. (2010). A modified Pareto genetic algorithm for multi-objective build-to-order supply chain planning with product assembly. *Advances in Engineering Software, 41*(7), 1011–1022. doi:10.1016/j.advengsoft.2010.04.001

Chiasson, J., & Vairamohan, B. (2005). Estimating the state of charge of a battery. *IEEE Transactions on Control Systems Technology, 13*(3), 465–470. doi:10.1109/TCST.2004.839571

Chisholm, G., & Hammons, M. J. (2006). Enabling Total Asset Visibility. *Defense Transportation Journal.* Retrieved from http://www.ndtahq.com/images/pdf/total_asset_visibility.pdf

Chis, M. (2010). Introduction: A survey of the evolutionary computation techniques for software engineering. In M. Chis (Ed.), *Evolutionary Computation amd Optimization Algorithms in Software Engineering: Applications and Techniques* (pp. 1–12). Hershey, PA: Information Science Reference; doi:10.4018/978-1-61520-809-8.ch001

Chitra, T. (2003). Life based maintenance policy for minimum cost. In *Proceedings of the annual reliability and maintainability symposium,* (pp. 470 – 474). doi:10.1109/RAMS.2003.1182034

Choi, J., Jeong, S., Bo, S., Park, J., El-Keib, A. A., & Watada, J. (2007). CO2, sox and NOx emissions constrained multi-criteria-best generation mix using fuzzy set theory. In *Proceedings, LESCOPE07- 2007 Large Engineering Systems Conference on Power Engineering.*

Choi, T. J., Ahn, C. W., & An, J. (2013). An Adaptive Cauchy Differential Evolution Algorithm for Global Numerical Optimization. *TheScientificWorldJournal, 2013,* 1–12. doi:10.1155/2013/969734 PMID:23935445

Chou, F. D., Chang, P. C., & Wang, H. M. (2006). A hybrid genetic algorithm to minimize makespan for the single batch machine dynamic scheduling problem. *International Journal of Advanced Manufacturing Technology, 31*(3-4), 350–359. doi:10.1007/s00170-005-0194-7

Chou, F. D., & Wang, H. M. (2008). A scheduling for a single semiconductor batch-processing machine to minimize total weighted tardiness. *Journal of the Chinese Institute of Industrial Engineers, 25*(2), 136–147. doi:10.1080/10170660809509079

Christie, R. (1993). *IEEE 118 bus test case.* College of Engineering, Electric Engineering, University of Washington. Available in: http://www.ee.washington.edu/research/pstca

Christopherson, D. G. (1941). A new mathematical method for the solution of film lubrication problems. *Proceedings - Institution of Mechanical Engineers, 146*(1), 126–135. doi:10.1243/PIME_PROC_1941_146_027_02

Christou, A., Lagodimos, A., & Lycopoulou, D. (2007). Hierarchical production planning for multi-product lines in the beverage industry: Production Planning and Control. *The Management of Operations, 18*(5).

Ciarlet, P. G. (1978). *The finite element method for elliptic problems.* Elsevier.

Civicioglu, P., & Besdok, E. (2013). A conceptual comparison of the Cuckoo-search, particle swarm optimization, differential evolution and artificial bee colony algorithms. *Artificial Intelligence Review, 39*(4), 315–346. doi:10.1007/s10462-011-9276-0

Clark, M. (2005). *JDepend*. Retrieved from http://www. clarkware. com/software/JDepend. html

Clerc, M. (1999). *The swarm and the queen: towards a deterministic and adaptive particle swarm optimization.* Paper presented at the 1999 Congress on Evolutionary Computation, Washington, DC. doi:10.1109/CEC.1999.785513

Cobb, S. H. (1997). *GPS Pseudolites: Theory, design, and applications.* (doctoral dissertation). Stanford University.

Cochran, D., Arinez, J., Duda, J., & Linck, J. (2002). A decomposition approach for manufacturing system design. *Journal of Manufacturing Systems*, *20*(6), 371–389. doi:10.1016/S0278-6125(01)80058-3

Coello, C. A. C., & Pulido, G. T. (2005). Multiobjective structural optimization using a micro genetic algorithm. *Structtural and Multidisciplinary Optimization*, *30*(5), 388–403. doi:10.1007/s00158-005-0527-z

Cook, W. D., & Kress, M. (1991). A multiple criteria decision model with ordinal preference data. *European Journal of Operational Research*, *54*(2), 191–198. doi:10.1016/0377-2217(91)90297-9

Cord'on, O., Herrera, F., Hoffmann, F., & Magdalena, L. (2001). Genetic Fuzzy Systems: Evolutionary Tuning and Learning of Fuzzy Knowledge Bases. World Scientific Publishing.

Cordon, O., Herrera, F., & Zwir, I. (2002). Francisco Herrera & Igor Zwir, Linguistic Modeling of Hierarchical Systems of Linguistic Rules. *IEEE Transactions on Fuzzy Systems*, *10*(1), 2–20. doi:10.1109/91.983275

Cormen, T. H., Leiserson, C. E., Rivest, R. L., & Stein, C. (2009). *Introduction to Algorithms (3rd ed.).* Cambridge, MA: The MIT Press.

Coşkun, S. B., & Atay, M. T. (2008). Fin efficiency analysis of convective straight fins with temperature dependent thermal conductivity using variational iteration method. *Applied Thermal Engineering*, *28*(17-18), 2345–2352. doi:10.1016/j.applthermaleng.2008.01.012

Cosser, E. (2004). *Bridge deformation monitoring with single frequency GPS augmented by Pseudolites.* (Doctoral Dissertation). The University of Nottingham.

Cosser, E., Roberts, G. W., Meng, X., & Dodson, A. H. (2003). The comparison of single and dual frequency GPS for bridge deflection and vibration monitoring. In *Proceedings of the 11th International Symposium on Deformation Measurements, International Federation Surveyors (FIG), Commission 6 - Engineering Surveys, Working Group 6.1.* Santorini, Greece.

Cottle, R. W. (1992). *The linear complementarity problem.* Siam.

Cottle, R. W., & Dantzig, G. B. (1968). Complementary pivot theory of mathematical programming. *Linear Algebra and Its Applications*, *1*(1), 103–125. doi:10.1016/0024-3795(68)90052-9

Cottle, R. W., Golub, G. H., & Sacher, R. S. (1977). On the solution of large, structured linear complementarity problems: The block partitioned case. *Applied Mathematics & Optimization*, *4*(1), 347–363. doi:10.1007/BF01442149

Cottle, R. W., & Sacher, R. S. (1976). On the solution of large, structured linear complementarity problems: The tridiagonal case. *Applied Mathematics & Optimization*, *3*(4), 321–340. doi:10.1007/BF01448184

Craig, D. & Locata Corporation. (2012). Truth on the range. *Inside GNSS*, *7*(3), 37-48.

Cryer, C. W. (1971). The method of Christopherson for solving free boundary problems for infinite journal bearings by means of finite differences. *Mathematics of Computation*, *25*(115), 435–443. doi:10.1090/S0025-5718-1971-0298961-7

Cryer, C. W. (1971). The solution of a quadratic programming problem using systematic overrelaxation. *SIAM Journal on Control*, *9*(3), 385–392. doi:10.1137/0309028

Csjáji, B., & Monostor, L. (2005). Stochastic Reactive Production Scheduling by Multi-agent Based Asynchronous Approximate Dynamic Programming. *CEEMAS LNAI*, 388-397.

Cui, Y., & He, B. (1999). A class of projection and contraction methods for asymmetric linear variational inequalities and their relations to Fukushima's descent method. *Computers & Mathematics with Applications (Oxford, England)*, *38*(11), 151–164. doi:10.1016/S0898-1221(99)00294-1

Current, J., & Marsh, M. (1993). Multiobjective transportation network design and routing problems: Taxonomy and annotation. *European Journal of Operational Research*, *65*(1), 4–19. doi:10.1016/0377-2217(93)90140-I

Current, J., & Min, H. K. (1986). Multiobjective design of transportation networks: Taxonomy and annotation. *European Journal of Operational Research*, *26*(2), 187–201. doi:10.1016/0377-2217(86)90180-3

Cvetković, L., & Kostić, V. (2014). A note on the convergence of the MSMAOR method for linear complementarity problems. *Numerical Linear Algebra with Applications*, *21*(4), 534–539. doi:10.1002/nla.1896

Cvetković, L., & Rapajić, S. (2005). How to improve MAOR method convergence area for linear complementarity problems. *Applied Mathematics and Computation*, *162*(2), 577–584. doi:10.1016/j.amc.2003.12.108

da Silva Maximiano, M., Vega-Rodríguez, M. A., Gómez-Pulido, J. A., & Sánchez-Pérez, J. M. (2013). A new Multi-objective Artificial Bee Colony algorithm to solve a real-world frequency assignment problem. *Neural Computing & Applications*, *22*(7-8), 1447–1459. doi:10.1007/s00521-012-1046-7

Dahiya, S. S., Chhabra, J. K., & Kumar, S. (2010). *Application of artificial bee colony algorithm to software testing.* Paper presented at the Software Engineering Conference (ASWEC), 2010 21st Australian. doi:10.1109/ASWEC.2010.30

Dai, Y. Q., & Xu, Z. S. (2007). Approach to multiple attribute decision making based on different intuitionistic preference structures. *J Southeast University*, *37*(4), 706–710.

Damodaran, P., Hirani, N. S., & Velez-Gallego, M. C. (2009). Scheduling identical parallel batch processing machines to minimise makespan using genetic algorithms. *European Journal of Industrial Engineering*, *3*(2), 187–206. doi:10.1504/EJIE.2009.023605

Damodaran, P., Manjeshwar, P. K., & Srihari, K. (2006). Minimizing makespan on a batch processing machine with non-identical job sizes using genetic algorithms. *International Journal of Production Economics*, *103*(2), 882–891. doi:10.1016/j.ijpe.2006.02.010

Dantzig, G. B., & Wolfe, P. (1960). Decomposition principle for linear programs. *Operations Research*, *8*(1), 101–111. doi:10.1287/opre.8.1.101

Darania, P., & Ebadian, A. (2007). A method for the numerical solution of the integro-differential equations. *Applied Mathematics and Computation*, *188*(1), 657–668. doi:10.1016/j.amc.2006.10.046

Darania, P., & Ivaz, K. (2008). Numerical solution of nonlinear Volterra–Fredholm integro-differential equations. *Computers & Mathematics with Applications (Oxford, England)*, *56*(9), 2197–2209. doi:10.1016/j.camwa.2008.03.045

Das, I., & Dennis, J. E. (1998). Normal-boundary intersection: A new method for generating the Pareto surface in nonlinear multicriteria optimization problems. *SIAM Journal on Optimization*, *8*(3), 631–657. doi:10.1137/S1052623496307510

Das, S., & Suganthan, P. N. (2011). Differential Evolution: A Survey of the State-of-the-Art. *IEEE Transactions on Evolutionary Computation*, *15*(1), 4–31. doi:10.1109/TEVC.2010.2059031

Davarynejad, M., Berg, J., & Rezaei, J. (2014). Evaluating center-seeking and initialization bias: The Case of Particle Swarm and Gravitational Search Algorithms. *Journal of Information Science*, *278*, 802–821. doi:10.1016/j.ins.2014.03.094

David, R. C., Precup, R., Petriu, E. M., Purcaru, C., & Preitl, S. (2012). PSO and GSA algorithms for fuzzy controller tuning with reduced process small time constant sensitivity. In *Proceedings of 16th International Conference on System Theory, Control and Computing (ICSTCC)*.

Davis, L. (1991). *Handbook of Genetic Algorithm*. New York, NY: Van Nostrand.

De Leone, R., & Mangasarian, O. L. (1988). Asynchronous parallel successive overrelaxation for the symmetric linear complementarity problem. *Mathematical Programming*, *42*(1-3), 347–361. doi:10.1007/BF01589410

De Leone, R., Mangasarian, O. L., & Shiau, T.-H. (1990). Multi-sweep asynchronous parallel successive overrelaxation for the nonsymmetric linear complementarity problem. *Annals of Operations Research*, *22*(1), 43–54. doi:10.1007/BF02023047

De los Cobos, S., Goddard, J., & Gutiérrez, M. A. (2011). Fuzzy regression vs. ordinary least squares regression: Case study. *Revista de Matemática: Teoría y Aplicaciones*, *18*(1), 33–48.

De Pierro, A. R. (1990). Nonlinear relaxation methods for solving symmetric linear complementarity problems. *Journal of Optimization Theory and Applications*, *64*(1), 87–99. doi:10.1007/BF00940024

De Pierro, A. R., & Iusem, A. N. (1991). On the convergence of SOR-and JOR-type methods for convex linear complementarity problems. *Linear Algebra and Its Applications*, *154*, 601–614. doi:10.1016/0024-3795(91)90396-E

De Pierro, A., & Lopes, J. (1994). Accelerating iterative algorithms for symmetric linear complementarity problems. *International Journal of Computer Mathematics*, *50*(1-2), 35–44. doi:10.1080/00207169408804240

De Sousa, L., Silvestre, B., & Bouchez, B. (2010). A combined multiphase electric drive and fast battery charger for electric vehicles. In *Vehicle Power and Propulsion Conference (VPPC)*. Lille, France: IEEE. doi:10.1109/VPPC.2010.5729057

De, M. H., Neto, G. F., Silva, A. M., Mariani, V. C., & Coelho, L. S. (2014). A Modified Gravitational Search Algorithm for Continuous Optimization. In *Proceedings of the conference companion on Genetic and evolutionary computation companion* (GECCO). ACM.

Deb, K., & Jain, S. (2002). *Running performance metrics for Evolutionary Multiobjective Optimization*. KanGAL Report No. 2002004. Kanpur Genetic Algorithms Laboratory, Indian Institute of Technology Kanpur.

Deb, K. (2001). *Multi-objective optimization using evolutionary algorithms* (Vol. 16). John Wiley & Sons.

Deb, K., Pratap, A., Agarwal, S., & Meyarivan, T. (2002). A Fast and Elitist Multiobjective Genetic Algorithm: NSGA-II. *IEEE Transactions on Evolutionary Computation*, *6*(2), 182–197. doi:10.1109/4235.996017

Deb, K., Pratap, A., & Moitra, S. (2000), Mechanical component design for multiple objectives using elitist non-dominatedsorting GA. In *Proceedings of the Parallel Problem Solving from Nature VI Conference*. doi:10.1007/3-540-45356-3_84

Degraeve, Z., & Jans, R. (2007). A New Dantzig- Wolfe Reformulation and Branch-and-Price Algorithm for the Capacitated Lot-Sizing Problem with Setup Times. *Operations Research*, *55*(5), 909–920. doi:10.1287/opre.1070.0404

Dehghan, M., & Hajarian, M. (2011). Two class of synchronous matrix multisplitting schemes for solving linear complementarity problems. *Journal of Computational and Applied Mathematics*, *235*(15), 4325–4336. doi:10.1016/j.cam.2011.03.021

Del Valle, Y., Hernandez, J. C., Venayagamoorthy, G. K., & Harley, R. G. (2006, August). Optimal STATCOM Sizing and Placement Using Particle Swarn Optimization. In Transmission & Distribution Conference and Exposition: Latin America, 2006. TDC'06. IEEE/PES (pp. 1-6). IEEE.

del Valle, Y., Venayagamoorthy, G. K., Mohagheghi, S., Hernandez, J.-C., & Harley, R. G. (2008). Particle swarm optimization: basic concepts, variants and applications in power systems. *Evolutionary Computation, IEEE Transactions on, 12*(2), 171-195.

Demarcke, P., Rogier, H., Goossens, R., & De Jaeger, P. (2009). Beamforming in the Presence of Mutual Coupling Based on Constrained Particle Swarm Optimization. *Antennas and Propagation. IEEE Transactions on, 57*(6), 1655–1666. doi:10.1109/TAP.2009.2019923

Dempe, S., & Dutta, J. (2012). Is bilevel programming a special case of a mathematical program with complementarity constraints? *Mathematical Programming, 131*(1-2), 37–48. doi:10.1007/s10107-010-0342-1

Dempe, S., & Zemkoho, A. B. (2012). On the Karush–Kuhn–Tucker reformulation of the bilevel optimization problem. *Nonlinear Analysis: Theory. Methods & Applications, 75*(3), 1202–1218.

Denholm, P., Kuss, M., & Margolis, R. M. (2013). Co-benefits of large scale plug-in hybrid electric vehicle and solar PV deployment. *Journal of Power Sources, 236*, 350–356. doi:10.1016/j.jpowsour.2012.10.007

Derghal, A., & Golea, N. (2014). Multi-Objective Generation Scheduling Using Genetic-Based Fuzzy Mathematical Programming Technique. In P. Vasant (Ed.), Handbook of Research on Novel Soft Computing Intelligent Algorithms: Theory and Practical Applications (pp. 450-474). IGI Global. doi:10.4018/978-1-4666-4450-2.ch015

Derigs, U., Li, B., & Vogel, U. (2009). Local search-based metaheuristics for the split delivery vehicle routing problem. *The Journal of the Operational Research Society, 61*(9), 1356–1364. doi:10.1057/jors.2009.100

Derriennic, Y., & Lin, M. (2003). The Central Limit Theorem for Markov Chains started at a point. *Probability Theory and Related Fields, 125*(1), 73–76. doi:10.1007/s004400200215

Desrochers, G., Blanchard, M., & Sud, S. (1986). A Monte Carlo simulation method for the economic assessment of the contribution of wind energy to power systems. *IEEE Transactions on Energy Conversion, 1*(4), 50–56. doi:10.1109/TEC.1986.4765774

Desrochers, M., Desrosiers, J., & Solomon, M. (2002). A new optimization algorithm for the vehicle routing problem with time windows. *Operations Research, 40*(2), 342–354. doi:10.1287/opre.40.2.342

Dhingra, A., & Chandna, P. (2010). Multi-objective flow shop scheduling using hybrid simulated annealing. *Measuring Business Excellence, 14*(3), 30–41. doi:10.1108/13683041011074191

Dhudshia, V. H. (1997). SEMI E10 – equipment reliability, availability and maintainability. *Semiconductor International, 20*(6), 167–174.

Diaby, M., Bahl, H. C., Karwan, M. H., & Zionts, S. (1992). A Lagrangean relaxation approach for very large scale capacitated lot-sizing. *Management Science, 38*(9), 1329–1340. doi:10.1287/mnsc.38.9.1329

Diamond, M. A. (1974). The solution of a quadratic programming problem using fast methods to solve systems of linear equations. *International Journal of Systems Science, 5*(2), 131–136. doi:10.1080/00207727408920083

Dieter, S., Fischer, C., & Menzel, W. (2011). Analysis of a folded reflect-array antenna using particle swarm optimization. *International Journal of Microwave and Wireless Technologies, 3*(3), 267–272. doi:10.1017/S1759078711000456

Dietrich, J., Yakovlev, V., McCartin, C., Jenson, G., & Duchrow, M. (2008). Cluster analysis of Java dependency graphs. In *Proceedings of the 4th ACM symposium on Software visualization*. doi:10.1145/1409720.1409735

Dieu, N. V., Peter, S., & Ongsakul, W. (2013). Cuckoo search algorithm for non-convex economic dispatch. *IET Generation. Transmission & Distribution, 7*(6), 645–654. doi:10.1049/iet-gtd.2012.0142

Dieu, V. N., & Ongsakul, W. (2009). Improved merit order and augmented Lagrange Hopfield network for short term hydrothermal scheduling. *Energy Conversion and Management*, *50*(12), 3015–3023. doi:10.1016/j.enconman.2009.07.021

Dijkstra, E. W. (1959). A note in two problems in connexion with graphs. *Numerische Mathematik*, *1*(1), 269–271. doi:10.1007/BF01386390

Diniz, A. L. (2010). *Test Cases for Unit Commitment and Hydrothermal Scheduling Problems*. IEEE Power and Energy Society General Meeting, Minneapolis, MN.

Dinu, S. & Bordea, G. (2011). A new genetic approach for transport network design and optimization. *Bulletin of the Polish Academy of Sciences - Technical Sciences, 59*, 263-272.

Domairry, G., & Fazeli, M. (2009). Homotopy analysis method to determine the fin efficiency of convective straight fins with temperature-dependent thermal conductivity. *Communications in Nonlinear Science and Numerical Simulation*, *14*(2), 489–499. doi:10.1016/j.cnsns.2007.09.007

Dong, J.-L. (2009). Inexact multisplitting methods for linear complementarity problems. *Journal of Computational and Applied Mathematics*, *223*(2), 714–724. doi:10.1016/j.cam.2008.02.013

Dong, Y., Tang, J., Xu, B., & Wang, D. (2005). An application of swarm optimization to nonlinear programming. *Computers & Mathematics with Applications (Oxford, England)*, *49*(11-12), 1655–1668. doi:10.1016/j.camwa.2005.02.006

Doraghinejad, M. & Nezamabadi-pour, H. (2014). Black Hole: A New Operator for Gravitational Search Algorithm. *Journal of Computational Intelligence Systems, 7*(5), 809-826.

Doraghinejad, M., Nezamabadi-pour, H., & Mahani, A. (2014). Channel assignment in multi-radio wireless mesh networks using an improved gravitational search algorithm. *Journal of Network and Computer Applications*, *38*, 163–171. doi:10.1016/j.jnca.2013.04.007

Doran, D. (2004). Rethinking the Supply Chain: An Automotive Perspective. *Supply Chain Management: An International Journal, 9*(1), 102 - 109.

Dorigo, M., Maniezzo, V., & Colorni, A. (1991). *Positive Feedback as a Search Strategy*. Technical Report no.91016. Politecnico di Milano.

Dorigo, M., Maniezzo, V., & Colorni, A. (1996). Ant System: Optimization by a Colony of Cooperating Agents. *IEEE Transactions on Systems, Man, and Cybernetics. Part B, Cybernetics*, *26*(1), 29–41. doi:10.1109/3477.484436 PMID:18263004

Doshi, P., Goodwin, R., Akkiraju, R., & Verma, K. (2004). Dynamic workflow composition using Markov decision processes.*IEEE International Conference on*, (pp. 576-582). doi:10.1109/ICWS.2004.1314784

Doucet, A., & Johansen, A. (2009). A tutorial on particle filtering and smoothing: Fifteen years later. In Handbook of Nonlinear Filtering, (pp. 656–704). Academic Press.

Doucet, A., & Tadic, V. (2003). Parameter estimation in general state-space models using particle methods. *Annals of the Institute of Statistical Mathematics*, *55*(2), 409–422. doi:10.1007/BF02530508

Doval, D., Mancoridis, S., & Mitchell, B. S. (1999). *Automatic clustering of software systems using a genetic algorithm*. Paper presented at the Software Technology and Engineering Practice. doi:10.1109/STEP.1999.798481

Doverspike, R. D. (1982). Some perturbation results for the linear complementarity problem. *Mathematical Programming*, *23*(1), 181–192. doi:10.1007/BF01583787

Dowlatshahi, M. B., Nezamabadi-pour, H., & Mashinchi, M. (2014). A discrete gravitational search algorithm for solving combinatorial optimization problems. *Information Sciences*, *258*, 94–107. doi:10.1016/j.ins.2013.09.034

Draper, N. R., & Smith, H. (1968). *Applied Regression Analysis*. New York: John Wiley & Sons.

Du Val, P. (1940). The unloading problem for plane curves. *American Journal of Mathematics*, *62*(1/4), 307–311. doi:10.2307/2371454

Dua, V. P. E., & Morari, M. (2004). *Hybrid Systems Modeling, Parametric Programming, and Model Predictive Control-Impact on Process Operations*. Paper presented at the Sixth International Conference on Foundations of Computer-Aided Process Design (FOCAPD), Austin, TX.

Duan, B. X., Zeng, W. Y., & Zhu, X. P. (2012). A Preconditioned Gauss-Seidel Iterative Method for Linear Complementarity Problem in Intelligent Materials System. *Advanced Materials Research*, *340*, 3–8. doi:10.4028/www.scientific.net/AMR.340.3

Dubey, H. M., Pandit, M., Panigrahi, B. K., & Udgir, M. (2013). Economic Load Dispatch by Hybrid Swarm Intelligence Based Gravitational Search Algorithm. *International Journal of Intelligent Systems and Applications*, *5*(8), 21–32. doi:10.5815/ijisa.2013.08.03

Dubois, D., & Prade, H. (1980). *Fuzzy Sets and Systems: Theory and Applications*. New York: Academic Press.

Dubois, D., & Prade, H. (1988). *Possibility Theory*. New York: Plenum. doi:10.1007/978-1-4684-5287-7

DuJulio, E. T., & Leet, J. H. (1988). Space station synergetic RAM - logistic analysis. In *Proceedings of Annual Reliability and Maintainability Symposium*. IEEE.

Duman, S., Yorukeren, N., & Altas, I. H. (2015). A novel modified hybrid PSOGSA based on fuzzy logic for non-convex economic dispatch problem with valve-point effect. *Journal of Electrical Power & Energy Systems*, *64*, 121–135. doi:10.1016/j.ijepes.2014.07.031

DuMouchel, W. H. (1973). On the asymptotic normality of the maximum-likelihood estimate when sampling from a stable distribution. *Annals of Statistics*, *1*(5), 948–957. doi:10.1214/aos/1176342516

Dunstall, S., & Wirth, A. (2005). A comparison of branch-and bound algorithms for a family scheduling problem with identical parallel machines. *European Journal of Operational Research*, *167*(2), 283–296. doi:10.1016/j.ejor.2004.03.022

Dupuy, M., Lamothe, J., Gaborit, P., & Dupont, L. (2005). Efficient neighbors in simulated annealing algorithm to optimize the lead time in a parallel multipurpose machine scheduling problem with setup and calendar constraints. In *Proceedings of the International Conference on Industrial Engineering and Systems Management*.

Durgun, İ., & Yildiz, A. R. (2012). Structural Design Optimization of Vehicle Components Using Cuckoo Search Algorithm. *Materials Testing*, *54*(3), 185–188. doi:10.3139/120.110317

Dusmez, S., Cook, A., & Khaligh, A. (2011). Comprehensive analysis of high quality power converters for level 3 off-board chargers. In *Vehicle Power and Propulsion Conference (VPPC)*. Chicago: IEEE. doi:10.1109/VPPC.2011.6043096

Dyson, F. J. (1969). 'Existence of a phase-transition in a one-dimensional Ising ferromagnet'. *Communications in Mathematical Physics*, *12*(2), 91–107. doi:10.1007/BF01645907

Dzielinski, B. P., & Gomory, R. E. (1965). Optimal programming of lot sizes, inventory, and labor allocations. *Management Science*, *11*(9), 874–890. doi:10.1287/mnsc.11.9.874

Eberhart, R. C., & Shi, Y. (2001). Particle swarm optimization: developments, applications and resources. In *Proceedings of the 2001 Congress on Evolutionary Computation*. Seoul, South Korea: IEEE. doi:10.1109/CEC.2001.934374

Edgar, R. C. (2004). MUSCLE: Multiple sequence alignment with improved accuracy and speed. In *Proceedings of the 2004 IEEE Computational Systems Bioinformatics Conference* (pp. 728-729). doi:10.1109/CSB.2004.1332560

EDINA, University of Edinburgh. (2000). *EDINA Digimap Ordinance Survey Service*. Retrieved from http://digimap.edina.ac.uk/digimap/

Eiben, A. E., Hinterding, R., & Michlewicz, Z. (1999). Parameter control in evolutionary algorithms. *IEEE Transactions on Evolutionary Computation*, *3*(2), 124–141. doi:10.1109/4235.771166

Ein-Dor, P., & Feldmesser, J. (1987). Attributes of the performance of central processing units: A relative performance prediction model. *Communications of the ACM*, *4*(4), 308–317. doi:10.1145/32232.32234

El Tayyan, A. A. (2011). Article. *Journal of Electron Devices*, *9*, 335.

Elamvazuthi, I., Ganesan, T., & Vasant, P. (2011). A comparative study of HNN and Hybrid HNN-PSO techniques in the optimization of distributed generation (DG) power systems. *International Conference on Advanced Computer Science and Information System (ICACSIS).*

El-Ashry, M. (2007). *REN21, Renewable energy policy network for the 21st century*. Renewables, Global Status Report, 51.

Ellson, J., Gansner, E., Koutsofios, L., North, S. C., & Woodhull, G. (2002). *Graphviz—open source graph drawing tools*. Paper presented at the Graph Drawing.

El-Naggar, K., AlRashidi, M., AlHajri, M., & Al-Othman, A. (2012). Simulated annealing algorithm for photovoltaic parameters identification. *Solar Energy*, *86*(1), 266–274. doi:10.1016/j.solener.2011.09.032

Emmerich, M., Beume, N., & Naujoks, B. (2005). An EMO Algorithm Using the Hypervolume Measure as Selection Criterion. In *Conference on Evolutionary Multi-Criterion Optimization (EMO 2005)*. doi:10.1007/978-3-540-31880-4_5

Ergezer, H., & Leblebicioglu, K. (2004). Multiplesequence alignment using hidden Markov model. In *Proceedings of the 2004 IEEE 12th Signal Processing and Communications Applications Conference* (pp.434-437). doi:10.1109/SIU.2004.1338556

Escobar, J. W., Linfati, R., Toth, P., & Baldoquin, M. G. (2014). A hybrid granular tabu search algorithm for the multi-depot vehicle routing problem. *Journal of Heuristics*, *20*(5), 483–509. doi:10.1007/s10732-014-9247-0

Eskandar, H., Sadollah, A., Bahreininejad, A., & Hamdi, M. (2012). Water cycle algorithm – a novel metaheuristic optimization method for solving constrained engineering optimization problems. *Computers & Structures*, *110-111*, 151–161. doi:10.1016/j.compstruc.2012.07.010

Espinoza, J. L. (2002). Obtaining reducts with a genetic algorithm. In K. Jajuga, A. Sokolowski, & H. H. Bock (Eds.), *Classification, Clustering and Analysis* (pp. 2119–2226). Berlin: Springer. doi:10.1007/978-3-642-56181-8_24

Evans, D., & Martins, M. (1992). On the convergence of the extrapolated AOR method. *International Journal of Computer Mathematics*, *43*(3-4), 161–171. doi:10.1080/00207169208804083

Evans, D., Martins, M. M., & Trigo, M. E. (2001). The AOR iterative method for new preconditioned linear systems. *Journal of Computational and Applied Mathematics*, *132*(2), 461–466. doi:10.1016/S0377-0427(00)00447-7

Eynan, A., & Rosenblat, M. (1996). Component commonality effects on inventory costs. *HE Transactions*, *28*(2), 93–104.

Facchinei, F., Jiang, H., & Qi, L. (1999). A smoothing method for mathematical programs with equilibrium constraints. *Mathematical Programming*, *85*(1), 107–134. doi:10.1007/s101070050048

Fahmy, S. A., ElMekkawy, T. Y., & Balakrishnan, S. (2008). Deadlock-free scheduling of flexible job shops with limited capacity buffers. *European Journal of Industrial Engineering*, *2*(3), 231–252. doi:10.1504/EJIE.2008.017685

Fan, Z. P., & Liu, Y. (2010). An approach to solve group decision making problems with ordinal interval Numbers. *IEEE Trans Syst Man Cybern Part B*, *40*(5), 1413–1423. doi:10.1109/TSMCB.2009.2039477 PMID:20172834

Fateen, S.-E. K., & Bonilla-Petriciolet, A. (2014). Gradient-Based Cuckoo Search for Global Optimization. *Mathematical Problems in Engineering, Article ID 493740*. doi:10.1155/2014/493740

Ferreira, J. A., & Condessa, B. (2012). Defining expansion areas in small urban settlements an application to the municipality of Tomar (Portugal). *Landscape and Urban Planning*, *107*(3), 283–292. doi:10.1016/j.landurbplan.2012.06.008

Ferris, M. C., & Pang, J.-S. (1997). Engineering and economic applications of complementarity problems. *SIAM Review*, *39*(4), 669–713. doi:10.1137/S0036144595285963

Field, A., Miles, J., & Field, Z. (2012). *Discovering Statistics Using R*. Thousand Oaks, CA: Sage Publications.

Fischer, A. (1995). A Newton-type method for positive-semidefinite linear complementarity problems. *Journal of Optimization Theory and Applications*, *86*(3), 585–608. doi:10.1007/BF02192160

Fishburn, P. C. (1967). *Additive Utilities with Incomplete Product Set: Applications to Priorities and Assignments*. Baltimore, MD: ORSA.

Fisher, M. L. (1981). The Lagrangian relaxation method for solving integer programming problems. *Management Science*, *27*(1), 1–18. doi:10.1287/mnsc.27.1.1

Fister, I., Fister, I., Jr., Brest, J., & Žumer, V. (2012). *Memetic artificial bee colony algorithm for large-scale global optimization*. Paper presented at the 2012 IEEE Congress on Evolutionary Computation, CEC 2012. doi:10.1109/CEC.2012.6252938

Fitouhi, M., & Nourelfath, M. (2014). Integrating non cyclical preventive maintenance scheduling and production planning for multi-state systems. *Reliability Engineering & System Safety*, *121*, 175–186. doi:10.1016/j.ress.2013.07.009

Fletena, S., & Kristoffersen, T. (2008). Short-term hydropower production planning by stochastic programming. *Computers & Operations Research*, *35*(8), 2656–2671. doi:10.1016/j.cor.2006.12.022

Floudas, C. A., & Lin, X. X. (2004). Continuous-time versus discrete-time approaches for scheduling of chemical processes: A review. *Computers & Chemical Engineering*, *28*(11), 2109–2129. doi:10.1016/j.compchemeng.2004.05.002

Floudas, C. A., & Lin, X. X. (2005). Mixed integer linear programming in process scheduling: Modeling, algorithms, and applications. *Annals of Operations Research*, *139*(1), 131–162. doi:10.1007/s10479-005-3446-x

Flug, P. (1996). *Optimization of Stochastic Models: The Interface Between Simulation and Optimization*. Kluwer Academic Publishers. doi:10.1007/978-1-4613-1449-3

Fogel, L. J., Owens, A. J., & Walsh, M. J. (1966). *Artificial intelligence thorough simulated evolution*. Chichester, UK: John Wiley&Sons, Ltd.

Fogel, L., Owens, A., & Walsh, M. (1966). *Artificial Intelligence Through Simulated Evolution*. New York: Wiley.

Fontana, R. J. (2004). Recent system applications of short-pulse ultra-wideband (UWB) technology. *IEEE Transactions on Microwave and Techniques*, *52*(9), 2087–2104. doi:10.1109/TMTT.2004.834186

Ford, T., Neumann, J., Toso, N., Petersen, W., Anderson, C., Fenton, P., . . . Barltrop, K. (1997, September). HAPPI- a high accuracy Pseudolite/GPS positioning integration. In *Proceedings of ION/GPS 1997, the 10th Technical Meeting of the Satellite Division of the Institute of Navigation* (pp. 1719-1728).

Foster, P. R., & Burberry, R. A. (1999). *Antenna problems in RFID systems.* Paper presented at the RFID Technology (Ref. No. 1999/123), IEE Colloquium on.

Fowler, J., & Rose, O. (2015). *Grand Challenges in modeling and simulation.* Retrieved from http://sim.sagepub.com/content/80/9/469.short

Franco, P. E. C., Carvalho, M. F., & Soares, S. (1994). A network flow model for short-term hydro-dominated hydrothermal scheduling problems. *IEEE Transactions on Power Systems*, *9*(2), 1016–1022. doi:10.1109/59.317642

Frangioni, A., Gentile, C., & Lacalandra, F. (2008). Solving unit commitment problems with general ramp constraints. *International Journal of Electrical Power & Energy Systems*, *30*(5), 316–326. doi:10.1016/j.ijepes.2007.10.003

Friesz, T. (1981). Multiobjective Optimization in Transportation: The Case of Equilibrium Network Design. *Lecture Notes in Economics and Mathematical Systems*, *190*, 116–127. doi:10.1007/978-3-642-45527-8_10

Friesz, T., Cho, H., Mehta, N., Tobin, R., & Anandalingam, G. (1992). A Simulated Annealing Approach to the Network Design Problem with Variational Inequality Constraints. *Transportation Science*, *26*(1), 18–26. doi:10.1287/trsc.26.1.18

Frommer, A., & Szyld, D. B. (1992). H-splittings and two-stage iterative methods. *Numerische Mathematik*, *63*(1), 345–356. doi:10.1007/BF01385865

Frutos, M., & Tohmé, F. (2012). A new approach to the optimization of the CVRP through Genetic Algorithms. *American Journal of Operations Research*, *2*(04), 495–501. doi:10.4236/ajor.2012.24058

Fu, M. C. (1994). Optimization via Simulation: A Review. *Annals of Operations Research*, *53*(1), 199–248. doi:10.1007/BF02136830

Fu, M. C. (2002). Optimization for simulation: Theory vs. practice. *INFORMS Journal on Computing*, *14*(3), 192–215. doi:10.1287/ijoc.14.3.192.113

Gale, D. (1960). *The theory of linear economic models*. University of Chicago press.

Gallien, L., & Wein, L. (2001). A simple and effective component procurement policy for stochastic assembly systems. *Queueing Systems*, *38*(2), 221–248. doi:10.1023/A:1010914600116

Gallo, M., D'Acierno, L., & Montella, B. (2012). A Meta-heuristic Algorithm for Solving the Road Network Design Problem in Regional Contexts. *Procedia: Social and Behavioral Sciences*, *54*, 84–95. doi:10.1016/j.sbspro.2012.09.728

Gandomi, A. H. (2014). Interior search algorithm (ISA): A novel approach for global optimization, ISA Transactions. *ISA Transactions*, *53*(4), 1168–1183. doi:10.1016/j.isatra.2014.03.018 PMID:24785823

Gandomi, A. H., & Alavi, A. H. (2011). Multi-stage genetic programming: A new strategy to nonlinear system modeling. *InfSci*, *181*, 5227–5239.

Ganesan, T., Elamvazuthi, I., Shaari, K. Z. K., & Vasant, P. (2013). An Algorithmic Framework for Multiobjective Optimization. The Scientific World Journal.

Ganesan, T., Elamvazuthi, I., Shaari, K. Z. K., & Vasan, P. (2013). Swarm Intelligence and Gravitational Search Algorithm for Multi-objective Optimization of Synthesis Gas Production. *Journal of Applied Energy*, *103*, 368–374. doi:10.1016/j.apenergy.2012.09.059

Ganesan, T., Vasant, P., & Elamvazuthi, I. (2014). Hopfield neural networks approach for design optimization of hybrid power systems with multiple renewable energy sources in a fuzzy environment. *Journal of Intelligent and Fuzzy Systems*, *2*(5), 2143–2154.

Ganesan, T., Vasant, P., & Elamvazuthy, I. (2012). A hybrid PSO approach for solving non-convex optimization problems. *Archives of Control Sciences*, *22*(1), 87–105. doi:10.2478/v10170-011-0014-2

Gan, L., Topcu, U., & Low, S. (2011). Optimal decentralized protocol for electric vehicle charging. In *Decision and Control and European Control Conference (CDC-ECC)*. IEEE. doi:10.1109/CDC.2011.6161220

Gao, X., & Wang, J. (2011). *A Neural Network for a Class of Horizontal Linear Complementary Problems*. Computational Intelligence and Security (CIS), 2011 Seventh International Conference on. doi:10.1109/CIS.2011.85

Gao, S., Todo, Y., Gong, T., Yang, G., & Tang, Z. (2014). Graph Planarization Problem Optimization Based on Triple-Valued Gravitational Search Algorithm. *Journal of IEEJ Transaction on Electrical and Electrical Engineering*, *9*(1), 39–48. doi:10.1002/tee.21934

Gao, X. (1999). Neural network for solving monotone linear complementary problems. *Communications in Nonlinear Science and Numerical Simulation*, *4*(2), 128–132. doi:10.1016/S1007-5704(99)90025-4

Garcia-Nieto, J., Olivera, A., & Alba, E. (2013). Optimal Cycle Program of Traffic Lights with Particle Swarm Optimization. *IEEE Transactions on Evolutionary Computation*, *17*(6), 823–839. doi:10.1109/TEVC.2013.2260755

Garg, H. (2015). A Hybrid GA-GSA Algorithm for Optimizing the Performance of an Industrial System by Utilizing Uncertain Data. In P. Vasant (Ed.), Handbook of Research on Artificial Intelligence Techniques and Algorithms (pp. 620-654). IGI Global. doi:10.4018/978-1-4666-7258-1.ch020

Garg, H. (2013). Reliability analysis of repairable systems using Petri nets and Vague Lambda-Tau methodology. *ISA Transactions*, *52*(1), 6–18. doi:10.1016/j.isatra.2012.06.009 PMID:22789401

Garg, H., & Rani, M. (2013). An approach for reliability analysis of industrial systems using PSO and IFS technique. *ISA Transactions*, *52*(6), 701–710. doi:10.1016/j.isatra.2013.06.010 PMID:23867122

Garg, H., Rani, M., & Sharma, S. P. (2013). Preventive maintenance scheduling of the pulping unit in a paper plant. *Japan Journal of Industrial and Applied Mathematics*, *30*(2), 397–414. doi:10.1007/s13160-012-0099-4

Garg, H., & Sharma, S. P. (2012). A two-phase approach for reliability and maintainability analysis of an industrial system. *International Journal of Reliability Quality and Safety Engineering*, *19*(3), 1250013. doi:10.1142/S0218539312500131

Garg, H., & Sharma, S. P. (2013). Multi-objective reliability-redundancy allocation problem using particle swarm optimization. *Computers & Industrial Engineering*, *64*(1), 247–255. doi:10.1016/j.cie.2012.09.015

Garg, H., Sharma, S. P., & Rani, M. (2012). Cost minimization of washing unit in a paper mill using artificial bee colony technique. *International Journal of System Assurance Engineering and Management*, *3*(4), 371–381. doi:10.1007/s13198-012-0128-3

Gavanidou, E. S., Bakirtzis, A. G., & Dokopoulos, P. S. (1992). A probabilistic method for the evalu- ation of the performance of wind-diesel energy systems. *IEEE Transactions on Energy Conversion*, *7*(3), 418–425. doi:10.1109/60.148561

Gayon, J., Benjaafar, S., & Vricourt, F. (2008). Using imperfect advance demand information in production-inventory systems with multiple customer classes. *Manufacturing & Service Operations Management*, *11*(1), 128–143. doi:10.1287/msom.1070.0201

Gen, B. (2005). *Reliability and Cost/Worth Evaluation of Generating Systems Utilizing Wind and Solar Energy*. (Ph.D. Thesis). Department of Electrical Engineering, University of Saskatchewan Askatoon.

Genetic Algorithm. (n.d.). Retrieved from http://en.wikipedia.org

Geng, J., Shi, C., Ge, M., Dodson, A., Lou, H. Y., Zhao, Q., & Liu, J. (2012). Improving the estimation of fractional-cycle biases for ambiguity resolution in precise point positioning. *Journal of Geodesy*, *86*(8), 579–589. doi:10.1007/s00190-011-0537-0

Gerchad, Y., & Henig, M. (1989). An inventory model with component commonality. *Naval Research Logistics*, 61-68.

Gerkensmeyer, C., Kintner-Meyer, M. C., & DeSteese, J. G. (2010). *Technical challenges of plug-in hybrid electric vehicles and impacts to the US power system: Distribution system analysis*. United States Department of Energy. doi:10.2172/974954

Gertsbakh, I. B. (1977). *Models of Preventive Maintenance*. New York: North-Holland Publishing Company.

Ghalambaz, M., Noghrehabadi, A. R., Behrang, M. A., Assareh, E., Ghanbarzadeh, A., & Hedayat, N. (2011). A Hybrid Neural Network and Gravitational Search Algorithm (HNNGSA) method to solve well known Wessinger's equation, Journal of World Academy of Science. *Engineering and Technology*, *73*, 803–807.

Ghanem, W. A. H. M., & Jantan, A. (2014). Using hybrid artificial bee colony algorithm and particle swarm optimization for training feed-forward neural networks. *Journal of Theoretical and Applied Information Technology*, *67*(3), 664–674.

Ghani, F., Duke, M., & Carson, J. (2013). Numerical calculation of series and shunt resistance of a photovoltaic cell using the Lambert W-function: Experimental evaluation. *Solar Energy*, *87*, 246–253. doi:10.1016/j.solener.2012.11.002

Giannelos, N. F., & Georgiadis, M. C. (2002). A simple new continuous-time formulation for short-term scheduling of multipurpose batch processes. *Industrial & Engineering Chemistry Research*, *41*(9), 2178–2184. doi:10.1021/ie010399f

Gil, E., Bustos, J., & Rudnick, H. (2003). Short-term hydrothermal generation scheduling model using a genetic algorithm. *IEEE Transactions on PWRS*, *18*, 1256–1264.

Gilland, W. G. (2002). A simulation study comparing performance of CONWIP and bottleneck-based release rule. *Production Planning and Control*, *13*(2), 211–219. doi:10.1080/09537280110069784

Glover, F. (1989). Tabu search - Part I. *ORSA Journal on Computing, 1*, 190-206.

Glover, F., & Kochenberger, G. A. (2003). *Handbook of Metaheuristics*. Kluwer Academic Publishers. doi:10.1007/b101874

Glover, F., & Laguna, M. (1997). *Tabu Search*. Boston, MA: Kluwer Academic Publishers. doi:10.1007/978-1-4615-6089-0

Glynn, P., & Iglehar, D. (1989). Importance sampling for stochastic simulations. *Management Science*, *35*(11), 1367–1392. doi:10.1287/mnsc.35.11.1367

Goel, S., Sharma, A., & Panchal, V. K. (2014). Multiobjective Cuckoo Search for Anticipating the Enemy's Movements in the Battleground. *International Journal of Applied Metaheuristic Computing*, *5*(4), 26–46. doi:10.4018/ijamc.2014100102

Goetschalckxa, M., Vidalb, C., & Dognac, K. (2002). Modeling and design of gloibal logistic systems: A review of integrated estrategic and tactical models and design algorithms. *European Journal of Operational Research*, *143*(1).

Golberg, D. (1989). *Genetic Algorithm in Search, Optimization and Machine Learning*. Cambridge, MA: Addison Wesley.

Goldberg, D. E. (2006). *Genetic algorithms*. Pearson Education India.

Goldberg, D. (1989). *Genetic Algorithms in Search, Optimisation and Machine Learning*. Addison–Wesley.

Goldberg, D. (1989). *Genetic Algorithms in Search, Optimization, and Machine Learning*. Addison Wesley.

Goldberg, D. E. (1989). *Genetic Algorithms in Search, Optimization and Machine Learning*. Reading, MA: Addison-Wesley.

Gold, R. (1963). *An iterative solution of the matrix representation of detection systems*. Argonne National Lab.

Gold, R. (1964). *An iterative unfolding method for response matrices*. Argonne National Lab. doi:10.2172/4634295

Gomathi, R., & Sharmila, D. (2014). A Novel Adaptive Cuckoo Search for Optimal Query Plan Generation. *TheScientificWorldJournal*, *7*. doi:10.1155/2014/727658 PMID:25215330

Gong, W., & Cai, Z. (2013). Parameter extraction of solar cell models using repaired adaptive differential evolution. *Solar Energy*, *94*, 209–220. doi:10.1016/j.solener.2013.05.007

Gong, W., Cai, Z., Ling, C., & Li, H. (2010). A real-coded biogeography-based optimization with mutation. *Applied Mathematics and Computation*, *216*(9), 2749–2758. doi:10.1016/j.amc.2010.03.123

Gong, W., Cai, Z., & Zhu, L. (2009). An efficient multiobjective differential evolution algorithm for engineering design. *Structural and Multidisciplinary Optimization*, *38*(2), 137–157. doi:10.1007/s00158-008-0269-9

Gonzalez, B., Valdez, F., Melin, P., & Prado-Arechiga, G. (2015). Fuzzy logic in the gravitational search algorithm for the optimization of modular neural networks in pattern recognition. *Journal of Expert Systems with Application*, *42*(14), 5839–5847. doi:10.1016/j.eswa.2015.03.034

Gonzalez-Feliu, J. (2013). Vehicle Routing in Multi-Echelon Distribution Systems with Cross-Docking: A Systematic Lexical-Metanarrative Analysis. *Computer and Information Science*, *6*(3), 28–47. doi:10.5539/cis.v6n3p28

Gopalakrishnan, M., Mohan, S., & He, Z. (2001). A tabu search heuristic for preventive maintenance scheduling. *Computers & Industrial Engineering*, *40*(1-2), 149–160. doi:10.1016/S0360-8352(01)00014-6

Gordini, N. (2014). Genetic Algorithms for Small Enterprises Default Prediction: Empirical Evidence from Italy. In P. Vasant (Ed.), Handbook of Research on Novel Soft Computing Intelligent Algorithms: Theory and Practical Applications (pp. 258-293). IGI Global. doi:10.4018/978-1-4666-4450-2.ch009

Gordini, N., & Veglio, V. (2014). Customer relationship management and data mining: a classification tree to predict customer purchasing behavior in global market. In P. M. Vasant (Ed.), *Handbook of Research on Novel Soft Computing Intelligent Algorithms: Theory and Practical Applications* (pp. 1–40). Hershey, PA: Information Science Reference; doi:10.4018/978-1-4666-4450-2.ch001

Gordon, G. (1988). *System simulation*. Englewood Cliffs, NJ: Prentice-Hall Inc.

Gottschalg, R., Rommel, M., Infield, D. G., & Kearney, M. J. (1999). The influence of the measurement environment on the accuracy of the extraction of physical parameters of solar cells. *Meas. Sci. Technol., 10*, 796–804.

Goudos, S. K., Siakavara, K., & Sahalos, J. N. (2013a). *Modified spiral RFID tag antenna optimal design using Artificial Bee Colony optimization*. Paper presented at the Microwave Conference (EuMC), 2013 European.

Goudos, S. K., Siakavara, K., & Sahalos, J. N. (2013b). *Novel spiral antennas design using swarm intelligence for passive UHF RFID tags*. Paper presented at the 7th European Conference on Antennas and Propagation, EuCAP 2013, Gothenburg.

Goudos, S. K., Siakavara, K., & Sahalos, J. N. (2013c). *Novel spiral antennas design using swarm intelligence for passive UHF RFID tags*. Paper presented at the Antennas and Propagation (EuCAP), 2013 7th European Conference on.

Goudos, S. K., Siakavara, K., & Sahalos, J. N. (2014). *Optimizing meandered spiral antennas for RFID tags using gbest-guided Artificial Bee Colony algorithm.* Paper presented at the Antennas and Propagation (EuCAP), 2014 8th European Conference on.

Goudos, S. K., Siakavara, K., & Sahalos, J. N. (2014). Novel spiral antenna design using artificial bee colony optimization for UHF RFID Applications. *IEEE Antennas and Wireless Propagation Letters*, *13*, 528–531. doi:10.1109/LAWP.2014.2311653

Goudos, S. K., Siakavara, K., & Sahalos, J. N. (2015). Design of load-ended spiral antennas for RFID UHF passive tags using improved artificial bee colony algorithm. *AEÜ. International Journal of Electronics and Communications*, *69*(1), 206–214. doi:10.1016/j.aeue.2014.09.008

Goudos, S. K., Siakavara, K., Samaras, T., Vafiadis, E. E., & Sahalos, J. N. (2011). Self-adaptive differential evolution applied to real-valued antenna and microwave design problems. *IEEE Transactions on Antennas and Propagation*, *59*(4), 1286–1298. doi:10.1109/TAP.2011.2109678

Gould, J. L., & Gould, C. G. (1988). *The Honey Bee*. New York: Scientific American Library.

Graves, S. C., & Willems, S. P. (2005). Optimizing the supply chain configuration for new products. *Management Science*, *51*(8), 1165–1180. doi:10.1287/mnsc.1050.0367

Grewal, M., & Andrews, A. (2008). *Kalman Filtering: Theory and Practice using MATLAB*. John Wiley and Sons. doi:10.1002/9780470377819

Groenen, P. J. F., Mathar, R., & Trejos, J. (2000). Global optimization methods for multidimensional scaling applied to mobile communications. In W. Gaul, O. Opitz, & M. Schader (Eds.), *Data Analysis. Scientific Modeling and Practical Application* (pp. 459–469). Berlin: Springer. doi:10.1007/978-3-642-58250-9_37

Groenen, P., Roosmalen, J., Trejos, J., & Castillo, W. (2008). Optimization strategies in two-mode partitioning. *Journal of Classification*, *26*(2), 155–181.

Grosan, C., & Abraham, A. (2011). *Intelligent Systems: A Modern Approach*. Berlin: Springer. doi:10.1007/978-3-642-21004-4

Grover, W. D. (1999). High availability path design in ring-based optical networks. *IEEE/ACM Transactions on Networking*, *7*(4), 558–574. doi:10.1109/90.793028

Guan, X. (1995). Nonlinear approximation method in Lagrangian relaxation-based algorithms for hydrothermal scheduling. *IEEE Transactions on Power Systems*, *10*(2), 772–778. doi:10.1109/59.387916

Gu, B., & Pan, F. (2013). Modified Gravitational Search Algorithm with Particle Memory Ability and its Application. *International Journal of Innovative Computing, Information, & Control*, *9*(11), 4531–4544.

Gunasekaran, A., & Ngai, E. W. T. (2004). Article. *Virtual Supply-Chain Management, 15*(6), 584–595. Retrieved from http://www.tandf.co.uk/journals

Gunawardena, A. D., Jain, S. K., & Snyder, L. (1991). Modified iterative methods for consistent linear systems. *Linear Algebra and Its Applications*, *154*, 123–143. doi:10.1016/0024-3795(91)90376-8

Günter H-O, & Veek P V. (2003). Advanced planning and scheduling in process industry. *Advances Planning and Scheduling Solutions in Process Industry*, 1-9.

Guo, Z., (2012). A Hybrid Optimization Algorithm Based on Artificial Bee Colony and Gravitational Search Algorithm. *International Journal of Digital Content Technology and its Applications, 6*(17), 620-626.

Gupta, A., & Joshi, S. (2006). A novel Least Mean Squares algorithm for tracking a discrete-time fBm Process. In *Proceedings of the IEEE Indian Conference 2006 (INDCON'06).* doi:10.1109/INDCON.2006.302790

Gupta, A., & Joshi, S. (2008). Some studies on the structure of covariance matrix of discrete-time fBm. *IEEE Transactions on Signal Processing*, *56*(10), 4635–4650. doi:10.1109/TSP.2008.927467

Gurtuna, F., Petra, C., Potra, F. A., Shevchenko, O., & Vancea, A. (2011). Corrector-predictor methods for sufficient linear complementarity problems. *Computational Optimization and Applications*, *48*(3), 453–485. doi:10.1007/s10589-009-9263-4

Gustafsson, F., Gunnarsson, F., Bergman, N., Forssell, U., Jansson, J., Karlsson, R., & Nordlund, P. (2002). Particle filters for positioning, navigation, and tracking. *Signal Processing, IEEE Transactions on*, 425–437.

Gustafsson, F., & Hendeby, G. (2012). Some relations between extended and unscented Kalman filters. *Signal Processing. IEEE Transactions on*, *60*(2), 545–555. doi:10.1109/TSP.2011.2172431

Gustavsson, M., & Wänström, C. (2009). Assessing information quality in manufacturing planning and control processes. *International Journal of Quality & Reliability Management*, *26*(4), 325–340. doi:10.1108/02656710910950333

Gu, Y., Xu, X., Du, J., Hou, R., & Qian, H. (2013). Anycast routing protocol for wireless sensor networks based on artificial bee colony. *Chinese Journal of Sensors and Actuators*, *26*(4), 564–569. doi:10.3969/j.issn.1004-1699.2013.04.024

Guzinski, J., Abu-Rub, H., Diguet, M., Krzeminski, Z. & Lewicki, A. (2010). Speed and load torque observer application in high-speed train electric drive. *Industrial Electronics, IEEE Transactions on*, 565–574.

Hackbusch, W., & Mittelmann, H. D. (1983). On multi-grid methods for variational inequalities. *Numerische Mathematik*, *42*(1), 65–76. doi:10.1007/BF01400918

Hacking, I. (1983). *Representing and Intervening. Introductory Topics in the Philosophy of Natural Science*. Cambridge University Press. doi:10.1017/CBO9780511814563

Hadjidimos, A. (1978). Accelerated overrelaxation method. *Mathematics of Computation*, *32*(141), 149–157. doi:10.1090/S0025-5718-1978-0483340-6

Hadjidimos, A., Lapidakis, M., & Tzoumas, M. (2012). On Iterative Solution for Linear Complementarity Problem with an H_+-Matrix. *SIAM Journal on Matrix Analysis and Applications*, *33*(1), 97–110. doi:10.1137/100811222

Hadjidimos, A., & Tzoumas, M. (2014). On the Solution of the Linear Complementarity Problem by the Generalized Accelerated Overrelaxation Iterative Method. *Journal of Optimization Theory and Applications*, 1–18.

Hadj, N., Kammoun, J., & Neji, R. (2014). Application of Evolutionary Algorithm for Triobjective Objective: Electric Vehicle. *International Journal of Energy Optimization and Engineering*, *3*(3), 1–19. doi:10.4018/ijeoe.2014070101

Han, K. H., & Kim, J. H. (2002). Quantum-inspired evolutionary algorithm for a class of combinatorial optimization. *IEEE Transactions on Evolutionary Computation*, *6*(6), 580–593. doi:10.1109/TEVC.2002.804320

Han, S.-P., & Mangasarian, O. L. (1983). A dual differentiable exact penalty function. *Mathematical Programming*, *25*(3), 293–306. doi:10.1007/BF02594781

Hansen, W. A., Edson, B. N., & Larter, P. C. (1992). Reliability, availability and maintainability expert system. In *Proceedings of Annual Reliability and Maintainability Symposium*. IEEE.

Han, X. H., Quan, L., Xiong, X. Y., & Wu, B. (2013). Facing the Classification of Binary Problems with a Hybrid System based on Quantum-Inspired Binary Gravitational Search Algorithm and K-NN Method. *Journal of Engineering Applications of Artificial Intelligence*, *26*(10), 2424–2430. doi:10.1016/j.engappai.2013.05.011

Han, X., & Chang, X. (2012). A Chaotic Digital Secure Communication based on a Modified Gravitational Search Algorithm Filter. *Journal of Information Science*, *208*, 14–27. doi:10.1016/j.ins.2012.04.039

Harcombe, P. (2012). *NSW Sydney satellite.* Oral Presentation from the NSW State Government. Retrieved from http://www.locata.com/wp-content/uploads/2012/12/

Haregeweyn, N., Fikadu, G., Tsunekawa, A., Tsubo, M., & Meshesha, D. T. (2012). The dynamics of urban expansion and its impacts on land use/land cover change and small-scale farmers living near the urban fringe: A case study of Bahir Dar, Ethiopia. *Landscape and Urban Planning*, *106*(2), 149–157. doi:10.1016/j.landurbplan.2012.02.016

Harikrishnan, K. K., & Ishii, H. (2005). Single machine batch scheduling problem with resource dependent setup and processing time in the presence of fuzzy due date. *Fuzzy Optimization and Decision Making*, *4*(2), 141–147. doi:10.1007/s10700-004-5870-9

Harjunkoski, I., & Grossmann, I. E. (2002). Decomposition techniques for multistage scheduling problems using mixed-integer and constraint programming methods. *Computers & Chemical Engineering*, *26*(11), 1533–1552. doi:10.1016/S0098-1354(02)00100-X

Harman, M. (2007). *The current state and future of search based software engineering.* Paper presented at the 2007 Future of Software Engineering. doi:10.1109/FOSE.2007.29

Harman, M., Hierons, R. M., & Proctor, M. (2002). *A New Representation And Crossover Operator For Search-based Optimization Of Software Modularization.* Paper presented at the GECCO.

Harter, A., Hopper, A., Steggles, P., Ward, A., & Webster, P. (1999). The anatomy of a context-aware application. In *Proceedings of the 5th annual ACM/IEEE international conference on Mobile computing and networking (MobiCom '99).* New York. doi:10.1145/313451.313476

Hassanzadeh, H. R., & Rohani, M. (2010). A multi-objective gravitational search algorithm. In *Proceedings of the Second International Conference on Computational Intelligence, Communication Systems and Networks.*

Hatamlou, A. (2013). Black hole: A new heuristic optimization approach for data clustering. *Information Sciences*, *222*, 175–184. doi:10.1016/j.ins.2012.08.023

Hatamlou, A., Abdullah, S., & Nezamabadi-pour, H. (2011a). Application of Gravitational Search Algorithm on Data Clustering. In *Proceedings of Rough Sets and Knowledge Technology* (pp. 337–346). Berlin: Springer. doi:10.1007/978-3-642-24425-4_44

Hatamlou, A., Abdullah, S., & Nezamabadi-pour, H. (2012). A Combined Approach for Clustering based on K-means and Gravitational Search Algorithms. *Journal of Swarm and Evolutionary Computation*, *6*, 47–52. doi:10.1016/j.swevo.2012.02.003

Hatamlou, A., Abdullah, S., & Othman, Z. (2011b). *Gravitational Search Algorithm with Heuristic Search for Clustering Problems. In Proceeding of 3rd IEEE on Data Mining and Optimization* (pp. 190–193). Selangor, Malaysia: DMO.

Haykin, S. (Ed.). (2001). *Adaptive filter theory* (4th ed.). Prentice Hall Upper Saddle River.

He, B. (1992). A projection and contraction method for a class of linear complementarity problems and its application in convex quadratic programming. *Applied Mathematics & Optimization*, *25*(3), 247–262. doi:10.1007/BF01182323

He, B. (1994). A new method for a class of linear variational inequalities. *Mathematical Programming*, *66*(1-3), 137–144. doi:10.1007/BF01581141

He, B. (1996). A modified projection and contraction method for a class of linear complementarity problems. *Journal of Computational Mathematics-International Edition, 14*, 54–63.

He, B. (1997). A class of projection and contraction methods for monotone variational inequalities. *Applied Mathematics & Optimization, 35*(1), 69–76. doi:10.1007/BF02683320

He, C., Okada, N., Zhang, Q., Shi, P., & Li, J. (2007). Modelling dynamic urban expansion processes incorporating a potential model with cellular automata. *Landscape and Urban Planning, 86*(1), 79–91. doi:10.1016/j.landurbplan.2007.12.010

He, C., Tian, J., Shi, P., & Hu, D. (2011). Simulation of the spatial stress due to urban expansion on the wetlands in Beijing, China using a GIS-based assessment model. *Landscape and Urban Planning, 101*(3), 269–277. doi:10.1016/j.landurbplan.2011.02.032

Hedar, A., & Fukushima, M. (2006). Derivative-Free Filter Simulated Annealing Method for Constrained Continuous Global Optimization. *JGlobalOptim, 35*, 521–549.

Hedley, M., Humphrey, D., & Ho, P. (2008). System and algorithms for accurate indoor tracking using low-cost hardware. In *Proceedings of the IEEE/ION Position, Location And Navigation Symposium (PLANS)*. doi:10.1109/PLANS.2008.4569983

Heinonen, E.W., Tapscott, R.E., Wildin, M.W., & Beall, A.N. (1996). *Assessment of antifreeze solutions for ground-source heat pumps systems*. New Mexico Engineering Research Institute NMERI 96/15/32580.

Hejazi, S. R., Memariani, A., Jahanshahloo, G., & Sepehri, M. M. (2002). Linear bilevel programming solution by genetic algorithm. *Computers & Operations Research, 29*(13), 1913–1925. doi:10.1016/S0305-0548(01)00066-1

Hendizadeh, S. H., ElMekkawy, T. Y., & Wang, G. G. (2007). 207. "Bi-criteria scheduling of a flowshop manufacturing cell with sequence dependent setup times. *European Journal of Industrial Engineering, 1*(4), 391–413. doi:10.1504/EJIE.2007.015388

Heo, J. S., Lee, K. Y., & Garduno-Ramirez, R. (2006). Multiobjective Control of Power Plants Using Particle Swarm Optimization Techniques. *IEEE Transactions on Energy Conversion, 21*(2), 552–561. doi:10.1109/TEC.2005.858078

Heragu, S. S., & Alfa, A. S. (1992). Experimental analysis of simulated annealing based algorithms for the facility layout problem. *European Journal of Operational Research, 57*(2), 190–202. doi:10.1016/0377-2217(92)90042-8

Heredia, F. J., & Nabona, N. (1995). Optimum short-term hydrothermal scheduling with spinning reserve through network flows. *IEEE Transactions on Power Systems, 10*(3), 1642–1651. doi:10.1109/59.466476

Herrera, F., Herrera-Viedma, E., & Chiclana, F. (2001). Multiperson decision-making based on multiplicative preference relations. *European Journal of Operational Research, 129*(2), 372–385. doi:10.1016/S0377-2217(99)00197-6

Hess, A., Malandrino, F., Reinhardt, M. B., Casetti, C., Hummel, K. A., & Barceló-Ordinas, J. M. (2012). Optimal deployment of charging stations for electric vehicular networks. In *Proceedings of the first workshop on urban networking*. doi:10.1145/2413236.2413238

He, X., Chen, A., Chaovalitwongse, W., & Liu, H. (2012). An improved linearization technique for a class of quadratic 0-1 programming problems. *Optimization Letters, 6*(1), 31–41. doi:10.1007/s11590-010-0249-z

Hide, C. (2003). *Integration of GPS and low cost INS measurements.* (doctoral dissertation). University of Nottingham, Nottingham, UK.

Hildreth, C. (1954). Point estimates of ordinates of concave functions. *Journal of the American Statistical Association, 49*(267), 598–619. doi:10.1080/01621459.1954.10483523

Hills, D. W. (1990). Co-evolving parasites improve simulated evolution as an optimization procedure. In Artificial Life II. Addison Wesley.

Hindawi Publishing Corporation. (n.d.a). Retrieved from http://www.hindawi.com/28921463/

Hindawi Publishing Corporation. (n.d.b). Retrieved from http://www.hindawi.com/journals/tswj/2015/964581/

Hirano, H., & Niki, H. (2001). Application of a preconditioning iterative method to the computation of fluid flow. *Numerical Functional Analysis and Optimization*, *22*(3), 405–418. doi:10.1081/NFA-100105110

Hirst, J. D., King, R. D., & Sternberg, M. J. (1994). Quantitative structure-activity relationships by neural networks and inductive logic programming: 2. The inhibition of dihydrofolate reductase by triazines. *Journal of Computer-Aided Molecular Design*, *8*(4), 405–420. doi:10.1007/BF00125375 PMID:7815092

Hirvonen, M., Pursula, P., Jaakkola, K., & Laukkanen, K. (2004). Planar inverted-F antenna for radio frequency identification. *Electronics Letters*, *40*(14), 848–850. doi:10.1049/el:20045156

Hismath, I., Wan Aida, W. M., & Ho, C. W. (2011). Optimization of extraction conditions for phenolic compounds from neem (Azadirachta indica) leaves. *International Food Research Journal*, *18*(3), 931–939.

Ho, S., Hsieh, C., Yu, F., & Huang, H. (2007). An intelligent two-stage evolutionary algorithm for dynamic pathway identification from gene expression profiles. *Computational Biology and Bioinformatics, IEEE/ACM Transactions on*, 648–704.

Hoerl, A. E. (1962). Application of ridge analysis to regression problems. *Chemical Engineering Progress*, *58*, 54–59.

Hofmann-Wellenhof, B., Lichtenegger, H., & Collins, J. (Eds.). (1993). *Global positioning system - theory and practice*. Springer-Verlag.

Holland, J. (1975). *A aptation in natural and artificial systems*. University of Michigan Press.

Holland, J. H. (1975). *Adaptation in Natural and Artificial Systems*. Ann Arbor, MI: The University of Michigan Press.

Holtz, J. (1995). The representation of ac machine dynamics by complex signal flow graphs. *Industrial Electronics, IEEE Transactions on*, 263–271.

Holtz-Eakin, D., & Selden, T. M. (1995). Stoking the fires? CO_2 emissions and economic growth. *Journal of Public Economics*, *57*(1), 85–101. doi:10.1016/0047-2727(94)01449-X

Holtz, J. (2002). Sensorless control of induction motor drives. *Proceedings of the IEEE*, 1359–1394.

Hong, D. H., & Choi, C. H. (2000). Multicriteria fuzzy decision making problems based on vague set theory. *Fuzzy Sets and Systems*, *114*(1), 103–113. doi:10.1016/S0165-0114(98)00271-1

Hopfield, J. J. (1982). Neural Networks and Physical Systems with Emergent Collective Computational Abilities. *Proceedings of the National Academy of Sciences of the United States of America*, *29*(8), 2554–2558. doi:10.1073/pnas.79.8.2554 PMID:6953413

Hopfield, J. J. (1984). Neurons with Graded Response have Collective Computational Properties like those of Two-State Neurons. *Proceedings of the National Academy of Sciences of the United States of America*, *81*(10), 3088–3092. doi:10.1073/pnas.81.10.3088 PMID:6587342

Horvat, A., & Tošic, A. (2012). Optimization of traffic networks by using genetic algorithms. *Elektrotehniški Vestnik*, *79*, 197–200.

Hosking, J. R. M. (1981). Fractional differencing. *Biometrika*, *68*(1), 165–176. doi:10.1093/biomet/68.1.165

Hota, A. R., Juvvanapudi, M., & Bajpai, P. (2014). Issues and solution approaches in PHEV integration to the smart grid. *Renewable & Sustainable Energy Reviews*, *30*(0), 217–229. doi:10.1016/j.rser.2013.10.008

Hotaa, P. K., Barisala, A. K., & Chakrabarti, R. (2009). An improved PSO technique for short-term optimal hydrothermal scheduling. *Electric Power Systems Research*, *79*(7), 1047–1053. doi:10.1016/j.epsr.2009.01.001

Hota, P. K., Chakrabarti, R., & Chattopadhyay, P. K. (1999). Short-term hydrothermal scheduling through evolutionary programming technique. *Electric Power Systems Research*, *52*(2), 189–196. doi:10.1016/S0378-7796(99)00021-8

Howard, R. (1971). *Dynamic probabilistic systems. Volume II: Semi Markov and decisión process.* John Wiley & Sons.

Hsieh, F., & Lin, J. (2014). A dynamic scheme for scheduling complex tasks in manufacturing systems based on collaboration of agents. *Applied Intelligence*, *41*(2), 366–382. doi:10.1007/s10489-014-0521-5

Hu, J., & Wang, C. (2011). Discrete Optimization Via Approximate Annealing Adaptative Search With Stochastic Averaging. In *Proceedings of the 2011 Winter Simulation Conference* (pp. 4206-4216). Phoenix, AZ.

Huang, J. Z., Yang, P. H., Chew, W. C., & Ye, T. T. (2010). A novel broadband patch antenna for universal UHF RFID tags. *Microwave and Optical Technology Letters*, *52*(12), 2653–2657. doi:10.1002/mop.25560

Huang, S.-J., & Liu, X.-Z. (2013). Application of artificial bee colony-based optimization for fault section estimation in power systems. *International Journal of Electrical Power & Energy Systems*, *44*(1), 210–218. doi:10.1016/j.ijepes.2012.07.012

Huang, W., Jiang, C., Xue, L., & Song, D. (2011). Extracting solar cell model parameters based on chaos particle swarm algorithm. In *2011 International Conference on Electric Information and Control Engineering (ICEICE)*. doi:10.1109/ICEICE.2011.5777246

Hunt, F. Y., Kearsley, A. J., & O'Gallagher, A. (2003). A linear programming based algorithm for multiple sequence alignments. In *Proceedings of the 2003 IEEE Bioinformatics Conference* (pp. 532-533). doi:10.1109/CSB.2003.1227398

Hurst, H. E., Black, R. P., & Sinaika, Y. M. (1966). Long term storage in reservoirs: An experimental study. *Journal of the Royal Statistical Society. Series A (General)*, *129*(4), 591–593. doi:10.2307/2982267

Hussein, W. A., Sahran, S., & Abdullah, S. N. H. S. (2014). Patch-Levy-based Initialization Algorithm for Bees Algorithm. *Applied Soft Computing*, *23*, 104–121. doi:10.1016/j.asoc.2014.06.004

Husseinzadeh Kashan, A., & Karimi, B. (2008). Scheduling a single batch-processing machine with arbitrary job sizes and incompatible job families: An ant colony framework. *The Journal of the Operational Research Society*, *59*(9), 1269–1280. doi:10.1057/palgrave.jors.2602448

Husseinzadeh Kashan, A., Karimi, B., & Jolai, F. (2006). Effective hybrid genetic algorithm for minimizing makespan on a single-batch-processing machine with nonidentical job sizes. *International Journal of Production Research*, *44*(12), 2337–2360. doi:10.1080/00207540500525254

Husseinzadeh Kashan, A., Karimi, B., & Jolai, F. (2010). An effective hybrid multi-objective genetic algorithm for bi-criteria scheduling on a single batch processing machine with non-identical job sizes. *Engineering Applications of Artificial Intelligence*, *23*(6), 911–922. doi:10.1016/j.engappai.2010.01.031

Hu, T., Guo, X., Fu, X., & Lv, Y. (2010). A neural network approach for solving linear bilevel programming problem. *Knowledge-Based Systems*, *23*(3), 239–242. doi:10.1016/j.knosys.2010.01.001

Huynh, V.-N., Nakamori, Y., Lawry, J., & Inuiguchi, M. (Eds.). (2010). *Integrated Uncertainty Management and Applications*. Berlin: Springer-Verlag.

Hvolby, H., & Jensen, K. (2010). Technical and industrial issues of Advanced Planning and Scheduling (APS) systems. *Computers in Industry*, *61*(9), 845–851. doi:10.1016/j.compind.2010.07.009

Hwang, C. L., & Yoon, K. (1981). *Multiple attribute decision methods and applications: A state of the art survey*. New York: Springer Verlag. doi:10.1007/978-3-642-48318-9

Ibrahim, A., Mohamed, A., & Shareef, H. (2012). Application Of Quantum-Inspired Binary GSA For Optimal Power Quality Monitor Placement. In *Proceedings of The 11th International Conference On Artificial Intelligence, Knowledge Engineering And Data Bases*. World Scientific and Engineering Academy and Society (WSEAS).

Ibrahim, I., Ibrahim, Z., Ahmad, H., Yusof, Z. MD., Shapiai, M.I., Nawawi, S.W. & Mubin, M., (2014). An Assembly Sequence Planning Approach with BGSA. *Journal of New Trends in Software Methodologies, Tools and Techniques*, 179-193.

Ibrahim, A., Mohamed, A., & Shareef, H. (2012). A novel quantum-inspired binary gravitational search algorithm in obtaining optimal power quality monitor placement. *Journal of Applied Science*, *12*(9), 822–830. doi:10.3923/jas.2012.822.830

Ibrahim, H. (1996). Metaheuristics: A bibliography, Institute of Mathematics and Statistics. *Annals of Operations Research*, *63*, 513–623.

Ibrahim, Z., Muhammad, B., Ghazali, K. H., Lim, K. S., Nawawi, S. W., & Yusof, Z. (2012). Vector Evaluated Gravitational Search Algorithm (VEGSA) for Multi-Objective Optimization Problems. In *Proceedings of the Fourth International Conference on Computational Intelligence, Modeling and Simulation*. doi:10.1109/CIMSim.2012.29

Ikura, Y., & Gimple, M. (1986). Efficient scheduling algorithms for a single batch processing machine. *Operations Research Letters*, *5*(2), 61–65. doi:10.1016/0167-6377(86)90104-5

Ilgina, M., & Guptab, S. (2010). Environmentally conscious manufacturing and product recovery (ECMPRO): A review of the state of the art. *Journal of Environmental Management*, *91*(3), 563–592. doi:10.1016/j.jenvman.2009.09.037 PMID:19853369

Illés, T., Nagy, M., & Terlaky, T. (2010). A polynomial path-following interior point algorithm for general linear complementarity problems. *Journal of Global Optimization*, *47*(3), 329–342. doi:10.1007/s10898-008-9348-0

Ingersoll, J. G., & Perkins, C. A. (1996). The 2.1 kW photovoltaic electric vehicle charging station in the city of Santa Monica, California. In *Photovoltaic Specialists Conference, Conference Record of the Twenty Fifth IEEE*. doi:10.1109/PVSC.1996.564423

Investopedia. (2015). *Supply Chain Management-SCM*.http://www.investopedia.com/terms/s/scm.asp

Ishii, H., Tada, M., & Masuda, T. (1992). Two scheduling problems with fuzzy due-dates. *Fuzzy Sets and Systems*, *46*(3), 339–347. doi:10.1016/0165-0114(92)90372-B

Iyengar, G. N. (2005). Robust dynamic programming. *Mathematics of Operations Research*, *30*(2), 257–280. doi:10.1287/moor.1040.0129

Jackson. (1991) *A User's Guide to Principal Components*. New York: John Wiley & Sons.

Janak, S. L., Lin, X. X., & Floudas, C. A. (2005). Enhanced continuous-time unit-specificevent-based formulation for short-term scheduling of multipurpose batch processes: Resource constraints and mixed storage policies (vol. 43, p. 2529, 2002). *Industrial & Engineering Chemistry Research*, *44*(2), 426. doi:10.1021/ie048866r

Jans, R., & Degraeve, Z. (2007). Meta-heuristics for dynamic lot sizing: A review and comparison of solution approaches. *European Journal of Operational Research*, *177*(3), 1855–1875. doi:10.1016/j.ejor.2005.12.008

Jans, R., & Degraeve, Z. (2008). Modeling industrial lot sizing problems: A review. *International Journal of Production Research*, *46*(6), 1619–1643. doi:10.1080/00207540600902262

Jayakumar, A., & Asagarpoor, S. (2004). Maintenance optimization of equipment by linear programming. In *International conference on probabilistic methods applied to power systems*, (pp. 145 – 149).

Jayaprakasam, S., Rahim, S. K. A., & Leow, C. Y. (2015). PSOGSA-Explore: A new hybrid metaheuristic approach for beam-pattern optimization in collaborative beam-forming. *Journal of Applied Soft Computing*, *30*, 229–237. doi:10.1016/j.asoc.2015.01.024

Jeya Mala, D., Mohan, V., & Kamalapriya, M. (2010). Automated software test optimisation framework-an artificial bee colony optimisation-based approach. *Software, IET*, *4*(5), 334–348. doi:10.1049/iet-sen.2009.0079

Jia, J. (2009). Parameter estimation for nonlinear biological system model based on global sensitivity analysis. In *3rd International Conference on Bioinformatics and Biomedical Engineering*. doi:10.1109/ICBBE.2009.5163168

Jiang, M., & Dong, J. (2005). On convergence of two-stage splitting methods for linear complementarity problems. *Journal of Computational and Applied Mathematics*, *181*(1), 58–69. doi:10.1016/j.cam.2004.11.018

Jiang, Y., Li, X., Huang, C., & Wu, X. (2013). Application of particle swarm optimization based on CHKS smoothing function for solving nonlinear bilevel programming problem. *Applied Mathematics and Computation*, *219*(9), 4332–4339. doi:10.1016/j.amc.2012.10.010

Jia-zhao, C., Yu-xiang, Z., & Yin-sheng, L. (2012). A Unified Frame of Swarm Intelligence Optimization Algorithm. *Knowledge Discovery and Data Mining*, 745-751.

Ji, B., Yuan, X., Li, X., Huang, Y., & Li, W. (2014). Application of Quantum-Inspired Binary Gravitational Search Algorithm for Thermal Unit Commitment with Wind Power Integration. *Journal of Energy Conversion and Management*, *87*, 589–598. doi:10.1016/j.enconman.2014.07.060

Jie, H. X., & Wei, C. (2008). Optimization of extraction process of crude polysaccharides from wild edible BaChu mushroom by response surface methodology. *Carbohydrate Polymers*, *72*(1), 67–74. doi:10.1016/j.carbpol.2007.07.034

Jin, M., Liu, K., & Bowden, R. (2007). A two-stage algorithm with valid inequalities for the split delivery vehicle routing problem. *International Journal of Production Economics*, *105*(1), 228–242. doi:10.1016/j.ijpe.2006.04.014

Jiun-Peng, C., & Powen, H. (2013). A Compact Strip Dipole Coupled Split-Ring Resonator Antenna for RFID Tags. *IEEE Transactions on Antennas and Propagation*, *61*(11), 5372–5376. doi:10.1109/TAP.2013.2278481

Jo, H. Y., Katsumi, T., Benson, C. H., & Edil, T. B. (2001). Hydraulic conductivity and swelling of non-prehydrated GCLs permeated with single- species salt solutions. *Journal of Geotechnical and Geoenvironmental Engineering*, *127*(7), 557–567. doi:10.1061/(ASCE)1090-0241(2001)127:7(557)

Joliffe, I. T. (2002). *Principal Component Analysis* (2nd ed.). New York: Springer.

Jones, N., & Pevzner, P. (2004). *An Introduction to Bioinformatics Algorithms*. Cambridge, MA: MIT Press.

Jonsson, P., & Mattsson, S.-A. (2003). The implications of fit between planning environments and manufacturing planning and control methods. *International Journal of Operations & Production Management*, *23*(8), 872–900. doi:10.1108/01443570310486338

Jubaer, A., & Zainal, S. A. (2014). Maximum Power Point Tracking (MPPT) for PV system using Cuckoo Search with partial shading capability. *Applied Energy*, *119*, 118–130. doi:10.1016/j.apenergy.2013.12.062

Julier, S., & Uhlmann, J. (1997). New extension of the kalman filter to nonlinear systems. *Proceedings of the Society for Photo-Instrumentation Engineers*, 182–193. doi:10.1117/12.280797

Jung, S. H. (2003). Queen-bee Evolution for Genetic Algorithms. *Electronics Letters*, *39*(6), 575–576. doi:10.1049/el:20030383

Jung, W., & Guziewicz, M. (2009). Schottky diode parameters extraction using LambertWfunction. *Materials Science and Engineering B*, *165*(1-2), 57–59. doi:10.1016/j.mseb.2009.02.013

Kalganova, T., Russell, G., & Cumming, A. (1999). Multiple Traffic Signal Control Using A Genetic Algorithm. In Artificial Neural Nets and Genetic Algorithms (pp. 220-228). Roanne, France.

Kallrath, J. (1999). Mixed-Integer nonlinear Programming Applications. In T. A. Ciriani, S. Gliozzi, E. l. Johnson, & R. Tadei (Eds.), *Operational Research in Industry*. Basingstoke: Macmillan Press.

Kalman, R. E. (1960). A new approach to linear filtering and prediction problem, *Trans. ASME, Ser. D. Journal of Basic Engineering*, 34–45.

Kamenetsky, M., & Widrow, B. (2004). A variable leaky LMS adaptive algorithm. In *Proceedings of the 38th Asilomar Conference on Signals, Systems and Computers*. doi:10.1109/ACSSC.2004.1399103

Kanli, M. O. (2005). Limitations of pseudolites systems using off-the-shelf GPS receivers. *Journal of Global Positioning Systems*, *3*(1-2), 154–166.

Kanzow, C. (1996). Global convergence properties of some iterative methods for linear complementarity problems. *SIAM Journal on Optimization*, *6*(2), 326–341. doi:10.1137/0806019

Kanzow, C. (1996). Some noninterior continuation methods for linear complementarity problems. *SIAM Journal on Matrix Analysis and Applications*, *17*(4), 851–868. doi:10.1137/S0895479894273134

Kao, Y., & Chen, M. (2013). Solving the CVRP Problem Using a Hybrid PSO Approach. In K. Madani, A. Dourado, A. Rosa, & J. Filipe (Eds.), *Computational Intelligence, Studies in Computational Intelligence* (Vol. 465, pp. 59–67). Heidelberg: Springer. doi:10.1007/978-3-642-35638-4_5

Kappel, N. W., & Watson, L. T. (1986). Iterative algorithms for the linear complementarity problem. *International Journal of Computer Mathematics*, *19*(3-4), 273–297. doi:10.1080/00207168608803522

Karaboga, D. (2005). *An Idea based on Honey Bee Swarm for Numerical Optimization*. Technical Report TR06. Computer Engineering Department, Engineering Faculty, Erciyes University.

Karaboga, D., & Akay, B. (2009). A comparative study of Artificial Bee Colony algorithm. *Applied Mathematics and Computation*, *214*(1), 108–132. doi:10.1016/j.amc.2009.03.090

Karaboga, D., Akay, B., & Ozturk, C. (2007) Artificial Bee Colony (ABC) optimization algorithm for training feed-forward neural networks.Vol. 4617*LNAI.4th International Conference on Modeling Decisions for Artificial Intelligence, MDAI 2007* (pp. 318-329). Kitakyushu. doi:10.1007/978-3-540-73729-2_30

Karaboga, D., & Basturk, B. (2007). A powerful and efficient algorithm for numerical function optimization: Artificial bee colony (ABC) algorithm. *Journal of Global Optimization*, *39*(3), 459–471. doi:10.1007/s10898-007-9149-x

Karaboga, D., & Basturk, B. (2008). On the performance of artificial bee colony (ABC) algorithm. *Applied Soft Computing*, *8*(1), 687–697. doi:10.1016/j.asoc.2007.05.007

Karaboga, D., Gorkemli, B., Ozturk, C., & Karaboga, N. (2014). A comprehensive survey: Artificial bee colony (ABC) algorithm and applications. *Artificial Intelligence Review*, *42*(1), 21–57. doi:10.1007/s10462-012-9328-0

Karaboga, D., Okdem, S., & Ozturk, C. (2012). Cluster based wireless sensor network routing using artificial bee colony algorithm. *Wireless Networks*, *18*(7), 847–860. doi:10.1007/s11276-012-0438-z

Karaki, S. H., Chedid, R. B., & Ramadan, R. (1999a). Probabilistic performance assessment of wind energy conversion systems. *IEEE Transactions on Energy Conversion*, *14*(2), 217–224. doi:10.1109/60.766986

Karaki, S. H., Chedid, R. B., & Ramadan, R. (1999b). Probabilistic performance assessment of autonomous solar-wind energy conversion systems. *IEEE Transactions on Energy Conversion*, *4*(3), 766–772. doi:10.1109/60.790949

Karimi, B., Fatemi Ghomi, S. M. T., & Wilson, J. M. (2003). The capacitated lot sizing problem: A review of models and algorithms. *Omega*, *31*(5), 365–378. doi:10.1016/S0305-0483(03)00059-8

Kaufmann, A., & Gupta, M. M. (1985). *Introduction to Fuzzy Arithmetic: Theory and Applications*. New York: Van Nostrand Reinhold.

Kaviani, A. K., Baghaee, H. R., & Riahy, G. H. (2007). Design and optimal sizing of a photovoltaic/wind- generator system using particle swarm optimization. In *Proceedings of the 22nd Power System Conference*. Tehran, Iran: PSC.

Kay, E. (1976). The effectiveness of preventive maintenance. *International Journal of Production Research*, *14*(3), 329–344. doi:10.1080/00207547608956606

Kay, S. (Ed.). (1993). Fundamentals of statistical signal processing: Vol. 1. *Estimation theory* (1st ed.). Prentice Hall Signal Processing Series, Prentice Hall PTR Upper Saddle River.

Kazak, N., & Duysak, A. (2012). Modified gravitational search algorithm. *International Symposium on Innovations in Intelligent Systems and Applications*.

Kefayati, M., & Caramanis, C. (2010). Efficient energy delivery management for PHEVs. In *Smart Grid Communications (SmartGridComm), 2010 First IEEE International Conference on*. doi:10.1109/SMARTGRID.2010.5621990

Kefayat, M., Lashkar Ara, A., & Nabavi Niaki, S. A. (2015). A hybrid of ant colony optimization and artificial bee colony algorithm for probabilistic optimal placement and sizing of distributed energy resources. *Energy Conversion and Management*, *92*, 149–161. doi:10.1016/j.enconman.2014.12.037

Kennedy, J., & Eberhart, R. (1995). Particle Swarm Optimization. In *Proceedings of the International Conference on Neural Networks*.

Kennedy, J., & Eberhart, R. (1995). *Particle swarm optimization.* Paper presented at the IEEE International Conference on Neural Networks, Piscataway, NJ.

Kennedy, J., & Eberhart, R. C. (1997). *Discrete binary version of the particle swarm algorithm.* Paper presented at the IEEE International Conference on Systems, Man and Cybernetics. doi:10.1109/ICSMC.1997.637339

Kennedy, J., & Ebemart, R. C. (1995). Particle swarm optimization. In *ProceedingsofIEEE International Conference*. IEEE.

Kennedy, J., & Eberhart, R. (1995). Particle swarm optimization.*IEEE International Joint Conference on Neural Networks* (vol. 4, pp. 1942-1948). Perth, Australia: IEEE.

Kennedy, J., & Eberhart, R. (1995, Nov/Dec). *Particle swarm optimization.* In *Proceedings of IEEE International Conference on Neural Networks*. doi:10.1109/ICNN.1995.488968

Kennedy, J., & Eberhart, R. C. (2000). *Intelligent Swarm Systems*. New York: Academic Press.

Keshavarz Ghorabaee, M., Amiri, M., & Azimi, P. (in press). Genetic algorithm for solving bi-objective redundancy allocation problem with k-out-of-n subsystems. *Applied Mathematical Modelling*.

Ke, Y.-F., & Ma, C.-F. (2014). On the convergence analysis of two-step modulus-based matrix splitting iteration method for linear complementarity problems. *Applied Mathematics and Computation*, *243*, 413–418. doi:10.1016/j.amc.2014.05.119

Khajehzadeh, M., Taha, M. R., El-Shafie, A., & Eslami, M. (2012). A Modified Gravitational Search Algorithm for Slope Stability Analysis. *Journal of Engineering Applications of Artificial Intelligence*, *25*(8), 1589–1597. doi:10.1016/j.engappai.2012.01.011

Khamseh, A. A., & Mahmoodi, M. (2014). A new fuzzy topsis-todim hybrid method for green supplier selection using fuzzy time function. *Advances in Fuzzy Systems, Article ID 841405*.

Khayyal, A. (1985). AL Minimizing a Quasi-concave Function Over a Convex Set: A Case Solvable by Lagrangian Duality. In *ProceedingsIEEE International Conference on Systems, Man, and Cybemeties*, (pp. 661-663).

Khebbache-Hadji, S., Prins, C., Yalaoui, A., & Reghioui, M. (2013). Heuristics and memetic algorithm for the two-dimensional loading capacitated vehicle routing problem with time windows. *Central European Journal of Operations Research*, *21*(2), 307–336. doi:10.1007/s10100-011-0204-9

Kilger, C., Staedtler, H., & Meyr, H. (2008). *Supply Chain Management and Advanced Planning, Concepts, Models, Software, and Case Studies* (4th ed.). Springer.

Kim, D. W., Kim, K. H., Jang, W., & Chen, F. F. (2002). Unrelated parallel machine scheduling with setup times using simulated annealing. *Robotics and Computer-integrated Manufacturing*, *18*(3-4), 223–231. doi:10.1016/S0736-5845(02)00013-3

Kim, Y., Sul, S., & Park, M. (1994). Speed sensorless vector control of induction motor using extended kalman filter. *IEEE Transactions on Industry Applications*, *30*(5), 1225–1233. doi:10.1109/28.315233

Kingham, M., Mohammadian, M., & Stonier, R. J. (1998). Prediction of Interest Rate using Neural Networks and Fuzzy Logic. In *Proceedings of ISCA 7th International Conference on Intelligent Systems*.

King, R. D., Hirst, J. D., & Sternberg, M. J. (1995). Comparison of artificial intelligence methods for modeling pharmaceutical QSARS. *Applied Artificial Intelligence an International Journal*, *9*(2), 213–233. doi:10.1080/08839519508945474

Kirkpatrick, S., Gelatt, D., & Vecchi, M. P. (1983). Optimization by simulated annealing. *Science*, *220*(4598), 671–680. doi:10.1126/science.220.4598.671 PMID:17813860

Kocvara, M., & Zowe, J. (1994). An iterative two-step algorithm for linear complementarity problems. *Numerische Mathematik*, *68*(1), 95–106. doi:10.1007/s002110050050

Koehn. (1994). *Combining Genetic Algorithms and Neural Networks: The Encoding Problem.* (Master of Science Degree Thesis). The University of Tennessee, Knoxville, TN.

Koh, S. G., Koo, P. H., Kim, D. C., & Hur, W. S. (2005). Scheduling a single batch processing machine with arbitrary job sizes and incompatible job families. *International Journal of Production Economics*, *98*(1), 81–96. doi:10.1016/j.ijpe.2004.10.001

Kohshelan & Wahid, N. (2014). Improvement of Audio Feature Extraction Techniques in Traditional Indian Musical Instrument. Recent Advances on Soft Computing and Data Mining, 507-516

Kojima, M., Mizuno, S., & Yoshise, A. (1989). A polynomial-time algorithm for a class of linear complementarity problems. *Mathematical Programming*, *44*(1-3), 1–26. doi:10.1007/BF01587074

Kojima, M., Mizuno, S., & Yoshise, A. (1991). AnO(\sqrt n L) iteration potential reduction algorithm for linear complementarity problems. *Mathematical Programming*, *50*(1-3), 331–342. doi:10.1007/BF01594942

Kolodziej, K. W., & Hjelm, J. (Eds.). (2006). *Local Positioning Systems: LBS applications and services*. Taylor & Francis Group.

Kononoff, P., Lehman, H., & Heinrichs, A. (2002). Technical note—a comparison of methods used to measure eating and ruminating activity in confined dairy cattle. *Journal of Dairy Science*, *85*(7), 1801–1803. doi:10.3168/jds.S0022-0302(02)74254-9 PMID:12201531

Korostensky, C., & Gonnet, G. (1999). Near optimal multiple sequence alignments using a traveling salesman problem approach. In *Proceedings of String Processing and Information Retrieval Symposium, and International Workshop on Groupware* (pp. 105–114). doi:10.1109/SPIRE.1999.796584

Kotecha, J., & Djuric, P. (2003). Gaussian particle filtering. *IEEE Transactions on Signal Processing*, *51*(10), 2592–2601. doi:10.1109/TSP.2003.816758

Koza, J. R. (1992). *Genetic Programming: On the Programming of Computers by Means of Natural Selection*. MIT Press.

Koza, J. R. (1992). *Genetic programming: on the programming of computers by natural selection*. Cambridge, MA: MIT Press.

Kozeny, V. (2015). Genetic algorithms for credit scoring: Alternative fitness function performance comparison, Expert Systems with Applications. *Expert Systems with Applications*, *42*(6), 2998–3004. doi:10.1016/j.eswa.2014.11.028

Krajzewicz, D., Erdmann, J., Behrisch, M., & Bieker, L. (2012). Recent Development and Applications of "SUMO - Simulation of Urban Mobility". *International Journal On Advances in Systems and Measurements*, *5*(4), 128–138.

Krauss, S. (1998). *Microscopic Modeling of Traffic Flow: Investigation of Collision Free Vehicle Dynamics*. (Unpublished doctoral dissertation). Hauptabteilung Mobilitt und Systemtechnik des DLR. Koln.

Kreyszig, E. (2009). *Advanced Engineering Mathematics* (9th ed.). New York: John Wiley & Sons.

Kristensen, A. (1996). *Textbook notes of herd management: Dynamic programming and Markov decision processes*. Retrieved from http://www.academia.edu/4892288/Dina,Notat,No.,49

Krzeminski, Z. (2008). Observer of induction motor speed based on exact disturbance model. In *13th IEEE on Power Electronics and Motion Control Conference*, (pp. 2294–2299). doi:10.1109/EPEPEMC.2008.4635605

Kubota, H., Matsuse, K. & Nakano, T. (1993). Dsp-based speed adaptive flux observer of induction motor. *Industry Applications, IEEE Transactions on*, 344–348.

Kulshrestha, P., Wang, L., Chow, M. Y., & Lukic, S. (2009). Intelligent energy management system simulator for PHEVs at municipal parking deck in a smart grid environment. In Power & Energy Society General Meeting, 2009. PES'09. IEEE (pp. 1-6). IEEE. doi:10.1109/PES.2009.5275688

Kumar, S., Kumar, A., Sharma, V. K., & Sharma, H. (2014). *A novel hybrid memetic search in Artificial Bee Colony algorithm*. Paper presented at the 2014 7th International Conference on Contemporary Computing, IC3 2014. doi:10.1109/IC3.2014.6897149

Kumar, S., Sharma, V. K., & Kumari, R. (2014). *Memetic search in Artificial Bee Colony algorithm with fitness based position update*. Paper presented at the International Conference on Recent Advances and Innovations in Engineering, ICRAIE 2014. doi:10.1109/ICRAIE.2014.6909301

Kumar, S., & Naresh, R. (2007). Efficient real coded genetic algorithm to solve the non-convex hydrothermal scheduling problem. *International Journal of Electrical Power & Energy Systems*, *29*(10), 738–747. doi:10.1016/j.ijepes.2007.06.001

Kumar, S., & Panneerselvam, R. (2012). A survey on the vehicle routing problem and its variants. *Intelligent Information Management*, *4*(03), 66–74. doi:10.4236/iim.2012.43010

Kunche, P., Rao, G. S. B., Reddy, K., & Maheswari, R. U. (2014). A new approach to dual channel speech enhancement based on hybrid PSOGSA. *International Journal of Speech Technology*, 1–12.

Kurz, A., & Zäpfel, G. (2013). Modeling cost-delivery trade-offs for distribution logistics by a generalized PVRP model. *Journal of Business Economics*, *83*(7), 705–726. doi:10.1007/s11573-013-0678-0

Kusiak, A., Xu, G., & Tang, F. (2010). Optimization of an HVAC System With A Strength Multi-Objective Particle-Swarm Algorithm. *Energy*, *36*(5), 184–192.

Kyriakidis, E., & Dimitrakos, T. (2006). Optimal preventive maintenance of a production system with an intermediate buffer. *European Journal of Operational Research*, *168*(1), 86–99. doi:10.1016/j.ejor.2004.01.052

Laarhoven, P., & Aarts, E. (1988). *Simulated Annealing: Theory and Applications*. Dordrecht: Kluwer Academic Publishers.

Lakshminarasimman, L., & Subramanian, S. (2006). Short-term scheduling of hydrothermal power system with cascaded reservoirs by using modified differential evolution. *IEE Proceedings. Generation, Transmission and Distribution*, *153*(6), 693–700. doi:10.1049/ip-gtd:20050407

Lam, S. (1999). *The role of CSP in reduced chemistry modeling*. IMA Symposium, Minnesota University.

Lambert, T. (2014). *A Brief History of Transport*. Retrieved from http://www.localhistories.org/transport.html

Lambert, D. M., Stock, J. R., & Ellram, L. M. (1998). *Fundamentals of Logistics Management*. Irwin/McGraw-Hill.

Lambiase, A., Mastrocinque, E., Miranda, S., & Lambiase, A. (2013). Strategic planning and design of supply chains: A literature review. *Int. J. Eng. Bus. Manag*, *5*(49), 1–11. doi:10.5772/56858

Langdon, W. B. (2008). Genetic Programming: An Introduction and Tutorial, with a Survey of Techniques and Applications. *Studies in Computational Intelligence, 115*, 927–1028. Retrieved from www.springerlink.com

Lan, K. M., Wen, U. P., Shih, H. S., & Lee, E. S. (2007). A hybrid neural network approach to bilevel programming problems. *Applied Mathematics Letters*, *20*(8), 880–884. doi:10.1016/j.aml.2006.07.013

Lapa, C. M. F., Pereira, C. M., & Barros, M. P. D. (2006). A model for preventive maintenance planning by genetic algorithms based on cost and reliability. *Reliability Engineering & System Safety*, *91*(2), 233–240. doi:10.1016/j.ress.2005.01.004

Laplaze, D., & Youm, I. (1985). Modélisation dune cellule photovoltaïque: Détermination des paramètres à partir de la caractéristique courant-tension sous e'clairement. *Solar Cells*, *14*(2), 167–177. doi:10.1016/0379-6787(85)90039-0

Laporte, G., & Osman, I. (1995). Routing problems: A bibliography. *Annals of Operations Research*, *61*(1), 227–262. doi:10.1007/BF02098290

Lasdon, L. S., & Terjung, R. C. (1971). An efficient algorithm for multi-item scheduling. *Operations Research*, *19*(4), 946–969. doi:10.1287/opre.19.4.946

Leach, L. F., & Henson, R. K. (2007). The use and impact of adjusted R^2. Effects in published regression research. *Multiple Linear Regression Viewpoints*, *33*(1), 1–11.

Leardi, R., Boggia, R., & Terrile, M. (2002). Genetic algorithms as a strategy for feature selection. *Journal of Chemometrics*, *6*(5), 267–281. doi:10.1002/cem.1180060506

Lee, K., Lee, G., Yoon, T., Choy, I., & Song, J. (2005). Speed and flux estimation for an induction motor using a parameter estimation technique. *International Journal of Control, Automation and Systems*, 79–86.

Lee, Y., Jeong, C., & Moon, C. (2002). Advanced planning and scheduling with outsourcing in manufacturing supply chain. *Computers & Industrial Engineering, 43*(1-2), 351-374.

Lee, C., Epelman, M., White, C. III, & Bozer, Y. (2006). A shortest path approach to the multiplevehicle routing problem with split pick-ups. *Transportation Research, 40*(4), 265–284. doi:10.1016/j.trb.2004.11.004

Lee, J., & Ricker, N. (1994). Extended kalman filter based nonlinear model predictive control. *Industrial & Engineering Chemistry Research, 33*(6), 1530–1541. doi:10.1021/ie00030a013

Lee, K. W. (2000). Stochastic models for random-request availability. *IEEE Transactions on Reliability, 49*(1), 80–84. doi:10.1109/24.855539

Lee, K. Y., Sode-Yome, A., & Park, J. H. (1998). Adaptive Hopfield Neural Networks for Economic Load Dispatch. *IEEE Transactions on Power Systems, 13*(2), 519–526. doi:10.1109/59.667377

Lee, S., & Grossmann, I. E. (2003). Global optimization of nonlinear generalized disjunctive programming with bilinear equality constraints: Applications to process networks. *Computers & Chemical Engineering, 27*(11), 1557–1575. doi:10.1016/S0098-1354(03)00098-X

Lee, Z. Y. (2006). Method of bilaterally bounded to solution Blasius equation using particle swarm optimization. *Applied Mathematics and Computation, 179*(2), 779–786. doi:10.1016/j.amc.2005.11.118

Lei, C., Shu, Y., & Hanhua, Y. (2009). *Study and design of a modified fractal antenna for RFID applications.* Paper presented at the Computing, Communication, Control, and Management, 2009. CCCM 2009. ISECS International Colloquium on.

Leick, A. (Ed.). (2004). *GPS satellite surveying* (3rd ed.). New York: John Wiley & Sons.

Leitao, P., Mendez, J., Bepperling, S., Cachapa, D., Colombo, A., & Reestivo, F. (2012). Integration of virtual and real environments for engineering service-oriented manufacturing systems. *Journal of Intelligent Manufacturing, 23*(6), 2551–2563. doi:10.1007/s10845-011-0591-8

Lemke, C. E. (1965). Bimatrix equilibrium points and mathematical programming. *Management Science, 11*(7), 681–689. doi:10.1287/mnsc.11.7.681

Lemke, C. E. (1967). *On complementary pivot theory.* Department of Mathematics, Rensselaer Polytechnic Institute.

Lemke, C. E., Howson, J. Jr, & Joseph, T. (1964). Equilibrium points of bimatrix games. *Journal of the Society for Industrial and Applied Mathematics, 12*(2), 413–423. doi:10.1137/0112033

Leou, R. (2006). A method for unit maintenance scheduling considering reliability and operation expense. *Electrical Power and Energy Systems, 28*(7), 471–481. doi:10.1016/j.ijepes.2006.02.009

Lesk, A. (2012). *Introduction to Genomics.* New York, NY: Oxford University Press.

Letendre, S. (2009). Solar Electricity as a Fuel for Light Vehicles. In *Proceedings of the 2006 American Solar Energy Society Annual Conference.*

Leung, S., & Wub, Y. (2004). A robust optimization model for stochastic aggregate production planning. *Production Planning and Control, 15*(5), 502–514. doi:10.1080/09537280410001724287

Levi, E. & Wang, M. (2002). A speed estimator for high performance sensorless control of induction motors in the field weakening region. *Power Electronics, IEEE Transactions on*, 365–378.

Levitin, G., & Lisnianski, A. (2000). Optimal replacement scheduling in multi – state series-parallel systems. *Quality and Reliability Engineering International*, *16*(2), 157–162. doi:10.1002/(SICI)1099-1638(200003/04)16:2<157::AID-QRE311>3.0.CO;2-J

Lévy, P. (1924). Théorie des erreurs la loi de Gauss et les lois exceptionelles. *Bulletin Soc. France.*, *52*, 49–85.

Li, J., Xu, L., & Zhang, Z. (2005), An adaptive sliding-mode observer for induction motor sensorless speed control. *Industry Applications, IEEE Transactions on*, 1039–1046.

Li, Z., Sahinoglu, Z., Tao, Z., & Teo, K. H. (2010). Electric vehicles network with nomadic portable charging stations. In Vehicular Technology Conference Fall (VTC 2010-Fall), 2010 IEEE 72nd (pp. 1-5). IEEE. doi:10.1109/VETECF.2010.5594437

Li12 Li, P., & Duan, H. B. (2012). Path Planning of Unmanned Aerial Vehicle based on Improved Gravitational Search Algorithm. *Science China*, *55*(10), 2712–2719. doi:10.1007/s11431-012-4890-x PMID:23108865

Liang, J. J., Qin, A. K., Suganthan, P. N., & Baskar, S. (2006). Comprehensive learning particle swarm optimizer for global optimization of multimodal functions. *IEEE Transactions on Evolutionary Computation*, *10*(3), 281–295. doi:10.1109/TEVC.2005.857610

Liang, K., Lingfu, K., & Peiliang, W. (2012, July). Adaptive Gaussian particle filter for nonlinear state estimation. In *Control Conference (CCC), 2012 31st Chinese* (pp. 2146-2150). IEEE.

Liao, L.-Z., & Wang, S. (2003). A self-adaptive projection and contraction method for linear complementarity problems. *Applied Mathematics & Optimization*, *48*(3), 169–180. doi:10.1007/s00245-003-0767-5

Liao, S. H., Hsieh, C. L., & Lai, P. J. (2011). An evolutionary approach for multi-objective optimization of the integrated location–inventory distribution network problem in vendor-managed inventory. *Expert Systems with Applications*, *38*(6), 6768–6776. doi:10.1016/j.eswa.2010.12.072

Li, C., Svoboda, A. J., Tseng, C. V., & Johnson, R. B. (1997). Hydro unit commitment in hydro-thermal optimization. *IEEE Transactions on Power Systems*, *12*(2), 764–769. doi:10.1109/59.589675

Li, C., & Zhou, J. (2011). Parameters identification of hydraulic turbine governing system using improved gravitational search algorithm. *Energy Conversion and Management*, *52*(1), 374–381. doi:10.1016/j.enconman.2010.07.012

Li, D.-H., Nie, Y.-Y., Zeng, J.-P., & Li, Q.-N. (2008). Conjugate gradient method for the linear complementarity problem with S-matrix. *Mathematical and Computer Modelling*, *48*(5), 918–928. doi:10.1016/j.mcm.2007.10.017

Lien, L. C., & Cheng, M. Y. (2014). Particle bee algorithm for tower crane layout with material quantity supply and demand optimization. *Automation in Construction*, *45*, 25–32. doi:10.1016/j.autcon.2014.05.002

Li, G., Niu, P., & Xiao, X. (2012). Development and investigation of efficient artificial bee colony algorithm for numerical function optimization. *Applied Soft Computing*, *12*(1), 320–332. doi:10.1016/j.asoc.2011.08.040

Li, H. B., Wong, C. C., Chen, K. W., & Chen, F. (2008). Antioxidant properties in vitro and total phenolic contents in methanol extracts from medicinal plants. *LWT Food Science and Technology*, *41*(3), 385–390. doi:10.1016/j.lwt.2007.03.011

Lillacci, G., & Khammash, M. (2010). Parameter estimation and model selection in computational biology. *PLoS Computational Biology*, *6*(3), e1000696. doi:10.1371/journal.pcbi.1000696 PMID:20221262

Lindenmeyer, D., Dommel, H., Moshref, A., & Kundur, P. (2001). An induction motor parameter estimation method. *International Journal of Electrical Power & Energy Systems*, *23*(4), 251–262. doi:10.1016/S0142-0615(00)00060-0

Lin, Y., & Cryer, C. (1985). An alternating direction implicit algorithm for the solution of linear complementarity problems arising from free boundary problems. *Applied Mathematics & Optimization*, *13*(1), 1–17. doi:10.1007/BF01442196

Liu, H., Darabi, H., Banerjee, P., & Liu, J. (2007). Survey of wireless indoor positioning techniques and systems. *IEEE Transactions on Systems, Man, and Cybernetics (Part C)*, *37*(6), 1067-1079.

Liu, B. (2002). *Theory and Practice of Uncertain Programming*. Heidelberg, Germany: Physica-Verlag. doi:10.1007/978-3-7908-1781-2

Liu, B. (2004). *Uncertainty Theory: An Introduction to its Axiomatic Foundations*. Berlin: Springer-Verlag. doi:10.1007/978-3-540-39987-2

Liu, C., Shang, Y., & Liu, H. (2015). An O (\ sqrt {n} L) iteration Mehrotra-type predictor-corrector algorithm for monotone linear complementarity problem. *Optimization Letters*, 1–16.

Liu, F., & Song, J. (2012). God and bad news about the (S,T) policy. *Manufacturing and Service Operations Management*, *14*(1), 42–49. doi:10.1287/msom.1110.0353

Liu, H. Liu, X. & Liu, Changhe. (2012). Mehrotra-type predictor–corrector algorithms for sufficient linear complementarity problem. *Applied Numerical Mathematics*, *62*(12), 1685–1700.

Liu, H., Ding, G., & Sun, H. (2012). An Improved Opposition-based Disruption Operation Gravitational Search Algorithm. In *Processing Fifth International Symposium on Computational Intelligence and Design*.

Liu, H., Weiss, R. E., Jennrich, R. I., & Wenger, N. S. (1999). PRESS model selection in repeated measures data. *Computational Statistics & Data Analysis*, *30*(2), 169–184. doi:10.1016/S0167-9473(98)00088-7

Liu, J., & Chen, R. (1998). Sequential monte carlo methods for dynamic systems. *Journal of the American Statistical Association*, *93*(443), 1032–1044. doi:10.1080/01621459.1998.10473765

Liu, N., Qiu, Y., & Zhang, B. (2014). An Improved Gravitation Search Algorithm forVehicle Routing Problem. *Journal of Modern Transportation*, *3*(2), 39–48.

Liu, S., Zheng, H., & Li, W. (2015). A general accelerated modulus-based matrix splitting iteration method for solving linear complementarity problems. *Calcolo*, 1–11.

Liu, X., & Potra, F. A. (2006). Corrector-predictor methods for sufficient linear complementarity problems in a wide neighborhood of the central path. *SIAM Journal on Optimization*, *17*(3), 871–890. doi:10.1137/050623723

Li, W., & Zheng, H. (2014). Some new error bounds for linear complementarity problems of H-matrices. *Numerical Algorithms*, *67*(2), 257–269. doi:10.1007/s11075-013-9786-8

Li, X., & Chen, X. (2014). Extension of the TOPSIS method based on prospect theory and trapezoidal intuitionistic fuzzy numbers for group decision making. *Journal of Systems Science and Systems Engineering*, *23*(2), 231–247. doi:10.1007/s11518-014-5244-y

Li, X., Wang, J., & Yin, M. (2014). Enhancing the performance of cuckoo search algorithm using orthogonal learning method. *Neural Computing & Applications*, *24*(6), 1233–1247. doi:10.1007/s00521-013-1354-6

Li, X., Yin, M., & Ma, Z. (2011). Hybrid Differential Evolution and Gravitation Search Algorithm for Unconstrained Optimization. *International Journal of the Physical Sciences*, *6*(25), 5961–5981.

Li, Y., & Dai, P. (2007). Generalized AOR methods for linear complementarity problem. *Applied Mathematics and Computation*, *188*(1), 7–18. doi:10.1016/j.amc.2006.09.067

Li, Y., & Sun, D. (2012). Microscopic car-following model for the traffic flow: The state of the art. *Journal of Control Theory and Applications*, *10*(2), 133–173. doi:10.1007/s11768-012-9221-z

Li, Y., Zhu, X., Sun, X., & Wang, F. (2010). Landscape effects of environmental impact on bay-area wetlands under rapid urban expansion and development policy: A case study of lianyungang, china. *Landscape and Urban Planning*, *94*(34), 218–227. doi:10.1016/j.landurbplan.2009.10.006

Ljung, L. (1979). Asymptotic behavior of the extended kalman filter as a parameter estimator for linear systems. IEEE Trans. on Automatic Control, pp. 36–50.

Lodree, J. Jr, & Uzochukwub, B. (2008). Production planning for a deteriorating ítem with stochastic demand and consumer choice. *International Journal of Production Economics*, *116*(2), 219–232. doi:10.1016/j.ijpe.2008.09.010

Longley, J. W. (1967). An appraisal of least-squares programs from the point of view of the user''. *Journal of the American Statistical Association*, *62*, 819–841. doi:10.1080/01621459.1967.10500896

Low, C. (2005). Simulated annealing heuristic for flow shop scheduling problems with unrelated parallel machines. *Computers & Operations Research*, *32*(8), 2013–2025. doi:10.1016/j.cor.2004.01.003

Lu, L. X., & Swaminathan, J. M. (2013). *Advances in Supply Chain Management.* Retrieved from http://public.kenan-flagler.unc.edu/faculty/lul/LuSwaminathan_SupplyChainManagement_2013.pdf

Luke, S. (2013). *Essentials of Metaheuristics.* Retrieved November 10, 2014, from http://cs.gmu.edu/~sean/book/metaheuristics/

Lu, N., & Huang, Z.-H. (2014). A smoothing Newton algorithm for a class of non-monotonic symmetric cone linear complementarity problems. *Journal of Optimization Theory and Applications*, *161*(2), 446–464. doi:10.1007/s10957-013-0436-z

Lund, H., & Kempton, W. (2008). Integration of renewable energy into the transport and electricity sectors through V2G. *Energy Policy*, *36*(9), 3578–3587. doi:10.1016/j.enpol.2008.06.007

Luo, L., Fan, Y., & Tondeur, D. (2007). Heat exchanger: From micro to multi- scale design optimisation. *International Journal of Energy Research*, *31*(13), 1266–1274. doi:10.1002/er.1298

Luo, L., Tondeur, D., Le Gall, H., & Corbel, S. (2007). Constructal approach and multi- scale components. *Applied Thermal Engineering*, *27*(10), 1708–1714. doi:10.1016/j.applthermaleng.2006.07.018

Luo, Z.-Q., & Tseng, P. (1991). On the convergence of a matrix splitting algorithm for the symmetric monotone linear complementarity problem. *SIAM Journal on Control and Optimization*, *29*(5), 1037–1060. doi:10.1137/0329057

Luyben, M. L., & Floudas, C. A. (1994). Analyzing the interaction of design and control. 1. A multiobjective framework and application to binary distillation synthesis. *Computers & Chemical Engineering*, *18*(10), 933–969. doi:10.1016/0098-1354(94)E0013-D

Lv, Y., Hu, T., Wang, G., & Wan, Z. (2007). A penalty function method based on Kuhn–Tucker condition for solving linear bilevel programming. *Applied Mathematics and Computation*, *188*(1), 808–813. doi:10.1016/j.amc.2006.10.045

Lv, Y., Hu, T., Wang, G., & Wan, Z. (2008). A neural network approach for solving nonlinear bilevel programming problem. *Computers & Mathematics with Applications (Oxford, England)*, *55*(12), 2823–2829. doi:10.1016/j.camwa.2007.09.010

Ma, Z., Callaway, D., & Hiskens, I. (2010). Decentralized charging control for large populations of plug-in electric vehicles. In *Decision and Control (CDC), 2010 49th IEEE Conference on* (pp. 206-212). IEEE.

Machida, N., Fukushima, M., & Ibaraki, T. (1995). A multisplitting method for symmetric linear complementarity problems. *Journal of Computational and Applied Mathematics*, *62*(2), 217–227. doi:10.1016/0377-0427(94)00103-2

Madireddy, M., Medeiros, D. J., & Kumara, S. (2011). An agent based model for evacuation traffic management. In *Proceedings of the 2011 Winter Simulation Conference* (pp. 222-233). doi:10.1109/WSC.2011.6147753

Madu, N. C., & Kuci, C. H. (1994). A group decision support system (GDSS) framework for adjusting system availability levels. *International Journal of Quality & Reliability Management*, *11*(9), 90–100. doi:10.1108/02656719410074314

Maes, J., & Van Wassenhove, L. (1988). Multi-item single-level capacitated dynamic lot- sizing heuristics: A general review. *The Journal of the Operational Research Society*, *39*(11), 991–1004. doi:10.1057/jors.1988.169

Magdalena, L. (1998). Hierarchical Fuzzy Control of a Complex System using Metaknowledge. In *Proceedings of the 7th International conference on Information Processing and Management of Uncertainty in Knowledge-based Systems.*

Ma, H., & Simon, D. (2011). Blended biogeography-based optimization for constrained optimization. *Engineering Applications of Artificial Intelligence*, *24*(3), 517–525. doi:10.1016/j.engappai.2010.08.005

Mahalakshmi, M; Kalaivani, P; Kiruba Nesamalar, E. (2013). A Review on Genetic Algorithm and its Applications. *International Journal of Computing Algorithm, 2*, 415-423.

Mahdavi, K., Harman, M., & Hierons, R. M. (2003). *A multiple hill climbing approach to software module clustering.* Paper presented at the Software Maintenance, 2003. doi:10.1109/ICSM.2003.1235437

Mahnam, M., Yadollahpour, M. R., Famil-Dardashti, V., & Hejazi, S. R. (2009). Supply chain modeling in uncertain environment with bi-objective approach. *Computers & Industrial Engineering*, *56*(4), 1535–1544. doi:10.1016/j.cie.2008.09.038

Majumder, A. (2015). Application of Standard Deviation Method Integrated PSO Approach in Optimization of Manufacturing Process Parameters In *Handbook of Research on Artificial Intelligence Techniques and Algorithms* (pp. 536-563). IGI Global.

Ma, L., Hu, K., Zhu, Y., & Chen, H. (2014). Cooperative artificial bee colony algorithm for multi-objective RFID network planning. *Journal of Network and Computer Applications*, *42*, 143–162. doi:10.1016/j.jnca.2014.02.012

Mala, D. J., & Mohan, V. (2009). ABC tester—artificial bee colony based software test suite optimization approach. *International Journal of Software Engineering*, *2*(2), 15–43.

Maleki, I. E., & Gharehchopogh, L. (2014). Hybrid Approach of Firefly and Genetic Algorithms in Software Cost Estimation. *MAGNT Research Report*, *2*(6), 372–388.

Malik, M., Ahsan, F., & Mohsin, S. (2014). Adaptive image denoising using cuckoo algorithm. *Soft Computing*. doi:10.1007/s00500-014-1552-x

Malin, N., & Alex, W. (2000). Ground-Source Heat Pumps: Are They Green?Environmental Building News, 9(7-8).

Mallick, S., Ghoshal, S. P., Acharjee, P., & Thakur, S. S. (2013). Optimal static state estimation using improved particle swarm optimization and gravitational search algorithm. *International Journal of Electrical Power & Energy Systems*, *52*(0), 254–265. doi:10.1016/j.ijepes.2013.03.035

Mallipeddi, R., Mallipeddi, S., & Suganthan, P. N. (2010). Ensemble strategies with adaptive evolutionary programming. *Information Sciences*, *180*(9), 1571–1581. doi:10.1016/j.ins.2010.01.007

Mammadova, M., & Jabrayilova, Z. (2014). Application of Fuzzy Optimization Method in Decision-Making for Personnel Selection. *Computer Science & Communications*, *5*(4), 190–204.

Mancoridis, S., Mitchell, B. S., Chen, Y., & Gansner, E. R. (1999). *Bunch: A clustering tool for the recovery and maintenance of software system structures*. Paper presented at the Software Maintenance, 1999. doi:10.1109/ICSM.1999.792498

Mancoridis, S., Mitchell, B. S., Rorres, C., Chen, Y., & Gansner, E. R. (1998). *Using automatic clustering to produce high-level system organizations of source code*. Paper presented at the International Conference on Program Comprehension. doi:10.1109/WPC.1998.693283

Mandal, B., & Roy, P. K. (2013). Optimal reactive power dispatch using quasi-oppositional teaching learning based optimization. *International Journal of Electrical Power & Energy Systems*, *53*, 123–134. doi:10.1016/j.ijepes.2013.04.011

Mandal, K. K., Basu, M., & Chakraborty, N. (2008). Particle swarm optimization technique based short-term hydrothermal scheduling. *Applied Soft Computing*, *8*(4), 1392–3399. doi:10.1016/j.asoc.2007.10.006

Mandelbrot, B. B., & Van Ness, J. W. (1968). Fractional Brownian motions, fractional noises and applications. *RE:view*, *10*(4), 422–437.

Mandel, J. (1984). A multilevel iterative method for symmetric, positive definite linear complementarity problems. *Applied Mathematics & Optimization*, *11*(1), 77–95. doi:10.1007/BF01442171

Mangasarian, O. L. (1977). Solution of symmetric linear complementarity problems by iterative methods. *Journal of Optimization Theory and Applications*, *22*(4), 465–485. doi:10.1007/BF01268170

Mangasarian, O. L. (1991). Convergence of iterates of an inexact matrix splitting algorithm for the symmetric monotone linear complementarity problem. *SIAM Journal on Optimization*, *1*(1), 114–122. doi:10.1137/0801009

Mangasarian, O. L., & De Leone, R. (1987). Parallel successive overrelaxation methods for symmetric linear complementarity problems and linear programs. *Journal of Optimization Theory and Applications*, *54*(3), 437–446. doi:10.1007/BF00940194

Mangasarian, O., & De Leone, R. (1988). Parallel gradient projection successive overrelaxation for symmetric linear complementarity problems and linear programs. *Annals of Operations Research*, *14*(1), 41–59. doi:10.1007/BF02186473

Mankad, K. (2015). An Intelligent Process Development Using Fusion of Genetic Algorithm with Fuzzy Logic. In P. Vasant (Ed.), Handbook of Research on Artificial Intelligence Techniques and Algorithms (pp. 44-81). IGI Global doi:10.4018/978-1-4666-7258-1.ch002

Manne, A. S. (1958). Programming of economic lot-sizes. *Management Science*, *4*(2), 115–135. doi:10.1287/mnsc.4.2.115

Mansouri, M., Nounou, H., & Nounou, M. (2013, March). State estimation of a chemical reactor process model-A comparative study. In Systems, Signals & Devices (SSD), 2013 10th International Multi-Conference on (pp. 1-6). IEEE. doi:10.1109/SSD.2013.6563998

Mansouri, M., Nounou, H., & Nounou, M. (2014, February). Improved particle filtering for state and parameter estimation-CSTR model. In Multi-Conference on Systems, Signals & Devices (SSD), 2014 11th International (pp. 1-6). IEEE. doi:10.1109/SSD.2014.6808794

Mansouri, M. M., Mohamed-Seghir, M. M., Nounou, H. N., Nounou, M. N., & Abu-Rub, H. A. (2014). Bayesian methods for time-varying state and parameter estimation in induction machines. *International Journal of Adaptive Control and Signal Processing*.

Mansouri, M. M., Nounou, H. N., Nounou, M. N., & Datta, A. A. (2014). Modeling of nonlinear biological phenomena modeled by S-systems. *Mathematical Biosciences*, *249*, 75–91. doi:10.1016/j.mbs.2014.01.011 PMID:24524881

Mansouri, M. M., Nounou, H. N., Nounou, M. N., & Datta, A. A. (2014). State and parameter estimation for nonlinear biological phenomena modeled by S-systems. *Digital Signal Processing*, *28*, 1–17. doi:10.1016/j.dsp.2014.01.012

Mansouri, M., Ilham, O., Snoussi, H., & Richard, C. (2011). Adaptive quantized target tracking in wireless sensor networks. *Wireless Networks*, *17*(7), 1625–1639. doi:10.1007/s11276-011-0368-1

Mantegna, R. N. (1994). Fast, accurate algorithm for numerical simulation of Levy stable stochastic processes. *Physical Review E: Statistical Physics, Plasmas, Fluids, and Related Interdisciplinary Topics*, *49*(5), 4677–4683. doi:10.1103/PhysRevE.49.4677 PMID:9961762

Mao, X., Stonier, R. J., & Thomas, P. (2001). An Effective Fuzzy Image Filter. In *Proceedings of the International Conference on Computational Intelligence for Modelling Control and Automation (CIMCA'01).*

Mares, M. (1989). Network anaysis of fuzzy technologies. In G. W. Evans, W. Karwowski, & M. R. Wilhelm (Eds.), *Applications of Fuzzy Set Methodologies in Industrial Engineering*. Elsevier.

Mares, M., & Horak, J. (1983). Fuzzy quantities in networks. *Fuzzy Sets and Systems*, *10*(1-3), 123–134. doi:10.1016/S0165-0114(83)80110-9

Mariappan, V. (2004). *Some studies on reliability centered maintenance*. (Ph.D. thesis). IIT Bombay.

Marichelvam, M. K., Prabaharan, T., & Yang, X. S. (2014). Improved cuckoo search algorithm for hybrid flow shop scheduling problems to minimize makespan. *Applied Soft Computing*, *19*, 93–101. doi:10.1016/j.asoc.2014.02.005

Marmolejo, J. A., Litvinchev, I., Aceves, R., & Ramirez, J. M. (2011). Multiperiod optimal planning of thermal generation using cross decomposition. *Journal of Computer and Systems Sciences International*, *50*(5), 793–804. doi:10.1134/S1064230711050145

Marrocco, G. (2003). Gain-optimized self-resonant meander line antennas for RFID applications. *IEEE Antennas and Wireless Propagation Letters*, *2*(1), 302–305. doi:10.1109/LAWP.2003.822198

Martil, I., & Diaz, G. G. (1992). Determination of the dark and illuminated characteristic parameters of a solar cell for IV characteristicsEur. *Journal of Physics*, 13193–13197.

Martín García, J. A., & Gil Mena, A. J. (2013). Optimal distributed generation location and size using a modified teaching–learning based optimization algorithm. *International Journal of Electrical Power & Energy Systems*, *50*, 65–75. doi:10.1016/j.ijepes.2013.02.023

Mastorakis, E. N. (2006). Unstable ordinary differential equations: solution via genetic algorithms and the method of Nelder-Mead. In *Proceeding of 6th WSEAS International Conference on Systems Theory & Scientific Computation* (vol. 119, pp. 1-6). Elounda, Greece.

Mastrocinque, E., Yuce, B., Lambiase, A., & Packianather, M. S. (2013). A multi-objective optimization for supply chain network using the Bees Algorithm. *Int. J. Eng. Bus. Manage*, *5*, 1–11. doi:10.5772/56754

Mastrocinque, E., Yuce, B., Lambiase, A., & Packianather, M. S. (2014). A System of Systems Approach to Supply Chain Design. *Applied Mechanics and Materials*, *496*, 2807–2814. doi:10.4028/www.scientific.net/AMM.496-500.2807

Mathar, R. (1997). A hybrid global optimization algorithm for multidimensional scaling. In R. Klar & O. Opitz (Eds.), *Classification and Knowledge Organization* (pp. 63–71). Berlin, Heidelberg: Springer. doi:10.1007/978-3-642-59051-1_7

Mathematical Optimization. (n.d.). Retrieved from http://en.wikipedia.org

Mathieu, R., Pittard, L., & Anandalingam, G. (1994). Genetic algorithm based approach to bi-level linear programming. *Revue française d'automatique, d'informatique et de recherche opérationnelle. Recherche opérationnelle, 28*(1), 1-21.

Mathirajan, M., Bhargav, V., & Ramachandran, V. (2010). Minimizing total weighted tardiness on a batch-processing machine with non-agreeable release times and due dates. *International Journal of Advanced Manufacturing Technology, 48*(9-12), 1133–1148. doi:10.1007/s00170-009-2342-y

Mathirajan, M., & Sivakumar, A. I. (2006). A literature review, classification and simple meta-analysis on scheduling of batch processors in semiconductor. *International Journal of Advanced Manufacturing Technology, 29*(9-10), 990–1001. doi:10.1007/s00170-005-2585-1

Ma, Y., Han, J. J., & Trivedi, K. S. (2001). Composite performance and availability analysis of wireless communication networks. *IEEE Transactions on Vehicular Technology, 50*(5), 1216–1223. doi:10.1109/25.950322

Mayfield, D. (2012). *Site Design For Electric Vehicle Charging Stations, ver. 1.0.* Sustainable Transportation Strategies.

McFadden, R. H. (1990). Developing a database for a reliability availability and maintainability improvement program for an industrial plant or commercial building. *IEEE Transactions on Industry Applications, 26*(4), 735–740. doi:10.1109/28.56001

Medhi, K. T. (1994). A two-stage successive overrelaxation algorithm for solving the symmetric linear complementarity problem. *Mathematical Programming, 65*(1-3), 365–380. doi:10.1007/BF01581703

Melouk, S., Damodaran, P., & Chang, P. Y. (2004). Minimizing makespan for single batchprocessing machine with non-identical job sizes using simulated annealing. *International Journal of Production Economics, 87*(2), 141–147. doi:10.1016/S0925-5273(03)00092-6

Mendes, A. S., Muller, F. M., Franca, P. M., & Moscato, P. (2002). Comparing meta-heuristic approach for parallel machine scheduling problems. *Production Planning and Control, 13*(2), 143–154. doi:10.1080/09537280110069649

Meng, X., Dodson, A. H., Moore, T., & Roberts, G. W. (2007). Ubiquitous positioning: anyone, anything: anytime, anywhere. *GPS World, 18*(6), 60–65.

Meng, X., Roberts, G. W., Dodson, A. H., Cosser, E., Barnes, J., & Rizos, C. (2004). Impact of GPS satellite and pseudolite geometry on structural deformation monitoring: Analytical and empirical studies. *Journal of Geodesy, 77*(12), 809–822. doi:10.1007/s00190-003-0357-y

Metaheuristic. (n.d.). Retrieved from http://en.wikipedia.org

Metropolis, M., Rosenbluth, A., Teller, A., & Teller, E. (1953). Equations of state calculation by fast computing machines. *Journal of Chemical Physics, 21*, 1087-1092.

Miandoabchi, E., & Farahani, R. Z. (2011). Optimizing Reserve Capacity of Urban Road Networks in a Discrete Network Design Problem. *Advances in Engineering Software, 42*(12), 1041–1050. doi:10.1016/j.advengsoft.2011.07.005

Michalewicz, Z. (1994). *Genetic Algorithms + Data Structures = Evolution Programs*. Springer Verlag. doi:10.1007/978-3-662-07418-3

Milaszewicz, J. (1987). Improving jacobi and gauss-seidel iterations. *Linear Algebra and Its Applications, 93*, 161–170. doi:10.1016/S0024-3795(87)90321-1

Miliotis, P. (1976). Integer Programming approaches to the Traveling Salesman Problem. *Journal of Mathematical Programming, 10*(1), 367–378. doi:10.1007/BF01580682

Millar, H. H., & Yang, M. (1994). Lagrangian heuristics for the capacitated multi-item lot-sizing problem with backordering. *International Journal of Production Economics, 34*(1), 1–15. doi:10.1016/0925-5273(94)90042-6

Minoux, M. (2007). Models and Algorithms for Robust PERT Scheduling with Time-Dependent Task Durations. *Vietnam Journal of Mathematics*, *35*(4), 387–398.

Minoux, M. (2008). Solving some multistage robust decision problems with huge implicitly-defined scenario trees. *Algorithmic O.R.*, *4*(1), 1–18.

Minoux, M. (2009). On robust maximum flow with polyhedral uncertainty sets. *Optimization Letters*, *3*(3), 367–376. doi:10.1007/s11590-009-0116-y

Minoux, M. (2010). Robust network optimization under polyhedral demand uncertainty is NP-Hard. *Discrete Applied Mathematics*, *158*(5), 597–603. doi:10.1016/j.dam.2009.09.025

Minoux, M. (2011). On 2-stage robust LP with RHS uncertainty: Complexity results and applications. *Journal of Global Optimization*, *49*(3), 521–537. doi:10.1007/s10898-010-9645-2

Minoux, M. (2014). Two-stage robust optimization, state-space representable uncertainty and applications. *RAIRO Oper. Res.*, *48*(4), 455–475. doi:10.1051/ro/2014017

Minoux, M. (2015). (to appear). Efficient Robust Multistage Optimization with State-Space Representation of Uncertainty and Applications. *IMA Journal of Management Mathematics*.

Mirjalili, S., & Mohd Hashim, S. Z. (2010). A New Hybrid PSOGSA Algorithm for Function Optimization. In *Proceeding of International Conference on Computer and Information Application* (ICCIA 2010).

Mirjalili, S., Mohd Hashim, S. Z., & Moradian Sardroudi, H. (2012). Training feedforward neural networks using hybrid particle swarm optimization and gravitational search algorithm. *Applied Mathematics and Computation*, *218*(22), 11125–11137. doi:10.1016/j.amc.2012.04.069

Mishra, S., Shaw, K., & Mishra, D. (2012). A new meta-heuristic bat inspired classification approach for microarray data. *Procedia Technology*, *4*, 802–806. doi:10.1016/j.protcy.2012.05.131

Mitchell, B. S., & Mancoridis, S. (2002). *Using Heuristic Search Techniques To Extract Design Abstractions From Source Code*. Paper presented at the GECCO.

Mitchell, B. S. (2002). *A heuristic search approach to solving the software clustering problem*. Drexel University.

Mitchell, B. S., & Mancoridis, S. (2006). On the automatic modularization of software systems using the bunch tool. *Software Engineering. IEEE Transactions on*, *32*(3), 193–208.

Mitchell, B. S., & Mancoridis, S. (2008). On the evaluation of the Bunch search-based software modularization algorithm. *Soft Computing*, *12*(1), 77–93. doi:10.1007/s00500-007-0218-3

Mitchell, M. (1996). *An Introduction to Genetic Algorithms*. MIT Press.

Mitra, P., & Venayagamoorthy, G. K. (2010). Intelligent coordinated control of a wind farm and distributed smartparks. In *Industry Applications Society Annual Meeting (IAS), 2010 IEEE* (pp. 1-8). IEEE. doi:10.1109/IAS.2010.5615930

Mittnik, S., Doganoglu, T., & Chenyao, D. (1999). Computing the probability density function of the stable Paretian distribution. *Mathematical and Computer Modelling*, *29*(10), 235–240. doi:10.1016/S0895-7177(99)00106-5

Mladenovic, N., (1995), A variable neighborhood algorithm: a new metaheuristic for combinatorial optimization. In *Abstracts of papers presented at Optimization Days: Montre´al.*

Mohammadi, S. A., Feizi Derakhshi, M. R., & Akbari, R. (2012). *An Adaptive Multi-Objective Artificial Bee Colony with crowding distance mechanism.* Paper presented at the AISP 2012 - 16th CSI International Symposium on Artificial Intelligence and Signal Processing. doi:10.1109/AISP.2012.6313720

Mohammadian, M., & Stonier, R. J. (1996). Fuzzy Rule Generation by Genetic Learning for Target Tracking. In *Proceedings of the 5th International Intelligent Systems Conference.*

Mohammadian, M. (2002). Designing customised hierarchical fuzzy systems for modelling and prediction. In *Proceedings of the International Conference on sinulated Evolution and Learning (SEAL'02).*

Mohammadian, M., & Kingham, M. (2004, January-March). An adaptive hierarchical fuzzy logic system for modelling of financial systems. *Journal of Intelligent Systems in Accounting, Finance and Management, Wiley Interscience*, *12*(1), 61–82. doi:10.1002/isaf.241

Mohammadian, M., & Stonier, R. J. (1995). Adaptive Two Layer Fuzzy Logic Control of a Mobile Robot System. In *Proceedings of IEEE Conference on Evolutionary Computing.*

Mohammadian, M., & Stonier, R. J. (1998). Hierarchical Fuzzy Control. In *Proceedings of the 7th International conference on Information Processing and Management of Uncertainty in Knowledge-based Systems.*

Mohideen Anver, M., & Russel, J. (2003). Parameter Optimization and Rule Base Selection for Fuzzy Impulse Filters using Evolutionary Algorithms. In *Proceedings of IEEE Tencon 2003*, Allied Publishers Private Ltd.

Mohideen Anver, M., & Russel, J. (2002). Stonier, Design of an Hierarchical Fuzzy Filter for Removal of Heavy Impulse Noise from Corrupted Images. In *Proceedings of the Seventh International Conference on Control, Automation, Robotics and Vision (ICARCV2002).*

Mohideen Anver, M., & Stonier, R. J. (2003). A Coevolutionary Approach for Optimization of Image Enhancement Filters. In *Proceedings of the International Congress on Evolutionary Computation (CEC2003).*

Mollinetti, M. A. F., Souza, D. L., & Teixeira, O. N. (2014). *ABC+ES: A novel hybrid Artificial Bee Colony Algorithm with evolution strategies.* Paper presented at the GECCO 2014 - Companion Publication of the 2014 Genetic and Evolutionary Computation Conference.

Moncayo-Martínez, L. A., & Recio, G. (2014). Bi-criterion optimization for configuring an assembly supply chain using Pareto ant colony meta-heuristic. *Journal of Manufacturing Systems*, *33*(1), 188–195. doi:10.1016/j.jmsy.2013.12.003

Moncayo-Martínez, L. A., & Zhang, D. Z. (2011). Multi-objective ant colony optimization: A meta-heuristic approach to supply chain design. *International Journal of Production Economics*, *131*(1), 407–420. doi:10.1016/j.ijpe.2010.11.026

Mönch, L., Unbehaun, R., & Choung, Y. I. (2006). Minimizing earliness–tardiness on a single burn-in oven with a common due date and maximum allowable tardiness constraint. *OR-Spektrum*, *28*(2), 177–198. doi:10.1007/s00291-005-0013-4

Mondal, T. K., & Samanta, S. K. (2002). Generalized intuitionistic fuzzy sets. *The Journal of Fuzzy Mathematics*, *10*(4), 839–862.

Monti, G., Catarinucci, L., & Tarricone, L. (2009). Compact microstrip antenna for RFID applications. *Progress In Electromagnetics Research Letters*, *8*, 191–199. doi:10.2528/PIERL09042803

Montillet, J.-P. (2008). *Precise positioning in urban canyons: Applied to the localisation of buried assets.* (doctoral dissertation). University of Nottingham.

Montillet, J.-P., & Yu, K. (2011). Leaky LMS algorithm and fractional Brownian motion model for GNSS receiver position estimation. In *Proceedings of the IEEE Vehicular Technology Conference (VTC'11 fall)*. doi:.10.1109/VETECF.2011.6092850

Montillet, J.-P., & Yu, K. (2014). Modified leaky LMS algorithms applied to satellite positioning. In *Proceedings of the IEEE Vehicular Technology Conference (VTC'14 fall)*. Vancouver, Canada.

Montillet, J.-P., & Yu, K. (2014b). Modelling geodetic processes with Levy alpha-Stable distribution and FARIMA. *Mathematical Geosciences*. doi:.10.1007/s11004-014-9574-6

Montillet, J.-P., Yu, K., & Oppermann, I. (2007). Location performance enhancement with recursive processing of time-of-arrival measurements. In *Proceedings of the IEEE 18th International Symposium on Personal, Indoor and Mobile Radio Communications (PIMRC 2007)*. doi:10.1109/PIMRC.2007.4394529

Montillet, J.-P., Roberts, G. W., Hancock, C., Meng, X., Ogundipe, O., & Barnes, J. (2009). Deploying a Locata network to enable precise positioning in urban canyons. *Journal of Geodesy*, *83*(2), 91–103. doi:10.1007/s00190-008-0236-7

Montillet, J.-P., Taha, A., Meng, X., & Roberts, G. W. (2007). Buried assets: Testing GPS and GSM in urban canyons. *GPS World*, *18*(3), 39–43.

Montillet, J.-P., Tregoning, P., McClusky, S., & Yu, K. (2012). Extracting white noise statistics in GPS coordinate time series. *IEEE Geoscience and Remote Sensing Letters*, *10*(3), 563–567. doi:10.1109/LGRS.2012.2213576

Montillet, J.-P., & Yu, K. (2012). On a new weighting matrix to enhance the accuracy of stand-alone GNSS positioning. In *Proceedings of the IEEE International Symposium on Communications and Information Technologies (ISCIT 12)*. Gold Coast, Australia. doi:10.1109/ISCIT.2012.6380845

Montillet, J.-P., & Yu, K. (2015). Covariance Matrix Analysis for Higher Order Fractional Brownian Motion Time Series. In *Proceedings of the 28th IEEE Canadian Conference on Electrical and Computer Engineering (CCECE)*. Halifax, Canada. doi:10.1109/CCECE.2015.7129488

Moona, C., & Seob, Y. (2005). Evolutionary algorithm for advanced process planning and scheduling in a multi-plant. *Computers & Industrial Engineering*, *48*(2), 311–325. doi:10.1016/j.cie.2005.01.016

Moon, C., Kim, J., Choi, G., & Seo, Y. (2002). An efficient genetic algorithm for the traveling salesman problem with precedence constraints. *European Journal of Operational Research, EJOR*, *140*(3), 606–617. doi:10.1016/S0377-2217(01)00227-2

Moore, T., Hill, C., Hide, C., Ochieng, W., Feng, S., Aguado, E., & Lau, L. et al. (2007). *End-to-End testing of an integrated centimetric positioning test-bed. In Proceedings of the ION GNSS 2007*. Fort Worth, TX: Fort Worth Convention Center.

Moreno, L., Poggi de Aragao, M., & Uchoa, E. (2010). Improved lower bounds for the split delivery vehicle routing problem. *Operations Research Letters*, *38*(4), 302–306. doi:10.1016/j.orl.2010.04.008

Morrow, K., Karner, D., & Francfort, J. (2008). *Plug-in hybrid electric vehicle charging infrastructure review*. US Department of Energy-Vehicle Technologies Program.

Mota, E., Campos, V., & Corbéran, A. (2007). A new metaheuristic for the vehicle routing problem with split demands. In J. Van Hemert & C. Cotta (Eds.), *Evolutionary Computation in Combinatorial Optimization* (pp. 121–129). Heidelberg: Springer. doi:10.1007/978-3-540-71615-0_11

Mount, D. (2004). Bioinformatics: Sequence and Genome Analysis (2nd ed.). New York, NY: Cold Spring Harbor Laboratory Press.

Muhammad, B., Ibrahim, Z., Ghazali, K. H., Ghazali, M. R., Saealal, M. S., Lim, K. Sh., . . . Mokhtar, N. (2014). Performance Evaluation of Vector Evaluated Gravitational Search Algorithm II. In New Trends in Software Methodologies, Tools and Techniques. IOS Press.

Mukherjee, I., & Ray, P. K. (2008). Artificial Neural Network and Metaheuristic Strategies: Emerging Tools for Metal Cutting Process Optimization. In D. Laha & P. Mandal (Eds.), *Handbook of Computational Intelligence in Manufacturing and Production Management* (pp. 366–397). Hershey, PA: Information Science Reference; doi:10.4018/978-1-59904-582-5.ch019

Mula, P. J., García Sabater, J., & Lario, F. (2006). Models for production planning under uncertainty: A review. *International Journal of Production Economics*, *103*(1), 271–285. doi:10.1016/j.ijpe.2005.09.001

Müller, H. A., Orgun, M. A., Tilley, S. R., & Uhl, J. S. (1993). A reverse-engineering approach to subsystem structure identification. *Journal of Software Maintenance: Research and Practice*, *5*(4), 181–204. doi:10.1002/smr.4360050402

Murty, K. G., & Yu, F.-T. (1988). *Linear complementarity, linear and nonlinear programming*. Berlin: Heldermann Verlag.

Nagakava, T. (1977). Optimum preventive maintenance policies for repairable system. *IEEE Transactions on Reliability*, *26*(3), 168–173. doi:10.1109/TR.1977.5220105

Nagao, Y., & Nagamochi, H. (2007). A DP-based heuristic algorithm for the discrete split delivery vehicle routing problem. *Journal of Advanced Mechanical Design, Systems and Manufacturing*, *1*(2), 217–226. doi:10.1299/jamdsm.1.217

Nagel, K., & Schreckenberg, M. (1992). A cellular automaton model for freeway traffic. *Journal de Physique. I*, *2*(2), 2221–2229. doi:10.1051/jp1:1992277

Nain, P.K.S., Giri, J.M., Sharma, S. & Deb, K., (2010). Multi-objective Performance Optimization of Thermo-Electric Coolers Using Dimensional Structural Parameters. In *Proceedings of SEMCCO 2010* (LNCS), (vol. 6466, pp. 607-614). Springer-Verlag Berlin Heidelberg.

Nainar, I., & Mohammadian, M. (1996). *Adaptive Control of Traffic Signals using Fuzzy Logic and Genetic Algorithms. Fourth International Conference on Control, Automation, Robotics and Vision*, Westin Stamford, Singapore.

Nainar, I., Mohammadian, M., Stonier, R. J., & Millar, J. (1996). An adaptive fuzzy logic controller for control fo traffic signals. In *Proceedings of the 4th International Conference on Control, Automation, Robotics and Computer Vision (ICARCV'96)*.

Nainar, I., Mohammadian, M., Stonier, R. J., & Millar, J. W. L. (1996). *An adaptive fuzzy logic controller for control of traffic signals. Fourth International Conference on Control, Automation, Robotics and Vision*, Westin Stamford, Singapore.

Najid, N. M., Alaoui-Selsouli, M., & Mohafid, A. (2011). An integrated production and maintenance planning model with time windows and shortage cost. *International Journal of Production Research*, *49*(8), 2265–2283. doi:10.1080/00207541003620386

Nakagawa, T. (1981a). A summary of periodic replacement with minimal repair at failure. *Journal of the Operations Research Society of Japan*, *24*, 213–228.

Nakagawa, T. (1981b). Modified periodic replacement with minimal repair at failure. *IEEE Transactions on Reliability*, *30*(2), 165–168. doi:10.1109/TR.1981.5221018

Nakashima, K., Arimitsua, H., Nosea, T., & Kuriyamab, S. (2004). Optimal control of a remanufacturing system. *International Journal of Production Research*, *42*(17), 3619–3625. doi:10.1080/00207840410001721772

Nanda, J., & Bijwe, P. R. (1981). Optimal Hydrothermal Scheduling with cascaded plants using progressive optimality algorithm. *IEEE Transactions on Power Apparatus and Systems*, *PAS-100*(4), 2093–2099. doi:10.1109/TPAS.1981.316486

Narang, N., Dhillon, J. S., & Kothari, D. P. (2012). Multiobjective fixed head hydrothermal scheduling using integrated predator-prey optimization and Powell search method. *Energy*, *47*(1), 237–252. doi:10.1016/j.energy.2012.09.004

Naval Studies Board. (1997). *Technology for the United States Navy and Marie Corps, 2000-2035. Becoming a 21st-Century Force*. Washington, DC: National Academy Press.

Nayagama, V. L. G., Muralikrishnan, S., & Sivaraman, G. (2010). Multi-criteria decision-making method based on interval-valued intuitionistic fuzzy sets. *Expert Systems with Applications*, *38*, 1464–1467.

Nayeem, S. M. A. (2012). An expected value model of quadratic clique problem on a graph with fuzzy parameters. In *Proceedings of 12th International Conference on Intelligent Systems Design and Applications* (pp. 592 – 597). IEEE, Doi:10.1109/ISDA.2012.6416604

Nayeem, S. M. A., & Pal, M. (2005). Shortest path problem on a network with imprecise edge weight. *Fuzzy Optimization and Decision Making*, *4*(4), 293–312. doi:10.1007/s10700-005-3665-2

Nayeem, S. M. A., & Pal, M. (2008). PERT on a network with imprecise edge weight. *Journal of Fuzzy Mathematics*, *16*(4), 853–874.

Neshat, M.A., Sepidnam, G.A., Sargolzaei, M.A. & Toosi, A.A. (2012). Artificial fish swarm algorithm: a survey of the state-of-the-art, hybridization, combinatorial and indicative applications. *Journal of Artificial Intelligence Review*, 1-33.

Neto, T., Constantino, M., Martins, I., & Pedroso, J. P. (2013). A branch-and-bound procedure for forest harvest scheduling problems addressing aspects of habitat availability. *International Transactions in Operational Research*, *20*(5), 689–709. doi:10.1111/itor.12003

Neumann, H. M., Schär, D., & Baumgartner, F. (2012). The potential of photovoltaic carports to cover the energy demand of road passenger transport. *Progress in Photovoltaics: Research and Applications*, *20*(6), 639–649. doi:10.1002/pip.1199

Neumann, J. V. (1955). *Mathematical Foundations of Quantum Mechanics*. Princeton University Press.

Nezamabadi-pour, H., Yazdani, S., Farsangi, M. M., & Neyestani, M. (2011). A Solution to Economic Dispatch Problem by Fuzzy Adaptive Genetic Algorithm. *Iranian Journal of Fuzzy System*, *8*(3), 1–21.

Nguyen, V. P., Prins, C., & Prodhon, C. (2012). Solving the two-echelon location routing problem by a GRASP reinforced by a learning process and path relinking. *European Journal of Operational Research*, *216*(1), 113–126. doi:10.1016/j.ejor.2011.07.030

Niculescu, D. (2001) Ad-hoc Positioning System (APS). In *Proceedings of the IEEE Global telecommunications Conference, 2001 (GLOBECOM'01)* (vol. 5, pp. 2926-2931). doi:10.1109/GLOCOM.2001.965964

Niculescu, D. (2003) Ad-hoc Positioning System (APS) using AoA. In *Proceedings of the twenty-second Annual Joint Conference of the IEEE Computer and Communications Societies (INFOCOM03)* (vol. 3, pp. 1734-1743). doi:10.1109/INFOCOM.2003.1209196

Niculescu, D. (2004). Positioning in Ad Hoc Sensor Networks. *IEEE Network*, *18*(4), 24–29. doi:10.1109/MNET.2004.1316758

Niethammer, W. (1993). A note on the implementation of the successive overrelaxation method for linear complementarity problems. *Numerical Algorithms*, *4*(1), 197–200. doi:10.1007/BF02142746

Nilsson, O., & Sjelvgren, D. (1996). Mixed-integer programming applied to short-term planning of a hydro-thermal system. *IEEE Transactions on Power Systems*, *11*(1), 281–286. doi:10.1109/59.486107

Nobahari, H., Nikusokhan, M., & Siarry, P. (2012). A Multi-Objective Gravitational Search Algorithm Based on Non-Dominated Sorting. *International Journal of Swarm Intelligence Research*, *3*(3), 32–49. doi:10.4018/jsir.2012070103

Nocedal, J., & Wright, S. J. (2006). Conjugate gradient methods. *Numerical Optimization*, 101-134.

Nolan, J. P. (1999). *Fitting Data and Assessing Goodness-of-fit with Stable Distributions*. Department of Mathematics and Statistics, American University. Available in: http://academic2.american.edu/~jpnolan/

Nolan, J. P. (2005). *Modeling financial data with stable distributions*. Department of Mathematics and Statistics, American University. Available in http://academic2.american.edu/~jpnolan/

Nolan, J. P. (1997). Numerical calculation of stable densities and distribution functions. *Communications in Statistics. Stochastic Models*, *13*(4), 759–774. doi:10.1080/15326349708807450

Noro, R., Dobashi, Y., Wang, Z. Y., Takahashi, T., Okamoto, T., Tanaka, T., & Watada, J. (2006). Real Options Approach to Life Cycle Management of Power Equipment. In *Proceedings, CMD2006, International Conference on Condition Monitoring and Diagnosis Changwon.*

Norros, I. (1995). On the use of fractional Brownian motion in the theory of connectionless networks. *IEEE Journal on Selected Areas of Telecommunications*, *13*(6). doi:10.1109/49.400651

Nourelfath, M., & Châtelet, E. (2012). Integrating production, inventory and maintenance planning for a parallel system with dependent components. *Reliability Engineering & System Safety*, *101*, 59–66. doi:10.1016/j.ress.2012.02.001

Nourelfath, M., Fitouhi, M., & Machani, M. (2010). A genetic algorithm for integrated production and preventive maintenance planning in multi state systems. *IEEE Transactions on Reliability*, *59*(3), 496–506. doi:10.1109/TR.2010.2056412

Novais, F. T., Batista, L. S., Rocha, A. J., & Guimaraes, F. G. (2013). *A multiobjective estimation of distribution algorithm based on artificial bee colony*. Paper presented at the 1st BRICS Countries Congress on Computational Intelligence, BRICS-CCI 2013. doi:10.1109/BRICS-CCI-CBIC.2013.75

Nowak, M., Ergun, O., & White, C. III. (2009). An empirical study on the benefit of split loads with the pickup and delivery problem. *European Journal of Operational Research*, *198*(3), 734–740. doi:10.1016/j.ejor.2008.09.041

Ntuen, C. (1991). An economic maintenance scheduling model with truncated gamma function. *Reliability Engineering & System Safety*, *31*(1), 31–38. doi:10.1016/0951-8320(91)90034-5

Odolinski, R., Teunissen, P. J. G., & Odijk, D. (2014). Combined BDS, Galileo, QZSS and GPS single-frequency RTK. *GPS Solutions*, *19*(1), 151–163. doi:10.1007/s10291-014-0376-6

Ojugo, A. A., Emudianughe, J., Yoro, R. E., Okonta, E. O., & Eboka, A. O. (2013). A Hybrid Artificial Neural Network Gravitational Search Algorithm for Rainfall Runoff Modeling and Simulation in Hydrology. *Journal of Intelligent Computing and Applications*, *2*(1), 22–33.

O'Leary, D. P., & White, R. E. (1985). Multi-splittings of matrices and parallel solution of linear systems. *SIAM Journal on Algebraic Discrete Methods*, *6*(4), 630–640. doi:10.1137/0606062

Olhager, J., & Selldin, J. (2007). Manufacturing planning and control approaches: Market alignment and performance. *International Journal of Production Research*, *45*(6), 1469–1484. doi:10.1080/00207540600635250

Oliva, D., Cuevas, E., & Pajares, G. (2014). Parameter identification of solar cells using artificial bee colony optimization. *Energy*, *72*, 93–102. doi:10.1016/j.energy.2014.05.011

Oliveira, G. G., & Soares, S. (1995). A second order network flow algorithm for hydrothermal scheduling. *IEEE Transactions on Power Systems, 10*(3), 1625–1641. doi:10.1109/59.466477

Omer, A. M. (2009c). Energy use and environmental impacts: a general review. *Journal of Renewable and Sustainable Energy, 1.*

Omer, A. M. (2008). Development of integrated bioenergy for improvement of quality of life of poor people in developing countries. In F. L. Magnusson & O. W. Bengtsson (Eds.), *Energy in Europe: Economics, Policy and Strategy- IB* (pp. 341–373). New York, USA: NOVA Science Publishers, Inc.

Omer, A. M. (2009a). Environmental and socio-economic aspect of possible development in renewable energy use. In *Proceedings of the 4th International Symposium on Environment.*

Omer, A. M. (2009b). Energy use, environment and sustainable development. In *Proceedings of the 3rd International Conference on Sustainable Energy and Environmental Protection (SEEP 2009).*

Omer, A. M. (2009d). Energy use, environment and sustainable development. In R. T. Mancuso (Ed.), *Environmental Cost Management* (pp. 129–166). New York: NOVA Science Publishers, Inc.

Ong, P. (2014). Adaptive Cuckoo Search Algorithm for Unconstrained Optimization. *TheScientificWorldJournal, 8.* doi:10.1155/2014/943403 PMID:25298971

Ong, Y. S., Lim, M. H., Neri, F., & Ishibichi, H. (2009). Emerging Trends in Soft Computing: Memetic Algorithms. *Journal of Soft Computing, 13*(8-9), 739–740. doi:10.1007/s00500-008-0353-5

Oosterlee, C. W. (2003). On multigrid for linear complementarity problems with application to American-style options. *Electronic Transactions on Numerical Analysis, 15*(165-185), 2-7.

Oracle Engineered Systems. (2015). Retrieved from http://www.oracle.com/index.html

Ordon˜ez, F., & Zhao, J. (2007). Robust capacity expansion of network flows. Networks, 50(2), 136–145.

Orero, S. O., & Irving, M. R. A. (1998). Genetic algorithm modeling framework and solution technique for short term optimal hydrothermal scheduling. *IEEE Transactions on PWRS, 13*, 501–518.

Ortiz-Conde, A., Ma, Y., Thomson, J., Santos, E., Liou, J. J., Garcia, F. J., & Layman, P. et al. (2000). Parameter extraction using lateral and vertical optimisation. In *Proc.22nd Int. Conf. on Microelectronics, MIEL 2000.*

Oshima, T., Sagara, K., Yoshida, T., Tong, Y. Y., Zhang, G., & Chen, Y. H. (1988). Determination of geniposide, gardeniside, geniposidic acid and genipin-1-β-gentiobioside in Gardenia jasminoides by high-performance liquid chromatography. *Journal of Chromatography. A, 455*, 410–414. doi:10.1016/S0021-9673(01)82148-8 PMID:3235620

Otri, S. (2011). *Improving The Bees Algorithm for Complex Optimization Problems.* (PhD Thesis). Cardiff University.

Ouaarab, A., Ahiod, B., & Yang, X. S. (2014). (in press). Random-key cuckoo search for the travelling salesman problem. *Soft Computing.*

Ouaarab, A., Ahiod, B., & Yang, X. S. (2014). Discrete cuckoo search algorithm for the travelling salesman problem. *Neural Computing & Applications, 24*(7-8), 1659–1669. doi:10.1007/s00521-013-1402-2

Oyola, J., & Løkketangen, A. (2014). GRASP-ASP: An algorithm for the CVRP with route balancing. *Journal of Heuristics, 20*(4), 361–382. doi:10.1007/s10732-014-9251-4

Ozdamar, L., & Bozyel, M. A. (2000). The capacitated lot sizing problem with overtime decisions and setup times. *IIE Transactions, 32*(11), 1043–1057. doi:10.1080/07408170008967460

Ozdemir, Y., Basligil, H., & Sarsenov, B. (2012). A Large Scale Integer Linear Programming to the Daily Fleet Assignment Problem: A Case Study in Turkey. *Procedia: Social and Behavioral Sciences*, *62*, 849–853. doi:10.1016/j.sbspro.2012.09.143

Packianather, M. S., Yuce, B., Mastrocinque, E., Fruggiero, F., Pham, D. T., & Lambiase, A. (2014, August). Novel Genetic Bees Algorithm applied to single machine scheduling problem. In *World Automation Congress (WAC)*, (pp. 906-911). IEEE. doi:10.1109/WAC.2014.6936194

Padhi, S. K., Karmakar, N. C. Sr, Law, C. L., & Aditya, S. Sr. (2003). A dual polarized aperture coupled circular patch antenna using a C-shaped coupling slot. *IEEE Transactions on Antennas and Propagation*, *51*(12), 3295–3298. doi:10.1109/TAP.2003.820947

Padhy, N. P. (2004). Unit Commitment - A bibliographical survey. *IEEE Transactions on Power Systems*, *19*(2), 1196–1205. doi:10.1109/TPWRS.2003.821611

Pal, B. B., Chakraborti, D., & Biswas, P. (2010, July). A genetic algorithm approach to fuzzy quadratic bilevel programming. In *Computing Communication and Networking Technologies (ICCCNT), 2010 International Conference on* (pp. 1-7). IEEE. doi:10.1109/ICCCNT.2010.5591820

Palli, N., Azram, S., McCluskey, P., & Sundararajan, R. (1999). An interactive multistage ε-inequality constraint method for multiple objectives decision making. *ASME Journal of Mechanical Design.*, *120*(4), 678–686. doi:10.1115/1.2829331

Pal, M., Khan, S. K., & Shyamal, A. K. (2002). Intuitionistic fuzzy matrices. *Notes on IntuitionisticFuzzy Sets*, *8*(2), 51–62.

Pang, C., Dutta, P., Kim, S., Kezunovic, M., & Damnjanovic, I. (2010). PHEVs as dynamically configurable dispersed energy storage for V2B uses in the smart grid.*IET Conference Proceedings*. doi:10.1049/cp.2010.0903

Pang, J. (1984). Necessary and sufficient conditions for the convergence of iterative methods for the linear complementarity problem. *Journal of Optimization Theory and Applications*, *42*(1), 1–17. doi:10.1007/BF00934130

Pang, J.-S. (1986). More results on the convergence of iterative methods for the symmetric linear complementarity problem. *Journal of Optimization Theory and Applications*, *49*(1), 107–134. doi:10.1007/BF00939250

Pang, J.-S. (1993). Convergence of splitting and Newton methods for complementarity problems: An application of some sensitivity results. *Mathematical Programming*, *58*(1-3), 149–160. doi:10.1007/BF01581264

Papoulis, A., & Unnikrishna Pillai, S. (Eds.). (2002). *Probability, random variables, and stochastic processes* (4th ed.). Mc Graw Hill.

Paredes, F., Zamora, G., Herraiz-Martinez, F. J., Martin, F., & Bonache, J. (2011). Dual-Band UHF-RFID Tags Based on Meander-Line Antennas Loaded With Spiral Resonators. *IEEE Antennas and Wireless Propagation Letters*, *10*, 768–771. doi:10.1109/LAWP.2011.2162716

Pasandideh, S. H. R., Niaki, S. T. A., & Asadi, K. (2015). Optimizing a bi-objective multi-product multi-period three echelon supply chain network with warehouse reliability. *Expert Systems with Applications*, *42*(5), 2615–2623. doi:10.1016/j.eswa.2014.11.018

Passino, K. M. (2002). Bio mimicry of Bacterial Foraging for Distributed Optimization and Control. *IEEE Control Systems Magazine*, *22*(3), 52–67. doi:10.1109/MCS.2002.1004010

Pavlyukevich, I. (2007). Lévy flights, non-local search and simulated annealing. *Journal of Computational Physics*, *226*(2), 1830–1844. doi:10.1016/j.jcp.2007.06.008

Pawlak, Z. (1991). *Rough Sets. Theoretical Aspects of Reasoning About Data*. Dordrecht: Kluwer Publ. Co.

Pearce, J. (2002). Open access "Photovoltaics – A Path to Sustainable Futures". *Futures, 34*(7), 663–674. doi:10.1016/S0016-3287(02)00008-3

Perboli, G., Tadei, R., & Vigo, D. (2011). The Two-Echelon Capacitated Vehicle Routing Problem: Models and Math-Based Heuristics. *Transportation Science Archive, 45*(3), 364-380.

Perez, H. D. (2015). *Supply Chain Strategies: Which One Hits The Mark?*. Retrieved from http://www.supplychainquarterly.com/topics/Strategy/20130306-supply-chain-strategies-which-one-hits-the-mark/

Pérez, L. G., Gress, E., Karelyn, A., & Orán, M. (2009). Some efficiency measures in the operation of flexible manufacturing systems: A stochastic approach.*Proceedings of the 2009 SIAM Conference on "Mathematics for Industry".*

Petcharaks, N., & Ongsakul, W. (2007). Hybrid enhanced Lagrangian relaxation and quadratic programming for hydrothermal scheduling. *Electric Power Components and Systems, 35*(1), 19–42. doi:10.1080/15325000600815449

Peteghem, V. V., & Vanhoucke, M. (2010). A genetic algorithm for the preemptive and non-preemptive multi-mode resource-constrained project scheduling problem. *European Journal of Operational Research, 201*(2), 409–418. doi:10.1016/j.ejor.2009.03.034

Peter, J. (2003). Evolutionary Learning of a 5I2O Fuzzy Controller Including Wheel Lift Constraint. In *Proceedings of the 2nd International Conference on Computational Intelligence, Robotics and Autonomous Systems (CIRAS2003).*

Pfeiffer, F., & Glocker, C. (2000). *Multibody dynamics with unilateral contacts.* Springer Science & Business Media.

Pham, D. T., Afify, A. A., & Koç, E. (2007b). *Manufacturing cell formation using the Bees Algorithm.* 3rd International Virtual Conference on Intelligent Production Machines and Systems (IPROMS 2007), Whittles, Dunbeath, Scotland.

Pham, D. T., Castellani, M., & Ghanbarzadeh. (2007e). Preliminary design using the Bees Algorithm. In *Proc Eighth International Conference on Laser Metrology, CMM and Machine Tool Performance, LAMDAMAP.*

Pham, D. T., Koc, E., Lee, J. Y., & Phrueksanant, J. (2007a). *Using the Bees Algorithm to schedule jobs for a machine.* LAMDAMAP, 8th international Conference on Laser Metrology, CMM and Machine Tool Performance, Cardiff, Euspen, UK.

Pham, D. T., Otri, S., Afify, A., Mahmuddin, M., & Al-Jabbouli, H. (2007c). *Data clustering using the Bees Algorithm.* 40th CIRP International Manufacturing Systems Seminar, Liverpool, UK.

Pham, D. T., Packianther, M. S., Imaguliyev, A., & Yuce, B. (2012). Early Neighborhood Search and Efficiency Based Improved The Bees Algorithm. IMS 2012 Conference, Sakarya, Turkey.

Pham, D. T., Castellani, M., & Fahmy, A. A. (2008). Learning the inverse kinematics of a robot manipulator using the Bees Algorithm. *Proceedings of INDIN, 2008*, 493–498.

Pham, D. T., Darwish, A. H., Eldukhri, E. E., & Otri, S. (2007d). Using the Bees Algorithm to tune a fuzzy logic controller for a robot gymnast. In *Proceedings of IPROMS 2007 Conference.*

Pham, D. T., Ghanbarzadeh, A., Koc, E., Otri, S., Rahim, S., & Zaidi, M. (2005). *The Bees Algorithm. Technical Report: MEC 0501.* Cardiff, UK: Manufacturing Engineering Centre, Cardiff University.

Pham, D. T., Ghanbarzadeh, A., Koc, E., Otri, S., Rahim, S., & Zaidi, M. (2006a). The bees algorithm-a novel tool for complex optimization problems. In *Proceedings of the 2nd Virtual International Conference on Intelligent Production Machines and Systems (IPROMS 2006)* (pp. 454-459). doi:10.1016/B978-008045157-2/50081-X

Pham, D. T., Soroka, A. J., Koç, E., Ghanbarzadeh, A., & Otri, S. (2007f). Some applications of the Bees Algorithm in engineering design and manufacture. In *Proc Int. Conference on Manufacturing Automation (ICMA 2007)*, Singapore.

Philappacopoulus, A. J., & Berndt, M. L. (2001). Influence of de-bonding in ground heat exchangers used with geothermal heat pumps. *Geothermics*, *30*(5), 527–545. doi:10.1016/S0375-6505(01)00011-6

Photovoltaics. (n.d.). Retrieved from http://en.wikipedia.org

Pierskalla, W. P., & Voelker, J. A. (1976). A survey of maintenance models: The control and surveillance of deteriorating systems. *Naval Research Logistics Quarterly*, *23*(2), 353–389. doi:10.1002/nav.3800230302

Piller, S., Perrin, M., & Jossen, A. (2001). Methods for state-of-charge determination and their applications. *Journal of Power Sources*, *96*(1), 113–120. doi:10.1016/S0378-7753(01)00560-2

Pinedo, M. L. (2008). *Scheduling: Theory, Algorithms, and Systems* (3rd ed.). New York: Springer.

Pinto-Varela, T., Barbosa-Póvoa, A. P. F., & Novais, A. Q. (2011). Bi-objective optimization approach to the design and planning of supply chains: Economic versus environmental performances. *Computers & Chemical Engineering*, *35*(8), 1454–1468. doi:10.1016/j.compchemeng.2011.03.009

Pishvaee, M. S., & Razmi, J. (2012). Environmental supply chain network design using multi-objective fuzzy mathematical programming. *Applied Mathematical Modelling*, *36*(8), 3433–3446. doi:10.1016/j.apm.2011.10.007

Pisica, I., Bulac, C., Toma, L., & Eremia, M. (2009, June). Optimal SVC placement in electric power systems using a genetic algorithms based method. In PowerTech, 2009 IEEE Bucharest (pp. 1-6). IEEE. doi:10.1109/PTC.2009.5281841

Piza, E., Trejos, J., & Murillo, A. (2000). Clustering with non-Euclidean distances using combinatorial optimisation techniques. In J. Blasius, J. Hox, E. de Leeuw, & P. Schmidt (Eds.), Science Methodology in the New Millennium. Academic Press.

Poelmans, L., & Rompaey, A. V. (2010). Complexity and performance of urban expansion models. *Computers, Environment and Urban Systems*, *34*(1), 17–27. doi:10.1016/j.compenvurbsys.2009.06.001

Pollaris, H., Braekers, K., Caris, A., Janssens, G. K., & Limbourg, S. (2015). Vehicle routing problems with loading constraints: State-of-the-art and future directions. *OR-Spektrum*, *37*(2), 297–330. doi:10.1007/s00291-014-0386-3

Popa, R. (2014). Melanocytic Lesions Screening through Particle Swarm Optimization. In P. Vasant (Ed.), Handbook of Research on Novel Soft Computing Intelligent Algorithms: Theory and Practical Applications (pp. 355-384). IGI Global doi:10.4018/978-1-4666-4450-2.ch012

Popescu, C., Soto, M., & Lastra, J. (2012). A Petri net-based approach to incremental modeling of flow and resources in service oriented manufacturing systems. *International Journal of Production Research*, *50*(2), 325–343. doi:10.1080/00207543.2011.561371

Porter, M. E. (2008). *Competitive Strategy: Techniques for Analyzing Industries and Competitors*. New York: The Free Press.

Potter, M. A., & De Jong, K. (1994). *A Cooperative Coevolutionary Approach to Function Optimization, Parallel Problem Solving from Nature - PPSN III*. Berlin: Springer.

Potts, C. N., & Kovalyov, M. Y. (2000). Scheduling with batching: A review. *European Journal of Operational Research*, *120*(2), 228–249. doi:10.1016/S0377-2217(99)00153-8

Praditwong, K., Harman, M., & Yao, X. (2011). Software module clustering as a multi-objective search problem. *Software Engineering. IEEE Transactions on*, *37*(2), 264–282.

Pramanik, S., & Roy, T. K. (2007). Fuzzy goal programming approach to multilevel programming problems. *European Journal of Operational Research*, *176*(2), 1151–1166. doi:10.1016/j.ejor.2005.08.024

Prince, B. E., & Haire, M. J. (1991). The influence of maintenance system reliability and maintainability characteristics on plant availability. In *Proceedings of Annual Reliability and Maintainability Symposium*. IEEE. doi:10.1109/ARMS.1991.154508

Priyantha, N. B., Chakraborty, A., & Balakrishnan, H. (2000). The Cricket location-support system. In *Proceedings of the 6th ACM International Conference on Mobile Computing and Networking (ACM MOBICOM)*. doi:10.1145/345910.345917

Proakis, J. G. (Ed.). (2000). Digital communications (4th ed.). Mc-Graw Hill.

Protter, M. H., & Protter, P. E. (1988). *Calculus with analytic geometry*. Jones & Bartlett Learning.

Puterman, L. (2005). *Markov Decision Processes: Discrete Stochastic Dynamic Programming*. John Wiley & Sons.

Qinggang, W., Jinling, Z., & Qingzhi, Y. (2010). Some non-interior path-following methods based on a scaled central path for linear complementarity problems. *Computational Optimization and Applications*, *46*(1), 31–49. doi:10.1007/s10589-008-9188-3

Quarteroni, A., & Valli, A. (1999). *Domain decomposition methods for partial differential equations numerical mathematics and scientific computation*. Oxford University Press New York.

R Core Team. (2013). *R: A language and environment for statistical computing*. R Foundation for Statistical Computing, Vienna, Austria. Retrieved from http://www.R-project.org/

Rabadi, G., Mollaghasemi, M., & Anagnostopoulos, G. C. (2004). A branch-and-bound algorithm for the early/tardy machine scheduling problem with a common due-date and sequence dependent setup time. *Computers & Operations Research*, *31*(10), 1727–1751. doi:10.1016/S0305-0548(03)00117-5

Racioppi, G., Monaci, G., Michelassi, C., Saccardi, D., Borgia, O., & Carlo, F. D. (2007). A methodology to assess the availability of a sour gas injection plant for the production of oil. In *Proceedings of the European safety and reliability conference 2007 – risk, reliability and societal safety*. Taylor and Francis.

Rahman, I., Vasant, P. M., Mahinder Singh, B. S., & Abdullah-Al-Wadud, M. (2014). Swarm Intelligence-Based Smart Energy Allocation Strategy for Charging Stations of Plug-In Hybrid Electric Vehicles. *Mathematical Problems in Engineering*, *2014*, 620425.

Rahman, I., Vasant, P. M., Singh, B. S. M., & Abdullah-Al-Wadud, M. (2014). Intelligent energy allocation strategy for PHEV charging station using gravitational search algorithm. *AIP Conference Proceedings*, *1621*, 52–59. doi:10.1063/1.4898445

Rahman, I., Vasant, P. M., Singh, B. S. M., & Abdullah-Al-Wadud, M. (2014). Optimisation of PHEV/EV charging infrastructures: A review. *Int. J. Energy Technology and Policy*, *10*(3/4), 280–296. doi:10.1504/IJETP.2014.066878

Rahman, I., Vasant, P., Singh, B., & Abdullah-Al-Wadud, M. (2015). Swarm intelligence based State-of-Charge optimization for charging Plug-in Hybrid Electric Vehicles. *Energy and Sustainability V. Special Contributions*, *206*, 261.

Rajpal, P. S., Shishodia, K. S., & Sekhon, G. S. (2006). An artificial neural network for modeling reliability, availability and maintainability of a repairable system. *Reliability Engineering & System Safety*, *91*(7), 809–819. doi:10.1016/j.ress.2005.08.004

Raju, G. V. S., & Zhou, J. (1993). Adaptive Hierarchical Fuzzy Controller. *IEEE Transactions on Systems, Man, and Cybernetics*, *23*(4), 973–980. doi:10.1109/21.247882

Raju, S., Zhou, J., & Kisner, R. A. (1990). Fuzzy logic control for steam generator feedwater control. In *Proceedings of American Control Conference*.

Ramli, A. A., & Watada, J. (2010). A novelty of real-time fuzzy switching regression analysis: a nuclear power plants case study. In *Proceedings of IUM2010*. Berlin: Springer-Verlag.

Randall, M., Lewis, A., Galehdar, A., & Thiel, D. (2007). *Using ant colony optimisation to improve the efficiency of small meander line RFID antennas.* Paper presented at the 3rd IEEE International Conference on E-Science and Grid Computing, Bangalore. doi:10.1109/E-SCIENCE.2007.82

Ranjbar, M. (2013). A branch-and-bound algorithm for scheduling of new product development projects. *International Transactions in Operational Research*, *20*(2), 251–266. doi:10.1111/j.1475-3995.2012.00873.x

Rao, K. V. S., Nikitin, P. V., & Lam, S. F. (2005). Antenna design for UHF RFID tags: A review and a practical application. *IEEE Transactions on Antennas and Propagation*, *53*(12), 3870–3876. doi:10.1109/TAP.2005.859919

Rao, R. V., Savsani, V. J., & Vakharia, D. P. (2011). Teaching–learning-based optimization: A novel method for constrained mechanical design optimization problems. *Computer Aided Design*, *43*(3), 303–315. doi:10.1016/j.cad.2010.12.015

Rao, R. V., Savsani, V. J., & Vakharia, D. P. (2012). Teaching–learning-based optimization: An optimization method for continuous non-linear large scale problems. *Information Sciences*, *183*(1), 1–15. doi:10.1016/j.ins.2011.08.006

Rao, R., & Patel, V. (2012). An elitist teaching-learning-based optimization algorithm for solving complex constrained optimization problems. *International Journal of Industrial Engineering Computations*, *3*(4), 535–560. doi:10.5267/j.ijiec.2012.03.007

Rashedi, E. (2007d). *Gravitational Search Algorithm.* (M.Sc. Thesis). Shahid Bahonar University of Kerman, Kerman, Iran. (in Farsi)

Rashedi, E., & Nezamabadi-pour, H. (2007a). Gravitational Search Algorithm. In *Proceeding of Fifth Iranian Conference Electrical Engineering*. (in Farsi)

Rashedi, E., Nezamabadi-pour, H., & Saryazdi, S. (2007c). Binary Gravitational Search Algorithm, In *Proceeding of First Joint Conference on Fuzzy and Intelligent Systems*. (in Farsi)

Rashedi, E., Nezamabadi-pour, H., Saryazdi, S., & Farsangi, M. M. (2007b). Allocation of Static Var Compensator Using Gravitational Search Algorithm. In *Proceeding of First Joint Conference on Fuzzy and Intelligent Systems*.

Rashedi, E., & Nezamabadi-pour, H. (2013). A Stochastic Gravitational Approach to Feature based Color Image Segmentation. *Journal of Engineering Applications of Artificial Intelligence*, *26*(4), 1322–1332. doi:10.1016/j.engappai.2012.10.002

Rashedi, E., & Nezamabadi-pour, H. (2014). Feature Subset Selection using Improved Binary Gravitational Search Algorithm. *Journal of Intelligent and Fuzzy Systems*, *26*(3), 1211–1221.

Rashedi, E., Nezamabadi-Pour, H., & Saryazdi, S. (2009). GSA: A gravitational search algorithm. *Information Sciences*, *179*(13), 2232–2248. doi:10.1016/j.ins.2009.03.004

Rashedi, E., Nezamabadi-pour, H., & Saryazdi, S. (2010). BGSA: Binary Gravitational Search Algorithm. *Journal of Natural Computing*, *9*(3), 727–745. doi:10.1007/s11047-009-9175-3

Rashedi, E., Nezamabadi-pour, H., & Saryazdi, S. (2013). A simultaneous feature adaptation and feature selection method for content-based image retrieval systems. *Knowledge-Based Systems*, *39*, 85–94. doi:10.1016/j.knosys.2012.10.011

Rashid, A., & Nor, K. M. (1991). An efficient method for optimal scheduling of fixed head hydro and thermal plants. *IEEE Transactions on Power Systems*, *6*(2), 632–636. doi:10.1109/59.76706

Rawson, M., & Kateley, S. (1999). *Electric vehicle charging equipment design and health and safety codes* (No. 1999-01-2941). SAE Technical Paper.

Rechenberg, I. (1973). *Evolutions Strategie: Optimierung Technischer Systemenach Prinzipien der biologischen Evolution*. Stuttgart: Frommann-Holzboog.

Reddy, M. J., & Kumar, D. N. (2007). An efficient multi-objective optimization algorithm based on swarm intelligence for engineering design. In Engineering Optimization. Taylor & Francis.

Reeves, C. (Ed.). (1995). *Modern Heuristic Techniques for Combinatorial Problems*. London: McGraw-Hill.

Riddalls, C., Bennett, S., & Tipi, N. (2000). Modelling the dynamics of supply chains. *International Journal of Systems Science*, *31*(8), 969–976. doi:10.1080/002077200412122

Ries, J., Beullens, P., & Wang, Y. (2013). Instance-specific parameter tuning for meta-heuristics. In P. N. Vasant (Ed.), *Meta-Heuristics Optimization Algorithms in Engineering, Business, Economics, and Finance* (pp. 136–170). Hershey, PA: IGI Global; doi:10.4018/978-1-4666-2086-5.ch005

Rios-Mercado, R. Z., & Bard, J. F. (1999). A branch-and-bound algorithm for permutation flow shops with sequence-dependent setup times. *IIE Transactions*, *31*(8), 721–731. doi:10.1080/07408179908969871

Rivera, J. C., Afsar, H. M., & Prins, C. (2015). A multistart iterated local search for the multitrip cumulative capacitated vehicle routing problem. *Computational Optimization and Applications*, *61*(1), 159–187. doi:10.1007/s10589-014-9713-5

Rizos, C. (2008). Multi-constellation GNSS/RNSS from the perspective of high accuracy users in Australia. *Journal of Spatial Sciences*, *53*(2), 29–63. doi:10.1080/14498596.2008.9635149

Roberts, G. W. (1997). *Real Time on-the-fly Kinematic GPS*. (Doctoral Dissertation). The University of Nottingham.

Roberts, G. W., Ashkenazi, V., & Dodson, A. H. (1999). Monitoring the height deflections of the Humber Bridge by GPS, GLONASS and finite element modeling. In *Proceedings of the International Association of Geodesy (IAG) Symposia: Geodesy Beyond 2000, the challenges of the first decade*.

Roberts, G. W., Meng, X., & Brown, C. (2004, July). From St Paul's to the Tate Modern; Overcoming problems in monitoring bridge deflections using GPS. In *Proceedings of the 1st FIG (Federation Internationale des Geometres – International Federation of Surveyors) International Symposium on Engineering Surveys for Construction Works and Structural Engineering, session 8: Frequencies and Cyclic Monitoring of Structures*.

Roberts, G. W., Meng, X., Brown, C. J., & Dallard, P. (2006). GPS measurements on the London Millennium Bridge. *The Institution of Civil Engineers: Bridge Engineering*, *159*(4), 153–161. doi:10.1680/bren.2006.159.4.153

Robino, G., & Sericola, S. (1995). Interval availability analysis using denumerable markov process: Application to multiprocessor subject to break-downs and repair. *IEEE Transactions on Computers*, *44*(2), 286–291. doi:10.1109/12.364539

Rodriguez, M., Pece, J. P., & Escudero, C. J. (2005) In-building location using Bluetooth. In *Proceedings of the International Workshop on Wireless Ad-hoc Networks (IWWAN)*. doi:10.1.1.126.2802

Rodríguez, R. & Ake, S. (2013). Valuación de opciones de tipo de cambio asumiendo distribuciones α-estables. *Contaduría y Administración*, *58*(4).

Romeijn, H., & Smith, L. (1994). Simulated annealing and adaptive search in global optimization. *Probability in the Engineering and Informational Sciences*, *8*(04), 571–590. doi:10.1017/S0269964800003624

Ross, S. M. (1983). *Introduction to stochastic dynamic programming*. New York, NY: Academic Press.

Roul, P., & Meyer, P. (2011). Numerical solutions of systems of nonlinear integro-differential equations by Homotopy-perturbation method. *Applied Mathematical Modelling*, *35*(9), 4234–4242. doi:10.1016/j.apm.2011.02.043

Rouyendegh, B. D., & Saputro, T. E. (2014). Supplier Selection Using Integrated Fuzzy TOPSIS and MCGP: A Case Study. *Procedia: Social and Behavioral Sciences*, *116*(21), 3957–3970. doi:10.1016/j.sbspro.2014.01.874

Roy, P. K. (2013). Teaching learning based optimization for short-term hydrothermal scheduling problem considering valve point effect and prohibited discharge constraint. *International Journal of Electrical Power & Energy Systems*, *53*, 10–19. doi:10.1016/j.ijepes.2013.03.024

Roy, P. K., & Bhui, S. (2013). Multi-objective quasi-oppositional teaching learning based optimization for economic emission load dispatch problem. *International Journal of Electrical Power & Energy Systems*, *53*, 937–948. doi:10.1016/j.ijepes.2013.06.015

Rubenstein, Y. (1980). Selecting the best stochastic stable system. Stochastic Processes and their Applications, 10, 75-85.

Rubenstein, Y., & Kroese, P. (2007). *Simulation and the Monte Carlo Method.* Wiley Series in Probability and Statistics. doi:10.1002/9780470230381

Rubio-Largo, Á., & Vega-Rodríguez, M. A. (2013). A Multi-objective Approach Based on the Law of Gravity and Mass Interactions for Optimizing Networks, Evolutionary Computation in Combinatorial Optimization. *Lecture Notes in Computer Science*, *7832*, 13–24. doi:10.1007/978-3-642-37198-1_2

Rubio-Largo, Á., Vega-Rodríguez, M. A., Gómez-Pulido, J. A., & Sánchez-Pérez, J. M. (2011). A Multi-objective Gravitational Search Algorithm Applied to the Static Routing and Wavelength Assignment Problem", Applications of Evolutionary Computation. *Lecture Notes in Computer Science*, *6625*, 41–50. doi:10.1007/978-3-642-20520-0_5

Rubio-Largo, Á., Vega-Rodríguez, M. A., Gómez-Pulido, J. A., & Sánchez-Pérez, J. M. (2013). A multiobjective approach based on artificial bee colony for the static routing and wavelength assignment problem. *Soft Computing*, *17*(2), 199–211. doi:10.1007/s00500-012-0887-4

Rudolph, G. (1994). Convergence analysis of canonical genetic algorithms. *IEEE Transactions on Neural Networks*, *5*(1), 96–101. doi:10.1109/72.265964 PMID:18267783

Rui, Y., & Chen, Y. (2001). Better proposal distributions: Object tracking using unscented particle filter. In *Computer Vision and Pattern Recognition, 2001. CVPR 2001.Proceedings of the 2001 IEEE Computer Society Conference on* (*Vol. 2*, pp. II-786). IEEE.

Ruiz, R., Sivrikaya-Serifoglu, F., & Urlings, T. (2006). *Evolutionary approach to realistic hybrid flexible flowshop schedulingproblems. Technical Report. Polytechnic University of Valencia.*

Ryua, K., & Jung, M. (2003). Agent-based fractal architecture and modelling for developing distributed manufacturing systems. *International Journal of Production Research*, *41*(17), 4233–4255. doi:10.1080/0020754031000149275

Saastamoinen, J. (1973). Contribution to the theory of atmospheric refraction. *Bulletin Géodésique*, *107*(1), 13–34. doi:10.1007/BF02522083

Saberi Najafi, H., & Edalatpanah, S. (2012). A kind of symmetrical iterative methods to solve special class of LCP (M, q). *International Journal of Applied Mathematics and Applications*, *4*(2), 183–189.

Saberi Najafi, H., & Edalatpanah, S. (2013a). Comparison analysis for improving preconditioned SOR-type iterative method. *Numerical Analysis and Applications*, *6*(1), 62–70. doi:10.1134/S1995423913010084

Saberi Najafi, H., & Edalatpanah, S. (2013b). Modification of iterative methods for solving linear complementarity problems. *Engineering Computations*, *30*(7), 910–923. doi:10.1108/EC-10-2011-0131

Saberi Najafi, H., & Edalatpanah, S. (2013c). On the convergence regions of generalized accelerated overrelaxation method for linear complementarity problems. *Journal of Optimization Theory and Applications*, *156*(3), 859–866. doi:10.1007/s10957-012-0135-1

Saberi Najafi, H., & Edalatpanah, S. A. (2014b). A new family of (I+S)-type preconditioner with some applications. *Computational & Applied Mathematics*, 1–15.

Saberi Najafi, H., Edalatpanah, S. A., & Sheikhani, A. H. (2014a). Convergence analysis of modified iterative methods to solve linear systems. *Mediterranean Journal of Mathematics*, *11*(3), 1019–1032. doi:10.1007/s00009-014-0412-3

Sabri, N. M., Puteh, M., & Mahmood, M. R. (2013). A Review of Gravitational Search Algorithm. *Int. J. Advance. Soft Comput. Appl*, *5*(3).

Sachdeva, A., Kumar, D., & Kumar, P. (2008). Planning and optimizing the maintenance of paper production systems in a paper plant. *Computers & Industrial Engineering*, *55*(4), 817–829. doi:10.1016/j.cie.2008.03.004

Sadollah, A., Eskandar, H., Yoo, D. G., & Kim, J. H. (2015). Approximate solving of nonlinear ordinary differential equations using least square weight function and metaheuristic algorithms. *Engineering Applications of Artificial Intelligence*, *40*, 117–132. doi:10.1016/j.engappai.2015.01.014

Saeidi-Khabisi, F.S. & Rashedi, E., (2012). Fuzzy Gravitational Search Algorithm. In *Proceeding of Computer and Knowledge Engineering* (ICCKE).

Sahab, M. G., Toropov, V. V., & Gandomi, A. H. (2014). Traditional and Modern Structural Optimization Techniques – Theory and Application. In Metaheuristic Applications in Structures and Infrastructures. Elsevier.

Saha, S., & Chakraborty, D. (2015). Improved Prediction Accuracy with Reduced Feature Set Using Novel Binary Gravitational Search Optimization, Computational Advancement in Communication Circuits and Systems. *Lecture Notes in Electrical Engineering*, *335*, 177–183. doi:10.1007/978-81-322-2274-3_22

Saigal, R. (1970). *Lemke's algorithm and a special linear complementarity problem*. Berkeley, CA: Center for Research in Management Science, University of California.

Sakawa, M., & Matsui, T. (2012). Stackelberg solutions for random fuzzy two-level linear programming through possibility-based probability model. *Expert Systems with Applications*, *39*(12), 10898–10903. doi:10.1016/j.eswa.2012.03.001

Sakawa, M., & Nishizaki, I. (2001). Interactive fuzzy programming for two-level linear fractional programming problems. *Fuzzy Sets and Systems*, *119*(1), 31–40. doi:10.1016/S0165-0114(99)00066-4

Sakawa, M., Nishizaki, I., & Uemura, Y. (1998). Interactive fuzzy programming for multilevel linear programming problems. *Computers & Mathematics with Applications (Oxford, England)*, *36*(2), 71–86. doi:10.1016/S0898-1221(98)00118-7

Salam, N. K. M., & Hamdan, A. R. (1998). Hydrothermal scheduling based Lagrangian relaxation approach to hydrothermal coordination. *IEEE Transactions on Power Systems*, *13*(1), 226–235. doi:10.1109/59.651640

Salhia, S., Imranb, A., & Wassana, N. A. (2014). The multi-depot vehicle routing problem with heterogeneous vehicle fleet: Formulation and a variable neighborhood search implementation. *Computers & Operations Research*, *52*, 315–325. doi:10.1016/j.cor.2013.05.011

Sallem, A., Kotti, M., Fakhfakh, M., Tlelo-Cuautle, E., & Siarry, P. (2015). Combination of Metaheuristics to the Optimal Design of Analog and RF Circuits. In M. Fakhfakh, E. Tlelo-Cuautle, & M. Fino (Eds.), *Performance Optimization Techniques in Analog, Mixed-Signal, and Radio-Frequency Circuit Design* (pp. 210–234). Hershey, PA: Engineering Science Reference; doi:10.4018/978-1-4666-6627-6.ch009

Salmi, S., & Toivanen, J. (2011). An iterative method for pricing American options under jump-diffusion models. *Applied Numerical Mathematics*, *61*(7), 821–831. doi:10.1016/j.apnum.2011.02.002

Samorodnitsky, G., & Taqqu, M. (1994). *Stable Non-Gaussian Random Processes: Stochastic models with infinite variance*. New York: Chapman and Hall.

Samrout, M., Chatelet, E., Kouta, R., & Chebbo, N. (2009). Optimization of maintenance policy using the proportional hazard model. *Reliability Engineering & System Safety*, *94*(1), 44–52. doi:10.1016/j.ress.2007.12.006

Sandgren, E. (1994). Multicriteria design optimization by goal programming. In Advances in Design Optimization. Chapman & Hall.

SAP para Pymes!-SAP Business One. (n.d.). Retrievedfrom http://evolve-it.com.mx/sap-para-pymes/?gclid=CKrKwLfetMMCFYM8aQod9XMAGw

Sarafrazi, S., Nezamabadi-pour, H. & Seydnejad, S.R., (2015). A Novel Hybrid Algorithm of GSA with Kepler Algorithm for Numerical Optimization. *Journal of King Saud University - Computer and Information Sciences*.

Sarafrazi, S., Nezamabadi-pour, H., & Brahman, M. (2010). A GSA-SVM Hybrid System for Classification of Binary Problems. In *Proceeding of 4th Global Conference on Power Control and Optimization* (PCO'2010).

Sarafrazi, S., & Nezamabadi-pour, H. (2012). A Hybrid Method of Gravitational Search Algorithm and an Adaptive Stochastic Local Search with Application to Function Optimization. *Journal of Soft Computing and Information Technology*, *1*(4), 13–28.

Sarafrazi, S., & Nezamabadi-pour, H. (2013). Facing the Classification of Binary Problems with a GSA-SVM Hybrid System. *Journal of Mathematical and Computer Modeling*, *57*(1-2), 270–278. doi:10.1016/j.mcm.2011.06.048

Sarafrazi, S., Nezamabadi-pour, H., & Saryazdi, S. (2011). Disruption: A new operator in gravitational search algorithm. *Journal of Scientia Iranica-D*, *18*(3), 539–548. doi:10.1016/j.scient.2011.04.003

Saraswat, S., & Yadava, G. (2008). An overview on reliability, availability, maintainability and supportability (RAMS) engineering. *International Journal of Quality & Reliability Management*, *25*(3), 330–344. doi:10.1108/02656710810854313

Sarkar, A., & Maulik, U. (2015). Cancer Biomarker Assessment Using Evolutionary Rough Multi-Objective Optimization Algorithm. In P. Vasant (Ed.), Handbook of Research on Artificial Intelligence Techniques and Algorithms (pp. 509-535). IGI Global doi:10.4018/978-1-4666-7258-1.ch016

Sarrafha, K., Rahmati, S. H. A., Niaki, S. T. A., & Zaretalab, A. (2015). A bi-objective integrated procurement, production, and distribution problem of a multi-echelon supply chain network design: A new tuned MOEA. *Computers & Operations Research*, *54*, 35–51. doi:10.1016/j.cor.2014.08.010

Sasikala, J., & Ramaswamy, M. (2010). Optimal gamma based fixed head hydrothermal scheduling using genetic algorithm. *Expert Systems with Applications*, *37*(4), 3352–3357. doi:10.1016/j.eswa.2009.10.015

Satapathy, J., & Subhashini, K. (2008). A tabu based Neural Network training algorithm for equalization of communication channels. In *Proceedings of International Conference on Artificial Intelligence* (pp.79-85).

Savvides, A., Han, C.-C., & Srivastava, M. (2001). Dynamic fine-grained localization in ad-hoc networks of sensors. In *Proceedings of the seventh Annual international Conference on Mobile Computing and Networking (ACM MOBICOM)*. Rome, Italy.

Sazvar, Z., Mirzapour Al-e-hashem, S. M. J., Baboli, A., & Akbari Jokar, M. R. (2014). A bi-objective stochastic programming model for a centralized green supply chain with deteriorating products. *International Journal of Production Economics*, *150*, 140–154. doi:10.1016/j.ijpe.2013.12.023

Scalas, E., & Kim, K. (2007). The art of Fitting Financial Time Series with Lévy Stable Distributions. *Journal of the Korean Physical Society*, *50*(1).

Schabe, H. (1995). A new approach to optimal replacement times for complex systems. *Microelectronics and Reliability*, *35*(8), 1125–1130. doi:10.1016/0026-2714(94)00166-L

Schaffer, J. D. (1987) Some Effects of Selection Procedures on Hyperplane Sampling by Genetic Algorithms. In Genetic Algorithms and Simulated Annealing. Colorado State University.

Scheffer, R., & Maciel Filho, R. (2001). The fractional Brownian motion as a model for an industrial airlift reactor. *Chemical Engineering Science*, *56*(2), 707–711. doi:10.1016/S0009-2509(00)00279-7

Schmidt, J. W. (1987). On tridiagonal linear complementarity problems. *Numerische Mathematik*, *51*(1), 11–21. doi:10.1007/BF01399692

Schoenherr, T., Tummala, V., & Harrison, T. (2008). Assessing supply chain risks with the analytic hierarchy process: Providing decision support for the offshoring decision by a US manufacturing company. *Journal of Purchasing and Supply Management*, *14*(2), 100–111. doi:10.1016/j.pursup.2008.01.008

Schwanke, R. W. (1991). *An intelligent tool for re-engineering software modularity*. Paper presented at the Software Engineering. doi:10.1109/ICSE.1991.130626

Schwefel, H. P. (1975). *Evolutionsstrategie undNumerische Optimierung*. (PhD dissertation). Technische Berlin, Germany.

Schwieger, V. (2003). Using handheld GPS receivers for precise positioning. In *Proceedings of the 2nd International Federation of Surveyors (FIG)Regional Conference*. Marrakech, Morocco.

Scientific Modelling. (n.d.). Retrieved from http://en.wikipedia.org

Seeley, T. D. (1995). *The Social Physiology of Honey Bee Colonies: The Wisdom of the Hive*. London: Hardvard University Press.

Selko, A. (2006). Nissan To Use RFID Enabled RTLS To Automate Supply Chain. *Industry Week: Advancing the Business of Manufacturing*. Retrieved from http://www.industryweek.com/planning-amp-forecasting/nissan-use-rfid-enabled-rtls-automate-supply-chain

Sengupta, A., & Pal, T. K. (2000). On comparing interval numbers. *European Journal of Operational Research*, *127*(1), 28–43. doi:10.1016/S0377-2217(99)00319-7

Senvar, O., Turanoglu, E., & Kahraman, C. (2013). Usage of Metaheuristics in Engineering: A Literature Review. In P. Vasant (Ed.), *Meta-Heuristics Optimization Algorithms in Engineering, Business, Economics, and Finance* (pp. 484–528). Hershey, PA: Information Science Reference; doi:10.4018/978-1-4666-2086-5.ch016

Sethi, S., Yan, H., Zhang, H., & Zhang, Q. (2002). Optimal and hierarchical controls in dynamic stochastic manufacturing systems: A review. *Manufacturing & Service Operations Management*, *4*(2), 133–170. doi:10.1287/msom.4.2.133.281

Sevaux, M., & Peres, S. D. (2003). Genetic algorithms to minimize the weighted number of late jobs on a single machine. *European Journal of Operational Research*, *151*(2), 296–306. doi:10.1016/S0377-2217(02)00827-5

Shafiei, A., & Williamson, S. S. (2010). Plug-in hybrid electric vehicle charging: Current issues and future challenges. In *Vehicle Power and Propulsion Conference (VPPC)* (pp. 1-8). IEEE. doi:10.1109/VPPC.2010.5729134

Shair, E. F., Khor, S. Y., Abdullah, A. R., Jaafar, H. I., Ali, N. M., & Abidin, A. Z. (2014). A Brief Review of Cuckoo Search Algorithm (CSA) Research Progression from 2010 to 2013. *Romania*, *1*, 1–4.

Shamsudin, H. C., Irawan, A., Ibrahim, Z., Abidin, A. F. Z., Wahyudi, S., Rahim, M. A. A., & Khalil, K. (2012). A Fast Discrete Gravitational Search Algorithm. In *Proceeding of Fourth International Conference on Computational Intelligence, Modeling and Simulation.*

Shang, Z. (2013). Neighborhood Crossover Operator: A new Operator in Gravitational Search Algorithm. *International Journal of Computer Science Issues*, *10*(2).

Shapiro, M., & Nemiroski, A. (2006). On complexity of stochastic programming problems. In V. Jeyakumar & A. Rubinov (Eds.), Continuous Optimization: Current Trends and Applications, (pp. 111-144). Berlin: Springer Verlag.

Shapiro, M. (2006). On complexity of multistage stochastic programs. *O.R. Letters*, *34*, 1–8.

Sharadindu, R., Uttam, P., & Nikhil, S. (2014). Efficient technique to solve Travelling Salesman Problem using Genetic Algorithm. *International Journal of Computer Engineering and Technology*, *5*(3), 132–137.

Sharma, R. K., & Kumar, S. (2008). Performance modeling in critical engineering systems using RAM analysis. *Reliability Engineering & System Safety*, *93*(6), 913–919. doi:10.1016/j.ress.2007.03.039

Sharp, I., Yu, K., & Sathyan, T. (2012). Positional accuracy measurement and error modeling for mobile tracking. *IEEE Transactions on Mobile Computing*, *11*(6), 1021–1032. doi:10.1109/TMC.2011.119

Shashi, D. K., & Katiyar, V. K. (2010). Multiobjective Extraction Optimization of Bioactive Compounds from Gardenia Using Real Coded Genetic Algorithm. In *Proceedings of 6th World Congress of Biomechanics* (WCB 2010).

Shaw, B., Mukherjee, V., & Ghoshal, S. P. (2012). A Novel Opposition-based Gravitational Search Algorithm for Combined Economic and Emission Dispatch Problems of Power Systems. *Journal of Electrical Power and Energy Systems*, *35*(1), 21–33. doi:10.1016/j.ijepes.2011.08.012

Sheikhpour, S., Sabouri, M., & Zahiri, S.-H. (2013). A Hybrid Gravitational Search Algorithm-Genetic Algorithm for Neural Network Training. In *Proceeding of 21st Iranian Conference of Electrical Engineering* (ICEE).

Shen, Z., Dessouky, M., & Ordóñez, F. (2009). A two-stage vehicle routing model for large scale bioterrorism emergencies. *Networks*, *54*, 255–269. doi:10.1002/net.20337

Sherwin, D. J. (2000). Steady-state series availability. *IEEE Transactions on Reliability*, *49*(2), 131–132. doi:10.1109/24.877327

Sheskin, D. J. (2003). *Handbook of parametric and nonparametric statistical procedures. CRC Press.* doi:10.1201/9781420036268

Shiau, T. (1988). An LP-based successive overrelaxation method for linear complementarity problems. *Journal of Optimization Theory and Applications*, *59*(2), 247–259.

Shih, H. S., Lai, Y. J., & Lee, E. S. (1996). Fuzzy approach for multi-level programming problems. *Computers & Operations Research*, *23*(1), 73–91. doi:10.1016/0305-0548(95)00007-9

Shlesinger, M. F. (2006). Mathematical physics: Search research. *Nature*, *443*(7109), 281–282. doi:10.1038/443281a PMID:16988697

Sholedolu, M. O. (2009). *Nature-inspired Optimization: Improvements to the Particle Swarm Optimization Algorithm and the Bees Algorithm.* (PhD Thesis). Cardiff University.

Shu-xian, L. (2013). An explicit approximate I–V characteristic model of a solar cell based on pade´ approximants. *Solar Energy*, *92*, 147–159.

Shyamal, A. K., & Pal, M. (2005). Distance between fuzzy matrices and its applications-I, *J. Nature. Physical Science (London)*, *19*(1), 39–58.

Simon, D. (2008). Biogeography - based optimization. *IEEE Transactions on Evolutionary Computation*, *12*(6), 702–713. doi:10.1109/TEVC.2008.919004

Simon, D., Ergezer, M., Du, D., & Rarick, R. (2011). Markov models for biogeography-based optimization. *IEEE Transactions on Systems, Man, and Cybernetics. Part B, Cybernetics*, *41*(1), 299–306. doi:10.1109/TSMCB.2010.2051149 PMID:20595090

Simopoulos, D. N., Kavatza, S. D., & Vournas, C. D. (2007). An enhanced peak shaving method for short term hydrothermal scheduling. *Energy Conversion and Management*, *48*(11), 3018–3024. doi:10.1016/j.enconman.2007.07.001

Sims, R. H. (2007). Not too late: IPCC identifies renewable energy as a key measure to limit climate change. *Renewable Energy World*, *10*(4), 31–39.

Singer, C. (1941). *A Short History of Science to the Nineteenth Century*. Clarendon Press.

Singh, C., & Lago-Gonzalez, A. (1985). Reliability modeling of generation systems including un- conventional energy sources. *IEEE Transactions on Power Apparatus and Systems*, *104*(5), 1049–1056. doi:10.1109/TPAS.1985.323455

Singh, S., Agrawal, S., Tiwari, A., Al-Helal, I. M., & Avasthi, D. V. (2015). Modeling and parameter optimization of hybrid single channel photovoltaic thermal module using genetic algorithms. *Solar Energy*, *113*, 78–87. doi:10.1016/j.solener.2014.12.031

Sinha, N., Chakrabarti, R., & Chattopadhaya, P. K. (2003). Fast evolutionary programming techniques for short-term hydrothermal scheduling. *Electric Power Systems Research*, *66*(2), 97–103. doi:10.1016/S0378-7796(03)00016-6

Sipper, D., & Bulfin, R. Jr. (1997). *Production Planning Control and Integration*. New York: McGraw Hill.

Sirjani, R., Mohamed, A., & Shareef, H. (2011). Optimal placement and sizing of Static Var Compensators in power systems using Improved Harmony Search Algorithm. *Przeglad Elektrotechniczny*, *87*(7), 214–218.

Sirjani, R., Mohamed, A., & Shareef, H. (2012). Optimal allocation of shunt Var compensators in power systems using a novel global harmony search algorithm. *International Journal of Electrical Power & Energy Systems*, *43*(1), 562–572. doi:10.1016/j.ijepes.2012.05.068

Sitek, P. (2014). A Hybrid Approach to the Two-Echelon Capacitated Vehicle Routing Problem. In R. Szewczyk, C. Zieliński, & M. Kaliczyńska (Eds.), *Recent Advances in Automation, Robotics and Measuring Techniques, Advances in Intelligent Systems and Computing* (Vol. 267, pp. 251–263). Heidelberg: Springer. doi:10.1007/978-3-319-05353-0_25

Sitek, P., & Wikarek, J. (2014). A novel integrated approach to the modelling and solving of the Two-Echelon Capacitated Vehicle Routing Problem. *Production & Manufacturing Research: An Open Access Journal*, *2*(1), 326–340.

Sitek, P., & Wikarek, J. (2015). A Hybrid Approach to the Optimization of multi-echelon Systems. *Mathematical Problems in Engineering*, *2015*, 1–12. doi:10.1155/2015/925675

Sitek, P., Wikarek, J., & Grzybowska, K. (2014). A Multi-Agent Approach to the Multi-Echelon Capacitated Vehicle Routing Problem. In J. M. Corchado, J. Bajo, J. Kozlak, & P. Pawlewski (Eds.), *Highlights of Practical Applications of Heterogeneous Multi-Agent Systems, The PAAMS Collection, Communications in Computer and Information Science* (Vol. 430, pp. 121–132). Heidelberg: Springer. doi:10.1007/978-3-319-07767-3_12

Sivasubramani, S., & Shanti Swarup, K. (2011). Hybrid DE–SQP algorithm for non-convex short term hydrothermal scheduling problem. *Energy Conversion and Management*, *52*(1), 757–761. doi:10.1016/j.enconman.2010.07.056

Smith, J. (2003). Survey on the use of simulation fr manufacturing systems design and operation. *Journal of Manufacturing Systems*, *22*(2), 157–171. doi:10.1016/S0278-6125(03)90013-6

Smith, L., & Ball, P. (2012). Steps towards sustainable manufacturing through modelling material, energy and waste flows. *International Journal of Production Economics*, *140*(1), 227–238. doi:10.1016/j.ijpe.2012.01.036

Soares, J., Sousa, T., Morais, H., Vale, Z., Canizes, B., & Silva, A. (2013). Application Specific Modified Particle Swarm Optimization for Energy Resource Scheduling Considering Vehicle-to-Grid. *Applied Soft Computing*, *13*(11), 4264–4280. doi:10.1016/j.asoc.2013.07.003

Soares, S., & Ohishi, T. (1995). Hydro-dominated short-term hydrothermal scheduling via a hybrid simulation-optimisation approach: A case study. *IEE Proceedings. Generation, Transmission and Distribution*, *142*(6), 569–575. doi:10.1049/ip-gtd:19952156

Sojoudi, S., & Low, S. H. (2011). Optimal charging of plug-in hybrid electric vehicles in smart grids. In *Power and Energy Society General Meeting, 2011 IEEE*. doi:10.1109/PES.2011.6039236

Solar Cell. (n.d.). Retrieved from www.wikipedia.org

Solar Cells. (n.d.). Retrieved from www.chemistryexplained.com

Solar Panel. (n.d.). Retrieved from wikipedia.org

Soleimanpour, M., & Nezamabadi-pour, H. (2013). A Modified Monkey Algorithm for Real-Parameter Optimization. *Journal of Multiple-Valued Logic and Soft Computing*, *21*, 453–477.

Soleimanpour-moghadam, M., & Nezamabadi-pour, H. (2012). An Improved Quantum Behaved Gravitational Search Algorithm. In *Proceeding of 20th Iranian Conference on Electrical Engineering* (ICEE2012).

Soleimanpour-moghadam, M., Nezamabadi-pour, H., & Farsangi, M. M. (2011). A quantum behaved gravitational search algorithm. *Journal of Intelligent Information Management*, *4*(6), 390–395. doi:10.4236/iim.2012.46043

Soleimanpour-moghadam, M., Nezamabadi-pour, H., & Farsangi, M. M. (2014). A Quantum Inspired Gravitational Search Algorithm for Numerical Function Optimization. *Journal of Information Science*, *267*, 83–100. doi:10.1016/j.ins.2013.09.006

Soman, C., Van Donk, D., & Gaalman, G. (2004). Combined make-to-order and make-tostock in a food production system. *International Journal of Production Economics, 90*(2), 223-235.

Son, Y., & Venkateswaran, J. (2007). Hierarchical supply chain planning architecture for integrated analysis of stability and performance. *Int. J. Simulation and Process Modelling, 3*(3).

Song, J., & Yao, J. (2002). Performance Analysis and Optimization of Assemble-to-Order Systems with Random Leadtimes. *Operations Research*, *50*(5), 889–903. doi:10.1287/opre.50.5.889.372

Soysal, M., Bloemhof-Ruwaard, J. M., & Bektas, T. (2014). The time-dependent two-echelon capacitated vehicle routing problem with environmental considerations. *International Journal of Production Economics*, 1-13.

Spall, J. C. (2003). *Introduction to Stochastic Search and Optimization*. Wiley. doi:10.1002/0471722138

Srinivas, N., & Deb, K. (1994). Multi-objective Optimization using Non-Dominated Sorting in Genetic Algorithms. *Journal of Evolutionary Computation*, *2*(3), 221–248. doi:10.1162/evco.1994.2.3.221

Stanilov, K., & Batty, M. (2011). Exploring the historical determinants of urban growth patterns through cellular automata. *Transactions in GIS*, *15*(3), 253–271. doi:10.1111/j.1467-9671.2011.01254.x

Statnikov, R. B., & Matusov, J. B. (1995). *Multicriteria Optimization and Engineering*. New York: Chapman and Hall. doi:10.1007/978-1-4615-2089-4

Steele, J. (1997). *Sustainable architecture: principles, paradigms, and case studies*. New York: McGraw-Hill Inc.

Steel, S. J., & Uys, D. W. (2006). Influential data cases when the *Cp* criterion is used for variable selection in multiple linear regression. *Computational Statistics & Data Analysis*, *50*(7), 1840–1854. doi:10.1016/j.csda.2005.02.003

Stefansson, H., Sigmarsdottir, S., Jensson, P., & Shah, N. (2011). Discrete and continuous time representations and mathematical models for large production scheduling problems: A case study from the pharmaceutical industry. *European Journal of Operational Research*, *215*(2), 383–392. doi:10.1016/j.ejor.2011.06.021

Stephens, M. A. (1977). Goodness of Fit for the Extreme Value Distribution. *Biometrika*, *64*(3), 583–588. doi:10.1093/biomet/64.3.583

Stickney, A., & Watson, L. (1978). Digraph models of Bard-type algorithms for the linear complementarity problem. *Mathematics of Operations Research*, *3*(4), 322–333. doi:10.1287/moor.3.4.322

Stoev, S., & Taqqu, M. S. (2003). Lecture Notes in Physics: Vol. 621. *Wavelet estimation for the Hurst parameter in stable processes*. doi:10.1007/3-540-44832-2_4

Stonier & Anver. (2003). A new multi-layered fuzzy image filter for removing impulse noise. *Journal of Systemics, Cybernetics and Informatics*, *1*(4).

Stonier, R. J., & Anver, M. (2002). *Multi-layered Fuzzy Image Filters for removing Impulse Noise*. ISCA 11th International Conference on Intelligent Systems on Emerging Technologies (ICIS-2002), Boston, MA.

Stonier, R. J., & Mohammadian, M. (1996). Intelligent Hierarchical Control for Obstacle- Avoidance. In Computational Techniques and Applications: CTAC95. World Scientific.

Stonier, R. J. (1999). Evolutionary learning of fuzzy logic controllers over a region of initial states. In *Proceedings of the 1999 Congress on Evolutionary Computation*. doi:10.1109/CEC.1999.785538

Stonier, R. J., Drumm, M. J., & Bell, J. (2003). Optimal boundary control of a tracking problem for a parabolic distributed system using hierarchical fuzzy control and evolutionary algorithms. In *Proceedings of the 18th International Conference on Computers and their Applications*.

Stonier, R. J., & Mohammadian, M. (1995). Self Learning Hierarchical Fuzzy Logic Controller in Multi-Robot Systems. In *Proceedings of the IEA Conference Control95*.

Stonier, R. J., & Mohammadian, M. (1998). Knowledge Acquisition for Target Capture. In *Proceedings of the International Conference on Evolutionary Computing ICEC'98*.

Stonier, R. J., Stacey, A. J., & Messom, C. (1998). Learning fuzzy controls for the inverted pendulum. In *Proceedings of ISCA 7th International Conference on Intelligent Systems*.

Stonier, R. J., Stacey, A., Mohammadian, M., & Smith, S. F. (1999). Application of evolutionary learning in fuzzy logic and optimal control. In *Proceedings of the International Conference on Computational Intelligence for Modelling, Control and Automation*.

Stonier, R. J., & Zajaczkowski, J. (2003). Hierarchical fuzzy controllers for the inverted pendulum. In *Proceedings of the 2nd International Conference on Computational Intelligence, Robotics and Autonomous Systems (CIRAS 2003)*.

Storn, R., & Price, K. (1995). Differential Evolution – A Simple and Efficient Heuristic for Global Optimization over Continuous Spaces. International Computer Science Institute.

Storn, R., & Price, K. V. (1997). Differential evolution, a simple and efficient heuristic for global optimization over continuous spaces. *Journal of Global Optimization*, *11*(4), 341–359. doi:10.1023/A:1008202821328

Stoye, J., Dress, A., & Perrey, S. (1997). Improving the Divide and conquer Approach to Sum-of-pairs Multiple Sequence Alignment. *Applied Mathematics Letters*, *10*(2), 67–73. doi:10.1016/S0893-9659(97)00013-X

Strang, G., & Borre, K. (1997). *Linear algebra, geodesy, and GPS*. Wellesley, MA: Wellesley Cambridge Press.

Su, W., & Chow, M. Y. (2011). Performance evaluation of a PHEV parking station using particle swarm optimization. In *Power and Energy Society General Meeting, 2011 IEEE* (pp. 1-6). IEEE. doi:10.1109/PES.2011.6038937

Subramanian, A., Garcia, M., Dominguez-Garcia, A., Callaway, D., Poolla, K., & Varaiya, P. (2012, June). Real-time scheduling of deferrable electric loads. In *American Control Conference (ACC)*, (pp. 3643-3650). IEEE.

Subramanian, P. (1988). A dual exact penalty formulation for the linear complementarity problem. *Journal of Optimization Theory and Applications*, *58*(3), 525–538. doi:10.1007/BF00939395

Suder, A., & Kahraman, C. (2015). Minimizing Environmental Risks Using Fuzzy TOPSIS: Location Selection for the ITU Faculty of Management, *Human and Ecological Risk Assessment. International Journal (Toronto, Ont.)*, *21*(5), 1326–1340.

Sun, J., Feng, B., & Xu, W. (2004). Particle swarm optimization with particles having quantum behavior. In Proceeding of Congress on Evolutionary Computation.

Sun, L. Y. (2006). Some extensions of the improved modified Gauss–Seidel iterative method for H-matrices. *Numerical Linear Algebra with Applications*, *13*(10), 869–876. doi:10.1002/nla.498

Surekha, P., & Sumathi, S. (2011). Solution To Multi-Depot Vehicle Routing Problem Using Genetic Algorithms. *World Applied Programming*, *1*(3), 118–131.

Survey, R. (2015). Evolutionary algorithms for de novo drug design, Applied Soft Computing. *Applied Soft Computing*, *27*, 543–552.

Su, W., & Chow, M. Y. (2010). An intelligent energy management system for PHEVs considering demand response. In *Proc. 2010 FREEDM Annual Conference.*

Su, W., & Chow, M.-Y. (2012a). Performance evaluation of an EDA-based large-scale plug-in hybrid electric vehicle charging algorithm. *Smart Grid. IEEE Transactions on*, *3*(1), 308–315.

Su, W., & Chow, M.-Y. (2012b). Computational intelligence-based energy management for a large-scale PHEV/PEV enabled municipal parking deck. *Applied Energy*, *96*, 171–182. doi:10.1016/j.apenergy.2011.11.088

Su, Z., & Wang, H. (2015). A Novel Robust Hybrid Gravitational Search Algorithm for Reusable Launch Vehicle Approach and Landing Trajectory Optimization. *Journal of Neurocomputing*, *162*, 116–127. doi:10.1016/j.neucom.2015.03.063

Tabatabaei, S. (2014). A new Gravitational Search Optimization Algorithm to Solve Single and Multiobjective Optimization Problems. *Journal of Intelligent and Fuzzy Systems*, *26*(2), 993–1006.

Taillard, E. (1993). *VRP benchmarks*. Retrieved from http://mistic.heigvd.ch/taillard/problemes.dir/vrp.dir/ vrp.html

Takriti, S., & Murty, K. G. (1997). On the convergence of the block principal pivotal algorithm for the LCP. *European Journal of Operational Research*, *102*(3), 657–666. doi:10.1016/S0377-2217(96)00243-3

Talbi, E. G. (2002). A Taxonomy of Hybrid Metaheuristic. *Journal of Heuristics*, *8*(5), 541–546. doi:10.1023/A:1016540724870

Talbi, E. G. (2009). *Metaheuristics: From Design to Implementation*. John Wiley and Sons. doi:10.1002/9780470496916

Tam, A. S. B., Chan, W. M., & Price, J. W. H. (2006). Optimal maintenance intervals for multi - component system. *Production Planning and Control*, *17*(8), 769–779. doi:10.1080/09537280600834452

Tan, K. C., Lee, T. H., & Khor, E. F. (2002). Evolutionary algorithms for multi-objective optimization: Performance assessments and comparisons. *Artificial Intelligence Review*, *17*(4), 251–290. doi:10.1023/A:1015516501242

Tank, D. W., & Hopfield, J. J. (1986). Simple Neural Optimization Network: An A\D Converter, Signal Decision Circuit and a Linear Programming Circuit. *IEEE Transactions on Circuits and Systems*, *CAS-33*(5), 533–541. doi:10.1109/TCS.1986.1085953

Tan, W. S., Hassan, M. Y., Rahman, H. A., Abdullah, M. P., & Hussin, F. (2013). Multi-distributed generation planning using hybrid particle swarm optimisation-gravitational search algorithm including voltage rise issue. *IET Generation. Transmission & Distribution*, *7*(9), 929–942. doi:10.1049/iet-gtd.2013.0050

Taqqu, M. S., Teverovsky, V., & Willinger, W. (1996). Estimators for long-range dependence: An empirical study. *Fractals*, *3*(4), 785–798. doi:10.1142/S0218348X95000692

Tavakkoli-Moghaddam, R., Rahimi-Vahed, A., & Mirzaei, A. H. (2007). A hybrid multi-objective immune algorithm for a flow shop scheduling problem with bi-objectives: Weighted mean completion time and weighted mean tardiness. *Information Sciences*, *177*(22), 5072–5090. doi:10.1016/j.ins.2007.06.001

Tavakkoli-Moghaddam, R., Safaei, N., Kah, M., & Rabbani, M. (2007). A new capacitated vehicle routing problem with split service for minimizing fleet cost by simulated annealing. *Journal of the Franklin Institute*, *344*(5), 406–425. doi:10.1016/j.jfranklin.2005.12.002

Tesla Motors. (2009). *Tesla roadster spec sheet 2009*. Author.

Teunissen, P. (1995). The Least-Squares AMbiguity Decorrelation Adjustment: A method for fast GPS integer ambiguity estimation. *Journal of Geodesy*, *70*(1-2), 65–82. doi:10.1007/BF00863419

Teunissen, P. J. G., Odolinski, R., & Odijk, D. (2014). Instantaneous BeiDou plus GPS RTK positioning with high cut-off elevation angles. *Journal of Geodesy*, *88*(4), 335–350. doi:10.1007/s00190-013-0686-4

Thamilselvan, R., & Balasubramanie, P. (2009). Integrating Genetic Algorithm, Tabu search Approach for Job Shop scheduling. *International Journal of Computer Science and Information Security*, *2*(1).

Thangiah, S., Fergany, A., & Awan, S. (2007). Real-time split-delivery pickup and delivery time window problems with transfers. *Central European Journal of Operations Research*, *15*(4), 329–349. doi:10.1007/s10100-007-0035-x

The Nature of Mathematical Programming. (n.d.). Mathematical Programming Glossary, INFORMS Computing Society.

The United Nations Framework Convention on Climate Change. (2005). Retrieved from http://unfccc.int/essentialbackground/convention/background/items/1353.php

Thoai, N. V., Yamamoto, Y., & Yoshise, A. (2005). Global optimization method for solving mathematical programs with linear complementarity constraints. *Journal of Optimization Theory and Applications*, *124*(2), 467–490. doi:10.1007/s10957-004-0946-9

Thomas & Stonier. (2003). Fuzzy control in robot-soccer, evolutionary learning in the first layer of control. *Journal of Systemics, Cybernetics and Informatics*, *1*(1).

Thomas, P. J., & Stonier, R. J. (2003). Hierarchical Fuzzy Control in Robot Soccer using Evolving Algorithms. In *Proceedings of the International Congress on Evolutionary Computation (CEC2003)*. doi:10.1109/CEC.2003.1299393

Thompson, K. M. (1987). *A two-stage successive overrelaxation algorithm for solving the linear complementarity problem*. University of Wisconsin-Madison, Computer Sciences Department.

Tibshirani, R. (1996). Regression shrinkage and selection via the Lasso. *Journal of the Royal Statistical Society. Series B. Methodological*, *58*(1), 267–288.

Tie, S. F., & Tan, C. W. (2013). A review of energy sources and energy management system in electric vehicles. *Renewable & Sustainable Energy Reviews*, *20*, 82–102. doi:10.1016/j.rser.2012.11.077

Tizhoosh, H. R. (2005). Opposition-Based Learning: A New Scheme for Machine Intelligence. In *Proceedings of the International Conference on Computational Intelligence for Modeling, Control and Automation, and International Conference on Intelligent Agents, Web Technologies and Internet Commerce*. doi:10.1109/CIMCA.2005.1631345

Toivanen, J. & C. W. Oosterlee (2011). *A projected algebraic multigrid method for linear complementarity problems*. Available at SSRN 1739902.

Tomassini. (n.d.). *Parallel and Distributed Evolutionary Algorithms: A Review*. Institute of Computer Science, University of Lausanne.

Tomassone, R., Audrain, S., Lesquoy, E., & Millier, C. (1992). *La Régression. Des Nouveaux Regards sur une Ancienne Méthode Statistique*. Paris: Masson.

Torabi, M., Yaghoobi, H., & Fereidoon, A. (2012). Application of differential transformation method and padé approximant for the Glauert-jet problem. *Recent Patents on Mechanical Engineering*, *5*, 150–155.

Toshio, N., & Shnji, O. (1976). Reliability analysis of a one unit system with unrepairable spare units and its optimization applications. *Operations Research*, *27*(1), 101–110. doi:10.1057/jors.1976.10

Totally Integrated Automation. (2011). *Benefits and Latest Advances of 2D Bar Codes for Consumer Packaged Goods Manufacturers*. Retrieved from http://www.totallyintegratedautomation.com/2011/06/benefits-and-latest-advances-of-2d-bar-codes-for-consumer-packaged-goods-manufacturers/

Towsend, T. U. A. (1989). Method for estimating the long-term performance of direct-coupled photovoltaic systems. (M.S. Thesis). Solar Energy Laboratory, University of Wisconsin.

Trejos, J. (1993). A simulated annealing implementation for oblique varimax rotations. In J. Janssen & C. H. Skiadas (Eds.), Applied Stochastic Models and Data Analysis (vol. 2, pp. 981-989). Singapore: World Scientific.

Trejos, J., & Piza, E. (2000). Partitioning of binary data applying a genetic algorithm (in Spanish). Technical Report, CIMPA, University of Costa Rica.

Trejos, J., Villalobos, M., Murillo, A., Chavarria, J., & Fallas, J. J. (2014). Evaluation of optimization metaheuristics in clustering. In C. Travieso, J. Arroyo & M. Ramírez (Eds.), *IEEE 3rd International Conference and Workshop on Bioinspired Intelligence* (pp. 154-161). doi:10.1109/IWOBI.2014.6913956

Trejos, J., & Castillo, W. (2000). Simulated annealing optimization for two-mode partitioning. In W. Gaul & R. Decker (Eds.), *Classification and Information at the Turn of the Millenium* (pp. 133–142). Berlin: Springer. doi:10.1007/978-3-642-57280-7_15

Trejos, J., Castillo, W., González, J., & Villalobos, M. (2000). Application of simulated annealing in some multidimensional scaling problems. In H. A. L. Kiers, J.-P. Rasson, P. J. F. Groenen, & M. Schader (Eds.), *Data Analysis, Classification and Related Methods* (pp. 297–302). Berlin: Springer. doi:10.1007/978-3-642-59789-3_47

Trejos, J., Murillo, A., & Piza, E. (1998). Global stochastic optimization for partitioning. In A. Rizzi, M. Vichi, & H. H. Bock (Eds.), *Advances in Data Science and Classification* (pp. 185–190). Berlin: Springer. doi:10.1007/978-3-642-72253-0_25

Trejos, J., Murillo, A., & Piza, E. (2004). Clustering by ant colony optimization. In D. Banks, L. House, F. R. McMorris, P. Arabie, & W. Gaul (Eds.), *Classification, Clustering, and Data Mining Applications* (pp. 25–32). Berlin: Springer. doi:10.1007/978-3-642-17103-1_3

Trejos, J., & Villalobos, M. (2007). Partitioning by particle swarm optimization. In P. Brito, P. Bertrand, G. Cucumel, & F. de Carvalho (Eds.), *Selected Contributions in Data Analysis and Classification* (pp. 235–244). Berlin: Springer. doi:10.1007/978-3-540-73560-1_22

Trelea, I. C. (2003). The particle swarm optimization algorithm: Convergence analysis and parameter selection. *Information Processing Letters*, *85*(6), 317–325. doi:10.1016/S0020-0190(02)00447-7

Trevor, T. (2007). Fridge recycling: Bringing agents in from the cold. *Waste Management World*, *5*, 43–47.

Triantaphyllou, E., (2000). *Multi-Criteria Decision Making: A Comparative Study*. Dordrecht, The Netherlands: Kluwer Academic Publishers.

Trigeiro, W., Thomas, L. J., & McLain, J. O. (1989). Capacitated lot sizing with setup time. *Management Science*, *35*(3), 353–366. doi:10.1287/mnsc.35.3.353

Truyen, B., & Cornelis, J. (1995). Adiabatic layering: A new concept of hierarchical multi-scale optimization. *Neural Networks*, *8*(9), 1373–1378. doi:10.1016/0893-6080(95)00076-3

Tsai, H. C., Tyan, Y. Y., Wu, Y. W., & Lin, Y. H. (2013). Gravitational Particle Swarm. *Journal of Applied Mathematics and Computation*, *219*(17), 9106–9117. doi:10.1016/j.amc.2013.03.098

Tsai, Y., Wang, K., & Tsai, L. (2004). A study of availability - centered preventive maintenance for multi-component systems. *Reliability Engineering & System Safety*, *84*(3), 261–270. doi:10.1016/j.ress.2003.11.011

Tsarouhas, P. H., & Arvanitoyannis, I. S. (2010). Reliability and maintainability analysis of bread production line. *Critical Reviews in Food Science and Nutrition*, *50*(4), 327–343. doi:10.1080/10408390802544488 PMID:20301018

Tsoulos, I. G., & Vasant, P. (2009). Product mix selection using an evolutionary technique.*Proceedings of AIP Conference* (Vol. 1159, pp. 240 – 247). doi:10.1063/1.3223936

Tulpule, P. J., Marano, V., Yurkovich, S., & Rizzoni, G. (2013). Economic and environmental impacts of a PV powered workplace parking garage charging station. *Applied Energy*, *108*, 323–332. doi:10.1016/j.apenergy.2013.02.068

Türkay, B., Mecitoğlu, F., & Baran, S. (2011). Application of a Fast Evolutionary Algorithm to Short-term Hydrothermal Generation Scheduling. *Energy Sources, Part B: Economics, Planning, and Policy*, *6*(4), 395–405. doi:10.1080/15567249.2010.489098

Tzerpos, V. (2001). *Comprehension-driven software clustering*. University of Toronto.

Ubaidullah, S., Upadhyay, R., Raut, S. A. & Rahman, I. (2012). Optimization of Continuous Extraction Column and Solvent Selection Using Differential Evolution Technique. *Journal of The Institution of Engineers (India): Series E*, *93*(1), 49-54.

Ukkusuri, S. V., Mathew, T. V., & Travis Waller, S. (2007). Robust Transportation Network Design Under Demand Uncertainty. *Computer-Aided Civil and Infrastructure Engineering*, *22*(1), 6–18. doi:10.1111/j.1467-8667.2006.00465.x

University of Nottingham (UoN). (2014). *Lecture notes in space geodesy. Material retrieved from lectures in GNSS.* Surveying and Space Geodesy at the University of Nottingham.

US Army, & Us Marine Corps. (1997). *Field Manual (FM)101-5-1/Marine Corps Reference Publication (MCRP)5-2A, Operational Terms and Graphics.* Retrieved from http://fas.org/man/dod-101/army/docs/fm101-5-1/f545con.htm

US Army. (1997). *Army Container Operations, Field Manual (FM) 55-80.* Retrieved from http://archive.org/stream/ost-military-doctrine-fm55_80/fm55_80_djvu.txt

US Government Accounting Office (GAO) National Security and International Affairs Division. (1999). *NSIAD.* Washington, DC: Defense Inventory DOD Could Improve Total Asset Visibility Initiative With Results Act Framework.

Usher, J. S., Kamal, A. H., & Hashmi, S. W. (1998). Cost optimal preventive maintenance and replacement scheduling. *IIE Transactions*, *30*(12), 1121–1128. doi:10.1080/07408179808966568

Uzsoy, R., Lee, C., & Martin-Vega, L. (1992). A review of production planning and scheduling models in the semiconductor industry, Part I: System characteristics, performance evaluation and production planning. *IIE Transactions*, *24*(4), 47–60. doi:10.1080/07408179208964233

Uzsoy, R., Lee, C., & Martin-Vega, L. (1994). A review of prod planning and scheduling models in the semiconductor industry, Part II: Shop floor control. *IIE Transactions*, *26*(5), 44–55. doi:10.1080/07408179408966627

Valente, J. M. S. (2007). Heuristics for the single machine scheduling problem with early and quadratic tardy penalties. *European Journal of Industrial Engineering*, *1*(4), 431–448. doi:10.1504/EJIE.2007.015391

Valian, E., Mohanna, S., & Tavakoli, S. (2011). Improved cuckoo search algorithm for feedforward neural network training. *International Journal of Artificial Intelligence & Applications*, *2*(3), 36–43. doi:10.5121/ijaia.2011.2304

Van Bokhoven, W. (1980). A class of linear complementarity problems is solvable in polynomial time. Unpublished Paper, Dept. of electrical engineering, university of technology, the Netherlands.

Van Breedam, A. (1994). *An Analysis of the Behavior of Heuristics for the Vehicle Routing Problem for a Selection of Problems with Vehicle Related, Customer-related, and Timerelated Constraints.* (PhD thesis). University of Antwerp, Belgium.

Van Calsteren, M. R., Bissonnette, M. C., Cormier, F., Dufresne, C., Ichi, T., LeBlanc, J. C. Y., & Roewer, I. et al. (1997). Spectroscopic characterization of crocetin derivatives from Crocus sativus and Gardenia jasminoides. *Journal of Agricultural and Food Chemistry*, *45*(4), 55–1061. doi:10.1021/jf9603487

Van Der Merwe, R., Doucet, A., De Freitas, N., & Wan, E. (2000, August). *The unscented particle filter.* NIPS.

Van Dierendonck, A. (Ed.). (1997). *Satellite radio navigation, in avionic navigation sytems.* John Wiley & Sons.

Van Eijndhoven, J. (1986). Solving the linear complementarity problem in circuit simulation. *SIAM Journal on Control and Optimization*, *24*(5), 1050–1062. doi:10.1137/0324062

Vanderbeck, F. (1998). Lot-sizing with start-up times. *Management Science*, *44*(10), 1409–1425. doi:10.1287/mnsc.44.10.1409

Varga, R. S. (2009). *Matrix iterative analysis.* Springer Science & Business Media.

Vasant, P. (2011). Solving Fuzzy Optimization Problems of Uncertain Technological Coefficients with Genetic Algorithms and Hybrid Genetic. In P. Vasant (Ed.), Innovation in Power, Control, and Optimization: Emerging Energy Technologies (pp. 344-368). IGI Global.

Vasant, P. (2012). Hybrid Linear Search, Genetic Algorithms, and Simulated Annealing for Fuzzy Non-Linear Industrial Production Planning Problems. In P. Vasant (Ed.), Meta-Heuristics Optimization Algorithms in Engineering, Business, Economics and Finance (pp. 87-109). IGI Global.

Vasant, P. (2014). Hybrid Optimization Techniques for Industrial ProductionPlanning: A Review. In P. Vasant (Ed.), Handbook of Research on Novel Soft Computing Intelligent Algorithms: Theory and Practical Applications (pp.1-40), IGI Global.

Vasant, P., Ganesan, T., & Elamvazuthi, I. (2012). An improved PSO approach for solving non-convex optimization problems. In *ICT and Knowledge Engineering (ICT & Knowledge Engineering), 2011 9th International Conference on* (pp. 80-87). IEEE. doi:10.1109/ICTKE.2012.6152418

Vasant, P. (2010). Hybrid simulated annealing and genetic algorithms for industrial production management problems. *International Journal of Computational Methods*, *7*(02), 279–297. doi:10.1142/S0219876210002209

Vasant, P. (2011). *Solving fuzzy optimization problems of uncertain technological coefficients with genetic algorithms and hybrid genetic algorithms pattern search approaches. In Innovation in Power, Control, and Optimization: Emerging Energy Technologies* (pp. 344–368). IGI Global.

Vasant, P. (2012). *Hybrid linear search, genetic algorithms, and simulated annealing for fuzzy nonlinear industrial production planning problems. In Meta-Heuristics Optimization Algorithms in Engineering, Business, Economics, and Finance* (pp. 87–109). IGI Global.

Vasant, P. (2014). *Hybrid Optimization Techniques for Industrial Production Planning: A Review. In Handbook of Research on Novel Soft Computing Intelligent Algorithms: Theory and Practical Applications* (pp. 41–68). IGI Global.

Vasant, P. (2015). *Handbook of Research on Artificial Intelligence Techniques and Algorithms*. Hershey, PA: IGI Global; doi:10.4018/978-1-4666-7258-1

Vasant, P. M. (2013). *Meta-Heuristics Optimization Algorithms in Engineering, Business, Economics, and Finance*. Hershey, PA: IGI Global; doi:10.4018/978-1-4666-2086-5

Vasant, P. M. (2014). *Handbook of Research on Novel Soft Computing Intelligent Algorithms: Theory and Practical Applications*. Hershey, PA: IGI Global; doi:10.4018/978-1-4666-4450-2

Vasant, P., & Barsoum, N. (2009). Hybrid genetic algorithms and line search method for industrial production planning with non-linear fitness function. *Engineering Applications of Artificial Intelligence*, *22*(4), 767–777. doi:10.1016/j.engappai.2009.03.010

Vasant, P., & Barsoum, N. (2010). Hybrid pattern search and simulated annealing for fuzzy production planning problems. *Computers & Mathematics with Applications (Oxford, England)*, *60*(4), 1058–1067. doi:10.1016/j.camwa.2010.03.063

Vasant, P., Barsoum, N., & Webb, J. (2012). *Innovation in Power, Control, and Optimization: Emerging Energy Technologies*. Hershey, PA: IGI Global; doi:10.4018/978-1-61350-138-2

Vasant, P., Elamvazuthi, I., Ganesan, T., & Webb, J. F. (2010). Iterative fuzzy optimization approach for crude oil refinery industry. *Scientific Annals of Computer Science*, *8*(2), 261–280.

Veldhuizen, D. A. V., & Lamont, G. B. (1998). *Multiobjective evolutionary algorithm research: A history and analysis.* Technical Report TR-98-03. Department of Electrical and Computer Engineering, Graduate School of Engineering, Air Force Institute of Technology, Wright-Patterson AFB, Ohio.

Vella, D. (2015). *Supply Chain Management: A Needed and Growing Opportunity.*http://www.inboundlogistics.com/cms/article/supply-chain-management-a-needed-and-growing-opportunity/

Venables, W. N., & Ripley, B. D. (1994). *Modern Applied Statistics with S-Plus.* New York: Springer-Verlag. doi:10.1007/978-1-4899-2819-1

Venayagamoorthy, G. K., & Mitra, P. (2011). SmartPark shock absorbers for wind farms. *Energy Conversion. IEEE Transactions on, 26*(3), 990–992.

Vermaak, J., Lawrence, N. D., & Pérez, P. (2003, June). Variational inference for visual tracking. In *Computer Vision and Pattern Recognition, 2003. Proceedings. 2003 IEEE Computer Society Conference on* (Vol. 1, pp. I-773). IEEE. doi:10.1109/CVPR.2003.1211431

Vicente, L., Savard, G., & Júdice, J. (1994). Descent approaches for quadratic bilevel programming. *Journal of Optimization Theory and Applications, 81*(2), 379–399. doi:10.1007/BF02191670

Vidal, T., Crainic, T. G., Gendreau, M., Lahrichi, N., & Rei, W. (2012). A hybrid genetic algorithm for multi-depot and periodic vehicle routing problems. *Operations Research, 60*(3), 611–624. doi:10.1287/opre.1120.1048

Villalobos, M., & Trejos, J. (2000a). Metric multidimensional scaling using tabu search[in Spanish]. *Revista de Matemática: Teoría y Aplicaciones, 7*(1-2), 71–76.

Villalobos, M., & Trejos, J. (2000b). Simulated annealing optimization in nonlinear regression: Algorithm and software[in Spanish]. *Investigación Operacional, 21*(3), 236–246.

Villalobos, M., Trejos, J., & de los Cobos, S. (2006). Application of tabu search in non linear regression[in Spanish]. *Revista de Matemática: Teoría y Aplicaciones, 13*(1), 79–92.

Viswanathan, G. M., Afanasyev, V., Buldyrev, S. V., Havlin, S., Luz, M., Raposo, E. P., & Stanley, H. E. (2000). Lévy flights in random searches. *Physica A: Statistical Mechanics and its Applications, 282*, 1-12.

Vitanov, V., Julkaa, N., Bainesb, T., Tjahjonob, B., & Lendermanna, P. (2007). A review of multi-factor capacity expansion models for manufacturing plants: Searching for a holistic decisions aid. *International Journal of Production Economics, 106*(2), 607–621. doi:10.1016/j.ijpe.2006.08.002

Von Frisch, K. (1955). *The dancing bees: An Account of the Life and Senses of the Honey Bee.* New York: Harcourt.

Vora, M., & Mirnalinee, T. (2015). From Optimization to Clustering: A Swarm Intelligence Approach. In P. Vasant (Ed.), Handbook of Research on Artificial Intelligence Techniques and Algorithms (pp. 594-619). IGI Global doi:10.4018/978-1-4666-7258-1.ch019

Voudouris, V. T., & Consulting, A. (1996). Mathematical programming techniques to debottleneck the supply chain of fine chemical industries. *Computers & Chemical Engineering, 20*, 1269–1274. doi:10.1016/0098-1354(96)00219-0

Wagner, H. M., & Whitin, T. M. (1985). A dynamic version of the economic lot size model. *Management Science, 5*(1), 89–96. doi:10.1287/mnsc.5.1.89

Walton, S., Hassan, O., Morgan, K., & Brown, M. R. (2011). Modified cuckoo search: A new gradient free optimisation algorithm. *Chaos, Solitons, and Fractals, 44*(9), 710–718. doi:10.1016/j.chaos.2011.06.004

Wan, Z., Wang, G., & Hou, K. (2008, October). An interactive fuzzy decision making method for a class of bilevel programming. In *Fuzzy Systems and Knowledge Discovery, 2008. FSKD'08. Fifth International Conference on* (Vol. 1, pp. 559-564). IEEE. doi:10.1109/FSKD.2008.537

Wang, X., Dai, H., & Thomas, R.J. (1984). Reliability modeling of large wind farms and associated electric utility interface system. *IEEE Transactions, 23*(3), 569-575.

Wang, Y. (2012). A hybrid approach based on ant colony system for the VRPTW. In Wu, Y. (Ed.), *Advanced Technology in Teaching,Proceedings of WTCS 2009, Advances in Intelligent and Soft Computing* (*vol. 117*, pp. 327-333). Springer. doi:10.1007/978-3-642-25437-6_46

Wang, Y. R., Watada, J., Yaakob, S. B., & Choi, J. (2010). Optimal Supply Balance of Power System Based on Sustainable Generators by Reliability Cost/Worth Method. In *Proceedings, Czeck-Japan Seminar.*

Wang. (1997). *A Course in Fuzzy Systems and Control.* Prentice-Hall Inc.

Wang, B., Li, Y., & Watada, J. (2010). Fuzzy power system reliability model based on Value-at-Risk. In *Proceedings of 14th International Conference on Knowledge-Based and Intelligent Information and Engineering Systems*, (pp. 445-453). doi:10.1007/978-3-642-15390-7_46

Wang, C. S., & Uzsoy, R. (2002). A genetic algorithm to minimize maximum lateness on a batch processing machine. *Computers & Operations Research, 29*(12), 1621–1640. doi:10.1016/S0305-0548(01)00031-4

Wang, G., Jianq, B., Zhu, K., & Wan, Z. (2010). Global convergent algorithm for the bilevel linear fractional-linear programming based on modified convex simplex method. *Systems Engineering and Electronics. Journalism, 21*(2), 239–243.

Wang, G.-Q., & Bai, Y.-Q. (2012). A class of polynomial interior point algorithms for the Cartesian P-matrix linear complementarity problem over symmetric cones. *Journal of Optimization Theory and Applications, 152*(3), 739–772. doi:10.1007/s10957-011-9938-8

Wang, G., Wan, Z., Wang, X., & Lv, Y. (2008). Genetic algorithm based on simplex method for solving linear-quadratic bilevel programming problem. *Computers & Mathematics with Applications (Oxford, England), 56*(10), 2550–2555. doi:10.1016/j.camwa.2008.05.006

Wang, G., Yu, C., & Teo, K. (2014). A full-Newton step feasible interior-point algorithm for P_*(\ kappa)-linear complementarity problems. *Journal of Global Optimization, 59*(1), 81–99. doi:10.1007/s10898-013-0090-x

Wang, H. (2002). A survey of maintenance policies of deteriorating systems. *European Journal of Operational Research, 139*(3), 469–489. doi:10.1016/S0377-2217(01)00197-7

Wang, H. M. (2011). Solving single batch-processing machine problems using an iterated heuristic. *International Journal of Production Research, 49*(14), 4245–4261. doi:10.1080/00207543.2010.518995

Wang, H. M., Chang, P. C., & Chou, F. D. (2007). A hybrid forward/backward approach for single batch scheduling problems with non-identical job sizes. *Journal of the Chinese Institute of Industrial Engineers, 24*(3), 191–199. doi:10.1080/10170660709509033

Wang, H., & Pham, H. (1999). Some maintenance models and availability with imperfect maintenance in production systems. *Annals of Operations Research, 91*, 305–318. doi:10.1023/A:1018910109348

Wang, J., Jin, S., Qin, S., & Jiang, H. (2014). Swarm Intelligence-Based Hybrid Models for Short-Term Power Load Prediction. *Mathematical Problems in Engineering, Article ID 712417.* doi:10.1155/2014/712417

Wang, L., Groves, P., & Ziebart, M. K. (2013). GNSS shadow matching: Improving urban positioning accuracy using a 3D city model with optimized visibility scoring scheme. *Journal of Navigation*, *60*(3), 195–207. doi:10.1002/navi.38

Wang, L., & Jiang, T. (1994). On the complexity of multiple sequence alignment. *Journal of Computational Biology*, *1*(4), 337–348. doi:10.1089/cmb.1994.1.337 PMID:8790475

Wang, L., & Lin, S. (2009). A multi-agent based agile manufacturing planning and control system. *Computers & Industrial Engineering*, *57*(2), 620–640. doi:10.1016/j.cie.2009.05.015

Want, R., Hopper, A., Falcao, V., & Gibbons, J. (1992). The active badge location system. *ACM Transactions on Information Systems*, *10*(1), 91–102. doi:10.1145/128756.128759

Wan, Z., Wang, G., & Sun, B. (2013). A hybrid intelligent algorithm by combining particle swarm optimization with chaos searching technique for solving nonlinear bilevel programming problems. *Swarm and Evolutionary Computation*, *8*, 26–32. doi:10.1016/j.swevo.2012.08.001

Watada, J., Yaakob, S. B., Takahashi, T., & Okamoto, T. (2010). The Diagnosis of Power Transformer Failures by Fuzzy Random Based Rough Sets Analysis. In *Proceedings, International Conference on Condition Monitoring and Diagnosis 2010 (CMD2010).*

Watson, L. T. (1978). An algorithm for the linear complementarity problem. *International Journal of Computer Mathematics*, *6*(4), 319–325. doi:10.1080/00207167808803148

Wei, J., Takahashi, T., Okamoto, T., Tanaka, T., & Watada, J. (2006). Mean-Variance Analysis for Replacing Unreliable Units in a Power Systems. In *Proceedings of CMD2006: International Conference on Condition Monitoring and Diagnosis*.

Weinbach, U., Raziq, N., & Collier, P. (2009). Mitigation of periodic GPS multipath errors using a normalised least mean square adaptive filter. *Journal of Spatial Science*, *54*(1), 1–13. doi:10.1080/14498596.2009.9635162

Weinstein, L., & Chung, C. (1999). Integrating maintenance and production decisions in a hierarchical production planning environment. *Computers & Operations Research*, *26*(10-11), 1059–1074. doi:10.1016/S0305-0548(99)00022-2

Wen, Z., & Tzerpos, V. (2005). *Software clustering based on omnipresent object detection*. Paper presented at the Program Comprehension.

Weng, W. T., & Wen, U. P. (2000). A primal-dual interior point algorithm for solving bilevel programming problem. *Asia-Pacific Journal of Operational Research*, *17*(2), 213.

Wiggerts, T. A. (1997). *Using clustering algorithms in legacy systems remodularization*. Paper presented at the Reverse Engineering. doi:10.1109/WCRE.1997.624574

Wilck, I. V. J. IV, & Cavalier, T. (2012). A construction heuristic for the split delivery vehicle routing. *American Journal of Operations Research*, *2*(02), 153–162. doi:10.4236/ajor.2012.22018

Wolpert, D. H., & Macready, W. G. (1997). No free lunch theorems for optimization. *IEEE Transactions on Evolutionary Computation*, *1*(1), 67–82. doi:10.1109/4235.585893

Wong, K. P., & Wong, Y. W. (1994). Short-term hydrothermal scheduling, part-I: Simulated annealing approach. *IEE Proceedings. Generation, Transmission and Distribution*, *141*(5), 497–501. doi:10.1049/ip-gtd:19941350

Wong, T. N., Zhang, S., Wang, G., & Zhang, L. (2012). Integrated process planning and scheduling – multi-agent system with two-stage ant colony optimisation algorithm. *International Journal of Production Research*, *50*(21), 6188–6201. doi:10.1080/00207543.2012.720393

Wood, A. (1989). Availability calculations with exhaustible spares. *IEEE Transactions on Reliability*, *38*(3), 388–391. doi:10.1109/24.44189

Wood, A. J., & Wollenberg, B. F. (1996). *Power generation, operation and control*. John Wiley & Sons.

Wooldridge, J. M. (Ed.). (2010). *Econometric analysis of cross section and panel data*. MIT Press.

Wu, A., Yu, H., Jin, S., Lin, K., & Schiavone, G. (2004). An incremental genetic algorithm approach to multiprocessor scheduling. *IEEE Transactions on Parallel and Distributed Systems*, *15*(9), 824–834. doi:10.1109/TPDS.2004.38

Wu, B., Lin, J., & Dong, M. (2013). Artificial Bee Colony Algorithm for three-dimensional loading capacitated vehicle routing problem. In E. Qi, J. Shen, & R. Dou (Eds.), *Proceedings of 20th International Conference on Industrial Engineering and Engineering Management* (pp. 815-825). Springer. doi:10.1007/978-3-642-40063-6_81

www.mathworks.in/products/matlab/

Xiadong, Y., Fernandez-G, E., & Fu, M. (2004). Optimal preventive maintenance scheduling in semiconductor manufacturing. *Semiconductor Manufacturing. IEEE Transactions on*, *17*(3), 345–356.

Xiang, W., Ma, S., & An, M. (2014). HABCDE: A hybrid evolutionary algorithm based on artificial bee colony algorithm and differential evolution. *Applied Mathematics and Computation*, *238*, 370–386. doi:10.1016/j.amc.2014.03.055

Xianming, Q., & Ning, Y. (2004). *A folded dipole antenna for RFID*. Paper presented at the Antennas and Propagation Society International Symposium, 2004. IEEE.

Xiaohui, Y., Bo, C., Bo, Y., & Yanbin, Y. (2008). Hydrothermal scheduling using chaotic hybrid differential evolution. *Energy Conversion and Management*, *49*(12), 3627–3633. doi:10.1016/j.enconman.2008.07.008

Xiao, J., Shen, Y., Ge, J., Tateishi, R., Tang, C., Liang, Y., & Huang, Z. (2006). Evaluating urban expansion and land use change in Shijiazhuang, China, by using GIS and remote sensing. *Landscape and Urban Planning*, *75*(12), 69–80. doi:10.1016/j.landurbplan.2004.12.005

Xie, S., Sun, Z., & Zeng, Y. (2013). A two-level domain decomposition algorithm for linear complementarity problem. *Journal of Inequalities and Applications*, (1): 1–9.

Xue, J. L., & Watada, J. (2010). Short-term Power Load Forecasting Method by Radial-basis-function Neural Net-work with Support Vector Machine Model. *ICIC Express Letters*, *5*(6B).

Xu, M., & Luan, G. (2007). A rapid algorithm for a class of linear complementarity problems. *Applied Mathematics and Computation*, *188*(2), 1647–1655. doi:10.1016/j.amc.2006.11.184

Xu, W. (2015). Modified modulus-based matrix splitting iteration methods for linear complementarity problems. *Numerical Linear Algebra with Applications*, *22*(4), 748–760. doi:10.1002/nla.1985

Xu, W.-, & Liu, H. (2014). A modified general modulus-based matrix splitting method for linear complementarity problems of H-matrices. *Linear Algebra and Its Applications*, *458*, 626–637. doi:10.1016/j.laa.2014.06.022

Xu, X., Hung, P.-F., & Ye, Y. (1996). A simplified homogeneous and self-dual linear programming algorithm and its implementation. *Annals of Operations Research*, *62*(1), 151–171. doi:10.1007/BF02206815

Xu, Z. S. (2007). Some similarity measure of intuitionistic fuzzy sets and their applications to multi attribute decision making. *Fuzzy Optimization and Decision Making*, *6*(2), 109–121. doi:10.1007/s10700-007-9004-z

Yaakob, S. B., Watada, J., Tsuguhiro, T., & Okamoto, T. (2010). A Hybrid Neural Network Approach for Solving Bi-level Programming Problems in a Power System Environment. In *Proceedings of CMD2010: International Conference on Condition Monitoring and Diagnosis 2010.*

Yaakob, S. B., Takahashi, T., Okamoto, T., Tanaka, T., Minh, T. D., Watada, J., & Zhu, X. J. (2008). Structural learning approach to replacing unreliable units in a power system, In Proceedings of International Conference on Condition Monitoring and Diagnosis. *CMD. Current Medical Dialog*, *2008*, 570–575.

Yaakob, S. B., & Watada, J. (2010). Power System Equipment Investment Decision-Making under U- certainty: A Real Options Approach. *Advances in Intelligent Decision Technologies*, *14*, 699–708. doi:10.1007/978-3-642-14616-9_68

Yaakob, S. B., Watada, J., & Liu, S. W. (2010). A Hybrid Neural Network Approach for Solving Bi-level Programming Problems and its Application in Power System. *In Proceedings of ISME2010.*

Yaghoobi, H., & Torabi, M. (2012). Novel solution for acceleration motion of a vertically falling non-spherical particle by VIM–Padé approximant. *Powder Technology*, *215-216*, 206–209. doi:10.1016/j.powtec.2011.09.049

Yalaoui, A., Chaabi, K., & Yalaoui, F. (2014). Integrated production planning and preventive maintenance in deteriorating production systems. *Information Sciences*, *278*, 841–861. doi:10.1016/j.ins.2014.03.097

Yang, H. & Bell, M.G. (1997). Traffic restraint, road pricing and network equilibrium. *Transportation Research, 31*(B), 303-314.

Yang, X. S., & Deb, S. (2009). *Cuckoo Search via Levy flights.* Paper presented at the World Congress on Nature & Biologically Inspired Computing (NaBIC 2009) doi:10.1109/NABIC.2009.5393690

Yang, X. S., & Deb, S. (2009, December). Cuckoo search via Lévy flights. In *Nature & Biologically Inspired Computing, 2009. NaBIC 2009. World Congress on* (pp. 210-214). IEEE.

Yang, B. S., Yeun, Y.-S., & Ruy, W.-S. (2002). Managing approximation models in multiobjective optimization. *Structural Multidiscsiplinary Optimization*, *24*(2), 141–156. doi:10.1007/s00158-002-0224-0

Yang, H., & Bell, M. G. (1998). Models and algorithms for road network design: A review and some new developments. *Transport Reviews*, *18*(3), 25–278. doi:10.1080/01441649808717016

Yang, H., & Li, Q. (2012). Overlapping restricted additive Schwarz method applied to the linear complementarity problem with an H-matrix. *Computational Optimization and Applications*, *51*(1), 223–239. doi:10.1007/s10589-010-9334-6

Yang, H., Li, Q., & Xu, H. (2009). A multiplicative Schwarz iteration scheme for solving the linear complementarity problem with an H-matrix. *Linear Algebra and Its Applications*, *430*(4), 1085–1098. doi:10.1016/j.laa.2008.10.005

Yang, L., & Iwamura, K. (2008). Fuzzy chance-constrained programming with linear combination of possibility measure and necessity measure. *Applied Mathematical Sciences*, *2*, 2271–2278.

Yang, P. C., Yang, H. T., & Huang, C. L. (1996). Scheduling short-term hydrothermal generation using evolutionary programming technique. *IEE Proceedings. Generation, Transmission and Distribution*, *143*(4), 371–276. doi:10.1049/ip-gtd:19960463

Yang, T., & Shen, Y. (2012). Heuristic algorithms for a practical-size dynamic parallel-machine scheduling problem: Integrated-circuit wire bonding. *Production Planning and Control*, *23*(1), 67–78. doi:10.1080/09537287.2011.572090

Yang, X. (2014). Diversity and Mechanisms in Swarm Intelligence.[IJSIR]. *International Journal of Swarm Intelligence Research*, *5*(2), 1–12. doi:10.4018/ijsir.2014040101

Yang, X. S. (2010). *Nature-Inspired Metaheuristic Algorithms* (2nd ed.). Luniver Press.

Yang, X. S. (2010a). Firefly algorithm. *Engineering Optimization*, 221–230.

Yang, X. S. (2010b). Firefly algorithm, Levy flights and global optimization. In *Research and Development in Intelligent Systems XXVI* (pp. 209–218). Springer London. doi:10.1007/978-1-84882-983-1_15

Yang, X. S., & Deb, S. (2010). Engineering optimisation by cuckoo search. *International Journal of Mathematical Modelling and Numerical Optimisation*, *1*(4), 330–343. doi:10.1504/IJMMNO.2010.035430

Yang, X., & Deb, S. (2015). Cuckoo Search for Optimization and Computational Intelligence. In M. Khosrow-Pour (Ed.), *Encyclopedia of Information Science and Technology* (3rd ed.; pp. 133–142). Hershey, PA: Information Science Reference; doi:10.4018/978-1-4666-5888-2.ch014

Yang, X.-S. (2010a). Firefly algorithm, stochastic test functions and design optimisation. *International Journal of Bio-inspired Computation*, *2*(2), 78–84. doi:10.1504/IJBIC.2010.032124

Yang, X.-S. (2010b). *A new metaheuristic bat-inspired algorithm Nature inspired cooperative strategies for optimization (NICSO 2010)*. Springer. doi:10.1007/978-3-642-12538-6_6

Yang, X.-S. (2011). Bat algorithm for multi-objective optimisation. *International Journal of Bio-inspired Computation*, *3*(5), 267–274. doi:10.1504/IJBIC.2011.042259

Yan, X., Zhu, Y., Zou, W., & Wang, L. (2012). A new approach for data clustering using hybrid artificial bee colony algorithm. *Neurocomputing*, *97*, 241–250. doi:10.1016/j.neucom.2012.04.025

Yao, X., Fernandez-Gaucherand, E., Fu, M. C., & Marcus, S. I. (2004). Optimal preventive maintenance scheduling in semiconductor manufacturing, Semiconductor Manufacturing. *IEEE Transactions on*, *17*(3), 345–356.

Yazdani, S., & Nezamabadi-pour, H. (2011). A new Gravitational Solution for Multimodal Optimization. In *Proceeding of the 18th Iranian Conference on Electrical Engineering* (ICEE20111).

Yazdani, S., Nezamabadi-pour, H., & Kamyab, S. (2014). A Gravitational Search Algorithm for Multimodal Optimization. *Journal of Swarm and Evolutionary Computation*, *14*, 1–14. doi:10.1016/j.swevo.2013.08.001

Yeh, C. H. (2002). A problem based selection of multi attribute decision making methods. *International Transactions in Operational Research*, *9*(2), 169–181. doi:10.1111/1475-3995.00348

Yeh, W. C., & Chuang, M. C. (2011). Using multi-objective genetic algorithm for partner selection in green supply chain problems. *Expert Systems with Applications*, *38*(4), 4244–4253. doi:10.1016/j.eswa.2010.09.091

Ye, Y. (1997). On homogeneous and self-dual algorithms for LCP. *Mathematical Programming*, *76*(1), 211–221. doi:10.1007/BF02614384

Yildirim, I., Tan, B., & Karaesmen, F. (2005). A multiperiod stochastic production planning and sourcing problem with service level constraints. *OR-Spektrum*, *27*(2-3), 471–489. doi:10.1007/s00291-005-0203-0

Yildiz, A. (2013). Cuckoo search algorithm for the selection of optimal machining parameters in milling operations. *International Journal of Advanced Manufacturing Technology*, *64*(1-4), 55–61. doi:10.1007/s00170-012-4013-7

Yilmaz, M., & Krein, P. T. (2012). Review of charging power levels and infrastructure for plug-in electric and hybrid vehicles. In *Electric Vehicle Conference (IEVC), 2012 IEEE International*. doi:10.1109/IEVC.2012.6183208

Yin, K., Yin, G., & Liu, H. (2004). Stochastic modeling for inventory and production planning in the paper industry. *AIChE Journal. American Institute of Chemical Engineers*, *50*(11), 2877–2890. doi:10.1002/aic.10251

Yin, M., Hu, Y., Yang, F., Li, X., & Gu, W. (2011). A Novel Hybrid K-harmonic Means and Gravitational Search Algorithm Approach for Clustering. *Journal of Expert Systems with Applications*, *38*(8), 9319–9324. doi:10.1016/j.eswa.2011.01.018

Yin, P. Y. (2013). *Trends in Developing Metaheuristics, Algorithms, and Optimization Approaches*. Hershey, PA: IGI Global; doi:10.4018/978-1-4666-2145-9

Young, N., & Stonier, R. J. (2003). Co-evolutionary learning and hierarchical fuzzy control for the inverted pendulum. In *Proceedings of the International Congress on Evolutionary Computation (CEC2003)*.

Yousefi, M., & Yusuff, R. M. (2013). Minimising earliness and tardiness penalties in single machine scheduling against common due date using imperialist competitive algorithm. *International Journal of Production Research*, *51*(16), 4797–4804. doi:10.1080/00207543.2013.774475

Yu, K., Saarnisaari, H., Montillet, J.-P., Rabbachin, A., Oppermann, I., & Thadeu Freitas de Abreu, G. (2006). Localization. In *Ultra-wideband wireless communications and networks*. John Wiley & Sons.

Yuan, D., & Song, Y. (2003). Modified AOR methods for linear complementarity problem. *Applied Mathematics and Computation*, *140*(1), 53–67. doi:10.1016/S0096-3003(02)00194-7

Yuan, X., & Yuan, Y. (2006). Application of cultural algorithm to generation scheduling of hydrothermal systems. *Energy Conversion and Management*, *47*(15-16), 2192–2201. doi:10.1016/j.enconman.2005.12.006

Yuce, B. (2012). *Novel Computational Technique for Determining DepthUsing Bees Algorithmand Blind Image Deconvolution*. (PhD Thesis). Cardiff University.

Yuce, B., Mastrocinque, E., Lambiase, A., Packianather, M. S., & Pham, D. T. (2014). A multi-objective supply chain optimization using enhanced Bees Algorithm with adaptive neighborhood search and site abandonment strategy. *Swarm and Evolutionary Computation*, *18*, 71–82. doi:10.1016/j.swevo.2014.04.002

Yuce, B., Mastrocinque, E., Packianather, M. S., Lambiase, A., & Pham, D. T. (2015). The Bees Algorithm and Its Applications. In P. Vasant (Ed.), *Handbook of Research on Artificial Intelligence Techniques and Algorithms* (pp. 122–151). Hershey, PA: Information Science Reference; doi:10.4018/978-1-4666-7258-1.ch004

Yuce, B., Packianather, M. S., Mastrocinque, E., Pham, D. T., & Lambiase, A. (2013). Honey bees inspired optimization method: The Bees Algorithm. *Insects*, *4*(4), 646–662. doi:10.3390/insects4040646 PMID:26462528

Yuce, B., Pham, D. T., Packianather, M. S., & Mastrocinque, E. (2015a). An enhancement to the Bees Algorithm with slope angle computation and Hill Climbing Algorithm and its applications on scheduling and continuous-type optimization problem. *Production & Manufacturing Research*, *3*(1), 3–19. doi:10.1080/21693277.2014.976321

Yu, K., & Dutkiewicz, E. (2012). Geometry and motion based positioning algorithms for mobile tracking in NLOS environments. *IEEE Transactions on Mobile Computing*, *11*(2), 254–263. doi:10.1109/TMC.2011.24

Yu, K., Sharp, I., & Guo, J. Y. (2009). *Ground-based wireless positioning*. Wiley-IEEE Press. doi:10.1002/9780470747988

Yu, Q., Huang, C., & Wang, X. (2006). A combined homotopy interior point method for the linear complementarity problem. *Applied Mathematics and Computation*, *179*(2), 696–701. doi:10.1016/j.amc.2005.12.010

Yu, Y., Chen, H., & Chu, F. (2008). A new model and hybrid approach for large scale inventory routing problems. *European Journal of Operational Research*, *189*(3), 1022–1040. doi:10.1016/j.ejor.2007.02.061

Yüzgeç, U., Türker, M., & Hocalar, A. (2009). On-line evolutionary optimization of an industrial fed-batch yeast fermentation process. *ISA Transactions*, *48*(1), 79–92. doi:10.1016/j.isatra.2008.09.001 PMID:18849027

Zadeh, L. A. (1965). Fuzzy sets. *Information and Control*, *8*(3), 338–353. doi:10.1016/S0019-9958(65)90241-X

Zadeh, L. A. (1978). Fuzzy sets as a basis for a theory of possibility. *Fuzzy Sets and Systems*, *1*(1), 3–28. doi:10.1016/0165-0114(78)90029-5

Zaghlool, M. F., & Trutt, F. C. (1988). Efficient methods for optimal scheduling of ﬁxed head hydrothermal power systems. *IEEE Transactions on Power Systems*, *3*(1), 24–30. doi:10.1109/59.43176

Zagrouba, M., Sellami, A., Bouaıcha, M., & Ksouri, M. (2010). Identification of PV solar cells and modules parameters using the genetic algorithms: Application to maximum power extraction. *Solar Energy*, *84*(5), 860–866. doi:10.1016/j.solener.2010.02.012

Zainuddin, Z., Wan Daud, W. R., Pauline, O., & Shafie, A. (2011). Wavelet neural networks applied to pulping of oil palm fronds. *Bioresource Technology*, *102*(23), 10978–10986. doi:10.1016/j.biortech.2011.09.080 PMID:21996481

Zanjania, M., Nourelfatha, M., & Kadia, D. (2010). A multi-stage stochastic programming approach for production planning with uncertainty in the quality of raw materials and demand. *International Journal of Production Research*, *48*(16).

Zeng, M.-L., & Zhang, G.-F. (2015). Modulus-based GSTS Iteration Method for Linear Complementarity Problems. *J. Math. Study*, *48*(1), 1–17. doi:10.4208/jms.v48n1.15.01

Zhang, X.-P., & Schaffner, C. S. (Eds.). (2010). *Minutes of the Working Group meeting on Test Systems for Economic Analysis, 2007, 2008, 2009 and 2010*. Available at http://www.ieee.tea.ethz.ch/wiki/WG_meetings

Zhang, Z., Li, Y., Xia, Y., & Luo, F. (2012). Immunity-based Gravitational Search Algorithm. In *Proceeding of 3rd International Conference on Information Computing and Applications*.

Zhang, C., Zhang, J., Hao, Y., Lin, Z., & Zhu, C. (2011). A simple and efficient solar cell parameter extraction method from a single current-voltage curve. *Journal of Applied Physics*, *110*(6), 064504. doi:10.1063/1.3632971

Zhang, G. (2011). Quantum-inspired evolutionary algorithms: A survey and empirical study. *Journal of Heuristics*, *17*(3), 303–351. doi:10.1007/s10732-010-9136-0

Zhang, G., Lu, J., & Xiang, Q. (2008). Application of Genetic Algorithm to Network Design Problem.*2008 International Conference on Intelligent Computation Technology and Automation (ICICTA) 1*, (pp. 26-29). doi:10.1109/ICICTA.2008.252

Zhang, J., Lam, W. H. K., & Chen, B. Y. (2013). A stochastic vehicle routing problem with travel time uncertainty: Trade-off between cost and customer service. *Networks and Spatial Economics*, *13*(4), 471–496. doi:10.1007/s11067-013-9190-x

Zhang, L.-L. (2014). Two-stage multisplitting iteration methods using modulus-based matrix splitting as inner iteration for linear complementarity problems. *Journal of Optimization Theory and Applications*, *160*(1), 189–203. doi:10.1007/s10957-013-0362-0

Zhang, L.-L., & Ren, Z.-R. (2013). Improved convergence theorems of modulus-based matrix splitting iteration methods for linear complementarity problems. *Applied Mathematics Letters*, *26*(6), 638–642. doi:10.1016/j.aml.2013.01.001

Zhang, L.-T., & Li, J.-L. (2014). The weaker convergence of modulus-based synchronous multisplitting multi-parameters methods for linear complementarity problems. *Computers & Mathematics with Applications (Oxford, England)*, *67*(10), 1954–1959. doi:10.1016/j.camwa.2014.04.018

Zhang, L.-T., Zhang, Y.-X., Gu, T.-X., Liu, X.-P., & Zhang, L.-W. (2015). New convergence of modulus-based synchronous block multisplitting multi-parameter methods for linear complementarity problems. *Computational & Applied Mathematics*, 1–12.

Zhang, L.-T., Zuo, X.-Y., Gu, T.-X., & Liu, X.-P. (2014). Improved convergence theorems of multisplitting methods for the linear complementarity problem. *Applied Mathematics and Computation*, *243*, 982–987. doi:10.1016/j.amc.2014.06.038

Zhang, Q., & Li, H. (2007). MOEA/D: A multiobjective evolutionary algorithm based on decomposition. *IEEE Transactions on Evolutionary Computation*, *11*(6), 712–731. doi:10.1109/TEVC.2007.892759

Zhang, Q., Tezuka, T., Ishihara, K. N., & Mclellan, B. C. (2012). Integration of PV power into future low-carbon smart electricity systems with EV and HP in Kansai Area, Japan. *Renewable Energy*, *44*, 99–108. doi:10.1016/j.renene.2012.01.003

Zhang, X., Yue, G., & Teng, S. Z. (2009). Possibility degree of interval-valued fuzzy numbers and its application. *International Symposium on Information Processing.*

Zhang, Y., Li, R., & Tsai, C. L. (2010). Regularization parameter selections via generalized information criterion. *Journal of the American Statistical Association*, *105*(489), 312–323. doi:10.1198/jasa.2009.tm08013 PMID:20676354

Zhang, Y., Wang, S., Ji, G., & Dong, Z. (2013). Genetic Pattern Search and Its Application to Brain Image Classification. *Mathematical Problems in Engineering*. doi:10.1155/2013/580876

Zhang, Y., Wu, L., & Wang, S. (2013). Solving Two-Dimensional HP Model by Firefly Algorithm and Simplified Energy Function. *TheScientificWorldJournal*. doi:10.1155/2013/398141

Zhao, S. Z., & Suganthan, P. N. (2010). Multi-objective evolutionary algorithm with ensemble of external archives. *International Journal of Innovative Computing, Information, & Control*, *6*(1), 1713–1726.

Zheng, H., & Zhou, Y. (2012). A novel cuckoo search optimization algorithm based on Gauss distribution. *Journal of Computer Information Systems*, *8*(10), 4193–4200.

Zheng, N., & Yin, J. (2013). Modulus-based successive overrelaxation method for pricing American options. *Appl Math Informatics*, *31*(5), 769–784. doi:10.14317/jami.2013.769

Zheng, N., & Yin, J.-F. (2013). Accelerated modulus-based matrix splitting iteration methods for linear complementarity problem. *Numerical Algorithms*, *64*(2), 245–262. doi:10.1007/s11075-012-9664-9

Zheng, N., & Yin, J.-F. (2014). Convergence of accelerated modulus-based matrix splitting iteration methods for linear complementarity problem with an H+-matrix. *Journal of Computational and Applied Mathematics*, *260*, 281–293.

Zhongbao, W., Shaojun, F., Shiqiang, F., & Shouli, J. (2011). Single-Fed Broadband Circularly Polarized Stacked Patch Antenna With Horizontally Meandered Strip for Universal UHF RFID Applications. *IEEE Transactions on Microwave Theory and Techniques*, *59*(4), 1066–1073. doi:10.1109/TMTT.2011.2114010

Zhongping, W., Guangmin, W., & Yibing, L. (2011). A dual-relax penalty function approach for solving nonlinear bilevel programming with linear lower level problem. *Acta Mathematica Scientia*, *31*(2), 652–660. doi:10.1016/S0252-9602(11)60265-8

Zhong, Y. B., Xiang, Y., & Liu, H. L. (2014). A multi-objective artificial bee colony algorithm based on division of the searching space. *Applied Intelligence*, *41*(4), 987–1011. doi:10.1007/s10489-014-0555-8

Zhou, J. K. (1986). *Differential transform and its applications for electrical circuits*. Wuhan: Huarjung University Press.

Zhou, Y., & Zheng, H. (2013). A Novel Complex Valued Cuckoo Search Algorithm. *TheScientificWorldJournal*, *2013*, 1–6. doi:10.1155/2013/597803 PMID:23766699

Zhu, G., & Kwong, S. (2010). Gbest-guided artificial bee colony algorithm for numerical function optimization. *Applied Mathematics and Computation*, *217*(7), 3166–3173. doi:10.1016/j.amc.2010.08.049

Zhu, L., Yu, F. R., Ning, B., & Tang, T. (2012). Optimal charging control for electric vehicles in smart microgrids with renewable energy sources. In *Vehicular Technology Conference (VTC Spring), 2012 IEEE 75th* (pp. 1-5). IEEE. doi:10.1109/VETECS.2012.6240169

Zhu, T., Xue-ming, D., & Can, L. (2013). T-S Model Identification Based on a New Hybrid GSAPSO Algorithm. *Journal of University of Shanghai for Science and Technology*, *4*, 351–354.

Zimmerman, R. D., Murillo-Sánchez, C. E., & Thomas, R. J. (2011). MATPOWER: Steady-state operations, planning, and analysis tools for power systems research and education. *Power Systems. IEEE Transactions on*, *26*(1), 12–19.

Zio, E., Baraldi, P., & Patelli, E. (2006). Assessment of the availability of an offshore installation by Monte Carlo simulation. *International Journal of Pressure Vessels and Piping*, *83*(4), 312–320. doi:10.1016/j.ijpvp.2006.02.010

Zitzler, E., Deb, K., & Thiele, L. (2000). Comparison of multiobjective evolutionary algorithms: Empirical results. *Evolutionary Computation*, *8*(2), 173–195. doi:10.1162/106365600568202 PMID:10843520

Zitzler, E., & Thiele, L. (1998). Multiobjective Optimization Using Evolutionary Algorithms - A Comparative Case Study. *In Conference on Parallel Problem Solving from Nature (PPSN V)*. doi:10.1007/BFb0056872

Zitzler, E., Thiele, L., Laumanns, M., Fonseca, C. M., & da Fonseca, V. G. (2003). Performance assessment of multiobjective optimizers: An analysis and review. *IEEE Transactions on Evolutionary Computation*, *7*(2), 117–132. doi:10.1109/TEVC.2003.810758

Zolotarev, V. M. (1986). One-dimensional Stable Distributions, Volume 65 of Translations of mathematical monographs. American Mathematical Society.

Zuo, J., Jia, Y., Chen, Y., & Zhang, Y. (2012, April). A New Particle Filter Guided by Iterated EKF. In *Proceedings of the 2012 Second International Conference on Electric Information and Control Engineering* (pp. 124-127). IEEE Computer Society.

About the Contributors

Pandian Vasant is an Executive Editor of Global Journal of Technology and Optimization and Co- Editor of International Journal of Energy Optimization & Engineering and International Journal of Computing & Optimization. His main research interests are in the areas of Optimization Methods and Applications to Decision Making and Industrial Engineering, Fuzzy Optimization, Computational Intelligence, and Hybrid Soft Computing. Pandian Vasant has co-authored 200 research papers and articles in national journals, international journals, conference proceedings, conference paper presentation, and special issues lead guest editor, lead guest editor for book chapters' project, conference abstracts, edited books and book chapters. In the year 2009, Pandian Vasant was professionally honored as top reviewer for the journal of Applied Soft Computing (Elsevier). He has been Co-editor for AIP Conference Proceedings of PCO (Power Control and Optimization) conferences since 2008 and editorial board member of international journals in the area of Soft Computing, Hybrid Optimization and Computer Applications. Currently he's a senior lecturer at Department of Fundamental & Applied Sciences, Universiti Teknologi PETRONAS. Google Scholar Citations: h-index is 18; i10-index is 32 (Citation Indices).

Gerhard-Wilhelm Weber is a Professor at IAM, METU, Ankara, Turkey. His research is on OR, financial mathematics, optimization and control, life sciences, data mining, education and development; he is involved in the organization of scientific life internationally. He received his Diploma and Doctorate in mathematics, and economics / business administration, at RWTH Aachen, and his Habilitation at TU Darmstadt. He held Professorships by proxy at University of Cologne, and TU Chemnitz, Germany. At IAM, he is in the programs of Financial Mathematics, Actuarial Sciences and Scientific Computing and Assistant to the Director, and he is a member of four further graduate schools, institutes and departments of METU. Further, he has affiliations at the universities of Siegen, Ballarat, Aveiro, North Sumatra, and Malaysia University of Technology, and he is "Advisor to EURO Conferences".

Vo Ngoc Dieu received his B.Eng. and M.Eng. degrees in electrical engineering from Ho Chi Minh City University of Technology, Ho Chi Minh city, Vietnam, in 1995 and 2000, respectively and his D.Eng. degree in energy from Asian Institute of Technology (AIT), Pathumthani, Thailand in 2007. He is Research Associate at Energy Field of Study, AIT and lecturer at Department of Power Systems, Faculty ofElectrical and Electronic Engineering, Ho Chi Minh City University of Technology, Ho Chi Minh city, Vietnam. His interests are applications of AI in power system optimization, power system operation and control, power system analysis, and power systems under deregulation.

* * *

M. Abdullah-Al-Wadud received his B.S. degree in computer science and M.S. in computer science and engineering from the University of Dhaka, Bangladesh in 2003 and 2004, respectively. In 2009, he completed his PhD in Computer Engineering from Kyung Hee University, South Korea. Afterwards, he served as a member of the faculty of the Department of Industrial and Management Engineering in Hankuk University of Foreign Studies, South Korea, from 2009 to 2014. He also served as a lecturer in Faculty of Sciences and Information Technology, Daffodil International University, Bangladesh, and in Faculty of Sciences and Engineering, East West University, Bangladesh, in 2003 and 2004, respectively. Currently, he is working as an associate professor in the Department of Software Engineering, King Saud University, Saudi Arabia. His research interest includes image enhancement, pattern recognition, optimization, sensor and ad hoc networks.

Amal Kumar Adak received his MSc degree in Mathematics and PhD from Vidyasagar University, West Bengal, India in 2006 and 2013, respectively. He is an Assistant Teachers in Department of Mathematics of Jafuly Deshpran School (H.S.), West Bengal, India since 2007. His research interest includes fuzzy and intuitionistic fuzzy sets, intuitionistic fuzzy matrices and operations research. His papers are published in several national and international journals. He has participated and presented papers in several national and international conferences.

Marouane Alaoui-Selsouli is a doctor engineer in ORTEMS SA.He received a Phd in computer science from Ecole des Mines of Nantes. His research interests are the development and application of methods and tools for modeling, optimization and decision support for the design, planning, operation scheduling logistics systems for goods and services.

Fatemeh Barani received the B.Sc. degrees in computer engineering from Ferdowsi University of Mashhad in 2008 and the M.Sc. degrees in computer engineering from Tarbiat Modares University. Her main research interests are network security, intrusion detection, evolutionary computation, and swarm intelligence. Currently, she is instructor at Department of Mathematics and computing, Higher Education Complex of Bam.

Enrique Gabriel Baquela is an Industrial Engineer (from Universidad Tecnológica Nacional, Argentina) an PhD Student (in Universidad Nacional de Rosario). His research is focused in the use of optimization and simulation methods in order to improve traffic systems.

Monoranjan Bhowmik received his M.Sc. in Mathematics from Indian Institute of Technology, Kharagpur, West Bengal, India and Ph.D. from Vidyasagar University, India in 1995 and 2010, respectively. Currently, he is an Assistant Professor of VTT College, Paschim Midnapore, West Bengal, India. He is Associate-Editor of International Journal of Fuzzy Mathematical Archive. He is a reviewer of several international journals. His main scientific interest concentrates on fuzzy and intuitionistic fuzzy sets, fuzzy and intuitionistic fuzzy matrices and fuzzy algebras, Math Education and Educational Statistics. He is author of more than 30 articles published in various scholarly journals and conference proceedings and also had organized several Seminar, Conference in his institution.

Lukasz Kosma Bonenberg is a Senior Experimental Officer at the NGI, with ten years industrial experience in civil and engineering surveying focusing mostly on precise applications, working in Europe and USA. He is a Corporate Member of Chartered Institution of Civil Engineering Surveyors. He completed his PhD, at the University of Nottingham, looking at augmentation of GNSS with ground transmitters (pseudolite, Locatalite) to improve accuracy and confidence of positioning in areas of limited sky visibility – such as urban canyons. His five years research experience include GNSS, sensor fusion (LSA, KF), hardware testing, workflow automation and data analysis. His current interests include navigation for transport, safety of life and indoor positioning, sensor fusion (LSA, KF), advanced aspects of deformation monitoring using multi-sensors and GNSS, data analysis.

Ibibia K. Dabipi research interests include Computer Security and Network Management, Parallel Computing and Algorithms Development, Performance Evaluation of Computer Networks, Optimization of Transportation Networks and Economic Analysis of Transportation Facilities. I. K. Dabipi holds the following degrees: Ph.D., in Electrical Engineering, 1987, and a Master of Science, in Electrical Engineering, 1981, Louisiana State University, Baton Rouge, Louisiana, Bachelor of Science, in Electrical Engineering, and Bachelor of Science in Physics/Mathematics, 1979 from Texas A&I University, Kingsville, Texas. I. K. Dabipi was Professor and Chair in the Department of Engineering and Aviation Sciences University of Maryland Eastern Shore from 2001-2006. Prior to coming to University of Maryland Eastern Shore, he was the Professor, Interim Chair, and Chair, Electrical Engineering Department, Southern University, Baton Rouge, Louisiana from 1997-2001. His experiences include working at Bell Communications Research and AT&T Bell Labs as member of technical staff during the summers of 1984 through 1987. He has authored or co-authored many technical articles for publications and presentations. He is currently a professor of Electrical Engineering in the Engineering and Aviation Sciences Department.

Renu Dhir is working as Associate professor and Head in the Department of Computer Science and Engineering National Institute of Technology, Jalandhar (Punjab), India. She has done M. Tech. in Computer Science and Engineering from TIET Patiala, India, in 1997 and Doctor of Philosophy (Ph.D.) in Computer Science and Engineering, Punjabi University, Patiala, India, in March 2008. She has published her research work in more than 40 International / National Conferences and Journals in various fields like Information Security, Image Processing, OCR, NLP and Pattern recognition.

S. A. Edalatpanah is an academic member of Islamic Azad University of Lahijan and University of Guilan. His research interests are numerical linear algebra, operation research and computational mathematics.He is on the editorial boards of more than 30 refereed journals.

I. Elamvazuthi obtained his PhD from the Department of Automatic Control & Systems Engineering, University of Sheffield, UK in 2002. Currently, he is an Associate Professor at the Department of Electrical and Electronic Engineering, Universiti Teknologi PETRONAS (UTP). Before joining UTP, he has worked at University Kuala Lumpur, Standards and Industrial Research Institute of Malaysia (SIRIM) and UMW Industrial Power. His research interests include Control & Systems Engineering with focus on Energy, Robotics and Bio-Medical Applications. He is a member of IEEE, IEEE Robotics & Automation Society (RAS), IEEE Control Systems Society.

Jose Luis Espinoza, M.Sc. in Computer Science, Technological Institute of Costa Rica (1998), Licence in Mathematics, University of Costa Rica (1988) and Bachelor in Mathematics, UCR (1987). Professor at the School of Mathematics, ITCR, since 1987.

Mariano Frutos is an Assistant Researcher of CONICET (National Research Council of Argentina) and Teaching Assistant at the Department of Engineering of the Universidad Nacional del Sur, in Bahía Blanca, Argentina. He got an undergraduate degree in Industrial Engineering as well as a Master's and a Ph.D. in Engineering from his home university. His research is centered on scheduling problems in production and its treatment through metaheuristic methods. He has published in Operational Research, Annals of Operations Research, Dyna, and American Journal of Operations Research among others. He participates actively in the Operations Research community in Latin-America.

Harish Garg is an Assistant Professor at Thapar University Patiala, India. Prior to joining Thapar University, Patiala, Dr. Garg received a PhD in Applied Mathematics with specialization of Reliability theory and soft computing techniques, from Indian Institute of Technology Roorkee, India in 2013 and an MSc in Mathematics from Punjabi University, Patiala, India in 2008. His research interest are in the field of Computational Intelligence, Reliability theory, Optimization techniques, various nature inspired algorithms (e.g. genetic algorithms, swarm optimization), fuzzy and intuitionistic fuzzy set theory. Application areas include wide range of industrial and structural engineering design problems. Dr. Garg is the author of over 50 scholarly publications in the top-tier reputed journals such as Applied Soft Computing, Expert Systems with Application, Computer and Industrial Engineering, ISA Transactions, Journal of Manufacturing Systems, International Journal of Uncertainty, Fuzziness and Knowledge-Based System, Journal of Intelligent and Fuzzy Systems, Journal of Industrial and Management Optimization, Mathematics and Computers in Simulation, Japan Journal of Industrial and Applied Mathematics etc. He has also published 4 book chapters.

Sotirios K. Goudos received the B.Sc. degree in Physics in 1991 and the M.Sc. of Postgraduate Studies in Electronics in 1994 both from the Aristotle University of Thessaloniki. In 2001, he received the Ph.D. degree in Physics from the Aristotle University of Thessaloniki and in 2005 the Master in Information Systems from the University of Macedonia, Greece. In 2011, he obtained the Diploma degree in Electrical and Computer Engineering from the Aristotle University of Thessaloniki. From 2004 to 2007, he was a part-time lecturer with the Department of Technology Management, University of Macedonia, Greece. Since 2013, he has been with the Department of Physics, Aristotle University of Thessaloniki. His research interests include antenna and microwave structures design, electromagnetic compatibility of communication systems, evolutionary computation algorithms, and mobile communications. Dr. Goudos is a Senior member of the IEEE, the Greek Physics Society, the Technical Chamber of Greece, and the Greek Computer Society.

Kawal Jeet is an Assistant Professor in Post-Graduate Department of Computer Science, D.A.V. College, Jalandhar, India. She received her Master's of Technology in Computer Science from Dr. B.R. Ambedkar National Institute of Technology, Jalandhar, India in 2012. Currently she is pursuing Ph.D from this institute. Her current research interest focuses on nature inspired computation, software modularization, Bayesian networks, software quality. She has published her research work in more than 15 international journals and conference proceedings. She is member of the IRED, UACEE and ACM India.

Joong Hoon Kim, a faculty of Korea University in the School of Civil, Environmental and Architectural Engineering, obtained his Ph.D. degree from the University of Texas at Austin in 1992 with the thesis title "Optimal replacement/rehabilitation model for water distribution systems". Prof. Kim's major areas of interest include: optimal design and management of water distribution systems, application of optimization techniques to various engineering problems, and development and application of evolutionary algorithms. His publication includes "A New Heuristic Optimization Algorithm: Harmony Search", Simulation, February 2001, Vol. 76, pp 60-68, which has been cited over 1,100 times by other journals of diverse research areas. Recently, he gave a distinguished lecture in the Annual Meeting of Asia Oceania Geosciences Society held in Brisbane, Australia in June 2013, under the title of "Optimization Algorithms as Tools for Hydrological Science".

S. Kohshelan received his Master in Information Technology from Universiti Tun Hussein Onn Malaysia (UTHM), Johor, Malaysia. He is a specialist in multimedia, mobile and web development. His research titles "Mobile ticket checker & trading for express bus (MTCTEB)" and "Mobile Hindu deity for daily prayers using augmented reality (IShiva)" has been awarded Silver medals in Research & Innovation Fest. His current research interests include Big Data, Optimization, Data Mining, Signal Processing, Algorithms, and Modeling.

Gilberto Pérez Lechuga is Industrial Engineering from the University Autonomous of the Hidalgo State, Mexico. He specialized in Systems Engineering and he holds the degree of Master in Science in Operations Research, by the National Polytechnic Institute, Mexico. Doctor Pérez Lechuga obtained the Doctor in Engineering degree in Operations Research, by the National Autonomous University of Mexico. He has done postdoctoral stays at the University of Campinas Brazil, in Bayesian Statistics; Havana Cuba in Reliability Engineering in Nuclear Plants; the University of Lleida Spain in Stochastic Optimization, the ULM University (Germany) in Dynamic of non Nonlinear systems, and in Montreal Canada in Mathematical Modeling of Supply Chain and Transport Systems. The research line of Dr. Pérez Lechuga focuses on the modeling and optimization of stochastic systems, design and application of biological algorithms and metaheuristics to emerging dynamic systems, modeling and optimization of stochastic flexible manufacturing systems and mathematical modeling of the supply chain, as well as applications in reliability.

Ali Mahmoodirad is a PHD candidate of applied mathematics (Operation Research) in Central Tehran Branch of Islamic Azad University. His research interests include: Fuzzy mathematical programming, Metaheuristics and hybrid methods, Decomposition methods. He has published research articles in international journals of Mathematics and industrial engineering.

Debashree Manna received hes MSc degree in Mathematics from Vidyasagar University, West Bengal, India in 2012. She is an Assistant Teachers in Department of Mathematics of Damda High School, West Bengal, India since 2010. Her research interest includes fuzzy and intuitionistic fuzzy matrices.

Majdi Mansouri has over seven years of research and practical experiences in the area of systems engineering and signal processing. His work focuses on the utilization of applied mathematics and statistics concepts to develop algorithms for modeling, estimation and prediction, which can help improve process operations. In academia, He worked extensively on the utilization and development of Bayesian inference to improve filtering, modeling, and state estimation, He worked also on process monitoring and faults detection for chemical and environmental systems. He is the author of over 70 papers (25 international journals, 7 book chapters and 38 international conferences).

Jose Antonio Marmolejo is Professor in the Faculty of Engineering of the Universidad Anahuac Mexico Norte. He is PhD in Engineering with specialization in Operations Research from the National Autonomous University of Mexico and currently a member of the National System of Researchers of the National Council of Science and Technology (CONACYT) of Mexico. He is on the Board of the Mexican Society of Operations Research. His areas of interest are large scale optimization and mathematical modeling in power systems and logistical problems in the supply chain.

Francisco Venegas Martínez has doctorates in mathematics and economics by Washington State University, master's degrees in mathematics and operations research by the National Autonomous University of Mexico. He has won the following awards: Economic Investigations Jesus Silva Herzog (2002), National Award of derivatives Mex-Der research (2005), by applied research at the National Polytechnic Institute (2011), medal Lazaro Cardenas (2012). He is a member of the Mexican Academy of Sciences. Doctor Venegas has published more than 200 scientific articles in national and international journals and has authored more than 20 research books.

Ernesto Mastrocinque is a Research Fellow in the Simulation and Modelling of Supply Chain at the School of Management of Royal Holloway, University of London. Before joining Royal Holloway's School of Management, he was a Research Fellow at the Department of Industrial Engineering of University of Salerno (Italy) and Visiting Researcher at the School of Engineering of Cardiff University (UK), where he worked on the development and application of swarm-based algorithm for supply chain optimisation. He holds a PhD in Mechanical Engineering from the University of Salerno (Italy). His research interests include: modelling, simulation and optimisation of supply chain, intelligent manufacturing systems, single machine and job-shop scheduling, lean manufacturing, swarm-based and metaheuristic optimisation algorithms. He is an author of papers in the field of operations management, technology and manufacturing systems, artificial intelligence. He also acts as a reviewer for international peer-reviewed journals and for book chapters.

Haydee Rocio Melo received the B.Sc. degree in Electro-mechanics engineering from National Polytechnic Institute (IPN), Mexico and the M.Sc. degree from the Graduate School of Information, Production and Systems, Waseda University, Japan. She is currently a Doctor student in Graduate School of Information, Production and Systems, Waseda University, Japan as a scholarship recipient student from the Ministry of Education, Culture, Sports, Science and Technology, Japan. Her research interests include soft computing, management engineering, reliability and Meta-heuristic algorithms.

Fabio Miguel is a Teaching Assistant at the Sede Alto Valle y Valle Medio of the Universidad Nacional de Río Negro, in Villa Regina, Argentina. He got an undergraduate degree in Economics from the Universidad Nacional de Córdoba. He is finishing his M.Sc. thesis in Management Science at the Universidad Nacional del Sur. His research is on scheduling problems in production and their solution through metaheuristic methods. His work has been presented at national and international conferences and published in their corresponding proceedings.

Michel Andre Minoux is a Professor 'Emeritus' at University of Paris-6. His research interests include linear and nonlinear optimization, combinatorial optimization and graph theory, optimization under uncertainty. He has authored or co-authored several books in these fields, in particular: 'Graphs and Algorithms' (Wiley, 1984), 'Mathematical Programming, theory and algorithms' (Wiley 1986), 'Graphs, Dioids and Semirings' (Springer 2008).

Abdelmoula Mohafid is Associate Professor at the University of Nantes Institut Universitaire de Technologie de Nantes & IRCCyN UMR CNRS 6597 Main research areas: - Modeling and optimization of maintenance policies - Integrating preventive maintenance scheduling and medium-term planning of the production.

Saber Molla-Alizadeh-Zavardehi is an assistant professor of Industrial Engineering at College of Engineering, Islamic Azad University (IAU), Masjed-Soleiman Branch in Iran. He is the Graduate Programs Manager of Industrial Engineering at the IAU, Masjed-Soleiman Branch, in Iran. His research interests are production scheduling, supply chain modeling, fuzzy set theory and metaheuristic algorithms. He has about 8 years of teaching and research experience and has authored 25 papers published in various international journals.

Jean-Philippe Montillet is a Senior IEEE member. He graduated of a PhD in civil engineering (mathematical geodesy) from the University of Nottingham (2008). He is a Research fellow working in GPS daily solutions and Real Time Kinematic for geophysical and civil engineering applications. A large amount of his work is in applied statistics on time series analysis (coloured noise, fBm, Bayesian statistics, MCMC), co-seismic offset estimation, navigation and tracking (Kalman filter). He is also interested in ubiquitous positioning, LBS and sensor networks (convex optimisation) and Locata technology (ground based positioning system). His recent work spans signal processing and operator theory (Energy Operators) with applications to wave theory.

Tierney Moore grew up on the North side of Houston and attended G.W. Carver High School where she majored in Architectural Engineering. While in high school Tierney remained on the honor roll and held various leadership positions including Student Council President and Programming Captain. Tierney currently attends Prairie View A&M University where she is double majoring in Civil Engineering and Construction Science. She has interned at GE (General Electric) and Shell during her summer breaks. While in school, Tierney has participated in various research projects under the guidance of Dr. Judy A. Perkins, PE. After graduation, Tierney would like to pursue a career in Environment Engineering.

H. Saberi Najafi, Associate Professor of Applied Mathematics at the University of Guilan and Islamic Azad University of Lahijan. His fields of interest are numerical linear algebra, numerical analysis, computational mathematics and numerical modeling of oceans.

Najib Mohamed Najid is a Professor in the department of Mechanical Engineering. He received a bachelor's Degree in in Electronic and Automatic from the University of Nantes, Master's Degree in Robotic from ENSAM of Paris and Phd in Computer science from Ecole Centrale of Nantes in 1988, 1990 and 1994 respectively. He joined the Department Of Mechanical Engineering, University Institute of Technology of Nantes, in 1994 where he is teaching production management, design and control of manufacturing systems courses. Since 2000, he is a member of Production and Logistics Systems (SLP) team of the Institute of Communications and Cybernetic of Nantes (UMR CNRS 6795) where he received his Habilitation in 2008. His research interests are the development and application of methods and tools for modeling, optimization and decision support for the design, planning, operations scheduling, monitoring and maintenance logistics systems for goods and services.

Sk. Md. Abu Nayeem has received his M.Sc. and Ph.D. in Applied Mathematics from Vidyasagar University, India in 2001 and 2007 respectively. He is currently working as an Assistant Professor of Mathematics at Aliah University, India. His research interest includes graph theory, optimization, computational intelligence, etc.

Hossein Nezamabadi-pour received his B.Sc. degree in Electrical Engineering from Shahid Bahonar University of Kerman in 1998, and his M.Sc. and Ph.D. degrees in Electrical Engineering from Tarbait Moderres University, Iran, in 2000 and 2004, respectively. In 2004, he joined the Department of Electrical Engineering at Shahid Bahonar University of Kerman, Kerman, Iran, as Assistant Professor, and was promoted to Full Professor in 2012. Dr. Nezamabadi-pour is author and co-author of more than 330 peer reviewed journal and conference papers. His research interests include Evolutionary Computation, Pattern Recognition, Soft Computing, and Image Processing.

Khai Phuc Nguyen received his B.Eng. and M.Eng. degrees in electrical engineering from Ho Chi Minh City University of Technology, Vietnam, in 2010 and 2012, respectively. From 2010, he worked as a lecturer at Department of Power System, Faculty of Electrical and Electronic Engineering, Ho Chi Minh City University of Technology. From 2014 until now, he is a Doctoral Candidate in Shibaura Institute of Technology, Tokyo, Japan. His interests are applications of AI in power system optimization, power system operation and control, power system analysis and power systems under deregulation.

Thang Trung Nguyen received his B.Eng. and M.Eng. degrees in Electrical Engineering from University of Technical education Ho Chi Minh City (UTE), Ho Chi Minh city, Vietnam in 2008 and 2010, respectively. Now, he is teaching at department of electrical and electronics engineering, Ton Duc Thang university and pursuing D.Eng. Degree at UTE, Ho Chi Minh city, Vietnam. His research interests include optimization of power system, power system operation and control and Renewable Energy.

Sadegh Niroomand is a PhD graduate from Eastern Mediterranean University. Currently is working as an invited lecturer in Shiraz University of Technology.

Hazem Numan Nounou (SM'08) received the B.S. degree (magna cum laude) from Texas A&M University, College Station, in 1995, and the M.S. and Ph.D. degrees from Ohio State University, Columbus, in 1997 and 2000, respectively, all in electrical engineering. In 2001, he was a Development Engineer for PDF Solutions, a consulting firm for the semiconductor industry, in San Jose, CA. Then, in 2001, he joined the Department of Electrical Engineering at King Fahd University of Petroleum and Minerals in Dhahran, Saudi Arabia, as an Assistant Professor. In 2002, he moved to the Department of Electrical Engineering, United Arab Emirates University, Al-Ain, UAE. In 2007, he joined the Electrical and Computer Engineering Program at Texas A&M University at Qatar, Doha, Qatar, where he is currently a full Professor. He published more than 80 refereed journal and conference papers. He served as an Associate Editor and in technical committees of several international journals and conferences. His research interests include intelligent and adaptive control, control of time-delay systems, system biology, and system identification and estimation. Dr. Nounou is a senior member of IEEE.

Mohamed Numan Nounou (SM'08) received the BS degree (magna cum laude) from Texas A&M University, College Station, in 1995, and the MS and PhD degrees from the Ohio State University, Columbus, in 1997 and 2000, respectively, all in chemical engineering. From 2000 to 2002, he was with PDF Solutions, a consulting company for the semiconductor industry, in San Jose, CA. In 2002, he joined the Department of Chemical and Petroleum Engineering at the United Arab Emirates University as a full professor. In 2006, he joined the Chemical Engineering Program at Texas A&M University at Qatar, Doha, Qatar, where he is currently an Associate Professor. He has published more than 80 refereed journal and conference papers and book chapters. He also served as an associate editor and in technical committees of several international journals and conferences. His research interests include process modeling and estimation, system biology, and intelligent control. He is a member of the American Institute of Chemical Engineers (AIChE) and a senior member of the IEEE.

Ana Carolina Olivera is an Assistant Researcher of CONICET (National Research Council of Argentina) stationed at the Department of Exact and Natural Sciences of the Universidad Nacional de la Patagonia Austral. She has a Ph.D. in Computer Science from the Universidad Nacional del Sur and held a post-doctoral position at the University of Malaga, in Spain. Her research focuses on urban traffic and production chain optimization problems using bio-inspired algorithms. She has published book chapters and several papers in indexed journals and proceedings of refereed international conferences.

Abdeen Mustafa Omer (BSc, MSc, PhD) is an Associate Researcher at Energy Research Institute (ERI). He obtained both his PhD degree in the Built Environment and Master of Philosophy degree in Renewable Energy Technologies from the University of Nottingham. He is qualified Mechanical Engineer with a proven track record within the water industry and renewable energy technologies. He has been graduated from University of El Menoufia, Egypt, BSc in Mechanical Engineering. His previous experience involved being a member of the research team at the National Council for Research/Energy Research Institute in Sudan and working director of research and development for National Water Equipment Manufacturing Co. Ltd., Sudan. He has been listed in the WHO'S WHO in the World 2005, 2006, 2007 and 2010. He has published over 200 papers in peer-reviewed journals, 50 review articles, 5 books and 50 chapters in books.

Pauline Ong received her Ph.D in applied mathematics from Universiti Sains Malaysia, Penang, Malaysia in year 2011, with specialiation in artificial neural networks. She is currently a senior lecturer with Universiti Tun Hussein Onn Malaysia, Johor, Malaysia. Her current research interests include artificial neural networks, mathematical modeling, and metaheuristic algorithm.

Madhumangal Pal is currently a Professor of Applied Mathematics, Vidyasagar University, India. He has received Gold and Silver medals from Vidyasagar University for rank first and second in M.Sc. and B.Sc. examinations respectively. Also he received jointly with Prof. G.P.Bhattacherjee, "Computer Division Medal" from Institute of Engineers (India) in 1996 for best research work. Prof. Pal has successfully guided 21 research scholars for Ph.D. degrees and has published more than 175 articles in international and national journals. His specializations include Algorithmic Graph Theory, Fuzzy Correlation & Regression, Fuzzy Game Theory, Fuzzy Matrices, Genetic Algorithms and Parallel Algorithms. Prof. Pal is the Editor-in-Chief of "Journal of Physical Sciences" and "Annals of Pure and Applied Mathematics", and member of the editorial Boards of several journals. He is also a reviewer of several international journals. Prof. Pal is the author of eight books published from Indian and Oxford, UK. He organized several national seminars/ conferences/ workshop. Also, visited China, London, Malaysia, Thailand, Hong Kong, Dubai and Bangladesh to participated, delivered invited talks and chaired in national and international seminars/ conferences.

Judy A. Perkins, PE holds a B.S., M.S., and Ph.D. in Civil Engineering from Southern University, University of Illinois (Urbana-Champaign), and Georgia Institute of Technology, respectively. She is a Regents Professor, and a licensed professional engineer in the state of Texas. Her research has focused on statewide intermodal transportation planning, asset management, emergency operations, transportation logistics, hurricane evacuation analysis, military logistics, engineering education, minority outreach, optimization of transportation infrastructure investments, and radio-frequency identification (RFID) technologies used by motor carriers and for locating underground utilities. Professor Perkins has accumulated extensive experience in the development of survey design, data collection, state- level transportation planning, and the refinement of economic development methodologies used to evaluate transportation-related activities. Her extensive record of publication extends into both the national and international transportation and engineering education arenas.

Imran Rahman received his Bachelor of Science in Engineering from Department of Electrical and Electronic Engineering, Islamic University of Technology (IUT), Gazipur, Bangladesh in the year of 2011 and Master of Science (by research) from the Department of Fundamental and Applied Sciences, Universiti Teknologi PETRONAS, Tronoh, Perak, Malaysia in 2016. His current research interests are smart grid optimization, optimization of plug-in hybrid electric vehicles (PHEVs), computational intelligence techniques.

Elvia Perez Ramirez has a degree in physics and master of sciences in mathematical modeling of complex systems with applications to geophysics, both, by the Autonomous National University of Mexico. She has directed several research projects related to their specialty for domestic companies as the National Petroleum Institute. She is the author of many research articles and, is currently is a doctoral student of mathematics applied and modeling of geophysical sciences.

Gethin Wyn Roberts is Professor of Geospatial Engineering at the University of Nottingham Ningbo China. He has been at the University of Nottingham since 1989, but on a seconded onto the UNNC campus since September 2010. His research focus is on engineering surveying, and in particular using GNSS for deformation monitoring, and integrating GNSS with other sensors. This includes monitoring the movements of large bridges using GNSS, such as the Severn, Humber and Forth bridges, using GNSS to control the movements of construction plant, integration of GNSS with pseudolites and Locatalites. He has authored and co-authored over 230 publications, as well as supervising 20 PhD students. He was the Chairman of the International Federation of Surveyors' Commission 6; Engineering Surveys, from 2010-2015. He is a Fellow of the Chartered Institution of Civil Engineering Surveyors, and a Fellow of the Higher Education Academy.

Román Rodríguez is PhD in Economics with specialization in mathematical finance from the School of Economics at the National Polytechnic Institute, Master in Engineering and statistician by the National Autonomous University of Mexico, Master in Public Policy Institute of Technology and Research superiors of Monterrey. Dr. Rodriguez is a member of the Mexican Mathematical Society, The National Association of Economists and Economic Modeling Network EcoMod. His areas of interest are stochastic optimization, mathematical finance and econometrics.

Ali Sadollah received his BS degree in mechanical engineering, solid states, from Azad University, Semnan Branch, Iran in 2007 and MS degree in mechanical engineering, applied mechanics from University of Semnan, Semnan, Iran in 2010. He is obtained his PhD at the Faculty of Engineering, University of Malaya, Kuala Lumpur, Malaysia in 2013. Currently, he is serving as a post-doctoral research fellow at Korea University, Seoul, South Korea. His research interests include algorithm development, optimization and metaheuristics, applications of soft computing methods in engineering, computational solid mechanics, artificial neural networks, cold roll forming process, and so forth.

John N. Sahalos received his B.Sc. degree in Physics, in 1967 and his Ph.D degree in Physics, in 1974, from the Aristotle University of Thessaloniki, (AUTH), Greece. Except of his PhD, during 1970-75, he studied at the School of Engineering of AUTH and he received the Diploma (BCE+MCE) in Civil Engineering, (1975). He also, during 1972-74, studied at the School of Science of AUTH and he received the professional Diploma of postgraduate studies in Radio-Electrology, (1975). In 1976, he worked at the ElectroScience Laboratory, the Ohio State University, Columbus, as a Postdoctoral University Fellow. From 1977 to 1986, he was a Professor in the Electrical Engineering Department, University of Thrace, Greece, and Director of the Microwaves Laboratory. From 1986 to 2010, he has been a Professor at the School of Science, AUTH, where he was the director of the postgraduate studies in Electronic Physics and the director of the Radio-Communications Laboratory (RCL). Dr. Sahalos is now Professor at the Electrical & Computer Engineering Department of the University of Nicosia, Cyprus and director of the Radio & Telecommunications Laboratory (RTeLab). During 1981-82, he was a visiting Professor at the Department of Electrical and Computer Engineering, University of Colorado, Boulder. During 1989-90, he was a visiting Professor at the Technical University of Madrid, Spain. He is the author of three books in Greeks, of nine book chapters and more than 430 articles published in the scientific literature. He also is the author of the book "The Orthogonal Methods of Array Synthesis, Theory and the ORAMA Computer Tool", Wiley, 2006. His research interests are in the areas of antennas, high frequency techniques, radio-communications, EMC/EMI, RFID, microwaves, and biomedical engineering. Dr. Sahalos is a Professional Engineer and a Consultant to industry. In 2002-04, he was in the Board of Directors of the OTE, the largest Telecommunications Company in Southeast Europe. He served, as a technical advisor, in several national and international committees, as well as, in several Mobile Communications Companies. Since 1992, he has been a member of commissions A and E of URSI. In 1998-2011, he was the president of the Greek committees of URSI. He was the president of the section of Informatics, Telecommunications and Systems of the National Committee of Research and Technology. Dr. Sahalos is a Life Fellow of the IEEE. He also is a member of the Greek Physical Society, an Honor Fellow of Radio-electrology and a member of the Technical Chamber of Greece. He is the creator and leader of an EMC network with five laboratories (3 from the academy and 2 from the industry). He has been honored with a special investigation fellowship of the Ministry of Education & Science, Spain. He also has been honored from several Institutes and Organizations. He has been in the editorial board of three scientific journals. Dr. Sahalos was elected by the department representatives of the Aristotle University of Thessaloniki as the Vice-Chairman of the Research Committee of AUTH for the period 2007-2010. Since 2010, he is the executive manager of the University of Nicosia Research Foundation (UNRF) and a member of the consulting committee of the GRNET S.A. Dr. Sahalos is a member of EuRAAP as a delegate of Cyprus, Greece, Israel & Turkey. He has supervised 31 PhDs and more than 100 postgraduate diploma theses. With his colleagues Prof. Sahalos designed several well known innovative products like the Electric Impedance Tomography, (EIT), the Microwave Landing System, (MLS), the ORAMA simulator and the SMS-K monitoring system.

Katherine Siakavara received the BSc degree in Physics in 1977 and the MSc degree in Electronics, in 1979, from the Aristotle University of Thessaloniki, Greece. She was awarded the PhD degree in Electrical Engineering from the same university in 1982. She is currently Associate Professor in the Department of Physics of the University of Thessaloniki, Greece. Dr. Siakavara, is author or co-author of 46 international journal papers, of 43 papers in proceedings of international conferences, of 2 book chapters in English and 2 book chapters in Greek. Her research interests are in the areas of a) Applied Electromagnetism b)Analysis and Design of Antenna Systems c)Microwaves and d)Radio-wave propagation and radio-communications.

Balbir Singh Mahinder Singh is from Malaysia, born on 6 September 1967. He graduated with B.Sc. (Ed) Hons (Physics) in 1991, obtained his MSc (Physics) in 1996 and PhD in 2004 from Universiti Sains Malaysia. He joined Universiti Teknologi PETRONAS as a lecturer in 1999. Currently he is an Associate Professor and Head of the Fundamental and Applied Sciences Department. His research interest is in the area of solar energy utilization for electricity generation, solar hydrogen production and hybrid renewable energy based electricity generating system.

Benmessaoud Mohammed Tarik received the engineer in electronics. Magister and Ph.D Magister and degrees in electrical engineering from the University of Sciences and Technology of Oran (USTO-MB) - Algeria, in 2000, 2006, and 2012, respectively. He is professor in the University of Sciences and Technology of Oran (USTO-MB) University - the electrical and electronics engineering faculty as well as a member of the materials and its applications in electrical engineering laboratory. His research interests include electronics, renewable energy, solar cells, fuel cells, desalination, materials and his applications in electrical engineering, theoretical and computational mathematical concepts with his applications to engineering electrical, neural networks, genetic algorithms and optimization methods.

Fernando Tohmé is a Principal Researcher of CONICET (National Research Council of Argentina) and Full Professor at the Department of Economics of the Universidad Nacional del Sur, in Bahía Blanca, Argentina. A former Fulbright Scholar he held visiting positions at U.C. Berkeley, Washington University in St. Louis and Endicott College. He holds an undergraduate degree in Mathematics and a Ph.D. in Economics. His research has focused on decision problems, game theory and optimization in the socio-economic settings. He has published in Theory and Decision, Mathematics of Social Sciences, Artificial Intelligence, Mathematical and Computational Modeling, and Annals of Operations Research among others.

Javier Trejos, Doctor in Applied Mathematics, Université Paul Sabatier – Toulouse III, Toulouse, France (1994); M.Sc. in Applied Mathematics, Université Paul Sabatier (1989), Licence in Mathematics, University of Costa Rica (1988), Bachelor in Mathematics, University of Costa Rica (1987). Professor at the University of Costa Rica since 1993. Director of Graduate Studies, UCR (1996-2000 and 2003-2004), Director of the Research Centre for Pure and Applied Mathematics, UCR (2006-2010), and Dean of the Faculty of Science, UCR (2012-present). Visiting professor in Université de Limoges, France (1999 and 2010), and the Metropolitan University, Mexico (2001).

John Tsiligaridis obtained his most recent PhD in CSE from the State University of NY at Buffalo, USA, in Sept 2003. Before that he had earned a PhD in Computer Networks from National Technical University of Athens, Greece and he holds a M.Sc. in OR and Informatics and a M.Phil. in Data Mining. He has worked as Software Engineer, Research Faculty Fellow, and he is currently working as Associate Professor of Computer Science at Heritage University, Toppenish, WA. His current interests include Scheduling, Algorithms, Optimizations, Mobile Databases, Wireless Networks, Data Mining, Networking, and Bioinformatics.

Mario A. Villalobos-Arias, Doctor in Applied Mathematics, CINVESTAV, Mexico (2004); M.Sc. in Applied Mathematics, University of Costa Rica (1999), Licence in Mathematics, University of Costa Rica (1993). Professor at the University of Costa Rica since 1993. Director of Graduate Studies, UCR (2008-2010), Director of the Research Centre for Pure and Applied Mathematics, UCR (2012-present).

Bela Vizvari, MS: Eotvos Lorand University of Budapest, 1973, operations research. PhD: Eotvos Lorand University of Budapest, 1979, mathematics Dr.sc.nat: TU "Carl Schorlemmer", Merseburg, Germany, 1987, methematics Dr.habil.: Eotvos Lorand University of Budapest, 2003, mathematics.

Junzo Watada received his B.S. and M.S. degrees in electrical engineering from Osaka City University Japan, and Dr. of Engineering on "Fuzzy Multivariant Analysis and Its Applications" from Osaka Prefecture University, Japan. His professional interests include soft computing, tracking system, knowledge engineering, and management engineering. Prof. Watada is a recipient of the Henri Coanda Gold Medal from Invention in Romania in 2002. He is a Life Fellow of Japan Society for Fuzzy Theory and intelligent Informatics (SOFT) and Biomedical Fuzzy System Association (BMFSA). He is the Principal Editor, Co-Editor, or Associate Editor of various international journals, including the International Journal of Biomedical Soft Computing and Human Sciences, ICIC Express Letters, the International Journal of Systems and Control Engineering, and Fuzzy Optimization and Decision Making.

Kegen Yu received the Ph.D. degree from the University of Sydney in 2003. He was with Jiangxi Geological and Mineral Bureau; Nanchang University; University of Oulu; CSIRO ICT Centre; Macquarie University where he has been an Adjunct Professor since 2011; and University of New South Wales. He is currently a Professor with the School of Geodesy and Geomatics, Wuhan University. He coauthored the book Ground-Based Wireless Positioning and made contributions to five other books. He has authored or coauthored over 80 journal and conference papers. His current research interests include GNSS reflectometry and radio positioning. Dr Yu is on the editorial boards of EURASIP Journal on Advances in Signal Processing, IEEE Transactions on Aerospace and Electronic Systems and IEEE

Transactions on Vehicular Technology. He was the Lead Guest Editor for two special issues: "indoor navigation and tracking" and "GNSS remote sensing". He is a senior member of IEEE.

Baris Yuce obtained his BSc and MSc of Engineering (Industrial) degree with First Class Honours from Sakarya University. He has then been awarded his PhD degree from Cardiff University in 2012. During his MSc, he has also worked as a teaching and research assistant at Industrial Engineering Department, Sakarya University between 2006-2008. After the completion of his PhD, he started to work as a Research Associate at Cardiff University, School of Computer Science in 2012. In 2013, he joined the BRE Centre for Sustainable Construction, School of Engineering, Cardiff University. His research interest focuses on intelligent systems, in particular swarm optimization algorithms such as the Bees Algorithm, Genetic Algorithm, Ant Colony Optimization, Hill Climbing, and Simulated Annealing. He currently conducts research in Intelligent Buildings, Energy Management Systems and Supply Chain Management. He has published several papers on intelligent systems and their application in the area of Optimization, Supply Chain Management, Building Management Systems, and manufacturing systems. He has completed several EU and (Building Research Establishment (BRE) funded projects, including SPORTE2, KnoholEM, and Estate Energy Optimization.

Index

T

U

V

W

www.ingramcontent.com/pod-product-compliance
Lightning Source LLC
LaVergne TN
LVHW080616300926
849162LV00005B/1181